DEVELOPMENTAL NEUROBIOLOGY

Fourth Edition

DEVELOPMENTAL NEUROBIOLOGY
Fourth Edition

Edited by

MAHENDRA S. RAO

National Institute on Aging
Bethesda, MD

and

MARCUS JACOBSON[†]

University of Utah
Salt Lake City, UT

[†]Deceased

Kluwer Academic / Plenum Publishers
New York, Boston, Dordrecht, London, Moscow

Library of Congress Cataloging-in-Publication Data

ISBN 0-306-48330-0

© 2005 by Kluwer Academic / Plenum Publishers, New York
233 Spring Street, New York, New York 10013

http://www.kluweronline.com

10 9 8 7 6 5 4 3 2 1

A C.I.P. record for this book is available from the Library of Congress

Permissions for books published in Europe: permissions@wkap.nl
Permissions for books published in the United States of America: permissions@wkap.com

Printed in Singapore

Marcus Jacobson

Marcus Jacobson, a prominent scholar of developmental neurobiology, died of cancer at his home in Torrey, Utah in November, 2001; he was 71.

Jacobson was born in South Africa and finished medical training at the University of Cape Town. He then completed graduate study at Edinburgh University, receiving a Ph.D. in 1960 for a dissertation concerning specificity of synaptic connections in the *Xenopus* retinotectal system. Over the next two decades, Jacobson exploited the experimental opportunities provided by this preparation to become one of the best-known researchers of nervous system development, first at Purdue University then at Johns Hopkins University and the University of Miami (Hunt and Jacobson, 1974). In 1977, Jacobson moved to the University of Utah to become chairman of the Department of Neurobiology & Anatomy; he expanded the department and refocused its research on developmental neurobiology, a field in which it maintains a strong reputation. Shortly after moving to Utah, Jacobson began using single-cell injection techniques and lineage tracing in *Xenopus* to study early patterning of the nervous system (Jacobson, 1985).

In 1970, Jacobson published *Developmental Neurobiology* (Jacobson, 1970), a landmark book that critically summarized the status of the core topics in the emerging field that thereafter became known as developmental neurobiology. In two subsequent editions of this leading reference text (published by Plenum Press in 1977 and 1991), Jacobson enlarged the book substantially to maintain comprehensive coverage of a field that was growing rapidly. Throughout his career, Jacobson showed a strong interest in the history of neuroscience and embryology.

His deep understanding of the history of the field was integral to all of his scientific publications but became more explicit and extensive in the third edition of *Developmental Neurobiology* and in his *Foundations of Neuroscience* (Jacobson, 1993), a consideration of historical, epistemological and ethical aspects of neuroscience research.

Jacobson was a man of formidable energy and intellect who was adept at provoking his colleagues to think deeply about the ideas underlying their work. Although he readily adopted new methods into his own research program, he warned against a preoccupation with techniques and observations at the expense of hypotheses and models (Jacobson, 1993). Jacobson was a connoisseur and collector of Chinese art and he amassed an important collection of modern Chinese paintings that, along with his large collection of rare books on the history of embryology and neuroscience, has been donated to the University of Utah. He is survived by his wife and three adult children.

REFERENCES

Hunt, R.K. and Jacobson, M., 1974, Neuronal specificity revisited, *Curr. Top. Dev. Biol.* 8:203–259.

Jacobson, M., 1985, Clonal analysis and cell lineages of the vertebrate central nervous system, *Ann. Rev. Neurosci.* 8:71–102.

Jacobson, M., 1970, *Developmental Neurobiology*, Holt Rinehart & Winston, New York.

Jacobson, M., 1993, *Foundations of Neuroscience*, Plenum Press, New York.

This book is dedicated to the memory of
Marcus and to graduate students everywhere.

Marcus wanted the book to serve as an
introduction to this fascinating field and it is our hope that we have retained the
spirit of Marcus's third edition in this new revised version of his book.

Contributors

Eva S. Anton
Department of Cell and Molecular
  Physiology
University of North Carolina
  School of Medicine
Chapel Hill, NC 27599

Clare Baker
Department of Anatomy
University of Cambridge
Cambridge, CB2 3DY,
United Kingdom

Robert W. Burgess
The Jackson Laboratory
Bar Harbor, ME 04609

Chi-Bin Chien
Department of Neurobiology and
  Anatomy
University of Utah, SOM
Salt Lake City, UT 84132

Maureen L. Condic
Department of Neurobiology and
  Anatomy
University of Utah, SOM
Salt Lake City, UT 84132

Diana Karol Darnell
Assistant Professor of Biology
Lake Forest College
Lake Forest, IL 60045

Jean de Vellis
Mental Retardation Research Center
University of California, Los Angeles
Los Angeles, CA 90024

Richard I. Dorsky
Department of Neurobiology and
  Anatomy
University of Utah, SOM
Salt Lake City, UT 84132

James E. Goldman
Department of Pathology and the
  Center for Neurobiology and
  Behaviors
Columbia University College of
  Physicians and Surgeons
New York, NY 10032

N. L. Hayes
Department of Neuroscience and Cell
  Biology
UMDNJ-Robert Wood Johnson
  Medical School
Piscataway, NJ 08854

Marcus Jacobson[†]
Department of Neurobiology and
  Anatomy
University of Utah, SOM
Salt Lake City, UT 84132

Raj Ladher
Laboratory of Sensory Development
RIKEN Center for Developmental
  Biology
Chuo-Ku, Kohe, Japan

Steven W. Levison
Department of Neurology and
  Neuroscience
UMDNJ-New Jersy Medical School,
Newark, NJ 07101.

[†] Deceased

Tobi L. Limke
Laboratory of Neurosciences
National Institute on Aging Intramural
  Research Program
Baltimore, MD 21224

Mark P. Mattson
Laboratory Chief-Laboratory of
  Neurosciences
National Institute on Aging Intramural
  Research Program
Baltimore, MD 21224
and
Department of Neuroscience
Johns Hopkins University School of
  Medicine
Baltimore, MD 21224

Margot Mayer-Pröschel
Department of Biomedical Genetics
University of Rochester Medical Center
Rochester, NY 14642

Robert H. Miller
Department of Neurosciences
Case Western Reserve University School
  of Medicine
Cleveland, OH 44106

Mark Noble
Department of Biomedical Genetics
University of Rochester Medical Center
Rochester, NY 14642

R. S. Nowakowski
Department of Neuroscience and
  Cell Biology
UMDNJ-Robert Wood Johnson
  Medical School
Piscataway, NJ 08854

Bruce Patton
Oregon Health and Science University
Portland, OR 97201

Franck Polleux
Department of Pharmacology
University of North Carolina School of
    Medicine
Chapel Hill, NC 27599

Kevin A. Roth
Division of Neuropathology
Department of Pathology
University of Alabama at
    Birmingham
Birmingham, AL 35294-0017

Gary Schoenwolf
Professor, Department of Neurobiology
    and Anatomy

Director, Children's Health Research
    Center
University of Utah
Salt Lake City, UT 84132

Monica L. Vetter
Department of Neurobiology and
    Anatomy
University of Utah, SOM
Salt Lake City, UT 84132

Contents

[†] Deceased.

Making a Neural Tube: Neural Induction and Neurulation

Raj Ladher and Gary C. Schoenwolf

INTRODUCTION

As subsequent chapters will describe, the vertebrate nervous system is necessarily complex. However, this belies its humble beginnings, segregating relatively early as a plate of cells in the dorsal ectoderm of the embryo. This process of segregation, termed neural induction, occurs as a result of instructive cues within the embryo and is described in this chapter. Once induced, the neural plate, in most vertebrates, rolls into a tube during a process known as neurulation. This tube is then later elaborated to form the central nervous system. In this chapter, we describe the model for how ectodermal cells become committed to a neural fate, and the studies that have led to this model. We will then review the mechanisms by which the induced neural ectoderm rolls up to form the neural tube.

SETTING THE SCENE

In this section, we describe some of the fundamental events that occur in embryogenesis prior to neural induction. We also introduce the main vertebrate model organisms used to investigate neural induction, and we discuss their strengths and appropriateness for various types of experimental studies.

Neural induction, the process by which a subset of the ectoderm is instructed to follow a pathway leading to the formation of the nervous system, has been studied in model systems comprising four classes of vertebrates. Despite obvious differences in the geometry of the embryos of these classes (e.g., the early frog embryo is spherical, whereas the early chick embryo is a flat disc), by and large their embryogenesis is comparable, and researchers can use the respective strengths of these models to address experimentally very specific research questions. By synthesizing data that have emerged from these studies, a model has been formulated of how neural tissue is induced.

Model Organisms

Four vertebrate model systems have been used extensively to study neural induction (Fig. 1). Two of these are classified as lower vertebrates—zebrafish and *Xenopus*—and two are classified as higher vertebrates—chick and mouse. Two major differences exist between lower and higher vertebrates. First, lower vertebrates lack an extraembryonic membrane called the amnion, which was developed by higher vertebrates as an adaptation to terrestrial life. Thus, lower vertebrates are anamniotes and higher vertebrates are amniotes. Second, true growth (i.e., cell division followed by an increase in cytoplasm in each daughter cell to an amount comparable to the parental cell—in contrast to cleavage, where cells get progressively smaller with division) is minimal during morphogenesis in lower vertebrates, but plays an integral role in morphogenesis of higher vertebrates. In addition to these differences between lower and higher vertebrates, another major difference exists among the four model organisms: the relationship between the formative cells of the embryo and their food source. Namely, *Xenopus* eggs contain a large internal store of yolk. With cleavage of the egg to form the spherical blastula, this yolk is incorporated into the forming blastomeres with the vegetal blastomeres being much larger than, and containing much more yolk than, the animal blastomeres. In both zebrafish and chick, a blastoderm forms as a disc on top of the yolk mass. Finally, in the mouse egg, yolk is sparse; rather the embryo receives its nourishment from the mother, initially by simple diffusion and later through the placenta. These differences in the amount and distribution of yolk in the eggs of the four vertebrate models result in very different geometries in the four organisms. Thus, during the early developmental stages of cleavage, gastrulation, neural induction, and neurulation, the four model organisms appear very different from one another, yet developmental mechanisms at the tissue, cellular, and molecular–genetic levels are highly conserved.

In *Xenopus* and zebrafish, early development is directed by maternal products laid down during oogenesis; at the

Raj Ladher • Laboratory of Sensory Development, RIKEN Center for Developmental Biology, Chuo-Ku, Kobe, Japan.   Gary C. Schoenwolf • Department of Neurobiology and Anatomy, and Children's Health Research Center, University of Utah, Salt Lake City, UT, 84132.

Developmental Neurobiology, 4th ed., edited by Mahendra S. Rao and Marcus Jacobson. Kluwer Academic / Plenum Publishers, New York, 2005.

mid-blastula transition, or MBT, zygotic transcription commences (Newport and Kirschner, 1982; Kane and Kimmel, 1993). Maternally provided products are important in axis formation and germ layer identity. In chicks and mice, "MBT," or the onset of zygotic transcription, occurs soon after fertilization; thus, the exact role of maternal products in early development has been difficult to decipher.

The *Xenopus* Embryo

A large body of literature exists on the development of the amphibian embryo. Indeed, two of the most important findings regarding the embryogenesis of the vertebrate nervous system—the discovery of the organizer and the elucidation of its role in neural induction (Spemann and Mangold, 1924, 2001) and the discovery of the molecular mechanisms of neural induction (Sasai and De Robertis, 1997; Nieuwkoop, 1999; Weinstein and Hemmati-Brivanlou, 1999)—were obtained using amphibian embryos. These will be discussed later in this chapter. The class itself can be split into the Anurans (frogs and toads) and the

Urodeles (newts and salamanders), and despite some differences in the details of their development, the many similarities make it possible to generalize the results and extend them to other organisms. Although the Anuran, *Xenopus*, is the model most used today, the starting point for most studies was the pivotal work performed in Urodeles by Spemann and Mangold in the course of discovering the organizer (Spemann and Mangold, 2001). For a summary of the differences between Anurans and Urodeles, see the excellent review by Malacinski *et al.* (1997). For a schematic view of key phases of early *Xenopus* development, see Fig. 2.

The amphibian embryo is large, easily obtained, readily accessible, and easily cultured in a simple salt solution. As all cells of the embryo have a store of yolk, pieces of the embryo and even single cells from the early embryo (i.e., blastomeres) can be cultured in simple salt solution. A recent advantage in the use of *Xenopus* is the ability to overexpress molecules of interest. Because early blastomeres are large, it is a simple matter to make RNA corresponding to a gene of interest and inject it into selected cells. The injected RNA is translated at high efficiency

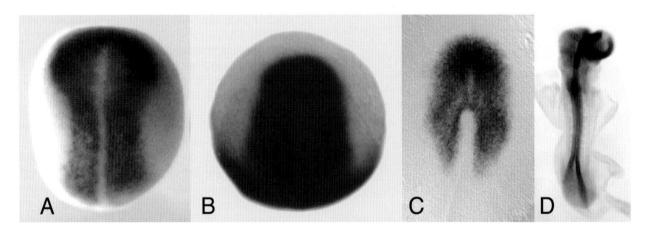

FIGURE 1. Photographs showing the locations of the neuroectoderm at neurula stages in (A) *Xenopus* (dorsal view, immunohistochemistry for N-CAM at stage 15; courtesy of Yoshiki Sasai); (B) zebrafish (dorsal view, *in situ* hybridization for *Sox-31* at tail bud stage; courtesy of Luca Caneparo and Corinne Houart); (C) chick (dorsal view, *in situ* hybridization for *Sox-2* at stage 6; courtesy of Susan Chapman); and (D) mouse (dorsolateral view, *in situ* hybridization for *Sox-2* at 8.5 dpc; courtesy of Ryan Anderson, Shannon Davis, and John Klingensmith).

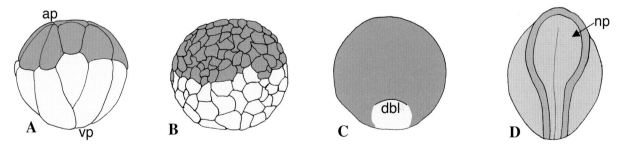

FIGURE 2. *Xenopus* development leading up to neurulation. Diagrams of embryos at the (A) morula, (B) blastula, (C) gastrula, and (D) neurula stages of development. Once the egg is fertilized, cleavage occurs, with the cells of the animal hemisphere darker and smaller than cells of the vegetal hemisphere. At blastula stages, mesoderm is induced. In particular, dorsal mesoderm is specified and at gastrula stages, this mesoderm starts to involute, forming the dorsal blastoporal lip and marking the site of the organizer. The organizer induces neural tissue in the overlying animal hemisphere. ap, animal pole; dbl, dorsal blastoporal lip; np, neural plate; vp, vegetal pole. Modified from Nieuwkoop and Faber (1967).

and is active. Indeed this technique has been used not only to assay a whole molecule, but also modified (i.e., systematically and selectively mutated) versions of the gene.

As most developmental biology research in amphibians is performed on the *Xenopus* embryo, we will consider its development. Smith (1989) provides an excellent synthesis of the early embryological events that occur prior to neural induction.

The *Xenopus* egg has an animal–vegetal polarity, with the darker (i.e., more heavily pigmented) animal hemisphere forming the ectoderm and mesoderm, and the lighter vegetal, yolk-rich hemisphere forming the endoderm. Fertilization imparts an additional asymmetry on the egg, with the sperm entering the animal hemisphere. The sperm entry point also determines the direction of rotation of the cortex of the egg in relation to the core cytoplasm, and this activates a specific pathway leading ultimately to the establishment of the dorsal pole of the embryo (Vincent and Gerhart, 1987; Moon and Kimelman, 1998). Specifically, the region of the vegetal hemisphere, the Nieuwkoop center, which is diametrically opposite the sperm entry point, is now conferred with the ability to induce the Spemann organizer in the adjacent animal hemisphere (Boterenbrood and Nieuwkoop, 1973). The Spemann organizer has the ability to induce dorsal mesoderm and pattern the rest of the mesoderm, as well as to direct the formation of the neuroectoderm (Gimlich and Cooke, 1983; Jacobson, 1984; see below and Box 1).

Following fertilization, mesoderm is induced in the equatorial region of the embryo, at the junction between the animal and vegetal poles (Nieuwkoop, 1969). Amazingly, this induction has been experimentally recreated to great effect in later assays for both mesoderm-inducing signals and neural-inducing signals. When challenged with the appropriate signal, an isolated piece of *Xenopus* animal tissue, which would normally form epidermal structures, will change its fate accordingly. This animal cap assay has, for years, provided researchers with a powerful assay for induction. One important caveat must be noted here though. Barth (1941) found that the animal cap of the amphibians *Ambystoma mexicanum* and *Rana pipiens*, amongst others, autoneuralizes; that is, the removal of the presumptive epidermis from its normal environment actually changes its fate to neural, a result supported and extended by Holtfreter (1944), who among other things showed that neural induction could occur even after the inducer had been killed (Holtfreter, 1947). This result could only be contextualized years later when the pathway for neural induction was worked out (see below). It should be noted here, however, that the animal cap of *Xenopus* does not show such auto-neuralization; indeed as we discuss below, the *Xenopus* animal cap is resistant to nonspecific neural induction by diverse agents (Kintner and Melton, 1987). This resistance to nonspecific neural induction strengthened the role of *Xenopus* embryos in the search for inducing signals.

Neural induction occurs during the process of gastrulation when the mesoderm and endoderm invaginate through the blastopore and, via a set of complex morphological movements (see Keller and Winklbauer, 1992, for details of this process), are internalized. This results in the ectoderm remaining on the surface and forming the crust, and the mesoderm and endoderm coming to lie deep to the ectoderm, forming the core. A fuller description of neural induction is given below.

The Zebrafish Embryo

Two large-scale mutagenesis screens propelled the zebrafish embryo to the forefront of developmental biology (Mullins and Nusslein-Volhard, 1993; Driever, 1995). The combination of

---

BOX 1. The Organizer

The discovery of the organizer in 1924 is one of the major milestones in developmental biology. This discovery has had a major influence on our thinking about the mechanisms underlying neural induction (Spemann and Mangold, 1924). The German scientists, Hans Spemann and Hilda Mangold, discovered that a region of the amphibian gastrula, the dorsal lip of the blastopore, had the ability to direct formation of the neural plate (Fig. 3A). By transplanting the dorsal lip from a donor embryo to the ventral side of a host embryo, they found that a second axis can be initiated. The experiment was performed using salamander embryos, not *Xenopus*, the current favorite amphibian model. By using two species of salamander, one pigmented and the other unpigmented, Spemann and Mangold could identify which structures in the duplicated axis were derived from the donor and which were derived from the host. Careful analysis showed that whereas the secondary notochord and parts of the somites were derived from the donor dorsal lip, the neural plate and other regions of the somites within the secondary axis were derived from the host. As host tissues should have been fated to form ventral derivatives, such as lateral mesoderm and epidermal ectoderm, Spemann and Mangold reasoned that the action of the donor dorsal tissue was not autonomous, and that a nonautonomous action induced the surrounding tissues to take on a dorsal fate. By using a classical definition of the word "induction"—the action of one tissue on another to change the latter's fate, Spemann and Mangold defined neural induction in vertebrate embryos and localized its center of activity.

As mentioned above, the action of an organizer is not just limited to amphibian embryos. A large number of studies have extended the findings of Spemann and Mangold to embryos of the fish, bird, and mammal (Waddington, 1934; Oppenheimer, 1936; Beddington, 1994; Fig. 3B). All of these studies have found that the organizer can induce the formation of a secondary axis. However, in the mouse, there is an important difference. Whereas in the fish, frog, and chick, transplantation of the organizer can induce a secondary axis with all rostrocaudal levels (i.e., from the forebrain to the caudal spinal cord), transplantation of the node in the mouse can induce only a supernumerary axis that begins rostrally at the level of the hindbrain (Beddington, 1994; Tam and Steiner, 1999). This has led to the identification of a second organizing center, the anterior visceral endoderm (Thomas and Beddington, 1996; Tam and Steiner, 1999). Using a series of transplants, it has been found that the anterior visceral endoderm, unlike the node of the mouse, cannot induce neural tissue. Instead, it provides a patterning activity, imparting rostral identity upon already induced neuroectoderm. As this is beyond the scope of this chapter, the anterior visceral endoderm will be more appropriately covered in greater detail in Chapter 3 on neural patterning.

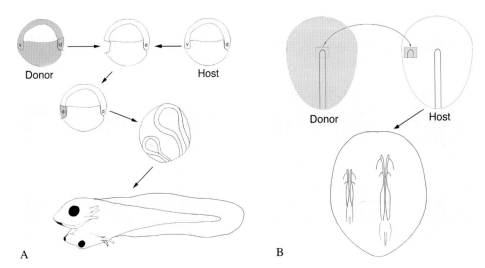

FIGURE 3. Axis duplication in (A) amphibians and (B) the chick after transplantation of the organizer regions of these embryos to ectopic locations. Details of the experiments are given in the main text. Transplantation of the dorsal lip (in amphibians) or Hensen's node (in chick) gives rise to a duplicated neuroaxis, derived from host tissue. This experiment mapped the site of neural induction to the organizer. d, dorsal; v, ventral. (A), modified from Spemann and Mangold (1924); (B), modified from Waddington (1932).

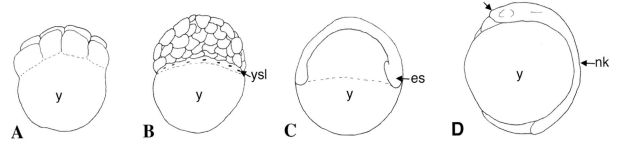

FIGURE 4. Zebrafish development leading up to neurulation. Diagrams of embryos at (A) morula, (B) blastula, (C) gastrula, and (D) neurula stages. The zebrafish embryo floats on top of the yolk (y), a situation that is not changed until gastrulation. At blastula stages, a belt of cells is formed at the junction between the embryo and the yolk; it is known as the yolk syncytial layer (ysl). This induces the formation of the mesoderm and also directs the formation of the embryonic shield (es), the organizer of the fish embryo. The embryo shield also induces the formation of neural ectoderm (i.e., the neural keel, nk). Arrow indicates the head end of the embryo. Modified from Langeland and Kimmel (1997).

generating mutants, cloning the affected genes and using traditional embryological techniques has made the zebrafish embryo especially attractive to researchers. For a schematic view of key phases of early zebrafish development, see Fig. 4.

Fertilization causes the segregation of the cytoplasm from the yolky matter in the egg, resulting in a polarity manifested by the presence of a transparent blastodisc on top of an opaque yolky, vegetal hemisphere (Langeland and Kimmel, 1997). Cell division increases the number of cells, forming the blastoderm, and at the 256-cell stage, the first overt specialization occurs within the blastoderm. The most superficial cells of the blastoderm form an epithelial monolayer, known as the enveloping layer, confining the deeper cells of the blastoderm. At around the tenth cell division, the cells at the vegetal edge of the enveloping layer of the blastoderm fuse with the underlying yolk cell. Interestingly, the tenth cell cycle marks the MBT for the zebrafish embryo. A belt of nuclei, the yolk syncytial layer (YSL), resides within the yolk cell cytoplasm just under the blastoderm. It provides a motive force for gastrulation, and it has been postulated also to function in establishing the dorsal–ventral axis of the zebrafish (Feldman *et al.*, 1998).

The initial phase of gastrulation is marked by the blastoderm flattening on top of the yolk. This causes the embryo to change from dome-shaped to spherical, and it results from the process of epiboly: the spreading of the blastoderm over the yolk hemisphere. The YSL drives epiboly, pulling the enveloping layer with it. The process has been likened to "pulling a knitted ski hat over one's head" (Warga and Kimmel, 1990). At about 50% epiboly, that is, when the blastoderm has covered half of the yolk hemisphere, the germ ring forms. This is a bilayered belt of cells: The upper layer is the "epiblast," whereas the lower layer is the "hypoblast." The lower layer forms by involution; that is, as the deeper cells of the blastoderm are driven superficially toward the vegetal margin, they fold back under and migrate toward the animal pole. At the same time, there is a movement of deep blastoderm cells toward the future dorsal side of the embryo. This creates a thicker region in the germ ring, marking the organizer of the zebrafish, a structure known as the embryonic shield. Similar to the situation in amphibia, this structure can be transplanted to the ventral side of a host fish embryo, where it induces the formation of a secondary axis (Oppenheimer, 1936; Box 1). As gastrulation proceeds and the body plan becomes clearer, the neural primordium becomes apparent as a thickened monolayer of cells. The mechanisms by which this happens will be discussed in detail later in this chapter.

The Chick Embryo

Chick eggs are readily available and embryos are easily accessible throughout embryogenesis. Embryos readily tolerate manipulation such as microsurgery. As a result of these attributes, the chick embryo has long been a favorite organism for experimental embryology. For a schematic view of key phases of early chick development, see Fig. 5.

After the egg is fertilized, which occurs within the oviduct of the hen, shell components are added during the day-long journey through the oviduct prior to laying. Cleavage begins immediately after fertilization, and by the time the egg is laid, it contains a bilaminar blastoderm floating on the surface of the yolk (Schoenwolf, 1997). The upper layer of the bilaminar blastoderm is termed the epiblast, whereas the lower layer (i.e., the one closest to the yolk) is termed the hypoblast. The epiblast gives rise to all of the tissue of the embryo proper, that is, the ectodermal, mesodermal, and endodermal derivatives. The hypoblast is displaced during embryogenesis and will contribute to extraembryonic tissue.

Like the fish embryo, the region of the chick egg that gives rise to the embryo proper floats on top of a yolky mass.

During cleavage, the blastoderm becomes 5–6 cells thick and is separated from the yolk by the subgerminal cavity. The deep cells in the central portion of the disc are shed, leaving the monolaminar area pellucida. This region of the blastoderm will give rise to the definitive embryo. The peripheral ring of cells, where the deeper cells have not been shed, is the area opaca. This region, in conjunction with the peripheral part of the area pellucida, will give rise to the extraembryonic tissues. Many of the extraembryonic tissues will eventually cover the entire yolk, providing the embryo with nourishment during development. At the border between the area opaqua and area pellucida at the time of formation of these two regions is a specialized ring of cells, the marginal zone. This zone plays an important role in establishing the body axis of the embryo (Khaner and Eyal-Giladi, 1986; Khaner, 1998; Lawson and Schoenwolf, 2001).

Shortly after the formation of the area pellucida, some of the cells in this region delaminate and form small polyinvagination islands beneath the outer layer (the epiblast). These cells flatten and join to form a structure known as the primary hypoblast. Within the caudal marginal zone, a sickle-shaped structure appears called Koller's sickle; it gives rise to a sheet of cells, called the secondary hypoblast, which migrates rostrally, joining the primary hypoblast. This results in an embryo with two layers—the uppermost layer epiblast and the lowermost hypoblast. These layers are separated from the yolk by a fluid-filled space called the blastocoel.

Once the egg is laid, further development requires incubation at about 38°C. After about 4 hr of incubation, the first signs of gastrulation become apparent. The cells of the hypoblast begin to reorganize in a swirl-like fashion, termed a Polinase movement. Viewed ventrally, that is, looking down on the surface of the hypoblast, the cells of the left side of the hypoblast move counterclockwise, whereas those on the right side move clockwise. Concomitantly, epiblast cells as they extend rostromedially

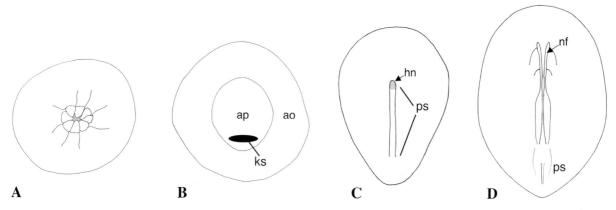

A   **B**   **C**   **D**

FIGURE 5. Chick development leading up to neurulation. Diagrams of embryos at (A) morula, (B) blastula, (C) gastrula, and (D) neurula stages; the blastoderm is shown removed from the yolk and viewed from its dorsal surface. At the time that the chick egg is laid, a multicellular blastoderm floats upon the yolk. The blastoderm is subdivided into an inner area pellucida (ap) and an outer area opaca (ao), with Koller's sickle (ks) marking the caudal end of the blastoderm. The ao forms the extraembryonic vasculature, providing nutrition for the growing embryo. By blastula stages, the central portion of the embryo is two cell layers thick: the upper epiblast will form all of the structures of the adult; the lower hypoblast will contribute to extraembryonic tissues. The primitive streak (ps) forms in the epiblast of the embryo, and the mesoderm and definitive endoderm ingress through it and into the interior. The primitive streak extends rostrally and once it has reached its maximal length, it forms a knot of cells known as Hensen's node (hn; shaded). This is the organizer of the chick embryo; it is responsible for neural induction. Shortly after neural induction, the embryo undergoes neurulation. nf, neural folds. Modified from Schoenwolf (1997).

from Koller's sickle begin to pile up at the caudal of the midline of the area pellucida. These cells accumulate as a wedge, with the base of the wedge at the caudal end and the apex pointing along the midline rostrally. This wedge-like structure is the initial primitive streak, the equivalent to the blastopore lip in the frog and the embryonic shield in fish, that is, the structure through which cells of the epiblast will ingress to give rise to mesoderm and definitive endoderm. It forms just rostral to Koller's sickle, and this has led to the belief that Koller's sickle acts in much the same way as the Nieuwkoop center in *Xenopus* (Callebaut and Van Nueten, 1994). As development progresses, the streak elongates reaching a maximal length at about 18 hr of incubation. As the streak reaches its maximal length, its rostral end forms a knot of cells called Hensen's node. Hensen's node is the embryological equivalent of the dorsal lip in *Xenopus* and the embryonic shield in zebrafish; that is, Hensen's node is the organizer of the avian embryo (Waddington and Schmidt, 1933; Waddington, 1934). The role of Hensen's node in neural induction is discussed further in Box 1.

The Mouse Embryo

The mouse, being a mammal, has an embryo that should be highly relevant for understanding development of the human embryo. Nevertheless, there are some caveats that make this model less than ideal. The fact that mouse development occurs within the maternal uterus and that the embryo is highly dependent upon its mother for respiration, nutrition, and the removal of its waste products makes the embryo relatively unsuitable for the kinds of embryological experimentation that have characterized research on the other three model systems discussed above. Early development of the mouse embryo also is peculiar in that unlike the other three model organisms, the gastrula stage of the mouse develops "inside-out"; that is, with its ectoderm on the "inside" and its endoderm on the "outside." For a schematic view of key phases of early mouse development, see Fig. 6.

Recent advances in whole-embryo culture have substantially increased the value of the mouse embryo for experimental embryology. Consequently, cutting- and pasting-type experiments in the mouse embryo are becoming increasingly common. However, it is in the realm of genetic analysis that the mouse embryo has excelled as a model organism. The ability to remove genes, to place genes into an unnatural context and to elucidate the genetic controls that genes are subject to, has advanced developmental biology considerably. These molecular genetic techniques are introduced in this chapter where necessary; for further information, the reader is directed to several excellent reviews (Capecchi, 1989; Rossant *et al.*, 1993; Soriano, 1995; St-Jacques and McMahon, 1996; Beddington, 1998; Osada and Maeda, 1998; Stanford *et al.*, 2001). In the subsequent section, we discuss development of the mouse up to the stage when neural induction occurs.

The mouse oocyte is released into the oviduct from the ovary and it is in the ampulla of the oviduct that fertilization occurs (Cruz, 1997). Cleavage begins as the oocyte passes down the oviduct toward the uterus. It should be noted that cleavage

occurs within the confines of the zona pellucida, the covering of the oocyte. The zona plays an important role in regulating the site (and time of) implantation in that until the embryo hatches from the zona pellucida, the embryo cannot implant. If the embryo hatches too early, then implantation can occur in the oviduct, resulting in an ectopic pregnancy.

After the third cleavage, that is, after the eight-cell stage, the conceptus transforms from a group of loosely arranged blastomeres called a morula (Latin for mulberry) to a mass of flattened and tightly interconnected cells. This change is referred to as compaction. As a result of compaction, the blastomeres flatten against each other at the surface of the morula, maximizing their contact with one another, and a blastocoel appears within the morula. As the blastocoel is forming, a small group of internal cells appears, known as the inner cell mass, surrounded by external cells, known as the trophoblast. With formation of the inner cell mass and trophoblast, the morula is converted into the blastocyst. Formation of these two cell types constitutes the first lineage restriction that occurs in mouse development, with cells of the trophoblast eventually forming the chorion—the embryonic portion of the placenta—and those of the inner cell mass forming the embryo proper and some associated extraembryonic tissue.

By the 64-cell stage, a large blastocoel has formed and the inner cell mass is displaced to one side of the blastocyst. There is now polarity to both the inner cell mass (a blastocoel-facing side and a trophoblast-facing side) and the trophoblast (the polar trophoectoderm in contact with the inner cell mass and the opposite side, not in contact with the inner cell mass, the mural trophoectoderm). This polarity plays an important role in subsequent development. The cells of the inner cell mass that face the blastocoel flatten and partition themselves from the remainder of the inner cell mass. These cells eventually form an epithelium and represent the murine hypoblast or primitive endoderm. The remaining cells within the inner cell mass become the primitive ectoderm or the epiblast. The cells of the primitive endoderm divide and some of the progeny migrate to cover the surface of the mural trophoectoderm, where they are known as the parietal endoderm. The cells of the primitive endoderm that remain in contact with the inner cell mass constitute the visceral endoderm.

By 5 days after fertilization (referred to as 5 days post coitum or 5 dpc), the blastocyst hatches from the zona pellucida and implants into the uterine wall. During this time the polar trophoectodermal cells have accumulated to form a pyramidal mass of cells. The outermost surface of the mass (i.e., the surface that faces the uterine wall) invades the uterine wall, forming the ectoplacental cone; the remainder of the polar trophoectoderm forms the extraembryonic ectoderm, namely, the ectoderm of the chorion. Cells of the mural trophoectoderm also invade the uterine walls, leaving behind the parietal endoderm. The latter becomes adherent to a thickened basement membrane called Reichart's membrane. At this stage in development, the endoderm of the embryo proper encases an epiblastic core; during subsequent turning of the embryo, this configuration is reversed, so that the ectoderm comes to lie on the outside of the embryo and the endoderm, on the inside, the typical situation present in the other vertebrate model organisms.

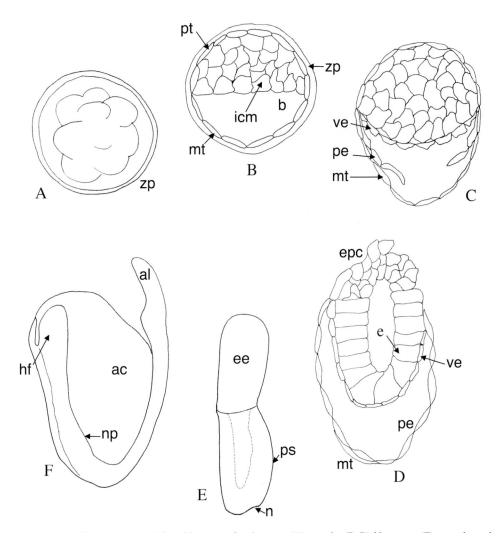

FIGURE 6. Mouse development leading up to neurulation. Diagrams of embryos at (A) morula, (B–D) blastocyst, (E) gastrula, and (F) neurula stages. Once fertilized, the mouse embryo cleaves within the confines of the zona pellucida (zp), an extracellular membrane important in preventing premature implantation and lost at the blastocyst stage (C). At the third cell division, the cells of the embryo undergo compaction to form the morula (A). With formation of the blastocyst (B), the inner cell mass (icm) and trophoblast can be identified; the latter becomes subdivided into mural trophectoderm (mt) and polar trophectoderm (pt). The inner cell mass will form the embryo proper, as well as contribute to the extraembryonic tissue. The cells of the inner mass that face the blastocoel (b) form the hypoblast or primitive endoderm. The latter gives rise to the visceral endoderm (ve) and parietal endoderm (pe; C). The remaining cells of the inner cell mass form the epiblast (D). By the late blastocyst stage (D), the epiblast has cavitated and now forms a cylindrical structure encased in visceral endoderm; the composite is known as the egg cylinder. The polar trophectoderm now forms a structure known as the ectoplacental cone (epc). The primitive streak (ps) of the mouse is initiated at the caudal end of the egg cylinder, and like the chick primitive streak, it is the site of ingression of cells that will form the mesoderm and definitive endoderm (E). The streak extends rostrally and eventually forms a knot of cells, known as the node (n), the organizer of the mouse embryo. To view embryos at this stage, the trophoblast is typically removed revealing the extraembryonic ectoderm (ee) and cup-shaped blastoderm containing epiblast on the inside of the cup and endoderm on the outside (E). At neurula stages (F), the neural plate (np) has formed and the body plan is apparent. The neural folds jut forward as the head folds (hf). Two extraembryonic membranes are visible at this stage: the amnion and allantois (al). The former encloses the developing embryo within the amniotic cavity (ac). Modified from Cruz (1997).

As implantation is occurring, the epiblast (i.e., the primitive ectoderm) cavitates to form the amniotic cavity, and growth transforms the conceptus into the egg cylinder. It is likely that the constraints of the uterine wall cause the epiblast (and adherent visceral endoderm) to assume this shape, reminiscent of a round-bottomed shot glass. During gastrulation, the epiblast will give rise to the embryo proper and also to the extraembryonic mesoderm (of the allantois and chorion).

Gastrulation of the mouse embryo commences with the formation of the primitive streak, at around 6 dpc, in the epiblast. It is during these stages that similarities with chick gastrulation become apparent. Like in the chick embryo, epiblast cells migrate through the primitive streak to form the mesoderm and definitive endoderm. As development proceeds, the streak elongates until, at 7.5 dpc, it reaches its maximal length. The distal tip of the streak is known as the node, the equivalent of Hensen's node in the chick, the dorsal lip in amphibians and the embryonic shield in fish; the node shares many of the same properties as the organizer in the other models and as such, it constitutes the murine organizer (Beddington, 1994; see also Box 1). The cells that migrate through the node become axial tissues, whereas those emanating from the rostral streak just caudal to the node give rise to paraxial mesoderm and endoderm. The definitive endoderm, as in the chick, displaces the hypoblast/visceral endoderm rostrally during its formation. The rostral displacement of the visceral endoderm plays an important role in the patterning of the embryo, which is more fully described in the subsequent chapter, with the anterior visceral endoderm acting in the generation of the forebrain (Thomas and Beddington, 1996), and the node acting in the induction of the neural plate caudal to the level of the midbrain.

NEURAL INDUCTION

The identification of the organizer prompted a vigorous search for the biochemical nature of the neural-inducing signal, a quest that has lasted over 75 years. In the intervening period, studies were undertaken to address the nature of the inducing signal. Unsurprisingly, virtually all of the work was performed in amphibian embryos; their heritage, ease of culture, and establishment (through the work of Spemann and Mangold) of a simple assay for neural induction made the choice straightforward.

One of the main controversies was whether the induction signal acted vertically, emanating from the involuted dorsal mesoderm and acting upon the overlying ectoderm, or whether the signal acted in the plane of the ectoderm, emanating from the dorsal ectoderm prior to its involution into the interior of the embryo during gastrulation. Spemann's subsequent experiments suggested that the vertical signaling predominated. Using the "einsteckung" method, he inserted the organizer into the blastocoel of the embryo, finding that a secondary axis could be induced (Geinitz, 1925). Extending these results, he found that whereas dorsal mesoderm was able to induce a secondary axis, dorsal ectoderm could not (Marx, 1925). In subsequent experiments, Holtfreter found that when the animal ectoderm was wrapped around pieces of notochord, neural tissue was induced (Holtfreter, 1933a). Similar experiments in the chick (Smith and Schoenwolf, 1989; van Straaten et al., 1989) showed that the notochord acts vertically on the overlying ectoderm. This strengthened the argument for vertical signals emanating from the dorsal axial tissue. Holtfreter also devised an experimental scheme unique to amphibian embryos (Holtfreter, 1933b). When blastulae are placed in a high salt solution, cells do not involute into the interior during gastrulation; instead, they expand outward to form what is known as an exogastrula—a mass of mesoderm and endoderm attached to an empty sac of ectoderm. In such cases, vertical signals cannot occur, as the two tissues are never juxtaposed vertically. Holtfreter found that no morphologically recognizable neural tissue was present in exogastrulae, indicative of the need for vertical signaling. This experiment has revisited using molecular markers. Kintner and Melton (1987),

using Xenopus embryos, found that although the neural tissue was not morphologically apparent, neural markers such as N-CAM could be detected. This led to the argument that a planar signal initiated neural induction. An alternative explanation is that the dorsomost mesoderm and endoderm of Xenopus is placed under the dorsal blastopore lip during pre-gastrula movements; thus, these cells are in a position to signal vertically even in exogastrulae (Jones et al., 1999). Unfortunately, there are currently little data distinguishing planar from vertical signaling in amniotes; however, the current thinking is that both modes of neural induction can occur.

Although much headway has been made into the identification of the tissues producing the neural-inducing signal, as well as the timing of neural induction, the identity of the inducing signal remained elusive. In early studies, it was discovered that neural induction could be initiated by a variety of tissues, ranging from the extract of a fish swim bladder to guinea pig bone marrow (Grunz, 1997). This proved quite exciting; perhaps, it would be easier to purify the signal from adult tissue, which was present in far greater mass and lacked yolk, which made amphibian tissues difficult for biochemical purification studies. Tiedemann showed that the phenol phase of an extract of an 11-day chick embryo was able to neuralize animal caps, demonstrating that proteins were the likely candidate for the inducing signal (Tiedemann and Tiedemann, 1956). Saxén (Saxén, 1961) and Toivonen (Toivonen and Wartiovaara, 1976) separated organizers juxtaposed to animal caps by using filters that excluded cell–cell contact. Their results showed that neuralization could still occur in the absence of direct cell–cell contact, indicating that the responsible protein was diffusible.

This is not quite the case in Xenopus. The Xenopus animal cap is resistant to induction by "nonspecific" neural inducers (Kintner and Melton, 1987), and it is also resistant to autoneuralization; however, these attributes have been more of an asset than a liability, as Xenopus tissues allow a more stringent test of candidate neural inducers. Thus, most modern studies on the molecular nature of the neural-inducing substance have used this amphibian and have relied heavily on the animal cap assay (Fig. 7).

The Default Pathway

As discussed below, neural fate is a default state, resulting from an inhibition of a non-neural fate within the ectoderm. There are some layers of complexity, but the majority data that have been gathered so far points to an inhibition of the inducing signal for the non-neural ectoderm. This is clearly true for amphibian (Xenopus) neural induction. However, the case for antagonistic signals inducing the nervous system of chickens and mice is less clear.

An indication that the neural fate may be a default one in the amphibian came from a number of studies where the Xenopus blastula animal cap was dissociated into single cells (Godsave and Slack, 1989; Grunz and Tacke, 1989; Sato and Sargent, 1989). By culturing the animal cap in media free of calcium and magnesium ions, the animal cap dissociates into a suspension of cells. If the ions are immediately added back, the animal cap cells

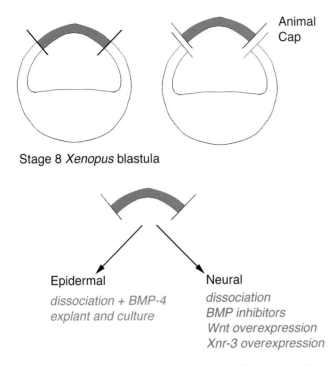

Animal
Cap

Stage 8 *Xenopus* blastula

Epidermal

*dissociation + BMP-4
explant and culture*

Neural

*dissociation
BMP inhibitors
Wnt overexpression
Xnr-3 overexpression*

FIGURE 7. Neuralization of the *Xenopus* animal cap. Shown are the effectors required to cause the isolated animal cap of a blastula-staged *Xenopus* embryo to change its fate from epidermal to neural. Modified from Wilson and Edlund (2001).

reassociate and form epidermis, similar to the intact cap. If the reassociation is delayed, the fate of the animal cap cells once they are reassociated is neural. These results suggested that intact blastula animal caps had an activity that maintained non-neural character, an activity that was diluted out during dissociation. Grunz also made the finding that this activity was located in the extracellular matrix (Grunz and Tacke, 1990).

Noggin was first isolated as an activity able to rescue dorsal development in *Xenopus* embryos that had been ventralized by UV irradiation of the vegetal pole (Smith and Harland, 1992). Using *in situ* hybridization, noggin was found to be expressed first in the dorsal mesoderm and later in the notochord of the embryo. Both places had already been defined as sites of the neural-inducing signal. That the molecule was secreted, made its involvement in neural induction more likely. This role was confirmed when Lamb and Harland incubated *Xenopus* animal caps in a simple salt solution containing purified noggin protein (Lamb *et al.*, 1993). These caps changed their fate from epidermis to neural. What made the activity of noggin unique was that it was able to directly induce the animal cap to become neural, without the concomitant induction of mesoderm. The induction of mesoderm and neural tissue had already been described for activin, a member of the TGF-β family (Box 2). In fact, the next neural inducer identified was a known inhibitor of activin activity, follistatin (Hemmati-Brivanlou *et al.*, 1994). Like noggin, it was able to directly induce neural tissue in animal caps. The fact

---

BOX 2. The BMP Signaling Pathway

BMP-2 and BMP-4 are members of the TGF-β superfamily, a group with a large number of members and with diverse functions during development. The transduction pathway of these genes has become well known and what follows is a simplified description of the components of the pathway. For a more in-depth review of the transduction pathway, the reader is directed to a number of excellent reviews on the subject (Massagué and Chen, 2000; von Bubnoff and Cho, 2001; Moustakas and Heldin, 2002; Fig. 8).

Transduction of the BMP signal involves two kinds of serine/threonine receptors, the type 1 and type 2. The ligand binds preferentially to the type 1 receptor, causing a conformational change that allows the association of the type 2 receptor. The juxtaposition of the type 2 receptors results in its phosphorylation of the type 1 receptor within the key glycine/serine (GS-rich) domain (Wrana *et al.*, 1994). The phosphorylation of the type 1 receptor causes the recruitment of Smad to the plasma membrane (Liu *et al.*, 1996). There are a number of Smad molecules in the cell, and they form two distinct classes (Attisano and Tuen Lee-Hoeflich, 2001). The receptor-regulated Smad or R-Smads, associate with the type 1 receptor via an adaptor protein, Smad Anchor for Receptor Activation (SARA) (Tsukazaki *et al.*, 1998). In fact, the R-Smads themselves can be split into two subclasses; Smad2 and Smad3 transduce responses elicited by activin or TGF-β signals, whereas Smad1, Smad5, and Smad8 generally transduce the BMP response (Attisano and Tuen Lee-Hoeflich, 2001). The association between Smad and the type 1 receptor results in the serine phosphorylation of the R-Smad, releasing it from the SARA/type 1 receptor complex. The phosphorylated R-Smad can now associate with the second class of Smads, the Co-Smad, usually Smad4, or additionally in *Xenopus*, Smad10. The R-Smad/Co-Smad complex results in the

nuclear translocation of these molecules (Lagna *et al.*, 1996). Once in the cytoplasm, the Smads complex acts as coordinators for the assembly of a number of transcription factors and thereby modulates the transcription of specific genes.

The BMP signal transduction pathway is also subjected to intracellular antagonism, an aspect that provides negative feedback for BMP activity. As well as the R-Smads that are responsible for activating BMP responsive genes, there are at least two inhibitory Smads (I-Smads), Smad6 and Smad7, which associate with the type 1 receptor to prevent the binding of the R-Smad/SARA complex (Imamura *et al.*, 1997; Tsuneizumi *et al.*, 1997; Inoue *et al.*, 1998; Souchelnytskyi *et al.*, 1998). It seems that the expression of I-Smad is induced by BMP activity itself (Nakao *et al.*, 1997; Afrakhte *et al.*, 1998). Another intracellular inhibitor is BMP and Activin Membrane Bound Inhibitor (BAMBI). BAMBI shows considerable sequence homology to the BMP receptors, but lacks the intracellular kinase domain, making it a naturally occurring dominant negative receptor (Onichtchouk *et al.*, 1999). Homologues have been identified in mouse (Grotewold *et al.*, 2001), humans (Degen *et al.*, 1996), and zebrafish (Tsang *et al.*, 2000). The expression pattern correlates well with the expression of BMP-2 and BMP-4, and indeed BAMBI is induced by BMP-4 expression and is lost in zebrafish mutant for bmp-2b (Tsang *et al.*, 2000).

Another feature of the BMP pathway is its ability to intersect with other signaling pathways (von Bubnoff and Cho, 2001). Particularly pertinent to this consideration of neural induction is the interaction, within the cell, with signaling from the fibroblast growth factor (FGF) family of molecules and the wingless/wnt group. Both can negatively influence BMP activity, and this is particularly germane to the role of these factors in the induction of the nervous system in amniotes.

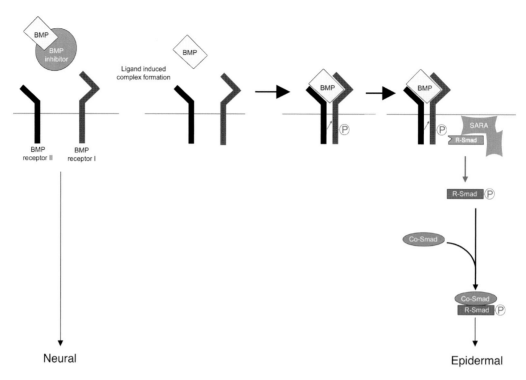

FIGURE 8. The BMP signal transduction pathway. BMP activity specifies the ectoderm as epidermal; its inhibition (e.g., by binding to a soluble inhibitor-like chordin) leads to neural induction. Ligand binding induces the type I and type II receptors to associate and causes the phosphorylation of the intracellular intermediate R-Smad, held in place by the adaptor molecule SARA. R-Smad is now free to associate with a Co-Smad, causing translocation into the nucleus, where the complex participates in the transcriptional modulation of a number of genes. Modified from von Bubnoff and Cho (2001).

that follistatin, an inhibitor of TGF-β signaling, was able to induce neural tissue suggested that inhibition of a pathway involving perhaps activin was responsible for the induction of neural ectoderm. These data were supported by studies using a truncated receptor for activin. RNA encoding the activin receptor lacking the transducing, cytosolic domain but with the extracellular and transmembrane domains, acts as a dominant negative, that is, although ligand binding can occur, it is unable to elicit a response (Hemmati-Brivanlou and Melton, 1992). As this modified molecule is present in far excess of the wild-type molecule, it has the effect of sequestering the ligand. Animal caps that express the dominant negative, truncated activin receptor follow a neural pathway of differentiation (Hemmati-Brivanlou and Melton, 1994).

This led to somewhat of a paradox. Though it seemed that neural induction was a result of activin inhibition, activin itself induced mesoderm and neural ectoderm. In actuality, the activin receptor used by Hemmati-Brivanlou and Melton was not specific for activin; rather it recognized other members of the TGF-β superfamily (Hemmati-Brivanlou and Melton, 1994). As the truncated receptor also induced dorsal mesoderm, rather than recognizing activin, another TGF-β family member active on the ventral side of the embryo could be the native ligand.

BMP-2 and BMP-4, members of the TGF-β superfamily, are both expressed in the ventral part of the embryo (Dale *et al.*, 1992; Jones *et al.*, 1992). Consequently, their potential role in neural induction was placed under scrutiny, which grew more

intense with the discovery of chordin, another secreted molecule capable of inducing neural tissue. Chordin was discovered by virtue of its expression in Spemann's organizer. Later, it is expressed in the axial tissue of the prechordal mesoderm and notochord, all structures capable of neural induction (Sasai *et al.*, 1994). Examination of the primary sequence of chordin provided further insight into the mechanism of neural induction. It was found that chordin shows considerable homology to the fruit fly *Drosophila* gene *short of gastrulation* (*sog*). Genetic analysis in *Drosophila* had already shown that *sog* acted as an antagonist to another gene, *decapentaplegic* (*dpp*), which is homologous to the vertebrate genes BMP-2 and BMP-4. The similarities with flies are not limited to the sequence (Holley *et al.*, 1995). In flies, eliminating *dpp* converts the epidermal cells of the fly into neuroectoderm. Overexpression of *dpp* changes the fate of neuroectodermal cells into epidermal (Biehs *et al.*, 1996). In the amphibian, BMP-4 is also expressed in the non-neural ectoderm, consistent with it being an epidermal inducer. Moreover, when BMP-4 is added to dissociated animal cap cells, neural induction is prevented regardless of how long reassociation is delayed (Wilson and Hemmati-Brivanlou, 1995). Overexpressing BMP-4 RNA on the dorsal side of the embryo results in an embryo with a loss of neural ectoderm. However, it should be noted that dorsal mesoderm, the primary neural-inducing tissue, is also missing (Dale *et al.*, 1992; Jones *et al.*, 1992). The data pointed to neural induction occurring by inhibition of the BMP pathway, and indicated that perhaps not only chordin, like its *Drosophila* counterpart *sog*, but also noggin and follistatin acted as antagonists of BMP activity. Indeed chordin, noggin, and follistatin bind to BMP-4 and the closely related BMP-2 (Piccolo *et al.*, 1996; Zimmerman *et al.*, 1996; Iemura *et al.*, 1998), and from genetic analysis in *Drosophila*, where chordin or noggin were ectopically expressed in various fly mutants in components of the BMP pathway, the site of action of chordin and noggin was placed upstream of the receptor, in the extracellular matrix (Holley *et al.*, 1995, 1996). An additional number of extracellular, secreted antagonists of BMP activity have been found. These molecules, such as Cerberus, Gremlin, and Xnr-3 (*Xenopus* nodal related-3), all induce neural fates in the animal cap of the *Xenopus* embryo (Smith *et al.*, 1995; Bouwmeester *et al.*, 1996; Hsu *et al.*, 1998).

Further support for the idea that BMP inhibition is germane to the induction of neural tissue came from inhibiting the intracellular components of the BMP signal-transduction pathway (Box 2). As well as the truncated activin receptors, acting as dominant negative forms of the endogenous receptor, which have been shown to bind BMP-2 and BMP-4, negative forms of the Smad molecules have been shown to promote neural differentiation in the animal cap (Liu *et al.*, 1996; Bhushan *et al.*, 1998). Indeed, even negative forms of the transcription factors that form the nuclear response to BMP signaling have been shown to neuralize the animal cap (Onichtchouk *et al.*, 1998; Trindade *et al.*, 1999). Many of these experiments have been repeated in the zebrafish embryo, with similar, if not identical, results (e.g., Imai *et al.*, 2001).

Complexities and Questions

That BMP inhibition, emanating from the organizer, is responsible for neural induction has been well demonstrated in anamniote (fish and frog) embryos. However, the data from the chick and mouse are confusing and challenge this idea.

Is the Organizer Responsible for Neural Induction?

The role of the chick and mouse equivalents of the organizer—Hensen's node and the node, respectively—in neural induction has been questioned over the years. In the chick, neural induction can occur even after the node is surgically ablated (Waddington, 1932; Abercrombie and Bellairs, 1954). This result was interpreted as showing that Hensen's node, though sufficient for neural induction, was not necessary. However, subsequent studies have shown that after extirpation, the node is reconstituted quickly owing to a series of complex inductive interactions (Yuan *et al.*, 1995; Psychoyos and Stern, 1996; Yuan and Schoenwolf, 1998, 1999; Joubin and Stern, 1999). Genetic ablation of the node and notochord in the mouse and fish also has little effect on the induction of neural tissue (Gritsman *et al.*, 1999; Klingensmith *et al.*, 1999). Recently, it has become clear that neural induction in all vertebrates occurs earlier than previously thought, beginning before the appearance of a morphologically distinct organizer. For example, in chick, neural induction begins before the appearance of Hensen's node, as determined by the

stage at which explants of *prospective* neural ectoderm first express neural markers (Darnell *et al.*, 1999; Wilson *et al.*, 2000). In *Xenopus*, neural induction is initiated before gastrulation. Using the clearance of the expression of components of the BMP signaling pathway as a marker for when neural induction is occurring, it has been shown that neural induction occurs during late blastula stages of *Xenopus* embryogenesis (Hemmati-Brivanlou and Thomsen, 1995; Faure *et al.*, 2000).

In fish containing the mutation one-eyed-pinhead (oep), the embryonic shield and dorsal mesoderm do not form. Despite this, these mutants still express chordin, indicating that some neural-inducing activity still persists (Gritsman *et al.*, 1999). The situation in the mouse HNF-3β mutant is more striking. Even in the absence of a node and axial mesoderm, and despite the lack of expression of many markers of the mouse organizer, the rostral streak, from which the node derives, is still capable of neural induction (Klingensmith *et al.*, 1999).

Is BMP Inhibition Sufficient for Neural Induction?

Experiments again in the chick first questioned the hypothesis that BMP inhibition mediates neural induction. Streit and coworkers showed that neural tissue could not be induced by clumps of noggin- or chordin-expressing cells, even though grafts of Hensen's node in parallel experiments induced neural tissue (Streit *et al.*, 1998). In the same study, Streit *et al.* (1998) showed that cells expressing BMP-2 or BMP-7 failed to inhibit neural plate formation. However, Wilson and coworkers showed that BMP-4 was able to induce epidermis in explants of the chick embryos fated to become neural ectoderm (Wilson *et al.*, 2000). The difference between these sets of data seem to be the stage at which the experiments were performed, with the experiments using expressing cells being done at mid-gastrula stages, and the explant-induction experiments being done at blastula to early-gastrula stages. In the mouse, null mutants of BMP-2 (Zhang and Bradley, 1996), BMP-4 (Winnier *et al.*, 1995), and BMP-7 (Dudley *et al.*, 1995) do not alter their pattern of neural induction. However, there is probably functional redundancy between these molecules, with one compensating for the loss of another (Dudley and Robertson, 1997). Compound mutants have not yet been established to address this issue.

The expression patterns in the chick of the BMP inhibitors noggin, follistatin, and chordin are not strictly correlated with tissues that contain neural-inducing ability (Connolly *et al.*, 1995, 1997; Streit *et al.*, 1998). Taken with the data from mice doubly mutant for noggin and chordin, which still have neural tissue (Bachiller *et al.*, 2000), this seems to indicate that BMP inhibition is not required for neural induction in amniotes. However, as discussed above, there are other inhibitors of BMP signaling, both extracellular and intracellular, which may account for neural induction (von Bubnoff and Cho, 2001; Muñoz-Sanjuan and Hemmati-Brivanlou, 2002). For example, support for the idea that BMP inhibition induces neural character in the chick embryo comes from an inspection of the localization of phosphorylated Smad1, -5, and -8. Using an antibody that recognizes the activated form of these Smads as an indication of BMP signaling,

Faure *et al.* (2002) showed that there is no BMP signaling activity in the forming neural plate. An argument has also been made that BMP inhibition merely stabilizes and reinforces neural cell fates, and that other families of signaling molecules are the primary neural inducers (Streit and Stern, 1999). Until the full complement of molecules that can induce neural tissue is known, and a full understanding of the signaling networks is understood, this question will not be fully resolved.

The Role of Other Signals in Neural Induction

Fibroblast Growth Factors (FGF)

Both the FGF family and the wnt family have been shown to play a role in the induction of neural tissue. This role is distinct from their roles in patterning of the neural tube, which are discussed in the subsequent chapter. In *Xenopus*, FGF can actually induce neuralization of animal cap cells that have undergone brief dissociation, a procedure that diminishes the amount of BMP activity (Kengaku and Okamoto, 1993). Furthermore, blocking FGF signaling using a truncated FGF receptor makes the animal cap refractory to neuralization by low amounts of chordin (Launay *et al.*, 1996). In chick, the role of FGF in neural induction has received considerable attention. Streit *et al.* (2000) reported that an FGF-responsive gene, Early Response to Neural Induction (ERNI), marks the territory in the chick epiblast fated to become neural, and it rapidly induced FGF expression. By using an FGF receptor antagonist, SU5402, Wilson *et al.* (2000) showed that neural differentiation could be blocked in chick epiblast explants normally fated to become neural ectoderm. The exact role of the FGF pathway in neural induction is unclear. Some of the data point to a role for FGF signaling in aiding the clearance of BMP activity from the neural plate; indeed, downstream effectors of the FGF pathway have been shown to inhibit the nuclear accumulation of the R-Smad/Co-Smad complex (Kretzschmar *et al.*, 1997, 1999). FGF may also induce neural tissue by a mechanism independent of BMP inhibition. An investigation of Smad10, a Co-Smad, in *Xenopus*, has yielded some relevant data (LeSeur *et al.*, 2002). Smad10, a component of the BMP signaling pathway, actually induces neural tissue within the animal cap. More surprisingly, by removing Smad10 protein using antisense oligonucleotides, neural tissue is never formed in the affected embryos. Using co-injection studies, it has been found that Smad10 cannot inhibit the BMP pathway, indicating some other mechanism for its function. One such mechanism is the identification of a site in the Smad10 protein that becomes phosphorylated and activated as a result of FGF signaling (LeSeur *et al.*, 2002).

An alternative view suggests that FGF signaling provides the ectoderm with competence to become defined as neural. There is precedence for this; Cornell *et al.* (1995) have shown that FGF signaling acts to define the competence of tissue to respond to mesoderm induction by TGF-β signals in *Xenopus*, the very same tissue that can respond to neural-inducing signals.

In fact, it is likely that both a competence-defining role early in development and a later neural-stabilizing role will be

shown for the FGF family. However, like many of the controversies surrounding neural induction, we will have to wait until all the players and the way they interact are known before adequate resolution can be achieved.

Wnts

The role of the wnt family of molecules has also been investigated during the induction of neural ectoderm. In the chick, wnt overexpression converts the epiblast fated to become neural to become epidermal (Wilson *et al.*, 2001). Conversely, in presumptive epidermal tissue fated to form epidermis, wnt inhibition causes the explant to take on a neural fate. In addition, at a sub-threshold concentration of wnt inhibitors, below the level required for neural induction in the epidermal epiblast explants, BMP inhibition and FGF signaling were able to induce neural ectoderm. One proposed mechanism is that wnt signaling causes an upregulation of BMP expression (Wilson *et al.*, 2001), and thereby induces epidermal fate, although in *Xenopus*, additional data suggest that wnt expression downregulates BMP expression (Baker *et al.*, 1999; Gomez-Skarmeta *et al.*, 2001). However, wnt signaling may also regulate the strength of the transduced BMP signal via activation of the calmodulin/Ca^{2+} pathway (Zimmerman *et al.*, 1998; Scherer and Graff, 2000). This may explain why BMP inhibition cannot induce neural tissue in epidermal epiblast explants. If the level of abrogation of BMP signaling is not complete, the sensitized transduction pathway can still receive an input, resulting in epidermal cell fates. If, however, wnt signaling is also inhibited, reception is desensitized and when combined with BMP inhibition, can lead to neural cell fates. Interestingly, two naturally occurring inhibitors of wnt signaling, FrzB and Sfrp-2, are expressed in the presumptive neural plate at around the stages that neural induction has been proposed to be occurring (Ladher *et al.*, 2000).

Insulin-Like Growth Factor

The insulin-like growth factor (IGF) family can also neuralize the *Xenopus* animal cap (Pera *et al.*, 2001). The necessity for IGF signaling has also been shown using a truncated IGF receptor. In these embryos, neural induction mediated by noggin is inhibited. The authors propose that the IGF pathway may act downstream of BMP inhibition during neural induction, and that as well as a passive role for BMP inhibition, neural induction may not be a default as previously thought. Instead, it may also require an active signal, induced as a result of BMP inhibition.

Summary of the Molecular Events of Neural Induction

As discussed above, the main mechanism by which the neural ectoderm is induced is via the inhibition of the BMP pathway. Other factors do play a role, namely the FGF family and the wnt family. As yet it is unclear what the exact roles of these molecules are, whether they are required as competence factors or whether they act to aid the clearing of BMP signals and their reception from the neural plate.

Once induced, the neural ectoderm—also known at this juncture as the neuroepithelium—still has a daunting journey ahead of it to form the central nervous system: it must roll up into a tube, which is subsequently patterned. We will describe in the next section the mechanism by which the specified neural ectoderm becomes a tube; other chapters later in this book deal with the elaboration of the neural tube into the adult central nervous system.

NEURULATION

The process of neural induction results in a plate of cells running along the rostrocaudal length of the embryo. The medial part of the neural plate will eventually form the ventral part of the neural tube, and the lateromost edges will be brought together to form the dorsal part of the tube during the process of neurulation. The end result of neurulation is a hollow nerve cord.

Neurulation can be subdivided into a number of events, each requiring different interactions. First, neurulation occurs in two phases called primary and secondary neurulation. When one speaks of neurulation, they are typically referring to primary neurulation, a process that occurs in four stages defined as formation, shaping, and bending of the neural plate, and closure of the neural groove (Figs. 9 and 10). Each stage will be described in turn. This discussion focuses primarily on the chick embryo, as most of the mechanistic studies have been performed on this embryo. For a more in-depth discussion, the reader is directed to several reviews (Schoenwolf and Smith, 1990; Smith and Schoenwolf, 1997; Colas and Schoenwolf, 2001).

Formation of the Neural Plate

The neural plate is a thickened region of the ectoderm located medially within the embryo. The thickening forms by an apicobasal elongation of ectodermal cells, an action known as cell pallisading. The thickening of the neural plate is not a result of an increase in the number of cell layers; the neuroepithelium remains pseudostratified (see Fig. 9A). It has been shown that thickening of the neural plate is an intrinsic property of the ectodermal cells once they have been induced as neural (Schoenwolf, 1988).

Shaping of the Neural Plate

During shaping, different cell behaviors convert the neural plate from a relatively short (in the rostrocaudal axis) and squat (wide in the mediolateral plane) structure to one that is long and narrow (see Fig. 10). This results from a combination of continued cell elongation, convergent extension, and cell division, as well as the caudalward regression of the primitive streak (Schoenwolf and Alvarez, 1989; Schoenwolf et al., 1989).

Neuroepithelial cells continue their apicobasal elongation during shaping, a process initiated shortly after neural induction and resulting in formation of the neural plate. As a result of cell elongation, a concomitant narrowing of the neural plate occurs, as neural plate cells maintain their individual volumes. Convergent extension movements further exaggerate the narrowing of the neural plate; that is, cells of the neural plate intercalate in the mediolateral plate, effectively causing the neural plate to lengthen rostrocaudally while narrowing simultaneously. Cell division also contributes to the lengthening of the neuroepithelium; about half of the division planes are oriented such that they place the daughter cells into the length of the neural plate rather than adding to its width (Sausedo et al., 1997). Isolation experiments have shown that the cell behaviors causing shaping of the neural plate are autonomous to the neural plate. In other words, such changes in cell behavior within the neural plate generate intrinsic forces for its shaping. However, for the shaping of the neural plate to occur completely normally, normal gastrulation movements also must occur, as the axis develops in the wake of the regressing Hensen's node.

Bending of the Neural Plate

Bending involves the establishment of localized deformations of the cells of the neuroepithelium and the subsequent elevation of the two flanks of the neuroepithelium, converting it from the neural plate to the neural groove. Bending is actually driven by two distinct types of movement: furrowing and folding (Colas and Schoenwolf, 2001). Furrowing is a behavior intrinsic to the hinge points within the neuroepithelium. There are three hinge points within the neural plate: a single median hinge point, found along the neuroaxis (except at the future forebrain level) and coincident with the floor plate of the neuroepithelium; and the paired (right and left) dorsolateral hinge points, found primarily at levels where the brain will form (Fig. 11; Schoenwolf and Franks, 1984). Neuroepithelial cells within the hinge points undergo wedging, that is, apical constriction with a concomitant basal expansion, driven in part by the basalward interkinetic movement of the nucleus (see Fig. 11; Smith and Schoenwolf, 1987, 1988). This acts not only to deform the neuroepithelium, creating a furrow, but it also provides points around which the neural plate can rotate during folding; that is to say, true to their nomenclature, the hinge points do act like hinges during neurulation.

Folding is a more complicated process and is driven by the non-neural ectoderm. The net result is rather like closing a pair of calipers. The easiest way to close calipers is to apply a force laterally at the tip of the calipers, and eventually the tips will meet, folding around the hinge. Like calipers, the neural plate elevates and folds by forces generated laterally in the non-neural ectoderm. This force results, in part, from cell shape changes in the non-neural ectoderm (Alvarez and Schoenwolf, 1992; Sausedo et al., 1997). These cells undergo apicobasal flattening, thus effectively increasing their surface area. Folding itself can be divided into three distinct events (Fig. 12), occurring while the lateral epithelium provides a medialward force (Lawson et al., 2001). The first is epithelial kinking, where cells at the interface between the neural and non-neural ectoderm deform, with each

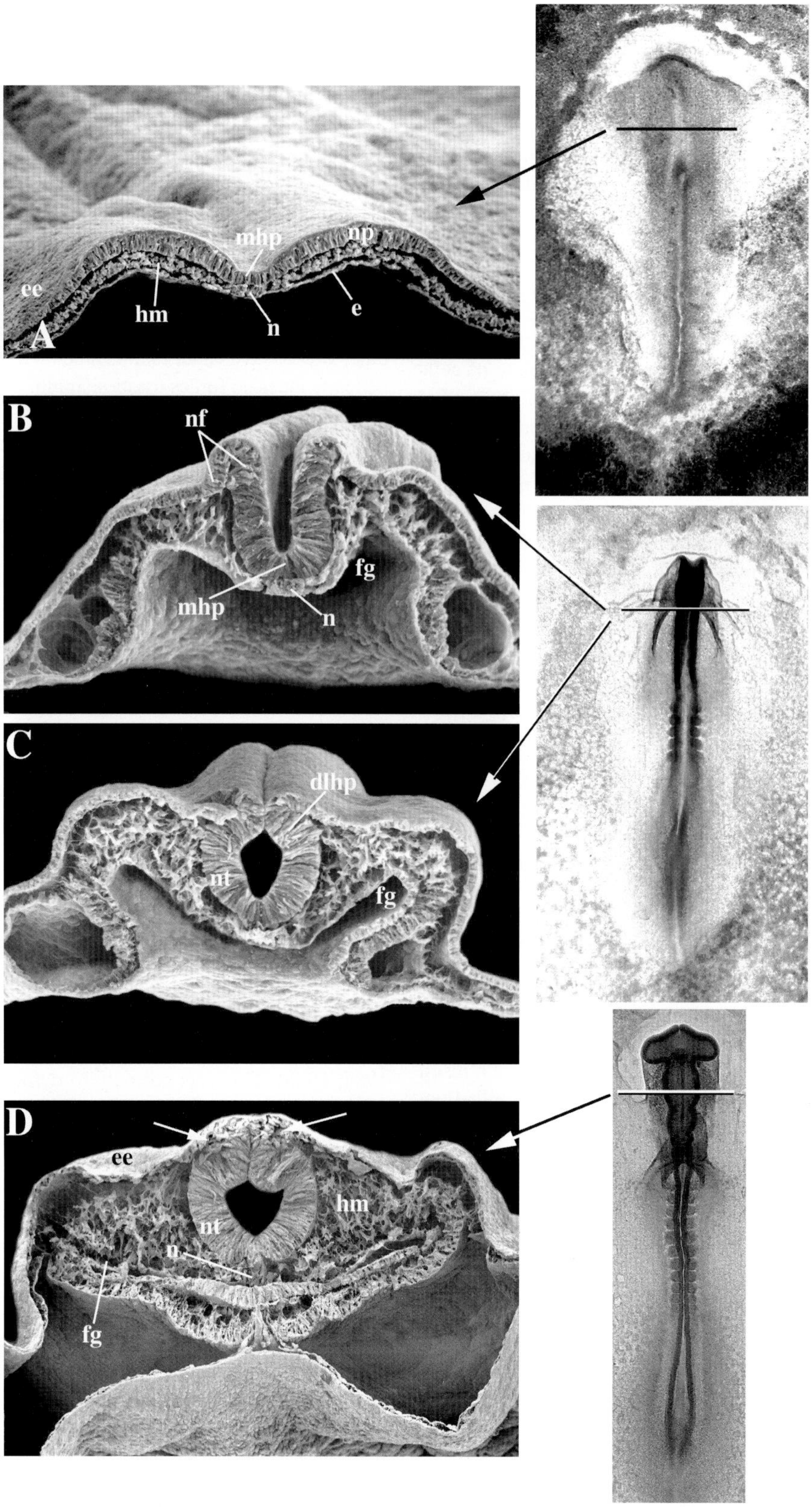

FIGURE 9. Whole mounts (for orientation; transverse lines indicate levels of cross-sections identified by long arrows) and scanning electron micrograph cross-sections of the neuroepithelium during neurulation. Shown are changes in the neuroepithelium that occur during the (A) formation, (B, C) shaping, and (A–C) bending of the neural plate, and closure of the neural groove (C, D). Details are provided in the text. dlhp, dorsolateral hinge point; e, endoderm; ee, epidermal ectoderm; fg, foregut; hm, head mesenchyme; mhp, median hinge point; n, notochord; nf, neural fold; nt, neural tube; arrows (D), neural crest cells. Modified from Schoenwolf (2001).

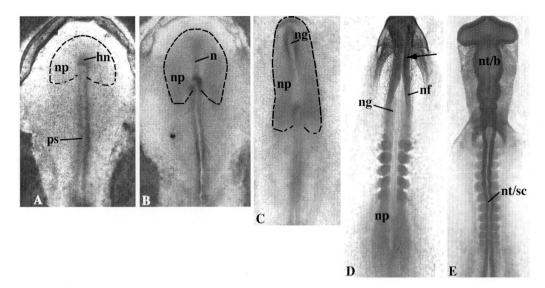

FIGURE 10. Whole mount embryos viewed from their dorsal side during neurulation. A–E indicate progressively older, yet partially overlapping, stages of neurulation, beginning with (A) formation of the neural plate, (B–E) shaping of the neural plate, (B–E) bending of the neural plate, and (D, E) closure of the neural groove. The neuroepithelium at the time of its formation is a relatively short and squat structure, as seen in surface view. However, during convergent extension movements that commence concomitant with regression of Hensen's node, the neural plate lengthens rostrocaudally and narrows mediolaterally. hn, Hensen's node; nf, neural fold; ng, neural groove; np, neural plate; ps, primitive streak; dashed lines, lateral borders of neural plate. Modified from Smith and Schoenwolf (1997).

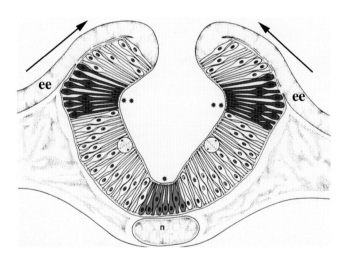

FIGURE 11. Cell behavior in the neural plate during its bending. Shown is a diagram of a cross-section through the neural tube during bending. Highlighted (darker shading) are the three hinge points: the median hinge point (asterisk), coincident with the floor plate of the neural tube, and the dorsolateral hinge points (double asterisks), found in the future brain level of the neuroaxis. ee, epidermal ectoderm; n, notochord; arrows, directions of expansion of the epidermal ectoderm. Modified from Schoenwolf and Smith (1990).

forming an inverted wedge that is apically expanded and basally constricted. The next step is epithelial delamination. This involves the deposition of extracellular matrix at the neural fold interface (i.e., the space between the two ectodermal layers of each neural fold) and a re-orientation of neural and non-neural cells around the interface, such that their basal surfaces abut. The final step is epithelial apposition, which occurs in the brain

region. This is essentially a rapid expansion of the neural folds, with extension of the area of epithelial delamination in the mediolateral plane and further deposition of extracellular matrix along the expanding width of the neural fold interface. Additionally, the non-neural ectoderm intercalates and undergoes oriented cell division, thereby contributing to the mediolateral forces generated in the epidermal ectoderm.

Tissue isolation experiments have been used to identify the cell types responsible for generating the forces of folding (Schoenwolf, 1988). Removal of the lateral, non-neural ectoderm results in the loss of folding, but furrowing of the neural plate within the hinge points still occurs (Hackett *et al.*, 1997). If the mesoderm and endoderm lateral to the neural plate are removed, but leaving the non-neural ectodermal layer intact, both folding and furrowing occur (Alvarez and Schoenwolf, 1992). Thus, the non-neural ectoderm is both necessary and sufficient for folding to occur.

Closure of the Neural Groove

Bending brings the tips of the neural folds into close contact at the site of the dorsal midline of the embryo. During closure, the two tips attach and fuse. Each component of the tip must fuse correctly, such that the non-neural epithelium forms a continuous sheet overlying the newly formed roof plate of the neural tube and the associated neural crest. The exact mechanism of this concluding step of neurulation is not well understood, and the molecules that mediate adhesion, epithelial breakdown, and fusion are not known.

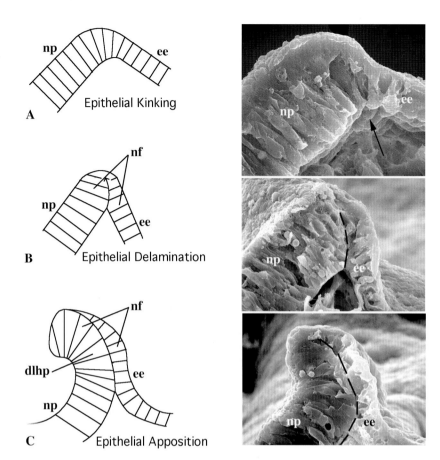

FIGURE 12. Formation of the neural folds. The scanning electron micrographs and accompanying diagrams highlight the formation and morphogenesis of the neural folds, in particular, (A) epithelial kinking, (B) delamination, and (C) (in the brain region) apposition. dlhp, dorsolateral hinge point; ee, epidermal ectoderm; nf, neural folds; np, neural plate; dashed lines, interface between the two ectodermal layers of the neural fold. Modified from Lawson *et al.* (2001).

Secondary Neurulation

At caudal levels of the neuraxis of birds and mammals (e.g., the lumbar and sacral regions), the neural tube develops in a manner distinct from more rostral regions. Caudal neural tube formation occurs through a process known as secondary neurulation. Rather than the rolling up of a flat plate of cells, as is the case in primary neurulation, secondary neurulation consists of the cavitation of a solid epithelial cord of cells in the tail of the embryo.

Secondary neurulation begins when cells within the tail bud condense to form an epithelial cord of cells, known as the medullary cord (Schoenwolf, 1979, 1984; Schoenwolf and DeLongo, 1980). The outer cells of the medullary cord then undergo elongation, forming a pseudostratified columnar epithelium similar to that of the neural plate during primary neurulation. This pseudostratified epithelium then becomes polarized, resulting in the formation and fusion of small lumina at the apices of the outer layer, around a central core of mesenchymal cells. These inner cells are removed during cavitation, by cell rearrangements and perhaps limited apoptosis. Cavitation results in the formation of a single, secondary lumen, which will join with the primary lumen of the rostral neural tube.

SUMMARY

The future central nervous system is derived from an unspecified sheet of ectoderm, with fate being instructed by signals emanating, in the main, from a specialized region of the early embryo, the organizer. The organizer secretes signals that have the net effect of inhibiting the BMP pathway, be it by extracellular antagonism or by intracellular modulation of the ability of the cell to perceive BMP signals. Other factors also play a role in neural induction, for example, the FGF family of molecules, but their exact role in neural induction remain unknown. As more players are identified in what undoubtedly will be a signaling network leading to neural induction, the exact molecular mechanism of neural induction can be established.

Once induced, the neuroepithelium rolls into the neural tube. One model, and one that has gained widespread acceptance, is the hinge point model. In this model, both extrinsic (i.e., outside the neural plate) and intrinsic forces cooperate and synergize in bending the neural plate. Although the cellular behaviors of much of this process have been well characterized, the molecular bases for these behaviors have so far proved elusive. The relationship between induction of the neuroepithelium and its subsequent morphological movements is of particular interest to the developmental neurobiologist.

ACKNOWLEDGMENTS

R.K.L. acknowledges the support of MEXT and the leading Projects of Japan. Results described from the Schoenwolf laboratory were obtained with support by grants from the National Institutes of Health. Support was also gratefully received from the Ministry of Science and Education of Japan and the Leading Projects. We thank past and present members of the Schoenwolf laboratory for their contributions.

REFERENCES

Abercrombie, M. and Bellairs, R., 1954, The effects in chick blastoderms of replacing the primitive node by a graft of posterior primitive streak. *J. Embryol. Exp. Morph.* 2:55–72.

Afrakhte, M., Moren, A., Jossan, S., Itoh, S., Sampath, K., Westermark, B. *et al.*, 1998, Induction of inhibitory Smad6 and Smad7 mRNA by TGF-beta family members, *Biochem. Biophys. Res. Commun.* 249: 505–511.

Alvarez, I.S. and Schoenwolf, G.C., 1992, Expansion of surface epithelium provides the major extrinsic force for bending of the neural plate, *J. Exp. Zool.* 261:340–348.

Attisano, L. and Tuen Lee-Hoeflich, S. 2001, The Smads, *Genome Biol.* 2:3010.1–3010.8.

Bachiller, D., Klingensmith, J., Kemp, C., Belo, J.A., Anderson, R.M., May, S.R. *et al.*, 2000, The organizer factors Chordin and Noggin are required for mouse forebrain development, *Nature* 403:658–661.

Baker, J.C., Beddington, R.S., and Harland, R.M., 1999, Wnt signaling in Xenopus embryos inhibits bmp4 expression and activates neural development, *Genes Dev.* 13:3149–3159.

Barth, L.G., 1941, Neural differentiation without organizer, *J. Exp. Zool.* 87:471–481.

Beddington, R.S., 1994, Induction of a second neural axis by the mouse node, *Development* 120:613–620.

Beddington, R., 1998, Mouse mutagenesis: From gene to phenotype and back again, *Curr. Biol.* 8:R840–R842.

Bhushan, A., Chen, Y., and Vale, W., 1998, Smad7 inhibits mesoderm formation and promotes neural cell fate in Xenopus embryos, *Dev. Biol.* 200:260–268.

Biehs, B., Francois, V., and Bier, E., 1996, The Drosophila short gastrulation gene prevents Dpp from autoactivating and suppressing neurogenesis in the neuroectoderm, *Genes Dev.* 10:2922–2934.

Boterenbrood, E.C. and Nieuwkoop, P.D., 1973, The formation of the mesoderm in urodelean amphibians. V. Its regional induction by the endoderm, *Wilhelm Roux' Arch. Dev. Biol.* 173:319–332.

Bouwmeester, T., Kim, S., Sasai, Y., Lu, B., and De Robertis, E.M., 1996, Cerberus is a head-inducing secreted factor expressed in the anterior endoderm of Spemann's organizer, *Nature* 382:595–601.

Callebaut, M. and Van Nueten, E., 1994, Rauber's (Koller's) sickle: The early gastrulation organizer of the avian blastoderm, *Eur. J. Morph.* 32:35–48.

Capecchi, M.R., 1989, The new mouse genetics: Altering the genome by gene targeting, *Trends Genet.* 5:70–76.

Colas, J.F. and Schoenwolf, G.C., 2001, Towards a cellular and molecular understanding of neurulation. *Dev. Dyn.* 221:117–145.

Connolly, D.J., Patel, K., Seleiro, E.A., Wilkinson, D.G., and Cooke, J., 1995, Cloning, sequencing, and expressional analysis of the chick homologue of follistatin. *Dev. Genet.* 17:65–77.

Connolly, D.J., Patel, K., and Cooke, J., 1997, Chick noggin is expressed in the organizer and neural plate during axial development, but offers no evidence of involvement in primary axis formation, *Int. J. Dev. Biol.* 41:389–396.

Cornell, R.A., Musci, T.J., and Kimelman, D., 1995, FGF is a prospective competence factor for early activin-type signals in Xenopus mesoderm induction, *Development* 121:2429–2437.

Cruz, Y.P., 1997, Mammals. In *Embryology: Constructing the Organism* (S.C. Gilbert and A.M. Raunio, eds.), Sinauer, Sunderland, MA, pp. 459–492.

Dale, L., Howes, G., Price, B.M., and Smith, J.C., 1992, Bone morphogenetic protein 4: A ventralizing factor in early Xenopus development, *Development* 115:573–585.

Darnell, D.K., Stark, M.R., and Schoenwolf, G.C., 1999, Timing and cell interactions underlying neural induction in the chick embryo, *Development* 126:2505–2514.

Degen, W.G., Weterman, M.A., van Groningen, J.J., Cornelissen, I.M., Lemmers, J.P., Agterbos, M.A. *et al.*, 1996, Expression of nma, a novel gene, inversely correlates with the metastatic potential of human melanoma cell lines and xenografts, *Int. J. Cancer* 65:460–465.

Driever, W., 1995, Axis formation in zebrafish, *Curr. Opin. Genet. Dev.* 5:610–618.

Dudley, A.T. and Robertson, E.J., 1997, Overlapping expression domains of bone morphogenetic protein family members potentially account for limited tissue defects in BMP7 deficient embryos, *Dev. Dyn.* 208:349–362.

Dudley, A.T., Lyons, K.M., and Robertson, E.J., 1995, A requirement for bone morphogenetic protein-7 during development of the mammalian kidney and eye, *Genes Dev.* 9:2795–2807.

Faure, S., de Santa Barbara, P., Roberts, D.J., Whitman, M., 2002, Endogenous patterns of BMP signaling during early chick development, *Dev. Biol.* 244(1):44–65.

Faure, S., Lee, M.A., Keller, T., ten Dijke, P., and Whitman, M., 2000, Endogenous patterns of TGFbeta superfamily signaling during early Xenopus development, *Development* 127:2917–2931.

Feldman, B., Gates, M.A., Egan, E.S., Dougan, S.T., Rennebeck, G., Sirotkin, H.I. *et al.*, 1998, Zebrafish organizer development and germ-layer formation require nodal-related signals, *Nature* 395:181–185.

Geinitz, B., 1925, Embryonale transplantation zwischen Urodelen und Anuren, *Roux Arch. Entwicklungsmech* 106:357–408.

Gimlich, R.L. and Cooke, J., 1983, Cell lineage and the induction of second nervous systems in amphibian development, *Nature* 306:471–473.

Godsave, S.F. and Slack, J.M., 1989. Clonal analysis of mesoderm induction in *Xenopus laevis*, *Dev. Biol.* 134:486–490.

Gomez-Skarmeta, J., de La Calle-Mustienes, E., and Modolell, J., 2001, The Wnt-activated Xiro1 gene encodes a repressor that is essential for neural development and downregulates Bmp4, *Development* 128:551–560.

Gritsman, K., Zhang, J., Cheng, S., Heckscher, E., Talbot, W.S., and Schier, A.F., 1999, The EGF-CFC protein one-eyed pinhead is essential for nodal signaling, *Cell* 97:121–132.

Grotewold, L., Plum, M., Dildrop, R., Peters, T., and Ruther, U., 2001, Bambi is coexpressed with Bmp-4 during mouse embryogenesis, *Mech. Dev.* 100:327–330.

Grunz, H., 1997, Neural induction in amphibians, *Curr. Top. Dev. Biol.* 35:191–228.

Grunz, H. and Tacke, L., 1989, Neural differentiation of *Xenopus laevis* ectoderm takes place after disaggregation and delayed reaggregation without inducer, *Cell. Diff. Dev.* 28:211–217.

Grunz, H. and Tacke, L., 1990, Extracellular matrix components prevent neural differentiation of disaggregated Xenopus ectoderm cells, *Cell. Diff. Dev.* 32:117–123.

Hackett, D.A., Smith, J.L., and Schoenwolf, G.C., 1997, Epidermal ectoderm is required for full elevation and for convergence during bending of the avian neural plate, *Dev. Dyn.* 210:1–11.

Hemmati-Brivanlou, A. and Melton, D.A., 1992, A truncated activin receptor inhibits mesoderm induction and formation of axial structures in Xenopus embryos, *Nature* 359:609–614.

Hemmati-Brivanlou, A. and Melton, D.A., 1994, Inhibition of activin receptor signaling promotes neuralization in Xenopus, *Cell* 77:273–281.

Hemmati-Brivanlou, A. and Thomsen, G.H., 1995, Ventral mesodermal patterning in Xenopus embryos: Expression patterns and activities of BMP-2 and BMP-4, *Dev. Genet.* 17:78–89.

Hemmati-Brivanlou, A., Kelly, O.G., and Melton, D.A., 1994, Follistatin, an antagonist of activin, is expressed in the Spemann organizer and displays direct neuralizing activity, *Cell* 77:283–295.

Holley, S.A., Jackson, P.D., Sasai, Y., Lu, B., De Robertis, E.M., Hoffmann, F.M. *et al.*, 1995, A conserved system for dorsal-ventral patterning in insects and vertebrates involving sog and chordin, *Nature* 376:249–253.

Holley, S.A., Neul, J.L., Attisano, L., Wrana, J.L., Sasai, Y., O'Connor, M.B. *et al.*, 1996, The Xenopus dorsalizing factor noggin ventralizes Drosophila embryos by preventing DPP from activating its receptor, *Cell* 86:607–617.

Holtfreter, J., 1933a, Nachweis der Induktionsfähigkelt abgetöteter Kiemteille, *Roux Arch. Entwicklungsmech* 129:584–633.

Holtfreter, J., 1933b, Die totale Exogastrulation, eine Selbstablösung des Ektoderms vom Entomesoderm, *Roux Arch. Entwicklungsmech* 129:669–793.

Holtfreter, J., 1944, Neural differentiation of ectoderm through exposure to saline solution, *J. Exp. Zool.* 98:169–209.

Holtfreter, J., 1947, Neural induction in explants that have passed through a sublethal cytolysis, *J. Exp. Zool.* 106:197–222.

Hsu, D.R., Economides, A.N., Wang, X., Eimon, P.M., and Harland, R.M., 1998, The Xenopus dorsalizing factor Gremlin identifies a novel family of secreted proteins that antagonize BMP activities, *Mol. Cell.* 1:673–683.

Iemura, S., Yamamoto, T.S., Takagi, C., Uchiyama, H., Natsume, T., Shimasaki, S. *et al.*, 1998, Direct binding of follistatin to a complex of bone-morphogenetic protein and its receptor inhibits ventral and epidermal cell fates in early Xenopus embryo, *Proc. Natl. Acad. Sci. USA* 95:9337–9342.

Imai, Y., Gates, M.A., Melby, A.E., Kimelman, D., Schier, A.F., and Talbot, W.S., 2001, The homeobox genes vox and vent are redundant repressors of dorsal fates in zebrafish, *Development* 128:2407–2420.

Imamura, T., Takase, M., Nishihara, A., Oeda, E., Hanai, J., Kawabata, M. *et al.*, 1997, Smad6 inhibits signalling by the TGF-beta superfamily, *Nature* 389:622–626.

Inoue, H., Imamura, T., Ishidou, Y., Takase, M., Udagawa, Y., Oka, Y. *et al.*, 1998, Interplay of signal mediators of decapentaplegic (Dpp): Molecular characterization of mothers against dpp, Medea, and daughters against dpp, *Mol. Biol. Cell.* 9:2145–2156.

Jacobson, M., 1984, Cell lineage analysis of neural induction: Origins of cells forming the induced nervous system, *Dev. Biol.* 102:122–129.

Jones, C.M., Lyons, K.M., Lapan, P.M., Wright, C.V., and Hogan, B.L., 1992, DVR-4 (bone morphogenetic protein-4) as a posterior-ventralizing factor in Xenopus mesoderm induction, *Development* 115:639–647.

Jones, C.M., Broadbent, J., Thomas, P.Q., Smith, J.C., and Beddington, R.S., 1999, An anterior signalling centre in Xenopus revealed by the homeobox gene XHex, *Curr. Biol.* 9:946–954.

Joubin, K. and Stern, C.D., 1999, Molecular interactions continuously define the organizer during the cell movements of gastrulation, *Cell* 98:559–571.

Kane, D.A. and Kimmel, C.B., 1993, The zebrafish midblastula transition, *Development* 119:447–456.

Keller, R. and Winklbauer, R., 1992, Cellular basis of amphibian gastrulation, *Curr. Top. Dev. Biol.* 27:39–89.

Kengaku, M. and Okamoto, H., 1993, Basic fibroblast growth factor induces differentiation of neural tube and neural crest lineages of cultured ectoderm cells from Xenopus gastrula, *Development* 119:1067–1078.

Khaner, O., 1998, The ability to initiate an axis in the avian blastula is concentrated mainly at a posterior site, *Dev. Biol.* 194:257–266.

Khaner, O. and Eyal-Giladi, H., 1986, The embryo-forming potency of the posterior marginal zone in stages X through XII of the chick, *Dev. Biol.* 115:275–281.

Kintner, C.R. and Melton, D.A., 1987, Expression of Xenopus N-CAM RNA in ectoderm is an early response to neural induction, *Development* 99:311–325.

Klingensmith, J., Ang, S.L., Bachiller, D., and Rossant, J., 1999, Neural induction and patterning in the mouse in the absence of the node and its derivatives, *Dev. Biol.* 216:535–549.

Kretzschmar, M., Doody, J., and Massague, J., 1997, Opposing BMP and EGF signalling pathways converge on the TGF-beta family mediator Smad1, *Nature* 389:618–622.

Kretzschmar, M., Doody, J., Timokhina, I., and Massague, J., 1999, A mechanism of repression of TGFbeta/ Smad signaling by oncogenic Ras, *Genes Dev.* 13:804–816.

Ladher, R.K., Church, V.L., Allen, S., Robson, L., Abdelfattah, A., Brown, N.A. *et al.*, 2000, Cloning and expression of the Wnt antagonists Sfrp-2 and Frzb during chick development, *Dev. Biol.* 218:183–198.

Lagna, G., Hata, A., Hemmati-Brivanlou, A., and Massague, J., 1996, Partnership between DPC4 and SMAD proteins in TGF-beta signalling pathways, *Nature* 383:832–836.

Lamb, T.M., Knecht, A.K., Smith, W.C., Stachel, S.E., Economides, A.N., Stahl, N. *et al.*, 1993, Neural induction by the secreted polypeptide noggin, *Science* 262:713–718.

Langeland, J. and Kimmel, C.B., 1997, Fishes. In *Embryology: Constructing the Organism* (S.C. Gilbert and A.M. Raunio, eds.), Sinauer, Sunderland, MA, pp. 383–408.

Launay, C., Fromentoux, V., Shi, D.L., and Boucaut, J.C., 1996, A truncated FGF receptor blocks neural induction by endogenous Xenopus inducers, *Development* 122:869–880.

Lawson, A. and Schoenwolf, G.C., 2001, New insights into critical events of avian gastrulation, *Anat. Rec.* 262:238–252.

Lawson, A., Anderson, H., and Schoenwolf, G.C., 2001, Cellular mechanisms of neural fold formation and morphogenesis in the chick embryo, *Anat. Rec.* 262:153–168.

LeSeur, J.A., Fortuno, E.S., 3rd, McKay, R.M., and Graff, J.M., 2002, Smad10 is required for formation of the frog nervous system, *Dev. Cell.* 2:771–783.

Liu, F., Hata, A., Baker, J.C., Doody, J., Carcamo, J., Harland, R.M. *et al.*, 1996, A human Mad protein acting as a BMP-regulated transcriptional activator, *Nature* 381:620–623.

Malacinski, G.M., Bessho, T., Yokota, C., Fukui, A., and Asashima, M., 1997, An essay on the similarities and differences between inductive interactions in anuran and urodele embryos, *Cell. Mol. Life. Sci.* 53:410–417.

Marx, A., 1925, Experimentelle Untersuchungen zur Frage der Determination der Medullarplatte, *Roux Arch. Entwicklungsmech* 105:20–44.

Massague, J. and Chen, Y.G., 2000, Controlling TGF-beta signaling, *Genes Dev.* 14:627–644.

Moon, R.T. and Kimelman, D., 1998, From cortical rotation to organizer gene expression: Toward a molecular explanation of axis specification in Xenopus, *Bioessays* 20:536–545.

Moustakas, A. and Heldin, C.H., 2002, From mono- to oligo-Smads: The heart of the matter in TGF-beta signal transduction, *Genes Dev.* 16:1867–1871.

Mullins, M.C. and Nusslein-Volhard, C., 1993, Mutational approaches to studying embryonic pattern formation in the zebrafish, *Curr. Opin. Genet. Dev.* 3:648–654.

Muñoz-Sanjuan, I. and Hemmati-Brivanlou, A., 2002, Neural induction, the default model and embryonic stem cells, *Nat. Rev. Neurosci.* 3:271–280.

Nakao, A., Afrakhte, M., Moren, A., Nakayama, T., Christian, J.L., Heuchel, R. *et al.*, 1997, Identification of Smad7, a TGFbeta-inducible antagonist of TGF-beta signalling, *Nature* 389:631–635.

Newport, J. and Kirschner, M., 1982, A major developmental transition in early Xenopus embryos: II. Control of the onset of transcription, *Cell* 30:687–696.

Nieuwkoop, P.D., 1969, The formation of mesoderm in Urodelean amphibians. I. Induction by the endoderm, *Wilhelm Roux' Arch. Dev. Biol.* 162:341–373.

Nieuwkoop, P.D., 1999, The neural induction process; its morphogenetic aspects, *Int. J. Dev. Biol.* 43:615–623.

Nieuwkoop, P. and Faber, J., 1967. Normal table of *Xenopus laevis*, North-Holland Publishing Company, Amsterdam, pp. 1–252.

Onichtchouk, D., Chen, Y.G., Dosch, R., Gawantka, V., Delius, H., Massague, J. *et al.* 1999, Silencing of TGF-beta signalling by the pseudoreceptor BAMBI, *Nature* 401:480–485.

Onichtchouk, D., Glinka, A., and Niehrs, C., 1998, Requirement for Xvent-1 and Xvent-2 gene function in dorsoventral patterning of Xenopus mesoderm, *Development* 125:1447–1456.

Oppenheimer, J., 1936, Transplantation experiments on developing teleosts (Fundulus and Perca), *J. Exp. Zool.* 72:409–437.

Osada, J., and Maeda, N., 1998, Preparation of knockout mice, *Meth. Mol. Biol.* 110:79–92.

Pera, E.M., Wessely, O., Li, S.Y., and De Robertis, E.M., 2001, Neural and head induction by insulin-like growth factor signals, *Dev. Cell* 1:655–665.

Piccolo, S., Sasai, Y., Lu, B., and De Robertis, E.M., 1996, Dorsoventral patterning in Xenopus: Inhibition of ventral signals by direct binding of chordin to BMP-4, *Cell* 86:589–598.

Psychoyos, D. and Stern, C.D., 1996, Restoration of the organizer after radical ablation of Hensen's node and the anterior end of the primitive streak in the chick embryo, *Development* 122:3263–3273.

Rossant, J., Bernelot-Moens, C., and Nagy, A., 1993, Genome manipulation in embryonic stem cells, *Philos. Trans. R. Soc. Lond. B. Biol. Sci.* 339:207–215.

Sasai, Y. and De Robertis, E.M., 1997, Ectodermal patterning in vertebrate embryos, *Dev. Biol.* 182:5–20.

Sasai, Y., Lu, B., Steinbeisser, H., Geissert, D., Gont, L.K., and De Robertis, E.M., 1994, Xenopus chordin: A novel dorsalizing factor activated by organizer-specific homeobox genes, *Cell* 79:779–790.

Sato, S.M. and Sargent, T.D., 1989, Development of neural inducing capacity in dissociated Xenopus embryos, *Dev. Biol.* 134:263–266.

Sausedo, R.A., Smith, J.L., and Schoenwolf, G.C., 1997, Role of non-randomly oriented cell division in shaping and bending of the neural plate, *J. Comp. Neurol.* 381:473–488.

Saxén, L., 1961, Transfilter neural induction of amphibian ectoderm, *Dev. Biol.* 3:140–152.

Scherer, A. and Graff, J.M., 2000, Calmodulin differentially modulates Smad1 and Smad2 signaling, *J. Biol. Chem.* 275:41430–41438.

Schoenwolf, G.C., 1979, Histological and ultrastructural observations of tail bud formation in the chick embryo, *Anat. Rec.* 193:131–147.

Schoenwolf, G.C., 1984. Histological and ultrastructural studies of secondary neurulation in mouse embryos. *Am. J. Anat.* 169:361–376.

Schoenwolf, G.C., 1988, Microsurgical analyses of avian neurulation: Separation of medial and lateral tissues, *J. Comp. Neurol.* 276:498–507.

Schoenwolf, G.C., 1997, Reptiles and birds. In *Embryology: Constructing the Organism* (S.C. Gilbert and A.M. Raunio, eds.), Sinauer, Sunderland, MA, pp. 437–458.

Schoenwolf, G.C., 2001, *Laboratory Studies of Vertebrate and Invertebrate Embryos. Guide and Atlas of Descriptive and Experimental Development*, Prentice Hall, Upper Saddle River, NJ, pp. 100, 101.

Schoenwolf, G.C. and Alvarez, I.S., 1989, Roles of neuroepithelial cell rearrangement and division in shaping of the avian neural plate. *Development* 106:427–439.

Schoenwolf, G.C. and DeLongo, J., 1980, Ultrastructure of secondary neurulation in the chick embryo, *Am. J. Anat.* 158:43–63.

Schoenwolf, G.C. and Franks, M.V., 1984, Quantitative analyses of changes in cell shapes during bending of the avian neural plate, *Dev. Biol.* 105:257–272.

Schoenwolf, G.C. and Smith, J.L., 1990. Mechanisms of neurulation: Traditional viewpoint and recent advances, *Development* 109: 243–270.

Schoenwolf, G.C., Everaert, S., Bortier, H., and Vakaet, L., 1989, Neural plate- and neural tube-forming potential of isolated epiblast areas in avian embryos, *Anat. Embryol.* 179:541–549.

Smith, J.C., 1989, Mesoderm induction and mesoderm-inducing factors in early amphibian development, *Development* 105:665–677.

Smith, J.L. and Schoenwolf, G.C., 1987, Cell cycle and neuroepithelial cell shape during bending of the chick neural plate, *Anat. Rec.* 218:196–206.

Smith, J.L. and Schoenwolf, G.C., 1988, Role of cell-cycle in regulating neuroepithelial cell shape during bending of the chick neural plate, *Cell. Tissue Res.* 252:491–500.

Smith, J.L. and Schoenwolf, G.C., 1989, Notochordal induction of cell wedging in the chick neural plate and its role in neural tube formation, *J. Exp. Zool.* 250:49–62.

Smith, J.L. and Schoenwolf, G.C., 1997, Neurulation: Coming to closure, *Trends Neurosci.* 20:510–517.

Smith, W.C. and Harland, R.M., 1992, Expression cloning of noggin, a new dorsalizing factor localized to the Spemann organizer in Xenopus embryos, *Cell* 70:829–840.

Smith, W.C., McKendry, R., Ribisi, S., Jr., and Harland, R.M., 1995, A nodal-related gene defines a physical and functional domain within the Spemann organizer, *Cell* 82:37–46.

Soriano, P., 1995, Gene targeting in ES cells, *Annu. Rev. Neurosci.* 18:1–18.

Souchelnytskyi, S., Nakayama, T., Nakao, A., Moren, A., Heldin, C.H., Christian, J.L. *et al.* 1998, Physical and functional interaction of murine and Xenopus Smad7 with bone morphogenetic protein receptors and transforming growth factor-beta receptors, *J. Biol. Chem.* 273:25364–25370.

Spemann, H. and Mangold, H., 1924, Über die Inducktion won Embryoanalagen durch Implantation artfremder Organisatoren, *Roux Arch. Entwicklungsmech* 100:599–638.

Spemann, H. and Mangold, H., 2001, Induction of embryonic primordia by implantation of organizers from a different species. 1923, *Int. J. Dev. Biol.* 45:13–38.

St-Jacques, B. and McMahon, A.P., 1996, Early mouse development: Lessons from gene targeting, *Curr. Opin. Genet. Dev.* 6:439–444.

Stanford, W.L., Cohn, J.B., and Cordes, S.P., 2001, Gene-trap mutagenesis: Past, present and beyond, *Nat. Rev. Genet.* 2:756–768.

Streit, A. and Stern, C.D., 1999, Establishment and maintenance of the border of the neural plate in the chick: Involvement of FGF and BMP activity, *Mech. Dev.* 82:51–66.

Streit, A., Lee, K.J., Woo, I., Roberts, C., Jessell, T.M., and Stern, C.D., 1998, Chordin regulates primitive streak development and the stability of induced neural cells, but is not sufficient for neural induction in the chick embryo, *Development* 125:507–519.

Streit, A., Berliner, A.J., Papanayotou, C., Sirulnik, A., and Stern, C.D., 2000, Initiation of neural induction by FGF signalling before gastrulation, *Nature* 406:74–78.

Tam, P.P. and Steiner, K.A., 1999, Anterior patterning by synergistic activity of the early gastrula organizer and the anterior germ layer tissues of the mouse embryo, *Development* 126:5171–5179.

Thomas, P. and Beddington, R., 1996, Anterior primitive endoderm may be responsible for patterning the anterior neural plate in the mouse embryo, *Curr. Biol.* 6:1487–1496.

Tiedemann, H. and Tiedemann, H., 1956, Versuche zur chemischen kennzeichnung von embryonale induktionsstoffen, *Z. Physiol. Chem.* 306:7–32.

Toivonen, S. and Wartiovaara, J., 1976, Mechanisms of cell interaction during primary embryonic induction studied in transfilter experiments, *Differentiation* 5:61–66.

Trindade, M., Tada, M., and Smith, J.C., 1999, DNA-binding specificity and embryological function of Xom (Xvent-2), *Dev. Biol.* 216:442–456.

Tsang, M., Kim, R., de Caestecker, M.P., Kudoh, T., Roberts, A.B., and Dawid, I.B., 2000, Zebrafish nma is involved in TGFbeta family signaling, *Genesis* 28:47–57.

Tsukazaki, T., Chiang, T.A., Davison, A.F., Attisano, L., and Wrana, J.L., 1998, SARA, a FYVE domain protein that recruits Smad2 to the TGFbeta receptor, *Cell* 95:779–791.

Tsuneizumi, K., Nakayama, T., Kamoshida, Y., Kornberg, T.B., Christian, J.L., and Tabata, T., 1997, Daughters against dpp modulates dpp organizing activity in Drosophila wing development, *Nature* 389: 627–631.

van Straaten, H.W., Hekking, J.W., Beursgens, J.P., Terwindt-Rouwenhorst, E., and Drukker, J., 1989, Effect of the notochord on proliferation and differentiation in the neural tube of the chick embryo, *Development* 107:793–803.

Vincent, J.P. and Gerhart, J.C., 1987, Subcortical rotation in Xenopus eggs: An early step in embryonic axis specification, *Dev. Biol.* 123:526–539.

von Bubnoff, A. and Cho, K.W., 2001, Intracellular BMP signaling regulation in vertebrates: Pathway or network?, *Dev. Biol.* 239:1–14.

Waddington, C.H., 1932. Experiments on the development of chick and duck embryos, cultivated in vitro, *Philos. Trans. R. Soc. Lond. B. Biol. Sci.* 221:179–230.

Waddington, C.H., 1934, Experiments on embryonic induction. *J. Exp. Biol.* 11:211–227.

Waddington, C.H. and Schmidt, G.A., 1933, Induction by heteroplastic grafts of primitive streak in birds, *Roux Arch. Entwicklungsmech* 128: 522–563.

Warga, R.M. and Kimmel, C.B., 1990, Cell movements during epiboly and gastrulation in zebrafish, *Development* 108:569–580.

Weinstein, D.C. and Hemmati-Brivanlou, A., 1999, Neural induction, *Ann. Rev. Cell. Dev. Biol.* 15:411–433.

Wilson, S.I. and Edlund, T., 2001, Neural induction: Toward a unifying mechanism, *Nat. Neurosci.* 4 (Suppl):1161–1168.

Wilson, P.A. and Hemmati-Brivanlou, A., 1995, Induction of epidermis and inhibition of neural fate by Bmp-4, *Nature* 376:331–333.

Wilson, S.I., Graziano, E., Harland, R., Jessell, T.M., and Edlund, T., 2000, An early requirement for FGF signalling in the acquisition of neural cell fate in the chick embryo, *Curr. Biol.* 10:421–429.

Wilson, S.I., Rydstrom, A., Trimborn, T., Willert, K., Nusse, R., Jessell, T.M. et al. 2001, The status of Wnt signalling regulates neural and epidermal fates in the chick embryo, *Nature* 411:325–330.

Winnier, G., Blessing, M., Labosky, P.A., and Hogan, B.L., 1995, Bone morphogenetic protein-4 is required for mesoderm formation and patterning in the mouse, *Genes Dev.* 9:2105–2116.

Wrana, J.L., Attisano, L., Wieser, R., Ventura, F., and Massague, J., 1994, Mechanism of activation of the TGF-beta receptor, *Nature* 370:341–347.

Yuan, S. and Schoenwolf, G.C., 1998, De novo induction of the organizer and formation of the primitive streak in an experimental model of notochord reconstitution in avian embryos, *Development* 125: 201–213.

Yuan, S. and Schoenwolf, G.C., 1999, Reconstitution of the organizer is both sufficient and required to re-establish a fully patterned body plan in avian embryos, *Development* 126:2461–2473.

Yuan, S., Darnell, D.K., and Schoenwolf, G.C., 1995, Identification of inducing, responding, and suppressing regions in an experimental model of notochord formation in avian embryos, *Dev. Biol.* 172:567–584.

Zhang, H. and Bradley, A., 1996, Mice deficient for BMP2 are nonviable and have defects in amnion/chorion and cardiac development, *Development* 122:2977–2986.

Zimmerman, L.B., De Jesus-Escobar, J.M., and Harland, R.M., 1996, The Spemann organizer signal noggin binds and inactivates bone morphogenetic protein 4, *Cell* 86:599–606.

Zimmerman, C.M., Kariapper, M.S., and Mathews, L.S., 1998, Smad proteins physically interact with calmodulin, *J. Biol. Chem.* 273:677–680.

2

Cell Proliferation in the Developing Mammalian Brain

R. S. Nowakowski and N. L. Hayes

BOX 1. Nomenclature

Stem cells—Cells that can produce neurons, glia, progenitor cells and also more stem cells.

Progenitor cells—Cells that can produce one lineage (e.g., neurons or glia) and more progenitor cells. In some systems the distinction between progenitor and stem cells may be a matter of degree of "stemness" (Blau *et al.*, 2001).

Neurogenesis—The production of the cells of the nervous system (both neurons and glia).

Neuronogenesis—The production of neurons.

Neuronogenetic interval (NI)—The period of time during which neurons arising in the PVE become permanently post-proliferative.

Pseudostratified ventricular epithelium (PVE)—A population of proliferating cells that lines the ventricles of the brain; the PVE is, for the most part, co-Extensive with the VZ.

Secondary proliferative population (SPP)—A population of proliferating cells that is adjacent to the ventricular zone; the SPP is, for the most part, coextensive with the SVZ.

Ventricular zone (VZ)—A cytoarchitectonically defined layer that is adjacent to the ventricles of the brain.

Subventricular zone (SVZ)—A cytoarchitectonically defined layer that is adjacent to the VZ and that appears after the VZ.

As the neural tube closes, the future brain consists of a single layer of cells that lines the lumen of the tube. The lumen is the developing ventricular system of the brain, and the layer of cells is the ventricular zone (VZ). The proliferating cells of the VZ will either directly or indirectly give rise to all of the cells of the developing central nervous system (CNS).

There are four major proliferative populations in the developing brain. The first of these to appear is the ventricular zone, which is the name agreed upon by the Boulder Committee (1970) as part of an effort to standardize and clarify a nomenclature that sometimes did not reflect accurately the known functions of the layers of the developing CNS. At the same time, the Boulder Committee recognized a second proliferative zone, the subventricular zone (SVZ) that develops later in much of the CNS. Two other proliferative populations arise in specific locations and give rise to specific populations of cells. These are the external granule cell layer of the cerebellum and the subhilar proliferative zone in the dentate gyrus. By the end of the developmental period, these four proliferative populations will give rise to all of the cells of the adult brain (with the exception of a small number that migrate into the brain from the periphery, and two of them, the SVZ and the subhilar zone in the dentate gyrus will continue to proliferate and produce neurons destined for limited areas of the nervous system into adulthood. The regulation of proliferation in these four proliferative zones is responsible for producing the right number of cells of the appropriate classes for all of the subdivisions of the CNS.

THE VENTRICULAR ZONE

The basic cellular organization of the VZ was recognized by both His (1889, 1897, 1904) and Ramon y Cajal (1894, 1909–1911). Both of these giants of the field recognized that the VZ was several cell diameters in thickness and that there were mitotic figures adjacent to the ventricular surface (Fig. 1). His thought that the mitotic figures were "germinal cells" ("*Keimzellen*") and that after mitotic division one daughter cell remained adjacent to the ventricular surface to divide again whereas the other became a postmitotic "neuroblast" that migrates away from the VZ and eventually develops into a neuron. His also thought that the remaining cells of the VZ were a syncytium of "spongioblasts" that give rise to the glial cells of the CNS. Cajal's views were similar to His' in that he also thought that the mitotic figures on the ventricular surface were a separate population of germinal cells that give rise to neuroblasts, but he did not think that the remaining cells of the epithelium,

R. S. Nowakowski and N. L. Hayes • Department of Neuroscience and Cell Biology, UMDNJ-Robert Wood Johnson Medical School, Piscataway, NJ 08854.

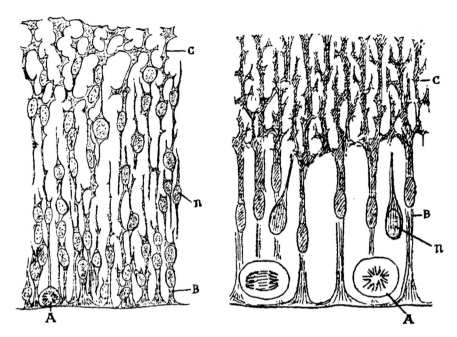

FIGURE 1. Early conceptions of the cellular organization of the early ventricular zone. On the left is a drawing from His (1889) and on the right is a drawing from Ramon y Cajal (1894). In both drawings the ventricular surface is at the bottom. His shows a germinal cell (A) in mitosis near the ventricular surface, the spongioblasts (B) forms a syncytium, and neuroblasts (n) migrating from the germinal zone at the ventricular surface to the marginal zone (C). Ramon y Cajal accepted the general conceptual framework put forward by His of the separate populations of germinal cells, neuroblasts, and spongioblasts, but he did not believe that the spongioblasts formed a syncytium (Jacobson, 1991). Thus, Ramon y Cajal's drawings of germinal cells and spongioblasts tend to look more like independent cellular entities, that is, more like our modern understanding of the structure of the ventricular zone.

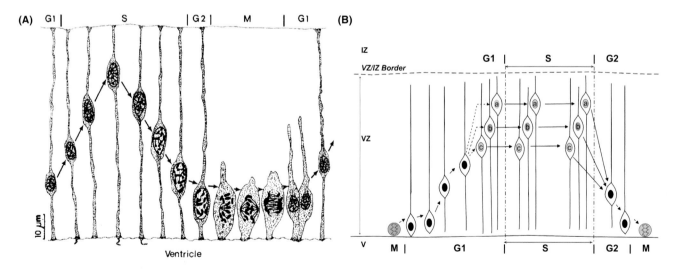

FIGURE 2. (A) A schematic diagram of Sauer's view of interkinetic nuclear migration (from Jacobson, 1991). During the cell cycle the cell nucleus moves to different levels of the ventricular zone. Mitosis occurs at the ventricular surface; the nucleus moves abventricularly during G1 and then again adventricularly before M. (B) A schematic diagram showing the correlation of interkinetic movements with the phases of the cell cycle (from Hayes and Nowakowski, 2000). During the cell cycle, both the direction and rate of movement of the nuclei are correlated with the phase of the cell cycle. Following mitosis, the nucleus of a PVE cells moves away from the ventricular surface. In the outer half of the VZ it enters the S-phase zone, where the nuclei labeled "a," "b," and "c" represent nuclei distributed throughout the thickness of the S-phase zone in the outer half of the VZ. During S, the nuclei do not seem to move, but as they finish S and enter G2, they move rapidly back to the ventricular surface.

that is, the spongioblasts, formed a syncytium but that they were independent cellular entities.

An alternative interpretation of the histological picture was presented by Vignal (1888), Schaper (1897a, b), and Koelliker (1896). These authors suggested that the so-called "germinal cells" adjacent to the ventricular surface were a transitional form of the "spongioblasts." Schaper (1897a, b) was most explicit about this; he suggested that the germinal cells of His "are not to be considered as a special type of cell in contrast to the main epithelial cells" but rather that they were part of the same population "in the process of continuous proliferation." In other words, Schaper suggested that the germinal cells and spongioblasts of His were really cells of the same type which move to different levels of the VZ during different phases of the cell cycle.

These issues were examined again in the late 1930s when F. Sauer (1935, 1936, 1937) undertook a careful cytological analysis of the VZ in the neural tube of chick and pig embryos. F. Sauer observed that the nuclei of the cells of the VZ were not identical in size or appearance and that a logical picture of the transitions of the cells through the cell cycle could be constructed from the distribution of these nuclei through the thickness of the VZ (Fig. 2(A)). With improved histological methods F. Sauer was also able to show convincingly that the VZ was not a syncytium, but that each cell had a distinct plasma membrane and that each cell was columnar in shape connected to both the ventricular and pial surfaces. He suggested, in essence, that after a mitotic division that the nuclei of the VZ cells move away from the ventricular surface during G1, that they remain in the outer half of the VZ during S, and that they return to the ventricular surface during G2 where they divide during M.

An important confirmation of these nuclear movements was made shortly after the introduction of the DNA precursor tritiated thymidine (^{3}H-thymidine) as a tracer. These key experiments showed that shortly after an exposure to ^{3}H-thymidine all of the labeled nuclei were in the outer half of the VZ but that a few hours later the labeled nuclei were adjacent to the ventricular surface (Sauer, 1959; Sauer and Chittenden, 1959; Sauer and Walker, 1959; Sidman *et al.*, 1959) which unequivocally showed that at least some of the cells of the VZ comprise a single population which have nuclear movements that correlate with the cell cycle. These nuclear movements which are collectively referred to as interkinetic nuclear migration have since been confirmed to occur in all columnar and pseudostratified columnar epithelia. The interkinetic movements of the nuclei are the hallmark of the VZ in the developing CNS and none of the other proliferating zones exhibit such nuclear movements. In the VZ, more recent experiments using two DNA tracers (bromodeoxyuridine [BUdR] and ^{3}H-thymidine, Fig. 3) simultaneously and clearly show the separation of cells in G2 vs S (Hayes and Nowakowski, 2000). In addition, by changing the interval between the exposure to the two tracers it was possible to show that both the speed and the direction of movements of the nuclei is closely correlated with the phase of the cells cycle (Fig. 2(B)). These results show, as original suggested by F. Sauer, that during G1 the nucleus moves outwards away from the ventricular surface, that during S it is stationary, and that during G2, the nucleus moves rapidly toward the ventricular surface. The inward movement of the nucleus during G2 is quite rapid, occurring within 20–40 min. In contrast, the outward movement of the nucleus during G1 is about 4–10 times slower.

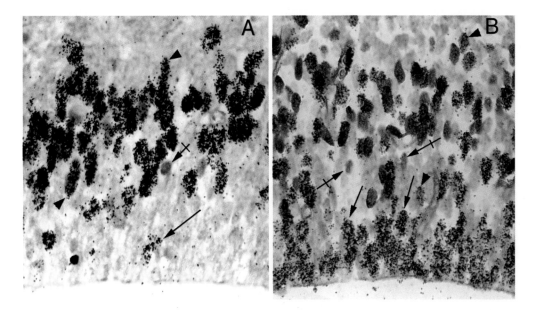

FIGURE 3. The labeling pattern in the ventricular zone (VZ) of an E14 mouse that received ^{3}H-TdR followed either 0.5 (A) or 2.0 (B) hr later by BUdR and was killed 0.5 hr after the BUdR injection. Some cells labeled only with ^{3}H-TdR are indicated by an arrow; cells labeled only with BUdR are indicated by a crossed-arrow. The cells labeled by the BUdR define the S-phase, and they are located in the S-phase zone in the outer half of the VZ. The ^{3}H-TdR-only labeled cells have left the S-phase during the period between the two injections. The 2 hr period (B) is long enough for many of the nuclei of the ^{3}H-TdR-only labeled cells to move towards the ventricular surface where they will divide.

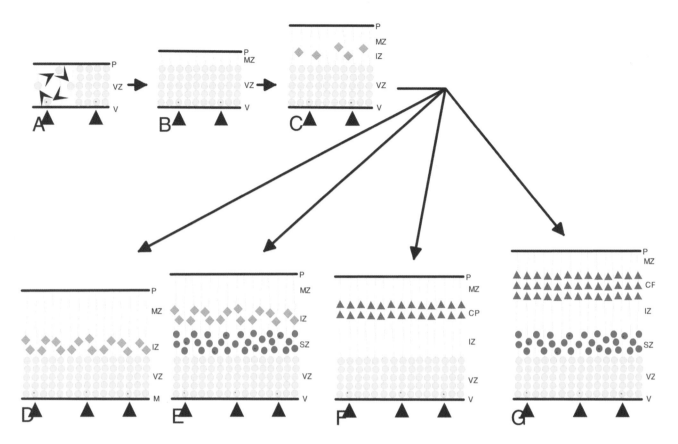

FIGURE 4. Radial differentiation of the neural tube. (A), (B), and (C) are schematic diagrams of the early stages of the radial differentiation of the neural tube through which every part of the CNS passes. (D), (E), (F), and (G) are schematic diagrams of various options for the later stages of the radial differentiation of the neural tube. Each of these options is characteristic of a different part of the neural tube. (A) At the time of closure of the neural tube its wall consists of a population of proliferating cells organized into a pseudostratified columnar epithelium, known as the ventricular zone (VZ). In this proliferative zone the nuclei of the cells are stratified, but each cell has processes that contact the ventricular (V) and pial (P) surfaces of the neural tube. As diagrammed on the right-hand side of the drawing, mitosis occurs at the pial surface (asterisks), and during the cell cycle the nucleus of each cell moves to a different level. DNA synthesis, for example, occurs in the outer half of the ventricular zone. This to-and-fro movement of the cell nuclei is known as interkinetic nuclear migration and means that all cells, even though they are apparently at different levels, are part of the proliferative population. (B) The next zone to appear during the radial differentiation of the neural tube is the marginal zone (MZ) which is an almost cell-free zone between the ventricular zone and the pial surface. (C) The intermediate zone (IZ), which contains the first postmitotic cells in the nervous system, is the next to form. This zone is located between the ventricular zone and the marginal zone. (D) In some parts of the neural tube, such as the spinal cord, the postmitotic cells derived from the ventricular zone aggregate and mature in a densely populated intermediate zone. (E) In some areas, such as the dorsal thalamus, a second proliferative zone, the subventricular zone (SVZ), is formed between the ventricular zone and the intermediate zone. In the subventricular zone, interkinetic nuclear migration does not occur; instead mitotic figures (asterisks) are found scattered throughout the thickness of the zone (DNA synthesis also occurs throughout the thickness of the subventricular zone). The postmitotic cells derived from both the ventricular and subventricular zones aggregate and mature in a densely populated intermediate zone. (Note, however, that any cells derived from the ventricular zone must cross the subventricular zone.) (F) In the hippocampus, the postmitotic cells derived from the ventricular zone migrate across a sparsely populated intermediate zone to form a cortical plate. (G) In the cerebral cortex, postmitotic cells derived from both the ventricular and subventricular zones migrate across a sparsely populated intermediate zone to form a cortical plate. Abbreviations: V: ventricular surface; VZ: ventricular zone; SZ: subventricular zone; IZ: intermediate zone; CP, cortical plate; MZ, marginal zone; P, pial surface.

The histological appearance of the VZ is remarkably constant, except for slight variations in thickness, with little variation regionally, as a function of time or in different species. However, the fact that different portions of the wall of the neural tube can develop into the highly different areas of the adult brain is *de facto* evidence that the output and capacity of the VZ must be remarkably variable. One early expression of this variability is the appearance of a second proliferative zone, in some regions of the CNS, but not in others. This second proliferative zone, which is known as the SVZ, appears adjacent to the VZ. This second

zone, known as the SVZ, differs in several ways from the VZ (Fig. 4). It is attractive to speculate that the SVZ is a phylogenetically recently acquired specialization. For example, the hippocampus is classified as an archicortical (i.e., "old" cortex) structure and the neurons of its major subdivisions (areas CA1, CA2, and CA3) are all derived from the VZ (Nowakowski and Rakic, 1981). In contrast, in the neocortex (i.e., "new" cortex) the SVZ is substantial and, although it is unlikely to contribute large numbers of neurons to the neocortex (Takahashi *et al.*, 1995a), it produces glial cells and also neurons in other parts of the telencephalon (Goldman, 1995; Garcia-Verdugo, 1998). A similar contrast occurs in the developing diencephalon in which the hypothalamus lacks a SVZ whereas other diencephalic subdivisions have both ventricular and SVZ (Rakic, 1977). The SVZ appears early and becomes greatly enlarged in the ganglionic eminence, a population of proliferating cells that produces the striatum, parts of the basal forebrain, and a population of interneurons that migrate into the neocortex (Corbin et al., 2001; Wichterle et al., 2001; Anderson et al., 2002; Nery et al., 2002; Powell et al., 2003). The cells of the SVZ, in contrast, neither maintain an attachment to the ventricular or pial surfaces nor do their nuclei move as they move through the cell cycle (Sidman, 1970). The contributions of the SVZ to the adult brain are important and most of the glia for most of the brain are produced there (Goldman, 1995). In addition, in some areas of the brain a significant number of neurons are also produced in the SVZ (Garcia-Verdugo, 1998).

The VZs and the SVZs are cytoarchitectonic entities; that is, they are defined by their appearance in histological sections. The secondary proliferative population (SPP) arises from the primordial pseudostratified ventricular epithelium (PVE) (Smart, 1972; Altman and Bayer, 1990; Halliday and Cepko, 1992; Takahashi et al., 1993) but comes to have a more diffuse and widespread distribution through the cerebral wall overlying the VZ although it overlaps the PVE at the outer fringe of the VZ (Takahashi et al., 1993). The distribution of SPP to the architectonically defined SVZ, in the depths of the intermediate zone abutting the VZ, was originally emphasized by the Boulder Committee. The SPP is a principal spawning ground for neuroglial cells (Smart, 1961; Smart and Leblond, 1961; Privat, 1975; Mares and Bruckner, 1978; Smart and McSherry, 1982; LeVine and Goldman, 1988a, b; Levinson and Goldman, 1993). Neurons of the olfactory bulb (Hinds, 1968a, b; Luskin, 1993; Lois and Alvarez-Buylla,1994; Luskin and McDermott, 1994) and possibly, a small number of neurons destined for the neocortex

(Reynolds and Weiss, 1992; Levinson and Goldman, 1993) may also undergo their terminal divisions in this proliferative population. The cells of the SPP, in contrast to those of the PVE, are not attached to each other as a pseudostratified epithelium (Rakic et al., 1974), and this population does not undergo interkinetic nuclear migration in the course of the cell cycle (Boulder Committee, 1970; Smart, 1972; Altman and Bayer, 1990).

The Cell Cycle in the Ventricular Zone

The use of the DNA precursors ^{3}H-thymidine and BUdR has also provided other insights into the behavior of the cells of the VZ. Notably, over the decades there have been a variety of methods used to measure the length of the cell cycle in various regions of the neural tube (for reviews see Sidman, 1970; Jacobson, 1991; Nowakowski et al., 2002). In particular, the dynamics of the cell cycle for both the PVE and the SPP, that is, the proliferating cells of the SVZ, are now well known (Caviness et al., 1995; Takahashi et al., 1995a, b, 1996a, 1997). These studies have examined the entire period of time for mouse neocortical during the neuronogenetic interval (NI), defined as the period of time during which neurons arising in the PVE become permanently postproliferative. The results show that the amount of time required for a single cell to pass through one cell cycle, that is, from the beginning of one G1 to the beginning of the next G1, varies systematically during the development of the neocortex (Fig. 5). At the time of the production of the first neurons in the mouse neocortex, measurements of the cell cycle using cumulative labeling with BUdR show that the total length of the cell cycle (T_c) is about 8 hr, with an S-phase of about 3 hr, G2+M of about 2 hr, and G1 of about 3 hr (Takahashi et al., 1995). As development proceeds, the cell cycle lengthens until at the end of the period of neuron production T_c reaches ~18 hr; S and G2+M remain approximately constant at 3–4 hr and 2 hr,

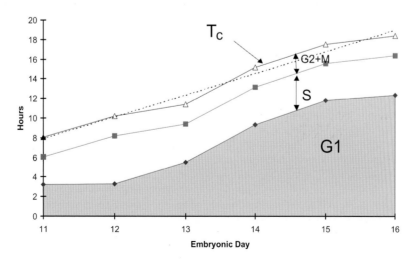

FIGURE 5. In the developing neocortex, the cell cycle lengthens systematically over the course of the six-day period during which neurons are produced. At the onset of E11, the cell cycle is ~8 hr, and by the end of the E16 it is over 18 hr. During this time the length of G2+M and S do not change systematically, and, hence, most of the lengthening is within G1.

respectively, and thus, virtually all of the change in T_c is due to an increase in the length of G1 (T_{G1}) (Takahashi *et al.*, 1995b). The increase in T_{G1} is dramatic, from 3 hr early in development to almost 12 hr late in development. As a result of the lengthening of the cell cycle, there are different numbers of cell cycles on each of the six days of production of neurons for the neocortex. For example, on the first day of the NI (E11) the cell cycle starts at 8 hr and lengthens to over 10 hr; thus, there is time for approximately 2.5 cell cycles, where as on the sixth and last day of the NI (E16) the cell cycle is over 18 hr, and thus, there is sufficient time to complete only just over one cell cycle. By integration under the linear fit to the T_c (Fig. 5), it can be calculated that there is sufficient time for 11 cell cycles during the entire six-day period (Takahashi *et al.*, 1995b).

The question of the range of cell cycle lengths present in the PVE during the developmental period has been addressed using different cell cycle measuring techniques. With cumulative labeling methods (Nowakowski *et al.*, 1989), the percentage of labeled cells rises linearly if the population is proliferatively homogeneous, that is, the length of the cell cycle and the S-phase are similar for all cells. Thus, from these measurements we have estimated that the PVE of the mouse is 80–90% homogeneous (Takahashi *et al.*, 1995a) with cell cycle parameters within 10% of the mean (Nowakowski *et al.*, 1989). This means, of course, that as many as 10–20% of the cells might have cell cycle parameters outside of this range. The cumulative labeling method gives an estimate for a maximum value of T_c because it is derived from the detection of an inflection point in the slope of the rising labeling index. This inflection point corresponds to the time required to label the entire proliferative population and occurs when the last (or slowest cycling) proliferating cell that was not labeled by the first injection enters the S-phase and becomes labeled (see Nowakowski *et al.*, 1989). In contrast, an alternative method for measuring cell cycle lengths, the percent labeled mitosis method, gives an estimate of the minimum value of T_c because it detects the time required for the first (or fastest cycling) proliferating cell to transit the entire cell cycle and enter M-phase for a second time (Kaufmann, 1968; Hoshino *et al.*, 1973; Hamilton and Dobbin, 1983a, b). The difference between the maximum and minimum estimates of T_c is an estimate of the range in T_c for the slowest vs fastest cycling cells. When both methods are used to identify the range of the cell cycle lengths for the neocortical PVE, it was found that approximately 99% of the cells have a cell cycle within 5–7% of the mean (Hamilton and Dobbin, 1983a, b). This means that if there is present in the PVE a population of proliferating cells with either a longer or a shorter cell cycle, it comprises only about 1% of the total. Interestingly, this proportion corresponds to estimates of the proportion of true stem cells made by van der Kooy and colleagues (for review see Seaberg and van der Kooy, 2003).

Extensive data for changes in the cell cycle length are not available for other species, but there is some evidence that in the neocortex of primates, the cell cycle is longer (Kornack and Rakic, 1998), and that there are about 28 cell cycles required to make the monkey neocortex. In the human, the comparable period of time during which neurons are produced is much longer, about 120 days (Caviness *et al.*, 1995). From this and other considerations, it has been estimated that about 34–35 cell cycles would be required to make all of the neurons of the human neocortex (Caviness *et al.*, 1995).

Overall, this extensive analysis of cell proliferation in the neocortex indicates that the proliferating cells of the VZ form a coherent group of cells that have a similar cell cycle length that lengthens as development proceeds. In addition, most of the lengthening of the cell cycle is due to an elongation of the G1-phase. On the surface, this seems reasonable as the G1-phase of the cell cycle is generally considered to be regulatory. Another region of the brain for which cell cycle data is available that covers the entire period of neurogenesis is the retina (Alexiades and Cepko, 1996). The developing retina differs from the developing neocortex, however, in that both G1 and S lengthen. The lengthening of the cell cycle in the retina is detectable even when measured over a period of a few days (Rachel *et al.*, 2002).

The Output from the Ventricular Zone

The output from the PVE is the population of neurons and other cell types that populate the mature brain and the cells that "seed" the SVZ. *A priori*, the mechanisms that control this output depend on four factors, the number of proliferating cells, the length of the cell cycle, the period of time that the proliferating population exists, and the proportion of daughter cells that exit vs remain in the proliferating population at each pass through the cell cycle. By definition the beginning of the NI coincides with the first cell cycle during which neurons are produced. Thus, for the first cell cycle of the NI and for each of the subsequent cell cycles, to a total of 11, some of the daughter cells of the proliferating population exit the cell cycle (Fig. 6). The daughter cells that exit from the cell cycle are called "Q" cells, signifying that they are proliferatively *q*uiescent (or that they *q*uit the cell cycle). The daughter cells that remain in the cell cycle are called "P" cells because they re-enter the S-phase and, hence, continue to proliferate. If, for the moment, the possibility of cell death within the proliferative population is ignored (considered in more detail below), it is clear that all of the daughter cells must select either a P or a Q fate and, hence, the proportions P and Q must add up to 1 (or $P + Q = 1$). Examination of the "old" tritiated thymidine birthday literature (e.g., Angevine and Sidman, 1961; Caviness and Sidman, 1973; Rakic, 1974; Stanfield and Cowan, 1979; Nowakowski and Rakic, 1981; Rakic and Nowakowski, 1981) shows that in various cortical structures different numbers of neurons are born on each of the various days of development. This means that P and Q must change dynamically as development proceeds (for review see Nowakowski *et al.*, 2002).

From first principles, it seems clear that prior to the onset of neuron production, P must be 1 and Q must be 0. Similarly, at the end of the NI, in order to account for the disappearance and involution of the PVE at the end of the NI, P must be decreased to 0 and Q must be increased to 1. This means that during the NI, P decreases from 1 to 0, and Q increases from 0 to 1. In the neocortex, measurements of Q made on each of the days of the NI in both dorsomedial and rostrolateral cortex (Miyama *et al.*, 1997)

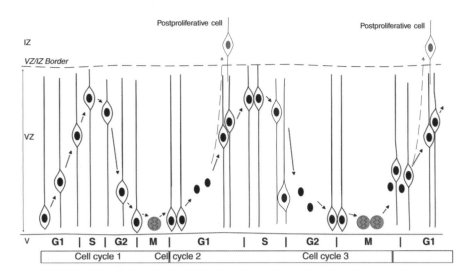

FIGURE 6. The cell cycle in the ventricular zone of the developing CNS: This schematic diagram illustrates the interkinetic movement of the nuclei of the cells comprising the proliferative ventricular epithelium of the ventricular zone (VZ). With each pass through the cell cycle the nucleus of a single cell moves from its starting position at the ventricular surface at the beginning of G1 to the border of the VZ where it enters S. During G2, the nucleus again moves down to the ventricular surface where it enters M and divides to form two cells. With each pass through the cell cycle some postmitotic neurons are produced. The postmitotic neurons migrate away from the VZ to produce the structures of the adult brain (in this case, the cerebral neocortex). During the production of the neocortex in the mouse, the cell cycle lengthens with each cell cycle and there are a total of 11 cell cycles.

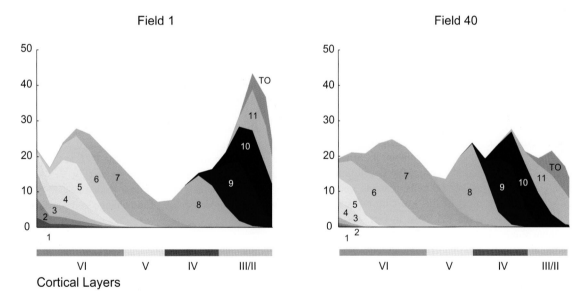

FIGURE 7. The laminar distribution of neurons produced during each of the 11 cell cycles of the Neuronogenetic Period for two nonadjacent cytoarchitectonic areas of the neocortex: Field 1, which is located dorsomedially, and Field 40, which is rostrolaterally. Each cell cycle produces neurons that are distributed in several layers, but there is a systematic change in the laminae of residence with each sequential cell cycle. Also, each layer receives neurons that are produced during more than one cell cycle. Note that the neurons of layers VI and V are produced during the first seven cell cycles, that is, during the period when the neopallium is still expanding (Fig. 8).

are shown in Fig. 7. In Fig. 7, the abscissa shows the 11 cell cycles of the NI, and the data for each of the embryonic days is plotted at its appropriate proportional position on this 11-cell-cycle scale. Thus, during this six-day NI of the neocortex, the nuclei of the PVE makes 11 round-trips through the cell cycle, and at each pass through the cell cycle the population produces an ever-increasing proportion of postmitotic neurons (Fig. 6)

(Caviness *et al.*, 1995; Takahashi *et al.*, 1997). The path of $Q = 0 \rightarrow Q = 1$ increases monotonically; P is the complement of Q, and thus, the path of $P = 1 \rightarrow P = 0$ decreases monotonically (Fig. 8). The P and Q curves intersect at $P = Q = 0.5$, which is between cell cycle 7 and 8. This divides the NI into two qualitatively different periods. During the first period, when $Q < 0.5$ and $P > 0.5$, the proliferative population expands. Since the

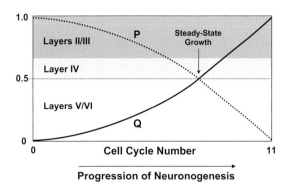

FIGURE 8. In the ventricular zone of the developing neocortex Q and P are complementary (i.e., $P = 1 - Q$), and as Q increases from 0 to 1, P decreases from 1 to 0. When P is > 0.5, more daughter cells re-enter the cell cycle than leave it, and the neopallium will expand. In the mouse, between cell cycle 7 and 8 a point is reached when this expansion stops, and both P and Q are equal to 0.5. This occurs approximately as the cells that will reside in the vicinity of the border between layers V and IV are produced.

VZ only increases slightly in thickness during this period, and the packing density of the cells remains constant (Takahashi *et al.*, 1996a), the bulk of this expansion results in an increase in the surface area of the VZ and, hence, in the surface area of the entire developing neocortex. As Q increases to 0.5 and P decreases to 0.5, a point is reached where "steady-state growth" is achieved transiently. At this time, the number of P cells produced is exactly enough to replenish the proliferative population; an equivalent number of Q cells are produced that leave the PVE. As development proceeds, however, a second period, when $Q > 0.5$ and $P < 0.5$, is reached. During this second period, the proliferative population contracts. This is because fewer daughter cells re-enter the S-phase than are needed to maintain it. Since the ventricular surface does not contract during the developmental period and the packing density of the cells remains constant (Takahashi *et al.*, 1996a, b), most of this must be reflected in a reduction of the thickness of the VZ. This, in fact, correlates with what is known about the development and involution of the VZ (Nowakowski *et al.*, 2002). It has been suggested that other species follow the same pathway of changes for P and Q except that they take more or fewer cell cycles than the 11 cell cycles needed to make mouse neocortex (Caviness *et al.*, 1995).

In the neocortex, the crossover point corresponding to the time when $P = Q = 0.5$, at which time the neocortical primordium ceases to expand, occurs as the NI passes through cell cycle 7. This point is important because it is when the expansion of the PVE stops. A cycle-by-cycle analysis of the laminar position of the neurons generated at each of the 11 cell cycles shows that this crossover point occurs as the last neurons of layer V are being produced (Fig. 7). This means that virtually all of the deep layers of the neocortex are produced during the first, expansion phase of the NI, and that virtually all of the superficial layers are produced during the second, extinction, phase. It is not known if the crossover point has similar significance in other regions of the developing CNS.

The pathway of changes in Q and P from $Q = 0 \rightarrow Q = 1$ and $P = 1 \rightarrow P = 0$ determines three properties of the proliferative population: (1) the life span of the PVE population, (2) the expansion of the PVE population and, hence, of the neocortical primordium, and (3) the output from the population, that is, the number of neurons produced both per cycle and also during the total NI. Each of these properties can be approached quantitatively (for review see Nowakowski *et al.*, 2002). The life span of the NI is most closely regulated by changes in Q as it changes from $Q = 0 \rightarrow Q = 1$. Neuron production begins as soon as Q becomes greater than 0 and the PVE disappears when $Q = 1$ because at this point both daughter cells would have to leave the cell cycle. Thus, the mechanisms which determine the changes in Q, that is, the changes in the probability at each successive cycle that postmitotic daughter cells will exit the proliferative population, determine the number of cycles in the NI. At present, there is no clear molecular explanation for the changes in Q.

The expansion of the PVE is also specified by the changes in Q and P. In this case, the expansion at each cell cycle is dictated by P, that is, the probability that the daughter cells will re-enter the cell cycle. The amount of expansion is twice the value of P at each cell cycle. In addition, the expansion is multiplicative at each cell cycle. For example, for the first three cell cycles of the NI, P is 0.99, 0.96, and 0.92; thus, the total expansion for a "unit volume" of the PVE during these first three cell cycles is the product of these three numbers ($1.982 \times 1.93 \times 1.846$) or 7.061. In other words, during the first three cell cycles of the NI (which occur in just over a day), the PVE expands over seven-fold.

The output (or the number of neurons formed) from a "single unit" of the PVE at each cell cycle is specified by the changes in both P and Q. At each cell cycle, the output is equal to twice the number of cells present in the PVE at the beginning of the cycle times Q. A graph of this series (Fig. 9) shows that the predicted output rises gradually to a peak and then falls. This is, of course, expected because of the change in the size of the PVE,

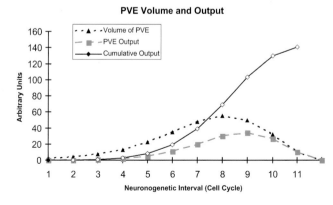

FIGURE 9. A graph of the changes in the PVE volume, PVE output per cell cycle and the cumulative PVE output as predicted by changes in P and Q (see Nowakowski et al., 2002). Note that the volume increases through cell cycle 8 when P first falls below 0.5.

which is controlled by P. It is also reassuring because it matches qualitatively the shape of the curves usually obtained from counts of the percentage of neurons born on sequential days during development (Angevine and Sidman, 1961; Caviness and Sidman, 1973; Rakic, 1974; Stanfield and Cowan, 1979; Nowakowski and Rakic, 1981; Rakic and Nowakowski, 1981). The cumulative total output at each cell cycle is simply the sum of the output for all of preceding cell cycles, and, thus, a graph of the cumulative output (Fig. 9) rises steadily to a total of over 140, indicating that in the mouse, on an average, a single PVE cell present at the beginning of the NI produces approximately 140 neurons.

The changes in the size of the PVE and developing cortical plate and the relationships of these changes can be seen more readily if results from a quantitative analysis (Nowakowski et al., 2002) are schematized (Fig. 10). At the beginning of the NI, the only cells present are the cells of the PVE; they are represented as a cube with a "unit volume" that is 1 unit high, 1 unit wide, and 1 unit deep. Since the PVE is about 6 cells high, such a unit volume would contain about 6^3 or 216 cells. During the first cell cycle, P is about 0.99 and Q is about 0.01. Thus, such a unit volume would produce only about two neurons on the first cell cycle and most of the daughter cells produced will remain in the proliferative population, and the PVE will expand. At the end of the third cell cycle the unit of the PVE has expanded to over seven times its original volume and produced about 1% of the neurons that will comprise the neocortex in the adult. Most of the cells of the neocortex are produced during the last few cell cycles (Fig. 10). This sequence of events corresponds at least qualitatively to histological observations. In principle, the neocortical VZ and, indeed, the VZ of the whole CNS contains many of these "units" arrayed across its surface. In the neocortex, it has been shown that the sequence of 11 cell cycles and the changes in P and Q in each are identical these "units," at least to the resolution that has been used so far (Miyama et al., 1997). The result of this arrangement is that different events in the sequence occur contemporaneously in different regions of the VZ (Fig. 11). The NI is initiated first in the rostrolateral cortex, and given the fact that there is more than a 24 hr difference between the rostrolateral cortex and the dorsomedial cortex (Miyama et al., 1997), when the NI is first initiated in the dorsomedial cortex, the rostrolateral cortex has already progressed into cell cycle 3 or even 4. This means that there is a gradient of maturation beginning in the rostrolateral cortex and spreading across the surface of the developing cortex. The gradient of maturation means that T_c, T_{G1}, Q, and P differ across the surface of the developing cortex (Fig. 11). Thus, at any given time the status of these proliferatively related parameters provides positional information. From the perspective of the cell cycle, this gradient divides the surface of the developing PVE into "cell cycle domains," that is, regions of the PVE in which all of the PVE cells are in the same cell cycle. As the developing cortex matures and each "unit" of the PVE progresses through the NI, these cell cycle domains "move" across the surface of the PVE defining a series of "waves" that radiate from the striatocortical fissure at the lateral edge of the neopallium (Fig. 11). In other regions of the CNS, the developmental progression of these "units" of the VZ presumably differ markedly with varied paths for $Q = 0 \rightarrow Q = 1$, and, hence, varied life spans of the VZ in different regions, and varied output and numbers of cells produced.

The Control of *P* and *Q* with Symmetric and Asymmetric Cell Divisions

The output of neurons by the PVE is controlled, in principle, by a variety of factors including the proportional representation of the three possible types of mitotic divisions: (1) symmetric nonterminal cell division (which produces two daughter cells that remain in the PVE and continue to proliferate), (2) symmetric terminal cell division (which produces two daughter cells that both migrate out of the PVE to become young neurons), and (3) asymmetric cell division (which produces one daughter cell that continues to proliferate and one that migrates out of the PVE). Changes in the proportions of these three types of mitotic divisions have been inferred from changes in the proportion of cells that enter vs leave the cell cycle (Takahashi et al., 1996a; Miyama et al., 1997), from time lapse cinematography (O'Rourke et al., 1992; Adams, 1996), from changes in the orientation of the mitotic apparatus (Smart, 1973; Chenn and McConell, 1995; Adams, 1996), and from immunohistochemistry (Chenn et al., 1995). How are such changes effected within single lineages? How might they be distributed among lineages making various cortical cell types? For example, it has been suggested that there are specific populations "reserved" in the PVE to produce either specific cell types or cells that occupy specific laminae (Dehay et al., 1993; Kennedy and Dehay, 1993; Luskin et al., 1993). Since all of the cells in a given neighborhood of the PVE are proliferating (Takahashi et al., 1995a, 1996a) with a similar cell cycle length (Cai et al., 1997a), in the absence of cell death such a "reserved population" would have a specific pattern of repeated symmetric nonterminal mitoses and would expand for several cell cycles to produce relatively large lineages of a specific and characteristic size (8, 16, 32, 32, etc.) containing only proliferating cells. Similarly, lineages following other specific patterns of proliferation, for example, repeated asymmetric divisions, would produce lineages of other specific characteristic sizes, for example, a preponderance of even-sized or odd-sized lineages, etc. The alternative to such repeated patterns of mitosis-type is the absence of pattern in the sequence of cell divisions within a lineage, in which case no specific and characteristic lineage size distribution will be produced. In general, the size distribution of lineages obtained during defined periods of development will reflect the dynamic changes in the proportions of these three types of cell divisions and will, thus, reveal any extant repeated patterns of mitosis. Note that the presence of cell death to any significant extent (cf. Blaschke et al., 1996; Thomaidou et al., 1997) would modify the specific and characteristic lineage sizes obtained, but would do so in a predictable way.

To estimate the frequency of occurrence of each of these distinct behaviors individual retrovirally labeled lineages were studied; each lineage consisting of proliferating cells in the PVE in the developing neocortex at known numbers of cell cycles after infection with a retrovirus. In contrast to most previously

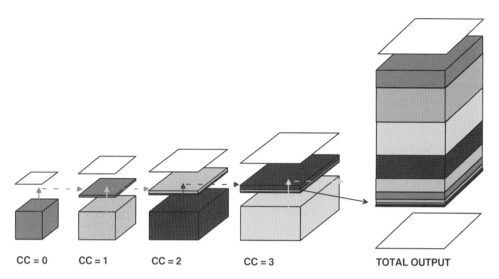

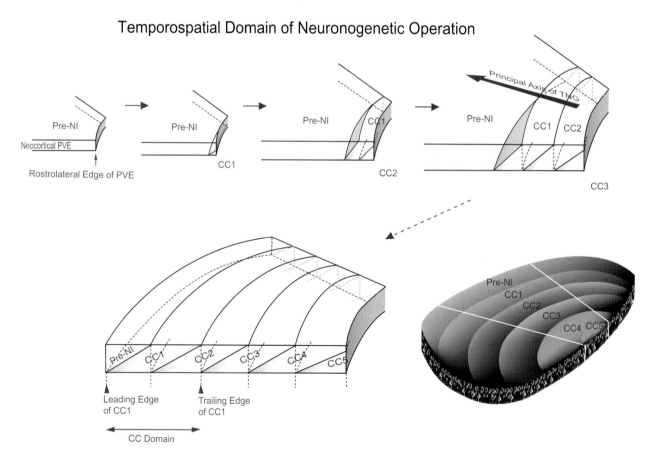

FIGURE 10. A visualization of the changes shown in the graphs of Fig. 12 and as given by the changes in *P* and *Q* per cell cycle (CC) (Nowakowski *et al.*, 2002). At the onset of the neuronogenetic interval (NI) (CC = 0), a single unit of the PVE is shown. At the next cell cycle (CC = 1) the PVE has an increased volume; the output from the first cell cycle is shown in the position of the cortical plate. At CC = 2, the PVE has increased in volume again, and now the output from the first two cell cycles is shown in the position of the cortical plate. At CC = 3, the process is repeated. In the right-hand side of the figure, the diagram shows the final Total Output of all of the 11 cell cycles of the NI. Note that the output from the first three cell cycles corresponds to only a small part of the Total Output, whereas the output of the last three cell cycles comprises about 50% of the Total Output.

Temporospatial Domain of Neuronogenetic Operation

FIGURE 11. The sequence of dynamic changes in the length of the cell cycle (and in *P* and *Q*) is initiated in the rostrolateral-most portions of the neopallium and then spreads as a gradient of maturation across the neopallial surface. This wave-like progression of maturation means that at any given time there are "domains" of the PVE that are in different states. This is, in theory, sufficient to provide a basis for cell cycle length to serve as positional information that could be involved in the development of cytoarchitectonic subdivisions.

published experiments using this method (Price, 1987; Luskin *et al.*, 1988; Walsh and Cepko, 1988, 1992; Williams *et al.*, 1991; Luskin, 1993; Mione *et al.*, 1994, 1997; Lavdas *et al.*, 1996), the resulting labeled lineages were examined after short survivals, that is, during the period that cell proliferation continues to occur, and have focused on the size of the proliferating population, that is, the cells that remain in the PVE, rather than on the cells that migrate to the cortical plate. There are three influences (Fig. 12) that could act at each cell cycle to reduce the number of cells per lineage from the maximum number that would be produced in a pure population of symmetric nonterminally dividing cells. First, some cells of the lineage could leave the cell cycle to migrate and become young neurons (Q-cells, *Q* in Fig. 1). Second, some PVE cells could die (*D* in Fig. 12). Cell death could, in theory, occur at any time during development and has been well studied in the maturing neocortex during the postnatal period (Leuba *et al.*, 1977; Finlay and Slattery, 1983; Heumann

and Leuba, 1983; Crandall and Caviness, 1984; Finlay and Pallas, 1989; Verney *et al.*, 2000). However, estimates of the magnitude of cell death occurring within the proliferative population and during the early period of cortical development vary greatly from <1.0% at any given time (Thomaidou *et al.*, 1997) to over 70% of the progenitor cells (Blaschke *et al.*, 1996). Thus, it remains unclear what role cell death in the proliferative population plays in the regulation of neuron number (Gilmore *et al.*, 2000). Third, some PVE cells could move tangentially (*T* in Fig. 12), that is, away from their sisters and cousins (Fishell *et al.*, 1993; Tan and Breen, 1993; Walsh, 1993). Such tangential movements would not affect the actual numbers of cells in the proliferative population, but would affect the apparent number of cells identified in a lineage and would concomitantly increase the putative number of lineages identified.

The cells in each retrovirally labeled lineage in the developing VZ reside in clusters (or clades) of varying size (Cai *et al.*, 1997a). The size of these clusters is dependent on the proliferative behavior of the cells in the labeled lineage, and depends on the mixture of symmetric nonterminal, symmetric terminal, and asymmetric cell divisions. There are three hypothetical mixtures of these three types of cell divisions that could occur (Fig. 13). The three Models differ only with respect to their composition of types of cell division, that is, they each have different ratios of asymmetric, symmetric nonterminal, and symmetric terminal cell divisions; however, all three Models are based on the same *P/Q* values measured using double S-phase labeling methods (Takahashi, 1996b; Miyama *et al.*, 1997). Importantly, the distribution of cluster sizes is best accounted for by the goodness-of-the-fit of the experimentally determined distribution with the distributions obtained from the model which assumes that all three types of cell divisions coexist during the entire NI, i.e., Model 1 of Fig. 13. Thus, these retroviral experiments: (1) provides evidence for the role of changes in *P/Q* in regulation of lineage size, (2) indicates that the amount of cell death and tangential movements in the PVE is low, and (3) indicate that the numbers of lineages that undergo a series of cell divisions with a repeated pattern is undetectable. In essence, these data suggest that the two daughter cells from a single cell division have their fate determined independently.

THE SUBVENTRICULAR ZONE

The SVZ was first recognized by Schaper and Cohen (1905) by the presence of mitotic figures in a location distant from the lateral ventricles. It was first shown definitively to have proliferating cells using ³H-thymidine label *in vitro* using slabs of human brain (Rakic and Sidman, 1968). The proliferating cells of the SVZ differ in two major ways from those of the VZ (Fig. 4). First, the nuclei of proliferating cells of the SVZ do not move during the cell cycle, but reflecting the fact that cells of the SVZ, in contrast to those of the VZ, are not attached to each other as a pseudostratified epithelium, this population does not undergo interkinetic nuclear migration in the course of the cell cycle (Boulder Committee, 1970; Smart, 1972; Altman and Bayer, 1990). Second, the cell bodies of the SVZ cells do not have long

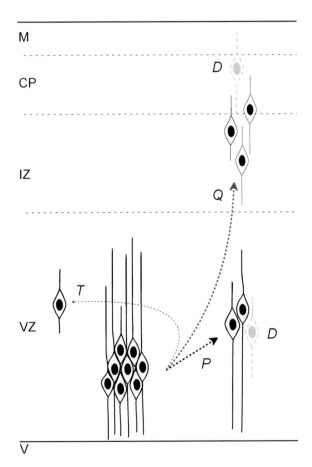

FIGURE 12. A schematic diagram depicting the influences in the ventricular zone that could affect lineage size in a single cell cycle. For a cluster of cells present at the beginning of G1 (in this example eight cells are shown) some cells could continue to proliferate (*P*), leave the proliferative population (*Q*), die (or lose the marker) in either the proliferative (*D*, VZ cell in gray with dashed lines) or postproliferative compartment (*D*, CP cell in gray with dashed lines), or move tangentially within the proliferative population (*T*). Abbreviations: M, marginal zone; CP, cortical plate; IZ, intermediate zone; VZ, ventricular zone; V, lateral ventricle.

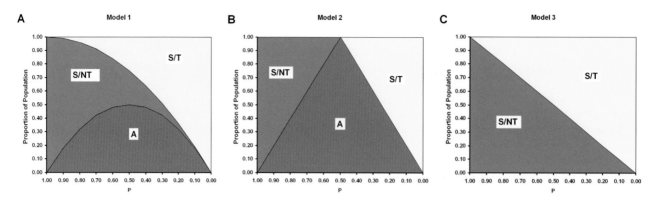

FIGURE 13. Schematic diagrams of the proportions of the changes in the proportions of symmetric non-terminal (S/NT), symmetric-terminal (S/T) and asymmetric cell divisions as a function of changes in P (abscissa) during the neuronogenetic interval. At any given time the sum of the proportions of the 3 types of cell divisions adds up to 1.0 (ordinate). The changes shown are the changes in the 3 types of cell divisions for the 3 different models developed for this study. For a detailed explanation of the assumptions of each model, see the text.

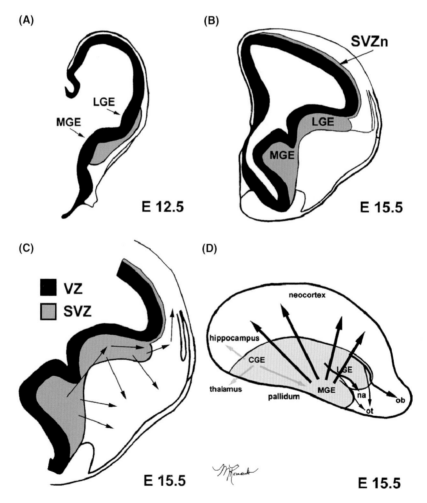

FIGURE 14. Contributions of the SVZ of the medial ganglionic eminence (MGE), lateral ganglionic eminence (LGE), and caudal ganglionic eminence (CGE) to early brain development. (A) A coronal view of the rodent forebrain germinal zones at E12.5. (B) The LGE and MGE are prominent structures in the E15.5 brain. By contrast, at this age the neocortical SVZ is unremarkable. (C) Directional movements of MGE and LGE cells as they migrate to the striatum, neocortex, and nucleus accumbens (na). Cells from the MGE also may migrate through the LGE en route to the neocortex. (D) Sagittal view of the rodent brain at E15.5 shows directional movements from the MGE, LGE, and CGE. Cells of the CGE migrate to the hippocampus, thalamus, pallidum, olfactory tract (ot), and olfactory bulb (ob). Panels (A), (B), (C) were adapted from Lavdas *et al.* (1999) and panel (D) was adapted from Wichterle *et al.* (2001). Figure modified from Brazel *et al.*, 2003.

radially oriented processes, but rather they have shorter processes that remain confined to the SVZ (Rakic *et al.*, 1974).

Regional Variation in the Subventricular Zone

Not all of the regions of the CNS have a SVZ. This and its distribution makes it attractive to speculate that the SVZ is a phylogenetically recently acquired specialization. For example, the hippocampus is classified as an archicortical (i.e., "old" cortex) structure and the neurons of its major subdivisions (areas CA1, CA2, and CA3) are all derived from the VZ (Nowakowski and Rakic, 1981). In contrast, in the neocortex (i.e., "new" cortex) the SVZ is substantial and may not contribute large numbers of neurons to the neocortex (Takahashi *et al.*, 1995a). A similar contrast occurs in the developing diencephalon in which the hypothalamus lacks a SVZ whereas other diencephalic subdivisions have both ventricular and SVZs (Rakic, 1977). The spinal cord, much of the brain stem, and the retinal also lack a SVZ. In the neocortex, at least the time of appearance of the SVZ is approximately coincident with the time of the production of the first neurons (Nowakowski and Rakic, 1981).

The regional variation in the SVZ is far more complex than simply whether or not it is present (Brazel *et al.*, 2003). Brazel and Levision (Brazel *et al.*, 2003) have recognized a set of geographically defined subdivisions of the SVZ which all differ not only in location but also in the types of cells that they produced. These subdivisions are SVZa, anterior SVZ; SVZdl, dorsolateral SVZ; SVZge, postnatal equivalent of the ganglionic eminences; SVZn, neocortical SVZ; SVZspt, septal SVZ. By far the largest of these subdivisions are the lateral and medial ganglionic eminences which appear quite early in development in the position of the future basal forebrain (Fig. 14). These two proliferative areas persist as a fairly large proliferative zone through the first postnatal week in a rodent (Sturrock and Smart, 1980; Bhide, 1996). In all of its subdivisions during the early part of its existence, the SVZ is intermixed with the VZ along their borders (Boulder Committee, 1970; Altman and Bayer, 1990; Takahashi *et al.*, 1993). Cells of the SVZ are also intermixed with nonproliferative cells including postmitotic neurons which arise from the VZ and intermingle with the proliferative cells of the SVZ in their ascent across the cerebral wall. In some regions of the brain the intermixed nonproliferating cells also include the somata of radial glial cells and probably other cells of glial lineage which have left the cell cycle during the epoch of neuronal migration but which may re-enter the cycle in the course of subsequent development of the cerebral wall (Schmechel and Rakic, 1979a; Schmechel and Rakic, 1979b). The SVZ in the lateral and medial ganglionic eminences is highly structured and the cells highly express members of the distal-less family in a pattern that is consistent with a maturation sequence (Fig. 15) (Panganiban and Rubenstein, 2002).

Output of the Subventricular Zone

The SVZ produces both neurons and glia, but the types of cells produced differ both regionally and temporally. For example,

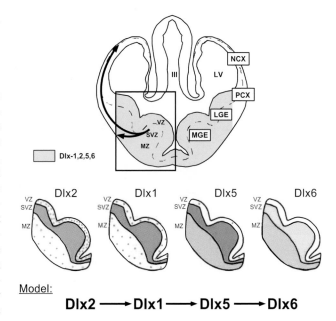

FIGURE 15. Expression domains of Dlx1, Dlx2, Dlx5, and Dlx6 during mouse brain development. (Top) Schema of a transverse section through the E12.5 mouse telencephalon showing the combined expression of Dlx transcripts. Most cells in the subpallial telencephalon express Dlx1, Dlx2, Dlx5, or Dlx6 at some stage of their differentiation. The arrows indicate the migration from the subpallium to the pallium (cortex) (Marin and Rubenstein, 2001). The boxed region on the left is used in the middle section to show the expression of Dlx2, Dlx1, Dlx5, and Dlx6. Dlx2 is primarily expressed in undifferentiated cells; it is expressed in scattered cells in the ventricular zone, in most cells in the subventricular zone and in scattered cells in the mantle zone. Dlx6 is primarily expressed in differentiated cells in the mantle zone. Dlx1 and Dlx5 are expressed in intermediate patterns. (Bottom) A hypothesized genetic and biochemical pathway that proposes the sequential role of Dlx2, Dlx1, Dlx5, and Dlx6 at different stages of differentiation. Telencephalic regions are as follows. Pallium: neocortex (NCX) and palliocortex (PCX). Subpallium: lateral ganglionic eminence (LGE). Medial ganglionic eminence (MGE). Stages of differentiation: ventricular zone (VZ); subventricular zone (SVZ); mantle zone (MZ). LV, lateral ventricle (ventricle of telencephalon); III, third ventricle (ventricle of the diencephalon). Figure modified from Panganiban and Rubenstein 2002.

although the neocortical SVZ coexists with the VZ for much of the time that neurons are produced for the neocortex (Takahashi *et al.*, 1995b), and, thus, it is possible that the SVZ may produce a small number of neurons destined for the neocortex (Reynolds and Weiss, 1992; Levinson and Goldman, 1993), during this time virtually all of the daughter cells of the SVZ re-enter the cell cycle, and, thus, the proportion of neurons produced is estimated to comprise only at most 5–10% of the total (Takahashi *et al.*, 1995b). The anterior part of the SVZ is, however, a major producer of the neurons of the olfactory bulb both during the prenatal and postnatal periods and also into adulthood (Hinds, 1968a, b; Luskin, 1993; Lois and Alvarez-Buylla, 1994; Luskin and McDermott, 1994). Neurons are also produced by the lateral and medial ganglionic eminence. Many of these neurons take up residence locally and comprise the future striatum

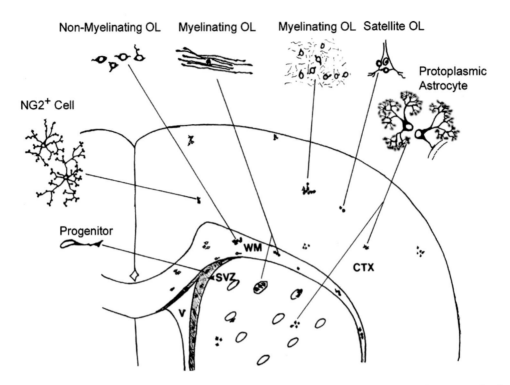

FIGURE 16. The descendants of the perinatal SVZdl. Depicted are the types of cells that are generated from postnatal day 2 SVZdl cells. Progenitors that leave the SVZdl and differentiate within the subcortical white matter become either myelinating or nonmyelinating oligodendrocytes. Few become astrocytes. Those progenitors that differentiate within the neocortex become myelinating oligodendrocytes as well as satellite oligodendrocytes and cells that label with the NG2 proteoglycan. Additionally, those progenitors that make contact with naked cerebral endothelial cells become protoplasmic astrocytes. Figure modified from Brazel *et al.*, 2003.

(Bhide, 1996). Another group of them have an interesting fate in that they migrate laterally and populate the neocortex (Marin and Rubenstein, 2001; Letinic *et al.*, 2002). This laterally migrating group produces many of the inhibitory interneurons (i.e., GABAergic) of the neocortex. An additional population of inhibitory interneurons (GABAergic) is produced in the ganglionic eminence of humans (but not in other primates or mice) that are destined for the dorsal thalamic nuclei in the thalamus (Rakic and Sidman, 1969; Letinic and Rakic, 2001). The migration of these telencephalic neurons into the dorsal thalamus forms a structure that is large enough to warrant a name, "corpus gangliothalamicum" (Rakic and Sidman, 1969). This special output population suggests that the SVZ proliferative population is available for recent evolutionary modification.

During perinatal life, the SVZ is a principal spawning ground for neuroglial cells (Smart, 1961; Smart and Leblond, 1961; Privat, 1975; Mares and Bruckner, 1978; Smart and McSherry, 1982; LeVine and Goldman, 1988b; Levinson and Goldman, 1993). The best studied of the gliogenic portions of the SVZ is the dorsal–lateral portion of the SVZ, that is, the SVZdl (Fig. 14). Mapping studies using retroviral markers (reviewed by Brazel and Levison, 2003) show that this zone produces a variety of glial cell types (Fig. 16). During postnatal life and into adulthood, parts of the SVZ persist as a population of stem/progenitor cells that seem to proliferate for the lifetime of the animal. These stem/progenitor cells produce both neurons and glia; the largest portion of these seems to be destined for the olfactory bulb,

which they reach through the rostral migratory stream (Alvarez-Buylla and Garcia-Verdugo, 2002).

THE DENTATE GYRUS

The subhilar region of the dentate gyrus is a specialized proliferative population that produces the granule cells of the dentate gyrus. The presence of a proliferating population of stem and progenitor cells in the dentate gyrus of mammals was first described in the mouse (Angevine, 1964, 1965). This proliferating population initially arises from the VZ of the medial wall of the lateral ventricle, that is, near the anlage of the dentate gyrus, and migrates into the future position of the dentate hilus (Nowakowski and Rakic, 1981). It persists there during the developmental period and even throughout adulthood in all mammals studied including rodents (Kaplan and Hinds, 1977; Bayer, 1982; Bayer *et al.*, 1982; Stanfield and Trice, 1988), monkeys (Kornack and Rakic, 1999), and humans (Eriksson *et al.*, 1998). Despite the persistence of this proliferative population into adulthood, the vast majority of the output of this proliferative population occurs between birth and P20, during which time approximately 80% of the neurons and glial cells of the murine dentate gyrus are born (Angevine, 1965; Bayer and Altman, 1975). However, there is also evidence that in the adult this proliferative population continues to give rise to neurons (and glia), some portion of which survive, migrate into the granule cell layer, form connections, and become a permanent part of the dentate gyrus granule cell layer (Bayer, 1982; Bayer et al., 1982; Crespo et al., 1986; Stanfield and Trice, 1988) and exhibit important functional properties (van Praag et al., 2002). Importantly, it has been shown that during the adult period the number of granule cells increases (Bayer, 1982; Bayer et al., 1982), the newly produced granule cells displace earlier generated granule cells (Crespo et al., 1986), and they grow an axon into the molecular layer of CA3 (Stanfield and Trice, 1988). In recent years, this proliferative population has been studied as an example of postnatal neurogenesis and stem cell proliferations. Proliferation in the subhilar region of the dentate gyrus has been shown to be affected by genetic differences (Kempermann et al., 1997; Hayes and Nowakowski, 2002), species differences (Kornack and Rakic, 1999), various treatments such as drugs (Eisch et al., 2000), stress (Tanapat et al., 1998; Gould and Tanapat, 1999), behavioral experiences (Kempermann et al., 1998a), hormones (Cameron et al., 1998; Tanapat et al., 1999), aging (Kempermann et al., 1998b), and exercise (van Praag et al., 1999).

Although proliferation in the dentate gyrus persists throughout the life span of the animal, there is a significant decline with age (Kuhn et al., 1996; Kempermann et al., 1998b); in mice at 18 months of age the reported number of BUdR labeled cells observed after 12 daily injections is only about 25% of the number observed after a similar labeling paradigm at 6 months of age (Kempermann et al., 1998b). This decline could be due to a decrease in the number of proliferating cells, an increase in the amount of cell death (in either the proliferating population or the output population) during the 12-day period during which the BUdR injections were given, or both. (However, as yet untested is the possibility that the difference could be a result of changes in T_c and/or T_s with age, for example, by a lengthening of G1 or a shortening of S.) What is significant, however, is that the proliferation continues even in aged animals and that even though there is a large decline over a one-year period, the decline is relatively small when considered with respect to the length of a single cell cycle, which is about 12–14 hr in mice (Hayes and Nowakowski, 2002) and about 24 hr in rats (Cameron and McKay, 2001). Using the longer cell cycle, that is, ~24 hr, the changes due to age would indicate that the size of the proliferating population declines at a rate of <0.15% per cell cycle. (Note that the converse also would hold; that is, if the proliferating population is in fact a constant size, then an increase in the length of the cell cycle of ~0.15% per cell cycle could account for the age changes.)

THE RHOMBIC LIP AND THE EXTERNAL GRANULE CELL LAYER OF THE CEREBELLUM

The external granule cell layer of the cerebellum is unique among the proliferating populations of the CNS in that it is adjacent to the pial surface rather than the ventricular surface (Fig. 17). The external granule cell layer was first recognized as the source of the granule cells of the cerebellum near the end of the 19th century (Obersteiner, 1883; Schaper, 1897a, b; Ramon y

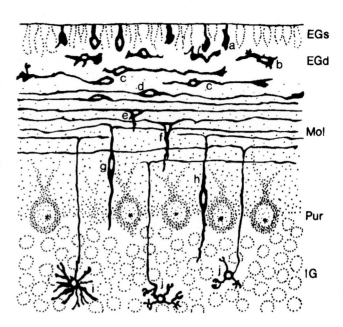

FIGURE 17. The external granule cell layer (EGL) lies beneath the pial surface of the developing cerebellum. These stem/progenitor cells divide in the EGL and migrate through the molecular layer (Mol), past the Purkinje cells into the internal granule cell layer (IG). Drawing from Jacobson (1991), modified from Ramon y Cajal (1909–1911).

Cajal, 1909–1911). The cells of the external granule cell layer originate from the rhombic lip and then migrate over the surface of the cerebellum. The rhombic lip also gives rise to neurons of the brain stem, chiefly of the inferior olivary nuclei but also of the cochlear and pontine nuclei (Harkmark, 1954; Taber-Pierce, 1973). In the human the cells migrating from the rhombic lip to the brain stem form a continuous band which was called the corpus pontobulbare by Essick (1907, 1909, 1912).

The external granule cell layer is present in every vertebrate that has been examined. It is a single layer of cells that is about 6–8 cell diameters thick. Importantly, mitotic figures are scattered throughout the external part of the layer indicating that there is no interkinetic nuclear migration. In this regard, the external granule cell layer is similar to the SVZ. The internal part of the external granule cell layer is not a proliferative zone, but instead it consists of cells that are "waiting" to migrate. The major output of the external granule cell layer is the many cells that comprise the internal granule cell, which are arguably the most numerous neurons in the brain. The life span of the external granule cell is long in comparison with the VZ that produces the Purkinje cells of the cerebellum. For example, in the mouse, the Purkinje cells are produced in a three-day period from E10 through E13 but the internal granule cells are produced over a much more extended period from late in the postnatal period through the third week after birth (Miale and Sidman, 1961). The relatively long period of neuron production in the external granule cell layer is similar in other species including humans (Zecevic and Rakic, 1976).

It is interesting to note that the two major cell classes of the cerebellum, the Purkinje cells and granule cells, are produced in two distinct proliferative zones, the VZ of the fourth ventricle and the external granule cell layer, respectively, at quite different times during development. Thus, it is clear that the final product, that is, the normal cerebellar cortex with a proper number of both types of cells, requires an elaborate regulatory system that would need to include some sort of feedback system through which the early developing cell (the Purkinje cell) could influence the production of the later developing cell (the granule cell). This interaction is hinted at by the changes in the thickness of the external granule cell layer in the reeler mutant mouse where it achieves normal thickness only in places where the Purkinje cell dendrites are normally oriented toward the pial surface (Caviness and Rakic, 1978). Recent evidence indicates that this interaction is mediated by sonic hedgehog which is released from the Purkinje cells and which then binds to the Patched1 receptor on the proliferating cells of the external granule cell layer (Corcoran and Scott, 2001). Mutations in the Patched1 receptor may be involved in the development of medulloblastoma, one of the most common brain tumors of childhood (Corcoran and Scott, 2001; Pomeroy et al., 2002).

OVERVIEW

The four major proliferative populations of the developing brain each have a specific role during the development of the brain. They have two important tasks which are to (1) produce the right number of cells for the particular brain region—either too many or too few will result in abnormalities—and (2) to produce the right class of cells (neurons vs glia, and subtypes of each). The delineation of the regulation of these two tasks is a major goal of developmental neuroscience. Progress toward some aspects of this are detailed in other chapters of this book.

REFERENCES

Adams, R.J., 1996, Metaphase spindles rotate in the neuroepithelium of rat cerebral cortex, *J. Neurosci.* 16:7610–7618.

Alexiades, M.R. and Cepko, C., 1996, Quantitative analysis of proliferation and cell cycle length during development of the rat retina, *Dev. Dyn.* 205:293–307.

Altman, J. and Bayer, S.A., 1990, Vertical compartmentation and cellular transformations in the germinal matrices of the embryonic rat cerebral cortex, *J. Exp. Neurol.* 107:23–35.

Alvarez-Buylla, A. and Garcia-Verdugo J.M., 2002, Neurogenesis in adult subventricular zone, *J. Neurosci.* 22: 629–634.

Anderson, S.A., Kaznowski, C.E., Horn, C., Rubenstein, J.L., and McConnell, S.K., 2002, Distinct origins of neocortical projection neurons and interneurons in vivo, *Cereb. Cortex* 12:702–709.

Angevine, J.B., Jr., 1965, Time of neuron origin in the hippocampal region: An autoradiographic study in the mouse, *Exp. Neurol. Suppl.* 2:1–71.

Angevine, J.B.J., 1964, Autoradiographic study of histogenesis in the area dentata of the cerebral cortex in the mouse, *Anat. Rec.* 148:255–.

Angevine, J.B.J. and Sidman, R.L., 1961, Autoradiographic study of cell migration during histogenesis of cerebral cortex in the mouse, *Nature* 192:766–768.

Bayer, S.A., 1982, Changes in the total number of dentate granule cells in juvenile and adult rats: A correlated volumetric and 3H-thymidine autoradiographic study, *Exp. Brain. Res.* 46:315–323.

Bayer, S.A. and Altman, J., 1975, Radiation-induced interference with postnatal hippocampal cytogenesis in rats and its long-term effects on the acquisition of neurons and glia, *J. Comp. Neurol.* 163:1–20.

Bayer S.A., Yackel J.W., and Puri P.S., 1982, Neurons in the rat dentate gyrus granular layer substantially increase during juvenile and adult life. *Science* 216:890-892.

Bhide, P.G., 1996, Cell cycle kinetics in the embryonic mouse corpus striatum, *J. Comp. Neurol.* 374:506–522.

Blau, H.M., Brazelton, T.R., and Weimann, J.M., 2001, The evolving concept of a stem cell: Entity or function? *Cell* 105:829–841.

Blaschke, A.J., Staley, K., and Chun, J., 1996, Widespread programmed cell death in proliferative and postmitotic regions of the fetal cerebral cortex, *Development* 122:1165–1174.

Boulder Committee, 1970, Embryonic vertebrate central nervous system: Revised terminology, *Anat. Rec.* 166:257–262.

Brazel, C.Y., Romanko, M.J., Rothstein, R.P., and Levison, S.W., 2003, Roles of the mammalian subventricular zone in brain development, *Prog. Neurobiol.* 69:49–69.

Cai, L., Hayes, N.L., and Nowakowski, R.S., 1997a, Synchrony of clonal cell proliferation and contiguity of clonally related cells: Production of mosaicism in the ventricular zone of developing mouse neocortex, *J. Neurosci.* 17:2088–2100.

Cai, L., Hayes, N.L., and Nowakowski, R.S., 1997b, Local homogeneity of cell cycle length in developing mouse cortex, *J. Neurosci.* 17:2079–2087.

Cameron, H.A. and McKay, R.D., 2001, Adult neurogenesis produces a large pool of new granule cells in the dentate gyrus, *J. Comp. Neurol.* 435:406–417.

Cameron, H.A., Tanapat, P., and Gould, E. 1998, Adrenal steroids and N-methyl-D-aspartate receptor activation regulate neurogenesis in the dentate gyrus of adult rats through a common pathway, *Neuroscience* 82:349–354.

Caviness, V., Takahashi, T., and Nowakowski, R., 1995, Numbers, time and neocortical neuronogenesis: A general developmental and evolutionary model, *Trends Neurosci.* 18:379–383.

Caviness, V.S. and Rakic, P., 1978, Mechanisms of cortical development: A view from mutations in mice, *Annu. Rev. Neurosci.* 1:297–326.

Caviness, V.S. and Sidman, R.L., 1973, Time of origin of corresponding cell classes in the cerebral cortex of normal and reeler mutant mice: An autoradiographic analysis, *J. Comp. Neurol.* 148:141–152.

Chenn, A. and McConnell, S., 1995, Cleavage orientation and the asymmetric inheritance of Notch1 immunoreactivity in mammalian neurogenesis, *Cell* 82:631–641.

Chenn, A., Zhang, Y.A., Chang, B.T., and McConnell, S.K., 1998, Intrinsic polarity of mammalian neuroepithelial cells, *Mol. Cell Neurosci.* 11:183–193.

Corbin, J.G., Nery, S., and Fishell, G., 2001, Telencephalic cells take a tangent: Non-radial migration in the mammalian forebrain, *Nat. Neurosci.* 4 Suppl:1177–1182.

Corcoran, R.B. and Scott, M.P., 2001, A mouse model for medulloblastoma and basal cell nevus syndrome, *J. Neurooncol.* 53:307–318.

Crandall, J.E. and Caviness, V.S.,1984, Axon strata of the cerebral wall in embryonic mice, *Dev. Brain. Res.* 14:185–195.

Crespo, D., Stanfield, B.B., and Cowan, W.M., 1986, Evidence that late-generated ganule cells do not simply replace earlier formed neurons in the rat dentate gyrus, *Exp. Brain. Res.* 62:541–548.

Dehay, C., Giroud, P., Berland, M., Smart, I., and Kennedy, H., 1993, Modulation of the cell cycle contributes to the parcellation of the primate visual cortex, *Nature* 366:464–466.

Eisch, A.J., Barrot, M., Schad, C.A., Self, D.W., and Nestler, E.J., 2000, Opiates inhibit neurogenesis in the adult rat hippocampus, *Proc. Natl. Acad. Sci. USA* 97:7579–7584.

Eriksson, P.S., Perfilieva, E., Bjork-Eriksson, T., Alborn, A.M., Nordborg, C., Peterson, D.A. *et al.*, 1998, Neurogenesis in the adult human hippocampus [see comments], *Nat. Med.* 4:1313–1317.

Essick, C.R., 1907, The corpus ponto-bulbare—A hitherto undescribed nuclear mass in the human hindbrain, *Am. J. Anat.* 7:119–135.

Essick, C.R., 1909, On the embryology of the corpus ponto-bulbare and its relationship to the development of the pons, *Anat. Rec.* 3:254–257.

Essick, C.R., 1912, The development of the nuclei pontis and the nucleus arcuatus in man, *Am. J. Anat.* 13:25–54.

Finlay, B.L. and Pallas, S.L., 1989, Control of cell number in the developing mammalian visual system, *Prog. Neurobiol.* 32:207–234.

Finlay, B.L. and Slattery, M., 1983, Local differences in the amount of early cell death in neocortex predict adult local specializations, *Science* 219:1349–1351.

Fishell, G., Mason, C.A., and Hatten, M.E.,1993, Dispersion of neural progenitors within the germinal zones of the forebrain [published erratum appears in *Nature* 1993 May 20; 363(6426):286] [see comments], *Nature* 362:636–638.

Garcia-Verdugo, J.M., Doetsch, F., Wichterle, H., Lim, D.A., and Alvarez-Buylla, A.,1998, Architecture and cell types of the adult subventricular zone: In search of the stem cells, *J. Neurobiol.* 36:234–248.

Gilmore, E.C., Nowakowski, R.S., Caviness, V.S., Jr., and Herrup, K., 2000, Cell birth, cell death, cell diversity and DNA breaks: How do they all fit together? *Trends. Neurosci.* 23:100–105.

Goldman, J.E., 1995, Lineage, migration, and fate determination of postnatal subventricular zone cells in the mammalian CNS, *J. Neurooncol.* 24:61–64.

Gould, E. and Tanapat, P., 1999, Stress and hippocampal neurogenesis, *Biol. Psychiatry* 46:1472–1479.

Halliday, A.L. and Cepko, C.L.,1992, Generation and migration of cells in the developing striatum, *Neuron* 9:15–26.

Hamilton, E. and Dobbin, J., 1983a, The percentage labeled mitosis technique shows the mean cell cycle time to be half its true value in Carcinoma TY. 1. [H3]thymidine and vincristine studies, *Cell Tissue Kinet.* 16: 473–482.

Hamilton, E. and Dobbin, J., 1983b, The percentage labeled mitoses technique shows the mean cell cycle time to be half its true value in Carcinoma NT. II. [3H]deoxyuridine studies, *Cell Tissue Kinet.* 16:483–492.

Harkmark, W., 1954, Cell migrations from the rhombic lip to the inferior olive, the nucleus raphe and the pons. A morphological and experimental investigation on chick embryos, *J. Comp. Neurol.* 100:115–209.

Hayes, N.L. and Nowakowski, R.S., 2000, Exploiting the dynamics of S-phase tracers in developing brain: Interkinetic nuclear migration for cells entering versus leaving the S-phase, *Dev. Neurosci.* 22:44–55.

Hayes, N.L. and Nowakowski, R.S., 2002, Dynamics of cell proliferation in the adult dentate gyrus of two inbred strains of mice, *Brain Res. Dev. Brain Res.* 134:77–85.

Heumann, D. and Leuba, G., 1983, Neuronal death in the development and aging of the cerebral cortex of the mouse, *Neuropathol. Appl. Neurobiol.* 9:297–311.

Hinds, J.W., 1968a, Autoradiographic study of histogenesis in the mouse olfactory bulb. I. Time of origin of neurons and neuroglia, *J. Comp. Neurol.* 134:287–304.

Hinds, J.W., 1968b, Autoradiographic study of histogenesis in the mouse olfactory bulb. II. Cell proliferation and migration. *J. Comp. Neurol.* 134:305–322.

His, W., 1889, Die Neuroblasten und deren Entstehung im embryonalen Mark, *Abh Kgl Sachs Ges Wiss Math-phys Cl* 15:313–372.

His, W., 1897, Address upon the development of the brain, *Trans. Roy. Acad. Med. Ireland* 15:1–21.

His, W., 1904, Die Entwicklung des Menschlichen Gehirns wahrend der ersten Monate, von S. Hirzel, Leipzig.

Hoshino, K., Matsuzawa, T., and Murakami, U., 1973, Characteristic of the cell cycle of matrix cells in the mouse embryo during histogenesis of telencephalon, *Exp. Cell Res.* 77:89–94.

Jacobson, M., 1991, *Developmental Neurobiology*, 3rd edn, Plenum Press: New York & London.

Kaplan, M.S. and Hinds, J.W., 1977, Neurogenesis in the adult rat: Electron microscopic analysis of light radioautographs, *Science* 197:1092–1094.

Kaufmann, S.L., 1968, Lengthening of the generation cycle during embryonic differentiation of the mouse neural tube, *Exp. Cell Res.* 49:420–424.

Kempermann, G., Brandon, E.P., and Gage, F.H., 1998a, Environmental stimulation of 129/SvJ mice causes increased cell proliferation and neurogenesis in the adult dentate gyrus, *Curr. Biol.* 8: 939–942.

Kempermann, G., Kuhn, H.G., and Gage, F.H., 1997, Genetic influence on neurogenesis in the dentate gyrus of adult mice, *Proc. Nat. Acad. Sci. USA* 94:10409–10414.

Kempermann, G., Kuhn, H.G., and Gage, F.H., 1998b, Experience-induced neurogenesis in the senescent dentate gyrus, *J. Neurosci.* 18:3206–3212.

Kennedy, H. and Dehay, C., 1993, The importance of developmental timing in cortical specification, *Perspect. Dev. Neurobiol.* 1:93–99.

Koelliker, R.A., 1896, *Handbuc der Gewebelchre des Menschen. Bd. 2 Nervensystem des Menschen und der Thiere,* 6th edn, W. Englemann, Leipzig.

Kornack, D.R. and Rakic, P., 1998, Changes in cell-cycle kinetics during the development and evolution of primate neocortex, *Proc. Nat. Acad. Sci. USA* 95:1242–1246.

Kornack, D.R. and Rakic, P., 1999, Continuation of neurogenesis in the hippocampus of the adult macaque monkey, *Proc. Natl. Acad. Sci. USA* 96: 5768–5773.

Kuhn, H.G., and Dickinson-Anson, H., and Gage, F.H., 1996, Neurogenesis in the dentate gyrus of the adult rat: Age-related decrease of neuronal progenitor proliferation, *J. Neurosci.* 16:2027–2033.

Lavdas, A.A., Grigoriou, M., Pachnis, V., and Parnavelas, J.G., 1999, The medial ganglionic eminence gives rise to a population of early neurons in the developing cerebral cortex, *J Neurosci* 19:7881-7888.

Lavdas, A.A., Mione, M.C., and Parnavelas, J.G., 1996, Neuronal clones in the cerebral cortex show morphological neurotransmitter heterogeneity during development, *Cerebral Cortex* 6:490–497.

Letinic, K. and Rakic, P., 2001, Telencephalic origin of human thalamic GABAergic neurons, *Nat. Neurosci.* 4:931–936.

Letinic, K., Zoncu, R., and Rakic, P., 2002, Origin of GABAergic neurons in the human neocortex, *Nature* 417:645–649.

Leuba, G., Heumann, D., and Rabinowicz, T., 1977, Postnatal development of the mouse cerebral neocortex. I. Quantitative cytoarchitectonics of some motor and sensory areas, *J. Hirnforsch.* 18:461–481.

LeVine, S.M. and Goldman, J.E., 1988a, Embryonic divergence of oligodendrocyte and astrocyte lineages in developing rat cerebrum, *J. Neurosci.* 8:3992–4006.

LeVine, S.M. and Goldman, J.E., 1988b, Spatial and temporal patterns of oligodendrocyte differentiation in rat cerebrum and cerebellum, *J. Comp. Neurol.* 277:441–455.

Levinson, S.W. and Goldman, J.E., 1993, Both oligodendrocytes and astrocytes develop from progenitors in the subventricular zone of the postnatal rat forebrain, *Neuron* 10:201–212.

Lois, C. and Alvarez-Buylla, A., 1994, Long-distance neuronal migration in the adult mammalian brain, *Science* 264:1145–1148.

Luskin, M.B., 1993, Restricted proliferation and migration of postnatally generated neurons derived from the forebrain subventricular zone, *Neuron* 11:173–189.

Luskin, M.B. and McDermott, K., 1994, Divergent lineages for oligodendrocytes and astrocytes originating in the neonatal forebrain subventricular zone, *Glia* 11:211–226.

Luskin, M.B., Parnavelas, J.G., and Barfield, J.A., 1993, Neurons, astrocytes, and oligodendrocytes of the rat cerebral cortex originate from separate progenitor cells: An ultrastructural analysis of clonally related cells, *J. Neurosci.* 13:1730–1750.

Luskin, M.B., Pearlman, A.L., and Sanes, J.R., 1988, Cell lineage in the cerebral cortex of the mouse studied in vivo and in vitro with a recombinant retrovirus, *Neuron* 1:635–647.

Mares, V. and Bruckner, G., 1978, Postnatal formation of neural cells in the rat occipital cerebrum: An autoradiographic study of the time and space pattern of cell division, *J. Comp. Neurol.* 177:519–528.

Marin, O. and Rubenstein, J.L., 2001, A long, remarkable journey: Tangential migration in the telencephalon, *Nat. Rev. Neurosci.* 2:780–790.

Miale, I. and Sidman, R.L., 1961, An autoradiographic analysis of histogenesis in the mouse cerebellum, *Exp. Neurol.* 4:277–296.

Mione, M.C., Cavanagh, J.F., Harris, B., and Parnavelas, J.G., 1997, Cell fate specification and symmetrical/asymmetrical divisions in the developing cerebral cortex, *J. Neurosci.* 17:2018–2029.

Mione, M.C., Danevic, C., Boardman, P., Harris, B., and Parnavelas, J.G., 1994, Lineage analysis reveals neurotransmitter, (GABA) or glutamate, but not calcium-binding protein homogeneity in clonally related cortical neurons, *J. Neurosci.* 14:107–123.

Miyama, S., Takahashi, T., Nowakowski, R.S., and Caviness, V.S., Jr., 1997, A gradient in the duration of the G1 phase in the murine neocortical proliferative epithelium, *Cereb. Cortex* 7:678–689.

Nery, S., Fishell, G., and Corbin, J.G., 2002, The caudal ganglionic eminence is a source of distinct cortical and subcortical cell populations, *Nat. Neurosci.* 5:1279–1287.

Nowakowski, R.S. and Rakic, P., 1981, The site of origin and route and rate of migration of neurons to the hippocampal region of the rhesus monkey, *J. Comp. Neurol.* 196:129–154.

Nowakowski, R.S., Caviness, V.S., Jr., Takahashi, T., and Hayes, N.L., 2002, Population dynamics during cell proliferation and neuronogenesis in the developing murine neocortex. In *Cortical Development: From Specification to Differentiation (Results and Problems in Cell Differentiation. Vol. 39)* (C. Hohmann, ed.), Springer-Verlag, New York, pp. 1–22.

Nowakowski, R.S., Lewin, S.B., and Miller, M.W., 1989, Bromodeoxyuridine immunohistochemical determination of the lengths of the cell cycle and the DNA-synthetic phase for an anatomically defined population, *J. Neurocytol.* 18:311–318.

O'Rourke, N.A., Dailey, M.E., Smith, S.J., and McConnell, S.K., 1992, Diverse migratory pathways in the developing cerebral cortex, *Science* 258:299–302.

Obersteiner, H., 1883, Der feinere Bau der Kleinhirnrinde beim Menschen und bie Tieren, *Biol. Zentralbl.* 3:145–155.

Panganiban, G. and Rubenstein, J.L., 2002, Developmental functions of the Distal-less/Dlx homeobox genes, *Development* 129:4371–4386.

Pomeroy, S.L., Tamayo, P., Gaasenbeek, M., Sturla, L.M., Angelo, M., McLaughlin, M.E. *et al.*, 2002, Prediction of central nervous system embryonal tumour outcome based on gene expression, *Nature* 415:436–442.

Powell, E.M., Campbell, D.B., Stanwood, G.D., Davis, C., Noebels, J.L., and Levitt, P., 2003, Genetic disruption of cortical interneuron development causes region- and GABA cell type-specific deficits, epilepsy, and behavioral dysfunction, *J. Neurosci.* 23:622–631.

Price, J., 1987, Retroviruses and the study of cell lineage, *Development* 101:409–419.

Privat, A., 1975, Postnatal gliogenesis in the mammalian brain, *Int. Rev. Cytol.* 40:281–323.

Rachel, R.A., Dolen, G., Hayes, N.L., Lu, A., Erskine, L., Nowakowski, R.S., and Mason, C.A., 2002, Spatiotemporal features of early neuronogenesis differ in wild-type and albino mouse retina, *J. Neurosci.* 22:4249–4263.

Rakic, P., 1974, Neurons in rhesus monkey visual cortex: Systematic relation between time of origin and eventual disposition. *Science* 183:425–427.

Rakic, P., 1977, Genesis of the dorsal lateral geniculate nucleus in the rhesus monkey: Site and time of origin, kinetics of proliferation, routes of migration and pattern of distribution of neurons, *J. Comp. Neurol.* 176:23–52.

Rakic, P., and Nowakowski, R.S.,1981, The time of origin of neurons in the hippocampal region of the rhesus monkey, *J. Comp. Neurol.* 196: 99–128.

Rakic, P. and Sidman, R.L., 1968, Supravital DNA synthesis in the developing human and mouse brain, *J. Neuropathol. Exp. Neurol.* 27:246–276.

Rakic, P. and Sidman, R.L., 1969, Telencephalic origin of pulvinar neurons in the fetal human brain, *Z Anat. Entwickl-Gesch.* 129:53–82.

Rakic, P., Stensaas, L.J., Sayer, E.P., and Sidman, R.L., 1974, Computer aided three-dimensional reconstruction and quantitative analysis of cells from serial electron microscopic montage of foetal monkey brain, *Nature* 250:31–34.

Ramon y Cajal, S., 1894, Les nouvelle idees sur la structure du systeme nerveuz chez l'homme et chez les vertebras, Reinwald, Paris.

Ramon y Cajal, S., 1909–1911, *Histologie du Systeme Nerveaux de l'Homme et des Vertebres*, Reprinted by Instituto Ramon y Cajal del CSIC, Madrid.

Reynolds, B.A. and Weiss, S., 1992, Generation of neurons and astrocytes from isolated cells of the adult mammalian central nervous system, *Science* 255:1707–1710.

Sauer, F.C., 1935, Mitosis in the neural tube, *J. Comp. Neurol.* 62:377–405.

Sauer, F.C., 1936, The interkinetic migration of embryonic epithelial nuclei, *J. Morphol.* 60:1–11.

Sauer, F.C., 1937, Some factors in the morphogenesis of vertebrate embryonic epithelia, *J. Morphol.* 61:563–579.

Sauer, M.E., 1959, Radioautographic study of the location of newly synthesized deoxyribonucleic acid in the neural tube of the chick embryo: Evidence for intermitotic migration of nuclei. *Anat. Rec.* 133:456.

Sauer, M.E. and Chittenden, A.C., 1959, Deoxyribonucleic acid content of cell nuclei in the neural tube of the chick embryo: Evidence for intermitotic migration of nuclei, *Exp. Cell. Res.* 16:1–6.

Schaper, A., 1897a, Die fruhesten Differenzierungsvorgange im Centralnervensystem, *Arch Entwicklungsmech Organ* 5:81–132.

Schaper, A., 1897b, The earliest differentiation in the central nervous system of vertebrates, *Science* 5:81–132.

Schaper, A. and Cohen, C., 1905, Beitraege zur Analyze des tierischen Wachstums. II. Teil: Ueber zellproliferatorische Wachstumszentren und deren Bezeihung zur Regeneration und Geschwulstbildung, *Arch. f Entwick-Mech.* 19:348–445.

Schmechel, D.E. and Rakic, P., 1979a, Arrested proliferation of radial glial cells during midgestation in rhesus monkey, *Nature* 277:303–305.

Schmechel, D.E. and Rakic, P., 1979b, A Golgi study of radial glial cells in developing monkey telencephalon: Morphogenesis and transformation into astrocytes, *Anat. Embryol. (Berl)* 156:115–152.

Seaberg, R.M. and van der Kooy, D., 2003, Stem and progenitor sells: The premature desertion of rigorous definitions, *Trends Neurosci.* 26:125–131.

Sidman, R.L., 1970, Autoradiographic methods and principles for study of the nervous system with thymidine-H3. In *Contemporary Research Methods in Neuroanatomy* (W.J.H. Nauta and S.O.E. Ebbesson, eds.), Springer, New York, pp. 252–274.

Sidman, R.L., Miale, I.L., and Feder, N., 1959, Cell proliferation and migration in the primitive ependymal zone: An autoradiographic study of histogenesis in the nervous system, *Exp. Neurol.* 1:322–333.

Smart, I., 1961, The subependymal layer of the mouse brain and its cell production as shown by autoradiography after thymidine-H3 injection, *J. Comp. Neurol.* 116:325–347.

Smart, I. and Leblond, C.P., 1961, Evidence for division and transformation of neuroglia cells in the mouse brain, as derived from radio-autography after injection of thymidine-H3, *J. Comp. Neurol.* 116:349–367.

Smart, I.H.M., 1972, Proliferative characteristics of the ependymal layer during the early development of the mouse diencephalon, as revealed by recording the number, location, and plane of cleavage of mitotic cells, *J. Anat.* 113:109–129.

Smart, I.H.M., 1973, Proliferative characteristics of the ependymal layer during the early development of the mouse neocortex: A pilot study based on recording the number, location and plane of cleavage of mitotic figures, *J. Anat.* 116:67–91.

Smart, I.H.M. and McSherry, G.M., 1982, Growth patterns in the lateral wall of the mouse telencephalon: II. Histological changes during and subsequent to the period of isocortical neuron production, *J. Anat.* 134:415–442.

Stanfield, B.B. and Cowan, W.M., 1979, The development of the hippocampus and dentate gyrus in normal and reeler mice, *J. Comp. Neurol.* 185: 423–459.

Stanfield, B.B. and Trice, J.E., 1988, Evidence that granule cells generated in the dentate gyrus of adult rats extend axonal projections, *Exp. Brain. Res.* 72:399–406.

Sturrock, R.R. and Smart, I.H., 1980, A morphological study of the mouse subependymal layer from embryonic life to old age, *J. Anat.* 130: 391–415.

Taber-Pierce, E., 1973, Time of origin of neurons in the brain stem of the mouse, *Prog. Brain Res.* 40:53–66.

Takahashi, T., Nowakowski, R.S., and Caviness, V.S., Jr., 1993, Cell cycle parameters and patterns of nuclear movement in the neocortical proliferative zone of the fetal mouse, *J. Neurosci.* 13:820–833.

Takahashi, T., Nowakowski, R.S., and Caviness, V.S., Jr., 1995a, The cell cycle of the pseudostratified ventricular epithelium of the embryonic murine cerebral wall, *J. Neurosci.* 15:6046–6057.

Takahashi, T., Nowakowski, R.S., and Caviness, V.S., Jr., 1995b, Early ontogeny of the secondary proliferative population of the embryonic murine cerebral wall, *J. Neurosci.* 15:6058–6068.

Takahashi, T., Nowakowski, R.S., and Caviness, V.S., Jr., 1996a, The leaving or Q fraction of the murine cerebral proliferative epithelium: A general model of neocortical neuronogenesis, *J. Neurosci.* 16:6183–6196.

Takahashi, T., Nowakowski, R.S., and Caviness, V.S., Jr., 1996b, Interkinetic and migratory behavior of a cohort of neocortical neurons arising in the early embryonic murine cerebral wall, *J. Neurosci.* 16:5762–5776.

Takahashi T., Nowakowski, R.S., and Caviness, V.S., Jr., 1997, The mathematics of neocortical neuronogenesis, *Dev. Neurosci.* 19:17–22.

Tan, S.S. and Breen, S., 1993, Radial mosaicism and tangential cell dispersion both contribute to mouse neocortical development [see comments], *Nature* 362:638–640.

Tanapat, P., Galea, L.A., and Gould, E., 1998, Stress inhibits the proliferation of granule cell precursors in the developing dentate gyrus, *Int. J. Dev. Neurosci.* 16:235–239.

Tanapat, P., Hastings, N.B., Reeves, A.J., and Gould, E., 1999, Estrogen stimulates a transient increase in the number of new neurons in the dentate gyrus of the adult female rat, *J. Neurosci.* 19:5792–5801.

Thomaidou, D., Mione, M.C., Cavanagh, J.F., and Parnavelas, J.G., 1997, Apoptosis and its relation to the cell cycle in the developing cerebral cortex, *J. Neurosci.* 17:1075–1085.

Van Praag, H., Christie, B.R., Sejnowski, T.J., and Gage, F.H., 1999, Running enhances neurogenesis, learning, and long-term potentiation in mice, *Proc. Natl. Acad. Sci. USA* 96:13427–13431.

Van Praag, H., Schinder, A.F., Christie, B.R., Toni, N., Palmer, T.D., and Gage, F.H., 2002, Functional neurogenesis in the adult hippocampus, *Nature* 415:1030–1034.

Verney, C., Takahashi, T., Bhide, P.G., Nowakowski, R.S., and Caviness, Jr., V.S., 2000, Independent controls for neocortical neuron production and histogenetic cell death, *Dev. Neurosci.* 22:125–138.

Vignal, W., 1888, Recherches sur le developpement des elements des couches corticales du cerveau et du cervelet chez l'homme et les mamiferes, *Arch. Physiol. Norm. Path. (Paris) Ser. IV* 2:228–254.

Walsh, C., 1993, Cell lineage and regional specification in the mammalian neocortex, *Perspect. Dev. Neurobiol.* 1:75–80.

Walsh, C. and Cepko, C.L., 1988, Clonally related cortical cells show several migration patterns, *Science* 242:1342–1345.

Walsh, C. and Cepko, C.L., 1992, Widespread dispersion of neuronal clones across functional regions of the cerebral cortex, *Science* 255:434–440.

Wichterle, H., and Turnbull, D.H., Nery, S., Fishell, G., and Alvarez-Buylla, A., 2001, In utero fate mapping reveals distinct migratory pathways and fates of neurons born in the mammalian basal forebrain, *Development* 128:3759–3771.

Williams, B.P., Read, J., and Price, J., 1991, The generation of neurons and oligodendrocytes from a common precursor cell, *Neuron* 7:685–693.

Zecevic, N. and Rakic, P., 1976, Differentiation of Purkinje cells and their relationship to other components of developing cerebellar cortex in man, *J. Comp. Neurol.* 167:27–47.

3

Anteroposterior and Dorsoventral Patterning

Diana Karol Darnell

PRINCIPLES AND MECHANISMS OF PATTERNING

If development is the process of reproducibly taking undifferentiated tissue and making it more complex in an organized way, then pattern formation is the mechanism for producing the organization in that complexity. This requires initiating differential gene expression within two or more apparently homogeneous cells. In some organisms this is initially done by segregating cytoplasmic determinants into specific daughter cells. These cytoplasmic determinants (proteins or RNAs) can result in the transcription of a restricted set of genes and begin the cascade that sets up tissues as different from one another in a coordinated pattern (Fig. 1). This is a totally cell autonomous mechanism and theoretically it could be the only mechanism for patterning the embryo. However, whereas this mechanism is well supported by evidence in the initiation of pattern formation in many invertebrates (e.g., *Drosophila*) and is probably invoked in vertebrates when asymmetrical cell division is the rule (e.g., stem cells), it does not appear to be the main method for embryonic pattern formation in vertebrates.

Vertebrate pattern formation, including the patterning of the nervous system, involves cellular responses to environmental asymmetries. Whereas embryonic cells initially may be a homogeneous population, they are not homogeneous in their relationship to asymmetrical environmental signals; by definition some are closer and some are further away. Thus, some receive a higher level of the signal and some a lower level or none at all. This difference gets translated into differential cellular response, which results in pattern formation within the field (Fig. 2).

Understanding pattern formation in the vertebrate nervous system means understanding this cascade of cellular and molecular interactions. The term **cascade** is often used to describe the events in development and pattern formation because one or more simple asymmetries initiate a pattern, which then becomes the foundation for the formation of a more complex pattern, which in turn forms the foundation for even finer patterning. The players in such a cascade are the cells and molecules of the early embryo. They include the *source* of the environmental asymmetry, which secretes the *signal*, which binds to the *receptors*, which initiate the

signal transduction pathway within the responding cells, which activates the *transcription factors*, which regulate the set of coordinated *downstream genes* whose expression is modulated (up or down) as a result. These downstream genes may code for new signals, receptors, signal transduction proteins, transcription factors, or extracellular-, membrane bound-, cytoplasmic-, or nuclear-*facilitators* or -*antagonists* to modulate the system (Fig. 3), adding the next layer to the cascade.

The asymmetrical environmental cues often come from neighboring embryonic tissues whose early differentiation has made them into signaling centers. If these signaling centers can both induce *differentiation and pattern* in an undifferentiated field, they are called **organizers**, after the first such center to be identified, the Spemann–Mangold Organizer in amphibians, which was observed to induce and pattern the neuraxis (Spemann and Mangold, 1924). The signaling molecules may be peptide growth factors, vitamin metabolites, or other soluble, transported, or tethered ligands. When they have different effects across a homogeneous field of responding cells depending on their concentration, these signaling molecules are called **morphogens**. Because they invest the cells within the field with information about their relative position, they are also called **positional signals**. Models involving differences in binding affinity have been offered to demonstrate how one signal could have differing affects at different concentrations (Fig. 4).

Regardless of mechanism, these signals activate or induce the expression of a specific set of transcription factors that are unique to responsive cells at a particular distance from the source, and thus at a particular location in the embryo. These transcription factors are called **positional identity** genes, and they are often used as markers to define a region. Before the molecular revolution, they were called **positional information**. These transcription factors regulate the expression of selected genes, which may code for a component in this or another patterning pathway, or for proteins involved in differentiation of these cells. In the nervous system, this could include proteins mediating neuronal migration, axon outgrowth and navigation, precise connections, specific neurotransmitter production, or receptors that characterize the neurons of this locale. In the event that these downstream genes are unique to this region, they

Diana Karol Darnell • Lake Forest College, Lake Forest, IL 60045.

Developmental Neurobiology, 4th ed., edited by Mahendra S. Rao and Marcus Jacobson. Kluwer Academic / Plenum Publishers, New York, 2005.

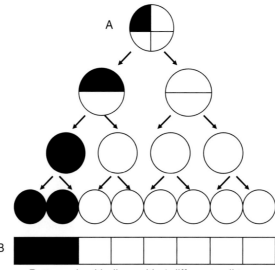

FIGURE 1. A patterned layer of cells can be achieved by localizing cytoplasmic determinants (shown here as various textures) within the parent cell. (A) Cell division segregates these determinants into different daughter cells, and they instruct their descendants (B) to acquire different phenotypes or fates. Cytoplasmic determinants are often RNAs for- or transcription factors themselves.

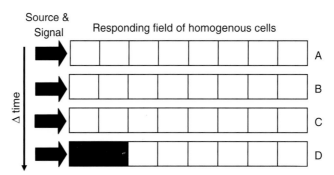

FIGURE 2. Asymmetric signaling (arrows) can change the fates of homogenous cells (white blocks) within the signal's reach. Cell fates can be specified in a stepwise pattern (as shown here, A > B > C > D) or all at once (A > D), depending on the timing of competence in the responding cells. This figure represents the formation of four different cell types (D) in response to a developing concentration gradient of a signaling molecule. Initially (A), the signal is low even near the source, but continued secretion yields a high concentration near the source and the possibility of inducing different cell types at several thresholds.

can also be used as **markers** when assessing the patterning or differentiation of the tissue.

The functions of various genes in these pathways are assessed through three types of experiments. First, candidate genes are identified because their expression shows a **correlation** with the timing and position of an observed patterning event. Second, the ectopic expression of the gene or presence of the protein causes a **gain of function**, showing that this gene product is **sufficient** to induce the observed pattern. Finally, failure to express the gene in the normal area results in a **loss of function**, indicating that the product is **necessary**. Evidence that a gene product is present, necessary, and sufficient is required to

demonstrate a cause and effect relationship between the gene expression and the patterning event.

Model Organisms

The current understanding of vertebrate neural pattern formation is due to research in a variety of model organisms including frog and other amphibians, chick, mouse, and zebrafish. Research with amphibians and birds has provided us with information on tissue interactions associated with patterning due to their accessibility to microsurgical manipulation, and more recently with specific localized protein function through

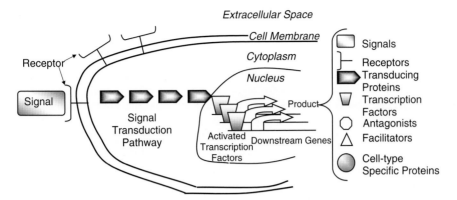

FIGURE 3. Pattern formation in vertebrates involves a signaling cascade that produces protein products, which can act in this cell or in the extracellular space to modify some aspect of a future signaling event. In addition, cell-type specific genes can be expressed leading to differentiation. Receptors may be membrane bound (as shown) for peptide ligands, or cytoplasmic as with RA and steroid ligands. Antagonists and facilitators can act in the extracellular space, in the membrane in conjunction with the receptor, with the signal transduction proteins or with a transcription factor. A transcription factor and its associated binding proteins can either up- or downregulate transcription of a given downstream gene.

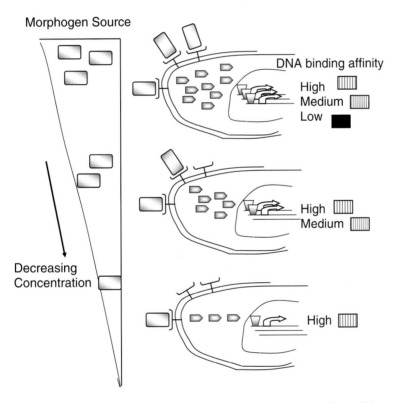

FIGURE 4. Model of morphogen action. Different concentrations of morphogen activate variable amounts of intracellular transcription factors. Downstream genes with variable affinity for these transcription factors are therefore activated at different concentrations of the morphogen. For example, at high levels of BMP (see Dorsal Patterning), high levels of nuclear SMAD activity would activate epidermal genes with low binding affinity (top cell), at intermediate levels neural crest genes would be activated (medium affinity, middle cell), and at low levels neural genes would be activated (high affinity, bottom cell). (Adapted from Wilson *et al.*, 1997, with permission from the Company of Biologists Ltd.)

injection (frog) or transfection (chick) with corresponding genes or mRNA. Mouse has allowed us to eliminate (or add) specific genes, individually or in combination, to understand their importance in specific pathways. Zebrafish has been useful for its ease of mutation, which has helped identify new players and reveal their importance in the signaling pathways.

In many cases, the molecular pathways and cellular responses that have been identified appear to be conserved between all vertebrates. In fact, for some molecular pathways, the conservation reaches back to our common ancestors with insects; the same pathways are used in *Drosophila*. In others, there appear to be differences in pattern regulation that are specific to classes of vertebrates. The best described of the general vertebrate central nervous system (CNS) patterning cascades include the anteroposterior (AP) patterning of the midbrain and hindbrain (reviewed by Lumsden and Krumlauf, 1996), and the dorsoventral (DV) patterning of the spinal cord (reviewed by Tanabe and Jessell, 1996; Lee and Jessell, 1999; Litingtung and Chiang, 2000). These will be discussed, and what is known about other regional CNS patterning pathways will be mentioned to highlight our current understanding of neural pattern formation.

Axes of the Nervous System

The vertebrate nervous system is initially induced as an apparently homogeneous epithelial sheet of ectoderm adjacent to its organizer (see Chapter 1). This **neural plate** has contact ventrally with the underlying dorsal mesoderm, and laterally with the epidermal ectoderm, and these two neighboring tissues assist the neural plate to form a **neural tube** in a generally rostral to caudal sequence. Subsequently, a number of broad, discrete regions will form, both anteroposteriorly and dorsoventrally, beginning the cascade of specialization that will ultimately give rise to the complex vertebrate CNS (Fig. 5). Traditionally we identify the prominent AP regions as forebrain, midbrain, hindbrain, and spinal cord, whereas in the DV plane (at least in the trunk) we recognize the dorsal sensory neurons and ventral motor neurons. In addition, from the lateral margins of the early neuroectoderm, the sensory placodes and neural crest form and generate the cranial nerves and the peripheral nervous system (PNS; Fig. 5, see also Chapter 4). At later stages, left vs right also becomes an important feature of the differentiated nervous system; however, virtually nothing is known at this time about the control of this patterning. The cellular and molecular mechanisms associated with the AP and DV cascades of patterning that give rise to distinctive regional development in the early vertebrate neuroectoderm is the focus of this chapter.

AP PATTERN

Early Decisions

At its inception, the neural plate has three axes, AP, mediolateral, and left–right. As it forms the neural tube, the AP axis comes to extend virtually the entire length of the dorsal embryo. Patterning in the AP plane proceeds from coarse to fine subdivisions and involves morphogens, receptors, internal and external regulators, signal transducers, transcription factors, and tissue specific target genes. The embryo matures in a head to tail direction, so more anterior structures are further along in their developmental cascade than are caudal structures. Thus, it is often not entirely meaningful to state the subdivisions as though they have formed concurrently. The AP cascade is much more complex than that. However, for simplicity's sake we say that the early neural plate begins its life in an anterior state (defined here as "head"), and the first step in patterning is to establish from

this a separate "trunk" region. Soon thereafter, beginning at the anterior end of the embryo, the neural plate forms a neural tube, which swells, extends, and further subdivides to form the **prosencephalon** or forebrain, the **mesencephalon** or midbrain, the **rhombencephalon** or hindbrain, and the narrow **spinal cord** (Fig. 6). Conventional embryology and anatomy include the forebrain, midbrain, and hindbrain with the head, and begin the trunk at the anterior spinal cord (either just caudal to the last rhombencephalic swelling at r7 and the first somite, or at the level of the fifth somite and first cervical vertebrae). However, evolutionarily, it appears that the hindbrain level of the AP axis may have come first in prevertebrate chordates, with structures anterior (new head) and posterior (trunk and tail) being added as vertebrates evolved. Within the realm of neural pattern formation, this "new head" including the forebrain and midbrain express Otx2 and other non-Hox transcription factors as positional information, and are dependent for their formation on several signaling factors called "head inducers" (see below), making this region of the head distinctly different from the hindbrain. In contrast, the spinal cord is clearly patterned as an extension of the hindbrain using *Hox* genes as positional information, and is dependent for its formation on several caudalizing factors, which are antagonistic to those involved in "new head" formation. Thus, for the purposes of discussing pattern formation, "head" will be defined as the neuroectoderm rostral to the midbrain/hindbrain boundary (site of the **isthmic organizer**), and "trunk" as the area caudal to it (including the future hindbrain and spinal cord). This "head–trunk" division represents a didactic effort to segregate major patterning differences.

Within the head and trunk further subdivisions are established in response to asymmetric signals through the expression of positional information genes (region specific transcription factors), and these regions in turn are also subdivided until the finely patterned detail of the fetal CNS is achieved. Details of our understanding of the pathways leading to these major and minor subdivisions appear below.

First Division

The longstanding models for AP patterning are founded on landmark experiments from the early part of the last century (Spemann and H. Mangold, 1924; Spemann, 1931; O. Mangold, 1933) and reconsidered in the 1950s by Nieuwkoop (Nieuwkoop *et al.*, 1952) and Saxen and Toivonen (reviewed by Saxen, 1989). Working with amphibian embryos, Spemann and H. Mangold discovered that the upper (dorsal) blastopore lip could induce a well-patterned ectopic neural axis. They called this region the organizer. Subsequently, Spemann (1931) determined that the organizer of younger embryos could induce a whole axis including head while older organizers could only induce the trunk neuraxis. Similarly, O. Mangold determined that the underlying mesendoderm having ingressed from the organizer at early stages induced the head, whereas the later mesoderm induced the trunk. Thus the concept of head and trunk as the first coarse AP division of the neuroectoderm was established.

Germ layer	Major division	Early subdivision	Later subdivisions	Mature derivatives	Cranial nerve (CN) associations	Classification
ectoderm	outer ectoderm	epidermal ectoderm	epidermis	skin	none	non-neural
				hair		
				nails		
				sebaceous glands		
				tooth enamel		
				anterior pituitary		
		placode ectoderm	lens placode	lens, cornea		non-neural
			nasal placode	olfactory epithelium	CN I	Peripheral Nervous System
			auditory (otic) placode	cochlea, vestibular ap.	CN VIII	
			epibranchial placodes	sensory ganglia	CN V, VII-X	
	neural crest	neural		Schwann cells	r1	
				neuroglial cells	r2, CN V	
				sympathetic NS	r3, CN V	
				Parasympathetic NS	r4, CN VI, VII, VIII	
		non-neural		facial cartilage	r5, none	
				dentine of teeth	r6, CN IX	
				melanocytes	r7, none	
				adrenal medula	r8, CN X, XI, XII	
	neural tube	prosen-cephalon	telencephalon	cerebral cortex	none	Central Nervous System
				basal ganglia		
				hippocampus		
			diencephalon	retina		
				thalamus	CN II	
				hypothalamus	none	
				infundibulum/post pit.		
				epiphysis/pineal		
		mesen-cephalon	mesencephalon	superior colliculus	none	
				inferior colliculus		
				tegmentum		
				cerebral peduncle		
		rhomben-cephalon	metencephalon	cerebellum		
				pons, r1	CN IV (motor)	
			myelencephalon	medula, r2–8	CN V–VII, X–XII	
		spinal cord	cervical, thoracic, lumbar & sacral nerves		none	

FIGURE 5. Chart showing developmental progression of ectodermal differentiation. CN, cranial nerves are: I Olfactory (special sensory), II Optic (special sensory), III Oculomotor (motor and autonomic), IV Trochlear (motor), V Trigeminal (sensory and motor), VI Abducens (motor), VII Facial (motor, sensory, and autonomic), VIII Auditory/Vestibulo-acoustic (special sensory), IX Glossopharyngeal (sensory, motor, and autonomic), X Vagus (autonomic, sensory, and motor), XI Accessory (motor and autonomic), XII Hypoglossal (motor). See also Fig. 12. Shading distinguishes major tissue classifications (CNS, PNS, Non-neural).

Head Neural Induction and Maintenance

The major similarity between the early models of AP patterning is the understanding that the initial neuroectoderm induced is rostral in character, either by default or due to primary rostralizing signals that are present as the neural ectoderm forms. This understanding has been supported at the molecular level by observations that the neural inducers chordin, noggin, and follistatin (all Bone Morphogenic Protein or BMP inhibitors) are able to induce forebrain but not neuroectoderm of more posterior character in amphibian animal caps (see Chapter 1), whereas in mouse double mutants for chordin and noggin, the forebrain does not form (Bachiller et al., 2000). These experiments indicate these factors are both sufficient and necessary to form the head.

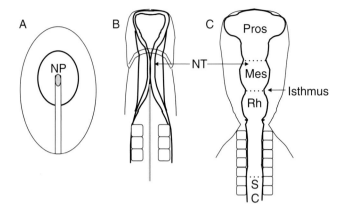

FIGURE 6. Drawings of avian embryos at various early stages. (A) At late stage 3, the neural plate (NP) (bold line) forms around the organizer (gray). (B) At stage 8, the neural plate rolls into a neural tube (NT) beginning at the future midbrain level. (C) At stage 11, the neural tube has formed its rostral vesicles, the prosencephalon (Pros) or forebrain, mesencephalon (Mes) or midbrain, and rhombencephalon (Rh) or hindbrain as well as the spinal cord (SC). Arrow shows the location of the isthmus, which forms an organizer between the mesencephalon and rostral rhombencephalon.

However, this one-step model of head formation appears to be an oversimplification because other proteins or tissues have been identified that are also sufficient and necessary for head formation. In mammals, there is a second signaling center, the anterior visceral endoderm (AVE) that secretes a TGFβ superfamily member (Nodal) and TGFβ and Wnt antagonist, Cerberus-like (cer1), that are involved in head formation. In many vertebrates cerberus and several other Wnt antagonists (Dickkopf-1 [Dkk1], Frzb1, and Crescent) are expressed in the rostral endoderm or cells in the early organizer, tissues which share head-forming qualities with the mammalian AVE. Ectopic expression of cerberus (in *Xenopus*; Cer, Bouwmeester *et al.*, 1996) and Dkk1 (in *Xenopus* and zebrafish; Kazanskaya *et al.*, 2000, Hashimoto *et al.*, 2000) show these proteins are sufficient to produce anterior neural ectoderm from ectodermal precursors. In addition, *Xenopus* embryos posteriorized experimentally (with bFGF, BMP4, or Smads: See below) are rescued by Dkk1 (Hashimoto *et al.*, 2000; Kazanskaya *et al.*, 2000). Conversely, overexpression of head inducers in caudal neuroectoderm results in the loss of caudal markers and the expansion of more rostral fates. All of these experiments indicate that these "head inducers" are sufficient to support rostral neural formation. These proteins are probably also necessary, because injections of anti-Dkk1 antibody resulted in loss of the telencephalon and diencephalon, and null mutation of *Dkk* in mouse leads to loss of all head structures anterior to the hindbrain (Mukhopadhyay *et al.*, 2001).

From these data we infer that these additional signaling factors induce head formation and this could be used to argue that anterior neuroectoderm is not the default state. On the other hand, rostral neural ectoderm could still be the default but undetermined state, and these factors could merely be required to protect it from transformation to more caudal fates in the presence of caudalizing signals. Because their function is the antagonism of Wnt action, and Wnts are caudalizing factors, it seems reasonable that anterior is the default and that "head inducers" like Cer and Dkk are required to override caudalizing factors to maintain (determine) the head in its original state (see below).

Trunk Neural Induction

Whereas the early modelers of AP pattern agreed that head neuroectoderm was primary, they differed in their ideas of how more caudal neuroectoderm was formed (Fig. 7). The Spemann/ Mangold model proposes that the cells in the early organizer induce and pattern the head, whereas at a later stage these cells are replaced with a population that induces the trunk neuroectoderm. Thus the organizer shifts from inducing the head to inducing the trunk over time (temporal separation) through the movement of cells (spatial separation). Nieuwkoop and coworkers proposed that signals (called transformers) from some other source could convert some of the rostral neuroectoderm into caudal neuroectoderm. Saxen and Toivonen proposed opposing gradients of morphogens whose relative levels would establish appropriate AP patterning separate from neural induction. One major difference between the models is whether a neural inducing and caudalizing signal is relayed through the organizer and coupled to induction or whether a caudalizing signal from a nonorganizer source transforms already-induced neuroectoderm directly by acting in a competitive or antagonistic manner. In the end, there is no reason that all of these pathways could not be used during AP patterning of the nervous system, and indeed, evidence indicates that they are (Kiecker and Niehrs, 2003).

Evidence in support of the Spemann/Mangold head- and trunk-organizer (Fig. 7A) model comes from several sources. First, classic amphibian and avian grafting experiments show that young organizers can induce a complete axis, whereas older organizers have lost the ability to induce the head. Second, "Keller sandwich" experiments, in which the amphibian neural ectoderm extends without underlying mesoderm, show that AP neural patterning can result from planar signals from the organizer (reviewed by Doniach, 1993; Ruiz i Altaba, 1993, 1994).

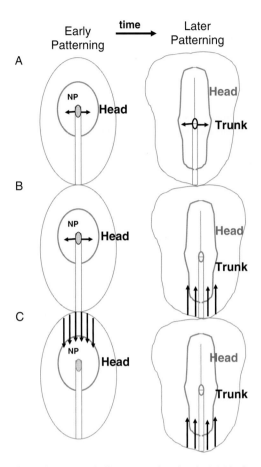

FIGURE 7. Three models of initial neural pattern formation. Arrows indicate patterning signals. (A) The Spemann/Mangold model wherein early signals from the organizer pattern the head and later signals from the organizer pattern the trunk. (B) The Nieuwkoop model wherein early signals from the organizer pattern the head and then later signals from other sources transform more caudal neuroectoderm into trunk. (C) The Saxen & Toivonen model wherein a rostral gradient of anteriorizing signals patterns the head and a caudal gradient of posteriorizing signals patterns the trunk.

Third, if the trunk organizer is going to exist with separate function from the head organizer, then one needs evidence that the organizer changes its secretory molecules over time and that the later ones can cause caudalization of the neuroectoderm. This has been demonstrated in mouse where retinoic acid (RA), a caudalizing agent, is produced by the older node but not the younger (Hogan *et al.*, 1992) and in *Xenopus*, where derivatives from the young node secrete chordin, which induces the head, whereas derivatives of older nodes secrete fibroblast growth factor (FGF), which induces the trunk (Tiara *et al.*, 1997). In addition, older chick nodes can induce *Xenopus* animal caps to express Pax3, a caudal marker, whereas younger nodes cannot (Bang *et al.*, 1997). Fourth, if the trunk organizer is going to be both inducing and patterning the trunk neuroectoderm in a single step, then a molecule that can both induce and caudalize must be identified. FGF is able to do both (Lamb and Harland, 1995). Fifth, there is evidence that trunk neuroectoderm is created *de novo* from later node and this generation requires FGF (Mathis *et al.*, 2001). Finally, recent experiments have implicated BMP-4 as a signal that acts directly on the *Xenopus* organizer to convert it from a head inducer to a trunk inducer (Sedohara *et al.*, 2002).

Thus tissue interactions appropriate for the Spemann/Mangold model of AP pattern play a role in AP neural patterning.

Significant evidence also exists in support of the Nieuwkoop model (Fig. 7B). This model is usually called activation/transformation for the initial activation (induction and patterning) of the head neuroectoderm by the organizer, followed by the subsequent transformation of the caudal cells in this head field into trunk neuroectoderm. Classic amphibian experiments demonstrate that vertical signaling from the mesoderm can directly pattern the neuroectoderm induced by the organizer (reviewed by Doniach, 1993; Ruiz i Altaba, 1993). Several secreted factors capable of caudalization have been identified including FGFs, RA, and vertebrate homologs of the *Drosophila* wingless protein (Wnts). FGFs (in *Xenopus*) are expressed in the posterior dorsal mesoderm during gastrulation. When anteriorized animal caps (which form anterior neural ectoderm expressing Otx-2 (forebrain and midbrain) and En2 midbrain–hindbrain boundary) were treated with bFGF both anterior and posterior markers (Krox-20/hindbrain and Hoxb-9/spinal cord) were expressed. When a later stage of the neural ectoderm was treated with bFGF it induced forebrain to express a hindbrain marker and hindbrain to express the spinal cord marker (Cox and Hemmati-Brivanlou, 1995). In another lab, Kengaku and Okamoto (1995) determined that progressively more posterior markers were induced when increasing concentrations of FGF were provided to neural ectoderm. Finally, recent work in zebrafish indicates that FGF3, through chordin (a BMP inhibitor), mediates expansion of the posterior- and suppression of the anterior neuroectoderm (Koshida *et al.*, 2002). Thus, FGFs would fit the role of Nieuwkoop's transforming signal. But they are not alone.

Retinoids can also serve this function. Retinoids are expressed at high levels in the posterior neuroectoderm and are involved in establishing the positional information for the hindbrain. RA and other retinoid derivatives of vitamin A act as signaling molecules much as steroid hormones do. They are able to pass through the plasma membrane of cells and bind to retinoic acid receptors called RARs and RXRs (retinoid X receptor peptides) in the cytoplasm. These translocate to the nucleus and act as transcription factors by binding to retinoic acid response elements (RAREs) within the promoters of certain genes. *Hox* genes contain RAREs and their expression is modified by levels of retinoids acting as morphogens. That is, *Hox* genes with rostral expression patterns (e.g., in the rostral hindbrain) are expressed at low levels of retinoids, while more caudal *Hox* genes are expressed only where the levels of retinoids are higher. Blocking RA signaling results in the loss of caudal rhombencephalic pattern and the transformation of this region into more rostral rhombencephalon (Dupe and Lumsden, 2001; see Hindbrain Patterning below). Artificially raising the concentration of RA in the environment results in changes in the expression patterns of some regionally expressed transcription factors including *Hox* genes, demonstrating the relationship between this morphogen and these positional information transcription factors. Phenotypically, increased RA results in a loss of anterior structures and markers (Fig. 8A). Distinct phenotypes are generated depending on the timing of exposure to RA (in mouse) indicating that RA can influence differentiation at several steps in the AP axis cascade (Fig. 8B; Simeone *et al.*, 1995).

Finally, a strong case can be made for Wnts as transformers in the caudalizing of the neuroectoderm. Overexpression of various Wnts, or of the elements in their canonical signal transduction pathway, or of lithium chloride, the artificial activator of this pathway, leads to loss of head structures and induction of posterior neural markers. Blocking Wnt activity leads to head gene expression, while mutations in various genes in this pathway lead to caudal truncations. Recently, Kiecker and Niehrs (2001) have shown that neuroectoderm associated with increasing concentrations of Wnt8 expresses genes associated with increasingly caudal levels of the neuraxis, demonstrating that Wnt, too, is a caudalizing morphogen. Thus, these three caudalizing morphogens, FGFs, RA, and Wnts, support the Nieuwkoop model of Activation and Transformation. By regulating the expression of positional identity genes within the already-formed anterior neuroectoderm, transforming signals can mediate posterior neural patterning.

Finally, the Saxen and Toivonen model (see Fig. 7C) seems to best express how the head is maintained in light of these transforming/caudalizing factors. But rather than a competition between two positive signaling gradients as originally proposed, we find the mechanism of head and trunk formation ultimately depends on antagonism gradients of inhibitors, comparable to the amphibian model for the induction of the neuroectoderm (Chapter 1; Fig. 9). In both cases, the default state is singular. In "neural induction" the default state of the ectoderm is neural (expressing transcription factors Sox1, 2, and 3). In "head induction" the default state is anterior ectoderm or head (expressing transcription factors Lim1, Otx2, and Anf). To increase complexity during development, secreted signals appear with the ability to transform this uniform tissue into another. For neural induction they are BMPs, and the secondary state is epidermal ectoderm. For AP neural pattern, these signals include RA, FGFs, Wnts, and BMPs (Glinka *et al.*, 1997; Piccolo *et al.*, 1999) and the secondary state is more caudal neuroectoderm. In order to protect the first state from this modification, antagonists of these signal(s) are generated. In neural induction, these are noggin, follistatin, and chordin expressed in the organizer and its derivatives. For AP patterning, these could be proteins such as cerberus, dickkopf, nodal, and lefty (reviewed by Perea-Gomez *et al.*, 2001), frzb, noggin, and crescent, which are secreted from the rostral mesendoderm and which are antagonists of Wnts, BMPs, and other signaling molecules involved in caudal specification. Successful protection of a subset of the original ectodermal region results in the formation of two separate potentials in each case (neural vs epidermal and "head" vs "trunk"). In addition, because the BMPs and caudalizers are morphogens, additional intermediate states can also be induced at the interface between these two states resulting in additional complexity. For neural induction, this begins the DV patterning cascade by inducing the neural crest, whereas for AP patterning the midbrain–hindbrain boundary or isthmus, appears to be the intermediate state. Thus, a three-step model of early AP pattern formation is supported: *Neural induction* (with anterior character), *caudalization* (new neural induction and transformation to generate trunk character), and *anterior maintenance* to protect two separate states, "head" and "trunk."

Although this three-step model is presented as a synthesis of the historical models that fits the current data, there are other ways of interpreting these data. One alternate interpretation still holds head induction to be the direct result of BMP and Wnt antagonism (an unmodified Saxen–Toivonen double-inhibitor model). This is supported by ectopic head induction using appropriate antagonists in *Xenopus* embryos (e.g., see Niehrs *et al.*, 2001). These antagonists are sufficient for head induction, but because they are also required for head maintenance and the neural state may be the default, it is difficult to demonstrate whether they are or are not actually required for induction of the head.

In addition, there may be some important differences between model animals in the caudalizer-antagonism step of this AP patterning. Specifically, the required source of the secreted caudalizing-factor antagonists ("head inducers") in mammals is the AVE (reviewed by Beddington and Robertson, 1998), although grafts to other species indicate the mouse node/organizer also produces the appropriate signals to induce and

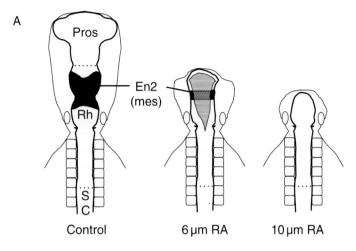

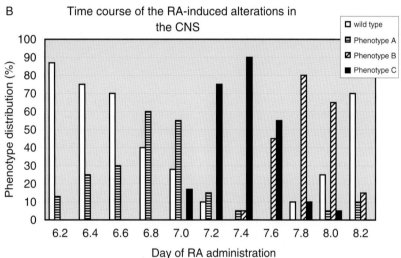

FIGURE 8. Effects of RA addition to developing CNS. (A) Diagrammatic representation of chick embryos treated with RA at stage 3 and cultured for 24 hr. Control embryos develop normal features and express En2 at the isthmus (solid black). Embryos treated with 6 μm RA express En2 in a smaller area and at lower levels. Embryos treated with 10 μm RA failed to express En2 or expressed it at levels undetectable with whole mount immunocytochemistry. Development of tissues rostral to the mesencephalon was not observed (Darnell, 1992). (B) 250–400 mouse embryos were analyzed for each time point and the percentage of each phenotype is shown on the graph. The wild-type phenotype dominates for RA treatment at both ends of the trial period, delineating the critical period for RA effect overall. The shifts in distribution between the other phenotypes indicates RA has different functions at different times during development. Phenotype A (mild: reduction in the olfactory pit and midbrain DV compression) reveals the structures most sensitive at 6.8 and 7 dpc. Phenotype B (severe, atelencephalic microcephaly: growth retardation; reduction or lack of anterior sense organs and neural vesicles back to the isthmus; branchial arches reduced or abolished and hindbrain disordered). Sensitive period 7.6–8.0 dpc. Phenotype C (moderate, anencephaly: hypertrophic obliteration of the ventricles, open neural roof for diencephalon through hindbrain, all anterior genes expressed but domains altered, for example, Hoxb1 expression expanded from normal r4, into presumptive r2–r3 territory). Sensitive period 7.2–7.6 dpc. (Redrawn after Simeone *et al.*, 1995, Fig. 1.)

maintain head (e.g., see Knoetgen *et al.*, 2000). Traditionally, in birds, fish, and amphibians the source of "head" inducers has been attributed solely to the early organizer/node and its derived prechordal plate mesendoderm, although this has been recently contested. In chick, the hypoblast, a tissue similar to the AVE, can transiently induce early head neural markers (Foley *et al.*, 2000) and the foregut endoderm is involved in forebrain patterning (Withington *et al.*, 2001). In fish, rostral endodermal cells are involved in anterior neural patterning through Wnt antagonism

(Houart *et al.*, 1998, 2002). And in *Xenopus*, endodermal expression of *Hex* (an AVE associated gene in mouse) is also involved in anterior patterning of the neuroectoderm (Jones *et al.*, 1999). Thus, it now seems less likely that the two-source localization of early head maintainers in mammals is due to mutations that occurred in the signals localizing the expression of these genes after mammals diverged from other vertebrates. Instead, it may be a more primitive pattern that has been maintained more robustly or localized differently in small embryos where the

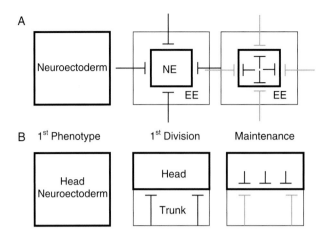

FIGURE 9. A comparison of the models for neuroectoderm "induction" and patterning. (A) The first phenotype of ectoderm is neuroectoderm. The first division of this tissue into two types occurs when inhibitory signals from the periphery (BMP) inhibit the neural signaling pathway and turn the outer area into epidermal ectoderm. The neural ectoderm is protected from these inhibitors by inhibitors from the organizer. (B) The first phenotype in patterning is head neuroectoderm. The first division of this tissue into two types occurs when signals from the caudal embryo transform closer neuroectoderm into trunk neuroectoderm. (These signals may either activate and/or inhibit certain gene expression.) The head is protected from these transforming signals by inhibitors expressed rostrally.

caudalizing signals would otherwise swamp out the rostral region. Experiments in diverse vertebrates with embryos of various sizes will be required to test this hypothesis.

Regional Patterning

Forebrain

The "head" is thus defined for pattern formation purposes as a region of anterior neuroectoderm that initially expresses the transcription factor Otx2 and extends from the anterior neural ridge at the rostral end of the embryo to the isthmus at the posterior margin of the future midbrain. Mouse mutants lacking Otx2 fail to form head structures (Acampora *et al.*, 2001), whereas in *Xenopus*, Otx2 is sufficient to induce anterior neural genes (Gammill and Sive, 2001). Thus, this transcription factor provides positional information for the head.

This Otx2 field subsequently subdivides within in the AP plane to generate the more complex pattern associated with the later forebrain and midbrain. These subdivisions result from responses to patterning signals from the underlying mesendoderm or prechordal plate and from new sources of environmental asymmetry, the anterior neural ridge in the anterior head and the isthmus in the posterior head. These signals could induce the appearance of active, region-specific transcription factors that could subdivide and further pattern the head. For example, Otx2 spans the head at the neural plate stage. Later, Otx1 is upregulated in all but the rostral region of Otx2-expression, then Emx2 is upregulated in the middle of the Otx2 region and Emx1 in the middle of this. The Otx2 pattern is followed by neural tube closure and the formation of anatomically identifiable pattern within the neural tube (36 hr in chick, 8–9.5 days in mouse, 4 weeks in human) correlated with the expression of these later genes (Fig. 10; Boncinelli *et al.*, 1993; Bell *et al.*, 2001).

Anatomically the prosencephalon (forebrain) forms the telencephalon (rostral forebrain) and diencephalon (caudal forebrain). The telencephalon, which ultimately forms the cerebral isocortex, olfactory cortex and bulbs, hippocampus, and basal ganglia (striatum and pallidum) expresses all of the head transcription factors mentioned previously, plus BF1. BF1 is upregulated in the telencephalon and retina by FGF8 (Shimamura and Rubenstein, 1997), a signaling molecule that is expressed in the anterior neural ridge and at the isthmus. Because the mesencephalic neuroectoderm does not upregulate BF1 in response to FGF8 (rather it upregulates the isthmic gene *En2*), it is clear that differential competence is established regionally within the head prior to the expression of these later marker genes.

The patterning of the diencephalon (in chick) has been described (Larsen *et al.*, 2001) but the signaling events required for this pattern formation have not been determined. The early diencephalon is subdivided into two functionally distinct regions: the anterior parencephalon and the posterior synencephalon. There is no cellular boundary (lineage or cell-mixing restriction) between the parencephalon and the telencephalon anterior to it; however, such a boundary does exist between the parencephalon and synencephalon (lineage restriction), and between the synencephalon and mesencephalon (lineage and cell-mixing restriction). Subsequently, the parencephalon is subdivided into ventral and dorsal thalamus by an anatomical feature called the zona limitans intrathalamica (zli), which is correlated with cells on either side becoming restricted to their compartment and with Gbx2 expression dorsally and Dlx2 and Pax6 expression ventrally.

Specific regulation of a number of other transcription factors has been correlated with the development of specific regions within the rostral head. For example, four POU-III transcription factor genes, *Brn-1*, *Brn-2*, *Brn-4*, and *Tst-1*, are expressed in the rat forebrain beginning on embryonic day 10 in a spatially and temporally complex pattern. The most restricted

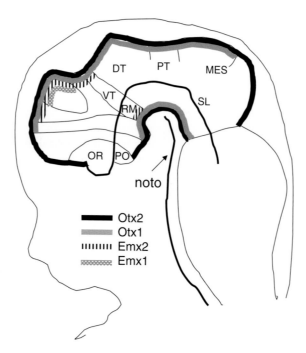

FIGURE 10. A diagram of the strong expression domains of four "head" genes in the mouse (E10). Internal lines correspond to locations where expression patterns change, indicating a possible functional boundary in AP patterning. Various anatomical subdivisions or precursor regions are labeled, including DT, dorsal thalamus; MES, mesencephalon; noto, notochord; OR, optic region; PO, post-optic; PT, pretectum; RM, retro-mammillary area; SL, sulcus limitans; and VT, ventral thalamus. (Redrawn after Boncinelli *et al.*, 1993.)

of these is *Brn-4*, which is expressed in the striatum of the telencephalon and parts of the thalamus and hypothalamus within the diencephalon (Alvarez-Bolado *et al.*, 1995). *Dlx-* and *Nkx2* gene families are regionally expressed in the diencephalon and other regions of the forebrain and their expression boundaries correlate with certain morphological boundaries (e.g., between isocortex and striatum within the telencephalon; Price, 1993). No clear boundaries of gene expression or cell-mixing restriction have been detected to subdivide the diencephalon into more restricted neuromeres, although the boundary between the diencephalon and mesencephalon is so defined (Larsen *et al.*, 2001).

Midbrain and Isthmus

Just caudal to the diencephalon, there is a bulge in the neural tube called the mesencephalon or midbrain. It is limited at its posterior margin by a constriction called the isthmus (see Fig. 6). The dorsal mesencephalon contributes to the superior and inferior colliculi (in mammals; equivalent to the optic tectum and torus semicircularis of birds), whereas the ventral mesencephalon (also known as tegmentum) generates structures such as the substantia nigra and the oculomotor nucleus. Otx2 is expressed broadly anterior to the isthmus, while the signaling molecule Wnt1 is expressed in a narrow band at the constriction. On the other side of the constriction, the transcription factors Pax2 and Gbx2 and signaling-molecule FGF8 are upregulated at the right time to be involved with the patterning of this region. Otx2 and Gbx2 appear to act as transcriptional repressors, each repressing the transcription of the other to generate a tight

boundary of gene expression at the isthmus, which is required for the appropriate expression of Fgf8, Pax2, and En2 (Glavic *et al.*, 2002). This boundary is not, however, a compartment boundary that limits cell movement across it (Jungbluth *et al.*, 2001). Another transcription factor, Xiro1, is expressed in a domain that overlaps the expression of Otx2, Gbx2, and FGF8 and is required for their correct spatial regulation (Glavic *et al.*, 2002).

Mouse mutants demonstrate that the signaling molecule Wnt1 and transcription factors En1/En2 expressed around this region are necessary for its development. Simultaneous knockouts of *En1* and *En2* result in failure of midbrain and cerebellar development. Knockouts of *Wnt1* show early expression of En1 and En2 but their increased expression is not maintained (McMahon *et al.*, 1992) and the mesencephalon and rostral rhombencephalon regions (cerebellar anlagen) subsequently fail to develop (McMahon and Bradley, 1990). Thus it appears that the transcription factors En1 and En2 are positional information genes required for the development of the midbrain and cerebellum and that they are initially expressed at the boundary between "head" and "trunk" neuroectoderm and maintained by Wnt1. So what turns on Wnt1 or En1 and En2?

Evidence showing that FGF8 secreted by the isthmus serves this function comes from bead implantation studies in the chick and mutation in zebrafish. Implanting FGF8 soaked beads in more rostral regions of the neuroectoderm induces several genes of the midbrain–rhombomere1 region in adjacent tissue including *Wnt1*, *En2*, and *FGF8*. FGF8 does this by binding to its receptor and initiating a signal transduction pathway that activates Pou2/Oct3/4 transcription factors (Reim and Brand, 2002).

Is FGF8 a morphogen? En2 is expressed in a gradient in the midbrain, an area that forms the optic tectum anterior to the isthmus (at low En2 levels) and the cerebellum posterior to the isthmus (at high En2 levels). This could be due to limited competence of these areas to respond, in which case they are prepatterned, or it could be a graded response to FGF concentration. To test this, the isthmus was grafted to either forebrain or hindbrain regions. When a part of the isthmus itself is grafted to the forebrain, a reversed gradient of En2 is induced nearby, with the higher concentrations near the graft (rostrally) and the lower concentration at a distance (caudally, Fig. 11). In these embryos, an ectopic cerebellar vesicle develops rostral to the ectopic optic tectum, supporting the conclusion that the concentration of the transcription factor En2 is differentially instructive within the development of the midbrain and hindbrain and thus that its inducer, FGF8, can act as a morphogen. However, in the hindbrain location, only cerebellum was induced, indicating that this tissue has received previous patterning information that limits its response to these inductive signals.

Thus the isthmus forms at a boundary between the midbrain (expressing Otx2) and the hindbrain (expressing Gbx2), which for patterning purposes we could say is between the "head" and the "trunk." This interface provides an asymmetrical source of signaling molecules that are involved in AP pattern of the cells both rostral and caudal to it. It is therefore referred to as the isthmic organizer.

Hindbrain

Just caudal to the isthmus, the neural swelling called the hindbrain or rhombencephalon develops (see Fig. 6). The rostral-most section of this vesicle (r1) expresses En2 in a gradient peaking at the rostral margin (the isthmus) and forms the cerebellum under the influence of FGF8 and Wnt1 (see above). The rhombencephalon is characterized early during development by its subdivision into anatomically identifiable rhombomeres. Rhombomeres 1–7 (r1–r7) form as identifiable bulges in the rhombencephalon proper, and the eighth metameric unit, r8, forms at the caudal end of the visible hindbrain, alongside the first five somites, and is similar in construction to the spinal cord. All eight rhombomeres constitute the rhombencephalon. At their dorsal margin, rhombomeres give rise to neural crest that forms the sensory component of the cranial nerves (along with contribution from ectodermal placodes, see Neural Crest and Placode). Laterally, interneurons form connecting sensory-motor reflex arcs and other inter-CNS connections. Ventrally, they produce motor neurons that contribute to the motor component of the IVth to XIIth cranial nerves. Specific cranial nerves arise from specific rhombomeres (Fig. 12) and cells within the

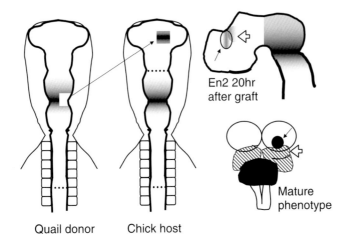

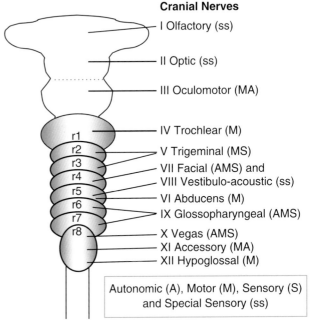

Cranial Nerves

I Olfactory (ss)

II Optic (ss)

III Oculomotor (MA)

IV Trochlear (M)

V Trigeminal (MS)

VII Facial (AMS) and
VIII Vestibulo-acoustic (ss)

VI Abducens (M)

IX Glossopharyngeal (AMS)

X Vegas (AMS)

XI Accessory (MA)

XII Hypoglossal (M)

Autonomic (A), Motor (M), Sensory (S)
and Special Sensory (ss)

FIGURE 11. Gain-of-Function experiment in chick showing the isthmus is sufficient to reestablish the mesencephalon and rostral rhombencephalon when grafted to an ectopic site. Shading indicates the gradient of En2 expression surrounding the isthmus. Neuroepithelium was taken from the isthmus region of a donor quail embryo (empty framed area) and grafted into the prosencephalon (stippled framed area) of a chick host. At 20 hr after grafting, the graft maintained En2 expression (small arrow) and induced En2 expression in the adjacent chick tissue. As with the normal expression, a gradient of En2 expression forms as the distance from the isthmus tissue increases. At later stages, the quail graft contributed directly to an ectopic cerebellum (thin arrow), and chick tissue just caudal to the graft formed an ectopic mesencephalon (open arrow) instead of dorsal thalamus (its normal fate). The ectopic mesencephalon/cerebellum is inverted in the AP plain relative to the host mesencephalon/cerebellum, indicating that their patterning is not influenced by a prepattern within the head neuroectoderm. (Redrawn after Alvarado-Mallart, 1993, Fig. 1.)

FIGURE 12. Cranial nerves: Diagram illustrating the AP origin of each cranial nerve in a d3 avian embryo. Motor and special sensory components come from the neural tube, whereas autonomic and sensory components come from the neural crest and placodes (see also Fig. 17). The motor branch of the trigeminal forms from axons of cell bodies in r2 and r3, and the glossopharyngeal from axons of cell bodies in r6 and r7. Axons contributing to the facial and auditory (vestibulo-acoustic) both exit at the same location in r4 (Lumsden and Krumlauf, 1996).

rhombomeres do not mix between rhombomeres beyond a certain stage. This demonstrates a new feature of patterning not yet addressed here: **segmentation**.

Most of what is known about segmentation and pattern formation was learned from the fruit fly, *Drosophila*. Fruit-fly body segmentation arises by a cascade of gene expression that subdivides a larger field. Large regions are specified by gap genes, and these are further subdivided into two-segment wide regions by the expression of pair-rule genes. Both gap and pair-rule genes are regulated by a morphogen gradient (bicoid) from one end of the embryo. These regions subdivide further under the influence of segment-polarity genes, which establish firm boundaries between the cells of each segment through negative-feedback circuits. As these boundaries are being established, the gap and pair-rule genes turn on specific sets of positional information transcription factors that will determine the later phenotype of each segment. In the fly, many of these positional information genes contain a conserved region called the homeobox. Homeobox-containing genes (*Hom* genes in flies) produce homeodomain proteins that are expressed in overlapping domains and establish positional information based on their rostral boundaries. The order of rostral expression of the *Hom* genes matches their 3′ to 5′ order within the *Hom* gene clusters on the chromosome, a feature called colinearity. *Hom* genes are assisted in their function of generating positional information by two other transcription factors, Extradenticle (Exd) and Homothorax (Hth). Segmentation of the vertebrate hindbrain shares some of these features.

No gap genes have been identified to define primordial subdivisions in the hindbrain as Otx2 and Gbx define the mesencephalic/rhombencephalic boundary and adjacent regions. So in vertebrates this first subdivision of the hindbrain may represent direct responsiveness to combinations of morphogen gradients. This has recently been shown for the normal development of r1, which is patterned by isthmic FGF8 and RA (Irving and Mason, 2000), and for r5 and r6, which depend on a different gradient of RA (Niederreither *et al.*, 2000) acting through RARα or RARγ (Wendling *et al.*, 2001). Within the posterior hindbrain many transcription factors are upregulated by the morphogen RA; however, the sources and directions of the RA gradients are a point of contention (Grapin-Botton *et al.*, 1998; Begemann and Meyer, 2001).

Although not necessarily involved in a primordial subdivision of the rhombencephalon, some "gaps" or shared qualities are observed between cells in the rostral rhombencephalon and are contrasted with other qualities shared by cells in the caudal rhombencephalon. For example, in humans, the rhombencephalon divides anatomically into metencephalon (which forms the cerebellum and pons and corresponds to the most rostral rhombomeres) and the myelencephalon (which forms the medulla and gives rise to cranial nerves VI–XII). However, this anatomical subdivision is not observed in other model animals. Instead there may be molecular differences between the rostral and caudal rhombencephalon. For example, the cells of r1–r3 differ in their cell division patterns from those in r4–r7/8 (Kulesa and Fraser, 1998) and r1–r4 have a different responsiveness to

RA than r5–r8 do (Niederreither *et al.*, 2000). Loss of RA signaling results in loss of r5–r8 character and their transformation to r4 identity (Dupe and Lumsden, 2001), whereas increases in RA result in expansion of r4–r8 at the expense of more rostral rhombomeres (e.g., Morriss-Kay *et al.*, 1991; Conlon and Rossant, 1992; Niederreither *et al.*, 2000). So, although gap genes have not been found in vertebrate hindbrain formation, the concept of larger pattern persists in this region.

In an approximation of the *Drosophila* pair-rule function, the hindbrain is initially subdivided into approximately two-segment units expressing transcription factors later associated with odd-numbered rhombomeres (e.g., Krox20, r3, and r5) and even-numbered rhombomeres (e.g., Hoxa2, r2; Hoxb1, r4; although Kreisler [kr] is expressed in both r5 and r6). At the interfaces between these two-segment regions, asymmetries provide positional information for full segmentation. For example, an analysis of *Krox20* mutant embryos indicates that Krox20 expression between even segments 2/4/6 and odd segments 3/5 is required for appropriate segment formation, cell segregation, and specification of regional identity. (Fig. 13; Voiculescu *et al.*, 2001).

The normal formation of boundaries between rhombomeres also depends on the expression of transcription factors Pou2/Oct4 (Burgess *et al.*, 2002), and bidirectional signaling mediated by Eph receptors (r3, r5) and their ligands (r2, r4, r6; Klein, 1999). In some ways this is similar to the action of the *Drosophila* segment polarity genes, although the Ephs/ephrins are **realizators** (revealing the cell's fate through their expression) whereas the crucial segment polarity genes are **selectors** (regulating the cell's fate through their expression). In any case, the juxtaposition of these alternating proteins restricts cell mixing *in vitro*, and likely generates the compartment boundaries observed *in vivo* (Lumsden, 1991). Ultimately, each rhombomere is well defined.

As with *Drosophila* segments, each rhombomere also expresses a different set of transcription factors that serve as its positional information (Fig. 14). In vertebrates, as in *Drosophila*, these genes frequently contain a homeobox (*Hox* genes in vertebrates). The order of the rostral boundaries of *Hox* gene expression in the nervous system shows colinearity with their position on the chromosomes. They are regulated by gradients of a morphogen (RA) or morphogens and their function depends on two other transcription factors, Pbx (the homolog of *Drosophila* Exd) and Meis (the homolog of *Drosophila* Hth; Waskiewicz *et al.*, 2001). As for being positional identity factors, ectopic expression or repression of these genes causes a shift in rhombomere identity to match the new code.

Thus the segmentation and segment identity cascade first determined in *Drosophila* is mirrored in the vertebrate hindbrain both at the mechanical and molecular level. It is generated through a cascade of signaling within the hindbrain and is autonomous from its surrounding mesoderm. This contrasts with the patterning of the hindbrain neural crest and the spinal cord, which are dependent on signals from the surrounding segmented mesoderm or branchial arches to determine their position.

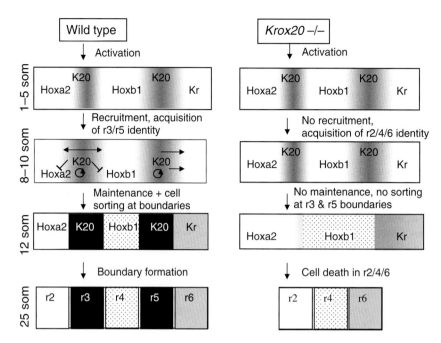

FIGURE 13. Model of hindbrain segmentation in mouse using wild-type and *Krox20* mutants. For wild-type embryos, at 1–5 somites, Krox20 is expressed in a few cells at two bands corresponding to prospective r3 and r5. The enhancers for *Hoxa2, -b1*, and *Kreisler (Kr)* are activated. Additional cells are recruited to express Krox20. At the 8–10 somite stage, prospective r3 and r5 express Krox20 homogeneously and recruit cells from adjacent regions (arrows). In addition, Krox20 regulates its own expression (circular arrows) and inhibits the expression of positional information genes from even numbered rhombomeres. By the 12 somite stage, r3 and r5 have acquired their identity. By the 25 somite stage, the rhombomere boundaries are well defined. In Krox20 mutants, the early stages look similar to wild-type embryos. However, the Krox20 regions do not expand or coalesce. Eventually these cells acquire an even numbered rhombomere identity and get incorporated into r2/4/6. By the 25 somite stage, significant cell death has reduced the size of the even-numbered rhombomeres leading to a reduction in the size of the hindbrain. (Adapted from Voiculescu *et al.*, 2001, with permission from the Company of Biologists Ltd.)

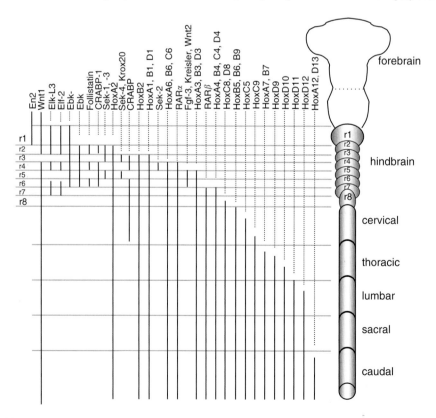

FIGURE 14. Diagram of localized gene expression in the developing "trunk." Rhombomere boundaries are specified by specific combinations of transcription factors. In the spinal cord, the rostral limit of *Hox* gene expression delineates positional information.

Spinal Cord

Colinear *Hox* gene expression is continuous from the hindbrain throughout the spinal cord, with genes located in more 3′ regions of the chromosomes being expressed more rostrally, and those at more 5′ regions in the clusters being expressed more caudally (Fig. 14). These transcription factors provide positional information within the neural tube and adjacent mesodermal somites that controls the development of cervical, thoracic, lumbar, and sacral development in the spine. Evidence in support of this comes from a comparison of the vertebrae of chick and mouse. These two species express similar *Hox* genes in their trunk, and the boundaries of expression of gene pairs match reproducibly with the division between cervical and thoracic (*Hoxc5* and *c6*) and between lumbar and sacral (*Hoxd9* and *d10*) even though these two points occur in different locations in mouse and chick (Fig. 15). In addition, grafting experiments that moved either neural tissue or paraxial mesoderm (somite) to another AP position in the embryo have demonstrated that neural positional information, as measured by AP-level specific motor neuron differentiation, tracks with the level of the adjacent paraxial mesoderm.

At a molecular level, it was anticipated that the mesoderm, which expresses *Hox* positional-information genes and directly underlies the trunk neuroectoderm, would pattern the overlying neuroectoderm directly. Unfortunately, the patterns of expression of the mesoderm and neuroectoderm do not line up. Three mechanisms have been suggested in chick and mouse to account for the observation that positional information genes in the spinal cord do not show the same rostral boundaries in ectoderm and mesoderm. The first possibility is that CNS position is regulated by adjacent paraxial mesoderm to express the same *Hox* genes, followed by differential growth or morphogenesis that would displace the rostral boundaries between these two tissues (e.g., Frohman *et al.*, 1990). Alternately, one *Hox* gene in the mesoderm could promote the secretion of signals that would induce another *Hox* gene in the CNS (e.g., Sundin and Eichele, 1992). Finally evidence also exists for the possibility that caudal sources secrete morphogens that form gradients that induce positional genes in the CNS and mesoderm independently, without the requirement for local signaling sources (e.g., Gaunt and Strachan, 1994). Again, it is possible that all of these mechanisms are functioning to regulate different parts of this complex cascade.

The point of establishing a specific *Hox* code within the neural tube is to regulate downstream genes appropriate to particular AP levels of the spinal cord. For example, although generally similar in function, the spinal cord sensory and motor neurons have specific targets depending of their AP level. For example, sensory and motor neurons from the brachial and lumbar regions target the arms and legs, whereas those of the cervical, thoracic, and sacral levels do not. Specific transcription factors, such as the LIM genes in motor neurons are expressed in a distinct pattern within the spinal cord in accordance with their projected targets and due to their *Hox* expression induced by patterning signals from the adjacent mesoderm (Ensini *et al.*, 1998).

Neural Crest

The neural crest cells (see Dorsal Patterning below) are induced at all AP levels of the neural tube except the rostral diencephalon and telencephalon. The regulation of their presence or absence in the AP plane is a function of the same caudalizing and caudal-antagonist signals that promote AP patterning in the CNS. Although no neural crest cells are formed at the boundary between the rostral-most CNS and epidermal ectoderm, treatment of rostral neural ectoderm in *Xenopus* with intermediate levels of BMP and either bFGF, Wnt8, or RA transforms this tissue into neural crest. This transformation can be blocked by expression of dominant negative forms of the appropriate receptor or dominant negative versions of the signal. Similar rostral crest induction can be achieved *in vivo* with the expression of a constitutively active RA receptor (Villanueva *et al.*, 2002). These data demonstrate elements of the patterning cascade regulating the no-crest/crest anterior boundary.

Within the crest-forming region, patterning also occurs (Fig. 16). Cells from the anterior crest (of the posterior diencephalon, mesencephalon, and rhombencephalon, down to the level of the fifth somite) form mesectoderm (non-neural cells forming the connective tissues of the cranial muscles and

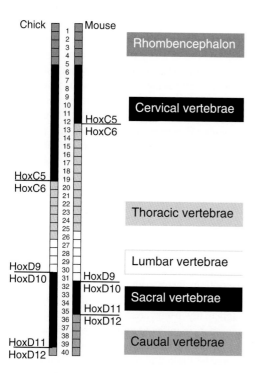

FIGURE 15. Specific anatomical boundaries in the mesoderm, for example, between the cervical and thoracic vertebrae, correlate with *Hox* gene expression in the mesoderm. Even though these anatomical transitions do not occur at the same level (somite number). In the chick there are many more cervical vertebrae than in the mouse, but *HoxC6* expression begins in the somite at the level of the first thoracic vertebrae in both species. Numbers down the middle of the figure represent somites.

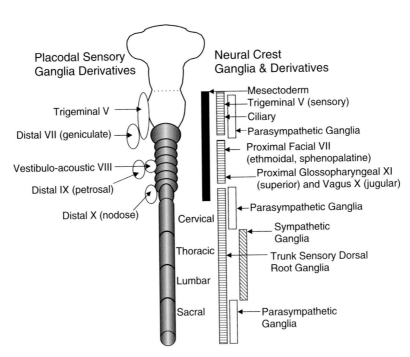

FIGURE 16. Placodal and neural crest contributions to the PNS (in part adapted from Le Douarin *et al.*, 1993, with permission from Academic Press, Orlando, FL).

the cartilage and membrane bone of the facial skeleton and skull vault), parasympathetic ganglia (cholinergic/Ach-secreting neurons from midbrain and rostral hindbrain levels [r1]), and sensory ganglia (also cholinergic). At spinal cord levels, parasympathetic ganglion cells give way to sympathetic ganglia cells (noradrenergic/noradrenaline-secreting neurons, T1-L2), whereas at the most caudal levels, parasympathetic ganglia reappear (second to fourth sacral segments). Sensory ganglia are formed at nearly all levels of the posterior cranial and spinal neural tube. Grafting studies using chick–quail chimeras, which allow tracking of heterotopically grafted cells to their new fates, demonstrate that all levels of the neural tube have the potential to produce sensory, sympathetic, and parasympathetic neurons from the crest. Therefore, limitations to the pattern must depend on signals independent of CNS patterning.

The understanding of the molecular mechanisms underlying neural crest positional identity is still limited. Many of these mechanisms, such as the involvement of cascades of certain types of transcription factors and lateral inhibition via the Notch-Delta system, have been conserved from our common ancestor with *Drosophila* (Ghysen *et al.*, 1993; Jan and Jan, 1993). For neural crest, the extracellular signaling tissues and molecules that control these cascades are still being elucidated. Within the hindbrain region, where crest forms specific cranial nerves associated both with particular rhombomeres and specific branchial arches (and pharyngeal pouches), one can ask if rhombomere positional identity or branchial arch positional identity determines the pattern of these crest cells. Zebrafish mutations that affect the mesendodermal patterning of the branchial arches through which these neural crest cells migrate without affecting the patterning of the rhombomeres indicate that the mesendoderm patterns

the crest and not vice versa as had previously been proposed (Piotrowski and Nusslein-Volhard, 2000). In a similar finding based on chick–quail grafting experiments (Couly *et al.*, 2002), Hox nonexpressing crest found rostral to the hindbrain were patterned by regional differences in the anterior endoderm (skeletal not neural structures were assessed). Crest from Hox-expressing regions failed to respond to similar signals, again indicating that a prepattern separates cells in the "head" from those in the "trunk."

Emerging evidence indicates that the neural crest choice between sensory and autonomic differentiation hinges on exposure to BMP2 expression in the peripheral tissues, perhaps from the dorsal aorta. *In vitro*, high concentrations of BMP2 initiates expression of the transcription factor MASH1 associated with autonomic differentiation. BMP2 acts instructively rather than selectively. Additional signals from specific AP locations that have not yet been identified could induce the expression of other transcription factors, which act in conjunction with MASH1 to specify the final phenotypes of the different autonomic neuron subtypes (sympathetic, parasympathetic, and enteric). In contrast, in the absence of BMP2, sensory neurons form and express several transcription factors including neurogenin 1 and 2, NeuroD, and NSCL1 and 2 (reviewed by Anderson, 1997).

Although many trunk crest cells are multipotent at the time their migration is initiated and can form either sensory or autonomic (sympathetic) neurons depending on their environment, others may be limited in their potential prior to migration. Trunk neural crest migrating from young neural tubes, which would normally form ventral structures, can differentiate into several cell types including catecholamine-positive (sympathetic) neuroblasts, whereas crest migrating from older neural tubes end up in the dorsal region (presynaptic-sympathetic or sensory ganglia) and never produce catecholamines. While young and older crest cells can be tricked into migrating to the dorsal or ventral locale that is inappropriate for them, old crest still cannot produce catacholamines (Artinger and Bronner-Fraser, 1992). This demonstrates that some DV pattern is not induced by the migratory environment but involves a cascade that includes changes to the crest that remain in the neuroectoderm layer longer. Perhaps neural crest cell differentiation has a dependence on birth order from a stem cell population as do the cells of the forebrain, where birth order determines the layering of the cerebral cortex.

Placodes

Placodes are neuroectodermal thickenings that form outside of the boundaries of the CNS and contribute to the paired specialized sense organs (olfactory/nose, optic/lens, otic or auditory/ear, and lateral line system) or to the anterior pituitary gland and cranial sensory ganglia (Fig. 16). Many early marker genes have been identified that are expressed in specific placodes such as *Pax6*, *Otx2*, and *Sox3* in the lens placodes, *Pax6* in the olfactory placodes; *Nkx5.1 Pax8*, and *Pax2* in the otic placodes; *Msx2* and *Dlx3* in the lateral line placodes; *Pax3*, *FREK*, and *neurogenin1* in the trigeminal placodes; and *Pax2* and *neurogenin2* in the epibranchial placodes that form the principal ganglia of the VIIth, IXth, and Xth cranial nerves (see Baker *et al.*, 1999 and references therein). However, how these regional specifications are patterned is still a work in progress (reviewed extensively by Baker and Bronner-Fraser, 2001).

In brief, a region of ectoderm competent to form the cranial placodes, the preplacodal domain, forms in the cranial neural plate border region. The expression of several *Pax* (paired-box transcription factor) genes in this ectoderm such that each placodal region expresses a different combination of *Pax* expression (see above). In *Drosophila*, Pax homologs (Ey and Toy) function synergistically with other transcription factors (so) and transcription factor facilitators (eya and dac). Various members of the vertebrate homologs of these transcription regulators, (Six, Eya, and Dach) are expressed with the various *Pax* genes in the placodes, suggesting that a conserved network of genetic regulation may be responsible for establishing specific placodal identity/pattern.

These transcription factors are regulated by signals from various sources. For olfactory placodes the anterior endoderm, prechordal mesoderm, and the anterior neural ridge all have been suggested as sources of inducing signal responsible for activating the appropriate set of transcription factors, although no signal has yet been identified that is either sufficient or necessary for olfactory placode induction. The hypophyseal placode is originally specified by BMP4 from the diencephalon. For lens placode induction, exposure to neural plate and anterior mesendoderm are sufficient, whereas exposure to the optic cup is both necessary and sufficient (via BMP4 and 7). For the trigeminal placodes, an interaction between the neural tube and the surface ectoderm is required to induce the placode but the signal and the method of restricting the placode to a certain location have not been determined. For the lateral line placode, neural plate, axial, and nonaxial mesoderm are each sufficient for induction, and no

signaling molecule has been identified. For otic placode formation, evidence points to mesendoderm as the source for an early signal, and to hindbrain as the source for a later signal in a two-step model of early ear patterning. For the epibranchial placodes, pharyngeal pouch endoderm expressing BMP7 is both necessary and sufficient. In summary, placodes are dependent on local environmental signals from various sources to initiate specific sets of highly conserved transcription regulators, which define their fate in the AP plane.

DV PATTERN

Ventral Patterning

As mentioned, the formation of the nervous system begins with the induction of a two-dimensional neural plate, which forms in an AP and mediolateral plane across the dorsal surface of the early embryo (Fig. 17). Along its mediolateral axis, polarity is established through asymmetrical signaling from neighboring tissues. At its midline, the neural plate contacts the dorsal mesoderm: the head process and notochord. These tissues formed as ingressed cellular derivatives of the Spemann–Mangold Organizer or node, which is responsible for neural

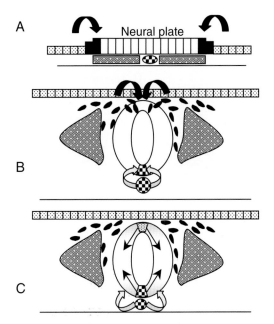

FIGURE 17. (A) Neural plate (white) has contact ventrally with the notochord (checkered) and somatic mesoderm (dark stipple), and laterally with the surface ectoderm (stipple). Black arrows indicate the morphogens, BMP-4 and -7 secreted from the surface ectoderm and altering the adjacent cells to form neural crest (black) at intermediate concentration and neuroectoderm (white) at low concentrations. (B) The neural tube is still under the influence of its adjacent tissues. Continued signaling has resulted in the migration of the neural crest (dorsally). Sonic hedgehog signaling (gray arrows) from the notochord induces the formation of a floor plate ventrally (checkered). (C) The induced roof plate (heavy stipple) becomes the new organizing center dorsally, and both the floor plate and notochord continue to secrete the morphogen Shh, which influences ventral patterning.

induction (see Chapter 1). This region becomes the ventral neural tube, but it starts out as the most dorsal (medial), and therefore most neuralized of all the neural ectoderm.

As one of the earliest parts of the neural patterning cascade following neural induction, these medial mesodermal tissues act as an asymmetrical signaling center to pattern the neural plate, and shortly thereafter the neural tube, by secreting a morphogen, Sonic hedgehog (Shh), which was induced in organizer cells and dorsal mesoderm by two transcription factors, goosecoid and HNF-3β (Fig. 18A). Shh sets up the ventral patterning center for the neuroectoderm (Fig. 18B) and orchestrates the specific development of three prospective cell types within the ventral neural tube: the floor plate, the motor neurons, and the ventral interneurons (Fig. 18C). Evidence supporting the cause and effect relationship between notochord, Shh, and ventral cell differentiation within the neural tube comes from several sources. For example, in the normal embryo, immediately adjacent to the floor plate, ventral interneurons (V3) and then motor neurons (MN) develop in response to decreasing levels of Shh signaling, and more lateral still, another type of ventral interneurons (V2) differentiate in response to the lowest levels of Shh (Fig. 18C). In chick, cutting the neural plate to segregate the ventral region from the floor plate or removing the notochord eliminates the formation of MN on the excised side. In addition, loss of the

Shh gradient in *Shh*−/−mutant mice results in an expansion of the dorsal phenotypes and a loss of ventral (Fig. 18D). These experiments show notochord and Shh signaling are necessary to induce the ventral pattern of cell phenotypes. In contrast, grafting of an additional notochord at a more dorsal position on the neural tube induces ectopic MN in more dorsal regions (Fig. 18E), showing that notochord is sufficient.

A controversy over the induction of floor plate by notochord has been raised (Le Douarin and Halpern, 2000) due to the observation that some notochordless or Shh deficient mutants nonetheless have a floor plate (e.g., see Halpern *et al.*, 1993, 1995; Schauerte *et al.*, 1998). Studies in zebrafish suggest that the floor plate may be two populations of cells (medial and lateral), the medial being independent of Shh signaling and derived directly from the organizer and lateral being Shh dependent and induced (Odenthal *et al.*, 2000). It is unclear if this represents a difference between teleosts and amniotes or is a constant feature of vertebrates. The later is a possibility, since floor plate cells in amniotes can derive directly from the node (Selleck and Stern, 1991; Schoenwolf *et al.*, 1992) or be induced in the neuroectoderm by Shh and thus may also represent two populations. Regardless, the floor plate cells express Shh, either through induction or as a direct derivative of the organizer.

The expression of Shh by the floor plate contributes to the morphogen gradient of Shh in the ventral neural tube and maintains it once the notochord has moved away from the ventral neural tube. But what is the molecular mechanism for ventral neural patterning? Studies of the hedgehog signal transduction pathway in *Drosophila* indicate hedgehog ligands work through a twelve-pass transmembrane receptor called Patched (Ptc) when it is bound to Smoothened (Smo), a G-protein-coupled transmembrane protein. Ptc constitutively inhibits signal transduction by Smo, and hedgehog binding lifts that inhibition. Smo uses a signal transduction pathway involving Protein Kinase A (PKA) and activates a Gli-family transcription factor (reviewed in Litingtung and Chiang, 2000). In vertebrates, two Ptc and three Gli homologs have been identified with appropriate expression localization. The Ptc homologs have high Shh binding affinity and the ability to form a complex with vertebrate Smo. Constituitively active vertebrate Smo mimics high Shh activity in the neural tube, and vertebrate Glis can be responsive to PKA. When activated, Gli proteins bind to Shh responsive promoters linked to a reporter gene and they ectopically induce ventral cell fates (reviewed in Litingtung and Chiang, 2000). This and other evidence indicates that this model pathway (Fig. 19) is probably conserved between flies and vertebrates.

What are the results of this signaling pathway to the patterning of the ventral neural tube? Shh signaling initially upregulates Pax6 and downregulates Pax3 and 7 in the ventral neural tube. Within this ventral Pax6 territory, Shh patterns the neural tube in two steps: first by inhibiting the transcription of certain transcription factors (Dbx1, Dbx2, Irx3, and Pax6; known as Class I transcription factors) in the ventral neuroectoderm in a concentration-dependent fashion, and second by inducing appropriate ventral transcription factors (Nks2.2 and Nkx6.1; known as Class II transcription factors) in these cells (Briscoe *et al.*,

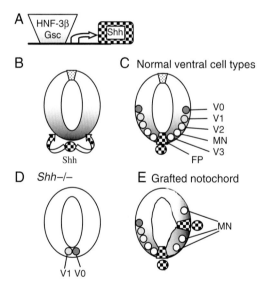

FIGURE 18. (A) In the spinal cord, cells of the Spemann–Mangold Organizer and its derivatives express the secreted protein Shh in response to transcription factors HNF-3β and Goosecoid (Gsc). (B) Shh secreted from the notochord and dorsal mesoderm (checkered) establishes a gradient along the ventral to dorsal axis of the neural tube. (C) This signal induces the differential differentiation of MN and four interneuron subtypes (V0–V3) in the ventral neural tube. Genes originally expressed throughout the neural tube, *Pax3* and *Pax7*, are now expressed only in the dorsal region. (D) In mice mutant for *Shh* (−/−), dorsal genes *Pax3* and *Pax7* expand into the ventral region, and ventral cell types are lost, with the exception of two lateral interneuron groups. (E) When a second notochord is grafted lateral to the forming neural tube, an ectopic floor plate and MN are induced nearby. Markers for interneuron were not assessed and the control side is presumed normal with regard to their expression.

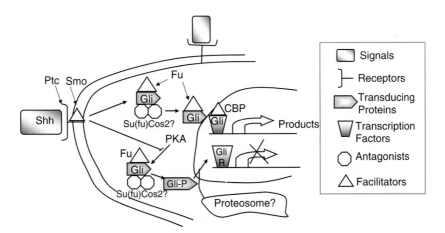

FIGURE 19. Signal transduction by Shh. Shh binds to receptor Patched (Ptc), which is associated with the membrane bound signal transduction facilitator Smo. Smo activates cytoplasmic Gli, which in conjunction with the cytoplasmic facilitator fu (if not antagonized by Su(fu)) can relocate to the nucleus. There Gli can cooperate with the nuclear facilitator CBP to activate ventral target genes including *Ptc*, *Gli* and *Shh*. A different pathway involving PKA allows Gli to act as a repressor (GliR) on dorsal genes such as *Pax3* and *Pax7*. Smo and Ptc activation by Shh can also block Gli repressor (GliR) formation, possibly by inhibiting the formation of phosphorylated forms of Gli (Gli-P). The efficient processing of Gli may require phosphoryletion and be proteosome dependent. The vertebrate homolog of Cos2 has not been identified; however, in *Drosophila*, Cos2 binds to antagonizes Shh signaling. (Adapted from Litingtung and Chiang, 2000, *Dev. Dynam.* Reprinted by permission of Wiley-Liss, Inc., a subsidiary of John Wiley & Sons, Inc.)

2000; Fig. 20). Class I and II transcription factors negatively regulate one another's gene expression to create clear boundaries between the progenitor domains. The specific combination of transcription factors in a given region then provides the DV positional information required to differentiate as the appropriate cell type for that region. For example, prospective MN neurons express Nkx6.1, while prospective V2 neurons express both Nkx6.1 and Irx3. Ectopic expression of Irx3 in prospective MN neurons changes their differentiation to V2 fate (Briscoe *et al.*, 2000). Similar gain and loss of function experiments indicate that the other cell types are also regulated by the specific combinatorial expression of these genes.

However, the Shh-neural story is more complicated than this. For example, through a pathway independent of the Ptc–Smo–Gli pathway, Shh may mediate adhesion of the neural tube and allow migration of neural crest from the dorsal neural tube where Shh concentration is minimal (Testaz *et al.*, 2001). Second, Shh has also been implicated as a mitogen in the neural tube (e.g., see Britto *et al.*, 2000), and differential growth is another aspect of pattern formation not considered here. Third, other intracellular and extracellular factors are known to facilitate or limit Shh activity or diffusion (reviewed by Capdevila and Belmonte, 1999; Robertson *et al.*, 2001), further regulating the activity of this morphogen. For example, Ptc the Shh receptor, in the absence of Smo acts as a Shh sink and limits its diffusion. Fourth, some ventral phenotypes do develop in the absence of Shh (V0, V1), and these can be induced in neural explant culture by RA; (Pierani *et al.*, 1999), a morphogen secreted by the paraxial mesoderm. Thus, other morphogens and signaling sources may also participate in the patterning of the ventral neural tube. Finally, the double *Shh:Gli3* mutant mouse has MN; thus there has to be some other induction path for MN that is normally inhibited by Gli3 in the absence of Shh (Litingtung and Chiang, 2000).

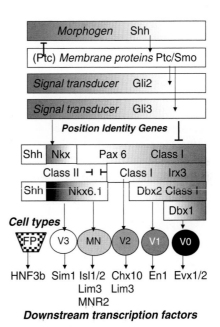

FIGURE 20. Shh induction of specific ventral cell fates. Shh, a morphogen, acts through receptor Ptc and its binding partner Smo to activate signal transducers Gli2 and Gli3. Gli activity gradients may result from differential transport of this protein into the nucleus. Gli may regulate both Class I (*Pax6*, *Irx3*, *Dbx2*, *Dbx1*) and Class II (*Nkx2.2*, *Nkx6*) genes. Class I genes are position identity genes expressed in a gradient with their highest level dorsally, whereas Class II gene gradients have their highest concentration ventrally. Thus Gli2 and 3, with their ventral gradient, could inhibit Class I genes and activate Class II genes. By combinatorial effect, the expression of these transcription factors establishes progenitor domains and results in the expression of specific downstream marker genes. In this case, these are also transcription factors that help determine the fate of these cells. (Adapted from Litingtung and Chiang, 2000, *Dev. Dynam.*, 2000. Reprinted by permission of Wiley-Liss, Inc. a subsidiary of John Wiley & Sons, Inc.)

Dorsal Patterning

Whereas Shh is a positive morphogen for the medial neural plate and ventral neural tube, several members of the TGFβ family of growth factors (including BMPs and GDFs) are positive morphogens for the lateral, and later dorsal, neural ectoderm, which includes neural crest, roof plate glia, and three dorsal interneuron types (D1A, D1B, and D2; Fig. 21; Lee *et al.*, 1998). BMPs (4 and 7) are expressed in the epidermal layer and initially act as morphogens such that the highest levels induce epidermal ectoderm, intermediate levels induce neural crest, and lower levels induce neural plate. BMPs are sufficient to upregulate the same genes (*Pax3, Pax7*) in the lateral neural plate and dorsal neural tube that Shh represses in the medial/ventral region. Thus it appears that these two morphogens are working as opposing gradients to pattern in this plane (like the Saxen–Toivonen model in AP pattern, see Fig. 7C). As these dorsalized *Pax* genes originally were expressed throughout the neural plate, their expression is not sufficient to initiate the cascade that will result in the patterning of the dorsal neuroectoderm. However, their expression is required for appropriate differentiation of the most dorsal cells, the neural crest.

Neural crest cells are the migratory founding cells of much of the PNS (see Fig. 5). They form at the margin of the epidermal ectoderm and neural plate (the neural fold) in response to required signals from both these tissues. However, they are not committed at the neural plate stage, as individual cells can contribute to crest, epidermis, or dorsal CNS. Induction of neural crest appears to be a multistep venture involving early Wnt signaling and later BMP signaling (Bronner-Fraser, 2002). Other positive factors involved include FGF (Lee and Jessell, 1999); Zic2, a zinc-finger transcription factor that promotes crest and inhibits neural differentiation (Brewster *et al.*, 1998); FoxD3, a winged-helix transcription factor that works with Zic2 in determining crest (Sasai *et al.*, 2001); and Noelin-1, a secreted factor

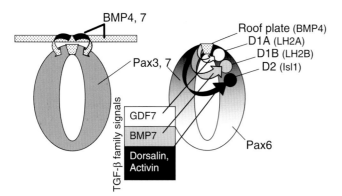

FIGURE 21. Dorsal neural patterning. Signals from the overlying surface ectoderm (BMPs) induce the roof plate (stipple), which also expresses BMPs. *Pax3* and *-7* shift from a pattern of expression throughout the neural tube to a concentration in the dorsal neural tube, whereas *Pax6* gets upregulated in the ventral neural tube. Several signaling molecules are secreted from the roof plate, and these induce the fates of several dorsal interneurons (D1A, D1B, and D2), which subsequently upregulate the expression of cell-type specific genes (*LH2A, LH2B,* and *Isl1*).

that is induced in a gradient in the dorsal neural tube and provides a competence factor to the dorsal cells, allowing them to differentiate as crest (Bronner-Fraser, 2002).

Just as Shh induces Shh secretion from the floor plate to assist in ventral patterning, exposure to secreted BMPs from the surface ectoderm also induces the expression of BMPs and other secreted factors in the adjacent ectoderm. This homeogenetic induction maybe necessary to maintain an accurate gradient within the neural tube as it grows and becomes separated from its neighboring tissues. The difference between the ventral and dorsal patterning is, rather than acting as morphogens, many of the dorsal signaling molecules (Noelin-1, several Wnts, Dsl1, BMP4, and BMP7) seem to regulate the differentiation of specific targets, and thus are secondarily involved in dorsal patterning (Fig. 21). For example, Noelin-1 is specifically involved in neural crest formation, whereas Wnt-1 and Wnt3a (Muroyama *et al.*, 2002) or TGF-family (BMP4, 5, 7, GDF6/7, and DSL1; Liem *et al.*, 1997; Lee *et al.*, 1998) expression in the roof plate provides the signal to induce the dorsal-most interneurons, D1A, and D1B. The more ventral D2 is induced by activin (Liem *et al.*, 1997).

Signal Transduction of BMPs in Dorsal Patterning

BMPs are the initial signaling molecules of dorsalization, and they require receptors and signal transduction pathways to initiate the expression of the other signaling molecules they induce. BMP 4 and 7 act as dimers and bind to serine/threonine kinase receptors (BMPRI, BMPRII, Alk8, tolloid, BMP2b/swirl, snailhouse, somitabun). These activate SMAD proteins intracellularly, which migrate to the nucleus and act as transcription factors for BMP-activated genes (Fig. 22). These are commonly used signaling pathways and a large number of signal transduction modifiers have been identified including other Smads (6 and 7), transcriptional activators (p300), and transcriptional repressors (Ski, Tob). These apparently allow this signal transduction pathway to regulate specific sets of target genes in a given tissue. In the zebrafish model system, reduction in BMP function through mutation of SMAD 5 (somitabun) causes an expansion of dorsal neural phenotypes into the epidermal ectoderm, while loss of one BMP directly from mutation results in loss of both epidermal ectoderm and dorsal neural phenotypes (Fig. 23), showing this pathway is both necessary and sufficient for dorsalization.

DV Pattern at Other AP Levels

Obviously, not all levels of the neural tube form the same types of neurons as the spinal cord, thus one must consider what mechanisms account for these differences in DV pattern at other AP levels. One could imagine that other responses in the DV plane might stem from intrinsic differences in the AP character of the neural plate. Alternately, one might imagine the differences stemming from differences in the localization of tissues that provide the morphogens and thus from morphogen availability. As usual, evidence exists for both.

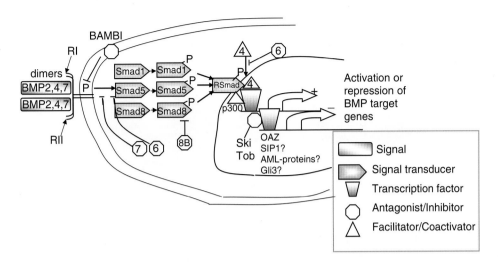

FIGURE 22. The main BMP–Smad signal transduction pathway. BMP2, 4, or 7 dimers bind to a receptor complex of type I and type II receptors (RI and RII). This leads RI to phosphorylate RII, which then phosphorylates Smad1, -5, or -8, depending on the cell. This allows this Smad to form a complex with the facilitator Smad4 and this complex enters the nucleus to bind to the MH1 domain and activate or repress target genes depending on which other facilitators or antagonist factors are present in the nucleus. Several nuclear factors also regulate this pathway by acting as transcription facilitators or antagonists. General facilitator p300 can bind to the MH2 domain of Smad1 and -4 and activate transcription through its histone acetylase activity. Ski and Tob act as antagonists to this pathway by binding to various Smads. Smads activate transcription by binding to other transcription factors. Transcription factors that can bind to Smads include OAZ, SIP1, AML, and Gli (C-terminally truncated). Although they are shown as heterodimers, the stoichiometry between Smad1, -5, -8, and Smad 4 is unknown. Smad6, -7, and -8B inhibit this transduction pathway cytoplasmically or in the nucleus, whereas BAMBI is a membrane bound antagonist. Smad6, -7, BAMBI, and Tob are all products of this pathway, leading to negative feedback loops. (Redrawn and adapted from von Bubnoff and Cho, 2001, with permission from Academic Press, Orlando Florida.)

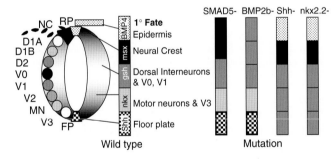

FIGURE 23. A summary of DV patterning in the vertebrate neural tube showing a schematic of the neural tube and the specific cell types that differentiate at various DV levels to the left, and the primary fate of various DV levels to the right. To the far right, the results of various mutations are shown. The *SMAD5*-mutation results in low BMP signaling, and the epidermis, which requires high BMP signaling, is replaced with neural crest cells. *BMP2b*-mutants (very low BMP signaling) replace both dorsal tissues with a dorsal interneuron phenotype. *Sonic hedgehog* mutations, in contrast, cause a shift in ventral cell types similar to the loss of *nks2.2*, which is required for MN-differentiation and some ventral interneurons. (Adapted from Cornell and Von Ohlen, 2000, with permission from Elsevier Science.)

Around the isthmus, Engrailed-2 (En2), a transcription factor required for normal cerebellar development, is expressed in the neural tube at all DV levels except the floor plate (Fig. 24A). When the notochord, which expresses Shh, is surgically removed, En2 expression expands into the ventral midline (Fig. 24B). When an ectopic Shh secreting notochord or floor plate is grafted adjacent to the neural tube at more dorsal levels, En2 expression is suppressed (Figs 24C, D). Also at the midbrain

level and immediately adjacent to the floor plate, dopaminergic rather than MN form. Dopaminergic neurons are not found at hindbrain or spinal cord levels. Why? Transplantation studies indicate that AP pattern limits the differential competence of ventral neuroepithelial precursors. The same is true for ventral genes in the forebrain. Shh upregulates *Nkx2.1*, which is needed in the diencephalon to form the hypothalamus, but only in the regions where AP patterning gene *Six3* is expressed (Kobayashi *et al.*, 2002). Shh also regulates *Pax* genes in the ventral brain, although their expression patterns differ somewhat from those in the spinal cord. In the head Pax3 and Pax7 are pushed far dorsally and are segregated from Pax6, which is also excluded from the most ventral region of the tube, restricting it to the area between the sulcus diencephalicus medius and ventralis. These examples all support the idea that Shh is a common ventral morphogen, but the response depends on previous changes wrought by the patterning cascade at each AP level of the neural tube.

The model in which the DV morphogens differ at various AP levels is also supported in some cases. For example, within the head a new dorsal signaling factor has recently been identified (in *Xenopus*), Tiarin, which dorsalizes the anterior neural plate, causing expansion of dorsal and suppression of ventral neural markers (Tsuda *et al.*, 2002). In addition, at the level of the rostral diencephalon an old morphogen, BMP7, is expressed in a new location: the *ventral* midline mesendoderm. At this AP level no floor plate forms and a different set of ventral marker genes including *Nkx2.1* is upregulated due to the inductive signals of both Shh and BMP7. Likewise, whereas an intermediate concentration of BMP induces neural crest as far rostrally as the mid

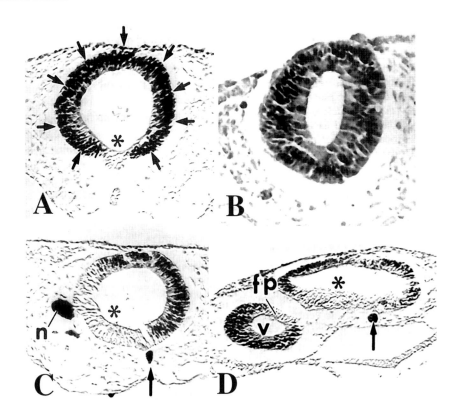

FIGURE 24. Engrailed-2 expression in the chick (stage 12). (A) Expression (arrows) in the normal embryo is high at all DV levels of the mesencephalon and rostral metencephalon, with the exception of the floor plate (asterisk). (B) In embryos in which the precursors of the notochord have been removed, no notochord and subsequently no floor plate form, and En2 is expressed uniformly at all DV levels of the neural tube. (C) Grafting of a notochord (n) to a position lateral to the neural tube results in a loss of En2 expression adjacent to the graft (asterisk), showing that notochord is sufficient to inhibit En2 expression. (D) Similar grafting of a fragment of the neural plate containing a floor plate (fp) results in similar suppression (asterisk). (Darnell and Schoenwolf, 1995; Darnell *et al.*, 1992) (Developmental Dynamics, and Journal of Neurobiology. Reprinted by permission of Wiley-Liss, Inc., a subsidiary of John Wiley & Sons, Inc.)

diencephalon, rostral to this, no neural crest is formed even though BMP is present in the epidermal ectoderm. A number of experiments have implicated caudalizing FGFs and Wnts as the requisite additional morphogens required to induce neural crest at appropriate levels, and a recent study has supported these and added RA to the list (Villanueva *et al.*, 2002). Thus, both AP differences in signal availability and AP differences in responding cell competence can shape the intersection between AP and DV pattern.

LEFT–RIGHT ASYMMETRY

Although significant recent progress has been made on the signaling cascade that confers left–right asymmetry on the early embryo, controlling heart looping and gut rotation (reviewed in Mercola and Levin, 2001), no connections have been made with the left–right asymmetries of the adult brain. These asymmetries do not correlate with known left–right (LR) patterning in the early embryo. For example, people with situs inversus, a condition in which the body LR axes are reversed such that their hearts are angled toward the right and their livers are on the left, still process language on the left side of their brains, as do 95% of

people with normal LR patterning. Studies of mirror-image identical twins (and conjoined twins) have lead to speculation that the mechanism for LR patterning in the head is separate from the patterning of the trunk just as many other aspect of head and trunk patterning are independent.

CONCLUSIONS

The predominant method of achieving a patterned vertebrate nervous system involves responses to signaling from an asymmetrical source. This instigates the activation of new transcription factors in the responding tissue, which leads to a cascade of cellular changes that generate additional asymmetry and cell differentiation. Several of the signaling sources have been identified, including the neural organizer/node, the anterior neural ridge, the isthmus, and the caudal mesoderm along the AP axis, and the notochord and epidermal ectoderm in the DV axis. Signals induce changes in target cells depending on concentration, and antagonists or distance protect other cells from responding, generating diverse cell types. These general concepts and in many cases the specific genetic networks for patterning have been conserved for hundreds of millions of years.

REFERENCES

Acampora, D., Gulisano, M., Broccoli, V., and Simeone, A., 2001, Otx genes in brain morphogenesis, *Prog. Neurobiol.* 64:69–95.

Alvarado-Mallart, R.-M., 1993, Fate and potentialities of the avian mesencephalic/metencephalic neuroepithelium, *J. Neurobiol.* 24(10): 1341–1355.

Alvarez-Bolado, G., Rosenfeld, M.G., and Swanson, L.W., 1995, Model of forebrain regionalization based on spatiotemporal patterns of POU-III homeobox gene expression, birthdates, and morphological features, *J. Comp. Neurol.* 355(2):237–295.

Anderson, D.J., 1997, Cellular and molecular biology of neural crest cell lineage determination, *Trends Genet.* 13(7):276–280.

Artinger, K.B. and Bronner-Fraser, M., 1992, Partial restriction in the developmental potential of late emigrating avian neural crest cells, *Dev. Biol.* 149(1):149–157.

Bachiller, D., Klingensmith, J., Kemp, C., Belo, J.A., Anderson, R.M., May, S.R. *et al.*, 2000, The organizer factors Chordin and Noggin are required for mouse forebrain development, *Nature* 403 (6770): 658–661.

Baker, C.V. and Bronner-Fraser, M., 2001, Vertebrate cranial placodes I. Embryonic induction, *Dev. Biol.* 232(1):1–61.

Baker, C.V., Stark, M.R., Marcelle, C., Bronner-Fraser, M., Iwasaki, M., Le, A.X., and Helms, J.A., 1999, Competence, specification and induction of Pax-3 in the trigeminal placode, *Development* 126(1): 147–156.

Bang, A.G., Papalopulu, N., Kintner, C., and Goulding, M.D., 1997, Expression of Pax-3 is initiated in the early neural plate by posteriorizing signals produced by the organizer and by posterior non-axial mesoderm, *Development* 124(10):2075–2085.

Beddington, R.S. and Robertson, E.J., 1998, Anterior patterning in mouse, *Trends Genet.* 14(7):277–284.

Begemann, G. and Meyer, A., 2001, Hindbrain patterning revisited: Timing and effects of retinoic acid signaling, *Bioessays* 23(11): 981–986.

Bell, E., Ensini, M., Gulisano, M., and Lumsden, A., 2001, Dynamic domains of gene expression in the early avian forebrain, *Dev. Biol.* 236(1): 76–88.

Boncinelli, E., Gulisano, M., and Broccoli, V., 1993, Emx and Otx homeobox genes in the developing mouse brain, *J. Neurobiol.* 24(10): 1356–1366.

Bouwmeester, T., Kim, S.H., Sasai, Y., Lu B., and De Robertis, E.M., 1996, Cerberus is a head-inducing secreted factor expressed in the anterior endoderm of Spemann's organizer, *Nature* 382:595–601.

Brewster, R., Lee, J., and Ruiz i Altaba, A., 1998, Gli/Zic factors pattern the neural plate by defining domains of cell differentiation, *Nature* 393(6685):579–583.

Briscoe, J., Pierani, A., Jessell, T.M., and Ericson, J., 2000, A homeodomain protein code specifies progenitor cell identity and neuronal fate in the ventral neural tube, *Cell* 101(4):435–445.

Britto, J.M., Tannahill, D., and Keynes, R.J., 2000, Life, death and Sonic hedgehog, *Bioessays* 22(6):499–502.

Bronner-Fraser, M., 2002, Molecular analysis of neural crest formation, *J. Physiol. Paris* 96(1–2):3–8.

Burgess, S., Reim, G., Chen, W., Hopkins, N., and Brand, M., 2002, The zebrafish spiel-ohne-grenzen (spg) gene encodes the POU domain protein Pou2 related to mammalian Oct4 and is essential for formation of the midbrain and hindbrain, and for pre-gastrula morphogenesis, *Development* 129(4):905–916.

Capdevila, J. and Belmonte, J.C., 1999, Extracellular modulation of the Hedgehog, Wnt and TGF-beta signalling pathways during embryonic development, *Curr. Opin. Genet. Dev.* 9(4):427–433.

Conlon, R.A. and Rossant, J., 1992, Exogenous retinoic acid rapidly induces anterior ectopic expression of murine Hox-2 genes in vivo, *Development* 116(2):357–368.

Cornell, R.A. and Ohlen, T.V., 2000, Vnd/nkx, ind/gsh, and msh/msx: Conserved regulators of dorsoventral neural patterning?, *Curr. Opin. Neurobiol.* 10(1):63–71.

Couly, G., Creuzet, S., Bennaceur, S., Vincent, C., and Le Douarin, N.M., 2002, Interactions between Hox-negative cephalic neural crest cells and the foregut endoderm in patterning the facial skeleton in the vertebrate head, *Development* 129(4):1061–1073.

Cox, W.G. and Hemmati-Brivanlou, A., 1995, Caudalization of neural fate by tissue recombination and bFGF, *Development* 121:4394–4358.

Darnell, D.K., 1992, The chick Engrailed-2 gene: Structure, expression and a marker for neural pattern, Doctoral Dissertation, University of California, San Francisco, 1992.

Darnell, D.K. and Schoenwolf, G.C., 1995, Dorsoventral patterning of the avian mesencephalon/metencephalon: Role of the notochord and floor plate in suppressing Engrailed-2, *J. Neurobiol.* 26(1):62–74.

Darnell, D.K., Schoenwolf, G.C., and Ordahl, C.P., 1992, Changes in dorsoventral but not rostrocaudal regionalization of the chick neural tube in the absence of cranial notochord, as revealed by the expression of Engrailed-2, *Dev. Dyn.* 193:389–396.

Doniach, T., 1993, Planar and vertical induction of anteroposterior pattern during the development of the amphibian central nervous system, *J. Neurobiol.* 24(10):1256–1275.

Dupe, V. and Lumsden, A., 2001, Hindbrain patterning involves graded responses to retinoic acid signaling, *Development* 128(12): 2199–2208.

Ensini, M., Tsuchida, T.N., Belting, H.G., and Jessell, T.M., 1998, The control of rostrocaudal pattern in the developing spinal cord: Specification of motor neuron subtype identity is initiated by signals from paraxial mesoderm, *Development* 125(6):969–982.

Foley, A.C., Skromne, I., and Stern, C.D., 2000, Reconciling different models of forebrain induction and patterning: A dual role for the hypoblast, *Development* 127(17):3839–3854.

Frohman, M.A., Boyle, M., and Martin, G.R., 1990, Isolation of the mouse *Hox-2.9* gene; Analysis of embryonic expression suggests that positional information along the anterior-posterior axis is specified by mesoderm, *Development* 110:589–607.

Gammill, L.S., and Sive, H., 2001, Otx2 expression in the ectoderm activates anterior neural determination and is required for *Xenopus* cement gland formation, *Dev. Biol.* 240(1):223–236.

Gaunt, S.J. and Strachan, L., 1994, Forward spreading in the establishment of a vertebrate Hox expression boundary: The expression domain separates into anterior and posterior zones, and the spread occurs across implanted glass barriers, *Dev. Dyn.* 199:229–240.

Ghysen, A., Dambly-Chaudiere, C., Jan, L.Y., and Jan, Y.N., 1993, Cell interactions and gene interactions in peripheral neurogenesis, *Genes Dev.* 7(5):723–733.

Glavic, A., Gomez-Skarmeta, J.L., and Mayor, R., 2002, The homeoprotein Xiro1 is required for midbrain–hindbrain boundary formation, *Development* 129(7):1609–1621.

Glinka, A., Wu, W., Onichtchouk, D., Blumenstock, C., and Niehrs, C., 1997, Head induction by simultaneous repression of Bmp and Wnt signalling in *Xenopus, Nature* 389(6650):517–519.

Grapin-Botton, A., Bonnin, M.A., Sieweke, M., and Le Douarin, N.M., 1998, Defined concentrations of a posteriorizing signal are critical for MafB/Kreisler segmental expression in the hindbrain, *Development* 125(7):1173–1181.

Halpern, M.E., Ho, R.K., Walker, C., and Kimmel, C.B., 1993, Induction of muscle pioneers and floor plate is distinguished by the zebrafish no tail mutation, *Cell* 75(1):99–111.

Halpern, M.E., Thisse, C., Ho, R.K., Thisse, G., Riggleman, B., Trevarrow, B. *et al.*, 1995, Cell-autonomous shift from axial to paraxial mesodermal development in zebrafish floating head mutants, *Development* 121: 4257–4264.

Hashimoto, H., Itoh, M., Yamanaka, Y., Yamashita, S., Shimizu, T., Solnica-Krezel, L. *et al.*, 2000, Zebrafish Dkk1 functions in forebrain specification and axial mesendoderm formation, *Dev. Biol.* 217(1): 138–152.

Hogan, B.L., Thaller, C., and Eichele, G., 1992, Evidence that Hensen's node is a site of retinoic acid synthesis, *Nature* 359(6392):237–241.

Houart, C., Caneparo, L., Heisenberg, C., Barth, K., Take-Uchi, M., and Wilson, S., 2002, Establishment of the telencephalon during gastrulation by local antagonism of Wnt signaling, *Neuron* 35(2):255–265.

Houart, C., Westerfield, M., and Wilson, S.W., 1998, A small population of anterior cells patterns the forebrain during zebrafish gastrulation, *Nature.* 391(6669):788–792.

Irving, C. and Mason, I., 2000, Signalling by FGF8 from the isthmus patterns anterior hindbrain and establishes the anterior limit of Hox gene expression, *Development* 127(1):177–186.

Jan, Y.N. and Jan, L.Y., 1993, Functional gene cassettes in development, *Proc. Natl. Acad. Sci. USA* 90(18):8305–8307.

Jones, C.M., Broadbent, J., Thomas, P.Q., Smith, J.C., and Beddington, R.S., 1999, An anterior signalling centre in Xenopus revealed by the homeobox gene XHex, *Curr. Biol.* 9(17):946–954.

Jungbluth, S., Larsen, C., Wizenmann, A., and Lumsden, A., 2001, Cell mixing between the embryonic midbrain and hindbrain, *Curr. Biol.* 11(3): 204–207.

Kazanskaya, O., Glinka, A., and Niehrs, C., 2000, The role of *Xenopus* dickkopf1 in prechordal plate specification and neural patterning, *Development* 127(22):4981–4992.

Kengaku, M. and Okamoto, H., 1995, bFGF as a possible morphogen for the anteroposterior axis of the central nervous system in *Xenopus, Development* 121(9):3121–3130.

Kiecker, C. and Niehrs, C., 2001, A morphogen gradient of Wnt/beta-catenin signalling regulates anteroposterior neural patterning in *Xenopus, Development* 128(21):4189–4201.

Kiecker, C. and Niehrs, C., 2003, The role of Wnt signalling in vertebrate head induction and the organizer-gradient model dualism. In *Wnt signalling in development* (Chapter 5). Michael Kuehl (ed.), Landis Biosciences Publishing, Georgetown, Texas, USA.

Klein, R., 1999, Bidirectional signals establish boundaries, *Curr. Biol.* 9(18): R691–694.

Knoetgen, H., Teichmann, U., Wittler, L., Viebahn, C., and Kessel, M., 2000, Anterior neural induction by nodes from rabbits and mice, *Dev. Biol.* 225(2):370–380.

Kobayashi, D., Kobayashi, M., Matsumoto, K., Ogura, T., Nakafuku, M., and Shimamura, K., 2002, Early subdivisions in the neural plate define distinct competence for inductive signals, *Development* 129(1): 83–93.

Koshida, S., Shinya, M., Nikaido, M., Ueno, N., Schulte-Merker, S., Kuroiwa, A. *et al.*, 2002, Inhibition of BMP activity by the FGF signal promotes posterior neural development in zebrafish. *Dev. Biol.* 244(1):9–20.

Kulesa, P.M., and Fraser, S.E., 1998, Segmentation of the vertebrate hindbrain: A time-lapse analysis, *Int. J. Dev. Biol.* 42(3 Spec No):385–392.

Lamb, T.M. and Harland, R.M., 1995, Fibroblast growth factor is a direct neural inducer, which combined with noggin generates anterior-posterior neural pattern, *Development* 121(11):3627–3636.

Larsen, C.W., Zeltser, L.M., and Lumsden, A., 2001, Boundary formation and compartition in the avian diencephalons, *J. Neurosci.* 21(13): 4699–4711.

Le Douarin, N.M. and Halpern, M.E., 2000, Discussion point. Origin and specification of the neural tube floor plate: Insights from the chick and zebrafish, *Curr. Opin. Neurobiol.* 10(1):23–30.

Le Douarin, N.M., Ziller, C., and Couly, G.F., 1993, Patterning of neural crest derivatives in the avian embryo: In vivo and in vitro studies, *Dev. Biol.* 159(1):24–49.

Lee, K.J., and Jessell, T.M., 1999, The specification of dorsal cell fates in the vertebrate central nervous system, *Annu. Rev. Neurosci.* 22:261–294.

Lee, K.J., Mendelsohn, M., and Jessell, T.M., 1998, Neuronal patterning by BMPs: A requirement for GDF7 in the generation of a discrete class of commissural interneurons in the mouse spinal cord, *Genes Dev.* 12(21):3394–3407.

Liem, K.F., Jr., Tremml, G., and Jessell, T.M., 1997, A role for the roof plate and its resident TGFbeta-related proteins in neuronal patterning in the dorsal spinal cord, *Cell.* 91(1):127–138.

Litingtung, Y. and Chiang, C., 2000, Control of Shh activity and signaling in the neural tube, *Dev. Dyn.* 219(2):143–154.

Lumsden, A., 1991, Cell lineage restrictions in the chick embryo hindbrain, *Philos. Trans. R. Soc. Lond. Biol.* 331(1261):281–286.

Lumsden, A. and Krumlauf, R., 1996, Patterning the vertebrate neuraxis, *Science* 274:1109–1123.

Mangold, O., 1933, Über die Induktionsfähigkeit der verschiedenen Bezirke der Neurula von Urodelen. *Naturwissenshaften* 4:761–766.

Mathis, L., Kulesa, P.M., and Fraser, S.E., 2001, FGF receptor signalling is required to maintain neural progenitors during Hensen's node progression, *Nat. Cell. Biol.* 3(6):559–566.

McMahon, A.P. and Bradley, A., 1990, The Wnt-1 (int-1) proto-oncogene is required for development of a large region of the mouse brain, *Cell* 62:1073–1085.

McMahon, A.P., Joyner, A.L., Bradley, A., and McMahon, J.A., 1992, The midbrain-hindbrain phenotype of *wnt-1-/wnt-1*-mice results from stepwise deletion of *engrailed*-expressing cells by 9.5 days *post-coitum, Cell* 69:581–595.

Mercola, M. and Levin, M., 2001, Left–right asymmetry determination in vertebrates, *Annu. Rev. Cell Dev. Biol.* 17:779–805.

Morriss-Kay, G.M., Murphy, P., Hill, R.E., and Davidson, D.R., 1991, Effects of retinoic acid excess on expression of Hox-2.9 and Krox-20 and on morphological segmentation in the hindbrain of mouse embryos, *EMBO* 10(10):2985–2995.

Mukhopadhyay, M., Shtrom, S., Rodriguez-Esteban, C., Chen, L., Tsukui, T., Gomer, L. *et al.*, 2001, Dickkopf1 is required for embryonic head induction and limb morphogenesis in the mouse, *Dev. Cell.* 1(3): 423–434.

Muroyama, Y., Fujihara, M., Ikeya, M., Kondoh, H., and Takada, S., 2002, Wnt signaling plays an essential role in neuronal specification of the dorsal spinal cord, *Genes Dev.* 16(5):548–553.

Niederreither, K., Vermot, J., Schuhbaur, B., Chambon, P., and Dolle, P., 2000, Retinoic acid synthesis and hindbrain patterning in the mouse embryo, *Development* 127(1):75–85.

Niehrs, C., Kazanskaya, O., Wu, W., and Glinka, A., 2001, Dickkopf1 and the Spemann–Mangold head organizer, *Int. J. Dev. Biol.* 45(1 Spec No): 237–240.

Nieuwkoop, P.D., Boterenbrood, E.C., Kremer, A., Bloemsma, F.F.S.N., Hoessels, E.L.M.J. *et al.*, 1952, Activation and organization of the central nervous system. I. Induction and activation. II. Differentiation and organization. III. Synthesis of a new working hypothesis, *J. Exp. Zool.* 120:1–108.

Odenthal, J., van Eeden, F.J., Haffter, P., Ingham, P.W., and Nusslein-Volhard, C., 2000, Two distinct cell populations in the floor plate of the zebrafish are induced by different pathways, *Dev. Biol.* 219(2): 350–363.

Perea-Gomez, A., Rhinn, M., and Ang, S.L., 2001, Role of the anterior visceral endoderm in restricting posterior signals in the mouse embryo, *Int. J. Dev. Biol.* 45(1 Spec No):311–320.

Piccolo, S., Agius, E., Leyns, L., Bhattacharyya, S., Grunz, H., Bouwmeester, T. *et al.*, 1999, The head inducer Cerberus is a multifunctional antagonist of Nodal, BMP and Wnt signals, *Nature* 397(6721): 707–710.

Pierani, A., Brenner-Morton, S., Chiang, C., and Jessell, T.M., 1999, A sonic hedgehog-independent, retinoid-activated pathway of neurogenesis in the ventral spinal cord, *Cell* 97(7):903–915.

Piotrowski, T. and Nusslein-Volhard, C., 2000, The endoderm plays an important role in patterning the segmented pharyngeal region in zebrafish (*Danio rerio*), *Dev. Biol.* 225(2):339–356.

Price, M., 1993, Members of the Dlx- and Nk x2-gene families are regionally expressed in the developing forebrain, *J. Neurobiol.* 24(10): 1385–1399.

Reim, G. and Brand, M., 2002, Spiel-ohne-grenzen/pou2 mediates regional competence to respond to Fgf8 during zebrafish early neural development, *Development* 129(4):917–933.

Robertson, C.P., Gibbs, S.M., and Roelink, H., 2001, cGMP enhances the sonic hedgehog response in neural plate cells, *Dev. Biol.* 238(1): 157–167.

Ruiz i Altaba, A., 1993, Induction and axial patterning of the neural plate: Planar and vertical signals, *J. Neurobiol.* 24(10):1276–1304.

Ruiz i Altaba, A., 1994, Pattern formation in the vertebrate neural plate, *TINS* 17(6):233–243.

Sasai, N., Mizuseki, K., and Sasai, Y., 2001, Requirement of FoxD3-class signaling for neural crest determination in *Xenopus*, *Development* 128(13):2525–2536.

Saxen, L., 1989, Neural induction, *Int. J. Dev. Biol.* 33(1):21–48.

Schauerte, H.E., van Eeden, F.J., Fricke, C., Odenthal, J., Strahle, U., and Haffter, P., 1998, Sonic hedgehog is not required for the induction of medial floor plate cells in the zebrafish, *Development* 125(15): 2983–2993.

Schoenwolf, G.C., Garcia-Martinez, V., and Dias, M.S., 1992, Mesoderm movement and fate during avian gastrulation and neurulation, *Dev. Dyn.* 193:235–248.

Sedohara, A., Fukui, A., Michiue, T., and Asashima, M., 2002, Role of BMP-4 in the inducing ability of the head organizer in *Xenopus* laevis, *Zoolog. Sci.* 19(1):67–80.

Selleck, M.A.J. and Stern, C.D., 1991, Fate mapping and cell lineage analysis of Hensen's node in the chick embryo, *Development* 112:615–626.

Shimamura, K. and Rubenstein, J.L., 1997, Inductive interactions direct early regionalization of the mouse forebrain, *Development* 124(14): 2709–2718.

Simeone, A., Avantaggiato, V., Moroni, M.C., Mavilio, F., Arra, C., Cotelli, F. et al., 1995, Retinoic acid induces stage-specific antero-posterior transformation of rostral central nervous system, *Mech. Dev.* 51(1): 83–98.

Spemann, H., 1931, Über den Anteil von Implantat und Wirtskeim an der Orientierung und Beschaffenheit der induzierten Embryonalanlage, *Roux' Arch. f. Entw. Mech.* 123:389–517.

Spemann, H. and Mangold, H., 1924, Über induktion von Embryonalanlagen durch Implantation artfremder Organisatoren. *Roux Arch*

Entwicklungsmech Org 100: 599–638. Translated into English by V. Hamburger (2001) In *Foundations of Experimental Embryology* (B.H. Willier and J.M. Oppenheimer, eds.), Prentice hall, Inc.; Englewood Cliffs, NJ, USA, pp. 146–184. And more recently (2001) reprinted in *Int. J. Dev. Biol.* 45(1):13–38.

Sundin, O. and Eichele, G., 1992, An early marker of axial pattern in the chick embryo and its respecification by retinoic acid, *Development* 114(4):841–852.

Tanabe, Y. and Jessell, T.M., 1996, Diversity and pattern in the developing spinal cord, *Science* 274(5290):1115–1123.

Testaz, S., Jarov, A., Williams, K.P., Ling, L.E., Koteliansky, V.E., Fournier-Thibault, C. *et al.*, 2001, Sonic hedgehog restricts adhesion and migration of neural crest cells independently of the Patched-Smoothened-Gli signaling pathway. *Proc. Natl. Acad. Sci. USA* 98(22):12521–12526.

Tiara, M., Saint-Jeannet, J.-P., and Davwid, I.B., 1997, Role of the Xlim-1 and Xbra genes in anteroposterior patterning of neural tissue by the head and trunk organizer, *PNAS* 94:895–900.

Tsuda, H., Sasai, N., Matsuo-Takasaki, M., Sakuragi, M., Murakami, Y., and Sasai, Y., 2002, Dorsalization of the neural tube by *Xenopus* tiarin, a novel patterning factor secreted by the flanking nonneural head ectoderm. *Neuron* 33(4):515–528.

Villanueva, S., Glavic, A., Ruiz, P., and Mayor, R., 2002, Posteriorization by FGF, Wnt, and retinoic acid is required for neural crest induction, *Dev. Biol.* 241(2):289–301.

Voiculescu, O., Taillebourg, E., Pujades, C., Kress, C., Buart, S., Charnay, P. *et al.*, 2001, Hindbrain patterning: Krox20 couples segmentation and specification of regional identity, *Development* 128(24):4967–4978.

von Bubnoff, A. and Cho, K.W., 2001, Intracellular BMP signaling regulation in vertebrates: Pathway or network? *Dev. Biol.* 239(1):1–14.

Waskiewicz, A.J., Rikhof, H.A., Hernandez, R.E., and Moens, C.B., 2001, Zebrafish Meis functions to stabilize Pbx proteins and regulate hindbrain patterning, *Development* 128(21):4139–4151.

Wendling, O., Ghyselinck, N.B., Chambon, P., and Mark, M., 2001, Roles of retinoic acid receptors in early embryonic morphogenesis and hindbrain patterning, *Development* 128(11):2031–2038.

Wilson, P.A., Lagna, G., Suzuki, A., and Hemmati-Brivanlou, A., 1997, Concentration-dependent patterning of the *Xenopus* ectoderm by BMP4 and its signal transducer Smad1, *Development* 124(16): 3177–3184.

Withington, S., Beddington, R. and Cooke, J., 2001, Foregut endoderm is required at head process stages for anteriormost neural patterning in chick, *Development* 128(3):309–320.

4

Neural Crest and Cranial Ectodermal Placodes

Clare Baker

GENERAL OVERVIEW AND CHAPTER LAYOUT

The entire peripheral nervous system (PNS) of vertebrates is derived from two transient embryonic cell populations: the neural crest (Hall, 1999; Le Douarin and Kalcheim, 1999) and cranial ectodermal placodes (Webb and Noden, 1993; Baker and Bronner-Fraser, 2001; Begbie and Graham, 2001a). Both originate from ectoderm at the border between the prospective neural plate and epidermis. Neural crest cells delaminate in a rostrocaudal wave and migrate through the embryo along specific migration pathways. They give rise to all peripheral glia, all peripheral autonomic neurons (postganglionic sympathetic and parasympathetic neurons; enteric neurons), all sensory neurons in the trunk, and some cranial sensory neurons, together with many nonneural derivatives such as pigment cells, endocrine cells, facial cartilage and bone, teeth, and smooth muscle. Cranial ectodermal placodes are paired, discrete regions of thickened cranial ectoderm that give rise to the paired peripheral sense organs (olfactory epithelium, inner ear, anamniote lateral line system plus the lens of the eye), most cranial sensory neurons, and the adenohypophysis (anterior pituitary gland). Neural crest, cranial ectodermal placodes, and their derivatives comprise many of the key defining characteristics of the craniates (vertebrates plus hagfish) within the chordate phylum (Gans and Northcutt, 1983; Northcutt and Gans, 1983; Maisey, 1986; Baker and Bronner-Fraser, 1997).

The neural crest and cranial ectodermal placodes share many similarities. Both arise from ectoderm at the neural plate border. Both give rise to multiple neuronal and non-neuronal cell types, including some overlapping derivatives, such as cutaneous sensory neurons in the trigeminal ganglion. Like cells in the central nervous system (CNS) (see Chapter 9), both placode-derived and neural crest cells have considerable migratory ability, although unlike CNS cells, they migrate in the periphery. There are also important differences between the neural crest and cranial ectodermal placodes. Neural crest cells form along the entire length of the neuraxis, except the rostral forebrain, while placode formation is restricted to the head. Neural crest cells give rise to various derivatives not formed by placodes, such as autonomic neurons, melanocytes, cartilage, and smooth muscle. Conversely, unlike neural crest cells, placodes form sensory ciliary receptor cells (sensory cells with a single modified cilium, e.g., olfactory receptor neurons, inner ear hair cells).

The neural crest and cranial ectodermal placodes were discovered independently toward the end of the 19th century; neural crest cells in chick embryos (His, 1868) and placodes in shark embryos (van Wijhe, 1883). They have been studied continuously ever since. What mechanisms and molecules control their formation in the embryo, their adoption of specific migration pathways, and their diversification into so many different cell types? This chapter summarizes our current understanding of these processes in both the neural crest and placodes.

After a brief description of the derivatives of the neural crest (section Neural Crest Derivatives), the chapter follows the order of neural crest cell development *in vivo*. The embryonic origin of neural crest cells at the border between the neural plate and epidermis is described, together with our current knowledge of the molecular nature of neural crest induction (sections Embryonic Origin of the Neural Crest; Neural Crest Induction). Neural crest cell migration pathways through the embryo are then outlined, including developments in our understanding of the molecular cues that guide migrating neural crest cells (section Neural Crest Migration). Finally, an overview is given of current hypotheses on how the diversity of neural crest cell derivatives is achieved (section Neural Crest Lineage Diversification), with particular emphasis on the formation of different cell types in the PNS (section Control of Neural Crest Cell Differentiation in the PNS).

The chapter then introduces the cranial ectodermal placodes (section Overview of Cranial Ectodermal Placodes). The evidence for a common "preplacodal field" at the anterior neural plate border is described (section A Preplacodal Field at the Anterior Neural Plate Border). Our current knowledge of the mechanisms of induction and neurogenesis within each individual placode is then discussed (sections Sense Organ Placodes; Trigeminal and Epibranchial Placodes). For the purposes of this part of the chapter, the placodes are divided into two groups: those that contribute to the paired sense organs (olfactory, lateral line, otic and lens placodes) (section Sense Organ Placodes), and those that only (or mainly) form sensory neurons (trigeminal and epibranchial placodes) (section Trigeminal and Epibranchial Placodes). The hypophyseal placode, which forms the endocrine

Clare Baker • Department of Anatomy, University of Cambridge, Cambridge, CB2 3DY, United Kingdom.

Developmental Neurobiology, 4th ed., edited by Mahendra S. Rao and Marcus Jacobson. Kluwer Academic / Plenum Publishers, New York, 2005.

cells of the adenohypophysis, falls outside the scope of this chapter and is not discussed.

NEURAL CREST DERIVATIVES

Neural crest cells form a startling array of different cell types, including cartilage and bone in the head, teeth, endocrine cells, peripheral sensory neurons, all peripheral autonomic neurons (enteric, postganglionic sympathetic, and parasympathetic neurons), all peripheral glia, and all epidermal pigment cells (Fig. 1). The neural crest origin of these cells has been determined by a variety of ablation and cell-labeling experiments, some of which are described in detail in the section on Experimental Approaches. Neural crest cells emigrating at different rostrocaudal levels along the neuraxis give rise to different but overlapping sets of derivatives (see Table 1). There are traditionally four rostrocaudal divisions of the neural crest along the neuraxis based on these differences: cranial (posterior diencephalon to rhombomere 6); vagal (axial level of somites 1–7); trunk (axial level of somites 8–28); and lumbosacral (axial level posterior to somite 28).

Cranial neural crest cells form a large amount of "mesectoderm," that is, ectodermal derivatives that are mesodermal in character, such as cartilage, bone, teeth, smooth muscle, and other connective tissues. Most of the vertebrate skull is derived from cranial neural crest cells (Fig. 1B). Cranial neural crest cells also form melanocytes (Fig. 1A), Schwann cells, all the satellite glia of the cranial ganglia, parasympathetic neurons, sensory neurons in some cranial sensory ganglia (see Fig. 11), and endocrine cells. Vagal and lumbosacral neural crest cells together provide all the neurons and glia of the enteric nervous system, plus sensory ganglia, parasympathetic ganglia, melanocytes, and endocrine cells (see Table 1). Trunk neural crest cells form the neurons and satellite glia of the sympathetic and dorsal root ganglia, together with Schwann cells, melanocytes, and endocrine cells in the adrenal medulla (Table 1; Figs. 1C and 5).

Most of the vagal neural crest is technically a subdivision of the cranial neural crest, since the boundary between the hindbrain and spinal cord falls at the level of somite 5 (Lumsden, 1990; Cambronero and Puelles, 2000). Vagal neural crest clearly also forms mesectoderm, including musculoconnective elements of the major arteries (Le Lièvre and Le Douarin, 1975; Etchevers et al., 2001) and the aorticopulmonary septum of the heart (Kirby et al., 1983). Although in birds, mesectoderm is only formed down to the level of the fifth somite (Le Lièvre and Le Douarin, 1975), corresponding precisely to the caudal boundary of the hindbrain, mesectoderm production cannot be used as a dividing line between cranial and trunk neural crest cells in all vertebrates. Trunk neural crest cells give rise to dorsal fin mesenchyme in fish and amphibians (Raven, 1931; DuShane, 1935; Collazo et al., 1993; Smith et al., 1994). They may contribute dermal bone to the fin rays of bony fish during normal development, although fish neural crest cells have not yet been followed late enough in development to prove this (Smith et al., 1994). When experimentally challenged with inducing tissues in culture, trunk neural crest cells from the level of the thoracic somites can form

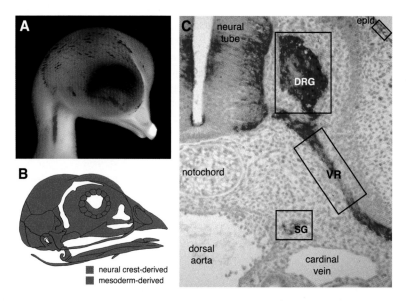

FIGURE 1. Diversity of neural crest derivatives. (A) Melanocytes, seen here as darkly pigmented feathers on the head of a quail–chick chimera. This 11 day-old chick embryo received a unilateral isotopic graft of migrating quail mesencephalic neural crest cells at the 10-somite stage. (B) Schematic to show that most of the vertebrate cranium derives from the neural crest. Redrawn from Couly et al. (1993). (C) Transverse section through the trunk of a 4 day-old chick embryo, stained with an anti-neurofilament antibody (dark staining), to show the location of trunk neural crest derivatives (boxes). These include all neurons and satellite cells of the dorsal root ganglia (DRG) and sympathetic ganglia (SG), Schwann cells along the ventral root (VR), and melanocytes in the epidermis (epid).

TABLE 1. Derivatives of the Neural Crest at Different Axial Levels

Axial level	Cell type	Tissues
Cranial (caudal diencephalon to rhombomere 6)	Mesectoderm	Most bones and cartilages of the neurocranium (brain capsule) and splanchnocranium (facial and pharyngeal)
		Tooth papilla; odontoblasts; dentine matrix
		Meninges of the brain
		Corneal "endothelium"
		Dermis of head and neck
		Tendons
		Non-endothelial components (pericytes, connective, and smooth muscle) of aortic arch-derived arteries
		Smooth muscle (feather arrector muscles in birds; in head blood vessels and aortic arch arteries)
		Connective component of striated muscles (facial and ocular)
		Subcutaneous adipose tissue
		Mesenchymal component of pituitary, salivary, thyroid and parathyroid glands, and the thymus
	Melanocytes	Epidermal pigment cells
	Neurons	
	Sensory	Proximal region of trigeminal (V) ganglion
		Proximal ganglia of cranial nerves VII, IX, and X
		Mesencephalic nucleus of the trigeminal nerve (inside brain)
	Parasympathetic	Postganglionic neurons in ciliary, ethmoidal (dorsal pterygopalatine), sphenopalatine (ventral pterygopalatine), submandibular, otic ganglia
	Glia	Schwann cells
		Satellite cells in cranial ganglia
	Endocrine	Calcitonin-producing cells of the ultimobranchial body (in mammals, parafollicular cells in the thyroid gland)
		Carotid body
Vagal (post-otic hindbrain: somite levels 1–7)	Mesectoderm	Aorticopulmonary septum of the heart
		Non-endothelial components (pericytes, connective, and smooth muscle) of aortic arch-derived arteries
	Melanocytes	Epidermal pigment cells
	Neurons	
	Sensory	Proximal ganglia of cranial nerves IX and X
		Dorsal root ganglia (somite levels 6–7 only)
	Parasympathetic	Postganglionic neurons of parasympathetic nerves innervating thoracic and abdominal visceral organs, including cardiac ganglia
	Sympathetic	Postganglionic neurons in superior cervical ganglion (somite levels 1–4 in the mouse)
	Enteric (sensory, motor, and interneurons)	Enteric ganglia
	Glia	Schwann cells
		Satellite cells in peripheral ganglia (including enteric)
	Endocrine	Calcitonin-producing cells of the ultimobranchial body (in mammals, parafollicular cells in the thyroid gland)
		Carotid body and groups of carotid cells in walls of large arteries arising from heart
Trunk (somite levels 8–28)	Mesectoderm	Fin mesenchyme in fish and amphibians
	Melanocytes	Epidermal pigment cells
	Neurons	
	Sensory	Dorsal root ganglia
	Sympathetic	Postganglionic neurons in sympathetic ganglia
	Glia	Schwann cells
		Satellite cells in peripheral ganglia
	Endocrine	Adrenal chromaffin cells (somite levels 18–24)
(caudal to somite 28)	Melanocytes	Epidermal pigment cells
	Neurons	
	Sensory	Dorsal root ganglia
	Parasympathetic	Remak's ganglion (birds); postganglionic neurons of pelvic splanchnic nerves
	Sympathetic	Postganglionic neurons in sympathetic ganglia
	Enteric (sensory, motor, and interneurons)	Enteric ganglic in post-umbilical gut

TABLE 1. (*Continued*)

Axial level	Cell type	Tissues
Lumbosacral (cont'd)	Glia	Schwann cells Satellite cells in peripheral ganglia (including post-umbilical enteric ganglia)
Pygostyle (birds only: somite levels 47–53)	Melanocytes Glia	Epidermal pigment cells Schwann cells

Source: Le Douarin and Kalcheim (1999); Etchevers *et al.* (2001); Durbec *et al.* (2001); Durbec *et al.* (1996); Smith *et al.* (1994); Collazo *et al.* (1993); Lim *et al.* (1987); Catala *et al.* (2000).

teeth and bone (Lumsden, 1988; Graveson *et al.*, 1997). Trunk neural crest cells can also form smooth muscle *in vitro* (e.g., Shah *et al.*, 1996). Results from both amphibian and chick embryos suggest that under the right circumstances, trunk neural crest cells can even form cartilage (Epperlein *et al.*, 2000; McGonnell and Graham, 2002; Abzhanov *et al.*, 2003). These experiments are examples of many such showing that restrictions in the fate of neural crest cell populations, at a given axial level (i.e., what they form during normal development), do not seem to result from restrictions in potential (the range of possible derivatives), at least at the population level. This will be discussed more fully in the section on Axial Fate-Restriction.

One proposed derivative of the neural crest has aroused controversy: The large sensory neurons that make up the mesencephalic nucleus of the trigeminal nerve (mesV) within the midbrain. These neurons were fate-mapped in the chick to mesencephalic neural crest cells that reenter into the brain immediately after delamination (Narayanan and Narayanan, 1978). Certainly, mesV precursors are not present in the migrating mesencephalic neural crest cell population that has moved away from the brain beneath the adjacent surface ectoderm (Baker *et al.*, 1997). The neural crest origin of mesV neurons has been challenged by a study of molecular marker expression (Hunter *et al.*, 2001), but the question will only be settled by combining a fate-mapping study with molecular markers. Similar large sensory neurons (Rohon-Beard neurons) in the dorsal neural tube in the trunk of fish and amphibian embryos were originally proposed to be a neural crest derivative (Du Shane, 1938; Chibon, 1966). Studies of different zebrafish mutants have shown that Rohon-Beard neurons share a lineage with neural crest cells (Artinger *et al.*, 1999; Cornell and Eisen, 2000, 2002). However, if neural crest cells are defined as cells that have delaminated from the neuroepithelium (section Neural Crest Induction), then Rohon-Beard neurons cannot be described as derivatives of the neural crest.

EMBRYONIC ORIGIN OF THE NEURAL CREST

Neural crest cells were first recognized in the neurula-stage chick embryo as a strip of cells lying between the presumptive epidermis and the neural tube (His, 1868). This area is already distinct at the open neural plate stage in amphibians (Brachet, 1907; Raven, 1931; Knouff, 1935; Baker and Graves, 1939) (see Fig. 13A). The prospective neural crest of urodele amphibians was fate-mapped in early gastrula stages, using vital dyes, to a narrow band of ectoderm between the presumptive neural plate and epidermis (Vogt, 1929). The prospective neural crest was also fate-mapped in the chick gastrula to a region between the prospective neural plate and epidermis, using isotopic grafts of tritiated-thymidine labeled epiblast tissue (Rosenquist, 1981). During neurulation, the neural plate border region forms the neural folds, which rise up and move together until they fuse to form the neural tube (Fig. 2). The prospective neural crest is thus brought from the lateral edges of the open neural plate to the dorsal midline, that is, the "crest" of the neural tube (although cranial neural crest cells are not always incorporated into the neural tube; see section Epithelial–Mesenchymal Transition). In fish, and in the tail region of tetrapods, the neural tube forms by secondary neurulation, in which the ectoderm thickens ventrally and the lumen of the neural tube forms by cavitation. However, the morphogenetic movements of secondary neurulation also involve infolding of the neural plate (Schmitz *et al.*, 1993; Papan and Campos-Ortega, 1994; Catala *et al.*, 1996). In the zebrafish, two bilaterally symmetrical thickenings form on either side of a medial thickening: These fuse to form the neural keel (Schmitz *et al.*, 1993). Prospective neural crest cells (as well as prospective neural and epidermal cells) are contained within the lateral thickenings; they subsequently converge toward the dorsal midline (Schmitz *et al.*, 1993; Thisse *et al.*, 1995). Neural crest cells, therefore, originate from the border between the neural plate and epidermis in all vertebrates.

Presumptive neural crest cells do not form a segregated population in the neural plate border region. When single cells in this region of open neural plate stage chick embryos were labeled and their progeny examined, it was found that individual cells within this field could form epidermis, neural crest and neural tube derivatives in the trunk (Selleck and Bronner-Fraser, 1995). Similarly, when small groups of cells were labeled at the cranial neural plate border, neural crest precursors were found to be intermingled with epidermal, placodal, and neural tube precursors (Streit, 2002). The epidermal lineage only segregates from the CNS and neural crest cell lineages when the neural tube closes (Selleck and Bronner-Fraser, 1995). Neural crest and CNS cell lineages do not seem to segregate at any stage within the neural tube: Single cells within the dorsal neural tube can form both neural tube and neural crest cell derivatives (Bronner-Fraser and Fraser, 1989). Dorsal root ganglion neurons and glia, and

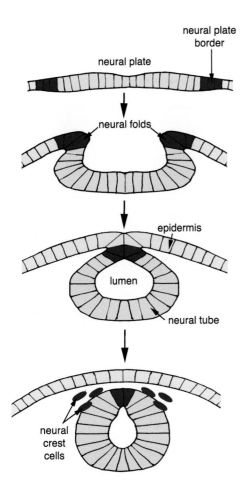

FIGURE 2. Schematic of neurulation in the trunk region of the vertebrate embryo, showing the location of prospective neural crest cells at the lateral borders of the neural plate. As the neural folds rise up and approximate to form the neural tube, prospective neural crest cells are brought dorsally to the "crest" of the neural tube. Cranial neural crest cells, however, are not always incorporated into the neural tube (section Embryonic Origin of the Neural Crest).

melanocytes, are generated by the dorsal neural tube as late as embryonic day 5 (E5) in the chick, several days after "classical" neural crest cell emigration has ceased (Sharma *et al.*, 1995). Furthermore, ventral neural tube cells grafted into neural crest cell migration pathways are able to form neural crest cell derivatives, although they eventually lose the potential to form neurons (Korade and Frank, 1996). Hence, neural crest cells do not constitute a separate population from the CNS until they delaminate from the neuroepithelium. Delamination, therefore, is a crucial defining characteristic of neural crest cells (section Neural Crest Induction).

As will be seen in the section on Evidence for Non-Neural Ecoderm Involvement, neural crest cells can be generated experimentally not only from the neural plate, but also from nonneural ectoderm (prospective epidermis), when these tissues are exposed to appropriate signals. Therefore, all ectodermal cells have the potential to form neural crest cells, at least during early stages of development. However, during normal development, neural crest cells only arise at the border between neural plate

and epidermis, which is underlain by nonaxial mesoderm. What mechanisms and molecules underlie the induction of neural crest cells in this region?

NEURAL CREST INDUCTION

Neural crest cells form at the border between prospective neural plate and prospective epidermis, above nonaxial (paraxial and lateral plate) mesoderm. The neural plate border itself is a recognizable domain, characterized by expression of various genes, including those encoding transcription factors such as *Pax3, Zic,* and *Snail* family members. Many of these genes are maintained in neural crest cells (see Table 1; LaBonne and Bronner-Fraser, 1999; Gammill and Bronner-Fraser, 2003). However, induction of the neural plate border is not equivalent to induction of the neural crest. The most rostral part of the neural plate border (prospective rostral forebrain) fails to produce neural crest (Adelmann, 1925; Knouff, 1935; Jacobson, 1959; Chibon, 1967a; Couly and Le Douarin, 1985; Sadaghiani and Thiébaud, 1987), except possibly for a few in the mouse (Nichols, 1981; Osumi-Yamashita *et al.*, 1994). In the head, the neural plate border also gives rise to cranial ectodermal placodes (section A Preplacodal Field at the Anterior Neural Plate Border). Furthermore, neural plate border markers and morphology can be induced experimentally without inducing neural crest cells (McLarren *et al.*, 2003).

The available evidence (reviewed in Kalcheim, 2000; Mayor and Aybar, 2001; Knecht and Bronner-Fraser, 2002; Gammill and Bronner-Fraser, 2003) suggests that neural crest induction can be divided into three main steps: (1) establishment of the neural plate border, which is initially anterior in character, via intermediate levels of bone morphogenetic protein (BMP) activity and Dlx transcription factor activity; (2) posteriorization of the neural plate border, and induction of neural crest cell precursors within it, by Wnt and/or FGF signaling; (3) epithelial–mesenchymal transition. Until a cell delaminates from the neuroepithelium into the periphery, it is not a *bona fide* neural crest cell. Indeed, failure to emigrate can lead to neural differentiation of neural crest precursors within the neuroepithelium (Borchers *et al.*, 2001). Hence, induction of delamination can be considered as the final step in neural crest induction.

Selected molecular markers of neural crest cells, many of which are used in assays for neural crest cell induction, are listed in Table 2 (also see Gammill and Bronner-Fraser, 2003). In *Xenopus*, induction of the genes encoding the zinc finger transcription factors, Slug and Twist (section Snail/Slug and FoxD3 Are Required for Neural Crest Precursor Formation), is commonly used as a proxy for neural crest cell induction. The HNK-1 epitope, a carbohydrate expressed on migrating neural crest cells, among other cell types, is frequently used in the chick to identify neural crest cells (see Table 2). The winged-helix transcription factor FoxD3 (sections Snail/Slug and FoxD3 Are Required for Neural Crest Precursor Formation; FoxD3 Promotes Neural Crest Cell Delimitation at All Axial Levels) and the HMG-box transcription factors Sox9 and Sox10 (section Sox10

TABLE 2. Some Genes Expressed in Premigratory and Migrating Neural Crest Cells

Molecule	Type	NC precursors	Migrating NC cells	Role in:	Selected references
Frizzled3	Wnt receptor	+	–	NC cell induction	Deardorff et al. (2001)
Pax3	Paired-domain transcr. factor	+	+ (early); reexpressed later	Postmigratory	Mansouri et al. (2001)
Zic family	Zinc finger transcr. factors	+	+	NC cell induction	Nakata et al. (2000)
AP-2α	Transcr. factor	+	+	NC precursor cell formation	Luo et al. (2003)
Sox9	HMG-domain transcr. factor	+	+	NC precursor cell formation and postmigratory	Spokony et al. (2002); Cheung and Briscoe (2003)
Sox10	HMG-domain transcr. factor	+	+	NC precursor cell formation and postmigratory	Britsch et al. (2001); Dutton et al. (2001); Honoré et al. (2003)
FoxD3 (=forkhead6)	Winged helix transcr. factor	+	+	NC cell induction	Dottori et al. (2001); Sasai et al. (2001)
Slug/Snail family	Zinc finger transcr. factors	+	+ (early)	NC cell induction, migration	LaBonne and Bronner-Fraser (2000); del Barrio and Nieto (2002)
Twist	bHLH transcr. factor	+ (cranial)	+ (cranial)	Unknown	Gitelman (1997)
Endothelin receptor B	Endothelin-3 receptor	+	+	Postmigratory	Nataf et al. (1996)
RhoB	GTP-binding protein	+	+ (early)	Emigration	Liu and Jessell (1998)
ADAM13	Metallo-protease	–	+	Emigration/ migration	Alfandari et al. (2001)
Cadherin7	Cell–cell adhesion	–	+	Migration	Nakagawa and Takeichi (1998)
p75[NTR]	Low-affinity neurotrophin receptor	–	+	Unknown	Stemple and Anderson (1992)
HNK1 epitope	Glucuronic acid-containing carbohydrate	–	+	Unknown	Le Douarin and Kalcheim (1999)

Note: References are selected to enable further reading: they are not comprehensive. bHLH, basic helix-loop-helix; HMG, high mobility group; NC, neural crest; transcr., transcription.

Is Essential for Formation of the Glial Lineage), which are expressed in neural crest precursors and migrating neural crest cells, are more recently identified neural crest cell markers.

Step 1: Establishment of the Neural Plate Border

Molecular signals involved in neural plate induction are discussed at length in Chapter 1 and will not be reviewed here. The classical "default" model for neural plate induction, whereby high levels of bone morphogenetic proteins (BMPs) specify epidermis, and low levels specify neural plate (see Chapter 1), led to the suggestion that intermediate levels of BMP activity specify the border between the two tissues (reviewed in Mayor and Aybar, 2001). Indeed, intermediate BMP activity levels are sufficient to induce some anterior neural plate border genes in *Xenopus* ectoderm *in vitro* (Wilson and Hemmati-Brivanlou, 1995; Knecht and Harland, 1997; Villanueva *et al.*, 2002). Importantly, however, no concentration of BMP antagonist is suf-

ficient to induce neural crest cells alone, that is, in the absence of neural and epidermal markers (Wilson *et al.*, 1997; LaBonne and Bronner-Fraser, 1998). This is consistent with the fact that the anterior neural plate border does not produce neural crest cells, and with the hypothesis that additional signals are required to induce neural crest cell precursors within the neural plate border region (see next section).

In *Xenopus*, overexpression of BMP antagonists *in vivo* leads to lateral expansion of neural crest markers, contiguous with their normal domain, at the expense of epidermal ectoderm (Mayor *et al.*, 1995; LaBonne and Bronner-Fraser, 1998). Conversely, overexpression of BMP4 has little effect on neural crest markers, but shifts the border medially at the expense of the neural plate (LaBonne and Bronner-Fraser, 1998). Zebrafish embryos carrying mutations in the BMP signaling pathway also show reduced or expanded domains of neural crest cell precursors, depending on the effect of the mutation on BMP activity levels (Nguyen *et al.*, 1998). In the chick, the balance between

BMP4 and its antagonists is important for establishing and/or maintaining the prospective neural plate border: This region, which itself expresses BMP4, is the only region responsive to changes in the level of BMP signaling at neural plate stages (Streit and Stern, 1999).

These results suggest that BMP signaling is required for neural plate border formation and maintenance, and that changes in BMP activity levels can affect neural crest cell formation, although they are not sufficient to induce neural crest cells.

Members of the Dlx family of transcription factors play an important role in positioning the neural plate border during gastrulation (McLarren et al., 2003; Woda et al., 2003). In the chick, gain-of-function experiments have shown that Dlx5, itself a marker of the neural plate border, represses neural fates without inducing epidermis (McLarren et al., 2003). Furthermore, Dlx5 acts non-cell autonomously (presumably by activating downstream signaling pathways) to promote the expression of other neural plate border markers in adjacent cells, such as the transcription factor Msx1, and BMP4 itself (McLarren et al., 2003). However, Dlx5 activity is not sufficient to induce either neural crest cells or placodes (McLarren et al., 2003). In Xenopus, gain-of-function and loss-of-function experiments have shown that Dlx3 and Dlx5 activity positions the neural plate border, and that Dlx protein function in non-neural ectoderm is required for the subsequent induction of both neural crest and placodes (Woda et al., 2003).

In summary, the activity of BMP signaling molecules and Dlx transcription factors appears to specify the neural plate border region. However, the activity of these molecules is insufficient to specify neural crest cells (or placode cells). Intermediate BMP activity levels induce neural plate border that is anterior in character. Hence, additional signals are required to posteriorize the neural plate border and induce neural crest precursor cells within it.

Step 2: Induction of Neural Crest Precursors

It is becoming increasingly evident that Wnt and/or FGF family members are involved both in posteriorizing the neural plate border and inducing neural crest precursor cells within it. These do seem to be separable processes, however, as neural crest induction can be experimentally uncoupled from the anterior–posterior patterning of the neural plate (e.g., Chang and Hemmati-Brivanlou, 1998; Monsoro-Burq et al., 2003).

Posteriorizing Signals (Wnts and FGFs)

A posteriorizing signal derived from the paraxial mesoderm enables rostral neural plate tissue to form neural crest cells in the chick (Muhr et al., 1997) and establishes Pax3 expression at the neural plate border in both chick and Xenopus embryos (Bang et al., 1997, 1999). In the chick, this posteriorizing activity is mediated by Wnt family members, in particular Wnt8c and Wnt11, in conjunction with permissive FGF signaling (Nordström et al., 2002). Paraxial mesoderm produces several other factors, including FGFs and retinoic acid, that are able to posteriorize the neural plate to induce posterior cell fates. In the

chick, though, FGFs and retinoic acid are insufficient to induce caudal character in neural cells in vitro: This requires Wnt activity from the caudal paraxial mesoderm (Muhr et al., 1997, 1999; Nordström et al., 2002).

Induction of Neural Crest Precursors (Wnts and FGFs)

In both Xenopus and chick embryos, Wnt family members are both sufficient to induce neural crest cells from neuralized ectoderm in vitro, and necessary for neural crest induction in vivo (reviewed in Wu et al., 2003). Wnts can induce neural crest markers in conjunction with BMP inhibitors in ectodermal explants in vitro (Saint-Jeannet et al., 1997; Chang and Hemmati-Brivanlou, 1998; LaBonne and Bronner-Fraser, 1998). Conversely, inhibiting Wnt function in vivo by overexpressing a dominant negative Wnt ligand prevents early neural crest cell marker expression (LaBonne and Bronner-Fraser, 1998). Morpholino oligonucleotide-mediated blockage of the translation of the Wnt receptor Frizzled3, or its proposed adaptor protein Kermit, both reduce Slug expression in Xenopus (Deardorff et al., 2001; Tan et al., 2001), again showing a requirement for Wnt signaling in neural crest cell formation. Furthermore, the Xenopus Slug promoter contains a functional binding site for a downstream effector of Wnt signaling (LEF/β-catenin) that is required to drive its expression in neural crest precursors, showing that the requirement for Wnt is direct (Vallin et al., 2001).

Wnt activity is also necessary and sufficient for neural crest cell induction in the chick (García-Castro et al., 2002). Overexpression of a dominant negative Wnt ligand inhibits Slug expression in vivo: This can be rescued by application of exogenous Wnt (García-Castro et al., 2002). Conversely, Drosophila Wingless (a Wnt1 homologue that triggers the Wnt signaling pathway in vertebrates) can induce neural crest cells from neural plate in a chemically defined medium that lacks any other growth factors and hormones (García-Castro et al., 2002). Importantly, BMP4, which was previously shown to induce neural crest cells from neural plate in vitro, in the presence of various additives (Liem et al., 1995), is unable to induce neural crest cells from the neural plate in their absence (García-Castro et al., 2002). Synergism with other factors present in the medium may also underlie the induction of neural crest cells by BMP2/4 from dissociated rat neural tube cells (Lo et al., 2002) or neuroepithelial stem cells (Mujtaba et al., 1998).

Wnt signaling seems to control the domain of expression of Iro1 and Iro7, homeodomain transcription factors homologous to the Iroquois family of factors that, in Drosophila, regulate the expression of proneural genes (section Proneural Genes: An Introduction) (Itoh et al., 2002). Functional knockdown of both Iro1 and Iro7 using morpholino antisense oligonucleotides leads to loss of FoxD3 expression (Itoh et al., 2002). This not only suggests that these transcription factors are upstream of FoxD3, but also provides indirect evidence that Wnt signaling regulates neural crest induction (Itoh et al., 2002). Furthermore, Wnt signaling is required for the induction of c-Myc, a basic helix-loop-helix zipper transcription factor whose expression is required for *Slug* and *FoxD3* expression and neural crest cell formation in *Xenopus* (Bellmeyer *et al.*, 2003).

The above results clearly show that Wnts are both necessary and sufficient to mediate neural crest cell induction from neuralized ectoderm. Several different models of neural crest induction have been proposed over the years, variously stressing the importance of nonaxial mesoderm and neural plate–epidermal interactions. (Some of the data supporting a role for both paraxial mesoderm and neural plate–epidermal interactions in neural crest induction are described in the following sections.) However, since both paraxial mesoderm and epidermis express Wnt family members, it is likely that both tissues are involved *in vivo*. Wnt8 is expressed in the paraxial mesoderm, and Wnt6 and Wnt7b are expressed in non-neural ectoderm (Chang and Hemmati-Brivanlou, 1998; García-Castro *et al.*, 2002).

Nonetheless, Wnts may not be the whole story. Work in *Xenopus* has suggested that not only Wnt8, but also retinoic acid and FGFs, are able to induce *Slug* expression, both in the anterior neural plate border, and in tissue transformed into anterior neural plate border by intermediate levels of BMP activity (Villanueva *et al.*, 2002; Monsoro-Burq *et al.*, 2003). Furthermore, FGF signaling is required for induction of neural crest markers by paraxial mesoderm in *Xenopus* (Monsoro-Burq *et al.*, 2003). Hence, although most of the evidence so far favors Wnts as the primary signals that induce neural crest cell precursors within the neural plate border (see Wu *et al.*, 2003), FGF involvement cannot be ruled out.

Evidence for Paraxial Mesoderm Involvement in Neural Crest Induction

Several lines of evidence have suggested a role for nonaxial mesoderm in neural crest cell induction. In 1945, Raven and Kloos showed in an amphibian model that fragments of lateral archenteron roof (prospective paraxial and lateral plate mesoderm) can induce neural crest cells from overlying ectoderm, in the absence of neural tissue, when grafted into the blastocoel (Raven and Kloos, 1945). Over fifty years later, prospective paraxial mesoderm was shown to induce neural crest marker expression and melanocytes from competent ectoderm in *Xenopus* explant cocultures (Bonstein *et al.*, 1998; Marchant *et al.*, 1998; Monsoro-Burq *et al.*, 2003). In the chick, paraxial mesoderm can induce neural plate explants to form melanocytes (though not neurons) (Selleck and Bronner-Fraser, 1995). Hence, paraxial mesoderm is sufficient to induce at least some neural crest cell markers and derivatives *in vitro*, both from non-neural ectoderm and neural plate. Importantly, removing prospective paraxial mesoderm at the start of gastrulation in *Xenopus* leads to a reduction in *Slug* expression and melanocyte formation *in vivo* (Bonstein *et al.*, 1998; Marchant *et al.*, 1998). This suggests that paraxial mesoderm is not only sufficient to induce neural crest cells *in vitro*, but also necessary for neural crest cell induction *in vivo*.

The molecular model of neural crest induction described thus far (i.e., intermediate BMP activity plus Wnt/FGF signaling) can explain the induction of neural crest cells by paraxial mesoderm. Paraxial mesoderm expresses both BMP inhibitors, such as Noggin and Follistatin (e.g., Hirsinger *et al.*, 1997; Marcelle *et al.*, 1997; Liem *et al.*, 2000), and Wnt and FGF family members (see previous section). The BMP inhibitors may induce intermediate levels of BMP activity in non-neural ectoderm, while the Wnt/FGF signals may subsequently induce neural crest cells from this neuralized ectoderm. However, this model has not been tested directly.

Evidence for Non-Neural Ectoderm Involvement in Neural Crest Induction

A role for non-neural ectoderm in neural crest cell induction was first proposed in the late 1970s and early 1980s. Rollhäuser-ter Horst used interspecific grafts between different species of urodele amphibians to follow the fate of gastrula ectoderm juxtaposed to different tissues (Rollhäuser-ter Horst, 1979, 1980). The ectoderm failed to form neural crest cells *in vitro* either when cultured alone, or when cocultured with neural-inducing tissue, but did form neural crest cells when both tissues were grafted to the belly of host embryos (Rollhäuser-ter Horst, 1979). This suggested a requirement for the host epidermis as well as neural-inducing tissue. When the gastrula ectoderm was grafted in place of the host neural folds, it also formed neural crest cells (Rollhäuser-ter Horst, 1980), again suggesting a role for interactions between neural and non-neural ectoderm in neural crest induction.

Moury and Jacobson similarly used pigmented and albino axolotl embryos as donors and hosts, respectively, to show that both neural folds and neural crest cells form at any newly created boundary between neural plate and epidermis (Moury and Jacobson, 1989, 1990). Under these circumstances, both epidermis and neural plate form neural crest cells. Interestingly, the neural plate forms melanocytes while the epidermis forms sensory neurons (Moury and Jacobson, 1990). In *Xenopus*, labeled neural plate grafted into epidermis *in vivo* leads to *Slug* upregulation in both donor and host tissues, at the interface between them (Mancilla and Mayor, 1996). Likewise, when quail neural plate is grafted into chick epidermis *in vivo*, both quail and chick tissue generate migratory HNK-1-positive cells (Selleck and Bronner-Fraser, 1995). *Slug* is also induced after similar experiments using unlabeled chick tissue (although in which tissues is unclear) (Dickinson *et al.*, 1995).

Although these *in vivo* experiments suggested a role for interactions between neural plate and epidermis in neural crest cell induction, all the grafted tissues were also exposed to signals from the underlying mesoderm. However, *in vitro* cocultures of neural plate and epidermis, in the absence of mesoderm, are sufficient to induce *Slug* expression in *Xenopus* (Mancilla and Mayor, 1996) and neural crest cells in the chick (*Slug* expression; formation of melanocytes and catecholaminergic neurons) (Dickinson *et al.*, 1995; Selleck and Bronner-Fraser, 1995). Hence, a local interaction between neural and non-neural ectoderm is sufficient to induce neural crest cells *in vitro*. This finding has been exploited in a subtractive hybridization screen of a macroarrayed chick cDNA library, in order to provide the first gene expression profile of newly induced neural crest cells (Gammill and Bronner-Fraser, 2002).

The interaction between neural and non-neural ectoderm seems to recapitulate all of the steps of neural crest induction seen *in vivo*, including induction of the neural plate border, since neural folds form at all experimentally generated neural/epidermal interfaces (Moury and Jacobson, 1989). Both epidermal and neural plate cells may contribute to the new neural plate border region, perhaps explaining why both tissues form neural crest cells after such interactions.

In summary, there is substantial evidence to implicate both paraxial mesoderm and non-neural epidermis in neural crest cell induction *in vivo*. Their involvement is probably due to their expression of Wnt (and/or FGF) family members, which can induce neural crest cell precursors within the neural plate border region.

AP2α and SoxE Transcription Factors Are Involved in the Earliest Steps of Neural Crest Precursor Formation

The transcription factor AP2α is expressed during early stages of neural crest development in all vertebrates, as well as in other tissues, such as the epidermis (see Luo *et al.*, 2003). In *Xenopus*, AP2α expression, which covers a broader territory than other early neural crest precursor markers such as *Sox9* (see next paragraph) and *Slug*, is upregulated by BMP and Wnt signaling (Luo *et al.*, 2003). Morpholino-mediated functional knockdown of AP2α results in failure of neural fold formation and the loss of *Sox9* and *Slug* expression (Luo *et al.*, 2003). These results suggest an important role for AP2α in the earliest stages of neural crest precursor formation. However, the broad expression pattern of AP2α, in particular in epidermis, implies that other factors must be involved in restricting neural crest precursor formation to the correct region.

Sox9 and Sox10 are members of the E subgroup of high-mobility-group (HMG) domain Sox transcription factors. Sox9 is one of the earliest markers of premigratory neural crest cell precursors within the neural plate border; its expression is maintained during early stages of neural crest migration (Spokony *et al.*, 2002; Cheung and Briscoe, 2003). Morpholino-mediated functional knockdown of either Sox9 or Sox10 in *Xenopus* blocks neural fold formation, as well as blocking expression of neural plate border markers and neural crest precursor markers, including *Slug* (Spokony *et al.*, 2002; Honoré *et al.*, 2003). Unlike Dlx activity (see section Establishment of the Neural Plate Border), Sox9 activity is sufficient to induce neural crest precursor markers, including *Slug* and *FoxD3*, in both dorsal and ventral regions of the chick neural tube (Cheung and Briscoe, 2003). However, Sox9-induced ectopic neural crest precursors rarely delaminate except in the most dorsal regions of the neural tube (Cheung and Briscoe, 2003). This suggests that additional signals are required for neural crest cell delamination, and that these signals are only present dorsally (see section Epithelial–Mesenchymal Transition).

Importantly, Sox9-mediated induction of neural crest precursor markers in the chick does not induce BMP or Wnt family members, nor require BMP activity, suggesting that, like AP2α, Sox9 lies downstream of these signaling pathways (Cheung and Briscoe, 2003). Blocking either FGF signaling or Wnt signaling in *Xenopus* also blocks Sox10 expression at the neural plate border (Honoré *et al.*, 2003), again suggesting that the SoxE transcription factors lie downstream of identifed neural crest precursor inducing signals.

Although morpholino-mediated functional knockdown of *Sox9a* in zebrafish does not affect neural crest precursors, it is possible that *Sox9b* may instead play this role in zebrafish (Yan *et al.*, 2002). Neural crest-specific knockout of *Sox9* in mice does not cause neural crest precursor defects (Mori-Akayama *et al.*, 2003), but it is possible that overlapping expression of the other SoxE subgroup members, Sox8 and Sox10, may compensate for the loss of Sox9.

In summary, it seems likely that AP2α, Sox9, and Sox10 may be crucial downstream target of BMP and Wnt/FGF signals in the formation of neural crest precursors. AP2α seems to lie upstream of Sox9, whose activity in turn induces the expression of multiple other markers of neural crest cell precursors, including *Slug* and *FoxD3* (see next section). However, delamination from the neuroepithelium (i.e., neural crest cell formation) requires additional signals that, at least in the chick, may only be present in the dorsal neural tube.

Snail/Slug and FoxD3 Are Required for Neural Crest Precursor Formation

The Snail superfamily of zinc finger transcriptional repressors contains two major families: Snail and Scratch (Nieto, 2002). In vertebrates, the Snail family is further subdivided into Snail and Slug subfamilies, both of which are essential during two stages of neural crest formation: (1) The formation of neural crest cell precursors within the neuroepithelium, and (2) delamination of cranial neural crest cells (section Snail Family Members Promote Cranial Neural Crest Cell Delamination). In *Xenopus, Slug* is first expressed at late gastrula stages, long before neural crest delamination occurs (Mayor *et al.*, 1995). Slug acts as a transcriptional repressor (LaBonne and Bronner-Fraser, 2000; Mayor *et al.*, 2000). *Slug* overexpression in *Xenopus* leads to an expansion of the neural crest domain at the expense of epidermis, and to overproduction of at least some neural crest derivatives (LaBonne and Bronner-Fraser, 1998). Conversely, other early neural crest precursor markers are lost after expression of a dominant negative *Slug* construct or antisense *Slug* RNA, showing that Slug function is necessary for the formation of neural crest precursors (Carl *et al.*, 1999; LaBonne and Bronner-Fraser, 2000). However, not all *Slug*-expressing cells delaminate to form neural crest cells (Linker *et al.*, 2000).

The winged-helix transcription factor FoxD3 (Forkhead6) is also important in early stages of neural crest cell formation (Dottori *et al.*, 2001; Kos *et al.*, 2001; Pohl and Knöchel, 2001; Sasai *et al.*, 2001). Like Slug, FoxD3 is a transcriptional repressor (Pohl and Knöchel, 2001; Sasai *et al.*, 2001) and is expressed both in premigratory neural crest cell precursors and migrating neural crest cells. In *Xenopus*, inhibiting FoxD3 function *in vivo* using a dominant negative *FoxD3* construct represses the expression of early neural crest precursor markers, including *Slug*, and leads to a corresponding expansion of the neural plate (Sasai *et al.*, 2001). Hence, like *Slug*, FoxD3 function is required for the formation of neural crest precursors. However, overexpression of *FoxD3* in the chick neural tube does not upregulate *Slug*, suggesting that *Slug* is not an obligate downstream target of FoxD3 (Dottori *et al.*, 2001). Instead, the two genes seem to act in concert, in partially overlapping pathways, to promote neural crest cell formation (Sasai *et al.*, 2001).

In addition to their importance for the formation of neural crest cell precursors, both FoxD3 and Slug can promote neural crest cell delamination (sections FoxD3 Promotes Neural Crest Cell Delamination at All Axial Levels; Snail Family Members Promote Cranial Neural Crest Cell Delamination).

Step 3: Epithelial–Mesenchymal Transition

The final step in neural crest induction is the activation of the epithelial–mesenchymal transition that leads to delamination from the neuroepithelium into the periphery. As described at the beginning of the section on Neural Crest Induction, a cell cannot be described as a *bona fide* neural crest cell until it emigrates from the neuroepithelium. Hence, induction of delamination is the final step in the induction of the neural crest.

In all vertebrates, neural crest cell precursors delaminate in a rostrocaudal wave along the neuraxis. Whether or not neural crest cell precursors are initially incorporated into the neural tube depends on the timing of neural crest cell delamination relative to the timing of fusion of the neural folds. This varies from species to species and on the axial level within the embryo. Cranial neural crest cells, in particular, which are the first to delaminate, may not be incorporated into the neural tube. In the mouse, cranial neural crest delamination begins in the midbrain/rostral hindbrain well before neural tube closure, when the neural folds are approaching one another in the cervical region (Nichols, 1981). In frogs, cranial neural crest cells form large masses that segregate from the neural tube prior to its closure; these masses do not take part in the morphogenetic movements of neurulation (Schroeder, 1970; Olsson and Hanken, 1996). In the chick, however, cranial neural crest cells delaminate as the neural folds meet or during early apposition, beginning at midbrain levels (Tosney, 1982). Trunk neural crest cells in the chick only emigrate after the epidermis and neural tube have separated (Tosney, 1978).

In the chick, the first sign of imminent neural crest cell delamination at cranial levels is that the neural crest cell precursor cell population becomes less tightly packed, and the cells extend long cellular processes into the intercellular spaces within the population (Tosney, 1982). As emigration starts, the basal lamina over the neural crest cells becomes fragmented, and the cells extend long processes into the adjacent cell-free space (Tosney, 1982). Clearly, major changes in cytoskeletal architecture, cell–cell and cell–matrix interactions occur during this epithelial–mesenchymal transition. Recent molecular evidence has given us a more detailed insight into the genes and signaling pathways controlling these processes.

The Basal Lamina Must Be Degraded before Delamination Can Occur

Neural crest cells do not seem to be able to penetrate an intact basal lamina (Erickson, 1987). The basal lamina clearly breaks down over neural crest cell precursors before they delaminate from the neuroepithelium (e.g., Tosney, 1982; Raible *et al.*, 1992) and this may be due to neural crest secretion of proteases, although it remains to be demonstrated. Neural crest cell precursors produce various proteolytic enzymes, including the serine protease plasminogen activator (Valinsky and Le Douarin, 1985; Agrawal and Brauer, 1996), BMP1/Tolloid metalloproteases (Martí, 2000), and members of the metalloprotease/disintegrin family (Alfandari *et al.*, 1997; Cai *et al.*, 1998). Some of these proteases are only found in cranial neural crest cell precursors and migrating cranial neural crest cells, for example, the metalloprotease/disintegrin ADAM13 in *Xenopus* (Alfandari *et al.*, 1997, 2001). However, a role for these proteases in neural crest cell delamination has not yet been shown.

Inhibiting Protein Kinase C Signaling Promotes Delamination

If avian neural tube explants are treated with protein kinase C inhibitors, cells immediately, and precociously, delaminate and migrate away from the neural tube (Newgreen and Minichiello, 1995, 1996). This occurs on both dorsal and ventral sides of the neural tube (although ventral cells are less sensitive than dorsal cells) (Newgreen and Minichiello, 1995, 1996). This stimulatory effect of protein kinase C inhibitors does not require protein synthesis (Newgreen and Minichiello, 1995). Similarly, protein kinase C inhibition triggers delamination, migration, and expression of the neural crest marker *Sox10*, in neuroectoderm cells produced from mouse embryonic stem cells in culture (Rathjen *et al.*, 2002). These results suggest that delamination can be induced by signals that modulate protein kinase C activity.

Delamination is Associated with Downregulation of Cadherin6B

Calcium-dependent cell–cell adhesions are required to prevent precocious emigration of neural crest cells (Newgreen and Gooday, 1985). In the chick, most neural tube cells express the calcium-dependent cell–cell adhesion molecule N-cadherin, while epidermal cells express E-cadherin; however, the dorsal neural tube, which contains neural crest cell precursors, expresses neither N-cadherin nor E-cadherin (Akitaya and Bronner-Fraser, 1992). In accordance with this, N-cadherin itself does not seem to be required for neural crest cell formation or migration, as pigmentation and cranial cartilages are normal in *N-cadherin* mutant zebrafish (Lele *et al.*, 2002). Instead, neural crest cell precursors within the neuroepithelium express *cadherin6B*; this expression is lost in emigrating neural crest cells (Nakagawa and Takeichi, 1995, 1998). Type II (atypical) cadherins are then upregulated in subpopulations of migrating neural crest cells, for example *cadherin7* and *cadherin 11*; these may be involved in controlling the rate of neural crest cell migration and/or in some aspects of fate specification (Nakagawa and Takeichi, 1995; Borchers et al., 2001).

FoxD3 Promotes Neural Crest Cell Delamination at All Axial Levels

FoxD3 is essential for the formation of neural crest cell precursors (section Snail/Slug and FoxD3 Are Required for Neural Crest Precursor Formation), and it may also play a role in neural crest cell delamination. Ectopic expression of FoxD3 in the chick neural tube promotes neural crest cell delamination at all axial levels (Dottori et al., 2001). This is achieved without upregulating Slug or, apparently, RhoB (section BMP4 Induces RhoB, Which Is Essential for Neutral Crest Cell Delamination), suggesting that FoxD3 and Slug function independently in regulating neural crest cell delamination (Dottori et al., 2001). The precise mechanism of action of FoxD3 in promoting delamination remains unclear.

Snail Family Members Promote Cranial Neural Crest Cell Delamination

Snail family transcription factors are required for the formation of neural crest cell precursors (section Snail/Slug and FoxD3 Are Required for Neural Crest Precursor Formation). Several different lines of evidence also support a role for Snail family genes in epithelial–mesenchymal transitions. Overexpression of mouse Slug in bladder carcinoma cells leads to desmosome dissociation at sites of cell–cell contact, a necessary prerequisite for epithelial–mesenchymal transition (Savagner et al., 1997). Overexpression of mouse Snail in epithelial cells represses transcription of the cell–cell adhesion molecule E-cadherin, and leads to epithelial–mesenchymal transition and migratory and invasive cell behaviors (Batlle et al., 2000; Cano et al., 2000). Since Snail and/or Slug genes are expressed in premigratory neural crest cell precursors in all vertebrates, a role in neural crest cell delamination from the neuroepithelium seems likely.

Early antisense experiments in chick embryos suggested a role for Slug in cranial neural crest cell migration (Nieto et al., 1994). Cranial neural crest cell migration is inhibited in Xenopus in the presence of antisense Slug RNA or a dominant negative Slug construct (Carl et al., 1999; LaBonne and Bronner-Fraser, 2000). Overexpression of Slug in the chick neural tube leads to an increase in the number of migrating cranial neural crest cells, although not of trunk neural crest cells (del Barrio and Nieto, 2002). Other experiments have also shown that, unlike FoxD3, increased Slug activity alone does not cause trunk neural crest cell delamination in the trunk (Sela-Donenfeld and Kalcheim, 1999). The basis of this difference between head and trunk is unknown.

BMP Signaling is Required for Delamination

In the trunk of the chick embryo, neural crest cells only begin to delaminate in areas adjacent to the epithelial somites: They do not emigrate at the level of the segmental plate mesoderm (Teillet et al., 1987). The timing of neural crest cell emigration in the trunk can be correlated with expression of the BMP2/4 antagonist Noggin (Sela-Donenfeld and Kalcheim, 1999). Noggin is strongly expressed in the dorsal neural tube opposite the segmental plate mesoderm, more weakly expressed opposite newly epithelial somites, and absent opposite fully dissociated somites, while BMP4 is expressed in the dorsal neural tube at all levels (Sela-Donenfeld and Kalcheim, 1999). Noggin overexpression (i.e., inhibition of BMP activity) inhibits neural crest cell delamination both in vivo and in vitro, and this can be rescued in vitro by BMP4 (Sela-Donenfeld and Kalcheim, 1999). This suggests that a balance between BMP4 and its antagonists plays a role in the onset of neural crest cell delamination in the trunk (Sela-Donenfeld and Kalcheim, 1999). This balance is now known to be controlled by the paraxial mesoderm itself: The dorsomedial region of developing somites produces a signal that downregulates noggin transcription in the dorsal neural tube (Sela-Donenfeld and Kalcheim, 2000). This enables the coordination of neural crest cell emigration with the formation of a suitable mesodermal substrate for migration (section Migration Pathways of Trunk Neural Crest Cells) (Sela-Donenfeld and Kalcheim, 2000).

BMP signaling is also essential for cranial neural crest cell migration in the mouse (Kanzler et al., 2000). When noggin is expressed in transgenic embryos under the control of a Hox2a enhancer, leading to noggin overexpression in the hindbrain, hindbrain-level neural crest cells fail to emigrate (Kanzler et al., 2000). Although Bmp4 is not expressed in the dorsal hindbrain in the mouse, Bmp2 is expressed there, and hindbrain neural crest cells fail to migrate in Bmp2 mutant embryos. Hence, it seems that BMP2 activity is necessary for cranial neural crest cell emigration in the mouse (Kanzler et al., 2000).

These results show that BMP signaling is essential not just to establish the neural plate border, but also at a later stage, to promote neural crest cell delamination.

BMP4 Induces RhoB, Which Is Essential for Neural Crest Cell Delamination

The small GTP-binding protein RhoB is expressed in neural crest precursors within the neuroepithelium and is downregulated shortly after delamination (Liu and Jessell, 1998). Rho proteins have been implicated in the assembly of the actin cytoskeleton required for motility (see Frame and Brunton, 2002). Treatment of chick neural tube explants with a Rho-specific inhibitor has shown that Rho function is essential for neural crest cell delamination, and that the actin cytoskeleton in neural crest cell precursors is perturbed (Liu and Jessell, 1998). RhoB also seems to be a downstream target of Slug activity, though whether direct or indirect is unknown (del Barrio and Nieto, 2002). It is not, however, detectably induced by FoxD3 (Dottori et al., 2001). Nor, interestingly, is RhoB detectably induced by Sox9, which induces neural crest precursor formation but is not sufficient to promote efficient delamination, except at the dorsalmost region of the neural tube (Cheung and Briscoe, 2003; section AP2α and SoxE Transcription Factors Are Involved in the Earliest Steps of Neural Crest Precursor Formation). However, RhoB is induced by BMP4: Indeed, it was originally identified in a PCR-based screen for genes induced by BMP4 in neural plate cells (Liu and Jessell, 1998). Since BMP4 is essential for delamination of neural crest cell precursors and induces RhoB, it seems that BMP4 activity is the most likely candidate for the dorsally located signal that induces neural crest cell formation from premigratory neural crest cell precursors. It will be important to establish whether all RhoB-expressing neural crest cell precursor cells do, in fact, emigrate from the neural tube.

Transition from G1 to S Phase of the Cell Cycle Is Required for Neural Crest Cell Delamination

In the chick, most trunk neural crest cells emigrate from the neural tube in the S phase of the cell cycle, when their nuclei are located at or near the basal margin of the neuroepithelium (Burstyn-Cohen and Kalcheim, 2002). Blocking the cell cycle transition from G1 to S phase blocks neural crest delamination, both *in vivo* and in explants (Burstyn-Cohen and Kalcheim, 2002). Thus, the cell cycle status of neural crest cell precursors is an essential prerequisite for the epithelial–mesenchymal transition that forms neural crest cells. It is possible that BMP signaling in the dorsal neural tube induces a cascade of signals that influence G1/S transition, perhaps by upregulating cyclin D1. Alternatively, independent pathways downstream of BMP signaling and the cell cycle may converge on common downstream targets to initiate delamination.

Summary of Neural Crest Induction

Neural crest induction is a multistep, multisignal process that can be divided into three distinct phases. Firstly, the neural plate border is induced during gastrulation, probably by intermediate levels of BMP activity, and with the involvement of Dlx transcription factors. Secondly, Wnt and/or FGF signals from surrounding tissues (paraxial mesoderm and non-neural ectoderm) posteriorize the neural plate border and induce neural crest cell precursors within it. Finally, BMP activity in the dorsal neural tube induces RhoB in a subset of neural crest cell precursors. After G1/S transition, these cells undergo an epithelial–mesenchymal transition, delaminate from the neuroepithelium as neural crest cells, and migrate into the periphery.

As neural crest cells delaminate from the neuroepithelium, they are faced with very different mesodermal environments depending on their axial level. In the head, they encounter the apparently disorganized cranial paraxial mesenchyme, while in the trunk, the paraxial mesoderm is segmented into repeating blocks, the somites. In both head and trunk, however, neural crest cells follow ordered pathways to their target sites, where they differentiate into an impressive array of different derivatives. The mechanisms underlying this migration are discussed in the following section.

NEURAL CREST MIGRATION

Experimental Approaches

Two main experimental approaches have been used to map the migration pathways and, concurrently, define the derivatives

of the neural crest. First, ablation studies have been performed, to determine what cell types and tissues are lacking as a result. Although such experiments yielded a wealth of information, particularly from fish and amphibians, drawbacks included the possibility of regulation to restore the missing cells, and indirect effects on other tissues. The second approach has been to label the neural folds, including premigratory neural crest cell precursors: Labeled neural crest cells delaminating into the periphery can be distinguished from surrounding unlabeled cells. Early studies in amphibian embryos employed vital dyes to label donor embryos, from which neural folds were explanted and grafted into unlabeled host embryos (e.g., Detwiler, 1937). Heterospecific grafts were also used extensively in amphibians as differences in pigmentation and/or cell size enabled donor and host tissues to be distinguished. Such grafts have also been combined with staining techniques that reveal differences in nuclear morphology (e.g., Sadaghiani and Thiébaud, 1987; Krotoski et al., 1988).

Tritiated thymidine labeling of the nuclei of donor embryos, followed by grafting of labeled neural folds into unlabeled hosts, was introduced in the 1960s for the chick (Weston, 1963) and immediately applied in amphibians (Chibon, 1964). This method was used in avian embryos for about 12 years (e.g., Johnston, 1966; Noden, 1975). It was superseded, however, by Le Douarin's discovery that the quail nucleolus is associated with a large mass of heterochromatin, enabling it to be distinguished clearly from chick nuclei after appropriate staining (Le Douarin, 1969, 1973). Hence, quail neural folds could be grafted into chick hosts, and the fate of the donor quail cells followed throughout development, up to and including hatching (although graft rejection occurs eventually). This technique was used in a series of elegant fate-mapping studies to define all the derivatives of the neural crest in the avian embryo along the length of the neuraxis (e.g., Le Douarin and Teillet, 1973, 1974; Teillet, 1978; Noden, 1978a, b) (reviewed in Le Douarin and Kalcheim, 1999). Today, a quail-specific antibody enables easier identification of grafted quail cells within the chick host, and the quail–chick chimera technique is still commonly used to study neural crest cell fate, migration, and potential (e.g., Baker et al., 1997; Catala et al., 2000; Etchevers et al., 2001).

Migrating neural crest cells have also been followed using monoclonal antibodies, such as the HNK1 antibody in chick and rat embryos (e.g., Rickmann et al., 1985; Bronner-Fraser, 1986; Erickson et al., 1989). Modern, nontoxic vital dyes have been extensively used to map neural crest cell migration pathways and derivatives *in situ*, avoiding any risk of artifacts introduced by invasive surgery or differences in behavior between donor and host cells. The lipophilic dye DiI can be injected into the lumen of the neural tube to label all neural tube cells, including premigratory neural crest cells, which can subsequently be followed as they migrate through the periphery (e.g., Serbedzija et al., 1989, 1990; Collazo et al., 1993). Time-lapse *in ovo* confocal microscopy, combined with DiI labeling, has also enabled migrating hindbrain neural crest cells to be followed *in vivo* at high resolution (e.g., Kulesa and Fraser, 2000). Membrane-impermeant dyes, such as lysinated rhodamine dextran, can be injected into individual neural crest cell precursors and migrating neural crest cells *in vivo*, allowing the progeny of single cells to be followed during development (Bronner-Fraser and Fraser, 1988, 1989; Fraser and Bronner-Fraser, 1991). Retroviral-mediated gene transfer has also enabled the clonal analysis of the progeny of single neural crest cells *in vivo* (Frank and Sanes, 1991). In mice, the fate of migrating cranial neural crest cells has been followed by using Cre–Lox transgenic technology to activate constitutive β-galactosidase expression under the control of the *Wnt1* promoter (Chai *et al.*, 2000).

Together, these different cell-labeling approaches have enabled a detailed picture to be drawn of the migration pathways followed by neural crest cells through the periphery.

Migration Pathways of Cranial Neural Crest Cells

Cranial neural crest cells migrate beneath the surface ectoderm, above the paraxial cephalic mesoderm (see Figs. 3 and 4B), although a few cells penetrate the paraxial mesoderm.

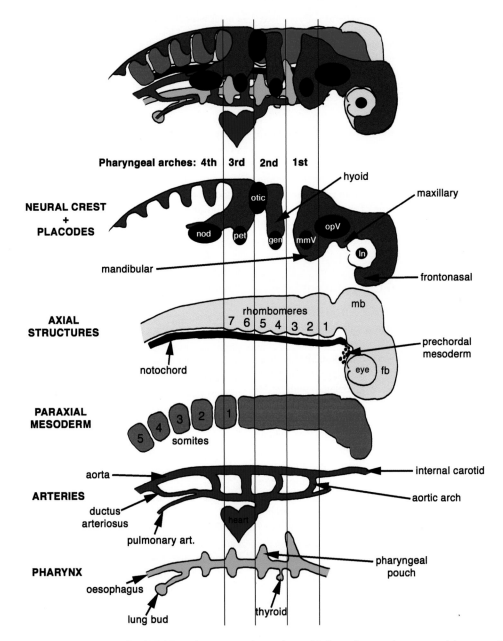

FIGURE 3. Schematic lateral views of a generalized 20–30 somite-stage amniote embryo with the surface ectoderm removed (except to show the positions of the cranial ectodermal placodes). Each tissue type from the embryo at the top is shown separately below, illustrating the relative positions of the migrating neural crest, placodes (filled black circles), axial structures, paraxial mesoderm, arteries, and pharyngeal endoderm. The olfactory placodes cannot be seen in this view. The vertical lines indicate which regions are in register with each pharyngeal arch. Redrawn from Noden (1991). art., artery; fb, forebrain; gen, geniculate; ln, lens; mb, midbrain; mmV, maxillomandibular trigeminal; nod, nodose; opV, ophthalmic trigeminal; pet, petrosal.

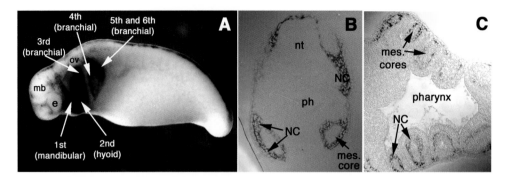

FIGURE 4. Cranial neural crest migration streams in the axolotl visualized by *in situ* hybridization for the *AP-2* gene. (A) Stage 29 (16-somite stage) axolotl embryo showing six *AP-2*[+] neural crest migration streams in the head (mandibular, hyoid, and four branchial streams). Premigratory trunk neural crest cell precursors can be seen as a dark line at the dorsal midline of the embryo. (B) Transverse section through a stage 26 (10–11 somite stage) axolotl embryo showing *AP-2*[+] neural crest cells (NC) moving out from the neural tube (nt) and down to surround the mesodermal core of the mandibular arch. (C) Horizontal section through the pharynx of a stage 34 (24–25 somite stage) axolotl embryo showing *AP-2*[+] neural crest cells (NC) around the mesodermal cores of each pharyngeal arch. e, eye; mb, midbrain; mes., mesodermal; NC, neural crest; nt, neural tube; ov, otic vesicle; ph, pharynx. Staging follows Bordzilovskaya *et al.* (1989). All photographs courtesy of Daniel Meulemans, California Institute of Technology, United States of America.

They migrate as coherent populations; indeed, at the hindbrain level, migrating neural crest cells are connected in chains by filopodia (Kulesa and Fraser, 1998, 2000). They populate the entire embryonic head and form much of the neurocranium (brain capsule) and all of the splanchnocranium (viscerocranium or visceral skeleton), that is, the skeleton of the face and pharyngeal arches. They also form neurons and satellite glia in cranial sensory and parasympathetic ganglia, Schwann cells, endocrine cells, and epidermal pigment cells (see Table 1).

Pharyngeal Arches and Neural Crest Streams

The patterning of cranial neural crest cell migration is intimately bound up with the segmental nature of both the hindbrain (rhombomeres; see Chapter 3) and the periphery (pharyngeal arches). Pharyngeal arches are also known as branchial arches, from the Latin *branchia* ("gill"), because in aquatic vertebrates the more caudal arches are associated with gills. However, "pharyngeal" is the more appropriate term, because all arches form in the pharynx, but not all arches support gills. Pharyngeal arches form between the pharyngeal pouches, which are outpocketings of the pharyngeal (fore-gut) endoderm that fuse with the overlying ectoderm to form slits in the embryo (see Fig. 3). The pharyngeal slits form the gill slits in aquatic vertebrates; the first pharyngeal slit in tetrapods forms the middle ear cavity. Paraxial mesoderm in the core of the pharyngeal arches (Figs. 4B, C) gives rise to striated muscles. Cranial neural crest cells migrate subectodermally to populate the space around the mesodermal core (Figs. 4B, C), where they give rise to all skeletal elements of the arches, and the connective component of the striated muscles.

The first pharyngeal arch is the mandibular, which forms the mandible (lower jaw). The second arch is the hyoid, which forms jaw suspension elements in fish but middle ear bones in tetrapods, together with parts of the hyoid apparatus/bone (supporting elements for the tongue and roof of the mouth). Varying numbers of arches follow more caudally. The third and fourth

arches also contribute to the hyoid apparatus and to laryngeal cartilages in tetrapods; in mammals, the fourth arch forms thyroid cartilages. More caudal arches in fish and aquatic amphibians support gills and form laryngeal cartilages in tetrapods. Importantly, pharyngeal arch formation *per se*, and the regionalization of gene expression patterns within them (excluding those of neural crest-derived structures) are both independent of neural crest cell migration (Veitch *et al.*, 1999; Gavalas *et al.*, 2001).

Cranial neural crest cells migrate in characteristic streams associated with the pharyngeal arches (Figs. 3 and 4A). There are three or more major migration streams in all vertebrates. The first stream, from the midbrain and rhombomeres 1 and 2 (r1,2), populates the first (mandibular) arch; the second stream, from r3–5, populates the second (hyoid) arch, and the third, from r5–7, populates the third arch (Fig. 4). In fish and amphibians, additional caudal streams populate the remaining arches: The axolotl, for example, has four branchial (gill) arches caudal to the mandibular and hyoid arches (Fig. 4A). How is the migrating neural crest cell population sculpted to achieve these different streams?

Separation of the First, Second, and Third Neural Crest Streams (Amniotes)

In chick and mouse embryos, there are neural crest cell-free zones adjacent to r3 and r5 (Fig. 3). It was suggested that neural crest cells at r3 and r5 die by apoptosis to generate adjacent neural crest-free zones (Graham *et al.*, 1993). However, both r3 and r5 give rise to neural crest cells during normal development in both chick and mouse, though r3 generates fewer neural crest cells than other rhombomeres (Sechrist *et al.*, 1993; Köntges and Lumsden, 1996; Kulesa and Fraser, 1998; Trainor *et al.*, 2002b). Neural crest cells from r3 and r5 migrate rostrally and caudally along the neural tube to join the adjacent neural crest streams; that is, r3-derived neural crest joins the r1,2 (first arch) and r4 (second arch) streams, while r5-derived neural crest joins the r4 (second arch) and r6,7 (third arch) streams (Sechrist *et al.*, 1993; Köntges and Lumsden, 1996; Kulesa and Fraser, 1998; Trainor *et al.*, 2002b). This deviation of the r3 and r5 neural crest generates the neural crest-free zones adjacent to r3 and r5, forming the three characteristic streams in birds and mice (Fig. 3). Hence, the first arch is populated by neural crest cells from the midbrain and r1–3, the second arch by neural crest cells from r3–5, and the third arch by neural crest cells from r5–7.

Neural crest cells leaving r5 are confronted by the otic vesicle (Fig. 3), which provides an obvious mechanical obstacle to migration. No such obstacle exists at r3; instead, paraxial mesoderm at the r3 level is inhibitory for neural crest cell migration, at least in amniotes (Farlie *et al.*, 1999). This inhibition is lost in mice lacking ErbB4, a high-affinity receptor for the growth factor Neuregulin1 (NRG1) (Golding *et al.*, 1999, 2000). ErbB4 is expressed in the r3 neuroepithelium, while NRG1 is expressed in r2; ErbB4 activation by NRG1 may somehow signal the production of inhibitory molecules in r3-level paraxial mesoderm (Golding *et al.*, 2000). A few hours after removing either r3 itself, or the surface ectoderm at the r3 level, r4 neural crest cells move aberrantly into the mesenchyme adjacent to r3, suggesting that both r3 itself and r3-level surface ectoderm are necessary to inhibit neural crest cell migration (Trainor *et al.*, 2002b).

Separation of the Third and Fourth Streams (Anamniotes)

Fish and amphibians also have additional cranial neural crest streams that populate the more caudal pharyngeal arches. In amphibians, at least, neural crest cells destined for different arches do not separate into different streams adjacent to the neural tube; instead, separation occurs at or just before entry into the arches (Robinson *et al.*, 1997). Another difference in *Xenopus*, in which the otic vesicle is adjacent to r4 rather than r5, is that all r5-derived neural crest cells seem to migrate into the third arch (Robinson *et al.*, 1997).

In *Xenopus*, migrating neural crest cells in the third and fourth cranial neural crest streams are separated by repulsive migration cues. These are mediated by the ephrin family of ligands, acting on their cognate Eph-receptor tyrosine kinases (Smith *et al.*, 1997; Helbling *et al.*, 1998; reviewed in Robinson *et al.*, 1997; for a general review of ephrins and Eph family members, see Kullander and Klein, 2002). The transmembrane ligand ephrinB2 is expressed in second arch neural crest cells and mesoderm. One ephrinB2 receptor, EphA4, is expressed in third arch neural crest cells and mesoderm, while a second ephrinB2 receptor, EphB1, is expressed in both third and fourth arch neural crest cells and mesoderm (Smith *et al.*, 1997). Inhibition of EphA4/EphB1 function using truncated receptors results in the aberrant migration of third arch neural crest cells into the second and fourth arches. Conversely, ectopic activation of EphA4/EphB1 (by overexpressing ephrinB2) results in the scattering of third arch neural crest cells into adjacent territories (Smith *et al.*, 1997). Hence, the complementary expression of ephrinB2 and its receptors in the second and third arches, respectively, is required to prevent mingling of second and third arch neural crest cells before they enter the arches. Since ephrinB2 is also expressed in second

arch mesoderm, it is also required to target third arch neural crest cells correctly away from the second arch and into the third arch. *EphrinB2*-null mice also show defects in cranial neural crest cell migration, particularly of second arch neural crest cells, which scatter and do not invade the second arch (Adams *et al.*, 2001).

Migrating *Xenopus* cranial neural crest cells also express EphA2; overexpression of a dominant negative (kinase-deficient) EphA2 receptor similarly leads to the failure of the third and fourth neural crest streams to separate, as neural crest cells from the third stream migrate posteriorly (Helbling *et al.*, 1998).

Neural Crest Streams and Cranial Skeleto-Muscular Patterning

Cranial neural crest cells form not only many of the skeletal elements of the head, but also the connective component of the striatal muscles that are attached to them (see Table 1). When the long-term fate of neural crest cells arising from the midbrain and each rhombomere was mapped using quail-chick chimeras, it was found that each rhombomeric population forms the connective components of specific muscles, together with their respective attachment sites on the neurocranium and splanchnocranium (Köntges and Lumsden, 1996). Cranial muscle connective tissues arising from a given rhombomere attach to skeletal elements arising from the same initial neural crest population, explaining how evolutionary changes in craniofacial skeletal morphology can be accommodated by the attached muscles (Köntges and Lumsden, 1996). Similar results have also been obtained in frog embryos, where connective tissue components of individual muscles of either of the first two arches originate from the neural crest migratory stream associated with that arch (Olsson *et al.*, 2001). Hence, the streaming of cranial neural crest cells into the different pharyngeal arches is important for patterning not only skeletal elements, but also their associated musculature.

Migration Pathways of Trunk Neural Crest Cells

The migration pathways of trunk neural crest cells have been most extensively studied in avian embryos (e.g., Weston, 1963; Rickmann *et al.*, 1985; Bronner-Fraser, 1986; Teillet *et al.*, 1987). As described in this section, neural crest cells only leave the neural tube opposite newly epithelial somites (Fig. 5A) (for reviews of somite formation and maturation, see Stockdale *et al.*, 2000; Pourquié, 2001). Here, they enter a cell-free space that is rich in extracellular matrix. They only migrate into the somites at a level approximately 5–9 somites rostral to the last-formed somite, where the somites first become subdivided into different dorsoventral compartments, the sclerotome and dermomyotome (Fig. 5B) (Guillory and Bronner-Fraser, 1986). The sclerotome is formed when the ventral portion of the epithelial somite undergoes an epithelial–mesenchymal transition to form loose mesenchyme. This mesenchyme will eventually form the cartilage and bone of the ribs and axial skeleton. The dorsal somitic compartment, the dermomyotome, remains epithelial, and will eventually form dermis, skeletal muscle, and vascular derivatives.

There are two main neural crest cell migration pathways in the avian trunk (Fig. 5C): (1) a ventral pathway between the neural tube and somites, followed by neural crest cells that eventually give rise to dorsal root ganglia, Schwann cells, sympathetic ganglia, and (at somite levels 18–24 in birds) adrenal chromaffin cells, and (2) a dorsolateral pathway between the somite and the overlying ectoderm, followed by neural crest cells that eventually form melanocytes.

Ventral Migration Pathway

In the chick, neural crest cells that delaminate opposite epithelial somites initially migrate ventrally between the somites. Once the sclerotome forms, they migrate exclusively through the rostral half of each sclerotome, leading to a segmental pattern of

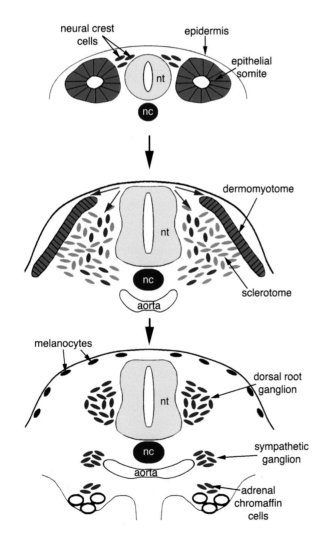

FIGURE 5. Schematic showing trunk neural crest cell migration pathways and derivatives (also see Fig. 1C). Neural crest cells migrate ventrally through the sclerotome to form neurons and satellite glia in the dorsal root ganglia and sympathetic ganglia, chromaffin cells in the adrenal gland (and Schwann cells on the ventral root; not shown). Neural crest cells also migrate dorsolaterally beneath the epidermis to form melanocytes. nc, notochord; nt, neural tube.

migration (Rickmann *et al.*, 1985; Bronner-Fraser, 1986). This pathway is almost identical to that followed by motor axons as they grow out from the neural tube, shortly after neural crest cells begin their migration (Rickmann *et al.*, 1985). Mouse neural crest cells are similarly restricted to the rostral sclerotome (Serbedzija *et al.*, 1990).

Neural crest cells that remain within the rostral sclerotome aggregate to form the dorsal root ganglia (primary sensory neurons and satellite glial cells), while those that move further ventrally form postganglionic sympathetic neurons (Fig. 8; section The Autonomic Nervous System: An Introduction) and adrenal chromaffin cells (Fig. 5C). The restriction of neural crest cells to the rostral half of each somite therefore leads to the segmental distribution of dorsal root ganglia; as will be seen in the section on Molecular Guidance Cues for Trunk Neural Crest Cell Migration, it results from the presence of repulsive migration cues in the caudal sclerotome.

Neural crest cells that delaminate opposite the caudal half of a somite migrate longitudinally along the neural tube in both directions. Once they reach the rostral half either of their own somite, or of the adjacent (immediately caudal) somite, they enter the sclerotome (Teillet *et al.*, 1987). Hence, each dorsal root ganglion is derived from neural crest cells emigrating at the same somite level and from one somite anterior to that level. In contrast, each sympathetic ganglion is derived from neural crest cells originating from up to six somite-levels of the neuraxis: This is approximately equal to the numbers of spinal cord segments contributing to the preganglionic sympathetic neurons that innervate each ganglion (see Fig. 8) (Yip, 1986).

There are some differences in the ventral neural crest migration pathway between different vertebrates. In fish and amphibians, the somites are mostly myotome, with very little sclerotome. In these animals, the ventral migration pathway is essentially a medial migration pathway, between the somites and the neural tube/notochord. In *Xenopus*, neural crest cells following this pathway give rise to dorsal root ganglia, sympathetic ganglia, adrenomedullary cells, and also pigment cells (Krotoski *et al.*, 1988; Collazo *et al.*, 1993). This is also a segmental migration, but in this case, the neural crest cells migrate between the neural tube and the caudal half of each somite (Krotoski *et al.*, 1988; Collazo *et al.*, 1993). The ventral pathway is the main pathway followed by pigment cell precursors in *Xenopus*; only a few pigment cells follow the dorsolateral pathway beneath the ectoderm (Krotoski *et al.*, 1988; Collazo *et al.*, 1993). In zebrafish, neural crest cells enter the medial pathway at any rostrocaudal location; however, they subsequently converge toward the middle of the somite so that their ventral migration is restricted to the region halfway between adjacent somite boundaries (Raible *et al.*, 1992). Rostral sclerotome precursors and motor axons also follow this pathway toward the center of the somite. However, rostral sclerotome cells are not required for this convergence of neural crest cells and motor axons, suggesting that unlike the situation in avian embryos (section Molecular Guidance Cues for Trunk Neural Crest Cell Migration), neural crest and motor axon guidance cues are not derived from the sclerotome (Morin-Kensicki and Eisen, 1997).

Dorsolateral Migration Pathway

Neural crest cells that migrate along the dorsolateral pathway, between the somites and the ectoderm, give rise to epidermal melanocytes in all vertebrates. In chick embryos, melanocytes only differentiate after they have invaded the ectoderm, while in amphibians, melanocytes often differentiate during migration (see, e.g., Keller and Spieth, 1984). In *Xenopus*, the subectodermal pathway is only a minor pathway for pigment cells, as most pigment cell precursors follow the ventral pathway (Krotoski *et al.*, 1988; Collazo *et al.*, 1993). However, in most amphibians, such as the axolotl, the dorsolateral pathway is a major pathway for pigment cell precursors (see, e.g., Keller and Spieth, 1984).

By injecting DiI into the lumen of the neural tube at progressively later stages, the fate of later-migrating neural crest cells can be specifically examined (Serbedzija *et al.*, 1989, 1990). The earliest injection labels all neural crest cells, while subsequent injections label neural crest cells leaving the neural tube at progressively later times. These experiments showed that neural crest cell derivatives are "filled" in a ventral–dorsal order, since the label is progressively lost first from sympathetic ganglia, and then from dorsal root ganglia, in both mouse and chick embryos (Serbedzija *et al.*, 1989, 1990). The last cells to leave the neural tube exclusively migrate along the dorsolateral pathway. (The same ventral–dorsal filling of derivatives is also seen in the head, where early-migrating mesencephalic neural crest cells form both dorsal and ventral derivatives, while late-migrating cells exclusively form dorsal derivatives; Baker *et al.*, 1997.)

Entry onto the dorsolateral pathway is delayed relative to entry onto the ventral pathway in the chick and zebrafish. In the chick, trunk neural crest cells only begin migrating dorsolaterally 24 hr after migration has begun on the ventral pathway (Erickson *et al.*, 1992; Kitamura *et al.*, 1992). This is concomitant with the dissociation of the epithelial dermomyotome to form a mesenchymal dermis. (In the vagal region of chick embryos, however, neural crest cells immediately follow the dorsolateral pathway, via which they reach the pharyngeal arches; Tucker *et al.*, 1986; Kuratani and Kirby, 1991; Reedy *et al.*, 1998.) In the zebrafish, there is also a delay of several hours before neural crest cells follow the dorsolateral pathway (Raible *et al.*, 1992; Jesuthasan, 1996). In contrast, neural crest cells follow both dorsolateral and ventral pathways simultaneously in the mouse (Serbedzija *et al.*, 1990), while in the axolotl, the dorsolateral pathway is followed before the ventral pathway (Löfberg *et al.*, 1980).

In the zebrafish, the lateral somite surface triggers collapse and retraction of neural crest cell protrusions but not Rohon-Beard growth cones, suggesting that the delay in entry onto the dorsolateral pathway is mediated by a repulsive cue on the dermomyotome that acts specifically on neural crest cells (Jesuthasan, 1996). In the chick trunk, inhibitory glycoconjugates, including peanut agglutinin-binding molecules and chondroitin-6-sulfate proteoglycans, are expressed on the dorsolateral pathway during the period of exclusion of neural crest cells; their expression decreases concomitant with neural crest cell entry (Oakley *et al.*, 1994). Dermomyotome ablation abolishes expression of these molecules and accelerates neural crest cell entry onto the dorsolateral pathway (Oakley *et al.*, 1994). Chondroitin-sulfate proteoglycans and the hyaluronan-binding proteoglycan aggrecan are also found in the perinotochordal space, which similarly excludes neural crest cells (see, e.g., Bronner-Fraser, 1986; Pettway *et al.*, 1996; Perissinotto *et al.*, 2000). It has also been suggested that, at least in the chick, only melanocyte precursors are able to enter the dorsolateral pathway (Erickson and Goins, 1995). However, this cannot be an absolute restriction, since multipotent neural crest cells (able to form not only melanocytes, but also sensory and autonomic neurons) have been isolated from the trunk epidermis of quail embryos (Richardson and Sieber-Blum, 1993).

Other Migration Pathways in the Trunk

In amphibians, neural crest cells also migrate dorsally to populate the dorsal fin (Löfberg *et al.*, 1980; Krotoski *et al.*, 1988; Collazo *et al.*, 1993). In *Xenopus*, DiI-labeling showed the existence of two migration pathways toward the ventral fin (Collazo *et al.*, 1993). One pathway leads along the neural tube and through the dorsal fin around the tip of the tail, while the other leads ventrally toward the anus and directly down the presumptive enteric region to the ventral fin (Collazo *et al.*, 1993).

Molecular Guidance Cues for Trunk Neural Crest Cell Migration

Various extracellular matrix molecules that are permissive for neural crest migration are prominent along neural crest migration pathways, including fibronectin, laminin, and collagen types I, IV, and VI (reviewed in Perris, 1997; Perris and Perissinotto, 2000). Function-blocking antibodies and antisense oligonucleotide experiments targeted against the integrin receptors for these molecules perturb neural crest cell migration (reviewed in Perris and Perissinotto, 2000). PG-M/versicans (major hyaluronan-binding proteoglycans) are expressed by tissues lining neural crest cell migration pathways and may be conducive to neural crest cell migration (Perissinotto *et al.*, 2000).

The most important guidance cues for neural crest cells seem to be repulsive. As discussed in the section on Dorsolateral Migration Pathway inhibitory extracellular matrix molecules such as chondroitin-sulfate proteoglycans and aggrecan are expressed in regions that do not permit neural crest cell entry, such as the perinotochordal space. Most molecular information is available about guidance cues that act to restrict neural crest cell migration to the rostral sclerotome in chick and mouse embryos (reviewed in Kalcheim, 2000; Krull, 2001). Microsurgical rotation of the neural tube or segmental plate mesoderm showed that the guidance cues responsible for the rostral restriction of neural crest cell migration, and also sensory and motor axon growth, reside in the mesoderm, not in the neural tube (Keynes and Stern, 1984; Bronner-Fraser and Stern, 1991). Similarly, when compound somites made up only of rostral somite-halves are surgically created, giant fused dorsal root ganglia form, while very small, irregular dorsal root ganglia form when only caudal halves are used (Kalcheim and Teillet, 1989). This also demonstrates the importance of the mesoderm in segmenting trunk neural crest cell migration. The presence of alternating rostral–caudal somite halves is also important for the correct formation of the sympathetic ganglionic chains (Goldstein and Kalcheim, 1991).

Many different molecules that are localized to the caudal sclerotome have been proposed as candidate repulsive cues for neural crest cells (see Krull, 2001). It is probable that multiple cues are present and act redundantly. Peanut agglutinin-binding molecules seem to be important, since application of peanut agglutinin leads to chick neural crest cell migration through both rostral and caudal half-sclerotomes; however, their identity is unknown (Krull et al., 1995). F-spondin, an extracellular matrix molecule originally isolated in the floor-plate, is also involved: Overexpression of F-spondin in the chick inhibits neural crest cell migration into the somite, while anti-F-spondin antibody treatment enables neural crest cell migration into previously inhibitory domains, including the caudal sclerotome (Debby-Brafman et al., 1999). Semaphorin 3A (Sema3A; collapsin1), a secreted member of the semaphorin family of proteins that act as (primarily) repulsive guidance cues for axon growth cones (reviewed in Yu and Bargmann, 2001), is also expressed in the caudal sclerotome (Eickholt et al., 1999). Migrating neural crest cells express the Sema3A receptor, Neuropilin1, and selectively avoid Sema3A-coated substrates in vitro (Eickholt et al., 1999). Mice mutant for either sema3A or neuropilin1 show normal neural crest migration through the caudal sclerotome (Kawasaki et al., 2002), but it is possible that other related molecules compensate for their loss.

Finally, as in the cranial neural crest (section Migration Pathways of Cranial Neural Crest Cells), ephrin–Eph interactions are also important (reviewed in Robinson et al., 1997; Krull, 2001). In the chick, trunk neural crest cells express the receptor EphB3, while its transmembrane ligand, ephrinB1, is localized to the caudal sclerotome (Krull et al., 1997). Neural crest cells enter both rostral and caudal sclerotomes in explants treated with soluble ephrinB1 (Krull et al., 1997). Similar ephrin–Eph interactions are also important in restricting rat neural crest cells to the rostral somite: Both ephrinB1 and ephrinB2 are expressed in the caudal somite, while neural crest cells express the receptor EphB2 and are repelled by both ligands (Wang and Anderson, 1997). Ephrin B ligands are also expressed in the dermomyotome in the chick: these seem to repel EphB-expressing neural crest cells from the dorsolateral pathway at early stages of migration, but promote entry onto the dorsolateral pathway at later stages, particularly of melanoblasts (Santiago and Erickson, 2002).

Importantly, ephrins do not simply block migration, but act as a directional cue. Eph[+] neural crest cells will migrate over a uniform ephrin[+] substrate, but when given a choice between ephrin[+] and ephrin-negative substrates, they preferentially migrate on the latter (Krull et al., 1997; Wang and Anderson, 1997).

Migration Arrest at Target Sites

Surprisingly little is known about the signals that control the arrest of neural crest cells at specific target sites.

FGF2 and FGF8 have been shown to promote chemotaxis of mesencephalic neural crest cells in vitro; both of these molecules are expressed in tissues in the pharyngeal arches, although an in vivo role has not been demonstrated (Kubota and Ito, 2000). Sonic hedgehog (Shh) in the ventral midline seems to act as a migration arrest signal for mesencephalic neural crest-derived trigeminal ganglion cells (Fedtsova et al., 2003). A local source of Shh blocks migration of these cells in chick embryos, while in Shh knockout mice, trigeminal precursors migrate toward the midline and condense to form a single fused ganglion (Fedtsova et al., 2003). Shh has also been shown to inhibit dispersal of avian trunk neural crest cells in vitro (Testaz et al., 2001), so it is possible that Shh may be a general migration arrest signal for neural crest cells.

Glial cell line-derived neurotrophic factor (GDNF), a ligand for the receptor tyrosine kinase Ret, has chemoattractive activity for Ret-expressing enteric neural crest cell precursors in the gut (Young et al., 2001). GDNF is expressed throughout the gut mesenchyme; it may promote neural crest cell migration through the gut and prevent neural crest cells leaving the gut to colonize other tissues, although this has not been proven (Young et al., 2001).

Sema3A, described in the last section as a potential repulsive guidance cue for neural crest cells migrating through the sclerotome (Eickholt et al., 1999), is required for the accumulation of sympathetic neuron precursors around the dorsal aorta (Kawasaki et al., 2002). In mice mutant either for sema3A or the gene encoding its receptor, neuropilin1, neural crest cells migrate normally through the caudal sclerotome, but sympathetic neuron precursors are widely dispersed, for example in the forelimb, where sema3A is normally expressed (Kawasaki et al., 2002). Sema3A also promotes the aggregation of sympathetic neurons in culture, suggesting a potential role for Sema3A in clustering sympathetic neuron precursors at the aorta (Kawasaki et al., 2002). Since sema3A is expressed in the somites (in the dermomyotome as well as in the caudal sclerotome) and in the forelimb, it is possible that secreted Sema3A forms a dorsoventral gradient, trapping sympathetic neuron precursors by the aorta, at the ventral point of the gradient (Kawasaki et al., 2002).

Summary of Neural Crest Migration

Neural crest cell migration pathways in the head and trunk are generally conserved across all vertebrates. Distinct streams of migrating cranial neural crest cells populate different pharyngeal arches. These streams are formed at least partly via the action of repulsive guidance cues from the mesoderm, including an unidentified ErbB4-regulated inhibitory cue in r3-level mesoderm in amniotes, and repulsive ephrin–Eph interactions between neural crest cells and pharyngeal arch mesoderm in amphibians. In the amniote trunk, the restriction of neural crest cell migration to the rostral sclerotome is mediated by multiple repulsive cues from the caudal sclerotome, including ephrins. This restriction is essential for the segmentation of the PNS in the trunk. Although relatively little is known about how migration arrest is controlled, a few potential molecular cues have been identified. These include Sema3A, which is required for the accumulation of sympathetic neuron precursors at the dorsal aorta.

NEURAL CREST LINEAGE DIVERSIFICATION

The astonishing diversity of neural crest cell derivatives has always been a source of fascination, and much effort has been devoted to understanding how neural crest lineage diversification is achieved (reviewed in Le Douarin and Kalcheim, 1999; Anderson, 2000; Sieber-Blum, 2000; Dorsky et al., 2000a; Sommer, 2001). The formation of different cell types in different locations within the embryo raises two distinct developmental questions (Anderson, 2000). First, how are different neural crest cell derivatives generated at distinct rostrocaudal axial levels? During normal development, for example, only cranial neural crest cells give rise to cartilage, bone, and teeth; only vagal and lumbosacral neural crest cells form enteric ganglia; and only a subset of trunk neural crest cells form adrenal chromaffin cells (see Table 1). Are these axial differences in neural crest cell fate determined by environmental differences or by intrinsic differences in the neural crest cells generated at different axial levels? Second, how are multiple different neural crest cell derivatives generated at the same axial level? For example, vagal neural crest cells form mesectodermal derivatives, melanocytes, endocrine cells, sensory neurons, and all three autonomic neuron subtypes (parasympathetic, sympathetic, and enteric). How is this lineage diversification achieved? These two questions will be examined in turn.

Axial Fate-Restriction Does Not Generally Reflect Restrictions in Potential

The restricted fate of different neural crest cell precursor populations along the neuraxis (see Table 1) has been extensively tested in avian embryos using the quail-chick chimera technique. Neural fold fragments from one axial level of quail donor embryos were grafted into different axial levels of chick host embryos (reviewed in Le Douarin and Kalcheim, 1999). These experiments revealed that, in general, neural crest cell precursors from all axial levels are plastic, as a population; that is, a premigratory population from one axial level can form the neural crest cell derivatives characteristic of any other axial level. For example, caudal diencephalic neural crest precursors, which do not normally form neurons or glia, will contribute appropriately to the parasympathetic ciliary ganglion and proximal cranial sensory ganglia after grafts to the mesencephalon or hindbrain (Noden, 1975, 1978b). Trunk neural crest precursors, which do not normally form enteric neurons, will colonize the gut and form enteric neurons, expressing appropriate neurotransmitters, when they are grafted into the vagal region (Le Douarin and Teillet, 1974; Le Douarin et al., 1975; Fontaine-Pérus et al., 1982; Rothman et al., 1986). Cranial and vagal neural crest cells, which do not normally form catecholaminergic derivatives, can form adrenergic cells both in sympathetic ganglia and the adrenal glands, when grafted to the "adrenomedullary level" (somites 18–24) of the trunk (Le Douarin and Teillet, 1974). These results suggest that axial differences in neural crest fate reflect axial differences in the environment, not intrinsic differences in the neural crest cells themselves, at least at the population level.

There are some exceptions to this general rule, however. For example, the most caudal neural crest cells in the chick embryo (those derived from the level of somites 47–53), only form melanocytes and Schwann cells during normal development (Catala et al., 2000). Furthermore, when tested both by in vitro culture and heterotopic grafting, they seem to lack the potential to form neurons (Catala et al., 2000).

Until very recently, it was accepted that trunk neural crest cells are intrinsically different from cranial neural crest cells in that they lack the potential to form cartilage. Trunk neural crest cells do not form cartilage when trunk neural folds are grafted in place of cranial neural folds in either amphibian or avian embryos (Raven, 1931, 1936; Chibon, 1967b; Nakamura and Ayer-Le Lièvre, 1982). One study suggested that trunk neural crest cells do not migrate into the pharyngeal arches after such grafts in the axolotl (Graveson et al., 1995) and hence are not exposed to cartilage-inducing signals from the pharyngeal endoderm. Even when trunk neural crest cells are cocultured in vitro with pharyngeal endoderm, however, under the same conditions that elicit cartilage from cranial neural crest cells, they do not form cartilage (Graveson and Armstrong, 1987; Graveson et al., 1995). Nonetheless, a study in the axolotl using DiI-labeled trunk neural folds found some aberrant migration by trunk neural crest cells in the head, and incorporation of a few trunk neural crest cells into cartilaginous skeletal elements (Epperlein et al., 2000).

Cervical and thoracic trunk neural crest cells isolated from avian embryos will eventually form both bone and cartilage when cultured for many days in a medium commonly used for growing these tissues (McGonnell and Graham, 2002; Abzhanov et al., 2003). Interestingly, this late differentiation in vitro correlates temporally with a downregulation of Hox gene expression in a subset of trunk neural crest cells in long-term culture (Abzhanov et al., 2003). This alteration in Hox expression may enable trunk neural crest cells to respond to chondrogenic signals (section Cranial Neural Crest Cells Are Not Prepatterned). Furthermore, when implanted as loosely packed aggregates directly into the mandibular and maxillary primordia, trunk neural crest cells were found scattered in multiple cartilaginous elements, including Meckel's cartilage and the sclera of the eyes (McGonnell and Graham, 2002). Hence, it appears that trunk neural crest cells do have the potential to form cartilage, although this is only expressed under particular experimental conditions. Notably, the formation of cartilage in vivo is only observed when the cells are scattered among host neural crest cells, rather than when they are present as a coherent mass (McGonnell and Graham, 2002). It is possible that these scattered cells alter their Hox gene expression pattern to accord with the surrounding host neural crest cells, enabling them to respond to chondrogenic signals (section Cranial Neural Crest Cells Are Not Prepatterned).

When trunk neural crest cell precursors are substituted for the rostral vagal region of the neural tube (somite levels 1–3), they are unable to supply connective tissue to the heart to form the aorticopulmonary septum (Kirby, 1989). It is possible that, were they implanted as loose aggregates of cells in the heart region in the same manner as for the cartilage induction experiments (McGonnell and Graham, 2002), they would be able to contribute to the aorticopulmonary septum; however, this remains to be tested.

Most current evidence, therefore, supports the idea that neural crest cells are largely plastic, at least at the population level. This plasticity was, until very recently, hard to reconcile with the classical "prepatterning" model of cranial neural crest cells, which is discussed briefly in the following section. The results that led to this model, though still valid, have been reinterpreted and the idea of prepatterning discarded.

Cranial Neural Crest Cells Are Not Prepatterned

Experiments carried out in the early 1980s led to the view that cranial neural crest cell precursors are extensively prepatterned before they delaminate from the neuroepithelium (Noden, 1983). When mesencephalic neural folds (prospective first arch neural crest) were grafted more caudally to replace hindbrain neural folds (prospective second arch neural crest) (see Fig. 3), a second set of jaw skeletal derivatives developed in place of the normal second (hyoid) arch derivatives (Noden, 1983). Moreover, anomalous first arch-type muscles were associated with the graft-derived first arch skeletal elements in the second arch (Noden, 1983). These experiments were interpreted as suggesting that patterning information for pharyngeal arch-specific skeletal and muscular elements is inherent in premigratory cranial neural crest cells (Noden, 1983).

This model has persisted until very recently. However, accumulating evidence suggests that although the results on which the model is based are valid, the original interpretation is incorrect. Given that this evidence pertains to skeletal patterning, rather than to the development of the PNS, there is insufficient space in this chapter to go into the evidence itself. The main thrust of the new results, however, is that cranial neural crest cells do not carry patterning information into the pharyngeal arches. Rather, they are able to respond to environmental cues from pharyngeal arch tissues, in particular pharyngeal endoderm (reviewed in Richman and Lee, 2003; Santagati and Rijli, 2003). After heterotopic grafts of mesencephalic neural folds to the hindbrain, *Hox* gene expression in the grafted neural crest cells is repatterned by signals from the isthmic organizer at the midbrain–hindbrain border (see Chapter 3), which is included in the graft (Trainor *et al.*, 2002a). The changes in *Hox* expression affect the response of neural crest cells to different patterning signals from pharyngeal endoderm in the different arches, resulting eventually in the jaw element duplication (Couly *et al.*, 2002).

The idea of a "prepattern" within the premigratory neural crest is now largely untenable, other than as a reflection of axial-specific *Hox* expression profiles that may alter the response of migratory neural crest cells to cranial environmental cues. How, then, can interspecies chimera experiments be explained, in which the size and shape of graft-derived skeletal elements are characteristic of the donor, not the host (e.g., Harrison, 1938; Wagner, 1949; Fontaine-Pérus *et al.*, 1997; Schneider and Helms, 2003)? In a striking recent example, interspecies grafts of cranial neural crest between quail and duck embryos resulted in donor-specific beak shapes (Schneider and Helms, 2003). At first sight

this may seem to indicate intrinsic patterning information within the grafted premigratory neural crest cells. However, it is clear that reciprocal signaling occurs between neural crest cells and surrounding tissues during craniofacial development. Environmental signals control the size and shape of neural crest-derived skeletal elements (e.g., Couly *et al.*, 2002), while skeletogenic neural crest cells regulate gene expression in surrounding tissues (e.g., Schneider and Helms, 2003). Species-specific differences are likely to exist in the interpretation both of environmental signals by neural crest cells, and of neural crest-derived signals by surrounding tissues. This is presumably due to species-specific differences in the upstream regulatory elements of the relevant genes. This may explain why donor-specific skeletal elements are seen in such interspecific chimeras (and also why murine neural crest cells form teeth in response to chick oral epithelium; Mitsiadis *et al.*, 2003). However, since our current knowledge of the molecular basis of morphogenesis is scanty, this hypothesis remains to be tested explicitly.

Summary

The general view gained from heterotopic grafting and culture experiments is that, given the right conditions, neural crest cell populations from every level of the neural axis are able to form the derivatives from every other. Hence, the normal restriction in fate that is observed along the neuraxis is not due to a restriction in potential, at least at the population level, but to differences in the environment encountered by the migrating neural crest cells. These experiments do not tell us, however, how the different neural crest lineages are formed at each axial level.

Lineage Segregation at the Same Axial Level

There are two main hypotheses to explain the lineage segregation of the neural crest at a given axial level: instruction and selection. The first (*instruction*) proposes that the emigrating neural crest is a homogeneous population of multipotent cells whose differentiation is instructively determined by signals from the environment. The second (*selection*) proposes that the emigrating neural crest is a heterogeneous population of determined cells (i.e., cells that will follow a particular fate regardless of the presence of other instructive environmental signals), whose differentiation occurs selectively in permissive environments, and which are eliminated from inappropriate environments.

Both of the above hypotheses are compatible with the heterotopic grafting experiments described in the preceding section. Although in their most extreme versions these hypotheses would appear to be mutually exclusive, there is evidence from *in vivo* and *in vitro* experiments to suggest that modified versions of both operate within the neural crest. Multipotent neural crest cells that adopt different fates in response to instructive environmental cues have been identified (reviewed in Anderson, 1997; Le Douarin and Kalcheim, 1999; Sommer, 2001). Conversely, fate-restricted subpopulations of neural crest cells have also been identified, either before or during early stages of migration, suggesting that the early-migrating neural crest cell population is indeed heterogeneous (reviewed in Anderson, 2000; Dorsky *et al.*, 2000a). Interestingly, there is evidence to suggest that at least some of the fate-restriction seen early in neural crest cell migration may result from interactions among neural crest cells themselves (e.g., Raible and Eisen, 1996; Henion and Weston, 1997; Ma *et al.*, 1999). However, a restriction in *fate* does not necessarily imply a restriction in *potential*, since the cell under consideration may only have encountered one particular set of differentiation cues. Latent potential to adopt different fates can only be revealed by challenging the cell with different environmental conditions. When isolated in culture in the absence of other environmental signals, a cell that follows its normal fate is defined as *specified* to adopt that fate. However, it may not be *determined*, that is, it may not have lost the potential to adopt a different fate when exposed to different environmental signals. Without knowing all the factors that a cell might encounter *in vivo*, it is difficult to know when the potential of a cell has been comprehensively tested *in vitro*. Hence, the most rigorous assays for cell determination involve grafting cells to different ectopic sites *in vivo*.

Evidence for Both Multipotent and Fate-Restricted Neural Crest Cells: (1) *In Vivo* Labeling

The fate of individual trunk neural crest cell precursors and their progeny has been analyzed *in vivo* by labeling single cells in the neural folds in chick (Bronner-Fraser and Fraser, 1988, 1989; Frank and Sanes, 1991; Selleck and Bronner-Fraser, 1995), mouse (Serbedzija *et al.*, 1994), and *Xenopus* (Collazo *et al.*, 1993). Two main methods have been used for these clonal lineage analyses. Lysinated rhodamine dextran, a fluorescent, membrane-impermeant vital dye of high molecular weight, can be iontophoretically injected into single cells; it is passed exclusively to the progeny of the injected cell. This technique was used in all the above-cited studies except that of Frank and Sanes (1991). These authors used retroviral-mediated transfection to introduce the gene for β-*galactosidase* (*lacZ*) into the genome of single cells in the dorsal neural tube; the gene is activated on cell division and is transmitted to the progeny of the infected cell (Frank and Sanes, 1991). Similar results were obtained using both marking techniques. In the chick, mouse, and *Xenopus*, many clones contained multiple derivatives, including both neural tube and neural crest derivatives. This showed that neural tube and neural crest cells share a common precursor within the neural folds. Multiple neural crest derivatives were often observed within the same clone, including both neuronal and non-neuronal derivatives, such as glial cells, melanocytes, and in *Xenopus*, dorsal fin cells.

These experiments suggested that individual neural crest precursors are multipotent, but left open the possibility that fate-restricted precursors are generated before the cells leave the neural tube. However, when the lineage of individual neural crest cells migrating through the rostral somite was similarly examined, most labeled clones were found to contain multiple derivatives, including both neuronal and non-neuronal cells (Fraser and Bronner-Fraser, 1991). In extreme cases, clones included both neurons and glia (neurofilament-negative cells) in both sensory and sympathetic ganglia, and Schwann cells along the ventral root (Fraser and Bronner-Fraser, 1991). Hence, at least some individual neural crest cells, early in their migration, are multipotent in the chick. However, some clones were also found that were fate-restricted with respect to a particular neural crest derivative. For example, clones that formed both neurons and glia (neurofilament-negative cells) were found only in the dorsal root ganglia, or only in sympathetic ganglia, while one clone only formed Schwann cells on the ventral root (Fraser and Bronner-Fraser, 1991).

The lineage of individual trunk and hindbrain neural crest cells has also been examined in the zebrafish, which has many fewer neural crest cells than tetrapods (only 10–12 cells per trunk segment) (Raible *et al.*, 1992). Trunk neural crest cells were labeled by intracellular injection of lysinated rhodamine dextran just after they segregated from the neural tube (Raible and Eisen, 1994). In contrast to the results in the chick (Fraser and Bronner-Fraser, 1991), most labeled clones in the zebrafish appeared to be fate-restricted; that is, all descendants of the labeled cell differentiated into the same neural crest derivative, for example, dorsal root ganglion neurons, or melanocytes, or Schwann cells (Raible and Eisen, 1994). Nonetheless, about 20% of clones produced multiple-phenotype clones, showing that at least some trunk neural crest cells are multipotent in the zebrafish (Raible and Eisen, 1994). Individual hindbrain neural crest cells in the most superficial 20% of the neural crest cell masses on either side of the neural keel were similarly labeled using fluorescent dextrans (Schilling and Kimmel, 1994). Strikingly, almost all clones were fate-restricted, giving rise to single identifiable cell types, such as trigeminal neurons, pigment cells, or cartilage; the remainder contained unidentified cell types (Schilling and Kimmel, 1994). Whether these results apply to the remaining, deeper 80% of neural crest cells in the cranial neural crest cell masses remains to be determined.

Similar analyses in the zebrafish trunk have also provided an excellent example of how fate-restriction in individual neural crest cells can be explained by regulative interactions between migrating neural crest cells, rather than by restrictions in potential (Raible and Eisen, 1996). Early-migrating neural crest cells along the medial pathway generate all types of trunk neural crest cell derivatives, including dorsal root ganglion neurons. Neural crest cells that migrate later along the same pathway form melanocytes and Schwann cells, but not dorsal root ganglion neurons (Raible *et al.*, 1992). When the early-migrating population was ablated, late-migrating cells contributed to the dorsal root ganglion, even when they migrated at their normal time (Raible and Eisen, 1996). This suggests that the fate-restriction of late-migrating cells in normal development is due neither to a restriction in potential, nor to temporal changes in, for example, mesoderm-derived environmental cues, but to regulative interactions between early- and late-migrating neural crest cells that restrict the fate choice of the latter (Raible and Eisen, 1996).

Evidence for Both Multipotent and Fate-Restricted Neural Crest Cells: (2) *In Vitro* Cloning

A wealth of data exists on the fate choices of single neural crest cells and their progeny *in vitro* (reviewed in Le Douarin and Kalcheim, 1999). Migrating neural crest cell populations can be cultured in low-density conditions, followed sometimes by serial subcloning of the primary clones (e.g., Cohen and Königsberg, 1975; Sieber-Blum and Cohen, 1980; Stemple and Anderson, 1992). Alternatively, single neural crest cells can be picked at random from a suspension of migrating neural crest cells and plated individually (e.g., Baroffio *et al.*, 1988; Dupin *et al.*, 1990). These clonal culture techniques have shown that both fate-restricted and multipotent neural crest cells can be isolated from avian and mammalian embryos. Most clones of migrating quail cranial neural crest cells gave rise to progeny that differentiated into 2–4 different cell types, that is, were multipotent (Baroffio *et al.*, 1991). Furthermore, single cells were found (at very low frequency, around 0.3%) that could give rise to neurons, glia, melanocytes, and cartilage, that is, all the major neural crest cell derivatives (Baroffio *et al.*, 1991). These highly multipotent founder cells were interpreted as stem cells, although self-renewal of these cells remains to be demonstrated. Self-renewing, multipotent neural crest stem cells have been isolated from the migrating mammalian trunk neural crest, based on their expression of the low-affinity neurotrophin receptor, p75^{NTR} (Stemple and Anderson, 1992). These cells are able to form autonomic neurons, Schwann cells and satellite glia, and smooth muscle cells, though they do not seem able to form sensory neurons (Shah *et al.*, 1996; White *et al.*, 2001).

As pointed out by Anderson (2000), it is difficult to be sure that the patterns and sequences of lineage restriction seen in these *in vitro* studies accurately reflect the composition of the migrating neural crest cell population *in vivo*. Although different founder cells might give rise to different subsets of neural crest cell derivatives *in vitro* (i.e., under the same culture conditions), this may not reflect intrinsic differences between the founder cells. It is possible that stochastic differences in their behavior, and/or the type and sequence of cell–cell interactions in each clone, might result in very different final outcomes, even if the initial founder cells were equivalent.

Single cell lineage analysis has also been performed on migrating neural crest cell explants *in vitro* (Henion and Weston, 1997). These authors injected lysinated rhodamine dextran intracellularly into random individual neural crest cells, migrating from trunk neural tubes placed in an enriched culture medium that supported the differentiation of melanocytes, neurons, and glia. Crucially, this method, unlike clonal culture, allows normal interactions between migrating neural crest cells to take place. The results showed that even during the first 6 hr of emigration, almost half of the labeled cells were fate-restricted, forming either neurons, glia, or melanocytes (Henion and Weston, 1997). Although the remaining clones formed more than one cell type, most formed neurons and glia, or glia and melanocytes, with only a few forming all three cell types (no cells formed only neurons and melanocytes) (Henion and Weston, 1997). Interestingly, neural crest cells sampled at later times (within a period

corresponding to one or two cell divisions) contained no neuronal-glial clones: Almost all the sampled cells that produced neurons were fate-restricted neuronal precursors (Henion and Weston, 1997). Since the medium remained unchanged, and random differentiation would not be expected reproducibly to produce or remove distinct sublineages, the authors suggested that interactions between the neural crest cells themselves are responsible for the sequential specification of neuron-restricted precursors (Henion and Weston, 1997). Again, fate-restriction may not reflect restriction in potential, but it is clear that the early-migrating neural crest cell population is heterogeneous, containing both fate-restricted (as assessed both *in vivo* and *in vitro*) and multipotent precursors.

Other Evidence for Heterogeneity in the Migrating Neural Crest

Some of the earliest evidence for heterogeneity in the migrating neural crest was based on antigenic variation within the migrating population. For example, various monoclonal antibodies raised against dorsal root ganglion cells also recognize early subpopulations of neural crest cells (e.g., Ciment and Weston, 1982; Girdlestone and Weston, 1985). The SSEA-1 antigen is expressed by quail sensory neuroblasts in dorsal root ganglia and in subpopulations of migrating neural crest cells that differentiate into sensory neurons in culture (Sieber-Blum, 1989). A monoclonal antibody raised against chick ciliary ganglion cells, associated with high-affinity choline uptake, also recognizes a small subpopulation of mesencephalic neural crest cells (which normally give rise to the cholinergic neurons of the ciliary ganglion) (Barald, 1988a, b). The progressive restriction of expression of the 7B3 antigen (transitin, a nestin-like intermediate filament) during avian neural crest cell development may reflect glial fate-restriction (Henion *et al.*, 2000). However, to show that expression of a particular antigen is related to the adoption of a particular fate, it must either be converted into a permanent lineage tracer, eliminated, or misexpressed ectopically, and this has not yet been achieved.

There is some evidence that late-migrating trunk neural crest cells in the chick may have reduced potential to form catecholaminergic neurons (see Fig. 9). Late-migrating chick trunk neural crest cells (i.e., those emigrating 24 hr after the emigration of the first neural crest cells at the same axial level) do not normally contribute to sympathetic ganglia (Serbedzija *et al.*, 1989). When transplanted into an "early" environment, these late-migrating cells are able to form neurons in sympathetic ganglia, but fail to adopt a catecholaminergic fate (Artinger and Bronner-Fraser, 1992). These results may not reflect a loss of all autonomic potential, however, as cholinergic markers were not examined in these embryos.

Neural Crest Cell Precursors are Exposed to Differentiation Cues within the Neural Tube

The dorsal neural tube expresses various signaling molecules known to promote different neural crest cell fates, including Wnt1, Wnt3a, and BMP4 (section Control of Neural Crest Cell

Differentiation in the PNS) (reviewed in Dorsky *et al.*, 2000a). Clearly, exposure of premigratory neural crest cell precursors to such factors could lead to at least some of the fate-restrictions and heterogeneity seen within the migrating neural crest cell population. For example, activation of the Wnt signaling pathway has been shown to be necessary and sufficient for melanocyte formation in both zebrafish and mouse (Dorsky *et al.*, 1998; Dunn *et al.*, 2000), via the direct activation of the *MITF/nacre* gene, which encodes a melanocyte-specific transcription factor (Dorsky *et al.*, 2000b). Continuous exposure to the neural tube stimulates melanogenesis in cultured neural crest cells (Glimelius and Weston, 1981; Derby and Newgreen, 1982), while Wnt3a-conditioned medium dramatically increases the number of melanocytes in quail neural crest cell cultures (Jin *et al.*, 2001). It is possible, therefore, that neural crest cell precursors exposed to Wnt3a in the dorsal neural tube for longer periods of time are more likely to generate progeny that will form into melanocytes, although this has not been directly tested. Wnts in the dorsal neural tube are not the only factors involved in melanocyte formation: For example, extracellular matrix from the subectodermal region specifically promotes neural crest cell differentiation into melanocytes (Perris *et al.*, 1988). Nonetheless, these results demonstrate that factors within the neural tube may play important roles in at least some fate decisions.

In summary, therefore, neural crest precursors within the neural tube are exposed to a variety of neural crest cell differentiation cues present within the neural tube (and overlying ectoderm). Although such exposure has not directly been shown to result in the formation of fate-restricted progeny, it may be relevant to at least some of the heterogeneity seen within the migrating neural crest cell population. It is possible that, for example, the early segregation of a subpopulation of sensory-biased progenitors (section Sensory-Biased Neural Crest Cells Are Present in the Migrating Population) and the loss of catecholaminergic potential in late-migrating cells (see preceding section) ultimately result from the exposure of neural crest cell precursors to environmental cues within the neural tube.

Molecular Control of Lineage Segregation: A Paradigm from the Immune System

Relatively little is known in the neural crest field about the downstream effects of transcription factors associated with particular neural crest lineages. The best characterized examples of the molecular control of lineage segregation from multipotent precursors are found in the immune system, for example, the transcriptional control of B-cell development from hematopoietic stem cells (reviewed in Schebesta *et al.*, 2002). Results from this field provide a paradigm for thinking about how lineage segregation might occur at the molecular level within the neural crest.

An emerging theme is that hematopoietic lineage segregation reflects not only the activation of lineage-specific genes, but also the suppression of alternative lineage-specific gene programs by negative regulatory networks of transcription factors (see Schebesta *et al.*, 2002). For example, the basic helix-loop-helix

transcription factors E2A and EBF coordinately activate the expression of B-cell-specific genes, but this is insufficient to determine adoption of a B-cell fate. For B-cell determination (commitment) to occur, the paired-domain homeodomain transcription factor Pax5 must also be present: This factor not only activates some genes in the B-cell program, but also represses lineage-inappropriate genes (Schebesta *et al.*, 2002). Indeed, continuous Pax5 expression is required in B-cell progenitors in order to maintain commitment to the B-cell lineage (Mikkola *et al.*, 2002).

Much less is known within the neural crest field about the downstream molecular effects of the expression of specific transcription factors. However, it is likely that similar networks of positive regulators activating transcription of lineage-appropriate genes, and negative regulators repressing transcription of lineage-inappropriate genes, are involved in neural crest cell lineage determination.

Segregation of Sensory and Autonomic Lineages

Postmigratory Trunk Neural Crest Cells Are Restricted to Forming Either Sensory or Autonomic Lineages

At postmigratory stages, distinct sensory-restricted and autonomic-restricted neural crest cells can be identified. When embryonic quail autonomic ganglia are "back-grafted" into early chick neural crest cell migration pathways, they are unable to contribute to dorsal root ganglion neurons and glia (reviewed by Le Douarin, 1986). Instead, they only form Schwann cells and autonomic derivatives (catecholaminergic sympathetic neurons, adrenal chromaffin cells, and sometimes enteric ganglia) (reviewed by Le Douarin, 1986). These results suggest that postmigratory neural crest cells in autonomic ganglia are restricted to an autonomic lineage. A similar autonomic restriction is seen in postmigratory neural crest cells in the gut, which normally form enteric ganglia. When these enteric neural precursor cells from rat embryos are grafted into chick neural crest migration pathways, they form neurons and satellite cells in sensory and sympathetic ganglia (White and Anderson, 1999). However, even in the sensory environment, the graft-derived neurons only express parasympathetic neuron markers, suggesting they are not able to form sensory neurons but are restricted to an autonomic lineage (White and Anderson, 1999).

Back-grafted dorsal root ganglia, in contrast, are additionally able to give rise to neurons and glia in the host dorsal root ganglia, provided that sensory neuroblasts are still mitotically active in the back-grafted ganglion (reviewed by Le Douarin, 1986). If sensory ganglia are back-grafted after all their sensory neuroblasts have withdrawn from the cell cycle, the postmitotic neurons die, and the non-neuronal cells within the ganglion differentiate into autonomic (sympathetic and enteric) but not sensory neurons (Ayer-Le Lièvre and Le Douarin, 1982; Schweizer *et al.*, 1983). Multipotent postmigratory neural crest progenitors have also been isolated from dorsal root ganglia: These are able to form autonomic neurons, glia, and smooth muscle, but not, apparently, sensory neurons (Hagedorn *et al.*, 1999, 2000a).

Hence, the potential to form dorsal root ganglion neurons and glia seems to be restricted, in postmigratory trunk neural crest cells, specifically to dividing sensory neuroblasts within sensory ganglia. Postmigratory neural crest cells in autonomic ganglia, and non-neuronal cells in sensory ganglia, are restricted to forming autonomic derivatives. These results point to a clear sensory vs autonomic lineage restriction within the postmigratory trunk neural crest, and also suggest that this decision occurs prior to any neuronal–glial lineage restriction.

A Model for Sensory–Autonomic Lineage Restriction

Based on the ganglion back-grafting experiments described above, Le Douarin put forward a model for the segregation of sensory and autonomic lineages within the neural crest (Le Douarin, 1986). The model proposed that (1) distinct sensory and autonomic neuronal progenitors are present in the migrating neural crest, as well as progenitors able to give rise to both lineages; (2) the sensory progenitors are only present until all sensory neurons have withdrawn from the cell cycle, while autonomic progenitors persist throughout development; (3) sensory progenitors only survive in sensory ganglia, while autonomic progenitors survive in all types of ganglia, suggesting different trophic requirements. Although the back-grafting data clearly support the existence of a sensory vs autonomic lineage restriction at postmigratory stages, the question of when this lineage restriction takes place has been much debated (see, e.g., Anderson, 2000).

The Le Douarin model proposes that some neural crest cells take the sensory–autonomic lineage decision early in their migration, while others retain the ability to form both lineages. The *in vivo* clonal analysis of migrating neural crest cells in the chick provides some support for this (Fraser and Bronner-Fraser, 1991). Some clones (which included both neurons and glia) were restricted either to dorsal root ganglia or sympathetic ganglia, while others gave rise to neurons and non-neuronal cells in both dorsal root and sympathetic ganglia (Fraser and Bronner-Fraser, 1991).

The ability to adopt a sensory fate may be rapidly lost, however. This is seen not only in postmigratory neural crest cells, as described above, but also in the migrating population. For example, self-renewing (re-plated) rat neural crest stem cells, which make up the bulk of the migrating neural crest cell population, seem to be unable to form sensory neurons, whether tested *in vitro* or *in vivo* (Shah *et al.*, 1996; Morrison *et al.*, 1999; White *et al.*, 2001). Given that neural crest-derived sensory neurons are only found proximal to the neural tube, in dorsal root ganglia and proximal cranial sensory ganglia, such a rapid loss of sensory potential may make some sense, but the underlying mechanism remains obscure.

Sensory-Biased Neural Crest Cells Are Present in the Migrating Population

No evidence as yet supports the existence of determined autonomic progenitors within the migrating neural crest cell population. However, sensory-determined and sensory-biased progenitors are present in the migrating mammalian neural crest (Greenwood *et al.*, 1999; Zirlinger *et al.*, 2002). When rat trunk neural crest cells are cultured in a defined medium that permits sensory neuron formation, sensory neurons develop from dividing progenitors even in the presence of a strong autonomic neurogenesis cue, BMP2 (section BMPs Induce Both Mash1 and Phox2b in Sympathetic Precursors) (Greenwood *et al.*, 1999). These results suggest that at least some dividing progenitors are already determined toward a sensory fate (Greenwood *et al.*, 1999).

In another work, an inducible-Cre recombinase system in mice was used to mark permanently a subpopulation of neural crest cells that expresses Neurogenin2 (Ngn2), a basic helix-loop-helix transcription factor required for sensory neurogenesis (sections Proneural Genes: An Introduction; Neurogenins Are Essential for the Formation of Dorsal Root Ganglia) (Zirlinger *et al.*, 2002). Ngn2[+] progenitors were four times as likely as the general neural crest cell population to contribute to dorsal root ganglia rather than sympathetic ganglia (Zirlinger *et al.*, 2002). Within the dorsal root ganglia, the Ngn2[+] cells were found to contribute to all the main sensory neuron subtypes, and to satellite glia, without any apparent bias toward a particular lineage (Zirlinger *et al.*, 2002). Since some Ngn2[+] precursors did contribute to sympathetic ganglia, these results suggest that while Ngn2 expression does not commit neural crest cells to a sensory fate, Ngn2 confers a strong bias toward a sensory fate. Ngn2 expression does not correlate with a bias toward any specific neuronal or glial subtype, however. These results therefore also support the idea that the restriction to sensory or autonomic lineages occurs before the decision to form neurons or glia.

Summary of Sensory/Autonomic Lineage Segregation

There is an autonomic vs sensory lineage restriction in postmigratory trunk neural crest cells in peripheral ganglia, and this seems to occur prior to the neuronal–glial decision. Some migrating neural crest cells may already be determined toward a sensory fate. Expression of the transcription factor Ngn2 in a subpopulation of migrating neural crest cells correlates with a strong bias, though not commitment, toward a sensory neural fate. Within dorsal root ganglia, Ngn2[+] cells are not restricted to a specific phenotype, but form multiple sensory neuronal subtypes and satellite glia. Although autonomic-restricted progenitors are found early in development (including, apparently, self-renewing neural crest stem cells), no autonomic-determined progenitors have yet been identified.

Sox10 Is Essential for Formation of the Glial Lineage

Neural crest cells give rise to all peripheral glia. These include satellite cells (glia that ensheathe neuronal cell bodies in peripheral ganglia) and Schwann cells (glia that ensheathe axonal processes of peripheral nerves). These can be distinguished molecularly: Satellite cells express the Ets domain transcription factor Erm (a downstream target of FGF signaling; Raible and Brand, 2001; Roehl and Nüsslein-Volhard, 2001) and do not express either the POU transcription factor Oct6 or the zinc finger transcription factor Krox20 (see Hagedorn et al., 2000b; Jessen and Mirsky, 2002). Schwann cells are Erm-negative, Oct6$^+$, Krox20$^+$, and also express, for example, the surface glycoprotein Schwann cell myelin protein (see Hagedorn et al., 2000b; Jessen and Mirsky, 2002). The satellite cell phenotype is maintained by the ganglionic microenvironment; when removed from this environment, satellite cells can adopt a Schwann cell fate, although the reverse does not seem to occur (Dulac and Le Douarin, 1991; Cameron-Curry et al., 1993; Murphy et al., 1996; Hagedorn et al., 2000b). Hence, satellite cells and Schwann cells are closely related.

The HMG-domain transcription factor Sox10 is essential for the formation of all neural crest-derived glia (and melanocytes) (Britsch et al., 2001; Dutton et al., 2001). In Sox10-null mice, all satellite cells and all Schwann cells are missing, leading to eventual degeneration of sensory, autonomic (including all enteric), and motor neurons (Britsch et al., 2001). Haploinsufficiency of Sox10 leads to neural crest defects that cause Waardenburg/Hirschsprung disease in humans (see McCallion and Chakravarti, 2001). Sox10 controls the expression of the ErbB3 gene (Britsch et al., 2001), which encodes one of the high-affinity receptors for the growth factor NRG1, a member of the epidermal growth factor superfamily. (For reviews of NRGs and their receptors, see Adlkofer and Lai, 2000; Garratt et al., 2000.)

Sox10 is expressed in migrating neural crest cells (also see section Ap2α and SoxE Transcription Factors), but is downregulated in all lineages except for glial cells and melanocytes. Sox10 function is required for the survival of at least a subpopulation of multipotent neural crest cells, at least in part by regulating their responsiveness to NRG1 (Paratore et al., 2001) (also see Dutton et al., 2001). Constitutive expression of Sox10 in migrating neural crest stem cells maintains both glial and neuronal differentiation potential, although an additional function of Sox10 is to delay neuronal differentiation (Kim et al., 2003). Hence, one role of Sox10 is to maintain multipotency of neural crest stem cells (Kim et al., 2003); thus Sox10 expression does not reflect determination toward the glial lineage.

Sox10 is essential for glial fate acquisition by neural crest stem cells in response to instructive gliogenic signals (Paratore et al., 2001). Such gliogenic cues include the type II isoform of NRG1 ("glial growth factor") and perhaps also NRG1 type III (sections Differentiation of DRG Satellite Cells; Neuregulin1 type III Is Essential for Schwann Cell Formation; Differentiation of Satellite Cells in Autonomic Ganglia; Shah et al., 1994; Shah and Anderson, 1997; Hagedorn et al., 1999, 2000b; Paratore et al., 2001; Leimeroth et al., 2002). Expression of the transmembrane receptor Notch1 is also missing from sensory ganglia in Sox10 mutant mice (Britsch et al., 2001): As will be seen in the section on Control of Neural Crest Cell Differentiation in the PNS, Notch activation is also a potent instructive cue for gliogenesis (Morrison et al., 2000b).

In summary, Sox10 is expressed in migrating neural crest cells and is maintained and required specifically in the glial lineage within the PNS. The early expression of Sox10 in migrating neural crest cells, as well as glial cells, may be consistent with the evidence (discussed in section Segregation of Sensory and Autonomic Lineages) suggesting that the sensory vs autonomic lineage decision occurs before the neuronal–glial decision.

Summary of Neural Crest Lineage Diversification

Two main hypotheses have been proposed to explain lineage segregation within the neural crest: (1) instruction, in which multipotent precursors are instructed by environmental cues to adopt particular fates, and (2) selection, in which determined cells, which are only able to adopt one fate, are selected in permissive environments. The available evidence suggests that the migrating population is heterogeneous, containing both highly multipotent cells and fate-restricted cells. However, there is little evidence to correlate fate-restriction with loss of potential to adopt other fates. Neural crest precursors are exposed to multiple environmental cues within the neural tube, and these may underlie at least some of the fate-restrictions seen within the migrating population. Ngn2 expression in a subset of migrating neural crest cells correlates with a strong bias (though not determination) toward a sensory fate. Apart from mitotic sensory neuroblasts in the DRG, postmigratory neural crest cells seem to be restricted to the autonomic lineage. The sensory–autonomic lineage decision seems to occur before the neuronal–glial decision. The transcription factor Sox10, expressed both in migrating neural crest cells and the glial lineage, is essential for, but does not determine, adoption of a glial fate.

CONTROL OF NEURAL CREST CELL DIFFERENTIATION IN THE PNS

A great deal of molecular information is now available concerning the signals and genetic machinery that underpin the differentiation of neural crest cells into specific cell types. Considerable progress has been made in understanding the molecular control of the differentiation of various non-neural and neural crest cell derivatives, for example, melanocytes (reviewed in Le Douarin and Kalcheim, 1999; Rawls et al., 2001), smooth muscle (see, e.g., Sommer, 2001), and even cartilage (Sarkar et al., 2001) (Fig. 6). However, any detailed discussion of the differentiation of these non-neural derivatives is beyond the scope of this chapter, which will concentrate on differentiation in the PNS. Numerous reviews provide additional information on this topic (e.g., Anderson, 1999; Le Douarin and Kalcheim, 1999; Anderson, 2000; Sieber-Blum, 2000; Morrison, 2001; Sommer, 2001). Chapter 5 should also be consulted for more general information on neuronal differentiation.

Within the PNS, it has become clear that vertebrate homologues of the invertebrate basic helix-loop-helix (bHLH) proneural transcription factors play essential roles in the differentiation of different neural crest cell types. Proneural genes are discussed in more detail in Chapter 5, but a brief introduction is given here for the purposes of this chapter.

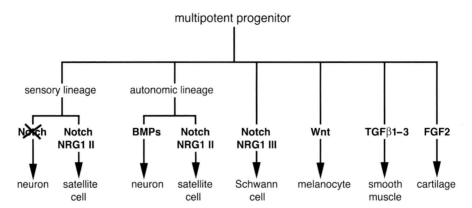

FIGURE 6. Schematic showing known signaling pathways involved in the differentiation of different cell types from multipotent neural crest cells. See the section on Contol of Neural Crest Cell Differentiation in the PNS for details. Modified from Dorsky *et al.* (2000a).

Proneural Genes: An Introduction

In both *Drosophila* and vertebrates, proneural bHLH transcription factors confer neuronal potential and/or specify neural progenitor cell identity (see Chapter 5) (reviewed in Bertrand *et al.*, 2002). They act in part by activating the expression of ligands of the Notch receptor, such as Delta. Cells with high levels of Notch activity downregulate Notch ligand expression and adopt a "secondary" (e.g., supporting) cell fate, while cells with low levels of Notch activity adopt a primary (e.g., neuronal) cell fate (see Chapter 5; Gaiano and Fishell, 2002). Two classes of proneural genes are active in the PNS of *Drosophila*: the *achaete-scute* complex and *atonal* (reviewed in Skaer *et al.*, 2002). Vertebrate homologues of the *achaete-scute* complex include *ash1* (*Mash1* in mice, *Cash1* in chick, etc.) and additional species-specific genes (e.g., *Mash2* in mice, *Cash4* in chick). The vertebrate *atonal* class contains many more genes, divided into various families based on the presence of specific residues in the bHLH domain (reviewed in Bertrand *et al.*, 2002). The *neurogenin*s (*ngn*s), which were briefly introduced in the section on Segregation of Sensory and Autonomic Lineages, make up one of these *atonal*-related gene families. In neural crest cells, the *atonal*-related *neurogenin* family is particularly important for the sensory lineage (section Neurogenins Are Essential for the Formation of Dorsal Root Ganglia), while the *achaete-scute* homologue *ash1* (*Mash1*) is important for aspects of autonomic neurogenesis (section *Mash1* Is Essential for Noradrenergic Differentiation).

Dorsal Root Gangliogenesis

Trunk neural crest cells that remain within the somite, in the vicinity of the neural tube, aggregate and eventually differentiate to form the sensory neurons and satellite glia of the dorsal root ganglia. Similar differentiation processes presumably occur within proximal neural crest-derived cranial sensory ganglia, but most information is available for dorsal root ganglia.

Neurogenins Are Essential for the Formation of Dorsal Root Ganglia

As described in the section Sensory-Biased Neural Crest Cells Are Present in the Migrating Population, Ngn2 expression biases (but does not determine) neural crest cells toward the sensory lineage, including both neurons and satellite glia (Zirlinger *et al.*, 2002). Ngn2 and a related factor, Ngn1, are expressed in complementary patterns in peripheral sensory neurons derived from neural crest and placodes (reviewed in Anderson, 1999) (sections Sense Organ Placodes; Trigeminal and Epibranchial Placodes). Knockout experiments in mice have shown that the Ngns are essential for the formation of sensory ganglia (Fode *et al.*, 1998; Ma *et al.*, 1998, 1999).

In the mouse, Ngn2 is expressed in cells in the dorsal neural tube, and in a subpopulation of migrating mammalian trunk neural crest cells, continuing into the early stages of dorsal root ganglion (DRG) condensation (Ma *et al.*, 1999). In contrast, Ngn1 is first expressed only after DRG condensation has begun (Ma *et al.*, 1999). In the chick, both Ngns are expressed in the dorsal neural tube, and in a subset of migrating neural crest cells (Perez *et al.*, 1999). Chick Ngn2 is transiently expressed during chick dorsal root gangliogenesis, while Ngn1 is maintained until late stages in non-neuronal cells and/or neuronal precursors at the DRG periphery (Perez *et al.*, 1999).

Normal Ngn2 expression in the mouse correlates with a strong bias toward the sensory lineage, but not toward any particular neuronal or glial phenotype within the sensory lineage (Zirlinger *et al.*, 2002) (section Sensory-Biased Neural Crest Cells Are Present in the Migrating Population).

In contrast, Ngn1 overexpression studies suggest that Ngn1 may act to promote a specifically sensory neuronal phenotype. Retroviral-mediated overexpression of mouse Ngn1 in premigratory neural crest precursors in the chick leads to a significant bias toward population of the DRG, and to ectopic sensory neuron formation in neural crest derivatives, and even in the somite (Perez *et al.*, 1999). Similar overexpression of Ngn1 in dissociated rat neural tube cultures, which are competent to form sensory and autonomic peripheral neurons, also leads to increased sensory neurogenesis (Lo *et al.*, 2002). However, permanent genetic labeling experiments, like those performed for Ngn2 (Zirlinger *et al.*, 2002), are needed to show whether this correlation holds true during normal development.

Differentiation of DRG Neurons Depends on Inhibition of Notch Signaling

There is accumulating evidence that the decision to follow a sensory vs autonomic lineage occurs before the neuronal–glial decision (section Segregation of Sensory and Autonomic Lineages). Hence, sensory precursors within the DRG give rise to both sensory neurons and satellite glia. How are both neurons and satellite glia produced from the same precursors within the same ganglionic environment? It is now clear that neuronal and glial differentiation within the DRG depend on inhibition and activation, respectively, of signaling by the transmembrane receptor Notch (see Chapter 5; Fig. 7) (Wakamatsu *et al.*, 2000; Zilian *et al.*, 2001).

Notch1 is expressed by most migrating chick trunk neural crest cells and is downregulated on differentiation of both neurons and glia. In the DRG, Notch1 is initially preferentially expressed by cycling cells in the periphery, while one of its ligands, Delta1, is expressed by differentiating neurons located

in the core of the ganglion (Wakamatsu *et al.*, 2000) (Fig. 7). If Notch signaling is activated in cultured quail trunk neural crest cells (by overexpression of the Notch1 cytoplasmic domain), neuronal differentiation is inhibited and cell proliferation is transiently increased, suggesting that in order for neurons to form, Notch activity must be inhibited (Wakamatsu *et al.*, 2000).

The Notch antagonist, Numb (see Chapter 5), is expressed asymmetrically in about 40% of the cycling cells at the periphery of the chick DRG (Wakamatsu *et al.*, 2000). It is not known how this asymmetrical expression is established, but, after these cells divide, Numb will be inherited in high concentrations by only one of the daughter cells. In the Numb-inheriting daughter cell, high levels of Numb will inhibit Notch signaling; Delta1 will be upregulated, and the cell will differentiate as a neuron. The daughter cell that does not inherit Numb will have high levels of Notch signaling, probably activated by Notch ligands (e.g., Delta1) expressed on differentiating neurons in the core. This daughter cell will therefore be able to divide again, and/or form a satellite cell (see the following section) (Fig. 7). In agreement with this model, knockout experiments in mice have shown that Numb is essential for the formation of DRG sensory neurons (but not for, e.g., sympathetic neurons, although it is expressed in sympathetic ganglia) (Zilian *et al.*, 2001).

As will be seen later, autonomic neuronal differentiation is promoted by instructive growth factors. Similar instructive sensory neuronal differentiation cues that act on multipotent progenitors have not been identified, although neural tube-derived neurotrophins, such as brain-derived neurotrophic factor (BDNF), are required for the survival and proliferation of DRG progenitors (reviewed in Kalcheim, 1996). Since the trigger for neuronal differentiation in the DRG seems to be the asymmetric expression of Numb in some of the cycling cells at the DRG periphery, understanding how this asymmetry is set up will shed light on how DRG neuronal differentiation is controlled.

Differentiation of DRG Satellite Cells Depends on Notch Activation and Instructive Gliogenic Cues

The above results give some insight into how neurogenesis occurs within the DRG. How, though, do satellite cells form in the same environment? Neuronal differentiation always occurs before glial differentiation in the DRG (Carr and Simpson, 1978), and it is likely that signals from the differentiating neurons instruct non-neuronal cells within the ganglion to form satellite cells. A model for how glial differentiation is controlled is emerging from studies of cultured neural crest stem cells and multipotent progenitors from cultured DRGs in the rat embryo (Hagedorn *et al.*, 1999, 2000b; Morrison *et al.*, 2000a; Leimeroth *et al.*, 2002). This model proposes a combinatorial action of Notch-mediated neurogenic repression and gliogenic instruction, triggered by Notch ligands on differentiating neurons, together with additional gliogenic growth factors expressed or secreted by differentiating neurons.

Notch activation, as well as inhibiting neurogenesis (Wakamatsu *et al.*, 2000), also instructively promotes a glial fate in cultured rat neural crest stem cells (Morrison *et al.*, 2000b;

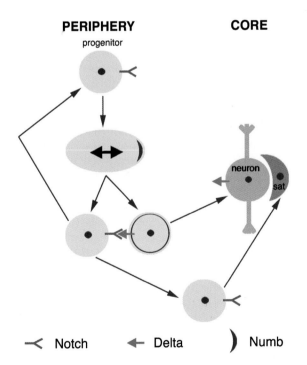

PERIPHERY   **CORE**

FIGURE 7. Schematic showing a model for neurogenesis within the dorsal root ganglion. The Notch inhibitor Numb is inherited asymmetrically by daughters of proliferating progenitors in the periphery of the ganglion. Cells with high levels of Numb have low levels of Notch activity: They upregulate the Notch ligand Delta, move to the core of the ganglion, and differentiate as neurons. Cells with low levels of Numb have high levels of Notch activity: They either divide again or differentiate into satellite cells (sat). Modified from Wakamatsu *et al.* (2000).

Kubu *et al.*, 2002). This is discussed more fully in the section on Notch Activation Leads to Gliogenesis by Neural Crest Stem Cells. Although these rat neural crest stem cells seem to lack sensory potential (Shah *et al.*, 1996; Morrison *et al.*, 1999; White *et al.*, 2001), it is likely that Notch activation is also involved in DRG satellite glial differentiation, probably in association with other instructive cues. Notch activation is presumably triggered by the Notch ligands, such as Delta1, expressed on differentiating neurons in the DRG core (Wakamatsu *et al.*, 2000). *Delta1*-null mice have reduced numbers of satellite glia and Schwann cells, providing some corroborating evidence for this (De Bellard *et al.*, 2002).

An independent instructive cue for satellite gliogenesis was also initially identified in studies of cultured rat neural crest stem cells (Shah *et al.*, 1994). These authors showed that the type II isoform ("glial growth factor") of the growth factor Neuregulin1 (NRG1) both inhibits neuronal differentiation and instructively promotes a glial fate in rat neural crest stem cells (Shah *et al.*, 1994; Shah and Anderson, 1997). Several NRG1 isoforms are expressed in DRG neurons (Meyer *et al.*, 1997; Wakamatsu *et al.*, 2000). NRG1 type II specifically induces the formation of satellite cells (as opposed to Schwann cells) in migrating neural crest stem cells and in DRG-derived progenitor cells *in vitro* (Hagedorn *et al.*, 2000b; Leimeroth *et al.*, 2002). However, knockout experiments in mice have failed to reveal a role either for NRG1 isoforms, or for one of their high-affinity receptors, ErbB3, in the DRG (Meyer *et al.*, 1997). Additional gliogenic signals, therefore, may also operate in the DRG.

Summary of Dorsal Root Gangliogenesis

Ngns are essential for the formation of sensory ganglia, including dorsal root ganglia. Mouse Ngn2 biases neural crest cells toward the sensory lineage, while Ngn1 may be involved in sensory neurogenesis within the DRG. Differentiation of DRG neurons requires inhibition of Notch signaling, mediated in part by asymmetric inheritance of Numb. Differentiation of satellite cells involves two instructive gliogenic cues: Notch activation, and gliogenic growth factors. Differentiating neurons in the core of the DRG express Notch ligands, which activate Notch signaling in cycling non-neuronal cells at the periphery of the DRG. Notch activation instructively promotes a glial cell fate. NRG1 type II, produced by differentiating DRG neurons, also instructively promotes a satellite cell fate.

Schwann Cell Differentiation

The differentiation of Schwann cells has been intensively studied (for reviews, see Le Douarin and Kalcheim, 1999; Jessen and Mirsky, 2002). As for satellite cells, Schwann cell differentiation may involve the combination of two independent pathways: Notch activation, and instructive gliogenic cues from neurons.

Notch Activation Leads to Gliogenesis by Neural Crest Stem Cells

Even transient activation of Notch signaling (using a soluble clustered form of its ligand, Delta) inhibits neuronal

differentiation and instructively promotes glial differentiation, in cultures of postmigratory neural crest stem cells isolated from fetal rat sciatic nerve (Morrison *et al.*, 2000b; Kubu *et al.*, 2002). While Notch activation also instructively promotes the glial differentiation of migrating neural crest stem cells, it is less efficient at inhibiting neuronal differentiation than in postmigratory cells, suggesting that glial promotion and neuronal inhibition are independent effects (Kubu *et al.*, 2002).

Neuregulin1 Type III Is Essential for Schwann Cell Formation

Knockout experiments in mice have shown that NRG1 type III, the major NRG1 isoform produced by sensory neurons and motor neurons, is essential for Schwann cell formation (Meyer *et al.*, 1997) (reviewed in Garratt *et al.*, 2000; Jessen and Mirsky, 2002). Migrating neural crest cells express ErbB3, a high-affinity NRG1 receptor that is downregulated in most lineages but maintained in glial lineages. As described in the section on Sox10 Is Essential for Formation of the Glial Lineage, *ErbB3* gene expression is at least partly controlled by Sox10, which is essential for the formation of all peripheral glia, including Schwann cells (Britsch *et al.*, 2001). Schwann cell precursors lining peripheral axons are missing in mice lacking NRG1 type III (see Meyer *et al.*, 1997). It was originally unclear whether this effect of NRG1 type III was solely due to its support of the survival and/or proliferation of Schwann cell precursors (reviewed in Garratt *et al.*, 2000; Jessen and Mirsky, 2002). However, membrane-bound NRG1 type III has now been shown to act as an instructive Schwann cell differentiation cue (Leimeroth *et al.*, 2002). Cultured rat neural crest stem cells and multipotent progenitors isolated from DRGs are specifically induced to form Schwann cells (as opposed to satellite cells) by membrane-bound NRG1 type III (Leimeroth *et al.*, 2002). Soluble NRG1 type III is unable to promote Schwann cell differentiation (Leimeroth *et al.*, 2002). Hence, locally presented NRG1 type III (e.g., on axons) may regulate Schwann cell differentiation. Signaling by membrane-bound NRG1 type III seems to be dominant over NRG1 type II, which induces satellite cell differentiation (see section Differentiation of DRG Satellite Cells) (Leimeroth *et al.*, 2002). This may underlie the apparent inability of Schwann cells to adopt a satellite cell fate (Hagedorn *et al.*, 2000b).

Differences in the Sensitivity of Different Neural Crest Stem Cells to Gliogenic Cues

In the rat, postmigratory neural crest stem cells from fetal sciatic nerves do not differentiate into neurons as readily as migrating neural crest stem cells, as shown by transplantations to chick neural crest cell migratory pathways (White and Anderson, 1999; White *et al.*, 2001). These fetal nerve neural crest stem cells express significantly higher levels of Notch1, and lower levels of the Notch antagonist Numb, than migrating neural crest stem cells (Kubu *et al.*, 2002). Postmigratory cells on the sciatic nerve are therefore more sensitive to Notch activation than migrating cells and hence more likely to differentiate into glia (Kubu *et al.*, 2002). The changes in Notch1 and Numb expression levels, and the sensitivity to Notch activation, require neural crest cell–cell interactions. These are probably mediated, at least in part, by Delta (or other Notch ligand) expression on differentiating neurons and peripheral nerves (Bixby *et al.*, 2002; Kubu *et al.*, 2002).

Similar intrinsic differences in the sensitivity of different neural crest stem cell populations to gliogenic signals have been observed in neural crest stem cells isolated from the rat gut (Bixby *et al.*, 2002; Kruger *et al.*, 2002). Fetal gut neural crest stem cells are highly resistant to gliogenic signals and form neurons, rather than glia, on chick peripheral nerves (probably in response to local BMPs; see the section BMPs Induce Both Mash1 and Phox2b in Sympathetic Precursors) (Bixby *et al.*, 2002). Conversely, postnatal gut neural crest stem cells are much more sensitive to gliogenic factors (including both NRG1 and Delta) than to neurogenic factors like BMPs and form glia on chick peripheral nerves (Kruger *et al.*, 2002). It remains to be seen whether differences in the expression levels of Notch and Numb also underlie these differences in sensitivity to gliogenic and neurogenic cues.

Summary of Schwann Cell Differentiation

Schwann cell differentiation, like satellite cell differentiation, involves two instructive gliogenic cues: activation of Notch signaling, and gliogenic growth factors. Notch activation, by Notch ligands present on differentiating neurons and axons, instructively promotes gliogenesis. Membrane-bound NRG1 type III, which is probably present on axons, instructively promotes Schwann cell differentiation. Different neural crest stem cell populations, isolated from different locations and developmental stages, show instrinsic differences in their sensitivity to gliogenic signals. These may be related to differences in the levels of expression of Notch and Numb, probably triggered by local neural crest cell–cell interactions involving Notch ligands. Such differences may help promote appropriate glial (or neuronal) fate decisions by multipotent neural crest progenitors.

Autonomic Gangliogenesis

The peripheral autonomic nervous system is by far the most complex division of the PNS. In order to aid the discussion of the control of differentiation of various autonomic cell types, the subdivisions of the autonomic nervous system are introduced below.

The Autonomic Nervous System: An Introduction

The autonomic nervous system has three major divisions: sympathetic, parasympathetic, and enteric. The sympathetic and parasympathetic subdivisions innervate smooth muscle, cardiac muscle, and glands (Fig. 8), and mediate various visceral reflexes. The enteric nervous system controls the motility and secretory function of the gut, pancreas, and gall bladder.

All peripheral autonomic neurons and glia are derived from the neural crest. These include the postganglionic motor neurons and satellite glia of the sympathetic and parasympathetic divisions, which are collected together in peripheral ganglia

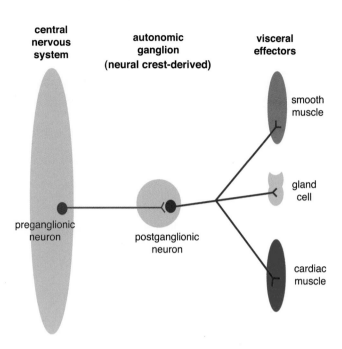

FIGURE 8. Schematic showing the structure of the autonomic nervous system. All peripheral autonomic neurons (sympathetic, parasympathetic, and enteric) are derived from the neural crest. Modified from Iversen *et al.* (2000).

(Fig. 8). The neurons in these ganglia are activated by preganglionic efferent neurons located in the brainstem and spinal cord (Fig. 8). Sympathetic ganglia are found in chains on either side of the spinal cord and hence are some considerable distance from their targets, while parasympathetic ganglia lie close to or are embedded in their target tissues. Enteric ganglia are located within the gut itself; they function relatively autonomously with respect to central nervous system input.

Preganglionic sympathetic neurons extend from the first thoracic spinal segment to upper lumbar segments; they innervate the bilateral chains of sympathetic ganglia. The postganglionic sympathetic neurons in these ganglia are derived from trunk neural crest cells that settle near the dorsal aorta to form the primary sympathetic chains. They innervate the glands and visceral organs, including the heart, lungs, gut, kidneys, bladder, and genitalia. Most of these neurons are noradrenergic, that is, release noradrenaline, a catecholamine derived from tyrosine via dopamine (Fig. 9). Some mature postganglionic sympathetic neurons, however, are cholinergic, that is, release acetylcholine. The endocrine (chromaffin) cells of the adrenal medulla, which are derived from a specific level of the trunk neural crest (somite levels 18–24 in the chick), are developmentally and functionally related to postganglionic sympathetic neurons (reviewed in Anderson, 1993). Adrenal chromaffin cells are adrenergic: They release adrenaline, another catecholamine, in turn derived from noradrenaline (Fig. 9).

Preganglionic parasympathetic neurons are found in various brain stem nuclei and in the sacral spinal cord. The brain stem nuclei innervate postganglionic neurons in cranial

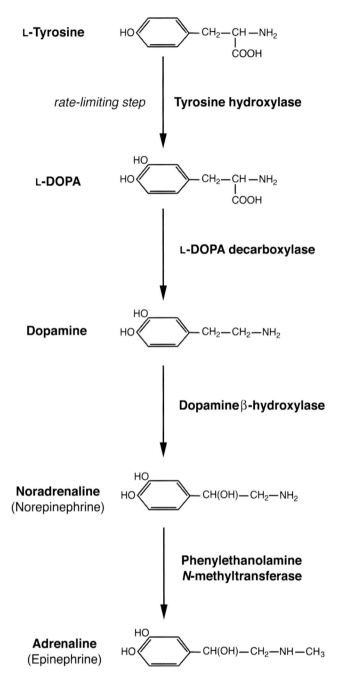

FIGURE 9. Catecholamine biosynthesis pathway: Intermediate stages in the formation of adrenaline. Redrawn from Blaschko (1973).

parasympathetic ganglia, including the ciliary, otic, sphenopalatine, and submandibular ganglia. These postganglionic neurons are derived from the cranial neural crest (Table 1), and innervate the eye, and lacrimal and salivary glands. Preganglionic parasympathetic axons exiting in the vagal nerve (cranial nerve X) innervate postganglionic neurons in cardiac ganglia and are embedded in the visceral organs of the thorax and abdomen. These postganglionic neurons are derived from vagal neural crest cells (Table 1). Preganglionic parasympathetic neurons in the

sacral spinal cord innervate the pelvic ganglion plexus, which is derived from sacral neural crest cells (Table 1). The neurons in this plexus innervate the colon, bladder, and external genitalia. Most of these postganglionic parasympathetic neurons are cholinergic, that is, release acetylcholine.

The enteric nervous system, which is entirely derived from vagal and sacral levels of the neural crest (Table 1), contains local sensory neurons (responding to specific chemicals, stretch, and tonicity), interneurons, and motor neurons, together with their associated glia. Enteric neurons innervate smooth muscle, local blood vessels, and mucosal secretory cells. They use a variety of neurotransmitters: Catecholaminergic, cholinergic, and serotonergic neurons can all be identified within the enteric nervous system.

Phox2b is Essential for the Formation of all Autonomic Ganglia

The paired-like homeodomain transcription factor Phox2b is expressed in all autonomic neural crest cell precursors (reviewed in Brunet and Pattyn, 2002; Goridis and Rohrer, 2002). *Phox2b* expression begins in prospective sympathetic neural crest cells as they aggregate at the aorta, and in enteric neural crest cells as they invade the gut (Pattyn *et al.*, 1997, 1999). In *Phox2b*-null mice, all these autonomic precursor cells die by apoptosis, so mutant animals lack all autonomic neurons and glia, that is, all sympathetic, parasympathetic, and enteric ganglia (Pattyn *et al.*, 1999).

Intriguingly, Phox2b is also expressed in and required for the development of visceral sensory neurons derived from the epibranchial placodes (Pattyn *et al.*, 1997, 1999) (Fig. 11; section Neurogenesis in the Epibranchial Placodes). These neurons provide autonomic afferent innervation to the visceral organs. Hence, Phox2b seems to be a pan-autonomic marker, despite the enormous variety of peripheral autonomic neural phenotypes. These include not only postganglionic neurons and satellite glia, but also autonomic sensory neurons, for example, enteric sensory neurons, and epibranchial placode-derived visceral sensory neurons. *Phox2b*-null mice lack the neural circuits underlying medullary autonomic reflexes (for a discussion of Phox2b in the CNS, see Brunet and Pattyn, 2002; Goridis and Rohrer, 2002).

Phox2b Is Required for Development of the Noradrenergic Phenotype

Within sympathetic and enteric precursors, Phox2b is required for expression of the *tyrosine hydroxylase* and *dopamine β-hydroxylase* (*DBH*) genes; these encode two enzymes in the catecholamine biosynthesis pathway (Fig. 9) (Pattyn *et al.*, 1999). Hence, Phox2b is an essential determinant of the catecholaminergic (particularly noradrenergic) phenotype. Several transcription factors that act downstream of Phox2b in sympathetic neurons to control noradrenergic differentiation have been identified. These include the closely related protein Phox2a (which functions upstream of Phox2b in epibranchial placode-derived neurons; see Brunet and Pattyn, 2002), the bHLH protein dHAND (HAND2), and the zinc finger protein Gata3 (reviewed

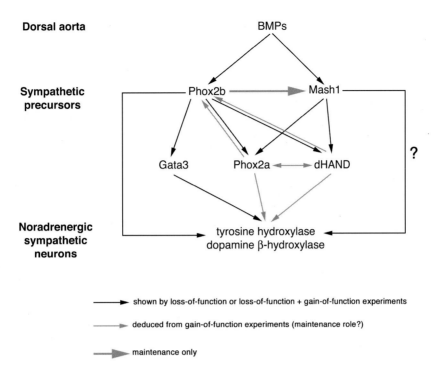

FIGURE 10. Regulatory network of transcription factors controlling sympathetic neuron development. See the section on Automatic Gangliogenesis for details. Question mark on arrow from Mash1 to *tyrosine hydroxylase* and *dopamine β-hydroxylase* indicates current uncertainty as to whether Mash1 acts on their promoters only through dHAND. BMPs, bone morphogenetic proteins. Modified from Goridis and Rohrer (2002).

in Brunet and Pattyn, 2002; Goridis and Rohrer, 2002). Although these factors are genetically downstream of Phox2b in sympathetic ganglia, together they form a complex regulatory network, in which most actions seem to be reciprocal (e.g., forced expression of dHAND can ectopically activate Phox2b) (Fig. 10) (reviewed in Brunet and Pattyn, 2002; Goridis and Rohrer, 2002).

Phox2b and Phox2a can each directly activate the *DBH* promoter, either alone or in conjunction with activation of the cyclic AMP second-messenger pathway (reviewed in Brunet and Pattyn, 2002; Goridis and Rohrer, 2002). There is some evidence that Phox2a can directly activate the *tyrosine hydroxylase* promoter, but again, cyclic AMP signaling may be required (see Goridis and Rohrer, 2002). Ectopic retroviral-mediated expression of either *Phox2b* or *Phox2a* in chick embryos promotes the formation of ectopic sympathetic neurons from trunk neural crest cells (Stanke *et al.*, 1999). These neurons express pan-neuronal markers, noradrenergic markers (tyrosine hydroxylase and *DBH*), and also cholinergic markers (e.g., *choline acetyltransferase*) (Stanke *et al.*, 1999). Hence, Phox2 proteins are sufficient to specify the differentiation of sympathetic neurons (including expression of both pan-neuronal and subtype-specific markers) *in vivo*.

In similar overexpression experiments in the chick, Phox2 proteins were found to be sufficient to induce expression of the bHLH transcription factor dHAND in trunk neural crest cells (Howard *et al.*, 2000). Expression of dHAND alone is likewise sufficient to elicit the formation of catecholaminergic sympathetic neurons, both *in vitro* and *in vivo* (Howard *et al.*, 1999,

2000). Indeed, dHAND and Phox2a act synergistically to enhance DBH transcription (Xu *et al.*, 2003).

The zinc finger transcription factor Gata3 is also genetically downstream of Phox2b (Goridis and Rohrer, 2002). In *Gata3*-null mice, sympathetic ganglia form but the neurons fail to express tyrosine hydroxylase and have reduced levels of DBH, suggesting that *Gata3* is also essential for the noradrenergic phenotype (Lim *et al.*, 2000).

This complex network of transcriptional regulation (Fig. 10) is perhaps the best characterized example of how neurotransmitter identity is controlled at the molecular level. One important gene in this network that has not yet been discussed, however, is the *achaete-scute* homologue *ash1* (*Mash1*) (sections Proneural Genes: An Introduction; *Mash1* Is Essential for Noradrenergic Differentiation). Although Phox2b is required to maintain *Mash1* expression, *Mash1* is induced independently of Phox2b in autonomic precursors, and itself induces a number of the same downstream genes (section *Mash1* Is Essential for Noradrenergic Differentiation).

Phox2b Is Required for Ret Expression in a Subset of Neural Crest Cells

Phox2b is required for expression of the receptor tyrosine kinase Ret in a subset of enteric precursors and in the most rostral sympathetic ganglion, the superior cervical ganglion (SCG) (Pattyn *et al.*, 1999). These cells are completely absent in Ret-deficient mice (Durbec *et al.*, 1996). One of the family of ligands that signal through Ret, glial cell line-derived neurotrophic factor (GDNF), is essential for the development of the entire enteric nervous system (Moore *et al.*, 1996) (section The Differentiation of Enteric Neurons; reviewed in Young and Newgreen, 2001; Airaksinen and Saarma, 2002).

Mash1 Is Essential for Noradrenergic Differentiation

Mash1 (mouse Ash1), a bHLH transcription factor related to the invertebrate proneural Achaete-Scute complex (section Proneural Genes: An Introduction; Chapter 5), was the first transcription factor found to be necessary for sympathetic development. Like Phox2b, Mash1 is expressed in all neural crest-derived autonomic precursors (sympathetic, parasympathetic, and enteric). Unlike Phox2b, however, it is not expressed in epibranchial placode-derived visceral sensory neurons. Mash1 is first expressed in sympathetic precursors shortly after they settle near the dorsal aorta. Like Phox2b, Mash1 is essential for *DBH* expression in all cell types except epibranchial placode-derived neurons; that is, Mash1 is a noradrenergic determinant, independent of Phox2b (Hirsch *et al.*, 1998).

In *Mash1*-null mice, sympathetic and parasympathetic ganglia form (and express Phox2b), but pan-neuronal markers, Phox2a, tyrosine hydroxylase, and *DBH* are all lacking, and most (but not all) sympathetic and parasympathetic neuroblasts subsequently degenerate (Guillemot *et al.*, 1993; Hirsch *et al.*, 1998). *dHAND* expression is also reported to be missing in these embryos (Anderson and Jan, 1997). If Mash1 is constitutively expressed in cultured neural crest stem cells, it induces both *Phox2a* and *Ret*, together with pan-neuronal markers and morphological neuronal differentiation (Lo *et al.*, 1998). Hence, *Phox2a*, *dHAND*, and *Ret* expression are induced not only by Phox2b, but also by Mash1. Mash1, like Phox2b, therefore, couples expression of pan-neuronal and neuronal subtype-specific markers (Fig. 10) (reviewed in Goridis and Rohrer, 2002). However, this linkage can be uncoupled experimentally: Floorplate ablation in the chick abolishes *Phox2a* and *tyrosine hydroxylase* expression, but not *Cash1* (chick Ash1) or pan-neuronal marker expression, in neural crest cells near the dorsal aorta (Groves *et al.*, 1995). This suggests that a floorplate-derived signal, in addition to Mash1, is required for noradrenergic identity in prospective sympathetic neurons (section Floorplate-Derived Signals). Hence, Mash1 expression is not sufficient, in all contexts, to promote noradrenergic identity. Indeed, Mash1 alone does not promote autonomic neurogenesis *in vitro* in the absence of BMP2; hence it must interact with other factors induced by BMP2, such as Phox2b (Lo *et al.*, 2002) (section BMPs Induce Both Mash1 and Phox2b in Sympathetic Precursors).

Interestingly, given the requirement of Gata3 for noradrenergic development (section *Phox2b* Is Required for Development of the Noradrenergic Phenotype), the *Drosophila* Gata factor Pannier can either activate or repress *achaete-scute* complex genes, in association with various transcriptional cofactors (Ramain *et al.*, 1993; Skaer *et al.*, 2002). This suggests

a mechanism whereby Gata3 might also interact with *Mash1*, as well as being downstream of Phox2b, although currently there is no evidence for this (Goridis and Rohrer, 2002).

A subset of enteric neurons, including apparently all serotonergic enteric neurons, is also missing in *Mash1*-null mice (Blaugrund *et al.*, 1996; Hirsch *et al.*, 1998). Since serotonergic enteric neurons seem to develop from tyrosine hydroxylase-expressing precursors, this loss is perhaps to be expected (Blaugrund *et al.*, 1996).

Mash1 Also Plays Roles in Sensory Neurogenesis

Mash1 is not only required for the development of autonomic neurons, and it does not always function by inducing *Phox2a*. The mesencephalic nucleus of the trigeminal nerve, which was introduced in the section on Neural Crest Derivatives as a (somewhat controversial) neural crest derivative within the brain, also depends on Mash1, but never expresses Phox2a (Hirsch *et al.*, 1998). Mash1 is also essential for the development of olfactory neuron progenitors in the olfactory placode, which likewise do not express Phox2a (Guillemot *et al.*, 1993; Cau *et al.*, 1997) (section A bHLH Transcription Factor Cascade Controls Olfactory Neurogenesis). Hence, different neuronal subtype-specific factors must cooperate with Mash1 in the formation of these cell types.

BMPs Induce Both Mash1 and Phox2b in Sympathetic Precursors

Neural crest cells that migrate past the notochord and stop in the vicinity of the dorsal aorta (section Migration Arrest at Target Sites) will form the neurons and satellite cells of the sympathetic ganglia. Transplantation, rotation, and ablation experiments in the chick suggest that catecholaminergic neuronal differentiation only occurs near the aorta/mesonephros and also requires the presence of either the ventral neural tube or the notochord (Teillet and Le Douarin, 1983; Stern *et al.*, 1991; Groves *et al.*, 1995).

As described above, both *Phox2b* and *Mash1* are first expressed shortly after neural crest cells arrive at the dorsal aorta. At this time, the dorsal aorta expresses *Bmp2*, *Bmp4*, and *Bmp7* (Reissmann *et al.*, 1996; Shah *et al.*, 1996). All three factors induce increased numbers of catecholaminergic cells in neural crest cell cultures, as does forced expression of a constitutively active BMP receptor (reviewed in Goridis and Rohrer, 2002). BMP2 induces *Mash1* and *Phox2a* in cultured neural crest stem cells (Shah *et al.*, 1996; Lo *et al.*, 1998). Overexpression of BMP4 near the developing sympathetic ganglia leads to the ectopic formation of catecholaminergic cells *in vivo* (Reissmann *et al.*, 1996). Conversely, when beads soaked in the BMP inhibitor Noggin are placed near the dorsal aorta in the chick, sympathetic ganglia initially form, but sympathetic neurons do not develop (Schneider *et al.*, 1999). In these Noggin-treated embryos, sympathetic ganglia lack expression of pan-neuronal markers, and of *Phox2b*, *Phox2a*, *DBH*, and tyrosine hydroxylase, while *Cash1* is strongly reduced (Schneider *et al.*, 1999). Together, these results provide overwhelming evidence that dorsal aorta-derived BMPs induce expression of both *Phox2b* and *Mash1*, thus initiating the regulatory network of transcription factors that leads eventually to sympathetic neuron differentiation. However, these cues may be insufficient for catecholaminergic differentiation *in vivo*, as discussed in the following section.

Floorplate-Derived Signals Are Also Required for Catecholaminergic Differentiation

In addition to signals from the dorsal aorta, the presence of floorplate and/or notochord is also required for catecholaminergic differentiation (Teillet and Le Douarin, 1983; Stern *et al.*, 1991; Groves *et al.*, 1995). In particular, although neurons differentiate in the sympathetic ganglia in the absence of floorplate, they do not express catecholaminergic markers (Groves *et al.*, 1995). This suggests that in addition to BMPs from the dorsal aorta (which induce *Phox2b* and *Mash1*), floorplate-derived signals are also required to induce or maintain subtype-specific markers in the sympathetic ganglia (Groves *et al.*, 1995). Sonic hedgehog (see Chapter 3) seems to have little effect on catecholaminergic differentiation (Reissmann *et al.*, 1996), and the molecular nature of the floorplate-derived signal(s) remains unclear. It may be relevant in this context that enhanced cyclic AMP signaling is required for efficient activation of the *tyrosine hydroxylase* promoter by Phox2a *in vitro* (reviewed in Goridis and Rohrer, 2002). Also, activation of the mitogen-activated protein (MAP) kinase signaling cascade in avian neural crest cells causes catecholaminergic differentiation independently of BMP4 (Wu and Howard, 2001). Clearly, there is still much to learn about the control of sympathetic neuron differentiation.

BMPs and Parasympathetic vs Sympathetic Differentiation

The differentiation of parasympathetic vs sympathetic autonomic neurons may be determined by local concentrations of BMPs at different neural crest target sites, as well as, perhaps, differential sensitivities of responding neural crest cells to BMPs (White *et al.*, 2001). Postmigratory rat neural crest stem cells, isolated from fetal sciatic nerve, are more likely to differentiate as cholinergic parasympathetic neurons than as catecholaminergic sympathetic neurons when back-grafted into chick neural crest migratory pathways (White *et al.*, 2001). After such grafts, they form cholinergic neurons in both sympathetic ganglia and parasympathetic ganglia, such as the pelvic plexus (White *et al.*, 2001). In culture, they respond to BMP2 by differentiating as both cholinergic and noradrenergic autonomic neurons. However, they are significantly less sensitive to the neuronal differentiation-inducing activity of BMP2 than are migrating neural crest stem cells (section Differences in the Sensitivity of Different Neural Crest Stem Cells to Gliogenic Cues), and differentiate as cholinergic neurons at lower BMP2 concentrations (White *et al.*, 2001). The molecular basis for this cholinergic bias is unknown.

BMPs are expressed at some sites of parasympathetic gangliogenesis. For example, the caudal cloaca, located proximal to the forming parasympathetic pelvic plexus, expresses BMP2 at an appropriate time to be involved in inducing parasympathetic neuronal differentiation (White *et al.*, 2001).

The Differentiation of Enteric Neurons

BMP2, which is expressed in gut mesenchyme, promotes the neuronal maturation of postmigratory enteric neural precursors isolated from the rat gut (Pisano *et al.*, 2000). However, several other growth factors have also been found to affect enteric neuronal differentiation.

Glial cell line-derived neurotrophic factor (GDNF) is the founding member of a family of ligands that act via a common signal transducer, the receptor tyrosine kinase Ret, complexed with ligand-specific receptors, the GDNF family receptor-α (GFRα) receptors (reviewed in Airaksinen and Saarma, 2002). GDNF is expressed in gut mesenchymal cells, and the entire enteric nervous system is missing in *GDNF*-deficient mice (Moore *et al.*, 1996). In *Ret*-deficient mice, all enteric neurons and glia are missing from the gut below the level of the esophagus and the immediately adjacent stomach (Durbec *et al.*, 1996). GDNF and Neurturin, another GDNF family ligand, promote the *in vitro* survival, proliferation, and neuronal differentiation of migrating and postmigratory Ret$^+$ enteric precursors from the rat gut (Taraviras *et al.*, 1999).

The growth factor Endothelin3 (Edn3), conversely, seems to inhibit the neuronal differentiation of enteric precursors, thus maintaining a sufficiently large pool of migratory, undifferentiated precursors to colonize the entire gut (Hearn *et al.*, 1998; Shin *et al.*, 1999). Endothelin3 prevents the neurogenic activity of GDNF on migrating enteric neural precursors isolated from the quail embryo gut (Hearn *et al.*, 1998).

Mutations that affect the Ret or Endothelin signaling pathways cause Hirschsprung's disease in humans, in which enteric ganglia are missing from the terminal colon (reviewed in Gershon, 1999; Manie *et al.*, 2001; McCallion and Chakravarti, 2001).

Differentiation of Satellite Cells in Autonomic Ganglia

Strong autonomic neurogenic cues, such as BMP2, are clearly present at sites of autonomic gangliogenesis. How, then, do satellite glia form within autonomic ganglia? Exposure to gliogenesis-promoting factors such as NRG1 type II (section Differentiation of DRG Satellite Cells) is insufficient. Cultured rat neural crest stem cells rapidly commit to an autonomic neuronal fate on exposure to BMP2, but only commit to a glial fate after prolonged exposure to NRG1 type II (Shah and Anderson, 1997). Furthermore, saturating concentrations of BMP2 are dominant over NRG1 type II (although at low BMP2 concentrations, NRG1 type II can attenuate *Mash1* induction by BMP2) (Shah and Anderson, 1997). These results may explain why, *in vivo*, neurons differentiate before glia in autonomic ganglia. What, then, prevents all autonomic progenitors from differentiating into neurons?

Activation of the Notch signaling pathway seems to be essential for adoption of a glial fate in the presence of BMP2

(Morrison *et al.*, 2000b). As discussed in the section on Notch Activation Leads to Gliogenesis by Neural Crest Stem Cells, even transient activation of Notch signaling inhibits neuronal differentiation and instructively promotes glial differentiation, in cultures of postmigratory neural crest stem cells isolated from fetal rat sciatic nerve (Morrison *et al.*, 2000b). This action of Notch is dominant over that of BMP2, blocking neurogenesis at a point upstream of *Mash1* induction (Morrison *et al.*, 2000b). It is likely that a similar mechanism of Notch activation acts within autonomic ganglia to promote satellite cell differentiation in the presence of BMP2. One model suggested by these results is that differentiating autonomic neurons express Notch ligands; these then activate Notch signaling in neighboring non-neuronal cells, which are then able to differentiate as glia (Morrison *et al.*, 2000b). Other gliogenesis-promoting factors, such as NRG1 type II, may also act in concert with, or reinforce, the gliogenic action of Notch in peripheral autonomic ganglia (Hagedorn *et al.*, 2000b). It is possible that once Notch is activated, preventing a neuronal fate and promoting a glial fate, NRG1 type II may then be able to promote a satellite cell fate (Hagedorn *et al.*, 2000b; Leimeroth *et al.*, 2002).

Summary of Autonomic Gangliogenesis

Phox2b is required for the formation of the entire peripheral autonomic nervous system. It is also necessary and sufficient for catecholaminergic (particularly noradrenergic) neuronal differentiation. Mash1 is necessary, but not sufficient, for noradrenergic differentiation. Phox2b and Mash1 interact in a complex regulatory network of transcription factors to induce noradrenergic differentiation. They are independently induced in sympathetic precursors by BMPs from the dorsal aorta; however, additional floorplate-derived signals are also required for catecholaminergic differentiation of sympathetic neurons. BMPs may also induce parasympathetic fates: The choice between parasympathetic and sympathetic fates may depend on local BMP concentrations and intrinsic differences in the sensitivity of different postmigratory neural crest cell populations to BMPs. BMPs and GDNF promote the differentiation of enteric neurons, while Edn3 may prevent enteric neuronal differentiation. Satellite cell differentiation requires Notch activation, which is dominant to the neurogenesis-promoting activity of BMPs. The gliogenic activity of NRG1 type II is subordinate to BMPs, but may be able to promote satellite cell differentiation once Notch has been activated.

Community Effects Alter Fate Decisions

A multipotent neural crest cell may adopt one fate in response to a given instructive growth factor when it is alone, but a different fate when it is part of a cluster ("community") of neural crest cells (reviewed in Sommer, 2001). Individual postmigratory multipotent cells isolated from embryonic rat DRG respond to BMP2 by forming both autonomic neurons and smooth muscle cells, while clusters of the same multipotent cells form significantly more autonomic neurons, at the expense of smooth muscle cells (Hagedorn *et al.*, 1999, 2000a).

This "community effect" (Gurdon *et al.*, 1993) may prevent neural crest cells in autonomic ganglia from adopting an aberrant (smooth muscle) fate in response to BMP2 *in vivo*.

Different concentrations of the same factor can also have different effects when local neural crest cell–cell signaling is allowed to occur. Individual postmigratory progenitors from rat DRG respond to TGFβ by adopting a predominantly smooth muscle fate; they never form neurons (Hagedorn *et al.*, 1999, 2000a). Although high doses of TGFβ cause some cell death, the predominant fate choice is still smooth muscle (Hagedorn *et al.*, 2000a). Clusters of these progenitors, in contrast, respond to high TGFβ doses by dying, and to low TGFβ doses by forming autonomic neurons (Hagedorn *et al.*, 1999, 2000a).

Similar community effects may underlie the results discussed in the section on Axial Fate-Restriction, in which individual trunk neural crest cells form cartilage in the head when surrounded by host cartilage cells, but coherent masses of trunk neural crest cells do not (McGonnell and Graham, 2002). Community effects also help to maintain neural crest cell regional identity: Individual neural crest cells will change their *Hox* gene expression patterns in response to environmental cues, while large groups of neural crest cells do not (e.g., Golding *et al.*, 2000; Trainor and Krumlauf, 2000; Schilling *et al.*, 2001).

In summary, local neural crest cell–cell interactions may reinforce fate choice in particular environmental contexts, and prevent inappropriate fate choices in response to environmental cues.

NEURAL CREST SUMMARY

Since the last edition of this book, in 1991, there has been an explosion of information about the genes and signaling pathways important for neural crest cell development. Molecular cues involved in neural crest cell induction at the neural plate border have now been identified. These include BMPs, which are important for setting up the neural plate border itself and, later, for neural crest cell delamination, and Wnts, which are both necessary and sufficient for neural crest precursor cell induction within the neural plate border. Numerous repulsive guidance cues, including ephrins, are now known to play essential roles in sculpting the migration pathways of both cranial and trunk neural crest cells, and some progress has been made in understanding migration arrest at target sites. The migrating neural crest cell population is heterogeneous, containing multipotent and fate-restricted cells; however, the latter do not seem to be determined; that is, they retain the potential to adopt other fates when challenged experimentally. There is a greater molecular understanding of lineage diversification, and it is becoming apparent that the sensory–autonomic lineage decision is taken before the neuronal–glial fate decision. Various transcription factors are known to be essential for the formation of particular neural crest lineages, including Phox2b for the autonomic lineage and Sox10 for the glial lineage. Several instructive differentiation cues that act on multipotent neural crest cells, including BMPs and NRGs, have been identified. Finally, an emerging theme is that neural crest cell–cell interactions, including community effects, are important in determining neural crest cell fate choice. Clearly, a great deal has been learned about neural crest cell induction, migration, and differentiation. However, many questions still remain, and there are many fruitful avenues for future research into the development of these fascinating cells.

OVERVIEW OF CRANIAL ECTODERMAL PLACODES

Cranial ectodermal placodes (Greek root πλακ, i.e., flat plate, tablet) are discrete patches of thickened ectoderm that appear transiently in the head of all vertebrate embryos (reviewed in Webb and Noden, 1993; Baker and Bronner-Fraser, 2001; Begbie and Graham, 2001a). They were discovered 120 years ago (van Wijhe, 1883) and given the name "placode" by von Kupffer (1894). Placodes give rise to the bulk of the peripheral sensory nervous system in the head. The olfactory, otic, and lateral line placodes give rise to the paired peripheral sense organs (olfactory epithelium, inner ears, and lateral line system of anamniotes) together with their afferent innervating neurons. The lens placodes give rise to the lenses of the eye. The trigeminal placodes form many of the cutaneous sensory neurons that innervate the head, including the jaws and teeth. The epibranchial placodes give rise to visceral sensory neurons that provide afferent innervation for tastebuds, and afferent autonomic innervation for the

visceral organs. Finally, the hypophyseal (or adenohypophyseal) placode gives rise to all of the endocrine cells and supporting cells of the adenohypophysis (anterior pituitary gland). Although the molecular mechanisms underlying the induction and development of the hypophyseal placode are perhaps the best understood of all the placodes, its development will not be discussed here (for detailed reviews of hypophyseal placode induction and development, see Baker and Bronner-Fraser, 2001; Dasen and Rosenfeld, 2001; Scully and Rosenfeld, 2002).

Like the neural crest, therefore, placodes give rise to a very diverse array of cell types, including sensory receptors, sensory neurons, supporting cells, secretory cells, glia, neuroendocrine, and endocrine cells (Table 3). Figure 11 shows a fate-map of the placode-forming ectoderm in the head of the 8-somite stage chick embryo, together with the respective neuronal contributions to cranial sensory ganglia of placodes and the neural crest. Figure 12 shows the location of the different placodes in the head of the 19-somite stage *Xenopus* embryo. It is evident from these schematics that a relatively large proportion of dorsal head ectoderm contributes to placodal tissue.

Also like the neural crest, cranial ectodermal placodes are usually considered to be a defining characteristic of the craniates (vertebrates plus hagfish) (Gans and Northcutt, 1983; Northcutt and Gans, 1983; Baker and Bronner-Fraser, 1997). However, molecular analyses suggest that at least some vertebrate placodes may have homologues in non-vertebrate chordates (e.g., Boorman and Shimeld, 2002; Christiaen *et al.*, 2002). Placodes

TABLE 3. Cell Types and Cells Derived from Cranial Ectodermal Placodes

Placode	General cell type	Cells
Olfactory	Sensory ciliary receptor	Chemoreceptive olfactory receptor neurons
	Sensory neurons	Olfactory receptor neurons
	Glia	Olfactory ensheathing glia
	Neuroendocrine cells	Gonadotropin-releasing hormone (GnRH)-producing neurons
	Secretory/support cells	Sustentacular cells (secrete mucus; provide support)
Otic	Sensory ciliary receptor	Mechanosensory hair cells
	Sensory neurons	Otic hair cell-innervating neurons, collected in vestibulo-cochlear ganglion of cranial nerve VIII
	Secretory cells	Cupula-secreting cells; endolymph-secreting cells; cells secreting biomineralized matrix of otoliths/otoconia
	Supporting cells	Hair cell support cells; non-sensory epithelia
Lateral line	Sensory ciliary receptor	Mechanosensory hair cells in neuromasts
		Electroreceptive cells in ampullary organs
	Sensory neurons	Lateral line hair cell-innervating neurons, collected in lateral line ganglia
	Secretory cells	Cupula-secreting cells in neuromasts
	Supporting cells	Hair cell support cells in neuromasts
Lens	Specialized epithelium	Lens fiber cells
Ophthalmic and maxillomandibular trigeminal	Sensory neurons	Cutaneous sensory neurons (pain, touch, temperature), collected in trigeminal ganglion of cranial nerve V
Epibranchial	Sensory neurons	Afferent neurons for taste buds and visceral organs, collected in geniculate, petrosal, and nodose ganglia (distal ganglia of cranial nerves VII, IX, and X, respectively)
Hypophyseal	Endocrine cells	All endocrine cells of the adenohypophysis (anterior pituitary gland)
	Supporting cells	Support cells of the adenohypophysis

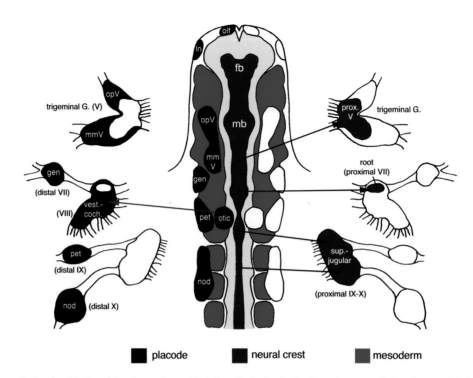

FIGURE 11. Fate-map of placodes (black ovals) and neural crest (dark blue) in the head of an 8-somite stage chick embryo, and their neuronal contribution to the sensory ganglia of the cranial nerves (Roman numerals). All satellite cells in cranial sensory ganglia are derived from the neural crest. fb, forebrain; G., ganglion; gen, geniculate; ln, lens; mb, midbrain; mmV, maxillomandibular trigeminal; nod, nodose; olf., olfactory; opV, ophthalmic trigeminal; pet, petrosal; prox., proximal; sup., superior; vest.-coch., vestibulocochlear. Redrawn from D'Amico-Martel and Noden (1983).

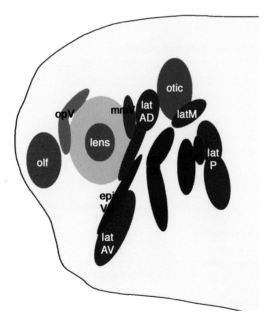

FIGURE 12. Location of placodes in the head of a 19-somite stage *Xenopus* embryo. epi, epibranchial placode; epi VII, facial/geniculate placode; epi IX, glossopharyngeal/petrosal placode; epi X, vagal/nodose placodes; lat, lateral line placode; latAD, anterodorsal lateral line placode; latAV, anteroventral lateral line placode; latM, middle lateral line placode; latP, posterior lateral line placode; mmV, maxillomandibular trigeminal placode; olf., olfactory placode; opV, ophthalmic trigeminal placode. Redrawn from Schlosser and Northcutt (2000).

have been studied in all craniate classes, including hagfish (e.g., Wicht and Northcutt, 1995) and lamprey (e.g., Bodznick and Northcutt, 1981; Neidert *et al.*, 2001; McCauley and Bronner-Fraser, 2002). Although most research has been done on the sense organ placodes, in particular the lens and otic placodes, as well as the hypophyseal placode, molecular information has also enabled closer investigation of the development of the trigeminal and epibranchial placodes. Here, a relatively brief summary is provided of the current state of knowledge of the induction and development of the different placodes. For a more detailed review of classical and modern research into placode induction, the reader is referred to Baker and Bronner-Fraser (2001).

A PREPLACODAL FIELD AT THE ANTERIOR NEURAL PLATE BORDER

All fate-mapping studies to date have shown that placodes arise from ectoderm at the neural plate border in the prospective head region (Baker and Bronner-Fraser, 2001). Older fate maps suggest that placodes originate from ectoderm lying lateral to the neural crest-forming area, except in the most rostral region, where no neural crest cells form and olfactory and hypophyseal placodes directly abut prospective neural plate territories (Baker and Bronner-Fraser, 2001). However, cell lineage analysis shows that placodal precursors, like neural crest precursors (section Embryonic Origin of the Neural Crest), do not exist as a segregated population (Streit, 2002). Although prospective placodal territory extends more laterally than prospective neural crest territory, placodal and neural crest precursors are mingled together more medially (Streit, 2002).

Molecular evidence supports some early morphological observations in suggesting that there is a preplacodal field, or panplacodal anlage, around the anterior neural plate. This field is morphologically visible in the frog, *Rana*, which has a continuous band of thickened ectoderm around the edge of the anterior neural plate, from which most placodes originate (Knouff, 1935) (Fig. 13A). Molecularly, this field seems to be characterized in multiple species by the expression of various genes in a horseshoe-shaped band around the anterior neural plate border (Figs 13B, C). These genes, which primarily encode transcription factors, are often subsequently maintained in all or multiple placodes. They include the homeodomain transcription factors Six1, Six4, Dlx3, Dlx5, and Dlx7, the HMG-domain transcription factor Sox3 (which is also expressed in the neural plate), and the transcription cofactors Eya1 and Eya2 (for original references, see Baker and Bronner-Fraser, 2001; also David *et al.*, 2001; Ghanbari *et al.*, 2001). See the section on Establishment of the Neural Plate Border for a discussion of the role of Dlx genes in positioning the neural plate border. In the chick, the expression domains of these genes are not coincident; rather, they are expressed in a series of overlapping domains that shift both spatially and temporally with the position of placodal precursors (Streit, 2002).

It is clear that several of these genes play important roles in the development of multiple placodes. For example, *dlx3*, acting in concert with *dlx7*, is necessary for the formation of both olfactory and otic placodes in the zebrafish (Solomon and Fritz, 2002). Ectopic expression of *Sox3* in another teleost fish, medaka, causes ectopic lens and otic vesicle formation in ectodermal regions relatively close to the endogenous lens and otic placodes (Köster *et al.*, 2000). Sox3 may, therefore, act as a competence factor, enabling ectopic ectoderm to respond to placode-inducing signals (Köster *et al.*, 2000).

However, the precise significance of the preplacodal domain of gene expression remains unclear: It does not seem to correlate either with the site of origin of all placodal precursor cells, or with determination toward a placodal fate. A cell lineage analysis in the chick showed that some otic placode precursors arise from ectoderm lying medial to the *Six4* expression domain (Streit, 2002). Hence, not all placodal precursors originate from the *Six4*$^+$ domain. Furthermore, cells within the *Six4*$^+$ domain form not only placodal derivatives, but also neural crest and epidermis (and neural tube until the 2-somite stage, at the level of the future otic placode) (Streit, 2002). Hence, cells within the preplacodal domain are not all determined toward a placodal fate.

Some insight into the function of the preplacodal domain may come from observations showing that there is a large degree of ectodermal cell movement in the neural plate border region (Whitlock and Westerfield, 2000; Streit, 2002). These studies combined cell lineage analysis (using DiI or fluorescent dextrans) with time-lapse analysis and *in situ* hybridization. Precursors of a particular placode, such as the olfactory placode in zebrafish (Whitlock and Westerfield, 2000) or the otic placode in chick (Streit, 2002), originate from a fairly large region of ectoderm at the anterior neural plate border, and subsequently converge to form the final placode. This may suggest a model whereby cells that move into the preplacodal gene expression domain upregulate the genes defining the domain, while cells that move out of the domain downregulate these genes. Cells that express the "preplacodal" genes may be rendered competent to respond to specific placode-inducing signals. However, the fate of a given cell within the preplacodal domain will depend on the precise combination of signals it subsequently receives. Hence, although it is competent to form placode, it may give rise to neural crest or epidermis instead.

The Pax/Six/Eya/Dach Regulatory Network

The overlapping expression of various *Six* and *Eya* genes in the preplacodal domain is of particular interest. Six and Eya

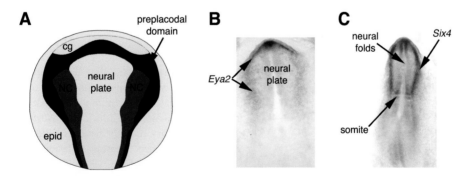

FIGURE 13. A preplacodal domain of ectoderm can be recognized around the anterior neural plate border, occasionally by morphology alone, more often by specific gene expression. (A) Fate map of open neural plate stage *Rana* embryo (dorsal view), showing the preplacodal domain, recognizable morphologically as a continuous strip of thickened ectoderm around the prospective neural crest (NC) domain. cg, prospective cement gland; epid, epidermis; NC, neural crest. Redrawn from Knouff (1935). (B) *Eya2* expression (dark staining) around the anterior neural plate border in a stage 6 (neurula-stage) chick embryo (dorsal view). (C) *Six4* expression (dark staining) around the anterior neural plate border in a 2-somite stage chick embryo (dorsal view). Both *Eya2* and *Six4* are subsequently maintained in most placodes (section A Preplacodal Field at the Anterior Neural Plate Border for details). Photographs courtesy of Dr. Andrea Streit and Anna Litsiou, King's College, London, United Kingdom. Chick staging after Hamburger and Hamilton (1951).

family members function in a complex cross-regulatory network with Pax transcription factors and the transcription cofactor Dachshund (Dach, also a multimember family in vertebrates), in a variety of developmental contexts. These include eye development in *Drosophila* (reviewed in Wawersik and Maas, 2000) and vertebrate muscle development (Heanue *et al.*, 1999). Dach family members are expressed in part of the preplacodal domain and in various placodes (Davis *et al.*, 2001; Hammond *et al.*, 2002; Heanue *et al.*, 2002; Loosli *et al.*, 2002). While *Pax* genes are not expressed in the preplacodal domain, most individual placodes are characterized by a particular combination of *Pax* gene expression. For example, *Pax6* is expressed in the olfactory and lens placodes, *Pax3* in the ophthalmic trigeminal placode, *Pax2/5/8* in the otic placode, and *Pax2* in the epibranchial placodes (reviewed in Baker and Bronner-Fraser, 2001). Mouse knockout studies have shown that *Pax* gene expression within the placodes is important for their proper development. For example, *Pax6* is essential for olfactory and lens placode formation, while *Pax2* is important for aspects of otic placode development (reviewed in Baker and Bronner-Fraser, 2001).

Given the above, it is possible that expression of *Six*, *Eya*, and perhaps also *Dach* genes within the preplacodal domain may represent a molecular framework common to all placodes. This network might then be able to interact with different *Pax* genes, induced in different regions of the preplacodal domain by specific placode-inducing signals, to specify individual placode identities. Although this model is attractive, further supporting evidence is required.

Models of Individual Placode Formation in the Preplacodal Domain

The active convergence of cells at the neural plate border to form specific placodes (Whitlock and Westerfield, 2000; Streit, 2002) suggests two possible models for the formation of individual placodes from ectoderm in this region (Streit, 2002). The first model proposes that cells in a large region of ectoderm receive a specific placode-inducing signal: Those cells that respond to the signal "sort out" from non-responding cells and actively migrate to the site of formation of the placode. The second model proposes that ectodermal cells move at random: Those that come within range of localized placode-inducing signals adopt a specific placodal fate. Evidence exists to support the existence of both localized and widespread placode-inducing signals (e.g., sections Induction of the Otic Placodes; Induction of the Trigeminal Placodes). However, it is currently unknown whether active sorting processes occur within the ectoderm to cause the aggregation of specific placode precursors.

Summary

All placodes originate from ectoderm at the anterior neural plate border. A horseshoe-shaped domain of ectoderm at the anterior neural plate border expresses numerous specific transcription factors, such as Six, Eya, and Dlx family members, all of which have roles in placode development. Cells expressing an appropriate combination of these genes may be competent to adopt a placodal fate in response to placode-inducing signals. Cells that respond to specific placode-inducing signals may sort out and aggregate to form the placode. Alternatively, cells may randomly move within range of specific placode-inducing signals, cease migrating, and differentiate accordingly. Further evidence is required to distinguish between these models.

In the following sections, the induction and some aspects of the development of individual placodes are discussed in turn, beginning with the sense organ placodes (olfactory, otic, lateral line, and lens), and ending with the trigeminal and epibranchial placodes.

SENSE ORGAN PLACODES

Olfactory Placodes

Olfactory Placode Derivatives

The paired olfactory placodes, which invaginate toward the telencephalon to form olfactory pits, give rise to the entire olfactory (odorant-sensing) and, where present, vomeronasal (pheromone-sensing) epithelia, together with the respiratory epithelium that lines the nasal passages. The olfactory and vomeronasal epithelia contain ciliated sensory receptor neurons, each of which bind odorants via a single member of an enormous family of G-protein-coupled, seven-transmembrane domain receptor molecules (reviewed in Mombaerts, 2001; Ronnett and Moon, 2002). The epithelia also contain basal cells, which generate olfactory sensory neurons throughout life (for a review on stem cells in the olfactory epithelium, see Calof *et al.*, 1998), and supporting sustentacular cells, which share some characteristics with glia (reviewed in Ronnett and Moon, 2002). All of these cells are derived from the olfactory placode.

The cell bodies of the olfactory receptor neurons remain in the placode, while their axons extend into the brain to form the olfactory, vomeronasal, and terminal nerves (for reviews of olfactory axon pathfinding, see Mombaerts, 2001; St. John *et al.*, 2002). These nerves are ensheathed by olfactory placode-derived glial cells (reviewed in Wewetzer *et al.*, 2002) that leave the placode and migrate along the nerves into the brain. In the zebrafish, pioneer neurons, distinct from olfactory receptor neurons, differentiate early within the placode and send their axons to the telencephalon (Whitlock and Westerfield, 1998). Axons from the olfactory receptor neurons follow this initial scaffolding, and the pioneer neurons subsequently die by apoptosis (Whitlock and Westerfield, 1998). Olfactory axons are the first peripheral input to reach the brain during development. The axons of pioneer neurons in the rat induce formation of the olfactory bulbs (Gong and Shipley, 1995), which fail to form if the olfactory placodes are missing or if olfactory axons fail to reach the brain (reviewed in Baker and Bronner-Fraser, 2001). The olfactory epithelium is also required for induction of the cartilaginous nasal capsule, which is derived from the neural crest.

The olfactory placode also forms neuroendocrine cells that migrate along the olfactory nerve into the forebrain and diencephalon. These neurons produce gonadotropin-releasing hormone (GnRH) and form the terminal nerve-septo-preoptic GnRH system (reviewed in Dubois *et al.*, 2002). This system regulates gonadotropin release from the adenohypophysis (anterior pituitary), another placodal derivative (section Overview of Cranial Ectodermal Placodes). Hence, the olfactory placode is not only essential for olfaction, but also for reproduction. This is seen clinically in Kallmann's syndrome, in which olfactory axons and GnRH neurons fail to migrate into the brain, resulting in anosmia and sterility (hypogonadism) (reviewed in MacColl *et al.*, 2002). An early-stage fate-map in zebrafish, however, challenges the olfactory placode origin of GnRH neurons, suggesting that terminal nerve GnRH neurons originate from the neural crest, and hypothalamic GnRH neurons from the hypophyseal placode (adenohypophysis) (Whitlock *et al.*, 2003). More early-stage fate-map data are needed from multiple species to resolve this controversy.

Olfactory Placode Formation Involves the Convergence of Cellular Fields

In the 4-somite stage zebrafish embryo, the olfactory placodes fate-map to bilateral regions of *Dlx3*$^+$ ectoderm at the lateral borders of the anterior-most neural plate, much longer in rostro-caudal extent than the final placodes, abutting prospective telencephalic territory rostrally and prospective neural crest caudally (Whitlock and Westerfield, 2000). *Dlx3*, acting in concert with *Dlx7*, is essential for the formation of the olfactory placode in the zebrafish, as shown both by mutant and knockdown analysis using antisense morpholino oligonucleotides (Solomon and Fritz, 2002). Each of these long bilateral *Dlx3*$^+$ cellular fields converges to form an olfactory placode (Whitlock and Westerfield, 2000).

In neurula-stage *Xenopus* embryos, and 3-somite stage chick and mouse embryos, the olfactory placodes fate-map to the outer edge of the anterior neural ridge (the rostral boundary of the neural plate) (Couly and Le Douarin, 1988; Eagleson and Harris, 1990; Osumi-Yamashita *et al.*, 1994). Future olfactory placode and olfactory bulb tissues are contiguous within the anterior neural plate. It is currently unknown whether olfactory placode formation in these species also involves cellular convergence, as in the zebrafish.

Induction of the Olfactory Placodes

Classical grafting and coculture experiments in amphibian embryos (reviewed in Baker and Bronner-Fraser, 2001) suggested that anterior mesendoderm is an important source of olfactory placode-inducing signals. This tissue is also important for forebrain induction (reviewed in Foley and Stern, 2001) (Chapter 3). Forebrain tissue is also important for olfactory placode induction and/or maintenance (reviewed in Baker and Bronner-Fraser, 2001). Nothing is currently known about which molecular signals from these tissues, or others, might be involved in the induction of the olfactory placode.

In the chick, FGF8 from the midfacial ectoderm is necessary and sufficient to induce the genes *erm* and *pea3* in the olfactory pits (Firnberg and Neubüser, 2002). (The induction and maintenance of these two Ets-domain transcription factors is generally tightly coupled to FGF signaling; Raible and Brand, 2001; Roehl and Nüsslein-Volhard, 2001.) FGFs stimulate the proliferation of olfactory receptor neuron progenitors *in vitro* (reviewed in Calof *et al.*, 1998) (see next section), hence, FGF8 may play a similar role in promoting olfactory neurogenesis *in vivo*. However, this remains to be demonstrated.

Experiments in mice and zebrafish have shown that the transcription factors Otx2, Pax6, Dlx5, and Dlx3 (acting in concert with Dlx7), all of which are expressed in the anterior neural ridge and in the olfactory placodes, are required for olfactory placode development (reviewed in Baker and Bronner-Fraser, 2001) (also Solomon and Fritz, 2002). However, their precise roles are currently undefined.

A bHLH Transcription Factor Cascade Controls Olfactory Neurogenesis

Mice lacking the Achaete-Scute homologue Mash1 (section *Mash1 Is Essential for Noradrenergic Differentiation*) have a drastically reduced number of olfactory receptor neurons, due to the death of most olfactory neuron progenitors (Guillemot *et al.*, 1993; Cau *et al.*, 1997). *Mash1* is required for the expression of the *atonal*-related bHLH genes *Neurogenin1* (*Ngn1*) (section Neurogenins Are Essential for the Formation of Dorsal Root Ganglia) and *NeuroD* (Cau *et al.*, 1997), and for activation of the Notch signaling pathway (Cau *et al.*, 2000, 2002). *Ngn1* is required for neuronal differentiation; it does not affect either *Mash1* expression or the Notch signaling pathway (Cau *et al.*, 2002).

Mash1 is not expressed in the earliest-differentiating neurons in the olfactory placode, whose formation is unaffected in *Mash1*-null mice (Cau *et al.*, 1997, 2002). These *Mash1*-independent progenitors express *Ngn2* as well as *Ngn1*, but their differentiation is blocked in *Ngn1*-mutant mice (Cau *et al.*, 1997, 2002). Given their early differentiation, these could represent the pioneer neurons whose axons set up the initial scaffold for the olfactory nerve (section Olfactory Placode Derivatives). Interestingly, vomeronasal sensory neurons are relatively unaffected in mice that are double mutant for both *Mash1* and *Ngn1* (Cau *et al.*, 2002). This suggests both that a different gene controls their development, and that vomeronasal and olfactory sensory neuron progenitors are molecularly distinct (Cau *et al.*, 2002).

As described in the previous section, FGF8, which is expressed in the epithelium around the placodes, may stimulate the proliferation of olfactory neuron stem cells and neuronal progenitors (reviewed in Calof *et al.*, 1998) (also see LaMantia *et al.*, 2000). Treatment of olfactory placode explants with a function-blocking FGF8 antibody causes a reduction in the numbers of neurons relative to controls, though not a complete loss, suggesting that FGF8 is not only sufficient but also necessary for olfactory neurogenesis (LaMantia *et al.*, 2000).

BMPs also play an important role in olfactory neurogenesis: In fact, they can both promote and inhibit neurogenesis in cultures of olfactory epithelium, depending on the concentration, the specific ligand, and the cellular context (Shou *et al.*, 1999, 2000). Exposure of Mash1$^+$ olfactory neuron progenitors to high concentrations of BMP4 or BMP7 in culture leads to the degradation of Mash1 protein via the proteasome pathway, and hence to inhibition of neuronal differentiation (Shou *et al.*, 1999). However, treatment of olfactory epithelium cultures with the BMP antagonist Noggin inhibits neuronal differentiation, showing a requirement for BMP signaling in neurogenesis (Shou *et al.*, 2000). This requirement is explained by the observation that low concentrations of BMP4, but not BMP7, promote the survival of newly born olfactory receptor neurons (Shou *et al.*, 2000). Hence, BMP4 inhibits the production of olfactory receptor neurons at high concentrations and promotes the survival of differentiated neurons at low concentrations. This may provide a feedback mechanism to maintain an appropriate number of olfactory receptor neurons in the epithelium, particularly as BMP4 may be produced by the olfactory receptor neurons themselves (Shou *et al.*, 2000).

Retinoic acid is produced by the neural crest-derived frontonasal mesenchyme between the olfactory placode and the ventrolateral forebrain (LaMantia *et al.*, 1993, 2000). *In vitro*, retinoic acid stimulates the maturation of olfactory receptor neurons from immortalized clonal cell lines derived from the mouse olfactory placode, suggesting a possible role in this process *in vivo* (Illing *et al.*, 2002). However, retinoic acid treatment of olfactory placode explants leads to reduced neuronal differentiation (LaMantia *et al.*, 2000). Further evidence is therefore required to establish the precise role of retinoic acid.

Lateral Line Placodes

Lateral Line Placode Derivatives

The lateral line system is a mechanosensory and electroreceptive sensory system in which individual sense organs are arranged in characteristic lines along the head and trunk of fish and amphibians (Fig. 14C) (Coombs *et al.*, 1989). The entire lateral line system seems to have been lost in amniotes, presumably in association with the transition to a primarily terrestrial lifestyle; it is also often lost in amphibians at metamorphosis. Lateral line electroreception was lost in most teleost fish and in anuran amphibians; indeed, different elements of the lateral line system have been lost independently in multiple vertebrate lineages (reviewed in Northcutt, 1997; Schlosser, 2002b).

There are two types of lateral line sense organs: Mechanosensory neuromasts (Fig. 14B) that respond to local disturbances in the water surrounding the animal, and electroreceptive ampullary organs that respond to weak electric fields. They are used in various behaviors, including schooling, obstacle avoidance, and prey detection. The sense organs themselves, and the neurons that provide their afferent innervation, are derived from a series of paired lateral line placodes on the head (reviewed in Winklbauer, 1989; Baker and Bronner-Fraser, 2001; Schlosser, 2002a) (Fig. 12). The same lateral line placode can form both mechanosensory neuromasts and electroreceptive ampullary organs (Northcutt *et al.*, 1995). Primitively, there were probably at least three pre-otic and three post-otic lateral line placodes (Northcutt, 1997). One pole of each lateral line placode gives rise to neuroblasts, which exit the placode and aggregate nearby to form the sensory neurons of a lateral line ganglion (the satellite

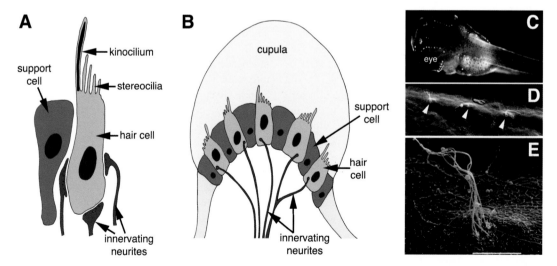

FIGURE 14. The mechanosensory lateral line system. (A) Schematic showing an individual unit of a lateral line neuromast: A mechanosensory hair cell bearing a true cilium (kinocilium) and stereocilia, innervated by lateral line neurites, with an adjacent support cell. Redrawn from Kardong (1998). (B) Schematic section through a neuromast organ: The cilia and stereocilia of each hair cell project into a gelatinous cupula, secreted by the supporting cells. Redrawn from Kardong (1998). (C) Lateral line neuromasts on the head (particularly visible in a ring around the eye) and along the trunk of a live stage 45 (4-day) *Xenopus* tadpole, visualized with the vital mitochondrial stain DASPEI. (D) Scanning electron micrograph of lateral line neuromasts along the trunk of a stage 49 (12-day) *Xenopus* tadpole. (E) High-power view of individual neuromast from (D), showing bundle of long kinocilia with smaller stereocilia at its base. *Xenopus* staging after Nieuwkoop and Faber (1967).

cells of the ganglion are derived from the neural crest). Neurites from these neurons, followed and ensheathed by neural crest-derived glial cells (Gilmour et al., 2002), track the remaining non-neurogenic part of the placode as it elongates to form a lateral line primordium and undergoes a remarkable migration through the epidermis, depositing clusters of cells in its wake. These cells give rise to the supporting and mechanosensory hair cells of the lateral line sense organ (neuromast). Hence, the placode that forms a given line of sense organs also typically forms their afferent innervating neurons. In the zebrafish, the atonal-related bHLH transcription factor Neurogenin1 (Ngn1; section Proneural Genes: An Introduction) is required for the formation of lateral line neurons, but its loss has no effect on the migration of the lateral line primordia, or on neuromast development (Andermann et al., 2002). Hence, as previously demonstrated in amphibian embryos (Tweedle, 1977), lateral line sense organ development is independent of innervation.

A single mechanosensory neuromast is composed of several hair cells, together with supporting cells; each hair cell has a single true cilium (kinocilium) with a bundle of stereocilia at its base (Figs. 14A, E). Lateral line hair cells are very similar in structure to inner ear hair cells derived from the otic placode (section Otic Placode Derivatives). The kinocilia and sterocilia of the hair cells in each lateral line neuromast are embedded in a gelatinous sheath, or cupula, which is secreted by the supporting cells of the neuromast (Fig. 14B). Water movements deflect the cupula. If the cupula movement bends the stereocilia toward the kinocilium, mechanosensitive ion channels in the hair cell open, depolarizing the hair cell and stimulating the afferent fibers of the lateral line nerve which synapses onto its basal surface. If the stereocilia are bent away from the kinocilium, this closes the few ion channels that are open at rest, causing hyperpolarization of the hair cell and neuronal inhibition (reviewed in Winklbauer, 1989; Pickles and Corey, 1992).

Electroreceptor cells are structurally similar to neuromast hair cells, although they are more variable across taxa (Bodznick, 1989). They are secondary sense cells (i.e., they require afferent innervation) with apical microvilli and/or a kinocilium. Interestingly, while teleost electroreceptors, where present, seem to have secondarily evolved from neuromast hair cells, nonteleost electroreceptors (i.e., those found in lamprey, nonteleost fish, and amphibia) are phylogenetically as old or older than neuromast hair cells (Bodznick and Northcutt, 1981; Bodznick, 1989). From the phylogenetic evidence, it is equally likely that neuromast hair cells evolved from electroreceptors, that electroreceptors evolved from mechanosensory hair cells, or that both evolved independently from a common ancestral ciliated cell type (Bodznick, 1989).

Induction of the Lateral Line Placodes

Relatively little is currently known about the sources of lateral line placode-inducing signals, and nothing of their molecular nature (reviewed in Baker and Bronner-Fraser, 2001; Schlosser, 2002a). Lateral line placodes are induced separately from the otic placodes (section Otic Placodes), despite their proximity and even (in some species) their apparent derivation from a common Pax2+ "dorsolateral placode area" (Schlosser and Northcutt, 2000; Schlosser, 2002a). Lateral line and otic placodes can be induced independently in grafting experiments, and ectodermal competence to form lateral line placodes persists much longer than that to form otic placodes (reviewed in Baker and Bronner-Fraser, 2001; Schlosser, 2002a). Furthermore, mutations in zebrafish can affect otic placodes but not lateral line placodes, and vice versa (Whitfield et al., 1996), and lateral line placodes have been lost multiple times in evolution with no effect on the otic placodes (e.g., Schlosser et al., 1999).

In the axolotl, lateral line placodes are determined (i.e., develop autonomously in ectopic locations) by late neural fold stages (Schlosser and Northcutt, 2001). Hence, lateral line placode-inducing signals must act before this time, although they persist until relatively late stages, as shown by grafts of non-placodal ectoderm to the placode-forming region of tailbud stage embryos (Schlosser and Northcutt, 2001). Grafting experiments in amphibians have implicated both mesoderm and neural plate as sources of lateral line placode-inducing signals (reviewed in Baker and Bronner-Fraser, 2001; Schlosser, 2002a). Their molecular nature is currently unknown. Migrating lateral line primordia in zebrafish express the FGF target genes, erm and pea3 (Münchberg et al., 1999; Raible and Brand, 2001; Roehl and Nüsslein-Volhard, 2001), the Wnt receptor gene Frizzled7a (Sumanas et al., 2002), and the BMP inhibitor follistatin (Mowbray et al., 2001). Hence, FGF, Wnt, and BMP signaling may all be involved in aspects of lateral line placode development; however, any role for these signaling pathways in lateral line placode induction remains to be demonstrated. Early development and migration of lateral line placodes in fgf8 mutant zebrafish appears normal, although the number of neuromasts formed is strongly reduced, suggesting an involvement of FGF8 at later stages (Léger and Brand, 2002).

Individual lateral line placodes differ in the number and type of sense organs that they form, and also in their gene expression patterns (e.g., only the middle lateral line placode expresses Hoxb3 in the axolotl; Metscher et al., 1997). Additional inducing signals are presumably involved, therefore, in specifying individual lateral line placode identity, but these are wholly unknown.

Migration of Lateral Line Primordia

Cells of the lateral line primordia in amphibian embryos actively migrate through and displace the inner cells of the bilayered epidermis. Cell division also occurs within the primordia as they migrate (Winklbauer, 1989; Schlosser, 2002a). In zebrafish, lateral line primordia migrate just beneath the epidermis; time-lapse analysis of living embryos shows that each cell migrates independently, rather than the whole primordium moving as a solid block of cells, but they generally retain their neighbor relationships (Gompel et al., 2001). The clusters of undifferentiated cells that will form the neuromasts are deposited when a group of cells at the trailing edge of the primordium progressively slows down relative to the rest of the primordium (Gompel et al., 2001).

During normal development, lateral line primordia migrate along invariant pathways. When pre-otic and post-otic lateral line placodes are exchanged, they migrate along the pathway appropriate for their new position (reviewed in Schlosser, 2002a), suggesting that they are following extrinsic guidance cues. Several such guidance cues have now been identified in zebrafish. The chemokine SDF1 is expressed in a trail along the migration pathway of the posterior lateral line primordium (David *et al.*, 2002). The migrating cells of the lateral line primordium express the SDF1 receptor, CXCR4, and inactivation of either the receptor or its ligand blocks migration (David *et al.*, 2002). Also, the posterior lateral line primordium migrates along the trunk at the level of the horizontal myoseptum that divides the axial muscles into dorsal and ventral halves. Semaphorin3A1 (Sema3A1) is expressed throughout the somites except in the horizontal myoseptum, and when the myoseptum is missing, as in certain mutant strains, the lateral line primordium migrates aberrantly (Shoji *et al.*, 1998). These results suggest that in addition to the SDF1-CSCR4 system, the primordium may be directed along the myoseptum by repulsive Sema3A1 migration cues from the dorsal and ventral somites (Shoji *et al.*, 1998). Finally, the posterior lateral line primordium also expresses *robo1* (Lee *et al.*, 2001), which encodes a receptor for the repulsive migration cue Slit (reviewed in Ghose and Van Vactor, 2002). Hence, Slit-Robo signaling may also be involved in guiding lateral line primordia along specific migration pathways.

The lateral line axons that track the migrating primordium are ensheathed in neural crest-derived glial cells, which lag slightly behind the axonal growth cones (Gilmour *et al.*, 2002). These are not required for the growth or pathfinding of the lateral line axons, but their genetic ablation leads to defasciculation of the lateral line nerves (Gilmour *et al.*, 2002). Hence, neural crest-derived glia are required for the organization of the lateral line nerves (Gilmour *et al.*, 2002).

Cell Fate Determination Within Lateral Line Neuromasts

The bHLH *atonal* homologue *zath1* (section Proneural Genes: An Introduction) is progressively restricted to prospective mechanosensory hair cells in lateral line neuromasts in the zebrafish, suggesting that its expression defines the cells with the potential to form hair cells (Itoh and Chitnis, 2001). The mouse *ath1* homologue, *Math1*, is Specifically required for inner ear (otic placode-derived) hair cell formation (Bermingham *et al.*, 1999) (section Hair Cell Specification Requires Notch Inhibition and *Math1*), so it is likely that *zath1* may similarly be required for lateral line hair cell formation. Determination of lateral line hair cell vs support cell fate probably involves Notch-mediated lateral inhibition (e.g., sections Differentiation of DRG Neurons Depends on Inhibition of Notch Signaling; Hair Cell Specification Requires Notch Inhibition and *Math1*): Expression of the Notch ligand DeltaB correlates with *zath1* expression in prospective hair cells, while *Notch3* expression is excluded from prospective hair cells (Itoh and Chitnis, 2001).

Otic Placodes

Otic Placode Derivatives

The paired otic placodes, which form in the ectoderm adjacent to the hindbrain (Figs. 3 and 10), invaginate to form closed otic vesicles (otocysts). In zebrafish, the placodes sink beneath the surface ectoderm and the vesicles form by cavitation (Whitfield *et al.*, 2002). Each simple hollow epithelial ball undergoes profound morphogenetic changes to produce the highly complex, three-dimensional structure of the inner ear, or vestibular apparatus (membranous labyrinth) (Fig. 15). A neurogenic region

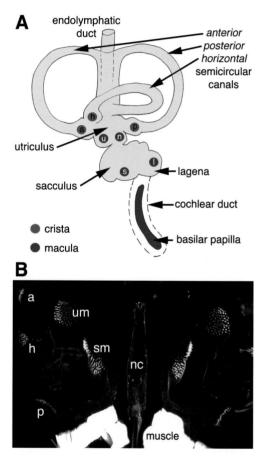

FIGURE 15. The complex structure of the inner ear. (A) Schematic showing a generalized vertebrate vestibular apparatus, with the three semicircular canals and major compartments: utriculus, sacculus, and lagena. The cochlear duct is an extension of the lagena that forms only in terrestrial vertebrates; hence it is shown as a dashed line. Specialized auditory hair cells are collected in a strip in the cochlear duct called the basilar papilla in birds, and the organ of Corti in the coiled mammalian cochlea. See section Otic Placode Derivatives for details. a, anterior crista; h, horizontal crista; l, lagenar macula; n, macula neglecta; p, posterior crista; s, saccular macula; u, utricular macula. Redrawn from Kardong (1998). (B) Confocal image of a 5-day zebrafish larva in dorsal view (rostral to the top), dissected to show the two ears on either side of the notochord (nc). The hair cells are visualized using fluorescent phalloidin, which binds the actin-rich stereocilia on the surface of each hair cell. Photograph courtesy of Dr. Tanya Whitfield, University of Sheffield, United Kingdom. a, anterior crista; h, horizontal (lateral) crista; nc, notochord; p, posterior crista; sm, saccular macula; um, utricular macula.

in each otic vesicle also gives rise to the sensory neurons that provide afferent innervation for the mechanosensory hair cells of the inner ear; these neurons are collected in the VIIIth cranial ganglion (vestibulocochlear/statoacoustic) (Fig. 11). The otic vesicle itself induces the formation of the cartilaginous otic capsule, which surrounds and protects the vestibular apparatus, from adjacent head mesenchyme (Frenz et al., 1994).

The vestibular apparatus contains three semicircular canals, oriented roughly in the three planes of space, and two or three relatively distinct chambers, the utriculus, sacculus and lagena (the latter being an extension of the sacculus) (Fig. 15A). These compartments all contain both non-sensory epithelium, and sensory vestibular epithelium containing neuromasts, that is, collections of mechanosensory hair cells and supporting cells. These are similar to those found in the lateral line system (section Lateral Line Placode Derivatives), except that otic neuromasts are usually much larger and contain many more hair cells than lateral line neuromasts. Each hair cell has a true cilium (kinocilium) with a bundle of stereocilia at its base; these are all embedded in a gelatinous cupula secreted by the supporting cells of the neuromast (Figs. 14A, B). Deflection of the cupula in a particular direction triggers depolarization of the hair cell, stimulating the afferent fibers of the sensory neurons that synapse onto the hair cell base. The high K^+/low Na^+ concentration of the endolymph filling the vestibular apparatus is essential for sensory tranduction by inner ear hair cells. Specialized epithelial cells within the vestibular apparatus (stria vascularis in mammals; tegmentum vasculosum in birds) secrete the endolymph.

Dilated ampullae at the base of each semicircular canal contain expanded neuromast organs, the cristae (Figs. 15A, B). When the head is turned, the semicircular canals are accelerated, but fluid inertia causes the endolymph to lag behind, relative to the semicircular canal itself, deflecting the cupula and stimulating the hair cells. The cristae therefore detect angular acceleration (rotation). Their afferent innervation is from otic placode-derived neurons in the vestibular part of the VIIIth ganglion (Fig. 11).

The utriculus and sacculus also contain large, modified neuromast organs, the maculae (Figs. 15A, B). The utricular macula in the adult zebrafish contains approximately 6,000 hair cells (Platt, 1993), which gives some idea of the size of these neuromasts. Each macula has dense crystals composed of protein and calcium carbonate, called otoliths or otoconia, embedded in the cupular surface (Riley et al., 1997). Otoliths intensify the displacements of the hair cells in response to linear acceleration. The maculae therefore detect gravity and linear acceleration. Like the cristae, their afferent innervation is provided by otic placode-derived vestibular ganglion neurons.

The maculae in the sacculus and lagena are also involved in hearing in all vertebrates, including fish, where compression waves cause movement of the maculae in relation to the otoliths resting on them. The lagena is lengthened in terrestrial vertebrates to form the cochlear duct, which is coiled in mammals (Fig. 15A). Cochlear auditory hair cells, which are often highly modified in structure, are collected in a specialized strip called the basilar papilla in birds and the organ of Corti in mammals. The afferent innervation for auditory hair cells is provided by otic placode-derived neurons in the auditory (cochlear) part of the VIIIth ganglion.

In summary, the entire vestibular apparatus, together with its afferent neurons, is derived from a simple epithelial ball, the otic vesicle, which in turn is derived from the otic placode. The formation of the inner ear is clearly a highly complex process, and only selected aspects will be discussed here. For more detailed analysis of otic morphogenesis, the reader is referred to several recent reviews (Torres and Giraldez, 1998; Rinkwitz et al., 2001; Whitfield et al., 2002).

Otic Placode Formation Involves Cell Movement and Convergence

Cell lineage analysis in the chick has shown that up to the 1-somite stage, otic placode precursors are found in a large region of ectoderm at the neural plate border, intermingled with precursors of neural tube, neural crest, epibranchial placodes (section Epibranchial Placodes), and epidermis (Streit, 2002). By the 4-somite stage, otic placode precursors extend from the level of the anterior hindbrain to the level of the first somite (Streit, 2002). By the 8-somite stage, a few hours before the otic placode is morphologically visible, quail-chick chimera analysis shows that prospective otic placode cells are found in a relatively small area adjacent to rhombomeres 5 and 6, just rostral to the first somite (Fig. 11) (D'Amico-Martel and Noden, 1983). A few hours later, at the 10-somite stage, the otic placode becomes morphologically visible.

Time-lapse video analysis of DiI-labeled ectodermal cells shows that the convergence of otic placode precursors to the final placode area results from extensive cell movement within the ectoderm (Streit, 2002). It is currently unclear whether this reflects active migration of otic-specified cells to the location of the future placode, or capture of randomly moving cells by progressively more localized otic placode-inducing signals (section Models of Individual Placode Formation in the Preplacodal Domain). Pax2, which is essential for proper otic placode development (Torres and Giraldez, 1998) is induced at the 4-somite stage in the broad region of ectoderm that contains otic precursors (Groves and Bronner-Fraser, 2000; Streit, 2002). However, not all cells within the $Pax2^+$ domain contribute to the otic placode: Even at the 7–10-somite stage, some cells in this domain contribute to epidermis, or to the epibranchial placodes, which also express Pax2 (Fig. 11; section Epibranchial Placodes) (Streit, 2002). Hence, Pax2 expression does not correlate with determination toward an otic fate (also see Groves and Bronner-Fraser, 2000). Nonetheless, it remains to be seen whether there is directed migration of otic precursor cells toward the site of the future otic placode.

Interestingly, a fate-map of different sensory areas within the otic placode in Xenopus suggests that extensive cell movement continues within the otic placode and vesicle until fairly late stages (Kil and Collazo, 2001).

Induction of the Otic Placodes

More detailed information on the induction of the otic placodes can be found in various reviews (Baker and Bronner-Fraser,

2001; Noramly and Grainger, 2002; Whitfield *et al.*, 2002; Riley and Phillips, 2003). Ablation and grafting experiments in a variety of species suggested that both mesendoderm and hindbrain are sources of otic placode-inducing signals. They led to a model in which the earliest otic placode-inducing signals are derived from mesendoderm, with later signals from the hindbrain. This model has been supported by experiments that have identified some of the signals involved in otic placode development.

Fgf3 is dynamically expressed in several vertebrates in rhombomeres 4–6 (r4–6), adjacent to the site of formation of the otic placodes (Fig. 11); however, otic vesicles form normally in *Fgf3*-null mice (Mansour *et al.*, 1993). Experiments in zebrafish and mouse suggest that FGF3 acts redundantly with a second FGF family member to induce the otic placodes (Phillips *et al.*, 2001; Maroon *et al.*, 2002; Wright and Mansour, 2003). In zebrafish, simultaneous knockdown of FGF3 and FGF8 function using antisense morpholino oligonucleotides results in the loss of early otic markers such as *pax2.1* and *dlx3* (see previous section and section A Preplacodal Field at the Anterior Neural Plate Border) and of the otic vesicles (Maroon *et al.*, 2002; Phillips *et al.*, 2001). *Fgf3* and *Fgf8* are co-expressed in r4 in the zebrafish, but *Fgf8* is not expressed in the hindbrain in chick or mouse. However, at neural plate stages in the chick, *Fgf4* is co-expressed with *Fgf3* in prospective hindbrain neuroectoderm (Maroon *et al.*, 2002), while *Fgf19* is expressed in paraxial mesoderm at same axial level (Ladher *et al.*, 2000) (see below). In the mouse, *Fgf10* is expressed in the mesenchyme underlying the prospective otic placode, and mice lacking both *Fgf3* and *Fgf10* fail to form otic vesicles and show abnormal otic placode marker expression patterns (Wright and Mansour, 2003). Hence, FGF3 may cooperate with other FGF family members in different vertebrate species to induce expression of early otic markers such as Pax2, and the otic vesicles themselves.

Nonetheless, abrogation of FGF3 and FGF8 function in zebrafish does not affect expression of *Pax8*, the earliest known specific marker for prospective otic placode ectoderm (Maroon *et al.*, 2002). *Pax8* expression, which normally appears at late gastrula/neural plate stages, is delayed in mutant zebrafish embryos that lack cranial mesendoderm (Mendonsa and Riley, 1999; Phillips *et al.*, 2001). It is possible, therefore, that early signals from cranial mesendoderm normally induce *Pax8*, but that in the mutant embryos, *Pax8* expression is rescued by later, as-yet unidentified hindbrain-derived signals.

Foxi1, a member of the forkhead family of winged-helix transcription factors, is expressed prior to *pax8* and is required for *pax8* expression in zebrafish (Solomon *et al.*, 2003). *Foxi1* mutant zebrafish show a severe reduction or loss both of *pax8* expression and the otic placodes, and *foxi1* misexpression induces ectopic *pax8* expression (Solomon *et al.*, 2003). However, while necessary for *pax8* expression, *foxi1* is not in fact sufficient, because *pax8* is not expressed in every cell that expresses *foxi1* (Solomon *et al.*, 2003). Hence, *pax8* expression requires additional regulatory factors besides Foxi1. Also, *foxi1* is expressed in and required for the development of the epibranchial placodes (section A Common Primordium for Epibranchial and Otic Placodes?), so *foxi1* is not specific to the otic placodes (Lee *et al.*, 2003).

In the chick, Wnt8c from the hindbrain and FGF19 from the paraxial mesoderm have been suggested to be involved in otic placode induction (Ladher *et al.*, 2000). However, loss of Wnt8c (via inhibition of retinoic acid signaling) does not affect otic vesicle formation in the chick, and FGF19 acts specifically through FGF receptor 4, whose loss does not affect otic placode development in the mouse (see discussion in Maroon *et al.*, 2002).

In summary, signals from cranial mesendoderm and the hindbrain are involved in induction of the otic placodes. FGF3 from the hindbrain, acting redundantly with another FGF family member, is required for otic vesicle formation: In their absence, Pax8 expression in prospective otic territory is unaffected, but Pax2 is not expressed and subsequent otic development is blocked. Wnts may also be involved in otic placode development, but a requirement for Wnt signaling has not yet been demonstrated.

Neurogenesis in the Otic Vesicle Requires *Neurogenin1* and Notch Inhibition

The neuroblasts that will form the neurons of the vestibulocochlear (VIIIth) ganglion delaminate from the ventromedial region of the otic vesicle and aggregate nearby to form the ganglion (Fig. 11). All satellite glia within the ganglion are derived from the neural crest.

In the mouse, Atonal-related neural bHLH factors such as Neurogenin1 (Ngn1; section Proneural Genes: An Introduction) and NeuroD are expressed in epithelial cells within the otic vesicle prior to delamination (Ma *et al.*, 1998). Vestibulocochlear neurons are entirely missing in *Ngn1*-null mice (Ma *et al.*, 1998). *Ngn1* is required prior to the delamination of otic neuroblasts from the otic vesicle, since *NeuroD* and the expression of the Notch ligand Delta-like1 are both missing from the otic epithelium in *Ngn1*-null mice (Ma *et al.*, 1998). Evidence that Notch inhibition (see section Proneural Genes: An Introduction) is involved in selection of neuronal cell fate within the otic vesicle comes from zebrafish carrying the *mindbomb* mutation: These embryos, in which Notch activation is blocked, have double the wildtype number of statoacoustic ganglion neurons (Haddon *et al.*, 1998).

Hair Cell Specification Requires Notch Inhibition and *Math1*

Cell fate specification in the inner ear is discussed in more detail in several recent reviews (Fekete and Wu, 2002; Whitfield *et al.*, 2002; Riley and Philllips, 2003). Cell fate specification within the sensory patches of the inner ear (the areas containing mechanosensory hair cells and supporting cells) depends on Notch signaling (section Proneural Genes: An Introduction). Well before hair cell differentiation occurs, prospective sensory patches are prefigured by their expression of Notch (initially expressed throughout the otic placode and later restricted to sensory epithelium) and its ligands Delta and Serrate. Delta expression eventually becomes restricted to nascent hair cells. Cells with low levels of Notch activity differentiate as hair cells, while

Notch activation leads to supporting cell differentiation (cf. glial differentiation, e.g., section Differentiation of DRG Satellite Cells Depends on Notch Activation and Instructive Gliogenic Cues) (reviewed in Eddison *et al.*, 2000; Fekete and Wu, 2002; Whitfield *et al.*, 2002). Numerous lines of evidence support this model. For example, in *mindbomb* mutant zebrafish (where Notch activation is blocked and cells are "deaf" to Delta signaling), sensory patch cells differentiate as hair cells at the expense of supporting cells (Haddon *et al.*, 1998). The Notch effector Hes1, a bHLH transcriptional repressor (Davis and Turner, 2001), negatively regulates hair cell production: *Hes1*-null mice have extra inner ear hair cells (Zheng *et al.*, 2000). Finally, the Notch antagonist Numb is expressed at high levels in hair cells in the chick (Eddison *et al.*, 2000).

The mouse Atonal homologue Math1 (section Proneural Genes: An Introduction) is both necessary and sufficient for hair cell differentiation, as shown by knockout and overexpression studies (Bermingham *et al.*, 1999; Zheng and Gao, 2000). Math1 expression first begins in nascent hair cells; it is not required to set up the area that will form a sensory patch (Chen *et al.*, 2002). Hence, Math1 may specify hair cell identity in the inner ear.

Lens Placodes

Lens Placode Derivatives

The lens placodes are unusual among the cranial ectodermal placodes, as they do not produce either sensory receptor cells or neurons. Where the evaginating optic vesicles approach the overlying surface ectoderm, it thickens to form the lens placodes; these invaginate and pinch off to form the eye lenses. The newly formed lenses have a distinct polarity, maintained throughout life, with proliferating cuboidal cells covering the anterior surface, and terminally differentiated lens fiber cells making up the bulk of the lens. Successive layers of lens fiber cells differentially accumulate highly stable, soluble proteins called crystallins, giving a smooth decreasing gradient of refractive index from the center to the periphery.

The Importance of Pax6 for Lens Placode Development

Pax6 has been implicated in eye and anterior head development in all major animal groups. The Pax/Six/Eya/Dach regulatory network (section The Pax/Six/Eya/Dach Regulatory Network) was first identified in studies of the *Pax6* homologue *eyeless* in *Drosophila* eye development (Wawersik and Maas, 2000). In vertebrates, Pax6 is essential for both lens placode and retinal development (reviewed in Ashery-Padan and Gruss, 2001; Baker and Bronner-Fraser, 2001). Pax6 is initially expressed in a broad region of head ectoderm and is eventually restricted to the lens placode itself. The homeobox transcription factors Meis1 and Meis2 are direct upstream regulators of Pax6 expression in lens ectoderm (Zhang *et al.*, 2002). Knockout experiments in mice have shown that Pax6 is required in head ectoderm both for competence to respond to a lens-inducing signal from the optic vesicle (see next section) and also for subsequent steps in lens placode development (reviewed in Ashery-Padan and Gruss, 2001; Baker and Bronner-Fraser, 2001). Pax6 is required for the upregulation of the HMG-domain transcription factor Sox2 (and/or Sox1, Sox3) in prospective lens ectoderm after it is contacted by the optic vesicles: These genes are essential for lens differentiation (Baker and Bronner-Fraser, 2001). Sox2 and Pax6 act together in subsequent steps of lens differentiation, by cooperatively binding *crystallin* gene enhancers and activating their expression (Kamachi *et al.*, 2001). Pax6 can induce ectopic lenses (and eyes) in *Xenopus* head ectoderm (Altmann *et al.*, 1997; Chow *et al.*, 1999), although in chick head ectoderm, ectopic lens induction requires both Pax6 and Sox2 (Kamachi *et al.*, 2001). Hence, Pax6 is necessary, though not sufficient, for lens formation.

Induction of the Lens Placodes

More detailed descriptions of lens placode induction can be found in recent reviews (Ogino and Yasuda, 2000; Ashery-Padan and Gruss, 2001; Baker and Bronner-Fraser, 2001). Classical grafting experiments demonstrated that anterior mesendoderm, anterior neural plate, and the optic vesicles may all play roles in lens placode induction (reviewed in Baker and Bronner-Fraser, 2001). Some of the signals from the optic vesicles have now been identified. Knockout experiments in mice have shown that optic vesicle-derived BMP7 is required for *Sox2* expression and *Pax6* maintenance in presumptive lens ectoderm (Wawersik *et al.*, 1999). FGF8 is expressed in the optic vesicles in the chick and can induce lens placode markers (Vogel-Höpker *et al.*, 2000), and genetic block of FGF signaling in prospective lens ectoderm in the mouse leads to defects in lens formation (Faber *et al.*, 2001). Presumptive lens ectoderm in the mouse also receives retinoic acid signals, as demonstrated by the activation of retinoic acid-responsive transgenes (Baker and Bronner-Fraser, 2001). Finally, optic vesicle-derived BMP4 is involved in a somewhat later phase of lens placode formation, downstream of Pax6 (Furuta and Hogan, 1998).

Lens Fiber Differentiation

Factors that induce lens fiber differentiation are found in the vitreous and aqueous humors of the eye, and several different families of growth factors have been implicated in lens fiber differentiation. FGFs can stimulate the differentiation of lens fiber cells from lens epithelial cells (Govindarajan and Overbeek, 2001), and transgenic expression of dominant negative FGF receptors in mouse lenses or eyes leads to delayed lens fiber differentiation and apoptosis (Robinson *et al.*, 1995; Govindarajan and Overbeek, 2001). Transgenic expression of dominant negative TGFβ receptors in the mouse lens also disrupts lens fiber differentiation, suggesting a role for TGFβ family members as well as FGFs (de Iongh *et al.*, 2001). Finally, retroviral-mediated overexpression of the BMP antagonist Noggin in chick eyes delays lens fiber development and results in lens cell death, suggesting that BMPs are also involved in lens fiber differentiation and survival (Belecky-Adams *et al.*, 2002).

Transgenic expression of a dominant negative BMP type I receptor in the mouse eye also leads to defects in lens fiber formation (Faber *et al.*, 2002). Hence, FGF, TGFβ, and BMP family members may all be involved in triggering lens fiber differentiation.

TRIGEMINAL AND EPIBRANCHIAL PLACODES

The trigeminal and epibranchial placodes (Figs. 10 and 11) do not contribute to the paired sense organs. However, trigeminal placode-derived neurons are important for touch, pain, and temperature sensations from the head, including the jaws and teeth, while epibranchial placode-derived neurons provide afferent innervation for taste buds, and autonomic afferent innervation for the visceral organs. The trigeminal placodes form in the surface ectoderm adjacent to the midbrain and rostral hindbrain, while the epibranchial placodes form above each pharyngeal (branchial) cleft (Figs. 3 and 10).

Trigeminal Placodes

Trigeminal Placode Derivatives

The sensory trigeminal ganglion complex of cranial nerve V is formed in most craniates by the fusion of two separate ganglia during development: the ophthalmic trigeminal (opV; sometimes called profundal) and maxillomandibular trigeminal (mmV; sometimes called gasserian) ganglia. The neurons in the trigeminal ganglion are of mixed origin, being derived both from neural crest and from the two separate trigeminal placodes (opV and mmV) (see Figs. 10 and 11). All satellite glial cells in the ganglion are derived from the neural crest. In the chick, both the opV lobe and the mmV lobe of the trigeminal ganglion contain large-diameter placode-derived neurons distally, and small-diameter neural crest-derived neurons proximally (Hamburger, 1961; D'Amico-Martel and Noden, 1983) (Fig. 11).

Trigeminal ganglion neurons are primary sensory neurons, like those in the dorsal root ganglia, transmitting cutaneous (touch, pain, and temperature) information from the skin and proprioceptive information from muscles. Neurons in the opV lobe/ganglion innervate the head, including the nose and eyeballs, while neurons in the mmV lobe/ganglion innervate the lower face, jaws, tongue, and teeth. Cutaneous neurons are derived from both the neural crest and the two placodes, while proprioceptive neurons seem only to be derived from the neural crest, at least in the chick (Noden, 1980). (Most of the proprioceptive neurons that innervate the jaws are found in the mesencephalic nucleus of the trigeminal nerve (mesV), which seems to be a neural crest-derived sensory ganglion within the brain; see section Neural Crest Derivatives.)

In fish and amphibians, trigeminal neurons are born very early and make up part of the primary nervous system that mediates swimming reflexes. Judging by their position, lateral to the *FoxD3*[+] neural crest domain in zebrafish (e.g., Kim *et al.*, 2000; Andermann *et al.*, 2002; Itoh *et al.*, 2002), these early-born trigeminal neurons are placode-derived. Like all other placode-derived neurons in the

zebrafish, they express the *atonal*-related proneural bHLH gene *ngn1* (section Proneural Genes: An Introduction) (Andermann *et al.*, 2002). Their early differentiation is consistent with the early birth of placode-derived trigeminal neurons relative to that of neural crest-derived neurons in other vertebrates, such as the chick (D'Amico-Martel and Noden, 1980).

Induction of the Trigeminal Placodes

The trigeminal placodes form in the surface ectoderm adjacent to the midbrain and rostral hindbrain (Figs. 3 and 10; D'Amico-Martel and Noden, 1983). For more detailed information about classical experiments on induction of the trigeminal placodes, see Baker and Bronner-Fraser (2001). Very little is known about the formation of the mmV placode. More information is available on induction of the opV placode in the chick, which begins to express Pax3 from the 4-somite stage (Stark *et al.*, 1997). Pax3 expression correlates with the determination of opV placode-derived cells to adopt a cutaneous sensory neuron fate (Baker and Bronner-Fraser, 2000; Baker *et al.*, 2002). The importance of Pax3 is shown by the severe reduction of the opV lobe of the trigeminal ganglion in mice carrying a mutated *Pax3* gene (Tremblay *et al.*, 1995). Barrier implantation and coculture experiments in the chick have shown that Pax3 is induced in head ectoderm by an unidentified neural tube-derived signal (Stark *et al.*, 1997; Baker *et al.*, 1999). The Pax3-inducing signal is produced along the entire length of the neuraxis; however, restriction of Pax3 expression to the forming opV placode may result, at least in part, from spatiotemporal changes in ectodermal competence to respond to this signal (Baker *et al.*, 1999).

Experiments in the zebrafish have shown that homologues of the Iroquois family of homeodomain transcription factors, which are required for the expression of proneural *achaete-scute* genes in *Drosophila* (section Proneural Genes: An Introduction), are involved in the formation of the trigeminal placodes (Itoh *et al.*, 2002). Zebrafish *iro1* and *iro7* are expressed at neural plate stages in a region of neuroectoderm extending from the midbrain to r4 (Itoh *et al.*, 2002). As somitogenesis begins, the expression of both genes expands into the ectoderm where the trigeminal placodes form, as defined by expression of the *atonal* homologue *neurogenin1* (*ngn1*) (Itoh *et al.*, 2002) (see next section). Functional knockdown of Iro7 (though not Iro1) using antisense morpholino oligonucleotides leads to loss of *ngn1* expression in the trigeminal placode (Itoh *et al.*, 2002). *Ngn1* in the mouse is essential for neurogenesis in the trigeminal placodes (see next section). Hence, *iro7* is required for trigeminal placode-derived neurogenesis.

The rostral border of *iro1* and *iro7* expression in the trigeminal placode ectoderm is expanded rostrally in zebrafish mutants with increased Wnt signaling (Itoh *et al.*, 2002); this correlates with the rostral expansion of *ngn1*[+] trigeminal placode-derived neurons seen in such mutants (Kim *et al.*, 2000; Itoh *et al.*, 2002). Wnt signaling may, therefore, be involved in trigeminal placode induction and/or neurogenesis. Several Wnt receptors are expressed broadly in rostral head ectoderm at appropriate stages in the chick (Stark *et al.*, 2000).

In the chick, the FGF receptor FREK is expressed in the opV placode, but only from the 10-somite stage, well after initial induction of Pax3 (Stark *et al.*, 1997). It continues to be expressed in delaminating neuroblasts, but is not maintained after gangliogenesis (Stark *et al.*, 1997). FGF family members may, therefore, play a role in trigeminal placode-derived cell migration.

Neurogenesis in the Trigeminal Placodes Requires *Neurogenin1*

In zebrafish, the trigeminal placodes are first detectable by *ngn1* expression in lateral patches of ectoderm at late gastrula stages; antisense morpholino-mediated functional knockdown of Ngn1 completely abrogates formation of the trigeminal ganglia (Andermann *et al.*, 2002). In the mouse, *Ngn1* is expressed in subsets of cells in the trigeminal placodes, in delaminating trigeminal neuroblasts, and in condensing trigeminal ganglion neurons in the mouse (Ma *et al.*, 1998). *Ngn2* is weakly expressed in the trigeminal ganglion well after *Ngn1* (Fode *et al.*, 1998; Ma *et al.*, 1998), and *Ngn2*-null mice have no trigeminal ganglion defects (Fode *et al.*, 1998). In contrast, the trigeminal ganglia are totally absent in *Ngn1*-null mice (Ma *et al.*, 1998). *Ngn1* is required in the trigeminal placodes for neuroblast delamination, and for expression of downstream neural bHLH genes, such as the *atonal*-related *NeuroD* family members *NeuroD* and *Math3*, the *achaete-scute*-related gene *NSCL1*, and *Ngn2* (Ma *et al.*, 1998) (for family relationships of proneural genes, see Bertrand *et al.*, 2002). *Ngn1* is also required for expression of the Notch ligand *Delta-like1* in the trigeminal placodes (Ma *et al.*, 1998); Delta–Notch signaling is presumably also involved in the selection of neuronal fate (section Proneural Genes: An Introduction). *Notch* expression within the trigeminal placodes is seen at the same time as *Ngn1* expression (Reaume *et al.*, 1992). However, abrogation of Notch signaling (in mice with mutations in a transcriptional effector of the Notch signaling pathway) has no effect on the initial expression of *Ngn1* in the trigeminal placodes (Ma *et al.*, 1998). Hence, the establishment of *Ngn1* expression in the trigeminal placodes is independent of Notch signaling.

The total absence of the trigeminal ganglion in *Ngn1*-null mice is due not only to trigeminal placode defects: Neural crest cells condense to form the trigeminal ganglionic primordium in *Ngn1*-null mice, but fail to form neurons (Ma *et al.*, 1998). The other proximal cranial sensory ganglia, whose neurons are all derived from the neural crest (Fig. 11), also fail to form in *Ngn1*-null mice (Ma *et al.*, 1998).

Interactions Between Neural Crest-Derived and Placode-Derived Trigeminal Cells in Gangliogenesis

Placode-derived neurons differentiate before neural crest-derived neurons in the trigeminal ganglion (D'Amico-Martel and Noden, 1980). However, the first ganglionic condensation is made up of neural crest cells, which are only later joined by

placode-derived neurons (Covell and Noden, 1989). Neural crest cells are not required for induction at least of the opV placodes (Stark *et al.*, 1997), and their ablation delays, but does not abolish, gangliogenesis and pathfinding by placode-derived trigeminal neurons (Hamburger, 1961; Moody and Heaton, 1983b). In the absence of neural crest cells, the placode-derived ganglia tend to remain as two separate ganglia, suggesting the neural crest cells act as an aggregation center for ganglionic fusion (Yntema, 1944; Hamburger, 1961). In contrast, when the trigeminal placodes are ablated, neural crest-derived trigeminal neurons do not make appropriate peripheral projections (Hamburger, 1961; Lwigale, 2001). Furthermore, the central projections of trigeminal placode-derived neurons are required for trigeminal motor neuron migration and axonal projection (Moody and Heaton, 1983a, b).

Epibranchial Placodes

Epibranchial Placode Derivatives

The epibranchial placodes form above the pharyngeal (branchial) clefts (section Pharyngeal Arches and Neural Crest Streams; Figs. 3 and 10). The first epibranchial placode (facial or geniculate) forms above the first pharyngeal cleft, and gives rise to sensory neurons in the distal (geniculate) ganglion of cranial nerve VII (facial) (Fig. 11). These neurons primarily provide afferent innervation for the taste buds. The second epibranchial placode (glossopharyngeal or petrosal) forms above the second pharyngeal cleft and gives rise to sensory neurons in the distal (petrosal) ganglion of cranial nerve IX (glossopharyngeal) (Fig. 11). These neurons provide afferent innervation for taste buds, and afferent autonomic innervation for visceral organs such as the heart. The third epibranchial placode (vagal or nodose) forms above the third pharyngeal cleft, and gives rise to sensory neurons in the distal (nodose) ganglion of cranial nerve X (vagal) (Fig. 11). These neurons primarily provide afferent autonomic innervation for the heart and other visceral organs. Additional vagal epibranchial placodes form above more posterior pharyngeal clefts and contribute neurons to the nodose ganglion or ganglia (see Baker and Bronner-Fraser, 2001). Satellite cells in all of these ganglia are derived from the neural crest (Narayanan and Narayanan, 1980).

The geniculate placode in nonteleost fish and birds has also been described as giving rise to a pouch-like sense organ associated with the first pharyngeal cleft, lined with mechanosensory hair cells (Vitali, 1926; Yntema, 1944; D'Amico-Martel and Noden, 1983; Baker and Bronner-Fraser, 2001). This organ appears to have been lost in teleosts, amphibians, reptiles, and mammals. In nonteleost fish, this "spiracular organ" is considered to be a specialized lateral line organ (reviewed in Barry and Bennett, 1989). However, if it is indeed derived from the geniculate placode and not from a lateral line placode, then it would appear that epibranchial placodes are able to form not only sensory neurons, but also mechanosensory hair cells like those of the inner ear and lateral line.

A Common Primordium for Epibranchial and Otic Placodes?

In the 10-somite stage chick embryo, the HMG-domain transcription factor Sox3, which labels the thickened ectoderm of the neural plate and cranial ectodermal placodes, is expressed in two narrow domains near the otic placode (Ishii *et al.*, 2001). One of these contains the otic placode itself plus prospective geniculate placode ectoderm; the other, more ventrocaudal domain, fate-maps to the petrosal and nodose placodes (Fig. 11) (Ishii *et al.*, 2001). Ectoderm that will eventually form the epibranchial placodes remains thickened and retains *Sox3* expression, while the ectoderm between the placodes thins and loses *Sox3* expression (Ishii *et al.*, 2001). These results suggest that a broad domain of thickened ectoderm is partitioned into the different epibranchial placodes in the chick.

Intriguingly, the broad domain of *Sox3*⁺ ectoderm that eventually forms the geniculate placode also contains the otic placode (Ishii *et al.*, 2001). Pax2 is also expressed in a broad region of ectoderm that includes precursors of both the otic and epibranchial placodes in the chick (Groves and Bronner-Fraser, 2000; Streit, 2002) (section Otic Placode Formation). It has been suggested in *Xenopus* that the Pax2⁺ "dorsolateral placode area," which includes otic and lateral line placodes (section Lateral Line Placode Derivatives), may also include the epibranchial placodes (Schlosser, 2002a), although this remains to be demonstrated. Furthermore, the winged-helix transcription factor Foxi1, which is required for otic placode formation (Solomon *et al.*, 2003; section Induction of the Otic Placodes) is also expressed in and required for epibranchial placode development (Lee *et al.*, 2003) (section Neurogenesis in the Epibranchial Placodes). The domain of *foxi1* expression in the zebrafish has been described as a "lateral cranial placodal domain" that encompasses otic and epibranchial placodes (Lee *et al.*, 2003). As described in the previous section, the geniculate placode may form mechanosensory hair cells, like otic and lateral line hair cells, during normal development in chick and nonteleost fish. It is possible, therefore, that the close spatial association of epibranchial placodes with the otic placodes, together with their shared expression of Pax2, might reflect previously unrecognized embryonic and, potentially, evolutionary relationships. However, additional evidence is required to support this hypothesis.

Induction of the Epibranchial Placodes

In all vertebrate species, epibranchial placode formation occurs in close spatiotemporal association with (1) contact between the outpocketing pharyngeal endoderm and the overlying surface ectoderm, and (2) migrating neural crest streams (Fig. 3; Baker and Bronner-Fraser, 2001). Mechanical and genetic ablation experiments have shown that neural crest cells are not required for the formation of the epibranchial placodes (Yntema, 1944; Begbie *et al.*, 1999; Gavalas *et al.*, 2001). Instead, signals from the pharyngeal endoderm seem to be important, at least for the induction of neurogenesis within the epibranchial placodes in the chick (Begbie *et al.*, 1999). Pharyngeal endoderm is sufficient to induce epibranchial neurons (*Phox2a*⁺; see next section) from non-placode-forming chick head ectoderm *in vitro* (Begbie *et al.*, 1999). BMP7, which is produced by pharyngeal endoderm, is also sufficient to induce epibranchial neurons from this ectoderm *in vitro* (Begbie *et al.*, 1999). Furthermore, the BMP7 inhibitor follistatin reduces neuronal induction by pharyngeal endoderm *in vitro*, suggesting that BMP7 might be the pharyngeal endoderm-derived signal *in vivo* (Begbie *et al.*, 1999). Nonetheless, pharyngeal endoderm cannot induce epibranchial neurons from trunk ectoderm (Begbie *et al.*, 1999), which is competent to make nodose placode neurons when grafted to the nodose placode region (Vogel and Davies, 1993). Hence, additional signals in the pharyngeal region must enable trunk ectoderm to form epibranchial neurons in response to signals from pharyngeal endoderm.

Neurogenesis in the Epibranchial Placodes Requires *Neurogenin2, Phox2b*, and *Phox2a*

The bHLH proneural transcription factor Neurogenin2 (Ngn2) (section Proneural Genes: An Introduction) is expressed in epibranchial placodes and delaminating cells prior to overt neuronal differentiation in the mouse (Fode *et al.*, 1998). In *Ngn2*-mutant mice, geniculate and petrosal placode-derived cells fail to delaminate, migrate, or differentiate (Fode *et al.*, 1998). In the nodose placode, which develops normally in *Ngn2* mutants, *Ngn2* may act redundantly with *Ngn1* (Fode *et al.*, 1998; Ma *et al.*, 1998). In all three epibranchial placodes, Ngn2 is required for *Delta-like1* expression, suggesting that Notch–Delta signaling is also involved in epibranchial placode-derived neurogenesis (Fode *et al.*, 1998).

In the zebrafish, Ngn1 seems to encompass all functions of murine Ngn1 and Ngn2, and *ngn1* is expressed in the epibranchial placodes (Andermann *et al.*, 2002). All peripheral ganglia, including the epibranchial placode-derived ganglia, are missing after antisense morpholino-mediated functional knockdown of Ngn1 (Andermann *et al.*, 2002). The winged-helix transcription factor Foxi1, which is expressed prior to *ngn1* in prospective epibranchial placode ectoderm, is required for *ngn1* expression in the epibranchial placodes (Lee *et al.*, 2003).

As described in the section Phox2b Is Essential for the Formation of All Autonomic Ganglia, the paired-like homeodomain transcription factor Phox2b is required for the development of all autonomic ganglia, including the epibranchial placode-derived ganglia (Pattyn *et al.*, 1999). The neurons in these ganglia provide autonomic afferent innervation to the visceral organs and transiently express the noradrenergic markers tyrosine hydroxylase and dopamine β-hydroxylase (DBH) (Fig. 9) (e.g., Katz and Erb, 1990; Morin *et al.*, 1997). As described in section *Phox2b* Is Required for Development of the Noradrenergic Phenotype, Phox2b and the related factor Phox2a directly activate the *DBH* promoter (reviewed in Brunet and Pattyn, 2002; Goridis and Rohrer, 2002). *Phox2b*-mutant mice show severe apoptotic atrophy of all three epibranchial placode-derived ganglia (Pattyn *et al.*, 1999).

In epibranchial placode-derived ganglia, unlike sympathetic ganglia (section *Phox2b* Is Required for Development of the Noradrenergic Phenotype), *Phox2a* lies genetically upstream of *Phox2b*, which is in turn required for *DBH* expression (Pattyn *et al.*, 1999, 2000). *Phox2a* is not required for delamination or aggregation of epibranchial placode-derived cells, or for the expression of certain neuronal markers, but is required for *DBH* and *Ret* expression (hence probably for neuronal survival in response to the Ret ligand GDNF) (Morin *et al.*, 1997). *Phox2a*-mutant mice show severe atrophy of the petrosal and nodose ganglia, while the geniculate ganglion is relatively unaffected (Morin *et al.*, 1997), perhaps via redundancy with Phox2b. Like sympathetic neurons (section BMPs Induce Both Mash1 and Phox2b in Sympathetic Precursors), a BMP family member, in this case BMP7, is able to induce *Phox2a* expression in head ectoderm (see previous section) (Begbie *et al.*, 1999).

Interactions Between Neural Crest-Derived and Epibranchial Placode-Derived Cells in Gangliogenesis

Although neural crest cells are not required for epibranchial placode formation or neurogenesis (Yntema, 1944; Begbie *et al.*, 1999; Gavalas *et al.*, 2001), they seem to play an important role in guiding the migration and projection patterns of epibranchial placode-derived neurons (Begbie and Graham, 2001b). After neural crest ablation, epibranchial placode-derived neurons remain subectodermal and make aberrant projections (Begbie and Graham, 2001b).

It is possible that some neural crest cells initially form neurons in the epibranchial ganglia in the chick (Kious *et al.*, 2002), although these presumably die, as they are not seen at later stages of development (D'Amico-Martel and Noden, 1983). Neural crest cells can also compensate to some extent for loss of the epibranchial placodes. Neural crest cells may form neurons in the geniculate ganglion in *Ngn2*-mutant mice, which lack epibranchial placode-derived neurons (Fode *et al.*, 1998). Also, neural crest cells from the same axial level as the nodose placode can form neurons in the nodose ganglion after the nodose placode is ablated (Harrison *et al.*, 1995). However, these neurons may not substitute functionally for nodose placode-derived neurons, as nodose placode-ablated embryos have abnormal cardiac function (Harrison *et al.*, 1995).

PLACODE SUMMARY

Cranial ectodermal placodes are, at first sight, a disparate collection of embryonic structures, united by their early-thickened morphology and association with the paired sense organs and/or cranial sensory ganglia. Each individual placode gives rise to very different derivatives, from mechanosensory hair cells to lens fibers to visceral sensory neurons. However, some early steps in placode induction may be common to all placodes. They share a common origin from a preplacodal field of ectoderm around the anterior border of the neural plate that can be identified molecularly and, in

some species, morphologically. The Pax/Six/Eya/Dach genetic regulatory network seems to be active in all placodes, with different combinations of Pax genes, in particular, expressed in different placodes and possibly serving to determine placode identity. Recent evidence suggests that there is a substantial degree of cell movement within the pre-placodal field. Individual placodes may form within this field either by differential cellular responses to widespread inducing signals and active convergence to the forming placode, or by the "trapping" of randomly moving cells by localized placode-inducing signals. Current evidence cannot distinguish between these two hypotheses. Each individual placode seems to be induced by a different combination of tissues (neural tube, pharyngeal endoderm, paraxial mesoderm, etc) and molecules: where identified, the latter include members of the BMP, FGF, and Wnt families. Neurogenesis within all neurogenic placodes involves one or both Ngns, and probably Delta–Notch signaling, showing clear parallels with sensory neurogenesis in the neural crest. As is the case for autonomic neural crest-derived neurons, Phox2a and Phox2b are required for the transient expression of the catecholaminergic phenotype within epibranchial placode-derived neurons, which provide afferent autonomic innervation to the visceral organs.

As should be evident from this section of the chapter, great strides have been made in our understanding of placode induction and development, particularly with the application of molecular techniques. However, there is much still to learn, from the earliest stages of placode induction at the neural plate border, to the final patterning and morphogenesis of their diverse derivatives.

OVERALL SUMMARY

Hopefully, this chapter has succeeded in giving a flavor of the complexity that underlies the induction and development of the neural crest and cranial ectodermal placodes. The neural crest forms the entire PNS in the trunk, while placodes are essential for the formation of the paired peripheral sense organs and most cranial sensory neurons. Although for the most part they have been treated separately, it is important to realize that neural crest and placodes do not develop in isolation from one another. As discussed in the preceding sections, placode-derived neurons in cranial sensory ganglia are supported by neural crest-derived satellite glia. Neural crest-derived trigeminal neurons need placode-derived trigeminal neurons in order to make appropriate peripheral projections. Migrating streams of cranial neural crest cells are required for proper migration of epibranchial placode-derived neurons. Hence, both the formation and interaction of placodes and neural crest cells are essential for the development of a fully functional peripheral nervous system. The mutual interdependence of these two cell populations reflects their long evolutionary history together: Both neural crest and placodes are present in hagfish, the most primitive extant craniate.

Since the last edition of this book, in 1991, our understanding of the induction and development of both neural crest and cranial ectodermal placodes has advanced in leaps and bounds. It is to be hoped that the next decade will prove similarly fruitful.

REFERENCES

Abzhanov, A., Tzahor, E., Lassar, A.B., and Tabin, C.J., 2003, Dissimilar regulation of cell differentiation in mesencephalic (cranial) and sacral (trunk) neural crest cells in vitro, *Development* 130:4567–4579.

Adams, R.H., Diella, F., Hennig, S., Helmbacher, F., Deutsch, U., and Klein, R., 2001, The cytoplasmic domain of the ligand ephrinB2 is required for vascular morphogenesis but not cranial neural crest migration, *Cell* 104:57–69.

Adelmann, H.B., 1925, The development of the neural folds and cranial ganglia of the rat, *J. Comp. Neurol.* 39: 19–171.

Adlkofer, K. and Lai, C., 2000, Role of neuregulins in glial cell development, *Glia* 29: 104–111.

Agrawal, M. and Brauer, P.R., 1996, Urokinase-type plasminogen activator regulates cranial neural crest cell migration *in vitro*, *Dev. Dyn.* 207:281–290.

Airaksinen, M.S. and Saarma, M., 2002, The GDNF family: Signalling, biological functions and therapeutic value, *Nat. Rev. Neurosci.* 3:383–394.

Akitaya, T. and Bronner-Fraser, M., 1992, Expression of cell adhesion molecules during initiation and cessation of neural crest cell migration, *Dev. Dyn.* 194:12–20.

Alfandari, D., Cousin, H., Gaultier, A., Smith, K., White, J.M., Darribère, T., and DeSimone, D.W., 2001, *Xenopus* ADAM 13 is a metalloprotease required for cranial neural crest-cell migration, *Curr. Biol.* 11:918–930.

Alfandari, D., Wolfsberg, T.G., White, J.M., and DeSimone, D.W., 1997, ADAM 13: A novel ADAM expressed in somitic mesoderm and neural crest cells during *Xenopus laevis* development, *Dev. Biol.* 182:314–330.

Altmann, C.R., Chow, R.L., Lang, R.A., and Hemmati-Brivanlou, A., 1997, Lens induction by Pax-6 in *Xenopus laevis*, *Dev. Biol.* 185:119–123.

Andermann, P., Ungos, J., and Raible, D.W., 2002, Neurogenin1 defines zebrafish cranial sensory ganglia precursors, *Dev. Biol.* 251:45–58.

Anderson, D.J., 1993, Molecular control of cell fate in the neural crest: The sympathoadrenal lineage. *Annu. Rev. Neurosci.* 16:129–158.

Anderson, D.J., 1997, Cellular and molecular biology of neural crest cell lineage determination, *Trends Genet.* 13:276–280.

Anderson, D.J., 1999, Lineages and transcription factors in the specification of vertebrate primary sensory neurons, *Curr. Opin. Neurobiol.* 9:517–524.

Anderson, D.J., 2000, Genes, lineages and the neural crest: A speculative review, *Phil. Trans. R. Soc. Lond. B*, 355:953–964.

Anderson, D.J. and Jan, Y.N., 1997, The determination of the neuronal phenotype, in *Molecular and Cellular Approaches to Neural Development* (W.M. Cowan, T.M. Jessell, and S.L. Zipursky, eds.), Oxford University Press, Oxford, pp. 26–63.

Artinger, K.B. and Bronner-Fraser, M., 1992, Partial restriction in the developmental potential of late emigrating avian neural crest cells, *Dev. Biol.* 149:149–157.

Artinger, K.B., Chitnis, A.B., Mercola, M., and Driever, W., 1999, Zebrafish *narrowminded* suggests a genetic link between formation of neural crest and primary sensory neurons, *Development* 126:3969–3979.

Ashery-Padan, R. and Gruss, P., 2001, Pax6 lights-up the way for eye development, *Curr. Opin. Cell Biol.* 13:706–714.

Ayer-Le Lièvre, C.S. and Le Douarin, N.M., 1982, The early development of cranial sensory ganglia and the potentialities of their component cells studied in quail-chick chimeras, *Dev. Biol.* 94:291–310.

Baker, C.V.H. and Bronner-Fraser, M., 1997, The origins of the neural crest. Part II: an evolutionary perspective, *Mech. Dev.* 69:13–29.

Baker, C.V.H. and Bronner-Fraser, M., 2000, Establishing neuronal identity in vertebrate neurogenic placodes, *Development* 127:3045–3056.

Baker, C.V.H. and Bronner-Fraser, M., 2001, Vertebrate cranial placodes I. Embryonic induction, *Dev. Biol.* 232:1–61.

Baker, C.V.H., Bronner-Fraser, M., Le Douarin, N.M., and Teillet, M.-A., 1997, Early- and late-migrating cranial neural crest cell populations have equivalent developmental potential in vivo, *Development* 124:3077–3087.

Baker, C.V.H., Stark, M.R., and Bronner-Fraser, M., 2002, Pax3-expressing trigeminal placode cells can localize to trunk neural crest sites but are committed to a cutaneous sensory neuron fate, *Dev. Biol.* 249:219–236.

Baker, C.V.H., Stark, M.R., Marcelle, C., and Bronner-Fraser, M., 1999, Competence, specification and induction of Pax-3 in the trigeminal placode, *Development* 126:147–156.

Baker, R.C. and Graves, G.O., 1939, The behaviour of the neural crest in the forebrain region of *Amblystoma*, *J. Comp. Neurol.* 71:389–415.

Bang, A.G., Papalopulu, N., Goulding, M.D., and Kintner, C., 1999, Expression of *Pax-3* in the lateral neural plate is dependent on a *Wnt*-mediated signal from posterior nonaxial mesoderm, *Dev. Biol.* 212:366–380.

Bang, A.G., Papalopulu, N., Kintner, C., and Goulding, M.D., 1997, Expression of *Pax-3* is initiated in the early neural plate by posteriorizing signals produced by the organizer and by posterior non-axial mesoderm, *Development* 124:2075–2085.

Barald, K.F., 1988a, Antigen recognized by monoclonal antibodies to mesencephalic neural crest and to ciliary ganglion neurons is involved in the high affinity choline uptake mechanism in these cells, *J. Neurosci. Res.* 21:119–134.

Barald, K.F., 1988b, Monoclonal antibodies made to chick mesencephalic neural crest cells and to ciliary ganglion neurons identify a common antigen on the neurons and a neural crest subpopulation, *J. Neurosci. Res.* 21:107–118.

Baroffio, A., Dupin, E., and Le Douarin, N.M., 1988, Clone-forming ability and differentiation potential of migratory neural crest cells, *Proc. Natl. Acad. Sci. USA* 85:5325–5329.

Baroffio, A., Dupin, E., and Le Douarin, N.M., 1991, Common precursors for neural and mesectodermal derivatives in the cephalic neural crest, *Development* 112:301–305.

Barry, M.A. and Bennett, M.V.L., 1989, Specialised lateral line receptor systems in elasmobranchs: The spiracular organs and vesicles of Savi, in *The Mechanosensory Lateral Line. Neurobiology and Evolution* (S. Coombs, P. Görner, and H. Münz, eds.), Springer-Verlag, New York, pp. 591–606.

Batlle, E., Sancho, E., Franci, C., Dominguez, D., Monfar, M., Baulida, J. *et al.*, 2000, The transcription factor Snail is a repressor of *E-cadherin* gene expression in epithelial tumour cells. *Nat. Cell Biol.* 2:84–89.

Begbie, J., Brunet, J.-F., Rubenstein, J.L., and Graham, A., 1999, Induction of the epibranchial placodes, *Development* 126:895–902.

Begbie, J. and Graham, A., 2001a, The ectodermal placodes: A dysfunctional family, *Phil. Trans. R. Soc. Lond. B* 356:1655–1660.

Begbie, J. and Graham, A., 2001b, Integration between the epibranchial placodes and the hindbrain, *Science* 294:595–598.

Belecky-Adams, T.L., Adler, R., and Beebe, D.C., 2002, Bone morphogenetic protein signaling and the initiation of lens fiber cell differentiation, *Development* 129:3795–3802.

Bellmeyer, A., Krase, J., Lindgren, J., and LaBonne, C., 2003, The protooncogene c-Myc is an essential regulator of neural crest formation in *Xenopus*, *Dev. Cell* 4:827–839.

Bermingham, N.A., Hassan, B.A., Price, S.D., Vollrath, M.A., Ben-Arie, N., Eatock, R.A. *et al.*, 1999, *Math1*: An essential gene for the generation of inner ear hair cells, *Science* 284:1837–1841.

Bertrand, N., Castro, D.S., and Guillemot, F., 2002, Proneural genes and the specification of neural cell types, *Nat. Rev. Neurosci.* 3:517–530.

Bixby, S., Kruger, G., Mosher, J., Joseph, N., and Morrison, S., 2002, Cell-intrinsic differences between stem cells from different regions of the peripheral nervous system regulate the generation of neural diversity, *Neuron* 35:643–656.

Blaschko, H., 1973, Catecholamine biosynthesis, *Br. Med. Bull.* 29:105–109.

Blaugrund, E., Pham, T.D., Tennyson, V.M., Lo, L., Sommer, L., Anderson, D.J. *et al.*, 1996, Distinct subpopulations of enteric neuronal progenitors defined by time of development, sympathoadrenal lineage markers and Mash-1-dependence, *Development* 122:309–320.

Bodznick, D., 1989, Comparisons between electrosensory and mechanosensory lateral line systems, in *The Mechanosensory Lateral Line. Neurobiology and Evolution* (S. Coombs, P. Görner, and H. Münz, eds.), Springer-Verlag, New York, pp. 655–678.

Bodznick, D. and Northcutt, R.G., 1981, Electroreception in lampreys: Evidence that the earliest vertebrates were electroreceptive, *Science* 212:465–467.

Bonstein, L., Elias, S., and Frank, D., 1998, Paraxial-fated mesoderm is required for neural crest induction in *Xenopus* embryos, *Dev. Biol.* 193:156–168.

Boorman, C.J. and Shimeld, S.M., 2002, *Pitx* homeobox genes in *Ciona* and amphioxus show left-right asymmetry is a conserved chordate character and define the ascidian adenohypophysis, *Evol. Dev.* 4:354–365.

Borchers, A., David, R., and Wedlich, D., 2001, *Xenopus* cadherin-11 restrains cranial neural crest migration and influences neural crest specification, *Development* 128:3049–3060.

Bordzilovskaya, N.P., Dettlaff, T.A., Duhon, S.T., and Malacinski, G.M., 1989, Developmental-stage series of axolotl embryos, in *Developmental Biology of the Axolotl* (J.B. Armstrong and G.M. Malacinski, eds.), Oxford University Press, Oxford, pp. 176–186.

Brachet, A., 1907, Recherches sur l'ontogénèse de la tête chez les Amphibiens, *Arch. Biol.* 23:165–257.

Britsch, S., Goerich, D.E., Riethmacher, D., Peirano, R.I., Rossner, M., Nave, K.A. *et al.*, 2001, The transcription factor Sox10 is a key regulator of peripheral glial development, *Genes Dev.* 15:66–78.

Bronner-Fraser, M., 1986, Analysis of the early stages of trunk neural crest migration in avian embryos using monoclonal antibody HNK-1, *Dev. Biol.* 115:44–55.

Bronner-Fraser, M. and Fraser, S., 1989, Developmental potential of avian trunk neural crest cells in situ, *Neuron* 3:755–766.

Bronner-Fraser, M. and Fraser, S.E., 1988, Cell lineage analysis reveals multipotency of some avian neural crest cells, *Nature* 335:161–164.

Bronner-Fraser, M. and Stern, C., 1991, Effects of mesodermal tissues on avian neural crest cell migration, *Dev. Biol.* 143:213–217.

Brunet, J.F. and Pattyn, A., 2002, *Phox2* genes—from patterning to connectivity, *Curr. Opin. Genet. Dev.* 12:435–440.

Burstyn-Cohen, T. and Kalcheim, C., 2002, Association between the cell cycle and neural crest delamination through specific regulation of G1/S transition, *Dev. Cell* 3:383–395.

Cai, H., Kratzschmar, J., Alfandari, D., Hunnicutt, G., and Blobel, C.P., 1998, Neural crest-specific and general expression of distinct metalloprotease-disintegrins in early *Xenopus laevis* development, *Dev. Biol.* 204:508–524.

Calof, A.L., Mumm, J.S., Rim, P.C., and Shou, J., 1998, The neuronal stem cell of the olfactory epithelium, *J. Neurobiol.* 36:190–205.

Cambronero, F. and Puelles, L., 2000, Rostrocaudal nuclear relationships in the avian medulla oblongata: A fate map with quail chick chimeras, *J. Comp. Neurol.* 427:522–545.

Cameron-Curry, P., Dulac, C., and Le Douarin, N.M., 1993, Negative regulation of Schwann cell myelin protein gene expression by the dorsal root ganglionic microenvironment, *Eur. J. Neurosci.* 5:594–604.

Cano, A., Pérez-Moreno, M.A., Rodrigo, I., Locascio, A., Blanco, M.J., del Barrio, M.G. *et al.*, 2000, The transcription factor snail controls epithelial–mesenchymal transitions by repressing E-cadherin expression, *Nat. Cell Biol.* 2:76–83.

Carl, T.F., Dufton, C., Hanken, J., and Klymkowsky, M.W., 1999, Inhibition of neural crest migration in *Xenopus* using antisense *slug* RNA, *Dev. Biol.* 213:101–115.

Carr, V.M. and Simpson, S.B., Jr., 1978, Proliferative and degenerative events in the early development of chick dorsal root ganglia. I. Normal development, *J. Comp. Neurol.* 182:727–739.

Catala, M., Teillet, M.-A., De Robertis, E.M., and Le Douarin, M.L., 1996, A spinal cord fate map in the avian embryo: While regressing, Hensen's node lays down the notochord and floor plate thus joining the spinal cord lateral walls, *Development* 122:2599–2610.

Catala, M., Ziller, C., Lapointe, F., and Le Douarin, N.M., 2000, The developmental potentials of the caudalmost part of the neural crest are restricted to melanocytes and glia, *Mech. Dev.* 95:77–87.

Cau, E., Casarosa, S., and Guillemot, F., 2002, Mash1 and Ngn1 control distinct steps of determination and differentiation in the olfactory sensory neuron lineage, *Development* 129:1871–1880.

Cau, E., Gradwohl, G., Casarosa, S., Kageyama, R., and Guillemot, F., 2000, *Hes* genes regulate sequential stages of neurogenesis in the olfactory epithelium, *Development* 127:2323–2332.

Cau, E., Gradwohl, G., Fode, C., and Guillemot, F., 1997, Mash1 activates a cascade of bHLH regulators in olfactory neuron progenitors, *Development* 124:1611–1621.

Chai, Y., Jiang, X., Ito, Y., Bringas, P., Han, J., Rowitch, D.H. *et al.*, 2000, Fate of the mammalian cranial neural crest during tooth and mandibular morphogenesis, *Development* 127:1671–1679.

Chang, C. and Hemmati-Brivanlou, A., 1998, Neural crest induction by *Xwnt7B* in *Xenopus*, *Dev. Biol.* 194:129–134.

Chen, P., Johnson, J.E., Zoghbi, H.Y., and Segil, N., 2002, The role of Math1 in inner ear development: Uncoupling the establishment of the sensory primordium from hair cell fate determination, *Development* 129:2495–2505.

Cheung, M. and Briscoe, J., 2003, Neural crest development is regulated by the transcription factor Sox9, *Development* 130:5681–5693.

Chibon, P., 1964, Analyse par la méthode de marquage nucléaire à la thymidine tritiée des dérivés de la crête neurale céphalique chez l'Urodèle *Pleurodeles waltlii*, *C. R. Acad. Sci., Ser. III* 259:3624–3627.

Chibon, P., 1966, Analyse expérimentale de la régionalisation et des capacités morphogénétiques de la crête neurale chez l'Amphibien Urodèle *Pleurodeles waltlii* Michah, *Mem. Soc. Fr. Zool.* 36:1–107.

Chibon, P., 1967a, Etude expérimentale par ablations, greffes et autoradiographie, de l'origine des dents chez l'Amphibien Urodèle *Pleurodeles waltlii* Michah, *Arch. Oral Biol.* 12:745–753.

Chibon, P., 1967b, Marquage nucléaire par la thymidine tritiée des dérivés de la crête neurale chez l'Amphibien Urodèle *Pleurodeles waltlii* Michah, *J. Embryol. Exp. Morphol.* 18:343–358.

Chow, R.L., Altmann, C.R., Lang, R.A., and Hemmati-Brivanlou, A., 1999, Pax6 induces ectopic eyes in a vertebrate, *Development* 126:4213–4222.

Christiaen, L., Burighel, P., Smith, W.C., Vernier, P., Bourrat, F., and Joly, J.S., 2002, *Pitx* genes in Tunicates provide new molecular insight into the evolutionary origin of pituitary, *Gene* 287:107–113.

Ciment, G. and Weston, J.A., 1982, Early appearance in neural crest and crest-derived cells of an antigenic determinant present in avian neurons, *Dev. Biol.* 93:355–367.

Cohen, A.M. and Königsberg, I.R., 1975, A clonal approach to the problem of neural crest determination, *Dev. Biol.* 46:262–280.

Collazo, A., Bronner-Fraser, M., and Fraser, S.E., 1993, Vital dye labelling of *Xenopus laevis* trunk neural crest reveals multipotency and novel pathways of migration, *Development* 118:363–376.

Coombs, S., Görner, P., and Münz, H., 1989, *The Mechanosensory Lateral Line. Neurobiology and Evolution*, Springer-Verlag, New York.

Cornell, R.A. and Eisen, J.S., 2000, Delta signaling mediates segregation of neural crest and spinal sensory neurons from zebrafish lateral neural plate, *Development* 127:2873–2882.

Cornell, R.A. and Eisen, J.S., 2002, Delta/Notch signaling promotes formation of zebrafish neural crest by repressing Neurogenin 1 function, *Development* 129:2639–2648.

Couly, G., Creuzet, S., Bennaceur, S., Vincent, C., and Le Douarin, N.M., 2002, Interactions between *Hox*-negative cephalic neural crest cells and the foregut endoderm in patterning the facial skeleton in the vertebrate head, *Development* 129:1061–1073.

Couly, G. and Le Douarin, N.M., 1988, The fate map of the cephalic neural primordium at the presomitic to the 3-somite stage in the avian embryo, *Development* 103:101–113.

Couly, G.F., Coltey, P.M., and Le Douarin, N.M., 1993, The triple origin of skull in higher vertebrates: A study in quail-chick chimeras, *Development* 117:409–429.

Couly, G.F. and Le Douarin, N.M., 1985, Mapping of the early neural primordium in quail-chick chimeras. I. Developmental relationships between placodes, facial ectoderm, and prosencephalon, *Dev. Biol.* 110:422–439.

Covell, D.A., Jr. and Noden, D.M., 1989, Embryonic development of the chick primary trigeminal sensory-motor complex, *J. Comp. Neurol.* 286:488–503.

D'Amico-Martel, A. and Noden, D.M., 1980, An autoradiographic analysis of the development of the chick trigeminal ganglion, *J. Embryol. Exp. Morphol.* 55:167–182.

D'Amico-Martel, A. and Noden, D.M., 1983, Contributions of placodal and neural crest cells to avian cranial peripheral ganglia, *Am. J. Anat.* 166:445–468.

Dasen, J.S. and Rosenfeld, M.G., 2001, Signaling and transcriptional mechanisms in pituitary development, *Annu. Rev. Neurosci.* 24:327–355.

David, R., Ahrens, K., Wedlich, D., and Schlosser, G., 2001, *Xenopus Eya1* demarcates all neurogenic placodes as well as migrating hypaxial muscle precursors, *Mech. Dev.* 103:189–192.

Davis, R.J., Shen, W., Sandler, Y.I., Heanue, T.A., and Mardon, G., 2001, Characterization of mouse *Dach2*, a homologue of *Drosophila dachshund*, *Mech. Dev.* 102:169–179.

Davis, R.L. and Turner, D.L., 2001, Vertebrate hairy and Enhancer of split related proteins: Transcriptional repressors regulating cellular differentiation and embryonic patterning, *Oncogene* 20:8342–8357.

De Bellard, M., Ching, W., Gossler, A., and Bronner-Fraser, M., 2002, Disruption of segmental neural crest migration and ephrin expression in *Delta-1* null mice, *Dev. Biol.* 249:121–130.

de Iongh, R.U., Lovicu, F.J., Overbeek, P.A., Schneider, M.D., Joya, J., Hardeman, E.D. *et al.*, 2001, Requirement for TGFβ receptor signaling during terminal lens fiber differentiation, *Development* 128:3995–4010.

Deardorff, M.A., Tan, C., Saint-Jeannet, J.P., and Klein, P.S., 2001, A role for frizzled 3 in neural crest development, *Development* 128:3655–3663.

Debby-Brafman, A., Burstyn-Cohen, T., Klar, A., and Kalcheim, C., 1999, F-Spondin, expressed in somite regions avoided by neural crest cells, mediates inhibition of distinct somite domains to neural crest migration, *Neuron* 22:475–488.

del Barrio, M.G. and Nieto, M.A., 2002, Overexpression of Snail family members highlights their ability to promote chick neural crest formation, *Development* 129:1583–1593.

Derby, M.A. and Newgreen, D.F., 1982, Differentiation of avian neural crest cells *in vitro*: Absence of a developmental bias toward melanogenesis, *Cell Tissue Res.* 225:365–378.

Detwiler, S.R., 1937, Observations upon the migration of neural crest cells, and upon the development of the spinal ganglia and vertebral arches in *Amblystoma*, *Am. J. Anat.* 61:63–94.

Dickinson, M.E., Selleck, M.A., McMahon, A.P., and Bronner-Fraser, M., 1995, Dorsalization of the neural tube by the non-neural ectoderm, *Development* 121:2099–2106.

Dorsky, R.I., Moon, R.T., and Raible, D.W., 1998, Control of neural crest cell fate by the Wnt signalling pathway, *Nature* 396:370–373.

Dorsky, R.I., Moon, R.T., and Raible, D.W., 2000a, Environmental signals and cell fate specification in premigratory neural crest, *Bioessays* 22:708–716.

Dorsky, R.I., Raible, D.W., and Moon, R.T., 2000b, Direct regulation of *nacre*, a zebrafish *MITF* homolog required for pigment cell formation, by the Wnt pathway, *Genes Dev.* 14:158–162.

Dottori, M., Gross, M.K., Labosky, P., and Goulding, M., 2001, The winged-helix transcription factor Foxd3 suppresses interneuron differentiation and promotes neural crest cell fate, *Development* 128:4127–4138.

Du Shane, G., 1938, Neural fold derivatives in the Amphibia: Pigment cells, spinal ganglia and Rohon-Beard cells, *J. Exp. Zool.* 78:485–503.

Dubois, E.A., Zandbergen, M.A., Peute, J., and Goos, H.J., 2002, Evolutionary development of three gonadotropin-releasing hormone (GnRH) systems in vertebrates, *Brain Res. Bull.* 57:413–418.

Dulac, C. and Le Douarin, N.M., 1991, Phenotypic plasticity of Schwann cells and enteric glial cells in response to the microenvironment, *Proc. Natl. Acad. Sci. USA* 88:6358–6362.

Dunn, K.J., Williams, B.O., Li, Y., and Pavan, W.J., 2000, Neural crest-directed gene transfer demonstrates Wnt1 role in melanocyte expansion and differentiation during mouse development, *Proc. Natl. Acad. Sci. USA* 97:10,050–10,055.

Dupin, E., Baroffio, A., Dulac, C., Cameron-Curry, P., and Le Douarin, N.M., 1990, Schwann-cell differentiation in clonal cultures of the neural crest, as evidenced by the anti-Schwann cell myelin protein monoclonal antibody, *Proc. Natl. Acad. Sci. USA* 87:1119–1123.

Durbec, P.L., Larsson-Blomberg, L.B., Schuchardt, A., Costantini, F., and Pachnis, V., 1996, Common origin and developmental dependence on c-ret of subsets of enteric and sympathetic neuroblasts, *Development* 122:349–358.

DuShane, G.P., 1935, An experimental study of the origin of pigment cells in amphibia, *J. Exp. Zool.* 72:1–31.

Dutton, K.A., Pauliny, A., Lopes, S.S., Elworthy, S., Carney, T.J., Rauch, J. *et al.*, 2001, Zebrafish *colourless* encodes *sox10* and specifies non-ectomesenchymal neural crest fates, *Development* 128:4113–4125.

Eagleson, G.W. and Harris, W.A., 1990, Mapping of the presumptive brain regions in the neural plate of *Xenopus laevis*, *J. Neurobiol.* 21:427–40.

Eddison, M., Le Roux, I., and Lewis, J., 2000, Notch signaling in the development of the inner ear: Lessons from *Drosophila*, *Proc. Natl. Acad. Sci. USA* 97:11,692–11,699.

Eickholt, B.J., Mackenzie, S.L., Graham, A., Walsh, F.S., and Doherty, P., 1999, Evidence for collapsin-1 functioning in the control of neural crest migration in both trunk and hindbrain regions, *Development* 126:2181–2189.

Epperlein, H.-H., Meulemans, D., Bronner-Fraser, M., Steinbeisser, H., and Selleck, M.A.J., 2000, Analysis of cranial neural crest migratory pathways in axolotl using cell markers and transplantation, *Development* 127:2751–2761.

Erickson, C.A., 1987, Behavior of neural crest cells on embryonic basal laminae, *Dev. Biol.* 120:38–49.

Erickson, C.A., Duong, T.D., and Tosney, K.W., 1992, Descriptive and experimental analysis of the dispersion of neural crest cells along the dorsolateral path and their entry into ectoderm in the chick embryo, *Dev. Biol.* 151:251–272.

Erickson, C.A. and Goins, T.L., 1995, Avian neural crest cells can migrate in the dorsolateral path only if they are specified as melanocytes, *Development* 121:915–924.

Erickson, C.A., Loring, J.F., and Lester, S.M., 1989, Migratory pathways of HNK-1-immunoreactive neural crest cells in the rat embryo, *Dev. Biol.* 134:112–118.

Etchevers, H.C., Vincent, C., Le Douarin, N.M., and Couly, G.F., 2001, The cephalic neural crest provides pericytes and smooth muscle cells to all blood vessels of the face and forebrain, *Development* 128:1059–1068.

Faber, S.C., Dimanlig, P., Makarenkova, H.P., Shirke, S., Ko, K., and Lang, R.A., 2001, Fgf receptor signaling plays a role in lens induction, *Development* 128:4425–4438.

Faber, S.C., Robinson, M.L., Makarenkova, H.P., and Lang, R.A., 2002, Bmp signaling is required for development of primary lens fiber cells, *Development* 129:3727–3737.

Farlie, P.G., Kerr, R., Thomas, P., Symes, T., Minichiello, J., Hearn, C.J. *et al.*, 1999, A paraxial exclusion zone creates patterned cranial neural crest cell outgrowth adjacent to rhombomeres 3 and 5, *Dev. Biol.* 213:70–84.

Fedtsova, N., Perris, R., and Turner, E.E., 2003, Sonic hedgehog regulates the position of the trigeminal ganglia, *Dev. Biol.* 261:456–469.

Fekete, D.M. and Wu, D.K., 2002, Revisiting cell fate specification in the inner ear, *Curr. Opin. Neurobiol.* 12:35–42.

Firnberg, N. and Neubüser, A., 2002, FGF signaling regulates expression of *Tbx2, Erm, Pea3,* and *Pax3* in the early nasal region, *Dev. Biol.* 247:237–250.

Fode, C., Gradwohl, G., Morin, X., Dierich, A., LeMeur, M., Goridis, C. *et al.*, 1998, The bHLH protein NEUROGENIN 2 is a determination factor for epibranchial placode-derived sensory neurons, *Neuron* 20:483–494.

Foley, A.C. and Stern, C.D., 2001, Evolution of vertebrate forebrain development: How many different mechanisms? *J. Anat.* 199:35–52.

Fontaine-Pérus, J., Halgand, P., Chéraud, Y., Rouaud, T., Velasco, M.E., Cifuentes Diaz, C. *et al.*, 1997, Mouse–chick chimera: A developmental model of murine neurogenic cells, *Development* 124:3025–3036.

Fontaine-Pérus, J.C., Chanconie, M., and Le Douarin, N.M., 1982, Differentiation of peptidergic neurones in quail–chick chimaeric embryos, *Cell Differ.* 11:183–193.

Frame, M.C. and Brunton, V.G., 2002, Advances in Rho-dependent actin regulation and oncogenic transformation, *Curr. Opin. Genet. Dev.* 12:36–43.

Frank, E. and Sanes, J.R., 1991, Lineage of neurons and glia in chick dorsal root ganglia: Analysis in vivo with a recombinant retrovirus, *Development* 111:895–908.

Fraser, S.E. and Bronner-Fraser, M., 1991, Migrating neural crest cells in the trunk of the avian embryo are multipotent, *Development* 112:913–920.

Frenz, D.A., Liu, W., Williams, J.D., Hatcher, V., Galinovic-Schwartz, V., Flanders, K.C. *et al.*, 1994, Induction of chondrogenesis: Requirement for synergistic interaction of basic fibroblast growth factor and transforming growth factor-beta, *Development* 120:415–424.

Furuta, Y. and Hogan, B.L.M., 1998, BMP4 is essential for lens induction in the mouse embryo, *Genes Dev.* 12:3764–3775.

Gaiano, N. and Fishell, G., 2002, The role of Notch in promoting glial and neural stem cell fates, *Annu. Rev. Neurosci.* 25:471–490.

Gammill, L.S. and Bronner-Fraser, M., 2002, Genomic analysis of neural crest induction, *Development* 129:5731–5741.

Gammill, L.S. and Bronner-Fraser, M., 2003, Neural crest specification: Migrating into genomics, *Nat. Rev. Neurosci.* 4:795–805.

Gans, C. and Northcutt, R.G., 1983, Neural crest and the origin of vertebrates: A new head, *Science* 220:268–274.

García-Castro, M.I., Marcelle, C., and Bronner-Fraser, M., 2002, Ectodermal Wnt function as a neural crest inducer, *Science* 297:848–851.

Garratt, A.N., Britsch, S., and Birchmeier, C., 2000, Neuregulin, a factor with many functions in the life of a schwann cell, *Bioessays* 22:987–996.

Gavalas, A., Trainor, P., Ariza-McNaughton, L., and Krumlauf, R., 2001, Synergy between *Hoxa1* and *Hoxb1*: The relationship between arch patterning and the generation of cranial neural crest, *Development* 128:3017–3027.

Gershon, M.D., 1999, Lessons from genetically engineered animal models. II. Disorders of enteric neuronal development: Insights from transgenic mice, *Am. J. Physiol.* 277:G262–G267.

Ghanbari, H., Seo, H., Fjose, A., and Brändli, A.W., 2001, Molecular cloning and embryonic expression of *Xenopus Six* homeobox genes, *Mech. Dev.* 101:271–277.

Ghose, A. and Van Vactor, D., 2002, GAPs in Slit-Robo signaling, *Bioessays* 24:401–404.

Gilmour, D.T., Maischein, H.M., and Nusslein-Volhard, C., 2002, Migration and function of a glial subtype in the vertebrate peripheral nervous system, *Neuron* 34:577–588.

Girdlestone, J. and Weston, J.A., 1985, Identification of early neuronal subpopulations in avian neural crest cell cultures, *Dev. Biol.* 109:274–287.

Gitelman, I., 1997, Twist protein in mouse embryogenesis, *Dev. Biol.* 189:205–214.

Glimelius, B. and Weston, J.A., 1981, Analysis of developmentally homogeneous neural crest cell populations *in vitro*. III. Role of culture environment in cluster formation and differentiation, *Cell Differ.* 10:57–67.

Golding, J.P., Tidcombe, H., Tsoni, S., and Gassmann, M., 1999, Chondroitin sulphate-binding molecules may pattern central projections of sensory axons within the cranial mesenchyme of the developing mouse, *Dev. Biol.* 216:85–97.

Golding, J.P., Trainor, P., Krumlauf, R., and Gassmann, M., 2000, Defects in pathfinding by cranial neural crest cells in mice lacking the neuregulin receptor ErbB4, *Nat. Cell Biol.* 2:103–109.

Goldstein, R.S. and Kalcheim, C., 1991, Normal segmentation and size of the primary sympathetic ganglia depend upon the alternation of rostro-caudal properties of the somites, *Development* 112:327–334.

Gompel, N., Cubedo, N., Thisse, C., Thisse, B., Dambly-Chaudière, C., and Ghysen, A., 2001, Pattern formation in the lateral line of zebrafish, *Mech. Dev.* 105:69–77.

Gong, Q. and Shipley, M.T., 1995, Evidence that pioneer olfactory axons regulate telencephalon cell cycle kinetics to induce the formation of the olfactory bulb, *Neuron* 14:91–101.

Goridis, C. and Rohrer, H., 2002, Specification of catecholaminergic and serotonergic neurons, *Nat. Rev. Neurosci.* 3:531–541.

Govindarajan, V. and Overbeek, P.A., 2001, Secreted FGFR3, but not FGFR1, inhibits lens fiber differentiation, *Development* 128:1617–1627.

Graham, A., Heyman, I., and Lumsden, A., 1993, Even-numbered rhombomeres control the apoptotic elimination of neural crest cells from odd-numbered rhombomeres in the chick hindbrain, *Development* 119:233–245.

Graveson, A.C. and Armstrong, J.B., 1987, Differentiation of cartilage from cranial neural crest in the axolotl (*Ambystoma mexicanum*), *Differentiation* 35:16–20.

Graveson, A.C., Hall, B.K., and Armstrong, J.B., 1995, The relationship between migration and chondrogenic potential of trunk neural crest cells in *Ambystoma mexicanum, Roux's Arch. Dev. Biol.* 204:477–483.

Graveson, A.C., Smith, M.M., and Hall, B.K., 1997, Neural crest potential for tooth development in a urodele amphibian: Developmental and evolutionary significance, *Dev. Biol.* 188:34–42.

Greenwood, A.L., Turner, E.E., and Anderson, D.J., 1999, Identification of dividing, determined sensory neuron precursors in the mammalian neural crest, *Development* 126:3545–3559.

Groves, A.K. and Bronner-Fraser, M., 2000, Competence, specification and commitment in otic placode induction, *Development* 127:3489–3499.

Groves, A.K., George, K.M., Tissier-Seta, J.-P., Engel, J.D., Brunet, J.-F., and Anderson, D.J., 1995, Differential regulation of transcription factor gene expression and phenotypic markers in developing sympathetic neurons, *Development* 121:887–901.

Guillemot, F., Lo, L.C., Johnson, J.E., Auerbach, A., Anderson, D.J., and Joyner, A.L., 1993, Mammalian achaete-scute homolog 1 is required for the early development of olfactory and autonomic neurons, *Cell* 75:463–476.

Guillory, G. and Bronner-Fraser, M., 1986, An *in vitro* assay for neural crest cell migration through the somites, *J. Embryol. Exp. Morphol.* 98:85–97.

Gurdon, J.B., Lemaire, P., and Kato, K., 1993, Community effects and related phenomena in development, *Cell* 75:831–834.

Haddon, C., Jiang, Y.J., Smithers, L., and Lewis, J., 1998, Delta–Notch signalling and the patterning of sensory cell differentiation in the zebrafish ear: Evidence from the *mind bomb* mutant, *Development* 125:4637–4644.

Hagedorn, L., Floris, J., Suter, U., and Sommer, L., 2000a, Autonomic neurogenesis and apoptosis are alternative fates of progenitor cell communities induced by TGFβ, *Dev. Biol.* 228:57–72.

Hagedorn, L., Paratore, C., Brugnoli, G., Baert, J.L., Mercader, N., Suter, U. *et al.*, 2000b, The Ets domain transcription factor Erm distinguishes rat satellite glia from Schwann cells and is regulated in satellite cells by neuregulin signaling, *Dev. Biol.* 219:44–58.

Hagedorn, L., Suter, U., and Sommer, L., 1999, P0 and PMP22 mark a multipotent neural crest-derived cell type that displays community effects in response to TGF-β family factors, *Development* 126:3781–3794.

Hall, B.K., 1999, *The Neural Crest in Development and Evolution*, Springer-Verlag, New York.

Hamburger, V., 1961, Experimental analysis of the dual origin of the trigeminal ganglion in the chick embryo, *J. Exp. Zool.* 148:91–117.

Hamburger, V. and Hamilton, H.L., 1951, A series of normal stages in the development of the chick embryo, *J. Morphol.* 88:49–92.

Hammond, K.L., Hill, R.E., Whitfield, T.T., and Currie, P.D., 2002, Isolation of three zebrafish *dachshund* homologues and their expression in sensory organs, the central nervous system and pectoral fin buds, *Mech. Dev.* 112:183–189.

Harrison, R.G., 1938, Die Neuralleiste Erganzheft, *Anat. Anz.* 85:3–30.

Harrison, T.A., Stadt, H.A., Kumiski, D., and Kirby, M.L., 1995, Compensatory responses and development of the nodose ganglion following ablation of placodal precursors in the embryonic chick (*Gallus domesticus*), *Cell Tissue Res.* 281:379–385.

Heanue, T.A., Davis, R.J., Rowitch, D.H., Kispert, A., McMahon, A.P., Mardon, G. *et al.*, 2002, *Dach1*, a vertebrate homologue of *Drosophila dachshund*, is expressed in the developing eye and ear of both chick and mouse and is regulated independently of *Pax* and *Eya* genes, *Mech. Dev.* 111:75–87.

Heanue, T.A., Reshef, R., Davis, R.J., Mardon, G., Oliver, G., Tomarev, S. *et al.*, 1999, Synergistic regulation of vertebrate muscle development by *Dach2, Eya2*, and *Six1*, homologs of genes required for *Drosophila* eye formation, *Genes Dev.* 13:3231–3243.

Hearn, C.J., Murphy, M., and Newgreen, D., 1998, GDNF and ET-3 differentially modulate the numbers of avian enteric neural crest cells and enteric neurons *in vitro*, *Dev. Biol.* 197:93–105.

Helbling, P.M., Tran, C.T., and Brändli, A.W., 1998, Requirement for EphA receptor signaling in the segregation of *Xenopus* third and fourth arch neural crest cells, *Mech. Dev.* 78:63–79.

Henion, P.D., Blyss, G.K., Luo, R., An, M., Maynard, T.M., Cole, G.J. *et al.*, 2000, Avian transitin expression mirrors glial cell fate restrictions during neural crest development, *Dev. Dyn.* 218:150–159.

Henion, P.D. and Weston, J.A., 1997, Timing and pattern of cell fate restrictions in the neural crest lineage, *Development* 124:4351–4359.

Hirsch, M.R., Tiveron, M.C., Guillemot, F., Brunet, J.F., and Goridis, C., 1998, Control of noradrenergic differentiation and Phox2a expression by MASH1 in the central and peripheral nervous system, *Development* 125:599–608.

Hirsinger, E., Duprez, D., Jouve, C., Malapert, P., Cooke, J., and Pourquié, O., 1997, Noggin acts downstream of Wnt and Sonic Hedgehog to antagonize BMP4 in avian somite patterning, *Development* 124:4605–4614.

His, W., 1868, *Untersuchungen über die erste Anlage des Wirbeltierleibes. Die erste Entwicklung des Hühnchens im Ei*, F.C.W. Vogel, Leipzig.

Honoré, S.M., Aybar, M.J., and Mayor, R., 2003, *Sox10* is required for the early development of the prospective neural crest in *Xenopus* embryos, *Dev. Biol.* 260:79–96.

Howard, M., Foster, D.N., and Cserjesi, P., 1999, Expression of HAND gene products may be sufficient for the differentiation of avian neural crest-derived cells into catecholaminergic neurons in culture, *Dev. Biol.* 215:62–77.

Howard, M.J., Stanke, M., Schneider, C., Wu, X., and Rohrer, H., 2000, The transcription factor dHAND is a downstream effector of BMPs in sympathetic neuron specification, *Development* 127:4073–4081.

Hunter, E., Begbie, J., Mason, I., and Graham, A., 2001, Early development of the mesencephalic trigeminal nucleus, *Dev. Dyn.* 222:484–493.

Illing, N., Boolay, S., Siwoski, J.S., Casper, D., Lucero, M.T., and Roskams, A.J., 2002, Conditionally immortalized clonal cell lines from the mouse olfactory placode differentiate into olfactory receptor neurons, *Mol. Cell. Neurosci.* 20:225–243.

Ishii, Y., Abu-Elmagd, M., and Scotting, P.J., 2001, *Sox3* expression defines a common primordium for the epibranchial placodes in chick, *Dev. Biol.* 236:344–353.

Itoh, M. and Chitnis, A.B., 2001, Expression of proneural and neurogenic genes in the zebrafish lateral line primordium correlates with selection of hair cell fate in neuromasts, *Mech. Dev.* 102:263–266.

Itoh, M., Kudoh, T., Dedekian, M., Kim, C.H., and Chitnis, A.B., 2002, A role for *iro1* and *iro7* in the establishment of an anteroposterior compartment of the ectoderm adjacent to the midbrain–hindbrain boundary, *Development* 129:2317–2327.

Iversen, S., Iversen, L., and Saper, C.B., 2000, The autonomic nervous system and the hypothalamus in *Principles of Neural Science* (E.R. Kandel, J.H. Schwartz, and T.M. Jessell, eds.), McGraw-Hill, New York, pp. 960–981.

Jacobson, C.-O., 1959, The localization of the presumptive cerebral regions in the neural plate of the axolotl larva, *J. Embryol. Exp. Morphol.* 7:1–21.

Jessen, K.R. and Mirsky, R., 2002, Signals that determine Schwann cell identity, *J. Anat.* 200:367–376.

Jesuthasan, S., 1996, Contact inhibition/collapse and pathfinding of neural crest cells in the zebrafish trunk, *Development* 122:381–389.

Jin, E.J., Erickson, C.A., Takada, S., and Burrus, L.W., 2001, Wnt and BMP signaling govern lineage segregation of melanocytes in the avian embryo, *Dev. Biol.* 233:22–37.

Johnston, M.C., 1966, A radioautographic study of the migration and fate of cranial neural crest cells in the chick embryo, *Anat. Rec.* 156:143–155.

Kalcheim, C., 1996, The role of neurotrophins in development of neural-crest cells that become sensory ganglia, *Phil. Trans. R. Soc. Lond. B* 351:375–381.

Kalcheim, C., 2000, Mechanisms of early neural crest development: From cell specification to migration, *Int. Rev. Cytol.* 200:143–196.

Kalcheim, C. and Teillet, M.A., 1989, Consequences of somite manipulation on the pattern of dorsal root ganglion development, *Development* 106:85–93.

Kamachi, Y., Uchikawa, M., Tanouchi, A., Sekido, R., and Kondoh, H., 2001, Pax6 and SOX2 form a co-DNA-binding partner complex that regulates initiation of lens development, *Genes Dev.* 15:1272–1286.

Kanzler, B., Foreman, R.K., Labosky, P.A., and Mallo, M., 2000, BMP signaling is essential for development of skeletogenic and neurogenic cranial neural crest, *Development* 127:1095–1104.

Kardong, K.V., 1998, *Vertebrates: Comparative Anatomy, Function, Evolution*, WCB/McGraw-Hill, Boston.

Katz, D.M. and Erb, M.J., 1990, Developmental regulation of tyrosine hydroxylase expression in primary sensory neurons of the rat, *Dev. Biol.* 137:233–242.

Kawasaki, T., Bekku, Y., Suto, F., Kitsukawa, T., Taniguchi, M., Nagatsu, I. *et al.*, 2002, Requirement of neuropilin 1-mediated Sema3A signals in patterning of the sympathetic nervous system, *Development* 129:671–680.

Keller, R.E. and Spieth, J., 1984, Neural crest cell behavior in white and dark larvae of *Ambystoma mexicanum*: Time-lapse cinemicrographic analysis of pigment cell movement *in vivo* and in culture, *J. Exp. Zool.* 229:109–126.

Keynes, R.J. and Stern, C.D., 1984, Segmentation in the vertebrate nervous system, *Nature* 310:786–789.

Kil, S.-H. and Collazo, A., 2001, Origins of inner ear sensory organs revealed by fate map and time-lapse analyses, *Dev. Biol.* 233:365–379.

Kim, C.H., Oda, T., Itoh, M., Jiang, D., Artinger, K.B., Chandrasekharappa, S.C. *et al.*, 2000, Repressor activity of Headless/Tcf3 is essential for vertebrate head formation, *Nature* 407:913–916.

Kim, J., Lo, L., Dormand, E., and Anderson, D.J., 2003, SOX10 maintains multipotency and inhibits neuronal differentiation of neural crest stem cells, *Neuron* 38:17–31.

Kious, B.M., Baker, C.V., Bronner-Fraser, M., and Knecht, A.K., 2002, Identification and characterization of a calcium channel gamma sub-unit expressed in differentiating neurons and myoblasts, *Dev. Biol.* 243:249–259.

Kirby, M.L., 1989, Plasticity and predetermination of mesencephalic and trunk neural crest transplanted into the region of the cardiac neural crest, *Dev. Biol.* 134:402–412.

Kirby, M.L., Gale, T.F., and Stewart, D.E., 1983, Neural crest cells contribute to normal aorticopulmonary septation, *Science* 220:1059–1061.

Kitamura, K., Takiguchi-Hayashi, K., Sezaki, M., Yamamoto, H., and Takeuchi, T., 1992, Avian neural crest cells express a melanogenic trait during early migration from the neural tube: Observations with the new monoclonal antibody, "MEBL-1", *Development* 114:367–378.

Knecht, A.K. and Bronner-Fraser, M., 2002, Induction of the neural crest: A multigene process, *Nat. Rev. Genet.* 3:453–461.

Knecht, A.K. and Harland, R.M., 1997, Mechanisms of dorsal-ventral patterning in noggin-induced neural tissue, *Development* 124: 2477–2488.

Knouff, R.A., 1935, The developmental pattern of ectodermal placodes in *Rana pipiens, J. Comp. Neurol.* 62:17–71.

Köntges, G. and Lumsden, A., 1996, Rhombencephalic neural crest segmentation is preserved throughout craniofacial ontogeny, *Development* 122:3229–3242.

Korade, Z. and Frank, E., 1996, Restriction in cell fates of developing spinal cord cells transplanted to neural crest pathways, *J. Neurosci.* 16:7638–7648.

Kos, R., Reedy, M.V., Johnson, R.L., and Erickson, C.A., 2001, The winged-helix transcription factor FoxD3 is important for establishing the neural crest lineage and repressing melanogenesis in avian embryos, *Development* 128:1467–1479.

Köster, R.W., Kühnlein, R.P., and Wittbrodt, J., 2000, Ectopic *Sox3* activity elicits sensory placode formation, *Mech. Dev.* 95:175–187.

Krotoski, D.M., Fraser, S.E., and Bronner-Fraser, M., 1988, Mapping of neural crest pathways in *Xenopus laevis* using inter- and intra-specific cell markers, *Dev. Biol.* 127:119–132.

Kruger, G., Mosher, J., Bixby, S., Joseph, N., Iwashita, T., and Morrison, S., 2002, Neural crest stem cells persist in the adult gut but undergo changes in self-renewal, neuronal subtype potential, and factor responsiveness, *Neuron* 35:657–669.

Krull, C.E., 2001, Segmental organization of neural crest migration, *Mech. Dev.* 105:37–45.

Krull, C.E., Collazo, A., Fraser, S.E., and Bronner-Fraser, M., 1995, Segmental migration of trunk neural crest: Time-lapse analysis reveals a role for PNA-binding molecules, *Development* 121:3733–3743.

Krull, C.E., Lansford, R., Gale, N.W., Collazo, A., Marcelle, C., Yancopoulos, G.D. *et al.*, 1997, Interactions of Eph-related receptors and ligands confer rostrocaudal pattern to trunk neural crest migration, *Curr. Biol.* 7:571–580.

Kubota, Y. and Ito, K., 2000, Chemotactic migration of mesencephalic neural crest cells in the mouse, *Dev. Dyn.* 217:170–179.

Kubu, C.J., Orimoto, K., Morrison, S.J., Weinmaster, G., Anderson, D.J., and Verdi, J.M., 2002, Developmental changes in Notch1 and Numb expression mediated by local cell–cell interactions underlie progressively increasing Delta sensitivity in neural crest stem cells, *Dev. Biol.* 244:199–214.

Kulesa, P.M. and Fraser, S.E., 1998, Neural crest cell dynamics revealed by time-lapse video microscopy of whole embryo chick explant cultures, *Dev. Biol.* 204:327–344.

Kulesa, P.M. and Fraser, S.E., 2000, In ovo time-lapse analysis of chick hindbrain neural crest cell migration shows cell interactions during migration to the branchial arches, *Development* 127:1161–1172.

Kullander, K. and Klein, R., 2002, Mechanisms and functions of Eph and ephrin signaling, *Nat. Rev. Mol. Cell Biol.* 3:475–486.

Kuratani, S.C. and Kirby, M.L., 1991, Initial migration and distribution of the cardiac neural crest in the avian embryo: An introduction to the concept of the circumpharyngeal crest, *Am. J. Anat.* 191:215–227.

LaBonne, C. and Bronner-Fraser, M., 1998, Neural crest induction in *Xenopus*: Evidence for a two-signal model, *Development* 125:2403–2414.

LaBonne, C. and Bronner-Fraser, M., 1999, Molecular mechanisms of neural crest formation, *Annu. Rev. Cell Dev. Biol.* 15:81–112.

LaBonne, C. and Bronner-Fraser, M., 2000, Snail-related transcriptional repressors are required in *Xenopus* for both the induction of the neural crest and its subsequent migration, *Dev. Biol.* 221:195–205.

Ladher, R.K. Anakwe, K.U., Gurney, A.L., Schoenwolf, G.C., and Francis-West, P.H., 2000, Identification of synergistic signals initiating inner ear development, *Science* 290:1965–1968.

LaMantia, A., Bhasin, N., Rhodes, K., and Heemskerk, J., 2000, Mesenchymal/epithelial induction mediates olfactory pathway formation, *Neuron* 28:411–425.

LaMantia, A.S., Colbert, M.C., and Linney, E., 1993, Retinoic acid induction and regional differentiation prefigure olfactory pathway formation in the mammalian forebrain, *Neuron* 10:1035–1048.

Le Douarin, N.M., 1969, Particularités du noyau interphasique chez la Caille japonaise (*Coturnix coturnix japonica*). Utilisation de ces particularités comme 'marquage biologique' dans les recherches sur les interactions tissulaires et les migrations cellulaires au cours de l'ontogenèse, *Bull. Biol. Fr. Belg.* 103:435–452.

Le Douarin, N.M., 1973, A biological cell labelling technique and its use in experimental embryology, *Dev. Biol.* 30:217–222.

Le Douarin, N.M., 1986, Cell line segregation during peripheral nervous system ontogeny, *Science* 231:1515–1522.

Le Douarin, N.M. and Kalcheim, C., 1999, *The Neural Crest*, Cambridge University Press, Cambridge.

Le Douarin, N.M., Renaud, D., Teillet, M.-A., and Le Douarin, G.H., 1975, Cholinergic differentiation of presumptive adrenergic neuroblasts in interspecific chimeras after heterotopic transplantations, *Proc. Natl. Acad. Sci. USA* 72:728–732.

Le Douarin, N.M. and Teillet, M.-A., 1973, The migration of neural crest cells to the wall of the digestive tract in avian embryo, *J. Embryol. Exp. Morphol.* 30:31–48.

Le Douarin, N.M. and Teillet, M.-A., 1974, Experimental analysis of the migration and differentiation of neuroblasts of the autonomic nervous system and of neurectodermal mesenchymal derivatives, using a biological cell marking technique, *Dev. Biol.* 41:162–184.

Le Lièvre, C.S. and Le Douarin, N.M., 1975, Mesenchymal derivatives of the neural crest: Analysis of chimaeric quail and chick embryos, *J. Embryol. Exp. Morphol.* 34:125–154.

Lee, J.S., Ray, R., and Chien, C.B., 2001, Cloning and expression of three zebrafish *roundabout* homologs suggest roles in axon guidance and cell migration, *Dev. Dyn.* 221:216–230.

Lee, S.A., Shen, E.L., Fiser, A., Sali, A., and Guo, S., 2003, The zebrafish forkhead transcription factor Foxi1 specifies epibranchial placode-derived sensory neurons, *Development* 130:2669–2679.

Léger, S. and Brand, M., 2002, Fgf8 and Fgf3 are required for zebrafish ear placode induction, maintenance and inner ear patterning, *Mech. Dev.* 119:91–108.

Leimeroth, R., Lobsiger, C., Lussi, A., Taylor, V., Suter, U., and Sommer, L., 2002, Membrane-bound neuregulin 1 type III actively promotes Schwann cell differentiation of multipotent progenitor cells, *Dev. Biol.* 246:245–258.

Lele, Z., Folchert, A., Concha, M., Rauch, G.J., Geisler, R., Rosa, F. *et al.*, 2002, *Parachute/n-cadherin* is required for morphogenesis and maintained integrity of the zebrafish neural tube, *Development* 129:3281–3294.

Liem, K.F., Jr., Jessell, T.M., and Briscoe, J., 2000, Regulation of the neural patterning activity of sonic hedgehog by secreted BMP inhibitors expressed by notochord and somites, *Development* 127:4855–4866.

Liem, K.F., Jr., Tremml, G., Roelink, H., and Jessell, T.M., 1995, Dorsal differentiation of neural plate cells induced by BMP-mediated signals from epidermal ectoderm, *Cell* 82:969–979.

Lim, K.C., Lakshmanan, G., Crawford, S.E., Gu, Y., Grosveld, F., and Engel, J.D., 2000, *Gata3* loss leads to embryonic lethality due to noradrenaline deficiency of the sympathetic nervous system, *Nat. Genet.* 25:209–212.

Lim, T.M., Lunn, E.R., Keynes, R.J., and Stern, C.D., 1987, The differing effects of occipital and trunk somites on neural development in the chick embryo, *Development* 100:525–533.

Linker, C., Bronner-Fraser, M., and Mayor, R., 2000, Relationship between gene expression domains of *Xsnail, Xslug*, and *Xtwist* and cell movement in the prospective neural crest of *Xenopus, Dev. Biol.* 224:215–225.

Liu, J.P. and Jessell, T.M., 1998, A role for rhoB in the delamination of neural crest cells from the dorsal neural tube, *Development* 125:5055–5067.

Lo, L., Dormand, E., Greenwood, A., and Anderson, D.J., 2002, Comparison of the generic neuronal differentiation and neuron subtype specification functions of mammalian *achaete-scute* and *atonal* homologs in cultured neural progenitor cells, *Development* 129:1553–1567.

Lo, L., Tiveron, M.C., and Anderson, D.J., 1998, MASH1 activates expression of the paired homeodomain transcription factor Phox2a, and couples pan-neuronal and subtype-specific components of autonomic neuronal identity, *Development* 125:609–620.

Löfberg, J., Ahlfors, K., and Fällström, C., 1980, Neural crest cell migration in relation to extracellular matrix organization in the embryonic axolotl trunk, *Dev. Biol.* 75:148–167.

Loosli, F., Mardon, G., and Wittbrodt, J., 2002, Cloning and expression of medaka *Dachshund, Mech. Dev.* 112:203–206.

Lumsden, A., 1990, The cellular basis of segmentation in the developing hindbrain, *Trends Neurosci.* 13:329–335.

Lumsden, A.G., 1988, Spatial organization of the epithelium and the role of neural crest cells in the initiation of the mammalian tooth germ, *Development* 103:155–169.

Luo, T., Lee, Y.-H., Saint-Jeannet, J.-P., and Sargent, T.D., 2003, Induction of neural crest in *Xenopus* by transcription factor AP2α, *Proc. Natl. Acad. Sci. USA* 100:532–537.

Lwigale, P.Y., 2001, Embryonic origin of avian corneal sensory nerves. *Dev. Biol.* 239:323–337.

Ma, Q., Chen, Z., del Barco Barrantes, I., de la Pompa, J.L., and Anderson, D.J., 1998, *Neurogenin1* is essential for the determination of neuronal precursors for proximal cranial sensory ganglia, *Neuron* 20:469–482.

Ma, Q., Fode, C., Guillemot, F., and Anderson, D.J., 1999, NEUROGENIN1 and NEUROGENIN2 control two distinct waves of neurogenesis in developing dorsal root ganglia, *Genes Dev.* 13:1717–1728.

MacColl, G., Bouloux, P., and Quinton, R., 2002, Kallmann syndrome: Adhesion, afferents, and anosmia, *Neuron* 34:675–678.

Maisey, J.G., 1986, Heads and tails: A chordate phylogeny, *Cladistics* 2:201–256.

Mancilla, A. and Mayor, R., 1996, Neural crest formation in *Xenopus laevis*:Mechanisms of *Xslug* induction, *Dev. Biol.* 177:580–589.

Manie, S., Santoro, M., Fusco, A., and Billaud, M., 2001, The RET receptor: Function in development and dysfunction in congenital malformation, *Trends Genet.* 17:580–589.

Mansour, S.L., Goddard, J.M., and Capecchi, M.R., 1993, Mice homozygous for a targeted disruption of the proto-oncogene *int-2* have developmental defects in the tail and inner ear, *Development* 117:13–28.

Mansouri, A., Pla, P., Larue, L., and Gruss, P., 2001, Pax3 acts cell autonomously in the neural tube and somites by controlling cell surface properties, *Development* 128:1995–2005.

Marcelle, C., Stark, M.R., and Bronner-Fraser, M., 1997, Coordinate actions of BMPs, Wnts, Shh and noggin mediate patterning of the dorsal somite, *Development* 124:3955–3963.

Marchant, L., Linker, C., Ruiz, P., Guerrero, N., and Mayor, R., 1998, The inductive properties of mesoderm suggest that the neural crest cells are specified by a BMP gradient, *Dev. Biol.* 198:319–329.

Maroon, H., Walshe, J., Mahmood, R., Kiefer, P., Dickson, C., and Mason, I., 2002, Fgf3 and Fgf8 are required together for formation of the otic placode and vesicle, *Development* 129:2099–2108.

Martí, E., 2000, Expression of chick BMP-1/Tolloid during patterning of the neural tube and somites, *Mech. Dev.* 91:415–419.

Mayor, R. and Aybar, M.J., 2001, Induction and development of neural crest in *Xenopus laevis, Cell Tissue Res.* 305:203–209.

Mayor, R., Guerrero, N., Young, R.M., Gomez-Skarmeta, J.L., and Cuellar, C., 2000, A novel function for the *Xslug* gene: Control of dorsal mesendoderm development by repressing BMP-4, *Mech. Dev.* 97:47–56.

Mayor, R., Morgan, R., and Sargent, M.G., 1995, Induction of the prospective neural crest of *Xenopus, Development* 121:767–777.

McCallion, A.S. and Chakravarti, A., 2001, EDNRB/EDN3 and Hirschsprung disease type II, *Pigment Cell Res.* 14:161–169.

McCauley, D.W. and Bronner-Fraser, M., 2002, Conservation of *Pax* gene expression in ectodermal placodes of the lamprey, *Gene* 287:129–139.

McGonnell, I.M. and Graham, A., 2002, Trunk neural crest has skeletogenic potential, *Curr. Biol.* 12:767–771.

McLarren, K.W., Litsiou, A., and Streit, A., 2003, DLX5 positions the neural crest and preplacode region at the border of the neural plate, *Dev. Biol.* 259:34–47.

Mendonsa, E.S. and Riley, B.B., 1999, Genetic analysis of tissue interactions required for otic placode induction in the zebrafish, *Dev. Biol.* 206:100–112.

Metscher, B.D., Northcutt, R.G., Gardiner, D.M., and Bryant, S.V., 1997, Homeobox genes in axolotl lateral line placodes and neuromasts, *Dev. Genes Evol.* 207:287–295.

Meyer, D., Yamaai, T., Garratt, A., Riethmacher-Sonnenberg, E., Kane, D., Theill, L.E. *et al.*, 1997, Isoform-specific expression and function of neuregulin, *Development* 124:3575–3586.

Mikkola, I., Heavey, B., Horcher, M., and Busslinger, M., 2002, Reversion of B cell commitment upon loss of Pax5 expression, *Science* 297:110–113.

Mitsiadis, T.A., Cheraud, Y., Sharpe, P., and Fontaine-Perus, J., 2003, Development of teeth in chick embryos after mouse neural crest transplantations, *Proc. Natl. Acad. Sci. USA* 100:6541–6545.

Mombaerts, P., 2001, How smell develops, *Nat. Neurosci.* 4 Suppl. 1:1192–1198.

Monsoro-Burq, A.H., Fletcher, R.B., and Harland, R.M., 2003, Neural crest induction by paraxial mesoderm in *Xenopus* embryos requires FGF signals, *Development* 130:3111–3124.

Moody, S.A. and Heaton, M.B., 1983a, Developmental relationships between trigeminal ganglia and trigeminal motoneurons in chick embryos. I. Ganglion development is necessary for motoneuron migration, *J. Comp. Neurol.* 213:327–343.

Moody, S.A. and Heaton, M.B., 1983b, Developmental relationships between trigeminal ganglia and trigeminal motoneurons in chick embryos. II. Ganglion axon ingrowth guides motoneuron migration, *J. Comp. Neurol.* 213:344–349.

Moore, M.W., Klein, R.D., Farinas, I., Sauer, H., *et al.*, 1996, Renal and neuronal abnormalities in mice lacking GDNF, *Nature* 382:76–79.

Mori-Akayama, Y., Akiyama, H., Rowitch, D., and de Crombrugghe, B., 2003, Sox9 is required for determination of the chondrogenic cell lineage in the cranial neural crest, *Proc. Natl. Acad. Sci. USA* 100:9360–9365.

Morin, X., Cremer, H., Hirsch, M.R., Kapur, R.P., Goridis, C., and Brunet, J.-F., 1997, Defects in sensory and autonomic ganglia and absence of locus coeruleus in mice deficient for the homeobox gene *Phox2a, Neuron* 18:411–423.

Morin-Kensicki, E.M. and Eisen, J.S., 1997, Sclerotome development and peripheral nervous system segmentation in embryonic zebrafish, *Development* 124:159–167.

Morrison, S.J., 2001, Neuronal potential and lineage determination by neural stem cells, *Curr. Opin. Cell Biol.* 13:666–672.

Morrison, S.J., Csete, M., Groves, A.K., Melega, W., Wold, B., and Anderson, D.J., 2000a, Culture in reduced levels of oxygen promotes clonogenic sympathoadrenal differentiation by isolated neural crest stem cells, *J. Neurosci.* 20:7370–7376.

Morrison, S.J., Perez, S.E., Qiao, Z., Verdi, J.M., Hicks, C., Weinmaster, G. *et al.*, 2000b, Transient Notch activation initiates an irreversible switch from neurogenesis to gliogenesis by neural crest stem cells, *Cell* 101:499–510.

Morrison, S.J., White, P.M., Zock, C., and Anderson, D.J., 1999, Prospective identification, isolation by flow cytometry, and in vivo self-renewal of multipotent mammalian neural crest stem cells, *Cell* 96:737–749.

Moury, J.D. and Jacobson, A.G., 1989, Neural fold formation at newly created boundaries between neural plate and epidermis in the axolotl, *Dev. Biol.* 133:44–57.

Moury, J.D. and Jacobson, A.G., 1990, The origins of neural crest cells in the axolotl, *Dev. Biol.* 141:243–253.

Mowbray, C., Hammerschmidt, M., and Whitfield, T.T., 2001, Expression of BMP signalling pathway members in the developing zebrafish inner ear and lateral line, *Mech. Dev.* 108:179–184.

Muhr, J., Graziano, E., Wilson, S., Jessell, T.M., and Edlund, T., 1999, Convergent inductive signals specify midbrain, hindbrain, and spinal cord identity in gastrula stage chick embryos, *Neuron* 23:689–702.

Muhr, J., Jessell, T.M., and Edlund, T., 1997, Assignment of early caudal identity to neural plate cells by a signal from caudal paraxial mesoderm, *Neuron* 19:487–502.

Mujtaba, T., Mayer-Proschel, M., and Rao, M.S., 1998, A common neural progenitor for the CNS and PNS, *Dev. Biol.* 200:1–15.

Münchberg, S.R., Ober, E.A., and Steinbeisser, H., 1999, Expression of the Ets transcription factors *erm* and *pea3* in early zebrafish development, *Mech. Dev.* 88:233–236.

Muñoz-Sanjuán, I. and Hemmati-Brivanlou, A., 2002, Neural induction, the default model and embryonic stem cells, *Nat. Rev. Neurosci.* 3:271–280.

Murphy, P., Topilko, P., Schneider-Maunoury, S., Seitanidou, T., Baron-Van Evercooren, A., and Charnay, P., 1996, The regulation of Krox-20 expression reveals important steps in the control of peripheral glial cell development, *Development* 122:2847–2857.

Nakagawa, S. and Takeichi, M., 1995, Neural crest cell-cell adhesion controlled by sequential and subpopulation-specific expression of novel cadherins, *Development* 121:1321–1332.

Nakagawa, S. and Takeichi, M., 1998, Neural crest emigration from the neural tube depends on regulated cadherin expression, *Development* 125:2963–2971.

Nakamura, H. and Ayer-Le Lièvre, C.S., 1982, Mesectodermal capabilities of the trunk neural crest of birds, *J. Embryol. Exp. Morphol.* 70:1–18.

Nakata, K., Koyabu, Y., Aruga, J., and Mikoshiba, K., 2000, A novel member of the *Xenopus* Zic family, Zic5, mediates neural crest development, *Mech. Dev.* 99:83–91.

Narayanan, C.H. and Narayanan, Y., 1978, Determination of the embryonic origin of the mesencephalic nucleus of the trigeminal nerve in birds, *J. Embryol. Exp. Morphol.* 43:85–105.

Narayanan, C.H. and Narayanan, Y., 1980, Neural crest and placodal contributions in the development of the glossopharyngeal-vagal complex in the chick, *Anat. Rec.* 196:71–82.

Nataf, V., Lecoin, L., Eichmann, A., and Le Douarin, N.M., 1996, Endothelin-B receptor is expressed by neural crest cells in the avian embryo, *Proc. Natl. Acad. Sci. USA* 93:9645–9650.

Neidert, A.H., Virupannavar, V., Hooker, G.W., and Langeland, J.A., 2001, Lamprey *Dlx* genes and early vertebrate evolution, *Proc. Natl. Acad. Sci. USA* 98:1665–1670.

Newgreen, D.F. and Gooday, D., 1985, Control of the onset of migration of neural crest cells in avian embryos. Role of Ca++-dependent cell adhesions, *Cell Tissue Res.* 239:329–336.

Newgreen, D.F. and Minichiello, J., 1995, Control of epitheliomesenchymal transformation. I. Events in the onset of neural crest cell migration are separable and inducible by protein kinase inhibitors, *Dev. Biol.* 170:91–101.

Newgreen, D.F. and Minichiello, J., 1996, Control of epitheliomesenchymal transformation. II. Cross-modulation of cell adhesion and cytoskeletal systems in embryonic neural cells, *Dev. Biol.* 176:300–312.

Nguyen, V.H., Schmid, B., Trout, J., Connors, S.A., Ekker, M., and Mullins, M.C., 1998, Ventral and lateral regions of the zebrafish gastrula, including the neural crest progenitors, are established by a *bmp2b/swirl* pathway of genes, *Dev. Biol.* 199:93–110.

Nichols, D.H., 1981, Neural crest formation in the head of the mouse embryo as observed using a new histological technique, *J. Embryol. Exp. Morphol.* 64:105–120.

Nieto, M.A., 2002, The Snail superfamily of zinc-finger transcription factors, *Nat. Rev. Mol. Cell Biol.* 3:155–166.

Nieto, M.A., Sargent, M.G., Wilkinson, D.G., and Cooke, J., 1994, Control of cell behavior during vertebrate development by *Slug*, a zinc finger gene, *Science* 264:835–839.

Nieuwkoop, P.D. and Faber, J., 1967, *Normal Table of* Xenopus laevis *(Daudin)*, North-Holland, Amsterdam.

Noden, D.M., 1975, An analysis of migratory behavior of avian cephalic neural crest cells, *Dev. Biol.* 42:106–130.

Noden, D.M., 1978a, The control of avian cephalic neural crest cytodifferentiation. I. Skeletal and connective tissues, *Dev. Biol.* 67:296–312.

Noden, D.M., 1978b, The control of avian cephalic neural crest cytodifferentiation. II. Neural tissues, *Dev. Biol.* 67:313–329.

Noden, D.M., 1980, Somatotopic and functional organization of the avian trigeminal ganglion: An HRP analysis in the hatchling chick, *J. Comp. Neurol.* 190:405–428.

Noden, D.M., 1983, The role of the neural crest in patterning of avian cranial skeletal, connective, and muscle tissues, *Dev. Biol.* 96:144–165.

Noden, D.M., 1991, Vertebrate craniofacial development: the relation between ontogenetic process and morphological outcome, *Brain Behav. Evol.* 38:190–225.

Noramly, S. and Grainger, R.M., 2002, Determination of the embryonic inner ear, *J. Neurobiol.* 53:100–128.

Nordström, U., Jessell, T.M., and Edlund, T., 2002, Progressive induction of caudal neural character by graded Wnt signaling, *Nat. Neurosci.* 5:525–532.

Northcutt, R.G., 1997, Evolution of gnathostome lateral line ontogenies, *Brain Behav. Evol.* 50:25–37.

Northcutt, R.G., Brändle, K., and Fritzsch, B., 1995, Electroreceptors and mechanosensory lateral line organs arise from single placodes in axolotls, *Dev. Biol.* 168:358–373.

Northcutt, R.G. and Gans, C., 1983, The genesis of neural crest and epidermal placodes: A reinterpretation of vertebrate origins, *Quart. Rev. Biol.* 58:1–28.

Oakley, R.A., Lasky, C.J., Erickson, C.A., and Tosney, K.W., 1994, Glycoconjugates mark a transient barrier to neural crest migration in the chicken embryo, *Development* 120:103–114.

Ogino, H. and Yasuda, K., 2000, Sequential activation of transcription factors in lens induction, *Dev. Growth Differ.* 42:437–448.

Olsson, L., Falck, P., Lopez, K., Cobb, J., and Hanken, J., 2001, Cranial neural crest cells contribute to connective tissue in cranial muscles in the anuran amphibian, *Bombina orientalis, Dev. Biol.* 237:354–367.

Olsson, L. and Hanken, J., 1996, Cranial neural-crest migration and chondrogenic fate in the Oriental fire-bellied toad *Bombina orientalis*: Defining the ancestral pattern of head development in anuran amphibians, *J. Morph.* 229:105–120.

Osumi-Yamashita, N., Ninomiya, Y., Doi, H., and Eto, K., 1994, The contribution of both forebrain and midbrain crest cells to the mesenchyme in the frontonasal mass of mouse embryos, *Dev. Biol.* 164:409–419.

Papan, C. and Campos-Ortega, J.A., 1994, On the formation of the neural keel and neural tube in the zebrafish *Danio (Brachydanio) rerio, Roux's Arch. Dev. Biol.* 203:178–186.

Paratore, C., Goerich, D.E., Suter, U., Wegner, M., and Sommer, L., 2001, Survival and glial fate acquisition of neural crest cells are regulated by an interplay between the transcription factor Sox10 and extrinsic combinatorial signaling, *Development* 128:3949–3961.

Pattyn, A., Goridis, C., and Brunet, J.-F., 2000, Specification of the central noradrenergic phenotype by the homeobox gene *Phox2b, Mol. Cell. Neurosci.* 15:235–243.

Pattyn, A., Morin, X., Cremer, H., Goridis, C., and Brunet, J.-F., 1997, Expression and interactions of the two closely related homeobox genes *Phox2a* and *Phox2b* during neurogenesis, *Development* 124:4065–4075.

Pattyn, A., Morin, X., Cremer, H., Goridis, C., and Brunet, J.F., 1999, The homeobox gene *Phox2b* is essential for the development of autonomic neural crest derivatives, *Nature* 399:366–370.

Perez, S.E., Rebelo, S., and Anderson, D.J., 1999, Early specification of sensory neuron fate revealed by expression and function of neurogenins in the chick embryo, *Development* 126:1715–1728.

Perissinotto, D., Iacopetti, P., Bellina, I., Doliana, R., Colombatti, A., Pettway, Z. *et al.*, 2000, Avian neural crest cell migration is diversely regulated by the two major hyaluronan-binding proteoglycans PG-M/versican and aggrecan, *Development* 127:2823–2842.

Perris, R., 1997, The extracellular matrix in neural crest-cell migration, *Trends Neurosci.* 20:23–31.

Perris, R. and Perissinotto, D., 2000, Role of the extracellular matrix during neural crest cell migration, *Mech. Dev.* 95:3–21.

Perris, R., von Boxberg, Y., and Lofberg, J., 1988, Local embryonic matrices determine region-specific phenotypes in neural crest cells, *Science* 241:86–89.

Pettway, Z., Domowicz, M., Schwartz, N.B., and Bronner-Fraser, M., 1996, Age-dependent inhibition of neural crest migration by the notochord correlates with alterations in the S103L chondroitin sulfate proteoglycan, *Exp. Cell Res.* 225:195–206.

Phillips, B.T., Bolding, K., and Riley, B.B., 2001, Zebrafish *fgf3* and *fgf8* encode redundant functions required for otic placode induction, *Dev. Biol.* 235:351–365.

Pickles, J.O. and Corey, D.P., 1992, Mechanoelectrical transduction by hair cells, *Trends Neurosci.* 15:254–259.

Pisano, J.M., Colon-Hastings, F., and Birren, S.J., 2000, Postmigratory enteric and sympathetic neural precursors share common, developmentally regulated, responses to BMP2, *Dev. Biol.* 227:1–11.

Platt, C., 1993, Zebrafish inner ear sensory surfaces are similar to those in goldfish, *Hear. Res.* 65:133–140.

Pohl, B.S. and Knöchel, W., 2001, Overexpression of the transcriptional repressor FoxD3 prevents neural crest formation in *Xenopus* embryos, *Mech. Dev.* 103:93–106.

Pourquié, O., 2001, Vertebrate somitogenesis, *Annu. Rev. Cell Dev. Biol.* 17:311–350.

Raible, D.W. and Eisen, J.S., 1994, Restriction of neural crest cell fate in the trunk of the embryonic zebrafish, *Development* 120:495–503.

Raible, D.W. and Eisen, J.S., 1996, Regulative interactions in zebrafish neural crest, *Development* 122:501–507.

Raible, D.W., Wood, A., Hodsdon, W., Henion, P.D., Weston, J.A., and Eisen, J.S., 1992, Segregation and early dispersal of neural crest cells in the embryonic zebrafish, *Dev. Dyn.* 195:29–42.

Raible, F. and Brand, M., 2001, Tight transcriptional control of the ETS domain factors Erm and Pea3 by Fgf signaling during early zebrafish development, *Mech. Dev.* 107:105–117.

Ramain, P., Heitzler, P., Haenlin, M., and Simpson, P., 1993, *pannier*, a negative regulator of *achaete* and *scute* in *Drosophila*, encodes a zinc finger protein with homology to the vertebrate transcription factor GATA-1, *Development* 119:1277–1291.

Rathjen, J., Haines, B.P., Hudson, K.M., Nesci, A., Dunn, S., and Rathjen, P.D., 2002, Directed differentiation of pluripotent cells to neural lineages: Homogeneous formation and differentiation of a neurectoderm population, *Development* 129:2649–2661.

Raven, C.P., 1931, Zur Entwicklung der Ganglienleiste. I: Die Kinematik der Ganglienleisten Entwicklung bei den Urodelen, *Wilhelm Roux Arch. EntwMech. Org.* 125:210–293.

Raven, C.P., 1936, Zur Entwicklung der Ganglienleiste. V: Uber die Differenzierung des Rumpfganglienleistenmaterials, *Wilhelm Roux Arch. EntwMech. Org.* 134:122–145.

Raven, C.P. and Kloos, J., 1945, Induction by medial and lateral pieces of the archenteron roof with special reference to the determination of the neural crest, *Acta. Néerl. Morph.* 5:348–362.

Rawls, J.F., Mellgren, E.M., and Johnson, S.L., 2001, How the zebrafish gets its stripes, *Dev. Biol.* 240:301–314.

Reaume, A.G., Conlon, R.A., Zirngibl, R., Yamaguchi, T.P., and Rossant, J., 1992, Expression analysis of a *Notch* homologue in the mouse embryo, *Dev. Biol.* 154:377–387.

Reedy, M.V., Faraco, C.D., and Erickson, C.A., 1998, Specification and migration of melanoblasts at the vagal level and in hyperpigmented Silkie chickens, *Dev. Dyn.* 213:476–485.

Reissmann, E., Ernsberger, U., Francis-West, P.H., Rueger, D., Brickell, P.M., and Rohrer, H., 1996, Involvement of bone morphogenetic protein-4 and bone morphogenetic protein-7 in the differentiation of the adrenergic phenotype in developing sympathetic neurons, *Development* 122:2079–2088.

Richardson, M.K. and Sieber-Blum, M., 1993, Pluripotent neural crest cells in the developing skin of the quail embryo, *Dev. Biol.* 157:348–358.

Richman, J.M. and Lee, S.-H., 2003, About face: Signals and genes controlling jaw patterning and identity in vertebrates, *Bioessays* 25:554–568.

Rickmann, M., Fawcett, J.W., and Keynes, R.J., 1985, The migration of neural crest cells and the growth of motor axons through the rostral half of the chick somite, *J. Embryol. Exp. Morphol.* 90:437–455.

Riley, B.B. and Phillips, B.T., 2003, Ringing in the new ear: Resolution of cell interactions in otic development, *Dev. Biol.* 261:289–312.

Riley, B.B., Zhu, C., Janetopoulos, C., and Aufderheide, K.J., 1997, A critical period of ear development controlled by distinct populations of ciliated cells in the zebrafish, *Dev. Biol.* 191:191–201.

Rinkwitz, S., Bober, E., and Baker, R., 2001, Development of the vertebrate inner ear, *Ann. N. Y. Acad. Sci.* 942:1–14.

Robinson, M.L., MacMillan-Crow, L.A., Thompson, J.A., and Overbeek, P.A., 1995, Expression of a truncated FGF receptor results in defective lens development in transgenic mice, *Development* 121:3959–3967.

Robinson, V., Smith, A., Flenniken, A.M., and Wilkinson, D.G., 1997, Roles of Eph receptors and ephrins in neural crest pathfinding, *Cell Tissue Res.* 290:265–274.

Roehl, H. and Nüsslein-Volhard, C., 2001, Zebrafish *pea3* and *erm* are general targets of FGF8 signaling, *Curr. Biol.* 11:503–507.

Rollhäuser-ter Horst, J., 1979, Artificial neural crest formation in amphibia, *Anat. Embryol.* 157:113–120.

Rollhäuser-ter Horst, J., 1980, Neural crest replaced by gastrula ectoderm in amphibia. Effect on neurulation, CNS, gills and limbs, *Anat. Embryol.* 160:203–211.

Ronnett, G.V. and Moon, C., 2002, G proteins and olfactory signal transduction, *Annu. Rev. Physiol.* 64:189–222.

Rosenquist, G.C., 1981, Epiblast origin and early migration of neural crest cells in the chick embryo, *Dev. Biol.* 87:201–211.

Rothman, T.P., Sherman, D., Cochard, P., and Gershon, M.D., 1986, Development of the monoaminergic innervation of the avian gut: Transient and permanent expression of phenotypic markers, *Dev. Biol.* 116:357–380.

Sadaghiani, B. and Thiébaud, C.H., 1987, Neural crest development in the *Xenopus laevis* embryo, studied by interspecific transplantation and scanning electron microscopy, *Dev. Biol.* 124:91–110.

Santagati, F. and Rijli, F.M., 2003, Cranial neural crest and the building of the vertebrate head, *Nat. Rev. Neurosci.* 4:806–818.

Saint-Jeannet, J.P., He, X., Varmus, H.E., and Dawid, I.B., 1997, Regulation of dorsal fate in the neuraxis by *Wnt-1* and *Wnt-3a*, *Proc. Natl. Acad. Sci. USA* 94:13,713–13,718.

Santiago, A. and Erickson, C.A., 2002, Ephrin-B ligands play a dual role in the control of neural crest cell migration, *Development* 129:3621–3632.

Sarkar, S., Petiot, A., Copp, A., Ferretti, P., and Thorogood, P., 2001, FGF2 promotes skeletogenic differentiation of cranial neural crest cells, *Development* 128:2143–2152.

Sasai, N., Mizuseki, K., and Sasai, Y., 2001, Requirement of *FoxD3*-class signaling for neural crest determination in *Xenopus*, *Development* 128:2525–2536.

Savagner, P., Yamada, K.M., and Thiery, J.P., 1997, The zinc-finger protein slug causes desmosome dissociation, an initial and necessary step for growth factor-induced epithelial-mesenchymal transition, *J. Cell Biol.* 137:1403–1419.

Schebesta, M., Heavey, B., and Busslinger, M., 2002, Transcriptional control of B-cell development, *Curr. Opin. Immunol.* 14:216–223.

Schilling, T.F. and Kimmel, C.B., 1994, Segment and cell type lineage restrictions during pharyngeal arch development in the zebrafish embryo, *Development* 120:483–494.

Schilling, T.F., Prince, V., and Ingham, P.W., 2001, Plasticity in zebrafish *hox* expression in the hindbrain and cranial neural crest, *Dev. Biol.* 231:201–216.

Schlosser, G., 2002a, Development and evolution of lateral line placodes in amphibians. I. Development, *Zoology* 105:119–146.

Schlosser, G., 2002b, Development and evolution of lateral line placodes in amphibians. II. Evolutionary diversification, *Zoology* 105:177–193.

Schlosser, G., Kintner, C., and Northcutt, R.G., 1999, Loss of ectodermal competence for lateral line placode formation in the direct developing frog *Eleutherodactylus coqui*, *Dev. Biol.* 213:354–369.

Schlosser, G. and Northcutt, R.G., 2000, Development of neurogenic placodes in *Xenopus laevis*, *J. Comp. Neurol.* 418:121–146.

Schlosser, G. and Northcutt, R.G., 2001, Lateral line placodes are induced during neurulation in the axolotl, *Dev. Biol.* 234:55–71.

Schmitz, B., Papan, C., and Campos-Ortega, J.A., 1993, Neurulation in the anterior trunk region of the zebrafish *Brachydanio rerio*, *Roux's Arch. Dev. Biol.* 203:250–259.

Schneider, R.A. and Helms, J.A., 2003, The cellular and molecular origins of beak morphology, *Science* 299:565–568.

Schneider, C., Wicht, H., Enderich, J., Wegner, M., and Rohrer, H., 1999, Bone morphogenetic proteins are required *in vivo* for the generation of sympathetic neurons, *Neuron* 24:861–870.

Schroeder, T.E., 1970, Neurulation in *Xenopus laevis*. An analysis and model based upon light and electron microscopy, *J. Embryol. Exp. Morphol.* 23:427–462.

Schweizer, G., Ayer-Le Lièvre, C., and Le Douarin, N.M., 1983, Restrictions of developmental capacities in the dorsal root ganglia during the course of development, *Cell Differ.* 13:191–200.

Scully, K.M. and Rosenfeld, M.G., 2002, Pituitary development: Regulatory codes in mammalian organogenesis, *Science* 295:2231–2235.

Sechrist, J., Serbedzija, G.N., Scherson, T., Fraser, S.E., and Bronner-Fraser, M., 1993, Segmental migration of the hindbrain neural crest does not arise from its segmental generation, *Development* 118:691–703.

Sela-Donenfeld, D. and Kalcheim, C., 1999, Regulation of the onset of neural crest migration by coordinated activity of BMP4 and Noggin in the dorsal neural tube, *Development* 126:4749–4762.

Sela-Donenfeld, D. and Kalcheim, C., 2000, Inhibition of noggin expression in the dorsal neural tube by somitogenesis: A mechanism for coordinating the timing of neural crest emigration, *Development* 127:4845–4854.

Selleck, M.A. and Bronner-Fraser, M., 1995, Origins of the avian neural crest: the role of neural plate-epidermal interactions, *Development* 121:525–538.

Serbedzija, G.N., Bronner-Fraser, M., and Fraser, S.E., 1989, A vital dye analysis of the timing and pathways of avian trunk neural crest cell migration, *Development* 106:809–816.

Serbedzija, G.N., Bronner-Fraser, M., and Fraser, S.E., 1994, Developmental potential of trunk neural crest cells in the mouse, *Development* 120:1709–1718.

Serbedzija, G.N., Fraser, S.E., and Bronner-Fraser, M., 1990, Pathways of trunk neural crest cell migration in the mouse embryo as revealed by vital dye labelling, *Development* 108:605–612.

Shah, N.M. and Anderson, D.J., 1997, Integration of multiple instructive cues by neural crest stem cells reveals cell-intrinsic biases in relative growth factor responsiveness, *Proc. Natl. Acad. Sci. USA* 94:11,369–11,374.

Shah, N.M., Groves, A.K., and Anderson, D.J., 1996, Alternative neural crest cell fates are instructively promoted by TGFβ superfamily members, *Cell* 85:331–343.

Shah, N.M., Marchionni, M.A., Isaacs, I., Stroobant, P., and Anderson, D.J., 1994, Glial growth factor restricts mammalian neural crest stem cells to a glial fate, *Cell* 77:349–360.

Sharma, K., Korade, Z., and Frank, E., 1995, Late-migrating neuroepithelial cells from the spinal cord differentiate into sensory ganglion cells and melanocytes, *Neuron* 14:143–152.

Shin, M.K., Levorse, J.M., Ingram, R.S., and Tilghman, S.M., 1999, The temporal requirement for endothelin receptor-B signalling during neural crest development, *Nature* 402:496–501.

Shoji, W., Yee, C.S., and Kuwada, J.Y., 1998, Zebrafish semaphorin Z1a collapses specific growth cones and alters their pathway in vivo, *Development* 125:1275–1283.

Shou, J., Murray, R.C., Rim, P.C., and Calof, A.L., 2000, Opposing effects of bone morphogenetic proteins on neuron production and survival in the olfactory receptor neuron lineage, *Development* 127:5403–5413.

Shou, J., Rim, P.C., and Calof, A.L., 1999, BMPs inhibit neurogenesis by a mechanism involving degradation of a transcription factor, *Nat. Neurosci.* 2:339–345.

Sieber-Blum, M., 1989, SSEA-1 is a specific marker for the spinal sensory neuron lineage in the quail embryo and in neural crest cell cultures, *Dev. Biol.* 134:362–375.

Sieber-Blum, M., 2000, Factors controlling lineage specification in the neural crest, *Int. Rev. Cytol.* 197:1–33.

Sieber-Blum, M. and Cohen, A.M., 1980, Clonal analysis of quail neural crest cells: they are pluripotent and differentiate *in vitro* in the absence of noncrest cells, *Dev. Biol.* 80:96–106.

Skaer, N., Pistillo, D., Gibert, J.M., Lio, P., Wulbeck, C., and Simpson, P., 2002, Gene duplication at the *achaete-scute* complex and morphological complexity of the peripheral nervous system in Diptera, *Trends Genet.* 18:399–405.

Smith, A., Robinson, V., Patel, K., and Wilkinson, D.G., 1997, The EphA4 and EphB1 receptor tyrosine kinases and ephrin-B2 ligand regulate targeted migration of branchial neural crest cells, *Curr. Biol.* 7:561–570.

Smith, M., Hickman, A., Amanze, D., Lumsden, A., and Thorogood, P., 1994, Trunk neural crest origin of caudal fin mesenchyme in the zebrafish *Brachydanio rerio, Proc. R. Soc. Lond. B* 256:137–145.

Solomon, K.S. and Fritz, A., 2002, Concerted action of two *dlx* paralogs in sensory placode formation, *Development* 129:3127–3136.

Solomon, K.S., Kudoh, T., Dawid, I.B., and Fritz, A., 2003, Zebrafish *foxi1* mediates otic placode formation and jaw development, *Development* 130:929–940.

Sommer, L., 2001, Context-dependent regulation of fate decisions in multipotent progenitor cells of the peripheral nervous system, *Cell Tissue Res.* 305:211–216.

Spokony, R.F., Aoki, Y., Saint-Germain, N., Magner-Fink, E., and Saint-Jeannet, J.-P., 2002, The transcription factor Sox9 is required for cranial neural crest development in *Xenopus, Development* 129:421–432.

St. John, J.A., Clarris, H.J., and Key, B., 2002, Multiple axon guidance cues establish the olfactory topographic map: How do these cues interact? *Int. J. Dev. Biol.* 46:639–647.

Stanke, M., Junghans, D., Geissen, M., Goridis, C., Ernsberger, U., and Rohrer, H., 1999, The Phox2 homeodomain proteins are sufficient to promote the development of sympathetic neurons, *Development* 126:4087–4094.

Stark, M.R., Biggs, J.J., Schoenwolf, G.C., and Rao, M.S., 2000, Characterization of avian *frizzled* genes in cranial placode development, *Mech. Dev.* 93:195–200.

Stark, M.R., Sechrist, J., Bronner-Fraser, M., and Marcelle, C., 1997, Neural tube-ectoderm interactions are required for trigeminal placode formation, *Development* 124:4287–4295.

Stemple, D.L. and Anderson, D.J., 1992, Isolation of a stem cell for neurons and glia from the mammalian neural crest, *Cell* 71:973–985.

Stern, C.D., Artinger, K.B., and Bronner-Fraser, M., 1991, Tissue interactions affecting the migration and differentiation of neural crest cells in the chick embryo, *Development* 113:207–216.

Stockdale, F.E., Nikovits, W., Jr., and Christ, B., 2000, Molecular and cellular biology of avian somite development, *Dev. Dyn.* 219:304–321.

Streit, A., 2002, Extensive cell movements accompany formation of the otic placode, *Dev. Biol.* 249:237–254.

Streit, A. and Stern, C.D., 1999, Establishment and maintenance of the border of the neural plate in the chick: involvement of FGF and BMP activity, *Mech. Dev.* 82:51–66.

Sumanas, S., Kim, H.J., Hermanson, S.B., and Ekker, S.C., 2002, Lateral line, nervous system, and maternal expression of Frizzled 7a during zebrafish embryogenesis, *Mech. Dev.* 115:107–111.

Tan, C., Deardorff, M.A., Saint-Jeannet, J.P., Yang, J., Arzoumanian, A., and Klein, P.S., 2001, Kermit, a frizzled interacting protein, regulates frizzled 3 signaling in neural crest development, *Development* 128:3665–3674.

Taraviras, S., Marcos-Gutierrez, C.V., Durbec, P., Jani, H., Grigoriou, M., Sukumaran, M. *et al.*, 1999, Signalling by the RET receptor tyrosine kinase and its role in the development of the mammalian enteric nervous system, *Development* 126:2785–2797.

Teillet, M.-A., 1978, Evolution of the lumbo-sacral neural crest in the avian embryo: Origin and differentiation of the ganglionated nerve of

Remak studied in interspecific quail-chick chimaerae, *Roux's Arch. Dev. Biol.* 184:251–268.

Teillet, M.-A., Kalcheim, C., and Le Douarin, N.M., 1987, Formation of the dorsal root ganglia in the avian embryo: Segmental origin and migratory behavior of neural crest progenitor cells, *Dev. Biol.* 120:329–347.

Teillet, M.A. and Le Douarin, N.M., 1983, Consequences of neural tube and notochord excision on the development of the peripheral nervous system in the chick embryo, *Dev. Biol.* 98:192–211.

Testaz, S., Jarov, A., Williams, K.P., Ling, L.E., Koteliansky, V.E., Fournier-Thibault, C. *et al.*, 2001, Sonic hedgehog restricts adhesion and migration of neural crest cells independently of the Patched-Smoothened-Gli signaling pathway, *Proc. Natl. Acad. Sci. USA* 98: 12,521–12,526.

Thisse, C., Thisse, B., and Postlethwait, J.H., 1995, Expression of *snail2*, a second member of the zebrafish snail family, in cephalic mesendoderm and presumptive neural crest of wild-type and spadetail mutant embryos, *Dev. Biol.* 172:86–99.

Torres, M. and Giraldez, F., 1998, The development of the vertebrate inner ear, *Mech. Dev.* 71:5–21.

Tosney, K.W., 1978, The early migration of neural crest cells in the trunk region of the avian embryo: An electron microscopic study, *Dev. Biol.* 62:317–333.

Tosney, K.W., 1982, The segregation and early migration of cranial neural crest cells in the avian embryo, *Dev. Biol.* 89:13–24.

Trainor, P. and Krumlauf, R., 2000, Plasticity in mouse neural crest cells reveals a new patterning role for cranial mesoderm, *Nat. Cell Biol.* 2:96–102.

Trainor, P.A., Ariza-McNaughton, L., and Krumlauf, R., 2002a, Role of the isthmus and FGFs in resolving the paradox of neural crest plasticity and prepatterning, *Science* 295:1288–1291.

Trainor, P.A., Sobieszczuk, D., Wilkinson, D., and Krumlauf, R., 2002b, Signalling between the hindbrain and paraxial tissues dictates neural crest migration pathways, *Development* 129:433–442.

Tremblay, P., Kessel, M., and Gruss, P., 1995, A transgenic neuroanatomical marker identifies cranial neural crest deficiencies associated with the *Pax3* mutant *Splotch, Dev. Biol.* 171:317–329.

Tucker, G.C., Ciment, G., and Thiery, J.P., 1986, Pathways of avian neural crest cell migration in the developing gut, *Dev. Biol.* 116:439–450.

Tweedle, C.D., 1977, Ultrastructure of lateral line organs in aneurogenic amphibian larvae (*Ambystoma*), *Cell Tissue Res.* 185:191–197.

Valinsky, J.E. and Le Douarin, N.M., 1985, Production of plasminogen activator by migrating cephalic neural crest cells, *EMBO J.* 4:1403–1406.

Vallin, J., Thuret, R., Giacomello, E., Faraldo, M.M., Thiery, J.-P., and Broders, F., 2001, Cloning and characterization of three *Xenopus Slug* promoters reveal direct regulation by Lef/β-catenin signaling, *J. Biol. Chem.* 276:30,350–30,358.

van Wijhe, J.W., 1883, Uber die Mesodermsegmente und die Entwicklung der Nerven des Selachierkopfes, *Verhandelingen der Koninklijke Akademie van Wetenschappen (Amsterdam)* 22(E):1–50.

Veitch, E., Begbie, J., Schilling, T.F., Smith, M.M., and Graham, A., 1999, Pharyngeal arch patterning in the absence of neural crest, *Curr. Biol.* 9:1481–1484.

Villanueva, S., Glavic, A., Ruiz, P., and Mayor, R., 2002, Posteriorization by FGF, Wnt, and retinoic acid is required for neural crest induction, *Dev. Biol.* 241:289–301.

Vitali, G., 1926, La façon de se comporter du placode de la première fente branchiale (placode épibranchiale) dans la série des vertébrés, *Arch. Ital. Biol.* 76:94–106.

Vogel, K.S. and Davies, A.M., 1993, Heterotopic transplantation of presumptive placodal ectoderm changes the fate of sensory neuron precursors, *Development* 119:263–276.

Vogel-Höpker, A., Momose, T., Rohrer, H., Yasuda, K., Ishihara, L., and Rapaport, D.H., 2000, Multiple functions of fibroblast growth factor-8 (FGF-8) in chick eye development, *Mech. Dev.* 94:25–36.

Vogt, W., 1929, Gestaltungsanalyse am Amphibienkeim mit örtlicher Vitalfärbung Vorwort über Wege une Ziele. II: Gastrulation und Mesodermbildung bei Urodelen und Anuren, *Wilhelm Roux Arch. EntwMech. Org.* 120:384–706.

von Kupffer, C., 1894, Ueber Monorhinie und Amphirhinie, *Sitzungsberichte der mathematisch-physikalischen Classe der k. Bayerischen Akademie der Wissenschaften zu München* 24:51–60.

Wagner, G., 1949, Die Bedeutung der Neuralleiste für die Kopfgestaltung der Amphibienlarven. Untersuchungen an Chimaeren von *Triton, Rev. Suisse Zool.* 56:519–620.

Wakamatsu, Y., Maynard, T.M., and Weston, J.A., 2000, Fate determination of neural crest cells by NOTCH-mediated lateral inhibition and asymmetrical cell division during gangliogenesis, *Development* 127:2811–2821.

Wang, H.U. and Anderson, D.J., 1997, Eph family transmembrane ligands can mediate repulsive guidance of trunk neural crest migration and motor axon outgrowth, *Neuron* 18:383–396.

Wawersik, S. and Maas, R.L., 2000, Vertebrate eye development as modeled in *Drosophila, Hum. Mol. Genet.* 9:917–925.

Wawersik, S., Purcell, P., Rauchman, M., Dudley, A.T., Robertson, E.J., and Maas, R., 1999, BMP7 acts in murine lens placode development, *Dev. Biol.* 207:176–188.

Webb, J.F. and Noden, D.M., 1993, Ectodermal placodes: Contributions to the development of the vertebrate head, *Amer. Zool.* 33:434–447.

Weston, J.A., 1963, A radioautographic analysis of the migration and localization of trunk neural crest cells in the chick, *Dev. Biol.* 6:279–310.

Wewetzer, K., Verdú, E., Angelov, D.N., and Navarro, X., 2002, Olfactory ensheathing glia and Schwann cells: Two of a kind? *Cell Tissue Res.* 309:337–345.

White, P.M. and Anderson, D.J., 1999, In vivo transplantation of mammalian neural crest cells into chick hosts reveals a new autonomic sublineage restriction, *Development* 126:4351–4363.

White, P.M., Morrison, S.J., Orimoto, K., Kubu, C.J., Verdi, J.M., and Anderson, D.J., 2001, Neural crest stem cells undergo cell-intrinsic developmental changes in sensitivity to instructive differentiation signals, *Neuron* 29:57–71.

Whitfield, T.T., Granato, M., van Eeden, F.J., Schach, U., Brand, M., Furutani-Seiki, M. *et al.*, 1996, Mutations affecting development of the zebrafish inner ear and lateral line, *Development* 123:241–254.

Whitfield, T.T., Riley, B.B., Chiang, M.Y., and Phillips, B., 2002, Development of the zebrafish inner ear, *Dev. Dyn.* 223:427–458.

Whitlock, K.E. and Westerfield, M., 1998, A transient population of neurons pioneers the olfactory pathway in the zebrafish, *J. Neurosci.* 18:8919–8927.

Whitlock, K.E. and Westerfield, M., 2000, The olfactory placodes of the zebrafish form by convergence of cellular fields at the edge of the neural plate, *Development* 127:3645–3653.

Whitlock, K.E., Wolf, C.D., and Boyce, M.L., 2003, Gonadotropin-releasing hormone (GnRH) cells arise from cranial neural crest and adenohypophyseal regions of the neural plate in the zebrafish, *Danio rerio, Dev. Biol.* 257:140–152.

Wicht, H. and Northcutt, R.G., 1995, Ontogeny of the head of the Pacific hagfish (*Eptatretus stouti*, Myxinoidea): Development of the lateral line system, *Phil. Trans. R. Soc. Lond. B* 349:119–134.

Wilson, P.A. and Hemmati-Brivanlou, A., 1995, Induction of epidermis and inhibition of neural fate by Bmp-4, *Nature* 376:331–333.

Wilson, P.A., Lagna, G., Suzuki, A., and Hemmati-Brivanlou, A., 1997, Concentration-dependent patterning of the *Xenopus* ectoderm by BMP4 and its signal transducer Smad1, *Development* 124:3177–3184.

Winklbauer, R., 1989, Development of the lateral line system in *Xenopus, Prog. Neurobiol.* 32:181–206.

Woda, J.M., Pastagia, J., Mercola, M., and Artinger, K.B., 2003, Dlx proteins position the neural plate border and determine adjacent cell fates, *Development* 130:331–342.

Wright, T.J. and Mansour, S.L., 2003, *Fgf3* and *Fgf10* are required for mouse otic placode induction, *Development* 130:3379–3390.

Wu, X. and Howard, M.J., 2001, Two signal transduction pathways involved in the catecholaminergic differentiation of avian neural crest-derived cells *in vitro, Mol. Cell. Neurosci.* 18:394–406.

Wu, J., Saint-Jeannet, J.-P., and Klein, P.S., 2003, Wnt-frizzled signaling in neural crest formation, *Trends Neurosci.* 26:40–45.

Xu, H., Firulli, A.B., Zhang, X., and Howard, M.J., 2003, HAND2 synergistically enhances transcription of dopamine-β-hydroxylase in the presence of Phox2a, *Dev. Biol.* 262:183–193.

Yan, Y.L., Miller, C.T., Nissen, R.M., Singer, A., Liu, D., Kirn, A. *et al.*, 2002, A zebrafish *sox9* gene required for cartilage morphogenesis, *Development* 129:5065–5079.

Yip, J.W., 1986, Migratory patterns of sympathetic ganglioblasts and other neural crest derivatives in chick embryos, *J. Neurosci.* 6:3465–3473.

Yntema, C.L., 1944, Experiments on the origin of the sensory ganglia of the facial nerve in the chick, *J. Comp. Neurol.* 81:147–167.

Young, H.M., Hearn, C.J., Farlie, P.G., Canty, A.J., Thomas, P.Q., and Newgreen, D.F., 2001, GDNF is a chemoattractant for enteric neural cells, *Dev. Biol.* 229:503–516.

Young, H.M. and Newgreen, D., 2001, Enteric neural crest-derived cells: Origin, identification, migration, and differentiation, *Anat. Rec.* 262:1–15.

Yu, T.W. and Bargmann, C.I., 2001, Dynamic regulation of axon guidance, *Nat. Neurosci.* 4 Suppl. 1:1169–1176.

Zhang, X., Friedman, A., Heaney, S., Purcell, P., and Maas, R.L., 2002, Meis homeoproteins directly regulate Pax6 during vertebrate lens morphogenesis, *Genes Dev.* 16:2097–2107.

Zheng, J.L. and Gao, W.Q., 2000, Overexpression of *Math1* induces robust production of extra hair cells in postnatal rat inner ears, *Nat. Neurosci.* 3:580–586.

Zheng, J.L., Shou, J., Guillemot, F., Kageyama, R., and Gao, W.Q., 2000, Hes1 is a negative regulator of inner ear hair cell differentiation, *Development* 127:4551–4560.

Zilian, O., Saner, C., Hagedorn, L., Lee, H.Y., Sauberli, E., Suter, U. *et al.*, 2001, Multiple roles of mouse Numb in tuning developmental cell fates, *Curr. Biol.* 11:494–501.

Zirlinger, M., Lo, L., McMahon, J., McMahon, A.P., and Anderson, D.J., 2002, Transient expression of the bHLH factor neurogenin-2 marks a subpopulation of neural crest cells biased for a sensory but not a neuronal fate, *Proc. Natl. Acad. Sci. USA* 99:8084–8089.

5

Neurogenesis

Monica L. Vetter and Richard I. Dorsky

INTRODUCTION

The function of the nervous system is controlled at the most basic level by individual cells—the neurons. In order to generate the enormous diversity of function and connectivity present in the mature nervous system, each neuron must be directed to differentiate at a particular time and place and to adopt a particular phenotype. The process of generating a neuron from a field of neurectodermal cells, known as neurogenesis, is the focus of this chapter. We will largely focus on neurogenesis in the vertebrate nervous system, but when appropriate will use examples from invertebrates to illustrate conserved aspects of nervous system development and in some cases demonstrate molecular mechanisms.

In every vertebrate nervous system, neural precursor cells initially occupy a uniform neuroepithelial sheet. The central nervous system (CNS) arises from a flat neural plate that is patterned along the rostral/caudal (RC) and dorsal/ventral (DV) axes by signals in the embryo beginning during gastrulation (see Chapter 3), while the neural crest and placodes, which are the source for cells of the peripheral nervous system (PNS), arise from the lateral border of this tissue (see Chapter 4). The neural plate eventually rolls (or intercalates in the case of fish) into a neural tube forming a lumen at the center, which defines the ventricular surface of the neural tube. At early stages of development the neural tube consists of proliferating neuroepithelial cells that are multipotent and give rise to all of the major cell populations of the CNS and much of the PNS (see Chapter 2). Throughout development, proliferating neuroepithelial cells remain in contact with the ventricular surface of the neural tube forming a ventricular zone (VZ—see Chapter 2). This zone contains the proliferating cells throughout CNS development, at all rostrocaudal levels of the embryo. As neuroepithelial cells begin the process of differentiation into CNS neurons they detach from the ventricular surface, exit the cell cycle, and migrate away from the VZ to their final location in the developing mantle layer (see Fig. 1A). Neuroepithelial cells also give rise to neural crest cells, which delaminate from the dorsal aspect of the neural tube, migrate away from the neural tube, and differentiate into

a variety of cell types, including neurons of the PNS (see Chapter 4).

The cellular process of neurogenesis can be generally considered as a progression from multipotent stem cells to fate-restricted neuronal precursors, through the gradual reduction of potential fates. Once a particular cell fate has been specified, neurons will withdraw from the cell cycle and differentiate. In this chapter we will illustrate the many steps of neurogenesis and provide examples that explain the genetic and molecular mechanisms behind each step. First, cells from the neuroectoderm acquire the competence to become neural, and these stem cells expand to provide the raw material for all subsequent cell generation. In the next step, neural progenitors are produced by asymmetric divisions of stem cells, lose the ability to self-renew, and begin to be restricted in potential. Cell number is tightly controlled at these early stages through regulation of both proliferation and survival of stem cells and progenitors. Third, neural progenitors express genes that promote differentiation, while negative regulators constrain the number of neurons that are generated at any given place and time. The fourth step of neurogenesis is the irreversible decision to leave the cell cycle and form a neuron. Fifth, neural precursors migrate to their final position in the nervous system and differentiate. Finally, neurons mature and adopt a particular phenotype by activating gene programs that direct their ultimate differentiation into functioning neurons. Many different subtypes of neurons exist in the mature nervous system. During development it is essential that the generation of these different classes of neurons be carefully orchestrated so that functionally integrated neuronal structures can assemble.

The two main processes that contribute to the generation of neuronal diversity are spatial patterning and temporal regulation of birthdates. Through the combination of these two events, each neural progenitor has a unique positional identity and history by virtue of being exposed to a different combination of inductive factors. This ultimately results in neural progenitors expressing a distinct combination of transcription factors that will regulate their differentiation into specific neuronal subtypes. In some cases the phenotype of a differentiating neuron can also be influenced as it migrates to its final position, or after innervation

Monica L. Vetter and Richard I. Dorsky • Department of Neurobiology and Anatomy, University of Utah, SOM, Salt Lake City, UT 84132.

Developmental Neurobiology, 4th ed., edited by Mahendra S. Rao and Marcus Jacobson. Kluwer Academic / Plenum Publishers, New York, 2005.

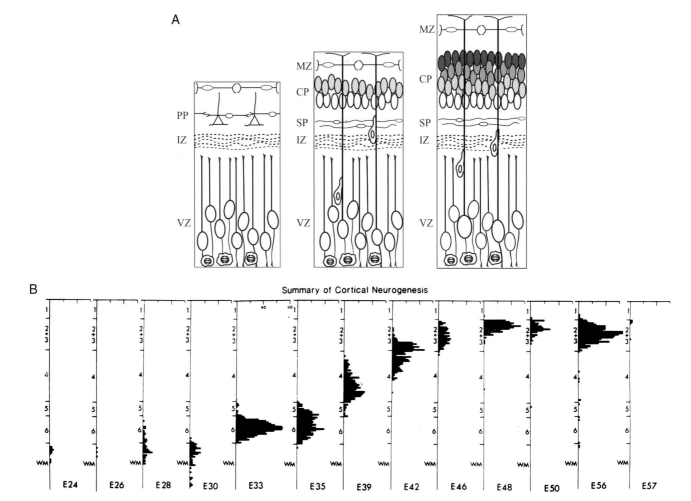

FIGURE 1. (A) Development of the cerebral cortex. The ventricular zone (VZ) contains proliferating progenitors that divide at the ventricular surface. The first neurons to differentiate are those forming the preplate (PP), which is separated from the VZ by PP axons and incoming thalamic axons in the intermediate zone (IZ). As development progresses the cortical plate (CP) forms from neurons which migrate out from the VZ along radial glial fibers, separating the PP into the subplate (SP) and superficial marginal zone (MZ). Within the CP, deep layer neurons are generated first and later-born neurons migrate past the early-born neurons to populate more superficial layers (dark grey). Ultimately, the SP neurons and VZ disappear and the MZ becomes layer I of the mature cortex. The CP neurons develop into the remaining cortical layers (II–VI) and overlay the white matter. Figure generated by Diana Lim. (B) Cortical neurons are born in an inside-out sequence. Each histogram shows the relative depth distribution of heavily labeled neurons in the developing visual cortex of the cat resulting from a single injection of [³H]thymidine given at the embryonic age shown underneath. Neurons of different cortical layers are generated in an inside-out sequence between E30 and E57. Modified from M.B. Luskin and C.J. Shatz, 1985, *J. Comp. Neurol.* 242:611–631.

of its target tissue. We will now consider in detail each of these steps in the process of neurogenesis, beginning with an overview of histogenesis, the cellular process of differentiation, in different parts of the developing nervous system.

HISTOGENESIS IN THE VERTEBRATE NERVOUS SYSTEM

Birthdating, Transplantation, and Lineage Analysis

The vertebrate nervous system is a highly organized tissue and its cellular organization is critical for its proper function.

In many parts of the nervous system the tissue is laminated; that is, neurons with similar structural and functional properties are organized into discrete layers. In other places, neurons assemble into nuclei or ganglia rather than layers. How are these patterns of tissue organization established? Historically, several techniques have been important for defining how neurons are generated and become organized within specific domains of the developing nervous system. The birthdating technique, developed by Richard Sidman in the late 1950s, can be used to label groups of neurons as they are born and then track them to their final position (Sidman *et al.*, 1959). This method involves labeling proliferating precursor cells within an embryo by pulsing with tritium-labeled thymidine, which incorporates into the DNA during replication. If the cell continues to divide then this label becomes diluted through subsequent rounds of DNA synthesis. However, if a cell becomes labeled during its final division and subsequently differentiates, then that cell remains heavily labeled and can be detected by autoradiography of histological sections. The "birthdate" of a cell is defined as the time when it undergoes its final division, and this can be assessed by pulsing with tritiated thymidine at various times in development and determining when that type of cell becomes heavily labeled. In addition, by analyzing the location of heavily labeled cells at progressively later times following a pulse of tritiated thymidine, it is possible to track the position of cells born at a particular time as they migrate to their final position.

The fate of cells can also be followed by transplanting cells from one species into another then using specific markers or cellular features to distinguish donor cells from host. For example, Nicole Le Douraïn used a heterochromatin marker in the nuclei of quail cells to track them after transplantation into chick embryos (Le Douarin, 1973, 1982). This approach has not only been valuable for tracking the migratory pathways of cells, particularly those derived from the neural crest, but has also made it possible to transplant cells into new environments to determine their developmental potential.

The third technique, called lineage analysis, made it possible to track all of the progeny from a single precursor cell and determine their phenotypes and their ultimate resting position. One approach to lineage analysis is to intracellularly inject a tracer such as a fluorescent dye or horseradish peroxidase that would be passed on to the progeny of that cell (Fig. 2; Weisblat *et al.*, 1978). This approach can be problematic since multiple rounds of cell division can dilute the tracer, so it is not always a reliable marker of lineage. Alternatively, retroviruses carrying a reporter gene can be used to stably label cells and their progeny (Cepko, 1988). Small amounts of retroviruses are injected so that only a few proliferating progenitor cells become infected and their progeny can be followed. One problem with this approach is that it is difficult to determine whether all labeled progeny in a given domain were derived from a single infected progenitor. To address this concern, libraries of retroviruses have been used carrying large numbers of individual tags that can be distinguished by amplifying specific tag sequences using the polymerase chain reaction (PCR; Walsh and Cepko, 1992). A single retrovirus will infect a progenitor and the labeled progeny will all carry the same tag, arguing for clonal origin.

Together these approaches have revealed a few general principles in nervous system development. First, the birthdate of a neuron is an important predictor of cell fate. In a given region, neurons born at a certain time generally adopt similar fates. Second, newborn neurons often migrate a considerable distance from their site of origin to their final resting place. Finally, within a given region of the nervous system, neurons of similar phenotype and birthdate cluster together in discrete layers, nuclei, or ganglia. We will consider several examples of histogenesis in the developing vertebrate nervous system to illustrate these points.

Cerebral Cortex

The mature cerebral cortex is a beautiful example of a laminated neuronal tissue. The mammalian neocortex consists of six layers that can be distinguished histologically based upon the morphology and density of neurons within each layer. This also reflects distinct functions for the neurons in each layer. Layer I is closest to the pial surface and contains relatively few neurons. Neurons in layers II/III provide connections between different cortical areas, while layer IV neurons receive inputs from subcortical structures such as the thalamus. Layer V and VI neurons send projections to subcortical structures, such as thalamus, brainstem, and spinal cord. The thickness of these layers varies depending upon whether a given cortical region serves largely sensory, motor, or association functions. This precise laminar organization is important for proper functioning of the neocortex. Developmental disorders that result in disruption of neurogenesis and lamination of the cortex are associated with severe mental retardation and epilepsy.

The cerebral cortex begins as a single layer of proliferating neuroepithelial cells in the walls of the telencephalon. At some point these neuroepithelial cells begin to divide asymmetrically generating first neurons and later glia. Birthdating studies have revealed a very tight correlation between birth order of neurons and their final laminar position (Angevine and Sidman, 1961). In the mammalian cortex, the earliest generated neurons migrate away from the VZ and form a layer of cells beneath the pial surface known as the preplate (Fig. 1A). Later-generated neurons then migrate into the preplate to form the cortical plate, thus splitting the preplate into a superficial marginal zone (future layer I) and a deeper zone called the intermediate zone that contains subplate neurons and incoming axons. Thus both

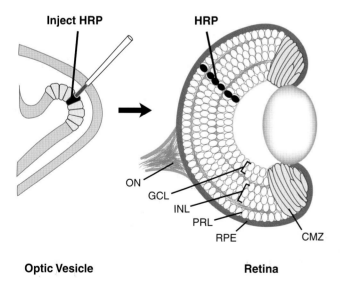

Inject HRP   **HRP**

ON
GCL
INL
PRL
RPE   CMZ

Optic Vesicle   **Retina**

FIGURE 2. Retinal progenitors are multipotent. Injection of HRP, a lineage tracer, into a single retinal progenitor at the optic vesicle stage in *Xenopus laevis* reveals that a single progenitor can generate multiple retinal cell types that span the layers of the mature retina (Holt *et al.*, 1988). HRP, horseradish peroxidase; ON, optic nerve; GCL, ganglion cell layer; INL, inner nuclear layer; PRL, photoreceptor layer; RPE, retinal pigment epithelium; CMZ, ciliary marginal zone. Figure generated by Diana Lim.

the marginal and intermediate zones contain neurons that were generated earliest. The marginal zone neurons include Cajal-Retzius cells, which provide important signals for later-born neurons as they migrate out and establish the cortical layers (see Chapter 8). The subplate neurons in the intermediate zone serve a transient developmental role as guideposts for incoming thalamic axons preparing to innervate the cortical layers.

Within the developing cortical plate, tritiated thymidine labeling reveals a very orderly pattern of generation, migration, and assembly of neurons in tangential strata (Fig. 1B; Angevine and Sidman, 1961). The emerging cortical layers are established in an inside-out sequence such that deep layer neurons are born first followed progressively by neurons that will migrate radially past the deep layer neurons to occupy more superficial layers (Fig. 1A). Thus, pulsing with thymidine at early stages of development results in labeling of neurons in deeper layers of the cortical plate, while pulsing at later stages of development results in labeling of more superficial layers. The older deep layer neurons have already begun to differentiate and send out axons as the later-born neurons migrate past them to populate the more superficial layers. In addition, there are spatial gradients across the cortex with respect to the timing of neurogenesis in different cortical regions. Even in three-layered allocortex, such as the hippocampus, deep neurons are generated before superficial neurons and the younger neurons migrate through previously formed layers to generate more superficial layers (Angevine, 1965).

In general, excitatory projection neurons follow this pattern of genesis and migration (Tan *et al.*, 1998). They are generated from progenitors in the VZ and then migrate radially to populate the emerging cortical layers in radial columns, although there is also evidence for non-radial tangential migration of developing cortical neurons (O'Rourke *et al.*, 1995, 1997; see Chapter 8). However, lineage analysis and studies of neuronal migration have revealed that most local circuit GABAergic inhibitory interneurons are generated from a distinct population of progenitors in subcortical ventral forebrain regions (Tan *et al.*, 1998). These interneurons are born in the VZ of the lateral and medial ganglionic eminences, then migrate dorsally and disperse through the cortical layers (Anderson *et al.*, 1997; Lavdas *et al.*, 1999; Parnavelas *et al.*, 2000).

At early stages of cortical development, neurons are generated from progenitors in the VZ, although the VZ diminishes as the cortex develops. At later stages of vertebrate development a second zone of proliferating cells known as the subventricular zone (SVZ) forms between the VZ and the intermediate zone. As the VZ disappears, the SVZ continues to proliferate and generate cortical neurons, as well as most of the glial cells in the cortex. The SVZ also gives rise to neurons that will migrate to the olfactory bulb along a specific migratory path known as the rostral migratory stream (Lois and Alvarez-Buylla, 1994). Although the SVZ also diminishes as development progresses, there is good evidence that the SVZ retains its capacity to generate new cells in the adult (Lois and Alvarez-Buylla, 1993), a topic that will be discussed in more detail later.

Retina

Like the cerebral cortex, the vertebrate retina is a laminated CNS structure consisting of three major cellular layers. The outermost layer closest to the non-neural retinal pigment epithelium is the photoreceptor layer and contains rod and cone photoreceptors. The middle layer, called the inner nuclear layer (INL), contains several classes of interneurons such as horizontal cells, bipolar cells, and amacrine cells. The innermost layer closest to the vitreal surface is the retinal ganglion cell layer, which consists of retinal ganglion cells, the projection neurons of the retina, and in some species considerable numbers of displaced amacrine cells. There is also one major type of glial cell in the retina, the Müller glial cell, which spans the width of the retina with the cell body being localized to the INL.

The retina begins as a single cell-wide epithelial sheet, and progenitors are attached to both the outer (ventricular) and inner limiting membranes, which are composed of neuroepithelial and eventually glial endfeet. As they proceed through the cell cycle, progenitor nuclei migrate from the outer surface (M-phase) to the inner surface (S-phase) in a process termed interkinetic migration (see Chapter 2). As progenitors continue to proliferate, the retinal thickness expands and dividing cells are split into inner and outer neuroblastic layers. The inner neuroblastic layer will eventually differentiate into ganglion, amacrine, and Müller cells, while the outer neuroblastic layer produces photoreceptor, horizontal, and bipolar cells. While there is no true "radial migration" of neural precursor cells in the retina, cells do detach from the retinal surfaces and move to their ultimate positions. As rod, bipolar, and Müller cells differentiate, neurons derived from the same region of neuroepithelium remain spatially associated. In contrast, cone, ganglion, horizontal, and amacrine cells undergo extensive tangential migration (Fekete *et al.*, 1994; Reese *et al.*, 1995).

Cell birthdating studies using the methods described previously have shown a generally conserved order of genesis for retinal cell types across all vertebrate species (Cepko *et al.*, 1996). Ganglion cells, the projection neurons of the retina, are the first cell type born, shortly followed by horizontal and amacrine interneurons, and cone photoreceptors. At the end of histogenesis, late-born cell types include rod photoreceptors, bipolar cells, and Müller glia. In rapidly developing vertebrates such as *Xenopus*, there is considerable overlap between the birthdates of these cell types, but the general order is preserved (Holt *et al.*, 1988). Importantly, this order suggests that some factor, either internal or external to the retinal progenitors, biases them toward particular fates at different times during development. Although cell fate in the retina is partially determined by temporal order of histogenesis, birth order does not correlate with laminar position, which is unlike the cerebral cortex. Instead, as progenitors withdraw from the cell cycle and differentiate, they migrate to the appropriate position for their function.

Interestingly, retinal histogenesis continues throughout the life of the animal in fish and frogs. As the eye continues to grow in these animals, new cells are added to the periphery from a structure called the ciliary marginal zone (CMZ) (see Fig. 2).

The CMZ has been studied as a model of retinal cell-fate determination because all the mature cell types are generated from this small region, and at any given time, all stages of progenitor development can be observed (Perron *et al.*, 1998). Furthermore, these characteristics of the CMZ suggest that extracellular signals influencing cell fate must be supplied very locally.

Spinal Cord

The spinal cord has become an important model system for studying neural cell-fate specification because it contains populations of anatomically and molecularly identifiable motoneurons and interneurons and a transient population of sensory neurons. In addition, the spinal cord is a relatively simple CNS structure in which histogenesis follows the same general rules as other regions of the nervous system. Proliferation takes place in the VZ, which, as in the cortex and retina, begins as a single cell-wide neuroepithelium. Progenitors undergo interkinetic nuclear migration then detach from the ventricular surface and migrate laterally through an intermediate zone into a mantle zone where they differentiate. In addition to radial migration, some differentiating precursors migrate tangentially in the intermediate zone, along dorsoventral and rostrocaudal pathways (Leber and Sanes, 1995). Therefore the final position of differentiated spinal neurons often does not correspond to the region from which they were generated.

The general order of histogenesis in the spinal cord is the same as in the brain—neurons are generated first, followed by astrocytes and oligodendrocytes. Within the neuronal population, there is also a conserved order of birth. Ventral motoneurons are born first, followed by more dorsal interneurons (Nornes and Carry, 1978). Single progenitors can give rise to multiple subtypes of neurons, and some produce both neurons and glia. As in the retina, it appears that both the timing and spatial localization of differentiation play important roles in ultimate cell fate. Particular types of neurons and glia arise from different dorsoventral positions in the VZ. In addition, progenitor fate appears to be restricted over time, to the point where some glial and neural-restricted precursors have been identified by clonal analysis both in culture and *in vivo* (Mayer-Proschel *et al.*, 1997; Rao *et al.*, 1998).

Different Classes of PNS Neurons Have Distinct Birthdates

Even in the PNS, different subtypes of neurons are born at different times and aggregate into discrete domains. For example, neurons in the dorsal root ganglia (DRG) are derived from neural crest precursor cells that have migrated away from the neural tube and aggregated into ganglia (see Chapter 4). Within the developing DRG, precursor cells proliferate then ultimately stop dividing and differentiate. The DRG contains several different classes of sensory neurons, such as proprioceptive and cutaneous neurons, which are born in an overlapping sequence (Carr and Simpson, 1978). These different types of DRG neurons then partially segregate within the DRG. For example, in chick, most proprioceptive neurons are born early and occupy the ventral half of the ganglia, while cutaneous neurons are, for the most part, born later than the proprioceptive neurons and are more broadly distributed within the ganglia, including the dorsal domain (Carr and Simpson, 1978; Henrique *et al.*, 1995). There is now evidence that early markers can distinguish these cell populations even before their axons reach their targets, suggesting that their fates are determined early (Guan *et al.*, 2003).

Conserved Role of Timing in Neurogenesis

In all these different regions of the vertebrate nervous system there is evidence of a strong link between birthdate and neuronal phenotype, suggesting that there is temporal regulation of the neuronal cell-fate decision. In fact, this appears to be a conserved feature of neurogenesis across animal phyla. We can use this conservation to help study the process of neurogenesis in simpler invertebrate organisms that are amenable to genetic manipulation. The most fruitful of these studies have taken place in *Drosophila*, where precise examination of neurogenesis has been undertaken throughout development. In the *Drosophila* embryonic CNS, precise numbers of neurons are generated from single neuroblasts in a defined temporal sequence. Individual neuroblasts arise from the ectoderm then divide to produce a series of ganglion mother cells (GMCs; see Fig. 3). These cells then divide to produce neuronal and glial siblings that undergo terminal differentiation. GMCs are produced sequentially and each successive GMC generates different progeny. If an individual GMC is ablated, its fate is skipped entirely and the next GMC goes on to produce daughters appropriate for its time of generation (Doe and Smouse, 1990). Thus there is a tight link between the birthdate of a GMC and the phenotype of the cells that it generates.

We will now step back and consider how neurogenesis is regulated, highlighting examples from both vertebrate and invertebrate nervous system development.

NEUROEPITHELIAL CELLS ARE MULTIPOTENT AND HAVE POSITIONAL IDENTITY

The vertebrate neural tube is initially formed of highly proliferative neuroepithelial cells that when isolated and placed in culture exhibit properties characteristic of neural stem cells: They are capable of long-term self-renewal and can generate the major cell types of the nervous system—neurons, astrocytes, and oligodendrocytes (see Chapter 2). In addition, infection of these early stem cells with retroviral lineage tracers *in vivo* shows that a single progenitor cell can give rise to all three major cell types (Kalyani and Rao, 1998). These neuroepithelial cells have long processes that span the width of the early neural tube; however, cell division occurs at the ventricular surface (see Chapter 2). Neuroepithelial cells initially divide symmetrically expanding the pool of early neural stem cells. In symmetric divisions

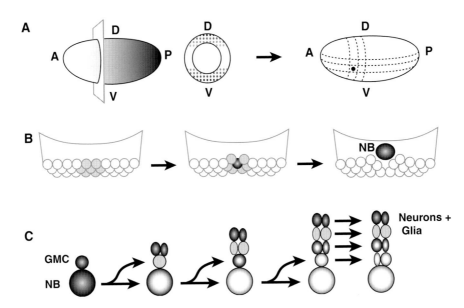

FIGURE 3. Neuroblast development in the *Drosophila* CNS. (A) Gradients of signaling molecules pattern the early *Drosophila* embryo along the anterior–posterior (AP) and dorsal/ventral (DV) axes. The embryo is thus subdivided by the expression of segment polarity genes (vertical stripes) and columnar genes (horizontal stripes), and each neuroblast within these segments (black dot—only one shown) has a positional identity that determines the phenotype of the cells that it generates. (B) Within the neuroectoderm a neuroblast (NB) is selected from a cluster of cells (light grey) through a process of lateral inhibition (see text) and delaminates from the ectoderm. All cells within the cluster (light grey) initially express equivalent levels of proneural genes. As the neuroblast is selected it expresses elevated levels of proneural genes, while the surrounding cells downregulate proneural gene expression and assume a non-neural ectodermal fate. (C) The neuroblast undergoes a series of divisions to generate ganglion mother cells (GMCs) which then divide and differentiate into neurons and glia of the ventral nerve cord. Figure generated by Diana Lim.

the plane of cell division is perpendicular to the ventricular surface generating two identical daughters (Chenn and McConnell, 1995). This mode of division is important for self-renewal and is prominent during the early expansion phase of neuronal development.

Coincident with neural induction, the nervous system becomes patterned along the RC and DV axes in response to gradients of signaling molecules from neighboring tissues (see Chapter 3). As a result, neuroepithelial cells at the earliest stages of development already have a positional identity and express genes appropriate for their region of origin even when isolated and grown in culture. This positional identity influences the types of neurons that arise from precursors in different parts of the nervous system. For example, neuroepithelial cells isolated from spinal cord can generate the complement of neuronal cell types appropriate for spinal levels (Kalyani *et al.*, 1997, 1998), while basal forebrain stem cells generate GABAergic interneurons similar to those that normally populate the cerebral cortex (He *et al.*, 2001). DV position is also important. For example, within the developing spinal cord, progenitors respond to gradients of signaling molecules, such as Sonic hedgehog (Shh) ventrally and BMPs dorsally, that define DV position within the spinal cord. These progenitors then have a unique positional identity that allows them to generate the appropriate types of neurons for that position in the spinal cord, such as ventral motoneurons and dorsal sensory interneurons (Lee and Pfaff, 2001).

As in vertebrates, positional identity is also a critical factor in insect nervous system development, arguing that this is an evolutionarily conserved mechanism for generating regional diversity in the nervous system. During insect CNS development,

neuroblasts arise at segmentally repeated positions in the ventral neurogenic region of the embryo in a precise spatiotemporal pattern. Within each hemisegment, around 30 neuroblasts delaminate from the epithelium and begin a series of cell divisions, generating first ganglion mother cells then post-mitotic neurons (Fig. 3). Neuroblasts in different positions within the hemisegment have distinct identities and generate a specific complement of neuronal and glial cell types. The gap and pair-rule genes act prior to neurogenesis to subdivide the embryo into segments along the anterior–posterior (AP) axis (Akam, 1987). Subsequently segment polarity genes, such as wingless (*wg*) and sonic hedgehog (*shh*), pattern the segments and have an important influence on the formation and identity of neuroblasts within a segment (Bhat, 1999). In addition, the dorsal–ventral position of neuroblasts is defined by signaling through NF-κB, BMP, and EGF pathways, which creates DV subdivisions of gene expression within the neuroectoderm (von Ohlen and Doe, 2000). Thus, the combination of AP and DV positional information provides each neuroblast in *Drosophila* with a positional identity and allows it to generate a unique complement of post-mitotic cell types appropriate for that position in the embryo.

NEURAL PROGENITORS ARE MULTIPOTENT BUT BECOME RESTRICTED IN COMPETENCE

Together with positional identity of the progenitors, the temporal birth order of post-mitotic cells from these progenitors is also a critical variable in determining the ultimate phenotype of the cells that result. In a given region of the vertebrate CNS neurons are generated first, followed by astrocytes then oligodendrocytes. This is also true if neural stem cells are isolated and grown in culture, although this can be influenced by addition of growth factors or other signaling molecules (Qian *et al.*, 2000). As development proceeds neuroepithelial cells begin to undergo asymmetric divisions, first generating progenitors for neurons, then for glia in a stage-dependent manner. When placed in culture, these progenitors have a limited capacity for self-renewal and are restricted in their potential, giving rise to a much more limited complement of cell types than the neuroepithelial stem cells (Rao, 1999). Thus, more restricted progenitors can divide to generate neurons that will exit the cell cycle, begin to differentiate, and migrate to their final position.

We know that in each region of the developing nervous system cells are born in a general order, but where do the individual cell types come from? More specifically, are there separate populations of progenitors that produce early and late neuronal cell types, or do they arise from a common pool? The fate of progenitor cells has been examined through a number of lineage-tracing methods, including direct label injection and retroviral infection. The results of these studies confirm that in many parts of the developing nervous system, progenitors are multipotent. For example, in the developing cerebral cortex, progenitor cells are multipotent, giving rise to clones of cells that will populate multiple cortical layers (Walsh and Cepko, 1988). At any given time in development cortical progenitors are biased towards generating cells of specific laminar fates. Deep layer neurons are generated early, while neurons in more superficial layers are generated later (Angevine and Sidman, 1961). Progenitors from older animals normally dedicated to making superficial layer neurons do not make early-born deep layer neurons, even when transplanted back into a younger environment; thus, their competence appears to be restricted over developmental time (Frantz and McConnell, 1996). However, progenitors isolated from the VZ at early stages of development can be transplanted into older animals, and these cells, if transplanted prior to their final division, will respond to their new environment and generate late-born cells appropriate for the later stage of development (McConnell, 1988; McConnell and Kaznowski, 1991). Thus, early cortical progenitors are competent to make both early and late cell types, while later progenitors appear to be restricted in their competence.

In the developing retina, individual retinal progenitors have the ability to produce many different combinations of retinal cells, including neurons and Müller glia (Fig. 2). Lineage analysis has revealed no predictable pattern to the cell composition of retinal clones, ruling out the idea of dedicated progenitors for specific neurons or combinations of neurons (Turner and Cepko, 1987; Holt *et al.*, 1988; Turner *et al.*, 1990). In many cases progenitors remain multipotent up until their final division generating two nonidentical daughters. Although retinal progenitors are multipotent, at any given stage of development they appear to be limited in their competence and generate only the subset of retinal cell types appropriate for that stage of development (Belliveau and Cepko, 1999; Belliveau *et al.*, 2000).

This competence appears to change over developmental time so that early retinal progenitors are biased toward making early-born cell types, such as retinal ganglion cells, while later progenitors are biased toward producing later-born fates, such as rod photoreceptors and Müller glia (Livesey and Cepko, 2001). An extreme case of this restriction occurs in the mature fish retina, where a population of dividing precursor cells generates only rods (Raymond and Rivlin, 1987).

Unlike in the cortex, retinal progenitors do not appear to change their intrinsic competence in response to new environments and appear to be restricted to a limited repertoire of fates at different times during development. For example, early progenitors grown in culture continue to generate retinal ganglion cells, an early-born cell type, even when cultured in the presence of older cells (Austin *et al.*, 1995). The mechanisms underlying changes in progenitor competence, both in the retina and cerebral cortex, remain to be defined. Although progenitors in many parts of the nervous system are multipotent, in a given region at any one time progenitors are not necessarily a uniform population. There is now good molecular evidence for progenitor diversity in the developing retina and cortex, and this may ultimately contribute to neuronal subtype diversity in the nervous system (Livesey and Cepko, 2001; Nieto *et al.*, 2001).

Up to this point, we have described cellular aspects of neuron formation, including the physical development of nervous system structures, and cellular histogenesis. We have also shown that progenitor cells are initially multipotent and become progressively restricted to a limited number of fates due to positional cues from their environment. Next, we will discuss the intrinsic and extrinsic molecular mechanisms by which these cells are driven down the pathway of neurogenesis.

THE PRONEURAL GENES

Like many developmental events, neurogenesis is regulated by a balance between positive regulators that promote neural competence or neuronal differentiation and negative regulators that constrain when and where differentiation occurs. There is evidence that these fundamental mechanisms, although they may vary in detail, are largely conserved during nervous system development of all animals. Subsets of cells within the neural ectoderm are selected to become neural precursors, which will then divide and differentiate to form post-mitotic neurons. How are these neural precursors specified?

Neurogenesis absolutely requires the function of proneural genes, which encode basic helix-loop-helix (bHLH) transcription factors (Bertrand *et al.*, 2002). The basic domain in this family of proteins mediates DNA binding to specific DNA sequences known as E boxes (CANNTG), while the helix-loop-helix motif allows heterodimerization with ubiquitously expressed bHLH partners or E proteins (Fig. 4A; Murre *et al.*, 1989a, b). Proneural bHLH genes were first described in *Drosophila* and include genes of the *achaete-scute* complex (*achaete, scute, lethal of scute*, and *asense*) and atonal-related genes (*atonal, amos,* and *cato*), which regulate the development of different classes of neurons in the fly PNS and CNS (Bertrand *et al.*, 2002). Multiple proneural bHLH

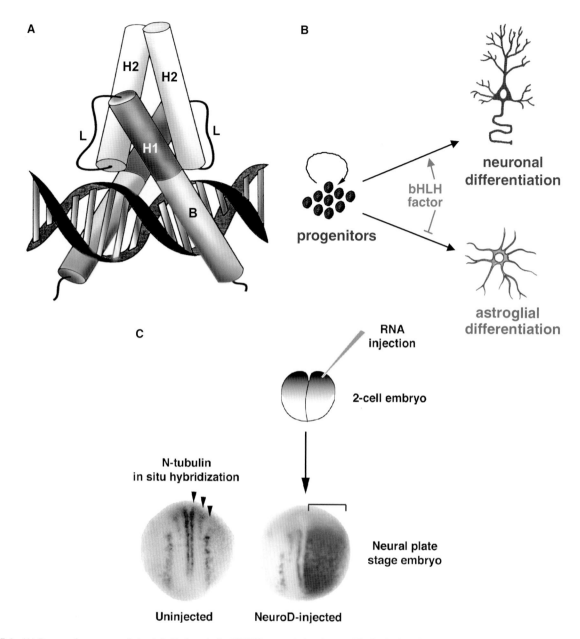

FIGURE 4. (A) Proneural genes encode basic helix-loop-helix (bHLH) transcription factors. The basic domain (B) mediates DNA binding. Helix 1 (H1) and helix 2 (H2) are joined by a loop (L) and mediate dimerization. Figure generated by Diana Lim. (B) Vertebrate proneural bHLH factors act in progenitors to promote the neuronal fate and suppress astroglial fate. (C) Three stripes of primary neurons (arrowheads) develop on either side of the midline in the neural plate of *Xenopus* embryos, as revealed by *in situ* hybridization for the neuronal marker N-tubulin (uninjected). Overexpression of NeuroD by RNA injection into a two-cell stage *Xenopus* embryo promotes ectopic neurogenesis throughout the ectoderm on the injected side (square bracket), showing that NeuroD is sufficient to convert ectodermal cells to a neuronal fate (Lee *et al.*, 1995).

genes have been identified in vertebrates and are expressed in distinct domains within the developing CNS. These can be classified into subfamilies based upon their homology to the *Drosophila* proneural genes. One vertebrate subfamily is most closely related to genes of the *achaete-scute* complex in *Drosophila* and includes genes such as *Mash1* (Guillemot and Joyner, 1993). The other subfamily shows stronger homology to *Drosophila atonal* and includes the *Ath* genes, *neurogenins* and *NeuroD*-related genes (Bertrand *et al.*, 2002). As in *Drosophila*, different vertebrate

proneural bHLH proteins are required for the development of different subpopulations of neurons, and in some cases act redundantly. For example, mice mutant for *neurogenin 1* (*ngn1*) or *ngn2* fail to develop complementary sets of cranial sensory ganglia, while mice mutant for both *ngn1* and *ngn2* lack both these populations of neurons and additionally lack neurons in the ventral spinal cord and DRG (Fode *et al.*, 1998; Ma *et al.*, 1998, 1999). During vertebrate CNS development, early multipotent stem cells will eventually give rise to neural precursors that generate solely neurons. Proneural bHLH factors such as Mash1 or Ngn1 are expressed in neural precursor cells in the ventricular zone and play an important role in promoting the neural fate and suppressing competence to make astroglia (Fig. 4B). For example, when Ngn1 is overexpressed in cortical progenitors in culture almost all of the cells differentiate into neurons and the astrocyte fate is suppressed (Sun *et al.*, 2001). Conversely, in mice mutant for *ngn2* and *mash1*, progenitors that would normally have differentiated into neurons fail to do so and instead are biased towards differentiating as astrocytes (Nieto *et al.*, 2001). Thus bHLH factors such as Ngn or Mash1 not only promote the neuronal fate but also act to suppress the astroglial fate.

The ability of proneural bHLH factors to promote neural competence was first demonstrated during nervous system development in *Drosophila*. The first step in *Drosophila* neurogenesis is to define a cluster of cells within the ectoderm with the potential to form neural precursors. This is achieved through the expression of proneural genes within a group of cells known as the proneural cluster (Cubas *et al.*, 1991; Skeath and Carroll, 1991, 1992). All cells within a proneural cluster express low levels of proneural genes and have equivalent potential to become a neuroblast. Cell–cell communication through the Notch pathway (discussed in detail below) causes one cell to be selected as the neuroblast and express elevated levels of the proneural genes while the other cells adopt a non-neural epidermal fate and downregulate proneural gene expression (Fig. 3; Skeath and Carroll, 1992). If a newly delaminating neuroblast is ablated with a laser, then a neighboring cell within the equivalence group can take its place. If all cells within the equivalence group are ablated then no neuroblast forms (Taghert *et al.*, 1984). Does a similar process happen in vertebrates? One important model system for understanding the function of proneural bHLH genes during vertebrate neurogenesis has been the neural plate of the amphibian embryo. Rather than being expressed in proneural clusters, early proneural bHLH genes in the *Xenopus* neural plate are expressed in broad stripes that ultimately give rise to more discrete sets of differentiated neurons within the stripes (see Fig. 4C). As discussed below, this refinement in the pattern of neurogenesis within the neural plate is mediated through the Notch signaling pathway. The first proneural bHLH gene expressed during primary neurogenesis in *Xenopus* is *X-Ngn-R1*, which is related to mammalian *ngn* (Ma *et al.*, 1996). *X-Ngn-R1* in turn regulates the expression of the downstream bHLH factor NeuroD and ultimately promotes cell cycle exit and terminal neuronal differentiation. Misexpression of *X-Ngn-R1* by RNA injection into cleavage stage *Xenopus* embryos is sufficient to promote the expression of downstream genes such as *NeuroD* and convert non-neural ectodermal cells into neurons (Ma *et al.*, 1996). *NeuroD* appears to be a critical regulator of the neuronal differentiation step and itself can promote the differentiation of ectopic neurons within the ectoderm when misexpressed (Fig. 4C; Lee *et al.*, 1995).

Similarly, in the developing mammalian nervous system, proneural bHLH factors appear to act in a cascade that reflects the progressive stages in the neuronal differentiation process. For example, in the developing neural tube, early proneural bHLH factors such as Ngn2 are expressed in subsets of proliferating neural precursors in the ventricular zone, while later acting bHLH factors, such as Ath3/NeuroM and NeuroD are expressed in differentiating neurons as they exit the cell cycle then migrate away from the ventricular zone toward the mantle layer and become post-mitotic neurons (Lee *et al.*, 1995; Roztocil *et al.*, 1997). In cranial sensory neurons, Ngn1 or Ngn2 is required for the expression of NeuroM and NeuroD, which are expressed in differentiating neurons (Fode *et al.*, 1998; Ma *et al.*, 1998).

Proneural bHLH genes are also required for the expression of genes that are involved in the differentiation of specific neuronal subtypes. For example, in sympathetic ganglia Mash1 regulates the expression of Phox2a, which is important for acquisition of a noradrenergic phenotype (Hirsch *et al.*, 1998; Lo *et al.*, 1998). Thus, in addition to regulating a core program of neuronal differentiation, proneural bHLH factors may also contribute to neuronal subtype decisions. This may be modulated through cooperation with region-specific patterning factors so that the same bHLH factor can regulate the development of distinct neuronal subtypes in different regions. In the developing forebrain, for example, Mash1 regulates the development of GABAergic neurons rather than noradrenergic neurons (Letinic *et al.*, 2002). As discussed below, differentiating neurons integrate multiple intrinsic and extrinsic signals to determine their ultimate phenotype.

REGULATION OF THE NUMBER OF NEURAL PROGENITORS—LATERAL INHIBITION

During vertebrate neurogenesis there is considerable spatial and temporal control over the differentiation of specific neuronal populations. Thus proneural bHLH factor activity must be constrained in some progenitors so that not all precursors differentiate simultaneously. The Notch signaling pathway plays an important role in regulating proneural bHLH factor activity and thus can control the pattern and timing of neurogenesis through a process known as lateral inhibition.

Study of invertebrates has given us much understanding of the molecular mechanisms of lateral inhibition, and these mechanisms are conserved in vertebrates. As described above, the selection of a neuroblast during *Drosophila* CNS development is governed by lateral inhibitory proteins that allow cells within an equivalence group to communicate with one another and essentially compete for the neuroblast fate. The core components of this pathway are the transmembrane Notch receptor and its transmembrane ligand Delta (Fig. 5). Activation of the Notch receptor by Delta initiates an intracellular signaling cascade that suppresses the neural fate within that cell (Artavanis-Tsakonas *et al.*, 1999). This signaling pathway begins with ligand-dependent cleavage of the Notch receptor and translocation of the intracellular domain of Notch to the nucleus. There it interacts with cofactors such as Suppressor of Hairless [Su(H)] and activates transcription of bHLH repressors such as Enhancer of Split proteins [E(Spl)]. These repressors in turn inhibit expression of proneural bHLH genes and prevent cells with active Notch

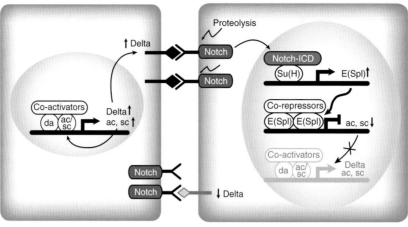

FIGURE 5. Lateral inhibition is mediated by Notch signaling between adjacent cells within a proneural cluster in the *Drosophila* neuroectoderm. Cells within the cluster express the proneural bHLH factors achaete (ac) and scute (sc), which dimerize with the bHLH partner daughterless (da), bind DNA, and regulate expression of the transmembrane ligand Delta. Delta activates the Notch receptor on adjacent cells, which initiates proteolysis of the Notch receptor and translocation of the intracellular domain (ICD) into the nucleus. Notch-ICD interacts with Suppressor of Hairless [Su(H)] and activates expression of Enhancer of Split [E(Spl)]. These repressors inhibit expression of the proneural bHLH factors causing suppression of the neuroblast fate within that cell. Loss of proneural gene expression also results in reduced Delta expression. Within a proneural cluster unknown mechanisms result in one cell (precursor cell) more strongly activating Notch signaling in neighboring cells. The neighboring cells downregulate ac/sc and Delta gene expression and ultimately differentiate into non-neural ectodermal cells. The selected precusor cell upregulates proneural gene expression through feedback autoregulation and becomes a neuroblast by Diana Lim.

signaling from adopting a neural fate. The expression of Delta in turn is positively controlled by proneural bHLH factors so that if proneural gene expression is inhibited by Notch signaling then Delta expression in that cell is also inhibited. The cell destined to become the neuroblast has slightly higher levels of Delta and thus activates Notch signaling more strongly in neighboring cells (Artavanis-Tsakonas *et al.*, 1990). The selected cell has reduced Notch signaling, upregulates proneural gene expression through feedback autoregulation and in turn maintains high levels of Delta expression (Heitzler *et al.*, 1996). The selected cell ultimately delaminates to become the neuroblast while the surrounding cells assume non-neural epidermal cell fates (Fig. 3). The process of lateral inhibition is fundamental to neural precursor selection throughout the developing nervous system.

Additional negative regulatory factors act outside of the proneural clusters to restrict proneural bHLH activity to only those cells within a cluster. These negative regulatory factors include bHLH factors that function as transcriptional repressors, such as Hairy (Van Doren *et al.*, 1991; Ohsako *et al.*, 1994), or HLH factors such as extramachrochaete (Emc) that lack a basic domain and antagonize proneural bHLH function by forming nonfunctional dimers and preventing DNA binding (Van Doren *et al.*, 1991). Elimination of these negative regulators results in ectopic neuroblast formation demonstrating that these negative regulators are important for constraining proneural bHLH activity to the proneural cluster.

Identical mechanisms have been shown to operate during vertebrate neurogenesis. For example, during primary neurogenesis in *Xenopus*, the proneural bHLH factor X-Ngn-R1 promotes Delta expression, which in turn activates the Notch receptor on adjacent cells (Ma *et al.*, 1996). Through the process of lateral inhibition, Notch signaling limits the number of cells that can activate expression of NeuroD and differentiate into neurons. Ectopic activation of the Notch signaling pathway inhibits primary neurogenesis, while interfering with Notch signaling results in expansion of the number of differentiating neurons within the normal domains of primary neurogenesis (Coffman *et al.*, 1993; Chitnis *et al.*, 1995).

Notch signaling is also important for regulating the timing of neurogenesis in the vertebrate nervous system. The components of the Notch signaling pathway in mammals are similar to *Drosophila*, with Notch receptor activation leading to upregulation of bHLH repressor genes called *Hairy/Enhancer of Split*-related genes or *Hes* genes (Davis and Turner, 2001). *Hes* genes in turn can repress the expression of proneural bHLH genes and prevent neurogenesis. *Hes1* and *Hes5* are expressed by progenitors in the VZ and mediate many effects of Notch in the developing nervous system (Kageyama and Ohtsuka, 1999). Disruption of *Hes1* causes premature neuronal differentiation (Lo *et al.*, 1998), while overexpression of *Hes1* can inhibit neurogenesis (Ishibashi *et al.*, 1994). Thus the *Hes* genes function as effectors of Notch activation and are important for limiting the number of neurons that differentiate at a given time.

REGULATION OF CELL NUMBER IN THE EARLY NERVOUS SYSTEM: MAINTENANCE OF A PROGENITOR POOL

Inhibition of Neuronal Differentiation

In order to generate appropriate numbers of neurons in the correct spatial and temporal patterns, it is critical to regulate progenitor cell number. This can be achieved by regulating the onset of differentiation, survival, and/or proliferation of progenitors. Stem cell and progenitor maintenance depends upon constraining the expression or function of proneural factors that act to promote neuronal differentiation. This is because proneural bHLH factors promote cell cycle exit of progenitors, which is an important step in the neuronal differentiation process. Overexpression of certain proneural bHLH factors in cell culture can promote neuronal differentiation and cell cycle exit (Farah et al., 2000). This may be achieved in part through upregulation of the cell cycle inhibitor p27^{Kip1}. Conversely, cortical progenitors isolated from *ngn2/mash1* mutant mice can proliferate much more extensively in culture than wild type progenitors, suggesting that these bHLH factors normally limit progenitor proliferation (Nieto et al., 2001).

Negative regulators that constrain bHLH factor expression or function are important regulators of the size of the progenitor pool since they act to prevent neuronal differentiation and cell cycle exit. In addition to coordinating the timing and pattern of neuronal differentiation, Notch signaling is also important for maintaining a population of proliferating progenitors within the VZ of the developing vertebrate neural tube. In many parts of the developing CNS, distinct neuronal subpopulations are born in the same region but at different times in development. As neurons begin to differentiate they activate Notch signaling in their neighbors, inhibit proneural gene expression or function, and thus prevent these neighboring cells from differentiating at the same time. If all progenitors were to differentiate early then the progenitor population would be depleted and later-born cell types would fail to be generated. In the developing vertebrate retina, interfering with Notch signaling by expressing a dominant negative form of the ligand Delta causes cells to preferentially adopt early-born cell fates at the expense of later-born populations (Dorsky et al., 1997).

A second class of negative regulators, Id proteins, can also inhibit the function of vertebrate bHLH factors and thus prevent neuronal differentiation. The *Id* genes encode HLH factors that, like Emc in *Drosophila*, lack a basic domain and antagonize proneural bHLH function by forming nonfunctional dimers with the partner E proteins, thus preventing DNA binding. *Ids* are expressed in the VZ of the developing neural tube and are important for promoting progenitor proliferation and preventing the onset of neurogenesis. For example, neural progenitors from mice mutant for both *Id1* and *Id3* show premature neuronal differentiation and cell cycle exit (Lyden et al., 1999). Thus Ids prevent neuronal differentiation by inhibiting proneural bHLH factor function.

Regulation of Cell Death and Proliferation

Another mechanism for regulating the size of the progenitor pool in the developing nervous system is regulation of progenitor survival. Although it has long been recognized that apoptosis is an important component of nervous system development, it was generally believed that the majority of deaths in the nervous system occurred in post-mitotic neurons as they compete for limiting amounts of trophic support from target tissue (see Chapter 11). More recently however, it has become clear that large numbers of progenitors normally die early in development, and that this is essential for regulating morphogenesis and cell number in the nervous system. This was revealed by generating mutant mice deficient for critical cell death regulators such as caspase 3, caspase 9, or Apaf1 (see Chapter 11). These mice all showed dramatic reductions in cell death in the early nervous system that resulted in severe malformations of the embryonic brain including protrusions and exencephaly of the forebrain, ventricular obstruction due to tissue hyperplasia, ectopic neural masses, and early lethality (Kuida et al., 1996, 1998; Yoshida et al., 1998). Thus, normal regulation of progenitor survival is a critical factor regulating the size of the progenitor pool during early development.

Proliferation in the early nervous system is also precisely regulated and is critical for controlling progenitor cell number. Proliferation and cell cycle exit are also intimately related to histogenesis and the neuronal cell-fate decision. Neural stem cells and progenitors respond to certain extrinsic factors with an increase in mitotic activity. For example, early neural stem cells are dependent upon FGF or EGF to proliferate and expand (Rao, 1999), while precursor cells in the cerebellum proliferate in response to Sonic hedgehog (Dahmane and Ruiz-i-Altaba, 1999; Wallace, 1999; Wechsler-Reya and Scott, 1999). Proliferation in all cell types depends upon the core cell cycle machinery, including cyclins, cyclin-dependent kinases (CDKs), CDK inhibitors, and Rb family proteins. However, it is now appreciated that these are large protein families and that different family members may play specialized roles in different tissues during development. For example, Cyclin D1 is the principal D-type cyclin regulating the transition to S-phase in the developing retina. In mice mutant for Cyclin D1, retinal progenitors show reduced proliferation (Sicinski et al., 1995). Conversely, CDK inhibitors such as p27^{Kip1} or p57^{Kip2} are expressed in retinal progenitors, and when these genes are mutated, retinal progenitors divide an extra round or two before exiting the cell cycle (Dyer and Cepko, 2000, 2001; Levine et al., 2000). In mice deficient for the retinoblastoma protein Rb, progenitor proliferation in the CNS is profoundly deregulated, resulting in excess dividing cells localized to normally post-mitotic regions (Dyer and Cepko, 2000, 2001; Levine et al., 2000). The extra cells that are generated in both these cases die by apoptosis, illustrating that cell number is regulated by the tight balance between proliferation and survival.

Asymmetric vs Symmetric Cell Division

Progenitor cell number is also dependent upon the ratio of asymmetric to symmetric cell divisions (Lu et al., 2000). At early stages of development cells have been observed to undergo symmetric divisions, that is, stem cells divide perpendicular to the ventricular surface generating two daughters that both remain in contact with the ventricular surface and continue to proliferate (Chenn and McConnell, 1995). As development progresses this mode of cell division becomes less common, and the plane of cell division is more often horizontal to the ventricular surface, generating daughters that are fundamentally different from each other. One daughter remains in contact with the ventricular surface and will continue to divide. The other daughter loses contact with the ventricular surface, will exit the cell cycle, and differentiate into a post-mitotic neuron that migrates away from the VZ (Chenn and McConnell, 1995). A neural progenitor can undergo repeated asymmetric divisions generating post-mitotic daughter neurons over a prolonged developmental period. Since neural progenitors have a limited capacity for self-renewal, the progenitor will ultimately undergo a final division, which can be asymmetric, generating two nonidentical, post-mitotic daughters.

The molecular basis for asymmetric cell division was first described in *Drosophila*, where it was shown that cell-fate determinants such as Numb and Prospero function as key components in this process. During asymmetric division in *Drosophila*, Numb and Prospero proteins are localized in a crescent to one half of a dividing cell and are then asymmetrically inherited, generating two nonequivalent daughters (Jan and Jan, 2001). For example, neuroblasts in the *Drosophila* CNS undergo a series of asymmetric divisions, in each case generating another neuroblast and a GMC (see above). As the neuroblast divides, Numb and Prospero become localized to one half of the cell and are inherited by the GMC (Hirata *et al.*, 1995; Knoblich *et al.*, 1995; Spana and Doe, 1995). The GMC in turn can divide asymmetrically producing two post-mitotic daughters that acquire distinct neuronal or glial fates. Loss of Numb results in both daughters adopting identical fates. In *Drosophila*, Numb acts in part by antagonizing the activity of Notch, which is also required for generating two nonidentical daughters (Frise *et al.*, 1996; Spana and Doe, 1996). Prospero is a homeodomain transcription factor that regulates the fate of the cell that inherits it. The localization of these determinants is regulated by a complex signaling pathway that controls the polarity of the dividing cell and the plane of cell division.

Vertebrate homologs of the Numb protein have been identified, and vertebrate Numb proteins can also be asymmetrically localized during cell division in the vertebrate CNS (Zhong *et al.*, 1996). Multiple Numb family members exist and may serve diverse functions; however, there is a clear requirement for these proteins in progenitor maintenance. Mice deficient for both vertebrate *numb* and *numb-like* exhibit a premature depletion of neural progenitors and early overproduction of neurons (Petersen *et al.*, 2002). These excess early-born neurons eventually die, once again demonstrating that cell number is tightly regulated. In vertebrates, the relationship between Numb and Notch remains to be fully defined.

In the preceding sections, we have shown how neuronal progenitors are specified and their numbers are regulated. Generating the correct number of progenitors is an important step in assembling the ultimate structure of the nervous system. Next, we will turn to the question of neuronal cell fate and examine how a single progenitor can give rise to multiple types of neurons.

CELL-FATE SPECIFICATION—INTRINSIC AND EXTRINSIC CUES

As a cell exits the cell cycle and becomes committed to becoming a neuron, it must also decide what type of neuron it is going to be. Although many neurons express the same genes early in their development, at some point they diverge and begin to express unique genes and proteins required for their ultimate fate. An individual neuron must express the correct neurotransmitters, receptors, and intracellular signaling molecules, and make the proper axonal and dendritic connections to other cells. All of these aspects of cellular phenotype require regulated gene expression that must be acquired over a relatively short developmental time. Previously in this chapter, we have shown that the timing of progenitor differentiation has a great influence on cell fate. A major unresolved question in the field of neurogenesis is whether the general neurogenic program and specific fate specification happen simultaneously, or as two successive steps. Evidence for both possibilities exists, and ultimately it may be more informative to explore the mechanisms by which fate specification occurs.

For many years, there have been two models for the specification of cell fate—intrinsic and extrinsic. In the intrinsic model, a cell's lineage is most important. When a progenitor cell divides, its daughters inherit "determinants" consisting of mRNA or proteins that result in a specific developmental program. These determinants could be divided asymmetrically, producing different fates from a single progenitor. Extrinsic specification instead depends on the environment, primarily through secreted or cell surface molecules. In this model, the time and place of differentiation play a greater role in cell fate than its parental lineage. Ultimately, the line between intrinsic and extrinsic specification becomes blurred, because extracellular signals can cause changes in a progenitor cell that are then passed on to its daughters. Whatever the mechanism, it is clear that all neuronal precursors begin with many possible fates and are progressively limited in potential until they differentiate.

In the following sections, we will give several examples of neuronal fate specification in different model systems, illustrating how both intrinsic and extrinsic factors contribute to cell fate. We provide examples from both vertebrates and *Drosophila*, but in each case focus on the system where the molecular factors that are required for fate specification are best understood. The examples presented here do not necessarily represent the extreme possibilities—completely intrinsic or extrinsic mechanisms. Each system seems to use a mechanism that is best suited for the final organization of its nervous system, taking into account the needs for control of precision in cell number, position, and plasticity. Importantly, all these systems use a similar hierarchy of gene expression to produce an ultimate phenotype, illustrating how a common developmental program has been adapted throughout evolution to produce specialized components of the nervous system.

MECHANISMS FOR CNS NEURONAL FATE SPECIFICATION—EXTRINSIC AND INTRINSIC CONTROL

Vertebrate Spinal Cord

The huge number of neurons generated in vertebrate nervous systems necessitates a strong role for extracellular signals in specification of neural precursor cells. During vertebrate spinal cord development, much of the positional information that goes into the process of cell-fate specification comes from environmental signals produced by surrounding tissues. As we have mentioned previously, the neural plate already has rostrocaudal and dorsoventral polarity by the time neurogenesis begins (see Chapter 3). For example, rostrocaudal identity is encoded in the CNS by the overlapping expression of Hox proteins, as a result of early patterning molecules. In addition, the secreted molecules BMP and Hedgehog, respectively, promote dorsal and ventral identity in the developing CNS at neural plate and neural tube stages (Fig. 6). These molecules appear to act as morphogens, such that cells respond differently to increasing concentrations in their environment (Liem et al., 1995; Roelink et al., 1995). Therefore, any given cell can "sense" its DV position based on relative levels of BMP and Hedgehog signaling. Importantly, cells that occupy a particular position in the neural tube, but have not yet begun to express region-specific genes, can be respecified by exposure to ectopic environmental signals.

In response to morphogen signals, region-specific transcription factors are expressed in subsets of spinal cord progenitors. Individual homeodomain and bHLH-class transcription factors are expressed in dividing cells at different DV positions, induced by BMP and Hedgehog in a dose-dependent manner (Fig. 6). In the ventral spinal cord, these genes can be divided into two classes—those that are repressed by Hedgehog and those that are activated (Briscoe et al., 2000). Pairs of genes comprising a member of each class of Hh-responsive factors set up mutually exclusive domains of expression by repressing each other's expression. Once each cell in the spinal cord expresses a set of region-specific transcription factors, it then exits the cell cycle and begins to express a new set of factors that control differentiation and ultimate fate (Fig. 6). One example of this process can be seen in the expression of the Mnx class of homeodomain factors in spinal motoneurons. The two homeodomain factors HB9 and MNR2 have been shown to be necessary and sufficient for motoneuron differentiation and are themselves directly regulated by Shh and region-specific homeodomain factor expression (Tanabe et al., 1998; Thaler et al., 1999). As they begin to differentiate, neurons express a complete program of cell type-specific factors that will be discussed below.

Precision in Neuronal Fate Specification

Thus, in the spinal cord a cell's position and exposure to environmental factors leads to the expression of a cascade of transcription factors that results in ultimate fate. How universal is this mechanism to the process of neurogenesis in all animals? The large number of neurons in the vertebrate CNS allows for a high degree of plasticity. Such a system is inherently "sloppy," but is also more adaptable—if a cell is incorrectly specified, the nervous system can still function. However, we have also learned much about different mechanisms to specify neural cell fate from the study of invertebrate models. In invertebrates, precise numbers of neurons must be generated in order to ensure proper connectivity and function.

In most invertebrate nervous systems, a regular array of neurons is generated during neurogenesis, each of which can be identified by position and morphology. However, even when precise organization is required, *Drosophila* has shown us that multiple mechanisms can be used to generate defined numbers of neuronal cell fates. Following are two examples from

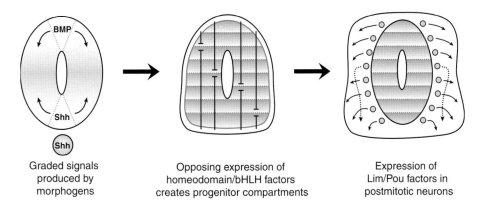

Graded signals produced by morphogens

Opposing expression of homeodomain/bHLH factors creates progenitor compartments

Expression of Lim/Pou factors in postmitotic neurons

FIGURE 6. In the vertebrate spinal cord, environmental signals are translated into discrete zones of transcription factor expression that produce distinct neuronal cell types. Gradients of BMP (dorsal) and Shh (ventral) signals give each position in the spinal cord a unique dorsal/ventral identity. This identity results in the expression of particular members of homeodomain and bHLH factors, which repress each others' expression. This mutual repression creates "compartments" of progenitor cells that will produce distinct neuronal types. As neurons are born, they express type-specific transcription factors from the Lim and Pou families, which in turn regulate their differentiation. Figure generated by Diana Lim.

Drosophila, illustrating how an intrinsic timing mechanism and lineage-independent local signals can both produce predictable numbers of individual cell types.

Drosophila CNS Neuroblasts

As described previously, individual *Drosophila* neuroblasts arise from the ectoderm as a result of proneural and lateral inhibitory gene function and have a distinct positional identity based upon AP and DV patterning information (Fig. 3). Once a neuroblast identity has been specified, it divides to produce a series of GMCs and each successive GMC generates different progeny. If an individual GMC is ablated, its fate is skipped entirely and the next GMC goes on to produce daughters appropriate for its time of generation (Doe and Smouse, 1990). This therefore represents an intrinsic mechanism of fate specification. Each GMC knows its identity internally and does not depend on outside signals to learn its fate. Possible mechanisms for this type of specification include asymmetric distribution of determinants inside the cell during division, or the molecular counting of cell cycles.

There is a distinct order of transcription factors expressed in successive GMCs. In order, Hunchback, Krüppel, Pdm, and Castor are expressed first in the neuroblast, then in the subsequently generated GMC (Fig. 7). These factors appear to be necessary and sufficient in the GMCs that express them for the correct progeny to be generated (Isshiki *et al.*, 2001). Interestingly, they convey a "temporal identity" on the GMC, instead of an absolute fate. As mentioned previously, neuroblasts in different positions generate different progeny, yet all their respective GMCs require these factors to produce neurons and glia appropriate for their lineage. In other words, Hunchback instructs a GMC to produce the primary fate for its position, whether that is a motoneuron or interneuron.

The *Drosophila* CNS is composed of relatively few neurons, and each makes a unique and specific connection with other neurons and muscles. Such an organization requires a high degree of precision to avoid the most serious potential problem— a missing neuron. Thus, although the initial pattern of neuroblast formation and specification is induced by environmental signals, the subsequent lineage-based system ensures that the correct number and type of each cell is produced. When a progenitor controls the fate of each of its progeny, high precision is possible.

Drosophila Retina

When many progenitor cells have the ability to produce neurons, clonally restricted lineage-dependent mechanisms are not required to generate defined numbers of mature cell types. An example of this is the *Drosophila* retina, often referred to as a "crystalline array" of ommatidia, the individual light-sensing units. Such a description is particularly illustrative of the process used to specify cell fate in this tissue. An initially uniform epithelial sheet must be patterned into a repeating array of differentiated cells, including eight photoreceptors and 12 accessory cells per ommatidium. In this case, the most important consideration is a cell's fate relative to its neighbors, rather than the presence or absence of a single cell. If one photoreceptor is missing, the fly can still see; however, if the array of ommatidia is disorganized, it cannot properly process visual information.

Differentiation proceeds across the eye imaginal disc as a wave, called the morphogenetic furrow. As this furrow moves across the disc from posterior to anterior, proneural gene activity results in a patterned array of the first photoreceptor to differentiate, R8 (Jarman *et al.*, 1994). The *atonal* gene is used to specify R8 cells that are spaced apart at a proper distance through lateral inhibition by Notch/Delta signaling. These R8 cells then recruit the entire ommatidium from their neighbors, through cell–cell interactions (Fig. 8). This is a lineage independent mechanism, and it is impossible to predict which progenitor will become which photoreceptor or accessory cell before they undergo specification.

General photoreceptor specification requires a common pathway, regardless of photoreceptor cell type. Extracellular factors from the EGF family signal through tyrosine kinase receptors to the intracellular Ras-MAPK pathway, which drives the expression of transcription factors that regulate differentiation. Elimination of any part of this pathway leads to a gain of accessory cells at the expense of photoreceptors. Thus, the process of general photoreceptor differentiation, but not fate specification of R1-8, is controlled by local EGF signaling.

Once the general photoreceptor pathway is activated, local signals from differentiated cells then drive the specification of cell fate in neighboring progenitors. Photoreceptors are recruited in an invariant order—R8, then R2/5, then R3/4, then R1/6, then R7 (Fig. 8). The outer photoreceptors, R2-6, form in pairwise fashion on either side of the R8 cell. Each successive pair of photoreceptors requires specific transcription factors for its

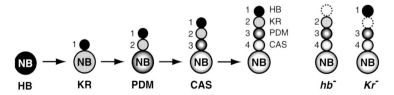

FIGURE 7. *Drosophila* CNS neurons are specified by a temporal progression of transcription factor expression. Neuroblasts express the transcription factors Hb, Kr, Pdm, and Cas at successively later timepoints during development. The neuronal progeny of these neuroblasts maintain expression of the factor that was expressed in the neuroblast when they were born. While the factors Hb and Kr are necessary and sufficient for the fates that express them, in different regions of the CNS these transcription factors drive different fates. (Modified from Isshiki *et al.*, 2001, with permission from Elsevier.)

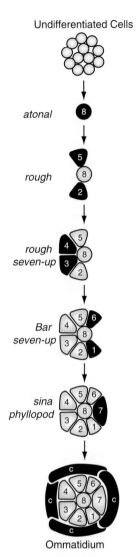

Undifferentiated Cells

atonal

rough

rough
seven-up

Bar
seven-up

sina
phyllopod

Ommatidium

FIGURE 8. *Drosophila* ommatidial cells are recruited in a lineage-independent manner from surrounding neuroepithelium. Newly recruited cells are depicted in black. The first photoreceptor to differentiate is R8, followed in order by R2/5, R3/4, R1/6, R7, and cone cells. Genes expressed in the photoreceptors at each step are listed on the left. These genes are required for the generation of the photoreceptors in which they are expressed. Figure generated by Diana Lim.

specification. R2/5 express and require the *rough* gene, and then signal with R8 to R3/4 which requires both *rough* and *seven-up*. R1/6, the last outer photoreceptors to form, require the *seven-up* and *BarI* genes. Cell contact is required for these factors to be induced at the correct time and place, allowing one cell to control the specification of the next.

The best studied cell induction in the fly eye is formation of R7. This cell requires a combination of signals from its neighbors, which result in the expression of the correct complement of transcription factors. The EGF-Ras-MAPK pathway is activated by the ligand Boss which is expressed by R8 and activates the receptor Sevenless. The Sevenless pathway activates the ETS domain factors Pnt and AP-1 and inhibits the factor Yan,

promoting general photoreceptor differentiation. Notch signaling from the neighboring R1/6 cells also plays a role in R7 specification so that Ras alone specifies the R1/6 fate, but high Ras with Notch specifies the R7 fate (Tomlinson and Struhl, 2001). Inside the R7 cell, the *lozenge* gene inhibits the expression of *seven-up*, thus preventing R1/6 differentiation. Conversely, signals from R1/6 and R8 activate genes that are required for R7 differentiation, such as *phyllopod* and *sevenless-in-absentia* (Daga *et al.*, 1996). In the fly eye, cell-fate specification is therefore controlled by the time and place of differentiation. Signals from neighboring cells regulate both general and cell type-specific gene expression. Thus, local cues result in reproducible, highly organized pattern.

PLASTICITY IN FATE—VERTEBRATE CNS NEURONS

When does a neuron become irreversibly committed to a particular phenotype? One would suspect that this step takes place upon the expression of cell type-specific genes, or axon outgrowth. In fact, neurons in different organisms develop with different degrees of plasticity. In some systems, cells cannot be respecified after they leave the cell cycle. In other cases, neuronal phenotype can be respecified until a cell begins its terminal differentiation. Here we will give examples of both cases.

Cerebral Cortex—Plasticity Until Final Cell Cycle

In the mammalian cerebral cortex, control of the cell cycle appears to correspond with cells' ability to be respecified. The environment plays a key role in determining how cells know where to migrate as development progresses, and this process is dependent on the state of the cell cycle. As mentioned previously, when younger cells are transplanted into an older cortex, a subset migrates into superficial layers, appropriate for the host age (McConnell, 1988). These early cells are therefore plastic and can be influenced by their environment to adopt new fates. In the converse experiment older cells do not migrate to deeper layers when transplanted into younger animals. Thus cortical plasticity is restricted over time, with older cells becoming limited to a small number of potential fates. However, further studies have shown that the plasticity of younger progenitor cells is itself limited. While the population as a whole shows evidence of respecification in an older environment, careful analysis of single cells has uncovered diverse responses to local signals.

Labeling of cells with tritiated thymidine shows that young progenitors that have yet to go through their final S-phase adopt the fates of their older hosts and migrate into superficial layers (McConnell and Kaznowski, 1991). However, cells that have completed their final S-phase remain committed to "younger" fates and migrate to deep layers even in older hosts. Therefore, sometime after a progenitor's final S-phase, it becomes irreversibly committed to the fate promoted by its local environment.

Zebrafish Spinal Cord—Plasticity Until Axonogenesis

In the zebrafish spinal cord, cell fate commitment appears to be coupled to terminal differentiation. Environmental cues during neural tube formation initially specify these cell fates, as described in the section above. In this system, 3–4 primary motoneurons form per spinal segment, and each has a stereotypical axon trajectory and target innervation. Additionally, each primary motoneuron expresses a unique subset of LIM-homeodomain transcription factors, whose function in cell differentiation will be discussed in the following section. However, experimental manipulations have shown that motoneuron identity is not fixed until the cells begin to put out axons.

If a zebrafish primary motor neuron is transplanted to a new location before axon outgrowth, it is respecified to express LIM genes appropriate for its new position (Appel *et al.*, 1995). Additionally, the axon projection of the transplanted cell follows a pathway equivalent to other neurons in the same location (Fig. 9). However, once the axon begins to grow, transplanted cells retain their original LIM gene expression and axon projection. For these cells, therefore, axonogenesis is the time when cells are irreversibly committed to a fate. From a developmental perspective, this timing makes sense because axon growth cones must express molecules on their surface to enable proper pathfinding. Once a cell switches fate, these molecules would have to be recycled and new ones expressed to allow for a new trajectory. Because all the primary motoneurons use the same neurotransmitters and function in similar circuits, gene expression before axonogenesis may be very similar between different cells and thus plasticity is possible.

Whenever extracellular signals play a role in cell-fate specification, one can measure the timing of commitment to a particular phenotype by challenging them with a new environment. By performing the above experiment *in vivo*, the researchers were able to determine the exact point at which signals in the embryo tell primary motoneurons which fate to produce. This could also be defined as the point at which extrinsic specification stops and intrinsic specification takes over, at least for some aspects of motoneuron phenotype. As we will see in a following section, other neuronal characteristics may still be plastic at this point and are regulated by target innervation. In all model systems described, this switch from extrinsic to intrinsic control happens at a slightly different point—but it happens nonetheless.

NEURONAL MATURATION

Once neurons have decided to exit the cell cycle and their fate has been specified, they undergo a process of maturation, which ultimately results in their final phenotype. As with every other event we have discussed so far, this process is controlled by gene expression. The complement of transcription factors expressed by a neural precursor cell as it differentiates will control its production of neurotransmitters and their receptors, axon guidance molecules that will regulate target innervation, and trophic dependence. The expression of these factors is a direct result of the specification process outlined in the previous section—the spatial and temporal history of each cell contributes to a "code" of transcription factors for each neuronal type that directly promotes all the above characteristics. We will give several examples of how these genes can ultimately regulate neuronal function by affecting maturation.

POU Genes Control Sensory Neurogenesis

Once they have been specified, there appears to be a conserved program of gene expression in all animal sensory neurons. Genes encoding transcription factors of the POU-homeodomain family are expressed in sensory neurons from worms to mammals. Functional analysis of these genes has demonstrated that they are necessary and sufficient to regulate sensory neurogenesis in both the CNS and PNS. In mouse, the three POU domain genes *Brn-3.0*, *Brn-3.1*, and *Brn-3.2* are expressed in and control

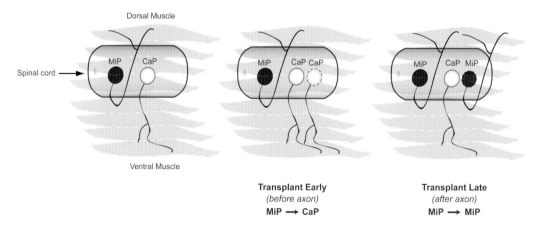

FIGURE 9. Some neurons exhibit plasticity in new environments after they are born. In the zebrafish spinal cord, the MiP primary motoneuron normally expresses Isl1 and projects dorsally, while the CaP motoneuron normally expresses Isl2 and projects ventrally. When MiP is transplanted to the CaP position before axonogenesis, it adopts a CaP phenotype. After axonogenesis, the MiP fate is fixed even when transplanted. Figure generated by Diana Lim.

the terminal differentiation of overlapping populations of sensory neuron populations. One of the clearest demonstrations of this role is in the retina, where deletion of *Brn-3.2* causes the loss of most retinal ganglion cells (Erkman *et al.*, 1996). In contrast, deletion of *Brn-3.1* results in a failure of inner ear hair cells to differentiate, leading to deafness. Simultaneous deletion of *Brn-3.1* and *Brn-3.2* results in additional losses of sensory neurons, indicating that these genes play redundant roles in some populations (Wang *et al.*, 2002).

What aspects of differentiation do POU-homeodomain factors regulate? Based on the phenotypes of knockouts, they act near the top of a hierarchy of gene expression that controls sensory axon formation and pathfinding. Cells in which *Brn-3* genes have been disrupted undergo cell death, rather than adopting inappropriate fates, perhaps due to a lack of trophic support from target tissues (Gan *et al.*, 1999). These cells appear to begin the differentiation process before they die, indicating that they are initially specified as neurons. Retinal ganglion cells lacking *Brn-3.2* are able to migrate to the ganglion cell layer and initially extend processes that are more characteristic of dendrites than of axons. Downstream of *Brn-3.2*, target genes include members of the LIM-homeodomain family, which in turn can regulate neuronal subtype specificity, as discussed below (Erkman *et al.*, 2000). These data indicate that *Brn-3.2* regulates aspects of a "projection neuron phenotype" including axon/dendrite polarity and axon guidance. While POU genes are required for terminal differentiation of sensory neurons, it is not currently clear whether there is a "code" of POU gene expression that defines each sensory subtype.

LIM Genes Regulate Subtype Specificity in CNS Motoneurons

Multiple members of the LIM homeodomain family of transcription factors are expressed in differentiating neurons of the vertebrate spinal cord. The first of these factors begins to be expressed as cells complete their final cell cycle, and others are only expressed after the final division. As mentioned in the previous section, in zebrafish primary motoneurons the expression of particular LIM factors corresponds with their axon projections. In the mouse and chick, it was also discovered that anatomically distinct pools of motoneurons express different members of this gene family, suggesting that they might contribute in some way to cellular diversity. Several of these genes are expressed in overlapping subpopulations of motoneurons: *isl1, isl2, lim1, lhx3*, and *lhx4*. One obvious way to test the role of these factors in regulating neuronal differentiation was to modulate their expression *in vivo* and examine the resulting effects on neurogenesis.

Multiple LIM genes have been knocked out in mouse, with very predictable effects on motoneuron development. *isl1* is first expressed by all motoneurons, suggesting that it activates a general program of motoneuron differentiation. When *isl1* function is removed, all motoneurons in the spinal cord are absent (Pfaff *et al.*, 1996), and the precursor cells appear to undergo programmed cell death. In contrast, *lhx3* and *lhx4* are expressed transiently in a subset of motoneurons with ventral projections (Fig. 10). When these two genes are simultaneously knocked out, motoneurons still develop, but ventrally projecting neurons are lost and appear to be converted into dorsally projecting cells (Sharma *et al.*, 1998). Therefore, some LIM factors may be generally required for motoneuron characteristics, while others control specific aspects of cell phenotype such as axon projection and target selection.

These same functions of LIM genes are mainly conserved in the insect nervous system as well, suggesting a common evolutionary history of neurogenesis pathways (Fig. 10). *Drosophila* CNS neurons express LIM homeodomain factors, which act to specify axon trajectories and neurotransmitter expression. In the fly, *isl*, the homologue of vertebrate *isl1* and *isl2*, is expressed by a subset of neurons in the ventral nerve cord including motoneurons. In contrast to the vertebrate spinal cord, *Drosophila* CNS neurons can still differentiate in the absence of *isl* expression, but they make errors in pathfinding and neurotransmitter expression (Thor and Thomas, 1997). This phenotype is more reminiscent of the *lhx3/4* knockout in mouse, suggesting that *isl* controls the final functional characteristics of *Drosophila* CNS neurons. In support of this role, *lim3*, the homologue of

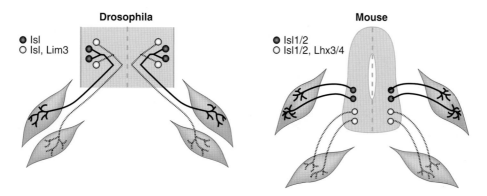

FIGURE 10. Similar LIM codes are used in fly and vertebrate motoneurons. In *Drosophila*, subsets of embryonic CNS motoneurons that express either Isl or Isl and Lim3 project to different muscles. In the vertebrate spinal cord, subsets of motoneurons express the homologues Isl1/2 or Isl1/2 and Lhx3/4 as well. As in the fly, these neurons project to different muscles depending on the combination of LIM factors they express. Figure generated by Diana Lim.

vertebrate *lhx3* and *lhx4*, is expressed in an subset of *isl*-expressing motoneurons (Fig. 10). When *lim3* expression is modified, neurons predictably adopt phenotypes characteristic of their new gene expression profile (Thor *et al.*, 1999). Therefore, the combination of LIM factors expressed by a neuron gives it a unique identity that allows the proper neural connections to be made.

ETS Genes Regulate Target Specificity

The process of neuronal differentiation is not complete by the time cells send out axons and connect to their final targets. During axon pathfinding, gene expression is carefully regulated to allow growth cones to appropriately respond to local environmental cues (see Chapter 9). Even if neurons make connections, retrograde signaling from their targets can affect aspects of phenotype such as neurotransmitter expression, synaptic maturation, and cell body migration. A classic example of this regulation by targets takes place during sympathetic innervation of sweat glands, during which the neurons switch their neurotransmitter from noradrenalin to acetylcholine. It has been demonstrated that this switch is directly promoted by the target tissue, because when these neurons are forced to innervate other targets, they maintain their adrenergic phenotype. While the molecular nature of this signal has not been identified, other systems have begun to give insight into mechanisms of retrograde signaling from targets.

A molecular pathway of retrograde signaling has been observed in developing spinal motoneuron circuits. Motoneurons that innervate different targets can be subdivided into electrically coupled "pools" with common anatomical localization, gene expression, and target arborization properties. Along with common expression of the homeodomain proteins discussed previously, motoneuron pools also express the same members of the ETS family of transcription factors. One of these factors, Pea3, has been shown to be necessary for the proper axonal arborization of a subset of motoneurons in their target muscles, as well as the final migratory position of the motoneuron cell bodies (Fig. 11) (Livet *et al.*, 2002). It may appear that ETS factors act in the same way as LIM factors in controlling differentiation.

However, ETS expression is in fact *regulated* by motoneuron targets. When expression of the trophic factor glial-derived neurotrophic factor (GDNF) in the muscle is disrupted, motoneurons fail to express Pea3 and differentiate incorrectly (Fig. 11; Haase *et al.*, 2002). Therefore, motoneurons that are genetically programmed to reach the same target subsequently receive a retrograde signal from the target that enhances their functional connectivity. Furthermore, Pea3 and a related gene ER81 have also been shown to function in proprioceptive sensory neurons that innervate muscle and connect to motoneurons in stretch reflex circuits. Because these factors regulate the connection of sensory neuron to their central targets, common expression of ETS genes in sensory and motor pools may define functional units that can be defined anatomically. What downstream targets of these factors might affect cellular connectivity? One candidate is the cadherin family of homotypic cell adhesion molecules, which are also coexpressed by common neuron pools and could regulate sorting of cell soma and axon arbors.

NEUROGENESIS IN THE ADULT VERTEBRATE NERVOUS SYSTEM

The majority of neurons in the vertebrate nervous system are generated during development through the mechanisms described above. However, there are examples of neurogenesis continuing beyond the initial embryonic period. In lower vertebrates, such as fish and amphibians, new neurons are generated during early stages of life as the animals grow. As mentioned previously, after the initial period of embryonic retinal development the retina grows in fish and amphibians by adding cells to the margins at the CMZ (Fig. 2; Straznicky and Gaze, 1971), and these new retinal cells integrate into the layers of the retina. A similar population of proliferative cells has been identified in the postnatal chick retina, but this population is lost in the adult (Fischer and Reh, 2000). In adult songbirds such as canaries, there is a seasonal replacement of neurons in specific brain nuclei involved in birdsong, in particular the HVC nucleus (Alvarez-Buylla and Kirn, 1997). New neurons are added to this

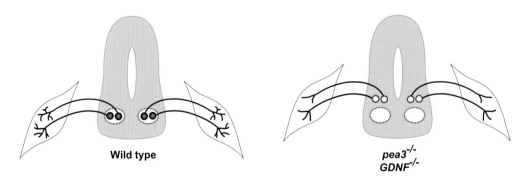

FIGURE 11. ETS factors play a role in neuronal differentiation following target innervation. Normally, a subset of vertebrate spinal motoneurons express the ETS factor Pea3 following limb muscle innervation. When Pea3 expression is lost, motoneuron cell bodies do not migrate to the correct final location, and the axons do not make appropriate synapses. The expression of Pea3 must in some part be driven by the target tissue, because loss of GDNF, expressed in the muscle, produces the same phenotype as loss of Pea3 in the neurons. Figure generated by Diana Lim.

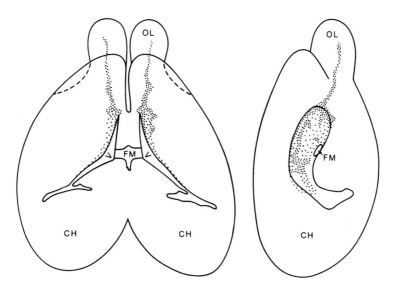

FIGURE 12. The subventricular zone (SVZ) in the telencephalon of the adult mouse. Dorsal view on the left; right lateral view on the right. The extent of the SVZ is shown by stippling and denser stippling shows where it is thickest. CH, cerebral hemisphere; FM, foramen of Monro; V, lateral ventricle; OL, olfactory lobe. From I. Smart, 1961, *J. Comp. Neurol.* 116:325–347.

nucleus in the spring, when birds learn a new song. These new neurons are born in the SVZ of the telencephalon and then migrate to populate the HVC nucleus.

Neurogenesis is much more restricted in the adult mammalian brain; however, there is now evidence that significant numbers of proliferating neural stem cells or progenitors exist within specific regions of the adult mammalian brain, and that these cells can give rise to new neurons. These regions are the subventricular zone, which lines the lateral ventricles (see Fig. 12) and the subgranular zone (SGZ) of the dentate gyrus (Taupin and Gage, 2002). SVZ progenitors give rise to interneurons that migrate along the rostral migratory stream to populate the olfactory bulb, while SGZ progenitors generate new granule cells that populate the dentate gyrus of the hippocampus. Specific stimuli can enhance neurogenesis in these brain regions. For example, exercise can stimulate the production of new neurons in the dentate gyrus (van Praag *et al.*, 1999). In addition, new neurons can be generated in response to injurious events, such as ischemia or seizure, potentially contributing to a repair response in the adult brain (see Chapter 12). The molecular mechanisms underlying differentiation of adult mammalian progenitors is only partially understood, but presumably there are fundamental similarities to neurogenesis during embryonic development.

SUMMARY

In this chapter, we have outlined the general steps of neurogenesis, and the mechanisms by which these steps occur. For over a hundred years, neurobiologists have described this process by making careful observations of the cellular events that occur as neuroepithelial cells mature into neurons. These observations

provided much of the groundwork for later studies and correctly predicted many of the mechanisms that were subsequently discovered. Recently, advances in molecular techniques have allowed us to understand how genetic and biochemical events control the progression from stem cells into differentiated neurons. At this point, we can observe a single neuron and know many of the genes that regulate every step of its differentiation. One important next step is to recapitulate this process *in vitro* to see if we can direct the differentiation of stem cells or progenitors in a carefully controlled environment. If successful, this work has the potential to provide therapy for human nervous system injury and disease.

REFERENCES

Akam, M., 1987, The molecular basis for metameric pattern in the *Drosophila* embryo, *Development* 101:1–22.

Alvarez-Buylla, A. and Kirn, J.R., 1997, Birth, migration, incorporation, and death of vocal control neurons in adult songbirds, *J. Neurobiol.* 33: 585–601.

Anderson, S.A., Eisenstat, D.D., Shi, L., and Rubenstein, J.L., 1997, Interneuron migration from basal forebrain to neocortex: Dependence on *dlx* genes, *Science* 278:474–476.

Angevine, J.B., 1965, Time of neuron origin in the hippocampal region: An autoradiographic study in the mouse, *Exp. Neurol. Suppl.* 2:1–70.

Angevine, J.B., Jr. and Sidman, R.L., 1961, Autoradiographic study of cell migration during histogenesis of cerebral cortex in the mouse, *Nature* 192:766–768.

Appel, B., Korzh, V., Glasgow, E., Thor, S., Edlund, T., Dawid, I.B., and Eisen, J.S., 1995, Motoneuron fate specification revealed by patterned LIM homeobox gene expression in embryonic zebrafish, *Development* 121:4117–4125.

Artavanis-Tsakonas, S., Delidakis, C., Fehon, R., Hartley, D., Herndon, V., Johansen, K. *et al.*, 1990, Notch and the molecular genetics of neuroblast segregation in *Drosophila*, *Mol. Reprod. Dev.* 27:23–27.

Artavanis-Tsakonas, S., Rand, M.D., and Lake, R.J., 1999, Notch signaling: Cell fate control and signal integration in development, *Science* 284: 770–776.

Austin, C.P., Feldman, D.E., Ida, J.A., Jr., and Cepko, C.L., 1995, Vertebrate retinal ganglion cells are selected from competent progenitors by the action of Notch, *Development* 121:3637–3650.

Belliveau, M.J. and Cepko, C.L., 1999, Extrinsic and intrinsic factors control the genesis of amacrine and cone cells in the rat retina, *Development* 126:555–566.

Belliveau, M.J., Young, T.L., and Cepko, C.L., 2000, Late retinal progenitor cells show intrinsic limitations in the production of cell types and the kinetics of opsin synthesis, *J. Neurosci.* 20:2247–2254.

Bertrand, N., Castro, D.S., and Guillemot, F., 2002, Proneural genes and the specification of neural cell types, *Nat. Rev. Neurosci.* 3:517–530.

Bhat, K.M., 1999, Segment polarity genes in neuroblast formation and identity specification during *Drosophila* neurogenesis, *Bioessays* 21: 472–485.

Briscoe, J., Pierani, A., Jessell, T.M., and Ericson, J., 2000, A homeodomain protein code specifies progenitor cell identity and neuronal fate in the ventral neural tube, *Cell* 101:435–445.

Carr, V.M. and Simpson, S.B.J., 1978, Proliferative and degenerative events in the early development of chick dorsal root ganglia. I. Normal development, *J. Comp. Neurol.* 182:727–739.

Cepko, C., 1988, Retroviruses and their applications in neurobiology, *Neuron* 1:345–353.

Cepko, C.L., Austin, C.P., Yang, X., Alexiades, M., and Ezzeddine, D., 1996, Cell fate determination in the vertebrate retina, *Proc. Natl. Acad. Sci. USA* 93:589–595.

Chenn, A. and McConnell, S.K., 1995, Cleavage orientation and the asymmetric inheritance of Notch1 immunoreactivity in mammalian neurogenesis, *Cell* 82:631–641.

Chitnis, A., Henrique, D., Lewis, J., Ish-Horowicz, D., and Kintner, C., 1995, Primary neurogenesis in *Xenopus* embryos regulated by a homologue of the *Drosophila* neurogenic gene *delta*, *Nature* 375:761–766.

Coffman, C.R., Skoglund, P., Harris, W.A., and Kintner, C.R., 1993, Expression of an extracellular deletion of xotch diverts cell fate in *Xenopus* embryos, *Cell* 73:659–671.

Cubas, P., de Celis, J.F., Campuzano, S., and Modolell, J., 1991, Proneural clusters of *achaete-scute* expression and the generation of sensory organs in the *Drosophila* imaginal wing disc, *Genes Dev.* 5: 996–1008.

Daga, A., Karlovich, C.A., Dumstrei, K., and Banerjee, U., 1996, Patterning of cells in the *Drosophila* eye by *lozenge*, which shares homologous domains with *aml1*, *Genes Dev.* 10:1194–1205.

Dahmane, N. and Ruiz-i-Altaba, A., 1999, Sonic hedgehog regulates the growth and patterning of the cerebellum, *Development* 126: 3089–3100.

Davis, R.L. and Turner, D.L., 2001, Vertebrate Hairy and Enhancer of split related proteins: Transcriptional repressors regulating cellular dif-ferentiation and embryonic patterning, *Oncogene* 20:8342–8357.

Doe, C.Q. and Smouse, D.T., 1990, The origins of cell diversity in the insect central nervous system, *Semin. Cell Biol.* 1:211–218.

Dorsky, R.I., Chang, W.S., Rapaport, D.H., and Harris, W.A., 1997, Regulation of neuronal diversity in the *Xenopus* retina by Delta signaling, *Nature* 385:67–70.

Dyer, M.A. and Cepko, C.L., 2000, P57(kip2) regulates progenitor cell proliferation and amacrine interneuron development in the mouse retina, *Development* 127:3593–3605.

Dyer, M.A. and Cepko, C.L., 2001, P27kip1 and p57kip2 regulate proliferation in distinct retinal progenitor cell populations, *J. Neurosci.* 21: 4259–4271.

Erkman, L., McEvilly, R.J., Luo, L., Ryan, A.K., Hooshmand, F., O'Connell, S.M. *et al.*, 1996, Role of transcription factors Brn-3.1 and Brn-3.2 in auditory and visual system development, *Nature* 381:603–606.

Erkman, L., Yates, P.A., McLaughlin, T., McEvilly, R.J., Whisenhunt, T., O'Connell, S.M. *et al.*, 2000, A POU domain transcription factor-dependent program regulates axon pathfinding in the vertebrate visual system, *Neuron* 28:779–792.

Farah, M.H., Olson, J.M., Sucic, H.B., Hume, R.I., Tapscott, S.J., and Turner, D.L., 2000, Generation of neurons by transient expression of neural bHLH proteins in mammalian cells, *Development* 127:693–702.

Fekete, D.M., Perez-Miguelsanz, J., Ryder, E.F., and Cepko, C.L., 1994, Clonal analysis in the chicken retina reveals tangential dispersion of clonally related cells, *Dev. Biol.* 166:666–682.

Fischer, A.J. and Reh, T.A., 2000, Identification of a proliferating marginal zone of retinal progenitors in postnatal chickens, *Dev. Biol.* 220:197–210.

Fode, C., Gradwohl, G., Morin, X., Dierich, A., LeMeur, M., Goridis, C. *et al.*, 1998, The bHLH protein Neurogenin 2 is a determination factor for epibranchial placode-derived sensory neurons, *Neuron* 20:483–494.

Frantz, G.D. and McConnell, S.K., 1996, Restriction of late cerebral cortical progenitors to an upper-layer fate, *Neuron* 17:55–61.

Frise, E., Knoblich, J.A., Younger-Shepherd, S., Jan, L.Y., and Jan, Y.N., 1996, The *Drosophila* Numb protein inhibits signaling of the Notch receptor during cell–cell interaction in sensory organ lineage, *Proc. Natl. Acad. Sci. USA* 93:11925–11932.

Gan, L., Wang, S.W., Huang, Z., and Klein, W.H., 1999, Pou domain factor Brn-3b is essential for retinal ganglion cell differentiation and survival but not for initial cell fate specification, *Dev. Biol.* 210: 469–480.

Guan, W., Puthenveedu, M.A., and Condic, M.L., 2003, Sensory neuron subtypes have unique substratum preference and receptor expression before target innervation, *J. Neurosci.* 23:1781–1791.

Guillemot, F. and Joyner, A.L., 1993, Dynamic expression of the murine achaete-scute homologue *mash-1* in the developing nervous system, *Mech. Dev.* 42:171–185.

Haase, G., Dessaud, E., Garces, A., de Bovis, B., Birling, M., Filippi, P. *et al.*, 2002, GDNF acts through Pea3 to regulate cell body positioning and muscle innervation of specific motor neuron pools, *Neuron* 35: 893–905.

He, W., Ingraham, C., Rising, L., Goderie, S., and Temple, S., 2001, Multipotent stem cells from the mouse basal forebrain contribute GABAergic neurons and oligodendrocytes to the cerebral cortex during embryogenesis, *J. Neurosci.* 21:8854–8862.

Heitzler, P., Bourouis, M., Ruel, L., Carteret, C., and Simpson, P., 1996, Genes of the enhancer of split and achaete-scute complexes are required for a regulatory loop between Notch and Delta during lateral signalling in *Drosophila*, *Development* 122:161–171.

Henrique, D., Adam, J., Myat, A., Chitnis, A., Lewis, J., and Ish-Horowicz, D., 1995, Expression of a Delta homologue in prospective neurons in the chick, *Nature* 375:787–790.

Hirata, J., Nakagoshi, H., Nabeshima, Y., and Matsuzaki, F., 1995, Asymmetric segregation of the homeodomain protein prospero during *Drosophila* development, *Nature* 377:627–630.

Hirsch, M.R., Tiveron, M.C., Guillemot, F., Brunet, J.F., and Goridis, C., 1998, Control of noradrenergic differentiation and *phox2a* expression by Mash1 in the central and peripheral nervous system, *Development* 125:599–608.

Holt, C.E., Bertsch, T.W., Ellis, H.M., and Harris, W.A., 1988, Cellular determination in the *Xenopus* retina is independent of lineage and birth date, *Neuron* 1:15–26.

Ishibashi, M., Moriyoshi, K., Sasai, Y., Shiota, K., Nakanishi, S., and Kageyama, R., 1994, Persistent expression of helix-loop-helix factor Hes-1 prevents mammalian neural differentiation in the central nervous system, *EMBO J.* 13:1799–1805.

Isshiki, T., Pearson, B., Holbrook, S., and Doe, C.Q., 2001, *Drosophila* neuroblasts sequentially express transcription factors which specify the temporal identity of their neuronal progeny, *Cell* 106:511–521.

Jan, Y.N. and Jan, L.Y., 2001, Asymmetric cell division in the *Drosophila* nervous system, *Nat. Rev. Neurosci.* 2:772–779.

Jarman, A.P., Grell, E.H., Ackerman, L., Jan, L.Y., and Jan, Y.N., 1994. Atonal is the proneural gene for *Drosophila* photoreceptors, *Nature* 369:398–400.

Kageyama, R. and Ohtsuka, T., 1999, The Notch-Hes pathway in mammalian neural development, *Cell Res.* 9:179–188.

Kalyani, A., Hobson, K., and Rao, M.S., 1997, Neuroepithelial stem cells from the embryonic spinal cord: Isolation, characterization, and clonal analysis, *Dev. Biol.* 186:202–223.

Kalyani, A.J., Piper, D., Mujtaba, T., Lucero, M.T., and Rao, M.S., 1998, Spinal cord neuronal precursors generate multiple neuronal phenotypes in culture, *J. Neurosci.* 18:7856–7868.

Kalyani, A.J. and Rao, M.S., 1998, Cell lineage in the developing neural tube, *Biochem. Cell Biol.* 76:1051–1068.

Knoblich, J.A., Jan, L.Y., and Jan, Y.N., 1995, Asymmetric segregation of Numb and Prospero during cell division, *Nature* 377:624–627.

Kuida, K., Haydar, T.F., Kuan, C.Y., Gu, Y., Taya, C., Karasuyama, H. *et al.*, 1998, Reduced apoptosis and cytochrome c-mediated caspase activation in mice lacking Caspase 9, *Cell* 94:325–337.

Kuida, K., Zheng, T.S., Na, S., Kuan, C., Yang, D., Karasuyama, H. *et al.*, 1996, Decreased apoptosis in the brain and premature lethality in Cpp32-deficient mice, *Nature* 384:368–372.

Lavdas, A.A., Grigoriou, M., Pachnis, V., and Parnavelas, J.G., 1999, The medial ganglionic eminence gives rise to a population of early neurons in the developing cerebral cortex, *J. Neurosci.* 19:7881–7888.

Le Douarin, N., 1973, A biological cell labeling technique and its use in experimental embryology, *Dev. Biol.* 30:217–222.

Le Douarin, N., 1982, *The Neural Crest*, Cambridge University Press, New York.

Leber, S.M. and Sanes, J.R., 1995, Migratory paths of neurons and glia in the embryonic chick spinal cord, *J. Neurosci.* 15:1236–1248.

Lee, J.E., Hollenberg, S.M., Snider, L., Turner, D.L., Lipnick, N., and Weintraub, H., 1995, Conversion of *Xenopus* ectoderm into neurons by NeuroD, a basic helix-loop-helix protein, *Science* 268:836–844.

Lee, S.K. and Pfaff, S.L., 2001, Transcriptional networks regulating neuronal identity in the developing spinal cord, *Nat. Neurosci.* 4(Suppl): 1183–1191.

Letinic, K., Zoncu, R., and Rakic, P., 2002, Origin of GABAergic neurons in the human neocortex, *Nature* 417:645–649.

Levine, E.M., Close, J., Fero, M., Ostrovsky, A., and Reh, T.A., 2000, P27(kip1) regulates cell cycle withdrawal of late multipotent progenitor cells in the mammalian retina, *Dev. Biol.* 219:299–314.

Liem, K.F., Jr., Tremml, G., Roelink, H., and Jessell, T.M., 1995, Dorsal differentiation of neural plate cells induced by BMP-mediated signals from epidermal ectoderm, *Cell* 82:969–979.

Livesey, F.J. and Cepko, C.L., 2001, Vertebrate neural cell-fate determination: Lessons from the retina, *Nat. Rev. Neurosci.* 2:109–118.

Livet, J., Sigrist, M., Stroebel, S., De Paola, V., Price, S.R., Henderson, C.E. *et al.*, 2002, Ets gene *pea3* controls the central position and terminal arborization of specific motor neuron pools, *Neuron* 35:877–892.

Lo, L., Tiveron, M.C., and Anderson, D.J., 1998, Mash1 activates expression of the paired homeodomain transcription factor Phox2a, and couples pan-neuronal and subtype-specific components of autonomic neuronal identity, *Development* 125:609–620.

Lois, C. and Alvarez-Buylla, A., 1993, Proliferating subventricular zone cells in the adult mammalian forebrain can differentiate into neurons and glia, *Proc. Natl. Acad. Sci. USA* 90:2074–2077.

Lois, C. and Alvarez-Buylla, A., 1994, Long-distance neuronal migration in the adult mammalian brain, *Science* 264:1145–1148.

Lu, B., Jan, L., and Jan, Y.N., 2000, Control of cell divisions in the nervous system: Symmetry and asymmetry, *Annu. Rev. Neurosci.* 23:531–556.

Lyden, D., Young, A.Z., Zagzag, D., Yan, W., Gerald, W., O'Reilly, R. *et al.*, 1999, Id1 and Id3 are required for neurogenesis, angiogenesis and vascularization of tumour xenografts, *Nature* 401:670–677.

Ma, Q., Chen, Z., del Barco Barrantes, I., de la Pompa, J.L., and Anderson, D.J., 1998, Neurogenin1 is essential for the determination of neuronal precursors for proximal cranial sensory ganglia, *Neuron* 20:469–482.

Ma, Q., Fode, C., Guillemot, F., and Anderson, D.J., 1999, Neurogenin1 and Neurogenin2 control two distinct waves of neurogenesis in developing dorsal root ganglia, *Genes Dev.* 13:1717–1728.

Ma, Q., Kintner, C., and Anderson, D.J., 1996, Identification of *neurogenin*, a vertebrate neuronal determination gene, *Cell* 87:43–52.

Mayer-Proschel, M., Kalyani, A.J., Mujtaba, T., and Rao, M.S., 1997, Isolation of lineage-restricted neuronal precursors from multipotent neuroepithelial stem cells, *Neuron* 19:773–785.

McConnell, S.K., 1988, Fates of visual cortical neurons in the ferret after isochronic and heterochronic transplantation, *J. Neurosci.* 8:945–974.

McConnell, S.K. and Kaznowski, C.E., 1991, Cell cycle dependence of laminar determination in developing neocortex, *Science* 254: 282–285.

Murre, C., McCaw, P.S., and Baltimore, D., 1989a, A new DNA binding and dimerization motif in immunoglobulin enhancer binding, Daughterless, MyoD, and Myc proteins, *Cell* 56:777–783.

Murre, C., McCaw, P.S., Vaessin, H., Caudy, M., Jan, L.Y., Jan, Y.N. *et al.*, 1989b, Interactions between heterologous helix-loop-helix proteins generate complexes that bind specifically to a common DNA sequence, *Cell* 58:537–544.

Nieto, M., Schuurmans, C., Britz, O., and Guillemot, F., 2001, Neural bhlh genes control the neuronal versus glial fate decision in cortical progenitors, *Neuron* 29:401–413.

Nornes, H.O. and Carry, M., 1978, Neurogenesis in spinal cord of mouse: An auto radiographic analysis, *Brain Res.* 159:1–16.

O'Rourke, N.A., Chenn, A., and McConnell, S.K., 1997, Postmitotic neurons migrate tangentially in the cortical ventricular zone, *Development* 124:997–1005.

O'Rourke, N.A., Sullivan, D.P., Kaznowski, C.E., Jacobs, A.A., and McConnell, S.K., 1995, Tangential migration of neurons in the developing cerebral cortex, *Development* 121:2165–2176.

Ohsako, S., Hyer, J., Panganiban, G., Oliver, I., and Caudy, M., 1994, Hairy function as a DNA-binding helix-loop-helix repressor of *Drosophila* sensory organ formation, *Genes Dev.* 8:2743–2755.

Parnavelas, J.G., Anderson, S.A., Lavdas, A.A., Grigoriou, M., Pachnis, V., and Rubenstein, J.L., 2000, The contribution of the ganglionic eminence to the neuronal cell types of the cerebral cortex, *Novartis Found. Symp.* 228:129–139; discussion 139–147.

Perron, M., Kanekar, S., Vetter, M.L., and Harris, W.A., 1998, The genetic sequence of retinal development in the ciliary margin of the *Xenopus* eye, *Dev. Biol.* 199:185–200.

Petersen, P.H., Zou, K., Hwang, J.K., Jan, Y.N., and Zhong, W., 2002, Progenitor cell maintenance requires *numb* and *numblike* during mouse neurogenesis, *Nature* 419:929–934.

Pfaff, S.L., Mendelsohn, M., Stewart, C.L., Edlund, T., and Jessell, T.M., 1996, Requirement for LIM homeobox gene *isl1* in motor neuron generation reveals a motor neuron-dependent step in interneuron differentiation, *Cell* 84:309–320.

Qian, X., Shen, Q., Goderie, S.K., He, W., Capela, A., Davis, A.A. *et al.*, 2000, Timing of CNS cell generation: A programmed sequence of neuron and glial cell production from isolated murine cortical stem cells, *Neuron* 28:69–80.

Rao, M.S., 1999, Multipotent and restricted precursors in the central nervous system, *Anat. Rec.* 257:137–148.

Rao, M.S., Noble, M., and Mayer-Proschel, M., 1998, A tripotential glial precursor cell is present in the developing spinal cord, *Proc. Natl. Acad. Sci. USA* 95:3996–4001.

Raymond, P.A. and Rivlin, P.K., 1987, Germinal cells in the goldfish retina that produce rod photoreceptors, *Dev. Biol.* 122:120–138.

Reese, B.E., Harvey, A.R., and Tan, S.S., 1995, Radial and tangential dispersion patterns in the mouse retina are cell-class specific, *Proc. Natl. Acad. Sci. USA* 92:2494–2498.

Roelink, H., Porter, J.A., Chiang, C., Tanabe, Y., Chang, D.T., Beachy, P.A. et al., 1995, Floor plate and motor neuron induction by different concentrations of the amino-terminal cleavage product of Sonic hedgehog autoproteolysis, *Cell* 81:445–455.

Roztocil, T., Matter-Sadzinski, L., Alliod, C., Ballivet, M., and Matter, J.M., 1997, NeuroM, a neural helix-loop-helix transcription factor, defines a new transition stage in neurogenesis, *Development* 124:3263–3272.

Sharma, K., Sheng, H.Z., Lettieri, K., Li, H., Karavanov, A., Potter, S. et al., 1998, Lim homeodomain factors Lhx3 and Lhx4 assign subtype identities for motor neurons, *Cell* 95:817–828.

Sicinski, P., Donaher, J.L., Parker, S.B., Li, T., Fazeli, A., Gardner, H. et al., 1995, Cyclin D1 provides a link between development and oncogenesis in the retina and breast, *Cell* 82:621–630.

Sidman, R.L., Miale, I.L., and Feder, N., 1959, Cell proliferation and migration in the primitive ependymal zone: An autoradiographic study of histogenesis in the nervous system, *Exp. Neurol.* 1:322–333.

Skeath, J.B. and Carroll, S.B., 1991, Regulation of achaete-scute gene expression and sensory organ pattern formation in the *Drosophila* wing, *Genes Dev.* 5:984–995.

Skeath, J.B. and Carroll, S.B., 1992, Regulation of proneural gene expression and cell fate during neuroblast segregation in the *Drosophila* embryo, *Development* 114:939–946.

Spana, E.P. and Doe, C.Q., 1995, The Prospero transcription factor is asymmetrically localized to the cell cortex during neuroblast mitosis in *Drosophila*, *Development* 121:3187–3195.

Spana, E.P. and Doe, C.Q., 1996, Numb antagonizes Notch signaling to specify sibling neuron cell fates, *Neuron* 17:21–26.

Straznicky, K. and Gaze, R.M., 1971, The growth of the retina in *Xenopus laevis:* An autoradiographic study, *J. Embryol. Exp. Morphol.* 26:67–79.

Sun, Y., Nadal-Vicens, M., Misono, S., Lin, M.Z., Zubiaga, A., Hua, X. et al., 2001, Neurogenin promotes neurogenesis and inhibits glial differentiation by independent mechanisms, *Cell* 104:365–376.

Taghert, P.H., Doe, C.Q., and Goodman, C.S., 1984, Cell determination and regulation during development of neuroblasts and neurones in grasshopper embryo, *Nature* 307:163–165.

Tan, S.S., Kalloniatis, M., Sturm, K., Tam, P.P., Reese, B.E., and Faulkner-Jones, B., 1998, Separate progenitors for radial and tangential cell dispersion during development of the cerebral neocortex, *Neuron* 21:295–304.

Tanabe, Y., William, C., and Jessell, T.M., 1998, Specification of motor neuron identity by the MNR2 homeodomain protein, *Cell* 95:67–80.

Taupin, P. and Gage, F.H., 2002, Adult neurogenesis and neural stem cells of the central nervous system in mammals, *J. Neurosci. Res.* 69:745–749.

Thaler, J., Harrison, K., Sharma, K., Lettieri, K., Kehrl, J., and Pfaff, S.L., 1999, Active suppression of interneuron programs within developing motor neurons revealed by analysis of homeodomain factor HB9, *Neuron* 23:675–687.

Thor, S., Andersson, S.G., Tomlinson, A., and Thomas, J.B., 1999, A LIM-homeodomain combinatorial code for motor-neuron pathway selection, *Nature* 397:76–80.

Thor, S. and Thomas, J.B., 1997, The *Drosophila islet* gene governs axon pathfinding and neurotransmitter identity, *Neuron* 18:397–409.

Tomlinson, A. and Struhl, G., 2001, Delta/Notch and Boss/Sevenless signals act combinatorially to specify the *Drosophila* R7 photoreceptor, *Mol. Cell* 7:487–495.

Turner, D.L. and Cepko, C.L., 1987, A common progenitor for neurons and glia persists in rat retina late in development, *Nature* 328:131–136.

Turner, D.L., Snyder, E.Y., and Cepko, C.L., 1990, Lineage-independent determination of cell type in the embryonic mouse retina, *Neuron* 4:833–845.

Van Doren, M., Ellis, H.M., and Posakony, J.W., 1991, The *Drosophila* Extramacrochaetae protein antagonizes sequence-specific DNA binding by Daughterless/Achaete-scute protein complexes, *Development* 113:245–255.

van Praag, H., Kempermann, G., and Gage, F.H., 1999, Running increases cell proliferation and neurogenesis in the adult mouse dentate gyrus, *Nat. Neurosci.* 2:266–270.

von Ohlen, T. and Doe, C.Q., 2000, Convergence of Dorsal, DPP, and EGFR signaling pathways subdivides the *Drosophila* neuroectoderm into three dorsal-ventral columns, *Dev. Biol.* 224:362–372.

Wallace, V.A., 1999, Purkinje-cell-derived sonic hedgehog regulates granule neuron precursor cell proliferation in the developing mouse cerebellum, *Curr. Biol.* 9:445–448.

Walsh, C. and Cepko, C.L., 1988, Clonally related cortical cells show several migration patterns, *Science* 241:1342–1345.

Walsh, C. and Cepko, C.L., 1992, Widespread dispersion of neuronal clones across functional regions of the cerebral cortex, *Science* 255:434–440.

Wang, S.W., Mu, X., Bowers, W.J., Kim, D.S., Plas, D.J., Crair, M.C. et al., 2002, Brn3b/Brn3c double knockout mice reveal an unsuspected role for Brn3c in retinal ganglion cell axon outgrowth, *Development* 129:467–477.

Wechsler-Reya, R.J. and Scott, M.P., 1999, Control of neuronal precursor proliferation in the cerebellum by Sonic hedgehog, *Neuron* 22:103–114.

Weisblat, D.A., Sawyer, R.T., and Stent, G.S., 1978, Cell lineage analysis by intracellular injection of a tracer enzyme, *Science* 239:1142–1145.

Yoshida, H., Kong, Y.Y., Yoshida, R., Elia, A.J., Hakem, A., Hakem, R. et al., 1998, Apaf1 is required for mitochondrial pathways of apoptosis and brain development, *Cell* 94:739–750.

Zhong, W., Feder, J.N., Jiang, M.M., Jan, L.Y., and Jan, Y.N., 1996, Asymmetric localization of a mammalian numb homolog during mouse cortical neurogenesis, *Neuron* 17:43–53.

6

The Oligodendrocyte

Mark Noble, Margot Mayer-Pröschel, and Robert H. Miller

INTRODUCTION

The oligodendrocyte is the cellular component of the brain and spinal cord responsible for the production of myelin, a fatty insulation composed of modified plasma membrane that surrounds axons and promotes the rapid and efficient conduction of electrical impulses along myelinated axons (Bunge, 1968). Myelination is essential for the normal functioning of the vertebrate central nervous system (CNS). As failures in myelination are associated with disruptions of normal impulse conduction, such alterations can interfere with neurological function as profoundly as the loss of neurons themselves. Thus, disruption of CNS myelin through injury, pathological degeneration, or genetic reasons leads to severe functional deficits and frequently a reduction in life span. Abnormalities related to myelination may represent the single largest category of neurological dysfunction in the CNS, being seen in multiple developmental syndromes, in traumatic injury of many varieties, and as a result of chronic degenerative processes.

Along with the importance associated with the function of the oligodendrocyte, this cell and its ancestors are of particular interest as a model system for the study of a large range of complex problems in cellular development. The developmental events leading to the generation of oligodendrocytes are currently among the best understood of all such processes (although not without controversy, as will be discussed). For this reason, studies on the precursor cells that give rise to oligodendrocytes have provided a fertile ground for the elucidation of general principles in developmental and cellular biology.

GENERATION OF OLIGODENDROCYTES FROM THEIR IMMEDIATE PRECURSOR CELLS

The cell that is believed to be the direct ancestor of the oligodendrocyte is a precursor cell that, at least *in vitro*, has the capacity to generate both oligodendrocytes and a particular subset of astrocytes (known as type-2 astrocytes). For this reason, these precursor cells originally were named oligodendrocyte-type-2 astrocyte (O-2A) progenitor cells (Raff *et al.*, 1983).

These cells constitutively differentiate into oligodendrocytes but require exposure to specific environmental signals in order to give rise to astrocytes (Hughes *et al.*, 1988; Lillien and Raff, 1990). Whether O-2A progenitors actually generate type-2 astrocytes during normal development has been a matter of considerable controversy and still remains unclear (Knapp, 1991; Espinosa de los Monteros *et al.*, 1993). Difficulties in identifying type-2 astrocytes as a derivative of these progenitor cells has led multiple laboratories to refer to this identical population of cells as oligodendrocyte precursor cells (OPCs). In recognition of this still unresolved controversy, the abbreviation O-2A/OPC will be used throughout this chapter.

Although O-2A/OPCs do not require the action of cell-extrinsic signaling molecules to generate oligodendrocytes, they do require the action of such molecules to undergo division. The most extensively characterized mitogen for these precursor cells is the platelet-derived growth factor AA homodimer (PDGF-A) (Noble *et al.*, 1988; Richardson *et al.*, 1988). In the intact CNS, PDGF-A is ubiquitously distributed, being synthesized by both astrocyte and neuronal populations (Yeh *et al.*, 1991; Hutchins and Jefferson, 1992). Overexpression of PDGF-A results in a dramatic increase in the number of spinal cord O-2A/OPCs (Calver *et al.*, 1998), while in PDGF-A knockouts, the number of these precursor cells is dramatically reduced (Fruttiger *et al.*, 1999).

The responsiveness of O-2A/OPCs to PDGF can be modified by synergistic interactions with a variety of other signaling molecules. For example, the chemokine CXCL1/GRO-α enhances the proliferation of spinal cord-derived O-2A/OPCs exposed to PDGF in a concentration-dependent manner (Robinson *et al.*, 1998; Wu *et al.*, 2000). Responsiveness to PDGF is also enhanced by co-exposure to neurotrophin-3 (NT-3; Barres *et al.*, 1994b; Ibarrola *et al.*, 1996) and basic fibroblast growth factor (FGF-2; Bogler *et al.*, 1990; McKinnon *et al.*, 1990).

In at least some cases, co-exposure to PDGF and other cytokines also alters the balance between self-renewal and differentiation in dividing O-2A/OPCs. Co-exposure to FGF-2, for example, causes these precursor cells to become trapped in a continuous program of self-renewal and appears to almost

Mark Noble and Margot Mayer-Pröschel • Department of Biomedical Genetics, University of Rochester Medical Center, Rochester, NY 14642. Robert H. Miller • Department of Neurosciences, Case Western Reserve University School of Medicine, Cleveland, OH 44106.

Developmental Neurobiology, 4th ed., edited by Mahendra S. Rao and Marcus Jacobson. Kluwer Academic / Plenum Publishers, New York, 2005.

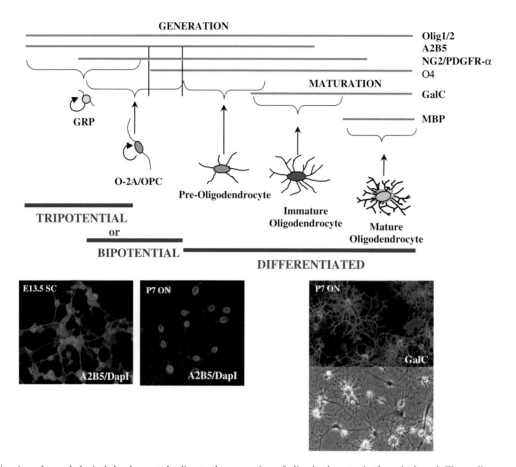

FIGURE 1. Antigenic and morphological development leading to the generation of oligodendrocytes in the spinal cord. The earliest progenitor cell that appears to be specialized for the generation of glial cells is the tripotential glial-restricted precursor (GRP) cell, which can be isolated from the rat spinal cord as early as E12.5. These cells are all A2B5⁺, and a subset of the ones derived from ventral spinal cord express the *Olig1* and *Olig2* genes. Beginning at ~E14 in the rat, some GRP cells (putatively from the Olig-expressing subset of GRP cells) express the NG2 proteoglycan and the alpha receptor for platelet-derived growth factor. The first cells that can be unambiguously defined as O-2A/OPCs are labeled with the O4 antibody and are seen at E18/19 in the rat spinal cord. O-2A/OPCs express a distinctive bipolar morphology. At E20/21, the first cells appear that express galactocerebroside (GalC), a myelin-specific galactolipid. GalC⁺ oligodendrocytes develop a multipolar morphology very different from that of the O-2A/OPC from which they are derived. These oligodendrocytes continue to mature to express such proteins as myelin basic protein (MBP). For more detailed discussion on the derivation of O-2A/OPCs from GRP cells, the reader is referred to Gregori *et al.* (2002).

It is not yet known if similar developmental progressions pertain outside of the spinal cord. As discussed in Noble *et al.* (2003), the cortex may be more complex than the spinal cord. Whether or not the O-2A/OPCs of the optic nerve are derived from GRP-like cells is also not known.

completely block the generation of oligodendrocytes (Bogler *et al.*, 1990). Co-exposure to NT-3 and PDGF also greatly enhances the extent of self-renewal (Ibarrola *et al.*, 1996). Such studies have revealed what has proven to be an important general principle in precursor cell biology, to whit, the mitotic capacity of a lineage-restricted precursor cell is regulated in part by the specific combination of signaling molecules to which the cell is exposed. The number of divisions such precursor cells can undergo can be greatly altered depending on exposure to different combinations and concentrations of cell-extrinsic signaling molecules.

The ability of combinations of growth factors to enhance precursor cell self-renewal is counterbalanced by the action of other signaling molecules that promote oligodendrocyte generation. The most extensively studied of such signaling molecules is thyroid hormone (TH), which has been of particular interest due to the severe deficiencies in myelination that occur in children or experimental animals that are hypothyroid for genetic, nutritional, or experimental reasons (Walters and Morell, 1981; Ibarrola *et al.*, 1996; Ibarrola and Rodriguez-Pena, 1997; Siragusa *et al.*, 1997; Jagannathan *et al.*, 1998; Rodriguez-Pena, 1999). Conditioned medium from cultured oligodendrocytes also inhibits the proliferation of O-2A/OPCs and promotes oligodendrocyte generation and this effect may be mediated in part by transforming growth factor-β (Louis *et al.*, 1992; McKinnon *et al.*, 1993). Still other cytokines, such as leukemia inhibitory effector and ciliary neurotrophic factor (CNTF), also have been found to enhance the generation of oligodendrocytes from O-2A/OPCs (Barres *et al.*, 1993; Louis *et al.*, 1993; Mayer *et al.*, 1994). It is important to note that these same factors, when applied to other precursor cells, appear to promote the generation of astrocytes (Johe *et al.*, 1996; Bonni *et al.*, 1997; Rao *et al.*, 1998; Mi and Barres, 1999; Park *et al.*, 1999; Aberg *et al.*, 2001). Moreover, if applied to O-2A/OPCs together with extracellular matrix (ECM) produced by endothelial cells, CNTF enhances astrocyte induction by the as yet unknown inducing factors present in the matrix (Lillien *et al.*, 1990; Mayer *et al.*, 1994). Thus, it quite clear that the effect of a signaling molecule on glial precursor cell differentiation may be very different for different precursor cells.

Along with soluble signaling molecules, contact-mediated signals also appear to modulate O-2A/OPC proliferation (Zhang and Miller, 1996; Nakatsuji and Miller, 2001). In cultures of embryonic rat spinal cord, oligodendrocyte lineage cells reach a steady-state density independent of the initial number of precursors and the presence of mitogens. This normalization of cell number, which reflects a feedback inhibition of precursor proliferation at high density, is cell-type specific and does not appear to be mediated through the release of a soluble factor (Zhang and Miller, 1996). The signaling pathways mediating density dependent inhibition of O-2A/OPC proliferation are unknown. The concept that cell proliferation may be regulated in a density-dependent fashion is not new (Wieser *et al.*, 1990) and some of the cellular mechanisms responsible for the decreases in O-2A/OPC proliferation are beginning to be defined. Increasing cell density of O-2A/OPCs, as in other cell types (Hengst and Reed, 1996; Kato *et al.*, 1997), is correlated with an increase in the expression levels of the cell cycle inhibitor p27^{Kip-1}, reductions in the expression levels of cyclins (Sherr, 1993), including cyclin A, and changes in the relative phosphorylation levels of Rb (Weinberg, 1995; Nakatsuji and Miller, 2001). Alterations in the expression of these components would all tend to inhibit the progression through the cell cycle and thus reduce proliferation (Nakatsuji and Miller, 2001). Also, it has been reported that increased cell density is associated with increased levels of oxidant production *in vitro*. Shifting of the intracellular redox balance to a more oxidized state is associated with differentiation of O-2A/OPCs (as discussed in more detail later in this chapter), as well as being associated (in at least some cell types) with increased expression of such cell cycle inhibitors as p21 (waf1/cip1) (Esposito *et al.*, 1997; Chen *et al.*, 1998; Kaneto *et al.*, 1999) and p27 (Kip1) (Hannken *et al.*, 2000).

In adult animals, O-2A/OPCs are arranged in a regular nonoverlapping distribution, which may be suggestive of the existence of similar regulatory mechanisms *in vivo* (Ong and Levine, 1999). Moreover, contact with CNS myelin prevents differentiation of perinatal O-2A/OPCs, possibly through Notch-1-mediated signaling (Wang *et al.*, 1998; Miller, 1999). Thus, the interesting possibility exists that differentiating oligodendrocytes send a signal back to their precursors that prevents the differentiation of the remaining cells. Such a hypothetical mechanism would help to insure development of a sufficient number of oligodendrocytes in addition to allowing for evolutionary (or stochastic) increases in the number of large diameter axons (Levine *et al.*, 2001).

Still another contributor to oligodendrocyte generation is the nerve cell itself. It seems likely that both soluble and cell-mediated signals from adjacent axons are integrated into the developmental profile of O-2A/OPCs resulting in cell differentiation, upregulation of myelin gene expression, and formation of the myelin organelle. Candidates for axonally derived soluble factors include FGFs (Bansal *et al.*, 1996; Qian *et al.*, 1997), while axonal cell surface molecules such as L1, myelin-associated glycoprotein (MAG), NCAM, and *N*-Cadherin may regulate formation of the myelin sheath (Trapp, 1990; Payne and Lemmon, 1993b). Oligodendrocyte development also is influenced by neuregulins that are expressed on many axons. Neuregulin exposure induces morphological changes in cultured oligodendrocytes (Vartanian *et al.*, 1994). Furthermore, in the absence of the neuregulin receptor ErbB2, many O-2A/OPCs develop, but few of these cells mature and those that do fail to interact with axons and do not produce myelin (Park *et al.*, 2001).

The generation of a complex structure such as the myelin sheath clearly requires a coordinated response in the myelinating cell. The synthesis and assembly of many myelin-specific components such as myelin basic protein (MBP), proteolipid protein (PLP), and the myelin-specific lipid galactocerebroside (GalC) have to be correctly orchestrated to give rise to myelin (Campignoni and Macklin, 1988; Campagnoni, 1995; Madison *et al.*, 1999; Campignoni and Skoff, 2001). The functional properties of different myelin proteins are highlighted in relevant mutant animals as discussed later in this chapter, and a detailed understanding of the regulation of assembly of the myelin sheath remains a major goal. Also, as discussed in more detail later, the interactions between the axon and myelinating glial cells are bidirectional and complex. Myelination leads to local changes in the cytoarchitecture of the axon (de Waegh *et al.*, 1992) as well as more systemic changes in the biology of myelinated neurons (Brady *et al.*, 1999).

THE ORIGIN OF O-2A/OPCs

Generation of Bipotential O-2A/OPCs from Tripotential Glial Restricted Precursors

Identifying the means by which neuroepithelial stem cells of the embryonic CNS give rise to O-2A/OPCs has been a matter of intensive investigation by multiple laboratories. At the time of writing this chapter, there are divergent views on how this occurs.

The only cell that arises earlier in development than O-2A/OPCs and has been shown to be able to give rise to O-2A/OPCs is a tripotential precursor cell that can be isolated from the embryonic spinal cord (Rao *et al.*, 1998) and also can be generated directly from neuroepithelial stem cells. This cell, which has been named as the glial-restricted precursor (GRP) cell, is restricted to the generation of glia and can generate oligodendrocytes, type-1 astrocytes and type-2 astrocytes. GRP cells do not give rise to neurons, either when transplanted into neurogenic zones of the CNS or when grown *in vitro* in conditions that promote generation of neurons from neuroepithelial stem cells or from neuron-restricted precursor (NRP) cells (Rao *et al.*, 1998; Herrera *et al.*, 2001). GRP cells represent one of the first two lineage-restricted populations to arise during differentiation of neuroepithelial stem cells (NSCs) *in vitro* (the other lineage-restricted population being NRP cells [Mayer-Pröschel *et al.*, 1997]). GRP cells arise early *in vivo* and can be isolated directly from the E12.5 rat spinal cord, a stage of development that precedes the appearance of any differentiated glia (Rao *et al.*, 1998; Liu *et al.*, 2002). In contrast, in the rat spinal cord, O-2A/OPCs (defined as cells that undergo bipotential differentiation into oligodendrocytes and type-2 astrocytes when examined in clonal culture) cannot be isolated from the rat spinal cord until at least E17 (Gregori *et al.*, 2002b).

GRP cells differ from O-2A/OPCs in a variety of ways (Fig. 2; Rao *et al.*, 1998; Gregori *et al.*, 2002b). Freshly isolated GRP cells are dependent upon exposure to FGF-2 for both their survival and their division, while division and survival of O-2A/OPCs can be promoted by PDGF and other growth factors. Consistent with this difference in growth factor–response patterns, GRP cells freshly isolated from the E13.5 spinal cord do not express receptors for PDGF, although they do express

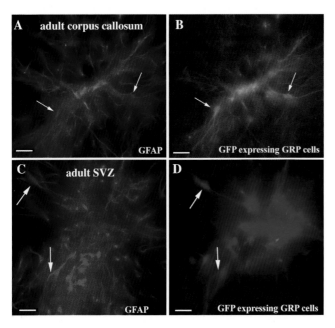

FIGURE 2. GRP cells generate astrocytes when transplanted into the normal CNS. Following isolation from the E13.5 spinal cord, stable green fluorescent protein (GFP) expressing GRP cells were generated *in vitro* using recombinant plnx-retrovirus encoding GFP (Herrera *et al.*, 2001). GRP cells were transplanted into the corpus callosum (A, B) or SVZ (C, D) of adult animals. As shown in this figure (and indicated by arrows), GFP-expressing GRP cells differentiated to yield GFAP[+] astrocytes *in vivo*. In other experiments, for which the reader is referred to Herrera *et al.* (2001) for details, transplantation into the neonatal SVZ or into spinal cords of myelin-deficient rats was associated with the generation of oligodendrocytes. In no instance did any of the GFP-labeled GRP cells express type III b-tubulin, a characteristic marker of neurons in the rat CNS.

such receptors with continued growth *in vitro* or *in vivo*. These populations also differ in their response to inducers of differentiation. For example, exposure of GRP cells to the combination of FGF-2 and CNTF induces these cells to differentiate into astrocytes (primarily expressing the antigenic phenotype of type-2 astrocytes) (Rao *et al.*, 1998). In contrast, exposure of O-2A/OPCs to FGF-2 + CNTF promotes the generation of oligodendrocytes (Mayer *et al.*, 1994).

A further striking difference between GRP cells and O-2A/OPCs is that GRP cells readily generate astrocytes following their transplantation into the adult CNS (Fig. 3; Herrera *et al.*, 2001). This is in striking contrast to primary O-2A/OPCs, which thus far only generate oligodendrocytes in such transplantations (Espinosa de los Monteros *et al.*, 1993) although it has been reported that O-2A/OPC cell lines will generate astrocytes if transplanted in similar circumstances (Franklin and Blakemore, 1995).

Antigenic and *in situ* analysis of development *in vivo* has confirmed that cells with the A2B5[+] antigenic phenotype of GRP cells arise in spinal development several days prior to the appearance of GFAP-expressing astrocytes, and also prior to the appearance of cells expressing markers of radial glia (Liu *et al.*, 2002). Thus, these cells can be isolated directly from the developing spinal cord, and cells with the appropriate antigenic phenotype have been found to exist *in vivo* at appropriate ages to play important roles in gliogenesis.

Thus far, analysis of A2B5[+] cells isolated from the early embryonic spinal cord reveals a great degree of homogeneity in their ability to generate oligodendrocytes and type-1 and type-2 astrocytes *in vitro* (Rao *et al.*, 1998; Gregori *et al.*, 2002b). In addition, GRP cells have been isolated from multiple species and by multiple means. For example, such cells have been isolated from the rat spinal cord, the mouse spinal cord, and from murine embryonic stem cells (Mujtaba *et al.*, 1999). In addition, A2B5[+] precursor cells restricted to the generation of astrocytes and oligodendrocytes have been derived from cultures of human embryonic brain cells (Dietrich *et al.*, 2002). Both mouse and human cells share the ability of rat GRP cells to generate oligodendrocytes and more than one antigenically defined population of astrocytes.

The GRP Cell As an Ancestor of the O-2A/OPC

A number or questions arise from the fact that it is possible to isolate two distinct precursor cell populations (i.e., GRP cells and O2A/OPCs) from the developing animal, each of which can generate oligodendrocytes. Is the relationship between these two populations one of lineage restriction or lineage convergence? If GRP cells and O2A/OPCs are related, what signals promote the generation of one from the other and how can the existence of both populations be integrated with existing studies on the generation of oligodendrocytes during spinal cord development?

In vitro studies have demonstrated that GRP cells can give rise to O-2A/OPCs if exposed to particular signaling molecules. In these experiments, cultures of GRP cells derived from spinal cords of E13.5 rats were grown in conditions known to induce the generation of oligodendrocytes (Gregori et al., 2002b). At the initiation of these experiments, no cells in the GRP cell cultures were labeled with the O4 antibody, which can be used to recognize O-2A/OPCs at a stage of development at which the generation of both oligodendrocytes and type-2 astrocytes in vitro is possible (Trotter and Schachner, 1989; Barnett et al., 1993; Grzenkowski et al., 1999). When GRP cells that originally had been grown in the presence of FGF for several days were exposed to a combination of PDGF and TH, however, O4$^+$ cells were generated in the cultures. Purification of cells that were O4$^+$ but did not express galactocerebroside (GalC, a marker of oligodendrocytes), and subsequent examination of the differentiation potential of these cells at the clonal level, confirmed that they behaved like O-2A/OPCs rather than like GRP cells (Gregori et al., 2002b). When O4$^+$GalC$^-$ cells were grown in conditions that induced the generation of astrocytes, the resulting clones contained only type-2 astrocytes. In contrast, O4$^-$ cells derived from the GRP cell cultures behaved as did freshly isolated GRP cells in these conditions and generated clones containing both type-1 and type-2 astrocytes. Moreover, experimental analysis suggested that no GalC$^+$ oligodendrocytes were generated in these cultures without prior passage through an O4$^+$GalC$^-$ stage of development, as has been seen in multiple other analyses.

Unlike PDGFR$^+$ cells, it does not appear that GRP cells are entirely restricted to the ventral spinal cord during early development. Even at E13.5 (up to a full day before the appearance of PDGFR$^+$ cells in the rat ventral cord; Hall et al., 1996), both dorsal and ventral regions of spinal cord contain A2B5$^+$ cells that, when analyzed at the clonal level, were found to be tripotential GRP cells (Gregori et al., 2002b). All clones contained both type-1 and type-2 astrocytes when exposed to fetal calf serum or bone morphogenetic protein (BMP), and all clones were capable of generating both O4$^+$GalC$^-$ cells and GalC$^+$ oligodendrocytes. Antigenic analysis in vivo also confirms that the domain of A2B5$^+$ cells in the spinal cord at E14.5 includes the domain of PDGFR$^+$ cells but extends further laterally, dorsally, and ventrally (Liu et al., 2002).

Despite the presence of GRP cells in both dorsal and ventral spinal cord of E13.5 rats, there appear nonetheless to be potentially interesting differences in these two populations. The frequency of A2B5$^+$ cells was greater ventrally than dorsally at E13.5 ($52 \pm 7\%$ vs $19 \pm 8\%$ of all cells, respectively), in agreement with immunohistochemical analysis of spinal cord development (Liu et al., 2002). While both dorsal and ventral GRP cells responded similarly to exposure to PDGF + TH in their ability to generate O4$^+$GalC$^-$ cells, only ventral-derived cells generated a significant number of oligodendrocytes over a five-day time period (Gregori et al., 2002b). Ventral-derived cells may be generally more inclined to differentiate at this stage, as they also showed a greater tendency to generate astrocytes in response to low concentrations (1 ng/ml) of BMP-4. Strikingly, in the ventral-derived cultures, exposure to BMP-4 was also associated with differentiation of over half of the cells into O4$^+$GalC$^-$ cells (although not further into GalC$^+$ oligodendrocytes), whereas only 12% of the cells in the dorsal-derived

cultures were O4$^+$GalC$^-$ in these conditions. Thus, it appears in general that although both dorsal- and ventral-derived GRP cells can generate oligodendrocytes, the ventral-derived populations exhibit a greater tendency to readily progress along this pathway.

That the overall population of GRP cells may contain subsets of cells with different properties is also indicated by analysis of patterns of antigen and mRNA expression in the developing spinal cord (Liu et al., 2002). For example, it appears that the domain of Nkx2.2-expressing cells in the E11.5–E13.5 spinal cord forms a subdomain within the population of A2B5$^+$ cells. Whether such heterogeneity in patterns of transcription factor expression is associated with heterogeneity of biological properties may only be revealed by applying techniques of quantitative analysis of what might be subtle differences in clonal properties, as have been developed for analysis of O-2A/OPCs (e.g., Yakovlev et al., 1998a, b; Boucher et al., 1999; Zorin et al., 2000).

At present, the simplest model of oligodendrocyte generation that appears to be consistent with the data discussed thus far would be that production of these cells requires the initial generation of GRP cells from NSCs followed by the generation of O2A/OPCs from GRP cells (Figure 3). Previous studies (Rao and Mayer-Pröschel, 1997; Rao et al., 1998) indicated strongly that GRP cells are a necessary intermediate between NSCs and differentiated glia, and subsequent experiments raise the possibility that O2A/OPCs are a necessary intermediate between GRP cells and oligodendrocytes, at least in the developing spinal cord (Gregori et al., 2002b).

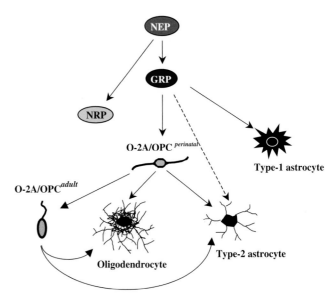

FIGURE 3. Theoretical lineage history of glia in the rat spinal cord. The only immediate ancestor that has thus far been demonstrated for the O-2A/OPC is the tripotential glial-restricted precursor (GRP) cell (Gregori et al., 2002). GRP cells appear to be directly descended from neuroepithelial (NEP) stem cells, which also generate neuron-restricted precursor (NRP) cells. It is possible that the generation of O-2A/OPCs is a necessary intermediate step in the generation of oligodendrocytes from GRP cells. The O-2A/OPC of the perinatal CNS then goes on to generate an adult-specific O-2A/OPC with quite distinct biological properties, as discussed in later sections of this chapter.

It is important to note that there are a number of claims that the GRP cell hypothesis explicitly does *not* make. Critically, this hypothesis does not state that all GRP cells give rise to both oligodendrocytes and astrocytes *in vivo*. It is clear that the differentiation fate of the progeny of a particular founder cell will be modulated by the microenvironment in which those progeny are localized. The generation of astrocytes and oligodendrocytes from a single founder cell requires that the progeny of the founder cell migrate into different microenvironments. What the hypothesis does predict, in contrast, is precisely what has been reported, that is, that A2B5$^+$ cells isolated from ventral regions (where oligodendrocytes will be generated) or dorsal regions (where astrocytes will be generated) of the early spinal cord will all be tripotential cells restricted to the generation of glia.

In addition, the GRP cell hypothesis does not require that all A2B5$^+$ cells derived from the embryonic spinal cord be alike in all ways. As discussed elsewhere in this chapter, O-2A/OPCs from different regions of the developing CNS express profound differences with respect to their tendency to undergo self-renewing division and in their responsiveness to inducers of differentiation (Power *et al.*, 2002). It has been suggested that these differences are reflective of the time courses of myelination in the tissues in which these cells are resident. Yet, these cells are all O-2A/OPCs as defined by their apparent restriction to the generation of oligodendrocytes and type-2 astrocytes *in vitro*. Similarly, although it is already clear that GRP cells derived from dorsal and ventral spinal cord of E13.5 rats may express some different properties, they are nonetheless thus far apparently identical with respect to their tripotentiality.

The Motor Neuron/Oligodendrocyte Precursor (MNOP) Cell Hypothesis

At the same time that studies have been ongoing on GRP cells, a wholly separate line of investigation has raised questions about whether oligodendrocytes are developmentally more closely related to motor neurons than they are to astrocytes (Fig. 4). These studies have pursued a hypothesis first suggested by Pringle *et al.* (1996) that motor neurons and oligodendrocytes shared a common precursor cell (i.e., an MNOP). The MNOP hypothesis was based initially on observations that both motor neurons and oligodendrocytes arise in a similar (and possibly identical) discrete zone of the ventral spinal cord (reviewed in Richardson *et al.*, 1997, 2000). Moreover, it was found that similar concentrations of Sonic hedgehog (Shh) are required for the induction of both cell types (Pringle *et al.*, 1996) and *in vitro* the induction of oligodendrocytes (e.g., by ectopic Shh presentation) is frequently accompanied by the induction of motor neurons (Pringle *et al.*, 1996; Orentas *et al.*, 1999).

The hypothesis that motor neurons and oligodendrocytes are developmentally related to each other has some intuitive attractiveness arising from the critical importance of this particular cell combination in evolution. The ensheathing of axons with myelin is associated with a large increase in conduction speed. In annelids and crustacea, pseudomyelin is preferentially associated with axons required for rapid escape responses (Roots, 1993; Davis *et al.*, 1999). It has been suggested that hagfish, which have no myelin, seem unable to accelerate to avoid capture (Richardson *et al.*, 2000). One way to ensure that motor neurons and the cells that ensheath them arise at the same place would be to derive both cells from the same precursor (as suggested in Richardson *et al.*, 1997, 2000).

The idea that motor neurons and oligodendrocytes might be developmentally related was given a further boost by the findings, from three separate laboratories, that compromising the function of members of the *Olig* gene family can prevent the generation of both motor neurons and oligodendrocytes (Lu *et al.*, 2002; Takebayashi *et al.*, 2002; Zhou *et al.*, 2002). These studies have been discussed in detail in a number of recent reviews (Richardson *et al.*, 2000; Rowitch *et al.*, 2002; Sauvageot and Stiles, 2002).

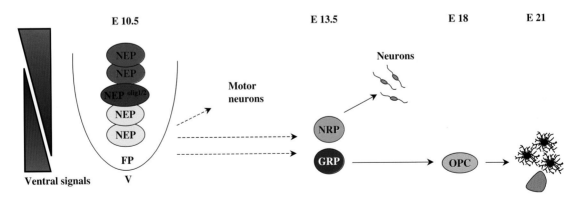

FIGURE 4. Theoretical scheme for generation of GRP and NRP cells in the ventral spinal cord. It appears that ventralizing signals (in particular, Sonic hedgehog) cause the specialization of NEP cells to generate particular cell types. Motor neurons are the first ventral-specific cell type to be generated. Motor neurons can be generated from NRP cells isolated from the spinal cord although, as discussed in the text, there are also arguments being made for the existence of a motor neuron/oligodendrocyte precursor (MNOP) cell. Due to this controversy, motor neuron derivation is indicated separately from GRP cells. Also generated are GRP cells, which go on to generate O-2A/OPCs. There are, however, alternative hypotheses as to how differentiation progresses in the early spinal cord, as summarized in Fig. 5.

Olig1 and *Olig2* genes are expressed in the developing mouse spinal cord within the specific region that appears to give rise to both oligodendrocytes and motor neurons (Lu *et al.*, 2000; Takebayashi *et al.*, 2000; Zhou *et al.*, 2000). Forced expression of *Olig1* or *Olig2* in neuroepithelial stem cells induces expression of early markers of the oligodendrocyte lineage (Lu *et al.*, 2000; Zhou *et al.*, 2001). Moreover, expression of *Olig2* in conjunction with *neurogenin2* appears to be critical for the generation of motor neurons (Mizuguchi *et al.*, 2001; Novitch *et al.*, 2001).

Among the most striking experiments that have been suggested to indicate the possible presence of a precursor cell committed to the generation of oligodendrocytes and motor neurons are experiments showing that targeted disruption of *Olig2* prevents oligodendrocyte and motor neuron specification in the spinal cord (Lu *et al.*, 2002; Takebayashi *et al.*, 2002; Zhou *et al.*, 2002). (Disruption of *Olig1*, in contrast, disrupted normal maturation of oligodendrocytes [Lu *et al.*, 2002].) Thus, the evidence is quite clear that expression of *Olig2* is required for the generation of both oligodendrocytes and motor neurons. Not all of the regulatory factors, however, that control motor neuron development also affect oligodendrocyte development, and *vice versa*. For example, disruption of function of the *Isl1* gene prevents the generation of motor neurons without apparently affecting oligodendrocyte generation (Pfaff *et al.*, 1996). Conversely, disruption of the neuregulin 1 signaling pathway (by knocking out the neuregulin 1 gene itself, or through disruption of the erbB receptors to which this protein binds) appears to disrupt normal oligodendrocyte development without impacting on the generation of motor neurons (Vartanian *et al.*, 1999; Park *et al.*, 2001).

The Ambiguous Case for a Restricted Motor Neuron/Oligodendrocyte Precursor Cell

Data derived from studies on the expression and function of *Olig1* and *Olig2* genes contain substantial ambiguities. As there is no evidence that has been presented (at the time of writing this chapter) of the derivation of both motor neurons and oligodendrocytes from a single lineage-restricted founder cell, the only conclusion that can be drawn from these experiments is that *Olig1* and *Olig2* genes are expressed in a population of ancestral cells of unknown heterogeneity. This population may well consist of separate *Olig1*$^+$/*Olig2*$^+$ precursors for neurons and for oligodendrocytes.

Critically, it is not yet possible to draw firm conclusions from existing studies on whether or not astrocytes are ever generated *in vivo* from cells that are at some point induced to express *Olig* genes. Claims that *Olig2*$^+$ cells do not generate astrocytes, and that disruption of *Olig* gene function does not alter astrocyte development (Lu *et al.*, 2002), are not without problems, as these suggestions have not been tested by determination of whether larger numbers of astrocytes are generated in the developing spinal cord of *Olig*-compromised animals (which would be a predicted consequence of such perturbations). It is also the case that in *Olig2*$^{-/-}$ mice in which the *Olig2* gene was disrupted by targeted replacement with tamoxifen-inducible Cre recombinase reveal at least some of the Cre expressing cells expressed the astrocyte marker S100β (Takebayashi *et al.*, 2002). Similarly, in *Olig1*$^{-/-}$ *Olig2*$^{-/-}$ mice in which GFP was expressed in the *Olig2* locus, half of the GFP-expressing cells differentiated into astrocytes *in vivo* (Zhou *et al.*, 2002). As these experiments are conducted in animals in which no functional *Olig* genes are expressed, no conclusions regarding *Olig* gene expression and lineage restriction can be drawn. The conclusion can be drawn, however, that the signals that induce *Olig* gene expression are not sufficient to cause restriction of the resultant precursor cells away from astrocytic pathways. If one believes, however, that disruption of the function of *Olig1* and *Olig2* genes *in vivo* is revealing of developmental plasticity, then the interpretation of the experiments of Takebayashi *et al.* (2002) and Zhou *et al.* (2002) would be that cells exposed to signals that induce expression of *Olig1/2* can readily generate astrocytes if *Olig* gene expression is disrupted. Such an interpretation is consistent with the observations that A2B5$^+$NG2$^+$PDGFR$^+$ cells derived from the ventral spinal cord of E16 rats readily generate oligodendrocytes and two populations of astrocytes *in vitro* when exposed to appropriate conditions, and thus appear to be GRP cells (Gregori *et al.*, 2002b).

Reconciling the GRP and MNOP Hypotheses of Oligodendrocyte Ancestry

One critical question that thus far appears to remain unanswered in studies on early spinal cord development concerns the heterogeneity of the *Olig2* expressing population found in the spinal cord at differing ages. At the time of writing this chapter, however, it is not clear if the population of *Olig1* and *Olig2* expressing cells of the embryonic spinal cord ever comprises an antigenically homogeneous population of cells at any stage beyond the NSC stage. The MNOP and GRP/NRP hypotheses predict very different outcomes of such experiments over a range of developmental stages.

One clear prediction of the GRP/NRP analysis of development is that the cells that express *Olig* genes in the E9.5–E10.5 are pluripotent NSCs. This prediction is made due to the failure to find PSA-N-CAM-expressing NRP or A2B5$^+$ GRP cells at this developmental stage (Mayer-Pröschel *et al.*, 1997; Liu *et al.*, 2002). The GRP/NRP hypothesis also predicts that once motor neuron generation is completed, the *Olig1/2*-expressing population should consist of A2B5$^+$ GRP cells, a prediction that appears thus far to be correct. This prediction appears to have been tested as a part of attempts to understand when and how the putative transition from GRP to O-2A/OPCs is regulated *in vivo*. Current data indicates that the "when" component of this transition occurs surprisingly late, and subsequent to the appearance of PDGFR$^+$ cells in the ventral region of the embryonic rat spinal cord at E14/14.5. Current data indicates that cells with the bipotential lineage restriction of O-2A/OPCs cannot be isolated from the rat spinal cord until at least E17 (Gregori *et al.*, 2002b; MMP *et al.*, in progress). Prior to this point, clonal analysis of A2B5$^+$ cells isolated from both dorsal and ventral rat spinal cord indicates that all of these cells appear to be tripotential GRP cells, including the ventral-derived ones that are PDGFR$^+$NG2$^+$.

As available data suggests a high level of overlap between *Olig*-expressing cells and PDGFR$^+$ cells (Lu *et al.*, 2000; Zhou *et al.*, 2000; Tekki-Kessaris *et al.*, 2001), it appears that the prediction that such cells are GRP cells is correct.

Nonetheless, it is very important—in the context of the MNOP hypothesis—to ask whether the outcomes of developmental studies *in vivo* are at least consistent with such a hypothesis. This appears not to be the case. The view that *Olig1/2* expression represents restriction to the motor neuron/oligodendrocyte pathways makes a very specific prediction that labeling of a founder cell and its clonal derivatives at stages of spinal cord development after *Olig1/2* expression occurs will reveal that motor neuron-containing clones contain oligodendrocytes but not astrocytes. Potentially relevant experiments appear to have been carried out over a decade ago, using the technique of injecting retroviral particles expressing bacterial β-galactosidase into the developing chick spinal cord, over a range of ages including up to one or two cell cycles before all motor neurons are born (Leber *et al.*, 1990). In these experiments, 82% of clones that contained motor neurons also had nonmotor neuron relatives. No evidence was found, however, for the occurrence of cell types in specific combinations. Forty-two percent of the multicellular clones that contained motor neurons contained cells that were clearly glial in both gray and white matter, with many of these glial cells being considered to be astrocytes by morphological criteria. Critically, with respect to the possible longevity of NSCs in the developing spinal cord, injection of retrovirus as late as one or two cell cycles before motor neurons are born revealed clones that contained motor neurons, interneurons, and glia as relatives, with putative astrocytes prominently represented among the glia. Although the focus of this study was on motor neuron development, the authors also noted the existence of other clones that appeared to contain both oligodendrocytes and astrocytes. The results of these studies are subject to multiple interpretations, but the reported frequency of motor neuron/astrocyte clones seems quite divergent from what one would expect were a restricted MNOP a critical contributor to spinal cord development.

While it is not an essential part of the GRP cell hypothesis that single precursor cells give rise to both oligodendrocytes and astrocytes *in vivo* (as this would require dispersion of the progeny of a single precursor cell into different microenvironments), it nonetheless would be of interest to know if such pairing does normally occur. Recent studies by Zerlin et al. (2004) demonstrate that this does indeed occur in the mammalian forebrain. In these experiments, retroviral labeling of cells of the neonatal (P0–P2) rat subventricular zone (SVZ) led to the later appearance of clonally related astrocytes and oligodendrocytes in the cortex. In white matter tracts, in contrast, the majority of progenitors become oligodendrocytes, although some astrocytes are also generated. Critically, as the GRP cell hypothesis predicts, it appears to be necessary for progeny to enter into different microenvironments in order for these clonally related cells to assume different fates. Moreover, if SVZ cells were labeled with retrovirus and then, after a further four or five days of *in vivo* growth, the cells that had migrated into the neocortex were isolated and allowed to develop *in vitro*, clones contained only glia and no neurons. Some clones were composed only of oligodendrocytes, some only of astrocytes, and some contained both kinds of glial cells. As dissection of these cells at earlier stages following retroviral labeling revealed mixed neuronal-glial clones (Levison and Goldman, 1997), it seems that continued development is associated with progressive fate restriction into glioblasts. Whether or not the cells that give rise to oligodendrocytes and astrocytes, but not to neurons, correspond to GRP cells remains to be determined.

There are two interpretations equally consistent with all of the above data, but which are fully compatible with the idea that lineage specialization in the CNS is initiated with the formation of GRP cells and NRP cells, and that do not require the evocation of a precursor cell specialized to generate oligodendrocytes and motor neurons. One interpretation is that the studies that have been discussed are indicative of localized influences on gene expression that cause expression of these genes in NSCs, but that such expression patterns occur before commitment to a particular lineage occurs. According to this hypothesis, *Olig1* and *Olig 2* are already expressed in NSCs prior to the generation of GRP and NRP cells. This hypothesis is consistent with the observations that these genes are expressed at a stage of neural tube development when all cells in the neural tube appear to be pluripotent NSCs able to generate all of the cell types of the CNS. In the ventral regions of the spinal cord, some NSCs coordinately express *neurogenin* genes with *Olig* genes, and such cells differentiate to generate motor neurons. Other NSCs that express *Olig1* and *Olig2*, but do not express neurogenins, differentiate to yield GRP cells and then become further restricted to yield O-2A/OPCs.

Another variant of the above hypothesis is that localized signals in this region cause NRP cells and GRP cells each to express *Olig1* and *Olig2* genes, which play different roles in these two lineages. When expressed in NRP cells in conjunction with neurogenins, they induce the formation of motor neurons. When expressed in GRP cells, they block the action of inducers of astrocytic differentiation (such as BMPs) and promote differentiation of these cells in O-2A/OPCs and, subsequently, into oligodendrocytes.

A common feature of both the above hypotheses is that, if correct, they would indicate that genes thought to be indicative of oligodendrocyte development are in fact expressed prior to terminal commitment to this differentiation path. It is possible, however, that none of the explanations of development that have been suggested thus far are correct. It may also be, as suggested in Fig. 5, that there exist both MNOP cells and GRP cells, each of which plays a contributory role in the generation of oligodendrocytes.

Regional Specialization of the First Appearance of Cells Expressing Oligodendrocyte Lineage Genes Is a Common Feature in CNS Development

Regardless of what emerges as the correct view of lineage specialization during oligodendrocyte development, it is clear that the first stages that can be identified in oligodendrocyte development *in vivo* consistently occur in a regionally specialized manner. As it is not clear whether the cells that have been

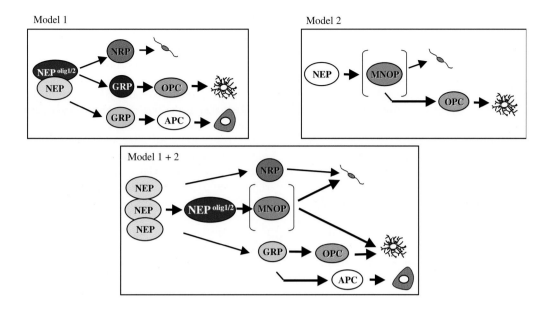

FIGURE 5. Theoretical models of lineage specification in the rat spinal cord between E11.5 and E13.5. The current state of knowledge does not allow any model to be decisively preferred over any other. For example, it could be that Olig1/2+ NEP cells generate NRP cells specialized to make motor neurons and GRP cells specialized to make O-2A/OPCs. It is also possible that NEP cells generate a motor neuron/oligodendrocyte precursor cell (MNOP) that makes only motor neurons and O-2A/OPCs. It is even possible that both of these processes occur, with the generation of GRP cells, NRP cells and MNOP cells.

defined in these studies are O-2A/OPCs, GRP cells, or as yet undefined lineage-restricted precursor cell populations, the generic term "oligodendrocyte precursor" will be used in this section to indicate cells thought to be participating in the eventual generation of oligodendrocytes.

Early oligodendrocyte development has been most extensively studied in caudal regions of the CNS such as the spinal cord and in the optic nerve. In the rat optic nerve, tissue culture studies suggested that the founder cells of the oligodendrocyte lineage originate in the brain or optic chiasm and migrate along the nerve during subsequent development (Small et al., 1987). The source of at least a subset of chick optic nerve oligodendrocytes was subsequently defined as a small foci of cells in the floor of the third ventricle (Ono et al., 1997). In the chick spinal cord, oligodendrocyte precursors are located along the entire rostral–caudal extent of the spinal cord as early as neural tube closure (Warf et al., 1991). By contrast, separation of dorsal and ventral regions of the spinal cord during early embryogenesis revealed that oligodendrocyte precursors were present only in ventral regions of the spinal cord (Warf et al., 1991; Ono et al., 1995). The ventral ventricular origin of spinal cord oligodendrocytes appears to be a common feature of vertebrate development and has been demonstrated in a broad range of species including *Xenopus* (Maier and Miller, 1995) and human as well as chick, mouse, and rat (Warf et al., 1991; Pringle and Richardson, 1993; Ono et al., 1995).

Whether the ventral spinal cord is the only source of oligodendrocytes is unresolved (Spassky et al., 1998). A dorsal source of oligodendrocyte precursors in caudal spinal cord regions was supported by initial chick–quail chimera studies in which dorsal

portions of the chick neural tube were replaced with equivalent quail tissue and the source of oligodendrocytes determined by species-specific labeling (Cameron-Currey and LeDouarin, 1995). This experimental outcome would be consistent with the observations that GRP cells able to generate oligodendrocytes *in vitro* can be isolated from both the ventral and dorsal spinal cords at times prior to the expression of such supposed markers of oligodendrocyte development as PDGFR-α (Gregori et al., 2002b). Nonetheless, subsequent analyses of similar chimeric spinal cords failed to substantiate these studies or provide evidence for a dorsal source for oligodendrocytes, but rather supported the notion that all spinal cord oligodendrocytes were derived from ventral regions (Pringle et al., 1998). At this stage, it is uncertain why studies that were theoretically identical in nature gave opposite results.

The spinal cord is not the only region where putative ancestors of oligodendrocytes arise in restricted locations. In more rostral areas of the CNS, the earliest cells thought to be oligodendrocyte precursors appear in defined domains of the ventricular zone and SVZ at particular stages of development (Ono et al., 1997). For example, a group of cells in the ventricular mantle zone of the ventral diencephalon of the E13 rat express mRNA for the PDGFR-α (Pringle and Richardson, 1993). During subsequent development, these cells appear to migrate into the developing thalamus and hypothalamus as well as to more dorsal regions including the developing cerebellum (Pringle and Richardson, 1993).

Not all the regions that initially generate oligodendrocyte precursors are ventrally located (Perez Villegas et al., 1999). In the chick metencephalon, while the earliest arising population of progenitors is adjacent to the floor plate in the ventral meten-cephalon (Ono *et al.*, 1997), a second more dorsal source of putative OPCs develops independently (Davies and Miller, 2001). The ancestors to oligodendrocytes that populate the telencephalon appear to be derived from both the medial and lateral ganglion eminence (LGE) and later migrate into the cortex (Spassky *et al.*, 1998; He *et al.*, 2001), as discussed in more detail later in this chapter. Whether both regions contribute equally to the eventual generation of forebrain oligodendrocytes is not clear. In mutants, in which the medial ganglion eminence (MGE) is converted to the LGE, there is a significant loss of oligodendrocytes suggesting that the MGE is the major source of oligodendrocyte precursors (Sussel *et al.*, 1999). Oligodendrocytes that populate the telencephalon also arise from alar regions such as the anterior entopeduncular area (Olivier *et al.*, 2001). The exact contribution of each area to the overall population of oligodendrocytes in the forebrain remains to be resolved. For example, it is unclear if individual domains generate oligodendrocytes that populate distinct regions of the forebrain, or if the different domains give rise to morphologically and biochemically distinct types of oligodendrocytes.

The Initial Appearance of Oligodendrocyte Precursors Is Regulated by Local Signals

In principle, two general mechanisms may account for the restricted geographical origin of oligodendrocyte precursors in the spinal cord and other regions of the CNS. Cells in dorsal regions may lack the intrinsic potential to generate oligodendrocytes or ventrally located signals may instruct neighboring cells to assume an oligodendrocyte fate while dorsal signals inhibit oligodendrocyte induction.

Several lines of evidence indicate that the localized appearance of oligodendrocyte precursors is a reflection of local signaling. For example, transplant studies indicate that the initial appearance of spinal cord oligodendrocytes is dependent on local influences from the adjacent notochord (Trousse *et al.*, 1995; Orentas and Miller, 1996; Pringle *et al.*, 1996). The notochord, a transient mesodermally derived structure, is located ventral to the developing neural tube and signals from the notochord have been shown to be involved in the formation of the dorsal/ventral axis in the developing CNS (van Straaten *et al.*, 1988). The establishment of dorsal and ventral polarity results in the subsequent specification of distinct populations of neurons found in the ventral spinal cord (van Straaten *et al.*, 1988, 1989; Jessell and Dodd, 1990). Transplantation of an additional notochord adjacent to the dorsal spinal cord resulted in the local induction of ventral neurons (Yamada *et al.*, 1991) and an ectopic cluster of oligodendrocyte precursors in chick and *Xenopus* embryos (Orentas and Miller, 1996; Maier and Miller, 1997). Likewise, coculture of dorsal spinal cord explants with isolated notochord is sufficient to induce generation of motor neurons (Yamada *et al.*, 1991) and oligodendrocytes in the spinal cord tissue (Orentas and Miller, 1996; Poncet *et al.*, 1996; Pringle *et al.*, 1996). The ability of the transplanted notochord to induce oligodendrocytes was restricted to a period during early embryonic chick development, which has

been suggested to reflect both a change in the signaling capacity of the notochord and a temporally dependent loss of responsiveness of the dorsal spinal cord cells (Orentas and Miller, 1996).

Not only is the notochord competent to induce ectopic oligodendrocytes, but it is also essential for the normal ventral appearance of spinal cord oligodendrocytes. In *Xenopus* embryos UV irradiated at the one-cell stage, oligodendrocytes failed to develop in spinal cord regions lacking a notochord (Maier and Miller, 1997). Likewise, oligodendrocytes did not develop in the spinal cord adjacent to the site of notochord ablation at embryonic or larval stages (Maier and Miller, 1997). Similarly, in the short-tailed Danforth mouse, the notochord is discontinuous along the length of the rostral–caudal axis, and while oligodendrocytes develop normally in regions of the spinal cord adjacent to the notochord, they are absent from those regions lacking a notochord (Pringle *et al.*, 1996). Thus, the notochord provides a local signal or signals that result in the subsequent appearance of spinal cord oligodendrocytes.

Many of the inductive properties of the notochord are due to production of the signaling molecule Shh (Echelard *et al.*, 1993; Roelink *et al.*, 1994). Shh, the vertebrate homologue of the *Drosophila* pattern forming gene *hedgehog*, is localized to the notochord and adjacent floor plate (Roelink *et al.*, 1994). *In vitro*, Shh induces the development of floor plate and different classes of motor neurons in a concentration-dependent manner (Roelink *et al.*, 1994, 1995), through the activation or repression of a series of homeodomain transcription factors (Jessell, 2000). Oligodendrocytes can be induced *in vitro* at similar concentrations of Shh required for the induction of motor neurons (Pringle *et al.*, 1996; Orentas *et al.*, 1999), suggesting that the development of these two cell types is closely linked (Richardson *et al.*, 1997, 2000). In the chick spinal cord, the generation of oligodendrocyte precursors requires continued Shh signaling after the formation of ventral–dorsal polarity and the generation of motor neurons (Orentas *et al.*, 1999). For example, inhibiting Shh signaling immediately prior to the appearance of oligodendrocyte precursors blocks their subsequent appearance, but has little effect on motor neuron pools (Orentas *et al.*, 1999). It seems likely that Shh contributes to the initial progression towards the oligodendrocyte lineage, possibly through induction of cell-type specific transcription factors such as the Olig genes (Lu *et al.*, 2000; Zhou *et al.*, 2000). In addition, recent *in vitro* studies suggest that the continued dependence of oligodendrocyte precursors on Shh signaling reflects a potent survival rather than proliferative influence on these cells (Davies and Miller, 2001).

In more rostral regions of the CNS, the expression of Shh and the appearance of oligodendrocytes also are spatially and temporally closely linked (Davies and Miller, 2001; Nery *et al.*, 2001; Tekki-Kessaris *et al.*, 2001). Furthermore, ectopic expression of Shh leads to concomitant local development of oligodendrocytes (Nery *et al.*, 2001). Whether Shh is essential for the development of all rostral populations of oligodendrocytes is less clear. In cell cultures derived from Shh knockout animals, considerable numbers of oligodendrocytes develop, indicating that oligodendrocytes can arise in the absence of Shh signaling (Nery *et al.*, 2001). It seems likely, however, that other members of the hedgehog family can substitute for Shh in its absence and blocking all hedgehog family member signaling with cyclopamine (Incardona *et al.*, 1998) appears to block all oligodendrocyte development (Tekki-Kessaris *et al.*, 2001).

In vitro, the development of oligodendrocyte precursors is inhibited by exposure to members of the TGF-β family (Mabie *et al.*, 1997). Specifically, BMP-2 and -4 appear to inhibit the development of oligodendrocytes, instead promoting the generation of astrocytes (Mabie *et al.*, 1997; Mehler *et al.*, 2000). Whether BMP signaling contributes to the spatial patterning of oligodendrocyte precursor induction in the developing intact CNS is currently unknown. For example, the failure of oligodendrocyte development in dorsal spinal cord may reflect active inhibition by BMPs, which is overcome in ventral regions by Shh (Mekki-Dauriac *et al.*, 2002). If the source of the BMPs was in dorsal tissue adjacent to the spinal cord, this hypothesis would explain why oligodendrocytes develop in isolated explants of dorsal spinal cord over time (Sussman *et al.*, 2000). Additional, as yet uncharacterized, inhibitors of oligodendrocyte precursor development may also exist. Indeed, dorsal spinal cord has been reported to contain an inhibitor of early oligodendrocyte development (Wada *et al.*, 2000) that is functionally distinct from any known BMP (Wada *et al.*, 2000).

Myelination of Developing White Matter Is Dependent on Oligodendrocyte Precursor Migration

Oligodendrocytes are widely distributed in the adult CNS, even though in early development their precursors arise in highly restricted ventricular domains (as discussed previously). The spatial separation between the location of origin of oligodendrocyte precursors and their final destination means that normal myelination is dependent on the long-distance migration of oligodendrocyte precursors. Although the migratory capacity of oligodendrocyte precursors has been clear for several years, the molecular mechanisms mediating this migration are only now becoming understood.

The earliest indications that oligodendrocyte precursors were capable of long-distance migration came from transplantation studies (Lachapelle *et al.*, 1984, 1994). Transplantation of fragments of normal CNS tissue into host animals lacking myelin proteins resulted in the substantial dispersal of normal oligodendrocytes throughout the host CNS (Lachapelle *et al.*, 1984), some of which may have occurred by trafficking through the ventricular system. Similar analyses using purified cell populations demonstrated that the capacity for long-distance cell migration through the neuropil of the CNS is predominantly a characteristic of less mature O-2A/OPCs and is lost as the cells mature (Warrington *et al.*, 1992, 1993). Likewise, retrospective analyses of the oligodendrocyte precursors that migrated into the cerebellum indicated they were highly immature (Goyne *et al.*, 1994), although in these studies (as in most studies of normal development), it is not yet clear if the migrating cell is a GRP cell, an O-2A/OPC, both, or something else entirely. Nonetheless, it seems very likely that the migration of immature oligodendrocyte

ancestors of some form is an essential component for normal myelination in the vertebrate CNS.

The extent of oligodendrocyte precursor migration during CNS development has been highlighted by analyses of the optic nerve (Small *et al.*, 1987; Ono *et al.*, 1997; Sugimoto *et al.*, 2001) and spinal cord (Warf *et al.*, 1991; Ono *et al.*, 1995). The cells that give rise to the oligodendrocytes that populate the optic nerve migrate into the nerve from the brain during late embryonic and early postnatal stages. This migration was first documented using cell culture (Small *et al.*, 1987). For example, cell cultures derived from brain or chiasmal regions of the optic nerve acquired the capacity to generate oligodendrocytes several days before cultures isolated from retinal regions of the nerve (Small *et al.*, 1987). The source of optic nerve oligodendrocytes and their dispersal along the nerve was directly visualized by labeling the cells at their origin in the floor of the third ventricle in developing chick embryos (Ono *et al.*, 1997). In contrast to rat, mouse, and human, where oligodendrocytes and myelin are restricted to the optic nerve, oligodendrocytes migrate into specific regions of the chick (Ono *et al.*, 1998) and rabbit retina (ffrench-Constant *et al.*, 1988) where they myelinate the proximal region of retinal ganglion cell axons. In the spinal cord, similar approaches have been used to document the migration of oligodendrocyte precursors from ventral to dorsal regions. Isolated cultures of ventral spinal cord generated oligodendrocytes from the time of neural tube closure (E12), while isolated cultures of dorsal spinal cord did not acquire the capacity to generate oligodendrocytes in a physiological time frame until E16 (Warf *et al.*, 1991). The acquisition of the capacity to generate oligodendrocytes in dorsal spinal cord correlated directly with the arrival of ventrally derived precursors (Ono *et al.*, 1995). Direct evidence of the ventral to dorsal migration of spinal cord oligodendrocyte precursors has been provided by cell tracking experiments using an *in vitro* preparation of rat spinal cord (Warf *et al.*, 1991).

In more rostral regions of the CNS, the migration of oligodendrocyte precursors is also pronounced (Fig. 6). During development of the cerebral cortex, immature OPCs have been suggested to migrate from the LGE as well as the MGE into the developing forebrain (Spassky *et al.*, 1998; He *et al.*, 2001; Marshall and Goldman, 2002). At later stages of development, glial precursor cells, including oligodendrocytes, migrate from the SVZ in radial and tangential directions toward the pial surface (Kakita and Goldman, 1999) to populate all regions of the cortex.

The nature of the cellular substrates utilized and the molecular mechanisms mediating oligodendrocyte precursor migration are not well understood. In the optic nerve, many of the migrating oligodendrocyte precursors are closely associated with retinal ganglion cell axons and it has been suggested that this migration is axophilic (Ono *et al.*, 1997). Migration along preexisting axon tracts would also provide a pathway for the ventral to dorsal migration in the spinal cord where migrating cells would utilize the earlier developed circumferential axon tracts. Not all oligodendrocyte precursor migration are dependent upon axons, however. In the rat optic nerve, removal of the retinal ganglion cell axons through eye enucleation or disruption of the neural retina in the postnatal animal (Ueda *et al.*, 1999) failed to

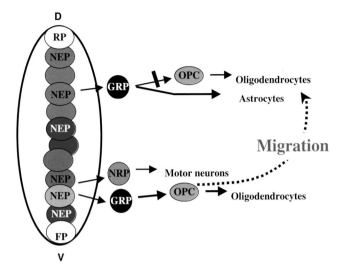

FIGURE 6. Migration plays a key role in the generation of white matter. Following the generation of ancestors of oligodendrocytes in highly specialized regions, these ancestral cells then migrate into regions where oligodendrocyte generation is required. For example, it may be that O-2A/OPCs are normally only generated from the GRP cells of the ventral spinal cord, and that these O-2A/OPCs migrate throughout this tissue to populate both dorsal and ventral white matter.

completely block the population of the nerve by oligodendrocytes although the number of cells was greatly reduced (Ueda *et al.*, 1999). It may be, however, that significant numbers of oligodendrocyte precursors had populated the nerve prior to the perturbation (Small *et al.*, 1987) or residual cues remained after the removal of the axons. Cell surface components such as adhesion molecules (Payne and Lemmon, 1993a; Wang *et al.*, 1994) and ECM receptors have also been proposed to play a role in regulating migration (Kiernan *et al.*, 1996; Garcion *et al.*, 2001). In explant studies, removal of neural cell adhesion molecule (NCAM)-associated polysialic acid (PSA) inhibits the dispersal of oligodendrocyte precursors (Wang *et al.*, 1994, 1996). However, in the developing chick, optic nerve removal of NCAM-associated PSA does not play a pivotal role in regulating precursor migration (Ono *et al.*, 1997). Oligodendrocyte precursors also express an array of integrin receptors that may play important role in regulation of both migration and cell differentiation (Milner and ffrench-Constant, 1994; Garcion *et al.*, 2001).

Guidance of Glial Precursor Migration

The migration of oligodendrocyte precursors throughout the developing CNS is likely to be mediated by specific directional and substrate cues. The immigration of oligodendrocyte precursors into the optic nerve seen in culture and labeling studies (Small *et al.*, 1987; Ono *et al.*, 1997) might simply have reflected the random movement of cells originating from a focal source within the brain. Likewise, the population of the dorsal spinal cord with ventrally derived precursors might occur by nondirectional dispersal (Ono *et al.*, 1995). Such a mechanism is unlikely. It would likely be very slow in dispersing cells and also fails to

accommodate the finding that different tracts of the spinal cord become populated with oligodendrocytes at defined times in development. The motility of immature oligodendrocyte precursors is promoted by PDGF, the same growth factor that promotes their proliferation and survival (Armstrong *et al.*, 1990b). In chemotaxis chambers, oligodendrocyte precursors migrate toward higher concentrations of PDGF indicating that this growth factor has chemotactic properties. *In vivo* PDGF appears to be ubiquitously distributed throughout the CNS, being made by populations of astrocytes and neurons (Richardson *et al.*, 1988; Yeh *et al.*, 1991) and so is unlikely to guide glial migration.

Recent studies suggest that the migration of oligodendrocyte precursors is guided by specific cues (Sugimoto *et al.*, 2001). For example, labeling of cells in a section of optic nerve demonstrated that the majority of the cell migration was unidirectional from the chiasm to the retina (Sugimoto *et al.*, 2001). This observation implied that the migration of cells along the nerve is guided by directional cues in the environment. Tissue culture studies indicated that the directional cues were located in the optic chiasm and appeared to be repellent signals produced by the optic chiasm. Without the optic chiasm, the migration of glial precursors was bidirectional with similar numbers of cells moving short distances in both directions along the nerve. By contrast, in the presence of the chiasm, OPCs in the optic nerve exhibited preferential migration away from the chiasm and toward the retina (Sugimoto *et al.*, 2001). Localization studies identified netrin-1 and semaphorin 3a as potential chiasm-derived chemorepellent signals, and consistent with this hypothesis, *in vitro* functional analyses confirmed that both netrin-1 and semaphorin 3a were chemorepellent for optic nerve glial precursors (Sugimoto *et al.*, 2001). It seems likely that the different chemorepellents act on different cell populations, with netrin being chemorepulsive for oligodendrocyte precursors and semaphorin 3a repulsive for astrocyte precursors in the optic nerve although this hypothesis awaits further confirmation.

One question raised by the above studies is how ubiquitous guided glial precursor migration is throughout the developing CNS. Studies in the spinal cord indicate that the dispersal of oligodendrocyte precursors from the ventral ventricular zone is in part mediated by localized expression of netrin-1 consistent with this protein being part of a common mechanism (Tsai *et al.*, 2003). Analyses of the migration paths of glial precursors in other regions of the developing CNS, however, suggest the situation may be more complex (Kakita and Goldman, 1999). For example, in striatum, glial precursors take two pathways that are almost perpendicular to each other. These discrete patterns of migration may reflect utilization of alternative cellular substrates of migration such as radial glial fibers or axon tracts (Ono *et al.*, 1997; Meintanis, 2001) rather than secreted guidance cues.

Stop Signals for Oligodendrocyte Precursors

One striking characteristic of the rodent optic system is the spatial restriction of myelination of retinal ganglion cell axons to the optic nerve, while more proximal portions of the same cell's axons in the retina are unmyelinated. Myelin is distributed evenly along the optic nerve but stops abruptly at the lamina cribosa of the optic nerve head in a number of species, because this region acts as a barrier for the migration of oligodendrocyte precursors (ffrench Constant et al., 1988). The mechanisms that inhibit the migration of oligodendrocyte precursors at the optic nerve head may involve localized expression of the ECM molecules such as Tenascin C in conjunction with other signals (Bartsch et al., 1994; Kiernan et al., 1996). During development, netrin-1 is transiently expressed at the optic nerve head (Deiner, 1997) and it may be that this localized expression of netrin serves as a repulsive cue to stop the migration of oligodendrocyte precursors into the retina. Another candidate stop signal for migrating oligodendrocyte precursors is the chemokine CXCL1, which appears to regulate spatial and temporal patterning of spinal cord myelination through inhibiting cell motility as well as promoting cell proliferation (Tsai et al., 2002). The cellular source of these stop signals is in part astrocytes and other glia.

Oligodendrocyte Development in the Embryonic Cortex: Similarities and Differences from the Spinal Cord

Studies on oligodendrocyte development in the cortex are currently indicating that important similarities and differences are seen from the spinal cord even during the earliest stages of brain formation.

As discussed earlier, in both the brain and spinal cord, it currently appears that the ancestors of oligodendrocytes are generated in discrete locations. Analysis of expression of PDGRα and plp/DM20 suggests the existence of a few localized ventral sites of origin (Spassky et al., 2000). In the early mouse forebrain, PDGFR-α expression is seen in the MGE and dorsal thalamus, and plp/DM20 is found in the basal (ventral) plate of the diencephalon, zona limitans intrathalamica, caudal hypothalamus, enteropeduncular area, amygdala, and olfactory bulb (Pringle and Richardson, 1993; Spassky et al., 1998; Nery et al., 2001), as is expression of the olig1 and olig2 genes (Lu et al., 2000; Zhou et al., 2000; Nery et al., 2001).

Two recent studies from the Temple (He et al., 2001) and Mehler (Yung et al., 2002) laboratories have analyzed several aspects of the cellular biology of early ancestors of oligodendrocytes. At least some of the stem cells that give rise to oligodendrocytes in the cortex appear to arise in the basal (ventral) forebrain and migrate into the overlying dorsal forebrain, including the ventricular zone, SVZ, and intermediate zones (Lavdas et al., 1999; Wichterle et al., 1999; Anderson et al., 2001; Marshall and Goldman, 2002). These basal progeny, which express the members of the dlx family of homeodomain transcription factors, migrate dorsally and intermix with other cells to form the dorsolateral SVZ (Marshall and Goldman, 2002).

Prior to the tangential migration of stem/progenitor cells from ventral to dorsal forebrain regions, it appears as if the early stages of specification are regionally biased. For example, when grown in medium supplemented with FGF-2 (He et al., 2001), or FGF-2 + Shh (Yung et al., 2002), MGE and LGE progenitors

of the E13.0 ventral forebrain are biased toward the generation of GABAergic neurons compared to stem cells and progenitors derived from dorsal cortex. In addition, prior to E12.5, few of the progenitor cells from dorsal or basal regions produce glia-only clones: most glia arise from stem cells at this stage, suggesting that divergence of glial lineages with the appearance of glial-restricted progenitors occurs predominantly at later stages. Indeed, it may be that the stem cells that are present in the dorsal forebrain prior to the period of tangential migration are not competent to make oligodendrocytes unless they are exposed to Shh (Yung et al., 2002), although this was not found by others (He et al., 2001; Qian et al., 1997).

The above results raise multiple questions. Are the stem cells or progenitor cells that make oligodendrocytes and/or GABAergic neurons truly migrating from the ventral to the dorsal cortex, or is instead the delayed appearance of such cells dorsally a reflection of a temporally regulated differentiation event that has a different timing in different regions of the CNS? To what extent is this specialization reflective of cell-intrinsic controlling mechanisms, and to what extent do cell-extrinsic signaling molecules contribute to this specification? The association of oligodendrocytes with a particular class of neuron (the GABAergic neuron of the cortex, the motor neuron of the spinal cord) is also of particular interest, particularly in light of the ongoing discussions about whether or not oligodendrocytes and motor neurons are derived from a single lineage-restricted progenitor cell (Pringle et al., 1996; Richardson et al., 1997, 2000; Orentas et al., 1999; Lu et al., 2000, 2002; Nery et al., 2001; Tekki-Kessaris et al., 2001; Mekki-Dauriac et al., 2002; Rowitch et al., 2002; Sauvageot and Stiles, 2002; Takebayashi et al., 2002; Zhou et al., 2002). But does this generation of oligodendrocytes within a single clone of cells reflect a lineage restriction of a stem/progenitor cell to the generation of only a limited subset of cell types?

Current evidence suggests strongly that the appearance of cells in the dorsal cortex that are able to generate clones containing both GABAergic neurons and oligodendrocytes is truly reflective of a migration of cells from ventral to dorsal regions. Consistent with these observations, the analysis of dlx2/tauLacZ knockin mouse (Corbin et al., 2000) also indicates that cells derived from subpallial dlx2-expressing progenitors migrate dorsally and intermix with other cells to form the dorsolateral SVZ (Marshall and Goldman, 2002). Moreover, in dlx1/2 −/− mice, in which there is a generalized defect in tangential migration and a reduction in cortical GABAergic neurons (Anderson et al., 2001), there is a failure of such cells to populate the dorsal cortex (He et al., 2001; Yung et al., 2002).

The consistent association of cell fate with position within a tissue raises the possibility that localized ventral and dorsal signals act on stem cells to make them generate particular, region appropriate, cell types. Hence, basal forebrain stem cells are biased early in development to generate GABAergic neurons that predominate in basal forebrain CNS areas (He et al., 2001; Yung et al., 2002). It has been suggested that initial ventral forebrain specification and tangential cortical migration would expose these bipotent progenitors to sequential ventral and dorsal gradient morphogens that normally mediate opposing developmental programs (Zhu et al., 1999; Yung et al., 2002).

Two of the factors thought to play important roles in inducing ventral cortical stem cells to be biased toward the generation of GABAergic neurons and oligodendrocytes are Shh and BMPs. It appears to be a common principle along the neuraxis that Shh and BMPs are ventral and dorsal gradient morphogens, respectively (Briscoe and Ericson, 1999; Miller et al., 1999; Thomas et al., 2000), and the role of these signaling molecules in development of the spinal cord has been discussed previously. The concentration of these molecules to which cells are exposed causes elaboration of specific sets of homeodomain and basic helix-loop-helix (bHLH) transcription factors that control the details of cell specification through their combinatorial interactions (Zhou et al., 2001, 2002; Rowitch et al., 2002). In dorsal domains of the spinal cord, BMP signaling is thought to promote the generation of astrocytes, while Shh promotes the localized generation of motor neurons and oligodendrocytes (Pringle et al., 1996; Mabie et al., 1997; Richardson et al., 1997, 2000; Orentas et al., 1999; Lu et al., 2000; Mehler et al., 2000; Zhou et al., 2000; Davies and Miller, 2001; Nery et al., 2001; Tekki-Kessaris et al., 2001; Mekki-Dauriac et al., 2002).

Despite the apparent role of Shh and BMP in directing differentiation of cortical stem/progenitor cells, as well as spinal cord stem/progenitor cells, there are important differences between these two tissues. This difference is already seen at the level of genes induced in cortical stem cells by exposure to Shh. For example, it currently appears that while both cortical and spinal cord stem cells are induced to express *olig2* by exposure to Shh, the cortical cells are induced to express *mash1* while spinal cord stem cells are induced to express *neurogenin2* (Mizuguchi et al., 2001; Novitch et al., 2001; Sun et al., 2001; Zhou et al., 2001; Yung et al., 2002).

Data reported thus far indicates that the role of BMP may be more complex in the cortex than has thus far been revealed in the spinal cord. Shh promotes generation of GABAergic neurons and oligodendrocytes, but the sequential elaboration of these cells requires spatial and temporal modulation of cortical BMP signaling by BMP and the BMP antagonist, noggin (Yung et al., 2002). For example, coincident with the establishment of the cortical SVZ, BMPs from the BMP2/4 factor subgroup now enhance the specification of late-born cortical (GABAergic) neurons. It seems that Shh promotes lineage restriction of ventral forebrain stem cells, in part, by upregulation of Olig2 and Mash1. BMP2 subsequently promotes GABAergic neuronal lineage elaboration by differential modulation of Olig2 and Mash1. Thus, when applied together with Shh, BMP2 potentiates the elaboration of GABAergic neurons from cortical stem/progenitor cells and suppresses oligodendrocyte generation (Mabie et al., 1999; Mehler et al., 2000), while the BMP antagonist noggin promotes the generation of oligodendrocytes (Li et al., 1998; Mehler et al., 2000).

How can the above results indicating BMP-promoted generation of neurons be integrated with experiments in the spinal cord (and also on cells derived from the developing brain) indicating that BMPs promote the generation of astrocytes and suppress the generation of oligodendrocytes (Gross et al., 1996; Mabie et al., 1997; Grinspan et al., 2000; Mehler et al., 2000; Nakashima et al., 2001; Mekki-Dauriac et al., 2002; Gregori et al., 2002b; Gomes et al., 2003)? It is possible that BMP, a potent anti-mitotic agent, is generally able to stimulate differentiation of progenitor cells but that the pathway of differentiation that is promoted is dependent upon as yet poorly understood changes in the target precursor cells themselves. One potentially interesting aspect of the studies of Yung et al. (2002), however, that may be relevant to BMP-mediated induction of neuron generation is that these studies address questions about what happens when cells are exposed to more than a single signaling molecule (i.e., Shh + BMP-2), a situation that seems likely to more closely resemble the realities of biology than exposure to a single agent. In this context, an attractive potential solution to this conundrum that needs to be explored is whether the combined exposure of cortical stem cells to BMP and Shh (the conditions applied in the studies of Yung et al., 2002) reveals an aspect of BMP signaling different from that which occurs when cells are exposed to BMP alone. Consistent with this possibility, continued Shh exposure also appears to suppress the generation of astrocytes in cortical stem/progenitor cells, which were only seen in cultures of these progenitor cells when expression of *olig2* and *Mash1* was ablated by exposure to antisense oligonucleotide constructs (Yung et al., 2002).

It will be of great interest to determine whether the correct paradigm for understanding the interactions between BMP-induced pathways and Shh-induced pathways might be that BMP always suppresses oligodendrocyte generation, but the directionality imposed by BMP is dependent upon the other signals to which the recipient cell is exposed, as well as on the differentiation potential of the target cell itself.

E13.5 Rat Cortex Contains A2B5⁺ Cells That Can Generate Oligodendrocytes, Two Different Astrocyte Populations and Neurons

As studies conducted by He et al. (2001) and Yung et al. (2002) indicate the existence of precursor cells that make neurons and glia, and other cells that are restricted to the generation of glia, it is of interest to know the identity of these precursor cells.

Analysis of the precursor cell populations in the cortex, although in their early stages, are revealing a level of complexity not seen in the spinal cord at this age (Noble et al., 2003). The E13.5 cortex contains abundant A2B5⁺ cells that do not express antigens associated with astrocytes or oligodendrocytes. In vitro characterization of the differentiation potential of these cells demonstrated that, in contrast with results in the spinal cord (Rao et al., 1998), at least some of the cortex-derived A2B5⁺ cells can generate neurons when grown in the presence of NT-3 and retinoic acid (Noble et al., 2003).

A more detailed analysis of the A2B5⁺ cell population isolated from E13.5 cortex indicates the presence of antigenically distinct subpopulations, only one of which thus far has been found to generate neurons in vitro. The subpopulation of cells that is competent to generate neurons also expresses PSA-NCAM, an antigen that has been found in several instances to be expressed by precursor cells able to generate neurons (Doetsch *et al.*, 1997; Mayer-Pröschel *et al.*, 1997; Weickert *et al.*, 2000). Removal of the PSA-NCAM⁺ cells from the A2B5⁺ population was associated with the loss of generation of neurons from this population (Noble *et al.*, 2003).

Although the A2B5⁺/PSA-NCAM⁻ cells derived from E13.5 cortex appear to be restricted to the generation of glia in their differentiation potential, this population is more heterogeneous than antigenically identical cells isolated from the spinal cord. Unlike the spinal cord, only 44% of clones derived from A2B5⁺ cells contained both type-1 and type-2 astrocytes when exposed to BMP-4. Many of the cortex-derived clones contained only one astrocyte population, with 16% of clones containing only type-2 astrocytes, and 17% containing type-1 astrocytes only, with no progenitor-like cells found in any of these clones. Virtually all cells appeared to be competent to generate oligodendrocytes, however, as 86% of clones contained at least one oligodendrocyte after being exposed to PDGF⁺ T3 for five days.

Thus, it appears that the E13.5 rat cortex contains cells with the same antigenic phenotype and differentiation potential of tripotential GRP cells isolated from the embryonic spinal cord. Further investigations are required to determine the degree of identity of these cells with GRP cells of the spinal cord, particularly due to the complexity of the A2B5⁺ populations isolated from the cortex. In addition, the embryonic cortex contains a further population of A2B5⁺ cells that co-express PSA-NCAM, an antigen not expressed by GRP cells of the spinal cord. These cells, but not the PSA-NCAM⁻/A2B5⁺ cells, are able to generate neurons *in vitro*. Moreover, the observations that ~16% of the clones derived from A2B5⁺/PSA-NCAM⁻ cells generated only type-2 astrocytes when exposed to BMP, and ~17% generated clones containing only type-1 astrocytes in these conditions, demonstrates further differences between the A2B5⁺ population of the E13.5 cortex and the E13.5 spinal cord. In the cord, in contrast with the cortex, this population shows a striking homogeneity with respect to the cell types generated in different conditions (Rao *et al.*, 1998; Gregori *et al.*, 2002b).

The full differentiation potential of the A2B5/PSA-NCAM double-positive cells that we have identified is still under study. Whether individual cells are capable of generating both neurons and glia is not yet. known. What seems clear, however, is that the embryonic rat cortex contains some A2B5+ precursor cell populations with properties quite different from the A2B5+ populations isolated from the embryonic spinal cord or the developing optic nerve.

Achieving a detailed understanding of the various PSA-NCAM⁺ populations in the cortex is going to require a considerable research effort. The studies of Noble *et al.* (2003) indicate that, at E13.5, there are both A2B5/PSA-NCAM double-positive cells and other cells that express PSA-NCAM only. It seems clear that the former cells can generate neurons, but it is not yet known whether either group of cells is restricted to the generation of neurons. Multiple previous studies have documented expression of PSA-NCAM on precursors of neurons,

including on neuron-restricted precursor cells of the spinal cord (Doetsch and Alvarez-Buylla, 1996; Doetsch *et al.*, 1997; Mayer-Pröschel *et al.*, 1997; Weickert *et al.*, 2000). It has also been previously reported that PSA-NCAM⁺ cells found in the perinatal SVZ differentiate into astrocytes and oligodendrocytes *in vivo* (Levison *et al.*, 1993; Keirstead *et al.*, 1999). Still other data suggest that PSA-NCAM⁺ cells may be able to generate neurons, oligodendrocytes, and astrocytes following transplantation *in vivo* (Vitry *et al.*, 2001), while *in vitro* studies have described a PSA-NCAM⁺/A2B5⁻ precursor cell that can give rise to A2B5⁺ O-2A/OPCs (Grinspan *et al.*, 1990; Grinspan and Franceschini, 1995; Ben-Hur *et al.*, 1998; Grinspan *et al.*, 2000).

The heterogeneity of the A2B5⁺ populations derived from the E13.5 cortex underscores the need for clonal analysis and detailed cell purification protocols in order to analyze successfully the developmental potential of a putative precursor cell population. Any studies on cortical development that do not separate these populations of cells from each other will be impossible to interpret unambiguously. As almost none of the previous studies conducted have combined antigenic characterization of precursor cells with clonal analysis, it is not possible to interpret data contained therein with regard to the lineage potential of particular precursor cell populations. For example, the analysis of purified A2B5⁺ cells from the E13.5 cortex would lead to the conclusion that cells with this antigenic phenotype can generate neurons. If one were to accept the conclusions of previous studies carried out in the developing rat CNS that A2B5⁺ cells are glial-restricted progenitor cells (whether O-2A/OPCs, GRP cells, or astrocyte progenitor cells [e.g., Raff *et al.*, 1983; Fok-Seang and Miller, 1992, 1994; Rao *et al.*, 1998; Mi and Barres, 1999; Power *et al.*, 2002]), one might then draw the conclusion that growth *in vitro* is associated with generation of neurons from glial progenitor cells (as in, e.g., Kondo and Raff, 2000). It has been suggested, at least in the case of the studies of Kondo and Raff, that a potential complicating issue in such studies is the presence of a low frequency of true multipotent NSCs in many regions of the perinatal CNS, including the perinatal optic nerve (D. Van der Kooy, unpublished observations). Another possibility is that the failure to distinguish between the PSA-NCAM-positive and negative subsets of A2B5⁺ cells would lead to a misinterpretation of the behavior of what appears from our analysis thus far to represent two distinct populations of cells.

One of the other potentially intriguing differences between cortical- and optic nerve-derived O-2A/OPCs that has been described is that only the cortical progenitor cells express members of the *dlx* family of transcriptional regulators (He *et al.*, 2001). While *dlx1/2* is not required for oligodendrocyte generation (He *et al.*, 2001), it is not known if such expression confers different properties on those precursor populations that are expression-positive. It is important to note, however, that just as generation of oligodendrocytes is an ongoing process in the cortex, so also is the generation of progenitor cells. For example, migration of cells from the LGE/MGE may continue after the earliest wave of tangential migration, as retroviral labeling of LGE/MGE cells in slice cultures harvested from E16 mice and grown *in vitro* for up to 72 hr demonstrates migration of cells into the perinatal SVZ of each slice (Marshall and Goldman, 2002). Nothing is known at this time as to whether O-2A/OPCs express different properties if they are generated from ancestral populations that differ in the spatial *or* their temporal origin, or whether the differences between O-2A/OPCs isolated from cortex and optic nerve discussed in the following section of this chapter are the results of exposure to tissue-specific instructive signals after this stage of lineage restriction has been achieved.

Is There More than One Path to an Oligodendrocyte and Is There More than One Kind of Oligodendrocyte?

All of the above discussions have been formulated as though there was only one path to generating an oligodendrocyte. It may well be that such an idea represents an oversimplification.

One of the striking aspects of CNS development is that different regions of this tissue develop according to different schedules, with great variations seen in the timing of both neurogenesis and gliogenesis. For example, neuron production in the rat spinal cord is largely complete by the time of birth, is still ongoing in the rat cerebellum for at least several days after birth, and continues in the olfactory system and in some regions of the hippocampus of multiple species throughout life. Similarly, myelination has long been known to progress in a rostral–caudal direction, beginning in the spinal cord significantly earlier than in the brain (e.g., Macklin and Weill, 1985; Kinney *et al.*, 1988; Foran and Peterson, 1992). Even within a single CNS region, myelination is not synchronous. In the rat optic nerve, for example, myelinogenesis occurs with a retinal-to-chiasmal gradient, with regions of the nerve nearest the retina becoming myelinated first (Skoff *et al.*, 1980; Foran and Peterson, 1992). The cortex itself shows the widest range of timing for myelination, both initiating later than many other CNS regions (e.g., Macklin and Weill, 1985; Kinney *et al.*, 1988; Foran and Peterson, 1992) and exhibiting an ongoing myelinogenesis that can extend over long periods of time. This latter characteristic is seen perhaps most dramatically in the human brain, for which it has been suggested that myelination may not be complete until after several decades of life (Yakovlev and Lecours, 1967; Benes *et al.*, 1994).

Variant time courses of development in different CNS regions could be due to two fundamentally different reasons. One possibility is that precursor cells are sufficiently plastic in their developmental programs that local differences in exposure to modulators of division and differentiation may account for these variances. Alternatively, it may be that the precursor cells, resident in particular tissues, express differing biological properties related to the timing of development in the tissue to which they contribute.

As has been discussed earlier, there is ample evidence for extensive plasticity in the behavior of O-2A/OPCs, which appear to be the direct ancestor of oligodendrocytes. O-2A/OPCs obtained from the optic nerves of seven-day-old (P7) rat pups and grown in the presence of saturating levels of PDGF exhibit an approximately equal probability of undergoing a self-renewing division or exiting the cell cycle and differentiating into an

oligodendrocyte (Yakovlev *et al.*, 1998b). The tendency of dividing O-2A/OPCs to generate oligodendrocytes is enhanced if cells are co-exposed to such signaling molecules as TH, CNTF, or retinoic acid (e.g., Barres *et al.*, 1994a; Mayer *et al.*, 1994; Ibarrola *et al.*, 1996). In contrast, co-exposure to NT-3 or basic FGF inhibits differentiation and is associated with increased precursor cell division and self-renewal (Bogler *et al.*, 1990; Barres *et al.*, 1994b; Ibarrola *et al.*, 1996). The balance between self-renewal and differentiation in dividing O-2A/OPCs can also be modified by the concentrations of the signaling molecules to which they are exposed, as well as by intracellular redox state (Smith *et al.*, 2000). Thus, the effects of the microenvironment could theoretically have considerable effects on the timing and extent of oligodendrocyte generation.

Recent experiments have raised the possibility that the differing timing of oligodendrocyte generation and myelination in different CNS regions is associated with the existence of regionally specialized O-2A/OPCs (Power *et al.*, 2002). Characterization of O-2A/OPCs isolated from different regions indicates that these developmental patterns are consistent with properties of the specific O-2A/OPCs resident in each region. In particular, cells isolated from optic nerve, optic chiasm, and cortex of identically aged rats show marked differences in their tendency to undergo self-renewing division and in their sensitivity to known inducers of oligodendrocyte generation. Precursor cells isolated from the cortex, a CNS region where myelination is a more protracted process than in the optic nerve, appear to be intrinsically more likely to begin generating oligodendrocytes at a later stage and over a longer time period than cells isolated from the optic nerve. For example, in conditions where optic nerve-derived O-2A/OPCs generated oligodendrocytes within 2 days, oligodendrocytes arose from chiasm-derived cells after 5 days and from cortical O-2A/OPCs only after 7–10 days. These differences, which appear to be cell-intrinsic, were manifested both in reduced percentages of clones producing oligodendrocytes and in a lesser representation of oligodendrocytes in individual clones. In addition, responsiveness of optic nerve-, chiasm-, and cortex-derived O-2A/OPCs to TH and CNTF, well-characterized inducers of oligodendrocyte generation, was inversely related to the extent of self-renewal observed in basal division conditions.

The above results indicate that the O-2A/OPC population may be more complex than initially envisaged, with the properties of the precursor cells resident in any particular region being reflective of differing physiological requirements of the tissues to which these cell contribute. For example, as discussed earlier, a variety of experiments have indicated that the O-2A/OPC population of the optic nerve arises from a germinal zone located in or near the optic chiasm and enters the nerve by migration (Small *et al.*, 1987; Ono *et al.*, 1995). Thus, it would not be surprising if the progenitor cells of the optic chiasm expressed properties expected of cells at a potentially earlier developmental stage than those cells that are isolated from optic nerve of the same physiological age. Such properties would be expected to include the capacity to undergo a greater extent of self-renewal, much as has been seen when the properties of O-2A/OPCs from optic nerves of embryonic rats and postnatal rats have been compared (Gao and Raff, 1997). With respect to the properties of cortical progenitor cells, physiological considerations also appear to be consistent with our observations. The cortex is one of the last regions of the CNS in which myelination is initiated, and the process of myelination can also continue for extended periods in this region (Macklin and Weill, 1985; Kinney et al., 1988; Foran and Peterson, 1992). If the biology of a precursor cell population is reflective of the developmental characteristics of the tissue in which it resides, then one might expect that O-2A/OPCs isolated from this tissue would not initiate oligodendrocyte generation until a later time than it occurs with O-2A/OPCs isolated from structures in which myelination occurs earlier. In addition, cortical O-2A/OPCs might be physiologically required to make oligodendrocytes for a longer time due to the long period of continued development in this tissue, at least as this has been defined in the human CNS (e.g., Yakovlev and Lecours, 1967; Benes et al., 1994).

The observation that O-2A/OPCs from different CNS regions express different levels of responsiveness to inducers of differentiation adds a new level of complexity to attempts to understand how different signaling molecules contribute to the generation of oligodendrocytes. This observation also raises questions about whether cells from different regions also express differing responses to cytotoxic agents, and whether such differences can be biologically dissected so as to yield a better understanding of this currently mysterious form of biological variability.

If there are multiple biologically distinct populations of O-2A/OPCs, it is important to consider whether similar heterogeneity exists among oligodendrocytes themselves. Evidence for morphological heterogeneity among oligodendrocytes is well established. Early silver impregnation studies identified four distinct morphologies of myelinating oligodendrocytes and this was largely confirmed by ultrastructural analyses in a variety of species (Bjartmar et al., 1968; Stensaas and Stensaas, 1968; Remahl and Hildebrand, 1990). Oligodendrocyte morphology is closely correlated with the diameter of the axons with which the cell associates (Butt et al., 1997, 1998). Type I and II oligodendrocytes arise late in development and myelinate many internodes on predominantly small diameter axons while type III and IV oligodendrocytes arise later and myelinate mainly large diameter axons. Such morphological and functional differences between oligodendrocytes are associated with different biochemical characteristics. Oligodendrocytes that myelinate small diameter fibers (type I and II) express higher levels of carbonic anhydrase II (CAII) (Butt et al., 1995, 1998), while those myelinating larger axons (type III and IV) express a specific small isoform of the MAG (Butt et al., 1998). Whether such differences represent the response of homogenous cells to different environments or distinct cell lineages is unclear. Transplant studies demonstrated that presumptive type I and II cells have the capacity to myelinate both small and large diameter axons suggesting that the morphological differences are environmentally induced (Fanarraga et al., 1998). By contrast, some developmental studies have been interpreted to suggest that the different classes of oligodendrocytes may be derived from biochemically distinct precursors

(Spassky et al., 2000) that differ in expression of PDGFR-α and PLP/Dm20, although more recent studies are not necessarily supportive of this hypothesis (Mallon et al., 2002).

Just as there is heterogeneity among O-2A/OPCs, it also seems likely that heterogeneity exists among earlier glial precursor cell populations. Separate analysis of GRP cell populations derived from ventral and dorsal spinal cord demonstrates that ventral-derived GRPs may differ from dorsal cells in such a manner as to increase the probability that they will generate O2A/OPCs and/or oligodendrocytes, even in the presence of BMP (Gregori et al., 2002b). Ventral-derived GRP cells yield several-fold larger numbers of oligodendrocytes over the course of several days of in vitro growth. When low doses of BMP-4 were applied to dorsal and ventral cultures, the dorsal cultures contained only a few cells with the antigenic characteristics of O-2A/OPCs. In contrast, over half of the cells in ventral-derived GRP cell cultures exposed to low doses of BMP differentiated into cells with the antigenic characteristics of O-2A/OPCs. Whether the O-2A/OPCs or oligodendrocytes derived from dorsal vs ventral GRP cells express different properties is not yet known.

OLIGODENDROCYTE PRECURSORS IN THE ADULT CNS

Once the processes of development ends, there is still a need for a pool of precursor cells for the purposes of tissue homeostasis and repair of injury. It is thus perhaps not surprising to find that the adult CNS also contains O-2A/OPCs. What is rather more remarkable is that current estimates are that these cells (or, at least cells with their antigenic characteristics) may be so abundant in both gray matter and white matter as to comprise 5–8% of all the cells in the adult CNS (Dawson et al., 2000). If such a frequency of these cells turns out to be accurate, then a strong argument can be made that they should be considered the fourth major component of the adult CNS, after astrocytes, neurons, and oligodendrocytes themselves. Moreover, as discussed later, it appears that these cells may represent the major dividing cell population in the adult CNS.

Studies *In Vitro* Reveal Novel Properties of Adult O-2A/OPCs

There are a variety of substantial biological differences between O-2A/OPCs of the adult and perinatal CNS (originally termed O-2A^{perinatal} and O-2A^{adult} progenitor cells, respectively) (Wolswijk and Noble, 1989, 1992; Wolswijk et al., 1990, 1991; Wren et al., 1992). For example, in contrast with the rapid cell cycle times (18 ± 4 hr) and migration (21.4 ± 1.6 μm hr^{-1}) of O-2A/OPCs^{perinatal}, O-2A/OPCs^{adult} exposed to identical concentrations of PDGF divide in vitro with cell cycle times of 65 ± 18 hr and migrate at rates of 4.3 ± 0.7 μm hr^{-1}. These cells are also morphologically and antigenically distinct. O-2A/OPCs^{adult} grown in vitro are unipolar cells, while O-2A/OPCs^{perinatal}

express predominantly a bipolar morphology. Both progenitor cell populations are labeled by the A2B5 antibody, but *adult* O-2A/OPCs share the peculiar property of oligodendrocytes of expressing no intermediate filament proteins. In addition, it appears thus far that *adult* O-2A/OPCs are always labeled by the O4 antibody, while *perinatal* O-2A/OPCs may be either O4$^-$ or O4$^+$ (although the O4$^+$ cells *perinatal* cells do express different properties than their O4$^-$ ancestors [Gard and Pfeiffer, 1993; Warrington *et al.*, 1993]).

One of the particularly interesting features of *adult* O-2A/OPCs is that when these cells are grown in conditions that promote the differentiation into oligodendrocytes of all members of clonal families of O-2A/OPCs^{perinatal}, O-2A/OPCs^{adult} exhibit extensive asymmetric behavior, continuously generating both oligodendrocytes and more progenitor cells (Wren *et al.*, 1992). Thus, even though under basal division conditions both *perinatal* and *adult* O-2A/OPCs undergo asymmetric division and differentiation, this tendency is expressed much more strongly in the *adult* cells. Indeed, it is not yet known if there is a condition in which *adult* progenitor cells can be made to undergo the complete clonal differentiation that occurs in *perinatal* O-2A/OPC clones in certain conditions (Ibarrola *et al.*, 1996).

Another feature of interest with regard to *adult* O-2A/OPCs is that these cells do have the ability to enter into limited periods of rapid division, which appear to be self-limiting in their extent. This behavior is manifested when cells are exposed to a combination of PDGF + FGF-2, in which conditions the *adult* O-2A/OPCs express a bipolar morphology and begin migrating rapidly (with an average speed of approximately 15 μm hr^{-1}. In addition, their cell cycle time shortens to an average of approximately 30 hr in these conditions (Wolswijk and Noble, 1992). These behaviors continue to be expressed for several days after which, even when maintained in the presence of PDGF + FGF-2, the cells re-express the typical unipolar morphology, slow migration rate and long cell cycle times of freshly isolated *adult* O-2A/OPCs. Other growth conditions, such as exposure to glial growth factor (GGF) can elicit a similar response (Shi *et al.*, 1998).

As can be seen from the above, *adult* O-2A/OPCs in fact express many of the characteristics that are normally associated with stem cells in adult animals. They are relatively quiescent, yet have the ability to rapidly divide as transient amplifying populations of the sort generated by many stem cells in response to injury. They also appear to be present throughout the life of the animal, and can even be isolated from elderly rats (which, in the rat, equals about two years of age). In this respect, the definition of a stem cell can be seen to be a complex one, for the *adult* O-2A/OPC would have to be classified as a narrowly lineage-restricted stem cell (in contrast with the pluripotent neuroepithelial stem cell).

The differing phenotypes of *adult* and *perinatal* O-2A/OPCs are strikingly reflective of the physiological requirements of the tissues from which they are isolated. O-2A/OPC^{perinatal} progenitor cells express properties that might be reasonably expected to be required during early CNS development (e.g., rapid division and migration, and the ability to rapidly generate large numbers of oligodendrocytes). In contrast, O-2A/OPC^{adult} progenitor cells express stem cell-like properties that appear to be more consistent with the requirements for the maintenance of a largely stable oligodendrocyte population, and the ability to enter rapid division as might be required for repair of demyelinated lesions (Wolswijk and Noble, 1989, 1992; Wren *et al.*, 1992).

It is of particular interest to consider the developmental relationship between *perinatal* and *adult* O-2A/OPCs in light of their fundamentally different properties. One might imagine, for example, that these two distinct populations are derived from different neuroepithelial stem cell populations, which produce lineage-restricted precursor cells with appropriate phenotypes as warranted by the developmental age of the animal. As it has emerged, the actual relationship between these two populations is even more surprising in its nature.

There are multiple indications that the ancestor of the O-2A/OPC^{adult} is in fact the *perinatal* O-2A/OPC itself (Wren *et al.*, 1992). This has been shown both by repetitive passaging of *perinatal* O-2A/OPCs, which yields over the course of a few weeks cultures of cells with the characteristics of *adult* O-2A/OPCs. Moreover, time-lapse microscopic observation of clones of *perinatal* O-2A/OPCs provides a direct demonstration of the generation of unipolar, slowly dividing and slowly migrating *adult* cells from bipolar, rapidly dividing and rapidly migrating *perinatal* ones. The processes that modulate this transition remain unknown, but appear to involve a cell-autonomous transition that can be induced to happen more rapidly if *perinatal* cells are exposed to appropriate inducing factors. Intriguingly, one of the inducing factors for this transition appears to be TH, which is also a potent inducer of oligodendrocyte generation (Tang *et al.*, 2000). How the choice of a *perinatal* O-2A/OPC to become an oligodendrocyte or an *adult* O-2A/OPC is regulated is wholly unknown.

The generation of *adult* O-2A/OPCs from *perinatal* O-2A/OPCs places the behavior of the *adult* cells exposed to PDGF + FGF-2 in an interesting context. It appears that the underlying genetic and metabolic changes that lead to expression of the *perinatal* phenotype are not irreversibly lost upon generation of the *adult* phenotype. Instead, they are placed under a different control so that very specific combinations of signals are required to elicit them (Wolswijk and Noble, 1992).

Studies *In Vivo*

Based upon the expression of such antigens as NG2 and PDGFR-α, a great deal has been learned regarding the biology of cells *in situ* that are currently thought to be *adult* O-2A/OPCs. Using these antibodies, and the O4 antibody, to label cells, it has been seen that the behavior of putative *adult* O-2A/OPCs *in vivo* is highly consistent with observations made *in vitro*. Adult OPCs do divide *in situ* but, as *in vitro*, they are not rapidly dividing cells in most instances. For example, the labeling index for cells of the adult cerebellar cortex is only 0.2–0.3%. Nonetheless, as there are few other dividing cells in the brain outside of those found in highly specialized germinal zones (such as the SVZ and the dentate gyrus of the hippocampus), the adult OPC appears to represent the major dividing cell population in the parenchyma of the adult brain (Levine *et al.*, 1993; Horner *et al.*, 2000). Indeed, of the cells of the uninjured adult brain and spinal cord, it appears that 70% or more of these cells express NG2 (and thus, by current evaluations, might be considered to be adult OPCs) (Horner *et al.*, 2000). That these cells are engaged in active division is also confirmed by studies in which retroviruses are injected into the brain parenchyma. As the retroviral genome requires cell division in order to be incorporated into a host cell genome, only dividing cells express the marker gene encoded in the retroviral genome. In these experiments, 35% of all the CNS cells that label with retrovirus are NG2-positive (Levison *et al.*, 1999). However, it must be stressed for all of these experiments that it is by no means clear that all of the NG2-expressing (or O4-expressing or PDGFR-α-expressing) cells in the adult CNS are adult O-2A/OPCs. In the hippocampus, for example, such cells may also be able to give rise to neurons (Belachew *et al.*, 2003).

One of the most likely functions of *adult* O-2A/OPCs is to provide a reservoir of cells that can respond to injury. As oligodendrocytes themselves do not appear to divide following demyelinating injury (Keirstead and Blakemore, 1997; Carroll *et al.*, 1998; Redwine and Armstrong, 1998), the O-2A/OPC^{adult} is of particular interest as a potential source of new oligodendrocytes following demyelinating damage.

Observations made *in vivo* are also consistent with *in vitro* demonstrations that *adult* O-2A/OPCs can be triggered to enter transiently into a period of rapid division. When lesions are created in the adult CNS by injection of anti-oligodendrocyte antibodies (Gensert and Goldman, 1997; Keirstead *et al.*, 1998; Redwine and Armstrong, 1998; Cenci di Bello *et al.*, 1999), division of NG2$^+$ cells is observed in the area adjacent to lesion sites. Rapid increases in the number of *adult* O-2A/OPCs are also seen following creation of demyelinated lesions by injection of ethidium bromide, viral infection, or production of experimental allergic encephalomyelitis (Armstrong *et al.*, 1990a; Redwine and Armstrong, 1998; Cenci di Bello *et al.*, 1999; Levine and Reynolds, 1999; Watanabe *et al.*, 2002). Most of the putative O-2A/OPCs^{adult} in the region of a lesion have the bipolar appearance of immature perinatal glial progenitors rather than the unipolar morphology that appears to be more typical of the *adult* O-2A/OPC, just as is seen *in vitro* when O-2A/OPCs^{adult} are induced to express a rapidly dividing phenotype by exposure to PDGF + FGF-2 (Wolswijk and Noble, 1992). It is also clear that cells that enter into division following injury are responsible for the later generation of oligodendrocytes (Watanabe *et al.*, 2002).

A variety of observations indicate that the *adult* O-2A/OPCs react differently depending upon the nature of the CNS injury to which they are exposed. Adult OPCs seems to respond to almost any CNS injury (Armstrong *et al.*, 1990a; Levine, 1994; Gensert and Goldman, 1997; Keirstead *et al.*, 1998; Redwine and Armstrong, 1998; Cenci di Bello *et al.*, 1999; Levine and Reynolds, 1999; Watanabe *et al.*, 2002). Response is rapid, and reactive cells (as determined by morphology) can be seen within 24 hr. Kainate lesions of the hippocampus produce the same kinds of changes in NG2+ cells. It appears, however,

that the occurrence of demyelination is required to induce *adult* O-2A/OPCs to undergo rapid division *in situ*, even though these cells do show evidence of reaction to other kinds of lesions. For example, *adult* O-2A/OPCs respond to inflammation by undergoing hypertrophy and upregulation of NG2 but, intriguingly, increases in cell division are only seen when inflammation is accompanied by demyelination or more substantial tissue damage (Levine, 1994; Nishiyama *et al.*, 1997; Redwine and Armstrong, 1998; Cenci di Bello *et al.*, 1999). It also appears that there is a greater increase in response to anti-GalC mediated damage if there is concomitant inflammation (Keirstead *et al.*, 1998; Cenci di Bello *et al.*, 1999), indicating that the effects of demyelination on these cells are accentuated by the occurrence of concomitant injury. In this respect, the ability of GRO-α to enhance the response of spinal cord–derived *perinatal* O-2A/OPCs to PDGF may be of particular interest (Robinson *et al.*, 1998), although it is not yet known if *adult* O-2A/OPCs show any similar responses to Gro-α. Also in agreement with *in vitro* characterizations of *adult* O-2A/OPCs are observations that the progression of remyelination in the adult CNS, however, is considerably slower than is seen in the perinatal CNS (Shields *et al.*, 1999).

The wide distribution of O-2A/OPCs *in situ* is also consistent with the idea that these cells are stem cells with a primary role of participating in oligodendrocyte replacement in the normal CNS and in response to injury. It is not clear, however, whether these cells might also express other functions. For example, it is not clear whether *adult* O-2A/OPCs contribute to the astrocytosis that occurs in CNS injury. Glial scars made from astrocytes envelop axons after most types of demyelination (Fok-Seang *et al.*, 1995; Schnaedelbach *et al.*, 2000). It is known that O-2A/OPCs produce neurocan, phosphacan, NF2, and versican, all of which are present in sites of injury (Asher *et al.*, 1999, 2000; Jaworski *et al.*, 1999) and can inhibit axonal growth (Dou and Levine, 1994; Fawcett and Asher, 1999; Niederost *et al.*, 1999). It is possible that much of the inhibitory chondroitin sulfate proteoglycans found at sites of brain injury are derived from *adult* O-2A/OPCS, or from astrocytes made by *adult* O-2A/OPCs. Whether still other possible functions also need to be considered is a matter of some interest. For example, glutaminergic synapses have been described in the hippocampus on cells thought to be *adult* O-2A/OPCs (Bergles *et al.*, 2000). What the cellular function of such synapses might be is not known.

If there are so many O-2A/OPCs in the adult CNS, then why is remyelination not more generally successful? It seems clear that remyelination of initial lesions is well accomplished (at least if they are small enough), but that repeated episodes of myelin destruction eventually result in the formation of chronically demyelinated axons. It seems that after the lesions are resolved, the O-2A/OPCs^{adult} return to pre-lesion levels, consistent with their ability to undergo asymmetric division (Wren *et al.*, 1992; Cenci di Bello *et al.*, 1999; Levine and Reynolds, 1999). It also seems clear that there are *adult* O-2A/OPCs within chronically demyelinated lesions (Nishiyama *et al.*, 1999; Chang *et al.*, 2000; Dawson *et al.*, 2000; Wolswijk, 2000). Thus, the stock of these does not appear to be completely exhausted.

However, the O-2A/OPCs that are found in such sites as the lesions of individuals with multiple sclerosis (MS) are remarkably quiescent, showing no labeling with antibodies indicative of cells engaged in DNA synthesis (Wolswijk, 2000). The reasons for such quiescent behavior are unknown. There are claims that electrical activity in the axon is involved in regulating survival and differentiation of *perinatal* O-2A/OPCs in development (Barres and Raff, 1993), and it is not known if similar principles apply in demyelinated lesions in which neuronal activity is perhaps compromised. It is also possible that lesion sites produce cytokines, such as TGF-β, that would actively inhibit O-2A/OPC division. At present, however, the reasons why the endogenous precursor pool is not more successful in remyelinating extensive, or repetitive, demyelinating lesions is not known.

The possibility must also be appreciated that there may exist heterogeneity within populations of *adult* O-2A/OPCs (analogous to that seen for *perinatal* O-2A/OPCs; Power *et al.*, 2002). Whether such heterogeneity exists, and what its biological relevance might be (e.g., with respect to sensitivity to damage and capacity for repair in the adult CNS), should prove a fruitful ground for continued exploration.

Oligodendrocytes and Their Precursors as Modulators of Neuronal Development and Function

There are multiple indications that oligodendrocytes not only myelinate neurons, but also provide a large variety of signals that modulate axonal function. It has long been known that association of axons with oligodendrocytes has profound physical effects on the axon, and is associated with substantial increases in axonal diameters. Animals in which oligodendrocytes are destroyed (e.g., by radiation) and defective (as in animals lacking PLP) show substantial axonal abnormalities (Colello *et al.*, 1994; Griffiths *et al.*, 1998). In addition, axonal damage, leading eventually to axonal loss, may also occur in MS (Trapp *et al.*, 1998).

One of the dramatic effects of O-2A/OPC lineage cells on axons is to modulate axonal channel properties. During early development, both Na^+ and K^+ channels are distributed uniformly along axons, but become clustered into different axonal domains coincident with the process of myelination (Peles and Salzer, 2000; Rasband and Shrager, 2000). Na^+ channels specifically become clustered into the nodes of Ranvier, the regions of exposed axonal membrane that lay between consecutive myelin sheaths. K^+ channels, in contrast, become clustered in the juxtaparanodal region.

It has become clear from multiple studies that Schwann cells in the peripheral nervous system (PNS), and oligodendrocytes in the CNS, play instructive roles in the clustering of axonal ion channels (Kaplan *et al.*, 1997, 2001; Peles and Salzer, 2000; Rasband and Shrager, 2000). These effects are quite specific in their effects on particular channels. Contact with oligodendrocytes, or growth of neurons in oligodendrocyte-conditioned medium, is sufficient to induce axonal clustering of $Na_v1.2$ and β2 subunits, but not of $Na_v1.6$ channels (Kaplan *et al.*, 2001).

It is not yet known what regulates $Na_v1.6$ clustering, but this may require myelination itself to proceed. Once clustering has occurred, *in vitro* analysis suggests that soluble factors produced by oligodendrocytes are not required to maintain the integrity of the channel clusters.

The ability of oligodendrocytes to modulate axonal channel clustering appears to depend on the age of both the oligodendrocytes and the neurons, with mature oligodendrocytes being more effective and mature axons being more responsive. This age-dependence is in agreement with *in vivo* observations that the increase in Na channel α and β subunit levels and their clustering on the cell surface do not reach the patterns of maturity until two weeks after birth in the rat (Schmidt *et al.*, 1985; Wollner *et al.*, 1988).

In vivo demonstrations of the importance of oligodendrocytes in the formation and maintenance of axonal nodal specializations come from studies of the jimpy mouse mutant and also of a mouse strain that allows controlled ablation of oligodendrocytes as desired by the experimenter. Jimpy mice have mutations in PLP that are associated with delayed oligodendrocyte damage and death, which occurs spontaneously during the first postnatal weeks (Knapp *et al.*, 1986; Vermeesch *et al.*, 1990). The timing of oligodendrocyte death in jimpy mice cannot be altered experimentally, as is possible through the study of transgenic mice in which a herpes virus thymidine kinase gene is regulated by the MBP promoter (Mathis *et al.*, 2001). Exposure of these animals to the nucleoside analogue FIAU causes specific death of oligodendrocytes; thus, application of FIAU at different time periods allows ablation of cells at any stage of myelination at which MBP is expressed. Killing of oligodendrocytes in the MBP-TK mice is associated with a failure to maintain nodal clusters of ion channels, although the levels of these proteins remained normal. In jimpy mice, a different picture emerges, in which nodal clusters of Na^+ channels remain even in the presence of ongoing oligodendrocyte destruction. K^+ channel clusters were also transiently observed along axons of jimpy mice, but they were in direct contact with nodal markers instead of in the juxtaparanodal regions in which they would normally be found. Thus, it appears that the effect of oligodendrocyte destruction on maintenance of nodal organization is to some extent dependent upon the specific means by which oligodendrocytes are destroyed (Mathis *et al.*, 2001).

Oligodendrocytes and O-2A/OPCs as Providers of Growth Factors

There are multiple indications that oligodendrocytes and/or O-2A/OPCs also can provide trophic support for neurons, with some studies indicating that such support may exhibit elements of regional specificity (reviewed in Du and Dreyfuss, 2002). Striatal O-2A/OPC lineage cells have been reported to enhance the survival of substantia nigra neurons through secreted factors (Takeshima *et al.*, 1994; Sortwell *et al.*, 2000), O-2A/OPC lineage cells from the optic nerve can enhance retinal ganglion cell survival *in vitro* (Meyer-Franke *et al.*, 1995), basal forebrain oligodendrocytes enhance the survival of cholinergic neurons from this same brain region (Dai *et al.*, 1998, 2003), and cortical O-2A/OPC lineage cells increase the *in vitro* survival of cortical neurons (Wilkins *et al.*, 2001). It is not yet known if the trophic effects that have been reported exhibit stringent regional specificities; if so, this will be indicative of a remarkable degree of specialization in cells of the oligodendrocyte lineage.

While the study of trophic support derived from O-2A/OPCs or oligodendrocytes is still in its infancy, an increasing number of interesting proteins have been observed to be produced by oligodendrocytes. For example, IGF-I, NGF, BDNF, NT-3, and NT-4/5 mRNAs and/or protein have been observed by *in situ* hybridization and via immunocytochemical studies in oligodendrocytes (Dai *et al.*, 1997, 2003; Dougherty *et al.*, 2000). Consistent with the idea that there might be trophism-related differences in oligodendrocytes from different CNS regions, it does appear that there is regional heterogeneity in the expression of these important proteins (Krenz and Weaver, 2000). Still other proteins that have been suggested to be produced by oligodendrocytes include neuregulin-1 (Vartanian *et al.*, 1994; Raabe *et al.*, 1997; Cannella *et al.*, 1999; Deadwyler *et al.*, 2000), GDNF (Strelau and Unsicker, 1999), FGF-9 (Nakamura *et al.*, 1999), and members of the TGF family (da Cunha *et al.*, 1993; McKinnon *et al.*, 1993). Many of the factors that oligodendrocytes appear to produce have been found to influence the development not only of neurons, but also of oligodendrocytes themselves. Thus, it may prove that one of the functions of oligodendrocytes is to produce factors that modulate their own functions. Such a notion is consistent with observations that oligodendrocytes produce factors that feedback to modulate the division and differentiation of O-2A/OPCs in a density-dependent manner (McKinnon *et al.*, 1993; Zhang and Miller, 1996).

O-2A/OPCs and oligodendrocytes also receive trophic support from both astrocytes and neurons. Astrocytes have long been known to produce such modulators of O-2A/OPC division and oligodendrocyte survival as PDGF and IGF-I (Ballotti *et al.*, 1987; Noble *et al.*, 1988; Raff *et al.*, 1988; Richardson *et al.*, 1988). Neurons appear to be a another source of PDGF (Yeh *et al.*, 1991), but also modulate the behavior of O-2A/OPC lineage cells by other means. For example, it has been reported that injection of tetrodotoxin into the eye, thus eliminating electrical activity of retinal ganglion cells, causes a decrease in proliferation of O-2A/OPCs (Barres and Raff, 1993). O-2A/OPCs and oligodendrocytes express K^+ channels (Barres *et al.*, 1990) and also express receptors for a variety of neurotransmitters, including glutamate and acetylcholine (Cohen and Almazan, 1994; Gallo *et al.*, 1994; Patneau *et al.*, 1994; Rogers *et al.*, 2001; Itoh *et al.*, 2002), thus enabling them to be responsive to the release of such transmitters in association with neuronal activity. Indeed, exposure to neurotransmitters can profoundly affect the proliferation and differentiation of O-2A/OPCs *in vitro* (Gallo *et al.*, 1996). Exposure to neurotransmitters can also alter the expression of neurotrophins (NTs) in oligodendrocytes (Dai *et al.*, 2001), raising the possibility that neuronal signaling to oligodendrocytes via neurotransmitter release can alter the trophic support that the oligodendrocyte may provide for the neuron. It is particularly intriguing that there appears to be a great

deal of specificity in the effects of different kinds of putative neuron-derived signals on trophic factor expression in oligodendrocytes. KCl has been reported to increase expression of BDNF mRNA, carbachol (an acetylcholine analogue) to increase levels of NGF mRNA, and glutamate specifically to decrease levels of BDNF expression (Dai *et al.*, 2001).

Functions of Myelin Components

As one might expect for such a highly specialized biological structure as myelin, there are a large number of proteins and lipids that are specifically produced by myelinating cells. It is therefore of considerable interest to understand the function of these myelin-specific molecules (as reviewed in more detail, e.g., in Campignoni and Macklin, 1988; Yin *et al.*, 1998; Campignoni and Skoff, 2001; Pedraza *et al.*, 2001; Woodward and Malcolm, 2001).

The two major structural proteins of myelin itself are PLP and MBP. PLP constitutes approximately 50% by weight of myelin proteins (Braun, 1984; Morell *et al.*, 1994). It appears to interact homophilically with other PLP chains from the surface of the myelin membrane in the next loop of the spiral (Weimbs and Stoffel, 1992). This ability of PLP to bind to PLP proteins in the next loop of the myelin spiral is thought to play an important role in leading to close apposition of the outer membranes of adjacent myelin spirals. The MBPs are actually a group of proteins that are the next most abundant myelin proteins, comprising 30–40% by weight of the proteins found in myelin (Braun, 1984; Morell *et al.*, 1994). In contrast with PLP, MBP is located on the cytoplasmic face of the myelin membrane. It is thought to stabilize the myelin spiral at the major dense line by interacting with negatively charged lipids at the cytoplasmic face of the lipid membrane (Morell *et al.*, 1994). Both PLP and MBP are critical in the creation of normal myelin.

The dependency on MBP for normal oligodendrocyte function has long been known due to studies of the shiverer mouse strain. Shiverer (*shi*) mice, which are neurologically mutant and exhibit incomplete myelin sheath formation, lack a large portion of the gene for the MBPs, have virtually no compact myelin in their CNS, and shiver, undergo seizures, and die early. Still another mouse mutant characterized by a deficiency of myelin is the *mld* mutation, which consists of two tandem MBP genes, with the upstream gene containing an inversion of its 3′ region. In these mice, MBP is expressed at low levels and on an abnormal developmental schedule (Popko *et al.*, 1988). Still another animal model of defective myelination associated with a mutation in the MBP gene is the Long Evans shaker (*les*) rat. Although scattered myelin sheaths are present in some areas of the CNS, most notably the ventral spinal cord in the young neonatal rat, this myelin is gradually lost, and by 8–12 weeks after birth, little myelin is present throughout the CNS. Despite this severe myelin deficiency, some mutants may live beyond 1 yr of age. Rare, thin myelin sheaths that are present early in development lack MBP. On an ultrastructural examination, these sheaths are poorly compacted and lack a major dense line. Many oligodendrocytes in these animals develop an accumulation of vesicles and membranous bodies, but no abnormal cell death is observed. Unlike *shi* and its allele, where myelin increases with time and oligodendrocytes become ultrastructurally normal, *les* oligodendrocytes are permanently disabled, continue to demonstrate cytoplasmic abnormalities, and fail to produce myelin beyond the first weeks of life (Kwiecien *et al.*, 1998). These various strains of MBP-defective animals also provide an opportunity for analyzing the function of individual MBP splice variants, of which there are at least five. Surprisingly, restoration of just the 17.2 kDa isoform (which is normally one of the minor myelin components) in the germline of transgenic shiverer mice is sufficient to restore myelination and nearly normal behavior (Kimura *et al.*, 1998).

Studies on the function of MBP are rendered more complex by the fact that the MBP gene also encodes a novel transcription unit of 105 Kb (called the Golli-mbp gene) (Campagnoni *et al.*, 1993). Three unique exons within the Golli gene are alternatively spliced to produce a family of MBP gene-related mRNAs that are under individual developmental regulation. These mRNAs are temporally expressed within cells of the oligodendrocyte lineage at progressive stages of differentiation. Golli proteins show a different developmental pattern than that of MBP, however, with the highest levels of *golli* mRNA expression being in intermediate stages of oligodendrocyte differentiation, and with levels being reduced in mature oligodendrocytes (Givogri *et al.*, 2001). Thus, the MBP gene is a part of a more complex gene structure, the products of which may play a role in oligodendrocyte differentiation prior to myelination (Campagnoni *et al.*, 1993). For these reasons, compromising the function of the MBP gene actually results in compromised expression of the Golli proteins, and attributing a particular developmental outcome selectively to either MBP transcripts or Golli transcripts is not possible.

Golli expression is also seen in cortical preplate cells, and targeting of herpes simplex thymidine kinase by the *golli* promoter allows selective ablation of preplate cells in the E11-12 embryo, leading to a dyslamination of the cortical plate and a subsequent reduction in short- and long-range cortical projection within the cortex and to subcortical regions (Xie *et al.*, 2002). Golli proteins, as well as PLP and DM-20 transcripts of the *plp* gene are also expressed by macrophages in the human thymus, which may be of relevance to the association between MS and immune response to MBP epitopes that are also expressed by *golli* gene products (Pribyl *et al.*, 1996).

There are also animal models of mutations in PLP, such as the jimpy mouse strain. In these mice, one sees delayed oligodendrocyte damage and death, which occurs spontaneously during the first postnatal weeks (Knapp *et al.*, 1986; Vermeesch *et al.*, 1990). PLP does not appear to be required for initial myelination, but is required for maintenance of myelin sheaths. In the absence of PLP, mice assemble compact myelin sheaths but subsequently develop widespread axonal swellings and degeneration (Griffiths *et al.*, 1998).

Along with analysis of myelin-specific proteins, it has also been possible to start dissecting the role of specific myelin lipids in oligodendrocyte function by examining CNS development in mice in which key enzymes required in lipid biosynthesis have been genetically disrupted. A particularly interesting

demonstration of the importance of the myelin-specific lipids has come from the study of mice that are incapable of synthesizing sulfatide due to disruption of the galactosylceramide sulfotransferase gene (Ishibashi *et al.*, 2002). Although compact myelin is itself preserved in these animals, abnormal paranodal junctions are found in both the PNS and CNS. Abnormal nodes are characterized by a decrease in Na^+ and K^+ channel clusters, altered nodal length, abnormal localization of K^+ channel localization, and a diffuse distribution of contactin-associated protein (Caspr) along the internode. This aberrant nodal organization arises despite the fact that the initial timing and number of Na^+ channel clusters are normal during development. The interpretation of these results is that sulfatide plays a critical role in maintaining ion channel organization but is not essential for establishing initial cluster formation. Similar results have been observed in mice lacking GalC (an essential precursor for sulfatide formation; Dupree *et al.*, 1998, 1999) and also in mice lacking Caspr (Bhat *et al.*, 2001) or contactin (Boyle *et al.*, 2001). Interestingly, sulfatide-deficient mice have a milder clinical phenotype than the animals deficient in both GalC and sulfatide, indicating that GalC may itself have other important roles that it plays. Whether the role of these lipids is to participate directly in interactions with components of the axonal membrane, to play a role in organizing oligodendrocyte membrane proteins that are themselves involved in oligodendrocyte–neuron interactions, or have still other unknown roles, is not yet known.

Other means by which oligodendrocyte function is disrupted, and the neurological consequences of such disruption are considered when we examine human genetic diseases that affect myelin.

MYELIN-RELATED DISEASES

Genetic Diseases of Oligodendrocytes and Myelin

A multitude of genetic diseases are associated with myelination defects. Experimental diseases of mice associated with structural mutations in important myelin proteins have been discussed earlier, such as seen in jimpy or shiverer mice, and human diseases associated with defects in myelin proteins are also known. In addition, there are a large number of metabolic diseases in humans in which myelination is abnormal, and white matter damage is even seen in individuals in which the underlying mutation affects proteins involved in RNA translation.

A myelin-related disease associated with a structural protein defect is the X-linked Pelizaeus–Merzbacher disease associated with mutations in the PLP gene (Woodward and Malcolm, 1999). Children with more severe symptoms tend to have severe abnormalities in protein folding in other structural aspects of the myelin, which would cause changes in the physical structure of the myelin. In addition, accumulation of misfolded proteins in the cell may trigger oligodendroglial apoptosis and consequent demyelination (Gow *et al.*, 1998). It is interesting that if the gene is completely deleted, affected children have a relatively mild form of the disease, despite the hypomyelination (Raskind *et al.*, 1991; Sistermans *et al.*, 1996).

Adrenoleukodystrophy is the most commonly occurring leukodystrophy in children. This X-linked disorder, caused by a mutation of the gene encoding a peroxisomal membrane protein, affects one in 20,000 boys (Dubois-Dalcq et al., 1999). The mutated protein (called ALD protein) is necessary for transferring very long-chain fatty acids into peroxisomes, where they are metabolized into shorter chain fatty acids for multiple purposes, including incorporation into the myelin membrane. ALD protein is found in all glial cells, but its expression in oligodendrocytes is limited to the locations that correlate well with locations of demyelination in affected children (Fouquet et al., 1997), such as corpus callosum, internal capsule, and anterior commissure. While it is not known why myelin breaks down in these children, it appears that the mutation somehow destabilizes the membrane. Then, in conjunction with inflammatory events in putatively dysfunctional microglia (in which the ALD protein is also expressed), this inherent weakness stimulates (or enables) consequent demyelination. MR imaging shows T2 prolongation during the early stages of disease, but whether this is primarily due to myelin breakdown or inflammation is not clear. The inflammation results in localized edema which itself is associated with imaging changes.

Metachromatic leukodystrophy (MLD) is an autosomal recessive disorder caused by deficient activity of the lysosomal enzyme arylsulfatase A. These patients may present at any age, have gait abnormalities, ataxia, nystagmus, hypotonia, diffuse spasticity, and pathologic reflexes (Barkovich, 2000). Myelin is usually formed normally in this condition, but the eventual membrane accumulation of sulfatide associated with this enzymatic defect results in an instability of the myelin membrane with ultimate demyelination. Damage may also occur due to progressive accumulation of sulfatides within oligodendroglial lysosomes, leading to eventual degeneration of the lysosomes themselves. There is extensive demyelination that develops, with complete or nearly complete loss of myelin in the most severely affected regions (van der Knaap and Valk, 1995).

Canavan's disease (CD) is another example of an autosomal recessive early-onset leukodystrophy, caused in this case by mutations in the gene for aspartoacylase. This is the primary enzyme involved in the catabolic metabolism of N-acetylaspartate (NAA), and its deficiency leads to a build-up of NAA in brain with both cellular and extracellular edema, as well as NAA acidemia and NAA aciduria. CD is characterized by loss of the axon's myelin sheath, while leaving the axons intact, and by spongiform degeneration, especially in white matter. The course of the illness can show considerable variation, and can sometimes be protracted. The mechanism by which a defect in NAA metabolism causes myelination deficits remains unknown, although it has been suggested that changes in osmotic balance due to buildup of NAA (which, even in the normal brain, is one of the most abundant single free amino acids detected) may be of importance (Baslow, 2000; Gordon, 2001; Baslow et al., 2002). It has also been suggested that NAA supplies acetyl groups for myelin lipid biosynthesis, a possibility consistent with known cellular expression of both NAA and its relevant enzymes (Urenjak et al., 1992, 1993; Bhakoo and Pearce, 2000; Bhakoo et al., 2001; Chakraborty et al., 2001).

Some of the most puzzling of genetic diseases in which myelin is affected are those in which the CNS initially undergoes normal development, and subsequently the individual is afflicted with a chronic and diffuse degenerative attack on the white matter. One of these disorders that has been genetically defined is a syndrome called vanishing white matter (VWM; MIM 603896) (Hanfield et al., 1993; van der Knaap et al., 1997), also called childhood ataxia with central hypomyelination (CACH; van der Knaap et al., 1997). VWM is the most frequent of the unclassified childhood leukoencephalopathies (van der Knaap et al., 1999). Onset is most often in late infancy or early childhood, but onset may occur at times ranging from early infancy to adulthood (Hanfield et al., 1993; van der Knaap et al., 1997, 2001; Francalanci et al., 2001; Prass et al., 2001). VWM is a chronic progressive disease associated with cerebellar ataxia, spasticity, and an initially, relatively mild mental decline. Death occurs over a very variable period, which may range from a few months to several decades. It has been suggested that oligodendrocyte dysfunction, leading to myelin destruction (and possibly associated with initial hypomyelination in cases with early onset) is the primary pathologic process in VWM (Schiffmann et al., 1994; Rodriguez et al., 1999; Wong et al., 2000).

VWM is an autosomal recessive disease, and it has been recently found that the underlying mutations may be in any of the five subunits of the eukaryotic translation initiation factor (eIF), eIF2B (Leegwater et al., 2001; van der Knaap et al., 2002). This discovery was quite surprising, as the widespread importance of initiation factors in cellular function makes it difficult to understand why a mutation in one of them should manifest itself so specifically as an abnormality in white matter. Indeed, despite the identification of the genetic basis of VWM, little is known about the biology of this disease, including the answers to such questions as: How can one have a disease in which oligodendrocyte function is apparently normal to begin with, and then at later stages—often after years of normal development and function— a chronic deterioration of myelin begins? And why would such a specific disease result from a mutation in a protein thought to be important in RNA translation throughout the body? Moreover, what function of initiation factors might explain the onset of the chronic white matter degeneration that characterizes this disease?

At the moment, one of the few clues to the underlying pathophysiology of VWM comes from observations that patients with this disease undergo episodes of rapid deterioration following febrile infections and minor head trauma. It has been suggested that mutations in eIF2B might be associated with an inappropriate response by oligodendrocytes to such stress (which would include within it febrile [thermal], oxidative, and chemical perturbations) (van der Knaap et al., 2002). Normally, mRNA translation is inhibited in such adverse circumstances, perhaps as a protective response against the capacity of such abnormal metabolic states to compromise normal folding of many proteins. Excessive accumulation of misfolded proteins then could lead to interference with normal cellular function, as has also been suggested earlier for Pelizaeus–Merzbacher disease. Attempts to understand the underlying pathophysiology of this disease remain speculative, however, in the absence of cellular and/or animal models suitable for detailed analysis. Moreover, it is difficult to reconcile such a hypothesis with observations that VWM disease is inherited as an autosomal recessive, rather than as a dominant trait, as a hypothesis invoking continued mRNA translation would be indicative of a dominant rather than a recessive function. Until such time as appropriate cellular tools (such as precursor cells from a patient with this disease) are available, it will remain unknown as to whether oligodendrocytes are particularly sensitive to alterations in the biology of mRNA translation, whether there is instead a failure in this disease to carry out the normal turning off of injury responses (thus leading to release of glutamate, secretion of tumor necrosis factor [TNF]-α, and other such responses as are associated with oligodendrocyte destruction), or whether other processes are involved in this tragic condition. Given only human autopsy tissue to study, one is limited to such observations as oligodendrocytes in the brains of VWM exhibiting an aberrant foamy cytological structure (Wong *et al.*, 2000), but it is wholly unknown whether this is a primary effect of the mutation in eIF2B or a secondary consequent of the extended period of destruction to which they have been subjected.

Studies on VWM also reveal another of the many areas in which our understanding of myelin function is incomplete. It is a striking feature of VWM that magnetic resonance imaging (MRI) reveals diffuse abnormalities of the cerebral white matter prior to the onset of symptoms (van der Knaap *et al.*, 1997). MRI and magnetic resonance spectroscopic analysis both indicate that as this disease progresses, increasing amounts of the cerebral white matter vanish and are replaced by cerebrospinal fluid (CSF), as is confirmed by examination of brains at autopsy (van der Knaap *et al.*, 1997, 1998; Rodriguez *et al.*, 1999). Still, it appears clear that damage to the white matter has already begun before clinical symptoms emerge.

The idea that one can have extensive loss of myelin without evidence of neurological abnormality seems extraordinarily counterintuitive. Yet, it has long been known that extensive demyelination is not always associated with clinical deficits in MS patients. The suggested explanations for this phenomena of "silent lesions" have generally been that they may be located in areas in which a loss of conduction does not manifest itself in a clinically detectable manner and/or that sufficient normally myelinated axons in these regions are spared to enable normal function. Such suggestions are consistent with multiple lines of evidence indicating functional redundancy in axonal pathways. Indeed, in such chronic neurodegenerative diseases as Parkinson's disease and Alzheimer's disease, it is clear that clinical symptoms are not seen until 50–70% of the relevant neurons have been destroyed. Still, it may be that there is a more complex biology that lies behind the situation in which loss of myelin is not associated with clinical manifestations. Such a possibility is indicated by experimental studies in which extensive demyelination was induced by infection of two different strains of mice with Theiler's virus (Rivera-Quinones *et al.*, 1998). Normal function was maintained in mice defective for expression of major histocompatibility complex (MHC) class I gene products, despite the presence of a similar distribution and extent of demyelinated

lesions as in other mouse strains in which neurological function was compromised. It has been proposed that the maintenance of normal neurological function in class I antigen-deficient mice with extensive demyelination results from increased sodium channel densities and the relative preservation of axons.

Nongenetic Diseases of Myelin

Aberrant myelination is also associated with a wide range of epigenetic physiological insults. Causes of such problems are so diverse as to include various nutritional deficiency disorders, hypothyroidism, fetal alcohol syndrome, treatment of CNS cancers of childhood by radiation, and treatment of even some non-CNS cancers of childhood by chemotherapy.

Hypothyroidism

A major cause of mental retardation and other developmental disorders is hypothyroidism, usually associated with iodine deficiency (e.g., Delange, 1994; Lazarus, 1999; Chan and Kilby, 2000; Thompson and Potter, 2000). It is well established in animal models that perinatal hypothyroidism is associated with defects in myelination and a reduced production of myelin-specific gene products, and that these defects can be at least partially ameliorated if TH therapy is initiated early enough in postnatal life (e.g., Noguchi *et al.*, 1985; Munoz *et al.*, 1991; Bernal and Nunez, 1995; Ibarrola and Rodriguez-Pena, 1997; Marta *et al.*, 1998). As for other deficiency disorders, however, application of hormonal replacement therapy after the appropriate critical period has been completed has relatively little effect.

The actions of TH to promote myelination are several. This hormone has been found to promote the generation of O-2A/OPCs from GRP cells, as well as promoting the generation of oligodendrocytes from dividing O-2A/OPCs (Barres *et al.*, 1994a; Ibarrola *et al.*, 1996; Gregori *et al.*, 2002a). TH also modulates the expression of multiple myelin genes (e.g., Oppenheimer and Schwartz, 1997; Jeannin *et al.*, 1998; Pombo *et al.*, 1999; Rodriguez-Pena, 1999). *In vivo*, reduction in TH levels are associated with an 80% reduction in the number of oligodendrocytes, which is the same degree of difference in oligodendrocyte prevalence observed in embryonic brain cultures grown in the presence or absence of TH (Ibarrola *et al.*, 1996).

Iron Deficiency

The most prevalent nutrient deficiency in the world is a lack of iron. It has been estimated that 35–58% of healthy women have some degree of iron deficiency (Fairbanks, 1994). Iron deficiency is particularly prevalent during pregnancy. Iron deficiency in children is associated with hypomyelination, changes in fatty acid composition, alterations to the blood brain barrier and behavioral effect (Pollitt and Leibel, 1976; Honig and Oski, 1978; Dobbing, 1990). It has been reported that the prevalence of iron deficiency may be as high as 25% for children under two years of age, as indicated by measurement of auditory brain responses as a measurement of conduction speed (Roncagliolo *et al.*, 1998).

That iron deficiency would be particularly important during specific developmental periods has been suggested by observations that there is a temporal correlation between the period in development when most oligodendrocytes are developing and a peak in iron uptake into the brain (Yu et al., 1986; Taylor and Morgan, 1990). In iron-deficient animals, where no such peak in iron uptake can occur, there is a relative lack of myelin lipids. The myelin isolated from these iron-deficient animals is normal in the ratios of its myelin components, however, suggesting that the reduced amount of myelin produced in these animals is normal in its biochemical composition.

The Role of Iron in Oligodendrocyte Generation

The role of iron in the myelination process is an emerging area of study in the development of the CNS. It has been noted that when the brains of many different species are histochemically labeled for iron, the cells with the highest iron levels are oligodendrocytes (Hill and Switzer, 1984; Dwork et al., 1988; Connor and Menzies, 1990; LeVine and Macklin, 1990; Morris et al., 1992; Benkovic and Connor, 1993). While the role of iron in oligodendrocytes is unknown, it has been suggested that a lack of iron might somehow interfere with the function of these cells (Connor and Menzies, 1996). The lack of myelination associated with iron deficiency has been measured in humans using auditory brainstem responses (ABRs). Changes in the latency of the ABRs have been related to the increased nerve conduction velocity that accompanies axonal myelination (Salamy and McKean, 1976; Hecox and Burkard, 1982; Jiang, 1995). A recent study has shown that there are measurable differences in ABR latency between normal and iron-deficient children (Roncagliolo et al., 1998), reflecting a myelination disorder.

Iron is taken up by cells predominantly when bound to transferrin, the mammalian iron transporter. Oligodendrocytes have the highest levels of transferrin mRNA and protein, and indeed seem to be responsible for transferrin production in the CNS (Connor and Fine, 1987; Dwork et al., 1988; Bartlett et al., 1991; Connor et al., 1993; Connor, 1994; Dickinson and Connor, 1995). These observations have led to the suggestion that oligodendrocytes are responsible for storing iron and for making it readily available to the environment, as well as suggestions that iron is important in critical—but currently unknown—steps in oligodendrocyte development (Connor and Menzies, 1996).

There is also a temporal correlation between the period in development when most oligodendrocytes are developing and a peak in iron uptake into the brain (Skoff et al., 1976a, b; Crowe and Morgan, 1992). In iron-deficient animals, where no such peak in iron uptake can occur, a reduction in myelin lipids can be measured (Connor and Menzies, 1990). The myelin isolated from these iron-deficient animals is normal in the ratios of its myelin components, suggesting that the myelin produced in iron-deficient rats is normal but that overall less myelin is being produced. The suggestion that it might be necessary to have adequate levels of bioavailable iron in order for normal myelination to occur is also supported by the observation that in myelin-deficient rats, in which oligodendrocytes fail to mature due to a genetic defect in the PLP, the levels of transferrin (bioavailable iron) in the brain are well below normal levels (Bartlett et al., 1991). Strikingly, exposure of myelin-deficient rats to transferrin can promote the production of myelin (Escobar Cabrera et al., 1997).

Despite the considerable evidence linking iron deficiency with defects in myelin production, it is still not clear how a defect in myelination might be established and at what timepoint during gliogenesis iron availability is important. As most data has been provided through descriptive studies in vivo, a mechanistic basis for iron-mediated myelin deficiency has not been established.

Cellular biological studies have indicated an importance of iron levels in the generation of oligodendrocytes from GRP cells (presumably through the intermediate generation of O-2A/OPCs, although this has not yet been confirmed) (Morath and Mayer-Proschel, 2001). In contrast, no effects of iron were found on oligodendrocyte maturation or survival in vitro, nor did increasing iron availability above basal levels increase oligodendrocyte generation from O-2A/OPCs. These results raise the possibility that iron may affect oligodendrocyte development at stages during early embryogenesis rather than during later development. This possibility is supported by in vivo studies demonstrating that iron deficiency during pregnancy affects the iron levels of various brain tissues in the developing fetus, and disrupts not only the proliferation of their glial precursor cells, but also disturbs the generation of oligodendrocytes from these precursor cells (Morath and Mayer-Proschel, 2002).

Selenium Deficiency

Still another syndrome associated with myelination defects is a deficiency in the essential trace element selenium. Selenium deficiency has been postulated to be associated with retarded intellectual development (Foster, 1993) and to neural tube defects (Guvenc et al., 1995). It has also been suggested that the incidence of MS is negatively correlated with selenium levels in the soil, suggesting that selenium deficiency may predispose oligodendrocytes to demyelinating injury (Foster, 1993).

In vitro studies have shown that normal selenium levels are required for both the normal morphological development and the survival of oligodendrocytes (Eccleston and Silberberg, 1984; Koper et al., 1984). Moreover, exposure to adequate levels of selenium is required for the normal upregulation of genes for PLP, MBP, and MAG. A deficiency of selenium in vitro is also associated with a reduction in the generation of oligodendrocytes from their precursor cells (Gu et al., 1997).

The mechanisms by which selenium deficiency may alter oligodendrocyte generation are far from clear. In vivo, it is known (Kohrle, 1996) that selenium is required for activity of the deiodinase that cleaves one iodine from T4 to make the bioactive T3 (triiodothyronine). Consistent with this role of selenium, deficiency in this trace element is known to cause further impairment of TH metabolism in iodine-deficient rats (Mitchell et al., 1998). Selenium also plays a critical role in redox regulation, however, particularly as many of the selenoproteins play critical roles in regulation of intracellular redox balance (Holben and Smith,

1999). In this regard, it may be that a lack of selenium leads to a more oxidized state in O-2A/OPCs, thus leading to their premature transition from dividing progenitor cells to nondividing oligodendrocytes (Smith *et al.*, 2000). As this would be associated with a reduction in oligodendrocyte number (secondary to a reduction in progenitor cell number), one would see associated reductions in myelin-specific genes when cultures were examined at the population level.

Nutrition and Oligodendrocyte Generation

We are not yet aware of any studies that have examined nutritional deficiency in a manner directly analogous to studies on TH or iron deficiency. Indeed, developing a model system for studying nutritional deficiency *in vitro* is problematic in a number of respects. Perhaps most importantly, true nutritional deficiency is associated with inadequate supplies of proteins, vitamins, and minerals and can itself lead to reduced production of normal hormonal supplies. This is a considerably more difficult syndrome to reproduce *in vitro* than TH deficiency, for example. Nonetheless, published data, from both *in vivo* and *in vitro* studies, are consistent with the possibility that oligodendrocyte generation is impaired in at least some models of undernourishment. *In vivo*, it is well established that the myelin deficits associated with undernutrition are even observed in animals in which oligodendrocyte number appears to be normal (Sikes *et al.*, 1981). In such animals, however, it has been reported (Royland *et al.*, 1993) that the mRNAs for three important myelin proteins (MAG, PLP, and MBP) do not undergo the normal increases seen in brains of well-nourished animals. Increases are delayed for several days beyond the normal time (i.e., day 7–9) at which they are observed, and the increases are lower in extent. In addition, still more severe malnutrition regimes have been reported to be associated with a clear reduction in glial cell number *in vivo* (Krigman and Hogan, 1976), although cell type specific markers were not utilized to determine whether this reduction preferentially effected oligodendrocytes rather than astrocytes.

In vitro studies on nutritional deficiency have largely focused on glucose deprivation as a means of mimicking caloric restriction. Such studies have raised the surprising possibility that transient caloric restriction at critical periods may lead to long-term effects on differentiated function (Royland *et al.*, 1993). In these experiments, mixed cultures were generated from newborn rat brain and exposed to different glucose concentrations, ranging from 0.55 to 10 mg/ml; the lower doses are within the range that occurs in clinical hypoglycemia. Low glucose concentrations were associated with markedly lower increases in levels of MAG, PLP, and MBP mRNA, and with a subsequent and abnormal downregulation in these mRNA levels. These effects were specific, in that total mRNA levels in the cultures were normal. Most importantly, these effects appeared to be irreversible if the glucose deprivation was applied over a time period that mirrors the critical period for nutritional deprivation *in vivo*. Deprivation coincident with the normal time of myelin gene activation and the period of rapid upregulation (6–14 DIV)

was irreversible. Deprivation at a later stage was instead associated with only transient depressing effects. It has also been previously reported that there is a relative reduction in the numbers of oligodendrocytes that are generated in glucose-deprived cultures (Zuppinger *et al.*, 1981).

Physiological Insults Associated with Developmental Abnormalities in Myelination

Still another means by which normal developmental processes may be thwarted is through the introduction of toxic substances into the developing organism.

Fetal Alcohol Syndrome

Evidence suggests that abnormal myelination is one factor contributing to the neuropathology associated with fetal alcohol syndrome. Studies on the expression of MBP and MAG, isoforms in experimental animals showed a considerable vulnerability to postnatal (but not prenatal) exposure to ethanol. These studies indicate that ethanol exposure during periods of rapid myelination (postnatal days 4–10) reduced the expression of specific MBP and MAG isoforms (Zoeller *et al.*, 1994). *In vitro* studies have also indicated that exposure to ethanol during early stages of oligodendrocyte development is associated with a specific repression of MBP expression, but not of the myelin-specific enzyme 2′,3′-cyclic nucleotide 3′-phosphodiesterase (CNPase). Delayed or decreased MBP expression could interfere with normal processes of myelination, as indicated by the adverse consequences of genetic interference with normal MBP expression or function (Bichenkov and Ellingson, 2001). In adult alcoholics, there are changes in expression of as many as 40% of superior frontal cortex-expressed genes (as determined from examination of postmortem samples). In particular, myelin-related genes were significantly downregulated in the brain specimens from alcoholics (Lewohl *et al.*, 2000).

Fetal Cocaine Syndrome

Abnormalities in myelination have also been associated with exposure to cocaine. The progeny born to pregnant rats treated daily with oral cocaine during gestation showed a 10% reduction in myelin concentrations in the brain. In contrast with the period of myelin vulnerability for undernourishment, which is thought to be largely postnatal, cross-fostering studies revealed that the fetal period of cocaine exposure presents a greater risk to postnatal myelination than exposure during the suckling period (Wiggins and Ruiz, 1990). As myelination in the human is not complete until the fourth decade (Yakovlev and Lecours, 1967), there has been some concern as to whether the ongoing processes of myelination might be disrupted in cocaine users. Indeed, in normal individuals, there is a continued increase in white matter volume in the frontal and temporal lobes that does not reach a maximum until age 47. In cocaine-dependent subjects, in contrast, this age-related expansion in white matter volume in the frontal and temporal cortex does not appear to occur (Bartzokis *et al.*, 2002).

Effects of Organic Mercury Compounds

Exposure to MeHg provides yet another example wherein exposure to toxic substances interferes with normal patterns of development. It is clear from unfortunate experiences with contaminated wheat in Iraq and contaminated fish in Japan that high levels of exposure to MeHg is associated with severe abnormalities in the developing brain, including neuronal migration disorders and diffuse gliosis of the periventricular white matter (Choi, 1989). Studies in the Faroe islands, the Seychelles Island, New Zealand, and the Amazon Basin have further found that children born from mothers exposed during pregnancy to moderate doses of MeHg showed significantly reduced performance on several neuropsychological tests (Crump et al., 1998, 2000; Grandjean et al., 1998, 1999; Dolbec et al., 2000). Children exposed to mercury during development may exhibit a range of neurological problems, including cerebral palsy (which includes failures in normal myelination), developmental delay, and white matter astrocytosis (Castoldi et al., 2001; Mendola et al., 2002).

The developing nervous system is more sensitive to MeHg neurotoxicity than the adult nervous system (Clarkson, 1997; Myers and Davidson, 1998). MeHg appears to have a wide range of toxic effects on the developing CNS. For example, developmental exposure to MeHg is associated with decreases in cell survival, myelination, and cerebral dysgenesis (Chang et al., 1977; Burbacher et al., 1990; Barone, Jr., et al., 1998), as well as decreased expression and/or activity of proteins involved in neurotrophic factor signaling (Barone, Jr., et al., 1998; Haykal-Coates et al., 1998; Mundy et al., 2000) and changes in neurotrophic factor expression (Lärkfors et al., 1991).

An organic mercury compound that has become of considerable recent interest as a potential inducer of developmental abnormalities is Thimerosal, a vaccine preservative that contains 49.6% ethylmercury (by weight) as its active ingredient. Concern has been raised that apparent increases in the prevalence of autism (from 1 in 2000 prior to 1970 up to 1 in 500 in 1996 (Gillberg and Wing, 1999)) have paralleled the increased mercury intake induced by mandatory inoculations. In 1999, the Food and Drug Administration (FDA) recorded Thimerosal usage in over 30 vaccine products (FDA, November 16, 1999). According to the classification of Thimerosal-containing vaccines provided by the Massachusetts Department of Public Health, as of June 2002, Thimerosal was still in use as a preservative in a significant number of vaccines, including diphtheria/tetanus, Hep B, Influenza, Meningococcus, and Rabies vaccines. The World Health Organization (WHO), the American Academy of Pediatrics, and the US Public Health Service have all voiced support for phasing out Thimerosal usage as a vaccine preservative, but the WHO has stressed that this may not be an option for developing countries. While a recent Danish study (Madsen et al., 2002) failed to find a link between autism and vaccination with the measles, mumps, rubella (MMR) vaccine, this is not a Thimerosal-containing vaccine and thus did not shed light on controversies related to autism and mercury exposure. The hypothesis that mercury exposure and autism are linked is discussed extensively in Bernard et al. (2001), including information on the multiple similarities between the neurological symptoms seen in mercury poisoning and those considered to typify autism.

The amount of mercury that would be delivered to a child born in the 1990s in association with vaccination over the first two years of life is not small, and is delivered in bolus form (as part of a vaccination). The amount of mercury injected at birth is 12.5 μg, followed by 62.5 μg at 2 months, 50 μg at 4 months, another 62.5 μg during the infant's 6-month immunizations, and a final 50 μg at about 15 months (Halsey, 1999). Concerns exist that infants under 6 months may be inefficient at mercury excretion, most likely due to their inability to produce bile, the main excretion route for organic mercury (Koos and Longo, 1976; Clarkson, 1993). More recent studies have challenged these concerns, reporting that blood mercury in Thimerosal-exposed 2-month-olds ranged from less than 3.75 to 20.55 parts per billion; in 6-month-olds, all values were lower than 7.50 parts per billion (Pichichero et al., 2002).

Ongoing studies on the effects of MeHg and Thimerosal on cells of the oligodendrocyte lineage have revealed a striking vulnerability of these cells to organic mercury compounds (MN, research in progress). Studies have thus far indicated that exposure of oligodendrocytes and O-2A/OPCs to doses of MeHg or Thimerosal in the ranges of 5–20 parts per billion is associated with significant cell death and inhibition of cell division. These are precisely the ranges of mercury levels that are routinely found in both infant and adult populations. Moreover, exposure to still lower levels of MeHg is sufficient to increase the sensitivity of O-2A/OPCs to killing by glutamate and of oligodendrocytes to killing by TNF. (Such vulnerabilities are discussed in more detail in the following section.) Thus, oligodendrocytes and their precursor cells may also be an important target of action of organic mercury compounds—and perhaps of many other environmental toxicants.

Neurotoxicity of Existing Cancer Treatments

It is becoming increasingly apparent that traditional approaches to cancer therapy are often associated with adverse neurological events, many of which affect the white matter tracts of the CNS. These neurological sequelae are seen in treatment regimes ranging from chemotherapy of primary breast carcinoma to radiation therapy of brain tumors. Even based on the figures available from recent publications (which represent only a beginning appreciation of this general problem), it seems likely that there are significant numbers of individuals for whom such neurotoxicity is a serious concern.

Even though there are still many cancer treatments for which cognitive changes and other neurological sequelae have not been noted in the literature, it appears that these adverse effects may be frequent. The Cancer Statistics Branch of NCI estimates a cancer prevalence in the United States for 1997 of nearly 9 million individuals. If cognitive impairment associated with treatment were to only effect 2.5% of this population, the total number of patients for whom this issue would be a concern is of similar size to the population of individuals with chronic spinal cord injury. As discussed in more detail later, recent studies raise the specter that such complications may occur in significantly more than 2.5% of individuals treated for cancer. Lowered IQ scores and other evidence of cognitive impairment are relatively frequent in children treated for brain tumors or leukemias, thus presenting survivors and their families with considerable challenges with respect to the ability of these children to achieve normal lives. Data for patients treated for non-CNS tumors are only beginning to emerge, and give grounds for further concern. For example, some studies suggest that as many as 30% of women treated with standard chemotherapy regimes for primary breast carcinoma show significant cognitive impairment 6 months after treatment (van Dam *et al.*, 1998; Schagen *et al.*, 1999). As the compounds used in the treatment for breast cancer (cyclophosphamide, methotrexate, and 5-fluorouracil) are used fairly widely, it would not be surprising to find problems emerging in other patient populations as more testing is conducted. Thus, current trends support the view that the number of individuals for whom cognitive impairment associated with cancer treatment is a problem may be as great as for many of the more widely recognized neurological syndromes.

Neurological complications have been most extensively studied with respect to radiation therapy to the brain, and these studies indicate the presence of a wide range of potential adverse effects. Radiation-induced neurological complications include radionecrosis, myelopathy, cranial nerve damage, leukoencephalopathy (i.e., white matter damage), and dopa-resistant Parkinsonian syndromes (Keime-Guibert *et al.*, 1998). Imaging studies have documented extensive white matter damage in patients receiving radiation to the CNS (Vigliani *et al.*, 1999). Cognitive impairment associated with radiotherapy also has been reported in many of these patients. For example, in examination of 31 children, aged 5–15 years, who had received radiotherapy for posterior fossa tumors, and who had been off therapy for at least 1 year, long-term cognitive impairment occurred in most cases (Grill *et al.*, 1999). Neurotoxicity also affects older patients, presenting as cognitive dysfunction, ataxia, or dementia as a consequence of leukoencephalopathy and brain atrophy (Schlegel *et al.*, 1999). In adults, "subcortical" dementia occurs 3–12 months after cerebral radiotherapy (Vigliani *et al.*, 1999).

Potential clues to the biological basis for cognitive impairment have come from studies on the effects of radiation on the brain, for which dose-limiting neurotoxicity has long been recognized (Radcliffe *et al.*, 1994; Roman and Sperduto, 1995). On a cellular basis, radiation appears to cause damage to both dividing and nondividing CNS cells. Recent studies have shown that irradiation causes apoptosis in precursor cells of the dentate gyrus subgranular zone of the hippocampus (Peissner *et al.*, 1999; Tada *et al.*, 2000) and in the subependymal zone (Bellinzona *et al.*, 1996), both of which are sites of continuing precursor cell proliferation in the adult CNS. Such damage is also associated with long-term impairment of subependymal repopulation. In addition, it seems to be clear that nondividing cells, such as oligodendrocytes, are killed by irradiation (Li and Wong, 1998). Damage to oligodendrocytes is consistent with clinical evidence, where radiation-induced neurotoxicity has been associated with diffuse myelin and axonal loss in the white matter, with tissue necrosis and diffuse spongiosis of the white matter characterized by the presence of vacuoles that displaced the normally stained myelin sheets and axons (Vigliani *et al.*, 1999). Although some damage *in vivo* may well be secondary consequences of vascular damage, evidence also has been provided that radiation is directly damaging to important CNS populations, such as OPCs (Hopewell and van der Kogel, 1999).

Although chemotherapy has been less well studied than radiation in terms of its adverse effects on the CNS, it is becoming increasingly clear that many chemotherapeutic regimens are associated with neurotoxicity. Multiple reports have confirmed cognitive impairment in children and adults after cancer treatment. In particular, improvements in survival for children with leukemias or brain tumors treated with radiotherapy and chemotherapy have led to increasing concerns on quality-of-life issues for long-term survivors, in which neuropsychological testing has revealed a high frequency of cognitive deficits (Appleton *et al.*, 1990; Glauser and Packer, 1991; Waber and Tarbell, 1997; Grill *et al.*, 1999; Riva and Giorgi, 2000). For example, Cetingul *et al.* recently reported that performance and total IQ scores were significantly reduced in children treated for acute lymphoblastic leukemia who had completed therapy at least a year before and survived more than five years after diagnosis (Cetingul *et al.*, 1999). Indeed, it is felt that neurotoxicity of chemotherapy is frequent, and may be particularly hazardous when combined with radiotherapy (Cetingul *et al.*, 1999; Schlegel *et al.*, 1999). For example, in CT studies of patients receiving both brain radiation and chemotherapy, all patients surviving a malignant glioma for more than 4 yrs developed leukoencephalopathy and brain atrophy (Stylopoulos *et al.*, 1988).

Studies on the effects of chemotherapeutic agents on normal CNS cells have revealed a significant vulnerability of oligodendrocytes to BCNU (carmustine, an alkylating agent widely used in the treatment of brain tumors, myeloma, and both Hodgkin and non-Hodgkin lymphoma) (Nutt *et al.*, 2000). BCNU was toxic for oligodendrocytes at doses that would be routinely achieved during treatment. More recent studies (MN *et al.*, research in progress) have revealed that such vulnerability extends to such widely used chemotherapeutic agents as cisplatin, and that O-2A/OPCs and GRP cells are as or more vulnerable to the effects of these compounds than are oligodendrocytes. Strikingly, it thus far appears that any dose of chemotherapeutic agents that kill cancer cells is sufficient to kill the cells of the oligodendrocyte lineage.

Myelin Destruction in the Adult

Loss of myelin in the adult is generally associated with chronic degenerative processes or with traumatic injury. As is the case in development, damage to myelin in the adult is a frequent event, associated with virtually all examples of traumatic injury (including spinal cord injury) and most examples of chronic degenerative processes. Even Alzheimer's disease appears to have myelin breakdown as one of its important components (Terry *et al.*, 1964; Chia *et al.*, 1984; Malone and Szoke, 1985; Englund *et al.*, 1988; de la Monte, 1989; Wallin *et al.*, 1989;

Svennerholm and Gottfries, 1994; Gottfries *et al.*, 1996; Bartzokis *et al.*, 2000, 2003; Braak *et al.*, 2000; Han *et al.*, 2002; Kobayashi *et al.*, 2002; Roher *et al.*, 2002). It has even been suggested that it is the breakdown of myelin that is the key precipitating event in the initiation of damage to neurons in this syndrome (Bartzokis, 2003).

The most widely known of demyelinating diseases of the adult, and the one that has been studied for the longest time, is that of multiple sclerosis (MS). The demyelination that characterizes the MS lesion, along with the variable amount of axonal destruction and scar formation, was first described in the mid-19th century by Rindfleisch (1863) and Charcot (1868).

Damage to oligodendrocytes in MS is thought to represent the outcome of an autoimmune reaction against myelin antigens. The number of antigens that have been found to be targets of immune attack in MS has continued to grow over the years. In most MS plaques, it is possible to visualize immunoglobulins and deposits of complement at the lesion site (Prineas and Graham, 1981; Gay *et al.*, 1997; Barnum, 2002). It has even been suggested that it is possible to observe deposition of antibodies against such specific antigens as myelin oligodendrocyte glycoprotein on dissolving myelin in active lesions (Genain *et al.*, 1999), although it is clear that MS patients produce antibodies against a variety of myelin antigens. Indeed, it seems clear that as this disease progresses, the continued destruction of myelin causes an auto-vaccination process that is associated with a phenomenon called epitope spreading, in which the number of antigens recognized continues to increase (Tuohy *et al.*, 1998; Goebels *et al.*, 2000; Tuohy and Kinkel, 2000; Vanderlugt and Miller, 2002).

The immune reaction that leads to myelin destruction is a complex one, with many components. Along with the clear presence of anti-oligodendrocyte antibodies in the serum and CSF of MS patients, there is also a T-cell mediated immune reaction, which secondarily leads to macrophage activation. Indeed, the range of possible immune-mediated destructive mechanisms that can lead to myelin destruction, and the substantial heterogeneity of the disease process itself, makes it seem likely that MS is more correctly viewed as a constellation of diseases which share certain characteristic features (see, e.g., Lassmann, 1999; Lassmann *et al.*, 2001 for review).

Protecting oligodendrocytes against further damage in the MS patient, and restoring the myelin that has been damaged, represent two of the main goals in MS treatment. It is important to note, however, that achieving these goals may be hindered by the presence of inhibitory substances in the MS lesion itself. Such a possibility is indicated by studies showing that MS lesions contain apparent O-2A/OPCs that exist in a condition of stasis, undergoing little or no cell division (Wolswijk, 1998, 2000; Chang *et al.*, 2000). In addition, even though there is a relative sparing of axons in MS lesion, there is nonetheless significant axonal loss. This was noted even in the earliest histological descriptions of MS pathology, and has been amply reconfirmed in more recent years (Fromman, 1878; Charcot, 1880; Marburg, 1906; Ferguson *et al.*, 1997; Trapp *et al.*, 1998; Bjartmar *et al.*, 2003). In lesions in which neurons also are lost, replacement of

oligodendrocytes (or treatment with 4-AP) is unlikely to provide clinical benefit.

For recent reviews on a variety of aspects of MS, the reader is referred to, for example, Bruck *et al.* (2003), Galetta *et al.* (2002), Hemmer *et al.* (2003), Neuhaus *et al.* (2003), Noseworthy (2003), Waxman (2002).

VULNERABILITIES OF OLIGODENDROCYTES AND THEIR PRECURSOR CELLS

The number of conditions in which oligodendrocytes and their precursors appear to be killed or otherwise compromised makes it of considerable importance to determine what are the mechanisms underlying the death of these cells. A variety of studies are revealing clues regarding such mechanisms.

It is well established that one of the major contributors to CNS damage following traumatic injury is excitotoxic death of neurons caused by exposure to supranormal levels of glutamate. In recent years, it has become apparent that such glutamate toxicity is also seen in cells of the O-2A/OPC lineage, an observation that may be of considerable importance in a variety of pathological conditions (Yoshioka *et al.*, 1996; Matute *et al.*, 1997; McDonald *et al.*, 1998). Glutamate toxicity has been demonstrated *in vitro*, and also has been shown to occur in isolated spinal dorsal columns (Li and Stys, 2000) and *in vivo* following infusion of AMPA/kainate agonists into the optic nerve (Matute *et al.*, 1997; Matute, 1998) or subcortical white matter (McDonald *et al.*, 1998).

The glutamate receptors expressed by oligodendrocytes and their precursors are of the AMPA-binding subclass, and have some peculiar features. AMPA receptors in differentiated oligodendrocytes lack the GluR2 subunit, thus rendering them permeable to Ca^{2+} (Burnashev, 1996). Moreover, the GluR6 subunit is edited in such a manner as to also result in receptors that are more permeable to Ca^{2+} (Burnashev, 1996). These features may be important in the sensitivity of oligodendrocytes to glutamate. Glutamate receptors have also been found in the myelin sheath (Li and Stys, 2000), and it is not known if local stimulation of sheaths with glutamate results in a localized pathology. As would be predicted from the types of glutamate receptors expressed by oligodendrocytes, it appears that AMPA antagonists can protect oligodendrocytes against ischemic damage, at least *in vitro* (Fern and Möller, 2000). Thus, once clinically useful AMPA antagonists become available, it may be that these agents will prove of use in protecting against damage to oligodendrocytes.

Glutamate may not only be intrinsically toxic, but it may also enhance the toxicity of other physiological insults. For example, ischemic injury is characterized by excessive release of glutamate into the extrasynaptic space (Choi, 1988; Lee *et al.*, 1999). Ischemia is also characterized by transient deprivation of oxygen and glucose, a physiological insult that is also toxic for oligodendrocytes. Strikingly, the toxicity associated with deprivation of oxygen and glucose is further enhanced by co-exposure to glutamate (Lyons and Kettenmann, 1998; McDonald *et al.*, 1998; Fern and Möller, 2000).

Glutamate mediated damage of oligodendrocytes could be of physiological importance in a variety of settings. One dramatic example of oligodendrocyte death in which these pathways have been invoked is that of ischemic injury occurring in birth trauma, which can be associated with periventricular leukomalacia and cerebral palsy (Kinney and Armstrong, 1997). It must also be considered whether glutamate contributes to the demyelination seen in MS, particularly as it has been observed that glutamate levels are increased in the CNS of patients with demyelinating disorders, with levels correlating with disease severity (Stover *et al.*, 1997; Barkhatova *et al.*, 1998). In this context, it is of potential interest that chronic infusion of kainate (an AMPA receptor agonist) into white matter tracts is associated with the generation of lesions that have many of the characteristics of MS lesions, including extensive regions of demyelination with plaque formation, massive oligodendrocyte death, axonal damage, and inflammation (Matute, 1998). Although acute infusion of kainate produces lesions that are repaired by endogenous cells, lesions induced by chronic kainate infusion are not spontaneously repaired.

Still other potential contributors to oligodendrocyte death are the inflammatory cytokine TNF-α and, surprisingly, the pro-form of nerve growth factor (proNGF). It is known from both *in vitro* and *in vivo* experiments that oligodendrocytes are vulnerable to killing by TNF-α (Louis *et al.*, 1993; Butt and Jenkins, 1994; Mayer and Noble, 1994). It has also been shown that glutamate-mediated activation of microglia induces release of TNF-α from these cells. As microglia can themselves release glutamate when they are activated (Piani *et al.*, 1991; Noda *et al.*, 1999), it is possible that inflammation elicits a set of responses that build upon each other with the eventual result of tissue destruction. The proNGF receptor p75 also is induced by various injuries to the nervous system. Recent studies have shown that p75 is required for the death of oligodendrocytes following spinal cord injury, and its action is mediated mainly by proNGF (Beattie *et al.*, 2002). Oligodendrocytes undergoing apoptosis expressed p75, and the absence of p75 resulted in a decrease in the number of apoptotic oligodendrocytes and increased survival of oligodendrocytes. ProNGF is likely responsible for activating p75 *in vivo*, since the proNGF from the injured spinal cord induced apoptosis among p75(+/+), but not among p75(−/−) oligodendrocytes in culture, and its action was blocked by proNGF-specific antibody.

In vivo, it is unlikely to ever be the case that single factors act alone, and in this regard, the interplay between glutamate and TNF-α is of particular interest with regard to induction of demyelination. The combination of glutamate and TNF-α shows a highly lethal synergy when applied together in the thoracic gray matter of the spinal cord (Hermann *et al.*, 2001). It is not yet known if similar synergies occur with respect to the killing of oligodendrocytes, either by TNF-α or by proNGF, but such combinatorial effects seem likely.

REPAIR OF DEMYELINATING DAMAGE

The enormous range of clinically important conditions in which myelination is not properly generated, or is destroyed, makes it of paramount importance to understand how to repair this damage. The extensive knowledge regarding myelin biology, and on O-2A/OPCs and other potential ancestors of oligodendrocytes, has made it possible to begin development of a variety of strategies for promoting such repair.

The development of approaches for the repair of demyelinating damage has several components, each of which needs to be successfully addressed to develop a clinically useful strategy. First, there needs to be a means of identifying individuals for whom remyelination therapy might be expected to provide clinical benefit. Second, there needs to be a means of evaluating the success of such therapy. The third and fourth considerations are whether one is going to use transplantation of exogenous precursor cells to generate new oligodendrocytes and myelin, or whether the preferred strategy will be to enhance recruitment of endogenous precursor cells.

Advance identification of individuals who have a high likelihood of benefiting from remyelination therapy is absolutely essential in evaluating the efficacy of the therapy under study. This is because the development of any novel therapy requires a positive outcome to warrant continued devotion of resources and effort to that therapeutic approach. Attempts to restore neurological function in individuals in which repair of abnormal myelination is not sufficient to improve function would fail for reasons that are not germane to evaluating the potential utility of such therapies. For example, the lesions of both spinal cord injury and MS may be associated with substantial axonal loss (Trapp *et al.*, 1998; Kakulas, 1999a, b; Dumont *et al.*, 2001; Doherty *et al.*, 2002), a problem that cannot be solved by remyelination therapies. As destruction of myelin can induce similar failures of impulse conduction as are associated with axonal transection, or with conduction block caused by pressure, a simple clinical examination may not provide unambiguous data regarding the contribution of demyelination to impulse failure. Examination of lesions with standard imaging tools also tends to reveal more information about inflammation and edema than about the local state of myelin.

At present, the most promising tool for identifying individuals who might benefit from remyelination therapy appears to be a blocker of voltage-gated potassium (K$^+$) channels called 4-aminopyridine (4-AP). Demyelinated axons show increased activity of 4-AP-sensitive K$^+$ channels (Blight and Gruner, 1987; Blight, 1989; Bunge *et al.*, 1993; Fehlings and Nashmi, 1996; Nashmi *et al.*, 2000). When myelin is intact, there is only an inward sodium (Na$^+$) current and little outward K$^+$ current (Chiu and Ritchie, 1980), but after disruption of the myelin sheath, there is an increased persistent outward K$^+$ current. 4-AP blocks the leak through the "fast" K$^+$ channels that are normally located underneath the myelin (Sherratt *et al.*, 1980; Bowe *et al.*, 1987; Rasband *et al.*, 1998). These channels have multiple properties that have been ascribed to them (Nashmi and Fehlings, 2001b), including roles in re-polarization (Kocsis *et al.*, 1986), stabilizing the node to prevent re-excitation after a single impulse (Chiu and Ritchie, 1984; Poulter *et al.*, 1989; David *et al.*, 1993; Poulter and Padjen, 1995), and thereby increasing the security of axonal conduction (Chiu and Ritchie, 1984), and limiting excessive axonal depolarization and inactivation of nodal Na$^+$ channels (David *et al.*, 1992).

A variety of clinical trials have indicated that administration of a sustained release formulation of 4-AP may provide significant benefit to a subset of individuals with MS and also to some individuals with incomplete spinal cord injury (wherein myelin destruction is a frequent event even in the presence of intact axons). Myelin destruction and oligodendrocyte death has been seen in both experimental and clinical injuries (Gledhill and McDonald, 1977; Griffiths and McCulloch, 1983; Bunge et al., 1993; Crowe et al., 1997; Li et al., 1999; Casha et al., 2001; Nashmi and Fehlings, 2001a; Koda et al., 2002).

If a given individual does not benefit from the utilization of 4-AP, then it may be very difficult to understand underlying reasons for a failure of functional gain associated with testing of a remyelination therapy. Would this be because there was insufficient remyelination to confer benefit, or because the axonal damage was itself sufficiently severe that remyelination was not sufficient to restore conduction? Despite some experimental evidence that 4-AP may also enhance synaptic transmission, separately from any effects on impulse conduction in unmyelinated axons, there thus far appears to be no better approach to the identification of suitable candidates for therapies targeted at enhancing remyelination.

The next critical distinction to be made in the development of remyelination therapies is that of distinguishing between repair by transplantation and repair by recruitment of endogenous precursor cells. As discussed below, these two options themselves segregate further into multiple strategic suboptions.

Attempts to repair demyelinated lesions by cell transplantation will necessarily be focused on instances in which most or all of the damage is found within a discrete lesion site and where there is a reasonable expectation that remyelination will provide functional benefit. There are several conditions that fulfil this requirement, including spinal cord injury, lacunar infarcts, and transverse myelitis. Although lesions in different patients may differ greatly in size, these different conditions nonetheless share the characteristic that successful repair within a single anatomical location has the highest probability of providing clear clinical benefit.

Once a decision is made to attempt to remyelinate lesions by cell transplantation, it is necessary to choose between the multitude of cellular populations that have emerged as candidates for such repair. In experimental animals, remyelination has been successful using O-2A/OPCs (Espinosa de los Monteros et al., 1993; Warrington et al., 1993; Groves et al., 1993a; Utzschneider et al., 1994; Duncan, 1996; Jeffery et al., 1999), GRP cells (Herrera et al., 2001), NSCs (Hammang et al., 1997), and embryonic stem cells that have been pretreated to bias differentiation toward a neural cell fate (Brustle et al., 1999; Liu et al., 2000). It has also been possible to isolate oligodendrocyte-competent glial precursor cells from embryonic stem cells ([Brustle et al., 1999; Liu et al., 2000], although it is not known whether these precursors are GRP cells, O-2A/OPCs, both, or neither). Precursor cells capable of making oligodendrocytes following transplantation can also be isolated from developing or from adult tissues. Moreover, many of the stem and progenitor cell populations of interest in the generation of new oligodendrocytes can be isolated from human tissues of different ages and sources (Roy et al., 1999; Dietrich et al., 2002; Windrem et al., 2002).

It is not presently known whether any individual population of cells capable of generating oligodendrocytes in vivo offers advantages over any other population, but there are reasons to be concerned that different populations may yield divergent outcomes. For example, if properties that cells express in vitro are indicative of their behavior in vivo, then O-2A/OPCs such as those isolated from the optic nerves of 7-day-old rats might be expected to generate a relatively restricted number of oligodendrocytes quite rapidly (Fig. 7). In contrast, O-2A/OPCs such as those isolated from cortices of the same animals might generate a far larger number of cells but may take a much longer time to generate oligodendrocytes (Power et al., 2002). GRP cells could also be used to generate both oligodendrocytes and astrocytes (Herrera et al., 2001), which may be beneficial. In contrast, O-2A/OPCs could be used to more selectively generate oligodendrocytes (Espinosa de los Monteros et al., 1993; Groves et al., 1993b; Warrington et al., 1993).

At this point in time, very little is known about the comparative utility of different precursor cell populations in lesion repair. Thus, an essential component of the development of remyelination therapies will be the determination of whether specific precursor populations are generally advantageous, or whether repair of different types of lesions will require transplantation of different types of cells.

In contrast with repair of focal lesions, the repair of the distributed lesions like those seen in MS patients seems more likely to be initially attempted by the application of strategies that recruit endogenous precursor cells. The most theoretically attractive strategy in this regard would be systemic administration of a therapeutic compound that specifically promotes division of glial precursor cells capable of generating oligodendrocytes.

At the time of writing this chapter, the only published approach to enhancing function of endogenous cells that seems

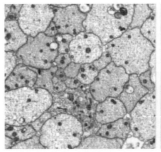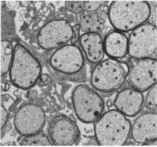

FIGURE 7. Remyelination by transplantation of O-2A/OPCs. In these experiments, O-2A/OPCs isolated from optic nerves of P7 rat pups and expanded in vitro for 3–4 weeks by being grown in the presence of PDGF + FGF-2. These cells were then transplanted into the spinal cord of rats that received a local injection of ethidium bromide to kill all cells with DNA in the injection site. Such an injection kills all glial cells while sparing the axons. In addition, the animals are irradiated so that host precursor cells cannot repair this damage. In the absence of cell transplantation, the tissue contains only axons running in a glial-free space (as shown in the left-hand electron micrograph). Following transplantation of O-2A/OPCs, >90% of the axons are remyelinated. For greater detail, the reader is referred to Groves et al. (1993a).

close to clinical evaluation is the application of antibodies that have been reported to promote remyelination. These antibodies were first identified in paradoxical studies indicating that monoclonal antibodies directed against myelin antigens could promote remyelination in a number of different circumstances (Asakura and Rodriguez, 1998; Warrington *et al.*, 2000). Effectiveness of these antibodies has been observed in the immune-mediated demyelination model of infection with Theiler's virus (Asakura and Rodriguez, 1998) as well as in the case of demyelination induced by injection of lysolecithin into white matter tracts (Pavelko *et al.*, 1998). Remyelination-promoting monoclonal antibodies also reduce relapse rates and prolong relapse onset in the autoimmune model of experimental allergic autoencephalomyelitis, an experimental model of MS (Miller *et al.*, 1997). The fact that many of the antibodies that have been found to be effective in this paradigm bind specifically to oligodendrocytes and/or their precursors provides an important potential for specificity of action of this strategy.

Antibodies that promote remyelination appear to work by physiologic stimulation of reparative systems. Intraperitoneal injection of remyelination-promoting antibodies labeled with radioactive amino acids has shown that these antibodies enter the CNS and bind primarily to cells in the demyelinated lesion (Hunter *et al.*, 1997). While the mechanism by which these antibodies promote remyelination remains uncertain, it is of potential interest that all remyelination-promoting antibodies tested evoke Ca^{++} transients in mixed glial cultures while isotype- and species-matched control antibodies do not. Thus, it may be that the ability of these antibodies to stimulate Ca^{++} fluxes activates a signal transduction cascade critical for myelinogenesis (Soldán *et al.*, 2003).

It is possible that growth factors will also be found that have the ability to beneficially stimulate specific precursor cell populations *in vivo* (McTigue *et al.*, 1998), but the ability of growth factors to modulate the biology of multiple cell types will make the careful elucidation of potential side effects of particular importance. Achieving adequate growth factor delivery is also a matter of concern. Although it is possible to infuse growth factors into CSF, many studies have shown that the extent to which such molecules can distribute into the CNS parenchyma due to diffusion is very limited (Bobo *et al.*, 1994; Lieberman *et al.*, 1995). Normal diffusion processes are intrinsically limited, with reductions in growth factor concentration being reduced according to the inverse square law that governs diffusion from a point source. Diffusion in the real setting of the CNS, moreover, is even more compromised. The fact that growth factors bind to cells and matrix in the diffusion path means that the distance of diffusion is reduced to an even greater extent than in a free diffusion system, and the reduction in growth factor concentration falls more sharply than in a simple inverse square relationship. Thus, successful growth factor delivery may require the utilization of convective delivery strategies (Bobo *et al.*, 1994; Lieberman *et al.*, 1995; Lonser *et al.*, 1999, 2002).

Successful application of strategies to recruit endogenous precursor cells will be dependent upon there being sufficient numbers of cells available to carry out repair and on the physiological condition of the patient being conducive to repair. At this point in time, little is known about whether there are limitations in precursor cell production that preclude extensive or repetitive repair, or whether the environment itself is refractory to repair. On the one hand, there are indications that there are such large numbers of putative *adult* O-2A/OPCs in the normal CNS as to potentially represent 5–8% of the total cells in the normal CNS (Nishiyama *et al.*, 1999; Dawson *et al.*, 2000; Levine *et al.*, 2001). On the other hand, we know little about the biological heterogeneity of this NG2$^+$ cell population, about the prevalence of cells following a lesion, or about the functional competence of those cells that are found in the post-lesioned CNS.

If it is the case that endogenous precursor cells are too depleted, or otherwise compromised, to allow effective repair, then usage of growth-promoting strategies in conjunction with cellular transplantation might provide an optimal approach to enhancing remyelination. If the CNS has become refractory to repair, for example, by generation of glial scar tissue that might inhibit O-2A/OPC migration into lesion sites (ffrench Constant *et al.*, 1988; Groves *et al.*, 1993b), then it will be essential to develop means of overcoming such inhibitory signals. That some form of refractory phenomena might occur is indicated by the apparent presence of nondividing O-2A/OPCs in lesions of MS patients (Chang *et al.*, 2000; Wolswijk, 2000). Moreover, it appears that although transplanted oligodendrocyte progenitor cells survive and remyelinate in acute lesion areas, normal white matter is inhibitory to the migration of these cells (O'Leary and Blakemore, 1997). Thus, there may well be *in vivo* constraints that limit the effectiveness of transplanted cells.

One of the most important and challenging ventures will be repair of myelination abnormalities that are diffusely distributed—or even globally distributed—throughout the CNS. Such a distributed failure of normal myelination occurs in many children with a variety of CNS diseases.

As indicated earlier in this chapter, the three general causes of diffuse, or global, abnormalities in myelination are (a) genetic disorders, (b) nutritional and hormonal deficiency disorders, and (c) exposure to any of a large variety of physiological insults. Different approaches may be required for each of these conditions.

A number of the genetic diseases that result in failures of normal myelination have been discussed previously in this chapter. They share the problem that recruitment of endogenous precursor cells is not a viable strategy in the absence of repair of the underlying genetic lesion, as it is clear that the genetically defective cells are themselves not capable of normal myelination. Thus, it is of paramount importance to develop strategies that allow the genetic lesion to be directly repaired, or for its effects to be overridden.

Two potential approaches to repair in the case of genetic diseases are to repair the genetic damage so that endogenous precursor cells can carry out repair or to transplant normal cells into the genetically abnormal environment. Promising progress has been made for both of these approaches. An example of the former approach has been the use of lentivirus vectors to obtain clear clinical improvement in adult beta-glucuronidase deficient (mucopolysaccharidosis type VII {MPS VII}) mice, an animal model of lysosomal storage disease (Brooks *et al.*, 2002). Lysosomal accumulation of glycosaminoglycans occurs in the brain and other tissues of individuals with this disease, causing a fatal progressive degenerative disorder, including mental retardation as one of its outcomes. Treatments are designed to provide a source of normal enzyme for uptake by diseased cells and thus can theoretically be treated by introduction of cells that express beta-glucuronidase. Improvement in this mouse model has also been obtained by transplantation of beta-glucuronidase-expressing neural stem cells into the cerebral ventricles of newborn animals. When these animals were examined at maturity, donor-derived cells were found to be present as normal constituents of diverse brain regions. β-Glucuronidase activity was expressed along the entire neuraxis, resulting in widespread correction of lysosomal storage in neurons and glia (Snyder *et al.*, 1995). A similar approach also has been applied in attempts to repair the global dysmyelination found in shiverer mice, in which myelin is not produced due to a genetic defect in the oligodendrocytes themselves. Transplantation of genetically normal NSCs in the ventricles of newborn shiverer mice was associated with widespread engraftment and generation of normal myelin in the shiverer brain (Yandava *et al.*, 1999).

Nutritional and hormonal deficiency disorders that compromise myelination may offer somewhat easier targets for repair than genetic myelination disorders in that there is a hope that existing cells are not compromised in their function. There is some reason to be optimistic about this possibility, due to the well-documented ability of myelination to return to normal levels in hypothyroid, or nutritionally-deprived, animals in which the underlying metabolic defect is corrected sufficiently early in development (Wiggins *et al.*, 1976; Wiggins, 1979, 1982; Wiggins and Fuller, 1979; Noguchi *et al.*, 1985; Munoz *et al.*, 1991; Bernal and Nunez, 1995; Ibarrola and Rodriguez-Pena, 1997; Marta *et al.*, 1998).

Despite the ability of endogenous precursor cells to correct myelination deficiencies if metabolic defects are corrected early enough in development, studies on nutritional and hormonal deficiency disorders have also demonstrated the critical importance of restoring normal metabolic function by an early enough time if one is going to achieve repair. For example, repair of dysmyelination associated with nutritional deprivation requires restoration of normal nutritional intake in order to achieve normal levels of myelination (Wiggins *et al.*, 1976; Wiggins, 1979, 1982; Wiggins and Fuller, 1979). Similarly, restoration of TH in the case of hypothyroidism only is associated with repair of dysmyelination if hormonal replacement therapy is initiated early enough in life (Noguchi *et al.*, 1985; Munoz *et al.*, 1991; Bernal and Nunez, 1995; Ibarrola and Rodriguez-Pena, 1997; Marta *et al.*, 1998). The existence of these critical developmental periods for enabling remaining CNS precursor cells to generate normal levels of myelination *in vivo* raises questions as to what is the underlying biology of such critical periods. One possible component of these periods of opportunity for successful repair could be the observed transition from the presence in the CNS by O-2A/OPCs of a *perinatal* phenotype to those with an *adult*-specific phenotype, a transition that occurs in the rat optic nerve largely during the period of 2–3 weeks after birth (Wolswijk *et al.*, 1990).

The existence of critical periods after which restoration of normal metabolism is no longer associated with an equivalent restoration of normal myelination suggests that it will also be necessary to apply strategies of enhancing function of endogenous precursor cells and/or transplanting additional precursor cells to achieve repair of these syndromes. It is important to stress, however, how little is known about the reasons for the failure of repair if metabolic repair is delayed too long. For example, it is not even known whether the CNS itself of older animals with metabolic disorders expresses properties that make it refractory to repair. This is a critical area for further study.

A further question that needs to be considered is whether there is a need to utilize more than one cell type for repair of tissue. For example, in global disorders of myelination, there may be value in transplanting O-2A/OPCs to achieve more rapid generation of oligodendrocytes, as well as transplanting NSCs in order to populate the germinal zones of the brain with cells capable of contributing glial precursor cells for a prolonged period. Or, in spinal cord injury or other forms of traumatic injury, there may be value in transplanting GRP cells to generate normal astrocytes together with O-2A/OPCs to increase the yield of oligodendrocytes. It is also not known whether successful remyelination will require multiple transplantations. And if so, then how many? With what interval between them? Will they need to be spread over particular physical distances?

While many questions remain to be answered to enable the application of our increasing knowledge about oligodendrocyte biology to the treatment of important medical problems, it is nonetheless extraordinary to consider the advances that have been made in a relatively short time. With such a rate of progress, it cannot be long before we are able to accomplish the remarkable feat of repairing damage to this vital component of the CNS. Moreover, it seems certain that the ongoing study of these fascinating cells will continue to provide insights relevant to a range of biological problems that extend far beyond the questions of how myelin is formed, maintained, and replaced.

REFERENCES

Aberg, M.A., Ryttsen, F., Hellgren, G., Lindell, K., Rosengren, L.E., MacLennan, A.J. *et al.*, 2001, Selective introduction of antisense oligonucleotides into single adult CNS progenitor cells using electroporation demonstrates the requirement of STAT3 activation for CNTF-induced gliogenesis, *Mol. Cell. Neurosci.* 17:426–443.

Anderson, S.A., Marin, O., Horn, C., Jennings, K., and Rubenstein J.L., 2001, Distinct cortical migrations from the medial and lateral ganglionic eminences, *Development* 128:353–363.

Appleton, R.E., Farrell, K., Zaide, J., and Rogers, P., 1990, Decline in head growth and cognitive impairment in survivors of acute lymphoblastic leukaemia, *Arch Dis. Child.* 65:530–534.

Armstrong, R., Friedrich, Jr., V.L., Holmes, K.V., and Dubois Dalcq, M., 1990a, In vitro analysis of the oligodendrocyte lineage in mice during demyelination and remyelination, *J. Cell Biol.* 111:1183–1195.

Armstrong, R.C., Harvath, L., and Dubois-Dalcq, M.E., 1990b, Type 1 astrocytes and oligodendrocyte-type 2 astrocyte glial progenitors migrate toward distinct molecules, *J. Neurosci. Res.* 27:400–407.

Asakura, K. and Rodriguez, M., 1998, A unique population of circulating autoantibodies promotes central nervous system remyelination, *Mult. Scler.* 4:217–221.

Asher, R.A. *et al.*, 1999, Versican is up-regulated in CNS injury and is a product of O-2A lineage cells, *Soc. Neurosci. Abstr.* 25:750.

Asher, R.A. *et al.*, 2000, Neurocan is upregulated in injured brain and in cytokine-treated astrocytes, *J. Neurosci.* 20:2427–2438.

Ballotti, R., Nielsen, F.C., Pringle, N., Kowalski, A., Richardson, W.D., Van Obberghen, E. *et al.*, 1987, Insulin-like growth factor I in cultured rat astrocytes: Expression of the gene, and receptor tyrosine kinase, *EMBO J.* 6:3633–3639.

Bansal, R., Kumar, M., Murray, K., and Pfeiffer, S.E., 1996, Developmental and FGF-2-mediated regulation of syndecans (1-4) and glypican in oligodendrocytes, *Mol. Cell. Neurosci.* 7:276–288.

Barkhatova, V.P. *et al.*, 1998, Changes in neurotransmitters in multiple sclerosis, *Neurosci. Behav. Physiol.* 28:341–344.

Barkovich, A.J., 2000, Toxic and metabolic brain disorders, In *Pediatric Neuroimaging*, 3rd edn. (A.J. Barkovich, ed.), Lippincott Williams & Wilkins, Philadelphia, PA.

Barnett, S.C., Hutchins, A.M., and Noble, M., 1993, Purification of olfactory nerve ensheathing cells from the olfactory bulb, *Dev. Biol.* 155:337–350.

Barnum, S.R., 2002, Complement in central nervous system inflammation, *Immunol. Res.* 26:7–13.

Barone, Jr., S., Haykal-Coates, N., Parran, D.K., and Tilson, H.A., 1998, Gestational exposure to methylmercury alters the developmental pattern of trk-like immunoreactivity in the rat brain and results in cortical dysmorphology, *Brain Res. Dev. Brain Res.* 109:13–31.

Barres, B.A., Koroshetz, W.J., Swartz, K.J., Chun, L.L., and Corey, D.P., 1990, Ion channel expression by white matter glia: The O-2A glial progenitor cell, *Neuron* 4:507–524, ISSN: 0896-6273.

Barres, B.A., Lazar, M.A., and Raff, M.C., 1994a, A novel role for thyroid hormone, glucocorticoids and retinoic acid in timing oligodendrocyte development, *Development* 120:1097–1108.

Barres, B.A. and Raff, M.C., 1993, Proliferation of oligodendrocyte precursor cells depends on electrical activity in axons, *Nature* 361:258–260.

Barres, B.A., Raff, M.C., Gaese, F., Bartke, I., Dechant, G., and Barde, Y.A., 1994b, A crucial role for neurotrophin-3 in oligodendrocyte development, *Nature* 367:371–375.

Barres, B.A., Schmidt, R., Sendtner, M., and Raff, M.C., 1993, Multiple extracellular signals are required for long-term oligodendrocyte survival, *Development* 118:283–295.

Bartlett, W.P., Li, X.S., and Connor, J.R., 1991, Expression of transferrin mRNA in the CNS of normal and jimpy mice, *J. Neurochem.* 57:318–322.

Bartsch, U., Faissner, A., Trotter, J., Dorries, U., Bartsch, S., Mohajeri, H. *et al.*, 1994, Tenascin demarcates the boundary between the myelinated and non-myelinated part of retinal ganglion cell axons in the developing and adult mouse, *J. Neurosci.* 14:4756–4768.

Bartzokis, G., 2003, Age-related myelin breakdown: A developmental model of cognitive decline and Alzheimer's disease, *Neurobiol. Aging* 25:5–8.

Bartzokis, G., Beckson, M., Lu, P.H., Edwards, N., Bridge, P., and Mintz, J., 2002, Brain maturation may be arrested in chronic cocaine addicts, *Biol. Psychiatry* 51:605–611.

Bartzokis, G., Cummings, J.L., Sultzer, D., Henderson, V.W., Nuechterlein, K.H., and Mintz, J., 2003, White matter structural integrity in healthy aging adults and patients with Alzheimer disease: A magnetic resonance imaging study, *Arch. Neurol.* 60:393–398.

Bartzokis, G., Sultzer, D., Cummings, J., Holt, L.E., Hance, D.B., Henderson, V.W. *et al.*, 2000, In vivo evaluation of brain iron in Alzheimer disease using magnetic resonance imaging, *Arch. Gen. Psychiatry* 57:47–53.

Baslow, M.H., 2000, Canavan's spongiform leukodystrophy: A clinical anatomy of a genetic metabolic CNS disease, *J. Mol. Neurosci.* 15:61–69.

Baslow, M.H., Kitada, K., Suckow, R.F., Hungund, B.L., and Serikawa, T., 2002, The effects of lithium chloride and other substances on levels of brain N-acetyl-L-aspartic acid in Canavan disease-like rats, *Neurochem. Res.* 27:403–406.

Beattie, M.S., Harrington, A.W., Lee, R., Kim, J.Y., Boyce, S.L., Longo, F.M. *et al.*, 2002, ProNGF induces p75-mediated death of oligodendrocytes following spinal cord injury, *Neuron* 36:375–386.

Belachew, S., Chittajallu, R., Aguirre, A.A., Yuan, X., Kirby, M., Anderson, S., and Gollo, V., 2003, Postnatal NGZ proteoglycan-expressing progenitor cells are intrinsically multipotent and generate functional neurons, *J. Cell Biol.* 161:169–186.

Bellinzona, M., Gobbel, G.T., Shinohara, C., and Fike, J.R., 1996, Apoptosis is induced in the subependyma of young adult rats by ionizing irradiation, *Neurosci. Lett.* 208:163–166.

Ben-Hur, T., Rogister, B., Murray, K., Rougon, G., and Dubois-Dalcq, M., 1998, Growth and fate of PSA-NCAM+ precursors of the postnatal brain, *J. Neurosci.* 18:5777–5788.

Benes, F.M., Turtle, M., Khan, Y., and Farol, P., 1994, Myelination of a key relay zone in the hippocampal formation occurs in the human brain during childhood, adolescence and adulthood, *Arch. Gen. Psychiatry* 51:477–484.

Benkovic, S.A. and Connor, J.R., 1993, Ferritin, transferrin, and iron in selected regions of the adult and aged rat brain, *J. Comp. Neurol.* 338:97–113.

Bergles, D.E. *et al.*, 2000, Glutaminergic synapses on oligodendrocyte precursor cells in the hippocampus, *Nature* 405:187–191.

Bernal, J. and Nunez, J., 1995, Thyroid hormones and brain development, *Eur. J. Endocrinol.* 133:390–398.

Bernard, S., Enayati, A., Redwood, L., Roger, H., and Binstock, T., 2001, Autism: A novel form of mercury poisoning, *Med. Hypotheses.* 56:462–471.

Bhakoo, K.K., Craig, T.J., and Styles, P., 2001, Developmental and regional distribution of aspartoacylase in rat brain tissue, *J. Neurochem.* 79:211–220.

Bhakoo, K.K. and Pearce, D., 2000, In vitro expression of N-acetyl aspartate by oligodendrocytes: Implications for proton magnetic resonance spectroscopy signal in vivo, *J. Neurochem.* 74:254–262.

Bhat, M.A., Rios, J.C., Lu, Y., Garcia-Fresco, G.P., Ching, W., St.Martin, M. *et al.*, 2001, Axon-glia interactions: The domain organization of myelinated axons requires Neurexin IV/Caspr/Paranodin, *Neuron* 30:369–383.

Bichenkov, E. and Ellingson, J.S., 2001, Ethanol exerts different effects on myelin basic protein and 2′,3′-cyclic nucleotide 3′-phosphodiesterase expression in differentiating CG-4 oligodendrocytes, *Brain Res. Dev. Brain Res.* 128:9–16.

Bjartmar, C., Hildebrand, C., and Loinder, K., 1968, Morphological heterogeneity of rat oligodendrocytes: Electron microscopic studies on serial sections, *Glia* 11:235–244.

Bjartmar, C., Wujek, J.R., and Trapp, B.D., 2003, Axonal loss in the pathology of MS: Consequences for understanding the progressive phase of the disease, *J. Neurol. Sci.* 206:165–171.

Blight, A.R., 1989, Effect of 4-aminopyridine on axonal conduction-block in chronic spinal cord injury, *Brain Res. Bull.* 22:47–52.

Blight, A.R. and Gruner, J.A., 1987, Augmentation by 4-aminopyridine of vestibulospinal free fall responses in chronic spinal-injured cats, *J. Neurol. Sci.* 82:145–159.

Bobo, R.H., Laske, D.W., Akbasak, A., Morrison, P.F., Dedrick, R.L., and Oldfield, E.H., 1994, Convection-enhanced delivery of macromolecules in the brain, *Proc. Natl. Acad. Sci. USA* 91:2076–2080.

Bogler, O., Wren, D., Barnett, S.C., Land, H., and Noble, M., 1990, Cooperation between two growth factors promotes extended selfrenewal and inhibits differentiation of oligodendrocyte-type-2 astrocytes (O-2A) progenitor cells, *Proc. Natl. Acad. Sci. USA* 87:6368–6372.

Bonni, A., Sun, Y., Nadal-Vicens, M., Bhatt, A., Frank, D.A., Rozovsky, I. et al., 1997, Regulation of gliogenesis in the central nervous system by the JAK-STAT signaling pathway, *Science* 278:477–483.

Boucher, K., Yakovlev, A., Mayer-Proschel, M., and Noble, M., 1999, A stochastic model of temporally regulated generation of oligodendrocytes in cell culture, *Math. Biosci.* 159:47–78.

Bowe, C.M., Kocsis, J.D., Targ, E.F., and Waxman, S.G., 1987, Physiological effects of 4-aminopyridine on demyelinated mammalian motor and sensory fibers, *Ann. Neurol.* 22:264–268.

Boyle, M.E.T., Berglund, E.O., Murai, K.K., Weber, L., Peles, E., and Ranscht, B., 2001, Contactin orchestrates assembly of the septate-like junctions at the paranode in myelinated peripheral nerve, *Neuron* 30:385–397.

Braak, H., Del Tredici, K., Schultz, C., and Braak, E., 2000, Vulnerability of select neuronal types to Alzheimer's disease, *Ann. NY Acad. Sci.* 924:53–61.

Brady, S.T., Witt, A.S., Kirkpatrick, L.L., de Waegh, S.M., Redhead, C., Tu, P.H. et al., 1999, Formation of compact myelin is required for maturation of the axonal cytoskeleton, *J. Neurosci.* 19:7278–7288.

Braun, P.E., 1984, Molecular organization of myelin. In *Myelin* (P. Morell, ed.), Plenum Press, New York, pp. 97–116.

Briscoe, J. and Ericson, J., 1999, The specification of neuronal identity by graded Sonic Hedgehog signaling, *Semin. Cell Dev. Biol.* 10:353–362.

Brooks, A.I., Stein, C.S., Hughes, S.M., Heth, J., McCray, P.M.J., Sauter, S.L. et al., 2002, Functional correction of established central nervous system deficits in an animal model of lysosomal storage disease with feline immunodeficiency virus-based vectors, *Proc. Natl. Acad. Sci. USA* 99:6216–6221.

Bruck, W., Kuhlmann, T., and Stadelmann, C., 2003, Remyelination in multiple sclerosis, *J. Neurol. Sci.* 206:181–185.

Brustle, O., Jones, K., Learish, R., Karram, K., Choudhary, K., Wiestler, O. et al., 1999, Embryonic stem cell-derived glial precursors: A source of myelinating transplants, *Science* 285:754–756.

Bunge, R.P., 1968, Glial cells and the central myelin sheath, *Physiol. Rev.* 48:197–251.

Bunge, R.P., Puckett, W.R., Becerra, J.L., Marcillo, A., and Quencer, R.M., 1993, Observations on the pathology of human spinal cord injury. A review and classification of 22 new cases with details from a case of chronic cord compression with extensive focal demyelination, *Adv. Neurol.* 59:75–89.

Burbacher, T.M., Rodier, P.M., and Weiss, B., 1990, Methylmercury developmental neurotoxicity: A comparison of effects in humans and animals, *Neurotoxicol. Teratol.* 12:191–202.

Burnashev, N., 1996, Calcium permeability of glutamate gated channels in the central nervous system, *Curr. Opin. Neurobiol.* 6:311–317.

Butt, A.M., Ibrahim, M., and Berry, M., 1997, The relationship between developing oligodendrocyte units and maturing axons during myelinogenesis in the anterior velum of neonatal rats, *J. Neurocytol.* 26:327–338.

Butt, A.M., Ibrahim, M., Gregson, N., and Berry, M., 1998, Differential expression of the L and S isoforms of myelin associated glycoprotein (MAG) in oligodendrocyte unit phenotypes in the adult anterior medullary velum, *J. Neurocytol.* 27:271–280.

Butt, A.M., Ibrahim, M., Ruge, F.M., and Berry, M., 1995, Biochemical subtypes of oligodendrocytes in the anterior velum of the rat revealed by the monoclonal antibody Rip, *Glia* 14:185–197.

Butt, A.M. and Jenkins, H.G., 1994, Morphological changes in oligodendrocytes in the intact mouse optic nerve following intravitreal injection of tumour necrosis factor, *J. Neuroimmunol.* 51:27–33.

Calver, A., Hall, A., Yu, W., Walsh, F., Heath, J., Betsholtz, C. et al., 1998, Oligodendrocyte population dynamics and the role of PDGF in vivo. *Neuron* 20:869–882.

Cameron-Currey, P. and LeDouarin, N.M., 1995, Oligodendrocyte precursors originate from both the dorsal and ventral parts of the spinal cord, *Neuron* 15:1299–1310.

Campagnoni, A.T., 1995, *Molecular Biology of Myelination*, Oxford University Press, New York.

Campagnoni, A.T., Pribyl, T.M., Campagnoni, C.W., Kampf, K., Amur-Umarjee, S., Landry, C.F. et al., 1993, Structure and developmental regulation of Golli-mbp, a 105-kilobase gene that encompasses the myelin basic protein gene and is expressed in cells in the oligodendrocyte lineage in the brain, *J. Biol. Chem.* 268:4930–4938.

Campignoni, A.T and Macklin, W.B., 1988, Cellular and molecular aspects of myelin gene expression, *Mol. Neurobiol.* 2:41–89.

Campignoni, A.T. and Skoff, R.P., 2001, The pathobiology of myelin mutants reveal novel biological functions of the MBP and PLP genes. *Brain Pathol.* 11:74–91.

Cannella, B., Pitt, D., Marchionni, M., and Raine, C.S., 1999, Neuregulin and erbB receptor expression in normal and diseased human white matter, *J. Neuroimmunol.* 100:233–242.

Carroll, W.M. et al., 1998, Identification of the adult resting progenitor cell by autoradiographic tracking of oligodendrocyte precursors in experimental CNS demyelination, *Brain* 121:293–302.

Casha, S., Yu, W.R., and Fehlings, M.G., 2001, Oligodendroglial apoptosis occurs along degenerating axons and is associated with FAS and p75 expression following spinal cord injury in the rat, *Neuroscience* 103:203–218.

Castoldi, A.F., Coccini, T., Ceccatelli, S., and Manzo, L., 2001, Neurotoxicity and molecular effects of methylmercury. *Brain Res. Bull.* 55:197–203.

Cenci di Bello, I. et al., 1999, Generation of oligodendroglial progenitors in acute inflammatory demyelinating lesions of the rat brain stem is associated with demyelination rather than inflammation, *J. Neurocytol.* 28:365–381.

Cetingul, N., Aydinok, Y., Kantar, M., Oniz, H., Kavakli, K., Yalman, O. et al., 1999, Neuropsychologic sequelae in the long-term survivors of childhood acute lymphoblastic leukemia, *Pediatr. Hematol. Oncol.* 16:213–220.

Chakraborty, G., Mekala, P., Yahya, D., Wu, G., and Ledeen, R.W., 2001, Intraneuronal N-acetylaspartate supplies acetyl groups for myelin lipid synthesis: Evidence for myelin-associated aspartoacylase, *J. Neurochem.* 78:736–745.

Chan, S. and Kilby, M.D., 2000, Thyroid hormone and central nervous system development, *J. Endocrinol.* 165:1–8.

Chang, A. et al., 2000, NG2+ oligodendrocyte progenitor cells in adult human brain and multiple sclerosis lesions. *J. Neurosci.* 20:6404–6412.

Chang, L.W., Reuhl, K.R., and Lee, G.W., 1977, Degenerative changes in the developing nervous system as a result of *in utero* exposure to methylmercury, *Environ. Res.* 14:414–423.

Charcot, J.M., 1868, Histologie de la sclerose en plaque, *Gaz. Hôpital (Paris).*

Charcot, J.M., 1880, *Lecons sur les maladies du systeme nerveux faites a la salpetriere* (A. Delahaye and E. Lecrosnier, eds.), Cerf et fils, Paris, pp. 189–220.

Chen, Q.M., Bartholomew, J.C., Campisi, J., Acosta, M., Reagan, J.D., and Ames, B.N., 1998, Molecular analysis of H_2O_2-induced senescent-like growth arrest in normal human fibroblasts: p53 and Rb control G1 arrest but not cell replication, *Biochem. J.* 332:43–50.

Chia, L.S., Thompson, J.E., and Moscarello, M.A., 1984, X-ray diffraction evidence for myelin disorder in brain from humans with Alzheimer's disease, *Biochim. Biophys. Acta* 775:308–312.

Chiu, S.Y. and Ritchie, J.M., 1980, Potassium channels in nodal and internodal axonal membrane of mammalian myelinated fibres, *Nature* 284:170–171.

Chiu, S.Y. and Ritchie, J.M.,1984, On the physiological role of internodal potassium channels and the security of conduction in myelinated nerve fibres, *Proc. R. Soc. Lond. B Biol. Sci.* 220:415–422.

Choi, B.H., 1989, The effects of methylmercury on the developing brain, *Prog. Neurobiol.* 32:447–470.

Choi, D.W., 1988, Calcium-mediated neurotoxicity: Relationship to specific channel types and role in ischemic damage, *Trends Neurosci.* 11:465–469.

Clarkson, T.W., 1993, Molecular and ionic mimicry of toxic metals, *Annu. Rev. Pharmacol. Toxicol.* 32:545–571.

Clarkson, T.W., 1997. The toxicology of mercury, *Crit. Rev. Clin. Lab. Sci.* 34:369–403.

Cohen, R.I. and Almazan, G., 1994, Rat oligodendrocytes express muscarinic receptors coupled to phosphoinositide hydrolysis and adenylyl cyclase, *Eur. J. Neurosci.* 6:1213–1224.

Colello, R.J., Pott, U., and Schwab, M.E., 1994, The role of oligodendrocytes and myelin on axon maturation in the developing rat retinofugal pathway, *J. Neurosci.* 14:2594–2605.

Connor, J.R., 1994, Iron regulation in the brain at the cell and molecular level. *Adv. Exp. Med. Biol.* 356:229–238.

Connor, J.R. and Fine, R.E., 1987, Development of transferrin-positive oligodendrocytes in the rat central nervous system, *J. Neurosci. Res.* 17:51–59.

Connor, J.R. and Menzies, S.L., 1990, Altered cellular distribution of iron in the central nervous system of myelin deficient rats, *Neuroscience* 34:265–271.

Connor, J.R. and Menzies, S.L., 1996, Relationship of iron to oligodendrocytes and myelination, *Glia* 17:83–93.

Connor, J.R., Roskams, A.J., Menzies, S.L., and Williams, M.E., 1993, Transferrin in the central nervous system of the shiverer mouse myelin mutant, *J. Neurosci. Res.* 36:501–507.

Corbin, J.G., Gaiano, N., Machold, R.P., Langston, A., and Fishell, G., 2000, The Gsh2 homeodomain gene controls multiple aspects of telencephalic development, *Development* 127:5007–5020.

Crowe, A. and Morgan, E.H., 1992, Iron and transferrin uptake by brain and cerebrospinal fluid in the rat, *Brain Res.* 592:8–16.

Crowe, M.J., Bresnahan, J.C., Shuman, S.L., Masters, J.N., and Beattie, M.S., 1997, Apoptosis and delayed degeneration after spinal cord injury in rats and monkeys, *Nat. Med.* 3:73–76.

Crump, K.S., Kjellstrom, T., Shipp, A.M., Silvers, A., and Stewart, A., 1998, Influence of prenatal mercury exposure upon scholastic and psychological test performance: Benchmark analysis of a New Zealand cohort, *Risk Anal.* 18:701–713.

Crump, K.S., Van Landingham, C., Shamlaye, C., Cox, C., Davidson, P.W., Myers, G.J. *et al.*, 2000, Benchmark concentrations for methylmercury obtained from the Seychelles Island Development Study, *Environ. Health Perspect.* 108:257–263.

da Cunha, A., Jefferson, J.A., Jackson, R.W., and Vitkovic, L., 1993, Glial cell-specific mechanisms of TGF-beta 1 induction by IL-1 in cerebral cortex, *J. Neuroimmunol.* 42:71–85.

Dai, X., Lercher, L.D., Clinton, P.M., Du, Y., Livingston, D., Vieira, C. *et al.*, 2003, Trophic role of oligodendrocytes in the basal forebrain, *J. Neurosci.* 23:5846–5863.

Dai, X., Lercher, L.D., Yang, L., Shen, M., Black, I.B., and Dreyfus, C.F., 1997, Expression of neurotrophins by basal forebrain (BF) oligodendrocytes, *Soc. Neurosci. Abstr.* 23:331.

Dai, X., Qu, P., and Dreyfus, C.F., 2001, Neuronal signals regulate neurotrophin expression in oligodendrocytes of the basal forebrain, *Glia* 34:234–239.

Dai, X., Vierira, C., Lercher, L.D., Black, I.B., and Dreyfus, C.F., 1998, The trophic role of basal forebrain oligodendrocytes on cholinergic neurons, *Soc. Neurosci. Abstr.* 24:1778.

David, G., Barrett, J.N., and Barrett, E.F., 1992, Evidence that action potentials activate an internodal potassium conductance in lizard myelinated axons, *J. Physiol.* 445:277–301.

David, G., Barrett, J.N., and Barrett, E.F., 1993, Activation of internodal potassium conductance in rat myelinated axons, *J. Physiol.* 472: 177–202.

Davies, J.E. and Miller, R.H., 2001, Local sonic hedgehog signaling regulates oligodendrocyte precursor appearance in multiple ventricular domains in the chick metencephalon, *Dev. Biol.* 233:513–525.

Davis, A.D., Weatherby, T.M., Hartline, D.K., and Lenz, P.H., 1999, Myelin-like sheaths in copepod axons, *Nature* 398:571–571.

Dawson, M.R., Levine, J.M., and Reynolds, 2000, NG2-expressing cells in the central nervous system: Are they oligodendroglial progenitors? *J. Neurosci. Res.* 61:471–479.

de la Monte, S.M., 1989, Quantitation of cerebral atrophy in preclinical and end-stage Alzheimer's disease, *Ann. Neurol.* 25:450–459.

de Waegh, S.M., Lee, V.M., and Brady, S.T., 1992, Local modulation of neurofilament phosphorylation, axonal caliber, and slow axonal transport by myelinating Schwann cells, *Cell* 68:451–463.

Deadwyler, G.D., Pouly, S., Antel, J.P., and DeVries, G.H., 2000, Neuregulins and erbB receptor expression in adult human oligodendrocytes, *Glia* 32:304–312.

Deiner, M.S., 1997, Netrin-1 and DCC mediate axon guidance locally at the optic disk: Loss of function leads to optic nerve hypoplasia, *Neuron* 19:575–589.

Delange, F., 1994, The disorders induced by iodine deficiency, *Thyroid* 4:107–128.

Dickinson, T.K. and Connor, J.R., 1995, Cellular distribution of iron, transferrin, and ferritin in the hypotransferrinemic (Hp) mouse brain, *J. Comp. Neurol.* 355:67–80.

Dietrich, J., Noble, M., and Mayer-Proschel, M., 2002, Characterization of A2B5+ glial precursor cells from cryopreserved human fetal brain progenitor cells, *Glia* 40:65–77.

Dobbing, J., 1990, *Brain, Behavior and Iron in the Infant Diet,* Springer-Verlag, London.

Doetsch, F. and Alvarez-Buylla, A., 1996, Network of tangential pathways for neuronal migration in adult mammalian brain, *Proc. Natl. Acad. Sci. USA* 93:14895–14900.

Doetsch, F., Garcia-Verdugo, J.M., and Alvarez-Buylla, A., 1997, Cellular composition and three-dimensional organization of the subventricular germinal zone in the adult mammalian brain, *J. Neurosci.* 17:5046–5061.

Doherty, J.G., Burns, A.S., O'Ferrall, D.M., and Ditunno, J.F.J., 2002, Prevalence of upper motor neuron vs lower motor neuron lesions in complete lower thoracic and lumbar spinal cord injuries, *J. Spinal Cord Med.* 25:289–292.

Dolbec, J., Mergler, D., Sousa-Passos, C.J., Sousa de Morais, S., and Lebel J., 2000, Methylmercury exposure affects motor performance of a riverine population of the Tapajos River, Brazilian Amazon, *Int. Arch. Occup. Environ. Health* 73:195–203.

Dou, C.L. and Levine, J.M., 1994, Inhibition of neurite growth by the NG2 chondroitin sulfate proteoglycan, *J. Neurosci.* 14:7616–7628.

Dougherty, K.D., Dreyfus, C.F., and Black, I.B., 2000, Brain-derived neurotrophic factor in astrocytes, oligodendrocytes, and microglia/macrophages after spinal cord injury, *Neurobiol. Dis.* 7:574–585.

Du, Y.L. and Dreyfuss, C.F., 2002, Oligodendrocytes as providers of growth factors, *J. Neurosci. Res.* 68:647–654.

Dubois-Dalcq, M., Feigenbaum, V., and Aubourg, P., 1999, The neurobiology of X-linked adrenoleukodystrophy, a demyelinating peroxisomal disorder, *Trends Neurosci.* 22:4–12.

Dumont, R.J., Okonkwo, D.O., Verma, S., Hurlbert, R.J., Boulos, P.T., Ellegala, D.B. *et al.*, 2001, Acute spinal cord injury, part I: pathophysiologic mechanisms, *Clin. Neuropharmacol.* 24:254–264.

Duncan, I.D., 1996, Glial cell transplantation and remyelination of the central nervous system, *Neuropathol. Appl. Neurobiol.* 22:87–100.

Dupree, J.L., Coetzee, T., Blight, T., Suzuki, K., and Popko, B., 1998, Myelin galactolipids are essential for proper node of Ranvier formation in the CNS, *J. Neurosci.* 18:1642–1649.

Dupree, J.L., Girault, J.A., and Popko, B., 1999, Axo-glial interactions regulate the localization of axonal paranodal proteins, *J. Cell Biol.* 147:1145–1151.

Dwork, A.J., Schon, E.A., and Herbert, J., 1988, Nonidentical distribution of transferrin and ferric iron in human brain, *Neuroscience* 27:333–345.

Eccleston, P.A. and Silberberg, D.H., 1984, The differentiation of oligodendrocytes in a serum-free hormone-supplemented medium, *Brain Res.* 318:1–9.

Echelard, Y., Epstein, D.J., St, J.B., Shen, L., Mohler, L., and McMahon, J.A., 1993, Sonic hedgehog, a member of a family of putative signaling molecules is implicated in the regulation of CNS polarity, *Cell* 75:1417–1430.

Englund, E., Brun, A., and Ailing, C., 1988, White matter changes in dementia of Alzheimer's type, *Brain* 111:1425–1439.

Escobar Cabrera, O.E., Zakin, M.M., Soto, E.F., and Pasquini, J.M., 1997, Single intracranial injection of apotransferrin in young rats increases the expression of specific myelin protein mRNA, *J. Neurosci. Res.* 47:603–608.

Espinosa de los Monteros, A., Zhang, M., and De Vellis, J., 1993, O2A progenitor cells transplanted into the neonatal rat brain develop into oligodendrocytes but not astrocytes, *Proc. Natl. Acad. Sci. USA*, 90:50–54.

Esposito, F., Cuccovillo, F., Vanoni, M., Cimino, F., Anderson, C.W., Appella, E. et al., 1997, Redox-mediated regulation of p21(waf1/cip1) expression involves a post-transcriptional mechanism and activation of the mitogen-activated protein kinase pathway, *Eur. J. Biochem.* 245:730–737.

Fairbanks, V.F., 1994, In *Modern Nutrition in Health and Disease* (M.E. Shils, J.A. Olson, and M. Shike, eds.), Lea and Febiger, Philadelphia, PA, pp. 185–213.

Fanarraga, M.L., Griffiths, I.R., Zhao, M., and Duncan, I.D., 1998, Oligodendrocytes are not inherently programmed to myelinate a specific size of axon, *J. Comp. Neurol.* 399:94–100.

Fawcett, J.W. and Asher, R.A., 1999, The glial scar and CNS repair, *Brain Res. Bull.* 49:377–391.

FDA, November 16, 1999, *Mercury Compounds in Drugs and Food*, 98N-1109.

Fehlings, M.G. and Nashmi, R., 1996, Changes in pharmacological sensitivity of the spinal cord to potassium channel blockers following acute spinal cord injury, *Brain Res.* 736:135–145.

Ferguson, B., Matyszak, M.K., Esiri, M.M., and Perry, V.H., 1997, Axonal damage in acute multiple sclerosis lesions, *Brain* 120:393–399.

Fern, R. and Möller, T., 2000, Rapid ischemic cell death in immature oligodendrocytes: A fatal glutamate release feedback loop, *J. Neurosci.* 20:34–42.

ffrench Constant, C., Miller, R.H., Burne, J.F., and Raff, M.C., 1988, Evidence that migratory oligodendrocyte-type-2 astrocyte (O-2A) progenitor cells are kept out of the rat retina by a barrier at the eye-end of the optic nerve, *J. Neurocytol.* 17:13–25, ISSN: 0300-4864.

Fok-Seang, J. et al., 1995, Migration of oligodendrocyte precurrocytes and meningeal cells, *Dev. Biol.* 171:1–15.

Fok-Seang, J. and Miller, H.R., 1992, Astrocyte precursors in neonatal rat spinal cord cultures, *J. Neurosci.* 12:2751–2764.

Fok-Seang, J. and Miller, R.H., 1994, Distribution and differentiation of A2B5+ glial precursors in the developing rat spinal cord, *J. Neurosci. Res.* 37:219–235.

Foran, D.R. and Peterson, A.C., 1992, Myelin acquisition in the central nervous system of the mouse revealed by an MBP-LacZ transgene, *J. Neurosci.* 12:4890–4897.

Foster, H.D., 1993, The iodine-selenium connection: Its possible roles in intelligence, cretinism, sudden infant death syndrome, breast cancer and multiple sclerosis, *Med. Hypotheses.* 40:61–65.

Fouquet, F., Zhou, J.M., Ralston, E. et al., 1997, Expression of the adrenoleukodystrophy protein in the human and mouse central nervous system, *Neurobiol. Dis.* 3:271–285.

Francalanci, P., Eynard-Pierre, E., Dionisi-Vici, C. et al., 2001, Fatal infantile leukodystrophy, a severe variant of CACH/VWM syndrome, allelic to chromosome 3q27, *Neurology* 57:265–270.

Franklin, R.J. and Blakemore, W.F., 1995, Glial-cell transplantation and plasticity in the O-2A lineage—implications for CNS repair, *Trends Neurosci.* 18:151–156.

Fromman, C., 1878, Untersuchungen über die Gewebsveränderungen bei der Multiplen Sklerose des Gehirns und Rückenmarks, Gustav Fischer, Jena, pp. 1–123.

Fruttiger, M., Karlsson, L., Hall, A., Abramsson, A., Calver, A., Bostrom, H. et al., 1999, Defective oligodendrocyte development and severe hypomyelination in PDGF-A knockout mice, *Development* 126:457–467.

Galetta, S.L., Markowitz, C., and Lee, A.G., 2002, Immunomodulatory agents for the treatment of relapsing multiple sclerosis: A systematic review, *Arch. Intern. Med.* 162:2161–2169.

Gallo, V., Wright, P., and McKinnon, R.D., 1994, Expression and regulation of a glutamate receptor subunit by bFGF in oligodendrocyte progenitors, *Glia* 10:149–153.

Gallo, V., Zhou, J.M., McBain, C.J., Wright, P., Knutson, P.L., and Armstrong, R.C., 1996, Oligodendrocyte progenitor cell proliferation and lineage progression are regulated by glutamate receptor-mediated K+ channel block, *J. Neurosci.* 16:2659–2670.

Gao, F. and Raff, M., 1997, Cell size control and a cell-intrinsic maturation program in proliferating oligodendrocyte precursor cells, *J. Cell Biol.* 138:1367–1377.

Garcion, E., Faissner, A., and ffrench-Constant, C., 2001, Knockout mice reveal a contribution of the extracellular matrix molecule tenascin-C to neural precursor proliferation and migration, *Development* 128:2485–2496.

Gard, A.L. and Pfeiffer, S.E., 1993, Glial cell mitogens bFGF and PDGF differentially regulate development of O4+GalC-oligodendrocyte progenitors, *Dev. Biol.* 159:618–630.

Gay, F.W., Drye, T.J., Dick, G.W., and Esiri, M.M., 1997, The application of multifactorial cluster analysis in the staging of plaques in early multiple sclerosis. Identification and characterization of primary demyelinating lesions, *Brain* 120:1461–1483.

Genain, C.P., Cannella, B., Hauser, S.L., and Raine, C.S., 1999, Autoantibodies to MOG mediate myelin damage in MS, *Nat. Med.* 5:170–175.

Gensert, J.M. and Goldman, J.E., 1997, Endogenous progenitors remyelinate demyelinated axons in the adult CNS, *Neuron* 19:197–203.

Gillberg, C. and Wing, L., 1999, Autism: Not an extremely rare disorder, *Acta Psychiatr. Scand.* 99:399–406.

Givogri, M.I., Bongarzone, E.R., Schonmann, V., and Campagnoni, A.T., 2001, Expression and regulation of golli products of myelin basic protein gene during in vitro development of oligodendrocytes, *J. Neurosci. Res.* 66:679–690.

Glauser, T.A. and Packer, R.J., 1991, Cognitive deficits in long-term survivors of childhood brain tumors, *Childs. Nerv. Syst.* 7:2–12.

Gledhill, R.F. and McDonald, W.I., 1977, Morphological characteristics of central demyelination and remyelination: A single-fiber study, *Ann. Neurol.* 1:552–560.

Goebels, N., Hofstetter, H., Schmidt, S., Brunner, C., Wekerle, H., and Hohlfeld, R., 2000, Repertoire dynamics of autoreactive T cells in multiple sclerosis patients and healthy subjects: Epitope spreading versus clonal persistence, *Brain* 123:508–518.

Gomes, W.A., Mehler, M.F., and Kessler, J.A., 2003, Transgenic overexpression of BMP4 increases astroglial and decreases oligodendroglial lineage commitment, *Dev. Biol.* 255:164–177.

Gordon, N., 2001, Canavan disease: A review of recent developments, *Eur. J. Paediatr. Neurol.* 5:65–69.

Gottfries, C.G., Karlsson, I., and Svennerholm, L., 1996, Membrane components separate early-onset Alzheimer's disease from senile dementia of the Alzheimer type, *Int. Psychogeriatr.* 8:363–372.

Gow, A., Southwood, C.M., and Lazzarini, R.A., 1998, Disrupted proteolipid protein trafficking results in oligodendrocyte apoptosis in an animal model of Pelizaeus–Merzbacher disease, *J. Cell Biol.* 140:925–934.

Goyne, G.E., Warrington, A.E., De Vito, J.A., and Pfeiffer, J.E., 1994, Oligodendrocyte precursor quantitation and localization in perinatal brain slices using a retrospective bioassay, *J. Neurosci.* 14:5365–5372.

Grandjean, P., Weihe, P., White, R.F., and Debes F., 1998, Cognitive performance of children prenatally exposed to "safe" levels of methylmercury, *Environ. Res.* 77:165–172.

Grandjean, P., White, R.F., Nielsen, A., Cleary, D., and de Oliveira Santos, E.C., 1999, Methylmercury neurotoxicity in Amazonian children downstream from gold mining, *Environ. Health Perspect.* 107:587–591.

Gregori, N., Proschel, C., Noble, M., and Mayer-Proschel, M., 2002a, The tripotential glial-restricted precursor (GRP) cell and glial development in the spinal cord: Generation of bipotential oligodendrocyte-type-2 astrocyte progenitor cells and dorsal-ventral differences in GRP cell function, *J. Neurosci.* 22:248–256.

Gregori, N., Proschel, C., Noble, M., and Mayer-Pröschel, M., 2002b, The tripotential glial-restricted precursor (GRP) cell and glial development in the spinal cord: Generation of bipotential oligodendrocyte-type-2 astrocyte progenitor cells and dorsal–ventral differences in GRP cell function, *J. Neurosci.* 22:248–256.

Griffiths, I., Klugmann, M., Anderson, T., Yool, D., Thomson, C., Schwab, M.H. *et al.*, 1998, Axonal swellings and degeneration in mice lacking the major proteolipid of myelin, *Science* 280:1610–1613.

Griffiths, I.R. and McCulloch, M.C., 1983, Nerve fibres in spinal cord impact injuries. Part 1. Changes in the myelin sheath during the initial 5 weeks, *J. Neurol. Sci.* 58:335–349.

Grill, J., Renaux, V.K., Bulteau, C., Viguier, D., Levy-Piebois, C., Sainte-Rose, C. *et al.*, 1999, Long-term intellectual outcome in children with posterior fossa tumors according to radiation doses and volumes, *Int. J. Radiat. Oncol. Biol. Phys.* 45:137–145.

Grinspan, J.B., Edell, E., Carpio, D.F., Beesley, J.S., Lavy, L., Pleasure, D. *et al.*, 2000, Stage-specific effects of bone morphogenetic proteins on the oligodendrocyte lineage, *J. Neurobiol.* 43:1–17.

Grinspan, J.B. and Franceschini, B., 1995, Platelet-derived growth factor is a survival factor for PSA-NCAM+ oligodendrocyte pre-progenitor cells, *J. Neurosci. Res.* 41:540–551.

Grinspan, J.B., Stern, J.L., Pustilnik, S.M., and Pleasure, D., 1990, Cerebral white matter contains PDGF-responsive precursors to O2A cells, *J. Neurosci.* 10:1866–1873.

Gross, R.E., Mehler, M.F., Mabie, P.C., Zang, Z., Santschi, L., and Kessler, J.A., 1996, Bone morphogenetic proteins promote astroglial lineage commitment by mammalian subventricular zone progenitor cells, *Neuron* 17:595–606.

Groves, A.K., Barnett, S.C., Franklin, R.J., Crang, A.J., Mayer, M., Blakemore, W.F. *et al.*, 1993a, Repair of demyelinated lesions by transplantation of purified O-2A progenitor cells, *Nature* 362:453–455.

Groves, A.K., Entwistle, A., Jat, P.S., and Noble, M., 1993b, The characterization of astrocyte cell lines that display properties of glial scar tissue, *Dev. Biol.* 159:87–104.

Grzenkowski, M., Niehaus, A., and Trotter, J., 1999, Monoclonal antibody detects oligodendroglial cell surface protein exhibiting temporal regulation during development, *Glia* 28:128–137.

Gu, J., Royland, J.E., Wiggins, R.C., and Konat, G.W., 1997, Selenium is required for normal upregulation of myelin genes in differentiating oligodendrocytes, *J. Neurosci. Res.* 47:626–635.

Guvenc, H., Karatas, F., Guvenc, M., Kunc, S., Aygun, A.D., and Bektas, S., 1995, Low levels of selenium in mothers and their newborns in pregnancies with a neural tube defect, *Pediatrics* 95:879–882.

Hall, A., Giese, N.A., and Richardson, W.D., 1996, Spinal cord oligodendrocytes develop from ventrally derived progenitor cells that express PDGF alpha receptors, *Development* 122:4085–4094.

Halsey, N.A., 1999, Limiting infant exposure to thimerosal in vaccines and other sources of mercury, *JAMA* 282:1763–1766.

Hammang, J., Archer, D., and Duncan, I., 1997, Myelination following transplantation of EGF-responsive neural stem cells into a myelin-deficient environment, *Exp. Neurol.* 147:84–95.

Han, X., Holtzman, D.M., McKeer, Jr., D.W., Kelley, J., and Morris, J.C., 2002, Substantial sulfatide deficiency and ceramide elevation in very early Alzheimer's disease: Potential role in disease pathogenesis, *J. Neurochem.* 82:809–818.

Hanfield, F., Holzbach, U., Kruse, B. *et al.*, 1993, Diffuse white matter disease in three children: An encephalopathy with unique features on magnetic resonance imaging and proton magnetic resonance spectroscopy, *Neuropediatrics* 24:244–248.

Hannken, T., Schroeder, R., Zahner, G., Stahl, R.A., and Wolf, G., 2000, Reactive oxygen species stimulate p44/42 mitogen-activated protein kinase and induce p27(Kip1): Role in angiotensin II-mediated hypertrophy of proximal tubular cells, *J. Am. Soc. Nephrol.* 11:1387–1397.

Haykal-Coates, N., Shafer, T.J., Mundy, W.R., and Barone Jr., S., 1998, Effects of gestational methylmercury exposure on immunoreactivity of specific isoforms of PKC and enzyme activity during postnatal development of the rat brain, *Brain Res. Dev. Brain Res.* 109:33–49.

He, W., Ingraham, C., Rising, L., Goderie, S., and Temple, S., 2001, Multipotent stem cells from the mouse basal forebrain contribute GABAergic neurons and oligodendrocytes to the cerebral cortex during embryogenesis, *J. Neurosci.* 21:8854–8862.

Hecox, K. and Burkard, R., 1982, Developmental dependencies of the human brainstem auditory evoked response, *Ann. NY Acad. Sci.* 388:538–556.

Hemmer, B., Kieseier, B., Cepok, S., and Hartung, H.P., 2003, New immunopathologic insights into multiple sclerosis, *Curr. Neurol. Neurosci. Rep.* 3:246–255.

Hengst, L., and Reed, S.I., 1996, Translational control of p27kip-1 accumulation during the cell cycle, *Science* 271:1861–1864.

Hermann, G.E., Rogers, R.C., Bresnahan, J.C., and Beattie, M.S., 2001, Tumor necrosis factor-α induces cFOS and strongly potentiates glutamate-mediated cell death in the rat spinal cord, *Neurobiol. Dis.* 8:590–599.

Herrera, J., Yang, H., Zhang, S.C., Proschel, C., Tresco, P., Duncan, I.D. *et al.*, 2001, Embryonic-derived glial-restricted precursor cells (GRP cells) can differentiate into astrocytes and oligodendrocytes in vivo, *Exp. Neurol.* 171:11–21.

Hill, J.M. and Switzer, R.C., 3rd, 1984, The regional distribution and cellular localization of iron in the rat brain, *Neuroscience* 11:595–603.

Holben, D.H. and Smith, A.M., 1999, The diverse role of selenium within selenoproteins: A review, *J. Am. Diet. Assoc.* 99:836–843.

Honig, A. and Oski, F., 1978, Developmental scores of the iron deficient infants and the effect of therapy, *Infant Behav. Dev.* 1:168–176.

Hopewell, J.W. and van der Kogel, A.J., 1999, Pathophysiological mechanisms leading to the development of late radiation-induced damage to the central nervous system, *Front Radiat. Ther. Oncol.* 33:265–75.

Horner, P.J. *et al.*, 2000, Proliferation and differentiation of progenitor cells throughout the intact adult rat spinal cord, *J. Neurosci.* 20:2218–2228.

Hughes, S.M., Lillien, L.E., Raff, M.C., Rohrer, H., and Sendtner, M., 1988, Ciliary neurotrophic factor induces type-2 astrocyte differentiation in culture, *Nature* 335:70–73, ISSN: 0028-0836.

Hunter, S.F., Miller, D.J., and Rodriguez, M., 1997, Monoclonal remyelination-promoting natural autoantibody SCH 94.03: Pharmacokinetics and in vivo targets within demyelinated spinal cord in a mouse model of multiple sclerosis, *J. Neurol. Sci.* 150:103–113.

Hutchins, J.B. and Jefferson, V.E., 1992, Developmental distribution of platelet-derived growth factor in the mouse central nervous system, *Brain Res. Dev. Brain Res.* 67:121–135.

Ibarrola, N., Mayer-Proschel, M., Rodriguez-Pena, A., and Noble, M., 1996, Evidence for the existence of at least two timing mechanisms that contribute to oligodendrocyte generation in vitro, *Dev. Biol.* 180:1–21.

Ibarrola, N. and Rodriguez-Pena, A., 1997, Hypothyroidism coordinately and transiently affects myelin protein gene expression in most rat brain regions during postnatal development, *Brain Res.* 752:285–293.

Incardona, J.P., Gassield, W., Kapur, R.P., and Roelink, H., 1998, The teratogenic veratrum alkaloid cyclopamine inhibits sonic hedgehog signal transduction, *Development* 125:3353–3562.

Ishibashi, T., Dupree, J.L., Ikenaka, K., Hirahara, Y., Honke, K., Peles, E. et al., 2002, A myelin galactolipid, sulfatide, is essential for mainenance of ion channels on myelinated axon but not essential for initial cluster formation, *J. Neuosci.* 22:6507–6514.

Itoh, T., Beesley, J., Itoh, A., Cohen, A.S., Kavanaugh, B., Coulter, D.A. et al., 2002, AMPA glutamate receptor-mediated calcium signaling is transiently enhanced during development of oligodendrocytes, *J. Neurochem.* 81:390–402.

Jagannathan, N.R., Tandon, N., Raghunathan, P., and Kochupillai, N., 1998, Reversal of abnormalities of myelination by thyroxine therapy in congenital hypothyroidism: Localized in vivo proton magnetic resonance spectroscopy (MRS) study, *Brain Res. Dev. Brain Res.* 109:179–186.

Jaworski, D.M. et al., 1999, Intracranial injury acutely induces the expression of the secreted isoform of the CNS-specific hyaluronan-binding protein BEHAB brevican, *Exp. Neurol.* 157:327–337.

Jeannin, E., Robyr, D., and Desvergnem, B., 1998, Transcriptional regulatory patterns of the myelin basic protein and malic enzyme genes by the thyroid hormone receptors α1 and β1, *J. Biol. Chem.* 273: 24239–24248.

Jeffery, N., Crang, A., O'leary, M., Hodge, S., and Blakemore, W., 1999, Behavioural consequences of oligodendrocyte progenitor cell transplantation into experimental demyelinating lesions in the rat spinal cord, *Eur. J. Neurosci.* 11:1508–1514.

Jessell, T.M., 2000, Neuronal specification in the spinal cord: Inductive signals and transcriptional codes, *Nat. Rev. Genet.* 1:20–29.

Jessell, T.M. and Dodd, J., 1990, Floor plate-derived signals and the control of neural cell pattern in vertebrates, *Harvey Lect.* 86:87–128.

Jiang, Z.D., 1995., Maturation of the auditory brainstem in low risk-preterm infants: A comparison with age-matched full term infants up to 6 years, *Early Hum. Dev.* 42:49–65.

Johe, K.K., Hazel, T.G., Muller, T., Dugich-Djordjevic, M.M., and McKay, R.D., 1996, Single factors direct the differentiation of stem cells from the fetal and adult central nervous system, *Genes Dev.* 10:3129–3140.

Kakita, A. and Goldman, J.E., 1999, Patterns and dynamics of SVZ cell migration in the postnatal forebrain: monitoring living progenitors in slice preparations, *Neuron* 23:461–472.

Kakulas, B.A., 1999a, The applied neuropathology of human spinal cord injury, *Spinal Cord* 37:79–88.

Kakulas, B.A., 1999b, A review of the neuropathology of human spinal cord injury with emphasis on special features, *J. Spinal Cord Med.* 22:119–124.

Kaneto, H., Kajimoto, Y., Fujitani, Y., Matsuoka, T., Sakamoto, K., Matsuhisa, M. et al., 1999, Oxidative stress induces p21 expression in pancreatic islet cells: Possible implication in beta-cell dysfunction, *Diabetologia* 42:1093–1097.

Kaplan, M.R., Cho, M.-H., Ullian, E.M., Isorn, L.L., Levinson, S.R., and Barres, B.A., 2001, Differential control of clustering of the sodium channels $Na_v1.2$ and $Na_v1.6$ at developing CNS nodes of Ranvier, *Neuron* 30:105–119.

Kaplan, M.R., Meyer-Franke, A., Lambert, S., Bennett, V., Duncan, I.D., Levinson, S.R. et al., 1997, Induction of sodium channel clustering by oligodendrocytes, *Nature* 386:724–728.

Kato, A., Takahashi, H., Takahashi, Y., and Matsushime, H., 1997, Inactivation of the cyclin D-dependent kinase in the rat fibroblast cell line 3Y1, induced by contact inhibition, *J. Biol. Chem.* 272:8065–8070.

Keime-Guibert, F., Napolitano, M., and Delattre, J.Y., 1998, Neurological complications of radiotherapy and chemotherapy, *J. Neurol.* 245:695–708.

Keirstead, H., Ben-Hur, T., Rogister, B., O'Leary, M., Dubois-Dalcq, M., and Blakemore, W., 1999, Polysialylated neural cell adhesion molecule-positive CNS precursors generate both oligodendrocytes and Schwann cells to remyelinate the CNS after transplantation, *J. Neurosci.* 19:7529–7536.

Keirstead, H., Hughes, H., and Blakemore, W., 1998, A quantifiable model of axonal regeneration in the demyelinated adult rat spinal cord, *Exp. Neurol.* 151:303–313.

Keirstead, H.S. and Blakemore, W.F., 1997, Identification of post-mitotic oligodendrocytes incapable of remyelination within the demyelinated adult spinal cord, *J. Neuropathol.* 56:1191–1201.

Kiernan, B.W., Gotz, B., Fassner, A., and ffrench-Constant, C., 1996, Tenascin-C inhibits oligodendrocyte precursor cell migration by both adhesion-dependent and adhesion-independent mechanism, *Mol. Cell Neurosci.* 7:322–335.

Kimura, M., M. Sato, Akatsuka, A., Saito, S., Ando, K., Yokoyama M. et al., 1998, Overexpression of a minor component of myelin basic protein isoform (17.2 kDa) can restore myelinogenesis in transgenic shiverer mice, *Brain Res.* 785:245–252.

Kinney, H.C. and Armstrong, D.D., 1997, Perinatal neuropathology. In *Greenfield's Neuropathology* (D.I. Graham and P.L. Lantos, eds.), Arnold, London, 535–599.

Kinney, H.C., Brody, B.A., Kloman, A.S., and Gilles, F.H., 1988, Sequence of central nervous system myelination in human infancy. II. Patterns of myelination in autopsied infants, *J. Neuropath. Exp. Neurol.* 47:217–234.

Knapp, P.E., 1991, Studies of glial lineage and proliferation in vitro using an early marker for committed oligodendrocytes, *J. Neurosci. Res.* 30:336–345.

Knapp, P.E., Skoff, R.P., and Redstone, D.W., 1986, Oligodendroglial cell death in jimpy mice: An explanation for the myelin deficit, *J. Neurosci.* 6:2813–2822.

Kobayashi, K., Hayashi, M., Nakano, H., Fukutani, Y., Sasaki, K., Shimazaki, M. et al., 2002, Apoptosis of astrocytes with enhanced lysosomal activity and oligodendrocytes in white matter lesions in Alzheimer's disease, *Neuropathol. Appl. Neurobiol.* 28:238–251.

Kocsis, J.D., Gordon, T.R., and Waxman, S.G., 1986, Mammalian optic nerve fibers display two pharmacologically distinct potassium channels, *Brain Res.* 383:357–361.

Koda, M., Murakami, M., Ino, H., Yoshinaga, K., Ikeda, O., Hashimoto, M. et al., 2002, Brain-derived neurotrophic factor suppresses delayed apoptosis of oligodendrocytes after spinal cord injury in rats, *J. Neurotrauma.* 19:777–785.

Kohrle, J., 1996, Thyroid hormone deiodinases—a selenoenzyme family acting as gate keepers to thyroid hormone action, *Acta Med. Austriaca.* 23:17–30.

Kondo, T. and Raff, M., 2000, Oligodendrocyte precursor cells reprogrammed to become multipotential CNS stem cells, *Science* 289:1754–1757.

Koos, B.J. and Longo, L.D., 1976, Mercury toxicity in the pregnant woman, fetus, and newborn infant, *Am. J. Obst. Gyn.* 126:390–406.

Koper, J.W., Lopes-Cardozo, M., Romijn, H.J., and van Golde, L.M., 1984, Culture of rat cerebral oligodendrocytes in a serum-free, chemically defined medium, *J. Neurosci. Meth.* 10:157–169.

Krenz, N.R. and Weaver, L.C., 2000, Nerve growth factor in glia and inflammatory cells of the injured rat spinal cord, *J. Neurochem.* 74: 730–739.

Krigman, M.R. and Hogan, E.L., 1976, Undernutrition in the developing rat: Effect upon myelination, *Brain Res.* 107:239–255.

Kwiecien, J.M., O'Connor, L.T., Goetz, B.D., Delaney, K.H., Fletch, A.L., and Duncan, I.D., 1998, Morphological and morphometric studies of the dysmyelinating mutant, the Long Evans shaker rat, *J. Neurocytol.* 27:581–591.

Lachapelle, F., Duhamel Clerin, E., Gansmuller, A., Baron Van Evercooren, A., Villarroya, H., and Gumpel, M., 1994, Transplanted transgenically marked oligodendrocytes survive, migrate and myelinate in the normal mouse brain as they do in the shiverer mouse brain, *Eur. J. Neurosci.* 6:814–24, ISSN: 0953-816x.

Lachapelle, F., Gumpel, M., Baulac, M., Jacque, C., Duc, P., and Baumann, N., 1984, Transplantation of CNS fragment into brain of shiverer mutant mice: Extensive myelination by implanted oligodendrocytes, *Dev. Neurosci.* 6:325–334.

Lärkfors, L., Oskarsson, A., Sundberg, J., and Ebendal, T., 1991, Methylmercury induced alterations in the nerve growth factor level in the developing brain, *Brain Res. Dev.* 62:287–291.

Lassmann, H., 1999, The pathology of multiple sclerosis and its evolution, *Phil. Trans. R. Soc. Lond. B.* 354:1635–1640.

Lassmann, H., Bruck, W., and Lucchinetti, C., 2001, Heterogeneity of multiple sclerosis pathogenesis: Implications for diagnosis and therapy, *Trends Mol. Med.* 7:115–121.

Lavdas, A.A., Grigoriou, M., Pachnis, V., and Parnavelas, J.G., 1999, The medial ganglionic eminence gives rise to a population of early neurons in the developing cerebral cortex, *J. Neurosci.* 19:7881–7888.

Lazarus, J.H., 1999, Thyroid hormone and intellectual development: A clinician's view, *Thyroid* 9:659–660.

Leber, S.M., Breedlove, S.M., and Sanes, J.R., 1990, Lineage, arrangement, and death of clonally related motoneurons in chick spinal cord. *J. Neurosci.* 10:2451–2462.

Lee, J.-M. *et al.*, 1999, The changing landscape of ischaemic brain injury mechanisms, *Nature* 399:A7–A14.

Leegwater, P.A.J., Vermeulen, G., Konst, A.A.M. *et al.*, 2001, Subunits of the translation initiation factor eIF2B are mutated in leukoencephalopathy with vanishing white matter, *Nat. Genet.* 29:383–388.

Levine, J.M., 1994, Increased expression of the NG2 chondroitin-sulfate proteoglycan after brain injury, *J. Neurosci.* 14:4716–4730.

Levine, J.M. and Reynolds, R., 1999, Activation and proliferation of endogenous oligodendrocyte precursor cells during ethidium bromide-induced demyelination, *Exp. Neurol.* 160:333–347.

Levine, J.M., Reynolds, R., and Fawcett, J.W., 2001, The oligodendrocyte precursor cell in health and disease, *TINS* 24:39–47.

Levine, J.M., Stincone, F., and Lee, Y.S., 1993. Development and differentiation of glial precursor cells in the rat cerebellum, *Glia* 7:307–321.

LeVine, S.M. and Macklin, W.B., 1990, Iron-enriched oligodendrocytes: A reexamination of their spatial distribution, *J. Neurosci. Res.* 26:508–512.

Levison, S.W., Chuang, C., Abramson, B.J., and Goldman, J.E., 1993, The migrational patterns and developmental fates of glial precursors in the rat subventricular zone are temporally regulated, *Development* 119:611–622.

Levison, S.W. and Goldman, J.E., 1997, Multipotential and lineage restricted precursors coexist in the mammalian perinatal subventricular zone, *J. Neurosci. Res.* 48:83–94.

Levison, S.W., Young, G.M., and Goldman, J.E., 1999, Cycling cells in the adult rat neocortex preferentially generate oligodendroglia, *J. Neurosci. Res.* 57:435–446.

Lewohl, J.M., Wang, L., Miles, M.F., Zhang, L., Dodd, P.R., and Harris, R.A., 2000, Gene expression in human alcoholism: Microarray analysis of frontal cortex, *Alcohol Clin. Exp. Res.* 24:1873–1882.

Li, G.L., Farooque, M., Holtz, A., and Olsson, Y., 1999, Apoptosis of oligodendrocytes occurs for long distances away from the primary injury after compression trauma to rat spinal cord, *Acta Neuropathol. (Berl.)* 98:473–480.

Li, S. and Stys, P.K., 2000, Mechanisms of ionotropic glutamate receptor-mediated excitotoxicity in isolated spinal cord white matter, *J. Neurosci.* 20:1190–1198.

Li, W., Cogswell, C.A., and LoTurco, J.J., 1998, Neuronal differentiation of precursors in the neocortical ventricular zone is triggered by BMP, *J. Neurosci.* 18:8853–8862.

Li, Y.Q. and Wong, C.S., 1998, Apoptosis and its relationship with cell proliferation in the irradiated rat spinal cord, *Int. J. Radiat. Biol.* 74:405–417.

Lieberman, D.M., Laske, D.W., Morrison, P.F., Bankiewicz, K.S., and Oldfield, E.H., 1995. Convection-enhanced distribution of large molecules in gray matter during interstitial drug infusion, *J. Neurosurg.* 82:1021–1029.

Lillien, L.E. and Raff, M.C., 1990, Differentiation signals in the CNS: Type-2 astrocyte development in vitro as a model system, *Neuron* 5:5896–6273.

Lillien, L.E., Sendtner, M., and Raff, M.C., 1990, Extracellular matrix-associated molecules collaborate with ciliary neurotrophic factor to induce type-2 astrocyte development, *J. Cell. Biol.* 111:635–644.

Liu, S., Qu, Y., Stewart, T.J., Howard, M.J., Chakrabortty, S., Holekamp, T.F. *et al.*, 2000, Embryonic stem cells differentiate into oligodendrocytes and myelinate in culture and after spinal cord transplantation, *Proc. Natl. Acad. Sci. USA* 97:6126–6131.

Liu, Y., Wu, Y., Lee, J.C., Xue, H., Pevny, L.H., Kaprielian, Z., and Rao, M.S., 2002, Oligodendrocyte and astrocyte development in rodents: An in situ and immunohistological analysis during embryonic development, *Glia* 40:25–43.

Lonser, R.R., Corthesy, M.E., Morrison, P.F., Gogate, N., and Oldfield, E.H., 1999, Convection-enhanced selective excitotoxic ablation of the neurons of the globus pallidus internus for treatment of parkinsonism in nonhuman primates, *J. Neurosurg.* 91:294–302.

Lonser, R.R., Walbridge, S., Garmestani, K., Butman, J.A., Walters, H.A., Vortmeyer, A.O. *et al.*, 2002, Successful and safe perfusion of the primate brainstem: In vivo magnetic resonance imaging of macromolecular distribution during infusion, *J. Neurosurg.* 97:905–913.

Louis, J.C., Magal, E., Takayama, S., and Varon, S., 1993, CNTF protection of oligodendrocytes against natural and tumor necrosis factor-induced death, *Science* 259:689–692, ISSN: 0036-8075.

Louis, J.C., Muir, D., and Varon, S., 1992, Autocrine inhibition of mitotic activity in cultured oligodendrocyte-type-2 astrocyte (O-2A) precursor cells. *Glia* 6:30–38.

Lu, Q.R., Sun, T., Zhu, Z., Ma, N., Garcia, M., Stiles, C.D. *et al.*, 2002, Common developmental requirement for Olig function indicates a motor neuron/oligodendrocyte connection, *Cell* 109:75–86.

Lu, Q.R., Yuk, D., Alberta, J.A., Zhu, Z., Pawlitzky, I., Chan, J. *et al.*, 2000, Sonic hedgehog-regulated oligodendrocyte lineage genes encoding bHLH proteins in the mammalian central nervous system, *Neuron* 25:317–329.

Lyons, S.A. and Kettenmann, H., 1998, Oligodendrocytes and microglia are selectively vulnerable to combined hypoxia and hypoglucemia injury in vitro, *J. Cereb. Blood Flow Metab. Brain Dis.* 18:521–530.

Mabie, P., Mehler, M., Marmur, R., Papavasiliou, A., Song, Q., and Kessler, J., 1997, Bone morphogenetic proteins induce astroglial differentiation of oligodendroglial-astroglial progenitor cells, *Neuroscience* 17:4112–4120.

Mabie, P.C., Mehler, M.F., and Kessler, J.A., 1999, Multiple roles of bone morphogenetic protein signaling in the regulation of cortical cell number and phenotype, *J. Neurosci.* 19:7077–7088.

Macklin, W.B. and Weill, C.L., 1985, Appearance of myelin proteins during development in the chick central nervous system, *Dev. Neurosci.* 7:170–178.

Madison, D., Krueger, W.H., Trapp, B.D., Cheng, D., and Pfeiffer, S.E., 1999, A model for vesicular transport in oligodendrocyte myelin biogenesis involving SNARE complex proteins VAMP-2 and syntaxin 4, *J. Neurochem.* 72:988–998.

Madsen, K.M., Hviid, A., Vestergaard, M., Schendel, D., Wohlfahrt, J., Thorsen, P. *et al.*, 2002, A population-based study of measles, mumps and rubella vaccination and autism, *N. Engl. J. Med.* 347:1477–1482.

Maier, C.E. and Miller, R.H., 1995, Development of glial cytoarchitecture in the frog spinal cord, *Dev. Neurosci.* 17:149–159.

Maier, C.E. and Miller, R.H., 1997, Notochord is essential for oligodendrocyte development in Xenopus spinal cord, *J. Neurosci. Res.* 47:361–371.

Mallon, B.S., Shick, H.E., Kidd, G.J., and Macklin, W.B., 2002, Proteolipid promoter activity distinguishes two populations of NG2-positive cells throughout neonatal cortical development, *J. Neurosci.* 22:876–885.

Malone, M.J. and Szoke, M.C., 1985, Neurochemical changes in white matter. Aged human brain and Alzheimer's disease, *Arch. Neurol.* 42:1063–1066.

Marburg, O., 1906, Die sogenannte "akute Multiple Sklerose." *Jahrb. Psychistrie* 27:211–312.

Marshall, C.A. and Goldman, J.E., 2002, Subpallial dlx2-expressing cells give rise to astrocytes and oligodendrocytes in the cerebral cortex and white matter, *J. Neurosci.* 22:9821–9830.

Marta, C.B., Adamo, A.M., Soto, E.F., and Pasquini, J.M., 1998, Sustained neonatal hyperthyroidism in the rat affects myelination in the central nervous system, *J. Neurosci. Res.* 53:251–259.

Mathis, C., Denisenko-Nehrbass, N., Girault, J.A., and Borrelli, E., 2001, Essential role of oligodendrocytes in the formation and maintenance of central nervous system nodal regions, *Development* 128:4881–4890.

Matute, C., 1998, Properties of acute and chronic kainate excitotoxic damage to the optic nerve, *Proc. Natl. Acad. Sci. USA* 95:10229–10234.

Matute, C., Sanchez-Gomez, M.V., Martinez-Millan, L., and Miledi, R., 1997, Glutamate receptor-mediated toxicity in optic nerve oligodendrocytes, *Proc. Natl. Acad. Sci. USA* 94:8830–8835.

Mayer, M., Bhakoo, K., and Noble, M., 1994, Ciliary neurotrophic factor and leukemia inhibitory factor promote the generation, maturation and survival of oligodendrocytes in vitro, *Development* 120:142–153.

Mayer, M. and Noble, M., 1994, N-acetyl-L-cysteine is a pluripotent protector against cell death and enhancer of trophic factor-mediated cell survival in vitro, *Proc. Natl. Acad. Sci. USA* 91:7496–7500.

Mayer-Pröschel, M., Kalyani, A., Mujtaba, T., and Rao, M.S., 1997, Isolation of lineage-restricted neuronal precursors from multipotent neuroepithelial stem cells, *Neuron* 19:773–785.

McDonald, J.W., Althomsons, S.P., Hyrc, K.L., Choi, D.W., and Goldberg, M.P., 1998, Oligodendrocytes from forebrain are highly vulnerable to AMPA/kainate receptor-mediated excitotoxicity, *Nat. Med.* 4:291–297.

McKinnon, R.D., Matsui, T., Dubois-Dalcq, M., and Aaronson, S.A., 1990, FGF modulates the PDGF-driven pathway of oligodendrocytic development, *Neuron* 5:603–614.

McKinnon, R.D., Piras, G., Ida, Jr., J.A., and Dubois Dalcq, M., 1993, A role for TGF-beta in oligodendrocyte differentiation, *J. Cell Biol.* 121:1397–407, ISSN: 0021-9525.

McTigue, D.M., Horner, P.J., Stokes, B.T., and Gage, F.H., 1998. Neurotrophin-3 and brain-derived neurotrophic factor induce oligodendrocyte proliferation and myelination of regenerating axons in the contused adult rat spinal cord, *J. Neurosci.* 18:5354–5365.

Mehler, M.F., Mabie, P.C., Zhu, G., Gokhan, S., and Kessler, J.A., 2000, Developmental changes in progenitor cell responsiveness to bone morphogenetic proteins differentially modulate progressive CNS lineage fate, *Dev. Neurosci.* 22:74–85.

Meintanis, S., 2001, The neuron–glia signal beta-neuregulin promotes Schwann cell motility via a MAPK pathway, *Glia* 34:39–51.

Mekki-Dauriac, S., Agius, E., Kan, P., and Cochard, P., 2002, Bone morphogenetic proteins negatively control oligodendrocyte precursor specification in the chick spinal cord, *Development* 129:5117–5130.

Mendola, P., Selevan, S.G., Gutter, S., and Rice, D., 2002, Environmental factors associated with a spectrum of neurodevelopmental deficits, *Ment. Retard. Dev. Disabil. Res. Rev.* 8:188–197.

Meyer-Franke, A., Kaplan, M.R., Pfrieger, F.W., and Barres, B.A., 1995, Characterization of the signaling interactions that promote the survival and growth of developing retinal ganglion cells in cultions regulate the localization of axonal paranodal proteins, *Neuron* 15:805–819.

Mi, H. and Barres, B.A., 1999, Purification and characterization of astrocyte precursor cells in the developing rat optic nerve, *J. Neurosci.* 19:1049–1061.

Miller, D.J., Bright, J.J., Sriram, S., and Rodriguez, M., 1997, Successful treatment of established relapsing experimental autoimmune encephalomyelitis in mice with a monoclonal natural autoantibody, *J. Neuroimmunol.* 75:204–209.

Miller, R.H., 1999, Contact with central nervous system myelin inhibits oligodendrocyte progenitor maturation, *Dev. Biol.* 216:359–368.

Miller, R.H., Hayes, J.E., Dyer, K.L., and Sussman, C.R., 1999, Mechanisms of oligodendrocyte commitment in the vertebrate CNS, *Int. J. Dev. Neurosci.* 17:753–763.

Milner, R. and Ffrench Constant, C., 1994, A developmental analysis of oligodendroglial integrins in primary cells: Changes in alpha v-associated beta subunits during differentiation, *Development* 120:3497–3506.

Mitchell, J.H., Nicol, F., Beckett, G.J., and Arthur, J.R., 1998, Selenoprotein expression and brain development in preweanling selenium- and iodine-deficient rats, *J. Mol. Endocrinol.* 20:203–210.

Mizuguchi, R., Sugimori, M., Takebayashi, H., Kosako, H., Nagao, M., Yoshida et al., 2001, Combinatorial roles of olig2 and neurogenin2 in the coordinated induction of panneuronal and subtype-specific properties of motoneurons, *Neuron* 31:757–771.

Morath, D.J. and Mayer-Proschel, M., 2001, Iron modulates the differentiation of a distinct population of glial precursor cells into oligodendrocytes, *Dev. Biol.* 237:232–243.

Morath, D.J. and Mayer-Proschel, M., 2002, Iron deficiency during embryogenesis and consequences for oligodendrocyte generation in vivo, *Dev. Neurosci.* 24:197–207.

Morell, P., Quarles, R.H., and Norton, W.T., 1994, Myelin formation, structure, and biochemistry, In *Basic Neurochemistry: Molecular, Cellular, and Medical Aspects*, 5th edn (G.J. Siegel, ed.), Raven Press, New York, pp. 117–143.

Morris, C.M., Candy, J.M., Bloxham, C.A., and Edwardson, J.A., 1992, Immunocytochemical localisation of transferrin in the human brain, *Acta Anat. (Basel)* 143:14–18.

Mujtaba, J., Piper, D., Groves, A., Kalyani, A., Lucero, M., and Rao, M.S., 1999, Lineage restricted precursors can be isolated from both the mouse neural tube and cultures ES cells, *Dev. Biol.* 214:113–127.

Mundy, W.R., Parran, D.K., and Barone, Jr., S., 2000, Gestational exposure to methylmercury alters the developmental pattern of neurotrophin- and neurotransmitter-induced phosphoinositide (PI) hydrolysis, *Neurotox. Res.* 1:271–283.

Munoz, A., Rodriguez-Pena, A., Perez-Castillo, A., Ferreiro, B., Sutcliffe, J.G., and Bernal, J., 1991, Effects of neonatal hypothyroidism on rat brain gene expression, *Mol. Endocrinol.* 5:273–280.

Myers, G.J. and Davidson, P.W., 1998, Prenatal methylmercury exposure and children: Neurologic, developmental, and behavioral research, *Environ. Health Perspect.* 106(Suppl. 3):841–847.

Nakamura, S., Todo, T., Motoi, Y., Haga, S., Aizawa, T., Ueki, A. et al., 1999, Glial expression of fibroblast growth factor-9 in rat central nervous system, *Glia* 28:53–65.

Nakashima, K., Takizawa, T., Ochiai, W., Yanagisawa, M., Hisatsune, T., Nakafuku, M. et al., 2001, BMP2-mediated alteration in the developmental pathway of fetal mouse brain cells from neurogenesis to astrocytogenesis, *Proc. Natl. Acad. Sci. USA* 98:5868–5873.

Nakatsuji, Y. and Miller, R.H., 2001, Control of oligodendrocyte precursor proliferation mediated by density-dependent cell cycle protein expression, *Dev. Neurosci.* 23:356–363.

Nashmi, R. and Fehlings, M.G., 2001a, Changes in axonal physiology and morphology after chronic compressive injury of the rat thoracic spinal cord, *Neuroscience* 104:235–251.

Nashmi, R. and Fehlings, M.G., 2001b, Mechanisms of axonal dysfunction after spinal cord injury: With an emphasis on the role of voltage-gated potassium channels, *Brain Res. Rev.* 38:165–191.

Nashmi, R., Jones, O.T., and Fehlings, M.G., 2000, Abnormal axonal physiology is associated with altered expression and distribution of Kv1.1 and Kv1.2 K+ channels after chronic spinal cord injury, *Eur. J. Neurosci.* 12:491–506.

Nery, S., Wichterle, H., and Fishell, G., 2001, Sonic hedgehog contributes to oligodendrocyte specification in the mammalian forebrain, *Development* 128:527–540.

Neuhaus, O., Archelos, J.J., and Hartung, H.P., 2003, Immunomodulation in multiple sclerosis: From immunosuppression to neuroprotection, *Trends Pharmacol. Sci.* 24:131–138.

Niederost, B.P. et al., 1999, Bovine CNS myelin contains neurite growth-inhibitory activity associated with chondroitin sulfate proteoglycans, J. Neurosci. 19:8979–8989.

Nishiyama, A. et al., 1997, Normal and reactive NG2+ glial cells are distinct from resting and activated microglia, J. Neurosci. Res. 48:299–312.

Nishiyama, A., Chang, A., and Trapp, B.D., 1999, NG2+ glial cells: A novel glial cell population in the adult brain, J. Neuropathol. Exp. Neurol. 58:1113–1124.

Noble, M., Arhin, A., Gass, D., and Mayer-Proschel, M., 2003, The cortical ancestry of oligodendrocytes: Common principles and novel features, Dev. Neurosci. 25:217–233.

Noble, M., Murray, K., Stroobant, P., Waterfield, M.D., and Riddle, P., 1988, Platelet-derived growth factor promotes division and motility and inhibits premature differentiation of the oligodendrocyte/type-2 astrocyte progenitor cell, Nature 333:560–562.

Noda, M. et al., 1999, Glutamate release from microglia via glutamate transporter is enhanced by amyloid-ß peptide. Neuroscience 92:1465–1474.

Noguchi, T., Sugisaki, T., Satoh, I., and Kudo, M., 1985, Partial restoration of cerebral myelination of the congenitally hypothyroid mouse by parenteral or breast milk administration of thyroxine, J. Neurochem. 45:1419–1426.

Noseworthy, J.H., 2003, Treatment of multiple sclerosis and related disorders: What's new in the past 2 years? Clin. Neuropharmacol. 26:28–37.

Novitch, B.G., Chen, A.I., and Jessell, T.M., 2001, Coordinate regulation of motor neuron subtype identity and pan-neuronal properties by the bHLH repressor Olig2, Neuron 31:773–789.

Nutt, C.L., Noble, M., Chambers, A.F., and Cairncross, J.G., 2000, Differential expression of drug resistance genes and chemosensitivity in glial cell lineages correlate with differential response of oligodendrogliomas and astrocytomas to chemotherapy, Cancer Res. 60:4812–4818.

O'Leary, M.T. and Blakemore, W.F., 1997, Oligodendrocyte precursors survive poorly and do not migrate following transplantation into the normal adult central nervous system, J. Neurosci. Res. 48:159–167.

Olivier, C., Cobos, I., Perez Villages, E.M., Spassky, N., Zalc, B., Martinez, S. et al., 2001, Monofocal origin of telencephalic oligodendrocytes in the anterior entopenduncular areas of the chick embryo, Development 128:1757–1769.

Ong, W.Y. and Levine, J.M., 1999, A light and electron microscopic study of NG2 chondroitin sulfate proteoglycan-positive oligodendrocyte precursor cells in the normal and kainate-lesioned rat hippocampus, Neuroscience 92:83–95.

Ono, K., Bansal, R., Payne, J., Rutishauser, U., and Miller, R.H., 1995, Early development and dispersal of oligodendrocyte precursors in the embryonic chick spinal cord, Development 121:1743–1754.

Ono, K., Tsumori, T., Kishi, T., Yokota, S., and Yasue, Y., 1998, Developmental appearance of oligodendrocytes in the embryonic chick retina, J. Comp. Neurol. 398:309–322.

Ono, K., Yasui, Y., Rutishauser, U., and Miller, R.H., 1997, Focal ventricular origin and migration of oligodendrocyte precursors into the chick optic nerve, Neuron 19:283–292.

Oppenheimer, J.H. and Schwartz, H.L., 1997, Molecular basis of thyroid hormone-dependent brain development, Endocr. Rev. 18:462–475.

Orentas, D.M., Hayers, J.E., Dyer, K.L., and Miller, R.H., 1999, Sonic hedgehog signaling is required during the appearance of spinal cord oligodendrocyte precursors, Development 126:2419–2429.

Orentas, D.M. and Miller, R.H., 1996, The origin of spinal cord oligodendrocytes is dependent on local influences from the notochord, Dev. Biol. 177:43–53.

Park, J.K., Williams, B.P., Alberta, J.A., and Stiles, C.D., 1999, Bipotent cortical progenitor cells process conflicting cues for neurons and glia in a hierarchical manner, J. Neurosci. 19:10383–10389.

Park, S.K., Miller, R., Krane, I., and Vartanian, T., 2001, The erbB2 gene is required for the development of terminally differentiated spinal cord oligodendrocytes, J. Cell Biol. 154:1245–1258.

Patneau, D.K., Wright, P.W., Winters, C., Mayer, M.L., and Gallo, V., 1994, Glial cells of the oligodendrocyte lineage express both kainate- and AMPA-preferring subtypes of glutamate receptor Neuron 12:357–371.

Pavelko, K.D., van Engelen, B.G., and Rodriguez, M., 1998, Acceleration in the rate of CNS remyelination in lysolecithin-induced demyelination, J. Neurosci. 18:2498–2505.

Payne, H.R. and Lemmon, V., 1993a, Glial cells of the O-2A lineage bind preferentially to N-cadherin and develop distinct morphologies, Dev. Biol. 159:595–607.

Payne, H.R. and Lemmon, V., 1993b, Glial cells of the O-2A lineage bind preferentially to N-cadherin and develop distinct morphologies, Dev. Biol. 159:595–607.

Pedraza, L., Huang, J.K., and Colman, D.R., 2001, Organizing principles of the axoglial apparatus, Neuron 30:335–344.

Peissner, W., Kocher, M., Treuer, H., and Gillardon, F., 1999, Ionizing radiation-induced apoptosis of proliferating stem cells in the dentate gyrus of the adult rat hippocampus, Brain Res. Mol. Brain Res. 71:61–68.

Peles, E. and Salzer, J.L., 2000, Molecular domains of myelinated fibers, Curr. Opin. Neurobiol. 10:558–565.

Perez Villegas, E.M., Olivier, C., Spassky, N., Poncet, C., Cochard, P., Zalc, B. et al., 1999, Early specification of oligodendrocytes in the chick embryonic brain, Dev. Biol. 216:98–113.

Pfaff, S.L., Mendelsohn, M., Stewart, C.L., Edlund, T., and Jessell, T.M., 1996, Requirement for LIM homeobox gene Isl1 in motor neuron generation reveals a motor neuron-dependent step in interneuron differentiation, Cell 84:309–320.

Piani, D. et al., 1991, Murine macrophages induce NMDA receptor mediated neurotoxicity in vitro by secreting glutamate, Neurosci. Lett. 133:159–162.

Pichichero, M.E., Cernichiari, E., Lopreiato, J., and Treanor, J., 2002, Mercury concentrations and metabolism in infants receiving vaccines containing thiomersal: A descriptive study, Lancet 360:1711–1712.

Pollitt, E. and Leibel, R., 1976, Iron deficiency and behavior, J. Pediatr. 88:732–781.

Pombo, P.M.G., Barettino, D., Ibarrola, N., Vega, S., and Rodriguez-Pena, A., 1999, Stimulation of the myelin basic protein gene expression by 9-cis-retinoic acid and thyroid hormone: Activation in the context of its native promoter, Mol. Brain Res. 64:92–100.

Poncet, C., Soula, C., Trousse, F., Kan, P., Hirsinger, E., Pourguié, O. et al., 1996, Induction of oligodendrocyte precursors in the trunk neural tube by ventralizing signals: Effects of notochord and floor plate grafts, and of sonic hedgehog, Mech. Dev. 60:13–32.

Popko, B., Puckett, C., and Hood, L., 1988, A novel mutation in myelin-deficient mice results in unstable myelin basic protein gene transcripts, Neuron 1:221–225.

Poulter, M.O., Hashiguchi, T., and Padjen, A.L., 1989. Dendrotoxin blocks accommodation in frog myelinated axons, J. Neurophysiol. 62:174–184.

Poulter, M.O. and Padjen, A.L., 1995, Different voltage-dependent potassium conductances regulate action potential repolarization and excitability in frog myelinated axon, Neuroscience 68:497–504.

Power, J., Mayer-Proschel, M., Smith, J., and Noble, M., 2002, Oligodendrocyte precursor cells from different brain regions express divergent properties consistent with the differing time courses of myelination in these regions, Dev. Biol. 245:362–375.

Prass, K., Brück, W., Schröder, N.J.W. et al., 2001, Adult onset leukoencephalopathy with vanishing white matter is located on chromosome 3q27, Am. J. Hum. Genet. 65:728–734.

Pribyl, T.M., Campagnoni, C., Kampf, K., Handley, V.W., and Campagnoni, A.T., 1996, The major myelin protein genes are expressed in the human thymus, J. Neurosci. Res. 45:812–819.

Prineas, J.W. and Graham, J.S., 1981, Multiple sclerosis: Capping of surface immunoglobulin G on macrophages involved in myelin breakdown, *Ann. Neurol.* 10:149–158.

Pringle, N., Guthrie, S., Lumsden, A., and Richardson, W., 1998, Dorsal spinal cord neuroepithelium generates astrocytes but not oligodendrocytes, *Neuron* 20:883–893.

Pringle, N.P. and Richardson, W.D., 1993, A singularity of PDGF alpha-receptor expression in the dorsoventral axis of the neural tube may define the origin of the oligodendrocyte lineage, *Development* 117:525–533.

Pringle, N.P., Wei-Ping, Y., Guthrie, S., Roelink, H., Lumsden, A., Peterson, A.C. *et al.*, 1996, Determination of neuroepithelial cell fate: Induction of the oligodendrocyte lineage by ventral midline cells and Sonic Hedgehog, *Dev. Biol.* 177:30–42.

Qian, X., Davis, A.A., Goderie, S.K., and Temple, S., 1997, FGF2 concentration regulates the generation of neurons and glia from multipotent cortical stem cells, *Neuron* 18:81–93.

Raabe, T.D., Clive, D.R., Wen, D., and DeVries, G.H., 1997, Neonatal oligodendrocytes contain and secrete neuregulins in vitro, *J. Neurochem.* 69:1859–1863.

Radcliffe, J., Bunin, G.R., Sutton, L.N., Goldwein, J.W., and Phillips, P.C., 1994, Cognitive deficits in long-term survivors of childhood medulloblastoma and other noncortical tumors: Age-dependent effects of whole brain radiation, *Int. J. Dev. Neurosci.* 12:327–334.

Raff, M.C., Lillien, L.E., Richardson, W.D., Burne, J.F., and Noble, M.D., 1988, Platelet-derived growth factor from astrocytes drives the clock that times oligodendrocyte development in culture. *Nature* 333:562–565.

Raff, M.C., Miller, R.H., and Noble, M., 1983, A glial progenitor cell that develops in vitro into an astrocyte or an oligodendrocyte depending on the culture medium, *Nature* 303:390–396.

Rao, M. and Mayer-Pröschel, M., 1997, Glial restricted precursors are derived from multipotent neuroepithelial stem cells, *Dev. Biol.* 188:48–63.

Rao, M., Noble, M., and Mayer-Pröschel, M., 1998, A tripotential glial precursor cell is present in the developing spinal cord, *Proc. Natl. Acad. Sci. USA* 95:3996–4001.

Rasband, M.N. and Shrager, P., 2000, Ion channel sequestration in central nervous system axons, *J. Physiol. (Lond.)* 525:63–73.

Rasband, M.N., Trimmer, J.S., Schwarz, T.L., Levinson, S.R., Ellisman, M.H., Schachner, M. *et al.*, 1998, Potassium channel distribution, clustering, and function in remyelinating rat axons, *J. Neurosci.* 18:36–47.

Raskind, W.H., Williams, C.A., Hudson, L.D., and Bird, T.D., 1991, Complete deletion of the proteolipid protein gene (PLP) in a family with X-linked Pelizaeus–Merzbacher disease, *Am. J. Hum. Genet.* 49:1355–1360.

Redwine, J.M. and Armstrong, R.C., 1998, In vivo proliferation of oligodendrocyte progenitors expressing PDGFalphaR during early remyelination, *J. Neurobiol.* 37:413–428.

Remahl, S. and Hildebrand, C., 1990, Relation between axons and oligodendroglial cells during myelination. 1. The glial unit, *J. Neurocytol.* 19:313–328.

Richardson, W.D., Pringle, N., Mosley, M.J., Westermark, B., and Dubois Dalcq, M., 1988, A role for platelet-derived growth factor in normal gliogenesis in the central nervous system, *Cell* 53:309–319.

Richardson, W.D., Pringle, N.P., Yu, W.-P., and Hall, A.C., 1997, Origins of spinal cord oligodendrocytes: Possible developmental and evolutionary relationships with motor neurons, *Dev. Neurosci.* 19:54–64.

Richardson, W.D., Smith, J.K., Sun, T., Pringle, N.P., Hall, A.C., and Woodruff, R., 2000, Oligodendrocyte lineage and the motor neuron connection, *Glia* 29:136–142.

Rindfleisch, E., 1863, Histologisches Detail zur grauen Degeneration von Gehirn und Rückenmark, *Arch. Pathol. Anat. Physiol. Klin. Med. (Virchow)*, 26:474–485.

Riva, D. and Giorgi, C., 2000, The neurodevelopmental price of survival in children with malignant brain tumours, *Childs Nerv. Syst.* 16:751–754.

Rivera-Quinones, C., McGavern, D., Schmelzer, J.D., Hunter, S.F., Low, P.A., and Rodriguez, M., 1998, Absence of neurological deficits following extensive demyelination in a class I-deficient murine model of multiple sclerosis, *Nat. Med.* 4:187–193.

Robinson, S., Tani, M., Strieter, R., Ransohoff, R., and Miller, R.H., 1998, The chemokine growth-regulated oncogene-alpha promotes spinal cord oligodendrocyte precursor proliferation, *J. Neurosci.* 18:10457–10463.

Rodriguez, D., Gelot, A., dellar Gaspera, B. *et al.*, 1999, Increased density of oligodendrocytes in childhood ataxia with diffuse central hypomyelination (CACH) syndrome: Neuropathological and biochemical study of two cases, *Acta Neuropathol.* 97:469–480.

Rodriguez-Pena, A., 1999, Oligodendrocyte development and thyroid hormone, *J. Neurobiol.* 40:497–512.

Roelink, H., Augsburger, A., Heemskerk, J., Korzh, V., Norin, S., Ruiz i Altaba, A. *et al.*, 1994, Floor plate and motor neuron induction by vhh-1, a vertebrate homolog of hedgehog expressed by the notochord, *Cell* 76:761–775.

Roelink, H., Porter, J.A., Chiang, D.T., Beachy, P.A., and Jessell, T.M., 1995, Floor plate and motor neuron induction by different concentrations of the amino-terminal cleavage product of sonic hedgehog autoproteolysis, *Cell* 81:445–455.

Rogers, S.W., Gregori, N.Z., Carlson, N., Gahring, L.C., and Noble, M., 2001, Neuronal nicotinic acetylcholine receptor expression by O2A/oligodendrocyte progenitor cells, *Glia* 33:306–313.

Roher, A.E., Weiss, N., Kokjohn, T.A., Kuo, Y.M., Kalback, W., Anthony, J. *et al.*, 2002, Increased A beta peptides and reduced cholesterol and myelin proteins characterize white matter degeneration in Alzheimer's disease, *Biochemistry* 41:11080–11090.

Roman, D.D. and Sperduto, P.W., 1995, Neuropsychological effects of cranial radiation: Current knowledge and future directions, *Int. J. Radiat. Oncol. Biol. Phys.* 31:983–998.

Roncagliolo, M., Garrido, M., Walter, T., Peirano, P., and Lozoff, B., 1998, Evidence of altered central nervous system development in infants with iron deficiency anemia at 6 mo: Delayed maturation of auditory brainstem responses, *Am. J. Clin. Nutr.* 68:683–690.

Roots, B.I., 1993, The evolution of myelin, *Adv. Neural Sci.* 1:187–213.

Rowitch, D.H., Lu, R.Q., Kessaris, N., and Richardson, W.D., 2002, An "oligarchy" rules neural development, *Trends Neurosci.* 25:417–422.

Roy, N.S., Wang, S., Harrison-Restelli, C., Benraiss, A., Fraser, R.A., Gravel, M. *et al.*, 1999, Identification, isolation, and promoter-defined separation of mitotic oligodendrocyte progenitor cells from the adult human subcortical white matter, *J. Neurosci.* 19:9986–9995.

Royland, J.E., Konat, G., and Wiggins, R.C., 1993, Abnormal upregulation of myelin genes underlies the critical period of myelination in undernourished developing rat brain, *Brain Res.* 607:113–116.

Salamy, A. and McKean, C.M., 1976, Postnatal development of human brainstem potentials during the first year of life, *Electroencephalogr. Clin. Neurophysiol.* 40:418–426.

Sauvageot, C.M. and Stiles, C.D., 2002, Molecular mechanisms controlling cortical gliogenesis. *Curr. Opin. Neurobiol.* 12:244–249.

Schagen, S.B., van Dam, F.S., Muller, M.J., Boogerd, W., Lindeboom, J., and Bruning, P.F., 1999, Cognitive deficits after postoperative adjuvant chemotherapy for breast carcinoma, *Cancer* 85:640–650.

Schiffmann, R., Moller, J.R., Trapp, B.D. *et al.*, 1994, Childhood ataxia with diffuse central nervous system hypomyelination, *Ann. Neurol.* 35:331–340.

Schlegel, U., Pels, H., Oehring, R., and Blumcke, I., 1999, Neurologic sequelae of treatment of primary CNS lymphomas, *J. Neurooncol.* 43:277–286.

Schmidt, J., Rossie, S., and Catterall, W.A., 1985, A large intracellular pool of inactive Na channel alpha subunits in developing rat brain, *Proc. Natl. Acad. Sci. USA* 82:4847–4851.

Schnaedelbach, O. *et al.*, 2000, N-Cadherin influences migration of oligodendrocytes on astrocyte monolayers, *Mol. Cell. Neurosci.* 15:288–302.

Sherr, C.J., 1993, Mammalian G1 cyclins, *Cell* 73:1059–1065.

Sherratt, R.M., Bostock, H., and Sears, T.A., 1980, Effects of 4-aminopyridine on normal and demyelinated mammalian nerve fibres, *Nature* 283:570–572.

Shi, J., Marinovich, A., and Barres, B., 1998, Purification and characterization of adult oligodendrocyte precursor cells from the rat optic nerve, *J. Neurosci.* 18:4627–4636.

Shields, S., Gilson, J., Blakemore, W., and Franklin, R., 1999, Remyelination occurs as extensively but more slowly in old rats compared to young rats following gliotoxin-induced CNS demyelination, *Glia* 28:77–83.

Sikes, R.W., Fuller, G.N., Colbert, C., Chronister, R.B., DeFrance, J., and Wiggins, R.C., 1981, The relative numbers of oligodendroglia in different brain regions of normal and postnatally undernourished rats, *Brain Res. Bull.* 6:385–391.

Siragusa, V., Boffelli, S., Weber, G., Triulzi, F., Orezzi, S., Scotti, G. *et al.*, 1997, Brain magnetic resonance imaging in congenital hypothyroid infants at diagnosis, *Thyroid* 7:761–764.

Sistermans, E.A., de Wijs, I.J., de Coo, R.F., Smit, L.M., Menko, F.H., and van Oost, B.A., 1996, A (G-to-A) mutation in the initiation codon of the proteolipid protein gene causing a relatively mild form of Pelizaeus–Merzbacher disease in a Dutch family. *Hum. Genet.* 97:337–339.

Skoff, R.P., Price, D.L., and Stocks, A., 1976a, Electron microscopic autoradiographic studies of gliogenesis in the rat optic nerve. II. Time of origin, *J. Comp. Neurol.* 169:313–334.

Skoff, R.P., Price, D.L., and Stocks, A., 1976b, Electron microscopic autoradiographic studies of gliogenesis in rat optic nerve. I. Cell proliferation, *J. Comp. Neurol.* 169:291–312.

Skoff, R.P., Toland, D., and Nast, E., 1980, Pattern of myelination and distribution of neuroglial cells along the developing optic system of the rat and rabbit, *J. Comp. Neurol.* 191:237–253.

Small, R.K., Riddle, P., and Noble, M., 1987, Evidence for migration of oligodendrocyte-type-2 astrocyte progenitor cells into the developing rat optic nerve, *Nature* 328:155–157.

Smith, J., Ladi, E., Mayer-Pröschel, M., and Noble, M., 2000, Redox state is a central modulator of the balance between self-renewal and differentiation in a dividing glial precursor cell, *Proc. Natl. Acad. Sci. USA* 97:10,032–10,037.

Snyder, E.Y., Taylor, R.M., and Wolfe, J.H., 1995, Neural progenitor cell engraftment corrects lysosomal storage throughout the MPS VII mouse brain, *Nature* 374:367–370.

Soldán, M.M.P., Warrington, A.E., Bieber, A.J., Ciric, B., Van Keulen, V., Pease, L.R. *et al.*, 2003, Remyelination-promoting antibodies activate distinct Ca2+ influx pathways in astrocytes and oligodendrocytes: Relationship to the mechanism of myelin repair, *Mol. Cell. Neurosci.* 22:14–24.

Sortwell, C.E., Daley, B.F., Pitzer, M.R., McGuire, S.O., Sladek, J.R.J., and Collier, T.J., 2000, Oligodendrocyte-type 2 astrocyte-derived trophic factors increase survival of developing dopamine neurons through the inhibition of apoptotic cell death, *J. Comp. Neurol.* 426:143–153.

Spassky, N., Goujet-Zalc, C., Parmantier, E., Olivier, C., Marinez, S., Ivanova, A. *et al.*, 1998. Multiple restricted origin of oligodendrocytes, *J. Neurosci.* 18:8331–8343.

Spassky, N., Olivier, C., Perez-Villegas, E., Goujet-Zalc, C., Martinez, S., Thomas, J. *et al.*, 2000, Single or multiple oligodendroglial lineages: A controversy, *Glia* 29:143–148.

Stensaas, L.J. and Stensaas, S.S., 1968, Astrocytic neuroglial cells, oligodendrocytes and microgliacytes in the spinal cord of the toad. 11. Electron microscopy. *Zeitschrift für Zellforschung und mikroskopische Anatomie* 86:184–213.

Stover, J.F. *et al.*, 1997, Neurotransmitters in cerebrospinal fluid reflect pathological activity, *Eur. J. Clin. Invest.* 27:1038–1043.

Strelau, J. and Unsicker, K., 1999, GDNF family members and their receptors: Expression and functions in two oligodendroglial cell lines representing distinct stages of oligodendroglial development, *Glia.* 26:291–301.

Stylopoulos, L.A., George, A.E., de Leon, M.J., Miller, J.D., Foo, S.H., Hiesiger, E. *et al.*, 1988, Longitudinal CT study of parenchymal brain changes in glioma survivors, *AJNR Am. J. Neuroradiol.* 9:517–22.

Sugimoto, Y., Taniguchi, M., Yagi, T., Akagi, Y., and Nojyo, Y., 2001, Guidance of glial precursor cell migration by secreted cues in the developing optic nerve, *Development* 128:3321–3330.

Sun, T., Echelard, Y., Lu, R., Yuk, D.I., Kaing, S., Stiles, C.D. *et al.*, 2001, Olig bHLH proteins interact with homeodomain proteins to regulate cell fate acquisition in progenitors of the ventral neural tube, *Curr. Biol.* 11:1413–1420.

Sussel, L., Marin, O., Kimura, S., and Rubinstein, J., 1999, Loss of Nkx2.1 homeobox gene function results in a ventral to dorsal molecular respecification with the basal telencephalon: Evidence for a transformation of the pallidum into the striatum, *Development* 126:3359–3370.

Sussman, C.R., Dyer, K.L., Marchionni, M., and Miller, R.H., 2000, Local control of oligodendrocyte developmental potential in the dorsal mouse spinal cord. *J. Neurosci. Res.* 59:413–420.

Svennerholm, L. and Gottfries, C.G., 1994, Membrane lipids, selectively diminished in Alzheimer brains, suggest synapse loss as a primary event in early-onset form (type I) and demyelination in later-onset form (type II). *J. Neurochem.* 62:1039–1047.

Tada, E., Parent, J.M., Lowenstein, D.G., and Fike, J.R., 2000, X-irradiation causes a prolonged reduction in cell proliferation in the dentate gyrus of adult rats. *Neuroscience* 99:33–41.

Takebayashi, H., Nabeshima, Y., Yoshida, S., Chisaka, O., Ikenaka, K., Nabeshima, Y. *et al.*, 2002, The basic helix-loop-helix factor olig2 is essential for the development of motoneuron and oligodendrocyte lineages, *Curr. Biol.* 12:1157–1163.

Takebayashi, H., Yoshida, S., Sugimori, M., Kosako, H., Kominami, R., Nakafuku, M. *et al.*, 2000, Dynamic expression of basic helix-loop-helix Olig family members: Implication of Olig2 in neuron and oligodendrocyte differentiation and identification of a new member, Olig3, *Mech. Dev.* 99:143–148.

Takeshima, T., Johnston, J.M., and Commissiong, J.W., 1994, Oligodendrocyte-type-2 astrocyte (O-2A) progenitors increase the survival of rat mesencephalic, dopaminergic neurons from death induced by serum deprivation, *Neurosci. Lett.* 166:178–182.

Tang, D.G., Tokumoto, Y.M. and Raff, M.C., 2000, Long-term culture of purified oligodendrocyte precursor cells: Evidence for an intrinsic maturation program that plays out over months, *J. Cell Biol.* 148:971–984.

Taylor, E.M. and Morgan, E.H., 1990, Developmental changes in transferrin and iron uptake by the brain in the rat, *Dev. Brain Res.* 55:35–42.

Tekki-Kessaris, N., Woodruff, R., Hall, A.C., Gaffield, W., Kimura, S., Stiles, C.D. *et al.*, 2001, Hedgehog-dependent oligodendrocyte lineage specification in the telencephalon, *Development* 128:2545–2554.

Terry, R.D., Gonatas, N.K., and Weiss, M., 1964, Ultrastructural studies in Alzheimer's presenile dementia, *Am. J. Pathol.* 49:269–281.

Thomas, J.L., Spassky, N., Perez Villegas, E.M., Olivier, C., Cobos, I., Goujet-Zalc, C. *et al.*, 2000, Spatiotemporal development of oligodendrocytes in the embryonic brain, *J. Neurosci. Res.* 59:471–476.

Thompson, C.C. and Potter, G.B., 2000, Thyroid hormone action in neural development, *Cereb. Cortex.* 10:939–945.

Trapp, B., Peterson, J., Ransohoff, R., Rudick, R., Mork, S., and Bo, L., 1998, Axonal transection in the lesions of multiple sclerosis, *N. Engl. J. Med.* 338:278–285.

Trapp, B.D., 1990, Myelin-associated glycoprotein. Location and potential functions, *Ann. NY. Acad. Sci.* 605:29–43.

Trotter, J. and Schachner, M., 1989, Cells positive for the O4 surface antigen isolated by cell sorting are able to differentiate into astrocytes or oligodendrocytes, *Brain Res. Dev. Brain Res.* 46:115–122.

Trousse, F., Giess, M.C., Soula, C., Ghandour, S., Duprat, A.M., and Cochard, P., 1995, Notochord and floor plate stimulate oligodendrocyte differentiating cultures of the chick dorsal neural tube, *J. Neurosci. Res.* 41:522–560.

Tsai, H.H., Frost, E., To, V., Robinson, S., ffrench-Constant, C., Geertman, R., 2002, The chemokine receptor CXCR2 controls positioning of oligodendrocyte precursors in developing spinal cord by arresting their migration, *Cell* 110:373–383.

Tsai, H.H., Tessier-Lavigne, M., and Miller, R.H., 2003, Netrin 1 mediates spinal cord oligodendrocyte precursor dispersal, *Development* 130:2095–2105.

Tuohy, V.K. and Kinkel, R.P., 2000, Epitope spreading: A mechanism for progression of autoimmune disease, *Arch. Immunol. Ther. Exp. (Warsz.)* 48:347–351.

Tuohy, V.K., Yu, M., Yin, L., Kawczak, J.A., Johnson, J.M., Mathisen, P.M. *et al.*, 1998, The epitope spreading cascade during progression of experimental autoimmune encephalomyelitis and multiple sclerosis, *Immunol. Rev.* 164:93–100.

Ueda, H., Levine, J., Miller, R.H., and Trapp, B.D., 1999, Rat optic nerve oligodendrocytes develop in the absence of viable retinal ganglion cell axons, *J. Cell Biol.* 146:1365–1374.

Urenjak, J., Williams, S.R., Gadian, D.G., and Noble, M., 1992, Specific expression of N-acetylaspartate in neurons, oligodendrocyte-type-2 astrocyte progenitors, and immature oligodendrocytes in vitro, *J. Neurochem.* 59:55–61.

Urenjak, J., Williams, S.R., Gadian, D.G., and Noble, M., 1993, Proton nuclear magnetic resonance spectroscopy unambiguously identifies different neural cell types, *J. Neurosci.* 13:981–989.

Utzschneider, D.A., Archer, D.R., Kocsis, J.D., Waxman, S.G., and Duncan, I.D., 1994, Transplantation of glial cells enhances action potential conduction of amyelinated spinal cord axons in the myelin-deficient rat, *Proc. Natl. Acad. Sci. USA* 91:53–57.

van Dam, F.S., Schagen, S.B., Muller, M.J., Boogerd, W., vd Wall, E., Droogleever Fortuyn, M.E. *et al.*, 1998, Impairment of cognitive function in women receiving adjuvant treatment for high-risk breast cancer: High-dose versus standard-dose chemotherapy, *J. Natl. Cancer. Inst.* 90:210–218.

van der Knaap, M.S., Barth, P.G., Gabreels, F.J.M. *et al.*, 1997, A new leukoencephalopathy with vanishing white matter, *Neurology* 48:845–855.

van der Knaap, M.S., Breiter, S.N., Naidu, S. *et al.*, 1999, Defining and categorizing leukoencephalopathies of unknown origin: MR imaging approach, *Radiology* 213:121–133.

van der Knaap, M.S., Kamphorst, W., Barth, P.G. *et al.*, 1998, Phenotypic variation in leukoencephalopathy with vanishing white matter, *Neurology* 51:540–547.

van der Knaap, M.S., Kamphorst, W., Barth, P.G. *et al.*, 2001, Adult onset leukoencephalopathy with vanishing white matter presenting with pre-senile dementia, *Ann. Neurol.* 50:665–668.

van der Knaap, M.S., Leegwater, P.A.J., Könst, A.A.M., Visser, A., Naidu, S., Oudegans, C.B.M. *et al.*, 2002, Mutations in each of the five subunits of translation initiation factor EIF2B can cause leukoencephalopathy with vanishing white matter, *Ann. Neurol.* 51:264–270.

van der Knaap, M.S. and Valk, J., 1995, *Magnetic Resonance of Myelin, Myelination and Myelin Disorders*, 2nd edn., Springer, Berlin.

van Straaten, H.W., Hekking, J.W., Beursgens, J.P., Terwindt, R.E., and Drukker, J., 1989, Effect of the notochord on proliferation and differentiation in the neural tube of the chick embryo, *Development* 107:793–803.

van Straaten, H.W., Hekking, J.W., Wiertz, H.E., Thors, F., and Drukker, J., 1988, Effect of the notochord on the differentiation of a floor plate area in the neural tube of the chick embryo, *Anat. Embryol. Berl.* 177:317–324.

Vanderlugt, C.L. and Miller, S.D., 2002, Epitope spreading in immune-mediated diseases: Implications for immunotherapy, *Nat. Rev. Immunol.* 2:85–95.

Vartanian, T., Corfas, G., Li, Y., Fischbach, G.D., and Stefansson, K., 1994, A role for the acetylcholine receptor-inducing protein ARIA in oligodendrocyte development, *Proc. Natl. Acad. Sci. USA* 91:11626–11630.

Vartanian, T., Fischbach, G., and Miller, R., 1999, Failure of spinal cord oligodendrocyte development in mice lacking neuregulin, *Proc. Natl. Acad. Sci. USA* 96:731–735.

Vermeesch, M.K., Knapp, P.E., Skoff, R.P., Studzinski, D.M., and Benjamins, J.A., 1990, Death of individual oligodendrocytes in jimpy brain precedes expression of proteolipid protein, *Dev. Neurosci.* 12:303–315.

Vigliani, M.C., Duyckaerts, C., Hauw, J.J., Poisson, M., Magdelenat, H., and Delattre, J.Y., 1999, Dementia following treatment of brain tumors with radiotherapy administered alone or in combination with nitrosourea-based chemotherapy: A clinical and pathological study, *J. Neurooncol.* 41:137–149.

Vitry, S., Avellana-Adalid, V., Lachapelle, F., and Evercooren, A.B., 2001, Migration and multipotentiality of PSA-NCAM+ neural precursors transplanted in the developing brain, *Mol. Cell. Neurosci.* 17:983–1000.

Waber, D.P. and Tarbell, N.J., 1997, Toxicity of CNS prophylaxis for childhood leukemia, *Oncology (Huntingt.)* 11:259–264; discussion 264–265.

Wada, T., Kagawa, T., Ivanova, A., Zalc, B., Shirasaki, R., Murkami, F. *et al.*, 2000, Dorsal spinal cord inhibits oligodendrocyte development, *Dev. Biol.* 227:42–55.

Wallin, A., Gottfries, C.G., Karlsson, I., and Svennerholm, L., 1989, Decreased myelin lipids in Alzheimer's disease and vascular dementia, *Acta. Neurol. Scand.* 80:319–323.

Walters, S.N. and Morell, P., 1981, Effects of altered thyroid states on myelinogenesis, *J. Neurochem.* 36:1792–1801.

Wang, C., Pralong, W.F., Schulz, M.F., Rougon, G., Aubry, J.M., and Pagliusi, S., 1996, Functional N-methyl-d-aspartate receptors in O-2A glial precursor cells: A critical role in regulating polysialic acid neural cell adhesion molecule expression and cell migration, *J. Cell Biol.* 135:1565–1581.

Wang, C., Rougon, G., and Kiss, J.Z., 1994, Requirement of polysialic acid for the migration of the O-2A glial progenitor cell from neurohypophyseal explants, *J. Neurosci.* 14:4446–4457.

Wang, S., Sdrulla, A.D., diSibio, G. Bush, G., Nofziger, D., Hicks, C. *et al.*, 1998, Notch receptor activation inhibits oligodendrocyte differentiation, *Neuron* 21:63–75.

Warf, B.C., Fok-Seang, J., and Miller, R.H., 1991, Evidence for the ventral origin of oligodendrocyte precursors in the rat spinal cord, *J. Neurosci.* 11:2477–2488.

Warrington, A.E., Asakura, K., Bieber, A.J., Ciric, B., Van Keulen, V., Kaveri, S.V. *et al.*, 2000, Human monoclonal antibodies reactive to oligodendrocytes promote remyelination in a model of multiple sclerosis, *Proc. Natl. Acad. Sci. USA* 97:6820–6825.

Warrington, A.E., Barbarese, E., and Pfeiffer, S.E., 1992, Stage specific, (04+GalC−) isolated oligodendrocyte progenitors produce MBP+ myelin in vitro, *Dev. Neurosci.* 14:93–97.

Warrington, A.E., Barbarese, E., and Pfeiffer, S.E., 1993, Differential myelinogenic capacity of specific developmental stages of the oligodendrocyte lineage upon transplantation into hypomyelinating hosts, *J Neurosci Res.* 34:1–13.

Watanabe, M., Toyama, Y., and Nishiyama, A., 2002, Differentiation of proliferated NG2-positive glial progenitor cells in a remyelinating lesion, *J. Neurosci. Res.* 69:826–836.

Waxman, S.G., 2002, Ion channels and neuronal dysfunction in multiple sclerosis, *Arch. Neurol.* 59:1377–1380.

Weickert, C.S., Webster, M.J., Colvin, S.M., Herman, M.M., Hyde, T.M., Weinberger, D.R. *et al.*, 2000, Localization of epidermal growth factor receptors and putative neuroblasts in human subependymal zone, *J. Comp. Neurol.* 423:359–372.

Weimbs, T. and Stoffel, W., 1992, Proteolipid protein (PLP) of CNS myelin: Positions of free, disulfide bonded, and fatty acid thioesterlinked cysteine residues: Implications for the membrane topology of PLP, *Biochemistry* 31:12289–12296.

Weinberg, R.A., 1995, The retinoblastoma protein and cell cycle control, *Cell* 81:323–330.

Wichterle, H., Garcia-Verdugo, J.M., Herrera, D.G., and Alvarez-Buylla, A., 1999, Young neurons from medial ganglionic eminence disperse in adult and embryonic brain, *Nat. Neurosci.* 2:461–466.

Wieser, R.J., Renauer, D., Schafer, A., Heck, R., Engel, R., Schutz, S. *et al.*, 1990, Growth control in mammalian cells by cell–cell contacts, *Environ. Health Perspect.* 88:251–253.

Wiggins, R.C., 1979, A comparison of starvation models in studies of brain myelination, *Neurochem. Res.* 4:827–830.

Wiggins, R.C., 1982, Myelin development and nutritional deficiency, *Brain Res. Rev.* 4:151–175.

Wiggins, R.C. and Fuller, G.N., 1979, Relative synthesis of myelin in different brain regions of postnatally undernourished rats, *Brain Res.* 162:103–112.

Wiggins, R.C., Miller, S.L., Benjamins, J.A., Krigman, M.R., and Morell, P., 1976, Myelin synthesis during postnatal nutritional deprivation and subsequent rehabilitation, *Brain Res.* 107:257–273.

Wiggins, R.C. and Ruiz, B., 1990, Development under the influence of cocaine. II. Comparison of the effects of maternal cocaine and associated undernutrition on brain myelin development in the offspring, *Metab. Brain Dis.* 5:101–109.

Wilkins, A., Chandran, S., and Compston, A., 2001, A role for oligodendrocyte-derived IGF-1 in trophic support of cortical neurons, *Glia* 36:48–57.

Windrem, M.S., Roy, N.S., Wang, J., Nunes, M., Benraiss, A., Goodman, R. *et al.*, 2002, Progenitor cells derived from the adult human subcortical white matter disperse and differentiate as oligodendrocytes within demyelinated lesions of the rat brain, *J. Neurosci. Res.* 69:966–975.

Wollner, D.A., Scheinman, R., and Catterall, W.A., 1988, Sodium channel expression and assembly during development of retinal ganglion cells, *Neuron* 1:727–737.

Wolswijk, G., 1998, Chronic stage multiple sclerosis lesions contain a relatively quiescent population of oligodendrocyte precursor cells, *J. Neurosci.* 18:601–609.

Wolswijk, G., 2000, Oligodendrocyte survival, loss and birth in lesions of chronic-stage multiple sclerosis, *Brain* 123:105–115.

Wolswijk, G. and Noble, M., 1989, Identification of an adult-specific glial progenitor cell, *Development* 105:387–400.

Wolswijk, G. and Noble, M., 1992, Cooperation between PDGF and FGF converts slowly dividing O-2Aadult progenitor cells to rapidly dividing cells with characteristics of O-2Aperinatal progenitor cells, *J. Cell Biol.* 118:889–900.

Wolswijk, G., Riddle, P.N., and Noble, M., 1990, Coexistence of perinatal and adult forms of a glial progenitor cell during development of the rat optic nerve, *Development* 109:691–698, ISSN: 0950-1991.

Wolswijk, G., Riddle, P.N., and Noble, M., 1991, Platelet-derived growth factor is mitogenic for O-2A^{adult} progenitor cells, *Glia* 4:495–503.

Wong, K., Armstrong, R.C., Gyure, K.A. *et al.*, 2000, Foamy cells with oligodendroglial phenotype in childhood ataxia with diffuse central nervous system hypomyelination syndrome, *Acta Neuropathol.* 2000:635–646.

Woodward, K. and Malcolm, S., 1999, Proteolipid protein gene: Pelizaeus-Merzbacher disease in humans and neurodegeneration in mice, *Trends Genet.* 15:125–129.

Woodward, K. and Malcolm, S., 2001, CNS myelination and PLP gene dosage, *Pharmacogenomics* 2:263–272.

Wren, D., Wolswijk, G., and Noble, M., 1992, In vitro analysis of the origin and maintenance of O-2A^{adult} progenitor cells, *J. Cell Biol.* 116:167–176.

Wu, Q., Miller, R.H., Ransohoff, R.M., Robinson, S., Bu, J., and Nishiyama, A., 2000, Elevated levels of the chemokine GRO-1 correlate with elevated oligodendrocyte progenitor proliferation in the jimpy mutant, *J. Neurosci.* 20:2609–2617.

Xie, Y., Skinner, E., Landry, C., Handley, V., Schonmann, V., Jacobs, E. *et al.*, 2002, Influence of the embryonic preplate on the organization of the cerebral cortex: A targeted ablation model, *J. Neurosci.* 22:8981–8991.

Yakovlev, A., Mayer-Proschel, M., and Noble, M., 1998a, A stochastic model of brain cell differentiation in tissue culture, *J. Math. Biol.* 37:49–60.

Yakovlev, A.Y., Boucher, K., Mayer-Pröschel, M., and Noble, M., 1998b, Quantitative insight into proliferation and differentiation of O-2A progenitor cells in vitro: The clock model revisited, *Proc. Natl. Acad. Sci. USA.* 95:14164–14167.

Yakovlev, P.L. and Lecours, A.R., 1967, The myelogenetic cycles of regional maturation of the brain. In *Regional Development of the Brain in Early Life* (A. Minkowski *et al.*, eds.), Blackwell, Oxford, pp. 3–70.

Yamada, T., Placzek, M., Tanaka, H., Dodd, J., and Jessell, T.M., 1991. Control of cell pattern in the developing nervous system: Polarizing activity of the floor plate and notochord, *Cell* 64:635–647.

Yandava, B., Billinghurst, L., and Snyder, E., 1999, "Global" cell replacement is feasible via neural stem cell transplantation: Evidence from the dysmyelinated shiverer mouse brain, *Proc. Natl. Acad. Sci. USA* 96:7029–7034.

Yeh, J.J., Ruit, K.G., Wang, Y.X., Parks, W.C., Snider, W.D., and Deuel, T.F., 1991, PDGF A-chain is expressed by mammalian neurons during development and in maturity, *Cell* 64:209–219.

Yin, X., Crawford, T.O., Griffin, J.W., Tu, P., Lee, V.M., Li, C. *et al.*, 1998, Myelin-associated glycoprotein is a myelin signal that modulates the caliber of myelinated axons, *J. Neurosci.* 18:1953–1962.

Yoshioka, A. *et al.*, 1996, Pathophysiology of oligodendroglial excitotoxicity, *J. Neurosci. Res.* 46:427–437.

Yu, G., Steinkirchner, T., Rao, G., and Larkin, E.C., 1986, Effect of prenatal iron deficiency on myelination in rat pups, *Am. J. Path.* 125:620–624.

Yung, S.Y., Gokhan, S., Jurcsak, J., Molero, A.E., Abrajano, J.J., and Mehler, M.F., 2002, Differential modulation of BMP signaling promotes the elaboration of cerebral cortical GABAergic neurons or oligodendrocytes from a common sonic hedgehog-responsive ventral forebrain progenitor species. *Proc. Natl. Acad. Sci. USA* 99: 16,273–16,278.

Zerlin, M., Milosevic, A., and Goldman, J.E., 2004, Glial progenitors of the neonatal subventricular zone differentiate asynchronously, leading to spatial dispersion of glial cones and to the persistence of immature glia in the adult mammalian CNS, *Devel. Biol.* 270:200–213.

Zhang, H. and Miller, R.H., 1996, Density-dependent feedback inhibition of oligodendrocyte precursor expansion, *J. Neurosci.* 16:6886–6895.

Zhou, Q., Anderson, D.J. *et al.*, 2002, The bHLH transcription factors olig2 and olig1 couple neuronal and glial subtype specification, *Cell* 109:61–73.

Zhou, Q., Choi, G., and Anderson, D.J., 2001, The bHLH transcription factor Olig2 promotes oligodendrocyte differentiation in collaboration with nkx2.2, *Neuron* 31:791–807.

Zhou, Q., Wang, S., and Anderson, D.J., 2000, Identification of a novel family of oligodendrocyte lineage-specific basic helix-loop-helix transcription factors. *Neuron* 25:331–343.

Zhu, G., Mehler, M.F., Zhao, J., Yu Yung, S., and Kessler, J.A., 1999, Sonic hedgehog and BMP2 exert opposing actions on proliferation and differentiation of embryonic neural progenitor cells, *Dev. Biol.* 215:118–129.

Zoeller, R.T., Butnariu, O.V., Fletcher, D.L., and Riley, E.P., 1994, Limited postnatal ethanol exposure permanently alters the expression of mRNAS encoding myelin basic protein and myelin-associated glycoprotein in cerebellum, *Alcohol Clin. Exp. Res.* 18:909–916.

Zorin, A., Mayer-Proschel, M., Noble, M., and Yakovlev, A.Y., 2000, Estimation problems associated with stochastic modeling of proliferation and differentiation of O-2A progenitor cells in vitro, *Math. Biosci.* 167:109–121.

Zuppinger, K., Wiesmann, U., Siegrist, H.P., Schafer, T., Sandru, L., Schwarz, H.P. *et al.*, 1981, Effect of glucose deprivation on sulfatide synthesis and oligodendrocytes in cultured brain cells of newborn mice, *Pediatr. Res.* 15:319–325.

7

Astrocyte Development

Steven W. Levison, Jean de Vellis, and James E. Goldman

INTRODUCTION

Astrocytes were first identified at the end of the 19th century as star-shaped glia within the CNS. Over the last century we have come to appreciate the myriad functions that they perform to maintain homeostasis in the brain (Andriezen, 1893; Lenhossék, 1895; Cajal, 1909). Great strides also have been made in understanding their origins from the neural ectoderm and the events that shape their development. We will begin by first reviewing the functions and types of astrocytes in the mature brain and then proceed to describe when they arise, where their precursors reside, how they migrate, and which intrinsic and extrinsic factors shape their development.

FUNCTIONS AND TYPES OF ASTROCYTES IN THE CNS

In this chapter, we will define an astrocyte as a mature cell that possesses specific attributes and performs specific functions in the mature CNS. The functions of astrocytes include regulating extracellular ion concentrations, detoxifying xenobiotics, modifying synaptic efficacy, inactivating neurotransmitters, inducing and maintaining the blood–brain barrier and glial limitans, and providing nutrients and trophic support for neurons and oligodendrocytes. In the adult CNS, astrocytes are positioned rather uniformly across the parenchyma of the CNS where they associate with blood vessels, the pial surface, neurons, oligodendrocytes, and other astrocytes (Fig. 1).

Astrocytes are not randomly dispersed across the CNS, but rather form a "matrix" in which their cell bodies are evenly separated in space by a distance of approximately 50 μm (Chan-Ling and Stone, 1991; Levison and Goldman, 1993; Tout et al., 1993). It has been proposed that astrocytes contact each other during development and establish individual domains (Bushong et al., 2002). Astrocytes that contact each other form gap junctions, which allow them to communicate. These gap junctions are formed by connexins which allow calcium and other small molecules to flow between the cells. As a consequence of these contacts, the astrocytes form a syncytium that enables intracellular signals or ions to flow across rather large regions of the CNS. In addition to contacting each other, astrocytes create functional connections with the endothelial cells that line the blood vessels of the brain. Indeed, the blood vessel end foot specialization is one of the distinguishing characteristics of a mature astrocyte. The basal lamina secreted by the endothelial cells, as well as the basal lamina formed by the fibroblasts located at the pial surfaces of the brain are covered by end feet processes of astrocytes (Peters et al., 1976). Astrocytes also invest neuronal cell bodies, their dendrites, and synapses (Kosaka and Hama, 1986; Hama et al., 1993; Ventura and Harris, 1999). At these contacts, there are bidirectional signals that regulate synaptic efficacy and enhance the metabolism of the astrocytes (Haydon, 2001). The formation of these various types of cell contacts must be accounted for in a developmental model for astrocyte genesis.

Differences in Astrocyte Types in Different Regions

Types of Astrocytes *In Vivo*

The broad category of cells referred to as astrocytes is traditionally subdivided into the subtypes of fibrous or protoplasmic on the basis of morphological criteria as well as by their localization to white matter or gray matter, respectively (Figs. 2 and 3). In the cerebellum, the Bergmann glia represent an additional astrocyte form. In brain regions other than forebrain, cerebellum, and spinal cord there are a number of other specialized CNS cells that share some characteristics with astrocytes. Included are pituicytes, tanycytes, ependymal cells, and Mueller

Steven W. Levison • Department of Neurology and Neuroscience, UMDNJ-New Jersey Medical School, Newark, NJ 07101.   Jean de Vellis • Director, Mental Retardation Research Center, University of California, Los Angeles, Los Angeles, CA 90024.   James E. Goldman • Department of Pathology and the Center for Neurobiology and Behavior, Columbia University College of Physicians and Surgeons, New York, NY 10032.

Developmental Neurobiology, 4th ed., edited by Mahendra S. Rao and Marcus Jacobson. Kluwer Academic / Plenum Publishers, New York, 2005.   **197**

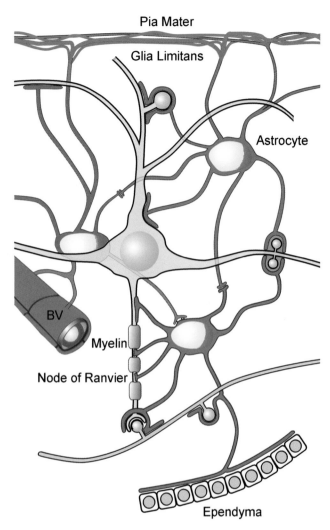

FIGURE 1. Astrocytes. In the adult CNS, astrocytes interact with multiple cell types. As depicted in this schematic, they form end feet on capillaries that induce the blood–brain barrier properties of the cerebral microvasculature. They interact with the cells along the pial surface, as well as at the ventricular lumen, and their processes interdigitate among the neurons, synapses, nodes of Ranvier, and oligodendrocytes. Additionally, they form gap junctions with other astrocytes.

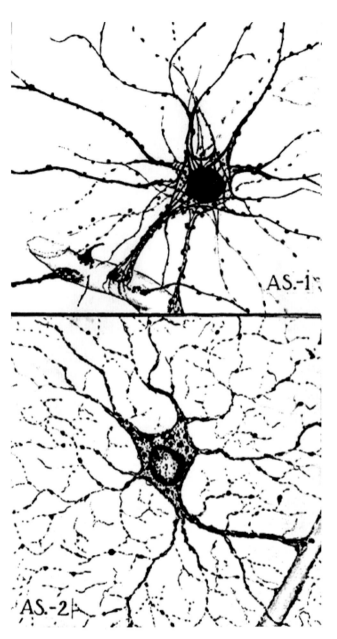

FIGURE 2. Types of astrocytes. Classical studies of astrocytes used metal impregnation techniques to reveal their structure. Histologists divided astrocytes into two main categories on the basis of their morphology. Reproduced here are two camera lucida drawings made by Wilder Penfield of a protoplasmic astrocyte and of a fibrous astrocyte (Penfield, 1932). Reproduced with permission.

glia. These glial cells will not be discussed at greater length; however, articles describing their development and functions can be found in the *Neuroglia* textbook (Kettenmann and Ransom, 1995).

    Fibrous astrocytes have a predominantly star-like morphology, with many cylindrical processes radiating symmetrically from the cell soma. These processes extend for long distances, branch infrequently, and contain abundant intermediate filaments. They frequently form end feet on capillaries and at the nodes of Ranvier. Fibrous astrocytes have oval nuclei containing evenly dispersed chromatin. In electron microscopic preparations, their cytoplasm is lightly tinted, with scattered glycogen granules and a relatively low density of organelles.

    Protoplasmic astrocytes have a more complex morphology than fibrous astrocytes. Their processes are highly branched and

form membranous sheets that enfold neuronal processes and cell bodies; they also form end feet on capillaries and at the pial surface. Protoplasmic astrocytes have spherical to oval nuclei, with slightly clumped chromatin. At the electron microscopic level their cytoplasm is lightly tinted; it contains glycogen and some microtubules. Compared to fibrous astrocytes, protoplasmic astrocytes have fewer intermediate filaments and a greater density of organelles. Whether the more complex morphology of protoplasmic astrocytes is an intrinsic property of the cell or

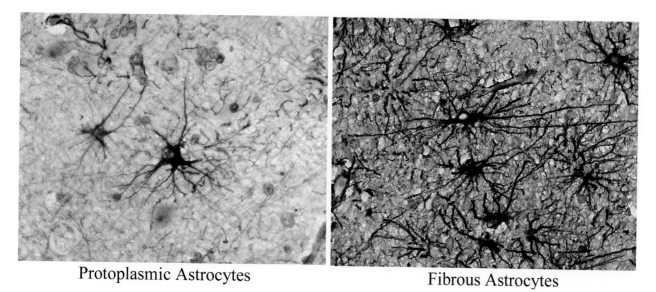

Protoplasmic Astrocytes   Fibrous Astrocytes

FIGURE 3. Human astrocytes stained for GFAP with a hematoxylin counterstain. Modern studies of astrocytes use immunohistochemical techniques to reveal their structure. Depicted here are protoplasmic astrocytes from the gray matter and fibrous astrocytes from the white matter stained for GFAP (brown) with a light hematoxylin counterstain (blue). Provided by S. W. Levison.

whether this form adopted as a consequence of the constraints of gray matter neuropil is not known. Since astrocytes can assume stellate shapes *in vitro* (see below), the process-bearing morphology of astrocytes is likely to reflect intrinsic biochemical and cytoskeletal properties.

Another astrocyte subtype known as the Bergmann glia, or Golgi epithelial cell, resides in the cerebellar cortex. The cell bodies of the Bergmann glia are present in the Purkinje cell layer, and these cells extend several long processes through the molecular layer, ending at the glial limitans of the pial surface and large blood vessels. Their processes ensheath Purkinje neurons and they send horizontal, lamellate expansions as they ascend through the molecular layer. These cells have a pale, bean-shaped nucleus that is usually oriented perpendicular to the pial surface. Their cytoplasm is typically pale, containing intermediate filaments, randomly oriented microtubules, glycogen, and scattered ribosomes.

The separation of astrocytes into the subcategories "fibrous" and "protoplasmic" has merit, but it is simplistic. For instance, Retzius (1894), drew images of at least six different types of astrocytes on the basis of Golgi metal impregnation staining, and Ramon-Molinar (1958) described cells stained with del Rio Hortega's modification of the Golgi method that displayed the star-like morphology of astrocytes, but had fewer processes, some of which were arranged in parallel, resembling those of oligodendrocytes. Additionally, there are astrocytes in white matter with a more protoplasmic topology, and astrocytes with mixed fibrous and protoplasmic features. Furthermore, classical histologists have described cells intermediate in form between oligodendrocytes and astrocytes that they have referred to as "transitional" neuroglia (Penfield, 1924; Wendell-Smith *et al.*, 1966). The terms "fibrous" and "protoplasmic" continue to be used because the morphological distinctions are generally sound,

and ultrastructural observations tend to confirm the separation of astrocytes into these two groups. However, the extent to which these differences are intrinsic, lineage-dependent properties, or are conferred upon the cells by the environments in which they reside is unknown. Whether "fibrous" and "protoplasmic" astrocytes are functionally distinct and whether they arise from the same progenitors are also unresolved issues.

Astrocyte Development Is Not Uniform Across Different Regions of the CNS

Studies carried out in a variety of regions in the mammalian CNS have revealed that glial development does not follow one pattern. There are notable anatomic differences among regions of the developing CNS and consequently there are multiple mechanisms of astrogliogenesis. For example, in most regions of the CNS, astrocytes arise from precursors that are direct descendents of the primitive neuroepithelium (the ventricular zone, VZ), such as in the spinal cord. Cells in the VZ can produce astrocytes directly. However, in other regions of the CNS where there is a subventricular zone (SVZ), some astrocytes arise from the precursors of the SVZ. In regions of the CNS where there is no SVZ, such as the optic nerve, there is strong evidence that the astrocytes are generated in a single wave, consistent with there being a single astrocyte lineage in these regions (Skoff, 1990). However, in regions of the CNS where there are multiple germinal zones, there may be two waves of astrogliogenesis. Thus, a conclusion that may be reached is there are more types of astrocytes in the forebrain (where there is an SVZ) than in the optic nerve. Below we will review in more depth the development of astrocytes within the optic nerve, cerebellum, spinal cord, and forebrain.

Molecular Markers for Astrocytes and Their Developmental Expression

In order to trace the development of an astrocyte from a precursor, one must have specific criteria that can be reliably used to classify different mature cell types. A number of immunological markers have been used over the years to identify specific cell types. Astrocyte intermediate filaments are composed of glial fibrillary acidic protein (GFAP), a protein enriched in astrocytes in the CNS, and vimentin, a less cell-specific filament protein (Bignami *et al.*, 1972; Antanitus *et al.*, 1975; Dahl *et al.*, 1981a). Thus, a positive immunohistochemical reaction for GFAP has often been used as a major criterion for identifying astrocytes. However, GFAP expression cannot be used as the sole criterion for identifying an astrocyte. For example, in early astrocyte development, vimentin can be the major or only intermediate filament expressed (Schnitzer *et al.*, 1981). Furthermore, some gray matter glia with the morphology and ultrastructural characteristics of astrocytes lack intermediate filaments (Herndon, 1964; Palay and Chan-Palay, 1974). Such astrocytes would be GFAP negative. Consistent with this observation, studies on GFAP protein levels and *in situ* hybridization for GFAP transcripts indicate that GFAP is expressed at lower levels in gray matter than in white matter (Kitamura *et al.*, 1987). Further restraint is recommended in using GFAP as the gold standard for astrocyte identification since some cells that may wander into the brain, such as a subset of lymphocytes, can express GFAP, and GFAP is also expressed by other cells outside of the CNS, such as myoepithelial cells, osteocytes, and chondrocytes (Neubauer *et al.*, 1996; Hainfellner *et al.*, 2001). Consequently, molecular markers other than intermediate filaments have been used to define astrocytes. For example, the enzyme glutamine synthetase (GS) is enriched in astrocytes, and fibrous and protoplasmic astrocytes are equally labeled by antibodies to GS (Norenberg and Martinez-Hernandez, 1979). The calcium-binding protein S-100β, and more recently the glutathione-*S*-transferase mu, are also useful markers for astrocytes (Boyes *et al.*, 1986; Cammer *et al.*, 1989).

Few antigenic markers are absolutely specific. Some gray matter oligodendrocytes and astrocytes, in fact, share a number of markers, consistent with the view that these cell populations might be closely related. For instance, the "oligodendrocytic" markers carbonic anhydrase II (CA) and another glutathione-*S*-transferase form, the π subunit, are expressed at low levels in some gray matter astrocytes in rodent CNS (Cammer and Tansey, 1988; Cammer *et al.*, 1989). Furthermore, the "astrocytic" marker GS has been demonstrated in some gray matter oligodendrocytes (D'Amelio *et al.*, 1990; Tansey *et al.*, 1991). Heterogeneous expression of these enzymes also illustrates the heterogeneity of astrocytes (also see below).

The macroglial cells of the CNS include many distinct types. Among these are several types of astrocytes, as well as progenitor cells. Due to this complexity, a full characterization of any given glial population should be based on a constellation of the attributes described above, including ultrastructure, the presence or absence of "astrocytic" markers (such as GFAP, S-100β, or GS) and the presence or absence of "oligodendrocytic" markers (such as 2′,3′ cyclic nucleotide 3′ phosphohydrolase [CNP] or myelin basic protein), and the absence of markers that are expressed by precursors. Examples of precursor markers could include nestin, vimentin, polysialic acid neural cell adhesion molecule (PSA-NCAM) and the NG2 proteoglycan.

A SUBSET OF ASTROCYTES ARE DIRECT DESCENDANTS OF THE VENTRICULAR ZONE

Radial Glia Arise from the VZ

Using metal impregnation staining, Lenhossék (1895) was the first to document the genesis of astrocytes. Using the Golgi method, which stains only a fraction of the cells in nervous tissues, cells were revealed that spanned the entire width of the developing neural tube, from ventricle to pial surface (Fig. 4).

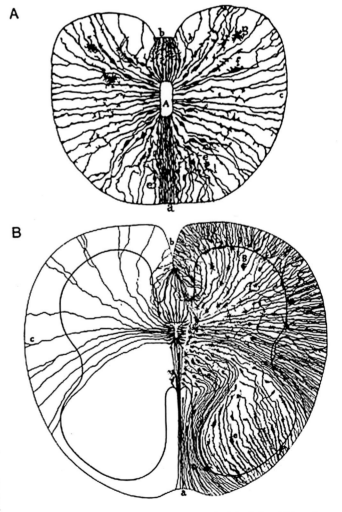

FIGURE 4. Radial glial cells in the spinal cord of the 9-day chick embryo (A) and 14 cm human embryonic spinal cord (B). These drawings depict golgi metal impregnation stains of radial glia, migrating precursors as well as radial glial transforming into astrocytes. (A) from Cajal (1894); (B) from Lenhossek (1893). Reproduced with permission.

Lenhossék provided a series of images that documented the transition of cells from these radial forms to cells that had the morphologies of mature astrocytes. The most primitive forms of these cells were classified using a number of terms and eventually given the name of radial glia by Rakic (1995). These radial cells had clearly evolved from the neuroepithelial cells of the VZ. Like earlier neuroepithelial cells these radial cells have processes that span the distance between the ventricular and pial surface, where they form pial end feet. As the CNS develops and more cells are added, the processes of the neuroepithelial cells elongate, but their cell bodies remain confined to the VZ. While these cells are referred to as radial glia, their processes may not, in fact, be strictly radial. Depending upon their location in the CNS, their processes may emerge and radiate perpendicular to the pial surface or they may curve substantially. As illustrated in Fig. 5, those radial glia that are located in the developing dorsomedial cerebral cortex extend processes that are perpendicular to the pial surface. Radial glial cells whose cell bodies reside in the lateral aspects of the VZ extend curvilinear process that first extend ventrolaterally, but then turn at right angles toward the pial surface. Based on both light and electron microscopic images of the developing brain, as well as time lapse video

microscopy of brain slices or of immature brain cells *in vitro*, it is clear that the radial glia act as substrates along which cortical neurons migrate during early forebrain development (Rakic, 1971, 1972; Gasser and Hatten, 1990). Migrating neurons adhere closely to the radial glia and the migrating neurons extend a leading process that wraps around the radial glial cell, forming a temporary adhesion to the radial glial cell which facilitates its movement (see Chapter 8).

The Radial Glial Cell Identity Is Actively Maintained by Neuron–Glial Cell Interactions

Radial glial cells receive signals from migrating neuroblasts that stabilize their status as radial glial cells. This type of signaling was first demonstrated using an *in vitro* paradigm where it was shown that immature granule neurons could induce immature cerebellar glia to adopt a radial glial morphology. To date, two ligand receptor signaling pairs have been discovered that likely mediate this effect; these are Delta/Notch and Neuregulin/erbB receptors (Fig. 6). Notch is a transmembrane receptor that is present on cells in the VZ and Notch is present on

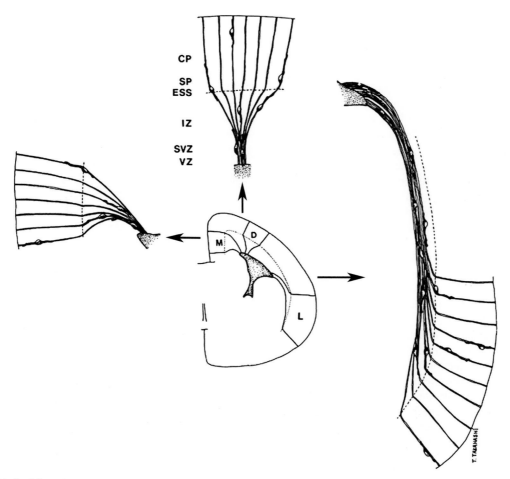

FIGURE 5. Radial glia. Schematic representation of patterns of radial glial fiber alignment and neuronal migration in medial (M), dorsal (D), and lateral (L) hemispheric regions. CP indicates cortical plate and SP indicates subplate. The somata of the radial glia are not depicted. From Misson *et al.* (1991), with permission.

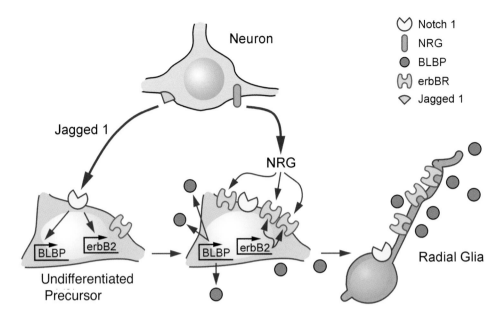

FIGURE 6. Neuronal induction of radial glia by sequential signaling through Notch and Erbb pathways. Initial contact by a Jagged1-expressing neuron activates Notch expression in the undifferentiated precursor. Notch signaling then induces expression of brain lipid-binding protein (BLBP) and erbB2 in the glial precursor. The increase in erbB receptor expression enhances the precursor's responsiveness to neuron-derived neuregulin (NRG), which subsequently induces the cell to adopt a radial morphology and to support neuronal migration.

radial glial cells. The ligands for Notch also are transmembrane proteins, and one of the ligands, Delta-1 is highly expressed by immature neurons. As the receptor is present on the radial glia, and the ligands are present on immature neurons, these molecules are appropriately situated to coordinate signals between migrating neurons and the radial glial cells they require to translocate to their appropriate destination. The introduction of an activated Notch receptor using replication deficient retroviruses increases the relative numbers of radial glia (Gaiano *et al.*, 2000). As activated Notch receptors increase the numbers of radial glia in the developing brain, this receptor system can provide a signal that maintains the numbers of radial glia needed during CNS development. However, as reviewed below, Notch signaling also can induce the radial glia to differentiate into astrocytes, which demonstrates that the context in which a precursor receives a signal is essential to how it interprets and responds to that signal. This example also demonstrates that no single ligand is responsible for the specification of an astrocyte from a precursor.

An additional ligand–receptor signaling system that has been shown to play a role in maintaining radial glial cell numbers is the neuregulin/erbB signaling pair. Like Notch, erbB receptors are transmembrane signaling receptors that are highly expressed by radial glial cells in the developing brain; however, erbBs activate disparate second messengers than Notch, and thus these receptors are not likely redundant. In the developing cerebellum, Bergmann glia cells express erbB4 whereas in the neocortex the radial glia express erbB2. Migrating neurons in both regions express the ligand for these receptors, neuregulin (Anton *et al.*, 1997; Rio *et al.*, 1997). Radial glial cells fail to develop normally in erbB2 genetically deficient mice and when antibodies against

neuregulin or against erbBs are introduced into radial glial/ immature neuron cell cultures, the length of the radial glial cells decreases and these antibodies perturb the migration of the immature neurons on the radial glia. Furthermore, adding neuregulin to these same cultures enhances the length of the radial glial fiber and increases the rate of neuronal migration. As the levels of neuregulin in the CNS decrease concurrent with neuronal maturation, which precedes the maturation of the radial glia (reviewed below), neuregulin/erbB likely play an important role in radial glial cell maintenance and function.

Radial Glia Express Markers of Immature Astrocytes

Radial glia express a number of molecules that are later shared by immature astrocytes. It is in part through the transient coexpression of these immature and mature cell markers that lineage relationships between radial glia and astrocytes have been deduced. The markers expressed by radial glia include (1) the intermediate filament proteins nestin and vimentin, and in primates, GFAP (Levitt *et al.*, 1981; Dahl *et al.*, 1981b; Benjelloun-Touimi *et al.*, 1985; Hockfield and McKay, 1985; Levine and Goldman, 1988; Tohyama *et al.*, 1992), (2) intermediate filament associated protein (IFAP) 300 (Yang *et al.*, 1993), (3) brain lipid-binding protein (BLBP) (Hartfuss *et al.*, 2001), (4) glutamate transporters, in particular GLAST (Shibata *et al.*, 1997), (5) the aldolase isoform, zebrin II (Staugaitis *et al.*, 2001; Marshall and Goldman, 2002), (6) the RNA-binding protein Musashi-1 (Sakakibara *et al.*, 1996), and (7) the radial cell markers RC1 and RC2 (Culican *et al.*, 1990; Edwards *et al.*, 1990).

Radial Glia Transform into Astrocytes at the End of Neurogenesis

The majority of radial glia undergo a transformation during the perinatal period to become astrocytes. As discussed earlier, Lenhossék used Golgi impregnations to reveal "intermediate" or "transitional" forms between radial glia and astrocytes in the early postnatal rodent brain and late gestational primate brain (Cajal, 1909; Choi and Lapham, 1978; Schmechel and Rakic, 1979b; Misson et al., 1988, 1991) (Fig. 7). Similar studies have been performed using antibodies to radial glial enriched molecules. These "intermediates" possess long processes that are

frequently radially oriented, but they gradually become more complex, as the cells become astrocytic. The accumulation of GFAP in cells with transitional forms was used as supportive evidence for this developmental transition. While these observations do not directly prove a transformation, they are consistent with it. More direct proof that radial glial differentiate into astrocytes comes from two types of experiments. First, radial glia were labeled in situ in the neonatal ferret brain by placing crystals of the fluorescent dye, DiI, onto the pial surface and then the development of the labeled cells was followed as they gradually transformed into astrocytes (Voigt, 1989). Second, the transformation of radial glial cells into astrocytes was followed using

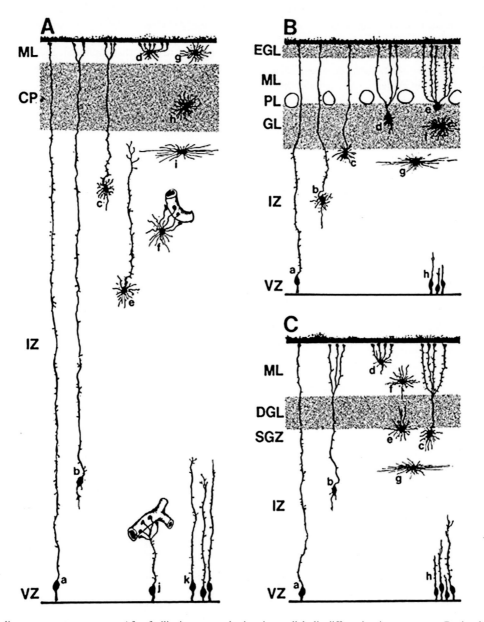

FIGURE 7. Radial glia are astrocyte precursors. After facilitating neuronal migration, radial glia differentiate into astrocytes. Depicted are radial glia from the cerebral cortex (A), cerebellar cortex (B), and dentate gyrus of the hippocampus (C). CP, cortical plate; EGL, external granule layer; GL, granular layer; DGL, dentate gyrus; IZ, intermediate zone; ML, molecular layer; PL, Purkinje cell layer; SGZ, subgranular zone; VZ, ventricular zone. From Cameron and Rakic, (1991), with permission.

replication deficient retrovirus infections. This type of analysis uses a retroviral vector to transfer a reporter gene, such as the gene encoding *Escherichia coli* beta-galactosidase, into the genome of dividing cells. Progeny of the infected cell will continue to express the transferred gene (Cepko, 1988; Sanes, 1989). In the case of beta-galactosidase, therefore, the descendants of an infected cell can be detected using either a histochemical stain for beta-galactosidase or by immunofluorescence with specific antibodies. When rat or chick embryos were injected with retroviruses and sacrificed at short intervals, the initial infected cells frequently had the morphologies of radial glial and they expressed markers of radial glia such as RC2 or vimentin. When infected animals were examined at later intervals after infection, many of the cells expressing β-galactosidase had the morphologies of astrocytes (Galileo *et al.*, 1990; Gray and Sanes, 1992). Thus, these linage-tracing experiments also established a precursor product relationship between radial glia and astrocytes. A third piece of evidence has been gleaned from cell culture studies where immature astrocytes can be transformed back to radial cells by the addition of neurons (which have neuregulin on their surface) or by the addition of neuregulin.

Radial Glial Are Bipotential Progenitors for Neurons and Astrocytes

Fate mapping studies using retroviruses have argued for a common progenitor for astrocytes and neurons in the retina (Turner and Cepko, 1987), optic tectum (Galileo *et al.*, 1990; Gray *et al.*, 1990; Gray and Sanes, 1992), cerebral cortex, and striatum (Halliday and Cepko, 1992; Walsh and Cepko, 1993). Studies that were performed when the retroviral lineage tracing technique was first applied to study CNS precursor differentiation reported that radial glia of the spinal cord and tectum could generate neurons and astrocytes, and the suggestion was made that the radial glia were self-renewing stem cells (Gray and Sanes, 1991) (Fig. 8). This latter hypothesis has been supported by more recent lineage studies and from time-lapse video microscopy to show that embryonic radial glial can divide asymmetrically to generate immature neurons, which then use the parent radial glial cell to ascend into the developing cortical plate (Malatesta *et al.*, 2000; Noctor *et al.*, 2001; Campbell and Gotz, 2002). When these radial guides are no longer required for neural precursor migration, most of the radial glia differentiate into astrocytes.

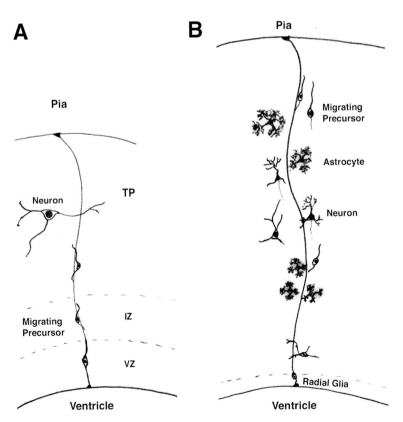

FIGURE 8. Radial glia are bipotential stem cells. Replication deficient retroviruses were injected into the tecta of early chick embryos to label dividing cells in the ventricular zone (VZ) and then examined over time to establish which cells were descended from the infected cells. (A) As depicted here, when analyzed within a few days of infection, clones of cells derived from the infected cells contained radial glia, cells migrating through intermediate zone (IZ) toward the tectal plate (TP) and the occasional neuron. (B) When analyzed more than a week after infection, the clones still contained one radial glial cell, but they also contained several types of neurons as well as astrocytes. Adapted from Gray and Sanes (1992).

Some Radial Glia Persist into Adulthood in Rodents and Human

Some glial cells with radial morphologies persist into adulthood. These include the Bergmann glia of the cerebellar cortex, radial glia in the hippocampus, and radially oriented glia found in the brainstem and spinal cord (King, 1966; Flament-Durand and Brion, 1985; Mori *et al.*, 1990; Reichenbach, 1990). Bergmann glia maintain their connections to the pial surface, while many of the radial glia in the brainstem and spinal cord maintain connections to the ventricular surface. These latter cells have been classically termed "tanycytes." Some of these radial cells guide migrating neurons during their development from the germinal zones to their final laminar locations. For example, cerebellar granule cells migrate from the external to the internal granule cell layer along Bergmann glial processes. Other radial glia in the adult brain function as conduits for ions or small proteins from the cerebral spinal fluid into the brain (Mori *et al.*, 1990). Recent studies have suggested that some of the residual radial glia, especially those in the hippocampus and related cells in the SVZ, may have stem cell properties. That is, they retain the developmental plasticity of radial glia of the embryonic CNS.

In an experiment to determine whether those cells in the SVZ that retained the features of radial glia are indeed neural stem cells, cytosine arabinoside (Ara-C), a potent antimitotic drug, was infused onto the surface of the brain of adult CD-1 mice for six days. This treatment effectively eliminates all mitotically active cells. However, within 10 days of the ceasing the Ara-C treatment, the SVZ fully regenerates, indicating that neural stem cells are resistant to Ara-C. Those cells that were resistant to Ara-C showed the ultrastructural and antigenic features of radial glia and they were the first cell type to divide after halting the Ara-C treatment (Doetsch *et al.*, 1999). Given their location and antigenic features, these stem cells are likely derivatives of the immature neuroepithelial cells of the VZ, although this lineage relationship has not been proven. Similarly, an immature astrocytic population that resides in the subgranular zone of the hippocampus generates neuroblasts that migrate into the dentate gyrus, where they differentiate into granule neurons (reviewed in Fabel *et al.*, 2003). These data indicate that the neural stem cells that persist in the mature brain retain properties of radial glia. It would be, however, a mistake to conclude that all of the immature astrocytes throughout the CNS can function as neural stem cells as those cells that leave their niche near the ventricles lose their stem cell properties. One question that remains unresolved is whether these persistent radial glia are bipotential precursors that produce neurons and astrocytes like their predecessors during development, or whether they are multipotent neural stem cells that are capable of generating all three major CNS cell types.

OTHER ASTROCYTES ARE DIRECTLY DESCENDED FROM THE SUBVENTRICULAR ZONE

The SVZs of the forebrain and cerebellum become prominent in late gestation and are highly productive. Indeed, during late gestation and during the first few weeks of postnatal life, the SVZs are the major sources of new astrocytes and oligodendrocytes. The SVZs contain large numbers of highly proliferative cells that are easily labeled by markers of DNA synthesis. Indeed, classic studies of forebrain gliogenesis inferred the migration of SVZ cells into white matter and gray matter and their differentiation into astrocytes and oligodendrocytes from thymidine pulse-labeling experiments (Altman, 1963; Paterson *et al.*, 1973). However, the conclusions reached from such DNA-labeling experiments were somewhat ambiguous, since dividing progenitors outside of the SVZ were also labeled as a consequence of systemically administered the ^{3}H-thymidine. Furthermore, interpretations of these experiments were confounded by the dilution of the label to undetectable levels with continued divisions of progenitor cells.

Lineage Tracing Experiments Demonstrate That Postnatal SVZ Cells Give Rise to Gray Matter, Protoplasmic Astrocytes

To more accurately trace the lineage of SVZ cells, dividing cells within the SVZ have been labeled with replication-deficient retroviruses (Fig. 9). When retroviruses are injected into the dorsolateral SVZ of the forebrain of two-day old rats, 95% of the labeled cells (when examined one day later) reside within the SVZ and they have immature morphologies (Levison and Goldman, 1993). The labeled cells typically have an ovoid or spindle shape, and the majority possess a single process that is oriented toward

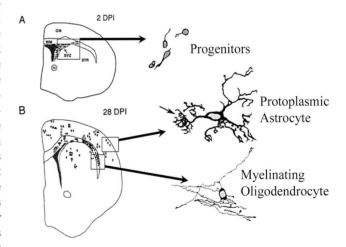

FIGURE 9. Schematic of postnatal SVZ descendants. Replication deficient retroviruses were injected into the SVZ of postnatal day two rats and the types of cells labeled were examined either 2 days post infection (DPI) or 28 DPI. The location of the labeled cells is indicated as well as the types of cells. At 2 DPI (A), the majority of labeled cells had simple unipolar morphologies and most of the labeled cells remained within the SVZ. At 28 DPI, labeled cells could be found dispersed throughout the forebrain, and these labeled cells could be classified as protoplasmic astrocytes (asterisks in B), myelinating oligodendrocytes (dark circles in B), or nonmyelinating oligodendrocytes (open circles in B). Camera lucida drawings of a typical protoplasmic astrocyte with an end foot on a blood vessel (arrow) and of a myelinating oligodendrocyte are provided. Adapted from Levison and Goldman (1993) and Levison *et al.* (1993), with permission.

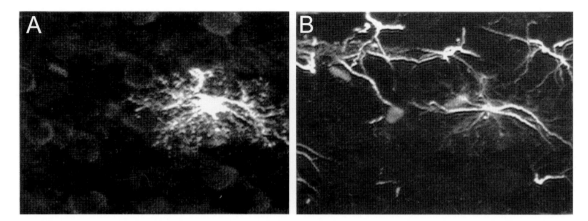

FIGURE 10. Protoplasmic astrocytes are descendants of the newborn SVZ. Panel A depicts immunofluorescence for the enzyme β-galactosidase encoded by the retroviral vector that had infected the precursor of this cell while it resided in the SVZ. The same cell is depicted in panel B stained for GFAP. From Levison and Goldman (1993), with permission.

the adjacent white matter. None of the labeled cells have the morphology of radial glia. Over time the labeled cells migrate out of the SVZ, generally moving laterally to colonize the adjacent striatum, or they migrate dorsally and laterally to reside in white matter and neocortical gray matter (Levison and Goldman, 1993; Levison *et al.*, 1993; Luskin and McDermott, 1994). SVZ cell migration also has been studied using time-lapse video microscopy in brain slices. These studies have demonstrated that migrating SVZ cells can use several means to migrate including radial glial cells and white matter tracts as well as glial tunnels (Suzuki and Goldman, 2003). Astrocytes generated from SVZ cells distribute throughout all cortical layers, up to the pial surface, and to the striatum. However, SVZ cells that migrate into the white matter preferentially differentiate into oligodendrocytes and not astrocytes (Levison and Goldman, 1993) (Fig. 9).

Contact with Blood Vessels Induces SVZ Precursors to Become Astrocytes

An interaction with blood vessels or the pial surface appears to be one of the first signs of astrocyte differentiation (Zerlin *et al.*, 1995; Zerlin and Goldman, 1997). After leaving the SVZ, many of the migrating SVZ cells still retain their simple, largely unipolar morphology. After contacting a blood vessel with its leading process, the SVZ cell begins to ensheath the blood vessel and it will form the classic end foot of an astrocyte (Zerlin *et al.*, 1995). During this ensheathment process, the cell slowly extends multiple fine processes into the surrounding parenchyma. The cell also acquires nestin and vimentin immunoreactivity—intermediate filaments characteristic of immature astrocytes that form the cytoskeleton of the cell (Zerlin *et al.*, 1995). Approximately half of the morphologically characterized astrocytes also express detectable levels of GFAP (Fig. 10). While this observation seems contradictory, it is, in fact not surprising, given that protoplasmic astrocytes have few intermediate filaments. Although half of the cells characterized as protoplasmic astrocytes do not express GFAP, they uniformly

express S-100β (Levison *et al.*, 1999). Interestingly, protoplasmic astrocyte differentiation is predominantly observed within the gray matter, with 65% of those cells from the P2 rat SVZ differentiating into these cells. Many of these newly generated astrocytes in the gray matter reside in tightly knit clusters, consistent with the view that astrocyte precursors continue to divide after they reach their final destination. By contrast, only 8.5% of the progeny of the P2 SVZ cells migrating into the white matter display astrocytic features, indicating that white matter tracts in the postnatal brain may not be permissive for astrocyte differentiation from SVZ cells.

This early interaction with blood vessels thus constitutes an early stage in astrocyte development. In fact, contact with endothelial cells induces astrocyte differentiation in astrocyte progenitors cultured from optic nerve (Mi *et al.*, 2001). That is, endothelial cells induce the expression of the astrocyte markers, GFAP and S-100β, in immature cells. This induction can be neutralized with antibodies to leukemia inhibitory factor (LIF), a growth factor expressed by endothelia (Mi *et al.*, 2001) (see below for discussion of molecular signals for astrocyte differentiation).

The Precursors in the SVZ That Generate Astrocytes Are Molecularly Distinct from VZ Derived Radial Glia

In contrast to the markers expressed by radial glia, precursors in the SVZ express a nonoverlapping set of molecular markers, which strongly suggests that the precursors in the SVZ that generate astrocytes are distinct from radial glia, which also give rise to astrocytes. As reviewed above, radial glia express a number of markers that include the intermediate filaments nestin, vimentin, and sometimes GFAP, IFAP-300, BLBP, GLAST, zebrin II, Musashi-1, and the radial cell markers RC1 and RC2. By contrast, the progenitors in the SVZ that give rise to glia do not express the majority of these markers. By contrast, a number of markers expressed by those SVZ cells that will go on to produce astrocytes are not expressed by radial glia. For example, when SVZ cells are dissociated and stained *in vitro* approximately two-thirds of the cells that would have gone on to generate astrocytes and oligodendrocytes label for PSA-NCAM and they express the ganglioside GD_3 (Levison and Goldman, 1997). These cell surface molecules are neither expressed by radial glia, nor are they expressed by type 1 astrocytes, which are lineally related to radial glia (Culican *et al.*, 1990). An additional molecule that is differentially expressed by SVZ cells and radial glia is the transcription factor Dlx2. Dlx2 is expressed by SVZ cells of the medial ganglionic eminence and it is not expressed by cells in the neocortical VZ (Anderson, 1997). Fate mapping studies performed using a Dlx2/tauLacZ knock-in mouse have shown that Dlx-2 expressing SVZ cells migrate into the striatum, white matter, and cerebral cortex where a subset differentiates into astrocytes (Marshall and Goldman, 2002).

SVZ Cells That Generate Astrocytes Can Also Produce Neurons or Oligodendrocytes

A small proportion of clones generated by single CNS cells from the postnatal SVZ contain astrocytes, neurons, and oligodendrocytes. On the basis of multiple lines of evidence we have argued that the retroviruses preferentially label progenitors, rather than stem cells (Levison and Goldman, 1997). Therefore, these experiments indicate that progenitors in the postnatal SVZ have the potential to give rise to both glia and neurons (Young and Levison, 1996; Levison and Goldman, 1997). Interpreting retroviral lineage studies *in vivo* can be difficult, depending on how one defines a "clone" of cells. Many studies have found that retroviral-labeled cells tend to congregate in homogenous clusters, with a small proportion of heterogenous clusters (astrocytes and oligodendrocytes) (Luskin *et al.*, 1988; Price and Thurlow, 1988; Grove *et al.*, 1993; Levison and Goldman, 1993; Luskin and McDermott, 1994; Parnavelas, 1999). Forebrain gliogenesis has been reexamined using a retroviral "library" (Walsh and Cepko, 1992), so that the proximity of cells to one another becomes irrelevant in judging clonality and two related cells that happen to be separated in space by some distance can be found to come from the same retrovirally infected progenitor. Most clones are indeed homogeneous, but about 15–20% are composed of both astrocytes and oligodendrocytes, sometimes appearing in the same cluster of glia (Zerlin and Goldman, submitted). Thus, not all SVZ cells are irrevocably committed to an astrocytic or oligodendrocytic fate before they emigrate from the SVZ. In fact, some do not make a final fate decision until they have stopped migrating, at which time they continue to divide to generate heterogeneous or homogenous clusters. An unresolved question is whether this difference in differentiation potential is not only a result of the actions of extrinsic inducers of astrocyte differentiation, but whether there also are intrinsic differences in the types of progenitors that exit the SVZ. While the majority of SVZ cells are molecularly distinct from VZ cells, a subset of SVZ cells share properties with VZ derived radial glia. The expression of the markers zebrin II and GLAST by a subset of SVZ cells may indicate that these cells are of direct VZ origin, whereas the PSA-NCAM population may be a separate precursor population. Thus, for the PSA-NCAM$^+$ cell population, their differentiation may be regulated by instructions that they receive once they cease migrating, whereas the zebrin II$^+$ population may be committed to an astrocyte fate prior to leaving the SVZ. Alternatively, it may turn out that the astrocyte fate decision is probabilistic, rather than strictly determined at one specific place and time.

Astrocyte Progenitors Emigrate from the SVZ along Radial Glial Guides

As progenitors emigrate from the SVZ, they usually leave in a radial orientation (Kakita and Goldman, 1999; Suzuki and Goldman, 2003). This suggests that they use radial glia to guide their exit from the SVZ. Indeed, examining progenitors as they leave the SVZ and as they migrate radially into white matter and cortex show that many contact radial glial processes (Suzuki and Goldman, 2003), which the CNS retains into early postnatal life. Glioblasts emigrating from the SVZ migrate radially, but once in the white matter they can move parallel to axons or radially into the cortex, where they can either continue or turn to migrate tangentially (Suzuki and Goldman, 2003). Furthermore, some SVZ cells migrate laterally along the white matter and then turn radially to enter the cortex, a pattern reminiscent of neuronal migration along radial glia (Bayer *et al.*, 1991; Misson *et al.*, 1991). (Fig. 11). Glioblasts migrate in a saltatory fashion at an average velocity of about 90 m per hr, but maximal speeds of up to 250 m per hr have been documented. By P14, however, the radial glia have transformed into astrocytes, and studies on the development of P14 SVZ cells have shown that progenitors at this stage either migrate into the subcortical white matter or the striatum, but they are unable to migrate into the neocortex. Thus, without radial glia, the SVZ cells are restricted from colonizing dorsal forebrain structures (Levison *et al.*, 1993).

DO DIFFERENT PATHWAYS OF ASTROCYTE DEVELOPMENT IN THE MAMMALIAN FOREBRAIN GIVE RISE TO DIFFERENT TYPES OF ASTROCYTES?

The studies summarized above indicate that there are two (at least) separate pathways through which the CNS provides astrocytes from immature cells: directly from radial glia and from SVZ cells. Is there any reason to think that these pathways generate different astrocyte types? Studies of glial development with retroviruses suggest that SVZ cells that migrate into gray matter preferentially generate astrocytes, whereas if they migrate into white matter they largely differentiate into oligodendrocytes (Levison and Goldman, 1993). This result suggests that many white matter astrocytes, at least in rodents, are not generated by SVZ cells, but from an earlier precursor pool, namely, the radial glia. Astrocytes constitute a heterogeneous group of cells, which vary in shape and molecular constituents, including growth factor expression, ion channels, levels of GFAP, and types of

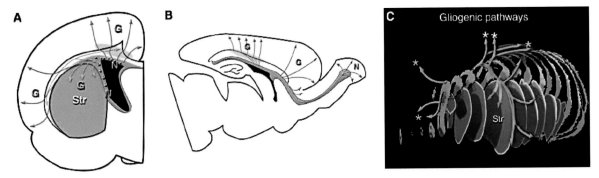

FIGURE 11. Models of migratory pathways of postnatal SVZ cells. (A) Coronal plane. Migratory pathways of SVZ cells into the white matter, cortex, and the striatum are shown by orange and yellow arrows, respectively. SVZ$_{DL}$ neuronal progenitor migration is depicted by purple dots and by short purple arrows. A region where cell accumulate between the corpus callosum and the subcortical white matter is indicated by light orange shades. (B) Sagittal plane. Representative migratory patterns of glial progenitors out of the anterior and posterior SVZ are shown by orange arrows. Migration of neuronal progenitors is shown by light purple arrows. (C) Reconstruction of the SVZ (green), the striatum (blue), and corresponding migration routes of glial progenitors. Migration toward the dorsal cortex (white asterisks), lateral cortex (red asterisks), and frontal cortex (pink asterisks) is shown. Yellow arrows indicate migration into the striatum. The orange ellipsoid shadow shows the cell accumulation layer between the corpus callosum and the subcortical white matter. Reproduced with permission from Suzuki and Goldman, (2003).

neurotransmitter transporters. How much of this heterogeneity is determined by lineage (radial glia vs SVZ), and how much is determined by local environmental factors (white matter vs gray matter) is not yet known.

Regardless of the source of a given astrocyte, the CNS must have a mechanism to generate large numbers of astrocytes in the perinatal period. It is in this time of rapid brain growth that the pial surface greatly enlarges, especially in larger mammals with gyriform brains, the vascular tree grows tremendously, and much of synaptogenesis peaks. As noted above, there seem to be constraints on astrocyte size; thus, as the brain enlarges there may be a requirement for more glia. It is likely that the need for additional astrocytes is satisfied by both the division of radial glia, the migration of astrocyte progenitors from the SVZ, and by the continued division of these progenitors within the parenchyma (Schmechel and Rakic, 1979a; Luskin and McDermott, 1994; Zerlin *et al.*, 1995; Zhang and Goldman, 1996; Levison *et al.*, 1999).

ASTROCYTE DEVELOPMENT IN THE CEREBELLUM

The cerebellum contains fibrous astrocytes in white matter, velate astrocytes in the internal granule cell layer, and Bergmann glia, whose cell bodies lie within the Purkinje cell layer (Palay and Chan-Palay, 1974). Bergmann glia send long complex processes dorsally that branch extensively as they project toward the pial surface. These processes resemble those of radial glia and guide immature granule cells as they migrate from the external to the internal granule cell layer (Grosche *et al.*, 1999). Early investigations using ^{3}H-thymidine uptake and morphological analyses suggested that Bergmann glia arise from other Bergmann glia or from some sort of immature cells, of unknown nature (Basco *et al.*, 1977; Moskovkin *et al.*, 1978; Choi and Lapham, 1980). Fate mapping studies using recombinant retroviruses show that progenitors that originate in the VZ at

the base of the cerebellum migrate from the base of the cerebellum through the white matter to reach the cortex to give rise to Bergmann glia and velate and white matter astrocytes (along with oligodendrocytes) (Miyake *et al.*, 1995; Zhang and Goldman, 1996; Milosevic and Goldman, 2002). At least some of the migratory progenitors in the white matter begin to express astrocyte characteristics, such as GLAST, during migration (Milosevic and Goldman, 2002). Since the viral-labeled astrocytes occur in clusters, either their immediate precursors or the immature astrocytes themselves continue to proliferate after they migrate from the VZ. The details of Bergmann glial growth can in general be inferred from images of Golgi or retroviral-labeled cells or immunocytochemical staining in the postnatal Purkinje cell layer (Choi and Lapham, 1980; Zhang and Goldman, 1996; Yamada *et al.*, 2000). The cell bodies arrest their transit from the VZ at the Purkinje cell layer and from there extend processes toward the pial surface. Over time, the numbers and complexity of their processes increase in dynamic synchrony with Purkinje cell dendritic growth and parallel fiber–Purkinje cell synapse formation. Interactions with neurons likely influence astrocyte morphology. Astrocytes derived from neonatal cerebellum assume Bergmann glial-like shapes, with elongated processes, when they are cocultured with migrating neurons, whereas they remain polygonal in the absence of neurons or in the presence of nonmigratory neurons (Hatten, 1985).

ASTROCYTE DEVELOPMENT IN THE SPINAL CORD

A number of immunocytochemical studies suggest that radial glia in the spinal cord, as elsewhere, generate astrocytes (Choi, 1981; Hirano and Goldman, 1988), although as noted above, it is difficult to infer lineage pathways from a chronological series of static images. Recent studies have linked the development of motor neurons and interneurons of the spinal cord to oligodendrocyte development, arguing that these three cell types arise from a single, ventral lineage, the specificity of which changes over time in a way correlated with changes in the expression of bHLH transcription factors that regulate patterning and cell fate (for review see Rowitch et al., 2002). Astrocytes, however, are not included in this lineage. Using fibroblast growth factor (FGF) receptor type 3 as a marker for astrocytes and their precursors in the developing cord, Pringle et al. (2003) have localized astrocyte precursors to both dorsal and ventral VZ, with the exception of the (ventral) pMN domain, which gives rise to oligodendrocytes and motor neurons. This is consistent with the earlier observation that astrocytes are generated from both dorsal and ventral parts of the neuroepithelium, while oligodendrocytes arise only from ventral parts (Pringle et al., 1998). Thus, the patterning of gliogenesis in the cord as reflected in the domains of the early neuroepithelium is different for astrocytes and oligodendrocytes and it is possible that cells in the pMN domain suppress the ability to generate astrocytes. In fact, in mice that are null for the bHLH factors Olig1 and Olig2, the pMN domain is converted to an adjacent homeodomain in the VZ, and that part of the VZ now generates interneurons followed by astrocytes (Takebayashi et al., 2002; Zhou and Anderson, 2002).

MAINTENANCE OF ASTROCYTES IN THE ADULT BRAIN

Astrocytes Are Maintained by Endogenous Precursors

Astrocytes continue to be generated in the adult CNS, albeit at an apparently very low rate. Studies that have counted astrocyte numbers or determined how many immature cells label in the adult brain with ^{3}H-thymidine conclude that immature cells persist within the parenchyma of the adult brain, but that there is little or no net accumulation of astrocytes with age (Altman, 1963; Hommes and Leblond, 1967; Korr et al., 1973; Ling and Leblond, 1973; Vaughan and Peters, 1974; Kaplan and Hinds, 1980; Paterson, 1983; Reyners et al., 1986; McCarthy and Leblond, 1988). For instance, in a study by McCarthy and LeBlond (1988), the authors administered ^{3}H-thymidine to nine-month-old "aged" male mice and then analyzed the astrocytes and oligodendrocytes of the corpus callosum. Some of the mice were given ^{3}H-thymidine as a 2 hr pulse to determine which cells were dividing, while in other mice ^{3}H-thymidine was infused for 30 days and the animals were sacrificed 60 or 180 days later. The ^{3}H-thymidine pulse paradigm revealed that there are cells resident within the white matter that label after an acute pulse of ^{3}H-thymidine. These cells have the morphological features of immature cells. After a 30-day infusion, approximately 12% of the astrocytes and 1% of the oligodendrocytes were labeled, indicating a net daily addition of 0.4% astrocytes and 0.04% oligodendrocytes per day. However, when the 30-day infusion was followed by 60 and 180 days without ^{3}H-thymidine, the labeled astrocytes decrease to 5% and 0% over time, respectively, whereas the number of labeled oligodendrocytes did not

change. These data indicate that there are resident immature cells in the corpus callosum that continue to divide across the life span to give rise to new astrocytes and oligodendrocytes. The production of new astrocytes appears to occur concurrent with cell turnover, so that the astrocyte population remains stable. By contrast, oligodendrocytes slowly accumulate in the adult brain, as confirmed by other studies (Ling and Leblond, 1973; Peters et al., 1991; Levison et al., 1999).

Several lines of evidence support the concept that the cells that are dividing in the mature brain are immature astrocytes rather than mature cells or bipotential glial precursors. For instance, when bulk fractionation techniques are used to isolate cells from the adult brain, a population of GD3 ganglioside[+] cells can be harvested from the white matter (a marker of immaturity). These GD3[+] cells from the adult brain are GFAP[-], divide in culture, and over time acquire mature astrocytic properties (such as GFAP), but not oligodendrocyte properties (Norton and Farooq, 1989). Using a retroviral strategy to label proliferating cells within the adult white matter, cells can be isolated that are able to generate oligodendrocytes in culture as well as a subset of cells that can generate astrocytes (Gensert and Goldman, 2001). The astrocyte "progenitor" subpopulation expresses the intermediate filament, vimentin, a cytoskeletal protein observed early in astrocyte differentiation.

Astrocyte Precursors Are Different from NG2 Cells

In addition to the vimentin[+] proliferating glial cells that are distributed throughout the adult CNS, there is another cell population that expresses the NG2 proteoglycan. As these cells are quite abundant and cycle slowly, the question has been raised as to whether these cells generate astrocytes in the adult brain. One reason that this question has been posed is that a cell that also expresses this antigen in vitro is the oligodendrocyte-type 2 astrocyte (O-2A) progenitor, which, as reviewed below, can generate oligodendrocytes or type 2 astrocytes. Despite this apparent overlap in the expression of NG2, the bulk of evidence suggests that adult NG2[+] cells do not produce astrocytes in the normal mature CNS. It is now well established that NG2[+] cells in vivo do not express GFAP under normal or reactive conditions (Levine and Card, 1987; Levine et al., 1993; Levine, 1994; Nishiyama et al., 1996; Bu et al., 2001). Since GFAP is not detectable in many protoplasmic astrocytes, the lack of GFAP immunoreactivity in NG2[+] cells itself does not rule out the possibility that NG2[+] cells might represent protoplasmic astrocytes as suggested by one study (Levine and Card, 1987). However, several observations indicate that this is not the case. Double immunolabeling studies using other antigens known to be expressed by protoplasmic astrocytes, such as the calcium-binding protein S-100β and GS, have shown that NG2 cells do not express these astrocyte markers (Nishiyama et al., 1996; Reynolds and Hardy, 1997). Furthermore, NG2 cells also lack several other characteristics of astrocytes. They are not coupled by gap junctions, they do not generate calcium waves, and they lack glutamate transporters (for review see Lin and Bergles, 2002).

NG2$^+$ cells are morphologically distinct from astrocytes. In a recent study, where fluorescent dyes were injected into hippocampal astrocytes, cells with bushy, highly arborized processes were labeled of which only one tenth are labeled for GFAP (Bushong *et al.*, 2002). Morphologically, these dye-filled cells showed striking similarity to classic images of protoplasmic astrocytes. By contrast, dye-filled cells that expressed NG2 had simpler morphologies with multiple thin processes emanating from the cytoplasm (Bergles *et al.*, 2000), similar to the pattern of cell surface NG2 immunolabeling. Protoplasmic astrocytes have small cell bodies, similar to NG2$^+$ cells, but astrocytes have a greater number of tertiary branches, and their primary branches are wider near the soma and taper distally. The extensive arborization of the processes and the presence of spines give them a bushy, fuzzy appearance. These cells, typically, have relatively weak strands of GFAP immunoreactivity along a subregion of their proximal processes. Interestingly, the territories of the CNS parenchyma occupied by each cell type do not overlap (for review see Nishiyama *et al.*, 2002).

CELL CULTURE STUDIES REVEAL MULTIPLE LINEAGES OF ASTROCYTES

Multipotent Neural Stem Cells Generate Astrocytes by Generating More Restricted Glial Precursors

The unambiguous demonstration of multipotent stem cells *in vitro* and *in vivo* suggests that differentiated astrocytes must be derived from an initially multipotent stem cell population. How these multipotent cells become restricted and specified toward neurons and glia is a focus of intense investigation. As reviewed in Chapter 4, the earliest multipotential precursors eventually become more restricted, so that over the course of development there emerge bipotent and eventually unipotent neural precursors. At the present time it is not clear how many intermediate precursors eventually form.

The first evidence for the existence of multiple glial precursors emerged from studies on the optic nerve. Martin Raff

and his colleagues published studies in 1983 characterizing two types of astrocytes in optic nerve cultures. Designated type 1 and type 2 astrocytes, these cells were delineated morphologically, antigenically, and by their responses to soluble growth factors (Raff *et al.*, 1983a) (Fig. 12). More recently, additional astrocyte types have been identified in forebrain and spinal cord cultures and these will be reviewed below.

Many studies on cell lineage have relied upon antibodies that react with either specific glial lineages or specific stages during the differentiation of cells within a lineage. Antibodies that have proven useful in studies of glial lineage *in vitro* include the antiganglioside antibodies A2B5, R24, and LB1 (the latter two react with GD3 ganglioside; Raff *et al.*, 1983a; Goldman *et al.*, 1986; Levi *et al.*, 1986), the rat neural antigen-2 (Ran-2) (Bartlett *et al.*, 1980), antibodies to fibronectin (Gallo *et al.*, 1987), anti-chondroitin sulfate or anti-NG2 antibodies (Gallo *et al.*, 1987; Levine and Stallcup, 1987), antibodies to the cell adhesion molecule J1 (ffrench-Constant *et al.*, 1986), and antibodies to GAP-43 (Deloulme *et al.*, 1990). All of these markers stain type 2 astrocytes and their progenitor, the O-2A progenitor, but they do not react with type 1 astrocytes, with the exception of Ran-2 and fibronectin, which stain subsets of type 1 and not type 2 astrocytes. Using markers alone is less than straightforward. For instance, it appears that type 2 astrocytes lose immunoreactivity for A2B5, LB1, and R24 with time in culture (Aloisi *et al.*, 1988; Lillien *et al.*, 1990; Levison and McCarthy, 1991) and they can become immunoreactive for Ran-2 after weeks in culture (Lillien and Raff, 1990). Furthermore, A2B5, LB1, and R24 are not specific for type 2 astrocytes, since some astrocytes that are clonally distinct from type 2 astrocytes express these markers (Vaysse and Goldman, 1990; Miller and Szigeti, 1991; Vaysse and Goldman, 1992).

The Lineage of the Type 1 Astrocyte

Type 1 astrocytes were originally defined as flat, polygonal cells that expressed GFAP but did not bind the monoclonal antibody, A2B5 (Raff *et al.*, 1983a). Optic nerve type 1 astrocytes can be distinguished from type 2 astrocytes by their

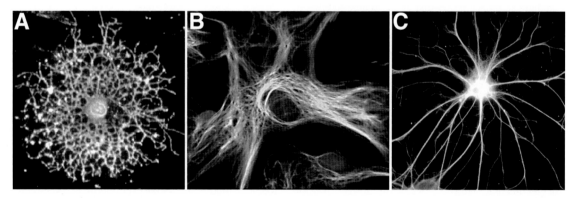

FIGURE 12. Three main types of macroglial cells in the CNS. Immunofluorescence micrographs of cells in culture. (A) Oligodendrocyte labeled on its surface with the O$_4$ monoclonal antibody. (B) Type 1 astrocyte labeled with antiserum against GFAP. (C) Type 2 astrocyte labeled with antiserum against GFAP. Photos provided by William Tyler and Phillip Albrecht.

immunoreactivity with the antibody Ran-2, by their absence of immunoreactivity with the other antibodies listed above, and by their separation from the oligodendrocyte lineage (Raff *et al.*, 1984). Unlike the O-2A lineage cells, type 1 astrocytes proliferate in response to epidermal growth factor (EGF) (Raff *et al.*, 1983a). Type 1 astrocytes develop early during gliogenesis. GFAP+/A2B5− astrocytes first appear in cell suspensions of developing rat optic nerve on embryonic day 16 (E16) (Miller *et al.*, 1985). Studies in forebrain cultures also support the early generation of astrocytes with a type 1 morphology and antigenic phenotype. For example, they are clonally distinct from the other glial lineages by E16 in rat forebrain cultures (Vaysse and Goldman, 1992). Culican *et al.* (1990) studied cultures from embryonic mouse forebrain and described cells with a radial glial-like morphology that bound the RC1 antibody, a monoclonal antibody that labels radial glia *in vivo* (Edwards *et al.*, 1990). While initially GFAP−, these cells became RC1+/GFAP+ with time, and eventually RC1−/GFAP+, a developmental and antigenic sequence that suggests type 1 astrocytes are generated *in vitro* from radial glia.

Applying the glial nomenclature derived from studies on optic nerve glia to other CNS regions can be problematic, since morphology and antigen expression can vary. For instance, studies of spinal cord astrocytes demonstrate that there is a greater variety of astrocyte types in the spinal cord than in optic nerve, and furthermore, that A2B5+ cells from the spinal cord give rise to "pancake"-shaped spinal cord astrocytes that are distinct from type 1 astrocytes (Miller and Szigeti, 1991; Fok-Seang and Miller, 1992). While clonally related cells tended to be morphologically similar, some are morphologically heterogeneous. Furthermore, the expression of A2B5 and Ran-2 varies even among clonally related cells. These and other observations illustrate astrocyte heterogeneity in different CNS regions and argue that antigen expression can be regulated by both lineage-dependent and lineage-independent factors.

Type 2 Astrocytes and the O-2A Lineage

Type 2 astrocytes were originally defined in optic nerve cultures (Raff *et al.*, 1983b), but type 2 astrocytes have been obtained from cultures of cerebellum (Levi *et al.*, 1986; Levine and Stallcup, 1987) and cerebral cortex (Goldman *et al.*, 1986; Behar *et al.*, 1988; Ingraham and McCarthy, 1989). As indicated above, a panel of additional cell markers is available that distinguish type 2 from type 1 astrocytes. In suspensions of developing brain, cells with the antigenic characteristics of type 2 astrocytes appear postnatally and derive from a bipotential O-2A progenitor (also referred to as an oligodendrocyte precursor cell or OPC) (Miller *et al.*, 1985; Williams *et al.*, 1985). O-2A progenitors differentiate into oligodendrocytes in a chemically defined medium, but into type 2 astrocytes in medium supplemented with fetal bovine serum (FBS) (Raff *et al.*, 1983b). Studies have characterized the molecules that induce type 2 astrocyte differentiation. Lillien *et al.* (1988) demonstrated that ciliary neurotrophic factor (CNTF) causes a transient commitment of the O-2A progenitor toward a type 2 astrocyte fate, but that the presence of an extracellular matrix-associated molecule derived from endothelial cells or fibroblasts is required for this phenotype to be expressed stably (Lillien *et al.*, 1990). Another stimulus that was partially characterized is the astrocyte-inducing molecule (AIM) that was isolated from the fetuin fraction of fetal bovine serum. Based on its biochemical properties, AIM may well turn out to be a member of the Galectins, since it has been recently demonstrated that Galectin-1, which is a fetuin-binding protein, can induce astrocyte differentiation from precursors (Sasaki *et al.*, 2003).

Direct evidence that the O-2A lineage is distinct from the type 1 astrocyte lineage was provided by an experiment where A2B5 and complement were combined to lyse the O-2A progenitor and its progeny. While the type 1 lineage was unaffected, the descendants of the O-2A progenitor failed to develop (Raff *et al.*, 1983b). Conversely, O-2A progenitors purified using fluorescence activated cell sorting (Williams *et al.*, 1985; Behar *et al.*, 1988), or grown as single cell microcultures (Temple and Raff, 1986) gave rise to oligodendrocytes or type 2 astrocytes, but not type 1 astrocytes. Furthermore, a retroviral analysis found that type 1 astrocytes are clonally distinct from oligodendrocytes in cultures from forebrain and spinal cord (Vaysse and Goldman, 1990). Whether type 2 astrocytes have a correlate *in vivo* has not yet been determined.

Other Astrocyte Types

Another astrocyte type has been identified *in vitro* (Vaysse and Goldman, 1992). In cultures of striatum, spinal cord, and cerebellum, these cells are very large, flat, and extend many fine cytoplasmic processes. They express both GFAP and GD3 ganglioside and remain GD3+ for at least eight weeks (the longest timepoint examined). Many, but not all of these cells, also stain with A2B5, but none express O4 or galactocerebroside (oligodendrocyte lineage markers). While these astrocytes antigenically resemble type 2 astrocytes, they are clonally distinct from type 1 astrocytes and from the O-2A lineage in the neonatal CNS. These astrocytes comprise a small percentage of the total cells and proliferate little, since the average clonal size is small. Whether these astrocytes have a correlate *in vivo* also has not yet been determined.

Heterogeneity within Astrocyte Lineages *In Vitro*

Subclasses of astrocytes with a type 1 phenotype have been revealed by analyses of cytoskeletal proteins, neuropeptide content, neuroligand receptors, secreted peptides, surface glycoproteins, release of prostaglandins, and by their influence on neuronal arborization patterns (for review, see Wilkin *et al.*, 1990). While many of these differences emerged by comparing cultures from different brain regions, subtypes have also been distinguished from the same brain region (McCarthy and Salm, 1991; Miller and Szigeti, 1991). Type 2 astrocytes also appear to be heterogeneous as revealed by receptor expression and class II MHC inducibility (Calder *et al.*, 1988; Sasaki *et al.*, 1989; Dave *et al.*, 1991; Inagaki *et al.*, 1991).

RECENT STUDIES PROVIDE EVIDENCE FOR THE SEQUENTIAL SPECIFICATION OF PRECURSORS FROM NEURAL STEM CELLS TO GLIAL-RESTRICTED PRECURSORS TO ASTROCYTE PRECURSOR CELLS

Glial-Restricted Precursors (GRPs) Are Cells That Can Differentiate into Type 1 Astrocytes, Oligodendrocytes, and Type 2 Astrocytes

In vitro experiments performed by several laboratories have identified a precursor that does not generate neurons, but which does produce type 1 astrocytes, oligodendrocytes, and under appropriate conditions, type 2 astrocytes. These precursors have been designated GRPs. Rao and colleagues have established that there are cells present in the developing spinal cord at E12 that are A2B5 and nestin immunoreactive (Rao and Mayer-Proschel, 1997; Rao *et al.*, 1998; Gregori *et al.*, 2002; Power *et al.*, 2002). Spinal cord GRPs lack PDGFR-alpha immunoreactivity and synthesize detectable levels of PLP/DM-20. Furthermore, they do not stain for ganglioside GD3 or for PSA-NCAM. Since GRPs are the earliest identifiable glial precursor and they generate two kinds of astrocytes *in vitro*, they are clearly at an earlier stage of restriction than type 1 astrocyte precursors and O-2A progenitors. This sequence of appearance of progressively more restricted precursors suggests, though does not prove, that a lineage relationship exists between them. A hypothetical relationship is schematized in Fig. 13, which is supported by *in vitro* studies.

Work performed by Rao and colleagues supports the model depicted where there is a gradual restriction in the developmental potential of neural precursors from a multipotential neuroepithelial precursor (NEP) to a cell-type specific neural progenitor (Mayer-Proschel *et al.*, 1997; Rao and Mayer-Proschel, 1997; Rao *et al.*, 1998). At least three intermediate precursors have been shown to arise from spinal cord neural stem cells. When A2B5$^+$/PSA-NCAM$^-$ precursors are generated from spinal cord NEPs and grown in serum-containing medium, they generate A2B5-negative, flat astrocytes. When these same precursors are stimulated with CNTF and FGF-2, they generate oligodendrocytes, but not neurons. The transition from an NEP to a GRP, and the subsequent production of more restricted glial cell types provides evidence for the transformation of multipotential precursors into more restricted glial precursors.

Analogous experiments conducted on precursors from the forebrain SVZ show that there are GRPs within the SVZ that are descended from multipotential neural stem cells. Clonal analyses have shown that precursors in the newborn rat SVZ can generate type 1 and type 2 astrocytes as well as oligodendrocytes (Levison *et al.*, 1993, 2003). In particular, when SVZ cells cultured under conditions that are permissive for neuronal differentiation, some SVZ derived progenitors generate astrocytes and oligodendrocytes, but they do not produce neurons. Thus, these cells can reasonably be called GRPs (Levison and Goldman, 1997). However, the markers expressed by GRPs from the SVZ appear

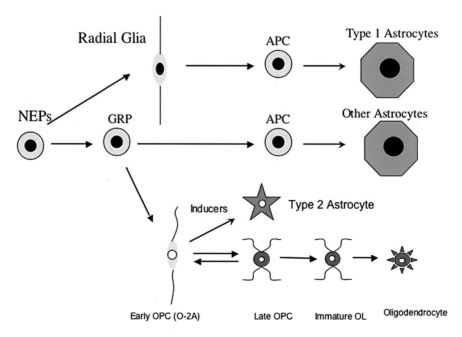

FIGURE 13. Model of astrocyte lineages. Depicted are several developmental pathways resulting in the production of a heterogeneous population of astrocyte types from neural epithelial precursors (NEPs). Depicted is the radial glia lineage which produces type 1 astrocytes through an intermediate astrocyte precursor cell (APC). Also depicted are the glial-restricted precursors (GRPs) such as those within the SVZ that produce both APCs as well as early oligodendrocytes progenitor cells (OPCs). These OPCs, *in vitro*, can be induced to produce type 2 astrocytes. Not depicted are other APCs, such as those in the optic nerve that are direct descendants of the NEPs without a radial glial intermediate.

to be different from the markers expressed by spinal cord GRPs in that SVZ GRPs express PSA-NCAM and ganglioside GD3 whereas these cell surface markers are not present on spinal cord GRPs (Levison *et al.*, 1993; Avellana-Adalid *et al.*, 1996; Ben-Hur *et al.*, 1998; Zhang *et al.*, 1999). Whether the properties and functional attributes of the astrocytes generated by spinal cord GRPs are different from the properties and functional attributes of the astrocytes generated by forebrain GRPs remains to be discerned.

Several Astrocyte-Restricted Precursors Have Been Isolated

There is clear evidence from *in vivo* studies that radial glia generate a subset of astrocytes, and these *in vivo* studies are supported by *in vitro* studies. For instance, in the study reported by Culican *et al.* (1990) the authors used the monoclonal antibody RC1, which recognizes an epitope present on radial glial, to follow the development of RC1-labeled cells *in vitro*. They observed that the cells from the E13 mouse brain that labeled with RC1 resembled radial glial cells *in vivo*. These cells possessed long, thin unbranched processes. After 3–4 days *in vitro* in the absence of neurons, these cells retained their RC1 epitope, acquired GFAP, and exhibited a polygonal shape reminiscent of type 1 astrocytes. In the presence of neurons, the RC1$^+$ cells acquired GFAP, but they possessed a more complex morphology, reminiscent of the stellate-shape typical of astrocytes *in vivo*. Unfortunately, these authors did not more fully characterize the antigenic phenotype of this astrocyte population, therefore, it is not entirely clear which type(s) of astrocytes were produced.

Other astrocyte-restricted precursors have been purified from the optic nerve using immunopanning. Mi *et al.* (2001) purified a population of cells from the E17 optic nerve that are Ran-2$^+$/A2B5$^+$/Pax-2$^+$/Vimentin$^+$ and they are S-100$^-$ and GFAP$^-$. Although A2B5$^+$, apparently, these cells express low levels of A2B5 when compared to O-2A progenitors. These astrocyte precursor cells (APCs) are clearly different from immature astrocytes and from O-2A progenitors. When maintained in a serum-containing medium, the APCs do not differentiate, but die, whereas immature astrocytes will differentiate and will readily divide. Moreover, when maintained in a culture medium that is permissive for oligodendrocyte differentiation, these APCs do not generate oligodendrocytes. Finally, when stimulated with either CNTF or LIF, APCs differentiate into A2B5$^-$/GFAP$^+$ polygonal astrocytes and not into type 2 astrocytes. Thus, on the basis of these studies, the authors conclude that these cells represent an astrocyte intermediate between the multipotential neural stem cell and a type 1 astrocyte. Unfortunately, these authors did not use markers of radial glia to determine whether these APCs might be similar to radial glia. However, these authors report that neither Pax-2 nor Ran-2 are expressed by forebrain APCs, suggesting that these optic nerve APCs are distinct from APCs in other regions of the CNS. Whether these different groups have identified slightly different precursors or whether the same precursor has been isolated multiple times remains to be determined.

MULTIPLE SIGNALS REGULATE ASTROCYTE SPECIFICATION

As alluded to earlier in this chapter, there are several sets of ligands and receptors that promote astrocyte differentiation: (1) the alpha helical family of cytokines and their receptors, (2) transforming growth factor beta (TGFβ) family members, particularly the bone morphogenetic proteins (BMPs) and BMP receptors, (3) Delta and Jagged ligands and Notch receptors, (4) FGFs and their receptors, (5) EGF family member ligands and the erbB family of receptors, and (6) Pituitary Adenylate Cyclase-Activating Polypeptide (PACAP) and the PAC1 receptor.

Members of the Alpha Helical Family of Cytokines Induce Astrocyte Specification through the LIF Receptor Beta and Activation of STATs

Hughes *et al.* (1988) initially found that CNTF would induce astrocyte differentiation in O-2A progenitors isolated from the postnatal optic nerve. Other members of the alpha helical cytokine family include leukemia inducing factor (LIF), interleukin-11, cardiotropin 1, and oncostatin M. The receptors for the alpha helical cytokines are expressed by cells in the VZ as well as by cells in the SVZ and CNTF has been shown to induce astrocytes from both cell populations (Johe *et al.*, 1996; Bonni *et al.*, 1997; Park *et al.*, 1999). However, CNTF deficient mice do not have a defect in astrocyte production, indicating that CNTF is not essential for astroglial differentiation (DeChiara *et al.*, 1995; Martin *et al.*, 2003). Whereas CTNF is dispensable for astrocyte differentiation, the LIF receptor may be important since LIF receptor deficient mice have reduced numbers of GFAP$^+$ cells at E19 (Koblar *et al.*, 1998).

Upon binding of alpha helical cytokines to their receptors, the janus kinases (JAKs) associated with those receptors become activated, whereupon they phosphorylate downstream signaling molecules such as the protranscription factors STAT3 and STAT1. Phosphorylating these protranscription factors enhances their ability to dimerize whereupon they form complexes with CBP/p300 (Bonni *et al.*, 1997; Kahn *et al.*, 1997) (Fig. 14). This transcriptional complex can then move into the nucleus where it can activate or repress genes that promote astrocyte differentiation as well as genes that are characteristic of astrocytes such as GFAP. Additionally, these cytokines will activate protein kinase B/AKT that will phosphorylate a transcriptional repressor known as N-CoR to keep that factor in the cytoplasm. When N-CoR is not phosphorylated it translocates into the nucleus, where it represses astrocyte differentiation. Indeed, astrocyte differentiation occurs prematurely in mice that lack N-CoR (Hermanson *et al.*, 2002).

Members of the TGF-β Family of Cytokines Induce Astrocyte Specification

During embryogenesis BMP signaling is essential for inducing mesoderm from ectoderm as well as for dorsal ventral

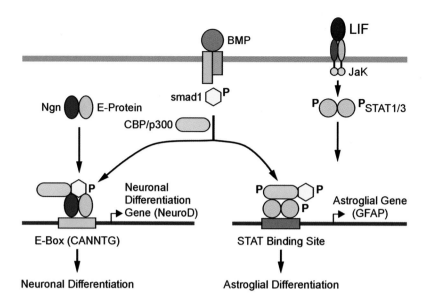

FIGURE 14. Model for developmental switch from neurogenesis to gliogenesis. The presence of neurogenin-1 in early VZ precursors inhibits glial differentiation by sequestering CBP–Smad1 away from glial-specific genes. When levels of neurogenin-1 decrease, CBP/p300 and Smad1, separately or together, are recruited to glial-specific genes (such as GFAP) by activated STAT1/ STAT3. Thus, neurogenin not only directly activates neuronal differentiation genes; it also inhibits glial gene expression.

patterning of the neural tube (Mehler, 1997). But later in development BMP homodimers and heterodimers potently induce astrocyte differentiation (D'Alessandro *et al.*, 1994; Gross *et al.*, 1996; Mabie *et al.*, 1997). The BMP receptors are expressed at high levels in the VZs and SVZs from as early as E12, and BMP-4 also is expressed in these regions (Gross *et al.*, 1996). *In vitro* studies have demonstrated that BMP ligands induce the differentiation of cells with the phenotypes of type 1 astrocytes or type 2 astrocytes depending upon which precursors are stimulated with ligand (Mabie *et al.*, 1997; Zhang *et al.*, 1998; Mehler *et al.*, 2000). BMPs also inhibit precursor proliferation (even in the presence of mitogens like EGF), and they also increase the maturation of astrocytes (D'Alessandro and Wang, 1994; D'Alessandro *et al.*, 1994). Comparative studies on the BMP ligands have shown that heterodimers comprised of BMP-2 and BMP-6, or BMP-2 and BMP-7, are potent at pico molar concentrations and that such heterodimers are more than three times more potent than homodimers of either ligand. Furthermore, they are much more potent than the related family member TGFβ1 which had been previously been implicated in astrocyte differentiation (Sakai *et al.*, 1990; Sakai and Barnes, 1991; D'Alessandro *et al.*, 1994; Gross *et al.*, 1996).

BMPs signal through a heterodimeric receptor composed of type 1 and type 2 subunits, which are serine/threonine kinases. The BMP bind to the type 2 receptor which then associates with the type 1 receptor resulting in the phosphorylation of the type 1 subunit. This activates the receptor leading to the phosphorylation of the protranscription factor Smad-1. The phosphorylated Smad-1 can then dimerize with another Smad, such as Smad-4, to produce a transcriptionally active complex that can induce or repress target genes. Several of the genes regulated by BMP

signaling are *Id1* and *Id3* which promote astrocytic differentiation and negatively regulate neuronal differentiation (Nakashima *et al.*, 2001). Another means by which BMP signaling inhibits neuronal differentiation is by sequestering CBP/p300, thus preventing neuronal specification (Fig. 14). Supporting these models, BMPs increase the percentage of astrocytes from neural stem cells while decreasing the production of neurons (as well as oligodendrocytes) without concurrent cell death, consistent with the concept that BMPs promote the specification of astrocyte-restricted precursors (Gross *et al.*, 1996; Nakashima *et al.*, 2001; Sun *et al.*, 2001).

Fibroblast Growth Factor-8b Promotes Astrocyte Differentiation

There are at least 21 FGFs, and these signaling molecules have long been known to affect astrocyte development. For instance, FGF-2 is a potent mitogen for type 1 astrocytes and their precursors and FGFs will increase GFAP and GS levels in cultured astrocytes (Morrison *et al.*, 1985; Perraud *et al.*, 1988). The FGFs exert their effects by stimulating one of four transmembrane tyrosine kinase FGF receptors and three of these receptors (FGFRs 1–3) are expressed by neural precursors in the VZ and SVZ (Bansal *et al.*, 2003). While the majority of studies have focused on FGF-2, a screen of nine FGF ligands (FGF-1, 4, 6, 7, 8a, 8b, 8c, 9, and 10) on embryonic rat neocortical precursors found that FGF-8b potently promoted the differentiation of a subpopulation of neocortical precursors toward astrocytes (Hajihosseini and Dickson, 1999). The other FGF8 ligands did not have this effect at the concentrations tested. As the precursors expressed FGFRs 1–3, it is not presently clear which FGFR is mediating this inductive effect. FGFR3 does not appear to be essential since FGFR-3 null mice have more astrocytes than their wild-type counterparts (Oh *et al.*, 2003). FGF-2 can have a similar effect to FGF8b, but at concentrations 10 times higher than are required for FGF8b (Qian *et al.*, 1997).

Signaling through the EGF Receptor Induces Astrocyte Specification

As discussed earlier, the ligand neuregulin, which binds to the erbB receptors, is produced and secreted by migrating neurons to prevent radial glia from differentiating into astrocytes (Anton *et al.*, 1997; Rio *et al.*, 1997). When the levels of neuregulin decrease, as they do during neuronal maturation, the radial glia become receptive to other astrocyte differentiating signals. As neural precursors become competent to generate astrocytes the levels of another receptor, the EGF receptor, increase, as does the level of one of its ligands, TGFα. In elegant experiments where the levels of the EGF receptor are experimentally increased, precursors that would not normally generate astrocytes do so precociously (Burrows *et al.*, 1997). This occurs because raising the levels of EGF receptor confers competence to these early progenitors to respond to LIF (Viti *et al.*, 2003). Indeed studies on early rat or mouse neural precursors or on precursors genetically deficient in EGF receptor show that LIF is incapable of inducing GFAP expression in cells lacking EGF receptors (Molne *et al.*, 2000; Viti *et al.*, 2003). In addition to providing competence to early progenitors to generate astrocytes, signaling through the EGF receptor has long been known to increase the proliferation of immature astrocytes (Leutz and Schachner, 1981). Thus, signaling through the EGF receptor coordinates several aspects of astrocytes development.

PACAP, Increases cAMP to Induce Astrocyte Differentiation

The neuropeptide PACAP and one of its receptors, PAC1, are expressed highly in the VZ during late gestation and the PAC1 receptor is expressed by E17 neocortical precursors *in vitro*. As this receptor is known to increase cAMP within cells, and as it had been shown previously that elevating cytosolic cAMP increases the expression of GFAP by immature astrocytes (Shafit-Zagardo *et al.*, 1988; Masood *et al.*, 1993; McManus *et al.*, 1999), Vallejo and Vallejo (2002) asked whether PACAP might induce astrocytic differentiation from fetal precursors. When they stimulated E17 forebrain precursors with PACAP, they observed increased levels of cAMP within 15 min, and the elevated levels of cAMP lead to phosphorylation of the transcription factor CREB. When examined 2 days later, PACAP exposed cells, or cells treated with a cAMP analog assumed a stellate shape, they had elevated levels of GFAP and they had decreased levels of nestin (McManus *et al.*, 1999). Prolonged treatment with PACAP was not necessary as a 30-min exposure was sufficient to induce GFAP expression and stellation. Finally, inhibiting the increase in cAMP is sufficient to inhibit the increased GFAP expression induced by PACAP. Thus, elevating cAMP by PACAP will induce astrocytic specification from fetal precursors (Fig. 15).

Notch Activation Can Promote Astrocyte Specification

The transmembrane signaling receptor Notch functions in a context dependent manner to regulate multiple aspects of neural development. The family of Notch transmembrane receptors control cell fate decisions by interaction with Notch ligands expressed on the surface of adjacent cells. As discussed earlier,

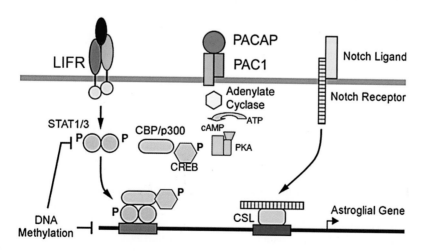

FIGURE 15. Signals regulating astrocyte specification. The LIF receptor (LIFR) activates the JAKs, and STATs, which can then combine with CBP/p300 to form a transcriptional regulator. Methylation of specific promotors will inhibit this complex from acting. The PAC1 receptor for PACAP increases levels of cAMP within the cell, which activates protein kinase A (PKA) to phosphorylate CREB, another transcription factor. Finally, cleavage of Notch receptors subsequent to binding by a Notch ligand releases the intracellular domain, which can combine with CSL to directly regulate genes involved in astrocyte specification.

Notch signaling promotes radial glial cell formation, and other studies have demonstrated that Notch inhibits differentiation at later stages in neural lineages as well. However, several recent studies show that Notch can instructively promote astrocytic differentiation. Studies by Tanigaki et al. (2001) and Ge et al. (2002) using either hippocampal-derived multipotent or E11 neocortical precursors, respectively, showed that introducing the signaling component of either the Notch1 or Notch3 receptors induces the expression of GFAP, increases the size of the cells and stimulates process formation. Moreover, activated Notch appears to act instructively as it reduces the number of neuronal and oligodendroglial cells while increasing the percentage of astrocytes. This effect of Notch on astroglial differentiation is not likely indirect, since the intracellular signaling domain of Notch forms a transcriptional complex with CSL and SKIP that binds to specific elements of the GFAP promotor to initiate transcription of GFAP. Notch signaling also induces the downstream target transcriptional regulator, Hes-1 (but not Hes-5). While Notch can clearly regulate GFAP expression, Hes-1 likely mediates some of Notch's effects on astrocyte differentiation. In experiments where the Hes transcription factors are overexpressed in glial-restricted precursors, overexpressing Hes-1, but not Hes-5, promotes astrocytic differentiation (as indicated by increased GFAP and CD44 expression) at the expense of oligodendrocyte differentiation (Wu et al., 2003). Importantly, this effect of Hes-1 is stage-specific because Hes-1 does not promote the astrocyte fate when overexpressed in neuroepithelial cells. Altogether, these experiments demonstrate that Notch can directly induce astroglial gene expression by forming a transcriptional complex with CSL and SKIP, and that this transcriptional complex also induces downstream signaling molecules like Hes-1 that also regulate astrocyte differentiation.

An Interplay of Multiple Pathways Contributes to Astrocyte Genesis

The competence of neural precursors to respond to extracellular signals is certainly one mechanism that regulates the onset of astroglial differentiation. One intrinsic feature that may determine whether a precursor will generate neurons or glia is the balance between "neurogenic" and "gliogenic" transcription factors. For instance, early neuroectodermal precursors express higher levels of Neurogenin 1, which correlates with the preference for these cells to differentiate into neurons rather than glia (Fig. 14). Overexpressing Neurogenin 1 in embryonic neuroepithelial cells not only promotes neurogenesis, but also decreases the ability of these cells to respond to astrocyte inducing signals, such as LIF (Sun et al., 2001). Sun et al. (2001) demonstrated that neurogenin 1 binds to the same CBP/p300, complex as the STATs. Furthermore, the Neurogenin-1-binding domain overlaps with the STAT-binding domain on CBP/p300; thus, Neurogenin 1 and STAT cannot physically bind to CBP/p300 simultaneously. Consequently, the relative levels of neurogenin 1 and STAT3 may in part determine whether an immature cell becomes a neuron or

an astrocyte. Furthermore, Neurogenin 1 inhibits STAT phosphorylation. Thus, competition between Ngn1 and STAT for these transcriptional coactivators as well as negative regulation of STAT phosphorylation provides a viable mechanism for determining a neocortical precursor's fate. However, merely overexpressing Neurogenins or Mash 1 by retroviral infection does not alter dramatically the numbers of neurons vs astrocytes that develop, suggesting that it is not just the levels of the transcription factor that determines cell fate in vivo. Similarly, knocking out both Neurogenin 2 and Mash 1 does not produce a dramatic decrease in neurons and increase astrocytes, although the cortices of these mice displayed marked disorganization of laminar patterning (Nieto et al., 2001).

DNA and histone methylation also regulate the intrinsic capacity of neural precursors to differentiate into astrocytes. A CpG dinucleotide within the STAT3-binding element of the GFAP promotor is highly methylated in early neuroepithelial cells, and the methylation of this site prevents STAT3 from binding. Consequently, the STATs cannot act as transcriptional activators of GFAP. This site is demethylated during CNS development, coincident with transcriptional activation by STATs and commensurate with astroglial differentiation (Takizawa et al., 2001). Furthermore, growth factors that have been shown to increase the competence of early precursors to generate astrocytes increase the methylation of Histone H3 at specific lysines which results in changes in chromatin conformation, again enabling specific genes involved in astroglial differentiation to be expressed (Song and Ghosh, 2004).

How might other extrinsic signaling molecules regulate astrocyte development in vivo? As discussed above, most of the soluble factors that can instructively drive astrocyte development are present in the developing CNS and some are present quite early. For instance, BMP-4 is present as early as E14, which is when neurons are produced, yet BMP-4 does not induce neuronal generation from early precursors. One reason is that the BMP antagonist, Noggin, is expressed in the developing cortex (Li and LoTurco, 2000) and in adult rodents, Noggin is found in ependymal cells (Lim et al., 2000). There it may function to counteract BMP-induced astrocytic development. LIF, which can induce astrocytes, also is present in the VZ quite early, and indeed, signaling through the LIF receptor is required to maintain the complement of neural stem cells. However, as reviewed above, in the absence of EGF receptor signaling, alpha helical cytokines cannot induce astrocyte differentiation. CNTF/LIF may be insufficient to induce astrocytes from SVZ cells later in development as factors present in the extracellular matrix may be required (Lillien et al., 1990). As discussed above, immature astrocytes derived from the SVZ interact with basal laminae at blood vessels and at the pial surface, and blood vessel interactions appear to be an early step in astrocyte differentiation (Zerlin and Goldman, 1997; Mi et al., 2001). Altogether, these examples demonstrate that astrocyte differentiation is coordinately regulated by the intrinsic properties of neural precursors as well as by the simultaneous signaling from multiple extrinsic signaling molecules.

CONCLUSION

We began this chapter by reviewing the types of astrocytes that populate the mature brain and then proceeded to discuss where and how astrocytes form. While there remain gaps in our knowledge, it is clear that there are multiple sources of astrocytes. In the forebrain, both the VZ and the SVZ produce astrocytes. The radial glia, which are direct descendants of the neuroepithelium, are one source of astrocytes. SVZ cells, which emerge later in development, are a second source, and they produce a subset of gray matter astrocytes. In the cerebellum, astrogliogenesis may proceed in a fashion similar to that established for the forebrain, but astrocyte generation in the spinal cord is different. Great strides continue to be made in defining the precursor product relationships between different types of phenotypically defined glial precursors and the cells that they produce. Moreover, elegant *in vitro* analyses are beginning to unravel the relative roles of the intrinsic competences of precursors at defined stages of development to respond to specific extrinsic signaling molecules. Multiple extrinsic signals have been identified that coordinate astrocyte differentiation. These include the alpha helical cytokines, BMPs, Notch ligands, FGF8b, EGF ligands, and PACAP, and as more is learned about the transcriptional regulators that they use, it may turn out that the internal signals used to establish an astrocytic fate are less complicated than the multiple signals that impinge upon their precursors. Clearly much has been learned over the last century when astrocytes were first discerned as a recognizable cell type, yet there are still many basic issues that remain to be addressed. We hope that this chapter has provided a conceptual framework onto which you, the reader, may incorporate the forthcoming answers.

REFERENCES

Aloisi, F., Agresti, C., and Levi, G., 1988, Establishment, characterization, and evolution of cultures enriched in type-2 astrocytes, *J. Neurosci. Res.* 21:188–198.

Altman, J., 1963, Autoradiographic investigation of cell proliferation in the brains of rats and cats, *Anat. Rec.* 145:573–591.

Anderson, S.A., 1997, Mutations of the homeobox genes *Dlx-1* and *Dlx-2* disrupt the striatal subventricular zone and differentiation of late born striatal neurons, *Neuron* 19:27–37.

Andriezen, W.L., 1893, The neuroglia elements in the human brain, *Br. J. Medicine* 2:227–230.

Antanitus, D.S., Choi, B.H., and Lapham, L.W., 1975, Immunofluorescence staining of astrocytes *in vitro* using antiserum to glial fibrillary acidic protein, *Brain Res.* 89:363–367.

Anton, E.S., Marchionni, M.A., Lee, K.F., and Rakic, P., 1997, Role of GGF/neuregulin signaling in interactions between migrating neurons and radial glia in the developing cerebral cortex, *Development* 124:3501–3510.

Avellana-Adalid, V., Nait-Oumesmar, B., Lachapelle, F., and Baron-Van Evercooren, A., 1996, Expansion of rat oligodendrocyte progenitors into proliferative "oligospheres" that retain differentiation potential, *J. Neurosci. Res.* 45:558–570.

Bansal, R., Lakhina, V., Remedios R., and Tole, S., 2003, Expression of FGF receptors 1, 2, 3 in the embryonic and postnatal mouse brain compared with Pdgfralpha, Olig2 and Plp/dm20: Implications for oligodendrocyte development, *Dev. Neurosci.* 25:83–95.

Bartlett, P.F., Noble, M.D., Pruss, R.M., Raff, M.C., Rattray, S., and Williams, C.A., 1980, Rat neural antigen-2 (RAN-2): A cell surface antigen on astrocytes, ependymal cells, Muller cells and lepto-meninges defined by a monoclonal antibody, *Brain Res.* 204:339–351.

Basco, E., Hajos, F., and Fulop, Z., 1977, Proliferation of Bergmann-glia in the developing rat cerebellum, *Anat. Embryol. (Berl.)* 151:219–222.

Bayer, S.A., Altman, J., Russo, R.J., Dai, X.F., and Simmons, J.A., 1991, Cell migration in the rat embryonic neocortex, *J. Comp. Neurol.* 307:499–516.

Behar, T., McMorris, F.A., Novotny, E.A., Barker, J.L., and Dubois-Dalcq, M., 1988, Growth and differentiation properties of O-2A progenitors purified from rat cerebral hemispheres, *J. Neurosci. Res.* 21:168–180.

Ben-Hur, T., Rogister, B., Murray, K., Rougon G., and Dubois-Dalcq, M., 1998, Growth and fate of PSA-NCAM+ precursors of the postnatal brain, *J. Neurosci.* 18:5777–5788.

Benjelloun-Touimi, S., Jacque, C.M., Derer, P., De Vitry, F., Maunoury, R., and Dupouey, P., 1985, Evidence that mouse astrocytes may be derived from the radial glia. An immunohistochemical study of the cerebellum in the normal and reeler mouse, *J. Neuroimmunol.* 9:87–97.

Bergles, D.E., Roberts, J.D., Somogyi, P., and Jahr, C.E., 2000, Glutamatergic synapses on oligodendrocyte precursor cells in the hippocampus, *Nature* 405:187–191.

Bignami, A., Eng, L.F., Dahl, D., and Uyeda, C.T., 1972, Localization of the glial fibrillary acidic protein in astrocytes by immunofluorescence, *Brain Res.* 43:429–435.

Bonni, A., Sun, Y., Nadal-Vicens, M., Bhatt, A., Frank, D.A., Rozovsky, I. et al., 1997, Regulation of gliogenesis in the central nervous system by the JAK-STAT signaling pathway, *Science* 278:477–483.

Boyes, B.E., Kim, S.U., Lee V., and Sung, S.C., 1986, Immunohistochemical co-localization of S-100b and the glial fibrillary acidic protein in rat brain, *Neuroscience* 17:857–865.

Bu, J., Akhtar, N., and Nishiyama, A., 2001, Transient expression of the NG2 proteoglycan by a subpopulation of activated macrophages in an excitotoxic hippocampal lesion, *Glia* 34:296–310.

Burrows, R.C., Wancio, D., Levitt, P., and Lillien, L., 1997, Response diversity and the timing of progenitor cell maturation are regulated by developmental changes in EGFR expression in the cortex, *Neuron* 19:251–267.

Bushong, E.A., Martone, M.E., Jones, Y.Z., and Ellisman, M.H., 2002, Protoplasmic astrocytes in CA1 stratum radiatum occupy separate anatomical domains, *J. Neurosci.* 22:183–192.

Cajal, S.R., 1909, *Histologie du systeme nerveux de l'homme et des vertebres*, Maloine, Paris.

Calder, V.L., Wolswijk, G., and Noble, M., 1988, The differentiation of O-2A progenitor cells into oligodendrocytes is associated with a loss of inducibility of Ia antigens, *Eur. J. Immunol.* 18:1195–1201.

Cameron, R.S. and Rakic, P., 1991, Glial cell lineage in the cerebral cortex: A review and synthesis, *Glia* 4:124–137.

Cammer, W., Tansey, F., Abramovitz, M., Ishigaki, S., and Listowsky, I., 1989, Differential localization of glutathione-S-transferase Yp and Yb subunits in oligodendrocytes and astrocytes of rat brain, *J. Neurochem.* 52:876–883.

Cammer, W. and Tansey, F.A., 1988, Carbonic anhydrase immunostaining in astrocytes in the rat cerebral cortex, *J. Neurochem.* 50:319–322.

Campbell, K. and Gotz, M., 2002, Radial glia: Multi-purpose cells for vertebrate brain development, *Trends in Neurosciences* 25:235–238.

Cepko, C.L., 1988, Retrovirus vectors and their applications in neurobiology, *Neuron* 1:345–353.

Chan-Ling, T. and Stone, J., 1991, Factors determining the morphology and distribution of astrocytes in the cat retina: A "contact-spacing" model of astrocyte interaction. *J. Comp. Neurol.* 303:387–399.

Choi, B.H., 1981, Radial glia of developing human fetal spinal cord: Golgi, immunohistochemical and electron microscopic study, *Brain Res.* 227:249–267.

Choi, B.H. and Lapham, L.W., 1978, Radial glia in the human fetal cerebrum: A combined Golgi, immunofluorescent and electron microscopic study, *Brain Res.* 148:295–311.

Choi, B.H. and Lapham, L.W., 1980, Evolution of Bergmann glia in developing human fetal cerebellum: A Golgi, electron microscopic and immunofluorescent study, *Brain Res.* 190:369–383.

Culican, S.M., Baumrind, N.L., Yamamoto, M., and Pearlman, A.L., 1990, Cortical radial glia: Identification in tissue culture and evidence for their transformation to astrocytes. *J. Neurosci.* 10:684–692.

D'Alessandro, J.S. and Wang, E.A., 1994, Bone morphogenetic proteins inhibit proliferation, induce reversible differentiation and prevent cell death in astrocyte lineage cells, *Growth Factors* 11:45–52.

D'Alessandro, J.S., Yetz-Aldape, J., and Wang, E.A., 1994, Bone morphogenetic proteins induce differentiation in astrocyte lineage cells. *Growth Factors* 11:53–69.

D'Amelio, F., Eng, L.F., and Gibbs, M.A., 1990, Glutamine synthetase immunoreactivity is present in oligodendroglia of various regions of the central nervous system, *Glia* 3:335–341.

Dahl, D., Bignami, A., Weber, K., and Osborn, M., 1981a, Filament proteins in rat optic nerves undergoing Wallerian degeneration: Localization of vimentin, the fibroblastic 100-A filament protein, in normal and reactive astrocytes, *Experimental Neurology* 73:496–506.

Dahl, D., Rueger, D.C., and Bignami, A., 1981b, Vimentin, the 57,000 molecular weight protein of fibroblast filaments, is the major cytoskeletal component of immature glia, *Eur. J. Cell Biol.* 24:191–196.

Dave, V., Gordon, G.W., and McCarthy, K.D., 1991, Cerebral type 2 astroglia are heterogeneous with respect to their ability to respond to neuroligands linked to calcium mobilization, *Glia* 4:440–447.

DeChiara, T.M., Vejsada, R., Poueymirou, W.T., Acheson, A., Suri, C., Conover, J.C. et al., 1995, Mice lacking the CNTF receptor, unlike mice lacking CNTF, exhibit profound motor neuron deficits at birth, *Cell* 83:313–322.

Deloulme, J.C., Janet, T., Au, D., Storm, D.R., Sensenbrenner, M., and Baudier, J., 1990, Neuromodulin (GAP43): A neuronal protein kinase C substrate is also present in O-2A glial cell lineage. Characterization of neuromodulin in secondary cultures of oligodendrocytes and comparison with the neuronal antigen. *J. Cell Biol.* 111:1559–1569.

Doetsch, F., Caille, I., Lim, D.A., Garcia-Verdugo, J.M., and Alvarez-Buylla, A., 1999, Subventricular zone astrocytes are neural stem cells in the adult mammalian brain, *Cell* 97:703–716.

Edwards, M.A., Yamamoto, M., and Caviness, V.S., Jr., 1990, Organization of radial glia and related cells in the developing murine CNS. An analysis based upon a new monoclonal antibody marker, *Neuroscience* 36:121–144.

Fabel, K., Toda, H., and Palmer, T., 2003, Copernican stem cells: Regulatory constellations in adult hippocampal neurogenesis, *J. Cell. Biochem.* 88:41–50.

ffrench-Constant, C., Miller, R.H., Kruse, J., Schachner, M., and Raff, M.C., 1986, Molecular specialization of astrocyte processes at nodes of Ranvier in rat optic nerve, *J. Cell Biol.* 102:844–852.

Flament-Durand, J. and Brion, J.P., 1985, Tanycytes: Morphology and functions: A review, *International Review of Cytology* 96:121–155.

Fok-Seang, J. and Miller, R.H., 1992, Astrocyte precursors in neonatal rat spinal cord cultures, *J. Neurosci.* 12:2751–2764.

Gaiano, N., Nye, J.S., and Fishell, G., 2000, Radial glial identity is promoted by Notch1 signaling in the murine forebrain, *Neuron* 26:395–404.

Galileo, D.S., Gray, G.E., Owens, G.C., Majors, J., and Sanes, J.R., 1990, Neurons and glia arise from a common progenitor in chicken optic tectum: Demonstration with two retroviruses and cell type-specific antibodies, *Proc. Natl. Acad. Sci. USA* 87:458–462.

Gallo, V., Bertolotto, A., and Levi, G., 1987, The proteoglycan chondroitin sulfate is present in a subpopulation of cultured astrocytes and in their precursors, *Dev. Biol.* 123:282–285.

Gasser, U.E. and Hatten, M.E., 1990, Neuron-glia interactions of rat hippocampal cells *in vitro*: Glial-guided neuronal migration and neuronal regulation of glial differentiation, *J. Neurosci.* 10:1276–1285.

Ge, W., Martinowich, K., Wu, X., He, F., Miyamoto, A., Fan, G. et al., 2002, Notch signaling promotes astrogliogenesis via direct CSL-mediated glial gene activation. *J. Neurosci. Res.* 69:848–860.

Gensert, J.M. and Goldman, J.E., 2001, Heterogeneity of cycling glial progenitors in the adult mammalian cortex and white matter, *J. Neurobiol.* 48:75–86.

Goldman, J.E., Geier, S.S., and Hirano, M., 1986, Differentiation of astrocytes and oligodendrocytes from germinal matrix cells in primary culture, *J. Neurosci.* 6:52–60.

Gray, G.E., Leber, S.M., and Sanes, J.R., 1990, Migratory patterns of clonally related cells in the developing central nervous system, *Experientia* 46:929–940.

Gray, G.E. and Sanes, J.R., 1991, Migratory paths and phenotypic choices of clonally related cells in the avian optic tectum, *Neuron* 6:211–225.

Gray, G.E. and Sanes, J.R., 1992, Lineage of radial glia in the chicken optic tectum, *Development* 114:271–283.

Gregori, N., Proschel, C., Noble, M., and Mayer-Proschel, M., 2002, The tripotential glial-restricted precursor (GRP) cell and glial development in the spinal cord: Generation of bipotential oligodendrocyte-type-2 astrocyte progenitor cells and dorsal-ventral differences in GRP cell function, *J. Neurosci.* 22:248–256.

Grosche, J., Matyash, V., Moller, T., Verkhratsky, A., Reichenbach, A., and Kettenmann, H., 1999, Microdomains for neuron–glia interaction: Parallel fiber signaling to Bergmann glial cells, *Nat. Neurosci.* 2:139–143.

Gross, R.E., Mehler, M.F., Mabie, P.C., Zang, Z., Santschi, L., and Kessler, J.A., 1996, Bone morphogenetic proteins promote astroglial lineage commitment by mammalian subventricular zone progenitor cells, *Neuron* 17:595–606.

Grove, E.A., Williams, B.P., Da-Qing, L., Hajihosseini, M., Friedrich, A., and Price, J., 1993, Multiple restricted lineages in the embryonic rat cerebral cortex, *Development* 117:553–561.

Hainfellner, J.A., Voigtlander, T., Strobel, T., Mazal, P.R., Maddalena, A.S., Aguzzi, A. et al., 2001, Fibroblasts can express glial fibrillary acidic protein (GFAP) *in vivo*, *J. Neuropathol. Exp. Neurol.* 60:449–461.

Hajihosseini, M.K. and Dickson, C., 1999, A subset of fibroblast growth factors (Fgfs) promote survival, but Fgf-8b specifically promotes astroglial differentiation of rat cortical precursor cells, *Mol. Cell. Neurosci.* 14:468–485.

Halliday, A.L. and Cepko, C.L., 1992, Generation and migration of cells in the developing striatum, *Neuron* 9:15–26.

Hama, K., Arii, T., and Kosaka, T., 1993, Three dimensional organization of neuronal and glial processes: High voltage electron microscopy, *Microscopy Research Techniques* 29:357–367.

Hartfuss, E., Galli, R., Heins, N., and Gotz, M., 2001, Characterization of CNS precursor subtypes and radial glia, *Dev. Biol.* 229:15–30.

Hatten, M.E., 1985, Neuronal regulation of astroglial morphology and proliferation *in vitro*, *J. Cell Biol.* 100:384–396.

Haydon, P.G., 2001, GLIA: Listening and talking to the synapse, *Nat. Rev. Neurosci.* 2:185–193.

Hermanson, O., Jepsen, K., and Rosenfeld, M.G., 2002, N-CoR controls differentiation of neural stem cells into astrocytes, *Nature* 419: 934–939.

Herndon, R.M., 1964, The fine structure of the rat cerebellum, II. The stellate neurons, granule cells and glia, *J. Cell Biol.* 23:277–293.

Hirano, M. and Goldman, J.E., 1988, Gliogenesis in rat spinal cord: Evidence for origin of astrocytes and oligodendrocytes from radial precursors, *J. Neurosci. Res.* 21:155–167.

Hockfield, S. and McKay, R.D.G., 1985, Identification of major cell classes in the developing mammalian nervous system, *J. Neurosci.* 5:3310–3328.

Hommes, O.R. and Leblond, C.P., 1967, Mitotic division of neuroglia in the normal adult rat, *J. Comp. Neurol.* 129:269–278.

Hughes, S.M., Lillien, L.E., Raff, M.C., Rohrer, H., and Sendtner, M., 1988, Ciliary neurotrophic factor induces type-2 astrocyte differentiation in culture, *Nature* 335:70–73.

Inagaki, N., Fukui, H., Ito, S., and Wada, H., 1991, Type-2 astrocytes show intracellular Ca2+ elevation in response to various neuroactive substances, *Neurosci. Lett.* 128:257–260.

Ingraham, C.A. and McCarthy, K.D., 1989, Plasticity of process-bearing glial cell cultures from neonatal rat cerebral cortical tissues, *J. Neurosci.* 9:63–71.

Johe, K.K., Hazel, T.G., Muller, T., Dugich-Djordjevic, M.M., and McKay, R.D.G., 1996, Single factors direct the differentiation of stem cells from the fetal and adult central nervous system, *Genes Dev.* 10:3129–3140.

Kahn, M.A., Ellison, J.A., Chang, R.P., Speight, G.J., and de Vellis, J., 1997, CNTF induces GFAP in a S-100 alpha brain cell population: The pattern of CNTF-alpha R suggests an indirect mode of action, *Brain Res. Dev. Brain Res.* 98:221–233.

Kakita, A. and Goldman, J.E., 1999, Patterns and dynamics of SVZ cell migration in the postnatal forebrain: Monitoring living progenitors in slice preparations, *Neuron* 23:461–472.

Kaplan, M.S. and Hinds, J.W., 1980, Gliogenesis of astrocytes and oligodendrocytes in the neocortical grey and white matter of the adult rat: Electron microscopic analysis of light radioautographs, *J. Comp. Neurol.* 193:711–727.

Kettenmann, H. and Ransom, B. (eds)., 1995, *Neuroglia.* Oxford University Press, Inc., New York.

King, J.S., 1966, A comparative investigation of neuroglia in representative vertebrates: A silver carbonate study, *J. Morphol.* 119:435–465.

Kitamura, T., Nakanishi, K., Watanabe, S., Endo, Y., and Fujita, S., 1987, GFA-protein gene expression on the astrocyte in cow and rat brains, *Brain Res.* 423:189–195.

Koblar, S.A., Turnley, A.M., Classon, B.J., Reid, K.L., Ware, C.B., Cheema, S.S. *et al.*, 1998, Neural precursor differentiation into astrocytes requires signaling through the leukemia inhibitory factor receptor. *Proc. Natl. Acad. Sci. USA* 95:3178–3181.

Korr, H., Schultze, B., and Maurer, W., 1973, Autoradiographic investigations of glial proliferation in the brain of adult mice. I. The DNA synthesis phase of neuroglia and endothelial cells, *J. Comp. Neurol.* 150:169–175.

Kosaka, T. and Hama, K., 1986, Three-dimensional structure of astrocytes in the rat dentate gyrus, *J. Comp. Neurol.* 249:242–260.

Lenhossék, M.v., 1895, Centrosum and sphäre in den spinalganglienzellen des frosches, *Arch. Mikr. Anat.* 46:345–369.

Leutz, A. and Schachner, M., 1981, Epidermal growth factor stimulates DNA-synthesis of astrocytes in primary cerebellar cultures, *Cell Tissue. Res.* 220:393–404.

Levi, G., Gallo, V., and Ciotti, M.T., 1986, Bipotential precursors of putative fibrous astrocytes and oligodendrocytes in rat cerebellar cultures express distinct surface features and "neuron-like" gamma-aminobutyric acid transport, *Proc. Natl. Acad. Sci. USA* 83:1504–1508.

Levine, J.M., 1994, Increased expression of the NG2 chondroitin-sulfate proteoglycan after brain injury, *J. Neurosci.* 14:4716–4730.

Levine, J.M. and Card, J.P., 1987, Light and electron microscopic localization of a cell surface antigen (NG2) in the rat cerebellum: Association with smooth protoplasmic astrocytes, *J. Neurosci.* 7:2711–2720.

Levine, J.M. and Stallcup, W.B., 1987, Plasticity of developing cerebellar cells *in vitro* studied with antibodies against the NG2 antigen, *J. Neurosci.* 7:2721–2731.

Levine, J.M., Stincone, F., and Lee, S.Y., 1993, Development and differentiation of glial precursor cells in the rat cerebellum, *Glia* 7:307–321.

Levine, S.M. and Goldman, J.E., 1988, Embryonic divergence of oligodendrocyte and astrocyte lineages in developing rat cerebrum, *J. Neurosci.* 8:3992–4006.

Levison, S.W., Chuang, C., Abramson, B.J., and Goldman, J.E., 1993, The migrational patterns and developmental fates of glial precursors in the rat subventricular zone are temporally regulated, *Development* 119:611–623.

Levison, S.W., Druckman, S., Young, G.M., Rothstein, R.P., and Basu, A., 2003, Neural stem cells in the subventricular zone are a source of astrocytes and oligodendrocytes, but not microglia, *Dev. Neurosci.* in press.

Levison, S.W. and Goldman, J.E., 1993, Both oligodendrocytes and astrocytes develop from progenitors in the subventricular zone of postnatal rat forebrain, *Neuron* 10:201–212.

Levison, S.W. and Goldman, J.E., 1997, Multipotential and lineage restricted precursors coexist in the mammalian perinatal subventricular zone, *J. Neurosci. Res.* 48:83–94.

Levison, S.W. and McCarthy, K.D., 1991, Characterization and partial purification of AIM: A plasma protein that induces rat cerebral type 2 astroglia from bipotential glial progenitors, *J. Neurochem.* 57:782–794.

Levison, S.W., Young, G.M., and Goldman, J.E., 1999, Cycling cells in the adult rat neocortex preferentially generate oligodendroglia, *J. Neurosci. Res.* 57:435–446.

Levitt, P., Cooper, M.L., and Rakic, P., 1981, Coexistence of neuronal and glial precursor cells in the cerebral ventricular zone of the fetal monkey: An ultrastructural immunoperoxidase analysis, *J. Neurosci.* 1:27–39.

Li, W. and LoTurco, J.J., 2000, Noggin is a negative regulator of neuronal differentiation in developing neocortex, *Dev. Neurosci.* 22:68–73.

Lillien, L.E. and Raff, M.C., 1990, Differentiation signals in the CNS: Type-2 astrocyte development *in vitro* as a model system, *Neuron* 5:111–119.

Lillien, L.E., Sendtner, M., and Raff, M.C., 1990, Extracellular matrix-associated molecules collaborate with ciliary neurotrophic factor to induce type-2 astrocyte development, *J. Cell Biol.* 111:635–644.

Lillien, L.E., Sendtner, M., Rohrer, H., Hughes, S.M., and Raff, M.C., 1988, Type-2 astrocyte development in rat brain cultures is initiated by a CNTF-like protein produced by type-1 astrocytes, *Neuron* 1:485–494.

Lim, D.A., Tramontin, A.D., Trevejo, J.M., Herrera, D.G., Garcia-Verdugo, J.M., and Alvarez-Buylla, A., 2000, Noggin antagonizes BMP signaling to create a niche for adult neurogenesis, *Neuron* 28:713–726.

Lin, S.C. and Bergles, D.E., 2002, Physiological characteristics of NG2-expressing glial cells, *J. Neurocytol.* 31:537–549.

Ling, E.A. and Leblond, C.P., 1973, Investigation of glial cells in semithin sections. II. Variation with age in the numbers of the various glial cell types in rat cortex and corpus callosum, *J. Comp. Neurol.* 149:73–82.

Luskin, M.B. and McDermott, K., 1994, Divergent lineages for oligodendrocytes and astrocytes originating in the neonatal forebrain subventricular zone, *Glia* 11:211–226.

Luskin, M.B., Pearlman, A.L., and Sanes, J.R., 1988, Cell lineage in the cerebral cortex of the mouse studied *in vivo* and *in vitro* with a recombinant retrovirus, *Neuron* 1:635–647.

Mabie, P.C., Mehler, M.F., Marmur, R., Papavasiliou, A., Song, Q., and Kessler, J.A., 1997, Bone morphogenetic proteins induce astroglial differentiation of oligodendroglial-astroglial progenitor cells, *J. Neurosci.* 17:4112–4120.

Malatesta, P., Hartfuss, E., and Gotz, M., 2000, Isolation of radial glial cells by fluorescent-activated cell sorting reveals a neuronal lineage, *Development—Supplement* 127:5253–5263.

Marshall, C.A. and Goldman, J.E., 2002, Subpallial dlx2-expressing cells give rise to astrocytes and oligodendrocytes in the cerebral cortex and white matter, *J. Neurosci.* 20:30–42.

Martin, A., Hofmann, H.D., and Kirsch, M., 2003, Glial reactivity in ciliary neurotrophic factor-deficient mice after optic nerve lesion, *J. Neurosci.* 23:5416–5424.

Masood, K., Besnard, F., Su, Y., and Brenner, M., 1993, Analysis of a segment of the human glial fibrillary acidic protein gene that directs astrocyte-specific transcription, *J. Neurochem.* 61:160–166.

Mayer-Proschel, M., Kalyani, A.J., Mujtaba, T., and Rao, M.S., 1997, Isolation of lineage-restricted neuronal precursors from multipotent neuroepithelial stem cells, *Neuron* 19:773–785.

McCarthy, G.F. and Leblond, C.P., 1988, Radioautographic evidence for slow astrocyte turnover and modest oligodendrocyte production in the corpus callosum of adult mice infused with 3H-thymidine, *J. Comp. Neurol.* 271:589–603.

McCarthy, K.D. and Salm, A.K., 1991, Pharmacologically-distinct subsets of astroglia can be identified by their calcium response to neuroligands, *Neuroscience* 41:325–333.

McManus, M.F., Chen, L.C., Vallejo, I., and Vallejo, M., 1999, Astroglial differentiation of cortical precursor cells triggered by activation of the cAMP-dependent signaling pathway, *J. Neurosci.* 19:9004–9015.

Mehler, M.F., Mabie, P.C., Zhang and Kessler, J.A., 1997, Bone morphogenetic proteins in the nervous system. *Trends Neurosci* 20:309–17.

Mehler, M.F., Mabie, P.C., Zhu, G., Gokhan, S., and Kessler, J.A., 2000, Developmental changes in progenitor cell responsiveness to bone morphogenetic proteins differentially modulate progressive CNS lineage fate, *Dev. Neurosci.* 22:74–85.

Mi, H. and Barres, B.A., 1999, Purification and characterization of astrocyte precursor cells in the developing rat optic nerve, *J. Neurosci.* 19: 1049–1061.

Mi, H., Haeberle, H., and Barres, B.A., 2001, Induction of astrocyte differentiation by endothelial cells, *J. Neurosci.* 21:1538–1547.

Miller, R.H., David, S., Patel, R., Abney, E.R., and Raff, M.C., 1985, A quantitative immunohistochemical study of macroglial cell development in the rat optic nerve: *In vivo* evidence for two distinct astrocyte lineages, *Dev. Biol.* 111:35–41.

Miller, R.H. and Szigeti, V., 1991, Clonal analysis of astrocyte diversity in neonatal rat spinal cord cultures, *Development* 113:353–362.

Milosevic, A. and Goldman, J.E., 2002, Progenitors in the postnatal cerebellar white matter are antigenically heterogeneous, *J. Comp. Neurol.* 452:192–203.

Misson, J.P., Austin, C.P., Takahashi, T., Cepko, C.L., and Caviness, V.S., Jr., 1991, The alignment of migrating neural cells in relation to the murine neopallial radial glial fiber system, *Cereb Cortex* 1:221–229.

Misson, J.P., Edwards, M.A., Yamamoto, M., and Caviness, V.S., Jr., 1988, Identification of radial glial cells within the developing murine central nervous system: Studies based upon a new immunohistochemical marker, *Brain Res. Dev. Brain Res.* 44:95–108.

Miyake, T., Fujiwara, T., Fukunaga, T., Takemura, K., and Kitamura, T., 1995, Glial cell lineage *in vivo* in the mouse cerebellum, *Dev. Growth. Differ.* 37:273–285.

Molne, M., Studer, L., Tabar, V., Ting, Y.T., Eiden, M.V., and McKay, R.D., 2000, Early cortical precursors do not undergo LIF-mediated astrocytic differentiation, *J. Neurosci. Res.* 59:301–311.

Mori, K., Ikeda, J., and Hayaishi, O., 1990, Monoclonal antibody R2D5 reveals midsagittal radial glial system in postnatally developing and adult brainstem, *Proc. Natl. Acad. Sci. USA* 87:5489–5493.

Morrison, R.S., de Vellis, J., Lee, Y.L., Bradshaw, R.A., and Eng, L.F., 1985, Hormones and growth factors induce the synthesis of glial fibrillary acidic protein in rat brain astrocytes, *J. Neurosci. Res.* 14:167–176.

Moskovkin, G.N., Fulop, Z., and Hajos, F., 1978, Origin and proliferation of astroglia in the immature rat cerebellar cortex. A double label autoradiographic study, *Acta Morphol. Acad. Sci. Hung.* 26:101–106.

Nakashima, K., Takizawa, T., Ochiai, W., Yanagisawa, M., Hisatsune, T., Nakafuku, M. *et al.*, 2001, BMP2-mediated alteration in the developmental pathway of fetal mouse brain cells from neurogenesis to astrocytogenesis, *Proc. Natl. Acad. Sci. USA* 98:5868–5873.

Neubauer, K., Knittel, T., Aurisch, S., Fellmer, P., and Ramadori, G., 1996, Glial fibrillary acidic protein—a cell type specific marker for Ito cells *in vivo* and *in vitro*, *J. Hepatol.* 24:719–730.

Nieto, M., Schuurmans, C., Britz, O., and Guillemot, F., 2001, Neural bHLH genes control the neuronal versus glial fate decision in cortical progenitors, *Neuron* 29:401–413.

Nishiyama, A., Lin, X.H., Giese, N., Heldin, C.H., and Stallcup, W.B., 1996, Co-localization of NG2 proteoglycan and PDGF alpha-receptor on O2A progenitor cells in the developing rat brain, *J. Neurosci. Res.* 43:299–314.

Nishiyama, A., Watanabe, M., Yang, Z., and Bu, J., 2002, Identity, distribution, and development of polydendrocytes: NG2-expressing glial cells, *J. Neurocytol.* 31:437–455.

Noctor, S.C., Flint, A.C., Weissman, T.A., Dammerman, R.S., and Kriegstein, A.R., 2001, Neurons derived from radial glial cells establish radial units in neocortex, *Nature* 409:714–720.

Norenberg, M.D. and Martinez-Hernandez, A., 1979, Fine structural localization of glutamine synthetase in astrocytes of rat brain, *Brain Res.* 161:303–310.

Norton, W.T. and Farooq, M., 1989, Astrocytes cultured from mature brain derive from glial precursor cells, *J. Neurosci.* 9:769–775.

Oh, L.Y., Denninger, A., Colvin, J.S., Vyas, A., Tole, S., Ornitz, D.M. *et al.*, 2003, Fibroblast growth factor receptor 3 signaling regulates the onset of oligodendrocyte terminal differentiation, *J. Neurosci.* 23:883–894.

Palay, S.L. and Chan-Palay, V., 1974, *Cerebellar Cortex, Cytology, and Organization*, Springer-Verlag, New York.

Park, J.K., Williams, B.P., Alberta, J.A., and Stiles, C.D., 1999, Bipotent cortical progenitor cells process conflicting cues for neurons and glia in a hierarchical manner, *J. Neurosci.* 19:10383–10389.

Parnavelas, J.G., 1999, Glial cell lineages in the rat cerebral cortex, *Exp. Neurol.* 156:418–429.

Paterson, J.A., 1983, Dividing and newly produced cells in the corpus callosum of adult mouse cerebrum as detected by light microscopic radioautography, *Anat. Anz.* 153:149–168.

Paterson, J.A., Privat, A., Ling, E.A., and Leblond, C.P., 1973, Investigation of glial cells in semithin sections III transformation of subependymal cells into glial cells as shown by radioautography after 3H-thymidine injection into the lateral ventricle of the brain of young rats, *J. Comp. Neurol.* 149:83–102.

Penfield, W., 1924, Oligodendroglia and its relation to classical neuroglia, *Brain* 47:430–450.

Penfield, W., 1932, *Neuroglia, Normal and Pathological, Cytology & Cellular Pathology of the Nervous System* (W. Penfield, ed.), Hoeber, New York, Vol. 2, pp. 421–480.

Perraud, F., Besnard, F., Pettmann, B., Sensenbrenner, M., and Labourdette, G., 1988, Effects of acidic and basic fibroblast growth factors (aFGF and bFGF) on the proliferation and the glutamine synthetase expression of rat astroblasts in culture, *Glia* 1:124–131.

Peters, A., Josephson, K., and Vincent, S.L., 1991, Effects of aging on the neuroglial cells and pericytes within area 17 of the rhesus monkey cerebral cortex, *Anat. Rec.* 229:384–398.

Peters, A., Palay, S.L., and Webster, H.F., 1976, *The Fine Structure of the Nervous System: The Neurons and Supporting Cells*, 1, Saunders, Philadelphia.

Power, J., Mayer-Proschel, M., Smith, J., and Noble, M., 2002, Oligodendrocyte precursor cells from different brain regions express divergent properties consistent with the differing time courses of myelination in these regions, *Dev. Biol.* 245:362–375.

Price, J. and Thurlow, L., 1988, Cell lineage in the rat cerebral cortex: A study using retroviral-mediated gene transfer, *Development* 104: 473–482.

Pringle, N.P., Guthrie, S., Lumsden, A., and Richardson, W.D., 1998, Dorsal spinal cord neuroepithelium generates astrocytes but not oligodendrocytes, *Neuron* 20:883–893.

Pringle, N.P., Yu, W.P., Howell, M., Colvin, J.S., Ornitz, D.M., and Richardson, W.D., 2003, Fgfr3 expression by astrocytes and their precursors: Evidence that astrocytes and oligodendrocytes originate in distinct neuroepithelial domains, *Development* 130:93–102.

Qian, X., Davis, A.A., Goderie, S.K., and Temple, S., 1997, FGF2 concentration regulates the generation of neurons and glia from multipotent cortical stem cells, *Neuron* 18:81–93.

Raff, M.C., Abney, E.R., Cohen, J., Lindsay, R., and Noble, M., 1983a, Two types of astrocytes in cultures of developing rat white matter: Differences in morphology, surface gangliosides, and growth characteristics, *J. Neurosci.* 3:1289–1300.

Raff, M.C., Abney, E.R., and Miller, R.H., 1984, Two glial cell lineages diverge prenatally in rat optic nerve, *Dev. Biol.* 106:53–60.

Raff, M.C., Miller, R.H., and Noble, M., 1983b, A glial progenitor cell that develops *in vitro* into an astrocyte or an oligodendrocyte depending on culture medium, *Nature* 303:390–396.

Rakic, P., 1971, Guidance of neurons migrating to the fetal monkey neocortex, *Brain Res.* 33:471–476.

Rakic, P., 1972, Mode of cell migration to the superficial layers of the fetal monkey neocortex, *J. Comp. Neurol.* 145:61–84.

Rakic, P., 1995, Radial glial cells: Scaffolding for brain construction. In *Neuroglia* (H. Kettenman, and B., Ransom, eds.), Oxford University Press, Inc., New York, pp. 746–762.

Ramon-Molinar, E., 1958, A study on neuroglia: The problem of transitional forms, *J. Comp. Neurol.* 110:157–171.

Ramon Y Caja, S. (1894) Les Novells Ideas sur la structure du system nerveaux chez l'homme et les vertebres. Reinwald, Paris.

Rao, M.S. and Mayer-Proschel, M., 1997, Glial-restricted precursors are derived from multipotent neuroepithelial stem cells, *Dev. Biol.* 188:48–63.

Rao, M.S., Noble, M., and Mayer-Proschel, M., 1998, A tripotential glial precursor cell is present in the developing spinal cord, *Proc. Natl. Acad. Sci. USA* 95:3996–4001.

Reichenbach, A., 1990, Radial glial cells are present in the velum medullare of adult monkeys, *J. Hirnforsch.* 31:269–271.

Retzius, G., 1894, Die neuroglia des gehirns beim menschen und bei säugethieren, *Biologische Untersuchungen. Neue Folge* 6:1–28.

Reyners, H., Gianfelici de Reyners, E., Regniers, L., and Maisin, J.R., 1986, A glial progenitor in the cerebral cortex of the adult rat, *J. Neurocytol.* 15:53–61.

Reynolds, R. and Hardy, R., 1997, Oligodendroglial progenitors labeled with the O4 antibody persist in the adult rat cerebral cortex *in vivo*, *J. Neurosci. Res.* 47:455–470.

Rio, C., Rieff, H.I., Qi, P., Khurana, T.S., and Corfas, G., 1997, Neuregulin and erbB receptors play a critical role in neuronal migration, *Neuron* 19:39–50.

Rowitch, D.H., Lu, Q.R., Kessaris, N., and Richardson, W.D., 2002, An "oligarchy" rules neural development, *Trends Neurosci.* 25:417–422.

Sakai, Y. and Barnes, D., 1991, Assay of astrocyte differentiation-inducing activity of serum and transforming growth factor beta, *Methods Enzymol.* 198:337–339.

Sakai, Y., Rawson, C., Lindburg, K., and Barnes, D., 1990, Serum and transforming growth factor beta regulate glial fibrillary acidic protein in serum-free-derived mouse embryo cells, *Proc. Natl. Acad. Sci. USA* 87:8378–8382.

Sakakibara, S., Imai, T., Hamaguchi, K., Okabe, M., Aruga, J., Nakajima, K. *et al.*, 1996, Mouse-Musashi-1, a neural RNA-binding protein highly enriched in the mammalian CNS stem cell, *Dev. Biol.* 176:230–242.

Sanes, J.R., 1989, Analyzing cell lineage with a recombinant retrovirus, *TINS* 12:21–28.

Sasaki, A., Levison, S.W., and Ting, J.P.Y., 1989, Comparison and quantitation of Ia antigen expression on cultured macroglia and amoeboid microglia from lewis rat cerebral cortex: Analyses and implications, *J. Neuroimmunol.* 25:63–74.

Sasaki, T., Hirabayashi, J., Manya, H., Kasai, K.I., and Endo, T., 2004, Galectin-1 induces astrocyte differentiation, which leads to production of brain-derived neurotrophic factor, *Glycobiology* 4:357–63.

Schmechel, D.E. and Rakic, P., 1979a, Arrested proliferation of radial glial cells during midgestation in rhesus monkey, *Nature* 277:303–305.

Schmechel, D.E. and Rakic, P., 1979b, A Golgi study of radial glial cells in developing monkey telencephalon: Morphogenesis and transformation into astrocytes, *Anat. Embryol. (Berl.)* 156:115–152.

Schnitzer, J., Franke, W.W., and Schachner, M., 1981, Immunocytochemical demonstration of vimentin in astrocytes and ependymal cells of developing and adult mouse nervous system, *J. Cell Biol.* 90:435–447.

Shafit-Zagardo, B., Kume-Iwaki, A., and Goldman, J.E., 1988, Astrocytes regulate GFAP mRNA levels by cyclic AMP and protein kinase C-dependent mechanisms, *Glia* 1:346–354.

Shibata, T., Yamada, K., Watanabe, M., Ikenaka, K., Wada, K., Tanaka, K. *et al.*, 1997, Glutamate transporter GLAST is expressed in the radial glia-astrocyte lineage of developing mouse spinal cord, *J. Neurosci.* 17:9212–9219.

Skoff, R.P., 1990, Gliogenesis in rat optic nerve: Astrocytes are generated in a single wave before oligodendrocytes, *Dev. Biol.* 139:149–168.

Song, M.R. and Ghosh, A., 2004, FGF2-induced chromatin remodeling regulates CNTF-mediated gene expression and astrocyte differentiation, *Nat. Neurosci.* 7:229–235.

Staugaitis, S.M., Zerlin, M., Hawkes, R., Levine, J.M., and Goldman, J.E., 2001, Aldolase C/zebrin II expression in the neonatal rat forebrain reveals cellular heterogeneity within the subventricular zone and early astrocyte differentiation, *J. Neurosci.* 21:6195–6205.

Sun, Y., Nadal-Vicens, M., Misono, S., Lin, M.Z., Zubiaga, A., Hua, X. *et al.*, 2001, Neurogenin promotes neurogenesis and inhibits glial differentiation by independent mechanisms, *Cell* 104:365–376.

Suzuki, S.O. and Goldman, J.E., 2003, Multiple cell populations in the early postnatal subventricular zone take distinct migratory pathways: A dynamic study of glial and neuronal progenitor migration, *J. Neurosci.* 23:4240–4250.

Takebayashi, H., Nabeshima, Y., Yoshida, S., Chisaka, O., and Ikenaka, K., 2002, The basic helix-loop-helix factor olig2 is essential for the development of motoneuron and oligodendrocyte lineages, *Curr. Biol.* 12:1157–1163.

Takizawa, T., Nakashima, K., Namihira, M., Ochiai, W., Uemura, A., Yanagisawa, M. *et al.*, 2001, DNA methylation is a critical cell-intrinsic determinant of astrocyte differentiation in the fetal brain, *Dev. Cell.* 1:749–758.

Tanigaki, K., Nogaki, F., Takahashi, J., Tashiro, K., Kurooka, H., and Honjo, T., 2001, Notch1 and Notch3 instructively restrict bFGF-responsive multipotent neural progenitor cells to an astroglial fate, *Neuron* 29:45–55.

Tansey, F.A., Farooq, M., and Cammer, W., 1991, Glutamine synthetase in oligodendrocytes and astrocytes: New biochemical and immunocytochemical evidence, *J. Neurochem.* 56:266–272.

Temple, S. and Raff, M.C., 1986, Clonal analysis of oligodendrocyte development in culture: Evidence for a developmental clock that counts cell divisions, *Cell* 44:773–779.

Tohyama, T., Lee, V.M.Y., Rorke, L.B., Marvin, M., McKay, R.D.G., and Trojanowski, J.Q., 1992, Nestin expression in embryonic human neuroepithelium and in human neuroepithelial tumor cells, *Lab. Invest.* 66:303–313.

Tout, S., Dreher, Z., Chan-Ling, T., and Stone, J., 1993, Contact-spacing among astrocytes is independent of neighbouring structures: *In vivo* and *in vitro* evidence, *J. Comp. Neurol.* 332:433–443.

Turner, D.L. and Cepko, C.L., 1987, A common progenitor for neurons and glia persists in rat retina late in development, *Nature* 328:131–136.

Vallejo, I. and Vallejo, M., 2002, Pituitary adenylate cyclase-activating polypeptide induces astrocyte differentiation of precursor cells from developing cerebral cortex, *Mol. Cell. Neurosci.* 21:671–683.

Vaughan, D.W. and Peters, A., 1974, Neuroglial cells in the cerebral cortex of rats from young adulthood to old age: An electron microscope study, *J. Neurocytol.* 3:405–429.

Vaysse, P.J.J. and Goldman, J.E., 1990, A clonal analysis of glial lineages in neonatal forebrain development *in vitro, Neuron* 5:227–235.

Vaysse, P.J.J. and Goldman, J.E., 1992, A distinct type of GD3+, flat astrocyte in rat CNS cultures, *J. Neurosci.* 12:330–337.

Ventura, R. and Harris, K.M., 1999, Three-dimensional relationships between hippocampal synapses and astrocytes, *J. Neurosci.* 19:6897–6906.

Viti, J., Feathers, A., Phillips, J., and Lillien, L., 2003, Epidermal growth factor receptors control competence to interpret leukemia inhibitory factor as an astrocyte inducer in developing cortex, *J. Neurosci.* 23:3385–3393.

Voigt, T., 1989, Development of glial cells in the cerebral wall of ferrets: Direct tracing of their transformation from radial glia into astrocytes, *J. Comp. Neurol.* 289:74–88.

Walsh, C. and Cepko, C.L., 1992, Widespread dispersion of neuronal clones across functional regions of the cerebral cortex, *Science* 255:434–440.

Walsh, C. and Cepko, C.L., 1993, Clonal dispersion in proliferative layers of developing cerebral cortex, *Nature* 362:632–635.

Wendell-Smith, C.P., Blunt, M.J., and Baldwin, F., 1966, The ultrastructural characterization of macroglial cell types, *J. Comp. Neurol.* 127:219–239.

Wilkin, G.P., Marriott, D.R., and Cholewinski, A.J., 1990, Astrocyte heterogeneity. *TINS* 13:43–46.

Williams, B.P., Abney, E.R., and Raff, M.C., 1985, Macroglial cell development in embryonic rat brain: Studies using monoclonal antibodies, fluorescence activated cell sorting, and cell culture, *Dev. Biol.* 112:126–134.

Wu, Y., Liu, Y., Levine, E.M., and Rao, M.S., 2003, Hes1 but not Hes5 regulates an astrocyte versus oligodendrocyte fate choice in glial restricted precursors, *Dev. Dyn.* 226:675–689.

Yamada, K., Fukaya, M., Shibata, T., Kurihara, H., Tanaka, K., Inoue, Y. *et al.*, 2000, Dynamic transformation of Bergmann glial fibers proceeds in correlation with dendritic outgrowth and synapse formation of cerebellar Purkinje cells, *J. Comp. Neurol.* 418:106–120.

Yang, H.Y., Lieska, N., Shao, D., Kriho, V., and Pappas, G.D., 1993, Immunotyping of radial glia and their glial derivatives during development of the rat spinal cord, *J. Neurocytol.* 22:558–571.

Young, G.M. and Levison, S.W., 1996, Persistence of multipotential progenitors in the juvenile rat subventricular zone, *Dev. Neurosci.* 18:255–265.

Zerlin, M. and Goldman, J.E., 1997, Interactions between glial progenitors and blood vessels during early postnatal corticogenesis: Blood vessel contact represents an early stage of astrocyte differentiation, *J. Comp. Neurol.* 387:537–546.

Zerlin, M., Levison, S.W., and Goldman, J.E., 1995, Early stages of dispersion and differentiation of glial progenitors in the postnatal mammalian forebrain, *J. Neurosci.* 15:7238–7249.

Zerlin M, Milosevic A, Goldman J.E., (2004) Gilal progenitors of the neonatal subventricular zone differentiate asynchronously, leading to spatial dispersion of glial clones and to the persistence of immature glia in the adult mammalian CNS. *Dev Biol.* 270:200–13.

Zhang, D., Mehler, M.F., Song, Q., and Kessler, J.A., 1998, Development of bone morphogenetic protein receptors in the nervous system and possible roles in regulating trkC expression, *J. Neurosci.* 18: 3314–3326.

Zhang, L. and Goldman, J.E., 1996, Developmental fates and migratory pathways of dividing progenitors in the postnatal rat cerebellum, *J. Comp. Neurol.* 370:536–550.

Zhang, S.C., Ge, B., and Duncan, I.D., 1999, Adult brain retains the potential to generate oligodendroglial progenitors with extensive myelination capacity, *Proc. Natl. Acad. Sci. USA* 96:4089–4094.

Zhou, Q. and Anderson, D.J., 2002, The bHLH transcription factors OLIG2 and OLIG1 couple neuronal and glial subtype specification, *Cell* 109:61–73.

8

Neuronal Migration in the Developing Brain

Franck Polleux and E. S. Anton

INTRODUCTION

The way in which a nervous system is constructed predisposes and constrains its functions. Thus the study of neuronal cell migration, an elementary step in the histogenesis of any nervous system, is critical if we are to understand how the structure and function of a nervous system come about. Specific neuronal networks emerge as a result of appropriate migration and final placement of neurons during development. In the developing nervous system, most, if not all, neurons undergo their terminal division and terminal differentiation in distinct locations. Specific neuronal populations have to migrate in distinct pathways and patterns over extensive distances to reach their final position. Two main types of migration predominate during the development of the central nervous system: radial vs tangential. Radial migration is characterized by close interactions between migrating neurons and the processes of radial glial cells, which constitute a scaffold bridging the proliferating neuroepithelium and the differentiating zone. Postmitotic neurons migrate radially from the ventricular zone toward the pial surface past previously generated neuronal layers (Rakic, 1971b, 1972a) to reach the top of the cortical plate (CP), where they terminate their migration and assemble into layers with distinct patterns of connectivity. Radial migration of cortical neurons can occur in two distinct modes: locomotion or somal translocation (Nadarajah *et al.*, 2001; Nadarajah and Parnavelas, 2002; Nadarajah *et al.*, 2002). In contrast, tangential migration is referred to as a nonradial mode of neuronal translocation that does not require specific interaction with radial glial cell processes. Observations of tangential dispersion of precursors or postmitotic neurons in the developing cortex suggested the possibility of nonradial migration in the cortex (O'Rourke *et al.*, 1992, 1995, 1997; Walsh and Cepko, 1992; Fishell *et al.*, 1993; Tan and Breen, 1993; Tan *et al.*, 1995; de Carlos *et al.*, 1996). Analysis of *Dlx1/2* double knock-out mice has demonstrated for the first time that subpopulations of GABAergic interneurons, originating from the ventral telencephalon (also called the ganglionic eminence [GE]), indeed migrate tangentially into the neocortex (Anderson *et al.*, 1997). Therefore, there is a tight correlation between neuronal subtype identity (glutamaergic vs GABAergic) and the mode of migration (radial vs tangential) in the developing cortex of mammals (Parnavelas, 2000).

Specific cell–cell recognition and adhesive interactions between neurons, glia, and the surrounding extracellular matrix (ECM) appear to modulate distinct patterns of neuronal migration, placement, and eventual differentiation within cortex. A fundamental challenge in the study of cortical development is the elucidation of mechanisms that determine how neurons migrate and coalesce into distinct layers or nuclei in the developing cerebral cortex. In this regard, several related questions need specific attention: (1) What are the cell-intrinsic and extracel-lular cues that trigger the onset of neuronal migration following last mitotic cell division? (2) What is the molecular basis and role of glial-independent and glial-guided neuronal migration in cortical development? (3) How do migrating neurons know where to end? and (4) What are the stage-specific genes that determine distinct aspects of neuronal migration in developing mammalian brain? In combination, analysis of these questions may elucidate some of the fundamental rules guiding the development of cerebral cortex.

PATTERNS OF NEURONAL MIGRATION

Extensive observations of neuronal migration in the past several decades in mammalian cerebral cortex and recent molecular characterization of migration deficits in mice and humans have raised the cerebral cortex as a prototype model for the analysis of migration in the developing mammalian central nervous system. Radial glial cells play a critical role in the construction of the mammalian brain by contributing to the formation of neurons and astrocytes and by providing a permissive and instructive scaffold for neuronal migration. The establishment of radial glial cells from an undifferentiated sheet of neuroepithelium precedes the generation and migration of neurons in the cerebral cortex. During early stages of corticogenesis, radial glial cells can give rise to neurons (reviewed in Fishell and Kriegstein, 2003; Rakic, 2003). Subsequent neuronal cell movement in the developing mammalian cerebral cortex occurs

Franck Polleux • Department of Pharmacology, University of North Carolina School of Medicine, Chapel Hill, NC 27599.   E. S. Anton • Department of Cell and Molecular Physiology, University of North Carolina School of Medicine, Chapel Hill, NC 27599.

Developmental Neurobiology, 4th ed., edited by Mahendra S. Rao and Marcus Jacobson. Kluwer Academic / Plenum Publishers, New York, 2005.

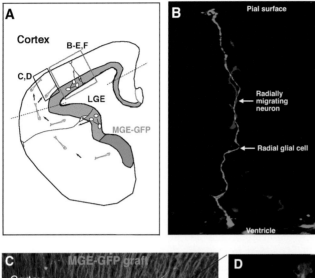

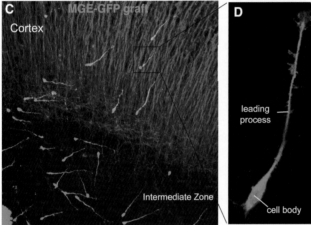

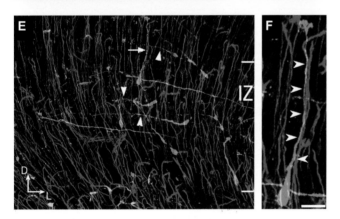

FIGURE 1. Radial vs tangential patterns of neuronal migration. In the developing embryonic cortex (A), radially (B) and tangentially (C) migrating neurons display a unipolor morphology characterized by a prominent leading process (D). These neurons are in intimate contact with either radia glial cells (B, green) or with neurites (C, red) within the developing cortex. Tangentially migrating neurons (arrowhead, E) eventually turn radially (arrow, E) in the intermediate zone and associate with radial fibers (F, red) during final stages of translocation to the cortical plate. Cells shown in B were electoporated with GFP, whereas tangentially migrating neurons (C–F) were isolated from GFP-expressing MGE graft on a slice co-culture assay (Polleux *et al.*, 2002).

mainly along radial glial fibers, though nonpyramidal neurons initially migrate into the cortex in a radial glial-independent manner (Fig. 1). Neurons migrate from the ventricular zone toward the pial surface past previously generated neuronal layers (Rakic, 1971a, b; 1972a, b) to reach the top of the CP, where they terminate their migration and assemble into layers with distinct patterns of connectivity. Radial migration of cortical neurons can occur in two distinct modes: locomotion or somal translocation (Nadarajah *et al.*, 2001; Nadarajah and Parnavelas, 2002; Nadarajah *et al.*, 2002). Somally translocating neurons, prevalent during early phases of corticogenesis, appear to move toward the pial surface by maintaining pial attachment while losing their ventricular attachments. In contrast, radial glial cell fibers serve as the primary migratory guides for locomoting neurons (Rakic, 1971a, b, 1972a, b, 1990; Gray *et al.*, 1990; Hatten and Mason, 1990; Misson *et al.*, 1991). These neurons form specialized membrane contacts variably referred to as junctional domains, interstitial junctions, or punctae adherentia with the underlying radial glial cell substrate (Gregory *et al.*, 1988; Cameron and Rakic, 1994; Anton *et al.*, 1996). Such specialized membrane contacts are hypothesized to be critical elements in the maintenance of directed neuronal cell migration along radial glial cell fibers (Rakic *et al.*, 1994). The radial movement of neurons stops abruptly at the interface between CP and cell sparse marginal zone. The signal to end neuronal cell migration is thought to be provided either by the afferent fibers that migrating neurons encounter near their target location or by the ambient neuronal cell population that had already reached its final position (Sidman and Rakic, 1973; Hatten, 1990, 1993; Hatten and Mason, 1990; D'Arcangelo *et al.*, 1995; Ogawa *et al.*, 1995). Alternately, a change in the cell surface properties of the radial glial substrate may signal a neuron migrating on it to stop, detach, and differentiate.

In contrast to radially migrating neurons, populations of GABAnergic interneurons, originating from the GE, migrate tangentially into the neocortex (Anderson *et al.*, 1997; Letinic and Rakic, 2001; Maricich *et al.*, 2001; Tamamaki *et al.*, 1997; Wichterle *et al.*, 2001; see Fig. 1). Some of these neurons migrate ventrally toward the cortical ventricular zone prior to radial migration toward the pial surface (Nadarajah *et al.*, 2002). These distinct patterns of neuronal migration enables several generations of neurons to reach their appropriate areal and laminar positions in the developing CP. Analysis of mutations in mice and humans have revealed several molecular cues controlling different aspects of neuronal migration. Evidently, a dynamic regulation of multiple cellular events such as cell–cell recognition, adhesion, transmembrane signaling, and cell motility events underlies the process of neuronal migration.

MECHANISMS UNDERLYING RADIAL MIGRATION

Initiation of Migration

Movement of neuronal cells from their site of birth in the ventricular areas to the specific laminar location involves a progressive unraveling of three interrelated cellular events: initiation of migration along appropriate pathways or substrates, maintenance of migration through a complex cellular milieu, and termination of migration in the CP at the appropriate laminar location.

In humans with periventricular heterotopia, mutations in actin-binding protein filamin 1 (or filamin-α [FLNA]) results in failure of neuronal migration and accumulation of neuroblasts in the ventricular zone of cerebral cortex (Eksioglu *et al.*, 1996; Fox *et al.*, 1998; Sheen *et al.*, 2001; Moro *et al.*, 2002). FLNA co-localizes to actin stress fibers, highly expressed by neural cells in the ventricular surface, is thought to crosslink F-actin network to facilitate cell motility (Fox *et al.*, 1998; Stossel *et al.*, 2001). The inability of neurons to initiate migration following FLNA mutations is indicative of the significance of actin dynamics in initiation of migration. Whether FLNA's interactions with cell surface integrin receptors (β1 or β2), presenilin, and small GTPase RalA is part of the cascade that conveys extracellular signals from the ventricular zone to initiate migration still needs further examination (Sharma *et al.*, 1995; Loo *et al.*, 1998; Zhang and Galileo, 1998; Ohta *et al.*, 1999). However, Filamin A interacting protein (FLIP) is expressed in the ventricular zone and degrades FLNA, thereby inhibiting premature onset of neuronal migration from the ventricular zone (Nagano *et al.*, 2002).

Maintenance of Migration

Once initiated, a dynamic regulation of multiple cellular events such as cell–cell recognition, adhesion, transmembrane signaling, and cell motility events underlies the process of neuronal migration (Lindner *et al.*, 1983; Grumet *et al.*, 1985; Antonicek *et al.*, 1987; Chuong *et al.*, 1987; Rutishauser and Jessell, 1988; Edmondson *et al.*, 1988; Sanes, 1989; Hatten and Mason, 1990; Stitt and Hatten, 1990; Takeichi, 1991; Misson *et al.*, 1991; Galileo *et al.*, 1992; Grumet, 1992; Komuro and Rakic, 1992, 1993, 1995; Shimamura and Takeichi, 1992; Fishman and Hatten, 1993; Hatten, 1993; Cameron and Rakic, 1994; Rakic *et al.*, 1994; Stipp *et al.*, 1994; Rakic and Komuro, 1995). A migrating neuron attaches itself to the radial glial substrate primarily by its leading process and cell soma. Only the actively migrating neurons form the specialized junctional domains or the interstitial densities with the apposing glial fibers (Gregory *et al.*, 1988; Cameron and Rakic, 1994), whereas the stationary neurons form desmosomes or puncta adherentia. The specialized subcellular accumulations of membrane proteins, such as radial glial based neuron–glial junctional protein 1 (NJPA1) or neuronal astrotactin, at the apposition of migrating neurons and radial glial cells may function in migration by orchestrating cell–cell recognition, adhesion, transmembrane signaling, and or motility. The homophilic or heterophilic nature of the junctional domain antigen interactions are unclear. However, the integrity of neuron–glial junctional complexes appears to depend on their association with microtubule cytoskeleton (Gregory *et al.*, 1988; Cameron and Rakic, 1994). Disruption of microtubules, but not actin filaments, adversely affect neuron–glial adhesion (Rivas and Hatten, 1995).

Junctional domain associated microtubules are thought to play a role in force generation during cell movement, in addition to being vital for the elaboration and maintenance of junctional domains (Gregory *et al.*, 1988). Furthermore, specific cell–cell interactions between migrating neurons and radial glial cells mediated by the junctional domain antigens may also modulate the properties of each other's cytoskeleton, akin to that observed between developing peripheral axons and Schwann cells (Kirkpatrick and Brady, 1994). It is argued that an increase in class III β-tubulin content leads to enhanced microtubule lability, thus allowing the continuous assembly and disassembly of microtubules needed to generate a forward force during cell movement (Falconer *et al.*, 1992; Moskowitz and Oblinger, 1995; Rivas and Hatten, 1995; Rakic *et al.*, 1996).

Significant deficits in neuronal migration were seen following mutations in genes regulating microtubule cytoskeleton (see Table 1 and Fig. 2). In humans, mutations in Lis1 (noncatalytic subunit of platelet-activating factor acetylhydrolase isoform 1b) Miller–Dieker syndrome, a severe form of lissencephaly (Reiner *et al.*, 1993; Hattori *et al.*, 1994). In mouse, truncation of Lis1 leads to slower neuronal migration and cortical plate disorganization characterized by unsplit preplate (Cahana *et al.*, 2001). Partial loss of Lis1 (i.e., mice with one inactive allele of Lis1) also results in retarded neuronal migration (Hirotsune *et al.*, 1998). Lis1 binds to microtubules, microtubule based motor protein, dyenin, and related microtubule interactors such as dynactin, NUDEL, and mNudE (Sapir *et al.*, 1997; Efimov and Morris, 2000; Faulkner *et al.*, 2000; Kitagawa *et al.*, 2000; Niethammer *et al.*, 2000; Sasaki *et al.*, 2000; Smith *et al.*, 2000). Loss of Lis1 leads to concentration of microtubules around the nucleus and failed dynein aggregation, whereas overexpression of Lis1 causes transport of microtubule to edges of the cell and aggregation of dynein and dynectin (Sasaki *et al.*, 2000; Smith *et al.*, 2000). NUDEL and mNudE appear to control cellular localization of dynein and the microtubule network around the microtubule-organizing centrosome, respectively (Feng *et al.*, 2000; Niethammer *et al.*, 2000; Sasaki *et al.*, 2000). How these associations of Lis1 modify the microtubule network to enable the nuclear translocation of neurons in the developing cortex remains to be elucidated.

Mutations in another microtubule-associated protein in migrating neurons, doublecortin (Dcx), leads to X-linked lissencephaly (double cortex syndrome) in humans (des Portes *et al.*, 1998; Gleeson *et al.*, 1998). In these patients, neurons that migrated aberrantly are deposited in a broad band in subcortical layers. Dcx is critical for the stabilization of microtubule network (Francis *et al.*, 1999; Gleeson *et al.*, 1999; Horesh *et al.*, 1999). Dcx can associate with Lis1 and promote tubulin polymerization *in vitro* (Gleeson *et al.*, 1999; Caspi *et al.*, 2000; Feng and Walsh, 2001). Overexpression of Dcx results in aggregates of thick microtubule bundles resistant to depolymerization (Gleeson *et al.*, 1999). Structural analysis of Dcx indicates that it contains a β-grasp superfold motif that can bind to tubulin and facilitate microtubule polymerization and stabilization (Taylor *et al.*, 2000). Point mutations within this tubulin-binding motif were seen in patients with double cortex syndrome (Gleeson *et al.*, 1999).

TABLE 1. Molecular Cues Affecting Radial and Tangential Neuronal Migration

Gene	Function and phenotype	References
Reelin	Essential for layer formation Inverted cortical layers	D'Arcangelo et al., 1995; Hirotsune et al., 1995; Ogawa et al., 1995; Hong et al., 2000
VLDLR/ ApoER2	Reelin receptors, critical for layer formation Inverted cortical layers	Trommsdorf et al., 1999
mDab1	Adapter protein component of reelin signaling cascade Inverted cortical layers	Howell et al., 1997; Sheldon et al., 1997; Ware et al., 1997
Astrotactin	Promotes neuron–radial glial adhesion Decreased rate of radial migration in mutants	Adams et al., 2002
Doublecortin	Critical for the stabilization of microtubule network Mutations lead to X-linked lissencephaly (double cortex syndrome) in humans	des Portes et al., 1998; Gleeson et al., 1998, 1999; Francis et al., 1999; Horesh et al., 1999
Lis1	Binds to microtubules, microtubule based motor protein, dyenin, and related microtubule interactors such as dynactin, NUDEL, and mNudE Mutations in Lis1 cause Miller–Dieker syndrome, a severe form of lissencephaly	Reiner et al., 1993; Hattori et al., 1994; Sapir et al., 1997; Efimov and Morris, 2000; Faulkner et al., 2000; Niethammer et al., 2000; Sasaki et al., 2000; Smith et al., 2000; Kitagawa et al., 2000
Cdk5	Facilitates neuronal migration to CP following splitting of the preplate Inverted layering in mutants	Oshima et al., 1996; Gilmore et al., 1998
p35/p39	Regulatory subunits of Cdk5 Similar in function to Cdk5	Chae et al., 1997; Kwon and Tsai, 1998; Ko et al., 2001
NRG1	Promotes radial migration Critical for the establishment of radial glial scaffold	Anton et al., 1999 Schmid et al., 2003
α_3 Integrin	Abnormal neuronal migration and laminar organization of cortex Neuron–glia interaction impaired Premature radial glial transformation	Anton et al., 1999; Kreidberg et al., 1996
α_6 Integrin	Lethal at birth Disorganized CP Ectopic neuroblasts in embryonic cortex Disorganized basal lamina assembly	Georges-Labouesse et al., 1996, 1998
α_3, α_6 Integrin	No neural tube closure Abnormal basal lamina assembly Multiple neuroblast ectopias in cortex	De Arcangelis et al., 1999
α_v Integrin	Survive until E14 or birth Intracortical hemorrhage Facilitates basic neuron–glial adhesion	Bader et al., 1998
β_1 Integrin (cond.)	Disrupted cortical laminar organization Radial glia end feet and pial basement membrane abnormalities	Graus-Porta et al., 2001
$\gamma1$ Laminin	Component of ECM in pial surface Misplaced neurons in CP	Halfter et al., 2002
Filamin α	Actin-binding protein, co-localizes to actin stress fibers, highly expressed by neural cells in the ventricular surface, crosslinks F-actin network to facilitate cell motility Needed to initiate migration from the ventricular zone Mutations cause periventricular heterotopia	Fox et al., 1998; Sheen et al., 2001; Moro et al., 2002
FILIP	Regulates degradation of Filamin α in the ventricular zone Prevents premature onset of migration	Nagano et al., 2002
Dlx1/2	Transcription factors regulating the differentiation of cortical and hippocampal interneurons from the subpallium Lateral ganglionic eminence (LGE)/medial ganglionic eminence (MGE)	Anderson et al., 1997; Pleasure et al., 2000

TABLE 1. (*Continued*)

Gene	Function and phenotype	References
Nkx 2.1	Transcription factor regulating the migration and differentiation of cortical interneurons from the MGE	Sussel *et al.*, 1999; Anderson *et al.*, 2001
TAG1	Neural cell-adhesion molecule expressed in corticofugal fibers Motogenic cue for tangentially migrating interneurons	Wolfer *et al.*, 1994; Denaxa *et al.*, 2001
HGF	Motogenic factor expressed in GE Promotes movement of cortical interneurons from MGE toward dorsal pallium	Powell *et al.*, 2001
u-PAR	Urokinase-type plasminogen activator receptor Enables HGF activation	Powell *et al.*, 2001
BDNF, NT-4	Motogenic factors for neuronal migration from MGE	Brunstrom *et al.*, 1997; Ringstedt *et al.*, 1998; Polleux *et al.*, 2002
TrkB	Receptor for BDNF/NT-4 Mutation leads to reduced interneuronal migration into cortex	Brunstrom *et al.*, 1997; Polleux *et al.*, 2002
Slit 1/2	Chemorepellent for GABAergic interneurons in the GE	Zhu *et al.*, 1999; Marin *et al.*, 2003
Sema3A/3F	Chemorepellent expressed in striatal mantle region Helps to channel cortical interneurons toward the cortex	Marin *et al.*, 2001; Tamamaki *et al.*, 2003
Nrp1/2	Receptors for class 3 secreted semaphorins Enables cortical interneurons to migrate away from striatum into the cortex	Marin *et al.*, 2001; Tamamaki *et al.*, 2003

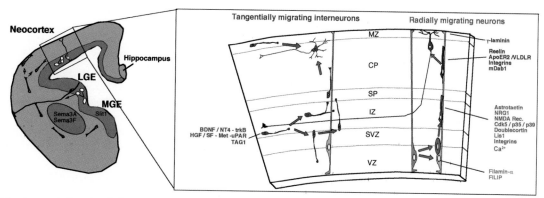

FIGURE 2. Molecular cues influencing distinct patterns of migration into the developing cerebral cortex. In the cerebral wall, neurons migrating tangentially into the cerebral cortex from ganglionic eminence and neurons migrating radially from the ventricular surface to the cortical plate are influenced by different sets of molecules on the right hand side panel. LGE-Lateral ganglionic eminence, MGE-medial ganglionic eminence, VZ-ventricular zone, SVZ-subventricular zone, IZ-intermediate, SP-subplate, CP-cortical plate, MZ-marginal zone.

Regulators of both actin and microtubule network associate with cyclin-dependent kinase-5 (Cdk5), expressed in migrating neurons and axon growth cones of the developing cortex (Nikolic *et al.*, 1998). Both filamin 1 and NUDEL are putative substrates for Cdk5 phosphorylation (Fox *et al.*, 1998; Niethammer *et al.*, 2000; Feng and Walsh, 2001). Mice deficient in Cdk5 and its activating subunits, p35 and p39, display abnormal neuronal migration and placement in cerebral cortex (Ohshima *et al.*, 1996; Chae *et al.*, 1997; Gilmore *et al.*, 1998; Kwon and Tsai, 1998; Ko *et al.*, 2001). Deficits in Brn1 and 2, class III POU domain transcription factors regulating p35 and p39 expression, also lead to cortical migrational abnormalities (McEvilly *et al.*, 2002). Interactions between p35 and β-catenin is thought to enable Cdk5 to regulate negatively *N*-cadherin-mediated adhesion and facilitate neuronal migration through the *N*-cadherin-rich developing cerebral wall up to the CP (Redies and Takeichi, 1993; Kwon *et al.*, 2000).

Transient, intracellular calcium fluxes that modulate neuronal migration *in vitro* can significantly influence the actin and microtubule network of neurons undergoing oriented migration (Rakic *et al.*, 1994). The link between extracelluar cues modulating migration and the internal cues involved in mechanics of migration is generated in a highly varied and redundant manner. For example, neurotransmitter receptors such as *N*-methyl-D-aspartate (NMDA) type glutamate receptors, and GABA receptors, or growth factors and their receptors such as neuregulins and its receptors erbB2, erbB3, and erbB4, or BDNF and its high-affinity receptor trkB, can promote radial-guided neuronal migration (Komuro and Rakic, 1993, 1996; Anton *et al.*, 1997; Rio *et al.*, 1997; Behar *et al.*, 2000, 2001). The most direct transmission of external cues via membrane receptors to cytoskeletal changes during migration is provided by integrins. Integrins are heterodimeric cell surface receptors that serve as structural links between the ECM and the internal cytoskeleton. Different integrin receptors display different adhesive properties, regulate different intracellular signal transduction pathways, and thus, different modes of adhesion-induced changes in cell physiology. Integrins are also capable of synergizing with other cell surface receptor systems to finely modulate a cell's behavior in response to multiple environmental cues. Developmental changes in the cell surface integrin repertoire and function may thus modulate distinct aspects of neuronal migration in the developing cerebral cortex by altering the strength and ligand preferences of cell–cell adhesion during development. Different α integrin subunits dimerize preferentially or exclusively with β_1 integrin, which is ubiquitously expressed in the developing cerebral cortex. The varied, yet distinct, cortical phenotypes of integrin subunit null mice provide striking insights into the distinct roles that cell–cell, cell–ECM adhesive interactions play in neuronal migration.

Mice homozygous for a targeted mutation in the α_3 integrin gene die during the perinatal period with severe defects in the development of the kidneys, lungs, skin, and cerebral cortex (Kreidberg *et al.*, 1996; Anton *et al.*, 1999). In the cerebral cortex, the normal laminar organization of neurons is lost, and neurons are positioned in a disorganized pattern. The α_3 integrin modulates neuron–glial recognition cues during neuronal migration and maintain neurons in a gliophilic mode until glial-guided neuronal migration is over and layer formation begins (Anton *et al.*, 1999). The gliophilic to neurophilic switch in the adhesive preference of developing neurons in the absence of α_3 integrin was hypothesized to underlie the abnormal cortical organization of α_3 integrin mutant mice. In contrast to α_3 integrin, α_v integrins appear to provide optimal levels of basic cell–cell adhesion needed to maintain neuronal migration and differentiation. Substantial disruption of cellular organization in cerebral wall and lateral ganglionic eminence (LGE) is seen at E11–12 in α_v null mice. Extensive intracerebral hemorrhage in α_v deficient mice, beginning at E12–13, prevents further evaluation of cortical development in late surviving (until birth) α_v null mice (Bader *et al.*, 1998). The α_v integrins expressed on radial glial cell surface can potentially associate with at least five different β subunits, β_1, β_3, β_5, β_6, and β_8. Adhesive interactions involving fibronectin, vitronectin, tenascin, collagen, or laminin, ECM

molecules that are found in the developing cerebral wall, can be mediated through these α_v-containing integrins (Cheresh *et al.*, 1989; Bodary and McLean, 1990; Moyle *et al.*, 1991; Hirsch *et al.*, 1994). Both transient cell-matrix interactions and cell-anchoring mechanisms that are mediated by different α_v-containing integrins and their respective ligands are likely to modulate the process of neuronal translocation in cerebral cortex.

In addition to α_3 integrin, some laminin isoforms in the developing cerebral cortex can also interact with α_6 integrin dimers (Georges-Labouesse *et al.*, 1998). The α_6 null mice die at birth (Georges-Labouesse *et al.*, 1996) with abnormal laminar organization of the cerebral cortex and retina (Georges-Labouesse *et al.*, 1998). Analysis of E13.5–E18.5 α_6 integrin-deficient embryos revealed ectopic neuronal distribution in the cortical plate, protruding out to the pial surface. The CP was further disorganized by wavy neurite outgrowth of ectopic neuroblasts. Coinciding abnormalities of laminin synthesis and deposition also occurs in the mutant brain. Persistence of glial laminin throughout development may have prevented neuroblasts from appropriately arresting their migration in the developing CP in α_6 null mice. Since cerebral cortex still formed in α_6 mutants, albeit abnormally, other integrin dimers may have overlapping functions with α_6 integrins during early cortical development. The similarities in the ligand preferences of α_3 and α_6 integrins are suggestive of potential functional overlap. The severe and novel cortical abnormalities in α_3, α_6 double knockout mutants, that is, disorganization of CP with large collection of ectopias, aberrant basal lamina organization, and abnormal choroid plexus, suggest a synergistic role for α_3 and α_6 integrins during cortical development (De Arcangelis *et al.*, 1999). Deficiency in β_4 integrin, which only associates with α_6, leads to an identical cortical phenotype. Mutations in α_6 or β_4 integrin in humans results in skin blistering (epidermolysis bullosa). However, the brain phenotype of the affected patients is unknown.

The β_1 integrin in the cerebral cortex can dimerize with at least 10 different α subunits; thus β_1 integrin deficiency leads to lethality from around E5.5 (Fassler and Meyer, 1995; Stephens *et al.*, 1995). Most of the cortical-specific α subunits seem to dimerize only with β_1 integrin. To study the role of β_1 integrin in the developing cortex, β_1 integrin-floxed mice were crossed with nestin-cre mice, resulting in widespread inactivation of β_1 integrins in cortical neurons and glia from around E10.5 (Graus-Porta *et al.*, 2001). Cortical layer formation is disrupted in these mice, in large part as a result of defective meningeal basement membrane assembly, marginal-zone formation, and glial end feet anchoring at the top of the cortex. BrdU birthdating studies suggest that glial-guided neuronal migration is not affected significantly. However, perturbed radial glial end feet development may contribute to the defective placement of neurons in the cortex. The varied cortical phenotypes of α_1, α_3, α_6, α_v, and β_1 null mice may reflect the transdominant, transnegative, or compensatory influences distinct integrin receptor dimers may exert over each other and the ECM ligands in the developing cerebral cortex. *In vitro*, binding of a ligand to a signal transducing integrin or inactivation of signaling through a particular integrin can initiate a unidirectional signaling cascade affecting the function of the target integrin in the same cell (Simon *et al.*, 1997; Hodivala-Dilke *et al.*, 1998; Blystone *et al.*, 1999). Elucidation of whether such integrin crosstalk regulates patterns of neuronal development and interactions with specific ECM molecules in the developing cortices of different integrin null mice will be informative in fully characterizing the role of integrins in neuronal migration.

Termination of Migration

Once neurons reach the top of the CP, the movement of neurons stops abruptly at the interface between the CP and the cell sparse marginal zone and cohorts of neurons begin to assemble into their respective layers. This final stage of neuronal migration is the least explored aspect of neuronal migration, in spite of its significance for genetic and acquired cortical malformations (Rakic, 1988; Rakic and Caviness, 1995; Olson and Walsh, 2002). The signal to terminate neuronal cell migration is thought to be provided either by the afferent fibers that migrating neurons encounter near their target location or by the ambient neuronal cell population that had already reached its final position (Sidman and Rakic, 1973; Hatten and Mason, 1990; D'Arcangelo *et al.*, 1995; Ogawa *et al.*, 1995). Alternatively, a change in the cell surface properties of the radial glial substrate at the top of the CP may signal a migrating neuron to stop, detach, and differentiate.

In the *reeler* mouse, deficits in this phase of migration have led to disorganized, inverted cortex, with early-born neurons occupying abnormally superficial positions and later-born neurons adopting abnormally deep positions (Caviness *et al.*, 1972; Caviness and Sidman, 1973; Lambert de Rouvroit and Goffinet, 1998). The inversion of final neuronal positions in the CP of the *reeler* mouse has made it a prototype model for the analysis of mechanisms controlling the final phase of neuronal migration, that is, how neurons disengage from a migratory mode to assemble into distinct layers. The *reeler* locus encodes Reelin, a 388 kDa secreted protein composed of a unique N-terminal sequence with similarity to F-spondin, followed by a series of eight 350–390 amino acid "Reelin repeats" each containing an EGF domain with homology to ECM proteins like Tenascin C (D'Arcangelo *et al.*, 1995; Hirotsune *et al.*, 1995). Reelin acts on noncell autonomously (Ogawa *et al.*, 1997), and the protein is synthesized and secreted in the cerebral cortex predominantly by the Cajal–Retzius (CR) cell of the marginal zone, the outermost layer of the developing cortex (D'Arcangelo *et al.*, 1995; Ogawa *et al.*, 1995).

Mutations in three molecules, VLDLR, ApoER2, and Dab1, have been found to phenocopy almost exactly the effects of the *reeler* gene mutation, suggesting that the corresponding proteins represent a reelin regulated biochemical pathway that mediates proper termination of neuronal migration and formation of cerebral cortical lamination (Gonzalez *et al.*, 1997; Howell *et al.*, 1997; Sheldon *et al.*, 1997; Ware *et al.*, 1997). The *dab1* gene encodes a cytoplasmic adapter protein (Dab1) expressed by neurons in the developing CP, suggesting that Dab1 represents a link in the signaling pathway that receives the Reelin

signal. This idea is confirmed by observation that Reelin expression is normal in the *dab1* mutant cortex (Gonzalez *et al.*, 1997) but Dab1 protein accumulates in the *reeler* mouse brain (Rice *et al.*, 1998) and Dab1 is phosphorylated in response to application of recombinant Reelin (Howell *et al.*, 1999a). Mammalian Dab1 was identified through a two-hybrid screen using the non-receptor tyrosine kinase Src as "bait" (Howell *et al.*, 1997) and found to have homology with Drosophila *disabled* (Gertler *et al.*, 1993). Dab1 has an N-terminal alpha helical structure and the critical amino acids of a protein interaction/phosphotyrosine-binding domain (PI/PTB) (Kavanaugh and Williams, 1994; Borg *et al.*, 1996; Margolis, 1996; Howell *et al.*, 1997). The PI/PTB domain of Dab1 binds proteins that contain an NPXY motif (Howell *et al.*, 1997, 1999b; Trommsdorff *et al.*, 1998) a motif that has been implicated in clathrin-meditated endocytosis (Chen *et al.*, 1990), and integrin signaling (Law *et al.*, 1999).

More recently, mice with compound mutations in both VLDLR and ApoER2 have been found to have a phenotype indistinguishable from *reeler* and *dab1* mutants (Trommsdorff *et al.*, 1999). VLDLR and ApoER2 are members of the low density lipoprotein (LDL) receptor superfamily that interacted with Dab1 in two-hybrid screens through the PI/PTB domain of Dab1 and the NPXY motif of LDL superfamily members (Trommsdorff *et al.*, 1998). The NPXY motif of LDL receptor family members is essential for clathrin-mediated endocytosis (Chen *et al.*, 1990). The implication of VLDLR and ApoER2 as potential Reelin receptors was surprising since LDL superfamily members are well characterized as mediating the endocytosis of specific ligands, but have never demonstrated a direct signaling function. Recent studies, however, have clearly demonstrated that both recombinant ApoER2 and the VLDLR bind Reelin and that this binding leads both to the tyrosine phosphorylation of Dab1 and in the case of VLDLR, the internalization of the receptor and Reelin (D'Arcangelo *et al.*, 1999; Hiesberger *et al.*, 1999). Thus there is compelling evidence that Reelin, VLDLR, ApoER2, and Dab1 function in a common signaling pathway between CR cells and CP neurons, but the downstream molecules that mediate Reelin signaling effect on either migration or adhesion of cortical neurons remains unclear.

Reelin's effect on cortical layering is hypothesized to result from three distinct cellular effects. First, reelin may regulate CP organization by initiating the splitting of preplate into marginal zone and subplate. Failure of this process in *reeler* mutants leads to the accumulation of cortical neurons underneath the preplate neurons. Second, a reelin gradient may act as an attractant for neurons to the top of the CP, thus enabling newly generated neurons to migrate past earlier generated ones in the developing CP. Third, reelin may induce detachment of neurons from their radial glial guides and thus end neuronal migration at the marginal zone-developing CP interface and initiate the differentiation of neurons into distinct layers.

Cortical neurons in β_1 integrin or laminin γ_1 nidogen-binding site (Halfter *et al.*, 2002) deficient mice invade the marginal zone in areas devoid of reelin producing CR cells, and in regions with CR cell ectopias, accumulate underneath them, within the CP. Invasion of neurons into areas devoid of reelin-producing CR cells supports a role for reelin in normal termination of neuronal migration. Furthermore, reelin appears to facilitate detachment of migrating neurons from glial guides *in vitro* and in the rostral migratory stream (RMS) (Hack *et al.*, 2002). The reelin-induced detachment of embryonic cortical neurons from glial guides *in vitro* depends on α_3 integrin signaling. It is hypothesized that during glial-guided migration to the CP neuronal α_3 integrin may interact with glial cell surface molecules such as fibronectin or laminin-2, and at the top of the CP, the ligand preference of α_3 integrins may change from radial glial cell surface ECM molecules to reelin. Different ligands or ligand concentration can determine the surface levels of integrins by regulating the rate at which integrin receptor is removed from the cell surface. Ligands can also regulate polarized flow of integrins toward or away from growth cone membranes. Reelin can also function as serine protease and degrade fibronectin and laminin normally used to maintain glial-based migration (Quattrocchi *et al.*, 2002). Thus changes in the availability, function, and ligand preference of α_3 integrins or reelin proteolytic activity may trigger the decrease in a migrating neuron's bias for gliophilic adhesive interactions and promote neurophilic interactions needed for neurons to detach from radial glial guides and organize into distinct layers. Interestingly, deficiencies in α_3 integrin ligands, laminin-2 and reelin lead to cortical anomalies such polymicrogyria or lissencephaly (Sunada *et al.*, 1995; Hong *et al.*, 2000).

TANGENTIAL MIGRATION IN THE FOREBRAIN

As introduced earlier in this chapter, two main types of migration are classically opposed during the development of the central nervous system: radial vs tangential migration. Radial migration is characterized by close interactions between migrating neurons and the processes of radial glial cells which constitute a scaffold bridging the proliferating neuroepithelium and the differentiating zone. By definition, tangential migration is referred to as a nonradial mode of neuronal translocation that does not require specific interaction with radial glial cell processes. Until recently, the predominant view was that the vast majority of neurons in the forebrain where generated through radial migration (Sidman and Rakic, 1973). The first evidence to suggest the need for a revised model came from observations of tangential dispersion of precursors or postmitotic neurons in the developing cortex (O'Rourke *et al.*, 1992a, 1995; Fishell *et al.*,

1993a; Tan and Breen, 1993; Tan *et al.*, 1995; de Carlos *et al.*, 1996). The widespread distribution of clonally related cells also suggested the possibility of non-radial migration in the cortex (Walsh and Cepko, 1992). In an elegant study, Parnavelas and collaborators coupled retroviral-mediated lineage-tracing studies with the determination of neuronal subtype identity and demonstrated a tight correlation between cell dispersion and neuronal subtype (Parnavelas *et al.*, 1991): most excitatory, glutamatergic pyramidal neurons are produced locally by a set of precursors migrating radially in the cortex, whereas most GABAergic, nonpyramidal neurons were produced by a set of progenitors migrating tangentially (Parnavelas, 2000).

Origin of Tangentially Migrating Cells in the Forebrain

The source and destination of these tangentially migrating cells, however, remained a mystery until experiments by Anderson *et al.* suggested that neurons migrated from the GE to the cortex where they gave rise preferentially to GABAergic interneurons (Anderson *et al.*, 1997; Tamamaki *et al.*, 1997). This conclusion is based mainly upon the observation that there are virtually no neocortical GABAergic neurons in *Dlx1/2* double knockout mice, two homeobox transcription factors expressed in the ventricular and subventricular zones of the GE (Anderson *et al.*, 1997). The GE is located in the ventral part of the telencephalon and is producing neurons of the basal ganglia (Fentress *et al.*, 1981; Qiu *et al.*, 1995). This ventral structure can be divided into three subregions using neuroanatomical and molecular criteria: the medial, the lateral, and the caudal parts (Corbin *et al.*, 2000). Several transcription factors are differentially expressed in these three regions (Table 2).

Recent *in utero* homotopic transplantation experiments performed in mice have revealed that these distinct regions give rise to specific neuronal populations displaying strikingly different patterns of cell migration (Fig. 3): the *medial* GE gives rise to the majority of GABAergic interneurons of the cortex and hippocampus (Lavdas *et al.*, 1999; Anderson *et al.*, 2001; Wichterle *et al.*, 2001; Polleux *et al.*, 2002) whereas precursors in the *lateral* GE generates projecting medium spiny neurons of the striatum, nucleus accumbens and olfactory tubercle and to the granule and periglomerular cells in the olfactory bulb (Wichterle *et al.*, 2001). The pattern of migration of neurons originating in the caudal GE is less well characterized but it has recently been shown that precursors in this region give rise to interneurons found in layer 5 of the neocortex, various regions of the limbic system and also neurons of the striatum (Nery *et al.*, 2002).

TABLE 2. Transcription Factors Expression in Different Subregions of the Ganglionic Eminence

	Mash1	*Dlx1/Dlx2*	*Nkx2.1*	*Lhx6*	*Gsh2*
MGE	+	+	+	+	−
LGE	+	+	−	−	+
CGE	+	+	?	?	?

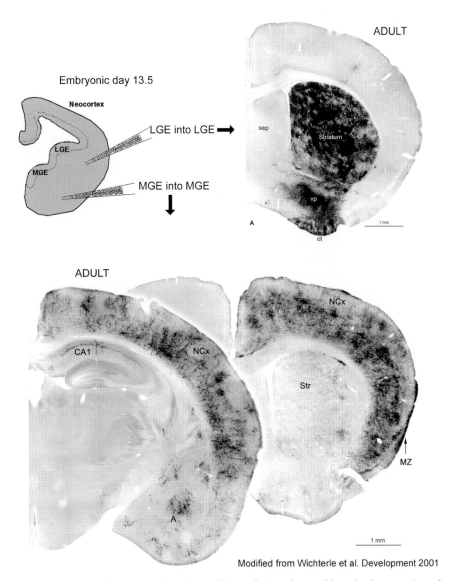

FIGURE 3. Generation and migration of cortical interneurons from the medial ganglionic eminence. Disssociated neurons (tagged with alkaline phosphatase) isolated from LGE or MGE were transplanted homotopically into LGE or MGE, respectively, at early stages of neuronal migration in cortex. Location and differentiation of transplanted neurons were analyzed in adult brains. Strikingly, MGE cells all migrated into cerebral cortex to become cortical interneurons, whereas LGE cells populated the striatum. LGE, lateral ganglionic eminence; MGE, medial ganglionic eminence. Modified with permission from Wichterle *et al.*, 2001.

Cellular and Molecular Substrates for Tangential Migration of Cortical Interneurons

Tangentially migrating interneurons display a characteristic unipolar morphology during translocation with a long leading process dragging behind their nucleus (Fig. 2) (Anderson *et al.*, 1997; Tamamaki *et al.*, 1997; Polleux *et al.*, 2002). Interneurons are migrating tangentially through the intermediate zone or the marginal zone, two axon-rich layers located, respectively, deep and superficial, relative to the CP, where all neurons accumulate in a layer-specific manner to undergo their terminal differentiation (O'Leary and Nakagawa, 2002).

During migration to the cortex, tangential migrating interneurons are not using radial glial cells processes as a scaffold during translocation and these cells do not appear to fasciculate along a specific cellular scaffold (Polleux *et al.*, 2002) although it has been proposed that they interact with corticofugal axons (Denaxa *et al.*, 2001). *In vitro*, the neural cell-adhesion molecule TAG-1 (also called contactin-2) expressed by corticofugal axons has been shown to play a role in the control of interneuron migration.

Extracellular Cues Regulating Tangential Migration in the Forebrain

The extracellular cues controlling the tangential migration of interneurons from the GE to the cortex can be classified in three categories: (1) extracellular cues regulating their motility (motogenic cues), (2) directional cues guiding their migration toward the appropriate territories, and (3) stop-signals abolishing their motility and therefore dictating where interneurons should terminally differentiate.

Cues Controlling the Motility of Tangentially Migrating Interneurons

Several factors expressed along the migrating pathway of cortical interneurons have recently been shown to be potent stimulators of interneurons motility. Both the hepatocyte growth factor (HGF, also called scatter factor) and the neurotrophin NT4/5 are expressed in the cortex during mouse embryogenesis and are potent stimulators of interneurons migration (Behar et al., 1997; Brunstrom et al., 1997; Powell et al., 2001; Polleux et al., 2002). Neurons migrating tangentially from the MGE to the cortex express c-Met and trkB, the high-affinity receptors for HGF and NT4, respectively. Furthermore, mice mutant for urokinase-type plasminogen activator receptor (u-PAR), a key component of HGF activation, exhibit reduced interneuron migration to the frontal and parietal cortex (Powell et al., 2001). This decreased number of interneurons in the cortex of u-PAR knockout mice has important behavioral consequences on the establishment of the normal cortical circuitry characterized by an imbalanced level of excitation and inhibition which leads to epilepsy (Powell et al., 2003). Mice presenting a targeted deletion of the tyrosine kinase receptor trkB, the high-affinity receptor of NT4, also present a significant reduction of the number of interneurons migrating from the MGE to the cortex (Polleux et al., 2002). The motogen activity resulting from the activitiation of these tyrosine kinase receptors (c-Met and trkB) is likely to be mediated through their ability to activate phosphoinositide 3-(PI3-)kinase (Polleux et al., 2002), a key regulator of cytoskeleton reorganization and cell motility in nonneuronal cell types (Iijima et al., 2002).

Guidance Cues (Sema 3A and Sema 3F; Slits)

Several axon guidance cues have been shown to play a role in directing interneuron migration from the GE to the cortex. The diffusible chemorepulsive Sema3A and Sema3F are expressed in the postmitotic mantle region of the developing striatum and migrating interneurons from the MGE express Neuropilin 1 (Npn1) and Neuropilin 2 (Npn2) (Marin et al., 2001; Tamamaki et al., 2003), Sema3A and -3F respective receptors (Chen et al., 1997; Kolodkin et al., 1997; Giger et al., 1998). In vitro experiments demonstrate that MGE-derived interneurons are repulsed by Sema3A and Sema3F in a cooperative manner. Furthermore, the in vivo analysis of mice presenting targeted deletion of Npn1 and Npn2 demonstrate that they are required for the selective avoidance of the striatum by cortical interneurons and therefore for the directed migration to the cortex (Marin et al., 2001; Tamamaki et al., 2003).

Slit1 and Slit2, another short-range chemorepulsive cue for axons expressed in the ventricular zone of the GE as well as in the medial part of GE, has been shown to repulse MGE-derived interneurons in vitro (Zhu et al., 1999). However, Slit 1/2 double knockout mice do not show any defect of guided migration toward the cortex but nevertheless show a defect in the position of specific interneuronal population within the basal telencephalon, close to the midline (Marin et al., 2003). The cortex exerts a chemoattractive activity on migrating interneurons but these cortex-derived cues remains to be identified.

Finally, membrane-bound cell-adhesion molecules, cadherins, delineate sharp territories of expression restricted to the dorsal telencephalon (R-Cadherin) and the LGE (Cadherin-6) in E10–11 developing mouse embryos. Evidence using both electroporation-mediated ectopic expression of cadherins or the in vivo analysis of Cadherin-6 knockout mice demonstrate its role in the appropriate sorting of striatal and cortical neuronal populations (Inoue et al., 2001).

Stop-Signals

Once migrating interneurons have reached the CP, they are targeting specific layers according to their birthdate just as radially migrating neurons do (Fairen et al., 1986). So far, few molecules have been characterized for their capacity to stop the motility of tangentially migrating interneurons and even less is known about the putative cues that coordinate the layer-specific targeting of these two populations of neurons. Interestingly, several studies have shown that tangentially migrating neurons are expressing functional calcium-permeable AMPA receptors (but not NMDA receptors) which could be activated by glutamate released from corticofugal axons (Metin et al., 2000) and/or GABA released from tangentially migrating interneurons themselves (Poluch and Konig, 2002). Both GABA and glutamate have been shown to control the motility of migrating neurons in the developing cortex (Behar et al., 1996, 1998, 1999, 2000) and AMPA receptor activation leads to neurite retraction and is sufficient to stop migration of cortical interneurons in embryonic slice cultures (Poluch et al., 2001). Because the neurotransmitter glutamate is expressed at high levels in the CP (Behar et al., 1999), it could trigger an AMPA-receptor-dependent calcium influx that could act as a stop-signal for tangentially migrating interneurons in their final cortical environment. Further work will be necessary to validate this model meanwhile the identity of the cues leading to the coordinated, layer-specific accumulation of interneurons and excitatory glutamaergic neurons remains mysterious and the center of a lot of attention.

Differences between the Pattern of Tangential Migration in Rodents and Humans

There might be important differences between the pattern of tangential migration of GE-derived interneurons between rodent and human brain. Recent work demonstrate that a contingent of GE-derived interneurons migrate medially from the ventral telencephalon to the diencephalon in the human developing brain but not in nonhuman primate or in mouse embryos (Letinic and Rakic, 2001). Moreover, in the human brain retroviral lineage studies performed in vitro suggest that a substantial proportion of cortical

GABAergic neurons are generated in the dorsal telencephalon (Letinic *et al.*, 2002) in contrast with what is observed in the embryonic mouse telencephalon (Anderson *et al.*, 1997a). In the human embryonic brain, dividing precursors located in the dorsal telencephalon are expressing *Mash1* and *Dlx1/2* and have been shown to be competent to generate GABAergic interneurons in the cortical neuroepithelium of human-primates (Letinic *et al.*, 2002) but not in rodents (Fode *et al.*, 2000). This suggests that modifications in the expression pattern of transcription factors in the forebrain could underlie species-specific programs for the generation of neocortical local circuit neurons (Letinic *et al.*, 2002).

Other Structures Displaying Nonradial Migration

Many other regions of the developing central nervous system are characterized by nonradial neuronal migration, including the RMS of olfactory interneurons and the tangential migration of granule cells in the cerebellum.

The Rostral Migratory Stream (RMS)

Precursors of the two populations of olfactory interneurons (periglomerular and granule cells) are not produced within the olfactory bulb but are generated by precursors located in the LGE during embryonic development (Altman, 1969; Lois and Alvarez-Buylla, 1993, 1994; Lois *et al.*, 1996; Dellovade *et al.*, 1998; Sussel *et al.*, 1999; Corbin *et al.*, 2000; Wichterle *et al.*, 2001). This migration is unique because it continues throughout adulthood in rodents providing a constant number of GABAergic neurons to the olfactory bulb. In the adult brain, olfactory interneurons are generated from the subependymal layer lining the lateral ventricles, a proliferative epithelium deriving from the subventricular zone of the embryonic GE (Doetsch and Alvarez-Buylla, 1996; Doetsch *et al.*, 1997, 1999a, b).

The interneurons migrating in the adult RMS are also unique with regard to their neurophilic rather than gliophilic mode of migration, requiring interactions between migrating interneurons (Lois and Alvarez-Buylla, 1994; Lois *et al.*, 1996). When explanted *in vitro*, these interneurons form chains by migrating along each other. This so-called *chain migration* is dependent of the expression of specific cell-adhesion molecules of the immunoglobulin superfamily such as the polysialylated form of neural cell-adhesion molecule (PSA-NCAM; reviewed in Marin *et al.*, 2003).

Cerebellar Granule Cell Migration

Another population of interneurons migrates nonradially in the cerebellum: the granule cells migrating from the external granule layer (EGL) to the internal granular layer (IGL) during early postnatal stages of rodent development (reviewed in Hatten, 1999). The rate of migration of cerebellar granule neurons also is modulated through the control of intracellular calcium levels by activation of NMDA-, AMPA-, and somatostatin receptors

(Komuro and Rakic, 1992, 1993, 1996; Yacubova and Komuro, 2002). Activation of somatostatin receptors increases the rate of granule cell migration near their site of birth in the EGL, but decreases their rate of migration near their final destination in the IGL. Correspondingly, the size and frequency of spontaneous $Ca2+$ fluctuations is enhanced by somatostatin in the early phase of migration, whereas spike-like $Ca2+$ transients are eliminated by somatostatin in the late phase (Yacubova and Komuro, 2002).

This mode of migration is characterized by a dynamic switch between tangential and radial mode of migration: after translocation in the superficial EGL, granule interneurons make a sharp 90° turn to migrate along the radial processes of Bergmann glia spanning the molecular layer, to reach the deep IGL where they will undergo terminal differentiation (Komuro and Rakic, 1995). This switch from tangential to radial mode of translocation is not unique to cerebellar granule cells but is also observed for cortical interneurons (Polleux *et al.*, 2002) and is likely to reflect a basic property of migrating interneurons.

REFERENCES

Altman, J., 1969, Autoradiographic and histological studies of postnatal neurogenesis. 3. Dating the time of production and onset of differentiation of cerebellar microneurons in rats, *J. Comp. Neurol.* 136:269–293.

Anderson, S.A., Eisenstat, D.D., Shi, L., and Rubenstein, J.L., 1997, Interneuron migration from basal forebrain to neocortex: Dependence on Dlx genes, *Science* 278:474–476.

Anderson, S.A., Marin, O., Horn, C., Jennings, K., and Rubenstein J.L., 2001, Distinct cortical migrations from the medial and lateral ganglionic eminences, *Development* 128:353–363.

Anton, E.S., Cameron, R.S., and Rakic, P., 1996, Role of neuron–glial junctional domain proteins in the maintenance and termination of neuronal migration across the embryonic cerebral wall, *J. Neurosci.* 16:2283–2293.

Anton, E.S., Kreidberg, J.A., and Rakic, P., 1999, Distinct functions of alpha3 and alpha(v) integrin receptors in neuronal migration and laminar organization of the cerebral cortex, *Neuron* 22:277–289.

Anton, E.S., Marchionni, M.A., Lee, K.F., and Rakic, P., 1997, Role of GGF/neuregulin signaling in interactions between migrating neurons and radial glia in the developing cerebral cortex, *Development* 124:3501–3510.

Antonicek, H., Persohn, E., and Schachner, M., 1987, Biochemical and functional characterization of a novel neuron–glia adhesion molecule that is involved in neuronal migration, *J. Cell. Biol.* 104:1587–1595.

Bader, B.L., Rayburn, H., Crowley, D., and Hynes, R.O., 1998, Extensive vasculogenesis, angiogenesis, and organogenesis precede lethality in mice lacking all alpha v integrins, *Cell* 95:507–519.

Behar, T.N., Schaffner, A.E., Scott, C.A., O'Connell, C., and Barker, J.L., 1998, Differential response of cortical plate and ventricular zone cells to GABA as a migration stimulus, *J. Neurosci.* 18:6378–6387.

Behar, T.N., Schaffner, A.E., Scott, C.A., Greene, C.L., and Barker, J.L., 2000, GABA receptor antagonists modulate postmitotic cell migration in slice cultures of embryonic rat cortex, *Cereb. Cortex* 10:899–909.

Behar, T.N., Smith, S.V., Kennedy, R.T., McKenzie, J.M., Maric, I., and Barker, J.L., 2001, GABA(B) receptors mediate motility signals for migrating embryonic cortical cells, *Cereb. Cortex* 11:744–753.

Behar, T.N., Li, Y.X., Tran, H.T., Ma, W., Dunlap, V., Scott, C. *et al.*, 1996, GABA stimulates chemotaxis and chemokinesis of embryonic cortical neurons via calcium-dependent mechanisms, *J. Neurosci.* 16: 1808–1818.

Behar, T.N., Scott, C.A., Greene, C.L., Wen, X., Smith, S.V., Maric, D. *et al.*, 1999, Glutamate acting at NMDA receptors stimulates embryonic cortical neuronal migration, *J. Neurosci.* 19:4449–4461.

Behar, T.N., Dugich-Djordjevic, M.M., Li, Y.X., Ma, W., Somogyi, R., Wen, X. *et al.*, 1997, Neurotrophins stimulate chemotaxis of embryonic cortical neurons, *Eur. J. Neurosci.* 9:2561–2570.

Blystone, S.D., Slater, S.E., Williams, M.P., Crow, M.T., and Brown, E.J., 1999, A molecular mechanism of integrin crosstalk: Alphavbeta3 suppression of calcium/calmodulin-dependent protein kinase II regulates alpha5beta1 function, *J. Cell. Biol.* 145:889–897.

Bodary, S.C. and McLean, J.W., 1990, The integrin beta 1 subunit associates with the vitronectin receptor alpha v subunit to form a novel vitronectin receptor in a human embryonic kidney cell line, *J. Biol. Chem.* 265:5938–5941.

Borg, J.P., Ooi, J., Levy, E., and Margolis, B., 1996, The phosphotyrosine interaction domains of X11 and FE65 bind to distinct sites on the YENPTY motif of amyloid precursor protein, *Mol. Cell. Biol* 16: 6229–6241.

Brunstrom, J.E., Gray-Swain, M.R., Osborne, P.A., and Pearlman, A.L., 1997, Neuronal heterotopias in the developing cerebral cortex produced by neurotrophin-4, *Neuron* 18:505–517.

Cahana, A., Escamez, T., Nowakowski, R.S., Hayes, N.L., Giacobini, M., von Holst, A. *et al.*, 2001, Targeted mutagenesis of Lis1 disrupts cortical development and LIS1 homodimerization, *Proc. Natl. Acad. Sci. USA.* 98:6429–6434.

Cameron, R.S. and Rakic, P., 1994, Identification of membrane proteins that comprise the plasmalemmal junction between migrating neurons and radial glial cells, *J. Neurosci.* 14:3139–3155.

Caspi, M., Atlas, R., Kantor, A., Sapir, T., and Reiner, O., 2000, Interaction between LIS1 and doublecortin, two lissencephaly gene products, *Hum. Mol. Genet.* 9:2205–2213.

Caviness, V.S., 1976, Patterns of cell and fiber ditribution in the neocortex of the reeler mutant mouse, *J. Comp. Neurol.* 170:435–448.

Caviness, V.S., Jr., So, D.K., and Sidman, R.L., 1972, The hybrid reeler mouse, *J. Heredity* 63:241–246.

Caviness, V.S., Jr. and Sidman, R.L., 1973, Time of origin or corresponding cell classes in the cerebral cortex of normal and *reeler* mutant mice: An autoradiographic analysis, *J. Comp. Neurol.* 148:141–151.

Chae, T., Kwon, Y.T., Bronson, R., Dikkes, P., Li, Ew., and Tsai, L.H., 1997, Mice lacking p35, a neuronal specific activator of Cdk5, display cortical lamination defects, seizures, and adult lethality, *Neuron* 18: 29–42.

Chazal, G., Durbec, P., Jankovski, A., Rougon, G., and Cremer, H., 2000, Consequences of neural cell adhesion molecule deficiency on cell migration in the rostral migratory stream of the mouse, *J. Neurosci.* 20:1446–1457.

Chen, H., Chedotal, A., He, Z., Goodman, C.S., and Tessier-Lavigne, M., 1997, Neuropilin-2, a novel member of the neuropilin family, is a high affinity receptor for the semaphorins Sema E and Sema IV but not Sema III, *Neuron* 19:547–559.

Chen, W.J., Goldstein, J.L., and Brown, M.S., 1990, NPXY, a sequence often found in cytoplasmic tails, is required for coated pit-mediated internalization of the low density lipoprotein receptor, *J. Biol. Chem.* 265: 3116–3123.

Cheresh, D.A., Smith, J.W., Cooper, H.M., and Quaranta, V., 1989, A novel vitronectin receptor integrin (alpha v beta x) is responsible for distinct adhesive properties of carcinoma cells, *Cell* 57:59–69.

Chuong, C.M., Crossin, K.L., and Edelman, G.M., 1987, Sequential expression and differential function of multiple adhesion molecules during the formation of cerebellar cortical layers, *J. Cell. Biol.* 104:331–342.

Colognato, H. and Yurchenco, P.D., 2000, Form and function: The laminin family of heterotrimers, *Dev. Dyn.* 218:213–234.

Corbin, J.G., Gaiano, N., Machold, R.P., Langston, A., and Fishell, G., 2000, The Gsh2 homeodomain gene controls multiple aspects of telencephalic development, *Development* 127:5007–5020.

Cousin, B., Leloup, C., Penicaud, L., and Price, J., 1997, Developmental changes in integrin beta-subunits in rat cerebral cortex, *Neurosci. Lett.* 234:161–165.

D'Arcangelo, G., Miao, G.G., Chen, S.C., Soares, H.D., Morgan, J.I., and Curran, T., 1995, A protein related to extracellular matrix proteins deleted in the mouse mutant *reeler*, *Nature* 374:719–723.

D'Arcangelo, G., Homayouni, R., Keshvara, L., Rice, D.S., Sheldon, M., and Curran, T., 1999, Reelin is a ligand for lipoprotein receptors, *Neuron* 24:471–479.

De Arcangelis, A., Mark, M., Kreidberg, J., Sorokin, L., and Georges-Labouesse, E., 1999, Synergistic activities of alpha3 and alpha6 integrins are required during apical ectodermal ridge formation and organogenesis in the mouse, *Development* 126: 3957–3968.

de Carlos, J.A., Lopez-Mascaraque, L., and Valverde, F., 1996, Dynamics of cell migration from the lateral ganglionic eminence in the rat, *J. Neurosci.* 16:6146–6156.

Dellovade, T.L., Pfaff, D.W., and Schwanzel-Fukuda, M., 1998, Olfactory bulb development is altered in small-eye (Sey) mice, *J. Comp. Neurol.* 402:402–418.

Denaxa, M., Chan, C.H., Schachner, M., Parnavelas, J.G., and Karagogeos, D., 2001, The adhesion molecule TAG-1 mediates the migration of cortical interneurons from the ganglionic eminence along the corticofugal fiber system, *Development* 128:4635–4644.

des Portes, V., Francis, F., Pinard, J.M., Desguerre, I., Moutard, M.L., Snoeck, I. *et al.*, 1998, Doublecortin is the major gene causing X-linked subcortical laminar heterotopia (SCLH), *Hum. Mol. Genet.* 7: 1063–1070.

Doetsch, F. and Alvarez-Buylla, A., 1996, Network of tangential pathways for neuronal migration in adult mammalian brain, *Proc. Natl. Acad. Sci. USA* 93:14895–14900.

Doetsch, F., Garcia-Verdugo, J.M., and Alvarez-Buylla, A., 1997, Cellular composition and three-dimensional organization of the subventricular germinal zone in the adult mammalian brain, *J. Neurosci.* 17: 5046–5061.

Doetsch, F., Garcia-Verdugo, J.M., and Alvarez-Buylla, A., 1999a, Regeneration of a germinal layer in the adult mammalian brain, *Proc. Natl. Acad. Sci. USA* 96:11619–11624.

Doetsch, F., Caille, I., Lim, D.A., Garcia-Verdugo, J.M., and Alvarez-Buylla, A., 1999b, Subventricular zone astrocytes are neural stem cells in the adult mammalian brain, *Cell* 97:703–716.

Edmondson, J.C., Liem, R.K., Kuster, J.E., and Hatten, M.E., 1988, Astrotactin: A novel neuronal cell surface antigen that mediates neuron–astroglial interactions in cerebellar microcultures, *J. Cell Biol.* 106:505–517.

Efimov, V.P. and Morris, N.R., 2000, The LIS1-related NUDF protein of *Aspergillus nidulans* interacts with the coiled-coil domain of the NUDE/RO11 protein, *J. Cell. Biol.* 150:681–688.

Eksioglu, Y.Z., Scheffer, I.E., Cardenas, P., Knoll, J., DiMario, F., Ramsby, G. *et al.*, 1996, Periventricular heterotopia: An X-linked dominant epilepsy locus causing aberrant cerebral cortical development, *Neuron* 16:77–87.

Fairen, A., Cobas, A., and Fonseca, M., 1986, Times of generation of glutamic acid decarboxylase immunoreactive neurons in mouse somatosensory cortex, *J. Comp. Neurol.* 251:67–83.

Falconer, M.M., Echeverri, C.J., and Brown, D.L., 1992, Differential sorting of beta tubulin isotypes into colchicine-stable microtubules during neuronal and muscle differentiation of embryonal carcinoma cells, *Cell Motil. Cytoskeleton* 21:313–325.

Fassler, R. and Meyer, M., 1995, Consequences of lack of beta 1 integrin gene expression in mice, *Genes Dev.* 9:1896–1908.

Faulkner, N.E., Dujardin, D.L., Tai, C.Y., Vaughan, K.T., O'Connell, C.B., Wang, Y. *et al.*, 2000, A role for the lissencephaly gene LIS1 in mitosis and cytoplasmic dynein function, *Nat. Cell. Biol.* 2:784–791.

Feng, Y. and Walsh, C.A., 2001, Protein-protein interactions, cytoskeletal regulation and neuronal migration *Nat. Rev. Neurosci.* 2:408–416.

Feng, Y., Olson, E.C., Stukenberg, P.T., Flanagan, L.A., Kirschner, M.W., and Walsh, C.A., 2000, LIS1 regulates CNS lamination by interacting with mNudE, a central component of the centrosome, *Neuron* 28: 665–679.

Fentress, J.C., Stanfield, B.B., and Cowan, W.M., 1981, Observation on the development of the striatum in mice and rats, *Anat. Embryol. (Berl)* 163:275–298.

Fishell, G., Mason, C.A., and Hatten, M.E., 1993a, Dispersion of neural progenitors within the germinal zones of the forebrain, *Nature* 362: 636–638.

Fishell, G. and Kriegstein, A.R., 2003, Neurons from radial glia: The consequences of asymmetric inheritance, *Curr. Opin. Neurobiol.* 13:34–41.

Fishman, R.B. and Hatten, M.E., 1993, Multiple receptor systems promote CNS neural migration, *J. Neurosci.* 13:3485–3495.

Fode, C., Ma, Q., Casarosa, S., Ang, S.L., Anderson, D.J., and Guillemot, F., 2000, A role for neural determination genes in specifying the dorsoventral identity of telencephalic neurons, *Genes. Dev.* 14: 67–80.

Fox, J.W., Lamperti, E.D., Eksioglu, Y.Z., Hong, S.E., Feng, Y., Graham, D.A. *et al.*, 1998, Mutations in filamin 1 prevent migration of cerebral cortical neurons in human periventricular heterotopia, *Neuron* 21: 1315–1325.

Francis, F., Koulakoff, A., Boucher, D., Chafey, P., Schaar, B., Vinet, M.C. *et al.*, 1999, Doublecortin is a developmentally regulated, microtubule-associated protein expressed in migrating and differentiating neurons, *Neuron* 23:247–256.

Franco, B., Guioli, S., Pragliola, A., Incerti, B., Bardoni, B., Tonlorenzi, R. *et al.*, 1991, A gene deleted in Kallmann's syndrome shares homology with neural cell adhesion and axonal path-finding molecules, *Nature* 353:529–536.

Galileo, D.S., Majors, J., Horwitz, A.F., and Sanes, J.R., 1992, Retrovirally introduced antisense integrin RNA inhibits neuroblast migration in vivo, *Neuron* 9:1117–1131.

Georges-Labouesse, E., Mark, M., Messaddeq, N., and Gansmuller, A., 1998, Essential role of alpha 6 integrins in cortical and retinal lamination, *Curr. Biol.* 8:983–986.

Georges-Labouesse, E., Messaddeq, N., Yehia, G., Cadalbert, L., Dierich, A., and Le Meur, M., 1996, Absence of integrin alpha 6 leads to epidermolysis bullosa and neonatal death in mice, *Nat. Genet.* 13:370–373.

Gertler, F.B., Hill, K.K., Clark, M.J., and Hoffmann, F.M., 1993, Dosage-sensitive modifiers of *Drosophila* abl tyrosine kinase function: Prospero, a regulator of axonal outgrowth, and disabled, a novel tyrosine kinase substrate, *Genes Dev.* 7:441–453.

Giger, R.J., Urquhart, E.R., Gillespie, S.K., Levengood, D.V., Ginty, D.D., and Kolodkin, A.L., 1998, Neuropilin-2 is a receptor for semaphorin IV: Insight into the structural basis of receptor function and specificity, *Neuron* 21:1079–1092.

Gilmore, E.C., Ohshima, T., Goffinet, A.M., Kulkarni, A.B., and Herrup, K., 1998, Cyclin-dependent kinase 5-deficient mice demonstrate novel developmental arrest in cerebral cortex, *J. Neurosci.* 18: 6370–6377.

Gleeson, J.G., Lin, P.T., Flanagan, L.A., and Walsh, C.A., 1999, Doublecortin is a microtubule-associated protein and is expressed widely by migrating neurons, *Neuron* 23:257–271.

Gleeson, J.G., Allen, K.M., Fox, J.W., Lamperti, E.D., Berkovic, S., Scheffer, I. *et al.*, 1998, Doublecortin, a brain-specific gene mutated in human X-linked lissencephaly and double cortex syndrome, encodes a putative signaling protein, *Cell* 92:63–72.

Gonzalez, J.L., Russo, C.J., Goldowitz, D., Sweet, H.O., Davisson, M.T., and Walsh, C.A., 1997, Birthdate and cell marker analysis of scrambler: A novel mutation affecting cortical development with a *reeler*-like phenotype, *J. Neurosci.* 17:9204–9211.

Graus-Porta, D., Blaess, S., Senften, M., Littlewood-Evans, A., Damsky, C., Huang, Z. *et al.*, 2001, Beta1-class integrins regulate the development of laminae and folia in the cerebral and cerebellar cortex, *Neuron* 31: 367–379.

Gray, G.E., Leber, S.M., and Sanes, J.R., 1990, Migratory patterns of clonally related cells in the developing central nervous system, *Experientia* 46: 929–940.

Gregory, W.A., Edmondson, J.C., Hatten, M.E., and Mason, C.A., 1988, Cytology and neuron–glial apposition of migrating cerebellar granule cells in vitro, *J. Neurosci.* 8:1728–1738.

Grumet, M., 1992, Structure, expression, and function of Ng-CAM, a member of the immunoglobulin superfamily involved in neuron-neuron and neuron–glia adhesion, *J. Neurosci. Res.* 31:1–13.

Grumet, M., Hoffman, S., Crossin, K.L., and Edelman, G.M., 1985, Cytotactin, an extracellular matrix protein of neural and non-neural tissues that mediates glia–neuron interaction, *Proc. Natl. Acad. Sci. USA* 82:8075–8079.

Hack, I., Bancila, M., Loulier, K., Carroll, P., and Cremer, H., 2002, Reelin is a detachment signal in tangential chain-migration during postnatal neurogenesis, *Nat. Neurosci.* 5:939–945.

Hagg, T., Portera-Cailliau, C., Jucker, M., and Engvall, E., 1997, Laminins of the adult mammalian CNS; laminin-alpha2 (merosin M-) chain immunoreactivity is associated with neuronal processes, *Brain. Res.* 764:17–27.

Halfter, W., Dong, S., Yip, Y.P., Willem, M., and Mayer, U., 2002, A critical function of the pial basement membrane in cortical histogenesis, *J. Neurosci.* 22:6029–6040.

Hatten, M.E., 1990, Riding the glial monorail: A common mechanism for glial-guided neuronal migration in different regions of the developing mammalian brain, *Trends Neurosci.* 13:179–184.

Hatten, M.E., 1993, The role of migration in central nervous system neuronal development, *Curr. Opin. Neurobiol.* 3:38–44.

Hatten, M.E., 1999, Central nervous system neuronal migration, *Annu. Rev. Neurosci.* 22:511–539.

Hatten, M.E. and Mason, C.A., 1990, Mechanisms of glial-guided neuronal migration in vitro and in vivo, *Experientia* 46:907–916.

Hattori, M., Adachi, H., Tsujimoto, M., Arai, H., and Inoue, K., 1994, Miller-Dieker lissencephaly gene encodes a subunit of brain platelet-activating factor acetylhydrolase [corrected], *Nature* 370:216–218.

Hiesberger, T., Trommsdorff, M., Howell, B.W., Goffinet, A., Mumby, M.C., Cooper, J.A. *et al.*, 1999, Direct binding of Reelin to VLDL receptor and ApoE receptor 2 induces tyrosine phosphorylation of disabled-1 and modulates tau phosphorylation, *Neuron* 24:481–489.

Hirotsune, S., Fleck, M.W., Gambello, M.J., Bix, G.J., Chen, A., Clark, G.D. *et al.*, 1998, Graded reduction of Pafah1b1 (Lis1) activity results in neuronal migration defects and early embryonic lethality, *Nat. Genet.* 19:333–339.

Hirotsune, S., Takahara, T., Sasaki, N., Hirose, K., Yoshiki, A., Ohashi, T. *et al.*, 1995, The *reeler* gene encodes a protein with an EGF-like motif expressed by pioneer neurons, *Nat. Genet.* 10:77–83.

Hirsch, E., Gullberg, D., Balzac, F., Altruda, F., Silengo, L., and Tarone, G., 1994, Alpha v integrin subunit is predominantly located in nervous tissue and skeletal muscle during mouse development, *Dev. Dyn.* 201:108–120.

Hodivala-Dilke, K.M., DiPersio, C.M., Kreidberg, J.A., and Hynes, R.O., 1998, Novel roles for alpha3beta1 integrin as a regulator of cytoskeletal assembly and as a trans-dominant inhibitor of integrin receptor function in mouse keratinocytes, *J. Cell. Biol.* 142:1357–1369.

Hong, S.E., Shugart, Y.Y., Huang, D.T., Shahwan, S.A., Grant, P.E., Hourihane, J.O. *et al.*, 2000, Autosomal recessive lissencephaly with cerebellar hypoplasia is associated with human RELN mutations, *Nat. Genet.* 26:93–96.

Horesh, D., Sapir, T., Francis, F., Wolf, S.G., Caspi, M., Elbaum, M. *et al.*, 1999, Doublecortin, a stabilizer of microtubules, *Hum. Mol. Genet.* 8: 1599–1610.

Howell, B.W., Herrick, T.M., and Cooper, J.A., 1999a, Reelin-induced tryosine phosphorylation of disabled 1 during neuronal positioning, *Genes Dev.* 13:643–648.

Howell, B.W., Hawkes, R., Soriano, P., and Cooper, J.A., 1997, Neuronal position in the developing brain is regulated by mouse disabled-1, *Nature* 389:733–737.

Howell, B.W., Lanier, L.M., Frank, R., Gertler, F.B., and Cooper, J.A., 1999b, The disabled 1 phosphotyrosine-binding domain binds to the internalization signals of transmembrane glycoproteins and to phospholipids, *Mol. Cell. Biol.* 19:5179–5188.

Hu, H., 2000, Polysialic acid regulates chain formation by migrating olfactory interneuron precursors, *J. Neurosci. Res.* 61:480–492.

Iijima, M., Huang, Y.E., and Devreotes, P., 2002, Temporal and spatial regulation of chemotaxis, *Dev. Cell.* 3:469–478.

Inoue, T., Tanaka, T., Takeichi, M., Chisaka, O., Nakamura, S., and Osumi, N., 2001, Role of cadherins in maintaining the compartment boundary between the cortex and striatum during development, *Development* 128:561–569.

Jacques, T.S., Relvas, J.B., Nishimura, S., Pytela, R., Edwards, G.M., Streuli, C.H. *et al.*, 1998, Neural precursor cell chain migration and division are regulated through different beta1 integrins, *Development* 125: 3167–3177.

Jankovski, A. and Sotelo, C., 1996, Subventricular zone–olfactory bulb migratory pathway in the adult mouse: Cellular composition and specificity as determined by heterochronic and heterotopic transplantation, *J. Comp. Neurol.* 371:376–396.

Kavanaugh, W.M. and Williams, L.T., 1994, An alternative to SH2 domains for binding tyrosine-phosphorylated proteins, *Science* 266: 1862–1865.

Kirkpatrick, L.L. and Brady, S.T., 1994, Modulation of the axonal microtubule cytoskeleton by myelinating Schwann cells, *J. Neurosci.* 14: 7440–7450.

Kirschenbaum, B., Doetsch, F., Lois, C., and Alvarez-Buylla, A., 1999, Adult subventricular zone neuronal precursors continue to proliferate and migrate in the absence of the olfactory bulb, *J. Neurosci.* 19: 2171–2180.

Kitagawa, M., Umezu, M., Aoki, J., Koizumi, H., Arai, H., and Inoue, K., 2000, Direct association of LIS1, the lissencephaly gene product, with a mammalian homologue of a fungal nuclear distribution protein, rNUDE, *FEBS Lett.* 479:57–62.

Kleinman, H.K., McGarvey, M.L., Liotta, L.A., Robey, P.G., Tryggvason, K., and Martin, G.R., 1982, Isolation and characterization of type IV procollagen, laminin, and heparan sulfate proteoglycan from the EHS sarcoma, *Biochemistry* 21:6188–6193.

Ko, J., Humbert, S., Bronson, R.T., Takahashi, S., Kulkarni, A.B., Li, E. *et al.*, 2001, p35 and p39 are essential for cyclin-dependent kinase 5 function during neurodevelopment, *J. Neurosci.* 21:6758–6771.

Kolodkin, A.L., Levengood, D.V., Rowe, E.G., Tai, Y.T., Giger, R.J., and Ginty, D.D., 1997, Neuropilin is a semaphorin III receptor, *Cell* 90: 753–762.

Komuro, H. and Rakic, P., 1992, Selective role of N-type calcium channels in neuronal migration, *Science* 257:806–809.

Komuro, H. and Rakic, P., 1993, Modulation of neuronal migration by NMDA receptors, *Science* 260:95–97.

Komuro, H. and Rakic, P., 1995, Dynamics of granule cell migration: A confocal microscopic study in acute cerebellar slice preparations, *J. Neurosci.* 15:1110–1120.

Komuro, H. and Rakic, P., 1996, Intracellular Ca2+ fluctuations modulate the rate of neuronal migration, *Neuron* 17:275–285.

Kornack, D.R. and Rakic, P., 2001, The generation, migration, and differentiation of olfactory neurons in the adult primate brain, *Proc. Natl. Acad. Sci. USA* 98:4752–4757.

Kreidberg, J.A., Donovan, M.J., Goldstein, S.L., Rennke, H., Shepherd, K., Jones, R.C. *et al.*, 1996, Alpha 3 beta 1 integrin has a crucial role in kidney and lung organogenesis, *Development* 122:3537–3547.

Kwon, Y.T. and Tsai, L.H., 1998, A novel disruption of cortical development in p35(−/−) mice distinct from *reeler*, *J. Comp. Neurol.* 395:510–522.

Kwon, Y.T., Gupta, A., Zhou, Y., Nikolic, M., and Tsai, L.H., 2000, Regulation of N-cadherin-mediated adhesion by the p35-Cdk5 kinase, *Curr. Biol.* 10:363–372.

Lambert de Rouvroit, C. and Goffinet, A.M., 1998, The *reeler* mouse as a model of brain development, *Adv. Anat. Embryol. Cell Biol.* 150:1–106.

Lavdas, A.A., Grigoriou, M., Pachnis, V., and Parnavelas, J.G., 1999, The medial ganglionic eminence gives rise to a population of early neurons in the developing cerebral cortex, *J. Neurosci.* 19:7881–7888.

Law, D.A., DeGuzman, F.R., Heiser, P., Ministri-Madrid, K., Killeen, N., and Phillips, D.R., 1999, Integrin cytoplasmic tyrosine motif is required for outside-in alphaIIbbeta3 signalling and platelet function, *Nature* 401:808–811.

Lei, Y. and Warrior, R., 2000, The Drosophila Lissencephaly1 (DLis1) gene is required for nuclear migration, *Dev. Biol.* 226:57–72.

Letinic, K. and Rakic, P., 2001, Telencephalic origin of human thalamic GABAergic neurons, *Nat. Neurosci.* 4:931–936.

Letinic, K., Zoncu, R., and Rakic, P., 2002, Origin of GABAergic neurons in the human neocortex, *Nature* 417:645–649.

Lindner, J., Rathjen, F.G., and Schachner, M., 1983, L1 mono- and polyclonal antibodies modify cell migration in early postnatal mouse cerebellum, *Nature* 305:427–430.

Liu, Z., Steward, R., and Luo, L., 2000, Drosophila Lis1 is required for neuroblast proliferation, dendritic elaboration and axonal transport, *Nat. Cell. Biol.* 2:776–783.

Lois, C. and Alvarez-Buylla, A., 1993, Proliferating subventricular zone cells in the adult mammalian forebrain can differentiate into neurons and glia, *Proc. Natl. Acad. Sci. USA* 90:2074–2077.

Lois, C. and Alvarez-Buylla, A., 1994, Long-distance neuronal migration in the adult mammalian brain, *Science* 264:1145–1148.

Lois, C., Garcia-Verdugo, J.M., and Alvarez-Buylla, A., 1996, Chain migration of neuronal precursors, *Science* 271:978–981.

Loo, D.T., Kanner, S.B., and Aruffo, A., 1998, Filamin binds to the cytoplasmic domain of the beta1-integrin. Identification of amino acids responsible for this interaction, *J. Biol. Chem.* 273:23304–23312.

Margolis, B., 1996, The PI/PTB domain: A new protein interaction domain involved in growth factor receptor signaling, *J. Lab. Clin. Med.* 128: 235–241.

Maricich, S.M., Gilmore, E.C., and Herrup, K., 2001, The role of tangential migration in the establishment of mammalian cortex, *Neuron* 31:175–178.

Marin, O., Yaron, A., Bagri, A., Tessier-Lavigne, M., and Rubenstein, J.L., 2001, Sorting of striatal and cortical interneurons regulated by semaphorin–neuropilin interactions, *Science* 293:872–875.

Marin, O., Plump, A.S., Flames, N., Sanchez-Camacho, C., Tessier-Lavigne, M., and Rubenstein, J.L., 2003, Directional guidance of interneuron migration to the cerebral cortex relies on subcortical Slit1/2-independent repulsion and cortical attraction, *Development* 130: 1889–1901.

McEvilly, R.J., de Diaz, M.O., Schonemann, M.D., Hooshmand, F., and Rosenfeld, M.G., 2002, Transcriptional regulation of cortical neuron migration by POU domain factors, *Science* 295:1528–1532.

Mechtersheimer, G., Barth, T., Hartschuh, W., Lehnert, T., and Moller, P., 1994, In situ expression of beta 1, beta 3 and beta 4 integrin subunits in non-neoplastic endothelium and vascular tumours, *Virchows Arch.* 425:375–384.

Metin, C., Denizot, J.P., and Ropert, N., 2000, Intermediate zone cells express calcium-permeable AMPA receptors and establish close contact with growing axons, *J. Neurosci.* 20:696–708.

Miner, J.H., Cunningham, J., and Sanes, J.R., 1998, Roles for laminin in embryogenesis: Exencephaly, syndactyly, and placentopathy in mice lacking the laminin alpha5 chain, *J. Cell. Biol.* 143:1713–1723.

Misson, J.P., Austin, C.P., Takahashi, T., Cepko, C.L., and Caviness, V.S., Jr., 1991, The alignment of migrating neural cells in relation to the murine neopallial radial glial fiber system, *Cereb. Cortex* 1: 221–229.

Miyagoe-Suzuki, Y., Nakagawa, M., and Takeda, S., 2000, Merosin and congenital muscular dystrophy, *Microsc. Res. Tech* 48:181–191. [pii].

Montgomery, A.M., Becker, J.C., Siu, C.H., Lemmon, V.P., Cheresh, D.A., Pancook, J.D. *et al.*, 1996, Human neural cell adhesion molecule L1 and rat homologue NILE are ligands for integrin alpha v beta 3, *J. Cell Biol.* 132:475–485.

Moro, F., Carrozzo, R., Veggiotti, P., Tortorella, G., Toniolo, D., Volzone, A., and Guerrini, R., 2002, Familial periventricular heterotopia: Missense and distal truncating mutations of the FLN1 gene, *Neurology* 58: 916–921.

Moskowitz, P.F. and Oblinger, M.M., 1995, Transcriptional and post-transcriptional mechanisms regulating neurofilament and tubulin gene expression during normal development of the rat brain, *Brain. Res. Mol. Brain. Res.* 30:211–222.

Moyle, M., Napier, M.A., and McLean, J.W., 1991, Cloning and expression of a divergent integrin subunit beta 8, *J. Biol. Chem.* 266: 19650–19658.

Nadarajah, B. and Parnavelas, J.G., 2002, Modes of neuronal migration in the developing cerebral cortex, *Nat. Rev. Neurosci.* 3:423–432.

Nadarajah, B., Alifragis, P., Wong, R.O., and Parnavelas, J.G., 2002, Ventricle-directed migration in the developing cerebral cortex, *Nat. Neurosci.* 5:218–224.

Nadarajah, B., Brunstrom, J.E., Grutzendler, J., Wong, R.O., and Pearlman, A.L., 2001, Two modes of radial migration in early development of the cerebral cortex, *Nat. Neurosci.* 4:143–150.

Nagano, T., Yoneda, T., Hatanaka, Y., Kubota, C., Murakami, F., and Sato, M., 2002, Filamin A-interacting protein (FILIP) regulates cortical cell migration out of the ventricular zone, *Nat. Cell. Biol.* 4:495–501.

Nery, S., Fishell, G., and Corbin, J.G., 2002, The caudal ganglionic eminence is a source of distinct cortical and subcortical cell populations, *Nat. Neurosci.* 5:1279–1287.

Niethammer, M., Smith, D.S., Ayala, R., Peng, J., Ko, J., and Lee, M.S. *et al.*, 2000, NUDEL is a novel Cdk5 substrate that associates with LIS1 and cytoplasmic dynein, *Neuron* 28:697–711.

Nikolic, M., Chou, M.M., Lu, W., Mayer, B.J., and Tsai, L.H., 1998, The p35/Cdk5 kinase is a neuron-specific Rac effector that inhibits Pak1 activity, *Nature* 395:194–198.

O'Leary, D.D. and Nakagawa, Y., 2002, Patterning centers, regulatory genes and extrinsic mechanisms controlling arealization of the neocortex, *Curr. Opin. Neurobiol.* 12:14–25.

O'Rourke, N.A., Chenn, A., and McConnell, S.K., 1997, Postmitotic neurons migrate tangentially in the cortical ventricular zone, *Development* 124:997–1005.

O'Rourke, N.A., Dailey, M.E., Smith, S.J., and McConnell, S.K., 1992a, Diverse migratory pathways in the developing cerebral cortex, *Science* 258:299–302.

O'Rourke, N.A., Sullivan, D.P., Kaznowski, C.E., Jacobs, A.A., and McConnell, S.K., 1995, Tangential migration of neurons in the developing cerebral cortex, *Development* 121:2165–2176.

Ogawa, M., Miyata, T., Nakajima, K., and Mikoshiba, K., 1997, [The action of reelin in the layering of cortical neurons in cerebrum], *Tanpakushitsu Kakusan Koso* 42:577–583.

Ogawa, M., Miyata, T., Nakajima, K., Yagyu, K., Seike, M., Ikenaka, K. *et al.*, 1995, The *reeler* gene-associated antigen on Cajal–Retzius neurons is a crucial molecule for laminar organization of cortical neurons, *Neuron* 14:899–912.

Ohshima, T., Ward, J.M., Huh, C.G., Longenecker, G., Veeranna, Pant, H.C. *et al.*, 1996, Targeted disruption of the cyclin-dependent kinase 5 gene results in abnormal corticogenesis, neuronal pathology and perinatal death, *Proc. Natl. Acad. Sci. USA* 93:11173–11178.

Ohta, Y., Suzuki, N., Nakamura, S., Hartwig, J.H., and Stossel, T.P., 1999, The small GTPase RalA targets filamin to induce filopodia, *Proc. Natl. Acad. Sci. USA* 96:2122–2128.

Olson, E.C. and Walsh, C.A., 2002, Smooth, rough and upside-down neocortical development, *Curr. Opinion. Gen. Dev.* 12:320–327.

Parnavelas, J.G., 2000, The origin and migration of cortical neurones: New vistas, *Trends. Neurosci.* 23:126–131.

Parnavelas, J.G., Barfield, J.A., Franke, E., and Luskin, M.B., 1991, Separate progenitor cells give rise to pyramidal and nonpyramidal neurons in the rat telencephalon, *Cereb. Cortex* 1:463–468.

Pencea, V., Bingaman, K.D., Freedman, L.J., and Luskin, M.B., 2001, Neurogenesis in the subventricular zone and rostral migratory stream of the neonatal and adult primate forebrain, *Exp. Neurol.* 172:1–16.

Peretto, P., Merighi, A., Fasolo, A., and Bonfanti, L., 1997, Glial tubes in the rostral migratory stream of the adult rat, *Brain. Res. Bull.* 42:9–21.

Pleasure, S.J., Anderson, S., Hevner, R., Bagri, A., Marin, O., Lowenstein, D.H. *et al.*, 2000, Cell migration from the ganglionic eminences is required for the development of hippocampal GABAergic interneurons, *Neuron* 28:727–740.

Polleux, F., Whitford, K.L., Dijkhuizen, P.A., Vitalis, T., and Ghosh, A., 2002, Control of cortical interneuron migration by neurotrophins and PI3-kinase signaling, *Development* 129:3147–3160.

Poluch, S. and Konig, N., 2002, AMPA receptor activation induces GABA release from neurons migrating tangentially in the intermediate zone of embryonic rat neocortex, *Eur. J. Neurosci.* 16:350–354.

Poluch, S., Drian, M.J., Durand, M., Astier, C., Benyamin, Y., and Konig, N., 2001, AMPA receptor activation leads to neurite retraction in tangentially migrating neurons in the intermediate zone of the embryonic rat neocortex, *J. Neurosci. Res.* 63:35–44.

Powell, E.M., Mars, W.M., and Levitt, P., 2001, Hepatocyte growth factor/scatter factor is a motogen for interneurons migrating from the ventral to dorsal telencephalon, *Neuron* 30:79–89.

Powell, E.M., Campbell, D.B., Stanwood, G.D., Davis, C., Noebels, J.L., and Levitt, P., 2003, Genetic disruption of cortical interneuron development causes region- and GABA cell type-specific deficits, epilepsy, and behavioral dysfunction, *J. Neurosci.* 23:622–631.

Qiu, M., Bulfone, A., Martinez, S., Meneses, J.J., Shimamura, K., Pedersen, R.A. *et al.*, 1995, Null mutation of Dlx-2 results in abnormal morphogenesis of proximal first and second branchial arch derivatives and abnormal differentiation in the forebrain, *Genes Dev.* 9: 2523–2538.

Quattrocchi, C.C., Wannenes, F., Persico, A.M., Ciafre, S.A., D'Arcangelo, G., Farace, M.G. *et al.*, 2002, Reelin is a serine protease of the extracellular matrix, *J. Biol. Chem.* 277:303–309.

Rakic, P., 1990. Principles of neural cell migration, *Experientia* 46:882–891.

Rakic, P., 1971a, Guidance of neurons migrating to the fetal monkey neocortex, *Brain. Res.* 33:471–476.

Rakic, P., 1971b, Neuron–glia relationship during granule cell migration in developing cerebellar cortex. A Golgi and electronmicroscopic study in Macacus Rhesus, *J. Comp. Neurol.* 141:283–312.

Rakic, P., 1971c, Neuron–glia relationship during granule cell migration in developing cerebellar cortex. A Golgi and electron microscopic study in Macacus Rhesus, *J. Comp. Neurol.* 141:283–312.

Rakic, P., 1972a, Mode of cell migration to the superficial layers of fetal monkey neocortex, *J. Comp. Neurol.* 145:61–83.

Rakic, P., 1972b, Extrinsic cytological determinants of basket and stellate cell dendritic pattern in the cerebellar molecular layer, *J. Comp. Neurol.* 146:335–354.

Rakic, P., 1972c, Mode of cell migration to the superficial layers of fetal monkey neocortex, *J. Comp. Neurol.* 145:61–83.

Rakic, P., 1988, Specification of cerebral cortical areas, *Science* 241:170–176.

Rakic, P., 1990, Principles of neural cell migration, *Experientia* 46:882–891.

Rakic, P. and Komuro, H., 1995, The role of receptor/channel activity in neuronal cell migration, *J. Neurobiol.* 26:299–315.

Rakic, P. and Caviness, V.S., Jr., 1995, Cortical development: View from neurological mutants two decades later, *Neuron* 14:1101–1104.

Rakic, P., Cameron, R.S., and Komuro, H., 1994, Recognition, adhesion, transmembrane signaling and cell motility in guided neuronal migration, *Curr. Opin. Neurobiol.* 4:63–69.

Rakic, P., Knyihar-Csillik, E., and Csillik, B., 1996, Polarity of microtubule assemblies during neuronal cell migration, *Proc. Natl. Acad. Sci. USA* 93:9218–9222.

Rakic, P., 2003, Elusive radial glial cells: Historical and evolutionary perspective, *Glia* 43:19–32.

Redies, C. and Takeichi, M., 1993, Expression of N-cadherin mRNA during development of the mouse brain, *Dev. Dyn.* 197:26–39.

Reiner, O., Carrozzo, R., Shen, Y., Wehnert, M., Faustinella, F. *et al.*, 1993, Isolation of a Miller-Dieker lissencephaly gene containing G protein beta-subunit-like repeats, *Nature* 364:717–721.

Rice, D.S., Sheldon, M., D'Arcangelo, G., Nakajima, K., Goldowitz, D., and Curran, T., 1998, Disabled-1 acts downstream of Reelin in a signaling pathway that controls laminar organization in the mammalian brain, *Development* 125:3719–3729.

Ringstedt, T., Linnarsson, S., Wagner, J., Lendahl, U., Kokaia, Z., Arenas, E. *et al.*, 1998, BDNF regulates reelin expression and Cajal–Retzius cell development in the cerebral cortex, *Neuron* 21:305–315.

Rio, C., Rieff, H.I., Qi, P., Khurana, T.S., and Corfas, G., 1997, Neuregulin and erbB receptors play a critical role in neuronal migration, *Neuron* 19:39–50.

Rivas, R.J. and Hatten, M.E., 1995, Motility and cytoskeletal organization of migrating cerebellar granule neurons, *J. Neurosci.* 15:981–989.

Rousselot, P., Lois, C., and Alvarez-Buylla, A., 1995, Embryonic (PSA) N-CAM reveals chains of migrating neuroblasts between the lateral ventricle and the olfactory bulb of adult mice, *J. Comp. Neurol.* 351: 51–61.

Rutishauser, U. and Jessell, T.M., 1988, Cell adhesion molecules in vertebrate neural development, *Physiol. Rev.* 68:819–857.

Sanes, J.R., 1989, Extracellular matrix molecules that influence neural development, *Annu. Rev. Neurosci.* 12:491–516.

Sapir, T., Elbaum, M., and Reiner, O., 1997, Reduction of microtubule catastrophe events by LIS1, platelet-activating factor acetylhydrolase subunit, *Embo. J.* 16:6977–6984.

Sarkisian, M.R., Frenkel, M., Li, W., Oborski, J.A., and LoTurco, J.J., 2001, Altered interneuron development in the cerebral cortex of the flathead mutant, *Cereb. Cortex* 11:734–743.

Sasaki, S., Shionoya, A., Ishida, M., Gambello, M.J., Yingling, J., Wynshaw-Boris, A. *et al.*, 2000, A LIS1/NUDEL/cytoplasmic dynein heavy chain complex in the developing and adult nervous system, *Neuron* 28:681–696.

Sharma, C.P., Ezzell, R.M., and Arnaout, M.A., 1995, Direct interaction of filamin (ABP-280) with the beta 2-integrin subunit CD18, *J. Immunol.* 154:3461–3470.

Sheen, V.L., Dixon, P.H., Fox, J.W., Hong, S.E., Kinton, L., Sisodiya, S.M. *et al.*, 2001, Mutations in the X-linked filamin 1 gene cause periventricular nodular heterotopia in males as well as in females, *Hum. Mol. Genet.* 10:1775–1783.

Sheldon, M., Rice, D.S., D'Arcangelo, G., Yoneshima, H., Nakajima, K., Mikoshiba, K. *et al.*, 1997, Scrambler and yotari disrupt the disabled gene and produce a *reeler*-like phenotype in mice, *Nature* 389: 730–733.

Shimamura, K. and Takeichi, M., 1992, Local and transient expression of E-cadherin involved in mouse embryonic brain morphogenesis, *Development* 116:1011–1019.

Sidman, R.L. and Rakic, P., 1973, Neuronal migration, with special reference to developing human brain: A review, *Brain Res.* 62:1–35.

Simon, K.O., Nutt, E.M., Abraham, D.G., Rodan, G.A., and Duong, L.T., 1997, The alphavbeta3 integrin regulates alpha5beta1-mediated cell migration toward fibronectin, *J. Biol. Chem.* 272:29380–29389.

Smith, D.S., Niethammer, M., Ayala, R., Zhou, Y., Gambello, M.J., Wynshaw-Boris, A. *et al.*, 2000, Regulation of cytoplasmic dynein behaviour and microtubule organization by mammalian Lis1, *Nat. Cell. Biol.* 2: 767–775.

Stephens, L.E., Sutherland, A.E., Klimanskaya, I.V., Andrieux, A., Meneses, J., Pedersen, R.A. *et al.*, 1995, Deletion of beta 1 integrins in mice results in inner cell mass failure and peri-implantation lethality, *Genes. Dev.* 9:1883–1895.

Stipp, C.S., Litwack, E.D., and Lander, A.D., 1994, Cerebroglycan: An integral membrane heparan sulfate proteoglycan that is unique to the developing nervous system and expressed specifically during neuronal differentiation, *J. Cell. Biol.* 124:149–160.

Stitt, T.N. and Hatten, M.E., 1990, Antibodies that recognize astrotactin block granule neuron binding to astroglia, *Neuron* 5:639–649.

Stossel, T.P., Condeelis, J., Cooley, L., Hartwig, J.H., Noegel, A., Schleicher, M. *et al.*, 2001, Filamins as integrators of cell mechanics and signalling, *Nat. Rev. Mol. Cell. Biol.* 2:138–145.

Sunada, Y., Edgar, T.S., Lotz, B.P., Rust, R.S., and Campbell, K.P., 1995, Merosin-negative congenital muscular dystrophy associated with extensive brain abnormalities, *Neurology* 45:2084–2089.

Sussel, L., Marin, O., Kimura, S., and Rubenstein, J.L., 1999, Loss of Nkx2.1 homeobox gene function results in a ventral to dorsal molecular respecification within the basal telencephalon: Evidence for a transformation of the pallidum into the striatum, *Development* 126: 3359–3370.

Sweet, H.O., Bronson, R.T., Johnson, K.R., Cook, S.A., and Davisson, M.T., 1996, Scrambler, a new neurological mutation of the mouse with abnormalities of neuronal migration, *Mamm. Genome* 7:798–802.

Takeichi, M., 1991, Cadherin cell adhesion receptors as a morphogenetic regulator, *Science* 251:1451–1455.

Tamamaki, N., Fujimori, K.E., and Takauji, R., 1997, Origin and route of tangentially migrating neurons in the developing neocortical intermediate zone, *J. Neurosci.* 17:8313–8323.

Tamamaki, N., Fujimori, K., Nojyo, Y., Kaneko, T., and Takauji, R., 2003, Evidence that Sema3A and Sema3F regulate the migration of GABAergic neurons in the developing neocortex, *J. Comp. Neurol.* 455:238–248.

Tan, S.-S. and Breen, S., 1993, Radial mosaicism and tangential cell dispersion both contribute to mouse neocortical development, *Nature* 362: 638–639.

Tan, S.-S., Faulkner-Jones, B., Breen, S.J., Walsh, M., Bertram, J.F., and Reese, B.E., 1995, Cell dispersion patterns in different cortical regions studied with an X-inactivated transgenic marker, *Development* 121:1029–1039.

Taylor, K.R., Holzer, A.K., Bazan, J.F., Walsh, C.A., and Gleeson, J.G., 2000, Patient mutations in doublecortin define a repeated tubulin-binding domain, *J. Biol. Chem.* 275:34442–34450.

Tian, M., Hagg, T., Denisova, N., Knusel, B., Engvall, E., and Jucker, M., 1997, Laminin-alpha2 chain-like antigens in CNS dendritic spines, *Brain Res.* 764:28–38.

Tomasiewicz, H., Ono, K., Yee, D., Thompson, C., Goridis, C., Rutishauser, U. and Magnuson, T., 1993, Genetic deletion of a neural cell adhesion molecule variant (N-CAM-180) produces distinct defects in the central nervous system, *Neuron* 11:1163–1174.

Trommsdorff, M., Borg, J.P., Margolis, B., and Herz, J., 1998, Interaction of cytosolic adaptor proteins with neuronal apolipoprotein E receptors and the amyloid precursor protein, *J. Biol. Chem.* 273: 33556–33560.

Trommsdorff, M., Gotthardt, M., Hiesberger, T., Shelton, J., Stockinger, W., Nimpf, J. *et al.*, 1999, *Reeler*/Disabled-like disruption of neuronal migration in knockout mice lacking the VLDL receptor and ApoE receptor 2, *Cell* 97:689–701.

Walsh, C. and Cepko, C.L., 1992, Widespread dispersion of neuronal clones across functional regions of the cerebral cortex, *Science* 255: 434–440.

Ware, M.L., Fox, J.W., Gonzalez, J.L., Davis, N.M., Lambert de Rouvroit, C., Russo, C.J. *et al.*, 1997 Aberrant splicing of a mouse disabled homolog, mdab1, in the scrambler mouse, *Neuron* 19:239–249.

Wichterle, H., Garcia-Verdugo, J.M., and Alvarez-Buylla, A., 1997, Direct evidence for homotypic, glia-independent neuronal migration, *Neuron* 18:779–791.

Wichterle, H., Turnbull, D.H., Nery, S., Fishell, G., and Alvarez-Buylla, A., 2001, In utero fate mapping reveals distinct migratory pathways and fates of neurons born in the mammalian basal forebrain, *Development* 128:3759–3771.

Yacubova, E. and Komuro, H., 2002, Stage-specific control of neuronal migration by somatostatin, *Nature* 415:77–81.

Yang, J.T., Rayburn, H., and Hynes, R.O., 1993, Embryonic mesodermal defects in alpha 5 integrin-deficient mice, *Development* 119: 1093–1105.

Yang, J.T., Rayburn, H., and Hynes, R.O., 1995, Cell adhesion events mediated by alpha 4 integrins are essential in placental and cardiac development, *Development* 121:549–560.

Yoneshima, H., Nagata, E., Matsumoto, M., Yamada, M., Nakajima, K., Miyata, T. *et al.*, 1997, A novel neurological mutant mouse, yotari, which exhibits *reeler*-like phenotype but expresses CR-50 antigen/reelin, *Neurosci. Res.* 29:217–223.

Zhang, Z. and Galileo, D.S., 1998, Retroviral transfer of antisense integrin alpha6 or alpha8 sequences results in laminar redistribution or clonal cell death in developing brain, *J. Neurosci.* 18:6928–6938.

Zhu, Y., Li, H., Zhou, L., Wu, J.Y., and Rao, Y., 1999, Cellular and molecular guidance of GABAergic neuronal migration from an extracortical origin to the neocortex, *Neuron* 23:473–485.

9

Guidance of Axons and Dendrites

Chi-Bin Chien

INTRODUCTION

How does the mechanical development of the nerve fibers occur, and wherein lies that marvelous power which enables the nerve fibers from very distant cells to make contact directly with certain other nerve cells of the mesoderm or ectoderm without going astray or taking a roundabout course?
Ramón y Cajal (1972), *The Structure of the Retina*, p. 146
Translated from the original, published 1892.

Wiring Complex Nervous Systems

As the nervous system develops, a key step is the directed outgrowth of axons and dendrites. The preceding events of early patterning (Chapters 1–3), neurogenesis and neural fate specification (Chapter 5), and neuronal migration (Chapter 8) produce a large number of neurons—of the order of 10^{12} in humans—located in their correct positions. Since each neuron can potentially synapse on thousands of others, the number of synapses is several orders of magnitude greater, perhaps 10^{15}. For these synapses to develop (see Chapter 10), the presynaptic axon and postsynaptic dendrite must grow out from their respective cell bodies, then rendezvous. As recognized by Santiago Ramón y Cajal over a century ago, this is no mean feat. Growing axons must navigate precisely over long distances to find their target region (up to several millimeters for some projection neurons). Once there, they must select from amongst thousands of neurons to find their target cells. Furthermore, each of the many types of neurons has its own individual program of axonal and dendritic development.

The complexity of this task demands a correspondingly complex developmental program. There are scores of genes known to be involved in axon guidance, and likely to be many more still to be discovered; furthermore, some of these genes give rise to multiple isoforms through alternative splicing or posttranslational modifications. This chapter addresses what is known about axonal and dendritic guidance, with particular emphasis on the axonal growth cone, which is much better studied than its dendritic counterpart.

Axons vs Dendrites

Mature axons and dendrites have quite different shapes. In general, neurons have only a single axon. This axon is relatively thin, often quite long, does not branch extensively except near the neuron's target, and bears presynaptic terminals. In contrast, neurons often have multiple dendrites. Dendrites are usually rather short and thick, branch elaborately, and bear postsynaptic densities and dendritic spines. Thus axons and dendrites face different developmental problems: axons must navigate over long distances through varied terrain and then select one of many possible target areas, while for dendrites navigation is not as difficult, but control of the exact pattern of branching is critical.

Studying the guidance of axons or dendrites requires methods for visualizing them. Axons are often easy to label using anterograde dye labeling, retrograde dye labeling, or antibody staining. Dendrites, lying close to the cell body, are often obscured by it. For this technical reason, and because of a historical emphasis on axons, much more is known about axonal than dendritic guidance.

The Growth Cone

Despite their differences, growing axons and dendrites have much in common. Most importantly, they grow by the movement of a specialized structure at their tip, the growth cone, which was first described by Ramón y Cajal in 1890 (Fig. 1). The growth cone is a motile structure which navigates through the developing brain, trailing the axon or dendrite behind it. (Sometimes, however, growth cones can arise from the axon shaft, forming an axon branch.) Axonal and dendritic growth cones are thought to be quite similar and are indistinguishable in cell culture. The growth cone responds to environmental cues by modulating its cytoskeleton to move appropriately: sometimes growing straight, sometimes zigging or zagging, slowing down at its target, and eventually stopping and initiating synaptogenesis. An example of a retinal axonal growth cone growing through the brain is shown in Fig. 2A, and an example of a dorsal root ganglion (DRG) axonal growth cone growing in culture is shown in Fig. 2B.

Chi-Bin Chien • Department of Neurobiology and Anatomy, University of Utah, SOM, Salt Lake City, UT 84132.

Developmental Neurobiology, 4th ed., edited by Mahendra S. Rao and Marcus Jacobson. Kluwer Academic / Plenum Publishers, New York, 2005.

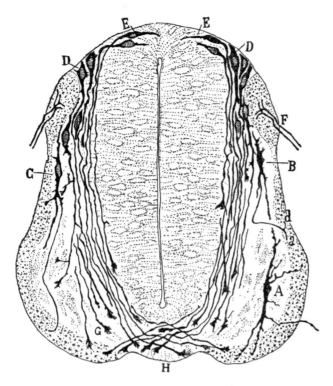

FIGURE 1. The first published drawing of growth cones, by Ramón y Cajal. In this transverse section of the spinal cord from a chick embryo, at day 4 of incubation, labeling with the rapid Golgi method reveals the structure of individual neurons, from cell bodies (A–E) to growth cones (G). Many of the growth cones are nearing or crossing the ventral midline (H). Dorsal up, ventral down. From Ramón y Cajal (1890).

The development of a growing axon can be broken into several distinct stages (Fig. 3): Axonogenesis, pathfinding, collateral formation in some cases, target recognition, target selection, branching, and synaptogenesis. Similarly, dendrite development consists of dendritogenesis, outgrowth and pathfinding, branching, and synaptogenesis. The importance of axonal and dendrite guidance continues even after initial development is finished: sprouting and branching of axons and dendrites are important for plasticity during normal adult learning and memory, and for repair after injury.

Key Questions about Axon and Dendrite Guidance

The three decades since 1970 have brought a wealth of knowledge both about the behavior of growing axons and dendrites, and about the molecules that control this behavior. Several basic questions have been important throughout:

1. What makes axons and dendrites different? Neurons usually have a single axon, but can have multiple dendrites. What controls how many there are, and what controls whether a process is an axon or a dendrite? Studies of this choice in culture have revealed some very interesting cell behaviors, but as yet, little of the underlying mechanisms.
2. How do axonal growth cones navigate over long distances? They seem to make not just one or two navigational choices, but instead a long series of decisions along their entire pathway. Not only do growth cones encounter

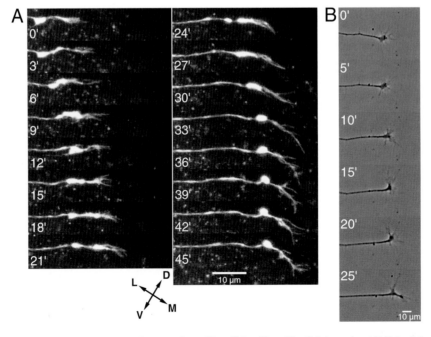

FIGURE 2. Time-lapse imaging of growth cones *in vivo* and *in vitro* shows filopodial and lamellipodial dynamics. (A) Zebrafish retinal growth cone labeled with the lipophilic fluorescent dye DiI, growing through the diencephalon of a living embryo at 36 hr after fertilization. The sequence starts after exiting the eye and ends before reaching the midline. D, dorsal; V, ventral; L, lateral; M, medial. (B) Chick dorsal root ganglion (DRG) growth cone viewed with phase contrast microscopy, growing in culture on a laminin substrate. (A) from L. Hutson and C.B. Chien; (B) courtesy of M. Lemons and M. Condic.

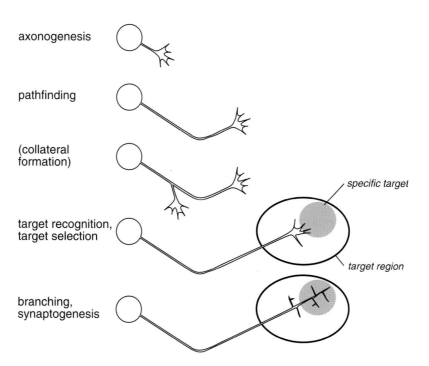

FIGURE 3. The life history of an axon. In order to connect to its appropriate target, an axon must go through a series of steps. During *axonogenesis*, the axon is generated; during *pathfinding*, the axon navigates through the brain; during *target recognition* and *target selection*, the axon recognizes its target region and in some cases finds a specific target within that region; finally, during *branching*, the axon starts to generate its terminal arbor and then begins *synaptogenesis*. For certain neurons, axonal branches are formed during *collateral formation*.

a large number of different guidance signals as they grow, but they are able to change their responses to particular molecules depending on their history.

3. How do growth cones move? Growth cone motility is similar to the movement of other cells, but is unique for three reasons: the cell body and nucleus stay behind; the growth cone can be very far from the nucleus; and the axonal shaft provides polarity and orientation. Much is known about the dynamics of the growth cone cytoskeleton from studies in culture, and a little bit is known even *in vivo*.

4. What are the signaling molecules used by growth cones? The advent of molecular biological and genetic techniques in the last decade has brought an explosion in our knowledge of growth cone signaling molecules, including ligands, receptors, and downstream signal transduction molecules. A striking finding is that these molecules are very highly conserved evolutionarily. Much of our knowledge about axon guidance in mammalian systems has its origins in studies of invertebrates, birds, amphibians, or fish.

5. How are axonal and dendritic branching controlled? This is a critical question that has not yet been extensively addressed. Axon branching responds to environmental guidance molecules, but also responds to activity-dependent interactions with target cells (see Chapter 10). Dendritic branching can be controlled

with exquisite precision, and is so far poorly understood.

ORGANIZATION OF THIS CHAPTER

This chapter discusses what we know of the answers to these questions, concentrating especially on the guidance of the axonal growth cone. It emphasizes the experimental approaches that have led to our rapidly growing understanding of axon and dendrite outgrowth. Lists of important molecules (Fig. 8) are included in order to provide an overview of the field, but the axon guidance literature has grown to the point where it is impossible to cite it comprehensively. Rather, some key systems and molecules have been chosen to illustrate the ideas and experimental strategies. This allows us to focus on principles, rather than the anatomical and molecular details of many different projections. A few references are given as jumping-off points into the broader literature.

We begin with a brief history of growth cone biology and a summary of experimental approaches and preparations. This is followed by a discussion of growth cone cell biology, primarily the control of the cytoskeleton. The major classes of guidance molecules are summarized briefly, and then the roles of a few molecules in well-studied model systems (the midline and the retinotectal system) are discussed in detail. We finish by discussing two newly emerging areas: growth cone signal transduction, and the control of dendrite outgrowth.

A BRIEF HISTORY OF GROWTH CONES

The Discovery of the Growth Cone

The history of the growth cone began at the end of the 19th century. At this time, the weight of evidence was just starting to favor the *neuron theory*, which posited that the brain is made up of a huge number of individual nerve cells, over the *reticular theory*, which held that the brain is a syncytium of continuously connected cells. Ramón y Cajal, one of the main proponents of the neuron theory, was carrying out exquisite anatomical analyses of the nervous systems of many different species using the recently invented method of Golgi staining. The ability of the Golgi method to stain individual cells in the brain lent strong support to the neuron hypothesis and also allowed Ramón y Cajal to see and describe the growth cone for the first time. In fixed embryonic brains, he saw cone-shaped structures at the ends of growing axons. He named these structures "growth cones," and based solely on observations of fixed specimens, was able to intuit that they are the growing ends of developing nerve fibers. Furthermore, he theorized that growth cones might navigate by using chemical signals in their environment:

> … it must be supposed that these cells are capable of amoeboid movement and are responsive to certain substances secreted by cells of the epithelium or mesoderm. The processes of the neuroblasts become oriented by chemical stimulation, and move toward the secretion products of certain cells.
>
> Ramón y Cajal (1972), *The Structure of the Retina*, p. 146
> Translated from the original, published 1892.

For over a century, neuroscientists have been trying to determine what "amoeboid movement" Cajal's growth cones use as they navigate through the developing brain, and what "chemical stimulation" controls this movement.

Classical Experiments

For most of the 20th century, advances in understanding axon and dendrite outgrowth came slowly, although there were several milestones along the way. Ross Harrison developed methods for tissue culture and observed living growth cones in culture (Harrison, 1907). Carl Speidel used the transparent tail of the tadpole to observe living growth cones *in vivo* (Speidel, 1933). Most significantly, Roger Sperry's landmark studies of the regenerating visual system, using surgical manipulations and classical anatomical tracing techniques, laid the intellectual groundwork for our modern understanding of axon guidance (Sperry, 1963; discussed in detail later in the chapter). Sperry's studies were followed by a plethora of studies of the amphibian visual system by Marcus Jacobson, Michael Gaze, and their collaborators. However, these early investigators were severely limited by the paucity of methods available for observing or manipulating growing axons.

Modern Experiments

This began to change in the 1970s, when it became possible to culture neurons routinely on defined substrates. Together with the development of video microscopy, this allowed the detailed observation of growth cone behavior using phase contrast or differential interference contrast. Time-lapse observations of cultured growth cones revealed the basic cytoskeletal movements associated with their growth. Now that researchers were able to watch how growth cones behaved, they started to test what stimuli would change this behavior. Over the next three decades, an increasingly sophisticated series of growth cone culture assays were developed. By careful manipulation, it is now possible to see how growth cones respond as they are challenged with other cells, other axons, purified membranes, or purified proteins. These stimuli can be presented in uniform layers, striped patterns, or precisely defined gradients. Results of such culture assays have been critical for our understanding of growth cone biology and are described in detail later in the chapter.

In parallel with advances in culture methods, there have been great strides in analyzing growing neurites *in vivo*. Again, methods for visualization were critical. The Golgi stain used by Ramón y Cajal gives beautiful staining, but is capricious and labels neurons at random, making it difficult to study a specific projection. Until the late 1970s, the only methods for specifically labeling particular projections were either to cut the axons and then stain for degenerating axons, or to test for electrophysiological connections using extracellular recordings. Direct tracing of axons became possible with the development of new anatomical techniques: first, labeling with reporters like tritiated proline or horseradish peroxidase that are transported anterogradely or retrogradely, then antibodies that specifically label particular axons, and most recently transgenic lines in which specific sets of neurons express reporters such as beta-galactosidase (lacZ) or green fluorescent protein (GFP). The development of fluorescent markers such as DiI (Honig and Hume, 1986) and GFP (Chalfie *et al.*, 1994) was especially important. Lipophilic dyes such as DiI can be used to trace axons or dendrites in fixed tissue, since they diffuse along the plasma membrane, which remains intact after aldehyde fixation. Fluorescent markers allow the visualization of very fine processes; furthermore, since lipophilic dyes and GFP are quite nontoxic, they can be used to observe the behavior of live growth cones *in vivo*. A wide array of methods for perturbing axons *in vivo* have also been developed. Molecular methods include antibody injection, pharmacological blockers, or misexpression of exogenous wild-type or dominant-negative constructs. Most importantly, there are now several genetic model systems in which genes critical to axon guidance can be mutated or knocked out. These *in vivo* techniques are critical for determining how the results from culture experiments apply to the behavior of growing axons or dendrites during normal development of the nervous system.

GROWTH CONE CELL BIOLOGY

The growth cone is the motile structure at the tip of a growing axon or dendrite. It is shaped much like a hand (Fig. 4A). Fingers (the filopodia) extend out from a palm (the body of the growth cone), which is attached to an arm (the axon). An internal skeleton (microfilaments and microtubules) supports

A

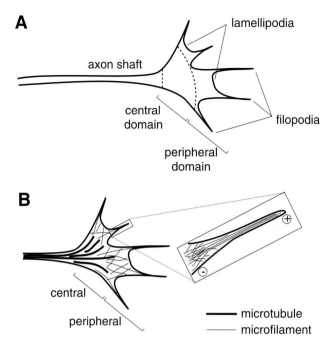

B

FIGURE 4. Structure of the growth cone. (A) Regions of the growth cone. The central domain is relatively thick and contains mitochondria and other organelles. The peripheral domain is relatively thin and consists of veil-like lamellipodia and spike-like filopodia. (B) The growth cone cytoskeleton. *Microtubules* are the primary structural elements of the axonal shaft, fill the central domain, and occasionally extend into the peripheral domain. Microtubules are oriented with (+) ends distal. *Microfilaments* are the primary structural elements of the peripheral domain. In lamellipodia, microfilaments form a meshwork, while in filopodia, they are aligned in tight bundles. In both lamellipodia and filopodia, microfilaments are oriented with their (+) ends distal.

the skin (the plasma membrane), which is filled with sensory organs (axon guidance receptors) (Fig. 4B).

The growth cone can be considered as a tiny autonomous path-finding device, which contains all of the internal machinery required to sense environmental signals and translate these into appropriate motions. Indeed, even when surgically separated from the cell body, growth cones continue to path-find normally for several hours (Harris *et al.*, 1987). Since there are multiple signals in the environment, some of which are present at very low concentrations, *integration* and *amplification* of external signals are both fundamental functions of the growth cone.

Growth Cone Behavior

As the growth cone crawls through the developing nervous system, it trails the axon behind it. Thus the path of the growth cone largely determines the final trajectory of the axon, although subsequent straightening, bending, or collateral branching can also play a role. Growth cones sometimes grow at a constant velocity, but more often their growth is *saltatory*, with periods of fast growth interrupted by periods of slow growth or pausing. As the growth cone advances, its filopodia constantly undergo cycles of extension and retraction, with the lamellipodia extending

and retracting between the filopodia (Fig. 2). The behavior of the growth cone is modulated as it encounters different stimuli in the culture dish or progresses to different parts of its pathway *in vivo*. Often, slowing or pausing is associated with an increase in filopodial complexity (i.e., more or longer filopodia). Such *in vivo* changes in behavior often occur at "decision points," where growth cones choose between alternate pathways (e.g., Hutson and Chien, 2002). To understand how growth cone behavior is controlled, we first consider the structure of the growth cone.

Growth Cone Structure

The growth cone cytoskeleton has many similarities to the leading edge of a migrating cell such as a keratinocyte or leukocyte. However, rather than being organized around the centriole as in the cell body, the cytoskeleton is organized around the end of the axon shaft. Figure 4A shows the three principal domains in the growth cone: the axon shaft; the central domain; and the peripheral domain.

In the *axon shaft*, the cytoskeleton consists largely of a bundle of parallel microtubules, oriented with their (+) ends distal (i.e., toward the growth cone). These microtubules tend to be acetylated and detyrosinated (two posttranslational modifications of tubulin indicative of long-lived, stable microtubules), and are decorated with the Tau microtubule-associated protein. The rigidity of these bundled microtubules makes the axon straight and stiff. Another function of these microtubules is to act as tracks for kinesin and dynein molecular motors, which actively transport membrane vesicles either anterogradely (toward the growth cone) or retrogradely (back toward the cell body).

In the *central domain*, the cytoskeleton still consists largely of microtubules with distal (+) ends. However, these microtubules are no longer so straight or tightly bundled as in the axon shaft and fan out from the axon's end. The central domain is relatively thick and contains mitochondria and other intracellular organelles, including ribosomes. It also contains reservoirs of membrane which can be added to the plasma membrane when the growth cone advances, or reinternalized when the growth cone retracts.

In the *peripheral domain*, the predominant cytoskeletal elements are microfilaments (F-actin). In the filopodia, microfilaments are bundled and oriented with their barbed or (+) ends distal, while in lamellipodia, microfilaments form a loose meshwork that is much less oriented, although their (+) ends still tend to be distal.

The *plasma membrane* of the growth cone contains the same classes of transmembrane proteins found in most neuronal membranes, including cell-adhesion molecules (CAMs), ion channels, and cell-surface receptors. The particular molecules expressed are specific to that growth cone; of special interest are its repertoire of axon guidance receptors. Some receptors are precisely localized within growth cones, for instance at the tips of filopodia (e.g., Grabham and Goldberg, 1997). There are also certain classes of membrane-associated proteins that are specifically expressed in growth cones, for instance the growth-associated protein GAP-43 (reviewed by Benowitz and Routtenberg, 1997).

Filopodia are especially interesting because they often react to signals encountered by the growth cone. They have the potential for both "motor" and "sensory" functions. Filopodia can exert force (Heidemann *et al.*, 1990), which may help to bias growth cone movement. They also bear axon guidance receptors and ion channels, and can change their internal concentration of second messengers such as calcium in response to external signals (Davenport *et al.*, 1993). Their long, narrow shape makes them particularly suited for amplifying external signals: Because of the high surface-to-volume ratio, activation of a few filopodial membrane receptors has a large effect on the concentration of downstream signaling molecules.

The key to growth cone function is to understand how events at the plasma membrane affect the cytoskeleton, eventually controlling the elongation of the axon shaft and the path taken by the axon. We discuss first the dynamics of microfilaments and microtubules (reviewed by Jay, 2000; Suter and Forscher, 2000).

Microfilament Dynamics

Microfilaments are clearly important for growth cone motility. Preventing actin polymerization with the drug cytochalasin removes growth cone filopodia and prevents axons from path-finding correctly *in vivo* (Bentley and Toroian-Raymond, 1986; Chien *et al.*, 1993). Microfilament dynamics have been best studied in culture, where cytoskeletal drugs can be easily

applied to perturb actin dynamics, and where the extension and retraction of filopodia can be observed with high-resolution microscopy. The picture that has emerged is shown in Fig. 5A. Even when the length of a filopodium is constant, its microfilaments are not static, but instead exist in a state of dynamic equilibrium, with polymerization at the (+) end being balanced by depolymerization at the (−) end. At the same time there is an overall retrograde flow of F-actin, driven by myosin motors, probably attached to the membrane. In a filopodium whose length is not changing, polymerization of microfilaments is balanced by this retrograde flow. Of course, actin monomers (G-actin) must diffuse outward in order to replenish the concentration at the filopodial tip. There must also be microfilament depolymerizing or severing activities in the lamellipodium or central domain (otherwise the microfilaments would eventually extend out the back of the growth cone!).

The filopodium can be controlled in three ways. First, by changing the rate of polymerization. Inhibiting actin polymerization pharmacologically causes filopodial retraction (Fig. 5B). Many actin-binding proteins are known to affect polymerization, offering a possible control mechanism. Second, by changing myosin function. Inhibiting myosin function pharmacologically causes filopodial extension (Fig. 5C), although little is known about how growth cone myosins are normally controlled. Third, by changing how actin or myosin are anchored. The "clutch hypothesis" (Mitchison and Kirschner, 1988) proposes that microfilaments

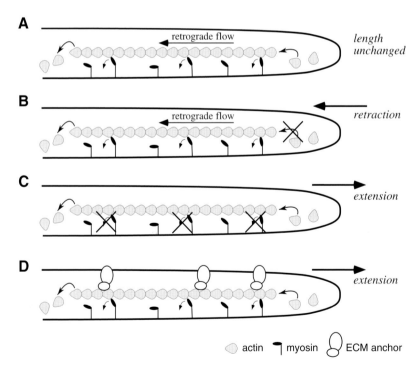

FIGURE 5. Actin dynamics in growth cone filopodia. Filopodial extension is controlled by a balance between the *polymerization* and *retrograde flow* of actin microfilaments. (A) In a filopodium whose length is constant, actin polymerization at the distal tip is balanced by retrograde flow, driven by myosin motors. The proximal ends of microfilaments undergo a net depolymerization. (B) Inhibiting actin polymerization causes filopodial retraction. (C) Inhibiting myosin function causes filopodial extension. (D) The *clutch hypothesis* proposes that microfilaments can become anchored to the extracellular matrix (ECM), so that myosin motors drive filopodial extension.

can be anchored reversibly to the extracellular matrix (ECM) via a transmembrane complex. When this "clutch" is engaged, the myosin motors, rather than pulling microfilaments backward, now push the filopodial membrane forward (Fig. 5D).

Microtubule Dynamics

Like microfilaments, microtubules exist in a state of dynamic equilibrium, with polymerization occurring preferentially at the (+) end (located peripherally), and depolymerization at the (−) end (located toward the axon shaft). They can also be actively transported by motor proteins. Time-lapse observations of fluorescently labeled microtubules show that they glide out into the growth cone, sometimes apparently bending under tension, then suddenly snapping straight (Tanaka and Kirschner, 1991). Microtubule dynamics are important for growth cone steering, since adding drugs that depolymerize or stabilize microtubules can prevent growth cone turning in response to external signals (e.g., Buck and Zheng, 2002).

Control of Cytoskeleton during Guidance

How does cytoskeletal dynamics come into play as growth cones navigate *in vivo*? The best-understood example is the behavior of the Ti1 pioneer neuron in the grasshopper limb bud as it encounters a guidepost cell. By injecting this large neuron with fluorescently labeled tubulin or actin, O'Connor, Bentley, and colleagues (Sabry *et al.*, 1991; O'Connor and Bentley, 1993) were able to directly visualize cytoskeletal rearrangements during this encounter (Fig. 6). Beforehand, the axon grows straight, and its filopodia and lamellipodia extend in random exploratory movements. When a single filopodium contacts the guidepost cell, F-actin accumulates in this filopodium, inflating it. Microtubules then invade this region, the growth cone reorients toward the guidepost cell, and eventually the contacting filopodium has been replaced by mature axonal shaft. These results suggest that the microfilament network is the first to react to external signals and guides the subsequent rearrangements of the microtubules. However, more recent work in culture suggests that there may also be reciprocal interactions: microtubules can also act to control the microfilaments (Buck and Zheng, 2002).

Transport of Components

As the axon elongates, its total content of membrane, cytoskeletal elements, and other proteins must increase. Where are these new components added? This question is not easy to test experimentally and therefore has been controversial over the years. There are three possible sites of insertion: At the cell body, along the axonal shaft, or at the growth cone. New *membrane* is synthesized as vesicles at the cell body and is transported distally by fast axonal transport (although there is some evidence for local lipid synthesis in the axon; reviewed by Futerman and Banker, 1996). It appears that these vesicles are exocytosed mainly at the growth cone, with a small contribution coming from addition

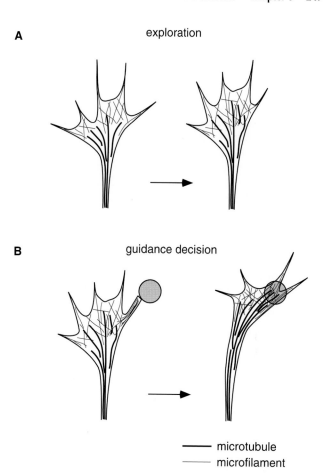

FIGURE 6. Cytoskeletal changes during growth cone guidance. As the growth cone advances, its microfilaments and microtubules constantly extend, rearrange, and retract. (A) When the growth cone is not actively responding to guidance signals (*exploration*), microfilaments and microtubules extend at random in all forward directions, so that on average the growth cone grows straight ahead. (B) A growth cone encountering an attractive guidance signal, in this case a guidepost cell (*shaded circle*) here. Microfilaments accumulate in the filopodium that contacts the guidepost cell, leading to preferential accumulation of microfilaments and microtubules on that side, and eventually causing the growth cone to steer toward that side and turn.

along the axonal shaft (Zakharenko and Popov, 1998, 2000). Microtubules seem to form in the cell body and then get actively transported toward the growth cone (reviewed by Baas, 2002). Indeed, most proteins are probably synthesized in the cell body and then actively transported toward the growth cone. For many years it was generally assumed that protein synthesis occurs *only* in the cell body. However, it has become quite clear that this assumption was wrong, and that some proteins are translated locally in the axon and the growth cone (reviewed by Giuditta *et al.*, 2002). This local translation is important because it could allow the growth cone to generate new proteins on a very rapid timescale, rather than relying solely on what the cell body decides to ship out. Indeed, it has now been shown that local protein translation is required for growth cones to respond to certain guidance signals (Campbell and Holt, 2001).

APPROACHES TO STUDYING AXON GUIDANCE

Criteria for Guidance Molecules

Understanding the mechanisms underlying growth cone guidance is essentially a problem of understanding the molecules involved. It is thus important to ask: When can we conclude that a particular molecule is involved in the guidance of a particular axon? There are two main criteria: expression and function. The molecule must be expressed in the right place (in or outside the growth cone); at the right time (when the axon is growing); in an active form; and at a biologically active concentration. At the same time, there must be some demonstration that the molecule has a biological function. Often it is asked whether a molecule is *sufficient* or *necessary* to guide axons. Sufficiency is most commonly demonstrated in culture, where a putative guidance signal can be applied to growing axons in the absence of confounding factors. Note however that showing sufficiency (what the molecule *can* do) may require a particular context: the right axons, presented with the right pattern of the molecule. Necessity is most stringently tested *in vivo*, by abrogating a molecule's function during normal development. Note that a molecule may act redundantly with others, so that even if not strictly necessary, it could still play an important role.

Culture Assays for Guidance Molecules

The advantages of culture assays are that neurons or growth cones are easily observed and can be presented with stimuli that are precisely defined molecularly, temporally, and spatially. Culture assays are often simple, fast, and easily quantified (especially important if they are to be used as assays for biochemical purification). The disadvantage of simple assays, however, is that some key element of the growth cone's natural environment may have been removed. The challenge is thus always to make the assay relevant to axon guidance. Many clever assays have been devised, of varying degrees of complexity. These are summarized in Fig. 7 and will be discussed in greater detail later in this chapter.

If a neuron is thought to interact with a particular cell type or substrate molecule *in vivo*, there must clearly be at least transient contact or adhesion involved. *Cell-adhesion assays* (Fig. 7A) are a simple test of whether a neuron can stick to other cells, or to a given substrate. The next question is whether a particular substrate (either a purified ECM molecule, or membranes isolated from a particular cell type) will allow axons to grow in an *axon outgrowth assay* (Fig. 7B). If so, it is said to be acting as a *permissive* substrate. Molecules can also be inhibitory for axon growth. In the *collapse assay* (Fig. 7C), a common test for inhibitory molecules, soluble or membrane-bound protein is bath-applied to a neuronal culture. When confronted with high

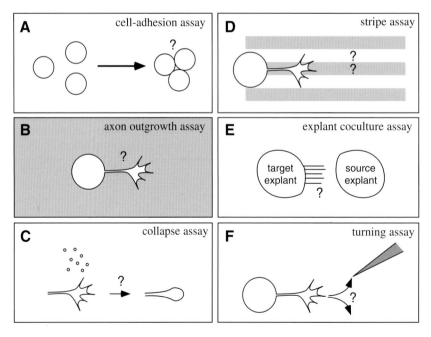

FIGURE 7. Culture assays for axon guidance signals: Several common culture assays used to test the guidance activities of particular molecules. (A) A *cell-adhesion assay* tests whether cells expressing a particular molecule will adhere to each other. (B) An *axon outgrowth assay* tests whether a uniform layer of a molecule will support axon outgrowth. (C) A *collapse assay* tests whether a soluble molecule, or a molecule found on membrane vesicles, causes a population of growth cones to *collapse*. (D) A *stripe assay* tests whether growth cones prefer to grow on one of two different substrates, applied in parallel stripes. (E) An *explant coculture assay* tests whether a diffusible molecule from the *source explant* will attract or repel axons from the *target explant*. Explants are embedded in a three-dimensional gel to stabilize diffusible gradients. (F) A *turning assay* tests whether a diffusible molecule delivered from a pipette attracts or repels a single growth cone.

concentrations of inhibitory molecule on all sides, growth cones typically "collapse" and pull in all their filopodia.

Testing whether a molecule can have an *instructive* function (i.e., can direct the path taken by an axon) requires an assay with a patterned substrate. The *stripe assay* (Fig. 7D) confronts growing axons with a choice between two sets of alternating stripes, each with a different substrate that allows axon outgrowth. If axons grow preferentially on one set of stripes, this shows that at least one of the substrates can be instructive; this may be due to either an attractive interaction (the axons prefer to grow on those stripes) or a repulsive interaction (they prefer *not* to grow on the other stripes).

To test whether a particular molecule can cause growing axons to turn, the growth cone must be presented with a local source from a defined direction. In the *explant coculture assay* (Fig. 7E), two explants are grown near each other in a three-dimensional gel. One (neuronal) explant is the source of the axons to be tested, while the other (often nonneuronal) is the source of a diffusible gradient of the candidate guidance molecule. The three-dimensional gel is critical for stabilizing this gradient, which otherwise would be washed away by fluid convection in the culture dish. It is then easy to see whether the axons turn up or down this concentration gradient. The *turning assay* (Fig. 7F) is a short-term assay, similar in principle to the explant coculture assay. A pipette containing a soluble candidate molecule is positioned about 100 μm from a growth cone, at 45° to its direction of travel. Pressure pulses are applied to slowly eject solution, which diffuses away from the tip and builds up a stable gradient within a few minutes. The growth cone is then observed, usually for 1 hr, to see whether it turns toward the pipette, turns away, or fails to respond. While the turning assay is quite labor intensive, it has many advantages: The gradient can be defined quantitatively, either by modeling or using a fluorescent tracer; it tests fast growth cone responses (in particular, the gradient can be turned on or off to test for changes in behavior), and other molecules can be added to the bath to test for interactions with the molecule in the pipette.

Examples will be given below of how these assays have been used to analyze the functions of different axon guidance molecules.

In Vivo Preparations for Axon Guidance

Despite the powerful culture assays that have been developed, which can show very beautifully that certain factors *can* guide growth cones, in the end we wish to know which factors actually *do* guide growth cones *in vivo*. During normal development growth cones encounter many signals, each of which is expressed in a spatiotemporally complex pattern. As seen later in the chapter, the history of the growth cone or the context in which it sees a particular signal can both be important determinants of the growth cone's responses. This complexity is impossible to reproduce fully in the culture dish. However, studying growth cones *in vivo* can be difficult. In contrast to culture systems, where it is easy to visualize growth cones, and the possible perturbations are limited largely by the experimenter's ingenuity, analysis of growth cone behavior *in vivo* is usually limited by the organism.

First, to see axons *in vivo* they must be labeled specifically, to pick them out from surrounding nervous tissue, and especially from the many other axons that are usually nearby. Axons can be labeled using tracers such as the lipophilic dye DiI; using specific antibodies; or using transgenes that drive reporters such as GFP or lacZ in specific sets of neurons. In the case of fluorescent labels such as DiI or GFP, it is possible with some care to watch the dynamic behavior of growth cones *in vivo* using time-lapse microscopy (see Fig. 2).

Second, perturbations are needed to test what factors can affect axon growth. These perturbations can be embryological: transplanting tissues from a different region of the brain (heterotopic transplants) or different developmental stage (heterochronic transplants) to test which tissues can guide axons, or simply removing pieces of tissue to see which are necessary for axon guidance. Perturbations can be molecular: using drugs, antibodies, or antisense oligos to block the function of a molecule. Finally, perturbations can be genetic: using *in vivo* expression systems to misexpress wild-type DNA constructs or express dominant-negative constructs, using transgenes for misexpression experiments, or using genetic mutants to eliminate or even replace the normal function of a gene.

Certain model systems are particularly easy to visualize or to perturb. The transparency of zebrafish and *Caenorhabditis elegans* makes them excellent for watching axons *in vivo*. Embryological "cut and paste" experiments are easy in chick, *Xenopus*, and zebrafish. Molecular biological techniques are rapidly improving in nearly all model systems, but are particularly powerful in conjunction with the genome sequencing projects of mouse, zebrafish, *Drosophila*, and *C. elegans*. Genetic experiments are the most powerful in mouse, zebrafish, *Drosophila*, and *C. elegans*. The widespread use of homologous recombination (knockout and knockin) in mice remains an especial advantage of this system. For all species, it is a general rule that sensory and motor axons are often the easiest to study. Since they run outside the central nervous system (CNS), they are physically accessible for visualization and manipulation.

In genetic model organisms, *genetic screens* are the most important strategy for analyzing axon guidance. Given a specific assay for testing the guidance of a particular set of axons, one screens through many potentially mutant individuals, looking for a *phenotype* indicating that a key axon guidance gene has been mutated. A main advantage of genetic screens is their lack of bias: They can find many classes of molecule, including novel ones. Furthermore, a mutant with a strong phenotype automatically indicates that the affected gene is required for axon guidance. Using "reverse genetic" approaches (generating a mutant in a known gene and looking for a phenotype), it is quite common to find that the targeted gene is not required for axon guidance, or plays only a minor role.

There are also disadvantages to genetic screens. Because of their lack of bias, the genes found may affect axon path-finding only very indirectly (e.g., transcription factors). Screens cannot find redundant genes, nor genes that are also required for the early development of the embryo (at least not without using sophisticated genetic tricks). Finally, since analyzing axon guidance phenotypes *in vivo* is hard, screens are a lot of work.

Summary

What, then, are the main strategies for finding new axon guidance genes? One is to start with a **functional effect**, either in culture or *in vivo*. Given a reliable effect in culture with extracts of a particular tissue, the relevant molecule(s) can be purified using biochemical or molecular biological methods. Given a mutant phenotype, the affected gene can be cloned. Another strategy is start with an **expression pattern**. Historically, this was done with *monoclonal antibody screens*: Generating panels of monoclonal antibodies against a tissue of interest, screening through for interesting expression patterns (labeling particular axons, for instance), and then isolating the antigen recognized. Similar screens can now be done with genetic techniques (enhancer or gene traps), or microarray experiments. Surprisingly often, one finds that proteins expressed in interesting patterns are indeed functionally important. The rest of the chapter will describe several examples in which these different strategies have been used to identify important axon guidance molecules.

GUIDANCE MOLECULES: OVERVIEW OF LIGANDS AND RECEPTORS

The Diversity of Guidance Signals

A large number of signaling molecules have been shown to act in axon or dendrite guidance. Some of these molecules are members of large gene families, while others are singletons. It now seems that most or all of these gene families play general roles in cell motility, not just in axon guidance. Furthermore, most of these families are important for axon guidance throughout the

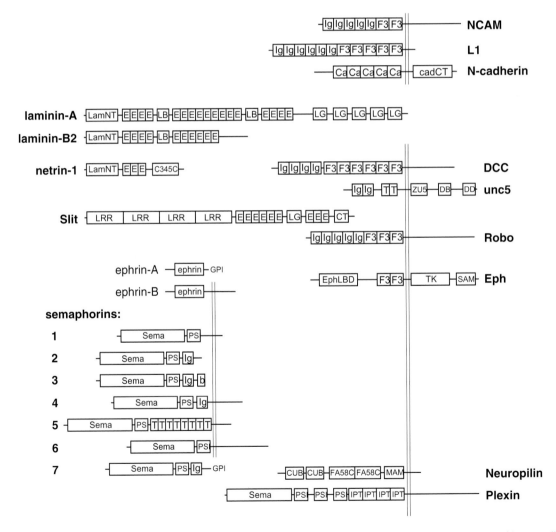

FIGURE 8. Domain structures of some well-studied axon guidance signals and receptors, showing that many domains are conserved between different axon guidance proteins. Structures given are for the mammalian genes. Transmembrane domains are indicated by a double line; glycophosphatidylinositol anchors are indicated as GPI. Abbreviations for domain names: b, basic domain; Ca, cadherin; cadCT, cadherin C-terminal; CT, cystine knot; F3, fibronectin type 3; Ig, immunoglobulin; lamNT, laminin N-terminal; LB, laminin B; LG, laminin G; E, EGF; Eph LBD, Eph ligand-binding domain; Sema, Semaphorin; T, thrombospondin type 1; TK, tyrosine kinase. Domain names are from the SMART database (http://smart.embl-heidelberg.de).

animal kingdom. Thus it seems that ancient mechanisms for the control of cell motility have been co-opted by evolution to control growth cones in many neuronal systems.

Figure 8 diagrams the primary structure of some of the best-studied axon guidance signals, and their cognate receptors where known. These molecules share many motifs: for instance, the immunoglobulin superfamily have extracellular domains consisting of several immunoglobulin (Ig) domains (also found in antibody molecules) and several fibronectin type 3 (FN3) domains (also found in the ECM protein fibronectin). Some receptors have cytoplasmic domains with known catalytic function (e.g., kinase or phosphatase domains); others have conserved cytoplasmic motifs likely to bind intracellular signaling partners; still others have no significant cytoplasmic domain, and presumably require coreceptors to signal intracellularly. Axon guidance receptors are necessarily associated with the plasma membrane of the growth cone. Axon guidance ligands, on the other hand, must be present in the environment encountered by the navigating growth cone, either on the surfaces of other cells or in the ECM.

Guidance ligands can therefore be either cell-surface or secreted. These two classes act at different ranges. Cell-surface ligands act only at short range: To receive a signal, the growth cone must contact the ligand-expressing cell. Secreted ligands can act at long range: Diffusion can carry them over a distance to reach the growth cone. Many secreted ligands, however, bind to the ECM with high affinity and in practice may not diffuse very far. Small molecules such as neurotransmitters or nitric oxide do not bind to the ECM, but may have limited range for other reasons (e.g., uptake mechanisms). Guidance signals can be further classified according to their activities: Positive (telling the growth cone that it is on the right track), negative (telling it that it is on the wrong track), or modulatory (modifying the effect of another signal). These classifications are summarized in Table 1. Some signals can be *either* cell-surface or secreted. Notice also that some signals can have different activities, depending on the particular growth cones and particular context being studied. Thus a blanket statement such as "netrin is a diffusible attractant" is likely to be oversimplified. A more careful statement would be "netrin is a diffusible attractant for commissural axons before they have reached the ventral midline of the spinal cord."

Cell-Surface Guidance Signals

When a growth cone encounters a cell expressing a particular cell-surface guidance signal, it can respond in one of two fundamental ways: It may prolong its contact with that cell (an *adhesive* signal), or shorten or terminate contact (a *repulsive* signal). The simplest mechanism for an adhesive action is to mediate mechanical binding between the growth cone and the cell that it has contacted. Indeed, some of the best-studied cell-surface molecules are CAMs such as the cadherins or neural cell-adhesion molecule (NCAM), which were originally characterized in mediating adhesion between cultured neuronal cells. These CAMs bind homophilically, meaning that they act in *trans* as their own receptors, and can mediate both cell–cell binding and axon–axon binding (fasciculation). The cadherins and

TABLE 1. Activities of Axon Guidance Signals

	Cell-surface	Secreted
Positive	*Adhesive* Cadherins NCAM	*Permissive* Laminin Fibronectin *Attractive* Netrin Semaphorins BDNF
Negative	*Repulsive* Ephrins Semaphorins Nogo MAG	*Inhibitory* CSPGs *Repulsive* Netrin Slit Semaphorins
Modulatory		HSPGs Laminin Slit

Notes: Axon guidance signals may be either cell-surface or secreted, and may have positive, negative, or modulatory activities. The table shows specific examples of molecules in each of these classes. Note that a given signal may be either cell-surface or secreted, and that many signals have multiple activities depending on context.

NCAM were originally isolated in experiments searching for the molecules that hold tissues together; subsequently, it has become clear that they function not only in mechanical binding, but also trigger intracellular signaling upon homophilic binding. Many other CAMs are known, with some binding homophilically and others heterophilically.

A *repulsive* cell-surface signal must first bind to a receptor on the growth cone, then trigger a cytoskeletal response causing the growth cone to pull away, and finally release from its receptor. There are many classes of such inhibitory molecules, including the ephrins, semaphorins, Nogo, and myelin-associated glycoprotein (MAG). In general, these bind heterophilically to their receptors.

Secreted Guidance Signals

Secreted axon guidance signals fall into three classes: Small molecules, major constituents of the ECM, and proteins that diffuse through or bind to the ECM, but are not considered part of it. Several small molecules, notably neurotransmitters and nitric oxide, can affect growth cone guidance in culture, but at present little is known about their actions *in vivo*.

ECM proteins can have *permissive* functions for axon guidance (laminin or fibronectin) or *inhibitory* functions (chondroitin sulfate proteoglycans or CSPGs). They may also function in concert with other guidance signals rather than on their own. ECM proteins can bind, stabilize, or shape the distribution of other guidance signals. In other cases, proteins such as laminin or heparan sulfate proteoglycans (HSPGs) can *modulate* the effects of other guidance signals. When neurons are cultured in an *axon outgrowth assay*, uniform layers of ECM proteins acting as *permissive* molecules can support axon outgrowth. Laminin and fibronectin are bound by integrin receptors on the surfaces of growth cones, providing substrate attachment. *In vivo*, laminin and fibronectin are widely distributed in basal laminae, and they are generally thought to define broad regions of the nervous system where axons can grow, rather than instructing axons to grow along specific pathways. When axons are challenged with patterns of laminin in culture, however, they can respond with specific turning behaviors, so a more instructive role for these molecules cannot be ruled out. *Inhibitory* ECM molecules are unfavorable for axon outgrowth in culture. Presented with a choice, axons generally prefer to avoid stripes of CSPG (although neurons plated on a uniform carpet of CSPG can extend axons). CSPGs are a diverse class of glycosaminoglycans with heterogeneous protein cores and long carbohydrate side chains. *In vivo*, CSPGs are thought to make certain regions "off-limits" for axon growth. Finally, HSPGs, another major component of the ECM, are primarily thought to affect axon guidance as cofactors for other proteins such as fibroblast growth factors.

Secreted proteins such as netrin, Slit, brain-derived neurotrophic factor (BDNF), and class 3 semaphorins, diffuse through and bind to the ECM, but are not considered part of the ECM because of their relatively low abundance and specific distribution patterns. They can have *attractive* actions (netrin, BDNF), *repulsive* actions (netrin, Slit, semaphorins), or *modulatory* actions (Slit), depending on the cell type and context. These are some of the most interesting guidance molecules because they are known to have very clear instructive roles in guiding axons.

Below are discussed specific examples of the roles of the best-characterized guidance ligands and receptors: netrin/DCC/unc5, Slit/Robo, ephrin/Eph, and semaphorin/neuropilin/plexin. There are of course many other classes of known axon guidance ligands and receptors (e.g., receptor tyrosine phosphatases, Nogo, MAG) which we do not have space to discuss.

ATTRACTION TO THE MIDLINE: NETRIN AND DCC

All vertebrates and many invertebrates have bilateral body plans, with a left and a right side, and have bilateral nervous systems to match. To coordinate the two sides of the body, some neurons must project contralaterally, across the midline; other neurons project ipsilaterally. Because the choice between ipsilateral and contralateral projections is such a fundamental one, understanding the control of midline crossing has been a major focus in the study of axon guidance. The main model systems for studying midline crossing have been the vertebrate spinal cord and the ventral nerve cords of *Drosophila* and *C. elegans*.

In order to cross the midline, axons must first reach it. It turns out that many axons are actively attracted to the midline, guided in part by the secreted ligand netrin.

Midline Attraction in the Vertebrate Spinal Cord

In the developing chick spinal cord, a particular class of commissural neurons has their cell bodies located dorsally (Fig. 9A). These neurons send their axons down toward the ventral midline of the spinal cord, where a specialized structure called the *floorplate* is located. The axons cross the floorplate upon reaching it, then project longitudinally. Tessier-Lavigne, Jessell, and colleagues set out to determine what factors were responsible for guidance toward the ventral midline, using collagen-gel cocultures as an assay (cf. Fig. 7). They dissected out explants from dorsal spinal cord which contained the commissural neurons and cocultured them with various other pieces of tissue. The commissural axons grew out preferentially toward floorplate tissue (Fig. 9B), but were unaffected by many other types of tissue. This implied that the floorplate specifically produces a diffusible factor. This factor might act in two different ways: It could be *attractive* for the commissural neurons, or it might merely stimulate their growth, acting with a limited range so that it only affected axons on the side of the explant nearest the floorplate. To distinguish between these possibilities, the assay was modified to place a strip of floorplate against one side of a dorsal spinal explant. Careful tracking of the explant's orientation

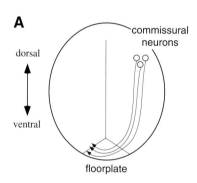

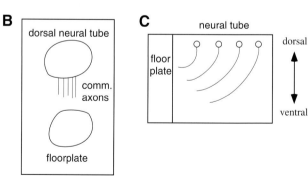

FIGURE 9. The floorplate attracts commissural neurons in the spinal cord. (A) Transverse section of the spinal cord. Commissural neurons are located dorsally and send their axons ventrally toward the floorplate, where they cross the midline. (B) Explant coculture of dorsal neural tube with floorplate shows that commissural axons are attracted by the floorplate. (C) Coculturing the floorplate against one side of an oriented neural tube explant shows that the floorplate can cause commissural axons to *turn* toward it. Based on Tessier-Lavigne *et al.* (1988) and Placzek *et al.* (1990).

predicted the direction (ventralward) in which the commissural axons would normally grow. Antibody staining showed that commissural axons now turned within the explant, toward the ectopic floorplate, thus proving that the floorplate factor was indeed truly attractive (Fig. 9C).

Biochemical Purification of Netrin

To biochemically purify the attractive floorplate factor, the Tessier-Lavigne group needed a rich source of protein and a simple bioassay for testing partially purified fractions. Floorplates are tiny, and the collagen-gel coculture assay is laborious. However, in addition to its ability to cause axon turning, the floorplate had a second biological effect: stimulation of axon outgrowth from spinal cord explants. Extracts of embryonic chick brain proved to have a similar stimulatory effect on axon outgrowth, and of course a whole brain provides much more material than a dissected floorplate. Under the assumption that the stimulatory factors from brain and floorplate were the same, the Tessier-Lavigne group undertook a biochemical purification starting with protein from 20,000 embryonic brains, using a simple axon outgrowth assay to test the biological activity of partially purified fractions. They were able to purify the floorplate factor to homogeneity and used protein microsequencing and cDNA cloning to identify two proteins (Serafini *et al.*, 1994) which they named netrin-1 and netrin-2 (after the Sanskrit for "one who guides").

Netrin Structure and Function

Both these netrins are similar in sequence to the terminal arm of the B2 subunit of laminin, fitting their presumed location in the ECM. The distribution of chick netrin-1 mRNA is just what would have been predicted: It is found specifically in the floorplate. Netrin-2 is found more broadly throughout the ventral two thirds of the spinal cord, but *not* in the floorplate. Thus a reasonable model is that commissural neurons are attracted to the floorplate by an increasing gradient of netrin concentration. Collagen-gel coculture assays with netrin-expressing cells and turning assays using pipettes filled with purified netrin both show that a netrin gradient is sufficient to cause growth cone turning (Kennedy *et al.*, 1994; de la Torre *et al.*, 1997).

To show that netrin is also necessary for axon pathfinding *in vivo*, the Tessier-Lavigne group used a genetic approach in mouse (Serafini *et al.*, 1996). They first cloned mouse netrin-1 and showed that its expression pattern in the spinal cord is a composite of the chick netrin-1 and -2 patterns: It is expressed both in the floorplate and in the ventral two-thirds of the cord. They then obtained a netrin-1 mutation that greatly reduced netrin function (a strong hypomorphic allele). Sure enough, in homozygous mutant embryos, the commissural axons are severely disrupted in their ventral migration, showing that netrin-1 is required for this guidance decision. Interestingly, when mutant floorplates were used in collagen-gel cocultures, they failed to stimulate axon outgrowth, but could still cause commissural axon turning. Thus, netrin-1 seems to account for all of the floorplate's outgrowth-stimulating activity, but only makes up part of its attractive activity.

Midline Attraction in *C. elegans*

One of the most interesting discoveries from the cloning of chick netrin-1 and -2 is that both genes are homologs of the *unc-6* gene from *C. elegans*. This gene had been shown to be required for the migration and axon guidance of many neurons (Hedgecock *et al.*, 1990; Ishii *et al.*, 1992; reviewed by Wadsworth, 2002). Since all of these guidance decisions occur in either a dorsal or a ventral direction, the UNC-6 protein had been hypothesized to be a guidance signal expressed in a dorsoventral gradient in the body of the nematode (just as the vertebrate netrins are thought to be expressed in spinal cord). It is now clear that the netrin/UNC-6 gene family is required for dorsoventral axon guidance in nematodes, flies, and vertebrates. This was one of the first examples of the striking conservation of axon guidance mechanisms across the animal kingdom.

Gene *unc-6* had originally been found by genetic analysis, together with two other genes, *unc-5* and *unc-40*. Wild-type *C. elegans* move forward by a smooth sinusoidal motion; in *unc* (uncoordinated) mutants, this movement is disrupted. Since this phenotype is easy to detect, and coordinated movement is a behavior that depends on many different genes, many *unc* mutants have been found. Such a behavioral phenotype can have many causes, ranging from early developmental defects to problems in synaptic transmission, but one possible cause is a disruption in neuronal wiring. After analyzing a large set of *unc* mutants, Hedgecock and colleagues (1990) found that *unc-5*, *unc-6*, and *unc-40* had related defects in axon guidance. In *unc-40*, certain axons failed to migrate ventrally, and others dorsally. In *unc-5*, axons failed to migrate dorsally. The *unc-6* phenotype was roughly a combination of the *unc-5* and *unc-40* phenotypes, with defects in both ventral and dorsal pathfinding. Based on genetic analysis of the three mutants, it was proposed that they were likely to be either guidance signals or receptors, acting in the same pathway. The cloning of *unc-6* suggested that it encoded a signal, later found to be expressed at the ventral midline. When *unc-5* and *unc-40* were cloned, they proved to encode transmembrane proteins, both now known to be receptors for UNC-6 (Leung-Hagesteijn *et al.*, 1992; Chan *et al.*, 1996).

DCC is an Attractive Netrin Receptor in Vertebrates

The cloning of UNC-40 revealed that it was a homolog of DCC (Deleted in Colorectal Carcinoma), a vertebrate gene named because the human version is often lost in certain cancers. It is still rather unclear what the role of DCC may be in cancer (although it can control apoptosis; Mehlen *et al.*, 1998), but it definitely acts as a netrin receptor for axon guidance (Keino-Masu *et al.*, 1996).

DCC is a member of the immunoglobulin superfamily (Fig. 8), with four Ig and six FN3 domains extracellularly, and several very highly conserved motifs in its intracellular domain. Several experiments show that DCC, like its homolog UNC-40, is an attractive netrin receptor. DCC binds netrin-1 with nanomolar affinity. It is expressed by commissural axons and their growth cones. An anti-DCC antibody can block the outgrowth-promoting effect of netrin on dorsal spinal cord explants. Furthermore, in the turning assay, anti-DCC antibody can block the attractive turning of retinal growth cones toward a source of netrin-1 (de la Torre *et al.*, 1997). Thus, DCC function is required for both the outgrowth-promoting and attractive effects of netrin in culture. When DCC is knocked out in the mouse, both spinal commissural axons and retinal axons show axon guidance defects similar to those seen in netrin mutants (Deiner *et al.*, 1997; Fazeli *et al.*, 1997). This combination of culture and *in vivo* studies makes a very strong case that DCC indeed acts as a netrin receptor.

Modulation of Netrin Signaling

While netrin was originally isolated as a chemoattractive signal, *C. elegans* genetics had suggested that UNC-6 acted both attractively and repulsively, since the *unc-6* mutant had defects in both dorsal and ventral guidance. In vertebrates, it is now also clear that for certain types of neurons, netrin can act as a chemorepellent; for others, netrin is chemoattractive at some times and has no effect at other times.

Netrin as a Repulsive Signal

The overall organization of the hindbrain is similar to that of the spinal cord, with longitudinally repeated segments, a floorplate, and a roofplate. In both, dorsally located commissural neurons generally project their axons ventralward to the floorplate and then across the midline, while ventrally located motoneurons send their axons out the ventral roots. However, the trochlear motoneurons in the hindbrain have an unusual projection pattern. Their cell bodies are located ventrally, but instead of projecting out on the ventral side, their axons navigate dorsalward, away from the midline, and then exit the hindbrain on the dorsal side. This route is similar to that taken by commissural axons, only reversed. Just as commissural axons are attracted by netrin, it turns out that trochlear motor axons are repelled by netrin (Colamarino and Tessier-Lavigne, 1995). In collagen-gel cocultures, trochlear axons grow *away* from floorplate or netrin-expressing COS cells. What is the significance of this finding *in vivo*? In the netrin-1 knockout mouse, trochlear axon trajectories are largely normal (Serafini *et al.*, 1996). Thus, in addition to netrin, the floorplate must also express another signal that repels trochlear axons.

UNC-5 Proteins are Repulsive Coreceptors for DCC

Genetic results from *C. elegans* led to a possible mechanism by which netrin can act as a repellent. The *unc-5* gene is required for guiding axons that normally migrate dorsally in the nematode, suggesting that UNC-5 protein acts as a repulsive receptor for the netrin homolog UNC-6. When this gene was cloned (Leung-Hagesteijn *et al.*, 1992), it was found to encode a transmembrane protein with just this function (Hamelin *et al.*, 1993). Homology cloning yielded two of the mammalian

homologs, named UNC5H1, UNC5H2, which can bind netrin (Leonardo *et al.*, 1997). Are the vertebrate UNC-5 proteins also netrin receptors? Genetic evidence in *C. elegans* had suggested that UNC-5 requires UNC-40 to function. Experiments using the turning assay on *Xenopus* spinal axons showed that vertebrate UNC-5 similarly needs DCC (Hong *et al.*, 1999). While these axons are normally attracted to netrin in a DCC-dependent manner, when they are also made to express UNC5H2, they are *repelled* by netrin. Biochemical experiments showed that UNC5H2 and DCC interact directly.

Silencing of Netrin after Crossing the Midline

Not only can the response to netrin vary between different axons, but it can change during the history of a single axon. Commissural axons in the hindbrain are attracted to netrin before reaching the midline, but then fail to respond to netrin after they cross the midline (Shirasaki *et al.*, 1998). This makes functional sense: If these axons were forever attracted to netrin, they might never leave the midline after reaching it. However, the axons still express DCC after crossing. How might the DCC/netrin interaction become "silenced"? One possibility is that this depends on the corepresentation of netrin with other guidance signals (Stein and Tessier-Lavigne, 2001).

Thus, netrin responses can be altered by differences between cell types; by expression of an UNC5/DCC heteromeric receptor; and when combined with other signals. A fourth route for modulating netrin responses, cyclic nucleotide signaling, is described below as part of the discussion of growth cone signal transduction.

CROSSING THE MIDLINE: Robo AND Slit

Once commissural axons reach the midline, guided in part by netrin attraction, they have further decisions to make. First, they must cross the midline, and afterwards, they must *stay* on the contralateral side, without crossing back. The best-understood mechanism for controlling midline crossing and recrossing is the Robo/Slit pathway, which was isolated using a genetic approach in *Drosophila*.

Drosophila Ventral Nerve Cord Screen

The *Drosophila* ventral nerve cord is analogous to the vertebrate spinal cord, being the division of the CNS that is located in the trunk. It is made up of longitudinally repeated segments which are bilaterally symmetric about the ventral midline, and so resembles a ladder (Fig. 10A): Each segment has an anterior and a posterior commissure (the rungs of the ladder), and the segments are connected by longitudinal fascicles (the uprights). To search for genes controlling the development of the ventral nerve cord, Corey Goodman's group carried out a large-scale screen (Seeger *et al.*, 1993), assaying with the BP102 monoclonal antibody. Since BP102 recognizes an epitope found on all axons in the CNS, this assay was not good for detecting defects in specific

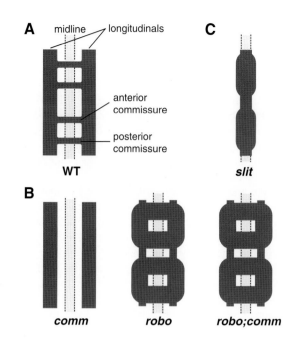

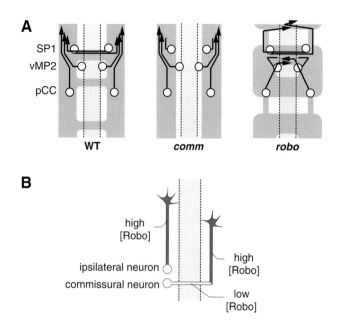

FIGURE 10. Crossing the midline in the *Drosophila* CNS. Diagrams showing axon pathways in the ventral nerve cord of wild type (WT) and mutant *Drosophila* embryos. (A) The normal nerve cord is a ladder-like structure composed of longitudinal and commissural axon bundles. Each repeated segment has an anterior and a posterior commissure. (B) In *commissureless* (*comm*) mutants, both commissures fail to form. In *roundabout* (*robo*) mutants, the longitudinals are greatly reduced, and the commissures are much thicker. *robo; comm* double mutants have the same phenotype as *robo*. (C) In *slit* mutants, all of the axons collapse onto the midline.

FIGURE 11. How behavior of single axons is controlled by Robo and Comm. (A) In wild type, SP1 axons cross the midline, while vMP2 and pCC axons stay ipsilateral. In *comm*, SP1 axons fail to cross the midline, while vMP2 and pCC project normally. In *robo*, SP1 axons can cross the midline more than once, while vMP2 and pCC axons now cross the midline abnormally. (B) Modulation of Robo protein levels controls axon crossing. Since ipsilateral axons express high levels of Robo constitutively, they are repelled by the midline and do not cross it. Commissural axons initially have low levels of Robo (due to downregulation by Comm), allowing them to cross the midline. After crossing, Comm function turns off, allowing Robo to be upregulated, so that the commissural growth cones are now repelled by the midline and inhibited from recrossing. In *robo* mutants, both types of axons can cross the midline freely; in *comm* mutants, Robo is never downregulated, so that both types of axons are always repelled by the midline.

axons (which would be hidden amongst all the other axons), but was rather a way to find defects in the overall pattern of the ventral nerve cord. This screen found some mutants in which the longitudinal fascicles were disrupted, and others in which the commissural axons were disrupted. The longitudinal mutants included *longitudinals gone* (*logo*), which has been little studied, and *longitudinals lacking* (*lola*), which proved to be a mutation in a transcription factor (Giniger *et al.*, 1994) and thus does not affect axon guidance directly. The commissural mutants included *roundabout* (*robo*) and *commissureless* (*comm*), which have proved to be key genes in the control of midline crossing.

robo, *comm,* and *slit*

As seen by BP102 staining, the *robo* and *comm* mutant phenotypes are opposites (Fig. 10B). In *robo*, the commissures are thickened and the longitudinals are somewhat reduced, so that the normal ventral nerve cord ladder now resembles a chain of traffic roundabouts. In *comm*, the commissures are completely absent. These phenotypes predict that normal *comm* gene function promotes midline crossing, while *robo* function discourages crossing. Furthermore, the double *robo; comm* phenotype looks like *robo*, showing that they act in the same pathway.

In understanding axon guidance phenotypes in mutants, it is critical to analyze the behavior of individual identified axons.

For *robo* and *comm*, this was done using antibodies that recognize the identified neurons pCC, vMP2, and SP1 (Kidd *et al.*, 1998b). In wild-type flies, the vMP2 and pCC axons both project ipsilaterally, while the SP1 axons cross the midline once, and not again (Fig. 11A). In mutants, single-cell analyses confirmed the predictions made from BP102 staining. *comm* shows reduced midline crossing: Normally noncrossing axons are unaffected, but the SP1 axons fail to cross the midline. *robo* shows enhanced crossing: Normally noncrossing axons (vMP2, pCC) now cross the midline, and axons such as SP1 that normally cross once, now cross more than once.

When these two genes were cloned, the structure of Comm was rather inscrutable—it had no recognizable motifs apart from a single transmembrane domain (Tear *et al.*, 1996). On the other hand, the structure of Robo immediately suggested its function (Kidd *et al.*, 1998a). Robo is a member of the immunoglobulin superfamily with five Ig domains, three FN3 domains, a single transmembrane domain, and several cytoplasmic motifs that are conserved in vertebrate Robo homologs (Fig. 8). This structure suggested that Robo was likely to be a receptor, with an extracellular domain that binds ligand, and an intracellular domain that communicates with downstream signaling components.

Subsequent experiments have shown that Robo indeed acts as a receptor, while Comm acts by regulating the levels of Robo protein.

The midline is ideally situated to be a source of attractive or repulsive guidance signals for commissural axons. (Indeed, it expresses fly *netrinA* and *netrinB*, which act as attractive signals.) The *robo* mutant phenotype suggested that Robo might act as a receptor for a repulsive signal, in whose absence axons would not be repelled and would thus cross more readily than in wild type. Such a receptor model makes two important predictions: Robo protein should be expressed on growing axons, and Robo should act cell-autonomously. Indeed, Robo is expressed on axons and growth cones, and expression of Robo in neurons can rescue the *robo* phenotype. Further, a mutation in the gene for the ligand should have a phenotype similar to the receptor mutant. What then is the Robo ligand? It proved to be a large secreted protein called Slit (Kidd *et al.*, 1999). The *slit* mutants had been isolated in the original CNS screen along with *comm* and *robo*, but have a different phenotype, in which all of the CNS axons are collapsed on the midline (Fig. 10C). As predicted by the repulsive receptor model, Slit protein is expressed by midline cells and binds to Robo *in vitro*. Genes *robo* and *slit* also interact genetically: *robo/+; slit/+* transheterozygotes have a *robo*-like phenotype, indicating that these genes function closely in the same pathway.

The difference between the *slit* and *robo* phenotypes is caused by redundancy in gene function: *Drosophila* has two more Robo homologs, *robo2* and *robo3*, which are expressed in many of the same neurons as *robo*. The *slit* mutants lack midline repulsion altogether, so that commissural axons are attracted to the midline and stay there. The *robo; robo2* double mutants have the same phenotype as *slit* mutants because their commissural axons cannot sense Slit (Rajagopalan *et al.*, 2000; Simpson *et al.*, 2000). However, in *robo* single mutants, axons lacking Robo reach the midline abnormally, but are then weakly repelled through the Robo2 that they do express and therefore exit the midline to reach the contralateral side.

What is the relationship between *comm* and *robo*? Double *robo; comm* mutants have exactly the same CNS phenotype as *robo*, showing that in the absence of *robo* function, *comm* function is unimportant. Antibody staining shows that Robo protein is expressed at high levels on axons in longitudinal fascicles, but only at low levels on axons in commissures (Kidd *et al.*, 1998b). Serial-section immuno-EM showed that this pattern reflects regulation within single axons: Commissural axons express low levels of Robo while crossing the midline, which allows them to cross the Slit barrier, then upregulate Robo after the midline, rendering them sensitive to Slit and preventing recrossing of the midline (Fig. 11B). The *comm* mutants show abnormally high levels of Robo, including in the midline. Conversely, driving ubiquitous overexpression of Comm from a transgene abolishes Robo expression, and yields a *robo*-like CNS phenotype. Comm regulates Robo levels by preventing Robo protein from reaching the cell-surface, apparently by triggering sorting into a degradation pathway (Keleman *et al.*, 2002; Myat *et al.*, 2002). Thus, Comm's function is to downregulate Robo protein on commissural growth cones as they cross the midline, rendering them insensitive to Slit repulsion. After crossing, Comm function turns off and Robo is upregulated, making the commissural axons sensitive to Slit and preventing them from recrossing.

Robo and Slit in Vertebrates

In mammals, cloning by homology to the fly genes revealed three Robos and three Slits (Brose and Tessier-Lavigne, 2000). The different Slit proteins seem to bind to all the Robos. Many culture studies have shown that vertebrate Slits can repel axons or migrating neurons, in cases where the axons or neurons express Robo endogenously.

In the vertebrate spinal cord, Robo/Slit signaling is likely to control midline crossing in a similar way to the fly ventral nerve cord. Slits are highly expressed in a stripe at the floorplate. As in flies, the responses of commissural axons to Slit are modulated in vertebrates: They are insensitive to Slit before reaching the midline and repelled by Slit after crossing the midline (Zou *et al.*, 2000). However, no Comm has been found to date either in vertebrates or in *C. elegans*, despite extensive searches. Thus the modulation of Slit responses in vertebrates is likely to be through a non-Comm mechanism. However, an *in vivo* function for Robo/Slit signaling in the spinal cord has yet to be demonstrated, and there is strong evidence that other molecules, particularly axonin, NrCAM, and NgCAM, are also involved in midline crossing (Stoeckli *et al.*, 1997).

The best-understood case of vertebrate Robo/Slit signaling is for retinal axons. Retinal ganglion cells express *robo2* as their axons grow across the optic chiasm, which is bounded rostrally and caudally by *slit* expressing cells. Optic chiasm formation is disrupted similarly in both *astray* (*robo2*) mutants in zebrafish (Fricke *et al.*, 2001; Hutson and Chien, 2002) and *slit1/slit2* double mutants in mouse (Plump *et al.*, 2002). The geometry of the chiasm differs from that of the spinal cord or fly ventral midline. Slits are not expressed in a midline stripe, but rather in bands parallel to the retinal axons. Thus Slit repulsion does not act as a gatekeeper at the midline, but instead seems to funnel the axons into their proper pathway. Similarly, Slit in *C. elegans* is not expressed at the ventral midline, and the Robo and Slit mutants *sax-3* and *slt-1* display axon guidance defects more complex than simple problems with midline crossing (Hao *et al.*, 2001).

THE SEMAPHORIN FAMILY OF GUIDANCE MOLECULES

The first identified axon repellent signals were members of the Semaphorin family, the largest known family of guidance molecules. Semaphorins and their receptors were discovered by the convergence of completely different experimental strategies in several model organisms.

Isolation of Collapsin (Sema3A)

The identification of collapsin arose from experiments in which Jonathan Raper and his colleagues grew different types of neurons together in culture. They noticed that axons from the same source would usually cross each other freely, while a growth cone encountering a "foreign" axon would often stop and pull back, repelled by the other axon (Kapfhammer and Raper, 1987). They then found that when DRG growth cones are presented with brain membrane vesicles instead of an intact axon, they exhibit "collapse," a behavior related to repulsion. The collapsing growth cone pulls in all its filopodia, pulls back slightly, and becomes a round bulb-like structure. Collapse is a response to a high uniform concentration of repellent—the growth cone would like to turn away, but has nowhere to go. Since this *collapse assay* (Fig. 7C) is simple, fast, and can test the activity of partially purified membrane-associated proteins, it is an ideal assay for a biochemical purification.

Purifying the DRG-collapsing activity from chick brain yielded collapsin-1 (Luo *et al.*, 1993), which was later renamed Semaphorin 3A when it was recognized as a member of a large family. Purified Sema3A can collapse DRG growth cones at low concentrations. It is a secreted, diffusible molecule, although it tends to bind to cell membranes. Structural analysis showed that in addition to a single Ig domain, Sema3A has a Sema domain, a type of domain first found in Sema1a (see below) and characteristic of all Semas (Fig. 8).

What is the normal function of Sema3A? Collapse is not known to occur frequently *in vivo*, but perhaps this is because growth cones usually encounter gradients rather than high uniform concentrations of Sema3A. Indeed, when DRGs are cocultured in collagen gels with Sema3A-expressing cells, the resulting gradient of diffusible Sema3A causes the DRG axons to turn away rather than collapse (Messersmith *et al.*, 1995). To test Sema3A's function *in vivo*, knockout mice were made. Mutant embryos show defasciculation of several peripheral nerves, and axons exit the DRGs laterally rather than via their normal ventral exit point (Taniguchi *et al.*, 1997).

Semaphorin Family

The first Semaphorin to be isolated was Sema1a from grasshopper (Kolodkin *et al.*, 1992). A monoclonal antibody screen had yielded the 6F8 monoclonal, which stained a subset of axon fascicles in the CNS, and specific bands of epithelial cells in the grasshopper limb bud. These bands coincided with the locations of specific turns made by the growing Ti1 axon. Certain antibodies can interfere with the functions of their ligands, but such "function-blocking" antibodies are the exception rather than the rule. Luckily, 6F8 proved to be such an exception. Culturing limb bud explants in the presence of 6F8 caused Ti1 axons to branch and extend into aberrant territories, thus proving that its antigen is somehow necessary for Ti1 guidance. This antigen was cloned and eventually named Sema1a.

The Semaphorin family is now known to comprise seven classes in animals (Fig. 8) plus one in viruses. Classes 1 and 2 are found in invertebrates, classes 3–7 in vertebrates, and class V in viruses (likely co-opted from their hosts long ago in evolution). Classes 2, 3, and V are secreted, while the other classes either have transmembrane domains or are linked to the membrane

through a glycophosphatidylinositol (GPI) linkage. Roles in axon guidance have been demonstrated for several vertebrate and many invertebrate Semas, but because of the size of the family, have yet to be studied in detail.

Isolation of Sema Receptors

The composition of Semaphorin receptors is complex, but the best-studied components are the neuropilins and the plexins. The founding members of these families were isolated from a monoclonal antibody screen carried out by Hajime Fujisawa's group to look for molecules expressed in specific patterns in the developing *Xenopus* visual system (Takagi *et al.*, 1987). Cloning the antigens identified them as novel transmembrane proteins with potential roles in cell adhesion (Takagi *et al.*, 1991; Ohta *et al.*, 1995), but their function as Sema receptors was discovered by a completely independent route.

The Kolodkin and Tessier-Lavigne groups were led to neuropilin while searching for a Sema3A receptor (He and Tessier-Lavigne, 1997; Kolodkin *et al.*, 1997). They reasoned that since DRG growth cones can be collapsed by Sema3A, DRGs must express the receptor. Fusing the Sema3A coding region to that of alkaline phosphatase (AP) yielded the "affinity reagent" Sema3A–AP—a fusion protein that should bind to Sema3A's receptor and can be visualized using a chromogenic AP reaction. They transfected cultured cells with a cDNA library made from rat DRGs. A few clones gave Sema3A–AP staining when expressed, and these proved to encode rat neuropilin-1. DRG axons express neuropilin-1, and an anti-neuropilin antibody can prevent their repulsion by Sema3A, strongly suggesting that neuropilin is a Sema3A receptor.

The first Plexin shown to be a Sema receptor was VESPR, a receptor for the *viral* semaphorins (Comeau *et al.*, 1998). This virologists' result prompted neurobiologists to test whether neural Plexins have similar roles, and Plexins were indeed found to act as axon guidance receptors for neural Semaphorins (Winberg *et al.*, 1998; Tamagnone *et al.*, 1999). There are two neuropilins and at least nine plexins known in vertebrates. Class 3 Semaphorins require both a plexin and a neuropilin as part of their receptors, while the other classes require plexin only.

The discovery of Semaphorins and their receptors from monoclonal antibody screens on the one hand, and culture assays for biochemical purification and expression cloning on the other hand, illustrates how fruitful it has been to study axon guidance in multiple systems, using multiple experimental approaches.

TARGET RECOGNITION AND TOPOGRAPHIC PROJECTIONS

Introduction

After navigating over long distances to reach their targets, growing axons still have two further tasks. First, they must *recognize* their targets and stop rather than growing past; second, they often must terminate *topographically* in order to preserve spatial information in a sensory or motor projection. Target recognition has been studied extensively in recent years, most notably in the mouse olfactory system (Mombaerts, 1999), the frog visual system (McFarlane *et al.*, 1996), the fly visual system (Clandinin and Zipursky, 2002), and the fly neuromuscular system (Rose and Chiba, 2000). Here, however, we will concentrate on the mechanisms of topographic projections in a classic model, the retinotectal system—the projection of the retina to the optic tectum, its principal target in lower vertebrates. This is the most intensively studied and best understood of all axonal projections.

Roger Sperry (Sperry, 1963) was the first to study the development of retinotectal topography, using fish and frogs as experimental systems. As in many sensory and motor systems, connections in the visual system are topographic in that neighboring neurons in the eye project to neighboring target neurons in the optic tectum. This projection is ordered along two orthogonal axes, dorsal–ventral (D–V) and anterior–posterior (A–P). The map is inverted along both axes. Axons from dorsal retina project to ventral tectum, and ventral retina projects to dorsal tectum; anterior retina projects to posterior tectum, and posterior retina projects to anterior tectum (Fig. 12A). This orderly projection produces a map of visual space on the tectum, allowing the animal to see a faithful representation of its visual world. Sperry surgically rotated the embryonic eye by 180°, and found that these rotated eyes still developed topographic projections to the tectum. Since retinal neurons projected according to their original positions rather than their rotated positions, these animals now saw the world upside-down. These results inspired Sperry's *chemospecificity hypothesis*, which proposed that chemical tags specify the positions of cells on both the retina and the tectum, and that the development of topography is a matching process between the tags expressed by retinal axons and the tags expressed on their target. It seemed implausible that there would be a distinct molecular tag for each of the many positions on the retina and the tectum. Therefore, Sperry proposed that there are only a few tags, but that each is expressed in a gradient across the retina or tectum, and that retinal or tectal position is specified by the *concentrations* of the tags. This model has been proven spectacularly correct by researchers following Sperry's footsteps.

Analyzing Retinotectal Topography *in Vitro*

To identify molecules that might act as chemospecificity cues along the anteroposterior axis, Friedrich Bonhoeffer's group took a functional approach in culture. They explanted tissue from different parts of the chick retina, growing it on carpets of membrane vesicles prepared from different parts of the tectum. Disappointingly, no differences were seen when nasal or temporal retinal explants were grown on uniform carpets of anterior (A) or posterior (P) tectal membranes. (In chick, anterior retina is called "nasal," and posterior retina, "temporal.") Reasoning that there might nevertheless be subtle differences between A and P membranes, the Bonhoeffer group then hit on the idea of presenting retinal axons with a *choice* between the two (Walter *et al.*, 1987). They designed an apparatus that could lay down alternating stripes

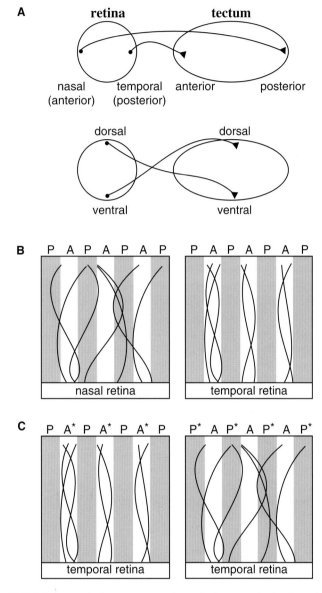

FIGURE 12. Analyzing anteroposterior retinal topography in the stripe assay. (A) Retinal axons project topographically to the tectum along two orthogonal axes, anterior–posterior and dorsal–ventral. (B) In a stripe assay using membranes from anterior (A) or posterior (P) tectum, axons from nasal retina show no preference, but axons from temporal retina prefer to grow on the A stripes. (C) Using heat or PI-PLC to inactivate A membranes (A*) does not affect the preference of temporal axons, but using either method to inactivate P membranes (P*) allows temporal axons to wander freely over A and P* stripes. This shows that P membranes contain a repellent activity.

of A and P membranes and placed strips of retinal tissue in such a way that retinal axons would grow out parallel to these stripes (Fig. 12B). Faced with this choice, axons from nasal retina pay no attention to the stripes. However, axons from temporal retina have a very clear preference for A membranes (which come from the region of the tectum to which these axons would normally project).

There are two possible explanations for this behavior: Either temporal axons could prefer A membranes, or they could be repelled by P membranes. Heat-inactivation of the A membranes had no effect on choice behavior, but heat-inactivation of P membranes abolished the choice (Fig. 12C). This showed that the axons were responding to a repulsive factor in P membranes, most likely a protein. Furthermore, choice was also abolished by pretreatment of the P membranes with phosphatidylinositol phospholipase C (PI-PLC), an enzyme that cleaves extracellular GPI linkages, suggesting that the repulsive factor on P membranes was likely GPI-linked. How does this repulsive factor cause the observed axon choice behavior? When a growth cone encounters the border of a P stripe, it sees repulsive cues only on that side and turns away, thus staying on the A stripe. The next step was to identify this repulsive molecule, which was done by biochemical purification from homogenates of chick brain.

A classical biochemical purification using the stripe assay would have been impractical because this assay is time consuming and requires a large amount of material. Instead, Uwe Drescher in the Bonhoeffer lab used two-dimensional protein gels to search for proteins that were expressed in posterior but not anterior tectum, and that were released by PI-PLC treatment (Drescher *et al.*, 1995). This approach isolated ephrin-A5. Ephrin-A5 mimicked P membranes both in the stripe assay and in the collapse assay. Just as Sperry had predicted long before, ephrin-A5 is expressed in a posterior > anterior gradient on the tectum (Fig. 13). At the same time, John Flanagan's group had been studying the ephrin genes and Eph receptors and trying to determine their function. They found that ephrin-A2 is expressed in a similar posterior > anterior gradient on the tectum, and that EphA receptors are expressed in temporal > nasal gradients in the retina (Cheng *et al.*, 1995). Based on these data, both groups proposed that ephrin-A/EphA signaling might be important for topography. Indeed, both ephrin-A2 and -A5 can guide retinal axons in the stripe assay; conversely, blocking EphA/ephrin-A interactions can abolish axon choice when the stripe assay is performed with P membranes (Monschau *et al.*, 1997; Ciossek *et al.*, 1998).

The ephrins are a family of proteins with very highly related extracellular domains, which are grouped into two subclasses, based on how they are attached to the membrane (reviewed in Kullander and Klein, 2002). The ephrin-As (ephrin-A1 through ephrin-A5) are GPI-linked, while the ephrin-Bs (ephrin-B1 through ephrin-B3) are transmembrane proteins with short intracellular domains. Their receptors are the Eph receptors, a family of receptor tyrosine kinases (RTKs), which are grouped into the EphAs (EphA1 through EphA8) and the EphBs (EphB1 through EphB6). In general, the EphAs preferentially bind the ephrin-As, with each EphA binding to most or all of the ephrin-As, though with differing binding affinities. Similarly, the EphBs bind the ephrin-Bs. As with other RTKs, Eph receptors become tyrosine-phosphorylated upon binding ligand (i.e., ephrin), triggering a signaling cascade within the Eph-expressing cell.

In addition to this *forward signaling*, it has recently been shown that binding of ephrins to Ephs can also trigger responses in the *ephrin*-expressing cell; this has been named *reverse signaling*. This is true for both ephrin-As and ephrin-Bs, and could be a more general phenomenon. Thus, when a membrane-bound "ligand" binds to a transmembrane "receptor," it must always be taken into account that signaling may be bidirectional (both forward and reverse).

A growth cone that encounters a repulsive signal on the surface of another cell, and binds the signal with a receptor on its own surface, now has a problem. The repulsive signal tells it to pull away, but the binding between the signal and its receptor physically links the growth cone and the other cell. Therefore there needs to be a release mechanism. In the case of ephrin-A signaling, the Flanagan lab has shown that ephrin-A can be cleaved extracellularly by a protease, which allows the growth cone to retract (Hattori *et al.*, 2000). When a mutated, uncleavable form of ephrin-A is used, EphA-expressing growth cones will respond to the signal, but are unable to pull away. Whether proteolytic cleavage is *generally* required for repulsive signaling is not yet known.

The Role of ephrin-A/EphA Signaling *in Vivo*

In the developing brain, the A–P distribution of ephrins on the tectum and Ephs on the retina is very much what Sperry had predicted in his chemoaffinity model. In both chicks and in mice, ephrin-A2 and ephrin-A5 are expressed in posterior > anterior gradients on the tectum. In the chick retinal ganglion cell layer, EphA3 is expressed in a temporal > nasal gradient, while EphA4 and A5 are expressed uniformly across the retina. While the mouse has a similar pattern of EphA expression in the retina, it deploys a different set of genes, with EphA5 and EphA6 expressed in temporal > nasal gradients, and with EphA4 expressed uniformly. Thus in both species, the total ephrin-A concentration is high in posterior tectum, and essentially zero in anterior tectum, while the total EphA concentration is high temporally, and lower (but not zero) nasally. Remembering that the stripe and collapse assays show that ephrin-As repel EphA-expressing axons, these expression patterns make functional sense. Temporal axons are most sensitive to ephrin-A repulsion since they express the highest levels of EphA and are thus confined to anterior tectum. Nasal axons are less sensitive to ephrin-A and therefore can reach posterior tectum.

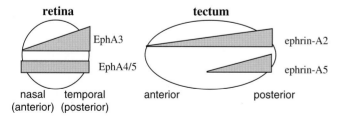

FIGURE 13. Distribution of ephrin-As and EphAs in the chick retinotectal system. Both ephrin-A2 and ephrin-A5 are expressed in high-posterior, low-anterior gradients in the tectum, explaining the repulsive activity of P but not A membranes. EphA3 is expressed in a high-temporal, low-nasal gradient in the retina, explaining why temporal but not nasal retinal axons are sensitive to the P-membrane activity. EphA4 and EphA5 are also expressed in the retina, but are expressed uniformly along the nasotemporal axis.

This model of the *in vivo* function of ephrin-As and EphAs has been tested in three ways. First, when ephrin-A2 is ectopically expressed in patches of the chick tectum using a retroviral vector, temporal axons are repelled by these patches (Nakamoto *et al.*, 1996). Thus, ephrin-A2 is *sufficient* to repel retinal axons *in vivo*. Second, in knockout mice that are doubly mutant for both ephrin-A2 and ephrin-A5, retinotectal topography along the A–P axis is almost completely abolished (Feldheim *et al.*, 2000), showing that the ephrin-A gradients are *necessary* for A–P topography. Finally, when a mouse transgene is used to misexpress EphA in a subset of retinal axons, these axons mistarget to more anterior parts of the colliculus, showing that increased EphA is sufficient to affect A–P targeting (Brown *et al.*, 2000).

Additional Mechanisms for A–P Topography

The experiments described above clearly show that signaling from ephrin-A in the tectum to EphA on retinal axons is critical for A–P topography. If this were the whole story, nasal axons would also be repelled by the tectal ephrin-A gradient, since they do after all express some EphA and would therefore get stuck at the anterior end of the tectum after entering. There are at least two other mechanisms that help retinal axons to spread out over the entire A–P axis.

1. *Ephrin-A expression in retina.* In the retina, where EphAs are expressed in a low-nasal to high-temporal gradient, ephrin-As are expressed in countervailing gradients, that is, high-nasal, low-temporal. Why should the retinal axons express ephrin-A? Removal of ephrin-As from nasal axons increases their sensitivity in the stripe assay, implying that ephrin-As normally antagonize the function of the EphAs (Hornberger *et al.*, 1999). Presumably, ephrin-As either bind to EphA on neighboring axons in *trans*, or to EphA on the same axon in *cis*, and cause habituation or downregulation of the EphA. Thus, nasal axons which express some EphA, but high ephrin-A, will have essentially no EphA function. On the other hand, temporal axons still have high EphA function since they express little ephrin-A. This masking by ephrin-As increases the effective steepness of the EphA gradient across the retina and means that although nasal axons do express EphAs, they should be relatively insensitive to tectal ephrin-A.

2. *Interaction between neighboring axons.* Retinal axons do not act independently when they select termination zones on the tectum. Instead, it is clear that they compete with one another for tectal space. The clearest evidence for this comes from an experiment that used an *islet-2*:*EphA3* mouse transgene to increase EphA levels in about 50% of retinal ganglion cells (Brown *et al.*, 2000). Axons of these cells projected more anteriorly than normal on the tectum, consistent with the expected increase in sensitivity to the tectal ephrin-A gradient. However, the other 50% of axons which express *normal* levels of EphA were also affected. Their axons projected more posteriorly than normal in the tectum, apparently having been pushed out of the anterior tectum by competition with the high-EphA axons. Similar competition is likely to occur during normal development, with the result that when temporal axons occupy anterior tectum, they help to force nasal axons into posterior territory.

A final complication in the A–P topography story is timing. In all vertebrates, retinotectal topography develops in two phases. During the initial termination phase, retinal axons enter their target and start to form arbors, whose size varies greatly with species. During the later refinement phase, arbors are resculpted by adding new branches and retracting old branches, yielding tightly focused termination zones and a very precise map. In rodents, initial termination is extremely imprecise. Axons enter the colliculus at its anterior end and project all the way to the posterior end, with no discernible topographical preference. Only during the refinement phase does topography become evident, as new branches are added specifically at the final termination zone. Thus, neither ephrin-A/EphA signaling nor competition seem to act during the initial phase; both then kick in during arbor refinement. In chicks, initial termination is somewhat more precise: Axons initially project most of the way across the tectum, but concentrate their branches at their eventual termination zone. Zebrafish are the acme of initial precision: initial arbors are already tightly focused in the correct location. Thus, ephrin-A/EphA signaling seems to act earlier in the development of birds and fish.

D–V Topography and Bidirectional Signaling

While a great deal is known about A–P topography, D–V topography is relatively poorly understood. One of the main reasons is that the D–V stripe assay does not work: Retinal axons do not seem to distinguish between membranes from dorsal and ventral tectum. It has long been known that ephrin-Bs and EphBs are expressed in D–V gradients on the retina and the tectum. On the retina, ephrin-B is dorsal > ventral, while EphB is ventral > dorsal. On the tectum, ephrin-B is again dorsal > ventral, and EphB is ventral > dorsal. Unlike ephrin-As and EphAs along the A–P axis, axons with high EphB project to areas of high ephrin-B, while axons with low EphB project to areas of low ephrin-B. This suggests that ephrin-B/EphB signaling might act *attractively* to help set up D–V topography. Indeed, recent experiments from *Xenopus* and mouse show that both ephrin-B to EphB forward signaling and EphB to ephrin-B reverse signaling are important for D–V topography (Hindges *et al.*, 2002; Mann *et al.*, 2002).

SIGNAL TRANSDUCTION

When guidance signals bind to receptors at the plasma membrane of the growth cone, this information must somehow be transmitted to the internal cytoskeleton. Intracellular signal transduction networks have four functions: To *distribute* information across the cell (e.g., by diffusion); to *amplify* small signals into large cytoskeletal effects; to *modulate* the effects of certain signals, controlling whether they act attractively or repulsively; and finally to *integrate* all of the signals the growth cone receives, turning these into well-defined path-finding events (there is no room in development for an indecisive growth cone!). Signal transduction in growth cones is a complex and rapidly growing field. Here we describe some of the most important results.

Classical Second Messengers: Calcium

The roles of calcium and cyclic nucleotides in growth cone guidance have been much studied because good pharmacological reagents have long been available. Calcium and cyclic nucleotides are known to act in so many signaling pathways that it would be surprising if they did not do *something*, and likely several different things, in growth cones. Changes in calcium have been shown to have several different effects. Laser-uncaging calcium in Ti1 growth cones in the grasshopper limb bud locally stimulates filopodial outgrowth (Lau *et al.*, 1999). Conversely, in the *Xenopus* spinal cord *in vivo*, spontaneous calcium transients inhibit axon outgrowth (Gomez and Spitzer, 1999). Uncaging calcium on one side of cultured *Xenopus* growth cones can cause the growth cone to turn either toward that side or away, depending on the resting calcium concentration (Zheng, 2000). Thus, intracellular calcium seems to have different and even opposite effects, depending on the cell type or the state of the cell. A netrin-1 gradient can induce local calcium increases in the growth cone, and whether these result in attraction or repulsion seems to depend on the precise pattern of calcium increase in the growth cone (Hong *et al.*, 2000).

Cyclic Nucleotides Modulate Responses to Guidance Signals

Over the years, cyclic AMP and GMP (cAMP and cGMP) have been shown to have a variety of effects on growth cones. However, recent compelling results suggest that a key role for cyclic nucleotides is to *modulate* the effects of different guidance signals (reviewed in Song and Poo, 1999). Using the *turning assay* on *Xenopus* spinal growth cones, Mu-Ming Poo's group found that certain guidance signals are modulated by cAMP levels (Fig. 14). Netrin-1, NGF, and BDNF, which normally attract these growth cones, will instead repel them when cAMP signaling is inhibited using a competitive antagonist of cAMP or an inhibitor of protein kinase A. On the other hand, MAG, which normally repels these growth cones, will attract them when cAMP signaling is activated. Other guidance signals are modulated by cGMP: NT-3 switches from attractive to repulsive when cGMP signaling is inhibited, while Sema3A switches from repulsive to attractive when cGMP signaling is activated. These results show very clearly that growth cones respond differently to different signals depending on their internal state, and furthermore, show that cyclic nucleotide levels are a key parameter. What then controls cAMP or cGMP levels? One possibility is the ECM protein laminin. Retinal axons grown on laminin have lowered cAMP levels, and furthermore, are repelled by netrin-1, in contrast to retinal axons grown on fibronectin or polylysine, which are attracted by netrin (Höpker *et al.*, 1999).

Small GTPases: Rho, Rac, cdc42

A large number of signaling proteins have been shown to be important in growth cone guidance, including cytoplasmic kinases such as Abl or Pak, which presumably contribute to signal amplification, and adapter proteins such as Dock (the fly homolog of Nck) and Ena (the fly Mena) which link receptors to

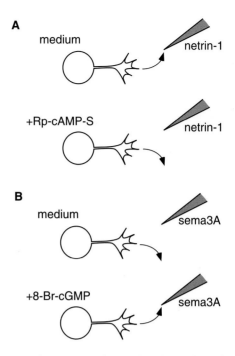

FIGURE 14. Turning assay experiments show that cyclic nucleotide levels can switch growth cone responses between attraction and repulsion. (A) A growth cone that is normally attracted to netrin-1 is now repelled when cAMP signaling is inhibited by bath application of the cAMP antagonist Rp-cAMP-S. (B) A growth cone that is normally repelled by Sema3A is now attracted when cGMP signaling is activated by the cGMP analog 8-Br-cGMP.

downstream signaling components. The best characterized are the small GTPases of the Rho family: Rho, Rac, and cdc42 (reviewed by Luo, 2000). These proteins function as molecular switches: In their GTP-bound form they are active; then over time they hydrolyze GTP to GDP and turn themselves off. They can then exchange GDP for a fresh GTP and turn back on again. They are regulated by specific GTPase-activating proteins (GAPs) and guanine nucleotide exchange factors (GEFs), which can turn them off (GAPs) or on (GEFs). These switches have powerful effects on the cytoskeleton. In fibroblasts and other non-neuronal cells, cdc42 induces filopodial formation, Rac induces lamellipodial activity, and Rho induces stress fiber formation; there is also evidence that cdc42 activates Rac, which in turn activates Rho. In growth cones, the Rho GTPases control analogous cytoskeletal changes, and therefore many investigators have tested whether Rho family members are involved in neuronal motility. One of the clearest examples comes not from axon guidance but from neural migration. Yi Rao's group studied SVZa cells, neurons whose migration in culture is repelled by Slit, acting through Robo (Wong *et al.*, 2001). A yeast two-hybrid screen for proteins that bind to the cytoplasmic tail of Robo isolated the srGAPs (for Slit–Robo GAPs). Slit increases the binding of srGAPs to Robo, and srGAPs specifically inactivate cdc42. In culture, repulsion of SVZa neurons by Slit requires both srGAP activity and cdc42 inactivation, which presumably inhibits filopodial formation. Thus, in this case cdc42 plays a key role downstream of Slit/Robo signaling, directly mediated by a specific GAP.

CONTROL OF DENDRITE OUTGROWTH

Initial Dendritic Development

Dendrites are just as critical to neuronal function as axons. Think of the Purkinje cell, whose hallmark is its baroque dendritic fan, receiving thousands of synaptic inputs. However, much less is known about the development of dendrites than that of axons, partly because of technical limitations (dendrites are harder to visualize), and partly for historical reasons (there are no dendrites at the neuromuscular junction!). However, work on dendrites has blossomed over the last decade and a half. Just as with axonal development, the development of dendrites can be separated into two phases: initial outgrowth (roughly speaking, before synaptogenesis), and later refinement (after synaptogenesis). While much recent interest has focused on activity-dependent refinement (reviewed in Cline, 2001; Wong and Ghosh, 2002), especially the dynamics of dendritic spines, we concentrate here on initial outgrowth. Three key questions are (1) how dendrites are generated, (2) how processes decide to be dendrites, and (3) what determines their direction of outgrowth. For each, the balance between intrinsic and extrinsic factors, and some of the molecules involved, are starting to be known.

Generation of Dendrites

When grown in pure neuronal cultures in serum-free medium, sympathetic neurons develop essentially no dendrites, which is very different from their behavior *in vivo*. This shows that extrinsic factors must play a role in inducing dendrites. Pamela Lein, Dennis Higgins, and colleagues have shown that bone morphogenetic proteins (BMPs) are likely one of these factors. Adding the growth factor BMP-7 (also known as OP-1) to sympathetic cultures leads to a normal number of dendrites (Lein *et al*., 1995). Adding glia derived from sympathetic ganglia also induces normal dendrites, and the glial effect can be blocked either by anti-BMP antibodies, or by the BMP antagonists follistatin and noggin (Lein *et al*., 2002). *In vivo*, BMPs and BMP receptors are expressed in sympathetic ganglia during perinatal ages (the normal period of rapid dendrite growth). Thus, a plausible working hypothesis is that glia upregulate BMP signaling in sympathetic neurons during perinatal ages, and that this leads to increased dendrite growth.

In addition to extrinsic factors such as BMPs, intrinsic factors must also be necessary for dendrite formation. Peter Baas and colleagues have shown that one such intrinsic factor is the motor protein CHO1/MKLP1, which slides oppositely oriented microtubules toward each other (Yu *et al*., 2000). The microtubules of axons are oriented with all their plus ends distal, while dendrites have a mixture of plus-end-distal and minus-end-distal. In culture, immature processes are "axon-like," with all plus ends distal. As dendrites mature, they gradually acquire a mixture of plus-end-distal and minus-end-distal microtubules. When CHO1/MKLP1 function is abrogated in cultured sympathetic neurons using antisense oligonucleotides, all of the neurites remain plus-end-distal; furthermore, dendrites fail to form. Thus,

rearrangement of the microtubule cytoskeleton by CHO1/MKLP1 seems to play a key role in dendrite formation.

Other known intrinsic factors are proteins that regulate cytoskeletal dynamics, which are likely to play similar roles in dendrites as they do in axons. The best-studied are the Rho GTPases: Rho, Rac, and Cdc42. Their function in dendrites has been studied in a variety of neurons, mostly using constitutively-active and dominant-negative forms, but also using genetic mutants in *Drosophila*. In general it seems that Rac and Cdc42 promote dendrite outgrowth, while Rho inhibits dendrite outgrowth (reviewed in Luo, 2000). However, there are several exceptions to this rule. For instance, when mutant forms of Rac and Cdc42 were misexpressed in embryonic *Drosophila* sensory neurons, Rac perturbation did not affect dendrites (though it did affect axons), while Cdc42 affected both dendrites and axons. As in axons, the regulation of Rho GTPases in dendrites is likely to be more complex than in fibroblasts, and to depend on the particular cell type being considered.

Dendritic vs Axonal Fate

In the brain, neurons generally have one axon, but multiple dendrites. How does each process know whether to be an axon or a dendrite? Hippocampal neurons are a good model, since (unlike sympathetic neurons) they reliably develop multiple dendrites when grown in culture. Thus, these neurons must have an intrinsic mechanism for generating one and only one axon. Extensive studies by Gary Banker, Carlos Dotti, and colleagues suggest that neurites compete with each other to decide which becomes the axon (reviewed in Bradke and Dotti, 2000).

The development of cultured hippocampal neurons has four characteristic stages (Fig. 15). In stage 1, the neuron initially attaches to the culture dish. In stage 2, the neuron starts to extend four or five neurites, all morphologically indistinguishable. At the end of stage 2, exactly one neurite becomes specified to become the axon. Its actin cytoskeleton becomes destabilized, its growth cone grows large, and the neuron's cytoplasmic flow becomes asymmetric, preferentially delivering mitochondria, ribosomes, and other organelles to this neurite. In stage 3, this chosen neurite becomes a *bona fide* nascent axon. It is much longer than the other neurites, grows much faster, and begins to acquire the normal complement of axon-specific proteins such as the microfilament-associated protein Tau. However, at this time neurite identity is still plastic: If the growing axon is cut, one of the nascent dendrites will take over and become the axon. In stage 4, neurite identity becomes determined: Cutting the axon no longer affects the nascent dendrites.

These results have led to a "tug of war" model. All neurites have an inherent tendency to become the axon, and each sends inhibitory, anti-axonal signals to the others. At the end of stage 2, one neurite starts to dominate. Its inhibition of the other neurites strengthens, while the inhibition it receives weakens. In stage 3 (but not stage 4), cutting the nascent axon removes the inhibition it sends to the other neurites, allowing one to become the new axon. Thus, a combination of positive feedback in the nascent axon and negative feedback to the other neurites ensures that

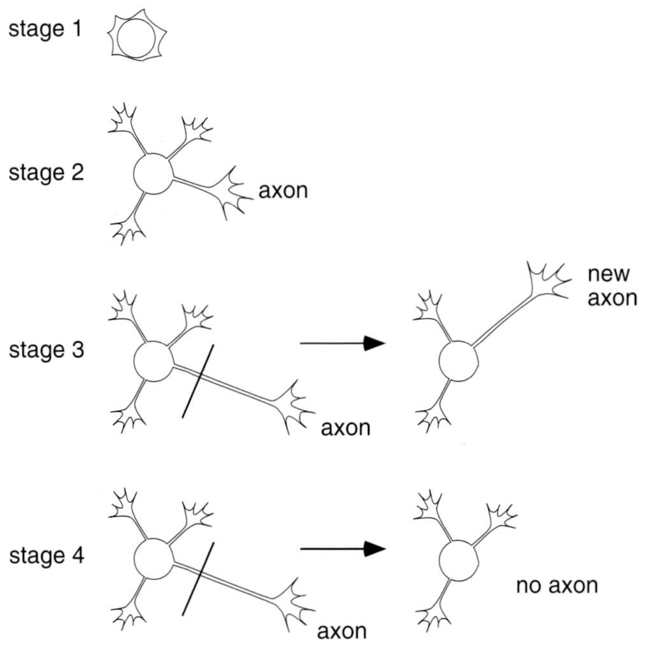

stage 1

stage 2

axon

stage 3

new
axon

axon

no axon

stage 4

axon

FIGURE 15. Development of dendrites in cultured hippocampal neurons. When hippocampal neurons are plated in culture, they go through four stages of neurite differentiation. In stage 1, the cell body attaches to the culture dish and spreads membrane ruffles in all directions. In stage 2, the neuron extends 4–5 processes. At the end of this stage, a single growth cone becomes slightly bigger and starts to grow faster; this process will become the axon, while the others will all become dendrites. In stage 3, the axon becomes much longer and begins to express axon-specific proteins. However, if the axon is severed, another process will take over and become a new axon. In stage 4, dendrite identity has become fixed, so that severing the axon yields a neuron with only dendrites.

there is always exactly one axon. The molecular basis of this competition remains to be determined.

Guidance of Dendrites

Once neurons have elaborated dendrites, what determines their orientation? The best example of a guidance signal for dendrites is Sema3A, which acts on the apical dendrites of pyramidal neurons in the cortex (Polleux *et al.*, 2000). These neurons' dendrites normally point up toward the marginal zone, near the pial surface, while their axons point down toward the ventricle. Anirvan Ghosh and collaborators used an overlay culture system to analyze the guidance of these dendrites. By plating GFP-expressing pyramidal neurons onto live or fixed cortical slices, they were able to visualize the behavior of individual dendrites in a normal or manipulated environment. They first showed that apical dendrites are attracted by a diffusible signal originating near the marginal zone (a region where Sema3A is expressed). In cultures that also contained a clump of cells transfected with Sema3A, dendrites were attracted toward the transfected cells, showing that Sema3A is *sufficient* to attract apical dendrites. Sema3A is also *necessary*: Apical dendrites lost their orientation when a Sema3A fusion protein was bath-applied (to swamp out the endogenous gradient) or when the GFP-labeled

neurons were grown on slices from *Sema3A−/−* mutant mice (which lack the endogenous gradient). Thus it is quite clear that apical dendrites of pyramidal neurons are normally attracted by Sema3A originating near the pial surface.

Intriguingly, these investigators had previously shown that the axons of these same pyramidal neurons are *repelled* by Sema3A (Polleux *et al.*, 1998). Both the axons and the dendrites express neuropilin-1, a component of the Sema3A receptor complex. What then makes the axon and dendrite of the same cell behave in opposite ways? A possible answer was suggested by the finding that raising cyclic GMP levels can change Sema3A from repulsive to attractive for *Xenopus* spinal growth cones (see above). Therefore Ghosh's group investigated a possible role for cGMP signaling in pyramidal neurons. They found that levels of soluble guanylate cyclase (sGC) are high in the apical dendrite and low in the soma (and, presumably, also low in the axon). Furthermore, pharmacological inhibitors of cGMP signaling abolished the orientation preference of apical dendrites in the overlay culture system. Thus, the different effects of Sema3A on the axon (repulsion) and dendrite (attraction) are likely due to greater cGMP signaling in the dendrite—a striking example of how a particular guidance signal can have different effects within the same neuron.

Future Questions

The branching of dendrites is one of their most notable properties. Many factors, both intrinsic (CAM kinase II, CPG15, Notch) and extrinsic (glutamate, neurotrophins, Slit) have been shown to affect dendritic branching either in culture or *in vivo* (reviewed in Whitford *et al.*, 2002), especially during activity-dependent refinement. An important future question will be what factors act *in vivo* during the *initial* formation of dendrite branches.

Certain areas of the nervous system are parceled out by *dendritic tiling*, so that the dendritic territories of neighboring neurons abut each other, but do not overlap. This tiling is important because it provides an efficient way to cover dendritic territory while maximizing spatial resolution. The best-studied examples are retinal ganglion cells in the vertebrate retina (Wässle *et al.*, 1981) and the dorsal arborization neurons of *Drosophila* embryos (Gao *et al.*, 1999). Genetic screens in the fly have started to elucidate the genes that control tiling by the latter, and this will be an area of intense interest in the future.

CONCLUSION

Many genes are now known to operate in the growth cone—that structure first identified by Ramón y Cajal so long ago. However, many questions remain about how the growth cone can integrate a large number of external and internal signals, and so guide the growing axon across varied terrain. Five broad questions are especially interesting.

1. What are *all* the important axon guidance genes? Many axon guidance molecules certainly remain to be identified, particularly those that make up intracellular signaling cascades, but also ligands and receptors.

2. How are guidance molecules regulated? It is clear that growth cones have much more interesting cell biology than we had realized. In the growth cone itself, the function of a guidance molecule can be regulated at the level of translation, insertion, recycling, internalization, or degradation. Back in the cell body, it can be regulated at the level of transcription, splicing, RNA targeting, or translation. The roles of each of these forms of regulation during *in vivo* guidance remains to be elucidated.

3. How do guidance signaling pathways interact? Intracellular signaling will be a rich subject for years to come, partly because it is so complex (sometimes it seems that all pathways interact with each other!). It will be particularly interesting to understand how the effects of particular axon guidance signals can be *modulated* (over time or across space), and how different signals *interact* (at the levels of the ligands themselves, their receptors, or downstream signaling cascades).

4. What is the difference between attraction and repulsion? Many guidance signals can be either attractive or repulsive, controlled by something as simple as a change in cyclic nucleotide levels. Does each signal have two downstream signaling pathways, one attractive and one repulsive, or do multiple signals somehow feed into a common machinery that is delicately balanced between attraction and repulsion?

5. How does each guidance molecule affect growth cone behavior? Improved *in vivo* imaging techniques are now making it possible to visualize growth cone behavior not just in fixed tissue, but as it takes place in real time. An especially interesting issue will be to see how the same molecules can have different roles, depending on the axon in which they are expressed.

Finally, the study of dendrite outgrowth and branching is just in its infancy, but will surely be just as interesting as axon guidance, with the additional twist that control by neural activity will be especially important.

ACKNOWLEDGMENTS

It is unfortunately not possible to cite comprehensively all the references in a field as broad and fast-moving as axon guidance. Apologies to those colleagues whose work could not be cited in detail. Thanks to Paul Garrity for comments on the manuscript, to Lara Hutson and Michele Lemons for growth cone time-lapse sequences, to Niki Hack and Molly Chien for making the writing of this chapter so interesting, and to Mahendra Rao for his patience. C-BC is supported by grants from the National Institutes of Health (National Eye Institute) and the National Science Foundation.

REFERENCES

Baas, P.W., 2002, Microtubule transport in the axon, *Int. Rev. Cytol.* 212:41–62.

Benowitz, L.I. and Routtenberg, A., 1997, GAP-43: An intrinsic determinant of neuronal development and plasticity, *Trends Neurosci.* 20:84–91.

Bentley, D. and Toroian-Raymond, A., 1986, Disoriented pathfinding by pioneer neurone growth cones deprived of filopodia by cytochalasin treatment, *Nature* 323:712–715.

Bradke, F. and Dotti, C.G., 2000, Establishment of neuronal polarity: Lessons from cultured hippocampal neurons, *Curr. Opin. Neurobiol.* 10:574–581.

Brose, K. and Tessier-Lavigne, M., 2000, Slit proteins: Key regulators of axon guidance, axonal branching, and cell migration, *Curr. Opin. Neurobiol.* 10:95–102.

Brown, A., Yates, P.A., Burrola, P., Ortuno, D., Vaidya, A., Jessell, T.M. *et al.* 2000, Topographic mapping from the retina to the midbrain is controlled by relative but not absolute levels of EphA receptor signaling, *Cell* 102:77–88.

Buck, K.B. and Zheng, J.Q., 2002, Growth cone turning induced by direct local modification of microtubule dynamics, *J. Neurosci.* 22:9358–9367.

Campbell, D.S. and Holt, C.E., 2001, Chemotropic responses of retinal growth cones mediated by rapid local protein synthesis and degradation, *Neuron* 32:1013–1026.

Chalfie, M., Tu, Y., Euskirchen, G., Ward, W.W., and Prasher, D.C., 1994, Green fluorescent protein as a marker for gene expression, *Science* 263:802–805.

Chan, S.S., Zheng, H., Su, M.W., Wilk, R., Killeen, M.T., Hedgecock, E.M. *et al.*, 1996, UNC-40, a *C. elegans* homolog of DCC (Deleted in Colorectal Cancer), is required in motile cells responding to UNC-6 netrin cues, *Cell* 87:187–195.

Cheng, H.J., Nakamoto, M., Bergemann, A.D., and Flanagan, J.G., 1995, Complementary gradients in expression and binding of ELF-1 and Mek4 in development of the topographic retinotectal projection map, *Cell* 82:371–381.

Chien, C.B., Rosenthal, D.E., Harris, W.A., and Holt, C.E., 1993, Navigational errors made by growth cones without filopodia in the embryonic *Xenopus* brain, *Neuron* 11:237–251.

Ciossek, T., Monschau, B., Kremoser, C., Loschinger, J., Lang, S., Muller, B.K. *et al.*, 1998, Eph receptor-ligand interactions are necessary for guidance of retinal ganglion cell axons in vitro, *Eur. J. Neurosci.* 10:1574–1580.

Clandinin, T.R. and Zipursky, S.L., 2002, Making connections in the fly visual system, *Neuron* 35:827–841.

Cline, H.T., 2001, Dendritic arbor development and synaptogenesis, *Curr. Opin. Neurobiol.* 11:118–126.

Colamarino, S.A. and Tessier-Lavigne, M., 1995, The axonal chemoattractant netrin-1 is also a chemorepellent for trochlear motor axons, *Cell* 81:621–629.

Comeau, M.R., Johnson, R., DuBose, R.F., Petersen, M., Gearing, P., VandenBos, T. *et al.*, 1998, A poxvirus-encoded semaphorin induces cytokine production from monocytes and binds to a novel cellular semaphorin receptor, VESPR, *Immunity* 8:473–482.

Davenport, R.W., Dou, P., Rehder, V., and Kater, S.B., 1993, A sensory role for neuronal growth cone filopodia, *Nature* 361:721–724.

de la Torre, J.R., Hopker, V.H., Ming, G.L., Poo, M.M., Tessier-Lavigne, M., Hemmati-Brivanlou, A. *et al.*, 1997, Turning of retinal growth cones in a netrin-1 gradient mediated by the netrin receptor DCC, *Neuron* 19:1211–1224.

Deiner, M.S., Kennedy, T.E., Fazeli, A., Serafini, T., Tessier-Lavigne, M., and Sretavan, D.W., 1997, Netrin-1 and DCC mediate axon guidance locally at the optic disc: Loss of function leads to optic nerve hypoplasia, *Neuron* 19:575–589.

Drescher, U., Kremoser, C., Handwerker, C., Loschinger, J., Noda, M., and Bonhoeffer, F., 1995, In vitro guidance of retinal ganglion cell axons by RAGS, a 25 kDa tectal protein related to ligands for Eph receptor tyrosine kinases, *Cell* 82:359–370.

Fazeli, A., Dickinson, S.L., Hermiston, M.L., Tighe, R.V., Steen, R.G., Small, C.G. *et al.*, 1997, Phenotype of mice lacking functional Deleted in colorectal cancer (Dcc) gene, *Nature* 386:796–804.

Feldheim, D.A., Kim, Y.I., Bergemann, A.D., Frisen, J., Barbacid, M., and Flanagan, J.G., 2000, Genetic analysis of ephrin-A2 and ephrin-A5 shows their requirement in multiple aspects of retinocollicular mapping, *Neuron* 25:563–574.

Fricke, C., Lee, J.S., Geiger-Rudolph, S., Bonhoeffer, F., and Chien, C.B., 2001, Astray, a zebrafish roundabout homolog required for retinal axon guidance, *Science* 292:507–510.

Futerman, A.H., and Banker, G.A., 1996, The economics of neurite outgrowth—the addition of new membrane to growing axons, *Trends Neurosci.* 19:144–149.

Gao, F.B., Brenman, J.E., Jan, L.Y., and Jan, Y.N., 1999, Genes regulating dendritic outgrowth, branching, and routing in *Drosophila*, *Genes Dev.* 13:2549–2561.

Giniger, E., Tietje, K., Jan, L.Y., and Jan, Y.N., 1994, lola encodes a putative transcription factor required for axon growth and guidance in Drosophila, *Development* 120:1385–1398.

Giuditta, A., Kaplan, B.B., van Minnen, J., Alvarez, J., and Koenig, E., 2002, Axonal and presynaptic protein synthesis: New insights into the biology of the neuron, *Trends Neurosci.* 25:400–404.

Gomez, T.M. and Spitzer, N.C., 1999, In vivo regulation of axon extension and pathfinding by growth-cone calcium transients, *Nature* 397:350–355.

Grabham, P.W. and Goldberg, D.J., 1997, Nerve growth factor stimulates the accumulation of beta1 integrin at the tips of filopodia in the growth cones of sympathetic neurons, *J. Neurosci.* 17:5455–5465.

Hamelin, M., Zhou, Y., Su, M.W., Scott, I.M., and Culotti, J.G., 1993, Expression of the UNC-5 guidance receptor in the touch neurons of *C. elegans* steers their axons dorsally, *Nature* 364:327–330.

Hao, J.C., Yu, T.W., Fujisawa, K., Culotti, J.G., Gengyo-Ando, K., Mitani, S. *et al.*, 2001, *C. elegans* slit acts in midline, dorsal–ventral, and anterior–posterior guidance via the SAX-3/Robo receptor, *Neuron* 32:25–38.

Harris, W.A., Holt, C.E., and Bonhoeffer, F., 1987, Retinal axons with and without their somata, growing to and arborizing in the tectum of Xenopus embryos: A time-lapse video study of single fibres in vivo, *Development* 101:123–133.

Harrison, R.G., 1907, Observations on the living developing nerve fiber, *Anat. Rec.* 1:116–118.

Hattori, M., Osterfield, M., and Flanagan, J.G., 2000, Regulated cleavage of a contact-mediated axon repellent, *Science* 289:1360–1365.

He, Z. and Tessier-Lavigne, M., 1997, Neuropilin is a receptor for the axonal chemorepellent Semaphorin III, *Cell* 90:739–751.

Hedgecock, E.M., Culotti, J.G., and Hall, D.H., 1990, The unc-5, unc-6, and unc-40 genes guide circumferential migrations of pioneer axons and mesodermal cells on the epidermis in *C. elegans*, *Neuron* 4:61–85.

Heidemann, S.R., Lamoureux, P., and Buxbaum, R.E., 1990, Growth cone behavior and production of traction force, *J. Cell. Biol.* 111:1949–1957.

Hindges, R., McLaughlin, T., Genoud, N., Henkemeyer, M., and O'Leary, D.D., 2002, EphB forward signaling controls directional branch extension and arborization required for dorsal-ventral retinotopic mapping, *Neuron* 35:475–487.

Hong, K., Hinck, L., Nishiyama, M., Poo, M.M., Tessier-Lavigne, M., and Stein, E., 1999, A ligand-gated association between cytoplasmic domains of UNC5 and DCC family receptors converts netrin-induced growth cone attraction to repulsion, *Cell* 97:927–941.

Hong, K., Nishiyama, M., Henley, J., Tessier-Lavigne, M., and Poo, M., 2000, Calcium signalling in the guidance of nerve growth by netrin-1, *Nature* 403:93–98.

Honig, M.G. and Hume, R.I., 1986, Fluorescent carbocyanine dyes allow living neurons of identified origin to be studied in long-term cultures, *J. Cell. Biol.* 103:171–187.

Höpker, V.H., Shewan, D., Tessier-Lavigne, M., Poo, M., and Holt, C., 1999, Growth-cone attraction to netrin-1 is converted to repulsion by laminin-1, *Nature* 401:69–73.

Hornberger, M.R., Dutting, D., Ciossek, T., Yamada, T., Handwerker, C., Lang, S. *et al.*, 1999, Modulation of EphA receptor function by coexpressed ephrinA ligands on retinal ganglion cell axons, *Neuron* 22:731–742.

Hutson, L.D. and Chien, C.B., 2002, Pathfinding and error correction by retinal axons: The role of astray/robo2, *Neuron* 33:205–217.

Ishii, N., Wadsworth, W.G., Stern, B.D., Culotti, J.G., and Hedgecock, E.M., 1992, UNC-6, a laminin-related protein, guides cell and pioneer axon migrations in *C. elegans*, *Neuron* 9:873–881.

Jay, D.G., 2000, The clutch hypothesis revisited: Ascribing the roles of actin-associated proteins in filopodial protrusion in the nerve growth cone, *J. Neurobiol.* 44:114–125.

Kapfhammer, J.P. and Raper, J.A., 1987, Interactions between growth cones and neurites growing from different neural tissues in culture, *J. Neurosci.* 7:1595–1600.

Keino-Masu, K., Masu, M., Hinck, L., Leonardo, E.D., Chan, S.S., Culotti, J.G. *et al.*, 1996, Deleted in Colorectal Cancer (DCC) encodes a netrin receptor, *Cell* 87:175–185.

Keleman, K., Rajagopalan, S., Cleppien, D., Teis, D., Paiha, K., Huber, L.A. *et al.*, 2002, Comm sorts robo to control axon guidance at the Drosophila midline, *Cell* 110:415–427.

Kennedy, T.E., Serafini, T., de la Torre, J.R., and Tessier-Lavigne, M., 1994, Netrins are diffusible chemotropic factors for commissural axons in the embryonic spinal cord, *Cell* 78:425–435.

Kidd, T., Brose, K., Mitchell, K.J., Fetter, R.D., Tessier-Lavigne, M., Goodman, C.S. *et al.*, 1998a, Roundabout controls axon crossing of the CNS midline and defines a novel subfamily of evolutionarily conserved guidance receptors, *Cell* 92:205–215.

Kidd, T., Russell, C., Goodman, C.S., and Tear, G., 1998b, Dosage-sensitive and complementary functions of roundabout and commissureless control axon crossing of the CNS midline, *Neuron* 20:25–33.

Kidd, T., Bland, K.S., and Goodman, C.S., 1999, Slit is the midline repellent for the robo receptor in Drosophila, *Cell* 96:785–794.

Kolodkin, A.L., Matthes, D.J., O'Connor, T.P., Patel, N.H., Admon, A., Bentley, D. *et al.*, 1992, Fasciclin IV: Sequence, expression, and function during growth cone guidance in the grasshopper embryo, *Neuron* 9:831–845.

Kolodkin, A.L., Levengood, D.V., Rowe, E.G., Tai, Y.T., Giger, R.J., and Ginty, D.D., 1997, Neuropilin is a semaphorin III receptor, *Cell* 90:753–762.

Kullander, K. and Klein, R., 2002, Mechanisms and functions of Eph and ephrin signalling, *Nat. Rev. Mol. Cell Biol.* 3:475–486.

Lau, P.M., Zucker, R.S., and Bentley, D., 1999, Induction of filopodia by direct local elevation of intracellular calcium ion concentration, *J. Cell. Biol.* 145:1265–1275.

Lein, P., Johnson, M., Guo, X., Rueger, D., and Higgins, D., 1995, Osteogenic protein-1 induces dendritic growth in rat sympathetic neurons, *Neuron* 15:597–605.

Lein, P.J., Beck, H.N., Chandrasekaran, V., Gallagher, P.J., Chen, H.L., Lin, Y. *et al.*, 2002, Glia induce dendritic growth in cultured sympathetic neurons by modulating the balance between bone morphogenetic proteins (BMPs) and BMP antagonists, *J. Neurosci.* 22:10377–10387.

Leonardo, E.D., Hinck, L., Masu, M., Keino-Masu, K., Ackerman, S.L., and Tessier-Lavigne, M., 1997, Vertebrate homologues of *C. elegans* UNC-5 are candidate netrin receptors, *Nature* 386:833–838.

Leung-Hagesteijn, C., Spence, A.M., Stern, B.D., Zhou, Y., Su, M.W., Hedgecock, E.M. *et al.*, 1992, UNC-5, a transmembrane protein with immunoglobulin and thrombospondin type 1 domains, guides cell and pioneer axon migrations in *C. elegans*, *Cell* 71:289–299.

Luo, L., 2000, Rho GTPases in neuronal morphogenesis, *Nat. Rev. Neurosci.* 1:173–180.

Luo, Y., Raible, D., and Raper, J.A., 1993, Collapsin: A protein in brain that induces the collapse and paralysis of neuronal growth cones, *Cell* 75:217–227.

Mann, F., Ray, S., Harris, W., and Holt, C., 2002, Topographic mapping in dorsoventral axis of the *Xenopus* retinotectal system depends on signaling through ephrin-B ligands, *Neuron* 35:461–473.

McFarlane, S., Cornel, E., Amaya, E., and Holt, C.E., 1996, Inhibition of FGF receptor activity in retinal ganglion cell axons causes errors in target recognition, *Neuron* 17:245–254.

Mehlen, P., Rabizadeh, S., Snipas, S.J., Assa-Munt, N., Salvesen, G.S., and Bredesen, D.E., 1998, The DCC gene product induces apoptosis by a mechanism requiring receptor proteolysis, *Nature* 395:801–804.

Messersmith, E.K., Leonardo, E.D., Shatz, C.J., Tessier-Lavigne, M., Goodman, C.S., and Kolodkin, A.L., 1995, Semaphorin III can function as a selective chemorepellent to pattern sensory projections in the spinal cord, *Neuron* 14:949–959.

Mitchison, T. and Kirschner, M., 1988, Cytoskeletal dynamics and nerve growth, *Neuron* 1:761–772.

Mombaerts, P., 1999, Molecular biology of odorant receptors in vertebrates, *Annu. Rev. Neurosci.* 22:487–509.

Monschau, B., Kremoser, C., Ohta, K., Tanaka, H., Kaneko, T., Yamada, T. *et al.*, 1997, Shared and distinct functions of RAGS and ELF-1 in guiding retinal axons, *Embo. J.* 16:1258–1267.

Myat, A., Henry, P., McCabe, V., Flintoft, L., Rotin, D., and Tear, G., 2002, Drosophila Nedd4, a ubiquitin ligase, is recruited by Commissureless to control cell surface levels of the roundabout receptor, *Neuron* 35:447–459.

Nakamoto, M., Cheng, H.J., Friedman, G.C., McLaughlin, T., Hansen, M.J., Yoon, C.H. *et al.*, 1996, Topographically specific effects of ELF-1 on retinal axon guidance in vitro and retinal axon mapping in vivo, *Cell* 86:755–766.

O'Connor, T.P. and Bentley, D., 1993, Accumulation of actin in subsets of pioneer growth cone filopodia in response to neural and epithelial guidance cues in situ, *J. Cell. Biol.* 123:935–948.

Ohta, K., Mizutani, A., Kawakami, A., Murakami, Y., Kasuya, Y., Takagi, S. *et al.*, 1995, Plexin: A novel neuronal cell surface molecule that mediates cell adhesion via a homophilic binding mechanism in the presence of calcium ions, *Neuron* 14:1189–1199.

Placzek, M., Tessier-Lavigne, M., Jessell, T., and Dodd, J., 1990, Orientation of commissural axons in vitro in response to a floor plate-derived chemoattractant, *Development* 110:19–30.

Plump, A.S., Erskine, L., Sabatier, C., Brose, K., Epstein, C.J., Goodman, C.S. *et al.*, 2002, Slit1 and Slit2 cooperate to prevent premature midline crossing of retinal axons in the mouse visual system, *Neuron* 33:219–232.

Polleux, F., Giger, R.J., Ginty, D.D., Kolodkin, A.L., and Ghosh, A., 1998, Patterning of cortical efferent projections by semaphorin–neuropilin interactions, *Science* 282:1904–1906.

Polleux, F., Morrow, T., and Ghosh, A., 2000, Semaphorin 3A is a chemoattractant for cortical apical dendrites, *Nature* 404:567–573.

Rajagopalan, S., Nicolas, E., Vivancos, V., Berger, J., and Dickson, B.J., 2000, Crossing the midline: Roles and regulation of Robo receptors, *Neuron* 28:767–777.

Ramòn y Cajal, S., 1890, A quelle époque apparaissent les expansions des cellules nerveuses de la moëlle épinière du poulet? *Anat. Anzeiger* 5:609–613.

Ramón y Cajal, S., 1972, *The Structure of the Retina* (Thorpe, S.A., and Glickstein, M., Trans.), Charles C. Thomas, Springfield, Illinois.

Rose, D. and Chiba, A., 2000, Synaptic target recognition at Drosophila neuromuscular junctions, *Microsc. Res. Tech.* 49:3–13.

Sabry, J.H., O'Connor, T.P., Evans, L., Toroian-Raymond, A., Kirschner, M., and Bentley, D., 1991, Microtubule behavior during guidance of pioneer neuron growth cones in situ, *J. Cell. Biol.* 115:381–395.

Seeger, M., Tear, G., Ferres-Marco, D., and Goodman, C.S., 1993, Mutations affecting growth cone guidance in Drosophila: Genes

necessary for guidance toward or away from the midline, *Neuron* 10:409–426.

Serafini, T., Kennedy, T.E., Galko, M.J., Mirzayan, C., Jessell, T.M., and Tessier-Lavigne, M., 1994, The netrins define a family of axon outgrowth-promoting proteins homologous to *C. elegans* UNC-6, *Cell* 78:409–424.

Serafini, T., Colamarino, S.A., Leonardo, E.D., Wang, H., Beddington, R., Skarnes, W.C. *et al.*, 1996, Netrin-1 is required for commissural axon guidance in the developing vertebrate nervous system, *Cell* 87:1001–1014.

Shirasaki, R., Katsumata, R., and Murakami, F., 1998, Change in chemoattractant responsiveness of developing axons at an intermediate target, *Science* 279:105–107.

Simpson, J.H., Kidd, T., Bland, K.S., and Goodman, C.S., 2000, Short-range and long-range guidance by slit and its Robo receptors. Robo and Robo2 play distinct roles in midline guidance, *Neuron* 28:753–766.

Song, H.J. and Poo, M.M., 1999, Signal transduction underlying growth cone guidance by diffusible factors, *Curr. Opin. Neurobiol.* 9:355–363.

Speidel, C.C., 1933, Studies of living nerves. II. Activities of ameboid growth cones, sheath cells, and myelin segments, as revealed by prolonged observation of individual nerve fibers in frog tadpoles, *Am. J. Anat.* 52:1–79.

Sperry, R.W., 1963, Chemoaffinity in the orderly growth of nerve fiber patterns and connections, *Proc. Natl. Acad. Sci. USA* 50:703–710.

Stein, E. and Tessier-Lavigne, M., 2001, Hierarchical organization of guidance receptors: Silencing of netrin attraction by slit through a Robo/DCC receptor complex, *Science* 291:1928–1938.

Stoeckli, E.T., Sonderegger, P., Pollerberg, G.E., and Landmesser, L.T., 1997, Interference with axonin-1 and NrCAM interactions unmasks a floorplate activity inhibitory for commissural axons, *Neuron* 18:209–221.

Suter, D.M. and Forscher, P., 2000, Substrate-cytoskeletal coupling as a mechanism for the regulation of growth cone motility and guidance, *J. Neurobiol.* 44:97–113.

Takagi, S., Tsuji, T., Amagai, T., Takamatsu, T., and Fujisawa, H., 1987, Specific cell surface labels in the visual centers of *Xenopus laevis* tadpole identified using monoclonal antibodies, *Dev. Biol.* 122:90–100.

Takagi, S., Hirata, T., Agata, K., Mochii, M., Eguchi, G., and Fujisawa, H., 1991, The A5 antigen, a candidate for the neuronal recognition molecule, has homologies to complement components and coagulation factors, *Neuron* 7:295–307.

Tamagnone, L., Artigiani, S., Chen, H., He, Z., Ming, G.I., Song, H. *et al.*, 1999, Plexins are a large family of receptors for transmembrane, secreted, and GPI-anchored semaphorins in vertebrates, *Cell* 99:71–80.

Tanaka, E.M. and Kirschner, M.W., 1991, Microtubule behavior in the growth cones of living neurons during axon elongation, *J. Cell. Biol.* 115:345–363.

Taniguchi, M., Yuasa, S., Fujisawa, H., Naruse, I., Saga, S., Mishina, M. *et al.*, 1997, Disruption of semaphorin III/D gene causes severe abnormality in peripheral nerve projection, *Neuron* 19:519–530.

Tear, G., Harris, R., Sutaria, S., Kilomanski, K., Goodman, C.S., and Seeger, M.A., 1996, Commissureless controls growth cone guidance across the CNS midline in Drosophila and encodes a novel membrane protein, *Neuron* 16:501–514.

Tessier-Lavigne, M., Placzek, M., Lumsden, A.G., Dodd, J., and Jessell, T.M., 1988, Chemotropic guidance of developing axons in the mammalian central nervous system, *Nature* 336:775–778.

Wadsworth, W.G., 2002, Moving around in a worm: Netrin UNC-6 and circumferential axon guidance in *C. elegans*, *Trends Neurosci.* 25:423–429.

Walter, J., Henke-Fahle, S., and Bonhoeffer, F., 1987, Avoidance of posterior tectal membranes by temporal retinal axons, *Development* 101:909–913.

Wässle, H., Peichl, L., and Boycott, B.B., 1981, Dendritic territories of cat retinal ganglion cells, *Nature* 292:344–345.

Whitford, K.L., Dijkhuizen, P., Polleux, F., and Ghosh, A., 2002, Molecular control of cortical dendrite development, *Annu. Rev. Neurosci.* 25:127–149.

Winberg, M.L., Noordermeer, J.N., Tamagnone, L., Comoglio, P.M., Spriggs, M.K., Tessier-Lavigne, M. *et al.*, 1998, Plexin A is a neuronal semaphorin receptor that controls axon guidance, *Cell* 95:903–916.

Wong, K., Ren, X.R., Huang, Y.Z., Xie, Y., Liu, G., Saito, H. *et al.*, 2001, Signal transduction in neuronal migration: Roles of GTPase activating proteins and the small GTPase Cdc42 in the Slit–Robo pathway, *Cell* 107:209–221.

Wong, R.O. and Ghosh, A., 2002, Activity-dependent regulation of dendritic growth and patterning, *Nat. Rev. Neurosci.* 3:803–812.

Yu, W., Cook, C., Sauter, C., Kuriyama, R., Kaplan, P.L., and Baas, P.W., 2000, Depletion of a microtubule-associated motor protein induces the loss of dendritic identity, *J. Neurosci.* 20:5782–5791.

Zakharenko, S. and Popov, S., 1998, Dynamics of axonal microtubules regulate the topology of new membrane insertion into the growing neurites, *J. Cell. Biol.* 143:1077–1086.

Zakharenko, S. and Popov, S., 2000, Plasma membrane recycling and flow in growing neurites, *Neuroscience* 97:185–194.

Zheng, J.Q., 2000, Turning of nerve growth cones induced by localized increases in intracellular calcium ions, *Nature* 403:89–93.

Zou, Y., Stoeckli, E., Chen, H., and Tessier-Lavigne, M., 2000, Squeezing axons out of the gray matter: A role for slit and semaphorin proteins from midline and ventral spinal cord, *Cell* 102:363–375.

10

Synaptogenesis

Bruce Patton and Robert W. Burgess

INTRODUCTION

The study of synapse formation requires an understanding of synaptic function, structure, and organization. This chapter, therefore, reviews the essential roles played by synapses in the nervous system, the basic mechanisms of synaptic transmission, and the presynaptic and postsynaptic specializations that support synaptic signaling, before considering the events that establish, maintain, and modulate synaptic connections.

Synapses are arbiters of information flow in the nervous system. Information is carried through the nervous system by distinct intracellular and intercellular processes. Within neurons, information is encoded in the patterns of electrochemical activity that pass in waves across neuronal surfaces. Neuronal activity is then transferred between neurons by means of specialized intercellular signaling structures, the synapses. Synapses with non-neuronal targets such as heart and skeletal muscle regulate most bodily functions. The term *synapse*, from the Greek for "connect," intimates a close physical proximity between the synaptic specializations in adjoining cells. Indeed, we now know that where the speed and fidelity of synaptic communication is critical, presynaptic and postsynaptic specializations are directly apposed and precisely aligned (Fig. 1). Originally, however, Sherrington coined "synapse" in a physiology textbook in order to designate the functional linkage between neurons whose activities are coupled (Foster, 1897). Although an anatomical substrate for Sherrington's functional synapse was separately anticipated by others, including Cajal, Held, and Langley, the precise cellular arrangement at synapses remained uncertain until synaptic connections were finally observed in the electron microscope (De Robertis and Bennett, 1955; Palay, 1956).

Studies in succeeding decades revealed the basic mechanisms of synaptic signaling, or *neurotransmission*. Most synapses transmit neuronal activity by means of an intercellular chemical messenger, the *neurotransmitter*. Chemical neurotransmission begins as electrical activity in the presynaptic cell triggers the secretion of neurotransmitter (Fig. 2). The released neurotransmitter diffuses within the fluids of the extracellular space and ultimately binds to specific receptor proteins embedded in the surface membrane of the postsynaptic cell. Synaptic transmission is completed as changes in the conformation of the receptor induced by transmitter binding alters postsynaptic electrochemical activity. The chemical nature of neurotransmission was initially predicted from the effect of nicotine on neural transmission through peripheral ganglia. Nicotine was eventually shown to act as a specific ligand for a subset of the receptors for acetylcholine (ACh), the first neurotransmitter identified in the peripheral nervous system (PNS) and the central nervous system (CNS) (Loewi, 1921; Dale *et al.*, 1936; Eccles *et al.*, 1956).

Two broad functional classes of chemical synapse differ principally in their speed of neurotransmission. Fast chemical synapses are composed of closely opposed presynaptic and postsynaptic elements and typically employ *ionotropic* neurotransmitter receptors (Figs. 2 and 3). Ionotropic receptors are ion channels whose conductance is directly regulated by neurotransmitter binding. In skeletal muscles, for example, ACh released from motor nerve terminals allosterically opens cation-selective pores formed by the subunits of nicotinic ACh receptors (AChRs), which are concentrated on the surfaces of muscle fibers opposite the nerve. The resulting influx of cations is immediate and large and rapidly stimulates muscle activity. In contrast, presynaptic and postsynaptic specializations at slow chemical synapses, which are common in the autonomic innervation of glands and organs, are diffusely organized and often are not closely opposed to each other. Slow chemical synapses also often employ *metabotropic* receptors, which regulate cell function indirectly, through intracellular second messengers. Thus, in the heart, parasympathetic axons from the vagus nerve release ACh that activates metabotropic AChRs on the surface of cardiac myocytes. These AChRs are pharmacologically distinguished by their sensitivity to muscarine rather than nicotine. Activation of muscarinic AChRs indirectly opens cardiac potassium channels (Sakmann *et al.*, 1983) through intermediary G-protein second messengers (reviewed by Brown and Birnbaumer, 1990). The resulting efflux of potassium from myocytes depresses cardiac excitability and gradually slows heart rate. Note that both the timing and strength of the response to ACh in cardiac muscle is muted compared to the immediate (millisecond), all-or-none contractile response in skeletal muscle. Regardless of synapse type and receptor mechanism, synaptic transmission ultimately ceases as transmitter is eliminated by re-uptake or catabolism, or the postsynaptic ion channels inactivate.

Bruce Patton • Oregon Health and Science University, Portland, OR 97201.   Robert W. Burgess • The Jackson Laboratory, Bar Habor, ME 04609.

Developmental Neurobiology, 4th ed., edited by Mahendra S. Rao and Marcus Jacobson. Kluwer Academic / Plenum Publishers, New York, 2005.

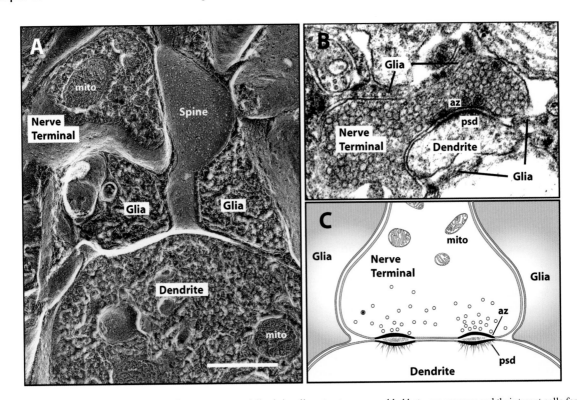

FIGURE 1. Cellular composition of the synapse. Synapses are specialized signaling structures assembled between neurons and their target cells for the accurate transmission of neural information. The location, speed, and strength of neurotransmission is dependent on the alignment of presynaptic specializations that control the secretion of neurotransmitter with postsynaptic specializations that transduce transmitter binding into changes in target cell activity. Synaptic features visible by microscopy include an enlarged presynaptic terminal (alternatively called a *bouton* or *varicosity*) containing mitochondria and high concentrations of small, clear "*synaptic vesicles.*" Nerve terminals at fast chemical synapses also contain *active zones* (az), membrane subdomains where transmitter secretion is enhanced; high concentrations of protein at active zones collect metal stains and appear dense in electron micrographs. The morphology of the postsynaptic cell often reflects the presynaptic terminal, and within the postsynaptic membrane, neurotransmitter receptors and signal transduction proteins are concentrated directly opposite the synaptic cleft from transmitter release sites. Glial cells typically surround synapses and provide metabolic support. (A) Scanning electron micrograph of a spine synapse on a pyramidal cell in the hippocampus of an adult rat, revealed by freeze-fracture methods. Spines are short protrusions from dendritic shafts, an anatomical arrangement which partially isolates many of the synaptic inputs to a single dendrite. Image kindly provided by Tom Reese; N.I.H. (B) Transmission electron micrograph of a synapse in the superior cervical ganglion of an adult mouse. Typical of many chemical synapses, the presynaptic terminal contains many clear synaptic vesicles concentrated opposite a dense region of postsynaptic membrane (the postsynaptic density, or PSD). Biochemical and immunochemical studies reveal PSDs are rich in cell adhesion proteins, transmitter receptors, and receptor-associated scaffolding proteins. Typical of excitatory synapses, the nerve terminal also contains a few dense core vesicles, which contain neuromodulatory peptides and/or components of the synaptic cleft, and a cluster of vesicles associated with a dense region of presynaptic membrane, known as an active zone (az). Notably, active zones and PSDs are precisely aligned. Most nerve terminals also have several mitochondria, not visible in this section. (C) Model chemical synapse, containing adherent pre- and postsynaptic elements with aligned sites of transmitter release and transmitter reception, surrounded by glial cells.

The properties of fast chemical synapses in particular have evolved beyond a simple means of exchanging neural information, to facilitate higher neural functions. By controlling the *timing*, the *location*, and the *strength* of neurotransmission, synapses act as gates to the flow of neural activity through the brain and body. Therefore, in most organisms, synaptic transmission is also a primary site of modulation of neural information. The strength and disposition of synaptic connections within the neural architecture so critically determine overall neural function that they comprise a secondary mechanism of encoding neural information. That is, an ability to change synaptic strength and location are most likely the biochemical and cellular substrates of learning and memory.

Fast synaptic transmission is promoted by an elaborate series of cellular and molecular mechanisms, which will remain the focus of this chapter. Neurotransmission is chiefly controlled at the steps where electrical and chemical signals are interconverted. Perhaps this should not be too surprising. The interconversion of electrochemical (ion flux) and chemical (transmitter) activity levels are the most complicated biochemical steps in the flow of neural information; many cellular processes are most heavily regulated at their slowest and most complex steps. The *timing* of neurotransmission is precisely controlled by tightly coupling presynaptic depolarization to neurotransmitter secretion. Coupling occurs through the use of calcium as a trigger for secretion, and by concentrating voltage-sensitive calcium channels at synaptic sites. *Location* is specified by tightly focusing neurosecretion and neuroreception at small sites on the pre- and postsynaptic cell surfaces. Importantly, these specialized signaling domains are co-localized at sites of adhesion between the

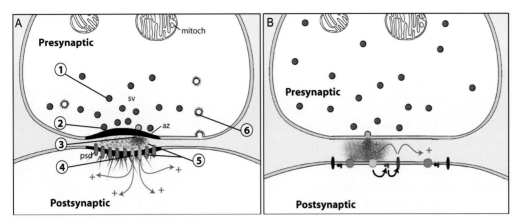

FIGURE 2. Neurotransmission. (A) (1) Neurotransmitter is initially concentrated in small lipid-walled vesicles within the presynaptic terminal. (2, 3) Transmission begins as presynaptic depolarization triggers the fusion of one or more synaptic vesicles with the synaptic membrane of the nerve terminal. Released transmitter diffuses across the synaptic cleft. (4) Binding of neurotransmitter to specific receptors in the postsynaptic membrane directly or indirectly changes the activity of postsynaptic ion channels. For example, excitatory transmitters such as glutamate open cation-selective ion channels and depolarize the postsynaptic cell. (5) Transmission ends as transmitter-induced currents are inactivated, either through clearance of transmitter from the synaptic cleft, or through biophysical properties intrinsic to the receptor or ion channels. (6) Excess presynaptic membrane is removed by endocytosis. At fast chemical synapses, pre- and postsynaptic specializations are directly apposed and precisely aligned. (B) In contrast, pre- and postsynaptic elements are loosely associated and minimally organized at slow chemical synapses. For example, presynaptic membranes lack active zones, and postsynaptic membranes have low concentrations of transmitter receptors. Transmission at slow synapses often relies on metabotropic receptors, which indirectly regulate membrane conductance through secondary messengers.

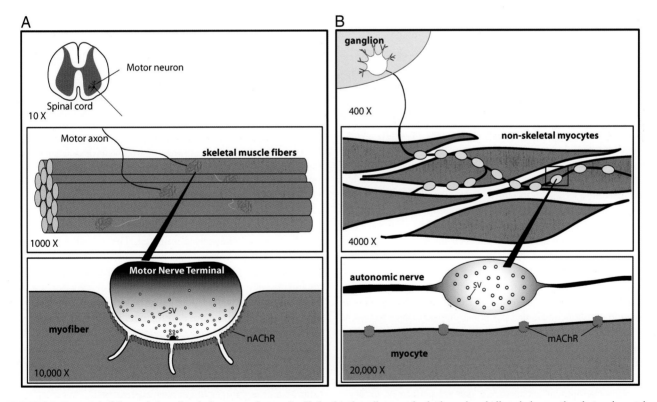

FIGURE 3. Prototypical fast and slow chemical synapses in muscle. Skeletal and cardiac muscles both receive cholinergic innervation, but each contains a different type of chemical synapse. (A) In skeletal muscle, axons from spinal motor neurons form fast chemical synapses at specific sites on each muscle fiber. Motor nerve terminals are located precisely opposite high concentrations of nicotine-sensitive ionotropic ACh receptors (nAChR) in the muscle fiber surface. Within the motor terminal, synaptic vesicles are polarized towards the synaptic surface, and further concentrated near active zones. (B) In cardiac muscle, sympathetic axons contain presynaptic varicosities which are widely distributed and which lack polarity or active zones. Cardiac myocytes contain metabotropic ACh receptors (mAChR), which are not concentrated near axons, and which are indirectly coupled to cardiac potassium channels through G-proteins. Approximate scales are provided at lower left corners of each figure.

axon and its target cell. Precise spatiotemporal control distinguishes synaptic transmission from the diffuse chemical signaling that coordinates the metabolic activity of an animal. Finally, the *strength* of synaptic transmission is dependent on the amount of neurotransmitter secreted in response to presynaptic activity, and the size of the postsynaptic response to a given amount of transmitter.

Fast chemical synapses have three cellular elements. First, the nerve terminal of the presynaptic cell contains specialized neurosecretory domains, which regulate the timing, location, and volume of neurotransmitter release. Since neurosecretion typically occurs far from the nucleus of the cell, presynaptic specializations also include mechanisms to locally synthesize and package transmitter into vesicles, and to recover synaptic vesicle materials following release. Second, the surface of the postsynaptic cell is specialized to recognize secreted neurotransmitter, and to transduce the chemical energy of binding into altered electrical activity. Excitatory transmitters often increase depolarizing conductances, as just described for ACh at the neuromuscular synapse. However, many variations on this mechanism have evolved. For example, neurotransmitters at some sensory synapses alter postsynaptic activity by closing ion channels. Third, most synapses are enshrouded by glial cell processes. Glial cells play important roles in supporting the metabolic activity of the pre- and postsynaptic elements. They also strongly influence the potential for growth and synaptogenesis by axons and dendrites.

Perhaps most importantly, fast synapses are sites of direct contact between the pre- and postsynaptic cells. The precise pairing of pre- and postsynaptic specializations is so fundamental to fast chemical neurotransmission that it may at first appear trivial. In fact, proximity is an essential mechanism underlying the speed and specificity of synaptic signaling and has profound consequences for neural function. A narrow synaptic cleft between the sites of transmitter release and reception means the neurotransmitter will diffuse only a few dozen nanometers to complete transmission. Just as importantly, restriction of synaptic transmission to small domains allows information to be distributed to specific subsets of cells and specific portions of those cell's surfaces, rather than willy-nilly between all potential matches. One consequence is that synapses often grossly outnumber the cell bodies they connect (Fig. 4). The resulting convergence and divergence of interneuronal signaling enables the nervous system to process and integrate information rather than merely relay it. A second consequence is that patterns of neural connectivity can be modified without wholesale cellular restructuring of the brain, by altering individual synaptic elements.

The coordinated assembly of pre- and postsynaptic specializations constitutes *synapse formation*. An initial phase of synaptic development establishes a general pattern of innervation, in which specific sets of cells are connected. The initial synaptic connections are then remodeled. Synaptic reorganization is influenced by fluctuating levels of activity among subsets of connections within the architecture, as well as by circulating humeral factors. In response to differing levels of activity, some synapses are selectively strengthened and maintained, while

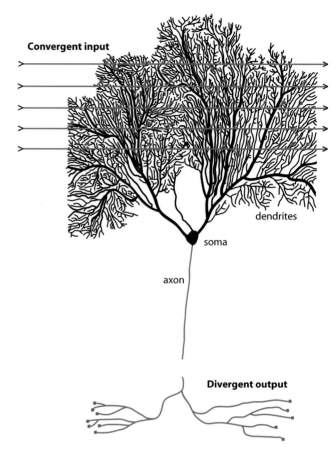

FIGURE 4. Convergence and divergence of neural information. Most neurons contain an array of dendrites, which receive hundreds of synaptic inputs. Dendritic processes ultimately converge at or near the neuronal soma. Conversely, a single axon typically emerges from the soma before branching to innervate many target cells. Synaptic activity at single dendritic sites is generally insufficient to bring the axon to the threshold of an action potential. Thus, activity in the axon represents the integrated synaptic activity in the dendritic arbor. The convergence and divergence of synaptic inputs allows neuronal systems to process information.

others are simultaneously eliminated. The overall effect is one of progressive restriction, narrowing initially broad patterns of innervation into functionally refined subpatterns. In some organisms, activity-dependent refinement of synaptic connections completes neural development. In many, however, the remodeling phase of neural development melds with processes of learning and memory and continues throughout life. A critical feature of vertebrate neural systems is that the capacity for computation, adaptation, and fine control in the adult animal depends as much on the specificity and plasticity of the synaptic connections as on the number of connected elements.

In principle, the precise colocalization of pre- and postsynaptic specializations could arise through cell-autonomous programs of development. Indeed, most synapsing cells independently express their synaptic components and can assemble functional pre- or postsynaptic elements alone, in the absence of a synaptic partner. Nevertheless, most synapse formation involves the coincident assembly of new pre- and postsynaptic specializations at sites where the two cells make contact. This indicates that neurons and their targets exchange *synaptogenic signals*, and that these signals act locally to promote the assembly of synapses from already synthesized components. In short, synapses are organized structures rather than induced programs of development.

THE NEUROMUSCULAR JUNCTION: MODEL SYNAPSE

Much of our understanding of synaptic organization is derived from studies of innervation in the skeletal muscles of vertebrate animals. Synapses between motor axons and muscle fibers are known as neuromuscular junctions (NMJs). Historically, innervation in muscle presented clear advantages for experimentalists. Compared to most interneuronal synapses, skeletal NMJs are large and physically isolated from each other (Fig. 5). They are also physiologically robust. The accessibility of this preparation led to the experiments that defined and confirmed the existence of chemical neurotransmitters, the vesicular hypothesis of neurotransmitter release, and the principal mechanisms of postsynaptic excitation.

An apparent disadvantage of NMJs is that they account for only a tiny fraction of the mass of a muscle. Ordinarily, this would prevent a straightforward analysis of the biochemical constitution of this synapse. Instead, biochemistry has been one of the NMJ's great advantages, due largely to an ontogenic relationship between the skeletal NMJ and the electric organ structures present in certain species of fish, such as the marine ray *Torpedo* (see Box 1). Fractionation of the electric organ led to the discovery of a number of key synaptic components. Some, like VAMP (vesicle associated membrane protein), turned out to be important components of virtually all chemical synapses; VAMP was later independently identified as synaptobrevin, in synaptosomal fractions of homogenized bovine brain. Others components were more specific to the neuromuscular synapse. For example, the

nicotinic AChR was the first neurotransmitter receptor (in fact, the first ion channel) to be molecularly characterized and cloned, due to its enrichment in electric organ membranes (Schmidt and Raftery, 1973; Noda *et al.*, 1982; Claudio *et al.*, 1983; Numa *et al.*, 1983). Another important example is agrin, the first synaptic organizing signal to be molecularly identified, which was also purified from Torpedo electric organ homogenates (Nitkin *et al.*, 1987). As a result, a good deal is known about how motor neurons direct synapse formation in skeletal muscles (described in a later section). In contrast, the identification of molecules that distinguish and organize the various types of chemical synapses in the brain has lagged, in no small measure because a homogeneous population of central synapses amenable to biochemistry has not been available. Instead, brain has proved to be good starting material for the identification of ubiquitous synaptic components. For example, SNAP-25, syntaxin, synapsin, synaptophysin, and munc18 are an ancient retinue of proteins discovered in extracts of mammalian brain that regulate presynaptic vesicle dynamics in nerve terminals throughout the body, in animals throughout the phylogenetic tree.

A further property of the neuromuscular system especially useful to developmental neurobiologists is that much of it is capable of regeneration. The ability of peripheral nerves and skeletal muscle fibers to regenerate has allowed processes of synapse assembly at the NMJ, which begins prenatally in mammals, to be reassessed following injury in adults. As a direct result, studies of reinnervation in skeletal muscle have played key roles in formulating and testing three fundamental concepts in neuroscience. The first is the essential notion that the synapse is the site of communication between nerves and their targets, which developed by the beginning of the last century. Second is the concept of synaptic specificity in neural development, as motor axons were found to reinnervate very specific sites on muscle fibers by Cajal and his students. Third is the molecular basis of synapse formation, conceived by Cajal early in the last century and pursued into the current one.

The NMJ possesses two final advantages for the current generation of neuroscientists. First, molecular information

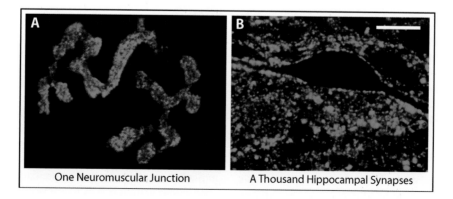

FIGURE 5. Neuromuscular synapses are much larger than most interneuronal synapses. (A) Motor nerve terminal at a single skeletal neuromuscular junction from an adult mouse. (B) Several hundred nerve terminals in the CA3 region of the hippocampal formation from a juvenile rat. Confocal images at similar scales show immunoreactivity for synapsin.

BOX 1. The Swimming Purified Acetylcholine Receptor

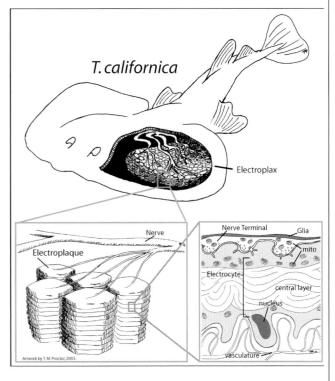

Strongly electric fish, including the Torpedinidae family of marine rays, contain specialized electrogenic organs. The Pacific marine ray (*Torpedo californica*) pictured above-left often reaches a meter in width and is capable of generating 50–100 V discharges. *T. cal*'s electric organs, or electroplax, generate moderate pulses as a defense against faster swimming predators such as sharks, and strong discharges to immobilize fast-swimming prey like salmon. Charles Darwin considered it "impossible to conceive by what steps these wonderous organs have been produced" (Darwin, 1981). Actually, the bilateral kidney-shaped organs are embryologically derived from the branchial musculature. Embryonic myotubes lose their skeletal muscle myosins and collapse longitudinally to form electroblasts. Electroblasts spread horizontally and intercalate to form stacks of disc-like differentiated electrocytes. One entire face of each electrocyte is then innervated by motor axons, but always on the same side (dorsally in *T. cal*), which orients all electrical activity in the same direction. Neural stimulation depolarizes the postsynaptic membrane of the electrocyte, producing an immediate 100 mV charge reversal. The electrocyte's central layer, which is a remnant of the sarcomeres, may transiently insulate the opposite side of the cell, polarizing the overall current flow. The depolarization of each electrocyte in the stack is synchronized through coordinated neural stimulation; their summed discharges peak at over 50 V, and repeat at more than 400 Hz, enough

to shock any adjacent sea creatures into compliance (either submission or avoidance). Thus, the voltage-generating electroplaques are hugely overgrown neuromuscular junctions, piled in series like a (very) tall stack of pancakes (above). There are typically about 400 electroplaques in each voltaic stack, and up to 400 stacks in each organ, which together make up to 30% of a ray's body mass. This represents an extraordinary (possibly even shocking) abundance of AChR-rich postsynaptic membrane in a tidy package. Indeed, to biophysicists and neurobiologists, "The torpedo ray … is essentially a swimming purified acetylcholine receptor" (Miller, 2000). (Photo by Howard Hall, used with permission; original artwork by Thomas M. Proctor.)

---

gained over the last several decades now permits sophisticated, mechanistic questions about synapse formation to be addressed. Second, the size, isolation, and regenerative capacity which attracted early students of the nervous system remain a distinct advantage to the advanced imaging methods that are now beginning to reveal the cellular and molecular dynamics involved in synapse formation and plasticity.

Innervation and Transmission in Muscle

A general principle of synaptic transmission at the vertebrate skeletal NMJ is that patterns of impulses in the nerve are highly correlated with contractile activity in the muscle. Reliable coupling of nerve activity to muscle activity ensures that, given adequate stimulation, every fiber in the muscle can be recruited

to heroic efforts, be it the sprint of a fieldmouse evading a hawk, or the strain of a paleo-hunter throwing a spear. Yet, in both predator and prey, the very same synaptic connections may also be employed in the performance of finely graded tasks. By moderating neural activity, the strength and timing of muscle activity can be exquisitely controlled to effect the fluid stroke of a cheek or a pen, the accurate movements in a throw and a catch, and the intricate labial, lingual, and laryngeal sequences of speech.

Two general features of neuromuscular innervation underlie the simultaneous robustness and fine control of neuromuscular coupling. First, the strength of individual synaptic connections in muscle is extraordinarily high. Second, each muscle fiber is innervated by a single motor axon, and each motor axon innervates a discrete number of muscle fibers. A single motor axon and the several muscle fibers it innervates are termed a *motor unit*. Combining these features ensures that a nerve impulse always generates a response in the muscle, but that the size of the response can be scaled, depending on how many motor neurons are activated, and the size of the motor units recruited. We consider these mechanisms in turn.

Synaptic Efficacy at the NMJ

Synaptic transmission at the NMJ rarely fails. A single action potential in the motor axon will ordinarily elicit a synaptic event capable of depolarizing the postsynaptic membrane in the muscle fiber to nearly 0 mV, which is well beyond the threshold for action potential propagation along the muscle fiber. The synaptic strength required to achieve the high fidelity of neuromuscular transmission is considerable. Not only must the postsynaptic current be large enough to overcome the low-input resistance that comes with the large diameter of the muscle fiber (often 50 μm), but signaling in most muscles occurs at levels that are several-fold above the minimum needed to gain full response to a single nerve impulse. In many muscles, more than 80% of the junctional receptors can be blocked before the muscle's response is detectably diminished (Fig. 6). This apparent excess capacity for transmission ensures that nerve and muscle activity remain tightly coupled during periods of intense demand and is known as the *safety factor* (Wood and Slater, 2001). The strength of a synaptic connection is a function of the amount of neurotransmitter secreted by the presynaptic cell in response to depolarization, and the amount of depolarization that occurs in the postsynaptic cell in response to neurotransmitter. A high safety factor for transmission depends in addition on specializations that sustain high levels of transmitter release and large postsynaptic responses during repetitive stimulation. These include both chemical and structural mechanisms, outlined below.

Presynaptic Mechanisms

The motor nerve terminal is highly specialized to promote and sustain high levels of neurotransmission. First, the extraordinary size of each motor nerve terminal (Fig. 5) accommodates hundreds of *active zones*, specialized membrane domains where neurotransmitter is preferentially released. Second, like other fast chemical synapses, motor terminals are specialized to speed both the release of neurotransmitter and the reconstitution of new transmitter-laden synaptic vesicles (Fig. 7).

Synaptic vesicles are initially concentrated near release sites along the presynaptic membrane. The molecular mechanisms that polarize the distribution of synaptic vesicles near release sites have not been confirmed, but likely depend in part on interactions with the actin cytoskeleton that permeates the terminal. Additional interactions with components of the active zone complex then recruit synaptic vesicles to docking sites along the terminal surface membrane. Vesicle docking is mediated by a SNARE complex, which includes the vesicle membrane protein VAMP/synaptobrevin and the plasma membrane proteins SNAP-25 and syntaxin. Docking effectively primes a subset of synaptic vesicles for immediate release. The SNARE complex also drives the fusion of vesicle and terminal surface

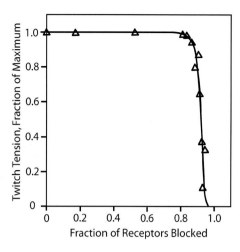

FIGURE 6. Safety factor for neurotransmission. At some synapses, the strength of neurotransmission dramatically exceeds the level required to guarantee a full postsynaptic response to evoked release of transmitter from the presynaptic terminal. For example, more than 80% of ACh receptors at the neuromuscular junction may be blocked by pharmacological antagonists before muscle contractions elicited by stimulation of the motor nerve are noticeably weakened. Thus, the safety factor for solitary synaptic events at the vertebrate neuromuscular junction is usually greater than 5-fold and may exceed 10-fold in muscles such as the diaphragm which are especially resistant to inhibition. The apparent excess signaling capacity is known as the safety factor in neurotransmission. It permits high fidelity neurotransmission to continue during periods of intense demand. The safety factor for neuromuscular transmission is significantly reduced in patients with the autoimmune disorder *myasthenia gravis*, in which antibodies to the muscle ACh receptor impair postsynaptic responsiveness. The extraordinarily high safety factor in diaphragm muscles allows neuromuscular blockers to be used in clinical care, as their proper titration will relax airway, limb, and axial muscle relaxants do not arrest breathing. In addition to levels of postsynaptic receptors, a high safety factor depends on elevated levels of neurotransmitter release from the presynaptic terminal, efficient coupling of transmitter binding to postsynaptic activity, and rapid clearance of spent neurotransmitter from the synaptic cleft.

membranes, but only when appropriately triggered (Sollner *et al.*, 1993). The trigger for fusion is calcium. Intracellular levels of calcium are maintained at very low concentrations in resting nerve terminals, but rise sharply upon depolarization of the nerve terminal membrane from influx through voltage-dependent calcium channels in the terminal surface. The voltage-dependence of the presynaptic calcium channels is critical; they open only when the nerve terminal membrane is strongly depolarized. The proposed calcium sensor is the calcium-binding protein synaptotagmin, which is concentrated in synaptic vesicle membranes. Calcium entry into the terminal is concentrated at vesicle docking sites by recruiting calcium channels to active zones, through interactions with presynaptic membrane proteins such as syntaxin. (In fact, it is entirely possible that active zone complexes are recruited to the location of calcium channels, which may themselves be anchored to extracellular substrates.) Together, these multiple features ensure that neurosecretion is targeted to specific sites on the neuronal surface, and tightly coupled to axonal activity.

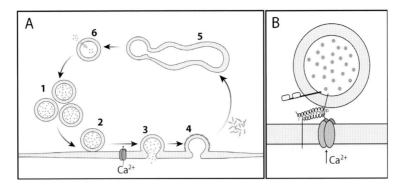

FIGURE 7. Synaptic function depends on regulated trafficking of synaptic vesicle components. Synaptic terminals far from the cell body employ signaling components that are locally synthesized or reused. Primary neurotransmitters are therefore simple biomolecules, such as amino acids or their metabolic relatives, and synaptic vesicles are reconstituted following synaptic activity. (A) The synaptic vesicle cycle. Prior to release of neurotransmitter, synaptic vesicles are concentrated near the synaptic surface of the nerve terminal (1), and dock at sites along the presynaptic membrane (2) through direct or indirect interactions with Ca^{2+} channels. Vesicle and surface membranes fuse in response to elevated intracellular Ca^{2+} concentrations following an action potential, releasing transmitter into the synaptic cleft (3). Vesicle membranes and proteins are internalized through clathrin-mediated endocytosis (4) at sites adjacent to the sites of fusion, and traffic through endosomal intermediates (5) before reforming small, clear synaptic vesicles. New vesicles are reloaded with neurotransmitter (6), by transporters powered by a pH gradient across the vesicle membrane. (B) Molecular model of vesicle docking. Docking is ultrastructurally defined by direct apposition of vesicle and plasma membranes, physiologically characterized by fusion in response to osmotic shock, and biochemically mediated by the formation of a SNARE complex. Synaptic vesicles contain the V-SNARE VAMP (synaptobrevin). Terminal membranes contain the target membrane T-SNAREs syntaxin and SNAP-25. Direct interactions between α-helical domains in each V- and T-SNARE produce a coiled-coil structure that holds vesicle and plasma membranes close together. Secondary interactions mediated by syntaxin link vesicles to voltage-activated Ca^{2+} channels. Synaptotagmin in the vesicle membrane likely mediates Ca^{2+}-induced fusion of vesicle and plasma membranes. Ca^{2+} binds to synaptotagmin at a pair of C2 domains, regulatory motifs first identified in the lipid- and Ca^{2+}-activated enzyme protein kinase C. Membrane fusion may be driven by conformational changes in the SNARE complex coiled-coil. How synaptotagmin drives fusion remains controversial.

The close proximity between sites of calcium entry and vesicle docking provides a three-fold benefit to rapid neurotransmission at the NMJ. First, the speed of neurosecretion is maximized by minimizing the delay between terminal membrane depolarization and vesicle fusion. Second, neurosecretion is topographically focused, as calcium levels rise fastest and highest where the channels are concentrated. Third, and for similar reasons, neurosecretion is chronologically focused and therefore synchronized at active zones throughout the motor terminal. Thus, active zones and calcium channels play a central role in fine-tuning the location and timing of neurosecretion. The concerted regulation of both the timing and location of transmitter release by calcium may ensure that depolarization does not cause release from errantly docked vesicles, and that properly docked vesicles release transmitter only in response to depolarization.

Postsynaptic Mechanisms

The postsynaptic region of the muscle fiber is called the *endplate*. The endplate's response to secreted neurotransmitter is determined principally by features of the nicotinic AChR. First, as described above, nicotinic AChRs directly translate neurotransmitter binding into membrane depolarization. The nicotinic AChR is composed of five homologous transmembrane subunits with a stoichiometry of 2α, 1β, 1γ, and 1δ. These are assembled as a ring around a membrane-spanning pore, which remains closed in the absence of ACh. By binding to specific sites on the extracellular surface of the ring, ACh allosterically opens, or

"gates," the central cation-selective pore: Upon binding, the channel bends slightly, the pore opens, and Na^+ (and some Ca^{2+}) ions flood into the muscle fiber. Importantly, the rate at which AChR channels open after ACh binds does not delay neurotransmission. A second way in which AChRs promote a postsynaptic response is through their relatively high ionic conductance, which speeds depolarization of the muscle fiber. Third, AChR channels close immediately upon ACh dissociation, but do not inactivate, and desensitize only slowly. Neurotransmission is therefore highly correlated to levels of ACh in the synaptic cleft. Fourth, the density of AChRs is maintained at extraordinarily high levels in the portion of the muscle membrane immediately subjacent to the nerve terminal (Fig. 8). The high density of transmitter-activated ion channels provides the endplate with the capacity to generate large postsynaptic currents in response to high levels of transmitter released from the nerve terminal. As discussed below, the clustering of transmitter receptors opposite the nerve terminal and the formation of a nerve terminal directly opposite clustered receptors are the fundamental events in the construction of a synapse.

One final high-performance feature of the nicotinic AChR is that channel activation is cooperatively dependent on ligand binding. The AChR channel opens only after two ACh molecules have bound. This makes it unlikely that low levels of ACh in the synaptic cleft will depolarize the postsynaptic membrane and thus improves the fidelity of neurotransmission by suppressing false alarms. The spontaneous release of neurotransmitter from the terminal is also suppressed, by mechanisms that remain incompletely understood, but which are likely to be directly

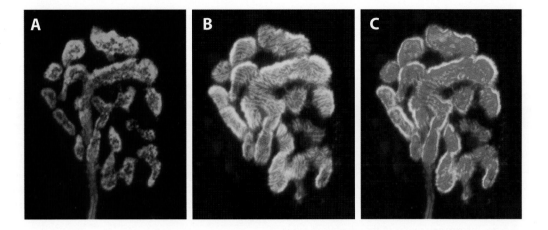

FIGURE 8. Morphological differentiation at the vertebrate neuromuscular junction. Staining skeletal muscle with antibodies to motor axons (neurofilaments and synaptophysin) and α-bungarotoxin, which binds tightly to ACh receptors, reveals that motor nerve terminal and endplate have matching conformations. In these confocal images of an adult mouse neuromuscular junction, antibody (blue) and toxin (yellow) were pseudocolored to visualize differences and overlap between nerve terminal and endplate. (A) The motor nerve terminal is branched and varicose compared to the pre-terminal axon. (B) ACh receptors are highly concentrated in postsynaptic membranes. Striations in toxin staining reveal the orientations of postsynaptic folds. (C) Pre- and postsynaptic domains show almost complete overlap.

incorporated into the transmitter-release machinery. These combined pre- and postsynaptic mechanisms not only reduce background noise in synaptic transmission, but also sharpen the postsynaptic response during *bona fide* synaptic events.

Single Innervation of Muscle Fibers

A second critical feature of neuromuscular innervation is that each muscle fiber is innervated by a single motor axon (Fig. 8). This facilitates control over the strength of muscle contraction. Other examples of monosynaptic input include the primary auditory relay synapse in the brain (the calyx of Held), and innervation in some autonomic ganglia, where information is rapidly and topographically exchanged without significant editing. In contrast, most neurons receive hundreds of convergent excitatory inputs, whose summary activity is required to depolarize the cell above threshold. Compared to the highly convergent and divergent patterns of innervation that complicate synaptic architecture in the CNS, innervation of muscle fibers appears simple. Indeed, the relative simplicity of muscle innervation has seduced several generations of neuroscientists, starting with Ramon y Cajal.

However, the simplicity of muscle innervation is deceptive. First, the exact match of innervating nerve terminals to muscle fiber number means that mechanisms of neuromuscular development must ensure complete innervation without polyinnervation. Stochastic methods of randomly plugging axons onto muscle fibers would leave some fibers without innervation, and some with more than one input. (Consider the likely result of playing 36 trials on a 36-slot roulette wheel.) Alternatively, it may be imagined that proper innervation would be most accurate if it followed a predetermined program of innervation, in which each neuron was genetically and hence, biochemically matched to a particular muscle fiber, perhaps through specific cell-surface

recognition molecules. In short, motor units would be molecularly defined. Evidence for a high level of cellular determinism has in fact been found in the neuromuscular systems of invertebrate animals, as in the genetic model organisms *Caenorhabditis* and *Drosophila*. In vertebrates, axon outgrowth is directed to target fields (see Chapter 9), and individual pools of spinal motor neurons innervate specific muscles groups. However, synapse formation in vertebrate muscles appears considerably less determined at the level of individual muscle fibers. For example, motor nerves still fully innervate skeletal muscles when the size of the target muscle is experimentally increased, or the pool of motor neurons is decreased. In adult muscles too, partial denervation leads to an increase in the size of the remaining motor units, as uninjured axons sprout collateral branches that innervate the denervated portion of the muscle. We understand, therefore, that muscles interact with their innervating population of motor axons, influencing their growth and propensity to form synapses. Muscles must supply signals that both promote and retard synapse formation in the nerve.

A second complication is that individual muscles usually contain a mixture of muscle fiber types, which are selectively innervated by subtypes of motor axons. Individual muscle fibers differ in their complement of myosin isoforms and levels of glycolytic enzymes—factors that determine the rates at which the fibers contract and subsequently fatigue. Similarly, motor axons supplying a given muscle vary in diameter, conduction velocity, and nerve terminal structure. In birds and mammals, which are best studied, direct mapping of connections and detailed recording of muscle contractions during graded stimulation of the nerve have shown that muscle fibers are not randomly assigned to motor units. Instead, motor units contain primarily muscle fibers of a similar type (fast- or slow-twitch; high- or low-activation threshold). Motor units also vary considerably in size, as each motor axon supplying a given muscle branches to innervate from tens to hundreds of muscle fibers. Large motor axons synapse with a greater number of large, more forceful muscle fibers. Small caliber motor axons innervate smaller groups of weaker, slow-twitch fibers. The resulting functional diversity in motor unit physiology enables individual muscles to shift among different use patterns by progressively activating motor units with differing contractile properties through increased levels of activity in the nerve. Progressive stimulation of larger proportions of a target cell population is called *recruitment*.

The selective organization of specific axons and muscle fiber types into particular motor units is one demonstration that the specificity of innervation extends to subtle differences among cells within a target population. Studies have established that one mechanism which sorts innervating axons within the target relies on graded differences in the display of target cell factors that selectively promote (or inhibit) axonal growth, combined with a graded susceptibility to these factors among the pool of innervating axons. Perhaps the best illustrated example of this mechanism is the topographically ordered projection of retinal axons onto the optic tectum, in birds and fish, which is regulated by tectal cell-derived ephrin signaling proteins and their cognate receptors differentially expressed in a topographically graded fashion by the retinal ganglion cells. The same molecular mechanism guides motor axons that have cell bodies in neighboring regions of the spinal cord to innervate different domains within multisegmental muscles, such as the diaphragm and intercostal (rib) musculature (Wigston and Sanes, 1982; Laskowski and Sanes, 1987; Feng *et al.*, 2000). However, for reasons of mechanical stability, the myofibers that comprise a motor unit are not fasiculated in a contiguous bundle. Rather, they are dispersed throughout the host muscle. This makes it impossible for a single motor axon to acquire a motor unit's worth of muscle fibers, or selectively innervate fibers of a particular type, simply by colonizing a small domain of the target muscle. Instead, axons and muscle fibers must exchange specific information during development that further biases the final outcome to favor certain matches.

A final complication in the apparent simplicity of muscle innervation is that most of the muscle is actually refractory to innervation. NMJs are much larger than most interneuronal synapses (Fig. 5). Nevertheless, they typically occupy only a small fraction of the muscle fiber's total surface area, leaving more than 99% uninnervated. Experimental attempts to form additional synapses in extrasynaptic portions have shown that existing synaptic sites actively suppress the formation and maintenance of novel synaptic sites. Mechanisms regulating the muscle's susceptibility to innervation are discussed in a later section. One consistent finding is that synaptic transmission and evoked activity in the muscle are important factors.

In summary, the apparent simplicity of neuromuscular innervation in mature muscles belies an underlying organizational complexity. Mature patterns of innervation arise through interactions between motor nerves and their target muscles that regulate the cumulative assembly and disassembly of individual synapses. It is worth noting that these conclusions resonate with initial studies of synaptogenesis in other systems, including the mammalian brain. We next review the structure of an individual neuromuscular synapse, before considering mechanisms that direct its development.

Synaptic Specializations at the Neuromuscular Junction

The NMJ is composed of three cell types: Motor neuron, skeletal muscle fiber, and Schwann cell (Fig. 9). The synaptic portion of each of these cells is morphologically and biochemically specialized to support neurotransmission. The extracellular matrix that fills the synaptic cleft is also specialized compared to the matrix that covers the extrasynaptic surfaces of nerve and muscle fibers.

Morphological Specialization

As previously noted, the skeletal muscle endplate occupies a very small region of the muscle fiber surface, usually located midway along the fiber's length. A central location allows action potentials generated at the synapse to spread most rapidly to the ends of the fiber, speeding and synchronizing contractions. The mature endplate is characterized by three morphological hallmarks. First, the postsynaptic surface is impressed with shallow channels and pits, sometimes called synaptic gutters. These postsynaptic depressions hold the branches of the nerve terminal and constitute the *primary synaptic cleft*. Second, the surface of the synaptic gutter is interrupted by a series of invaginations, prosaically named *folds*, which extend several microns into the subsynaptic sarcoplasm and thus constitute a set of *secondary synaptic clefts*. Folds are unique features of the neuromuscular synapse. One possible benefit of forming secondary clefts is to increase the fidelity of synaptic transmission at high firing frequencies by speeding the clearance of spent neurotransmitter from the primary synaptic cleft. Third, the postsynaptic membrane is thickened by an extremely high concentration of AChRs and a coterie of receptor-associated proteins. Synaptic AChRs are concentrated 1,000-fold above levels in extrasynaptic regions of the muscle membrane. Moreover, synaptic AChRs are asymmetrically distributed between the primary and secondary postsynaptic membranes. AChRs are concentrated in the primary postsynaptic membrane, at the crests of the folds. In contrast, the secondary postsynaptic membrane (in the depths of the folds) contains high concentrations of voltage-gated sodium (Na_V) channels.

The motor nerve terminal is also characterized by three morphological hallmarks. First, the motor axon ends in a series of branches, known as a terminal arbor. Unlike the long slender axon, the terminal branches are relatively short and swollen ("varicose"). Second, terminal varicosities are loaded with synaptic vesicles and mitochondria. Third, the synaptic membrane of the nerve terminal contains a large population of active zones, which appear in the electron microscope as thickened regions of the presynaptic membrane associated with several synaptic vesicles.

Within the nerve terminal, the distributions of synaptic vesicles and mitochondria are polarized. The mitochondria,

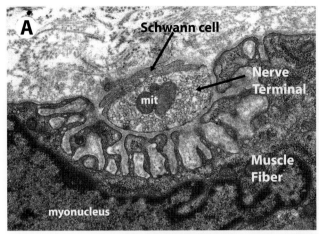

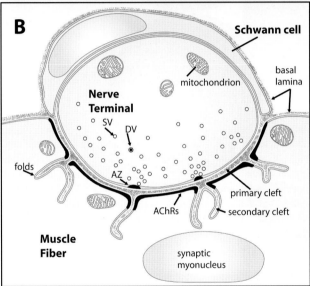

FIGURE 9. Organization of synaptic specializations at the neuromuscular junction (NMJ). Neuromuscular synapses comprise primary and secondary specializations in three cells. Primary specializations include a terminal Schwann cell that caps rather than wraps the motor nerve; a varicose nerve terminal, which accumulates mitochondria (mit) and synaptic vesicles (SV); and a high concentration of ACh receptors (AChRs) and scaffolding proteins in the postsynaptic membrane, which therefore appears thickened in electron micrographs. Secondary specializations develop postnatally and enhance neurotransmission. In the nerve terminal, active zones (AZ) appear along the junctional surface, and the distributions of synaptic vesicles and intra-terminal organelles become polarized with respect to the synaptic cleft. In the muscle, secondary synaptic clefts create folds in the postsynaptic membrane; postsynaptic membrane proteins are distributed asymmetrically in postsynaptic membranes, with AChRs concentrated in the primary postsynaptic membrane, and voltage-gated sodium channels in secondary postsynaptic membranes. Myofiber nuclei in the subsynaptic sarcoplasm express genes for postsynaptic proteins, such as AChR subunits, at higher levels than extrasynaptic myonuclei. A basal lamina (BL) covers the surfaces of muscle fiber and Schwann cell, and fills the primary and secondary synaptic clefts. Although separated by the synaptic BL, active zones in the nerve terminal surface are accurately aligned with postsynaptic folds. Thus, each cell adopts synapse specific behaviors that reflect specializations in its partners. (A) Electron micrograph of an adult mouse neuromuscular junction. (B) Interpreted view of cellular specializations at the vertebrate neuromuscular junction. DV, dense core vesicle.

microtubules, neurofilaments, and other components shared with the axon are concentrated in the center or abjunctional parts of the terminal, away from the synaptic cleft. In contrast, the synaptic vesicles are concentrated in the portion of the terminal nearest the synaptic cleft, and even further concentrated near active zones. The active zones themselves are spaced at intervals along the synaptic surface of the nerve terminal. The most striking example of active zone organization may be at frog NMJs, where they form a set of parallel stripes evenly spaced at 1 μm intervals across the straight terminal branches.

The geometry of the endplate and nerve terminal match precisely (Figs. 8–10). Nerve terminal branches conform exactly to the width and length of the endplate's gutters. Postsynaptic membranes concentrate AChRs directly opposite the nerve terminal branches. In addition, the polarized distribution of AChRs and Na$_v$ channels to the primary and secondary postsynaptic membranes, respectively, reflects the polarized distribution of synaptic vesicles in the nerve terminal. Note that those components closest to the primary cleft are directly involved in transmission; those in arrears augment transmission capacity. Perhaps most remarkable, the locations of the active zone complexes in the nerve terminal membrane are maintained in precise register with the mouths of the secondary synaptic clefts that pattern the postsynaptic surface (Fig. 9). Extraordinary size allows the subsynaptic organization of the NMJ to be recognized and imaged. However, apposition of pre- and postsynaptic specializations occurs to some extent at all synapses and is similarly precise at fast chemical synapses throughout the nervous system.

A wealth of data from physiological, molecular, genetic, and imaging experiments supports the notion that the morphological specializations of the NMJ foster high capacity neurotransmission. The large presynaptic area enables the motor axon to release enough ACh neurotransmitter to produce a postsynaptic current that overcomes the muscle fiber's low input resistance. The branching of motor nerve terminal also serves to spread neurotransmission over enough of the muscle fiber surface so that the resulting depolarization in the surrounding muscle membrane reliably generates a myofiber action potential. Active zones ensure that individual neurosecretory events are distributed, speeded, and synchronized throughout the enlarged synaptic area. Such outcomes are fostered by pre-docking synaptic vesicles preferentially at active zones and extended by concentrating and polarizing the distribution of synaptic vesicles and mitochondria within the terminal cytosol. The formation of a postsynaptic gutter increases the synaptic contact area between nerve and muscle, while postsynaptic folds lacking receptors speed the clearance of transmitter from the primary cleft following muscle fiber activation. It may be interesting to consider whether the structural specializations that support synaptic transmission at this (or any other) synapse could have taken another, significantly different form. Could coordinated control of movement have been achieved through alternative signaling mechanisms?

The importance of terminal morphology to synaptic function is illustrated by its highly stereotyped organization across vertebrate species and in different types of muscles. For example, the precise location of presynaptic active zones opposite postsynaptic

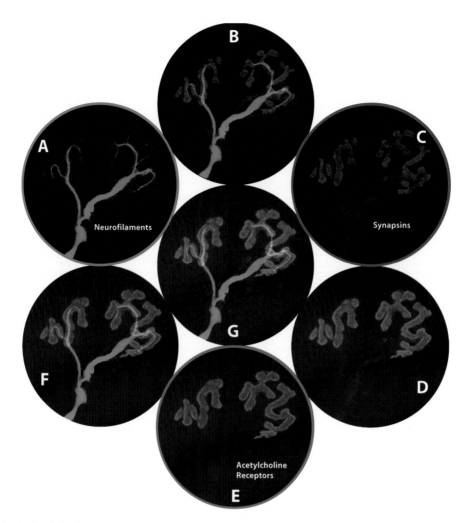

FIGURE 10. Molecular differentiation between nerve terminal and axon. Synaptic structure and function ultimately depend on molecular specializations. In one example, neurofilaments that fill the motor axon are restricted to the core regions of primary branches in nerve terminals (A, and green in other panels). In contrast, synapsins are concentrated throughout the nerve terminal branches and largely absent from the axon (C, and blue in other panels). In the muscle, ACh receptors in postsynaptic membranes (E, and red in other panels) are concentrated 1000-fold above levels in extrasynaptic portions of the myofiber surface. The molecular differentiation between axon and nerve terminal accurately reflects the location of postsynaptic membrane.

folds, described above, occurs in virtually all species examined (from snakes to *sapiens*). Similarly, the distribution of AChRs in the postsynaptic membrane precisely matches the arborization of the nerve terminals, regardless of the shape those terminals may take in a particular muscle or species. Interestingly, terminal arbors and their matching AChR-rich endplate have stereotypical shapes that vary among different muscle types. For example, terminals in jumping muscles in frogs form a series of parallel branches that extend several hundred microns up and down the muscle fiber. In contrast, axons innervating snake muscles that undergo slow and sustained contractions typically form *en grappe* terminals; these contain a cluster of spherical boutons that look like a bunch of grapes and cover only a few tens of microns of the myofiber surface. Similarly in birds and mammals, terminals on fast-twitch fibers form a set of curled branches, like a misshapen pretzel (termed *en plaque*), while terminals on slow-twitch muscles form *en grappe* bouton clusters.

In general, long-branched synapses are present where synaptic transmission is strongest and contraction is most vigorous. Smaller, bouton-like terminals are present on muscle fibers that contract more weakly but more sustainably. These distinctive synaptic morphologies have been maintained through hundreds of millions of years of evolutionary divergence, arguing that they impart significant functional advantages. One possible explanation is that an action potential spreads throughout the arbor of an *en plaque* nerve terminal, but spreads unevenly into a subset of the boutons of an *en grappe* nerve terminal. Active zones throughout the *en plaque* terminal branches would be synchronously recruited to neurotransmission, increasing synaptic strength and avoiding failures at periods of maximum muscle contraction. In contrast, active zones in subsets of *en grappe* boutons could be recruited in response to successive action potentials, potentially reducing the synapse's susceptibility to fatigue during extended periods of activity.

Molecular Specialization

A maxim of engineering is that form follows function. The demands of synaptic transmission at the NMJ are supported by structural specializations in pre- and postsynaptic elements, and these in turn are accompanied by biochemical specializations. Some molecular specializations at the synapse, like the concentration of AChRs in the postsynaptic membrane, are so closely allied to synaptic function that they serve as inviolate markers of synaptic sites in the muscle. Other constituents of the synapse have more subtle roles; although they may be concentrated at synaptic sites in all vertebrate phyla, their absence (as, e.g., in genetically engineered mutant mice) causes little perceptible defect in synaptic structure. Nevertheless, their evolutionary conservation suggests that they play essential functional roles in wild animal populations.

The nerve terminal lacks proteins, such as neurofilaments that are concentrated in the axon, and is enriched instead with proteins devoted to the control of synaptic vesicle dynamics (Fig. 10). These include vesicle membrane-associated proteins from the rab, rabphilin, and synapsin families, which regulate intra-terminal synaptic vesicle trafficking; membrane and cytosolic proteins which subserve the docking and fusion of synaptic vesicles with the terminal surface, including synaptobrevin/VAMP, synaptotagmin, SNAP-25, syntaxin, and munc18; voltage-gated calcium channels, which transduce terminal depolarization into the biochemical signal for transmitter release; additional proteins that promote the recovery of vesicle membrane from the terminal surface membrane (clathrin, AP2, dynamin, intersectin), and target the retrieved membrane packets to endosomal compartments (rab 5), where a final complement of enzymes and pumps reconstitute ACh-loaded synaptic vesicles (choline acetyltransferase—ChAT, vesicular acetylcholine transporter—VAChT).

Postsynaptic specializations in the muscle include proteins that establish and maintain the high concentration of AChRs directly opposite the nerve (Fig. 11). Central players in the initiation of AChR clustering opposite the nerve include MuSK (Muscle Specific Kinase), a receptor tyrosine kinase that initiates intracellular signaling in response to agrin, and rapsyn, a receptor-associated scaffolding protein. Their roles are elaborated in a further section. Also concentrated in the postsynaptic membrane is a second large transmembrane protein complex known as the DGC (dystroglycan-associated glycoprotein complex). The DGC is a multifunctional receptor composed of dystroglycan (α and β), the sarcoglycans (α, β, and γ), and sarcospan. It serves as the primary membrane-spanning link between the extracellular matrix in the synaptic cleft and the intracellular cytoskeleton. Primary extra-cellular ligands include the glycoprotein laminin-4 (the α2β2γ1 isoform of the laminin heterotrimer), and the heparan sulfate proteoglycans, agrin and perlecan. Each of these synaptic basal lamina components contains an LG-domain that binds to the α-dystroglycan component of the DGC. (LG-domains are protein modules first identified in *l*aminins and possessing a *g*lobular tertiary structure.) Perlecan appears to be the primary binding partner for AChE in the synaptic cleft, such that interaction between the DGC and perlecan appears to be especially important in maintaining the high concentration of AChE in the synaptic cleft.

DGC interactions with the cytoskeleton include the submembranous cytoskeletal proteins dystrophin and utrophin, which have similar structures and overlapping functions.

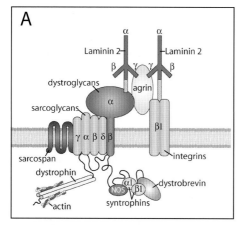 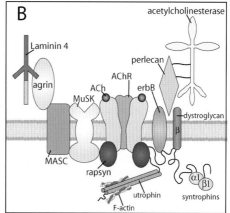

FIGURE 11. Postsynaptic scaffolds at the neuromuscular junction. The dystroglycan–glycoprotein complex in synaptic and extrasynaptic portions of the muscle. (A) In extrasynaptic portions of muscle fibers, the extracellular matrix is comprised of collagens, laminins, and Z(0) forms of agrin. These molecules interact with the cell surface by interactions with α-dystroglycan and β1-integrins. Within the membrane, β-dystroglycan and the sarcoglycans form a complex that binds to dystrophin intracellularly. Dystrophin then interacts with the actin cytoskeleton and with signaling molecules such as dystrobrevins, syntrophins, and nitric oxide synthase. (B) At the synapse, distinct isoforms of laminin and Z(+) agrin interact with different receptors at the cell surface. Agrin activates the tyrosine kinase receptor MuSK via a hypothetical accessory protein MASC. This interaction results in the clustering of pentameric acetylcholine receptors. Intracellularly, the scaffolding protein rapsyn aggregates AChRs and also binds to utrophin. Like dystrophin, utrophin links this complex to the actin cytoskeleton and to syntrophins for signal transduction. In addition to AChRs, erbB receptors and β-dystroglycan are also linked to this complex via utrophin, thus allowing the transduction of signals such as neuregulin in addition to acetylcholine synaptic transmission.

One function of the dystrophin-rich cytoskeleton is to anchor the specialized protein complexes of the postsynaptic membrane to the intracellular cytoskeleton (Fig. 11A). This function has in fact been best studied in extrasynaptic regions of the muscle, where interactions between dystrophin and the DGC stabilize the myofiber membrane during muscle contractions. In extrasynaptic muscle, dystrophin and the DGC are concentrated at costameres. Costameres are membrane-bound protein complexes that tether the sarcolemma to the contractile apparatus of the sarcomere. Mutations to dystrophin or to components of the DGC destabilize this linkage and cause severe forms of muscular dystrophy (Durbeej and Campbell, 2002). A similar set of interactions with postsynaptic membranes likely stabilizes the endplate during muscle activity.

In addition, at synaptic sites, the DGC serves as the membrane platform for utrophin-dependent interactions with the AChR-complex, and with a family of proteins called syntrophins (Fig. 11B). Utrophin specifically associates with MuSK and is concentrated in the AChR-rich primary postsynaptic membrane, along the crests of the junctional folds. (Dystrophin is concentrated in complementary fashion along the secondary postsynaptic membrane deep in the junctional folds, where Na_v channels are concentrated.) Syntrophins are co-concentrated with utrophin at synaptic sites. Little utrophin accumulates in the postsynaptic membrane in the absence of syntrophin, in mice bearing targeted syntrophin gene deletions. The mature structure of the endplate is adversely affected, as few postsynaptic folds are formed in syntrophin-deficient mice (Adams et al., 2000). An additional, abbreviated homologue of dystrophin called dystrobrevin is also concentrated at postsynaptic sites. Interestingly, postsynaptic sites in dystrobrevin-deficient mice are initially well formed, but begin to fragment during postnatal development (Grady et al., 2000). These and other results suggest that the DGC and its associated cytoskeletal partners promote the growth and maturation of the muscle's postsynaptic specialization (Cote et al., 1999; Albrecht and Froehner, 2002).

Postsynaptic differentiation also includes changes in gene expression by synaptic myonuclei. Each muscle fiber is a syncytium, containing hundreds of myonuclei from the fusion of myoblasts during embryonic development. Most myonuclei are spread more or less evenly through the fiber, but several nuclei are prominently clustered beneath the synaptic endplate. Muscle-specific proteins, such as skeletal muscle myosins and muscle creatine kinase, are highly expressed by myonuclei throughout the muscle. In contrast, several synapse-specific proteins are only expressed at high levels by synaptic myonuclei. These include the genes for subunits of the AChR and Na_v channels, for acetylcholine esterase (AChE) and for utrophin. High-level expression of synaptic components may be necessary to support the large size of the NMJ. Conversely, local synthesis may also be a mechanism to ensure that synaptic components are properly targeted to the synaptic sites along the muscle fiber. At interneuronal synapses, as well, there is growing evidence that local synthesis of synaptic components contributes to synapse formation and plasticity (reviewed by Steward and Schuman, 2001).

In these cases, however, where the postsynaptic cell has one nucleus, it is specific RNAs rather than specialized nuclei that are sequestered at synaptic sites.

Schwann Cells

Most synapses are surrounded by glial cell processes. At the NMJ, the motor nerve terminal is covered by the processes of one or a few Schwann cells. Although Schwann cells do not have a primary role in synaptic transmission, they play critical roles in supporting nerve terminal metabolism and influence overall levels of innervation in muscle. Like motor nerves and muscle fibers, Schwann cells at synaptic sites are distinguished from their extrasynaptic counterparts by structural and molecular differences.

The most obvious difference is the behavior of the Schwann cell's processes. Schwann cells located in the nerve myelinate motor axons by wrapping sheet-like processes around axon segments, in compact, concentric circles. Schwann cells at synapses do not wrap axon terminals. Rather, they extend short processes that cover the non-synaptic surface of the terminal (Fig. 9). Synaptic Schwann cell processes tend to follow the course of the terminal branches; they avoid the synaptic cleft and rarely extend into extrasynaptic muscle.

Molecular differences accompany the morphological differences between preterminal and terminal Schwann cells. For example, myelinating Schwann cells express specific transcription factors, including krox-20 and Oct-6/SCIP/Tst-1, and their membranes contain a unique complement of membrane glycolipids and glycoproteins that includes protein-zero, peripheral myelin protein 22, periaxin, and myelin-associated glycoprotein. In contrast, synaptic Schwann cells express little or no myelin-associated proteins, but do express high levels of krox-24 (also called EGR-1, zif/268, and NGFI-A), the intracellular calcium-binding protein S100, the cell-surface adhesion molecule NCAM, and the extracellular matrix component laminin α4. It is unlikely that these differences represent fully differentiated cell fates, for although synaptic and myelinating Schwann cells derive from common neural crest progenitors, their final state depends on which portion of which axon they contact and is reversible (Garbay et al., 2000; Lobsiger et al., 2002). Signals associated with large caliber motor and sensory axons cause their associated Schwann cells to form myelin. Following axonal degeneration, the myelinating Schwann cells revert to a pre-myelinating phenotype, a process that includes the downregulation of myelin proteins and upregulation of S100, NCAM, and laminin α4. One possibility is that synaptic Schwann cells are permanently pre-myelinating. Alternatively, unknown synaptic signaling factors may induce a uniquely differentiated Schwann cell state. In any event, factors concentrated in the synaptic cleft prevent Schwann cells from myelinating the terminal portion of the axon. One factor contributing to exclusion of Schwann cell processes from the synaptic cleft is laminin-11, a component of the synaptic cleft material that inhibits the motility of Schwann cell processes (Patton et al., 1998).

The Synaptic Cleft

At most chemical synapses, the gap between the pre- and postsynaptic elements is filled with a matrix of glycoproteins and proteoglycans. At interneuronal synapses, where the cleft is only 20 nm across, the synaptic matrix most likely consists of the extracellular domains of membrane proteins. In the synaptic cleft at the NMJ, the synaptic matrix contains secreted glycoproteins and proteoglycans assembled into a tightly woven sheet of material known as a basal lamina. The thickness of this structured matrix largely determines the 50–70 nm width of the neuromuscular synaptic cleft. As we shall see, the structure and activity of the synaptic basal lamina plays an essential role in the formation and maintenance of the NMJ, as well as its physiology (Patton, 2003).

Basal laminae are present in many tissues in the body, serving as substrates for cell adhesion and movement, imparting structural integrity, and organizing cell-signaling domains through membrane receptors such as integrins. In muscle, the synaptic matrix is part of a continuous basal lamina that covers the entire surface of each myofiber, and which has been structurally likened to a nylon stocking on a leg. A similar basal lamina made by Schwann cells covers each peripheral nerve fiber. Extrasynaptically, the myofiber basal lamina is closely bound to the surface of the myofiber; in the synaptic cleft, the basal lamina is bound directly to the surfaces of both nerve terminal and endplate. Although structurally similar, synaptic and extrasynaptic portions of the myofiber basal lamina contain distinct molecular components. In particular, the synaptic cleft contains distinct isoforms of the main structural components of the basal lamina, and in addition, incorporates a number of unique accessory factors.

The principal components of all basal laminae are laminin, type IV collagen, entactin (also called nidogen), and the heparan sulfate proteoglycan, perlecan (Fig. 12). Laminins and collagens IV are families of long, rope-like glycoproteins. Their structure comes from a trimeric composition of homologous subunits (here referred to as "chains") that are entwined along much of their length. Laminins and the type IV collagens self-polymerize, forming supramolecular networks cross-linked by entactin. Together, they account for much of the structural integrity of the basal lamina. Interestingly, synaptic and extrasynaptic basal laminae contain different isoforms of these components, which differ in chain composition. The extrasynaptic basal lamina primarily contains collagen IV trimers composed of $\alpha1(IV)$ and $\alpha2(IV)$ chains, and laminin-2 (the $\alpha2\beta1\gamma1$ heterotrimer). Synaptic basal lamina is more complicated. It contains collagen IV composed of the $\alpha3(IV)$, $\alpha4(IV)$, and $\alpha5(IV)$ chains and also contains three distinct laminin heterotrimers: Laminin-4 ($\alpha2\beta2\gamma1$), laminin-9 ($\alpha4\beta2\gamma1$), and laminin-11 ($\alpha5\beta2\gamma1$). The synaptic matrix also contains an uncharacterized variant of entactin, possibly nidogen-2 (Chiu and Ko, 1994).

Synaptic laminins principally differ from extrasynaptic laminins by virtue of their $\beta2$-chain and differ from each other by their α-chain component, discussed in more detail below. The laminin-$\beta2$ chain was originally named s-laminin, by Hunter, Merlie, and Sanes, for its *synaptic* concentration (Sanes *et al.*, 1990). Most synaptic basal lamina components are present throughout the synaptic cleft. One exception is laminin-9, containing the $\alpha4$ chain, which is absent from postsynaptic folds, and is concentrated in small patches within the primary cleft (Patton *et al.*, 2001). The several synaptic laminins have

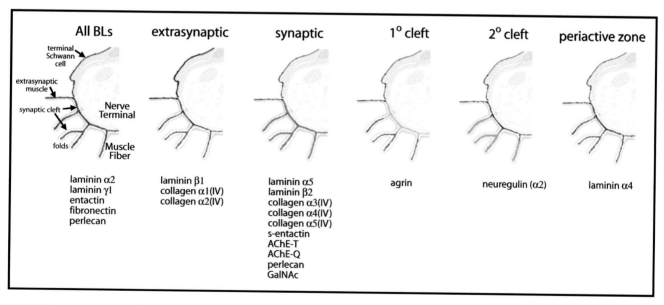

FIGURE 12. Specialization of the synaptic basal lamina. The basal lamina (BL) that ensheaths each myofiber is molecularly specialized at synapses. All BLs contain type IV collagen, laminin, and entactin, which together provide structural integrity. Synaptic and extrasynaptic BLs differ in specific isoforms of these ubiquitous components. In addition, synaptic BLs are enriched by a series of accessory components. These include the catalytic and collagen-like tail subunits of acetylcholine esterase (AChE-T and AChE-Q, respectively), the heparan sulfate proteoglycans agrin and perlecan, and the growth and differentiation factor neuregulin. Agrin, neuregulin, and the laminin $\alpha4$ subunit have restricted distributions within the synaptic BL. (Figure reprinted from Patton, 2003, with permission.)

important roles in the maturation of the nerve terminal and the terminal Schwann cell, which are discussed later.

The synaptic basal lamina also incorporates a number of accessory constituents, including agrin, neuregulin, a glycosyl-transferase, and a collagen-tailed form of AChE. Agrin and neuregulin are signaling molecules that act through membrane receptor tyrosine kinases to control the expression and distribution of synaptic components in the muscle fiber. Agrin is concentrated in the primary synaptic cleft, along the crests of the AChR-rich junctional folds (Trinidad *et al.*, 2000). In contrast, at least one isoform of neuregulin is concentrated in the troughs of the postsynaptic folds and is absent from the primary cleft. The roles of agrin and neuregulin in postsynaptic differentiation are addressed below.

Although our understanding of the role of glycosylation at the synapse is rudimentary, several components of the synaptic cleft bear unique and evolutionarily conserved glycosylation patterns. For example, VVA-b4 isolectin (isolated from the hairy vetch, *vicia villosa*) recognizes a carbohydrate group (terminally sialylated *N*-acetyl galactosamine) that is specifically concentrated at NMJs in nearly all classes of vertebrates: fish, amphibians, reptiles, birds, and mammals. This suggests that a role for specific glycosylation patterns developed early in vertebrate evolution and was important enough to be preserved for hundreds of millions of years in all viable offspring. One possibility suggested by recent studies is that glycosylation of the dystroglycan receptor alters its affinity for variants of laminin, agrin, and perlecan, fostering their concentration in the synaptic cleft (Xia *et al.*, 2002).

Perlecan plays a specific and important role in binding AChE into the synaptic cleft (Peng *et al.*, 1999). Perlecan is a long, multidomain molecule, with binding sites for a number of extracellular matrix proteins and receptors. One likely partner is ColQ, a collagen-like subunit of AChE present in the synaptic cleft. Mutations in the genes for perlecan and ColQ each prevent the accumulation of AChE at synaptic sites (Donger *et al.*, 1998, Feng *et al.*, 1999, Arikawa-Hirasawa *et al.*, 2002).

AChE hydrolyzes acetylcholine following synaptic transmission at the NMJ. Because nicotinic AChR channels in muscle do not rapidly inactivate, AChE effectively terminates neurotransmission by eliminating ACh as it dissociates from the AChRs. Two specializations support this role. First, AChE's rate of catalysis is extraordinarily fast. The measured k_{cat} is 14,000 s^{-1}, and the k_{cat}/K_M value is calculated to be 1.6×10^8 (Ms)$^{-1}$, which is near the diffusion limited rate for enzymatic reactions. AChE is essentially catalytically perfect. Second, AChE is concentrated to extraordinarily high levels in the synaptic basal lamina, up to 3,000 per μm^2 (Rogers *et al.*, 1969; Salpeter, 1969; Salpeter *et al.*, 1978; Anglister *et al.*, 1998). The dependence of synaptic function on AChE activity is shown by the potency with which anticholinesterase drugs affect neuromuscular synaptic transmission. Anticholinesterases are administered at low doses to myasthenic patients, to bolster weak neuromuscular transmission. At moderately higher doses, however, anticholinesterase exposure is lethal. AChE is a primary physiological target of organophosphate and carbamate insecticides used on crops and livestock, and military nerve gases stored for use against humans.

Collectively, the components of the synaptic cleft reflect the organization of the pre- and postsynaptic cells. While most components of the synaptic basal lamina are synthesized by the target muscle fiber, including the laminins, collagens, perlecan, and cholinesterase, the nerve supplies components as well, most notably agrin. In principle, specialization of the synaptic cleft could act primarily during maturation of the maturation factors, to strengthen transmission and mechanically stabilize the synaptic site during activity. In this case, the first evidence of specialization in the synaptic cleft might appear well after the establishment of initial synaptic contacts. Direct observations show, however, that most components of the synaptic cleft are present at early stages of synapse formation, and several, such as AChE and laminin-β2, are highly concentrated at synaptic sites shortly after their formation. This raises the alternative possibility that the restricted distributions of synaptic cleft components may reflect a direct role in organizing nerve terminals and endplates. Indeed, agrin, laminin β2, and neuregulin were originally identified by attempts to define the molecular signals that promote synaptic differentiation. These possibilities are explored more fully in the next section, where the molecular mechanisms underlying synaptic development and maturation are considered.

SYNAPSE FORMATION

From our discussion of synaptic structure and function, we conclude that there are two essential aspects to the formation of a chemical synapse. Small regions of the axon and its target cell become specialized to support effective neurotransmission, and these synaptic specializations are directly apposed to each other across a synaptic cleft. In broad terms, it is possible to imagine two alternative processes by which synaptic specializations in the motor axon and muscle fiber are formed and colocalized. In one case, cell autonomous programs of development in the nerve and muscle first produce synaptic specializations in each cell, independently; these specializations then become oriented with respect to each other through intercellular interactions. Alternatively, synaptic differentiation could be initiated by local signaling interactions between nerve and muscle. In this case, the location of the synaptic site could reflect sites of initial contact, or regions of especial susceptibility along the cell surfaces.

In support of the preprogrammed model of synaptic differentiation, muscle fibers and motor neurons do independently express most of the components of the mature NMJ and will organize primitive synaptic structures when cultured separately. For example, cultured myotubes (immature muscle fibers) express functional AChRs on their surface and spontaneously cluster receptors in small patches, similar to AChR plaques that form *in vivo*. Cultured myotubes also respond to application of neurotransmitter with weak contractions. Similarly, the neurites of motor neurons cultured in the absence of muscle cells are capable of spontaneous and evoked release of neurotransmitter. Nerves and muscles are therefore prepared to form rudimentary synaptic specializations without direction. Synaptic connections might then be established by either guided or random intersections of pre- and postsynaptic specializations. For the mature pattern of innervation observed in adult muscles to emerge by this mechanism, *bona fide* synapses would require stabilization, while ectopic (unconnected) synaptic specializations would require disassembly. In fact, synapse formation in muscle does include a partial misalignment of initial specializations, an enhancement of well-formed connections, and the elimination of weakly matched specializations. These events are discussed in further detail, below. In effect, mature synapses are sculpted from an initial population of specializations that are more crudely aligned. This model recalls earlier stages of neural development in vertebrates, wherein the final population of neurons and axonal projections represent a subset of those initially formed.

In the alternative scenario, synapsing cells directly organize each other's synaptic specializations, through diffusible and/or cell-surface cues. In fact, several direct observations support this signaling model of synapse formation at the NMJ. First, in studies of nerve–muscle cocultures, developing myotubes assembled a new AChR-rich postsynaptic apparatus at sites where they were contacted by axons (Fig. 13) (Anderson and Cohen, 1977; Anderson *et al.*, 1977). Preexisting postsynaptic specializations, which had formed spontaneously (without innervation), were not preferentially innervated by axons and were often *disassembled* in response to novel innervation of the myotube (Kuromi and Kidokoro, 1984). Second, presynaptic specializations preferentially formed where axons contact muscle surfaces, in nerve–muscle cocultures (Fig. 14), during normal

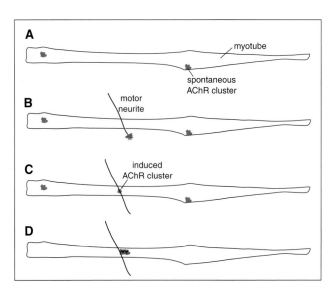

FIGURE 13. Motor nerves organize postsynaptic specializations in cultured muscle cells. Experiments in which spinal motor neurons were cocultured with differentiated muscle myotubes and then stained with α-bungarotoxin to show the distribution of ACh receptors (AChR), showed that (A) muscle cells are capable of clustering AChRs "spontaneously," independent of innervation; (B) motor neurites do not target spontaneous AChR clusters for preferential innervation; (C) neurites induce new AChR clusters where they contact the muscle fiber; and (D) maturation of nerve-induced AChR clusters is accompanied by disassembly of spontaneous clusters. These results, obtained in studies by Cohen and Anderson, and Frank and Fischbach, suggested that motor nerves provide signals that organize the local differentiation of the postsynaptic site.

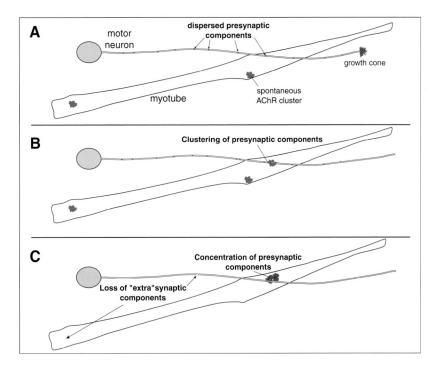

FIGURE 14. Myotubes promote presynaptic differentiation in cultured motor axons. In nerve–muscle coculture experiments by Lupa and Hall, developing motor neurites preferentially concentrated synaptic vesicles at sites of contact with muscle cells. Presynaptic sites coincided spatially and temporally with postsynaptic sites induced by the neurite in the muscle cell. The results indicated that muscle fibers provide local, retrograde signals that promote presynaptic differentiation in the axon.

development, and during reinnervation of muscles following nerve injury *in vivo* (Cajal, 1928; Marshall *et al.*, 1977; Lupa and Hall, 1989; Lupa *et al.*, 1990). Third, nerve terminals and muscle endplates develop in concert as synaptogenesis proceeds during normal development, *in vivo*. The initial axonal projection predicts the final location of synapses in the muscle, and early errors in matching terminals and endplates are rapidly corrected. Indeed, it is difficult to imagine how the highly stereotyped geometry of the mature nerve terminal could reliably match the AChR-rich endplate membranes, or how the position of terminal active zones could so accurately align with postsynaptic folds, through completely independent programs of development in nerve and muscle.

Historically, these observations provided strong *a priori* evidence that the organization of synaptic specializations is a concerted process regulated by locally produced nerve- and muscle-derived factors. Considerable molecular evidence has been added over the last 25 years. Nerve-derived factors that organize the muscle endplate have been identified and muscle-derived factors have been identified as candidate cues to guide nerve terminal formation and regulate Schwann cell motility. As a result, the signaling model of synapse formation has come to dominate our thinking of how most chemical synapses are established and maintained.

Recently, the signaling model has been challenged to accommodate earlier observations that nerve terminals and AChR-rich postsynaptic specializations are often misaligned during the earliest stages of synapse formation in muscle (Braithwaite and Harris, 1979; Lupa and Hall, 1989; Dahm and Landmesser, 1991). Most dramatically, it has recently been discovered that nearly normal patterns of postsynaptic differentiation in the muscle occur in the complete absence of a nerve (Yang *et al.*, 2000, 2001; Lin *et al.*, 2001). Discussed more fully in a later section, the implication of this surprising result is that innervation is strongly influenced by a preestablished pattern of synaptic differentiation in the muscle, but is initially imprecise. Synaptic specializations in the nerve and muscle then rapidly align with each other through an exchange of synaptogenic signals. Signals from the nerve may either stabilize the initial synaptic template in the muscle, or disassemble and replace it. Thus, the exact roles of signaling and preestablished programs in synapse formation remain uncertain, even at the NMJ. It seems quite possible that elements of both models of synapse formation, preprogrammed and signal-mediated, may be required to fully explain synapse formation as it occurs during development.

Presynaptic Differentiation

The first motor axons to grow into developing muscles arrive at or before the time myoblasts fuse to form myotubes (Bennett, 1983). Individual motor axons branch many times within the muscle to innervate the many fibers that eventually comprise a motor unit. Axons likely respond directly to guidance cues provided by developing muscles, although guidance to synaptic sites after injury in adults is largely mediated by Schwann cells (Son and Thompson, 1995; Riethmacher *et al.*, 1997; Nguyen *et al.*, 2002).

Developing motor axons release ACh at their growth cones and are capable of activity-dependent neurotransmission within minutes of contact with the myotube (reviewed in Sanes and Lichtman, 1999). Nevertheless, synaptic coupling at nascent synapses is very weak, due in part to the small number of synaptic vesicles present in the primitive terminal, and the low density of AChRs in the postsynaptic membrane. The first overt sign of presynaptic differentiation is the formation of small branches and varicosities at the end of the axon, which occurs after nerve–muscle contact *in vivo*. Terminal varicosities initially form in the vicinity of concentrated "plaques" of AChRs located on the surface of the muscle fibers. (Whether these AChR plaques are targets of, or are induced by, the innervating nerve is not yet clear.) Within a day or two, nerve terminals and receptor plaques are topographically matched: Each postsynaptic plaque of receptors is covered by a cluster of terminal varicosities, and few terminal branches stray beyond the edges of a receptor plaque. During the following week, nerve terminals concentrate synaptic vesicles at high densities, lose the microtubule structures of the axon, and widen to cover more of the postsynaptic surface. Finally, mitochondria accumulate, the terminal becomes polarized, and active zones form along the synaptic portion of the terminal membrane. Maturation of the terminal is accompanied by a large increase in the number of vesicles released per depolarization and thus a dramatic increase in the strength of the synaptic connection.

One conclusion from these observations is that presynaptic differentiation is largely a process of organization, with progressively higher levels of detail in successive steps. Presynaptic differentiation proceeds from the segregation of terminal and axonal compartments, through increasing levels of structural complexity, with relatively little change in the molecular components of the nerve terminal. A second lesson is that nerve terminal organization progresses gradually, in steps. Primitive synaptic connections form as presynaptic terminals and postsynaptic receptors become colocalized. Immature nerve terminals form as the levels of primary synaptic components increase. Mature nerve terminals appear as the synaptic components are further organized to concentrate neurotransmission at active zones. An emerging theme is that each of these steps may be separately regulated by factors derived from the muscle and/or the terminal Schwann cell.

Presynaptic Differentiation Factors

Nearly a century ago, Cajal surmised that muscles supply factors which cause motor axons to form nerve terminals, and that these factors are concentrated at postsynaptic sites. Fernando Tello, a student of Cajal, observed axons of crushed peripheral nerves as they grew back into the denervated muscle (Cajal, 1928). Tello noted that the reinnervating axons stopped growing and formed terminal-like structures at places that appeared to be the original sites of innervation on the muscle fibers. This result was later confirmed when histological stains for cholinesterase were developed and used to show that postsynaptic sites remain identifiable and at least partly intact during extended periods of denervation (McMahan *et al.*, 1978). Regenerating motor axons do in fact demonstrate remarkable synaptic specificity, faithfully reinnervating the tiny fraction (0.1%) of each muscle fiber's surface that was previously occupied by an original nerve terminal.

The implications of Tello's observations were explored more fully by U.J. McMahan and his colleagues, in a series of experiments now 25 years old (Fig. 15). They first noted that the surface of the muscle fiber, including the synaptic endplate, is entirely covered by a basal lamina. Contact between the nerve and muscle is therefore mediated by the myofiber basal lamina, and reinnervating axons form new synaptic contacts over the original synaptic basal laminae. At the time, the cleft material was known to be molecularly specialized, containing acetylcholinesterase (McMahan *et al.*, 1978). They therefore supposed that the synaptic basal lamina harbored additional, unidentified components, which controlled the growth and synaptic differentiation of reinnervating axons. To test this idea, they took advantage of an old observation that crushed muscle fibers retract and degenerate within their basal lamina sheath, which remains largely intact for a time. Then, like Tello before them, they observed the growing ends of cut axons during reinnervation of the muscle, although in this case after the muscle fibers had degenerated. Remarkably, not only did motor axons faithfully reinnervate the empty tubes of basal lamina, but new nerve terminals were located immediately adjacent to the original synaptic basal lamina, which was identified by stains for cholinesterase (Marshall *et al.*, 1977; Sanes *et al.*, 1978). The nerve terminals even assembled active zones in proper alignment with the stems of the basal lamina that previously lined postsynaptic folds

(Glicksman and Sanes, 1983). Most importantly, these experiments revealed that the underlying muscle fiber is not required for axons to accurately reinnervate the original synaptic basal lamina. This result strongly implicated the synaptic basal lamina itself as a reservoir of molecular cues that arrest the growth of the motor axon and organize the formation of the motor nerve terminal.

Several components of the synaptic basal lamina have been proposed to regulate presynaptic differentiation of the motor axon. The best understood are synaptic isoforms of the laminin heterotrimer ($\alpha\beta\gamma$), made by the muscle fiber (Fig. 16). Three synaptic laminin heterotrimers have been identified, as described earlier (laminin-4, -9, and -11). Each contains the $\beta2$-chain, but differ from each other in their α-chain ($\alpha2$, $\alpha4$, and $\alpha5$, respectively). Synapses form abnormally in mutant mice lacking the laminin $\beta2$-chain (Noakes *et al.*, 1995), a genetic modification that prevents muscles from synthesizing any of the synaptic laminin trimers. The normal synaptic differentiation of all three cells is perturbed (Fig. 17). Motor nerve terminal organization stalls at an immature stage, with no polarity and very few active zones; the formation of postsynaptic folds is grossly retarded; and synaptic Schwann cells extend processes into the synaptic cleft, nearly isolating the pre- and postsynaptic elements. Not surprisingly, mice lacking the $\beta2$-laminins move poorly and typically die at weaning.

In principle, the loss of any or all of synaptic isoforms could cause the synaptic defects seen in the laminin $\beta2$-deficient animals. However, comparisons of synaptic defects in mice lacking the individual laminin α-chains, combined with observations

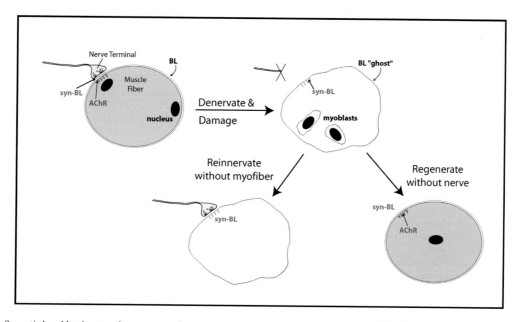

FIGURE 15. Synaptic basal lamina contains synaptogenic cues. Motor axons accurately reinnervate the original synaptic sites on muscle fibers, following axotomy in adult animals. Similarly, new muscle fibers concentrate AChRs at original synaptic sites, following injury-induced degeneration and regeneration. In a series of experiments by McMahon and his group, the ability of motor axons and muscle fibers to accurately target synaptic differentiation to original sites was determined in the absence of synaptic partners. Importantly, the myofiber basal lamina (BL) remains structurally intact during the degeneration and regeneration of a muscle fiber, and synaptic BL could be located by stains for acetylcholine esterase. When muscle fiber regeneration was prevented by irradiation of myoblasts, regenerating axons formed new nerve terminals opposite original synaptic BLs, even in the absence of a muscle fiber. When reinnervation was prevented by ligature, regenerating muscle fibers concentrated AChRs opposite original synaptic BLs, even in the absence of a nerve terminal. The results implied that synaptogenic cues were stably incorporated into the synaptic BL, and that these were sufficient to organize both pre- and postsynaptic differentiation.

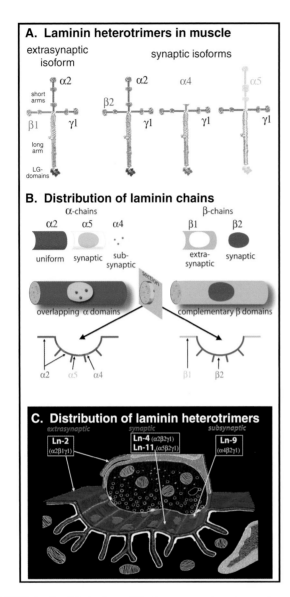

FIGURE 16. Laminin isoforms differ in extrasynaptic and synaptic muscle basal laminae (BLs). Laminins are large (*c.* 800 kDal) heterotrimers of related α, β, and γ chains. (A, B) Muscle fibers express three laminin α-chains: α2 is present in all BLs; α5 is restricted to synaptic BLs; α4 is further restricted to subsynaptic domains within the primary synaptic cleft, adjacent to junctional folds. Two laminin β-chains are expressed in muscle: β1 is restricted to extrasynaptic BL; β2 is concentrated in synaptic BL. All muscle laminins contain the γ1 chain (not shown). (C) Based on the distribution of the α, β, and γ chains: Extrasynaptic BLs contain primarily laminin-2 (α2β1γ1); synaptic BLs contain a mixture of laminins-4, -9, and -11 (α2β2γ1, α4β2γ1, and α5β2γ1, respectively). Specific associations between the distribution of laminin-9 and the location of presynaptic active zones is based on abnormal active zone placement in laminin α4-deficient mice (Fig. 17) and remains speculative. (Reprinted from Patton, 2003, with permission.)

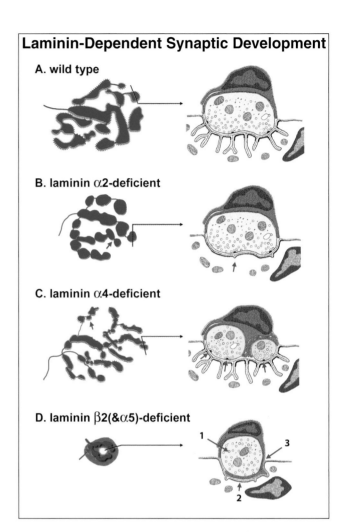

FIGURE 17. Neuromuscular junctions in laminin-deficient mice. Cellular and ultrastructural organization of the neuromuscular junction in normal mice (A) and mice with mutations in specific laminin chains (B–D). (A) In normal muscles, pre- and postsynaptic specializations are precisely aligned. Nerve terminals are embedded in ACh receptor-rich gutters in the muscle surface, synaptic vesicles and ACh receptors are concentrated near the synaptic cleft, and active zones are aligned with secondary synaptic clefts. (B) Loss of laminin α2 prevents the proper formation of postsynaptic folds. Nerve terminal branches are smaller, possibly in compensation for the paucity of folds. (C) Loss of laminin α4 prevents the proper alignment of active zones and secondary clefts, although active zones and folds form at normal frequencies. Terminal varicosities are markedly smaller than in controls. (D) (1) Loss of laminin β2 causes severe synaptic defects. Presynaptic terminals have few branches and lack polarity and active zones; (2) AChR-rich postsynaptic membranes do not match nerve terminal branches and have few folds; (3) Schwann cell processes invade the synaptic cleft, nearly isolating nerve terminal from endplate. The β2-deficient cleft also lacks laminin α5, but contains the laminin β1-chain, which is normally restricted to extrasynaptic muscle. The results indicate that individual synaptic laminins organize postnatal maturation at the neuromuscular junction, including the precise registration of pre- and post-synaptic specializations. (Reprinted from Patton, 2004, with permission.)

of the effects of purified synaptic laminins on the behavior of cultured cells, suggests that each isoform may promote a different aspect of synaptic development.

Laminin-4, which contains the α2-chain, is important for the development of normal postsynaptic folds, but otherwise appears redundant to laminin-9 and -11. Postsynaptic folds form poorly in the absence of laminin-4, in mutant mice lacking the α2-chain, but nerve terminal differentiation and Schwann cell processes appear nearly normal (Fig. 17).

Laminin-9, which contains the α4 chain, is required for the proper alignment of presynaptic active zones and postsynaptic folds (Patton *et al.*, 2001). In normal mice and indeed nearly all vertebrate animals, active zones and folds are precisely aligned across the neuromuscular synaptic cleft. In contrast, mice lacking the laminin α4-chain fail to registrate active zones and folds across the synaptic cleft (Fig. 17). Although lacking in alignment, the active zones and folds form at normal densities in the absence of the α4-laminin chain. Therefore, laminin-9 is not required for presynaptic maturation *per se*. Rather, it acts as a target-derived synaptic organizing factor.

Two unique features of laminin-9 may play a role in guiding the colocalization of active zones and folds. First, laminin-9 is concentrated in small patches in the synaptic basal lamina patches whose distribution reflects the orientation of the postsynaptic folds (Patton *et al.*, 2001). Second, presynaptic calcium channels form stable complexes with laminin-9 but not other endogenous laminins, in extracts of Torpedo electric organs (Sunderland *et al.*, 2000). Although the directness of their interaction remains uncertain, the concentration of calcium channels at active zones in vertebrates raises the possibility that laminin-9 orients the location of active zones through interactions with presynaptic calcium channels. In support of this idea, targeted disruption of the presynaptic calcium channel in mice causes a similar dislocalization of active zones and folds (Nishimune *et al.*, 2002). It could, of course, be the other way round, with active zones directing the location of folds, via laminin-9. In either case, the receptors that mediate selective interactions between synaptic laminins and the postsynaptic membrane are not known.

The remaining synaptic isoform of laminin (laminin-11, containing the α5-chain) might play a more fundamental role in organizing formation of the NMJ. First, synaptic defects in mice lacking laminin-4 and laminin-9 are mild compared to the defects in laminin β2-deficient mice. Second, purified preparations of laminin-11 arrest neurite outgrowth from cultured motor neurons, and recombinant preparations of the β2-chain are capable of inducing the morphological, biochemical, and functional properties of nerve terminals (Patton *et al.*, 1997; Son *et al.*, 1999). Third, nerve terminal formation is aberrant at embryonic stages in mutant mice specifically lacking laminin-11 through mutation of the α5-chain (Bierman *et al.*, 2003).

Laminin-11 appears to play a second important role at the NMJ, one that reveals complex interrelationships between nerve, muscle, and Schwann cell. Normally, Schwann cells cap the nerve terminal. In mutant mice lacking the laminin β2-chain, the terminal Schwann cell invades the synaptic cleft, interrupting neurotransmission (Noakes *et al.*, 1995). *In vivo* and *in vitro* experiments implicate laminin-11 as an inhibitory substrate to Schwann cell processes that directly prevents Schwann cell entry into the cleft (Patton *et al.*, 1998). Poor nerve terminal differentiation may also contribute to the Schwann cell's misbehavior at laminin β2-deficient synapses.

The synaptic laminins appear to act in concert to organize the behavior of all three synaptic cells. However, their identified roles are largely directed at mid and late stages of synaptic development. What factors regulate the initial transformation of growth cone into nerve terminal?

Candidate factors include agrin, fibroblast growth factors, neurotrophins, and the cell-adhesion molecule NCAM. Most of these have been proposed to regulate aspects of nerve terminal differentiation based on cell culture assays of motor neurons. For example, agrin inhibits motor neurite outgrowth and promotes the clustering of synaptic vesicles, neuronal behaviors that preferentially occur during nerve terminal formation *in vivo* (Campagna *et al.*, 1995, 1997). A role for agrin in presynaptic differentiation has been difficult to discern *in vivo*, however. Loss of agrin in mice grossly perturbs postsynaptic differentiation. Moreover, an absence of postsynaptic differentiation by other means, as by loss of the agrin-transducing receptor MuSK, causes a similar absence of nerve terminal formation and increased axonal growth, despite apparently normal levels of agrin.

Similarly, several neurotrophins increase neurosecretion by motor neurons in culture, but their potential role in nerve terminal differentiation *in vivo* is obscured by their strong roles in modulating postsynaptic differentiation. Growth factors, as well, have been implicated in both motor neuron survival and synaptic development. For example, FGF5 accounts for a major fraction of the muscle-derived survival activity when assayed on cultured motor neurons and also increases the expression of choline acetyltransferase by cholinergic neurons *in vitro* (Hughes *et al.*, 1993; Lindholm *et al.*, 1994). However, mice lacking FGF5 have modest synaptic defects, suggesting that their roles may be limited or may overlap with other unidentified factors (Moscoso *et al.*, 1998). Interestingly, FGF2-coated beads cause axonal swelling and synaptic vesicle accumulation at sites of contact along developing motor neurites in culture (Dai and Peng, 1995). Although roles for FGFs have not been established at synapses *in vivo*, their effects on cultured neurons indicate that sustained increases in intracellular calcium may be a critical intracellular determinant of presynaptic differentiation.

NCAM is concentrated in the neuromuscular synaptic cleft *in vivo* and regulates neurite outgrowth in culture. Functional synapses form in NCAM-deficient mice. Thus, NCAM does not play a dominant role in the establishment of this synapse (Moscoso *et al.*, 1998). However, additional studies revealed that synapses in adult NCAM-deficient mice retain functional and biochemical features of embryonic synapses (Rafuse *et al.*, 2000; Polo-Parada *et al.*, 2001). Compared to normal controls, neurotransmission at synapses in mutant muscles was markedly depressed and prone to fail entirely in response to repetitive stimulation. In pursuit of the underlying molecular events mediating NCAM's effects on synaptic transmission, Landmesser and colleagues confirmed in normal mice that synaptic vesicle cycling is regulated differently in immature and mature nerve terminals. Immature terminals use nifedipine-sensitive L-type Ca^{2+} channels to regulate release of neurotransmitter, and synaptic vesicle reformation is inhibited by brefeldin A. Transmitter release in mature terminals relies instead on P/Q-type Ca^{2+} channels, which are blocked by ω-conotoxin TK. Mature terminals have few L-type channels, and their clathrin-mediated vesicle recycling is relatively insensitive to brefeldin A treatment. The immature release mechanisms are fully replaced by the mature apparatus during postnatal development in normal mice.

However, in NCAM-deficient mice, nerve terminals retained the immature transmission components throughout the presynaptic terminal, as well as in nearby regions of the preterminal axon. Furthermore, although the mature components of vesicle cycle appeared in NCAM-deficient terminals, they were not organized around active zones, as occurs in normal terminals. NCAM therefore appears dispensable for proper initiation of synapse formation, but plays an important role in organizing nerve terminal maturation.

Postsynaptic Differentiation

The primary function of the muscle endplate is to translate neurotransmitter binding into a large postsynaptic depolarization. To this end, postsynaptic differentiation at the NMJ involves the creation of subcellular domains and morphological features that enhance the muscle's response to neurotransmitter.

The cardinal feature of postsynaptic differentiation at the NMJ is the clustering of AChRs into a high-density plaque, located in the sarcolemma opposite the nerve terminal. Ultimately, more than a dozen additional synaptic proteins become co-clustered with AChRs in the postsynaptic membrane, in the overlying extracellular matrix, and in the underlying cytoskeletal matrix. Nevertheless, AChR clustering is the earliest definitive postsynaptic specialization that can be identified, consistent with the central role of AChRs in mediating the postsynaptic response to neurotransmitter. AChR clustering is also one of the easiest synaptic features to detect experimentally, as AChRs are specifically and almost irreversibly labeled by α-bungarotoxin (see Box 2). Studies of postsynaptic differentiation have therefore focused on the mechanisms by which motor nerves and muscles control AChR clustering.

Postsynaptic differentiation is dependent on signals secreted by the nerve. Indeed, it has long been thought that sites of postsynaptic differentiation in muscle are determined by extrinsic signals secreted by axons where they contact developing muscle fibers. While this view continues to have great merit, recent studies indicate that muscle fibers possess an intrinsic program of postsynaptic differentiation, which is capable of forming rudimentary postsynaptic specializations without signals from motor neurons. To complicate matters further, innervation also provides a second, apposing signal that causes the disassembly of postsynaptic specializations. This inhibitory signal likely serves to eliminate secondary sites of postsynaptic differentiation along the muscle surface. We consider the central features of these mechanisms, below. One of the main tasks in the future will be to reconcile these separate programs of postsynaptic organization.

Extrinsic Control of Postsynaptic Differentiation

By monitoring the clustering of AChRs in cocultures of motor neurons and muscle fibers, Anderson and Cohen, and Frank and Fischbach, found that motor neurites promote postsynaptic differentiation at sites where they contact myotubes (Fig. 13) (Anderson and Cohen, 1977; Anderson *et al.*, 1977; Frank and Fischbach, 1977, 1979). In contrast, sites of postsynaptic differentiation preestablished by the myotube are

BOX 2. Taiwan Banded Krait—*Bungarus multicinctus*

Taiwan Banded Krait – *Bungarus multicinctus*

C.C. Chang, C.Y. Lee, and their colleagues identified and characterized α-bungarotoxin as a major bioactive component in the venom of the many-banded krait (*Bungarus multicinctus*; above) indigenous to Taiwan (Chang, 1963, Lee, 1972). In retrospect, this discovery profoundly influenced progress in molecular, cellular, and developmental neurobiology. Due to the toxin's specificity and extremely high affinity for the nicotinic AChR expressed in vertebrate skeletal muscle, α-bungarotoxin became a key tool in its isolation and characterization. The nicotinic AChR was, therefore, the first neurotransmitter receptor and ion channel to be molecularly dissected. Fluorochrome-conjugated α-bungarotoxin has remained the primary means of identifying and monitoring the molecular and cel-lular differentiation of the neuromuscular synapse since its introduction in the mid-1970s (Anderson, 1974).

disassembled (Kuromi and Kidokoro, 1984). These observations provided strong evidence that neurons present signals that cue postsynaptic differentiation in the muscle, but did not provide an easy means of identifying them. In search of the source of the signals, Burden, Sargent, and McMahan studied the clustering of AChRs in regenerating muscle fibers, *in vivo* (Burden *et al.*, 1979). They found that the synaptic basal lamina was capable of directing postsynaptic differentiation, even in the absence of a nerve terminal (Fig. 15). These experiments were similar to the empty basal lamina experiments used earlier to investigate presynaptic differentiation, but in this case, the muscle was allowed to regenerate within the original sheath of myofiber basal lamina, while the nerve was ligated to prevent reinnervation. Importantly, focus was newly directed at the composition of the synaptic basal lamina as a potential source of synaptogenic cues. McMahan and his colleagues identified a component that organized AChR clustering, which they called agrin. In a parallel series of studies, Fischbach and his colleagues identified a distinct basal lamina component that increased the levels of AChRs synthesized by the muscle fiber, which they named ARIA (for *a*cetylcholine *r*eceptor-*i*nducing *a*ctivity), and which was later identified as an isoform of neuregulin. Together, these studies demonstrated that nerve-derived signals can play dominant roles in the control of postsynaptic differentiation.

Lines of investigation leading from these discoveries ultimately identified three nerve-to-muscle signaling axes that control post-synaptic differentiation: Agrin-dependent activation of the muscle specific receptor tyrosine kinase, MuSK; neuregulin activation of ErbB receptors; and acetylcholine-mediated neurotransmission.

Agrin

McMahan and his colleagues hypothesized that axons induced local postsynaptic differentiation in the muscle by means of a secreted molecule, which was stably incorporated into the synaptic basal lamina at the mature NMJ. They identified this active component through biochemical purification, using Torpedo electric organ as a starting material rich in synaptic basal lamina, and the clustering of AChRs on cultured myotubes as a bioassay of postsynaptic differentiation. At each step of the purification, a sample of each fraction was added to the myotube culture medium; after a few hours, the cultures were stained with rhodamine-conjugated α-bungarotoxin. Active fractions caused a

significant increase in the number of "hot spots" on the myotube surface, clusters of stained AChRs that appeared similar to the plaques of AChRs present at nascent postsynaptic sites *in vivo*. The name agrin given to the purified factor recalls the aggregation of surface AChRs that occurs with its addition to myotube culture medium.

Agrin is produced from a single gene as a large secreted heparan sulfate proteoglycan molecule (Rupp *et al.*, 1991; Tsim *et al.*, 1992). The polypeptide core contains nearly 2,000 amino acids arranged into distinct domains (Fig. 18). The N-terminal portion of the sequence contains conserved sites for attachment of the long glycosaminoglycan chains that make agrin a heparan sulfate proteoglycan, as well as several sites for asparagine-linked glycosylation. The N-terminus of agrin also contains follistatin repeats that are homologous to Kazal protease inhibitor domains. The C-terminal half of agrin contains three LG-domains, which are homologous to globular domains first recognized in the laminin α-subunit. Agrin's interaction with cells is likely dominated by its G-domains, which contain the AChR-inducing

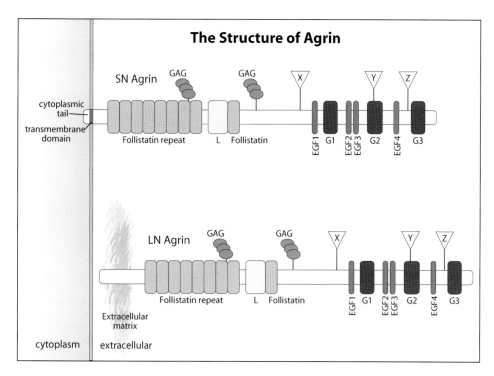

FIGURE 18. Agrin structure/function. Agrin is a large (*ca.* 400 kDal) heparan sulfate proteoglycan. The core polypeptide contains nearly 2,000 amino acids, most of which are present in recognizable domains. There are nine follistatin-like repeats in the N-terminal half of the protein; these bind growth factors and have protease inhibitor activity, but have uncertain biological roles at synapses. Heparan sulfate glycosaminoglycan chains are attached to serine or threonine residues near the middle of the protein. The C-terminal portion of agrin contains four epidermal growth factor (EGF)-like repeats, and three laminin-like G-domains that serve as ligands for the multifunctional matrix receptor dystroglycan. Among three conserved sites of alternative splicing (X, Y, Z) in the C-terminus, one dominates agrin's biological activity in postsynaptic differentiation. Variants with exon inserts at the Z site are up to 100-times more potent at causing ACh receptor aggregation on cultured muscle cells than variants lacking Z exons, and mouse embryos engineered to lack only the agrin Z-site exons fail to maintain postsynaptic sites during innervation. Agrin transcripts incorporating Z exons are found only in the nervous system; their absence in muscle cells explains why muscle-derived agrin is impotent at clustering ACh receptors. An additional transcriptional mechanism that may regulate agrin signaling is variation at the N-terminus. The SN form of the protein contains an N-terminal transmembrane domain, which is restricted to the nervous system and presumably tethers agrin to the neuronal surface (as a type II membrane protein). The LN form of agrin, which is produced from an alternative start site, contains an N-terminal laminin-binding domain in place of the transmembrane domain. The LN isoform of agrin is secreted by motor neurons and is concentrated in the synaptic basal lamina at the neuromuscular junction (NMJ), presumably through interactions with synaptic laminins.

activity, and which are ligands for the two major classes of matrix receptors in muscle cells, α-dystroglycan and β1-integrins. The incorporation of agrin into the synaptic basal lamina may rely on an identified interaction between N-terminal agrin domains and the laminin γ1-chain (Denzer *et al.*, 1997, 1998).

McMahan's "agrin hypothesis" (McMahan, 1990) led to several predictions, including that motor nerves selectively synthesize agrin and secrete it during synapse formation, that agrin-induced signaling in the muscle fiber is sufficient to direct postsynaptic differentiation, that agrin and its intramuscular

effectors are required for postsynaptic differentiation *in vivo*, and that agrin is concentrated in the synaptic basal lamina. These predictions have been tested and found accurate, although not without some interesting surprises.

Immunoreactivity for agrin is indeed concentrated in the synaptic cleft at the mature NMJ. However, an early quandary for the agrin hypothesis was that both neurons and muscle fibers synthesize and secrete agrin. In fact, agrin is abundant throughout the extrasynaptic basal lamina in developing muscle, a region that is not normally explored by the nerve. If muscles supply

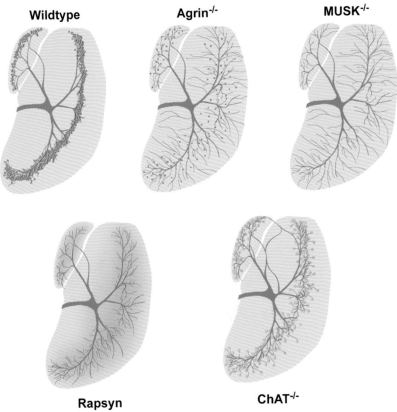

FIGURE 19. Comparision of synaptic defects in mutant mice lacking agrin, MuSK, rapsyn, and choline acetyltransferase (ChAT). Schematic diagrams show innervation of the left hemi-diaphragm, dorsal end up, in mutant mouse embryos genetically engineered to test the agrin pathway and the role of synaptic transmission in postsynaptic differentiation at the neuromuscular junction. In wild-type mice, the phrenic nerve contacts the developing diaphragm near its midpoint, and branches ventrally and dorsally; a medial branch extends to the crus, which attaches to the spine. Motor nerve terminals are located over AChR-rich postsynaptic sites, midway along each fiber, forming a so-called "endplate band." In agrin-deficient mice, the nerve initially enters and branches properly within the diaphragm, but ultimately grows far beyond the normal endplate. Few clusters of AChRs are present at perinatal ages, and these are broadly distributed and mostly unapposed by nerve terminals. The result is consistent with the hypothesis that agrin is required to establish stable postsynaptic sites. According to this view, axon overgrowth is secondary to the inability to establish synaptic contacts. No postsynaptic differentiation occurs in mice lacking the MuSK receptor tyrosine kinase. As in the absence of agrin, initial growth of the nerve is correct, but failure to terminate in the endplate region leads axons far into extrasynaptic muscle. The similarity of the MuSK and agrin phenotypes provides the strongest evidence that MuSK transduces the activity of nerve-derived agrin. Similarly in rapsyn-deficient mice, no clusters of AChRs form, and the nerve grows beyond the normal endplate region. However, AChRs are relatively enriched along the central portion of each fiber. The results support the idea that rapsyn mediates agrin- and MuSK-activated clustering of AChRs, but indicates that agrin/MuSK signaling (intact in these mice) promotes additional aspects of postsynaptic specialization that are rapsyn-independent, including elevated expression of AChRs. Additional studies support this notion, as there is a greater degree of postsynaptic transcriptional specialization in rapsyn mutants than in agrin- and MuSK-mutant mice. Correlation between the severity of nerve overgrowth and defects in postsynaptic differentiation supports the conclusion that axon outgrowth is normally inhibited by a retrograde signal associated with postsynaptic differentiation. Motor neurons in ChAT-deficient mice cannot synthesize ACh, preventing neurotransmission. AChR clusters on the muscle are individually larger, and collectively more broadly distributed. In part, this likely reflects an overabundance of motor axons, as there is less developmental apoptosis in the motor pools of ChAT mutant mice than in normal controls. The result suggests transmission, per se, is not required to establish neuromuscular junctions, but that nerve-evoked activity regulates the early pattern of synaptic connections in muscle.

agrin along their entire length, then how could axons employ agrin to specify the location of AChR clustering and postsynaptic differentiation? Would not muscle-derived agrin cause AChR clustering without contribution by the nerve?

Resolution came with the discovery that motor neurons produce an especially active isoform of agrin (Ruegg *et al.*, 1992; Ferns *et al.*, 1993; Hoch *et al.*, 1993). The active isoform is encoded by specific mRNA splice variants, which are made by motor neurons and not muscle fibers. The locations of the alternative splice site sequences in the structure of the agrin polypeptide are shown in Fig. 18. The most important splice site has been named Z (or B, in avian transcripts). Splicing at the Z site involves two exons, encoding 0, 8, 11, or (the 8 + 11 combination) 19 amino acid residues in one of the G-domains. The Z+ isoforms (containing the 8, 11, or 19 residue inserts) are at least 100-fold more potent at clustering AChRs than the Z0 isoforms. Importantly, while neurons produce transcripts for active Z+ agrin, muscles produce only the inactive Z0 isoform of agrin. Thus, the paradox of agrin's distribution in developing muscle is resolved by tissue-specific expression and differential activity of alternative splice variants.

Targeted mutagenesis of the agrin gene in mice has established that agrin is essential for postsynaptic development *in vivo* (Gautam *et al.*, 1996). Mutant mouse pups born without agrin are unable to move voluntarily, or even breath. Embryonic muscles and nerves develop normally without agrin, but fail to establish normal synaptic connections. At birth, muscle fibers in agrin-deficient mice have few AChR clusters, and little other molecular evidence of postsynaptic differentiation (Fig. 19). The motor axons also fail to form stable nerve terminals, and instead grow along the lengths of the muscle fibers, far beyond the normal region of axonal growth through the center of the muscle. Two observations suggest that defects in presynaptic differentiation in agrin-deficient mice are secondary to abnormal postsynaptic differentiation. First, nearly identical abnormalities are seen in mice lacking only the agrin gene's Z+ exons (Burgess *et al.*, 1999). That is, despite an abundance of Z0 agrin produced by developing muscle fibers, muscle fibers fail to cluster AChRs, and motor axons fail to form stable nerve terminals, when Z+ agrin is specifically absent. Second, motor axons fail to form stable nerve terminals in mice lacking postsynaptic sites by other genetic mutations, despite completely normal levels of agrin. Thus, defects in presynaptic differentiation correlate strongly with absence of postsynaptic differentiation, but weakly or not at all with overall levels of agrin. While establishing a primary role for agrin in postsynaptic differentiation, these studies support the idea that postsynaptic sites are associated with a retrograde signal (not yet identified) that promotes presynaptic differentiation in the motor axon.

Agrin's activity in clustering AChRs is transduced by MuSK (Glass *et al.*, 1996). MuSK is a prototypic member of the tyrosine kinase receptor family of transmembrane proteins. The single polypeptide contains an extracellular N-terminal domain with homology to the immunoglobulin superfamily, a single transmembrane segment, an intracellular protein tyrosine kinase domain, and several protein/protein interaction domains near the C-terminus. MuSK is concentrated at NMJs *in vivo* and co-clustered with AChRs on cultured myotubes. The application of agrin to myotubes in culture induces MuSK autophosphorylation. In other well-studied receptor tyrosine kinases, such as the EGF and PDGF receptors, autophosphorylation follows from ligand-induced dimerization and precedes binding and activation of downstream scaffolding and signaling components. These steps have not been well characterized in the case of MuSK. However, a number of experiments performed *in vivo* and *in vitro* have firmly established that MuSK is required for agrin-induced postsynaptic differentiation at the NMJ. First, MuSK and agrin are similarly required for postsynaptic development in embryonic mice (Fig. 19) (Dechiara *et al.*, 1996). Second, myotubes cultured from embryonic MuSK-deficient mice are unable to cluster AChRs or other postsynaptic proteins in response to purified agrin. Third, activation of MuSK kinase activity by independent methods causes AChR clustering, and interference of MuSK catalytic activity blocks agrin-induced AChR clustering. These several criteria firmly establish that MuSK is an essential component of the agrin signal transduction pathway (Fig. 20).

One uncertainty in this model is how directly agrin interacts with MuSK. While full-length agrin binds and activates

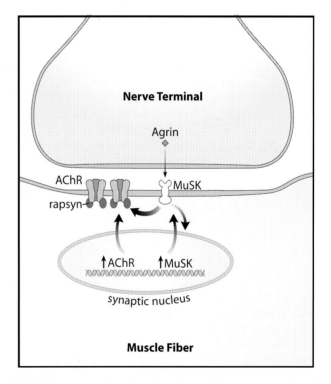

FIGURE 20. Agrin-induced signaling pathways. Motor terminals release Z+ splice variants of agrin (see Fig. 18). Z+ agrin activates MuSK concentrated in postsynaptic membranes. Autophosphorylation of MuSK is coincident with agrin-induced activation and may be prerequisite to the downstream activation of additional kinases leading to the clustering of acetylcholine receptors (AChRs). MuSK activation also increases expression of synapse-specific genes by subsynaptic myonuclei. Aggregation of AChRs on the cell surface occurs via the intracellular scaffolding protein rapsyn, likely through direct rapsyn–AChR binding.

MuSK (Parkhomovskiy *et al.*, 2000), truncated versions of agrin activate MuSK and cause AChR clustering without detectably binding to MuSK. Thus, the binding and activation of MuSK by agrin have not been tightly linked, leading to speculation that an accessory factor may promote productive interactions between agrin and MuSK.

Signal transduction downstream of MuSK also remains unresolved. MuSK activation stimulates several intracellular signaling pathways, and a large number of known intracellular signal transduction kinases have been implicated, including Abl/Arg, Cdk5, FYN, GSK3, Src, and YES (Burden *et al.*, 2002, 2003). Some are likely to activate small GTP-binding effector proteins, which promote actin remodeling and facilitate the structural changes that accompany postsynaptic differentiation. Others may mediate activation of the ras-mediated signaling pathway by which MuSK is known to regulate nuclear transcription. Several synapse-specific genes, including subunits of the acetylcholine receptor and MuSK itself, contain specific promoter sequences that mediate MuSK-induced upregulation. Auto-activation of MuSK expression provides a potential mechanism for positive feedback to concentrate MuSK signaling at developing synaptic sites. Finally, agrin/MuSK signaling also promotes the aggregation of ErbB receptors at the synapse, further influencing postsynaptic gene expression by mechanisms discussed below.

The best documented effector molecule downstream of MuSK activation is the 43 kDal AChR-associated protein known as rapsyn (Frail *et al.*, 1988). Rapsyn is closely associated with the intracellular portion of AchR subunits and serves as a scaffold for receptor aggregation. Rapsyn itself has a complicated structure with several protein–protein interaction domains. Evidence in support of rapsyn's role include the observation that rapsyn and AChRs will spontaneously co-cluster when expressed together in fibroblasts (Phillips *et al.*, 1991), and also the absence of AChR plaques or clusters in the muscles of rapsyn-deficient mutant mice (Gautam *et al.*, 1995). Muscles in rapsyn-deficient mice also lack most other postsynaptic features, including synapse-specific gene expression.

The agrin/MuSK signaling axis, therefore, acts through rapsyn-mediated AChR clustering to scaffold the development of a postsynaptic apparatus (Fig. 20). Additional studies suggest that MuSK itself forms a primary scaffold on which AChRs and other postsynaptic components co-aggregate. While the discrete steps between the interactions of agrin, MuSK, and rapsyn remain unresolved, their definitive involvement in postsynaptic differentiation establishes a pathway by which motor axons play an essential role in regulating the time and place of synapse formation in the muscle.

Neuregulin

Postsynaptic differentiation includes transcriptional changes in gene expression. In muscle fibers, several nuclei are clustered immediately adjacent to the AChR-rich postsynaptic membrane. Genes for AChR subunits are transcribed at much higher rates by subsynaptic nuclei than by the nuclei that populate the rest of the muscle fiber. Fischbach and colleagues found that cultured muscle cells synthesized more AChRs when the culture media was supplemented with extracts of brain and spinal cord. They named the activity of the extracts ARIA, for *a*cetylcholine *r*eceptor *i*nducing *a*ctivity. In contrast to agrin, which aggregates AChRs already synthesized and present in the membrane, ARIA increased the levels of AChRs in the membrane but did not affect their distribution. The protein responsible for the ARIA activity was later purified and cloned, revealing an isoform of the intercellular signaling protein neuregulin. Purified and recombinant neuregulin has the same activity as ARIA, and antibodies to neuregulin and their erbB receptors selectively stain NMJs in skeletal muscles. Neuregulin upregulates the expression of AChR genes, one of the transcriptional hallmarks of postsynaptic differentiation. These results support the idea that neuregulin is a second postsynaptic differentiation signal provided by motor nerve terminals. Neuregulin's role in increasing AChR expression levels is complimentary to that established for agrin, which clusters AChRs already present in the membrane (Fig. 21).

ARIA activity is encoded by neuregulin-1 (NRG-1), one of four related neuregulin genes in mammals. NRG-1 is alternatively spliced to produce a set of related growth and differentiation factors, including glial growth factor, heregulin, and the neu differentiation factor, as well as ARIA (Lemke and Brockes, 1984; Holmes *et al.*, 1992; Wen *et al.*, 1992; Falls *et al.*, 1993; Marchionni *et al.*, 1993). NRG-1 signaling activity is associated with the epidermal growth factor (EGF) domain, which is present once in each isoform. This domain activates three members of the EGF family of membrane receptors, ErbB2, ErbB3, and ErbB4, which regulate cellular activities through intracellular protein tyrosine kinase domains. Receptor activation likely occurs through ligand-induced dimerization and autophosphorylation, as occurs with the closely related EGF receptor.

Motor neurons express multiple isoforms of neuregulin, which arise through transcriptional mechanisms. Muscle fibers express multiple erbB receptor subtypes, which are encoded by separate genes, but which form functional receptors as heterodimers. Confusingly, muscle fibers also express neuregulins, and Schwann cells express both neuregulins and ErbB receptors, offering a bewildering array of potential signaling interactions at the NMJ. The benefit underlying this high level of transcriptional complexity in NRG/ErbB signaling is not well understood.

One likely possibility is that the non-EGF polypeptide domains in neuregulin limit the range and specificity of neuregulin signaling. These domains vary considerably among NRG-1 isoforms. Some isoforms (types I and III) contain a transmembrane domain, which tethers the EGF-signaling domain to the cell surface. Neuregulin signaling in these cases may require direct cell–cell membrane interactions, as between the axon and Schwann cell. Alternatively, proteolytic cleavage at conserved sites proximal to the membrane can release the active EGF-domain into the extracellular space and across the synaptic cleft. Proteolytic release would provide a means of coupling neuregulin secretion to remodeling of the extracellular space. An additional complication is that alternative transcription start sites generate two N-terminal variants of NRG-1. One variant contains an extracellular Ig-superfamily homology domain; ARIA was

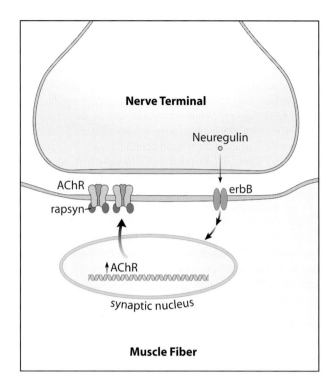

for NRG at synaptic sites. Nevertheless, four lines of evidence implicate NRG-1 in promoting the concentration of AChRs opposite the nerve terminal at the NMJ. First, treatment of cultured myotubes with purified and recombinant NRG selectively increases the levels of AChR subunit mRNAs, similar to the increased levels of subsynaptic receptor mRNAs observed *in vivo* (Harris *et al.*, 1988; Martinou *et al.*, 1991; Chu *et al.*, 1995). Neuregulin also upregulates the expression of utrophin and voltage-gated sodium channels, which are concentrated at mature postsynaptic sites. Second, motor neurons synthesize, axonally transport, and secrete active NRG (Corfas *et al.*, 1993). Third, NRG is present at developing NMJs, *in vivo*, and is concentrated in the synaptic cleft at mature NMJs (Goodearl *et al.*, 1995). Neuregulin is stably associated with the synaptic basal lamina, possibly through interactions with heparan sulfate proteoglycans such as perlecan and agrin (Holmes *et al.*, 1992; Loeb and Fischbach, 1995; Meier *et al.*, 1998). This likely explains the observation that the synaptic basal lamina is able to direct synapse specific transcription in denervated muscles (Jo and Burden, 1992). Fourth, and most importantly, mice with reduced levels of one subset of NRG-1 isoform (those containing the Ig-domain) have strongly reduced levels of AChRs at the NMJ and are myasthenic (Sandrock *et al.*, 1997).

Together, these studies suggest that neuregulin/ErbB signaling sustains the maturation of the postsynaptic apparatus by increasing the levels of synaptically abundant proteins as the size of the developing muscle fiber and the strength of the synaptic connection increase. An additional possibility is that neuregulin counteracts the effect of electrical activity in the muscle fiber; as we shall see in the next section, electrical activity reduces the synthesis of AChRs in extrasynaptic regions of the muscle.

Neuregulin signaling and agrin signaling neatly complement each other. Agrin appears to play a primary role in establishing the initial location and organization of postsynaptic differentiation. In contrast, neuregulin does not appear to be essential for establishing the site or pattern of synaptic gene expression, but rather supports the growth of the postsynaptic apparatus by amplifying the synthesis of postsynaptic components.

FIGURE 21. Neuregulin-induced postsynaptic signaling. Like agrin, neuregulin is concentrated in the synaptic basal lamina and is produced from alternatively spliced transcripts in neurons and muscles. In muscle fibers, neuregulin signaling activates ErbB receptors concentrated in the postsynaptic membrane. ErbB activation leads to an upregulation of AChR gene expression in subsynaptic myonuclei. Unlike agrin, neuregulin does not directly cause aggregation of AChRs.

originally identified in this form. A second N-terminal variant lacks the Ig-domain and instead encodes a cysteine-rich domain (CRD) coupled to a second likely transmembrane domain. Although the EGF domain remains extracellular in both N-terminal variants, two cleavage events are presumably required for the signaling domain to be released from the CRD-isoforms. The CRD-isoforms may therefore act primarily in cell-attached fashion. Consistent with this idea, studies in mutant mice indicate that the CRD-isoforms made by motor neurons act through ErbB2 receptors on Schwann cells to regulate Schwann cell survival (Wolpowitz *et al.*, 2000). The requirement for neuron : Schwann cell contact in this signaling interaction may ensure that the number of Schwann cells required to myelinate the nerve matches the number and length of the developing axons; in this model, supernumerary Schwann cells lacking axonal contact fail to receive a neuregulin signal and die. Thus, through its varying domain structure, neuregulin may sometimes act as a paracrine factor signaling over short extracellular distances, or as a juxtacrine factor requiring cell–cell contact to transmit its signal.

Tests of neuregulin's role in synaptic development *in vivo* have produced mixed results. As mentioned, neuregulin produced by neurons is essential for the survival and proliferation of Schwann cells. In turn, developing motor neurons die in the absence of Schwann cells, hindering attempts to establish roles

Acetylcholine

The third well-established signal at the NMJ is the neurotransmitter ACh. While the primary role of ACh is to open the cation-selective ion channel in the AChR and depolarize the muscle fiber membrane, ACh carries a second, longer lasting signal into the cytoplasm of the muscle fiber, in the form of calcium. The concentration of calcium inside the muscle fiber increases markedly during periods of electrical activity, as it does in most electrically excitable cells. In immature muscle fibers, this calcium comes principally from the extracellular medium through voltage-gated calcium channels in the sarcolemma. At synaptic sites, some calcium enters through AChR channels. In mature fibers, cytosolic calcium is rapidly infused through channels in the sarcoplasmic reticulum—an elaborate intracellular calcium storage/release system specific to muscle.

Calcium is well-established as a multifunctional second messenger. In muscle, of course, calcium triggers muscle fiber contraction by activating the myofibrillar actin–myosin complex. As in most cells, calcium regulates signal transduction pathways in muscle fibers, through the activation of protein kinases and phosphatases such as mitogen-activated protein kinase (MAPK), protein kinase C, calmodulin-dependent (CaM) kinases II and IV, and protein phosphatase 2B (calcineurin). These pathways, in turn, modulate the expression of specific genes, by regulating the function of transcriptional regulatory proteins, such as CREB and the basic helix–loop–helix (bHLH) factors. Synaptic transmission and muscle activity thereby have short-term and long-term effects on the development of the muscle fiber. Some of these effects contribute to the differentiation of postsynaptic and extrasynaptic regions of the muscle.

Muscle activity suppresses the expression of AChR subunits and other synaptic components in myonuclei throughout the extrasynaptic regions of the muscle (reviewed by Fromm and Burden, 1998). For example, cholinergic activity inactivates the gene for the AChR delta subunit through an E-box (CAnnTG) in the 5′-regulatory sequence. The E-box is a binding site for the muscle bHLH proteins MyoD, myogenin, MRF4, and myf5. This regulation is readily reversible, explaining why paralysis is accompanied by an upregulation of AChR levels in the extrasynaptic regions of muscle.

Since much of the electrical activity in muscles is driven by synaptic transmission, defects in synaptic function likely have secondary consequences for synaptic differentiation. One example where this seems very likely is in mutant mice lacking the gene for choline acetyltransferase (ChAT) (Misgeld *et al.*, 2002; Brandon *et al.*, 2003). Because ChAT is the sole enzyme responsible for the biosynthesis of acetylcholine from choline and acetyl-CoA, motor terminals in ChAT-deficient mice are unable to release ACh. Loss of synaptic transmission in ChAT-deficient mice does not prevent AChRs from clustering in the muscle membrane opposite the nerve terminal, which is consistent with the notion that agrin and neuregulin are secreted independent of ACh. However, loss of ChAT profoundly affects the distribution of synaptic sites in the muscle (Fig. 19). At birth, each muscle fiber normally contains one or a few centrally located synaptic sites. In ChAT-deficient mice, fibers appear to maintain as many as five or more AChR-rich postsynaptic sites, and these are spread along a much wider span of the muscle's length. The broader distribution of postsynaptic specializations is matched by a broader expression of synaptic transcripts in the muscle, and by increased formation of nerve terminals by the innervating axons. Qualitatively similar defects in the normal distribution of synapses occur in the muscles of chick embryos paralyzed by curare or bungarotoxin (Loeb *et al.*, 2002). These observations appear to confirm the notion formulated during studies of nerve and muscle paralysis in polyinnervated adult muscle, that effective synaptic transmission suppresses the formation of secondary synaptic specializations and is required during development to restrict synapse formation to a single site in the muscle.

The primary conclusion to be drawn from these studies is that the muscle fiber responds to successful innervation in a way that makes it refractory to additional innervation. One attractive possibility is that secreted factors produced by uninnervated fibers to attract innervation by the nerve are downregulated upon initiation of successful cholinergic synaptic transmission. However, an equally plausible idea is that synaptic transmission increases the expression of retrograde inhibitory factors in extrasynaptic regions. While retrograde factors that promote or impede additional innervation have not been identified, it seems likely that their production is tightly coupled to cholinergic transmission. An additional surprise in ChAT-deficient mice is that muscles have significantly fewer muscle fibers, suggesting that ACh and/or muscle activity has an important role in myogenesis.

In summary, motor nerves secrete three factors that together serve to organize postsynaptic differentiation in skeletal muscle (Fig. 22). Agrin plays a primary role in establishing postsynaptic differentiation opposite the nerve terminal. Neuregulin sustains postsynaptic specializations as they mature. ACh, likely acting through changes in intracellular calcium in myofibers, suppresses ectopic innervation and thereby fosters the high degree of synaptic specificity present in mature muscle.

Intrinsic Control of Postsynaptic Differentiation

One prediction of the preceding model of neurally controlled postsynaptic differentiation is that uninnervated muscles should bear little evidence of postsynaptic organization. It was therefore surprising to find that robust postsynaptic differentiation occurs in muscles lacking nerve contact. Historically, this question has been addressed in mutant animals lacking proper innervation, such as the peroneal muscular atrophy mutation in mice (Ashby *et al.*, 1993), in which a branch of the sciatic nerve fails to form and a muscle group develops without innervation. However, the patterns of innervation and synapse formation at embryonic ages are more readily addressed in the diaphragm, which develops earlier than many other muscles, and which is thinner and more readily stained as a wholemount preparation. Fortuitously, mutations in topoisomerase IIβ and the motor neuron transcription factor HB9 have been found to prevent the phrenic nerve from forming. The diaphragm in such mutants is never contacted by motor neurons (Yang *et al.*, 2000, 2001; Lin *et al.*, 2001).

The most intriguing observation in aneural embryonic muscles is that the AChRs aggregate in plaques, and the plaques are located in the central portion of the muscle (Fig. 23). Moreover, the uninnervated AChR-rich plaques contain many of the molecular specializations of normally innervated postsynaptic sites, including an AChE-rich synaptic basal lamina, and increased expression of AChR subunit genes by the underlying myonuclei. Clearly, muscle possesses an *intrinsic* program of postsynaptic differentiation.

At first glance, properly organized postsynaptic specialization in the absence of a nerve is contrary to the notion that nerve-derived agrin organizes postsynaptic differentiation. Indeed, the findings in aneural muscles are in striking contrast to the postsynaptic defects initially reported in mice missing either agrin or MuSK, where there is little or no postsynaptic

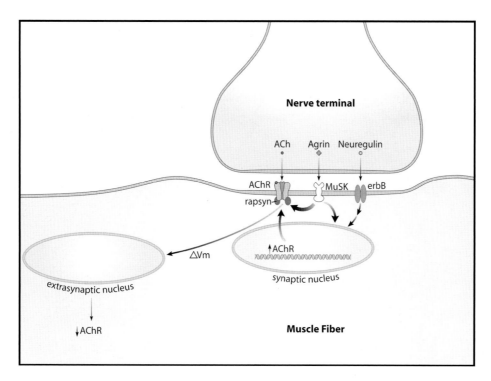

FIGURE 22. Agrin, neuregulin, and cholinergic transmission combine to control postsynaptic differentiation in muscle. Agrin promotes the concentration of AChRs and scaffolding proteins in the postsynaptic membrane. Neuregulin and agrin both amplify gene expression in postsynaptic nuclei, to enhance postsynaptic responses to ACh during growth and reorganization of the synapse. Nerve-evoked activity, mediated by the neurotransmitter ACh, suppresses the expression of synaptic genes by myonuclei in extrasynaptic regions of the muscle. Effects of muscle activity occur largely through increases in intracellular Ca^{2+}, which regulate a myriad of signaling pathways regulating gene expression and cellular activity.

differentiation. How are these observations to be reconciled? Close comparison of postsynaptic differentiation in normal, aneural, agrin-deficient, and MuSK-deficient animals yields several clear suggestions.

First, MuSK is required for all postsynaptic differentiation. No AChR plaques form in MuSK-deficient mice, with or without innervation. In fact, intrinsic postsynaptic differentiation is very sensitive to the levels of MuSK expression and is acutely affected by gene dosage. In HB9 mutant mice also heterozygous for a defective MuSK gene, the number of AChR clusters was reduced by 95% compared to HB9 mutants that carried two normal alleles of MuSK. It is not known whether this sensitivity represents cooperativity in MuSK activation, or a threshold in forming a MuSK-dependent scaffold for postsynaptic assembly. However, as developing muscle fibers grow preferentially at their ends, the central portion of the muscle is the oldest and has had the longest time to accumulate MuSK receptors. It seems likely that MuSK auto-activation, a phenomenon common among tyrosine kinase receptors, would be concentrated where its levels are highest, in the center of the muscle. Thus, MuSK autoactivation might initiate cell-autonomous postsynaptic differentiation in this part of the muscle. This explanation would account for the location of the endplate band in uninnervated muscles, as well as its dependence on and dosage sensitivity to MuSK.

Second, nerves provide two competing signals for postsynaptic differentiation: one (agrin) promotes postsynaptic specialization immediately adjacent to the nerve terminal; a second signal (most likely ACh) promotes muscle activity and causes degeneration of postsynaptic specializations located away from the nerve terminal. Reexamination of agrin-deficient mice at very early stages of innervation (at 13 days of embryogenesis, or about 6 days before birth) found that muscles initially contain a near-normal number and density of AChR-rich postsynaptic sites. These are presumably the muscle's intrinsic sites, now observed for the first time in an innervated (but agrin-less) muscle. The key observation is that these AChR-rich sites persist in muscles lacking a nerve (as e.g., in HB9-deficient mice) but disappear within a couple of days when the host muscle is innervated by agrin-deficient nerves. To be clear: In the absence of agrin, the nerve promotes *disassembly* of intrinsic postsynaptic sites.

Therefore, a more accurate assessment of neural-agrin's role in normal development may be that it acts to stabilize the intrinsic postsynaptic sites from activity-induced disassembly. Alternatively, neural-agrin could organize entirely new sites where it is secreted by the nerve, while the muscle's intrinsic sites are simultaneously disassembled. In fact, this replacement model is most consistent with original observations of nerve-induced postsynaptic differentiation, in nerve–muscle cocultures (Anderson and Cohen, 1977; Frank and Fischbach, 1979). As described earlier, cultured myotubes form spontaneous clusters of AChRs, which are eschewed by the neurites of cocultured motor neurons. Rather, motor axons organize new AChR clusters

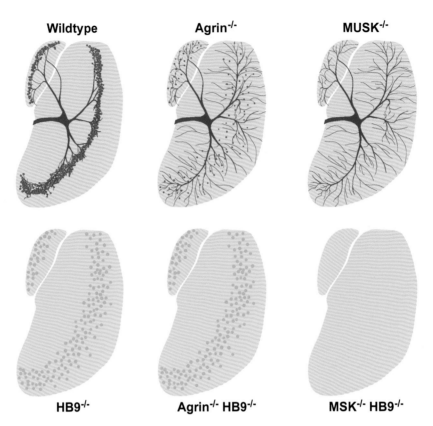

FIGURE 23. Antagonistic effects of agrin and innervation on postsynaptic differentiation in muscle. Schematic diagrams of innervation of the left hemidiaphragm in normal and mutant mice, as described in Fig. 19. Top panels reiterate dependence of postsynaptic specializations on agrin/MuSK signaling. Bottom panels illustrate postsynaptic differentiation occurring in the absence of innervation. In mice lacking the transcription factor HB9, cervical motor neurons that should innervate the diaphragm are absent. Interestingly, diaphragm myofibers in HB9$^{-/-}$ mice form plaques of AChRs near their midpoints, creating a rudimentary endplate band of postsynaptic sites in the absence of innervation. A similar distribution of postsynaptic sites is present in agrin-deficient mice, at very early stages of synaptic development. At late embryonic ages, postsynaptic differentiation in HB9$^{-/-}$ mice actually exceeds that in agrin$^{-/-}$ mice. The comparisons reveal that (1) muscle fibers have an agrin-independent, autonomous program of postsynaptic differentiation; (2) innervation provides a dispersal factor, which eliminates autonomous AChR clusters when agrin is absent; and (3) agrin/MuSK signaling is required to stabilize postsynaptic sites against nerve-induced dispersal. Muscle activity stimulated by release of ACh from motor axons may promote dispersal of any AChR clusters that are not stabilized by agrin. As autonomous AChR clustering is absent in the MuSK$^{-/-}$ background (i.e., in HB9$^{-/-}$; MuSK$^{-/-}$ double-mutant mice), it seems likely that autonomous clustering involves the activation of the MuSK signaling pathway at some point downstream of agrin. The results raise some uncertainty about the precise role of agrin, and whether motor axons or muscle fibers determine synaptic sites. If nerves innervate autonomous sites of AChR clusters provided by the muscle fibers, then agrin's role may be more of a maintenance factor. Alternatively, if nerves ignore autonomous clusters, then agrin both initiates and maintains permanent synaptic sites. This view is most consistent with observations in nerve–muscle cocultures, described in Fig. 13. Regardless, the complementary effects of agrin-induced clustering and activity-dependent dispersal of ACh receptors ensure that mature nerve terminals and endplates are aligned.

at sites of contact with the myotube, while the spontaneous AChR clusters are disassembled.

There remains, then, some uncertainty about the precise role played by the muscle's intrinsic program of postsynaptic differentiation. In principle, intrinsic postsynaptic sites may be selectively innervated *in vivo*, or they may be ignored and fully replaced. An intermediate possibility is that intrinsic sites are stabilized if the axon happens to arrive directly at their location, but that they are otherwise disassembled by activity-dependent mechanisms as new agrin-induced specializations are established at adjacent sites on the myofiber. A fourth possibility is that the initial set of postsynaptic specializations are not innervated immediately, but are associated with the production of retrograde signals that attract and induce the growth and presynaptic

differentiation of the motor axon. Attempts to distinguish these possibilities by close observation in wild-type animals, during the very earliest stages of neuromuscular innervation (at about E13 in a mouse, or six days before birth), reveal that a large percentage of the initial complement of AChR clusters have in fact no direct nerve contact. Perhaps intrinsic postsynaptic sites are not directly targeted by motor axons. Nevertheless, because many nerve endings do co-localize with early AChR plaques in the very same muscles, these possibilities are not currently resolvable. In the absence of a dynamic view of initial innervation *in vivo*, the replacement model initially identified *in vitro* remains the most likely mechanism.

The vertebrate NMJ is the best studied synapse. We have focused here on a few well-established molecular events that direct and define its formation, but myriad other signals undoubtedly contribute. Examples include growth factors and matrix proteases. Glial-derived neurotrophic factor (GDNF) perturbs NMJ formation when overexpressed in transgenic mice (Nguyen *et al.*, 1998). Matrix metalloprotease 3 (MMP3) is concentrated at synaptic sites, where it is capable of releasing agrin from the synaptic basal lamina, and could play a role in synaptic remodeling or the dispersal of uninnervated postsynaptic sites (Vansaun and Werle, 2000). Despite years of study and real progress at the NMJ, a great deal remains to be learned about the complex interactions of axon, target, and glial cell at this best understood synapse.

CNS SYNAPSES

Compared to the NMJ, synapse formation in the CNS is poorly understood, for several reasons. CNS synapses vary considerably in function and specificity, but relatively little in size and structure. In addition, the complex anatomical architecture of the brain has hindered the ability to identify either a single axon's presynaptic terminals, or the postsynaptic specializations associated with a single dendrite. Even within topographically mapped populations there are numerous functional subtypes, such as the "On" and "Off" retinal ganglion cells in the eye, which so far lack molecular or anatomical features of distinction. Next, it is hard to observe one CNS synapse even twice, in search of changes that occur with development or use. Finally, there has been no CNS ortholog of the Torpedo electroplaques that would allow the unique molecular signature of a specific type of CNS synapse to be identified by biochemical means. Perhaps it should not be surprising that no clear kingpin of CNS synapse formation has been identified. Nevertheless, while the mechanisms of CNS synaptogenesis are relatively unknown, there are many functional analogies and some direct commonalities between neuromuscular and central synapses. One emerging theme is that synapse formation in the CNS includes a higher degree of functional redundancy and overlap than found at the NMJ, possibly reflecting the fact that any given neuron in the brain is a target for many hundreds of other neurons, often of several subtypes employing different transmitters.

To understand the requirements of synaptogenesis in the CNS, we first consider how synaptic transmission in the CNS resembles and differs from the NMJ. We then review mechanisms of synaptogenesis in the CNS, insofar as data support their role. Points of significant homology to or departure from well-understood events at the NMJ will be considered in course.

Structure and Function at Central Synapses

As at the NMJ, the control of neurotransmitter release at interneuronal synapses relies on presynaptic morphological and biochemical specializations in the axon, usually concentrated in small domains located at an axonal branch tip. Release of transmitter is commonly focused by active zone complexes, which are visible in electron micrographs as thickened (electron dense) segments of the presynaptic membrane that accumulate synaptic

vesicles. SNARE complexes mediate docking and fusion of synaptic vesicles with the nerve terminal plasma membrane and trigger neurotransmitter release in response to elevated intracellular calcium. Fusion is followed by recovery and recycling of vesicle membrane components, enabling nerve terminals to function far from the cell nucleus. The molecular specializations supporting these functions (e.g., synaptotagmin, synaptobrevin, SNAP25, munc18, dynamin, rab5, voltage-gated calcium channels) are often identical or nearly identical to those at the NMJ. Thus, central and peripheral synapses rely on similar cellular and molecular presynaptic specializations.

The essential postsynaptic features of CNS synapses are also familiar. Neurotransmitter receptors are highly concentrated in the postsynaptic membrane directly opposite the presynaptic active zones. Additional voltage-gated ion channels are often concentrated in the membrane adjacent to the neurotransmitter receptor density, amplifying neurotransmitter-induced currents in the same way Na_v^+ channels concentrated in postsynaptic folds augment ACh-induced postsynaptic currents at the neuromuscular synapse. CNS transmitter receptors are co-concentrated with an array of primary scaffolding proteins and secondary signal transduction components that help co-concentrate the postsynaptic components and likely translate the recent history of synaptic activity into changes in synaptic strength and structure. A further parallel with the NMJ is that ribosomal complexes are found at postsynaptic sites in neurons. These may allow synaptic activity to regulate the synthesis of the postsynaptic components by translating synaptically localized mRNAs, analogous to the proposed role for transcriptional specialization of synaptic nuclei in skeletal muscle. CNS synapses also employ neurotransmitter clearance and re-uptake mechanisms to terminate synaptic signaling. Finally, the nerve terminal and postsynaptic specializations are maintained in precise register across a narrow synaptic cleft, through interactions between cell-surface adhesion receptors. As emphasized at the NMJ, proximity between sites of neurosecretion and reception is required for specific and effective neurotransmission. In many fundamental respects, therefore, interneuronal and neuromuscular synapses are alike.

One of the most notable features of synaptic transmission in the CNS, and one of the most obvious differences with skeletal NMJs, is the remarkable heterogeneity in inter-neuronal synaptic chemistry. The majority of inter-neuronal synapses use neurotransmitters other than acetylcholine, such as glutamate, GABA, or glycine. As there are few exceptions to Dale's hypothesis that each neuron employs a single primary neurotransmitter, each nerve terminal contains a restricted set of biosynthetic enzymes and transporters appropriate to the neurotransmitter. The variety of transmitters and neuromodulators used among interneuronal synapses is supported by an even greater variety of postsynaptic signal transduction mechanisms. These include ligand-gated ion channels, heterotrimeric G-protein coupled receptors, and peptidergic receptors.

A second, relatively obvious feature of most CNS synapses is their comparatively small size (Fig. 5). Most interneuronal synapses encompass a few square microns, rather than hundreds, and successful synaptic transmission in the CNS typically involves the release of transmitter from one or a few synaptic vesicles, instead of hundreds, and detection by a few dozen postsynaptic receptors, instead of tens of thousands. At many inter-neuronal synapses, nerve terminal depolarization fails to release transmitter more often than it succeeds. Some of these synapses could represent the persistence of immature synapses in the adult CNS. Alternatively, the stochastic nature of transmission at such synapses may be their fully developed form. Indeed, just as the certainty of synaptic transmission at the NMJ relies on elaborate pre- and postsynaptic specializations, the tuning of central synapses to successfully transmit with a certain probability rather than with uniformity seems likely to depend on a high order of synaptic specialization.

To be sure, the weakness of individual synaptic connections in the CNS is typically counterbalanced by a high density of synaptic sites; the surfaces of neurons are often almost entirely covered by nerve terminals. The postsynaptic neuron thus integrates many synaptic inputs, each small, some excitatory, and others inhibitory. One consequence of this convergence is that the contribution of each synapse to postsynaptic activity is weighted by its proximity to the site of action potential generation, usually the target cell's axon hillock. Thus, excitatory glutamatergic transmission at a synapse on a distal dendritic spine will ordinarily have less of an effect on the membrane voltage at the axon hillock than a similar synapse located downstream on a dendritic shaft, whose activity in turn can be readily nullified by inhibitory synaptic input to the perikaryon. Therefore, the degree of neuronal arborization and the number and distribution of synaptic connections are especially critical aspects of synaptic development in the CNS.

A final CNS departure is the synaptic cleft, which contains a proteinaceous material but lacks the basal lamina present in the synaptic cleft at the NMJ. Typically 20 nm apart, the pre- and postsynaptic membranes at interneuronal synapses are close enough to involve direct interactions between adhesion molecules in the opposed membranes. Thus, signals that promote and/or maintain synaptic differentiation may be integral components of the synaptic membranes, rather than secreted extracellular matrix components. Interneuronal synapses also lack postsynaptic folds. If folds are neuromuscular specializations that allow the massive release of ACh to rapidly dissipate, then their absence at interneuronal synapses may reflect the relatively small synaptic area and low level of transmitter release.

Development of CNS Synapses

The lessons of synaptic organization at the NMJ suggest that synaptic differentiation between neurons is dependent on an exchange of molecular cues. However, as CNS synapses are sites of direct contact between the membranes of their pre- and postsynaptic cells and lack the basal lamina that stably incorporates agrin, neuregulin, and laminin at the NMJ, it has seemed more likely that homo- and heterophilic cell-adhesion molecules play roles in establishing, aligning, and/or maintaining synaptic specializations in the brain. Important roles have been proposed

for cadherins and the neurexin:neuroligin complex. Certainly, soluble secreted factors may also play roles, and several have been suggested to play important roles in establishing or modulating synaptic connections. We consider each in turn.

ADHESION PROTEINS

Cadherins and Protocadherins

Cadherins are a large class of cell-surface membrane proteins, originally named for their dominant role in mediating calcium-dependent cell–cell adhesion. Four subgroups are identified: classical cadherins, protocadherins, desmosomal cadherins, and atypical cadherins. Each member contains at least one extracellular "cadherin" domain, and most are single-pass type I transmembrane proteins. Of these four types, we will discuss below the CNS roles of classical cadherins and protocadherins (Fig. 24), which are the best characterized.

Classical cadherins contain five extracellular "cadherin" repeats, and a relatively small intracellular domain. Classical cadherins mediate intercellular adhesion through homophilic interactions, such that among mixed populations of cells expressing different cadherins, cells expressing the same cadherin self-associate. The classical cadherin intracellular domain interacts with catenins, linking cadherin-rich membrane domains to actin cytoskeletal dynamics, and gene expression.

In the CNS, cadherins are concentrated at synapses. They have received special interest as mediators of synaptic connectivity, in part because homoselective binding offers a possible explanation for how axons select appropriate postsynaptic targets (Fannon and Colman, 1996; Uchida et al., 1996; Takeichi et al., 1997; Shapiro et al., 1999; Yagi et al., 2000). The "labeled line" model for synaptic connectivity in the CNS suggests that synapses preferentially form between pre- and postsynaptic cells that express complimentary adhesion molecules, as an electrician would splice a red wire to another red wire. In principle, homophilic cadherin interactions could serve as adhesive "labels" to instruct proper connectivity. However, while neurons in common circuits do express the same cadherins, they often express multiple cadherins, and synaptic connections do form between neurons that express different cadherins. This does not rule out an important role for cadherins in CNS circuitry, but suggests that whatever codes may exist are not simply reliant on cadherins.

Additional studies suggest that cadherins impart some of the specificity of synaptic connections in the CNS. One such example is in the avian optic tectum, a laminated region of the brain that receives multiple axonal projections from the eye and other brain regions. Retinal ganglion cell axons terminate in three of seven tectal cell layers. The laminar specificity of retinal innervation is directed by molecular cues that variously attract or repel the ingrowing retinal axons. Cadherins are among the cell-surface proteins differentially expressed between retino-recipient and non-recipient layers. Experiments designed to selectively perturb cadherin function altered the normal lamina-specific

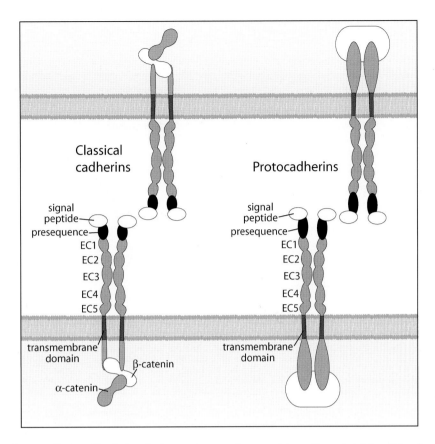

FIGURE 24. Cadherins and protocadherins. Classical cadherins are transmembrane proteins with modest intracellular domains and a series of five extracellular cadherin-specific domains. Cadherins play a significant role in promoting selective cell:cell interactions, through homophilic binding of specific cadherin isoforms. Intracellularly, classical cadherins bind to β-catenin, an important regulatory protein with links to both the actin cytoskeleton and to transcriptional regulation of gene expression. Protocadherins are similar to classical cadherins, but contain additional cadherin repeats. Intracellular interactions of protocadherins are less defined.

pattern of retinal innervation in the tectum (Inoue and Sanes, 1997). Other studies suggest that cadherins support cellular adhesion and molecular organization at synaptic sites. Detailed imaging found that cadherins are concentrated along the periphery of the synaptic densities, forming an adherens junction that surrounds the site of neurotransmission (Togashi *et al.*, 2002) (Fig. 25). Thus, cadherins may act somewhat like a molecular zipper to bind initial pre- and postsynaptic specializations in precise registration.

Despite the attractiveness of these models for cadherin function in synaptogenesis, there is considerable uncertainty regarding the contributions of specific cadherin isoforms. Genetic perturbation studies in mice so far indicate that the formation of most synapses does not depend on an individual form of cadherin. For example, mice lacking cadherin-11 have mild abnormalities in CNS function, and no obvious morphological defects. In contrast, approaches that simultaneously inhibit multiple cadherins do alter synaptic structure. For example, dominant negative cadherin constructs that mimic the conserved intracellular domain of classical cadherins, and thereby compete for downstream intracellular cadherin-binding proteins, cause defects in the formation of dendritic spines (which are postsynaptic structures) in cultured

hippocampal neurons (Togashi *et al.*, 2002). These constructs presumably interfere with the downstream signaling from all of the classical cadherins expressed in these cells and thus have a broader effect than the inhibition of individual cadherins. One implication of the enhanced effect of interfering with multiple cadherins is that there is a significant degree of functional overlap between cadherins expressed in the CNS, or that specific not-yet-tested versions play dominant roles. It has not yet been possible to test some of the most obvious candidates for dominant roles, such as N-cadherin, which is expressed by many neurons. N-cadherin-deficient mice die from cardiac defects at mid-gestational ages, prior to the normal period of synaptogenesis. However, synaptic defects similar to those caused by dominant-negative cadherin expression result from loss of the adaptor protein αN-catenin, which mediates interactions with the intracellular domain of classical cadherins.

The protocadherins are a large family of cadherin-like cell-adhesion proteins, composed of dozens of related cell-adhesion proteins. Typical members possess six or more extracellular cadherin repeats, a single transmembrane domain, and an intracellular domain that is less well conserved than in classical cadherins (Fig. 24). The large number of protocadherin proteins is

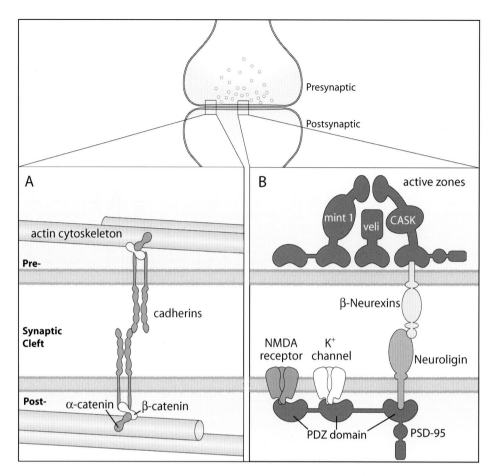

FIGURE 25. Synaptic adhesion complexes. (A) Cadherin complexes mediate homophilic adhesion. Cadherins are present at the borders of the presynaptic and postsynaptic densities, and interact with cytoskeletal elements within pre- and postsynaptic cells. (B) A second adhesion complex is formed by the interaction of β-neurexin with neuroligin, within the portions of the synapse involved in neurotransmission. Intracellular domains of both β-neurexin and neuroligin interact PDZ domains in synaptic scaffolding proteins. Presynaptically, β-neurexin interacts with the PDZ domain of CASK, which in turn interacts with veli and mint in the presynaptic density. Postsynaptically, neuroligin interacts with the PDZ domain of PSD-95, an integral component of the postsynaptic density. PSD-95 contains multiple PDZ domains, enabling it to link neuroligin to PDZ-binding neurotransmitter receptors and ion channels. Cadherins may serve to stabilize the adhesion of pre- and postsynaptic surfaces, and neuroligin/β-neurexin binding may serve to align the pre- and postsynaptic apparatus for neurotransmission.

partly a consequence of the genomic organization of their genes (Wu and Maniatis, 2000). Protocadherins are collected in three tandem gene clusters, termed α, β, and γ (Fig. 26). Within each cluster, the use of exons encoding the extracellular cadherin repeats, the transmembrane domain, and part of the cytoplasmic domain is highly variable; in contrast, exons encoding the remainder of the cytoplasmic domain are shared by all transcripts. This arrangement is generally similar to the arrangement of immunoglobulin genes and allows for a tremendous degree of diversity in the protein products. Such diversity would presumably be of tremendous value as a molecular array regulating synaptic specificity in the brain. However, the variable exon usage that produces individual protocadherins also hinders the study of individual variants. Moreover, deletion of the entire γ-protocadherin complex in mice results in neonatal lethality, and a great deal of apoptotic cell death in the nervous system (Wang *et al.*, 2002). While neurons cultured from these animals form an initial set of synapses before rapidly dying, more refined

perturbations will be required to understand whether synaptic abnormalities contribute to the excessive neuronal cell death.

Neurexin and Neuroligin

Neurexin and neuroligin are neuronal cell-surface proteins present at central synapses (Figs. 27 and 28). Unlike the cadherins, their interactions are heterophilic. Neurexins on the presynaptic cell bind to neuroligins and dystroglycan on the postsynaptic cell. Neuroligins preferentially bind β-neurexins, forming an especially tight complex. Much like cadherins, however, these interactions likely serve multiple roles in the CNS, quite possibly including the organization of new synapses and the stabilization of mature synapses.

Neurexins were identified in a search for the neuronal receptor for α-latrotoxin, a component of black widow spider venom (Ushkaryov *et al.*, 1992). The α-Latrotoxin causes massive exocytosis of neurotransmitter by stimulating the unregulated

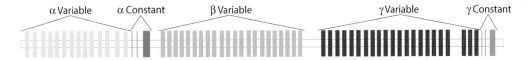

α Variable   α Constant   β Variable   γ Variable   γ Constant

FIGURE 26. Genetic organization of protocadherin diversity. Synaptic membrane proteins with hypervariable domains are attractive candidates to mediate the specificity of synaptic connections. Variability among protocadherins depends primarily on alternative splicing. The α-protocadherins are produced from a single gene containing fourteen "variable" exons, which are spliced to form the five or six extracellular cadherin repeats found in these isoforms, and three "constant" exons, which encode the transmembrane and intracellular domains present in all α-protocadherins. The β-protocadherins are produced from twenty-two variable exons. The γ-protocadherins are produced from 3 constant exons, and 22 variable exons. Given the possible number of exon combinations, these genes are capable of generating an astounding array of protein isoforms. The arrangement of protocadherin genes in clusters is similar to immunoglobulins.

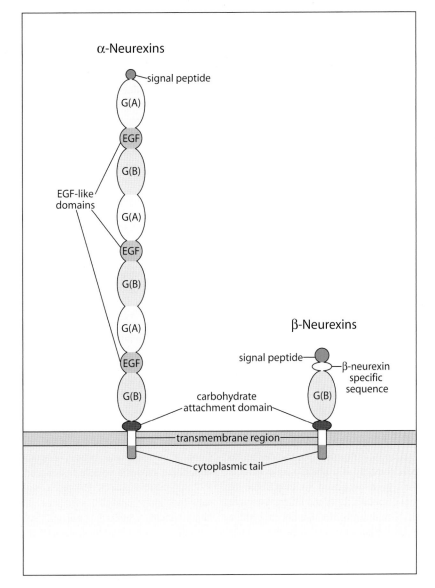

FIGURE 27. Neurexin structure. Neurexins are type I membrane proteins. Each contains a short cytoplasmic domain and a single transmembrane domain. The majority of neurexin mass is extracellular. The α-Neurexins contain 6 laminin-G domains and 3 EGF domains. Sequence similarities between the G-domains in α-neurexins suggest evolutionary triplication of an ancestral pair of G-domains across an EGF-like domain [i.e., G(A)-EGF-G(B)]. The β-neurexins contain a single G-domain and may represent a beneficial truncation of the ancestral α-neurexin G-domain pair. Considerable diversity in neurexin isoforms arises through a conserved splice site present in each G(B) domain. G-domains were originally named on their discovery in the α1-chain of laminin and have also been called LNS domains for their common appearance in laminins, neurexins, and the soluble hormone-binding S-protein. G-domains in agrin, perlecan, and laminin α-chains are ligands for receptors at the neuromuscular junction. The Z-splice site in agrin that regulates ACh receptor clustering is located within an agrin G-domain. Thus, through genetic duplication and alternative splicing, G-domains may have provided a common protein platform for organizing multiple aspects of pre- and postsynaptic differentiation across the synaptic cleft.

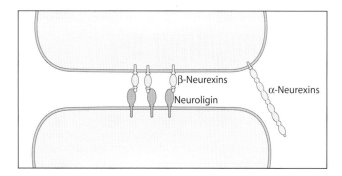

FIGURE 28. β-Neurexins, but not α-neurexins, interact with neuroligin across the synaptic cleft.

fusion of synaptic vesicles with the nerve terminal surface. The neurexin interaction with α-latrotoxin initially indicated that neurexin was not only present on presynaptic terminals, but in intimate association with the vesicle fusion machinery. This distribution has been difficult to confirm by conventional immunological methods, as antibodies to neurexins are poor. Nevertheless, transgenic mice concentrate neurexin-fusion protein epitopes at nerve terminals.

Neuroligins were identified by biochemical methods, as they bind directly and specifically to the β-neurexins. Antibodies specific for neuroligins readily label synaptic sites in brain, and staining with immunogold-labeled antibodies shows neuroligins specifically localize to the postsynaptic surface of the synaptic cleft.

The neurexin family is highly polymorphic. Gene duplication, multiple promoter elements, and alternative splicing produce a large number of potential neurexin isoforms. Neurons express neurexins from at least three genes (*Nrxn1*, *Nrxn2*, *Nrxn3*) (Missler *et al.*, 1998). A fourth neurexin gene encodes a more distantly related protein, which is selectively expressed by glia. The *Nrxn1–3* genes each contain two independent promoters, which generate longer α-neurexins and shorter β-neurexins (Fig. 27). Five conserved splice sites decorate the α-neurexins; two of these sites are included in β-neurexins. As a result, there are nearly 3,000 potential neurexin isoforms. Like the cadherins and protocadherins, neurexin diversity is a tantalizingly diverse molecular resource and has been proposed to contribute to the molecular basis of synaptic specificity in the brain. Analyses of neuronal transcripts indicate that a considerable number of the possible neurexin variants are actually expressed in the mature nervous system.

Variability in the neurexin gene transcription is targeted to the extracellular polypeptide domains. Each *Nrxn* gene encodes a major extracellular domain, a single transmembrane domain, and a modest intracellular domain. The extracellular domain is dominated by regions of homology to the LG-domain. The α-Neurexins contain six LG-domains. The β-Neurexins are initiated from a second, downstream promoter, and include only the final LG-domain, nearest the transmembrane domain. The tertiary structure of the LG-domain has been determined (Hohenester *et al.*, 1999;

Rudenko *et al.*, 1999; Timpl *et al.*, 2000). Of the five conserved alternative splice sites, three are specifically targeted to exposed loops of the LG-domain.

Interestingly, there is a notable precedent where alternative splicing in the LG-domain is critically important to synapse formation. Laminin G-domains are relatively common structural elements in extracellular matrix proteins and are concentrated in the synaptic basal lamina of the NMJ. Five LG-domains are present in tandem at the C-terminus of the laminin α2-, α4-, and α5-chains, and three G-domains are present in agrin (Figs. 16 and 18). They often (but not always) serve as binding sites for dystroglycan (Fig. 11), a matrix receptor concentrated at synaptic sites in both the PNS and CNS. However, LG-domains are also associated with neuronal signaling properties. The G-domains in the eponymous laminin-1 heterotrimer contribute to neurite adhesion and growth cone motility. Moreover, the AChR clustering activity of agrin is due to an alternative splice variation in a loop of the third LG-domain in agrin. LG-domains have a 14 β-strand structure, in which two antiparallel β-sheets are layered against each other, like an empty sandwich. Loops connecting the β-strands rim the margins (like a sandwich's crusts). The loops are relatively unconstrained and readily accommodate sequence variations. Accordingly, the Y- and Z-splice sites in agrin alter small peptide elements in adjacent LG-domain loops; both variations control agrin's ability to activate the MuSK receptor kinase. Possibly, splicing in neurexin's LG-domains mimics that in agrin. Moreover, it varies among brain regions, raising the possibility that neurexin LG-domain splicing has functional relevance to the organization of synaptic circuits. It remains uncertain whether documented differences represent cell-specific splice variation, or how many isoforms may be expressed at synaptic sites. There is also little notion of how variation in neurexin splice isoforms is recognized by postsynaptic receptors, as neuroligins do not appear to present a similar diversity. Nevertheless, functional studies suggest neurexins are important elements of nerve terminal differentiation.

Brain function in mice lacking individual neurexin genes is mildly or little affected. In contrast, mice lacking two or three of the α-neurexin genes are strongly affected and most die within one week, with disruptions to the rhythms of breathing (REF). Loss of α-neurexins causes a marked decrease in calcium-dependent synaptic vesicle fusion and evokes neurotransmission at both inhibitory (GABA-releasing) and excitatory (AMPA-sensitive glutamatergic) synapses. Importantly, while calcium channels are expressed at normal levels and have normal intrinsic conductances in the absence of α-neurexins, the calcium channel current density decreases precipitously during the period of synapse formation, compared to normal controls. There is no detectable defect in synaptic structure in the absence of α-neurexins, although there is a selective loss of brainstem GABA-releasing nerve terminals, which could account for the defects in breathing. Together, the results demonstrate an important functional role for the α-neurexins and indicate that α-neurexins are target-derived signals that regulate the location and/or activity of presynaptic calcium channels at sites of neurotransmitter release. They do not, however, discriminate functional differences between potential neurexin splice variants. These results also recall the previously described role of laminin-9 at the NMJ, which interacts specifically with presynaptic calcium channels and organizes the position of active zones in the nerve terminal membrane.

Mice lacking α-neurexins appear to express β-neurexins at normal levels. Additional studies suggest β-neurexins have important, but distinct functions at central synapses. First, the β-neurexins (one from each *Nrxn* gene) are specific trans-synaptic binding partners for neuroligins. Neuroligins are members of a gene family with at least three members in mammals. They are type I single-pass transmembrane proteins, with a single large extracellular domain that selectively binds β-neurexins. Alternative splicing of neurexin may alter this interaction, as incorporation of additional amino acid residues into the β-neurexin extracellular domain abolishes neuroligin binding. There also appears to be specificity through neuroligin expression; for example, neuroligin1 is excluded from GABAergic synapses. The extracellular domain bears strong sequence homologies to cholinesterases, but is catalytically inactive.

Second, *in vitro* studies have found that cultured neurons form presynaptic structures on non-neuronal cells that are transfected with constructs for recombinant neuroligins (Scheiffele *et al.*, 2000). Little or no nerve terminal formation occurred on neuroligin-expressing cells when soluble β-neurexin fusion proteins were added to the culture medium. The results suggest that neuroligin interactions with axon-associated β-neurexins promote the formation of presynaptic specializations, including terminal varicosities, synaptic vesicle accumulations, biochemical differentiation, and active zone localization.

The mechanisms by which neurexin/neuroligin bindings are transduced into synaptic organization are not yet known. One possibility is that they serve primarily as synaptic adhesives, tying pre- and postsynaptic membranes together, with additional membrane protein interactions driving synapse assembly. Alternatively, the neurexins and neuroligins could serve as platforms for signaling or scaffolding proteins and thus play more active roles in directing or stabilizing synapse formation. In support of this latter idea, the cytoplasmic domains of neurexins interact with the PDZ domain protein CASK (PDZ domains are described in detail later), which ultimately links to the presynaptic release apparatus (Fig. 25). In a blessed fact of simplicity, each α- and β-neurexin isoform encoded by a given gene (*Nrxn1, 2,* or *3*) has a common, invariant cytoplasmic domain. This could provides a mechanism to allow neurexins to directly connect diverse extracellular ligands (binding to the hypervariable neurexin LG-domains) to machinery of neurotransmitter release, which is shared at synapses throughout the nervous system. Similarly, neuroligins interact with the PDZ domain protein PSD95, which provide a direct link to the glutamate receptors and potassium channels concentrated at postsynaptic sites. Thus, by virtue of their localization, diversity, and extracellular adhesive properties, neurexins and neuroligins are attractive synaptogenic candidates at central synapses. Is summary, by simultaneously anchoring the anterograde and retrograde organization of synaptic protein complexes, neurexin/neuroligin interactions may promote the coincident formation of pre- and postsynaptic specialization.

Cadherin homophilic interactions and neurexin/neuroligin heterophilic interactions represent the best current view of CNS synapse formation. First, both are adhesion-based mechanisms that link extracellular interactions to intracellular signaling and protein localization. Second, each includes the potential for considerable molecular diversity, and they are therefore plausible candidate substrates underlying specificity in synaptic connections. Each may also play important roles in the nervous system beyond synaptogenesis. Cadherins are certainly involved in cell migration and the growth of axons and may be involved in neuronal survival as well. Neurexins and neuroligins seem well suited to regulate similar events before and after synaptogenesis. It is worth noting, however, that both sets of interactions are calcium dependent, while synaptic adhesion is not. Additional calcium-independent mechanisms of adhesion, such as immunoglobulin superfamily adhesion molecules, may therefore be essential components of synaptic interactions in the CNS.

SIGNALING FACTORS

Agrin and Neuregulin Play Uncertain Roles

Synaptogenesis at the NMJ relies on locally secreted cues passed between nerve and muscle. While agrin and neuregulins are obvious starting points in the search for similar controlling factors in the CNS, their roles there remain unclear. Several observations suggest agrin may promote the organization of synaptic specializations in the brain. Agrin is broadly expressed in the CNS, by many neuronal cell types in addition to cholinergic neurons. Much of the agrin expressed in the CNS is the Z+ isoform, which is "active" in clustering AChRs at the NMJ. Interestingly, unlike the NMJ, much of the agrin in the CNS is the product of an alternative transcriptional start site that creates an N-terminal transmembrane domain. This produces agrin as a type II transmembrane protein, in which the AChR-clustering signaling domain remains extracellular. Presumably, tethering agrin to the neuronal membrane represents a mechanism to anchor agrin to specific extracellular sites in the CNS, which lacks the semiautonomous form of extracellular matrix (the basal lamina) that pervades the PNS (Neumann *et al.*, 2001; Burgess *et al.*, 2002). Neurons are also capable of responding to agrin. In neuronal cultures, the addition of soluble agrin causes an increase in CREB phosphorylation and cFOS expression and alters neuronal morphology (Ji *et al.*, 1998; Hilgenberg *et al.*, 1999; Smith *et al.*, 2002). More provocatively, antiagrin antibodies and transfection with agrin-specific antisense oligonucleotides perturb synapse formation between neurons in culture; synapse formation is restored by application of exogenous agrin to the culture medium (Ferreira, 1999; Bose *et al.*, 2000; Mantych and Ferreira, 2001). Despite these supportive results, CNS development in agrin mutant mice appears relatively normal, and primary neurons cultured from these mice display few or no detectable defects in synaptogenesis (Li *et al.*, 1999; Serpinskaya *et al.*,

1999). How can these disparate *in vivo* and *in vitro* results be reconciled? One possibility is that the *in vitro* environment for synapse formation is artificially simple, allowing a minor, modulatory role for CNS agrin to be magnified. A second, common explanation for the lack of a "knockout" phenotype is redundancy among related factors. While no other agrin-like genes have been identified, it could be that the relevant signaling domain in agrin is reduplicated in other gene products. Indeed, the LG-domains which incorporate agrin's synaptogenic activity at the NMJ are present (as inactive isoforms) in a broad array of extracellular proteins in the CNS as well as the PNS. One of these, of course, is neurexin, described in the previous section.

A specific role for neuregulins in synapse formation in the CNS is even more obscure than that for agrin. Neuregulin is a multifunctional signaling factor in the nervous system, with significant roles in the fate and migration of neural crest derivatives. These events are especially crucial to the development of the brain's cellular architecture. Thus, defects in other neuronal behaviors may obscure specific roles for neuregulins in synapse formation. While agrin and neuregulin have uncertain roles in synapse formation in the CNS, other secreted signaling molecules have received more direct experimental support. These include the WNT/wingless signaling pathway, and NARP.

WNT Signaling

WNTs are a family of vertebrate proteins with homology to wingless (Wg), a secreted cell signaling glycoprotein in *Drosophila*. As the *Drosophila* name implies, wingless was identified through mutations that disrupt wing development. In the best characterized function of WNTs, Wg is a *Drosophila* morphogenetic factor that establishes polarity in developing anatomical elements, such as the segments of the embryonic body and the imaginal discs that produce the adult body structures. Vertebrate WNT proteins act in similar fashion, as short range signaling factors. They play critical roles in neural and axonal development (Burden, 2000; Patapoutian and Reichardt, 2000).

WNT signaling activities are mediated by Frizzled (Fz) receptors, a family of membrane proteins also first identified in *Drosophila* (Fig. 29). Fz receptors have a domain structure related to the seven-transmembrane domain, G-protein coupled receptors. Low-density lipoprotein receptor-related proteins (LRPs), a family of single-pass membrane proteins, serve as essential co-receptors for WNTs. WNTs also bind to heparan sulfate proteoglycans, which may be important for establishing gradients of WNT in the extracellular space. The WNT downstream signal pathway is best studied in non-neuronal cells. Activation of Fz receptors leads to the phosphorylation of Disheveled (Dsh). Phosphorylated Dsh prevents ubiquitin-dependent degradation of β-catenin, a protein that promotes the expression of WNT-responsive genes. Phosphorylated Dsh stabilizes β-catenin indirectly, by disrupting the formation of a complex between glycogen synthase kinase 3β (GSK3β), the adenomatous polyposis coli protein (APC), and the scaffolding protein Axin. The assembled complex phosphorylates β-catenin, promoting its ubiquitination and degradation. Stabilized β-catenin is required

for specific transcription factors (Lef/Tcf) to activate gene expression. In addition to affecting β-catenin, WNTs inhibit GSK3β-catalyzed phosphorylation of microtubules, thereby influencing cytoskeletal dynamics by increasing the stability of microtubule bundles.

Several studies indicate that WNT/Fz signaling is important during synaptogenesis. First, Wg/Fz signaling occurs at the *Drosophila* NMJ, and mutations in Wg cause defects in synaptic structure and function in *Drosophila* muscles (Packard *et al.*, 2002, 2003). The *Drosophila* NMJ is branched and varicose, like the vertebrate NMJ, but uses glutamate as neurotransmitter, like most excitatory synapses in the vertebrate CNS. Wg is secreted from motor neurons during synapse formation at *Drosophila* NMJs, where it activates myofiber Fz2 receptors. Mutations in Wg disrupt the normal postsynaptic aggregation of glutamate receptors and scaffolding proteins, as well as the elaborate structure of the postsynaptic membrane. Retrograde defects are also seen in Wg-deficient presynaptic boutons, which concentrate vesicles but lack their normal complement of mitochondria and presynaptic densities. It is attractive to consider that the presynaptic defects are a direct result of impaired microtubule-based trafficking in the absence of Wg. However, presynaptic defects could be secondary to impaired postsynaptic differentiation. For example, similar presynaptic defects arise at the vertebrate neuromuscular synapse, when postsynaptic differentiation is prevented by disrupting the agrin/MuSK/rapsyn pathway.

WNTs have been implicated in synapse formation in the vertebrate CNS, as well (Salinas *et al.*, 2003). WNT7a is produced by cerebellar granule cells and influences the presynaptic morphology of mossy fiber axons, which ascend from the brainstem (Hall *et al.*, 2000). Mossy fiber synapses on granule cells typically form elaborate multisynaptic structures, called glomerular rosettes. The morphology of these rosettes is controlled by WNT7a signaling. The formation of glomerular rosettes is delayed in WNT7a knockout mice, and direct application of WNT7a to mossy fiber axons causes an accumulation of synapsin 1, an early molecular marker of synapse formation. The effects of WNT7a on terminal remodeling are blocked by a secreted Fz-related protein, which antagonizes WNT signaling, and are inhibited by lithium, which antagonizes GSK activity downstream of Fz receptor activation. Since WNT7a is made primarily by the postsynaptic cell, in this case, it appears to act as a retrograde factor for presynaptic differentiation.

Similar retrograde signaling by WNTs has also been observed in the spinal cord (Krylova *et al.*, 2002). In the lateral column of the ventral horn, neurotrophin 3 (NT3)-responsive primary muscle afferents form monosynaptic connections with spinal motor neurons. These motor neurons produce WNT3 during the development of these connections. Application of WNT3 to the NT3-responsive sensory axons decreases axonal growth, but increases axonal branching and growth cone size. These effects are blocked by secreted Fz-related protein and are mediated by GSK interaction with the microtubule cytoskeleton. Although these studies lack the *in vivo* genetic analysis performed for WNT7a in the cerebellum, together they represent a consistent picture of WNTs as retrograde signals for presynaptic

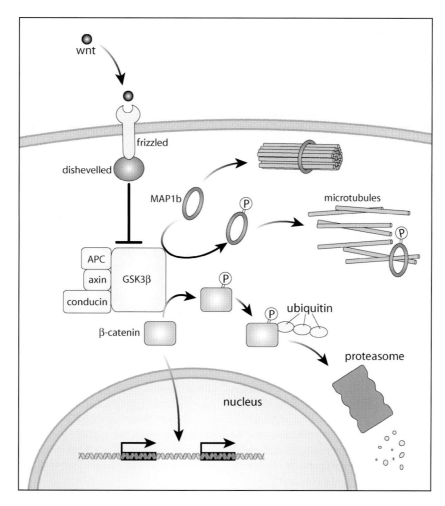

FIGURE 29. The wnt/frizzled pathway. WNT binding activates frizzled receptors, which leads to phosphorylation of dishevelled. Phosphorylated dishevelled inhibits GSK3β by promoting its association with APC. In the absence of WNT, active GSK3β phosphorylates MAP1b, which promotes dissociation of micro-tubule bundles. GSK3β also phosphorylates β-catenin, leading to its polyubiquitination and degradation. With WNT, phosphorylated dishevelled inhibits GSK3β, which stabilizes the microtubule cytoskeleton and allows levels of β-catenin to rise and regulate gene expression.

development in the vertebrate CNS. If WNTs prove to play roles in promoting presynaptic differentiation throughout the CNS, it will be important in determining how the specificity of synaptic connections is superimposed. The redundancy and complexity of the WNT/Fz signaling pathway represent an additional challenge.

Narp (Neuronal Activity-Regulated Pentraxin)

Narp was identified as an immediate early gene whose expression is induced by synaptic activity. Initially, activity-dependent regulation of Narp expression was taken as evidence that Narp functions after the initial steps in synaptogenesis, possibly to stabilize or refine initial connections (Tsui *et al.*, 1996). More recent studies suggest that Narp may also play an important role at nascent synapses (O'Brien *et al.*, 1999; Mi *et al.*, 2002). Narp is selectively concentrated at glutamatergic synapses, which have been best studied in the hippocampus and spinal cord. Overexpression of Narp in cultured spinal neurons

causes a substantial increase in the number of excitatory synapses present in the cultures. Narp co-aggregates with AMPA-type glutamate receptors after co-expression in non-neuronal cells, suggesting that it has a direct role in clustering glutamate receptors. However, Narp likely acts as a secreted factor to cluster receptors. For example, application of recombinant Narp to neuronal cultures causes cell-surface AMPA receptors to cluster. Thus, the activities of Narp on neuronal AMPA receptors are analogous to the activities of agrin on AChRs in cultured myotubes.

Several features of Narp deserve mention. First, the Narp polypeptide has homology to the pentraxin family of secreted proteins. Pentraxins form pentamers with a lectin-like three-dimensional structure. Lectins are plant proteins that bind with high avidity to carbohydrates. This and other biochemical features of Narp raise the interesting possibility that Narp acts as an extracellular bridge between carbohydrate moieties on neurotransmitter receptor or on receptor-associated proteins. Narp is secreted and could signal in anterograde fashion to promote postsynaptic differentiation *in vivo*. Second, Narp is associated with glutamatergic synapses and is absent from inhibitory synapses. Narp may therefore promote the specificity of synaptic connections. Third, Narp acts at both spiny synapses in the hippocampus, and aspiny synapses in the spinal cord. The notion that one factor may influence two morphologically distinct classes of synapses is a refreshing bit of simplicity for the CNS. Fourth, as mentioned at the start, Narp expression is regulated by synaptic activity. This most interesting observation suggests Narp may play roles in maintaining or remodeling connections in the mature CNS.

Mechanisms of Postsynaptic Specialization

Effective neurotransmission at chemical synapses depends critically on the density of neurotransmitter receptors in the postsynaptic membrane. Mechanisms underlying the concentration of postsynaptic receptors were first identified at the NMJ. The importance of rapsyn to AChR clustering at the NMJ had seemed to argue that receptor-associated clustering agents would likely play a dominant role at all fast chemical synapses. This concept has received considerable support from subsequent studies, although it now appears that CNS synapses use different molecular components to similar ends, even at cholinergic synapses. Rapsyn, which clusters AChRs at the NMJ, is apparently a muscle-specific postsynaptic scaffolding component, as it is not significantly expressed in the CNS (even at cholinergic synapses). AChR clustering mechanisms at interneuronal cholinergic synapses have not been indentified. However, an analogous component, gephyrin, appears to cluster receptors at inhibitory synapses in the brain (Fig. 30; Kneussel and Betz, 2000).

Much like rapsyn, gephyrin is an intracellular protein that interacts directly and specifically with pentameric neurotransmitter receptors, in this case glycine and GABA receptors (Fig. 30). Gephyrin also anchors receptor complexes with intracellular cytoskeletal elements, much like rapsyn. However, gephyrin interacts with microtubules instead of the actin cytoskeleton. Genetic experiments support gephyrin's role in sustaining postsynaptic

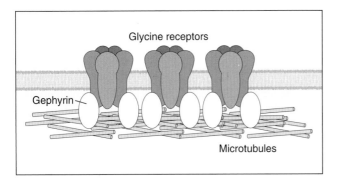

FIGURE 30. Glycine receptor clustering in the central nervous system is mediated by gephyrin. Gephyrin binds to the intracellular portion of the pentameric glycine receptors and also to the microtubule cytoskeleton. The role of gephyrin at inhibitory interneuronal synapses is analogous to the role of rapsyn at the neuromuscular junction.

receptor clustering. Targeted genetic deletion of gephyrin by homologous recombination in mice results in a failure to cluster glycine and GABA receptors, and an absence of glycinergic and GABAergic synapses (Feng *et al.*, 1998; Kneussel *et al.*, 1999). Not surprisingly, the mutant mice cannot survive beyond birth. In humans, as well, autoimmune reactions directed against gephyrin cause "Stiff-Man Syndrome," a human disorder caused by a lack of inhibitory synaptic transmission in the CNS (Butler *et al.*, 2000). These consistent series of observations were the first to definitively identify a specific receptor-clustering component in the CNS. Together, rapsyn and gephryn provide tangible evidence that tethering of postsynaptic receptors is a common mechanism of postsynaptic differentiation.

Postsynaptic specializations in the CNS contain a large number of additional scaffolding proteins. One broad class is known by a particular element of protein tertiary structure involved in protein : protein interactions, the PDZ domain (reviewed in Nourry *et al.*, 2003). PDZ domains were first identified in the tight junction protein ZO-1, the adherens junction protein Discs large (Dlg), and the 95 kDal postsynaptic density protein (PSD-95) concentrated at synaptic junctions in the vertebrate CNS (Kennedy, 1995). PDZ domains are present in all members of the PSD and SAP (synapse *a*ssociated *p*rotein) families, along with a catalytically inactive guanylate kinase homology domain. PDZ domains form hydrophobic pockets, which bind C-terminal amino acid motifs present on a number of transmembrane proteins. There is a loose consensus peptide sequence capable of interacting with PDZ domains. Most terminate with a valine residue, but differences at other positions promote preferential interactions with different PDZ domains.

The beauty of PDZ domain proteins is their modular structure. Multiple PDZ domains are typically present within a given polypeptide, in combinations with each other and additional protein interaction domains. PDZ domains are known to interact with glutamate receptors, potassium channels, and adhesion molecules, including neurexin and neuroligin discussed above. PSD-95, with three distinct PDZ domains, is able to interact with a neurotransmitter receptor, an ion channel, and a cell-adhesion molecule simultaneously. Thus, PDZ-proteins appear well-designed to link together multiple transmembrane and submembranous proteins. In this way, PDZ-proteins may serve to co-localize several functionally distinct membrane proteins that are fundamental to proper synaptic function. In this example, adhesion maintains proximity between pre- and postsynaptic elements, the neurotransmitter receptor responds to presynaptic exocytosis, and the ion channel propagates the depolarization into the neuron beyond. Although postsynaptic interactions involving PDZ-proteins are perhaps best described, PDZ domain proteins are also concentrated nerve terminals, where they may serve similar roles in linking presynaptic receptors, ion channels, and cell-adhesion molecules.

Glia-Derived Signals

Glial cells appear to be required for normal synaptogenesis. *In vivo*, synaptogenesis is concurrent with glial proliferation and maturation. Where specific loss of glial cells has been induced, neurons are observed to withdraw their synaptic connections. *In vitro*, the number and strength of synaptic connections among, for example, cultured retinal ganglion cell neurons increases many-fold in the presence of astrocytes, or astrocyte-conditioned medium (Ullian *et al.*, 2001; Slezak and Pfrieger, 2003). Despite these observations, a direct role for glial cells in promoting synapse formation is difficult to separate from their role in providing metabolic and trophic support to neurons. Glia absorb spent neurotransmitter and ions, which leach out of the synaptic cleft following transmission. They also provide trophic support to neurons. The ability of astrocytes to promote synapse formation *in vitro*, mentioned above, is associated with the ability of astrocytes to synthesize and supply cholesterol to the neurons (Mauch *et al.*, 2001; Pfrieger, 2003). Neurons are especially rich in cholesterol, and cholesterol is especially concentrated in "rafts" in the plasma membrane, which are domains rich in signaling receptors. It is attractive to speculate that interneuronal signaling is regulated by a glia-derived supply of cholesterol, although there is little *in vivo* evidence to support this idea at present, and no clear evidence that cholesterol is present at limiting levels in normal neurons.

SYNAPTIC REMODELING

Throughout the nervous system, the initial pattern of innervation undergoes significant remodeling during postnatal development. This has been particularly well-studied in muscle, where serial images of single synapses can be obtained over the course of days, weeks, and months. The most significant changes in innervation result from modifications at the synaptic sites themselves (Fig. 31). Three main changes take place. First, a majority of the initial synaptic connections are eliminated. Second, the strength of individual connections is enhanced through structural changes that increase synaptic territory, in part to accommodate growth of the muscle fiber. Third, changes in the structure and geometry of the synapse are accompanied by upgrades to the molecular composition of the synapse. Some of these molecular alterations are known to require altered patterns of gene expression.

Synapse Elimination

In sharp contrast to the single axon that innervates each adult muscle fiber, at birth, each neonatal muscle fiber is

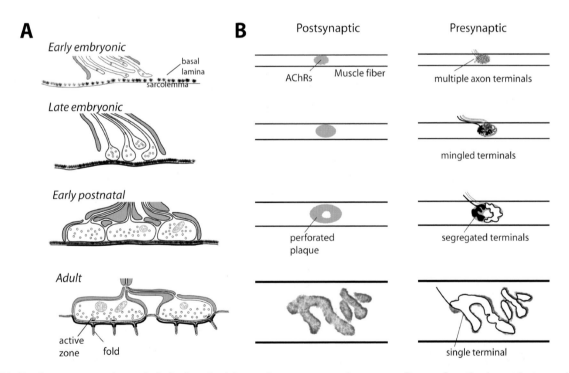

FIGURE 31. Development, maturation, and elimination of polyinnervation at neuromuscular synapses. Synapse formation in vertebrate muscles occurs in stages. (A) Embryonic myofibers are initially contacted by multiple motor axons, whose nerve terminals lack organized specializations and are loosely confederated within Schwann cell processes; postsynaptic sites have low concentrations of ACh receptors (AChRs) and a sparse basal lamina. Presynaptic terminals enlarge and concentrate synaptic vesicles, and postsynaptic membranes concentrate ACh receptors and specialized basal lamina components such as laminin β2 and ACh esterase, during embryonic and perinatal development. Secondary specializations, including active zones and folds, appear during postnatal maturation. (B) Pre- and postsynaptic development are spatially as well as temporally synchronized. Multiple axon terminals are initially co-mingled opposite a single AChR-rich plaque of postsynaptic membrane. As multiple inputs are eliminated through activity-dependent competition, surviving terminals become segregated, and holes appear in the postsynaptic plaque. Upon completion of synapse eliminated, the enlarged branches from a single nerve terminal innervate a matching series of postsynaptic gutters, which contain AChRs, a synapse-specific basal lamina, and secondary synaptic clefts.

innervated by multiple axons, typically from three to five (Fig. 31). Each of the axons innervating a single embryonic muscle fiber originates from a different motor neuron in the spinal cord. Nevertheless, their terminals interdigitate at a single, contiguous postsynaptic site on the muscle fiber (in twitch muscles; multiple postsynaptic sites appear on tonic muscle fibers). The apparent elevation in the number of synaptic connections present in neonatal muscle is real, compared to adult muscle. Neonates possess a mature number of spinal motor neurons, and a nearly complete number of muscle fibers. However, each spinal neuron has more intramuscular branches and innervates more muscle fibers in neonatal muscles than in adults. The large, overlapping motor units present at birth partly explain the exaggerated, uncoordinated movements of newborns.

Synapse elimination refers to the period during early postnatal development when all but one of the initial synaptic inputs to each fiber is disassembled. In a given muscle, most fibers become singly innervated within a few days of each other, although it may take several weeks to progress from the first to the final eliminated nerve terminal. Synapse elimination does not involve neuronal cell death, which is completed earlier. Instead, synapses are eliminated through the withdrawal of individual presynaptic terminals from the postsynaptic site (Bernstein and Lichtman, 1999). However, retraction of the preterminal axonal branch does bear similarity to the axonal atrophy that accompanies loss of trophic support, as if synapse elimination was a sort of subcellular, or subaxonal, demise.

The application of the term "synapse elimination" may at first seem confusing. Certainly, no neuron becomes targetless, and no muscle fiber becomes denervated. Rather, the number of neuron–muscle connections decreases by winnowing out the weakest connections from each hyperinnervated postsynaptic site. In addition, it is worth noting that the term "synapse elimination" does not describe the mechanism so much as the result. The term elimination might at first appear thoroughly myocentric, as it implies that the muscle fiber is the final arbiter in the decision of which of its inputs are rejected. As we shall see, current studies indicate that the muscle fiber does play a central role in *mediating* synapse elimination. However, the final outcome depends primarily on relative synaptic strengths and, therefore, relies as much on competitive interactions between the nerve terminals as on any controlling influence from the target itself. Synapse elimination must also be considered from the motor neuron's perspective, which selectively withdraws a majority of its embryonic nerve terminals, but necessarily succeeds in maintaining a substantial number as well.

The molecular mechanisms by which supernumerary nerve terminals are selectively eliminated from the muscle's postsynaptic site are not known. However, there are two known requirements to guide the ongoing search (Sanes and Lichtman, 1999). One requirement is postsynaptic activity in the muscle fiber. Simple paralysis of the muscle is in fact sufficient to prevent synapse elimination, strongly supporting the idea that retrograde factors play a role in eliminating connections. Polyinnervation persists on slowly contracting tonic muscles fibers, consistent with an absence of action potentials in these fibers.

A second requirement for normal synapse elimination is synaptic transmission. In particular, there must be *disparity* between both the strength and timing among the multiple axonal inputs whose terminals co-mingle on a given fiber. Synapse elimination in mice begins at neonatal ages, as motor neurons begin to lose gap-junctional coupling in the spinal cord, and electrical activity in the motor axons becomes temporally uncorrelated (Personius and Balice-Gordon, 2001). Complete neuromuscular blockade through pre- or postsynaptic mechanisms delays the elimination of polyinnervation; elimination proceeds when the block is released and neurotransmission is restored. Similarly, genetic perturbations that prevent release of neurotransmitter produce hyperinnervation, as in mice lacking the gene for choline acetyltransferase. In contrast, experimentally manipulating the levels of activity in a subset of axons supplying a muscle can accelerate the rate of elimination.

Recent genetic studies solidify support for the idea that it is the relative differences in synaptic activity that lead to the elimination of the weaker synapse. In one striking example, mice carrying a conditional mutation in choline acetyltransferase were used to eliminate ACh release from a subset of motor neurons, at neonatal ages (Buffelli *et al.*, 2003). In competitions for synaptic territory between wild-type and ChAT-deficient axons, the "silent" axon always lost, despite equal conditions during development. Presumably, the absence of neurotransmitter did not affect axonal activity, rates of synaptic vesicle fusion, or access to other target- or Schwann cell-derived factors.

Similar results have come from studies of adult NMJs, in which neurotransmission through a small portion of the synapse was selectively blocked. Normally, the adult neuromuscular synapse is a model of stoic persistence, enduring with little structural change for the life of the animal (and hopefully a century in all of us). Nevertheless, these studies found that an adult NMJ will readily eliminate an entire lobe of the synapse, including pre- and postsynaptic elements, after focal blockade of neurotransmission in that lobe. Focal blockade was performed using a micropipette to flow a stream of irreversible AChR-antagonist (such as α-bungarotoxin) across one end of the target NMJ (Balice-Gordon and Lichtman, 1994). Repeated imaging of the same synapse over the ensuing days showed that the inactive portion of the synapse is always eliminated, without altering the structure of the active remainder of the junction. As noted above, blockade of the entire junctional area has the opposite effect, suppressing the elimination of differing inputs. Thus, by all tests, disparity in synaptic transmission is critical for synapse elimination.

The hypothesis that synapse elimination is driven by competitive interactions between neighboring synaptic inputs on a single target cell is now generally accepted. One axiom of this thesis is that competition is fueled by differences in the levels of synaptic activity, with active sites displacing inactive sites. A second axiom is that the target cell (here, the muscle fiber) mediates the competition between its synaptic inputs. Several major questions remain. What factor(s) serve as the molecular substrate of synaptic competition? How is synaptic activity coupled to the activity of putative maintenance/elimination factors? What postsynaptic mechanism(s) in the muscle interpret different levels of synaptic activity between inputs and selectively eliminate the weakest? The answers to these questions are avidly sought, in part because they seem likely to apply to remodeling of synaptic connections throughout the nervous system, including the refinement of connections in the brain.

One possible mechanism for elimination at the NMJ is that motor nerve terminals compete for a retrograde trophic substance. Although none has been convincingly identified, it would presumably be available in limited amounts, and supplied in activity-dependent fashion by the muscle. In support of this idea, overexpression of glia-derived neurotrophic factor (GDNF) in the muscles of transgenic mice prevents synapse elimination and produces dramatic hyperinnervation (Nguyen et al., 1998). There is no direct evidence that GDNF or its like actually participate in regulating synapse elimination during normal development. Moreover, there is still no clear understanding of how a retrograde trophic factor could be differentially applied to terminals that vary only slightly in their temporal patterns of activity, or their physical location on the target. An alternative molecular mechanism posits that an alter ego to the retrograde trophic factor could provide the same competitive substrate. In this scenario, active terminals would be less susceptible to a toxic substance, such as a protease released by the muscle in response to synaptic activity. These putative activities have been dubbed "synaptotrophins" and "synaptotoxins," respectively (Sanes and Lichtman, 1999).

The mechanism by which muscles selectively couple differences in synaptic activity to differences in synaptic maintenance is an especially intriguing mystery. Careful observations of single NMJs show that the initial synaptic site is partially disassembled as synapse elimination proceeds. Repetitive observations of single NMJs during the period of synapse elimination show that nerve terminals undergoing elimination lose territory one branch at a time, starting in subregions of the synapse where they are especially underrepresented (Balice-Gordon and Lichtman, 1993; Balice-Gordon et al., 1993; Gan and Lichtman, 1998). Elimination of the terminal accelerates as the disparity in territory and efficacy increases (Colman et al., 1997; Kopp et al., 2000). Consistent with this accelerating disparity, synaptic sites that start out with evenly matched inputs are the last to complete the process of elimination. In addition, local disassembly of the postsynaptic apparatus beneath the losing terminal begins before the terminal completely withdraws. It seems likely that pre- and postsynaptic specializations that are destined for removal become molecularly distinguished from those that will be preserved. For example, activity in one region of the synapse could effectively "tag" adjacent regions, destabilizing them or marking them for disassembly. The nature of such a tag, the subsynaptic signals that would mediate differential tagging, and the mechanisms that could coordinate the removal of pre- and postsynaptic elements across the synaptic cleft remain speculative.

Structural Maturation

Paradoxically, synapse elimination occurs at the same time when the overall size and complexity of the NMJ is increasing

(Fig. 30). First, the branches and varicosities of the nerve terminal thicken and fuse to form the mature terminal arbor. In parallel, the AChR-rich regions of the endplate are sculpted to precisely match the profile of the overlying nerve. One mechanism that likely helps maintain the precise colocalization of AChRs opposite the nerve terminal varicosities is incorporation of nerve-derived isoforms of agrin into the synaptic basal lamina. Agrin is required early in synaptic development to maintain AChR clusters at synaptic sites. At mature synapses, agrin is concentrated in the synaptic basal lamina immediately adjacent to high concentrations of AChRs in the postsynaptic membrane. Agrin is localized to the basal lamina of the primary synaptic cleft, between the nerve terminal and endplate, and immediately adjacent to the AChR-rich tops of the postsynaptic folds (Trinidad et al., 2000). Agrin is absent from basal lamina that lines postsynaptic folds, whose membranes lack AChRs. Agrin has been found to bind directly and avidly to laminin in the basal lamina. This interaction tethers agrin to its site of secretion, and thereby serves as a "blueprint" of the nerve terminal's dimensions for the developing AChR-rich endplate.

Second, the synapse begins to adopt a complex geometry. Regions of the synaptic area are subtracted, as competing nerve terminals are eliminated. However, the size of the surviving synaptic area increases in absolute size, as the muscle fiber grows in length and caliber, continuing into adulthood. Growth of the synaptic area occurs by intercalary addition since, like a child's hand, the overall geometry of each individual endplate retains its basic shape through development. Synaptic growth need not have been accomplished this way; for example, neuromuscular synapses in *Drosophila* larval muscles increase in size by budding new varicosities from the edge of previous ones.

Third, the muscle forms postsynaptic gutters beneath the terminal branches, and postsynaptic folds beneath the active zones. These modifications further enhance the strength of the synaptic connection by increasing the postsynaptic surface area, and by isolating sites of high-efficiency synaptic transmission. The mechanisms that promote the formation of gutters and folds are not understood. One possibility is that adhesive interactions between the nerve terminal and muscle endplate pull their shapes into conformity. Similar interactions between regions of the postsynaptic membrane could sustain the tightly formed folds, if not their formation. One possibility suggested by Jeff Lichtman is that folds reflect the constraints on the addition of membrane to a region of the muscle surface that is tightly bound to another surface, in this case the nerve terminal (Marques et al., 2000). If expansion of the postsynaptic surface is laterally constrained, it can only increase by puckering, like a bunched blanket. Consistent with this idea, postsynaptic folds are typically absent in mutant mice that lack synaptic isoforms of laminin, the basal lamina component which serves as a primary anchor to dystroglycan in the muscle membrane. According to this puckered-blanket model of the endplate, the postsynaptic membranes in mice lacking tight linkage to (and within) the synaptic basal lamina may be free to slide laterally as new postsynaptic membrane components are intercalated during growth of the muscle fiber.

Molecular Maturation

The molecular machinery that supports synaptic transmission and which maintains the integrity of the synaptic connection is modified as the synapse reaches maturity. One prominent modification is the substitution of the γ-subunit of the AChR present during embryonic development for the ε-subunit present in adult muscle. This switch in receptor composition is accompanied by changes in channel kinetics, which may be required for efficient signaling in large, adult muscle fibers (Missias *et al.*, 1997). Voltage-dependent Na^+ channels become concentrated in the depths of the postsynaptic folds at this time. Additional molecular changes increase the stability of the endplate receptors (Salpeter and Loring, 1985; Shyng *et al.*, 1991), possibly through increased interactions with the submembranous cytoskeleton and the overlying synaptic basal lamina. For example, postnatal maturation of the dystrophin-associated protein complex, including dystrobrevin, is required to maintain the integrity of the AChR-rich postsynaptic domains (Grady *et al.*, 1997, 1999, 2000; Adams *et al.*, 2000). Similarly, the synaptic basal lamina undergoes a transition in composition during postnatal development; these changes are important to synaptic structure, as postnatal development of the synapse in the absence of basal lamina components, such as the laminin β2, α5, and α4 chains, leads to major structural defects (Patton, 2003).

Synaptic maturation at the NMJ has functional counterparts in the maturation of synaptic connections in the brain, which are beyond the scope of this review (Cowan *et al.*, 2001). During the postnatal development of the mammalian brain, for example, new synaptic territory is added as new neurons are added and dendritic fields enlarge. Broadly distributed projections are narrowed by increasing the number of synaptic connections in some areas and simultaneously eliminating connections in others. Finally, changes in synaptic strength are accompanied by alterations in the molecular composition of their pre- and postsynaptic elements. A common mechanism for change throughout the peripheral and central nervous systems is that structural changes are driven by relative levels of activity in neighboring synaptic connections. Activity-dependent changes in the synaptic architecture allow the initially crude, genetically determined pattern to adapt to best accommodate the host animal's interaction with the environment. Thus, remodeling likely represents an obligate solution to a fundamental problem in the development of any complex neural architecture: How to allocate synaptic connections in patterns which best fit the needs of each new member of the species.

REFERENCES

Adams, M.E., Kramarcy, N., Krall, S.P., Rossi, S.G., Rotundo, R.L., Sealock, R. *et al.*, 2000, Absence of alpha-syntrophin leads to structurally aberrant neuromuscular synapses deficient in utrophin, *J. Cell. Biol.* 150: 1385–1398.

Albrecht, D.E. and Froehner, S.C., 2002, Syntrophins and dystrobrevins: Defining the dystrophin scaffold at synapses, *Neurosignals* 11:123–129.

Anderson, M.J. and Cohen, M.W., 1977, Nerve-induced and spontaneous redistribution of acetylcholine receptors on cultured muscle cells, *J. Physiol.* 268:757–773.

Anderson, M.J. and Cohen, M.W., 1974, Fluorescent staining of acetylcholoine receptors in vertebrate skeletal muscle, *J Physiol* 237(2):385–400.

Anderson, M.J., Cohen, M.W., and Zorychta, E., 1977, Effects of innervation on the distribution of acetylcholine receptors on cultured muscle cells, *J. Physiol.* 268:731–756.

Anglister, L., Eichler, J., Szabo, M., Haesaert, B., and Salpeter, M.M., 1998, 125I-labeled fasciculin 2: A new tool for quantitation of acetylcholinesterase densities at synaptic sites by EM-autoradiography, *J. Neurosci. Methods* 81:63–71.

Arikawa-Hirasawa, E., Rossi, S.G., Rotundo, R.L., and Yamada, Y., 2002, Absence of acetylcholinesterase at the neuromuscular junctions of perlecan-null mice, *Nat. Neurosci.* 5:119–123.

Ashby, P.R., Wilson, S.J., and Harris, A.J., 1993, Formation of primary and secondary myotubes in aneural muscles in the mouse mutant peroneal muscular atrophy, *Dev. Biol.* 156:519–528.

Balice-Gordon, R.J., Chua, C.K., Nelson, C.C., and Lichtman, J.W., 1993, Gradual loss of synaptic cartels precedes axon withdrawal at developing neuromuscular junctions, *Neuron* 11:801–815.

Balice-Gordon, R.J. and Lichtman, J.W., 1993, *In vivo* observations of pre- and postsynaptic changes during the transition from multiple to single innervation at developing neuromuscular junctions, *J. Neurosci.* 13:834–855.

Balice-Gordon, R.J. and Lichtman, J.W., 1994, Long-term synapse loss induced by focal blockade of postsynaptic receptors, *Nature* 372:519–524.

Bennett, M.R., 1983, Development of neuromuscular synapses, *Physiol. Rev.* 63:915–1048.

Bernstein, M. and Lichtman, J.W., 1999, Axonal atrophy: The retraction reaction, *Curr. Opin. Neurobiol.* 9:364–370.

Bose, C.M., Qiu, D., Bergamaschi, A., Gravante, B., Bossi, M., Villa, A. *et al.*, 2000, Agrin controls synaptic differentiation in hippocampal neurons, *J. Neurosci.* 20:9086–9095.

Braithwaite, A.W. and Harris, A.J., 1979, Neural influence on acetylcholine receptor clusters in embryonic development of skeletal muscles, *Nature* 279:549–551.

Brandon, E.P., Lin, W., D'Amour, K.A., Pizzo, D.P., Dominguez, B., Sugiura, Y. *et al.*, 2003, Aberrant patterning of neuromuscular synapses in choline acetyltransferase-deficient mice, *J. Neurosci.* 23:539–549.

Brown, A.M. and Birnbaumer, L., 1990, Ionic channels and their regulation by G protein subunits, *Annu Rev Physiol* 52:197–213.

Buffelli, M., Burgess, R.W., Feng, G., Lobe, C.G., Lichtman, J.W., and Sanes, J.R., 2003, Genetic evidence that relative synaptic efficacy biases the outcome of synaptic competition, *Nature* 424:430–434.

Burden, S.J., 2000, Wnts as retrograde signals for axon and growth cone differentiation, *Cell* 100:495–497.

Burden, S.J., 2002, Building the vertebrate neuromuscular synapse, *J. Neurobiol.* 53:501–511.

Burden, S.J., Fuhrer, C., and Hubbard, S.R., 2003, Agrin/MuSK signaling: Willing and Abl, *Nat. Neurosci.* 6:653–654.

Burden, S.J., Sargent, P.B., and McMahan, U.J., 1979, Acetylcholine receptors in regenerating muscle accumulate at original synaptic sites in the absence of the nerve, *J. Cell. Biol.* 82:412–425.

Burgess, R.W., Dickman, D.K., Nunez, L., Glass, D.J., and Sanes, J.R., 2002, Mapping sites responsible for interactions of agrin with neurons, *J. Neurochem.* 83:271–284.

Burgess, R.W., Nguyen, Q.T., Son, Y.J., Lichtman, J.W., and Sanes, J.R. *et al.*, 1999, Alternatively spliced isoforms of nerve- and muscle-derived agrin: Their roles at the neuromuscular junction, *Neuron* 23:33–44.

Butler, M.H., Hayashi, A., Ohkoshi, N., Villmann, C., Becker, C.M., Feng, G. *et al.*, 2000, Autoimmunity to gephyrin in Stiff-Man syndrome, *Neuron* 26:307–312.

Cajal, S.R.Y., 1928, *Degeneration and Regeneration of the Nervous System*, Oxford University Press, London.

Campagna, J.A., Ruegg, M.A., and Bixby, J.L., 1995, Agrin is a differentiation-inducing "stop signal" for motoneurons *in vitro*, *Neuron* 15: 1365–1374.

Campagna, J.A., Ruegg, M.A., and Bixby, J.L., 1997, Evidence that agrin directly influences presynaptic differentiation at neuromuscular junctions *in vitro*, *Eur. J. Neurosci.* 9:2269–2283.

Chang, C.C. and Lee, C.Y., 1963, Isolation of neurotoxins from the venom of burgarus multicinctus and their modes of neuromuscular blocking action. *Arch. Int. Pharmacodyn. Ther.* 144:241–257.

Chiu, A.Y. and Ko, J., 1994, A novel epitope of entactin is present at the mammalian neuromuscular junction, *J. Neurosci.* 14:2809–2817.

Chu, G.C., Moscoso, L.M., Sliwkowski, M.X., and Merlie, J.P., 1995, Regulation of the acetylcholine receptor epsilon subunit gene by recombinant ARIA: An *in vitro* model for transynaptic gene regulation, *Neuron* 14:329–339.

Claudio, T., Ballivet, M., Patrick, J., and Heinemann, S., 1983, Nucleotide and deduced amino acid sequence of *Torpedo californica* acetylcholine receptor γ subunit, *Proc. Natl. Acad. Sci. USA* 80: 1111–1115.

Colman, H., Nabekura, J., and Lichtman, J.W., 1997, Alterations in synaptic strength preceding axon withdrawal, *Science* 275:356–361.

Corfas, G., Falls, D.L., and Fischbach, G.D., 1993, ARIA, a protein that stimulates acetylcholine receptor synthesis, also induces tyrosine phosphorylation of a 185-kDa muscle transmembrane protein, *Proc. Natl. Acad. Sci. USA* 90:1624–1628.

Cote, P.D., Moukhles, H., Lindenbaum, M., and Carbonetto, S., 1999, Chimaeric mice deficient in dystroglycans develop muscular dystrophy and have disrupted myoneural synapses, *Nat. Genet.* 23: 338–342.

Cowan, W.M., Sudhof, T.C., and Stevens, C.F., eds., 2001, *Synapses*, The Johns Hopkins University Press, Baltimore and London.

Dahm, L.M. and Landmesser, L.T., 1991, The regulation of synaptogenesis during normal development and following activity blockade, *J. Neurosci.* 11:238–255.

Dai, Z. and Peng, H.B., 1995, Presynaptic differentiation induced in cultured neurons by local application of basic fibroblast growth factor, *J. Neurosci.* 15:5466–5475.

Dale, H.H., Feldberg, W. and Vogt, M., 1936, Release of acetylcholine at voluntary motor nerve endings, *Journal of Physiology* 86:353–380.

Darwin, C., 1981, Origin of species. Cambridge University Press. 120p. (Originally published 1859).

De Robertis, E. and Bennett, H.S., 1955, Some features of the submicroscopic morphology of synapses in frog and earthworm. *J. Biochem. Biophys. Cytol.*, 1:47–58.

Dechiara, T.M., Bowen, D.C., Valenzuela, D.M., Simmons, M.V., Poueymirou, W.T., Thomas, S. *et al.*, 1996, The receptor tyrosine kinase MuSK is required for neuromuscular junction formation *in vivo*, *Cell* 85:501–512.

Denzer, A.J., Brandenberger, R., Gesemann, M., Chiquet, M., and Ruegg, M.A., 1997, Agrin binds to the nerve–muscle basal lamina via laminin, *J. Cell Biol.* 137:671–683.

Denzer, A.J., Schulthess, T., Fauser, C., Schumacher, B., Kammerer, R.A., Engel, J. *et al.*, 1998, Electron microscopic structure of agrin and mapping of its binding site in laminin-1, *Embo J.* 17:335–343.

Donger, C., Krejci, E., Serradell, A.P., Eymard, B., Bon, S., Nicole, S. *et al.*, 1998, Mutation in the human acetylcholinesterase-associated collagen gene, COLQ, is responsible for congenital myasthenic syndrome with end-plate acetylcholinesterase deficiency (Type Ic), *Am. J. Hum. Genet.* 63:967–975.

Durbeej, M. and Campbell, K.P., 2002, Muscular dystrophies involving the dystrophin–glycoprotein complex: An overview of current mouse models, *Curr. Opin. Genet. Dev.* 12:349–361.

Falls, D.L., Rosen, K.M., Corfas, G., Lane, W.S., and Fischbach, G.D., 1993, ARIA, a protein that stimulates acetylcholine receptor synthesis, is a member of the neu ligand family, *Cell* 72:801–815.

Eccles, J.C., Eccles, R.M. and Fatt, P., 1956, Pharmacological investigations on a central synapse operated by acetylcholine, *Journal of Physiology* 131:154–169.

Fannon, A.M. and Colman, D.R., 1996, A model for central synaptic junctional complex formation based on the differential adhesive specificities of the cadherins, *Neuron* 17:423–434.

Feng, G., Krejci, E., Molgo, J., Cunningham, J.M., Massoulie, J., and Sanes, J.R., 1999, Genetic analysis of collagen Q: Roles in acetylcholinesterase and butyrylcholinesterase assembly and in synaptic structure and function, *J. Cell. Biol.* 144:1349–1360.

Feng, G., Laskowski, M.B., Feldheim, D.A., Wang, H., Lewis, R., Frisen, J. *et al.*, 2000, Roles for ephrins in positionally selective synaptogenesis between motor neurons and muscle fibers, *Neuron* 25:295–306.

Feng, G., Tintrup, H., Kirsch, J., Nichol, M.C., Kuhse, J., Betz, H. *et al.*, 1998, Dual requirement for gephyrin in glycine receptor clustering and molybdoenzyme activity, *Science* 282:1321–1324.

Ferns, M.J., Campanelli, J.T., Hoch, W., Scheller, R.H., and Hall, Z., 1993, The ability of agrin to cluster AChRs depends on alternative splicing and on cell surface proteoglycans, *Neuron* 11:491–502.

Ferreira, A., 1999, Abnormal synapse formation in agrin-depleted hippocampal neurons, *J. Cell Sci.* 112:4729–4738.

Frail, D.E., McLaughlin, L.L., Mudd, J., and Merlie, J.P., 1988, Identification of the mouse muscle 43,000-dalton acetylcholine receptor-associated protein (RAPsyn) by cDNA cloning, *J. Biol. Chem.* 263:15602–15607.

Frank, E. and Fischbach, G.D., 1977, ACh receptors accumulate at newly formed nerve–muscle synapses *in vitro*, *Soc. Gen. Physiol. Ser.* 32:285–291.

Frank, E. and Fischbach, G.D., 1979, Early events in neuromuscular junction formation *in vitro*: Induction of acetylcholine receptor clusters in the postsynaptic membrane and morphology of newly formed synapses, *J. Cell Biol.* 83:143–158.

Fromm, L. and Burden, S.J., 1998, Transcriptional pathways for synapse-specific, neuregulin-induced and electrical activity-dependent transcription, *J. Physiol. Paris* 92:173–176.

Gan, W.B. and Lichtman, J.W., 1998, Synaptic segregation at the developing neuromuscular junction, *Science* 282:1508–1511.

Garbay, B., Heape, A.M., Sargueil, F., and Cassagne, C., 2000, Myelin synthesis in the peripheral nervous system, *Prog. Neurobiol.* 61:267–304.

Gautam, M., Noakes, P.G., Moscoso, L., Rupp, F., Scheller, R.H., Merlie, J.P. *et al.*, 1996, Defective neuromuscular synaptogenesis in agrin-deficient mutant mice, *Cell* 85:525–535.

Gautam, M., Noakes, P.G., Mudd, J., Nichol, M., Chu, G.C., Sanes, J.R. *et al.*, 1995, Failure of postsynaptic specialization to develop at neuromuscular junctions of rapsyn-deficient mice, *Nature* 377:232–236.

Glass, D.J., Bowen, D.C., Stitt, T.N., Radziejewski, C., Bruno, J., Ryan, T.E. *et al.*, 1996, Agrin acts via a MuSK receptor complex, *Cell* 85:513–523.

Glicksman, M.A. and Sanes, J.R., 1983, Differentiation of motor nerve terminals formed in the absence of muscle fibres, *J. Neurocytol.* 12:661–671.

Goodearl, A.D., Yee, A.G., Sandrock, A.W., Jr., Corfas, G., and Fischbach, G.D., 1995, ARIA is concentrated in the synaptic basal lamina of the developing chick neuromuscular junction, *J. Cell Biol.* 130:1423–1434.

Grady, R.M., Grange, R.W., Lau, K.S., Maimone, M.M., Nichol, M.C., Stull, J.T. *et al.*, Role for alpha-dystrobrevin in the pathogenesis of dystrophin-dependent muscular dystrophies, *Nat. Cell Biol.* 1:215–220.

Grady, R.M., Merlie, J.P., and Sanes, J.R., 1997, Subtle neuromuscular defects in utrophin-deficient mice, *J. Cell Biol.* 136:871–882.

Grady, R.M., Zhou, H., Cunningham, J.M., Henry, M.D., Campbell, K.P., and Sanes, J.R., 2000, Maturation and maintenance of the neuromuscular synapse: Genetic evidence for roles of the dystrophin–glycoprotein complex, *Neuron* 25:279–293.

Hall, A.C., Lucas, F.R., and Salinas, P.C., 2000, Axonal remodeling and synaptic differentiation in the cerebellum is regulated by WNT-7a signaling, *Cell* 100:525–535.

Harris, D.A., Falls, D.L., Dill-Devor, R.M., and Fischbach, G.D., 1988, Acetylcholine receptor-inducing factor from chicken brain increases the level of mRNA encoding the receptor alpha subunit, *Proc. Natl. Acad. Sci. USA* 85:1983–1987.

Hilgenberg, L.G., Hoover, C.L., and Smith, M.A., 1999, Evidence of an agrin receptor in cortical neurons, *J. Neurosci.* 19:7384–7393.

Hoch, W., Ferns, M., Campanelli, J.T., Hall, Z.W., and Scheller, R.H., 1993, Developmental regulation of highly active alternatively spliced forms of agrin, *Neuron* 11:479–490.

Hohenester, E., Tisi, D., Talts, J.F., and Timpl, R., 1999, The crystal structure of a laminin G-like module reveals the molecular basis of alpha-dystroglycan binding to laminins, perlecan, and agrin, *Mol. Cell* 4:783–792.

Holmes, W.E., Sliwkowski, M.X., Akita, R.W., Henzel, W.J., Lee, J., Park, J.W. *et al.*, 1992, Identification of heregulin, a specific activator of p185erbB2, *Science* 256:1205–1210.

Hughes, R.A., Sendtner, M., Goldfarb, M., Lindholm, D., and Thoenen, H., 1993, Evidence that fibroblast growth factor 5 is a major muscle-derived survival factor for cultured spinal motoneurons, *Neuron* 10:369–377.

Inoue, A. and Sanes, J.R., 1997, Lamina-specific connectivity in the brain: regulation by N-cadherin neurotrophins, and glycoconjugates. *Science.* 276:1428–1431.

Ji, R.R., Bose, C.M., Lesuisse, C., Qiu, D., Huang, J.C., Zhang, Q. *et al.*, 1998, Specific agrin isoforms induce cAMP response element binding protein phosphorylation in hippocampal neurons, *J. Neurosci.* 18:9695–9702.

Jo, S.A. and Burden, S.J., 1992, Synaptic basal lamina contains a signal for synapse-specific transcription, *Development* 115:673–680.

Kennedy, M.B., 1995, Origin of PDZ (DHR, GLGF) domains, *Trends Biochem. Sci.* 20:350.

Kneussel, M. and Betz, H., 2000, Clustering of inhibitory neurotransmitter receptors at developing postsynaptic sites: The membrane activation model, *Trends Neurosci.* 23:429–435.

Kneussel, M., Brandstatter, J.H., Laube, B., Stahl, S., Muller, U., and Betz, H., 1999, Loss of postsynaptic GABA(A) receptor clustering in gephyrin-deficient mice, *J. Neurosci.* 19:9289–9297.

Kopp, D.M., Perkel, D.J., and Balice-Gordon, R.J., 2000, Disparity in neurotransmitter release probability among competing inputs during neuromuscular synapse elimination, *J. Neurosci.* 20:8771–8779.

Krylova, O., Herreros, J., Cleverley, K.E., Ehler, E., Henriquez, J.P., Hughes, S.M. *et al.*, 2002, WNT-3, expressed by motoneurons, regulates terminal arborization of neurotrophin-3-responsive spinal sensory neurons, *Neuron* 35:1043–1056.

Kuromi, H. and Kidokoro, Y., 1984, Nerve disperses preexisting acetylcholine receptor clusters prior to induction of receptor accumulation in *Xenopus* muscle cultures, *Dev. Biol.* 103:53–61.

Laskowski, M.B. and Sanes, J.R., 1987, Topographic mapping of motor pools onto skeletal muscles, *J. Neurosci.* 7:252–260.

Lee, C.Y., 1972, Chemistry and pharmacology of polypeptide toxins in snake venoms. *Annu. Rev. Pharmacol.* 12:265–286.

Lemke, G.E. and Brockes, J.P., 1984, Identification and purification of glial growth factor, *J. Neurosci.* 4:75–83.

Li, Z., Hilgenberg, L.G., O'Dowd, D.K., and Smith, M.A. 1999, Formation of functional synaptic connections between cultured cortical neurons from agrin-deficient mice, *J. Neurobiol.* 39:547–557.

Lin, W., Burgess, R.W., Dominguez, B., Pfaff, S.L., Sanes, J.R., and Lee, K.F., 2001, Distinct roles of nerve and muscle in postsynaptic differentiation of the neuromuscular synapse, *Nature* 410:1057–1064.

Lindholm, D., Harikka, J., da Penha Berzaghi, M., Castren, E., Tzimagiorgis, G., Hughes, R.A. *et al.*, 1994, Fibroblast growth factor-5 promotes differentiation of cultured rat septal cholinergic and raphe seroto-

nergic neurons: Comparison with the effects of neurotrophins, *Eur. J. Neurosci.* 6:244–252.

Lobsiger, C.S., Taylor, V., and Suter, U., 2002, The early life of a Schwann cell, *Biol. Chem.* 383:245–253.

Loeb, J.A. and Fischbach, G.D., 1995, ARIA can be released from extracellular matrix through cleavage of a heparin-binding domain, *J. Cell Biol.* 130:127–135.

Loeb, J.A., Hmadcha, A., Fischbach, G.D., Land, S.J., and Zakarian, V.L., 2002, Neuregulin expression at neuromuscular synapses is modulated by synaptic activity and neurotrophic factors, *J. Neurosci.* 22:2206–2214.

Loewi, O., 1921, Uber humorale ubertragbarkeit der herznerven-wirkung, *Pflugers Archive* 189:239–242.

Lupa, M.T., Gordon, H., and Hall, Z.W., 1990, A specific effect of muscle cells on the distribution of presynaptic proteins in neurites and its absence in a C2 muscle cell variant, *Dev. Biol.* 142:31–43.

Lupa, M.T. and Hall, Z.W., 1989, Progressive restriction of synaptic vesicle protein to the nerve terminal during development of the neuromuscular junction, *J. Neurosci.* 9:3937–3945.

Mantych, K.B. and Ferreira, A., 2001, Agrin differentially regulates the rates of axonal and dendritic elongation in cultured hippocampal neurons, *J. Neurosci.* 21:6802–6809.

Marchionni, M.A., Goodearl, A.D., Chen, M.S., Bermingham-Mcdonogh, O., Kirk, C., Hendricks, M. *et al.*, 1993, Glial growth factors are alternatively spliced erbB2 ligands expressed in the nervous system, *Nature* 362:312–318.

Marques, M.J., Conchello, J.A., and Lichtman, J.W., 2000, From plaque to pretzel: Fold formation and acetylcholine receptor loss at the developing neuromuscular junction, *J. Neurosci.* 20:3663–3675.

Marshall, L.M., Sanes, J.R., and McMahan, U.J., 1977, Reinnervation of original synaptic sites on muscle fiber basement membrane after disruption of the muscle cells, *Proc. Natl. Acad. Sci. USA* 74:3073–3077.

Martinou, J.C., Falls, D.L., Fischbach, G.D., and Merlie, J.P., 1991, Acetylcholine receptor-inducing activity stimulates expression of the epsilon-subunit gene of the muscle acetylcholine receptor, *Proc. Natl. Acad. Sci. USA* 88:7669–7673.

Mauch, D.H., Nagler, K., Schumacher, S., Goritz, C., Muller, E.C., Otto, A. *et al.*, 2001, CNS synaptogenesis promoted by glia-derived cholesterol, *Science* 294:1354–1357.

McMahan, U.J., 1990, The agrin hypothesis, *Cold Spring Harb. Symp. Quant. Biol.* 55:407–418.

McMahan, U.J., Sanes, J.R., and Marshall, L.M., 1978, Cholinesterase is associated with the basal lamina at the neuromuscular junction, *Nature* 271:172–174.

Meier, T., Masciulli, F., Moore, C., Schoumacher, F., Eppenberger, U., Denzer, A.J. *et al.*, 1998, Agrin can mediate acetylcholine receptor gene expression in muscle by aggregation of muscle-derived neuregulins, *J. Cell Biol.* 141:715–726.

Mi, R., Tang, X., Sutter, R., Xu, D., Worley, P., and O'Brien, R.J., 2002, Differing mechanisms for glutamate receptor aggregation on dendritic spines and shafts in cultured hippocampal neurons, *J. Neurosci.* 22:7606–7616.

Miller, C., 2000, Ion channel surprises: prokaryotes do it again! Neuron. 25:7–9.

Misgeld, T., Burgess, R.W., Lewis, R.M., Cunningham, J.M., Lichtman, J.W., and Sanes, J.R., 2002, Roles of neurotransmitter in synapse formation: Development of neuromuscular junctions lacking choline acetyltransferase, *Neuron* 36:635–648.

Missias, A.C., Mudd, J., Cunningham, J.M., Steinbach, J.H., Merlie, J.P., and Sanes, J.R., 1997, Deficient development and maintenance of postsynaptic specializations in mutant mice lacking an "adult" acetylcholine receptor subunit, *Development* 124:5075–5086.

Missler, M., Fernandez-Chacon, R., and Sudhof, T.C., 1998, The making of neurexins, *J. Neurochem.* 71:1339–1347.

Moscoso, L.M., Cremer, H., and Sanes, J.R., 1998, Organization and reorganization of neuromuscular junctions in mice lacking neural cell adhesion molecule, tenascin-C, or fibroblast growth factor-5, *J. Neurosci.* 18:1465–1477.

Neumann, F.R., Bittcher, G., Annies, M., Schumacher, B., Kroger, S., and Ruegg, M.A., 2001, An alternative amino-terminus expressed in the central nervous system converts agrin to a type II transmembrane protein, *Mol. Cell Neurosci.* 17:208–225.

Nguyen, Q.T., Parsadanian, A.S., Snider, W.D., and Lichtman, J.W., 1998, Hyperinnervation of neuromuscular junctions caused by GDNF overexpression in muscle, *Science* 279:1725–1729.

Nguyen, Q.T., Sanes, J.R., and Lichtman, J.W., 2002, Pre-existing pathways promote precise projection patterns, *Nat. Neurosci.* 5:861–867.

Nitkin, R.M., Smith, M.A., Magill, C., Fallon, J.R., Yao, Y.M., Wallace, B.G. *et al.*, 1987, Identification of agrin, a synaptic organizing protein from Torpedo electric organ, *J. Cell Biol.* 105:2471–2478.

Noakes, P.G., Gautam, M., Mudd, J., Sanes, J.R., and Merlie, J.P., 1995, Aberrant differentiation of neuromuscular junctions in mice lacking s-laminin/laminin beta 2, *Nature* 374:258–262.

Noda, M., Takahashi, H., Tanabe, T., Toyosato, M., Furutani, Y., Hirose, T. *et al.*, 1982, Primary structure of α-subunit precursor of *Torpedo californica* acetylcholine receptor deduced from cDNA sequence, *Nature* 299:793–797.

Nourry, C., Grant, S.G., and Borg, J.P., 2003, PDZ domain proteins: Plug and play! *Sci STKE* 179:RE7.

Numa, S., Noda, M., Takahashi, H., Tanabe, T., Toyosato, M., Furutani, Y. *et al.*, 1983, Molecular structure of the nicotinic acetylcholine receptor, *Cold Spring Harb. Symp. Quant. Biol.* 48:57–69.

O'Brien, R.J., Xu, D., Petralia, R.S., Steward, O., Huganir, R.L., and Worley, P., 1999, Synaptic clustering of AMPA receptors by the extracellular immediate-early gene product Narp, *Neuron* 23:309–323.

Packard, M., Koo, E.S., Gorczyca, M., Sharpe, J., Cumberledge, S., and Budnik, V., 2002, The *Drosophila* Wnt, wingless, provides an essential signal for pre- and postsynaptic differentiation, *Cell* 111:319–330.

Packard, M., Mathew, D., and Budnik, V., 2003, Wnts and TGF beta in synaptogenesis: Old friends signalling at new places, *Nat. Rev. Neurosci.* 4:113–120.

Palay, S.L., 1956, Synapses in the central nervous system, *Journal of Biophysical and Biochemical Cytology* 2(Suppl.):193–202.

Parkhomovskiy, N., Kammesheidt, A., and Martin, P.T., 2000, N-acetyl-lactosamine and the CT carbohydrate antigen mediate agrin-dependent activation of MuSK and acetylcholine receptor clustering in skeletal muscle, *Mol. Cell Neurosci.* 15:380–397.

Patapoutian, A., and Reichardt, L.F. 2000, Roles of Wnt proteins in neural development and maintenance, *Curr. Opin. Neurobiol.* 10:392–399.

Patton, B.L., 2003, Basal lamina and the organization of neuromuscular synapses, *J. Neurocytol.* 32:883–903.

Patton, B.L., Chiu, A.Y., and Sanes, J.R., 1998, Synaptic laminin prevents glial entry into the synaptic cleft, *Nature* 393:698–701.

Patton, B.L., Cunningham, J.M., Thyboll, J., Kortesmaa, J., Westerblad, H., Edstrom, L. *et al.*, 2001, Properly formed but improperly localized synaptic specializations in the absence of laminin alpha4, *Nat. Neurosci.* 4:597–604.

Patton, B.L., Miner, J.H., Chiu, A.Y., and Sanes, J.R., 1997, Distribution and function of laminins in the neuromuscular system of developing, adult, and mutant mice, *J. Cell Biol.* 139:1507–1521.

Peng, H.B., Xie, H., Rossi, S.G., and Rotundo, R.L., 1999, Acetylcholinesterase clustering at the neuromuscular junction involves perlecan and dystroglycan, *J. Cell Biol.* 145:911–921.

Personius, K.E. and Balice-Gordon, R.J., 2001, Loss of correlated motor neuron activity during synaptic competition at developing neuromuscular synapses, *Neuron* 31:395–408.

Pfrieger, F.W., 2003, Role of cholesterol in synapse formation and function, *Biochim. Biophys. Acta* 1610:271–280.

Phillips, W.D., Kopta, C., Blount, P., Gardner, P.D., Steinbach, J.H., and Merlie, J.P., 1991, ACh receptor-rich membrane domains organized in fibroblasts by recombinant 43-kilodalton protein, *Science* 251:568–570.

Polo-Parada, L., Bose, C.M., and Landmesser, L.T., 2001, Alterations in transmission, vesicle dynamics, and transmitter release machinery at NCAM-deficient neuromuscular junctions, *Neuron* 32:815–828.

Rafuse, V.F., Polo-Parada, L., and Landmesser, L.T., 2000, Structural and functional alterations of neuromuscular junctions in NCAM-deficient mice, *J. Neurosci.* 20:6529–6539.

Riethmacher, D., Sonnenberg-Riethmacher, E., Brinkmann, V., Yamaai, T., Lewin, G.R., and Birchmeier, C., 1997, Severe neuropathies in mice with targeted mutations in the ErbB3 receptor, *Nature* 389:725–730.

Rogers, A.W., Darzynkiewicz, Z., Salpeter, M.M., Ostrowski, K., and Barnard, E.A., 1969, Quantitative studies on enzymes in structures in striated muscles by labeled inhibitor methods. I. The number of acetylcholinesterase molecules and of other DFP-reactive sites at motor endplates, measured by radioautography *J. Cell Biol.* 41:665–685.

Rudenko, G., Nguyen, T., Chelliah, Y., Sudhof, T.C., and Deisenhofer, J., 1999, The structure of the ligand-binding domain of neurexin Ibeta: Regulation of LNS domain function by alternative splicing, *Cell* 99:93–101.

Ruegg, M.A., Tsim, K.W., Horton, S.E., Kroger, S., Escher, G., Gensch, E.M. *et al.*, 1992, The agrin gene codes for a family of basal lamina proteins that differ in function and distribution, *Neuron* 8:691–699.

Rupp, F., Payan, D.G., Magill-Solc, C., Cowan, D.M., and Scheller, R.H., 1991, Structure and expression of a rat agrin, *Neuron* 6:811–823.

Sakmann, B., Noma, A. and Trautwein, W., 1983, Acetylcholine activation of single muscarinic K+ channels in isolated pacemaker cells of the mammalian heart, *Nature* 303(5914):250–253.

Salinas, P.C., 2003, Synaptogenesis: Wnt and TGF-beta take centre stage, *Curr. Biol.* 13:R60–62.

Salpeter, M.M., 1969, Electron microscope radioautography as a quantitative tool in enzyme cytochemistry. II. The distribution of DFP-reactive sties at motor endplates of a vertebrate twitch muscle, *J. Cell Biol.* 42:122–134.

Salpeter, M.M. and Loring, R.H., 1985, Nicotinic acetylcholine receptors in vertebrate muscle: Properties, distribution and neural control, *Prog. Neurobiol.* 25:297–325.

Salpeter, M.M., Rogers, A.W., Kasprzak, H., and McHenry, F.A., 1978, Acetylcholinesterase in the fast extraocular muscle of the mouse by light and electron microscope autoradiography, *J. Cell Biol.* 78:274–285.

Sandrock, A.W., Jr., Dryer, S.E., Rosen, K.M., Gozani, S.N., Kramer, R., Theill, L.E. *et al.*, 1997, Maintenance of acetylcholine receptor number by neuregulins at the neuromuscular junction *in vivo*, *Science* 276:599–603.

Sanes, J.R., Hunter, D.D., Green, T.L., and Merlie, J.P., 1990, S-laminin, *Cold Spring Harb. Symp. Quant. Biol.* 55:419–430.

Sanes, J.R. and Lichtman, J.W., 1999, Development of the vertebrate neuromuscular junction, *Annu. Rev. Neurosci.* 22:389–442.

Sanes, J.R., Marshall, L.M., and McMahan, U.J., 1978, Reinnervation of muscle fiber basal lamina after removal of myofibers. Differentiation of regenerating axons at original synaptic sites, *J. Cell Biol.* 78:176–198.

Scheiffele, P., Fan, J., Choih, J., Fetter, R., and Serafini, T., 2000, Neuroligin expressed in nonneuronal cells triggers presynaptic development in contacting axons, *Cell* 101:657–669.

Schmidt, J. and Raftery, M.A., 1973, Purification of acetylcholine receptors from *Torpedo californica* electroplax by affinity chromatography, *Biochemistry* 12:852–856.

Serpinskaya, A.S., Feng, G., Sanes, J.R., and Craig, A.M., 1999, Synapse formation by hippocampal neurons from agrin-deficient mice, *Dev. Biol.* 205:65–78.

Shapiro, L. and Colman, D.R., 1999, The diversity of cadherins and implications for a synaptic adhesive code in the CNS, *Neuron* 23:427–430.

Shyng, S.L., Xu, R., and Salpeter, M.M., 1991, Cyclic AMP stabilizes the degradation of original junctional acetylcholine receptors in denervated muscle, *Neuron* 6:469–475.

Slezak, M. and Pfrieger, F.W., 2003, New roles for astrocytes: Regulation of CNS synaptogenesis, *Trends Neurosci.* 26:531–535.

Smith, M.A., Hilgenberg, L.G., Hoover, C.L., Li, Z., and O'Dowd, D.K., 2002, Agrin in the CNS: A protein in search of a function? Evidence of an agrin receptor in cortical neurons, *Neuroreport* 13: 1485–1495.

Sollner, T., Bennett, M.K., Whiteheart, S.W., Scheller, R.H., and Rothman, J.E., 1993, A protein assembly–disassembly pathway *in vitro* that may correspond to sequential steps of synaptic vesicle docking, activation, and fusion, *Cell* 75:409–418.

Son, Y.J., Patton, B.L., and Sanes, J.R., 1999, Induction of presynaptic differentiation in cultured neurons by extracellular matrix components, *Eur. J. Neurosci.* 11:3457–3467.

Son, Y.J. and Thompson, W.J., 1995, Schwann cell processes guide regeneration of peripheral axons, *Neuron* 14:125–132.

Steward, O. and Schuman, E.M., 2001, Protein synthesis at synaptic sites on dendrites, *Annu. Rev. Neurosci.* 24:299–325.

Sunderland, W.J., Son, Y.J., Miner, J.H., Sanes, J.R., and Carlson, S.S., 2000, The presynaptic calcium channel is part of a transmembrane complex linking a synaptic laminin (alpha4beta2gamma1) with non-erythroid spectrin, *J. Neurosci.* 20:1009–1019.

Takeichi, M., Uemura, T., Iwai, Y., Uchida, N., Inoue, T., Tanaka, T. *et al.*, 1997, Cadherins in brain patterning and neural network formation, *Cold Spring Harb. Symp. Quant. Biol.* 62:505–510.

Timpl, R., Tisi, D., Talts, J.F., Andac, Z., Sasaki, T., and Hohenester, E., 2000, Structure and function of laminin LG modules, *Matrix Biol.* 19:309–317.

Togashi, H., Abe, K., Mizoguchi, A., Takaoka, K., Chisaka, O., and Takeichi, M., 2002, Cadherin regulates dendritic spine morphogenesis, *Neuron* 35:77–89.

Trinidad, J.C., Fischbach, G.D., and Cohen, J.B., 2000, The Agrin/MuSK signaling pathway is spatially segregated from the neuregulin/ErbB receptor signaling pathway at the neuromuscular junction, *J. Neurosci.* 20:8762–8770.

Tsim, K.W., Ruegg, M.A., Escher, G., Kroger, S., and McMahan, U.J., 1992, cDNA that encodes active agrin, *Neuron* 8:677–689.

Tsui, C.C., Copeland, N.G., Gilbert, D.J., Jenkins, N.A., Barnes, C., and Worley, P.F., 1996, Narp, a novel member of the pentraxin family,

promotes neurite outgrowth and is dynamically regulated by neuronal activity, *J. Neurosci.* 16:2463–2478.

Uchida, N., Honjo, Y., Johnson, K.R., Wheelock, M.J., and Takeichi, M., 1996, The catenin/cadherin adhesion system is localized in synaptic junctions bordering transmitter release zones, *J. Cell Biol.* 135:767–779.

Ullian, E.M., Sapperstein, S.K., Christopherson, K.S., and Barres, B.A., 2001, Control of synapse number by glia, *Science* 291:657–661.

Ushkaryov, Y.A., Petrenko, A.G., Geppert, M., and Sudhof, T.C., 1992, Neurexins: Synaptic cell surface proteins related to the alpha-latrotoxin receptor and laminin, *Science* 257:50–56.

Vansaun, M. and Werle, M.J., 2000, Matrix metalloproteinase-3 removes agrin from synaptic basal lamina, *J. Neurobiol.* 43:140–149.

Wang, X., Weiner, J.A., Levi, S., Craig, A.M., Bradley, A., and Sanes, J.R., 2002, Gamma protocadherins are required for survival of spinal interneurons, *Neuron* 36:843–854.

Wen, D., Peles, E., Cupples, R., Suggs, S.V., Bacus, S.S., Luo, Y. *et al.*, 1992, Neu differentiation factor: A transmembrane glycoprotein containing an EGF domain and an immunoglobulin homology unit, *Cell* 69:559–572.

Wigston, D.J. and Sanes, J.R., 1982, Selective reinnervation of adult mammalian muscle by axons from different segmental levels, *Nature* 299:464–467.

Wolpowitz, D., Mason, T.B., Dietrich, P., Mendelsohn, M., Talmage, D.A., and Role, L.W., 2000, Cysteine-rich domain isoforms of the neuregulin-1 gene are required for maintenance of peripheral synapses, *Neuron* 25:79–91.

Wood, S.J. and Slater, C.R., 2001, Safety factor at the neuromuscular junction, *Prog. Neurobiol.* 64:393–429.

Wu, Q. and Maniatis, T., 2000, Large exons encoding multiple ectodomains are a characteristic feature of protocadherin genes, *Proc. Natl. Acad. Sci. USA* 97:3124–3129.

Xia, B., Hoyte, K., Kammesheidt, A., Deerinck, T., Ellisman, M., and Martin, P.T., 2002, Overexpression of the CT GalNAc transferase in skeletal muscle alters myofiber growth, neuromuscular structure, and laminin expression, *Dev. Biol.* 242:58–73.

Yagi, T. and Takeichi, M., 2000, Cadherin superfamily genes: Functions, genomic organization, and neurologic diversity, *Genes Dev.* 14:1169–1180.

Yang, X., Arber, S., William, C., Li, L., Tanabe, Y., and Jessell, T.M., 2001, Patterning of muscle acetylcholine receptor gene expression in the absence of motor innervation, *Neuron* 30:399–410.

Yang, X., Li, W., Prescott, E.D., Burden, S.J., and Wang, J.C., 2000, DNA topoisomerase IIbeta and neural development, *Science* 287: 131–134.

11

Programmed Cell Death

Kevin A. Roth

INTRODUCTION

Programmed cell death is critical for normal nervous system development. From the initial sculpting of the size and shape of the developing brain through establishment of the number of cells in specific neuronal populations, programmed cell death ensures that nervous system development proceeds in an orderly and regulated fashion. Although neuronal programmed cell death research has historically focused on synapse-bearing neurons and their competition for limited supplies of target-derived neurotrophic molecules, recent studies have revealed a significant role for programmed cell death in neural precursor cells and immature neurons, prior to the establishment of synaptic contacts. Rapid advances in molecular biology and the use of gene-targeting approaches have led to tremendous progress in our understanding of the molecular regulation of programmed cell death. Recent studies have also revealed unexpected complexity in neuronal cell-specific death pathways and raised questions about the intrinsic and extrinsic triggers of programmed cell death.

HISTORICAL PERSPECTIVE

Programmed cell death refers to the reproducible, spatially- and temporally-restricted death of cells during organismal development (Burek and Oppenheim, 1996). As such, the term is synonymous with "physiological," "naturally-occurring," or "developmental" cell death. The original discovery of programmed cell death in the vertebrate nervous system has been attributed to Beard who described the degeneration of neurons in the skate nervous system over 100 years ago (Beard, 1896; Jacobson, 1991). In 1926, Ernst described three types of developmental cell death involving the regression of vestigial organs, the cavitation, folding or fusion of organ anlage, and the elimination of cells during tissue remodeling (Ernst, 1926). These three functional types of programmed cell death were termed phylogenetic, morphogenetic, and histogenetic cell death, respectively, by Glücksmann approximately 50 years ago (Glücksmann,

1951). Examples of phylogenetic death in the mammalian nervous system include degeneration of the paraphysis, vomeronasal nerve, and nervus terminalis. Morphogenetic cell death occurs during formation of the mammalian optic and otic vesicles and during maturation of the neural tube and neural plate. Histogenetic cell death in the mammalian nervous system is fairly widespread and numerous neuronal cell populations throughout the central and peripheral nervous systems have been reported to undergo this type of degeneration (Jacobson, 1991). The extent of histogenetic death varies between neuronal populations but has been estimated to range between 20% and 80% of neurons in some populations (Oppenheim, 1991). Prominent among these are motor neurons in the spinal cord and neurons in the sensory and sympathetic nervous systems.

Histogenetic cell death was the focus of many studies of neuronal programmed cell death during the second half of the 20th century (Hamburger, 1992). Histogenetic neuron death is typically triggered by insufficient trophic factor support. Following initial neurogenesis, immature neuron migration, and synaptogenesis, many neuronal populations enter a period of competition for target-derived trophic factors. Neurons obtaining inadequate trophic support during this period are eliminated through activation of a cell autonomous death program. This competitive process has been thought to ensure the proper matching of the size of each newly generated neuronal population with that of its target field. Nerve growth factor (NGF) was the first neurotrophic factor to be isolated and characterized and much of the recent progress in defining the molecular regulation of neuronal cell death can be attributed to investigations of NGF and related molecules. Despite the historical significance of NGF-related research, the concept that neuronal programmed cell death serves largely to match neuron numbers with post-synaptic target size is overly simplistic (Kuan *et al.*, 2000). Many neuronal populations exhibit no obvious requirement for target-derived neurotrophic molecules and a significant degree of cell death may occur in neural precursor cells and immature neurons prior to the elaboration of neuritic processes and formation of synaptic contacts. Ongoing research on neural precursor cells and immature neurons is extending our understanding of the role of programmed cell death in developmental neurobiology.

Kevin A. Roth • Division of Neuropathology, Department of Pathology, University of Alabama at Birmingham, Birmingham, AL 35294-0017.

Developmental Neurobiology, 4th ed., edited by Mahendra S. Rao and Marcus Jacobson. Kluwer Academic / Plenum Publishers, New York, 2005.

MORPHOLOGICAL TYPES OF PROGRAMMED CELL DEATH

Cells undergoing programmed cell death in the developing nervous system may exhibit various morphological appearances (Clarke, 1990). The most common morphological type of programmed cell death is type 1 or apoptotic cell death and many authors have erroneously equated apoptosis with programmed cell death (Häcker, 2000). The term "apoptosis" was originally used to describe a unique type of "non-necrotic" cell death in which the degenerating cells displayed a specific set of morphological features (Kerr *et al.*, 1972). Apoptotic cells exhibit chromatin condensation and margination, nuclear fragmentation, cytoplasmic membrane blebbing and convolution, and cell shrinkage (Fig. 1). These features are best appreciated by ultrastructural examination but can also be observed at the light microscopic level (Roth, 2002). The vast majority of cells that die during nervous system development, including neurons lost during competition for target-derived neurotrophic factor support, show apoptotic features.

The second major type of programmed cell death is type 2 or autophagic degeneration (Clarke, 1990). Autophagic cell death is characterized by the presence of numerous autophagic vacuoles and degradation of cytoplasmic elements in the degenerating cell (Fig. 2). The nucleus may become pyknotic and, in some cases, may exhibit typical "apoptotic-like" nuclear features. Autophagic vacuoles originate as double membrane sheets derived from the endoplasmic reticulum which engulf intracellular organelles and cytoplasmic materials followed by delivery of the vacuolar contents to lysosomes (Seglen and Bohley, 1992; Dunn, 1994). Unlike apoptotic cell death which typically involves single or scattered cells within normal parenchyma, autophagic cell death typically involves contiguous groups of degenerating cells (Lee and Baehrecke, 2001). Autophagic cell death has been observed at several sites in the vertebrate nervous system including the isthmo-optic nucleus where neurons undergo programmed cell death secondary to insufficient target-derived trophic factor support.

Less common morphological forms of programmed cell death have been described including type 3A, "non-lysosomal disintegration" and type 3B, "cytoplasmic type" (Clarke, 1990). These forms of death bear some resemblance to necrotic cell death in that swelling of intracellular organelles and fragmentation of the cell membrane are prominent. Type 3A and 3B programmed cell death are rarely observed in the mammalian nervous system. However, distinguishing between forms of cell death may be difficult, and in some cases, degenerating cells may exhibit features of multiple morphological death types.

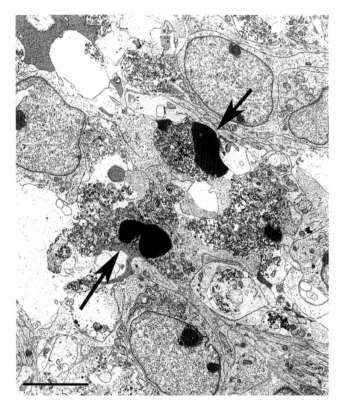

FIGURE 1. Ultrastructural examination of the embryonic day 12 mouse spinal cord shows several degenerating neurons (indicated by arrows) with apoptotic features including chromatin condensation and margination and cell shrinkage. Scale bar equals 5 μm.

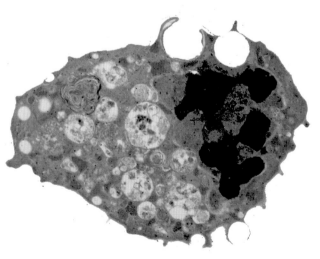

FIGURE 2. The morphological features of autophagic cell death are illustrated in this electron micrograph of a telencephalic neuron that was exposed to 20 μM chloroquine, a lysosomotropic agent, for 18 hr *in vitro*. The cell contains numerous membrane delimited autophagic vacuoles of various sizes, some containing osmophilic debris, a decreased number of cytoplasmic organelles, and degenerative nuclear features (clumped and fragmented chromatin) similar to those observed in cells undergoing apoptotic death. Scale bar equals 2 μm.

MOLECULAR REGULATION OF PROGRAMMED CELL DEATH

Apoptotic Cell Death

Apoptosis is the most common type of programmed cell death in the developing nervous system. Much of our understanding of the molecular pathways regulating mammalian cell apoptosis was anticipated by investigations of programmed cell death in the nematode *Caenorhabditis elegans* (Horvitz, 1999). In *C. elegans*, approximately 10% of the organism's cells undergo programmed cell death in a highly stereotyped, cell autonomous fashion. Four genes, *egl-1* (egg-laying defective), *ced-9* (ced, cell death abnormal), *ced-4*, and *ced-3*, act in a coordinated fashion to cause *C. elegans* cell death. Studies suggest that EGL-1 binds to CED-9, releasing CED-4 from a CED-9/CED-4 complex, and CED-4 in turn activates CED-3 which represents the commitment point to *C. elegans* cell death.

This basic pattern of apoptotic death regulation is recapitulated in mammals. Structural homologs of EGL-1, CED-9, CED-4, and CED-3 exist in mammals and consist of several multigene families. Mammalian EGL-1-like molecules are members of the BH3 domain-only Bcl-2 subfamily and include Bid, Bim, Bad, and Noxa (Korsmeyer, 1999). These molecules are thought to interact with multidomain, CED-9-like, Bcl-2 family members to regulate mitochondrial cytochrome c release and function. Multidomain Bcl-2 family members are divided into anti- (e.g., Bcl-2 and Bcl-X_L) and pro-apoptotic (e.g., Bax and Bak) subgroups. Bcl-2 and Bcl-X_L can block apoptotic stimulus-induced cytochrome c redistribution and Bax and Bak promote mitochondrial cytochrome c release. Apaf-1, the best-defined mammalian homolog of CED-4, binds cytosolic cytochrome c, and in the presence of dATP or ATP, assists in the conversion of caspase-9 into an active enzyme (Zou *et al.*, 1997, 1999). Caspases are the mammalian homologs of CED-3 and consist of approximately 15 cysteine-containing, aspartate-specific proteases (Nicholson, 1999). Caspases exist at baseline as inactive zymogens and are converted into active enzymes via cleavage of the proenzyme form into large and small subunits which together form the active caspase. This processing occurs at specific aspartic residues which are themselves caspase cleavage sites. Caspase-9 is an initiator caspase and upon its activation cleaves caspase-3, one of three effector caspases (caspase-3, caspase-6, and caspase-7). In most cell types, including neurons, caspase-3 is the predominant caspase effector and its activity is responsible for producing many of the morphological features that define apoptotic cell death (Zheng *et al.*, 1998; D'Mello *et al.*, 2000). Targeted gene disruptions of *apaf-1*, *bcl-2*, and *caspase* family members have revealed an important role for apoptotic cell death regulators in neuronal programmed cell death (see below).

Despite the many parallels between programmed cell death in *C. elegans* and mammals, recent studies have revealed increased complexity in mammalian cell death regulation (Joza *et al.*, 2002). For example, cytochrome c release from mitochondria plays an import role in mammalian apoptosome formation and caspase activation, yet cytochrome c is uninvolved in *C. elegans* cell apoptosis. An intriguing family of endogenously expressed mammalian caspase inhibitors has emerged as possible key regulators of mammalian programmed cell death. The inhibitors of apoptosis protein (IAP) family consists of multiple molecules including XIAP, cIAP-1, cIAP-2, and a subfamily of neuronal apoptosis inhibitory proteins (NAIPs), which in mice consist of multiple members (Deveraux and Reed, 1999). IAPs are characterized by the presence of one or more baculovirus inhibitory repeat (BIR) homologous domains and although IAPs may have other functions, they appear to affect apoptosis by potently inhibiting caspase enzymatic activity (Deveraux and Reed, 1999). IAP family members exhibit selective caspase inhibitory activity, and endogenous inhibition of activated caspases 2, 3, 7, and 9 has been reported (Chai *et al.*, 2001; Huang *et al.*, 2001; Riedl *et al.*, 2001). NAIP was originally reported to lack caspase inhibitory activity; however, more recent studies have shown that NAIP is a potent group II caspase (caspases 2, 3, and 7) inhibitor (Robertson *et al.*, 2000). Several IAPs have been reported to be expressed in the nervous system and a variety of studies suggest that IAPs may regulate neuronal apoptosis (Robertson *et al.*, 2000). Overexpression of XIAP, cIAP-1, or cIAP2 can prevent or delay cell death in both *in vitro* and *in vivo* neuronal apoptosis paradigms (Götz *et al.*, 2000; Kügler *et al.*, 2000; Mercer *et al.*, 2000; Perrelet *et al.*, 2000). XIAP-deficient mice have been generated but showed no obvious nervous system abnormalities (Harlin *et al.*, 2001). The potential role of NAIPs in regulating neuronal programmed cell death is particularly intriguing since partial deletions in the human *NAIP* gene have been found in patients with spinal muscular atrophy (Roy *et al.*, 1995) and targeted gene disruption of *naip-1* in mice resulted in increased susceptibility to kainic-acid-induced neuronal apoptosis (Holcik *et al.*, 2000). Recently, two molecules have been identified that can bind IAPs and block their caspase inhibitory effects. The first of these molecules, Smac/Diablo is released from mitochondria following an apoptotic stimulus and can promote apoptosis by displacing IAPs from caspase-9 (Verhagen *et al.*, 2000; Zheng *et al.*, 2000; Srinivasula *et al.*, 2001). The second molecule, XAF1, has been reported to bind to XIAP and antagonize its caspase inhibitory activity (Liston *et al.*, 2001). This multilevel regulation of caspase activity underscores the importance of caspases in apoptosis and suggests a possible role for IAPs in the developing nervous system.

Recent studies also suggest that apoptotic cell death may occur, in at least some cell types, independently of caspase activation (Nicotera, 2000; Cheng *et al.*, 2001). Apoptosis-inducing factor (AIF) is a mitochondrial localized flavoprotein that undergoes nuclear translocation in response to certain death stimuli (Susin *et al.*, 1999). AIF can produce cell death in the absence of caspase activation and it may mediate the death-promoting effects of poly(ADP-ribose)polymerase-1 in several models of neuronal cell death (Yu *et al.*, 2002). The significance of AIF in neuronal programmed cell death regulation remains to be determined.

Autophagic Cell Death

In contrast to apoptotic cell death, autophagic cell death has received relatively scant attention. However, recent studies are beginning to provide insights into the molecules involved in autophagic cell death (Bursch, 2001; Tolkovsky *et al.*, 2002). Several genes, including *beclin 1* and oncogenic *ras*, have been demonstrated to play a role in caspase-independent autophagic death and APG5, a molecule involved in the targeting of proteins for autophagic destruction, is upregulated in degenerating cells (Chi *et al.*, 1999; Liang *et al.*, 1999; Saeki *et al.*, 2000). Similarly, mRNA and protein for the lysosomal protease cathepsin D are upregulated in apoptotic cells indicating possible crosstalk between lysosomal-dependent autophagic death and caspase-dependent apoptotic death (Deiss *et al.*, 1996; Wu *et al.*, 1998). This concept is further supported by the finding that Bax deficiency significantly inhibited cell death in an *in vitro* model of neuronal autophagic cell death and the observation that lysosomal extracts were capable of cleaving the pro-apoptotic Bcl-2 family member Bid; providing a possible pathway for lysosomal mediated caspase activation (Stoka *et al.*, 2001; Zaidi *et al.*, 2001). Interestingly, trophic factor withdrawal induced neuronal death, which is typically considered a trigger of apoptotic death, may produce extensive autophagic vacuole formation and be attenuated by 3-methyladenine, an inhibitor of autophagic vacuole formation (Shibata *et al.*, 1998; Xue *et al.*, 1999; Uchiyama, 2001). Together, these observations indicate extensive crosstalk in the molecular pathways regulating these two major morphological types of programmed cell death (Fig. 3).

TARGETED GENE DISRUPTIONS

The generation of mice deficient in one or more cell death-associated molecule(s) has proven a powerful tool for investigating mammalian neuronal programmed cell death (Snider, 1994; Zheng, 2000). These "knockout" mice provide both an unambiguous assessment of the role of specific genes in neuronal development and an *in vivo* test of the epigenetic relationship between apoptosis-associated molecules (Kuan *et al.*, 2000; Roth *et al.*, 2000). Two caveats to the interpretation of the results obtained in such transgenic mouse studies deserve mention. First, a negative result, that is, the targeted gene disruption fails to affect neuronal programmed cell death, does not exclude a role for the disrupted gene in programmed cell death regulation. Compensatory changes in other genes or alternative cell death pathways may minimize the effects of single gene disruptions (Zheng, 2000). A positive result, however, implies that the disrupted gene has a noncompensatable, nonredundant function in programmed cell death regulation. Second, the results obtained with targeted gene disruptions may be incompletely penetrant and/or dramatically affected by mouse strain-specific genetic factors. For example, we have observed markedly different neurodevelopmental abnormalities in C57BL/6J and 129X1/SvJ caspase-3-deficient mice, and mouse strain-specific effects of other targeted gene disruptions have been reported (Lomaga *et al.*, 2000; Leonard *et al.*, 2002). Similarly, species-specific effects of gene disruption cannot be easily excluded and may limit extrapolation of results from mouse studies to human nervous system development. Despite these caveats, significant

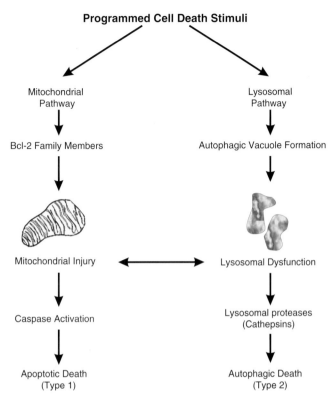

FIGURE 3. Programmed cell death stimuli trigger degeneration by activation of either mitochondrial or lysosomal-dependent pathways. The molecules involved in the apoptotic death pathway are fairly well defined and include Bcl-2 and caspase family members. Less is known about the autophagic death pathway, but lysosomal proteases, including cathepsins, are likely to play an important role in cellular destruction.

insights into the molecular regulation of neuronal programmed cell death have been obtained from transgenic mouse studies.

Bcl-2

Bcl-2 is the prototypical anti-apoptotic Bcl-2 family member (Korsmeyer, 1999). Bcl-2 is a 26 kDa protein that is localized to the outer mitochondrial membrane, nuclear envelope, and portions of the endoplasmic reticulum (Krajewski *et al.*, 1993). Bcl-2 immunoreactivity is present in the developing nervous system in relatively high amounts in neural precursor cells in the ventricular and subventricular zones and in neurons in the developing cortical plate (Merry *et al.*, 1994). Bcl-2 immunoreactivity decreases significantly in the postnatal central nervous system. High levels of Bcl-2 expression are retained, however, in sensory and sympathetic ganglion neurons in the adult (Merry and Korsmeyer, 1997).

The function of *bcl-2* in the nervous system has been explored in a variety of experimental paradigms. Overexpression of *bcl-2* in cultured sympathetic neurons prevents apoptosis in NGF-deprived cells (Garcia *et al.*, 1992; Allsopp *et al.*, 1993). *In vivo* neuronal overexpression of *bcl-2* in transgenic mice indicates a potential role for Bcl-2 in programmed cell death in the nervous system. Compared to nontransgenic littermates, transgenic mice have a 12% increase in brain size, and some brain regions show up to a 50% increase in neuron number (Martinou *et al.*, 1994). The targeted disruption of *bcl-2* indicates a particularly important role for Bcl-2 in the maintenance, during postnatal life, of motoneurons, sympathetic neurons, and sensory neurons, populations of neurons normally expressing high levels of Bcl-2 (Veis *et al.*, 1993; Nakayama *et al.*, 1994; Michaelidis *et al.*, 1996). Surprisingly, the pro-apoptotic effect of Bcl-2 deficiency occurs largely after the peak period of naturally occurring cell death in these neuronal populations. This finding does not imply that *bcl-2* is functionless in the embryonic nervous system since Bcl-2 may be involved in other developmental processes besides apoptosis (Chen *et al.*, 1997); rather, it may suggest functional redundancy for *bcl-2* and other anti-apoptotic *bcl-2* gene family members in regulating neuronal programmed cell death.

Bcl-X$_L$

Bcl-X$_L$ is an anti-apoptotic member of the Bcl-2 family (Boise *et al.*, 1993). The *bcl-x* gene (*Bcl2l1*) can be alternatively spliced to produce two major protein isoforms Bcl-X$_L$ and Bcl-X$_S$, a pro-apoptotic molecule (González-García *et al.*, 1994). Bcl-X$_L$ is the predominant isoform in mammals and Bcl-X$_S$ is present at only low levels, if at all, in the nervous system (González-García *et al.*, 1995). Bcl-X$_L$ immunoreactivity is present at relatively high levels in neurons in the adult nervous system. In the embryonic brain, Bcl-X$_L$ immunoreactivity is not detected in neural precursor cells in the ventricular zone but is abundant in immature neurons in the intermediate and marginal zones and in more mature neurons in the developing cortical plate (Motoyama *et al.*, 1995; Roth *et al.*, 2000). The targeted gene disruption of *bcl-x* failed to affect apoptosis of neural precursor cells but produced a dramatic increase in apoptotic neurons throughout the developing nervous system (Motoyama *et al.*, 1995). Degenerating cells with apoptotic nuclear features and TUNEL positivity were numerous in the embryonic day (E)11–12 mouse spinal cord, brainstem, and dorsal root ganglia (Fig. 4). These degenerating neurons were abundant in the intermediate zone of the developing brain suggesting that Bcl-X$_L$ plays a particularly important role in promoting the survival of newly post-mitotic immature neurons. Since Bcl-X$_L$ deficiency is lethal at approximately E13.5, secondary to hematopoietic

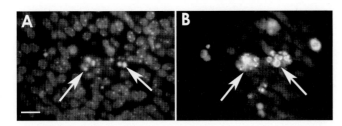

FIGURE 4. Apoptotic nuclei (examples indicated by arrows) are frequent in the Bcl-X$_L$-deficient embryonic spinal cord and can be detected by condensed nuclear labeling with (A) Hoechst 33,258 and/or (B) TUNEL staining. Scale bar equals 10 μm.

apoptosis, the effects of Bcl-X$_L$ deficiency on telencephalic neurons, which are only sparsely present prior to E13, was examined in primary E12 telencephalic cells that were cultured for several days in medium promoting neuronal differentiation. Bcl-X$_L$-deficient telencephalic neurons were found to be markedly susceptible to a variety of *in vitro* apoptotic insults including trophic factor deprivation and genotoxic injury indicating that Bcl-X$_L$ is a key regulator of neuron survival throughout the nervous system (Roth *et al.*, 1996; Shindler *et al.*, 1998; Zaidi *et al.*, 2001).

The findings cited above indicate that Bcl-X$_L$ is the most important Bcl-2 anti-apoptotic protein during the period of neuronal programmed cell death since its deficiency cannot be compensated for by other endogenously expressed family members. Bcl-2 does, however, complement Bcl-X$_L$'s action in immature neurons as mice lacking both Bcl-X$_L$ and Bcl-2 show even more substantial neuron apoptosis than mice lacking Bcl-X$_L$ alone (Shindler *et al.*, 1998). It is also possible that other anti-apoptotic Bcl-2 family members such as Bcl-W (Bcl2l2) and Bfl1/A1 (Bcl2a1) will be found to play a role in neuronal programmed cell death; however, such data are currently lacking.

Bax

The pro-apoptotic Bcl-2 subfamily consists of approximately 15 molecules divided into two subgroups (Korsmeyer, 1999). The EGL-1-like subgroup consists of molecules with a single Bcl-2 homology domain, the so called "BH3 domain-only" group, and the multidomain subgroup whose members contain two or more Bcl-2 homology domains. Targeted gene disruption studies have revealed a particularly important role for the pro-apoptotic multidomain family member Bax in neuronal programmed cell death (Knudson *et al.*, 1995; Deckwerth *et al.*, 1996). Bax is capable of heterodimerizing with Bcl-X$_L$ or Bcl-2 and it is expressed at relatively high levels in the embryonic and adult nervous system (Oltvai *et al.*, 1993; Krajewski *et al.*, 1994; Sedlak *et al.*, 1995). Bax-deficient mice exhibited markedly decreased neuronal programmed cell death and increased neuron numbers in neurotrophic factor dependent neuronal subpopulations (Deckwerth *et al.*, 1996). Bax-deficient neurons were resistant to trophic factor withdrawal induced apoptosis both *in vivo* and *in vitro*. Regionally dependent decreases in TUNEL positive cells were also observed in Bax-deficient embryos as early as gestational day 11.5, suggesting that Bax may also regulate immature neuron apoptosis (White *et al.*, 1998). No significant expansion in the embryonic neural precursor cell population or gross brain abnormalities were observed in Bax-deficient mice suggesting that Bax may not play a significant role, by itself, in regulating morphogenetic cell death in the developing brain.

A critical interaction between Bax and Bcl-X$_L$ in regulating immature neuron death was demonstrated in *bax$^{-/-}$/bcl-x$^{-/-}$* embryos (Shindler *et al.*, 1997). Bax/Bcl-X$_L$ dual deficient embryos were completely protected from the increased neuronal apoptosis observed in the Bcl-X$_L$-deficient embryonic nervous system (Fig. 5). This finding indicates that the mechanism of Bcl-X$_L$'s anti-apoptotic action in immature neurons is through its

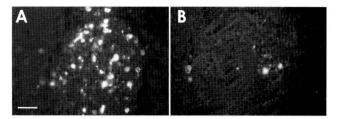

FIGURE 5. (A) Numerous apoptotic neurons with activated caspase-3 immunoreactivity are detected in the dorsal root ganglia of a Bcl-X$_L$-deficient embryo. (B) In contrast, an embryo lacking both Bcl-X$_L$ and Bax shows only rare caspase-3 immunoreactive dorsal root ganglion neurons. Scale bar equals 20 μm.

ability to inhibit the pro-apoptotic effects of *bax* expression. In many cell types and apoptotic paradigms, Bax and Bcl-X$_L$ exhibit additional interactions with BH3 domain-only molecules and the multidomain pro-apoptotic molecule Bak. For example, the anti-apoptotic effects of Bcl-2 and Bcl-X$_L$ are modulated by BH3-domain molecules such as Bad, Bim, and Noxa, and Bak interacts with Bax to regulate mitochondrial cytochrome *c* release and function following many different death-promoting stimuli (Bouillet *et al.*, 2001; Wei *et al.*, 2001).

BH3 domain-only proteins may affect neuronal apoptosis as evidenced by the fact that sympathetic neurons derived from Bim-deficient mice showed significantly reduced apoptosis in response to NGF deprivation *in vitro* (Putcha *et al.*, 2001). However, Bim deficiency had little effect on neuronal programmed cell death *in vivo* and targeted disruption of other BH3 domain-only genes, for example, *bid* and *bad*, have not resulted in significant neurodevelopmental abnormalities (Shindler *et al.*, 1998; Leonard *et al.*, 2001). Since several BH3 domain-only genes are expressed in the developing nervous system, the targeted disruption of a single BH3 domain-only gene may be compensated for by other family members resulting in an underestimation of the importance of this Bcl-2 subfamily in programmed cell death regulation.

The complementary effects of multidomain pro-apoptotic Bcl-2 family members is strikingly demonstrated in *bax*$^{-/-}$/*bak*$^{-/-}$ mice. Bax-deficient mice showed decreased neuronal programmed cell death, male infertility, and reduced female fertility, but were otherwise healthy, while Bak-deficient mice showed no obvious pathology in either the nervous system or other organs. In contrast, mice lacking both Bax and Bak typically died during the perinatal period and exhibited multiple developmental defects that were not observed in mice deficient in either Bax or Bak alone (Lindsten *et al.*, 2000). In the nervous system, *bax*$^{-/-}$/*bak*$^{-/-}$ mice showed decreased programmed cell death of neurons, similar to that observed in *bax*$^{-/-}$/*bak*$^{+/+}$ animals. In addition, dual-deficient mice exhibited a significantly increased periventricular neural precursor cell population which was more striking than that observed in either Bax- or Bak-deficient mice. The *bax*$^{-/-}$/*bak*$^{-/-}$ embryos did not, however, exhibit the massive expansion of neural precursor cells or display exencephaly and cranial bone defects that have been

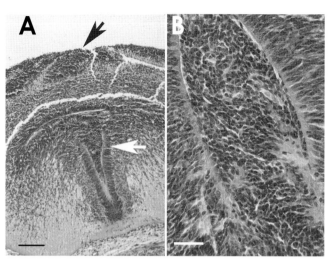

FIGURE 6. (A) An embryonic day 12.5 caspase-3-deficient mouse shows an expanded neural precursor cell population in the developing cerebellum (black arrow) and supernumerary cells in the intraventricular space (white arrow). Higher magnification of (B) the intraventricular infiltrate shows densely packed neural precursor cells within the ependymal lined ventricle. Scale bars in (A) and (B) equal 100 and 25 μm, respectively.

observed in *caspase-9*$^{-/-}$ or *apaf-1*$^{-/-}$ mice (see below). In total, these results suggest that Bax is the predominant pro-apoptotic multidomain Bcl-2 family member in post-mitotic neurons, but that Bax and Bak combine to regulate programmed cell death in at least a subpopulation of neural precursor cells.

Apaf-1, Caspase-9, and Caspase-3

In *C. elegans*, the downstream mediators of CED-9 action are CED-4 and CED-3. In the mammalian nervous system, the immediate downstream mediators of Bcl-2 family effects are Apaf-1, caspase-9, and caspase-3 (Joza *et al.*, 2002). The targeted disruptions of *apaf-1*, *caspase-9*, and *caspase-3* resulted in similar pathological changes consistent with the functional relationship between Apaf-1-dependent apoptosome formation, caspase-9, and caspase-3 activation (Kuida *et al.*, 1996, 1998; Cecconi *et al.*, 1998; Hakem *et al.*, 1998). Apaf-1-, caspase-9-, and caspase-3-deficient mice exhibited extensive perinatal lethality and gross structural neuropathology which included cranial facial abnormalities, exencephaly, neural precursor cell hyperplasia, and ectopic neural masses (Fig. 6). Examination of gene-disrupted embryos revealed a striking decrease in the number of cells undergoing neuronal programmed cell death. Neither *apaf-1*$^{-/-}$- nor *caspase-9*$^{-/-}$-deficient cells exhibited caspase-3 activation *in vivo* or *in vitro* indicating that these three molecules participate in a linear death pathway (Kuan *et al.*, 2000). This conclusion is further supported by data demonstrating that the increased neuronal cell death observed in Bcl-X$_L$-deficient embryos could be completely prevented by concomitant deficiency in any one of these three genes (Kuan *et al.*, 2000; Roth *et al.*, 2000; Zaidi *et al.*, 2001). Thus, in the developing mouse brain, Bcl-X$_L$ regulates immature neuron programmed cell death through its ability to inhibit Bax-dependent apoptosome formation and caspase-3 activation. This pathway, however, cannot account for all programmed cell death in the developing nervous system for several reasons. First, the striking expansion of neural precursor cells observed in *apaf-1*$^{-/-}$, *caspase-9*$^{-/-}$, and *caspase-3*$^{-/-}$ mice is not observed in *bax*$^{-/-}$ mice or *bax*$^{-/-}$/*bak*$^{-/-}$ mice. Thus, the upstream molecular mediators of caspase-3 activation in neural precursor cells are different from those in post-mitotic neurons. Second, the neurodevelopmental effects of Apaf-1, caspase-9, and caspase-3 deficiency are incompletely penetrant and are influenced by strain-specific genetic factors. For example, caspase-3-deficient 129X1/SvJ mice uniformly die during the perinatal period and have severe neurodevelopmental pathology, while caspase-3-deficient C57BL/6J mice survive into adulthood and have minimal neuropathological complications (Leonard *et al.*, 2002). Intercrosses of these two strains suggest the presence of strain-dependent genetic modifiers that influence the significance of caspase-3 activation in nervous system development cell death. Identification of these genes should provide new insights into the molecular regulation of nervous system programmed cell death.

DEVELOPMENTAL PHASES OF PROGRAMMED CELL DEATH

Programmed cell death in the mammalian nervous system involves an orderly progression of developmental phases (Fig. 7). As cells transition from one phase to the next, the stimuli triggering cell death change and the critical intracellular molecular regulators of death undergo revision.

Neural Precursor Cell

One of the most exciting and controversial areas of modern developmental neurobiology research is determination of the extent and significance of programmed cell death in neural precursor cells (Voyvodic, 1996; Sommer and Rao, 2002). Several techniques are available to identify degenerating and/or apoptotic cells in the developing nervous system and these have been applied by various investigators to determine the frequency of programmed cell death in mammalian neural precursor cells. Unfortunately, the resultant estimates have varied widely. Investigators using a technique called *in situ* end labeling plus (ISEL+) to detect DNA double strand breaks in apoptotic cells have estimated that between 50% and 70% of the proliferating cells in the embryonic mouse brain are destined to die (Blaschke *et al.*, 1996, 1998). In contrast, other investigators using the TUNEL method to detect apoptotic cells reported a death rate of 0.3–1.7% in the proliferative zone of the E14 rat forebrain (Thomaidou *et al.*, 1997). Low rates of neural precursor cell death were also identified by detection of activated caspase-3 immunoreactivity, *in vivo* infusion of annexin V, and ultrastructural examination of the developing mammalian brain (Ferrer *et al.*, 1992; Srinivasan *et al.*, 1998; van den Eijnde *et al.*, 1999; Rakic and Zecevic, 2000). These disparate results have led to interesting discussions on cell death detection methodologies and the possible role of DNA strand breaks in nervous system development (Chun and Schatz, 1999; Chun, 2000; de la Rosa and de Pablo, 2000; Gilmore *et al.*, 2000).

Regardless of the extent of cell death in the neural precursor cell population, the dramatic expansion of this population in mice lacking specific apoptotic effector molecules (e.g., *apaf-1*$^{-/-}$ or *caspase-9*$^{-/-}$ mice) demonstrates that a reduction in neural precursor cell programmed cell death significantly affects

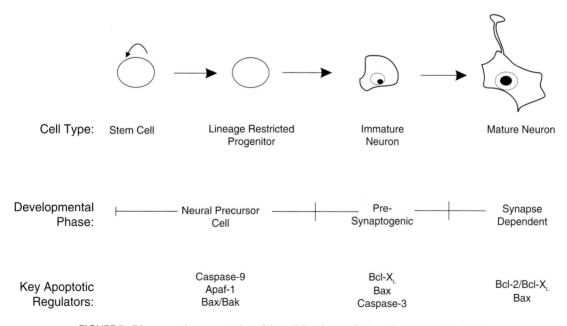

FIGURE 7. Diagrammatic representation of the cellular phases of neuronal programmed cell death.

nervous system development. In formulating hypotheses about neural precursor cell death and its role in mammalian nervous system development, it is important to recognize that neural precursor cells are heterogeneous, consisting of self-replicating multipotent stem cells and several types of lineage-restricted progenitors (Sommer and Rao, 2002). In addition, the neural precursor cell population size is affected not only by cell death but by cell proliferation. A recent study demonstrated that transgenic overexpression of β-catenin in neural precursor cells produced a marked expansion of this cell population (Chenn and Walsh, 2002). This effect was mediated by β-catenins ability to promote cell cycle reentry in dividing neural precursor cells and occurred despite a two-fold increase in apoptosis in the β-catenin transgenic brains. In contrast with the neural precursor cell expansion observed in $apaf-1^{-/-}$, $caspase-9^{-/-}$, and $caspase-3^{-/-}$ embryos, which results in thickening of the developing forebrain and expansion of the neural precursor cell population into the lateral ventricles, β-catenin transgenic mice exhibited no increase in cortical thickness or ventricular ingrowth, but a striking "horizontal expansion" of the cerebral hemispheres resulting in numerous cortical convolutions. These results with genetically modified mice indicate at least two mechanisms by which final cortical size can be regulated by the neural precursor cell population. First, increased numbers of neural stem cells ("founder" cells), as seen in β-catenin transgenic mice, can produce dramatic horizontal expansion of the cerebral cortical surface; and second, decreased death of neural stem cells and/or "daughter cells" (lineage-restricted progenitors), as seen in $apaf-1^{-/-}$, $caspase-9^{-/-}$, and $caspase-3^{-/-}$ mice, can produce a similarly dramatic increase in vertical expansion and cortical thickness. The balance between these two processes ultimately determines the size and shape of the developing brain.

A great deal remains to be learned about the stimuli triggering programmed cell death in neural precursor cells. Likely, a combination of cell intrinsic signals, for example, DNA damage or abnormal cell cycle progression, and extrinsic signals; for example, limited trophic factor support, combine to regulate neural precursor cell programmed cell death (Voyvodic, 1996; de la Rosa and de Pablo, 2000). Recent studies have revealed a significant degree of DNA aneuploidy in embryonic cortical neuroblasts which might trigger an apoptotic response in neural precursor cells similar to that observed when these cells are exposed to DNA damaging agents *in vivo* or *in vitro* (D'Sa-Eipper *et al.*, 2001; Rehen *et al.*, 2001). Defective cytokinesis of neuronal precursor cells may also be a stimulus for programmed cell death since mutations in the gene-encoding Citron-Kinase, which is normally expressed at high levels in the ventricular zone and is involved in neural precursor cell division, resulted in extensive apoptosis in the developing nervous system (Di Cunto *et al.*, 2000). It is likely that additional neural precursor cell death signals will be identified as developmental neurobiologists focus their attention on this intriguing cell population.

Pre-Synaptogenic

One of the most revealing outcomes of recent investigations of neuronal programmed cell death is the remarkable sensitivity of newly post-mitotic neurons to apoptosis. In addition to Bcl-X_L deficiency, targeted gene disruptions in *rb*, *DNA ligase IV*, and *XRCC4* all produced extensive immature neuron death (Jacks *et al.*, 1992; Lee *et al.*, 1992; Frank *et al.*, 1998; Gao *et al.*, 1998; Slack *et al.*, 1998). The protein product of the retinoblastoma tumor suppressor gene is an important regulator of the cell cycle and XRCC4 and DNA ligase IV are involved in DNA double strand break repair. Rb-deficient mice show numerous ectopic mitotic figures and apoptotic neurons in the developing central and peripheral nervous systems. Similarly, XRCC4- and DNA ligase IV-deficient embryos show abundant death of immature neurons suggesting that both cell cycle dysregulation and inadequate repair of DNA double strand breaks are apoptotic stimuli for immature neurons. The response of immature neurons to DNA damage and cell cycle dysregulation is regulated by p53 since p53 deficiency completely blocked the increased neuronal apoptosis observed in XRCC4- and DNA ligase IV-deficient mice and inhibited the death of Rb-deficient neurons in the central nervous system (Morgenbesser *et al.*, 1994; Macleod *et al.*, 1996; Frank *et al.*, 2000; Gao *et al.*, 2000). In contrast, the increased death of immature neurons in Bcl-X_L-deficient embryos was unaffected by concomitant p53 deficiency (Klocke *et al.*, 2002). These findings suggest that multiple death pathways exist in immature neurons and that different apoptotic stimuli may selectively engage different components of the death machinery. The relative contribution of different programmed cell death stimuli, including DNA damage, cell cycle dysregulation, and inadequate trophic factor support, to immature neuron death and normal nervous system development remains to be determined.

Synapse Dependent

Neurons that have survived earlier phases of programmed cell death extend neurites, form synapses, and enter a phase of development that is characterized by extensive death. Neuron death at this stage is typically triggered by inadequate neurotrophic support but other death stimuli, such as a lack of electrical activity and/or death receptor signaling, may also kill synapse-bearing neurons (Cowan *et al.*, 1984; Burek and Oppenheim, 1996; Raoul *et al.*, 2000). The importance of target-derived neurotrophic factors during this phase is well documented and has been the focus of intense investigation (Bibel and Barde, 2000); the significance of the latter two stimuli during normal development requires additional investigation.

During the maturation of neurons, a change occurs in the molecular regulation of programmed cell death. Unlike neural precursor cells and immature neurons which typically require Apaf-1-dependent caspase-9 and caspase-3 activation for programmed cell death, neuronal death due to neurotrophin deprivation, as well as programmed cell death of motor neurons *in vivo*, becomes Apaf-1- and caspase-independent (Honarpour *et al.*, 2001; Oppenheim *et al.*, 2001). These alterations emphasize the fact that programmed cell death pathways are remarkably cell-specific and differentiation-dependent. Mature neurons remain dependent on Bcl-2 family members for regulating viability and it is clear that Bax- and Bak-mediated mitochondrial dysfunction is a potent death stimulus independent of downstream caspase activation (Wei *et al.*, 2001).

GLIAL CELL DEATH

Like neurons, glial cells also undergo programmed cell death in the developing mammalian nervous system (Raff *et al.*, 1993). Many parallels exist between neuronal cell death and that of oligodendrocytes and astrocytes. Glial cell death is similarly extensive; for example, over half of the oligodendrocytes present in the rat optic nerve undergo programmed cell death as do numerous astrocytes in the developing cerebellum (Barres *et al.*, 1992; Krueger *et al.*, 1995). Glial cell death may also be regulated by trophic molecules, is typically apoptotic in appearance, may involve competition for limited quantities of trophic molecules, and is regulated by Bcl-2 and caspase family members (Barres *et al.*, 1993; Burne *et al.*, 1996; Jacobson *et al.*, 1997). Thus, the maturation-dependent phases of programmed cell death and the molecular regulation of death described earlier in this chapter can be equally applied to cells of glial lineage as to those of neuronal lineage.

SUMMARY

Programmed cell death research has had an arguably unparalleled impact on the field of developmental neurobiology. For over one hundred years, investigations of neuronal cell death have challenged our assumptions about nervous system development and its cellular and molecular regulation. Current studies on programmed cell death, including investigations of neural stem cell death, hold tremendous promise for elucidating the complex processes involved in normal mammalian nervous system development.

REFERENCES

Allsopp, T.E., Wyatt, S., Paterson, H.F., and Davies, A.M., 1993, The proto-oncogene *bcl-2* can selectively rescue neurotrophic factor-dependent neurons from apoptosis, *Cell* 73:295–307.

Barres, B.A., Hart, I.K., Coles, H.S.R., Burne, J.F., Voyvodic, J.T., Richardson, W.D. *et al.*, 1992, Cell death and control of cell survival in the oligodendrocyte lineage, *Cell* 70:31–46.

Barres, B.A., Jacobson, M.D., Schmid, R., Sendtner, M., and Raff, M.C., 1993, Does oligodendrocyte survival depend on axons? *Curr. Biol.* 3:489–497.

Beard, J., 1896, The history of a transient nervous apparatus in certain Ichthyopsida. An account of the development and degeneration of ganglion-cells and nerve fibres, *Zool. Jahrbüchér Abt. Morphol.* 9:1–106.

Bibel, M. and Barde, Y.-A., 2000, Neurotrophins: Key regulators of cell fate and cell shape in the vertebrate nervous system, *Genes Dev.* 14:2919–2937.

Blaschke, A.J., Staley, K., and Chun, J., 1996, Widespread programmed cell death in proliferative and postmitotic regions of the fetal cerebral cortex, *Development* 122:1165–1174.

Blaschke, A.J., Weiner, J.A., and Chun, J., 1998, Programmed cell death is a universal feature of embryonic and postnatal neuroproliferative regions throughout the central nervous system, *J. Comp. Neurol.* 396:39–50.

Boise, L.H., González-García, M., Postema, C.E., Ding, L., Linsten, T., Turka, L.A. *et al.*, 1993, *bcl-x*, a *bcl-2*-related gene that functions as a dominant regulator of apoptotic cell death, *Cell* 74:597–608.

Bouillet, P., Cory, S., Zhang, L.-C., Strasser, A., and Adams, J.M., 2001, Degenerative disorders caused by Bcl-2 deficiency prevented by loss of its BH3-only antagonist Bim, *Dev. Cell* 1:645–653.

Burek, M.J. and Oppenheim, R.W., 1996, Programmed cell death in the developing nervous system, *Brain Pathol.* 6:427–446.

Burne, J.F., Staple, J.K., and Raff, M.C., 1996, Glial cells are increased proportionally in transgenic optic nerves with increased numbers of axons, *J. Neurosci.* 16:2064–2073.

Bursch, W., 2001, The autophagosomal-lysosomal compartment in programmed cell death, *Cell Death Differ.* 8:569–581.

Cecconi, F., Alvarez-Bolado, G., Meyer, B.I., Roth, K.A., and Gruss, P., 1998, Apaf1 (CED-4 Homolog) regulates programmed cell death in mammalian development, *Cell* 94:727–737.

Chai, J., Shiozaki, E., Srinivasula, S.M., Wu, Q., Dataa, P., Alnemri, E.S. *et al.*, 2001, Structural basis of Caspase-7 inhibition by XIAP, *Cell* 104:769–780.

Chen, D.F., Schneider, G.E., Martinou, J.-C., and Tonegawa, S., 1997, Bcl-2 promotes regeneration of severed axons in mammalian CNS, *Nature* 385:434–439.

Cheng, E.H. Y., Wei, M.C., Weiler, S., Flavell, R.A., Mak, T.W., Lindsten, T. *et al.*, 2001, BCL-2, BCL-X$_L$ sequester BH3 domain-only molecules preventing BAX- and BAK-mediated mitochondrial apoptosis, *Mol. Cell* 8:705–711.

Chenn, A. and Walsh, C.A., 2002, Regulation of cerebral cortical size by control of cell cycle exit in neural precursors, *Science* 297:365–369.

Chi, S., Kitanake, C., Noguchi, K., Mochizuki, T., Nagashima, Y., Shirouzu, M. *et al.*, 1999, Oncogenic ras triggers cell suicide through the activation of a caspase-independent cell death program in human cancer cells, *Oncogene* 18:2281–2290.

Chun, J., 2000, Cell death, DNA breaks and possible rearrangements: An alternative view, *Trends Neurosci.* 23:407–408.

Chun, J. and Schatz, D.G., 1999, Rearranging views on neurogenesis: Neuronal death in the absence of DNA end-joining proteins, *Neuron* 22:7–10.

Clarke, P.G.H., 1990, Developmental cell death: Morphological diversity and multiple mechanisms, *Anat. Embryol.* 181:195–213.

Cowan, W.M., Fawcett, J.W., O'Leary, D.D.M., and Stanfield, B.B., 1984, Regressive events in neurogenesis, *Science* 225:1258–1265.

D'Mello, S.R., Kuan, C.-Y., Flavell, R.A., and Rakic, P., 2000, Caspase-3 is required for apoptosis-associated DNA fragmentation but not for cell death in neurons deprived of potassium, *J. Neurosci. Res.* 59:24–31.

D'Sa-Eipper, C., Leonard, J.R., Putcha, G., Zheng, T.S., Flavell, R.A., Rakic, P. *et al.*, 2001, DNA damage-induced neural precursor cell apoptosis requires p53 and caspase-9 but neither Bax nor caspase-3, *Development* 128:137–146.

de la Rosa, E.J. and de Pablo, F., 2000, Cell death in early neural development: Beyond the neurotrophic theory, *Trends Neurosci.* 23:454–458.

Deckwerth, T.L., Elliott, J.L., Knudson, C.M., Johnson, Jr., E.M., Snider, W.D., and Korsmeyer, S.J., 1996, Bax is required for neuronal death after trophic factor deprivation and during development, *Neuron* 17:401–411.

Deiss, L.P., Galinka, H., Berissi, H., Cohen, O., and Kimchi, A., 1996, Cathepsin D protease mediates programmed cell death induced by interferon-γ, Fas/APO-1 and TNF-α, *EMBO J.* 15:3861–3870.

Deveraux, Q.L. and Reed, J.C., 1999, IAP family proteins—suppressors of apoptosis, *Genes Dev.* 13:239–252.

Di Cunto, F., Imarisio, S., Hirsch, E., Broccoli, V., Bulfone, A., Migheli, A. *et al.*, 2000, Defective neurogenesis in citron kinase knockout mice by altered cytokinesis and massive apoptosis, *Neuron* 28:115–127.

Dunn, Jr., W.A., 1994, Autophagy and related mechanisms of lysosome-mediated protein degradation, *Trends Cell Biol.* 4:139–143.

Ernst, M., 1926, Über Untergang von Zellen während der normalen Entwicklung bei Wirbeltieren, *Z. Anat. Entwicklungsgesch* 79: 228–262.

Ferrer, I., Soriano, E., Del Rio, A., Alcántara, S., and Auladell, C., 1992, Cell death and removal in the cerebral cortex during development, *Progress in Neurobiol.* 39:1–43.

Frank, K.M., Sekiguchi, J.M., Seidl, K.J., Swat, W., Rathbun, G.A., Cheng H.-L. *et al.*, 1998, Late embryonic lethality and impaired V(D)J recombination in mice lacking DNA ligase IV, *Nature* 396:173–177.

Frank, K.M., Sharpless, N.E., Gao, Y., Sekiguchi, J.M., Ferguson, D.O., Zhu, C. *et al.*, 2000, DNA ligase IV deficiency in mice leads to defective neurogenesis and embryonic lethality via the p53 pathway, *Mol. Cell* 5:993–1002.

Gao, Y., Ferguson, D.O., Xie, W., Manis, J.P., Sekiguchi, J., Frank, K.M. *et al.*, 2000, Interplay of p53 and DNA-repair protein XRCC4 in tumorigenesis, genomic stability and development, *Nature* 404:897–900.

Gao, Y., Sun, Y., Frank, K.M., Dikkes, P., Fujiwara, Y., Seidl, K.J. *et al.*, 1998, A critical role for DNA end-joining proteins in both lymphogenesis and neurogenesis, *Cell* 95:891–902.

Garcia, I., Marinou, I., Tsujimoto, Y., and Martinou, J.-C., 1992, Prevention of programmed cell death of sympathetic neurons by the *bcl-2* proto-oncogene, *Science* 258:302–304.

Gilmore, E.C., Nowakowski, R.S., Caviness, Jr., V.S., and Herrup, K., 2000, Cell birth, cell death, cell diversity and DNA breaks: How do they all fit together? *Trends Neurosci.* 23:100–105.

Glücksmann, A., 1951, Cell deaths in normal vertebrate ontogeny, *Bio. Rev.* 26:59–86.

González-García, M., Garcia, I., Ding, L., O'Shea, S., Boise, L.H., Thompson, C.B. *et al.*, 1995, *bcl-x* is expressed in embryonic and postnatal neural tissues and functions to prevent neuronal cell death, *Proc. Natl. Acad. Sci. USA* 92:4304–4308.

González-García, M., Thompson, C.B., Ding, L., Duan, L., Boise, L.H., and Nuñez, G., 1994, *bcl-x*$_L$ is the major *bcl-x* mRNA form expressed during murine development and its product localizes to mitochondria, *Development* 120:3033–3042.

Götz, R., Karch, C., Digby, M.R., Troppmair, J., Rapp, U.R., and Sendtner, M., 2000, The neuronal apoptosis inhibitory protein suppresses neuronal differentiation and apoptosis in PC12 cells, *Hum. Mol. Genet.* 9:2479–2489.

Häcker, G., 2000, The morphology of apoptosis, *Cell Tissue Res.* 301: 5–17.

Hakem, R., Hakem, A., Duncan, G.S., Henderson, J.T., Woo, M., Soengas, M.S. *et al.*, 1998, Differential requirement for caspase-9 in apoptotic pathways *in vivo*, *Cell* 94:339–352.

Hamburger, V., 1992, History of the discovery of neuronal death in embryos, *J. Neurobiol.* 23:1116–1123.

Harlin, H., Reffey, S.B., Duckett, C.S., Lindsten, T., and Thompson, C.B., 2001, Characterization of XIAP-deficient mice, *Mol. Cell. Biol.* 21:3604–3608.

Holcik, M., Thompson, C.B., Yaraghi, Z., Lefebvre, C.A., MacKenzie, A.E., and Korneluk, R.G., 2000, The hippocampal neurons of neuronal apoptosis inhibitory protein 1 (NAIP1)-deleted mice display increased vulnerability to kainic acid-induced injury, *Proc. Natl. Acad. Sci. USA* 97:2286–2290.

Honarpour, N., Tabuchi, K., Stark, J.M., Hammer, R.E., Südhof, T.C., Parada, L.F. *et al.*, 2001, Embryonic neuronal death due to neuro-trophin and neurotransmitter deprivation occurs independent of Apaf-1, *Neuroscience* 106:263–274.

Horvitz, H.R., 1999, Genetic control of programmed cell death in the nematode *Caenorhabditis elegans*, *Cancer Res.* 59:1701s–1706s.

Huang, Y., Park, Y.C., Rich, R.L., Segal, D., Myszka, D.G., and Wu, H., 2001, Structural basis of caspase inhibition by XIAP: Differential roles of the Linker versus the BIR domain, *Cell* 104:781–790.

Jacks, T., Fazeli, A., Schmitt, E.M., Bronson, R.T., Goodell, M.A., and Weinberg, R.A., 1992, Effects of an *Rb* mutation in the mouse, *Nature* 359:295–300.

Jacobson, M., 1991, *Developmental Neurobiology*, 3rd edn, Plenum Press, New York.

Jacobson, M.D., Weil, M., and Raff, M.C., 1997, Programmed cell death in animal development, *Cell* 88:347–354.

Joza, N., Kroemer, G., and Penninger, J.M., 2002, Genetic analysis of the mammalian cell death machinery, *Trends Genet.* 18:142–149.

Kerr, J.F.R., Wyllie, A.H., and Currie, A.R., 1972, Apoptosis: A basic biological phenomenon with wide-ranging implications in tissue kinetics, *Br. J. Cancer* 26:239–257.

Klocke, B.J., Latham, C.B., C. D'Sa, and Roth, K.A., 2002, p53 deficiency fails to prevent increased programmed cell death in the Bcl-XL-deficient nervous system, *Cell Death Differ* 9:1063–1068.

Knudson, C.M., Tung, K.S.K., Troutellotte, W.G., Brown, G.A.J., and Korsmeyer, S.J., 1995, Bax-deficient mice with lymphoid hyperplasia and male germ cell death, *Science* 270:96–99.

Korsmeyer, S.J. 1999, *BCL-2* gene family and the regulation of programmed cell death, *Cancer Res.* 59:1693s–1700s.

Krajewski, S., Krajewska, M., Shabaik, A., Miyashita, T., Wang, H.-G., and Reed, J.C., 1994, Immunohistochemical determination of *in vivo* distribution of Bax, a dominant inhibitor of Bcl-2, *Am. J. Path.* 145:1323–1328.

Krajewski, S., Tanaka, S., Takayama, S., Schibler, M.J., Fenton, W., and Reed, J.C., 1993, Investigations of the subcellular distribution of the Bcl-2 oncoprotein residence in the nuclear envelope, endoplasmic reticulum, and other mitochondrial membranes, *Cancer Res.* 53:4701–4714.

Krueger, B.K., Burne, J.F., and Raff, M.C., 1995, Evidence for large-scale astrocyte death in the developing cerebellum, *J. Neurosci.* 15: 3366–3374.

Kuan, C.-Y., Roth, K.A., Flavell, R.A., and Rakic, P., 2000, Mechanism of programmed cell death in the developing brain, *Trends Neurosci.* 23: 287–293.

Kügler, S., Straten, G., Kreppel, F., Isenmann, S., Liston, P., and Bähr, M., 2000, The X-linked inhibitor of apoptosis (XIAP) prevents cell death in axotomized CNS neurons *in vivo*, *Cell Death Differ.* 7:815–824.

Kuida, K., Haydar, T.F., Kuan, C.-Y., Gu, Y., Taya, C., Karasuyama, H. *et al.*, 1998, Reduced apoptosis and cytochrome c-mediated caspase activation in mice lacking caspase-9, *Cell* 94:325–337.

Kuida, K., Zheng, T.S., Na, S., Kuan, C.-Y., Yang, D., Karasuyama, H. *et al.*, 1996, Decreased apoptosis in the brain and premature lethality in CPP32-deficient mice, *Nature* 384:368–372.

Lee, C.-Y. and Baehrecke, E.H., 2001, Steroid regulation of autophagic programmed cell death during development, *Development* 128: 1443–1455.

Lee, E.Y.H.P., Chang, C.-Y., Hu, N., Wang, Y.-C.J., Lai, C.-C., Herrup, K. *et al.*, 1992, Mice deficient for Rb are nonviable and show defects in neurogenesis and haematopoiesis, *Nature* 359:288–294.

Leonard, J.R., D'Sa, C., Cahn, R., Korsmeyer, S., and Roth, K.A., 2001, Bid regulation of neuronal apoptosis, *Dev. Brain Res.* 128:187–190.

Leonard, J.R., Klocke, B.J., D'Sa, C., Flavell, R.A., and Roth, K.A., 2002, Strain-dependent neurodevelopmental abnormalities in caspase-3-deficient mice, *J. Neuropathol. Exp. Neurol.* 61:673–677.

Liang, X.H., Jackson, S., Seaman, M., Brown, K., Kempkes, B., Hibshoosh, H. *et al.*, 1999, Induction of autophagy and inhibition of tumorigenesis by *beclin 1*, *Nature* 402:672–676.

Lindsten, T., Ross, A.J., King, A., Zong, W.-X., Rathmell, J.C., Shiels, H.A. *et al.*, 2000, The combined functions of proapoptotic Bcl-2 family members Bak and Bax are essential for normal development of multiple tissues, *Mol. Cell* 6:1389–1399.

Liston, P., Fong, W.G., Kelly, N.L., Toji, S., Miyazaki, T., Conte, D. *et al.*, 2001, Identification of XAF1 as an antagonist of XIAP anti-caspase activity, *Cell Biol. Nat.* 3:128–133.

Lomaga, M.A., Henderson, J.T., Elia, A.J., Robertson, J., Noyce, R.S., Yeh, W.-C. *et al.*, 2000, Tumor necrosis factor receptor-associated factor 6 (TRAF6) deficiency results in exencephaly and is required for apoptosis within the developing DNS, *J. Neurosci.* 20:7384–7393.

Macleod, K.F., Hu, Y., and Jacks, T., 1996, Loss of *Rb* activates both *p53*-dependent and independent cell death pathways in the developing mouse nervous system, *EMBO J.* 15:6178–6188.

Martinou, J.-C., Dubois-Dauphin, M., Staple, J.K., Rodriguez, I., Frankowski, H., Missotten, M. *et al.*, 1994, Overexpression of Bcl-2 in transgenic mice protects neurons from naturally occurring cell death and experimental ischemia, *Neuron* 13:1017–1030.

Mercer, E.A., Korhonen, L., Skoglösa, Y., Olsson, P.-A., Kukkonen, J.P., and Linkholm, D., 2000, NAIP interacts with hippocalcin and pro-tects neurons against calcium-induced cell death through caspase-3-dependent and -independent pathways, *EMBO J.* 19: 3597–3607.

Merry, D.E. and Korsmeyer, S.J., 1997, Bcl-2 gene family in the nervous system, *Annu. Rev. Neurosci.* 20:245–267.

Merry, D.E., Veis, D.J., Hickey, W.F., and Korsmeyer, S.J., 1994, *bcl-2* protein expression is widespread in the developing nervous system and retained in the adult PNS, *Development* 120:301–311.

Michaelidis, T.M., Sendtner, M., Cooper, J.D., Airaksinen, M.S., Holtmann, B., Meyer, M. *et al.*, 1996, Inactivation of *bcl-2* results in progressive degeneration of motoneurons, sympathetic and sensory neurons during early postnatal development, *Neuron* 17:75–89.

Morgenbesser, S.D., Williams, B.O., Jacks, T., and DePinho, R.A., 1994, *p53*-dependent apoptosis produced by *Rb*-deficiency in the developing mouse lens, *Nature* 371:72–74.

Motoyama, N., Wang, F., Roth, K.A., Sawa, H., Nakayama, K.-I., Nakayama, K. *et al.*, 1995, Massive cell death of immature hematopoietic cells and neurons in Bcl-x-deficient mice, *Science* 267:1506–1510.

Nakayama, K., Nakayama, K.-I., Negishi, I., Kuida, K., Sawa, H., and Loh, D.Y., 1994, Targeted disruption of Bcl-2 αβ in mice: Occurrence of gray hair, polycystic kidney disease, and lymphocytopenia, *Proc. Natl. Acad. Sci. USA* 91:3700–3704.

Nicholson, D.W., 1999, Caspase structure, proteolytic substrates, and function during apoptotic cell death, *Cell Death Differ.* 6:1028–1042.

Nicotera, P., 2000, Caspase requirement for neuronal apoptosis and neuro-degeneration, *IUBMB Life* 49:421–425.

Oltvai, Z.N., Milliman, C.T., and Korsmeyer, S.J., 1993, Bcl-2 hetero-dimerizes *in vivo* with a conserved homolog, Bax, that accelerates programmed cell death, *Cell* 74:609–619.

Oppenheim, R.W., 1991, Cell death during development of the nervous system, *Annu. Rev. Neurosci.* 14:453–501.

Oppenheim, R.W., Flavell, R.A., Vinsant, S., Prevette, D., Kuan, C.-Y., and Rakic, P., 2001, Programmed cell death of developing mammalian neurons after genetic deletion of caspases, *J. Neurosci.* 21: 4752–4760.

Perrelet, D., Ferri, A., MacKenzie, A.E., Smith, G.M., Korneluk, R.G., Liston, P. *et al.*, 2000, IAP family proteins delay motoneuron cell death *in vivo*, *Eur. J. Neurosci.* 12:2059–2067.

Putcha, G.V., Moulder, K.L., Golden, J.P., Bouillet, P., Adams, J.A., Strasser, A. *et al.*, 2001, Induction of BIM, a proapoptotic BH3-only BCL-2 family member, is critical for neuronal apoptosis, *Neuron* 29: 615–628.

Raff, M.C., Barres, B.A., Burne, J.F., Coles, H.S., Ishizaki, Y., and Jacobson, M.D., 1993, Programmed cell death and the control of cell survival: Lessons from the nervous system, *Science* 262:695–700.

Rakic, S. and Zecevic, N., 2000, Programmed cell death in the developing human telencephalon, *Eur. J. Neurosci.* 12:2721–2734.

Raoul, C., Pettmann, B., and Henderson, C.E., 2000, Active killing of neurons during development and following stress: A role for p75[NTR] and Fas? *Curr. Opin. Neurobiol.* 10:111–117.

Rehen, S.K., McConnell, M.J., Kaushal, D., Kingsbury, M.A., Yang, A.H., and Chun, J., 2001, Chromosomal variation in neurons of the devel-oping and adult mammalian nervous system, *Proc. Natl. Acad. Sci. USA* 98:13361–13366.

Riedl, S.J., Renatus, M., Schwarzenbacher, R., Zhou, Q., Sun, C., Fesik, S.W. *et al.*, 2001, Structural basis for the inhibition of caspase-3 by XIAP, *Cell* 104:791–800.

Robertson, G.S., Crocker, S.J., Nicholson, D.W., and Schulz, J.B., 2000, Neuroprotection by the inhibition of apoptosis, *Brain Pathol.* 10: 283–292.

Roth, K.A., 2002, *In situ* detection of apoptotic neurons. In *Neuromethods, Vol. 37: Apoptosis Techniques and Protocols* (A.C. LeBlanc, ed.), Humana Press, Inc., Totowa, NJ, pp. 205–224.

Roth, K.A., Kuan, C.-Y., Haydar, T.F., D'Sa-Eipper, C., Shindler, K.S., Zheng, T.S. *et al.*, 2000, Epistatic and independent apoptotic func-tions of Caspase-3 and Bcl-X$_L$ in the developing nervous system, *Proc. Natl. Acad. Sci. USA* 97:466–471.

Roth, K.A., Motoyama, N., and Loh, D.Y., 1996, Apoptosis of *bcl-x*-deficient telencephalic cells *in vitro*, *J. Neurosci.* 16:1753–1758.

Roy, N., Mahadevan, M.S., McLean, M., Shutler, G., Yaraghi, Z., Farahani, R. *et al.*, 1995, The gene for neuronal apoptosis inhibitory protein is partially deleted in individuals with spinal muscular atrophy, *Cell* 80: 167–178.

Saeki, K., You, A., Okuma, E., Yazaki, Y., Susin, S.A., Kroemer, G. *et al.*, 2000, Bcl-2 down-regulation causes autophagy in a caspase-independent manner in human leukemic HL60 cells, *Cell Death Differ.* 7:1263–1269.

Sedlak, T.W., Oltvai, Z.N., Yang, E., Wang, K., Boise, L.H., Thompson, C.B. *et al.*, 1995, Multiple Bcl-2 family members demonstrate selective dimerizations with Bax, *Proc. Natl. Acad. Sci. USA* 92:7834–7838.

Seglen, P.O. and Bohley, P., 1992, Autophagy and other vacuolar protein degradation mechanisms, *Experientia* 48:158–172.

Shibata, M., Kanamori, S., Isahara, K., Ohsawa, Y., Konishi, A., Kametaka, S. *et al.*, 1998, Participation of cathepsins B and D in apoptosis of PC12 cells following serum deprivation, *Biochem. Biophys. Res. Commun.* 251:199–203.

Shindler, K.S., Latham, C.B., and Roth, K.A., 1997, *bax* deficiency prevents the increased cell death of immature neurons in *bcl-x*-deficient mice, *J. Neurosci.* 17:3112–3119.

Shindler, K.S., Yunker, A.M.R., Cahn, R., Zha, J., Korsmeyer, S.J., and Roth, K.A., 1998, Trophic support promotes survival of *bcl-x*-deficient telencephalic cells *in vitro*, *Cell Death Differ.* 5: 901–910.

Slack, R.S., El-Bizri, H., Wong, J., Belliveau, D.J., and Miller, F.D., 1998, A critical temporal requirement for the retinoblastoma protein family during neuronal determination, *J. Cell Biol.* 140:1497–1509.

Snider, W.D., 1994, Functions of the neurotrophins during nervous system development: What the knockouts are teaching us, *Cell* 77: 627–638.

Sommer, L. and Rao, M., 2002, Neural stem cells and regulation of cell number, *Progress in Neurobiol.* 66:1–18.

Srinivasan, A., Roth, K.A., Sayers, R.O., Shindler, K.S., Wong, A.M., Fritz, L.C., and Tomaselli, K.J., 1998, *In situ* immunodetection of activated caspase-3 in apoptotic neurons in the developing nervous system, *Cell Death Differ.* 5:1004–1016.

Srinivasula, S.M., Hegde, R., Saleh, A., Datta, P., Shiozaki, E., Chai, J. *et al.*, 2001, A conserved XIAP-interaction motif in caspase-9 and Smac/DIABLO regulates caspase activity and apoptosis, *Nature* 410: 112–116.

Stoka, V.V., Turk, B., Schendel, S.L., Kim, T.W., Cirman, T., Snipas, S.J. *et al.*, 2001, Lysosomal protease pathways to apoptosis: Cleavage of bid, not pro-caspases is the most likely route, *J. Biol. Chem.* 276:3149–3157.

Susin, S.A., Lorenzo, H.K., Zamzami, N., Marzo, I., Snow, B.E., Brothers, G.M. *et al.*, 1999, Molecular characterization of mitochondrial apoptosis-inducing factor, *Nature* 397:441–446.

Thomaidou, D., Mione, M.C., Cavanagh, J.F.R., and Parnavelas, J.G., 1997, Apoptosis and its relation to the cell cycle in the developing cerebral cortex, *J. Neurosci.* 17:1075–1085.

Tolkovsky, A.M., Xue, L., Fletcher, G.C., and Borutaite, V., 2002, Mitochondrial disappearance from cells: A clue to the role of autophagy in programmed cell death and disease? *Biochemie* 84: 233–240.

Uchiyama, Y., 2001, Autophagic cell death and its execution by lysosomal cathepsins, *Arch. Histol. Cytol.* 64:233–246.

van den Eijnde, S.M., Lips, J., Boshart, L., Marani, E., Reutelingsperger, C.P.M., and De Zeeuw, C.I., 1999, Spatiotemporal distribution of dying neurons during early mouse development, *Eur. J. Neurosci.* 11:712–724.

Veis, D.J., Sorenson, C.M., Shutter, J.R., and Korsmeyer, S.J., 1993, Bcl-2-deficient mice demonstrate fulminant lymphoid apoptosis, polycystic kidneys, and hypopigmented hair, *Cell* 75:229–240.

Verhagen, A.M., Ekert, P.G., Pakusch, M., Silke, J., Connolly, L.M., Reid, G.E. *et al.*, 2000, Identification of DIABLO, a mammalian protein that promotes apoptosis by binding to and antagonizing IAP proteins, *Cell* 102:43–53.

Voyvodic, J.T., 1996, Cell death in cortical development: How much? Why? So what? *Neuron* 16:693–696.

Wei, M.C., Zong, W.-X., Cheng, E.H.Y., Lindsten, T., Panoutsakopoulou, V., Ross, A.J., Roth, K.A. *et al.*, 2001, Proapoptotic BAX and BAK: A requisite gateway to mitochondrial dysfunction and death, *Science* 292:727–730.

White, F.A., Keller-Peck, C.R., Knudson, C.M., Korsmeyer, S.J., and Snider, W.D., 1998, Widespread elimination of naturally occurring neuronal death in *Bax*-deficient mice, *J. Neurosci.* 18:1428–1439.

Wu, G.S., Saftig, P., Peters, C., and El-Deiry, W.S., 1998, Potential role for cathepsin D in p53-dependent tumor suppression and chemosensitivity, *Oncogene* 16:2177–2183.

Xue, L., Fletcher, G.C., and Tolkovsky, A.M., 1999, Autophagy is activated by apoptotic signalling in sympathetic neurons: An alternative mechanism of death execution, *Mol. Cell. Neurosci.* 14: 180–198.

Yu, S.-W., Wang, H., Poitras, M.F., Coombs, C., Bowers, W.J., Federoff, H.J. *et al.*, 2002, Mediation of poly(ADP-Ribose) polymerase-1-dependent cell death by apoptosis-inducing factor, *Science* 297: 259–263.

Zaidi, A.U., D'Sa-Eipper, C., Brenner, J., Kuida, K., Zheng, T.S., Flavell, R.A. *et al.*, 2001, Bcl-XL-Caspase-9 interactions in the developing nervous system: Evidence for multiple death pathways, *J. Neurosci.* 21: 169–175.

Zaidi, A.U., McDonough, J.S., Klocke, B.J., Latham, C.B., Korsmeyer, S.J., Flavell, R.A. *et al.*, 2001, Chloroquine-induced neuronal cell death is p53 and Bcl-2 family-dependent but caspase-independent, *J. Neuropathol. Exp. Neurol.* 60:937–945.

Zheng, T.S. and Flavell, R.A., 2000, Divinations and surprises: Genetic analysis of caspase function in mice, *Exp. Cell Res.* 256:67–73.

Zheng, T.S., Hunot, S., Kuida, K., Momoi, T., Srinivasan, A., Nicholson, D.W. *et al.*, 2000, Deficiency in caspase-9 or caspase-3 induces compensatory caspase activation, *Nat. Med.* 6:1241–1247.

Zheng, T.S., Schlosser, S.F., Dao, T., Hingorani, R., Crispe, I.N., Boyer, J.L. *et al.*, 1998, Caspase-3 controls both cytoplasmic and nuclear events associated with Fas-mediated apoptosis *in vivo*, *Proc. Natl. Acad. Sci. USA* 95:13618–13623.

Zou, H., Henzel, W.J., Liu, X., Lutschg, A., and Wang, X., 1997, Apaf-1, a human protein homologous to *C. elegans* CED-4, participates in cytochrome c-dependent activation of caspase-3, *Cell* 90:405–413.

Zou, H., Li, Y., Liu, X., and Wang, X., 1999, An APAF-1 cytochrome c multimeric complex is a functional apoptosome that activates procaspase-9, *J. Biol. Chem.* 274:11549–11556.

12

Regeneration and Repair

Maureen L. Condic

The central nervous system (CNS) of adult vertebrate animals is capable of considerable plasticity, both in the course of normal adult function and in response to injury. Patients suffering even severe injuries to the brain frequently achieve substantial, if not complete, recovery of function. Despite the ability of the nervous system to functionally compensate for injury, regenerative *repair* of neural injury is quite limited. Cells of the nervous system are exceptionally sensitive to ischemic insult (Goldberg and Barres, 2000; Allan and Rothwell, 2001) and cell death can be extensive following even mild injury. Neurons lost to injury or disease are rarely replaced in the adult brain (Garcia-Verdugo *et al.*, 2002; Lim *et al.*, 2002; Parent and Lowenstein, 2002; Turlejski and Djavadian, 2002). Even when injured neurons do not die, they are largely unable to regenerate axons and dendrites in order to reestablish functional connections with their normal synaptic partners. Thus, while significant functional recovery is often possible, the adult CNS does not appear to be capable of regenerative repair following injury. What are the possible mechanisms by which nervous system function is restored following injury?

RESTORATION OF FUNCTION VS REGENERATIVE REPAIR

Functional recovery following CNS injury in the adult can occur by at least three distinct mechanisms: restitution, substitution, and compensation (Singer, 1982). In restitutive or restorative recovery, the original system responsible for the function is repaired or its efficacy enhanced as a means of regaining normal behavior. This kind of recovery may not be complete, but the behavior observed following restitution is always qualitatively similar to the original behavior. Restitution may depend on redundancy of parallel or distributed systems that normally mediate a particular function. Alternatively, restitution of function may be based on neuronal sprouting and the generation of new synapses within the damaged system to replace those lost to injury or cell death.

Substitution, in contrast, involves the adoption of function by a related system that imperfectly replaces the failed or damaged system. Substitutive recovery is never complete, and always involves some qualitative change in the behavior. For example, patients with damage to the primary cortical visual centers ("cortical blindness") can recover significant visual behavior (avoiding obstacles, orienting toward light sources, detection of moving objects, etc.) without any conscious perception of sight, presumably by recruitment of visual processing pathways that are not normally involved in conscious experience of visual stimuli (Poppel *et al.*, 1973).

Finally, compensation for CNS injury involves the recovery of function due to adaptation of the undamaged components of the normal system, so as to minimize the effects of a partial loss of function. In compensatory recovery, changes in the gain or attenuation of a system's components can result in improved functional output of the system as a whole, without strictly restoring the aspect of normal function that was lost as a consequence of damage. Gradual recovery of balance following unilateral damage to the vestibular system is an example of a compensation. Recovery of balance occurs through changes in the normal reciprocal inhibition between the two vestibular nuclei (Dieringer and Precht, 1979) that balance the firing rates of these paired groups of neurons. Neither the damaged neurons nor their connections are replaced, but the system compensates for the lost function of these neurons to restore the output of the system as a whole.

Therapeutic approaches to CNS injury attempt to exploit all three of these naturally occurring mechanisms of functional recovery. Rehabilitative medicine works to enhance the efficacy of any residual function using physical training and biofeedback techniques. Recovery mediated through physical therapy is likely to reflect both compensatory and restitutive changes in the damaged system. Pharmacologic agents that increase conduction velocity of demyelinated axons are a compensatory treatment designed to increase the gain of circuitry that remains intact following injury. The nervous system can also be trained to utilize intact, local circuitry to mediate a function normally controlled by descending cortical activity (substitution). For example, the walking function can be imperfectly recovered following complete spinal transection by a substitutive mechanism that recruits local spinal circuits normally used for maintenance of balance and foot placement to generate walking behavior (Barbeau *et al.*, 1999). Entraining this "spinal walking circuit" by evoking spinal

Maureen L. Condic • Department of Neurobiology and Anatomy, University of Utah, SOM, Salt Lake City, Utah 84132.

Developmental Neurobiology, 4th ed., edited by Mahendra S. Rao and Marcus Jacobson. Kluwer Academic / Plenum Publishers, New York, 2005.

pattern generators takes considerable practice, but can ultimately result in reasonable, albeit imperfect, walking behavior in cats, rodents, and even human patients (Edgerton *et al.*, 1992; Chau *et al.*, 1998; Fouad *et al.*, 2000; Harkema, 2001). Notably absent from the current repertoire of therapeutic approaches to CNS injury are manipulations that strictly induce or enhance the regeneration of damaged cells or axons.

FUNCTIONAL RECOVERY: RECAPITULATION OF DEVELOPMENT?

There has long been a bias in the field of regeneration research that functional recovery following CNS injury involves a recapitulation or reactivation of the processes underlying embryonic development of the brain and spinal cord. In the majority of cases, however, recovery bears little resemblance to development, appearing instead to more closely mimic the normal adaptive processes of the adult CNS that are likely to underlie learning and memory (see Chapter 10). The same flexibility that enables the mature nervous system to learn and adapt appears to provide a mechanism by which functional deficits can be circumvented, and in many cases overcome. With very few exceptions, true regenerative recovery (due to either cell or axon replacement) does not contribute to functional recovery in humans.

The minimal contribution of cell and axon regeneration to functional recovery of the CNS in humans is by no means a universal phenomenon. The failure to reactivate developmental processes following injury appears to be a limitation that is largely restricted to higher vertebrate species (avians and mammals). In reptiles and amphibians, there is extensive regeneration in the adult, in addition to the functional recovery mediated through more adaptive mechanisms (Chernoff *et al.*, 2002). In most cases where the adult CNS regenerates, the process does indeed appear to mimic development. For example, the spinal cord of newts undergoes complete functional regeneration following ablation of spinal segments several millimeters in length. Spinal regeneration occurs through a process quite reminiscent of embryonic development. Following injury, specialized cells lining the ventricle (ependymal cells) dedifferentiate and migrate into the site of injury. Once there, these primitive cells proliferate to fill the ablated cavity and subsequently redifferentiate into mature spinal cord cells (Chernoff *et al.*, 2002).

The failure of the adult mammalian and avian CNS to fully reinitiate a developmental program as a means of repairing injury has led to extensive investigation into the underlying reasons for this failure. Curiously, in many mammalian and avian species, the CNS exhibits robust regeneration up until roughly the last third of prenatal development (Forehand and Farel, 1982; Shimizu *et al.*, 1990; Bates and Stelzner, 1993; Bandtlow and Loschinger, 1997; Sholomenko and Delaney, 1998; Wang *et al.*, 1998a,b). Changes occurring in both the neurons and the environment of the CNS during development have been implicated in the shift from regeneration competency to regeneration failure. CNS regenerative failure has also been extensively compared to

the relatively robust regeneration observed from adult peripheral neurons following injury. To understand regeneration failure in adult mammalian and avian CNS, it is useful to consider both what is known regarding peripheral nervous system (PNS) regeneration and what is known about the factors that prevent regenerative replacement of either cells or axons in the adult CNS.

PNS REGENERATION

The ability of peripheral nerve to regenerate following injury depends largely on the severity of the injury. Sir Sydney Sunderland (1965) defined five degrees of peripheral nerve injury from mild to severe, with the likelihood of spontaneous regeneration quite poor for all injuries above the third degree (Fig. 1). In addition, the proximity of the injury to the cell body greatly affects the likelihood of recovery, with neurons that sustain injuries close to the cell body being far less likely to regenerate than those subjected to more distal injuries. A similar correlation is seen in regenerating CNS neurons (Sunderland, 1965, 1970, 1990). The age of the individual at the time of injury also greatly affects the likelihood of peripheral regeneration, with younger individuals far more likely to regenerate peripheral nerve compared to more aged individuals.

The process of regeneration, whether it be for central or peripheral neurons, involves several distinct stages: Surviving the initial insult, initiating outgrowth (sprouting), traversing the region of injury, navigating back to the original targets, reestablishing appropriate synaptic contacts at those targets, and restoring normal myelination of regenerated axons (Fig. 2). At many of these stages, the response of peripheral nerve is distinct from that of CNS tissue (Fu and Gordon, 1997). While both CNS and peripheral nerve undergo an initial inflammatory response to injury, the damage to the CNS neurons is exacerbated by the swelling of the tissue against the rigid constraints of the vertebral column and cranium. In peripheral nerve and in the CNS, damaged neurons and glial cells undergo apoptosis and are cleared in a process known as Wallerian degeneration. Clearing of cellular debris occurs much more rapidly in peripheral nerve (within weeks) than it does following CNS damage (months). In response to cytokines released by infiltrating immune cells, quiescent peripheral nerve Schwann cell-precursors are activated, reenter the cell cycle, and actively migrate into the region of damage. A similar activation of microglia and astrocytes occurs in the CNS. However, proliferating Schwann cells are a rich source of trophic support to injured neurons, producing a wide range of beneficial factors including neurotrophins as well as other factors that enhance neuronal survival (Frostick *et al.*, 1998; Yin *et al.*, 1998; Terenghi, 1999). Schwann cells also produce a highly growth-promoting environment, both through their physical alignment into conduits that guide regenerating fibers as well as through the production of a specialized extracellular matrix (ECM) that stimulates regeneration. In contrast, activated CNS glia produce a highly nonpermissive environment in regions of CNS injury (see below). Finally, denervated peripheral targets participate in nerve cell regeneration by producing trophic and

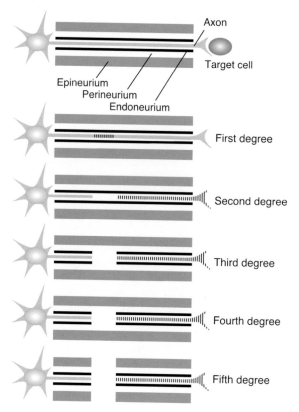

FIGURE 1. Regeneration of peripheral nerve depends on severity of the injury. Peripheral axons of mature nerves are surrounded by three layers of connective tissue: The epineurium surrounding the entire nerve, consisting of fibroblasts, fat cells, small blood vessels, and collagenous matrix; the perineurium surrounding individual nerve fascicles, consisting of a collagen matrix and specialized perineural cells (this layer also forms the blood–nerve barrier); the endoneurium surrounding individual myelinated axons, which is largely an acellular collagen layer separated from the axon itself by the Schwann cell and its basal lamina. Injuries that damage axons (by mild compression or temperature extremes) without disrupting the extracellular matrix are considered first degree injuries and are readily repaired. Second degree injuries that sever or crush the nerve result in Wallerian degeneration distal to the injury, but are also readily repaired, so long as connective tissue layers remain intact. Third through fifth degree injuries involve increasing disruption of the connective tissue of the nerve and are rarely repaired without surgical intervention to reconnect distal and proximal nerve stumps.

possibly tropic factors, a response that is generally not observed from denervated CNS targets.

The most significant challenge presented to regenerating peripheral and central neurons is the site of injury itself. If peripheral nerve injuries are relatively mild (Fig. 1; degrees 1–3), proliferating Schwann cells will successfully fill the gap between distal and proximal nerve, allowing a continuous pathway for the extension of regenerating fibers. Regenerating axons are guided to their original targets by the denervated nerve sheath that serves as a conduit for growing fibers. If injuries are more severe, however, Schwann cells form a dense mass at the injury site that regenerating axons are unable to traverse. Regeneration aborts at the injury site, with nerve fibers often forming a dense neuroma. Surgical interventions, either to ligate the severed ends of the

nerve or to provide an artificial conduit that bridges the gap between distal and proximal nerve stumps, are required to promote regeneration for severe peripheral nerve injuries.

CNS REGENERATION

Following injury, CNS neurons are faced with many of the same challenges presented by injured peripheral nerve. Neurons must survive the initial insult, initiate new axons and dendrites, navigate up to and beyond the injury site, extend to appropriate targets, arrest growth, reestablish synaptic contacts, and establish normal myelination for regenerated axons (Fig. 2). In contrast to peripheral nerve, however, regeneration in the CNS is rarely, if ever, accomplished. The reasons for CNS regenerative failure have been subject to considerable debate and interpretation. For many years, it was generally accepted that, in contrast to peripheral neurons, adult neurons of the mammalian CNS were intrinsically incapable of regeneration. This pessimistic conclusion was radically altered in the early 1980s by several convincing demonstrations that adult neurons could re-extend axons over long distances if provided with the permissive environment of the adult peripheral nerve (Richardson *et al.*, 1980; David and Aguayo, 1981). Over the next 20 years, the focus of regeneration research was largely the environment of the injured CNS, in an attempt to define what factors present in this environment prevent the reestablishment of contacts disrupted by injury. The dominant view was that adult CNS neurons are fully capable of regeneration, but that this intrinsic ability is somehow suppressed by the poorly supportive or actively inhibitory environment of the adult CNS.

In recent years, the pendulum of scientific opinion on the topic of adult CNS regeneration has begun to swing yet again: Away from a strict focus on the environment and toward a more nuanced and complex view of adult regeneration. It has become increasingly clear that the intrinsic ability of adult neurons to extend axons and dendrites is compromised relative to immature neurons. Even under optimal conditions, outgrowth of processes from adult neurons is weak relative to that observed from embryonic or fetal neurons. When fetal neurons are transplanted into injured adult brain, their regeneration is always superior to that of injured adult neurons in the same environment (Wictorin and Bjorklund, 1992; Nogradi and Vrbova, 1994; Lindvall, 1998; Broude *et al.*, 1999; Ito *et al.*, 1999), indicating that changes in cell-autonomous properties of neurons contribute to adult regenerative failure. The cell-intrinsic factors that contribute to poor adult regeneration are poorly understood, and such factors have increasingly become the topic of research.

Lastly, although restoration of function through regenerating axons and synaptic contacts has been the primary focus of research, recent work has begun to investigate the possibility of replacing damaged neurons entirely, either by supplying embryonic counterparts or by stimulating the proliferation of quiescent neuronal precursors present in the adult CNS. The use of neurons generated from either fetal or adult stem cells to replace adult neurons lost to injury or disease is an active area of research.

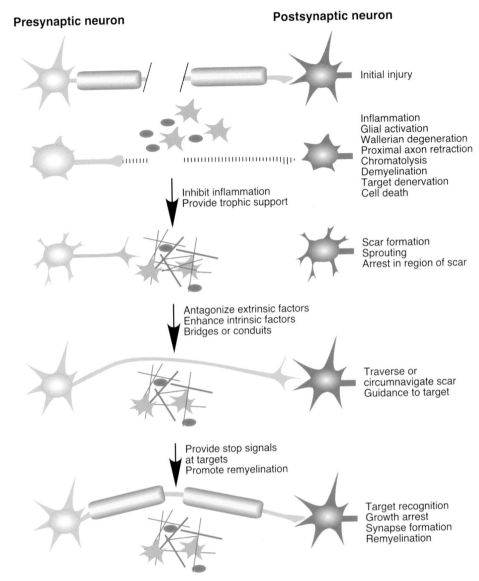

Presynaptic neuron   **Postsynaptic neuron**

Initial injury

Inflammation
Glial activation
Wallerian degeneration
Proximal axon retraction
Chromatolysis
Demyelination
Target denervation
Cell death

Inhibit inflammation
Provide trophic support

Scar formation
Sprouting
Arrest in region of scar

Antagonize extrinsic factors
Enhance intrinsic factors
Bridges or conduits

Traverse or
circumnavigate scar
Guidance to target

Provide stop signals
at targets
Promote remyelination

Target recognition
Growth arrest
Synapse formation
Remyelination

FIGURE 2. Response of the nervous system to injury involves distinct stages. Events occurring at each stage are listed at the right. Possible interventions designed to promote progression to the next stage are given adjacent to the arrows. Following initial injury, the damaged axon rapidly degenerates distal to the lesion. The proximal axon retracts from the injury site somewhat and becomes demyelinated. Cell bodies of damaged neurons swell (chromatolysis) and damaged neurons often undergo apoptosis. There is infiltration of inflammatory cells and activated glia into the injury site. Inhibiting inflammation and providing trophic support for damaged neurons at this stage can greatly enhance functional recovery by limiting cell death and scar formation. Following the acute phase of injury, surviving neurons initiate sprouts and re-extend toward the region of injury. Growth cones arrest upon encountering the glial scar associated with the injury site. At this stage, manipulations designed either to decrease scar-mediated inhibition of regeneration, to stimulate the intrinsic regenerative capability of neurons or to provide bridges or conduits that traverse the scar can improve functional recovery. Once growth cones have extended beyond the region of injury, they must be correctly guided to their original targets and must recognize those targets appropriately (i.e., arrest growth and reestablish synaptic contacts). Regenerated axons must be remyelinated to fully restore normal function. Providing stop signals at targets and introducing factors or cells that promote remyelination can improve functional recovery at this stage.

The current understanding of extrinsic environmental CNS factors, intrinsic changes in adult CNS neurons that limit their regenerative potential, and the possible mechanisms for replacing neurons lost to injury or disease will be discussed in turn below.

Extrinsic Factors: The Importance of the Glial Scar

During embryonic development, the nervous system expresses a plethora of molecules that promote the survival and differentiation of neurons (see Chapters 5, 9, 11). Many of these molecules are further believed to play a role in guiding growth cones to their appropriate targets, enabling them to recognize those targets and to respond by ceasing outgrowth and initiating synapse formation (see Chapter 1). Generally, the expression of molecules that promote neuronal survival, outgrowth, and synapse formation declines over developmental time to low levels in the adult. Consequently, the environment of the adult CNS is believed to be relatively nonpermissive for the generation of new neurons, for the establishment of new long-distance projections and even, perhaps, for the large-scale establishment of new synapses.

The generally poor environment of the adult CNS takes a rapid turn for the worse following injury. In response to CNS injury, a large number of factors that are not normally expressed in the mature CNS are induced in regions of injury. Many of these factors are produced by astrocytes and microglia that are activated in response to injury. Activated glial cells, in particular astrocytes, migrate into the region of injury and produce a wide range of factors that influence regenerating neurons (Table 1). Ultimately, the injury site transforms into a scar composed of dense glial networks and a complex ECM (Fig. 2). Recent data strongly suggests that the glial scar is critical to regeneration failure. When labeled sensory neurons are transplanted into degenerating white matter tracts rostral to a spinal lesion, the transplanted neurons are capable of extensive and rapid regeneration (behavior that is never normally observed from sensory neurons in the injured CNS). This regeneration abruptly ceases once the growth cones encounter the glial scar (Davies *et al.*, 1999). In the region of injury, regenerating growth cones arrest and assume characteristic "dystrophic" morphology (Fig. 3), originally described by Cajal over 100 years ago (Ramon y Cajal, 1991). These striking observations strongly suggest that, at least for sensory neurons, factors associated with the regions of scaring, rather than factors associated with degenerating white matter tracts are the critical components of regenerative failure.

Scar-associated molecules can have both positive and negative effects on growth cone extension *in vitro* (Table 2, Fig. 4). While it is important to understand the functions that specific molecules are *capable* of mediating under well-controlled, experimental circumstances, it is equally important to appreciate that the relationship between regeneration failure and the function(s) of scar-associated molecules is unlikely to be simple (Fig. 5). Importantly, molecules present in CNS scars that *do not function as intrinsically negative regulators* of neurite extension can contribute to regeneration failure. Factors that permit cell adhesion as well as those that prevent it can both contribute to regeneration failure by establishing an attachment state that is not conducive to growth cone motility (Fig. 5A). Mechanical barriers composed *entirely* of permissive molecules can inhibit regeneration by physically blocking the re-extension of axons. Trophic and growth factors may promote neuronal survival, and yet contribute to regenerative failure by stimulating the expression of negative growth regulators or by arresting growth cones in regions of high trophic support (Fig. 5B). Lastly, molecules with intrinsically negative functions may directly suppress growth

cone migration. The complex mechanisms likely to underlie regeneration failure and the ways in which specific molecules may contribute to these mechanisms are considered below.

Positive Regulators: The Dual Role of Cell Adhesion

Molecules with a positive influence on growth cone extension fall into three general classes: Those that *permit* (i.e., allow) extension by interacting with the cytoskeletal machinery underlying growth cone migration; those that *promote* (i.e., encourage) extension by stimulating growth cone migration without directly mediating motility; and those that do not directly interact with growth cone receptors but *enhance* outgrowth by modulating the function of other positive factors (Table 2, Fig. 4). Many discussions of growth cone guidance and regeneration make a further distinction between molecules that are *instructive* (i.e., those capable of guiding the direction of growth cone extension) and those that are "merely" permissive, promoting, or enhancing. It is important to recognize, however, that *all* growth-influential molecules (both positive and negative factors) can be instructive if they are spatially distributed in a manner that directs growth cone extension. For example, sensory growth cones are robustly instructed to extend on low concentrations of laminin when confronted with alternating lanes containing different laminin concentrations (Fig. 6A). Under some circumstances, permissive molecules may be avoided for a more "preferred" permissive molecule. For example, growth cones turn away from borders between laminin and fibronectin (Fig. 6B), although both molecules are strongly permissive for neurite extension (Gomez and Letourneau, 1994). Thus, growth cone preference and growth cone turning indicate that specific molecules *can be* instructive under some circumstances, yet such behaviors do not necessarily provide information regarding the *intrinsic functions* of guidance molecules or how (in general) they influence growth cone behavior. Simple experimental criteria, such as whether receptors are expressed and whether those receptors mediate motility (Table 2), define molecular function far more accurately than do observations of growth cone behavior under limited experimental situations.

Many of the factors expressed in regions of CNS injury are considered permissive factors for neuronal migration and axon extension (see Chapters 8 and 9), yet regeneration invariably aborts precisely in regions of scaring where expression of growth permissive molecules is highest. Regeneration failure, despite the expression of positive growth regulators, may simply reflect the preponderance of negative regulators present in the same region (Table 1). Alternatively, permissive factors may themselves contribute to regenerative failure due to their ability to promote strong cell adhesion when present at high concentrations (Fig. 5A).

For nonneuronal cells, both theoretical calculations and empirical data indicate that cell migration only occurs over a narrow range of matrix concentrations, where cells are adhered strongly enough to generate traction, without being so strongly adhered that they are unable to change position (Palecek *et al.*,

TABLE 1. Growth-Influential Molecules That Have Altered Expression Following CNS Injury

Class	Name	Function	GAGs	Growth factors	Molecular interactions (partial list)
Proteoglycan Decorin/SLRP	Biglycan	Pro	CS/DS	TGFβ	
	Decorin	B	CS	TGFβ	Collagen, fibronectin, thrombospondin
Hyalectin/lectican	Aggrecan	R/B	CS/KS		
	Brevican[a]	B	CS		
	Neurocan	B/T	CS	bFGF	Cadherins, integrins, galactosyl transferase, tenascin-C, NCAM, tenascin-R, L1, TAG-1, contactin
	Versican[b]	B	CS		
Transmembrane	NG2	B	CS		
	Neuroglycan[c]	B	CS		
	RPTPβ[d]	B/T	CS/KS	bFGF	Tenascin-C, tenascin-R, NCAM, L1, TAG-1, contactin
Other	Appican[e]	Per/T	CS		Laminin
	Perlecan	Per/T	HS/CS	TGFβ FGF	Fibronectin, laminin, tenascin
	Phosphacan[d]	B/T		bFGF	Tenascin-C, tenascin-R, NCAM, L1, TAG-1, contactin
Non-proteoglycan	Collagen I, collagen IV	Per			SPARC, Decorin, Fibronectin, integrins
	Fibronectin	Per			Collagen, perlecan, integrins
	Laminin	Per			Entactin/nidogen, integrins
	SC1/SPARC	B		PDGF, VEGF bFGF, TGFβ	Thrombospondin, vitronectin, entactin/nidogen, collagens
	Semaphorin	C			
	Tenascin-C	B			Perlecan, phosphacan, neurocan, integrins
	Tenascin-R	B			Perlecan, phosphacan, neurocan, RPTPβ, integrins
Growth and trophic	Neurotrophin	Pro			
	CNTF, FGF	Pro			
	TGFβ	Pro			
	Cytokines	?			
Cell-associated	L1	Per			Phosphacan, neurocan, integrins, RPTPβ
	NCAM	Per			Phosphacan, neurocan, RPTPβ
	Eph-Ephrins	C/Pro			
	Trks	Pro			
Myelin-associated	MAG	C			
	Nogo	C			

Note: Molecules that have been molecularly cloned are considered. Functions proposed correspond to Table 2; Pro = promoting, Per = permissive, T = trapping, B = blocking, C = collapsing, R = repressing.

[a]BEHAB is a cleavage product representing approximately the N-terminal half of Brevican.

[b]GHAP is a proteolytic fragment of Versican.

[c]CALEB is likely to be the chick homolog of rat neuroglycan.

[d]Phosphacan, also known as DSD-1 and 6B4, represents the cleaved ectodomain of RPTPβ.

[e]The core protein of Appican is a splice variant of amyloid precursor protein (APP). Appican exists in transmembrane and secreted forms.

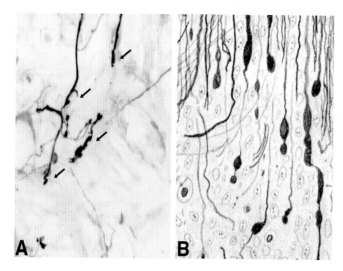

FIGURE 3. Axonal regeneration in the injured CNS is robust through degenerating white matter tracts, yet fails at the site of injury. (A) Darkly labeled, GFP-expressing sensory neurons transplanted into degenerating white matter tracts rostral to a spinal lesion regenerate robustly (1 mm/day), but form dystrophic endings (arrows) at the site of injury (S. Davies, unpublished image; Davies *et al.*, 1999). Dim background staining reflects labeling for activated astrocytes (GFAP) and scar-associated CSPGs. (B) Similar dystrophic endings ("terminal end bulbs") of endogenous spinal neurons were initially described at the site of adult CNS injury by Cajal (Ramon y Cajal, 1991).

1997, 1999). At high concentrations of adhesive molecules, cells become "trapped" or "stalled" by the strong attachments they establish with the substratum (Table 2; Fig. 5A). Surprisingly, this well-established relationship between motility and adhesion has not been rigorously applied to the study of growth-cone migration.

Embryonic neurons extending on the molecule laminin may be an exception to the general rule that only intermediate levels of adhesive proteins will support motility. Growth cones of embryonic neurons efficiently migrate on laminin concentrations that vary over several orders of magnitude (McKenna and Raper, 1988; Buettner and Pittman, 1991; Condic and Letourneau, 1997). The ability of neurons to migrate over a wide range of laminin concentrations is due to an unusual regulation of neuronal receptors for laminin (Condic and Letourneau, 1997). Laminin receptors are downregulated in response to high laminin concentrations, thereby reducing adhesion and allowing an intermediate level of attachment to be maintained over a wide range of ligand concentrations.

While embryonic neurons are able to compensate for a wide range of laminin concentrations, the response of growth cones to other molecules that utilize different receptors is unknown. There is evidence that the receptor molecule L1 (also known as Ng-CAM) can be efficiently removed from the surface of embryonic growth cones (Kamiguchi *et al.*, 1998; Kamiguchi and Lemmon, 2000; Long *et al.*, 2001), but whether the absolute levels of L1 compensate for availability of ligand to maintain a constant level of L1-mediated attachment is unknown. Nothing is known about the regulation of other receptors that promote neuronal adhesion to components of scar matrix. Moreover, very

little is known about the ability of *adult neurons* to regulate receptor levels in response to the molecular composition of the environment.

Whether the adhesive molecules expressed in regions of scaring promote or inhibit regeneration is unclear, although both functions are certainly possible. For example, appican is a chondroitin sulfate proteoglycan (CSPG) that contains an alternatively spliced version of the amyloid precursor protein as its core (Table 1). Appican expression increases following brain damage (Salinero *et al.*, 1998) and is also increased in the brains of patients with Alzheimer's disease (Salinero *et al.*, 1998, 2000). *In vitro* studies indicate that appican acts as a strongly adhesive matrix protein that promotes neurite extension at low concentrations (Coulson *et al.*, 1997; Wu *et al.*, 1997). In patients with Alzheimer's disease, appican has been proposed to trap growth cones during the formation of plaques and neural tangles (Coulson *et al.*, 1997; Wu *et al.*, 1997; Salinero *et al.*, 1998, 2000). Whether appican plays a similar role in regeneration failure is unknown.

Mechanical Barriers: Stability and Crosslinking of the Scar Matrix

Regeneration failure at the site of injury may be strongly influenced by the physical characteristics of the scar, as well as by its molecular composition. For example, several forms of collagen are upregulated at CNS injury sites and become structurally organized into a basal lamina (BL) surrounding the wound. Most collagens are permissive molecules that mediate the extension of neurites in culture. Nonetheless, recent work suggests the collagen-BL constitutes a physical barrier to regeneration following injury. Preventing formation of the scar-associated BL reduces the expression of a number of positive (permissive, promoting, and enhancing) molecules in the region of injury and, counterintuitively, improves regeneration across the injury site (Stichel *et al.*, 1999). This finding is controversial, given that other groups have reported no correlation between the formation of a collagen BL and regeneration failure (Weidner *et al.*, 1999; Joosten *et al.*, 2000), a discrepancy that may be due to anatomical differences between brain and spinal cord (Hermanns *et al.*, 2001). These experiments strongly suggest, however, that structural aspects of the scar matrix affect the ability of neurons to regenerate through the region of injury, *independent* of the positive or negative functions mediated by the molecules composing those structures.

The ability to structurally organize the extracellular environment of the scar is not restricted to collagens. A large number of molecules expressed in regions of injury have complex interactions with other scar-associated molecules (Table 1). Structurally diverse molecules such as tenascins, perlecan, appican, neurocan, phosphacan, and decorin all exhibit high-affinity interactions with other scar components as well as with a variety of neuronal receptors that are themselves upregulated following injury (reviewed in Condic and Lemons, 2002). The role of such complex molecular interactions in regenerative failure is poorly understood. Nonetheless, it seems likely that expression of such a large number of highly interactive molecules in regions of

TABLE 2. Experimentally Distinguishing Functions for "X" and Its Receptor "R$_x$"

Function	Definition	Example	Requires that	Criteria R_x	A/S[a]	2°[b]	Mechanisms[c]
Positive							
Permissive	X supports extension	Laminin	• X is substratum-bound • R$_x$ interacts with cytoskeleton to generate force and motility	Yes	Yes	No	R$_x$ interacts with cytoskeleton or recruits receptors that do
Promoting	Extension is improved by X; X is not permissive	NGF	• R$_x$ enhances extension without directly mediating migration or attachment	Yes	No	Yes	R$_x$ enhances actin polymerization or enhances function of permissive pathway
Enhancing	Extension is improved by X; R$_x$ is not expressed	Nidogen/entactin	• X neither promotes nor permits extension	No	No	Yes	X changes function of permissive or promoting factors or their receptors
Negative							
Collapsing	X induces growth cone collapse	Semaphorin	• X neither promotes nor permits extension • R$_x$ induces de-adhesion and collapse	Yes	No	No	R$_x$ depolymerizes actin or antagonizes permissive receptor function
Repressing/silencing	X blocks extension without inducing collapse	Aggrecan (?)	• X neither promotes nor permits extension	Yes	No	Yes	R$_x$ inhibits the signaling pathway downstream from a permissive or promoting receptor
Blocking	X blocks extension or induces collapse	CSPGs (?)	• X neither promotes nor permits extension	No	No	Yes	X changes the function of permissive or promoting factors
Trapping/stalling	High [X] blocks extension without collapse	Appican (?)	• X is permissive at low concentrations	Yes	Yes	No	R$_x$ mediates strong adhesion and does not desensitize to high [X]
Neutral							
Neutral	No response to X	Silicon	• R$_x$ does not exist or is not expressed	No	No	n/a	Neurons do not express R$_x$
Nonpermissive	No neurites form on X	NGF	• R$_x$ does not support neurite extension	Yes	No	No	R$_x$ does not interact with cytoskeleton
Nonpromoting	X does not enhance extension	Substance P	• R$_x$ does not alter neurite extension	Yes	No	No	R$_x$ does not affect motility pathways

Notes:

[a]Promotes adhesion and spreading.

[b]Does the effect on neurite extension require a specific secondary factor?

[c]Possible mechanisms are not intended to provide an exhaustive list, but rather to illustrate how functional terms *limit* the possible mechanisms and inform the direction of future experiments.

injury will contribute to the stability and crosslinking of the scar matrix in a manner that may impede the advance of axons.

Adding to the dense structural environment provided by scar-associated extracellular molecules is the dense accumulation of glial cells in regions of injury. Glia are the major source of the scar matrix and are found in great numbers associated with injury sites. Evidence suggests that both astrocytes and microglia actively migrate into regions of injury in response to cytokines and chemokines released during the inflammatory response (Goldberg and Barres, 2000; Allan and Rothwell, 2001). Glial densities remain high in the region of scarring for long periods following injury, perhaps indefinitely. Whether or not high glial densities themselves constitute a mechanical barrier to regeneration is difficult to determine, although some evidence suggests that high densities of astrocytes do not, in and of themselves, constitute a barrier to the advance of regenerating axons (Davies *et al.*, 1999).

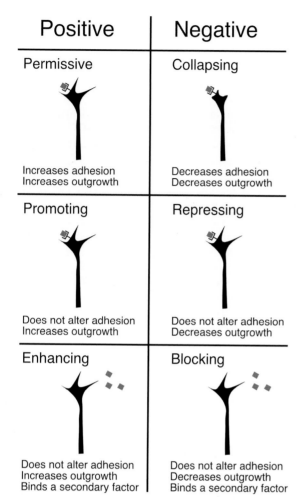

FIGURE 4. Positive and negative functions of growth influential molecules (see also Table 2). Molecules with permissive functions work through neuronal receptors to "permit" or allow growth cone migration by mediating attachment and interacting with cytoskeletal components required for cell motility. Collapsing molecules function through cellular receptors to oppose the activity of permissive factors. Molecules with collapsing function decrease cell attachment and suppress motility. Promoting molecules do not mediate cell adhesion directly, but activation of their receptors stimulates the rate of neurite extension. Analogously, repressing factors do not alter cell adhesion, but decrease motility without inducing collapse. Finally, enhancing and blocking molecules do not directly interact with neuronal receptors, but bind to secondary factors to increase (i.e., enhance) or decrease (i.e., block) neurite extension.

Trophic Molecules: Effects on Neurons and Glia

The role of trophic molecules in regeneration and regeneration failure is far from simple. Adult CNS neurons are highly sensitive to insult and die in large numbers following even mild injury (Allan and Rothwell, 2001). Extensive evidence suggests that providing trophic support can improve the survival of neurons following CNS injury, and in some cases, stimulate regeneration as well (Goldberg and Barres, 2000). Both neurotrophins and other growth factors, such as TGFβ and FGF, can promote neuronal survival and regeneration. However, in many cases,

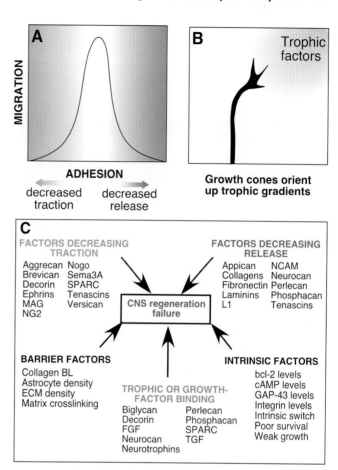

FIGURE 5. Molecules with both positive and negative functions may prevent migration of growth cones in regions of injury. (A) For nonneuronal cells, adhesion to the substratum increases linearly with the amount of bound matrix protein, yet peak motility is only supported by a narrow range of adhesion states (Palecek *et al.*, 1997, 1999). Low levels of attachment/matrix inhibit motility by reducing traction while high levels of attachment/matrix inhibit motility by reducing the ability of cells to release from the substratum. (B) Trophic and growth factors can function as attractive signals for growth cones (McFarlane, 2000). High concentration of such factors in regions of injury may attract and subsequently retain regenerating axons at the scar. (C) Growth cone motility may be reduced by more than a single mechanism following CNS injury. Cellular mechanisms and specific molecules that may contribute to those mechanisms are listed. Factors providing mechanical barriers to migration may physically block growth cone advance. Factors that antagonize the function of cell-matrix receptors or directly inducing growth cone collapse will prevent migration by reducing traction. Extracellular matrix molecules and receptors that increase growth cone adhesion will prevent migration by preventing release from the substratum. High levels of neurotrophic molecules or molecules that recruit growth factors may retain growth cones in regions of injury due to high trophic support. Finally, cell autonomous limitations to regeneration may prevent CNS neurons from regenerating efficiently, independent of environmental factors.

regeneration observed following exogenous application of trophic factors is disorganized and perhaps even maladaptive. Animals expressing NT3 in the spinal cord exhibit extensive invasion of sensory afferents into the dorsal horn following dorsal root crush. Yet many of these animals become excessively sensitive to painful stimuli (Romero *et al.*, 2000), suggesting

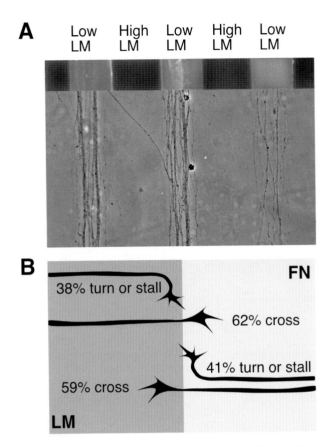

A Low LM High LM Low LM High LM Low LM

B FN LM

38% turn or stall
62% cross
41% turn or stall
59% cross

FIGURE 6. Growth-influential molecules are "instructive" for growth cone guidance if they are presented in discontinuous or graded distributions. (A) Sensory neurons preferentially elect to extend on lanes containing low levels of laminin and avoid lanes containing high levels of the same molecule (M. Condic, unpublished). (B) Some sensory growth cones will turn away from a border of two permissive molecules (Laminin and Fibronectin), although neither molecule is a negative regulator of sensory neurite extension (data taken from Gomez and Letourneau, 1994).

that the strong but diffuse response to exogenously supplied neurotrophins may induce the formation of inappropriate or inordinate synaptic contacts.

There is evidence for endogenous upregulation of trophic factors and growth factors in regions of CNS injury (Goldberg and Barres, 2000; Allan and Rothwell, 2001). The contribution of these factors to regeneration is likely to be complex. Neurotrophins and other growth factors can clearly promote both neuronal survival and neuronal regeneration, yet the neuroprotective effect of growth factors is often transient or specific for only subclasses of neurons. In some cases, neurotrophins may actually potentiate neural injury (Behrens *et al.*, 1999). The same factors that promote neuronal survival and regeneration also contribute to the activation of glia and the subsequent formation of the glial scar. Whether the positive effects of growth factors on neurons can be experimentally separated from the deleterious effects these same factors have in promoting scar formation is not known.

A number of molecules expressed at injury sites are able to bind growth factors with high affinity (Table 1). Expression of growth-factor binding proteins is likely to increase the localization of these factors to regions of injury. However, it is not clear whether increased growth factor function will have a positive or a negative effect on regeneration. Moreover, growth factor expression does not necessarily translate into increased activity. For example, the scar-associated proteoglycan decorin binds and inactivates TGFβ (Ruoslahti *et al.*, 1992; Hildebrand *et al.*, 1994). Exogenous application of decorin (presumably, inactivating TGFβ) attenuates scar formation (TGFβ induces many scar-associated molecules) and enhances regeneration (Logan *et al.*, 1999), suggesting that TGFβ has a net negative effect on regeneration.

A final role for scar-associated growth factors may be chemoattraction (Fig. 5B). Numerous studies have indicated that growth cones in culture will orient up growth factor gradients (McFarlane, 2000). Localization of such factors to the region of injury may induce arrest of growth cones at the high point of such chemoattractive gradients. Indeed, neurotrophic factors are believed to contribute to the arrest and terminal differentiation of growth cones in their appropriate target tissues during development (Chapter 10). Whether high expression of growth factors and trophic factors contribute to growth cone arrest is currently unknown.

Negative Regulators: Collapse, Repression, and Blocking

There are numerous proteins expressed in the CNS following injury that are believed to negatively regulate axonal regeneration (Tables 1, 2). Most of these molecules can be assigned to one of three categories: CSPGs, non-proteoglycan molecules, and factors associated with myelin. CSPGs are of particular interest due to their highly localized expression in regions of CNS scarring, their roles during development, and their influence on neuronal growth in culture (Bovolenta and Fernaud-Espinosa, 2000; Asher *et al.*, 2001; Condic and Lemons, 2002). Myelin-associated factors are expressed generally in the CNS and do not appear to play a critical role in regeneration; regenerating axons extend up to 1 mm/day in CNS regions expressing high concentrations of myelin-associated factors (Davies *et al.*, 1999).

Molecules that have a negative influence on growth cones can function in more than one manner (Table 2; Fig. 5C). As noted above, molecules that permit neurite extension at low concentrations can potentially suppress extension at high concentrations, due to trapping or stalling (Fig. 5A). Whether this form of negative regulation applies to neurons has not been tested, but appears likely given the precedent from non-neural cells.

Molecules can also have a negative impact on growth cone extension by inducing growth cone *collapse* (Table 2, Fig. 4). The morphology of dystrophic endings *in vivo* (Fig. 3) is similar to that of growth cones undergoing collapse and retraction *in vitro* (see Chapter 9), suggesting that factors present in regions of injury may induce growth cone collapse. Collapsing molecules directly promote cytoskeletal depolymerization and release of growth cones from the substratum. Collapsing molecules are not dependent on the context or on the presence of specific positive factors to mediate their effects on growth cones.

For example, the Eph-family tyrosine kinase receptors are a large class of cell surface molecules that interact with both cell-surface and matrix-associated ephrin ligands (see Chapter 9). In development, Eph–ephrin interactions most commonly mediate inhibitory or repulsive growth cone responses, independent of the substratum on which neurites are extending (reviewed in Wilkinson, 2001). Recent work has shown that several Eph receptors, including EphB3 (Miranda *et al.*, 1999), EphA4, EphA5, and EphB2 (Moreno-Flores and Wandosell, 1999) are upregulated following either traumatic or excitotoxic injury to the CNS, suggesting that these collapsing factors may contribute to regenerative failure.

Similar to ephrins, the collapsing factor semaphorin 3A is upregulated following CNS injury in regions of scarring (Pasterkamp *et al.*, 1998, 1999). Semaphorins are a large family of cell surface and secreted proteins that are believed to act as repulsive and/or stop signals in neural development (Nakamura *et al.*, 2000; Pasterkamp and Verhaagen, 2001). Thus far, there is no direct evidence for a role of semaphorins in CNS regenerative failure, yet by analogy to the role of semaphorins in development (Nakamura *et al.*, 2000) and based on the ability of semaphorins to inhibit the regeneration of adult sensory neurons in culture (Tanelian *et al.*, 1997), semaphorins could readily contribute to adult regenerative failure by collapsing growth cones in regions of injury.

Negative regulators can also act *via* receptors to *repress* neurite outgrowth without inducing collapse (Table 2, Fig. 4). Repressing factors do not directly antagonize the function of receptors that mediate adhesion, but rather inhibit the signaling pathway downstream of such receptors such that growth cone motility is suppressed. For example, growth cones extending on laminin arrest but do not collapse when they encounter the CSPG aggrecan (Snow *et al.*, 1994; Challacombe *et al.*, 1996, 1997). Aggrecan induces a rapid and sustained increase in growth cone calcium, suggesting the effects of this molecule are mediated by an uncharacterized neuronal receptor (Snow *et al.*, 1994). Thus, aggrecan does not disrupt growth cone attachment to laminin (although it can reduce whole cell attachment; Condic *et al.*, 1999), but suppresses growth cone motility that normally results from such attachment.

In contrast to factors that have an intrinsic negative effect on growth cones, *blocking* molecules act predominantly by suppressing the positive effects of growth-promoting or growth-permissive molecules (Table 2; Fig. 4). Molecules with blocking functions have often been described as "inhibitory," because such factors inhibit the function of something else. Yet, the term "inhibitory" is inherently imprecise as a descriptor of molecular function, due to the fact that "inhibition" can describe *either* the effect of a molecule on the rate of neurite extension *or* the impact of one molecule on the function of another. *All* molecules that slow or abolish neurite extension (i.e., all negative regulators) inhibit outgrowth, yet *only* molecules that antagonize the function of other factors have inhibitory (i.e., blocking) molecular function.

Molecules that alter neurite outgrowth *via* a blocking mechanism would be predicted to have *no direct effect on neurons when presented alone* (Table 2). Some of the CSPGs found in regions of injury are believed to inhibit growth cone extension by blocking the effects of growth-permissive molecules present in the CNS. Importantly, molecules that block neurite extension are entirely dependent on the context in which they are encountered and will only inhibit the positive functions of specific promoting or permissive molecules.

It is interesting to note that under this definition, blocking factors need not exclusively mediate negative effects on growth cones. While thus far the growth cone equivalent of a "derepressor" (i.e., an antagonist of a molecule that normally mediates a negative function) has not yet been described, it is possible that both positive and negative blocking molecules exist.

Molecules with More than One Function

A further complication in the study of regeneration failure is that a large number of growth-influential factors can have more than a single effect on neurite extension (Fig. 7). Growth-influential molecules can work through more than one receptor to mediate opposing effects on neurite extension (Fig. 7A). For example, netrin-1 can act through the receptor DCC to mediate attraction (Vielmetter *et al.*, 1994; Keino-Masu *et al.*, 1996; Kolodziej *et al.*, 1996) and through Unc-5 to mediate repulsion (Hedgecock *et al.*, 1990; Leonardo *et al.*, 1997; Colavita and Culotti, 1998). Differential expression of these receptors in particular populations of neurons or in the same neurons at different times results in netrin-1 having widely varying effects on growth cones. For example, commissural neurons of the spinal cord (that normally extend toward a source of netrin-1 during development) are attracted to netrin-1 *in vitro* (Kennedy *et al.*, 1994) while trochlear motor axons (that normally extend down a netrin-1 gradient) are repelled (Colamarino and Tessier-Lavigne, 1995). These findings indicate that subpopulations of CNS neurons are likely to have widely differing responses to the same scar-associated molecule, depending on which receptors are expressed.

In addition, the response of a neuron to the same factor can be modified by the internal state of the growth cone (see Chapter 9), most notably the levels of cyclic nucleotides (Fig. 7B; reviewed in McFarlane, 2000). The fact that the state of the neuron can critically alter its response suggests that the recent history of the growth cone can influence behavior. Indeed, while commissural axons of the spinal cord are initially attracted to netrin-1, they lose responsiveness to this molecule after having crossed the ventral midline (Shirasaki *et al.*, 1998), indicating that recent encounters can alter growth cone response. Similarly, sensory neurites extending on fibronectin will accelerate and change morphology in response to a single encounter with a laminin-coated bead (Kuhn *et al.*, 1995), yet a second encounter with laminin within a narrow time window following the first causes the growth cone to completely arrest (Diefenbach *et al.*, 2000). The differing response of growth cones to the *same molecule* depending on the recent history of the growth cone argues against a simple view of any specific molecule as having a strictly positive or negative function.

The effects of specific molecules on regeneration can be further complicated by the ability of one factor to modify or even eliminate the response of a growth cone to a second molecule

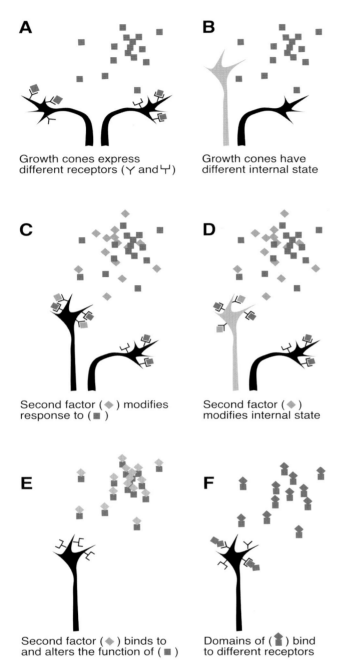

A Growth cones express different receptors (Y and ⊥)

B Growth cones have different internal state

C Second factor (◆) modifies response to (■)

D Second factor (◆) modifies internal state

E Second factor (◆) binds to and alters the function of (■)

F Domains of (⬆) bind to different receptors

FIGURE 7. Molecules that influence regeneration can have more than a single function. (A) Different classes of neurons can express receptors that mediate different responses to the same factor. (B) The internal state of the growth cone at the time the factor is encountered can alter growth cone response. (C) Secondary factors can modify the response of the growth cone through interaction of distinct receptors or their downstream signaling pathways. Examples include "silencing" of growth cone response to netrin-1 by Robo-slit signaling or inactivation of integrin signaling as a consequence of Eph–ephrin signaling. (D) Secondary factors can modify the internal state of the growth cone, thereby altering growth cone response, without a direct interaction between the receptors. Promoting and repressing factors work in this manner. (E) Secondary factors can modify the function of other growth influential factors. Enhancers, such as nidogen/entactin and blockers (possibly scar-associated CSPGs), work in this manner. (F) Single molecules can possess different functional domains that interact with distinct receptors.

(Figs. 7C–E). Such "context-dependent" effects can be either transient (i.e., dependent on the presence or absence of the modifying factor) or long lasting, thereby contributing to the "history-dependent" effects described above. In addition to factors that act as classical agonists or antagonists (i.e., activating or inactivating ligands for the same receptor) there are at least three ways in which secondary factors can alter a growth cone's response. First, direct interactions between the receptors or signaling pathways downstream from different molecules can modify the response a growth cone shows to either molecule in isolation (Fig. 7C). For example, a positive turning response to netrin-1 can be "silenced" via a physical interaction of the netrin-receptor DCC and an unrelated receptor robo, when robo is bound by its ligand, the extracellular protein slit (Stein and Tessier-Lavigne, 2001). In this situation, presence or absence of slit (itself a negative guidance cue) can radically alter the effects of netrin-1 on growth cone behavior.

In addition to mechanisms that depend directly or indirectly on receptor–receptor interactions, growth cone response can be altered by secondary factors that alter the internal state of the growth cone (Fig. 7D). For example, retinal growth cones are attracted to a soluble gradient of netrin-1 when they are extending on a fibronectin substratum, but repelled by the same gradient when extending on laminin (Hopker *et al.*, 1999), despite the absence of evidence for a physical interaction between receptors for netrin and receptors for either laminin or fibronectin. Similarly, the scar-associated molecule versican strongly inhibits the extension of sensory neurites on laminin (Schmalfeldt *et al.*, 2000), but does not inhibit regeneration on fibronectin (Braunewell *et al.*, 1995), despite the lack of evidence for a direct interaction between versican and either laminin or fibronectin receptors. In these cases, it is likely that independent receptor-ligand pathways alter the internal state of the growth cone, which in turn alters growth cone response (Fig. 7D).

Independent of cellular receptors, factors themselves can interact to modify function in both positive (Table 2, Fig. 4; enhancing) and negative (Table 2, Fig. 4; blocking) ways. The effects of enhancing and blocking factors do not require specific receptors or downstream signaling pathways, but rather occur via direct action on other growth influential molecules (Fig. 7E). The presence or absence of blocking or enhancing factors can alter or reverse the function of specific molecules that in isolation have only a single effect on neurons.

Finally, the effects of some growth-influential molecules are complicated by the multidomain structure of the molecules themselves (Fig. 7F). For example, members of the tenascin gene family have complex molecular structure (Faissner, 1997; Jones and Jones, 2000; Meiners *et al.*, 2000) and diverse effects on neurite extension. Tenascins can possess adhesive, counteradhesive, growth-promoting, and inhibitory activities all on the same molecule (Jones and Jones, 2000). Different tenascin domains interact with a wide range of cellular receptors expressed by neurons as well as with a large number of scar-associated molecules (Jones and Jones, 2000). How a particular neuron responds to tenascin will reflect both the receptors expressed by that neuron (Fig. 7F) and the crosslinking of tenascin to other molecules that may modify (or eliminate) the function of specific tenascin domains (Fig. 7E). Given the diversity of high-affinity interactions tenascin is able to maintain, it is difficult to predict what the functional properties of tenascin would be in a complex molecular environment and what net affect this molecule would have on a neuron expressing multiple tenascin receptors.

Dissecting the Contributions of Specific Molecules to Regeneration Failure

Understanding how specific molecules contribute to regeneration failure clearly goes well beyond a simple matter of understanding the functions those molecules are capable of mediating *in vitro* (Table 2, Fig. 4). While molecular function is clearly relevant to regeneration failure, in complex environments, the net effect of a specific factor can be hard to predict. In light of the multiple ways that molecules can function to prevent regeneration (Table 2; Figs. 4 and 5) and the multiple functions any given molecule can mediate (Fig. 7), sorting out the contribution of individual factors is extraordinarily complicated. While it is clear that the net effect of the scar matrix is to inhibit regeneration, the precise manner in which a specific molecule participates in this net function is hard to discern. Is scar-associated laminin a positive factor whose beneficial effect is masked by the numerous inhibitory molecules present in regions of injury or is it present in sufficiently high concentrations that it acts as a stalling factor, thereby itself contributing to regenerative failure? How can the net effect of a specific molecule in such a complex environment be determined?

One approach has been to generate animals deficient for a molecule believed to contribute to regenerative failure and then challenge those animals with CNS injury. This approach cannot be applied to the majority of molecules believed to play a role in regeneration failure, due to the essential functions these molecules play in development and normal physiology. However, in a small number of cases, genetic knockouts have proven informative. For example, myelin-associated glycoprotein (MAG) is a prominent component of CNS myelin that can induce growth cone collapse when applied to neurons *in vitro* (Li *et al.*, 1996). Animals deficient for MAG are viable and show subtle defects in myelin ultrastructure and in axonal conduction velocity (Bartsch, 1996). Surprisingly, animals deficient for MAG show no significant improvement in regeneration following CNS injury (Bartsch *et al.*, 1995), suggesting that MAG is not critical to regeneration failure. This conclusion has recently been supported by experiments demonstrating robust regeneration in intact (Davies and Silver, 1998) and degenerating white matter tracts (Davies *et al.*, 1999), sites where MAG is present in high concentrations. Thus, while MAG is clearly able to collapse axons *in vitro*, it appears either to play a minor role in regeneration failure or (more likely) to be merely one of a large number of players *in vivo*.

As for all genetic approaches, the interpretation of the MAG knockout experiment is compromised by the fact that it is difficult to know how removal of a major component of myelin may have altered the normal development of the CNS or its response to injury. While proteins such as MAG could play an important role in regenerative failure in wild-type animals, this role may be compensated for by other molecules in animals deficient in MAG. Thus, while genetic ablation indicates that MAG does not make a *critical* contribution to regenerative failure, the actual role of MAG in regeneration failure in wild-type animals is difficult to access from such an experiment.

An alternative approach has been to acutely antagonize the function of specific molecules or classes of molecules either *in vivo* or in an *in vitro* model. This approach has the advantage of studying regeneration failure in a genetically wild-type individual and asking what role does a specific molecule or combination of molecules normally play. For example, the contribution of CSPGs to regeneration failure has been examined using chondroitinase ABC, a glycolytic enzyme that removes chondroitin sulfate side chains from proteoglycan cores and greatly reduces the inhibition mediated by this class of proteins. Both in the cortex (Yick *et al.*, 2000; Moon *et al.*, 2001) and in the spinal cord (Bradbury *et al.*, 2002), chondroitinase treatment significantly improves regeneration. In an *in vitro* model system, McKeon *et al.* (1995) have similarly shown that CSPG associated with CNS scars inhibits axon outgrowth; as CSPG accumulates in the region of injury, axon outgrowth is progressively inhibited. Treatment of the scar-associated matrix with chondroitinase results in a significant increase in neurite extension over scar tissue *in vitro*. Interestingly, co-treatment with both chondroitinase and function-blocking anti-laminin antibodies reverses the growth-promoting effect of chondroitinase alone, suggesting that scar-associated laminin plays a positive role in regeneration that is antagonized by CSPGs co-expressed in regions of injury (McKeon *et al.*, 1995).

Acute manipulations allow for the contribution of particular molecules to be accessed, but are often limited by the specificity and reliability of the available reagents. In the example given above, the positive effects of chondroitinase treatment suggest that this class of molecules plays an important role in regeneration failure, yet it is not possible to determine the contribution of specific CSPGs from these experiments. Moreover, it is difficult to interpret negative findings of acute manipulations. Does a failure to improve regeneration indicate that a particular factor is not involved or merely that the manipulation did not sufficiently reduce the function of that factor? As noted above for manipulations of collagen deposition in injury models (Stichel *et al.*, 1999; Weidner *et al.*, 1999; Joosten *et al.*, 2000; Hermanns *et al.*, 2001), subtle differences in technique or even anatomical differences between CNS regions can potentially alter the experimental outcome.

Intrinsic Factors

In addition to the critical role of scar-associated factors in regeneration failure, the intrinsic state of neurons can play an important role in regeneration (Figs. 7B, D). The intrinsic ability of neurons to regenerate changes over developmental time and in response to injury itself. Although the cell-autonomous factors controlling (and limiting) adult regeneration are poorly understood, such factors present attractive targets for therapeutic intervention. To promote the maximum level of regeneration with the lowest level of side effects on undamaged regions of the nervous system, one would ideally like to control both the temporal and spatial extent of the manipulation. In most cases, it is quite difficult to control the effective temporal and spatial extent of manipulations designed to alter the extracellular environment of the CNS. Even with genetic or molecular manipulations, once molecules are secreted into the extracellular space, the time course over which they persist and the distances over which they diffuse are difficult to regulate.

Targeting cell intrinsic factors as a means of promoting adult regeneration presents several technical advantages. Inducible promoters can be used to regulate the temporal expression of intrinsic factors: Turning genes on to promote regeneration and turning them off once regeneration is accomplished. The ability to return adult gene expression to normal adult levels once connections have been reestablished is a strong advantage of approaching CNS regeneration from the perspective of cell-intrinsic factors. In addition, gene expression can be *locally* altered in the region of injury using microinjection techniques. Viral gene-delivery systems can be selected for high neuronal affinity that results in a very minimal spread of the viral agent away from the site of injection. Spatially restricting the manipulation to the region of injury minimizes any unintended effects on undamaged neurons distant from the site of injury. Damaged neurons take up factors from the environment (including viral vectors) more readily than do undamaged cells, an effect that serves to further enhance the "targeting" of gene manipulation to the cells actually affected by the injury. Thus, while relatively little is currently known regarding the contribution of cell-autonomous factors to regeneration failure, such factors are an active area of investigation and attractive targets for therapeutic intervention.

Changes in the Intrinsic Properties of CNS Neurons with Maturation

It is abundantly clear that there are developmental changes in the ability of neurons to regenerate axons. Both retinal (Cohen *et al.*, 1986, 1989; Neugebauer and Reichardt, 1991; Bates and Meyer, 1997) and sensory (Sango *et al.*, 1993; Golding *et al.*, 1999) neurons in culture show decreased rates of axon extension at progressively older stages. Several studies suggest that young neurons transplanted into injured adult CNS tissue show more extensive regeneration than do the adult neurons of the host (Wictorin and Bjorklund, 1992; Nogradi and Vrbova, 1994; Lindvall, 1998; Broude *et al.*, 1999; Ito *et al.*, 1999). The basis for this age-dependent decline in intrinsic regenerative potential is unknown.

A large number of intrinsic factors that may contribute to regeneration show altered expression over developmental time (reviewed in Caroni, 1997; Rossi *et al.*, 1997), but in most cases, the contribution of these factors to adult regeneration failure is unclear. For example, while GAP43 expression is associated with regeneration in some adult neurons (Vaudano *et al.*, 1995), many regenerating axons do not express GAP43 (Schreyer and Skene,

1991; Andersen and Schreyer, 1999). Overexpression of GAP43 in transgenic animals does not stimulate adult neuronal regeneration (Buffo *et al.*, 1997; Mason *et al.*, 2000), while overexpression of GAP-43 in combination with a related protein, CAP-23, does improve adult performance (Bomze *et al.*, 2001). Similarly, increased expression of other genes expressed at high levels in embryonic stages but at low levels in adults (e.g., the anti-apoptotic gene *bcl2* [Chen *et al.*, 1997] or receptors of the integrin class [Condic, 2001]) can improve the regeneration of adult neurons *in vitro*. Potentially relevant cell-intrinsic factors are not limited to molecules classically associated with axon extension and guidance. For example, as neurons mature, cAMP levels decline, and pharmacologically increasing cAMP can stimulate regeneration both *in vitro* (Cai *et al.*, 2001) and *in vivo* (Neumann *et al.*, 2002; Qiu *et al.*, 2002).

The mechanism underlying the age-dependent decline in regenerative potential is not known. One possibility is that molecules required for efficient regeneration are incompatible with normal adult CNS function and are therefore downregulated at adult stages in order to promote stabile adult CNS function. For example, receptors of the integrin class are the primary receptors for growth-promoting matrix proteins and are expressed at high levels in embryonic neurons during periods of axon extension (Reichardt and Tomaselli, 1991; Letourneau *et al.*, 1994). In contrast, adult CNS neurons express low levels of integrin message (Jones and Grooms, 1997; Pinkstaff *et al.*, 1999) and the majority of integrin protein expressed in the adult brain is found at synapses (Grotewiel *et al.*, 1998; Nishimura *et al.*, 1998; Chavis and Westbrook, 2001). Consistent with these observations, the major phenotype associated with loss of integrin function in the adult CNS is a learning defect (Grotewiel *et al.*, 1998). These results suggest that integrin expression is constitutively suppressed in the adult CNS and that integrins have distinct functions at adult and embryonic stages. Low levels of integrin protein appear to be required for the maintenance or formation of synaptic connections in the adult. Quite possibly, low levels of integrin expression also serve to explicitly *prevent* the large-scale formation of new neuronal connections in the adult, "crippling" adult neurons, to promote a stable pattern of circuitry in the brain.

An alternative (albeit, not mutually exclusive) possibility is that age-dependent changes in regeneration potential reflect a stable developmental switch from production of axons to production of dendrites (Fig. 8). In this view, molecules required for efficient long-distance regeneration of axons are downregulated once long-distance axonal projections have been established. This switch reflects a normal developmental program that promotes dendrite formation at the expense of axons. Due to the stability of this developmental switch, adult neurons are unable to revert to production of axons following injury, although they are able to produce short, dendrite-like processes. In retina, factors associated with amacrine cell membranes are able to induce a stable shift from production of axons to production of dendrites from retinal ganglion cell neurons (Goldberg *et al.*, 2002). Whether such a developmental switch could be reversed or inactivated to promote the regeneration of axons will depend on the precise nature of the switch.

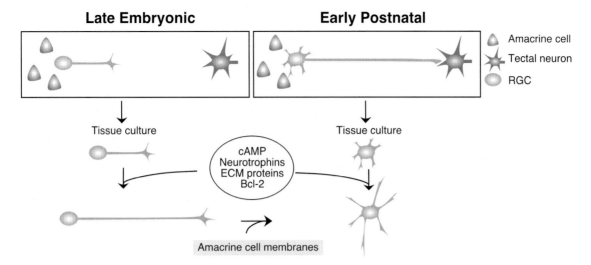

Late Embryonic   **Early Postnatal**

Amacrine cell
Tectal neuron
RGC

Tissue culture   Tissue culture

cAMP
Neurotrophins
ECM proteins
Bcl-2

Amacrine cell membranes

FIGURE 8. Maturing neurons may undergo a cell autonomous switch from production of axons to production of dendrites. Retinal ganglion cells (RGCs) *in vivo* (boxes) extend axons to innervate targets in the brain during late embryonic stages, and extend dendrites during postnatal stages. RGCs placed in tissue culture at embryonic or postnatal stages regenerate processes that are similar to the ones they generate *in vivo*; young neurons re-extend a single axon while older neurons extend multiple short dendrites. Factors that stimulate neurite extension (oval) can increase the length of the regenerated processes, but do alter the axonal vs dendritic nature of the process, suggesting that RGCs have undergone a stable, cell-intrinsic switch from production of axons to production of dendrites. Contact with cell membranes derived from postnatal amacrine cells is sufficient to switch embryonic RGCs to a postnatal pattern of growth in culture, suggesting that amacrine-associated factors may mediate this maturational switch in retina (Goldberg *et al.*, 2002). Figure adapted from Condic (2002).

Changes in Intrinsic Properties of CNS Neurons in Response to Injury

Independent of maturational changes in neuronal gene expression, the intrinsic state of adult neurons can be a key factor in CNS regeneration. For example, adult sensory neurons that have sustained a "conditioning" peripheral lesion regenerate more readily into the CNS following dorsal root injury (Neumann and Woolf, 1999). How such conditioning lesions enhance the ability of neurons to regenerate into the CNS is unknown, but it is possible that peripheral injuries indirectly promote expression of genes that are not upregulated in response to CNS injuries (Frostick *et al.*, 1998; Terenghi, 1999; Kury *et al.*, 2001). For example, activated Schwann cells may supply trophic factors to sensory neurons that are not supplied by activated central glia. Consequently, neurons that have been appropriately "conditioned" may have a distinct state of gene activation that enhances their ability to regenerate.

In the absence of a beneficial conditioning lesion, injured adult CNS neurons exhibit altered patterns of gene expression that can both improve and detract from their ability to regenerate. Following injury, CNS neurons express higher levels of cell adhesion molecules, such as NCAM (Becker *et al.*, 2001; Tzeng *et al.*, 2001) and L1 (Jung *et al.*, 1997), both of which interact with components of the scar matrix as well as with the surfaces of other neurons. The net effect of increased cell-adhesion molecule expression is hard to predict. Enhanced axon–axon interactions may promote regeneration along axon scaffolds. However, increased adhesion to the scar ECM may contribute to regenerative failure by stalling growth cones in the region of injury

(Fig. 5A). Adult neurons also upregulate receptors for collapsing factors, including members of the Eph-family (Miranda *et al.*, 1999; Moreno-Flores and Wandosell, 1999). Lastly, neurotrophin receptor expression is upregulated following injury, suggesting that the response of neurons to growth factors may be enhanced (Goldberg and Barres, 2000). The effect of such enhanced responsiveness on regeneration is unclear, with some evidence suggesting that neurotrophins may potentiate rather than reduce neuronal injury (Behrens *et al.*, 1999).

Adult CNS neurons are as much characterized by their *failure* to respond to injury as by their response. In the PNS, for example, numerous beneficial genes are upregulated in response to injury, including growth-associated molecules, neurotrophin receptors, and matrix receptors (Frostick *et al.*, 1998; Yin *et al.*, 1998; Terenghi, 1999). In many cases, these genes fail to increase in expression following CNS injury. Whether the failure to adaptively regulate gene expression reflects some suppressing property of the CNS environment or an intrinsic limitation of CNS neurons appears to vary depending on the cell type. For example, injured adult Purkinje neurons *in vivo* fail to upregulate the growth-associated molecule GAP-43 and do not express this gene even when provided with a permissive environment for regeneration (Gianola and Rossi, 2002). In contrast, adult retinal neurons only weakly upregulate GAP-43 *in vivo*, yet respond to permissive environments *in vitro* with a strong upregulation (Meyer *et al.*, 1994). While there may not be general rules that apply to *all* CNS neurons, it appears that failure to respond adaptively to injury can contribute to the limited intrinsic regenerative capability of some CNS neurons.

Cell Replacement: Endogenous or Transplanted Neuronal Stem Cells

Following CNS injury, there is extensive death of injured neurons. Replacing neurons lost to injury has long been considered an attractive option for the repair of CNS injury, particularly in light of the superior ability of young transplanted neurons to extend axons in the damaged adult CNS. Attempts to restore CNS function by replacing damaged or dead neurons have taken two general approaches; stimulating the division and differentiation of endogenous neuronal stem cells and transplanting stem cells or their derivatives into the injured CNS.

In most areas of the CNS, new neurons are not born in adult animals. Until quite recently, it was believed that all neurogenesis was completed during development and that new neurons were never added to the adult CNS. Recent work has modified this view somewhat. It is clear that in limited areas of the brain, there is ongoing neurogenesis during adult life (Garcia-Verdugo et al., 2002; Turlejski and Djavadian, 2002). It is likely that new neurons are generated throughout the CNS, albeit in very small numbers for most regions. The source of new neurons in the adult brain and spinal cord appears to be a resident population of adult neural stem cells. The existence of an adult stem cell population is in many ways quite surprising. What function do these cells normally serve, and why do they fail to repair the CNS following injury? The factors that stimulate and suppress the generation of mature neurons from endogenous stem cells are clearly of great scientific and therapeutic interest, yet remain poorly understood (Lim et al., 2002). It is also unclear whether stem cells derived from adult CNS tissue are capable of forming all, or only some of the neurons found in the mature nervous system. A significant advantage of stimulating endogenous cell replacement mechanisms or utilizing stem cells derived from patients is that autologous stem cell transplants would not be subject to immune rejection (Subramanian, 2001).

In contrast to adult CNS tissue, neural stem cells are abundant in fetal and embryonic CNS. Transplantation of fetal-derived stem cells and/or neurons into adult injury models has thus far had mixed results (Temple, 2001; Cao et al., 2002; Rossi and Cattaneo, 2002). In some cases, fetal tissue improves recovery following CNS injury. Typically this improvement is not due to fetal stem cells generating neurons, but rather due to fetal-derived astrocytes or other nonneuronal cells providing unknown factors that enhance the survival and regenerative performance of injured adult neurons. It is possible that the environment of the adult CNS promotes the differentiation of bipotential stem cells along a glial pathway. Alternatively, it is possible that newly generated fetal neurons are unable to survive or to integrate into existing adult CNS tissues. One beneficial aspect of the propensity of transplanted neural stem cells to form glia has been the generation of oligodendrocytes that are capable of myelinating axons. Much of the functional deficit experienced following CNS injury is attributable to reduced conduction velocities as a consequence of demyelination. Oligodendrocytes derived from transplanted stem cells readily migrate into areas of injury and can participate in myelination of existing axon tracts (Lundberg et al., 1997).

A significant concern for the use of cell-replacement strategies is the long-term survival and fate of such transplanted cells. Very few experiments have been done testing the function of stem cells or their derivatives over the long survival times (Temple, 2001; Cao et al., 2002; Rossi and Cattaneo, 2002). Little is known regarding the functional properties of replacement cells in vivo and the stability of those properties over time. It is critical to determine whether tissue differentiated in culture from stem cells remains stable and functional once transplanted into the CNS. The stability and normalcy of transplanted cells is of particular concern for derivatives of embryonic stem cells (ESCs). ESCs form teratomas in adult tissue with high frequency (Kirschstein and Skirboll, 2001). Whether ESCs can be safely differentiated into stable cell types that do not form teratomas is largely unknown. Lastly, immune rejection of allografts is also a concern for potential cell replacement therapies (Subramanian, 2001). Although the CNS enjoys a certain degree of "immune privilege," replacement cells would nonetheless be rejected by the immune system over the long term if immunosupression is not employed.

SUMMARY

1. In mammals and in avians, restoration of function is unlikely to be due to recapitulation of developmental mechanisms, but rather appears to come about through recruitment of the normal mechanisms underlying adult plasticity and learning. Restitution, substitution, and compensation can all contribute to recovery of function.

2. In lower vertebrates and during the embryonic life of most mammals, the CNS is capable of extensive regenerative repair that occurs largely through the dedifferentiation and redifferentiation of damaged CNS tissue.

3. In both the CNS and the PNS of adult mammals, regeneration involves distinct, sequential challenges: Surviving the initial insult, initiating new axons and dendrites, circumnavigating the region of injury, guidance back to original targets, recognition of appropriate synaptic partners, reestablishment of synaptic contacts, and reestablishment of myelination.

4. In the PNS, the effects of inflammation, the response of glia, and the ability of the nerve to serve as a permissive conduit for regeneration and guidance all contribute to superior performance.

5. In the CNS, regeneration is limited by both the intrinsic properties of CNS neurons and the extracellular environment of the CNS that suppresses regeneration.

6. CNS regeneration failure is largely due to factors present at the site of CNS injury. While factors that inhibit axon extension are expressed throughout CNS white matter, regeneration can be nonetheless robustly accomplished in degenerating white matter tracts. Regeneration abruptly fails once growth cones encounter the glial scar at the region of injury.

7. Numerous factors with both positive and negative effects on axon extension in culture are associated with CNS scar tissue. Regeneration is likely to be inhibited by a number of

distinct mechanisms, including mechanical barriers, growth cone collapse, inhibition of outgrowth, and growth cone trapping.

8. Specific molecules expressed in regions of CNS scarring have complex and changing effects on regeneration, depending on the type on neuron encountering the factor, the internal state of the growth cone at the time the factor is encountered, and the molecular context in which the factor is encountered. Dissecting the role of individual molecules in regeneration failure is a task of exceptional difficulty.

9. Adult CNS regeneration failure reflects maturational changes in the intrinsic properties of CNS neurons and the maladaptive response of these neurons to injury.

10. Cell replacement therapy may prove to be a means of restoring function lost due to death of CNS neurons, either by stimulating the division of endogenous neural stem cells or by transplanting fetal or ESCs into the CNS. Very little is known regarding the long-term survival and function of transplanted stem cell or their derivatives, due in part to the immune rejection of these cells and the tendency of ESCs to form teratomas in adult tissue.

ACKNOWLEDGMENT

This work was supported by grant R01 NS382138.

REFERENCES

Allan, S.M. and Rothwell, N.J., 2001, Cytokines and acute neurodegeneration, *Nat. Rev. Neurosci.* 2:734–744.

Andersen, L.B. and Schreyer, D.J., 1999, Constitutive expression of GAP-43 correlates with rapid, but not slow regrowth of injured dorsal root axons in the adult rat, *Exp. Neurol.* 155:157–164.

Asher, R.A., Morgenstern, D.A., Moon, L.D., and Fawcett, J.W., 2001, Chondroitin sulphate proteoglycans: Inhibitory components of the glial scar, *Prog. Brain. Res.* 132:611–619.

Bandtlow, C.E. and Loschinger, J., 1997, Developmental changes in neuronal responsiveness to the CNS myelin-associated neurite growth inhibitor NI-35/250, *Eur. J. Neurosci.* 9:2743–2752.

Barbeau, H., McCrea, D.A., O'Donovan, M.J., Rossignol, S., Grill, W.M., and Lemay, M.A., 1999, Tapping into spinal circuits to restore motor function, *Brain Res. Brain Res. Rev.* 30:27–51.

Bartsch, U., 1996, Myelination and axonal regeneration in the central nervous system of mice deficient in the myelin-associated glycoprotein, *J. Neurocytol.* 25:303–313.

Bartsch, U., Bandtlow, C.E., Schnell, L., Bartsch, S., Spillmann, A.A., Rubin, B.P. *et al.*, 1995, Lack of evidence that myelin-associated glycoprotein is a major inhibitor of axonal regeneration in the CNS, *Neuron* 15:1375–1381.

Bates, C.A. and Meyer, R.L., 1997, The neurite-promoting effect of laminin is mediated by different mechanisms in embryonic and adult regenerating mouse optic axons in vitro, *Dev. Biol.* 181:91–101.

Bates, C.A. and Stelzner, D.J., 1993, Extension and regeneration of corticospinal axons after early spinal injury and the maintenance of corticospinal topography, *Exp. Neurol.* 123:106–117.

Becker, C.G., Becker, T., and Meyer, R.L., 2001, Increased NCAM-180 immunoreactivity and maintenance of L1 immunoreactivity in injured optic fibers of adult mice, *Exp. Neurol.* 169:438–448.

Behrens, M.M., Strasser, U., Lobner, D., and Dugan, L.L., 1999, Neurotrophin-mediated potentiation of neuronal injury, *Microsc. Res. Tech.* 45:276–284.

Bomze, H.M., Bulsara, K.R., Iskandar, B.J., Caroni, P., and Pate Skene, J.H., 2001, Spinal axon regeneration evoked by replacing two growth cone proteins in adult neurons, *Nat. Neurosci.* 4:38–43.

Bovolenta, P. and Fernaud-Espinosa, I., 2000, Nervous system proteoglycans as modulators of neurite outgrowth, *Prog. Neurobiol.* 61:113–132.

Bradbury, E.J., Moon, L.D., Popat, R.J., King, V.R., Bennett, G.S., Patel, P.N. *et al.*, 2002, Chondroitinase ABC promotes functional recovery after spinal cord injury, *Nature* 416:636–640.

Braunewell, K.H., Pesheva, P., McCarthy, J.B., Furcht, L.T., Schmitz, B., and Schachner, M., 1995, Functional involvement of sciatic nerve-derived versican- and decorin-like molecules and other chondroitin sulphate proteoglycans in ECM-mediated cell adhesion and neurite outgrowth, *Eur. J. Neurosci.* 7:805–814.

Broude, E., McAtee, M., Kelley, M.S., and Bregman, B.S., 1999, Fetal spinal cord transplants and exogenous neurotrophic support enhance c-Jun expression in mature axotomized neurons after spinal cord injury, *Exp. Neurol.* 155:65–78.

Buettner, H.M. and Pittman, R.N., 1991, Quantitative effects of laminin concentration on neurite outgrowth in vitro, *Dev. Biol.* 145:266–276.

Buffo, A., Holtmaat, A.J., Savio, T., Verbeek, J.S., Oberdick, J., Oestreicher, A.B. *et al.*, 1997, Targeted overexpression of the neurite growth-associated protein B-50/GAP-43 in cerebellar Purkinje cells induces sprouting after axotomy but not axon regeneration into growth-permissive transplants, *J. Neurosci.* 17:8778–8791.

Cai, D., Qiu, J., Cao, Z., McAtee, M., Bregman, B.S., and Filbin, M.T., 2001, Neuronal cyclic AMP controls the developmental loss in ability of axons to regenerate, *J. Neurosci.* 21:4731–4739.

Cao, Q., Benton, R.L., and Whittemore, S.R., 2002, Stem cell repair of central nervous system injury, *J. Neurosci. Res.* 68:501–510.

Caroni, P., 1997, Intrinsic neuronal determinants that promote axonal sprouting and elongation, *Bioessays* 19:767–775.

Challacombe, J.F., Snow, D.M., and Letourneau, P.C., 1996, Actin filament bundles are required for microtubule reorientation during growth cone turning to avoid an inhibitory guidance cue, *J. Cell Sci.* 109: 2031–2040.

Challacombe, J.F., Snow, D.M., and Letourneau, P.C., 1997, Dynamic microtubule ends are required for growth cone turning to avoid an inhibitory guidance cue, *J. Neurosci.* 17:3085–3095.

Chau, C., Barbeau, H., and Rossignol, S., 1998, Effects of intrathecal alpha1- and alpha2-noradrenergic agonists and norepinephrine on locomotion in chronic spinal cats, *J. Neurophysiol.* 79:2941–2963.

Chavis, P. and Westbrook, G., 2001, Integrins mediate functional pre- and postsynaptic maturation at a hippocampal synapse, *Nature* 411:317–321.

Chen, D.F., Schneider, G.E., Martinou, J.C., and Tonegawa, S., 1997, Bcl-2 promotes regeneration of severed axons in mammalian CNS, *Nature* 385:434–439.

Chernoff, E., Sato, K., Corn, A., and Karcavich, R., 2002, Spinal cord regeneration: Intrinsic properties and emerging mechanisms, *Semin. Cell. Dev. Biol.* 13:361.

Cohen, J., Burne, J.F., Winter, J., and Bartlett, P., 1986, Retinal ganglion cells lose response to laminin with maturation, *Nature* 322:465–467.

Cohen, J., Nurcombe, V., Jeffrey, P., and Edgar, D., 1989, Developmental loss of functional laminin receptors on retinal ganglion cells is regulated by their target tissue, the optic tectum, *Development* 107:381–387.

Colamarino, S.A. and Tessier-Lavigne, M., 1995, The axonal chemoattractant netrin-1 is also a chemorepellent for trochlear motor axons, *Cell* 81:621–629.

Colavita, A. and Culotti, J.G., 1998, Suppressors of ectopic UNC-5 growth cone steering identify eight genes involved in axon guidance in Caenorhabditis elegans, *Dev. Biol.* 194:72–85.

Condic, M.L., 2001, Adult neuronal regeneration induced by transgenic integrin expression, *J. Neurosci.* 21:4782–4788.

Condic, M.L., 2002, Neural development: Axon regeneration derailed by dendrites, *Curr. Biol.* 12:R455–R457.

Condic, M.L. and Lemons, M.L., 2002, Extracellular matrix in spinal cord regeneration: Getting beyond attraction and inhibition, *NeuroReport* 13:A37–48.

Condic, M.L. and Letourneau, P.C., 1997, Ligand induced changes in integrin expression regulate neuronal adhesion and neurite outgrowth, *Nature* 389:852–856.

Condic, M.L., Snow, D.M., and Letourneau, P.C., 1999, Embryonic neurons adapt to the inhibitory proteoglycan aggrecan by increasing integrin expression, *J. Neurosci.* 19:10036–10043.

Coulson, E.J., Barrett, G.L., Storey, E., Bartlett, P.F., Beyreuther, K., and Masters, C.L., 1997, Down-regulation of the amyloid protein precursor of Alzheimer's disease by antisense oligonucleotides reduces neuronal adhesion to specific substrata, *Brain Res.* 770: 72–80.

David, S. and Aguayo, A.J., 1981, Axonal elongation into peripheral nervous system "bridges" after central nervous system injury in adult rats, *Science* 214:931–933.

Davies, S.J., Goucher, D.R., Doller, C., and Silver, J., 1999, Robust regeneration of adult sensory axons in degenerating white matter of the adult rat spinal cord, *J. Neurosci.* 19:5810–5822.

Davies, S.J. and Silver, J., 1998, Adult axon regeneration in adult CNS white matter, *Trends Neurosci.* 21:515.

Diefenbach, T.J., Guthrie, P.B., and Kater, S.B., 2000, Stimulus history alters behavioral responses of neuronal growth cones, *J. Neurosci.* 20:1484–1494.

Dieringer, M. and Precht, W., 1979, Synaptic mechanisms involved in compensation of vestibular function following hemilabrinthectomy, *Brain Res.* 50:607–615.

Edgerton, V.R., Roy, R.R., Hodgson, J.A., Prober, R.J., de Guzman, C.P., and de Leon, R., 1992, Potential of adult mammalian lumbosacral spinal cord to execute and acquire improved locomotion in the absence of supraspinal input, *J. Neurotrauma.* 9:S119–S128.

Faissner, A., 1997, The tenascin gene family in axon growth and guidance, *Cell Tissue Res.* 290:331–341.

Forehand, C.J. and Farel, P.B., 1982, Anatomical and behavioral recovery from the effects of spinal cord transection: Dependence on metamorphosis in anuran larvae, *J. Neurosci.* 2:654–662.

Fouad, K., Metz, G.A., Merkler, D., Dietz, V., and Schwab, M.E., 2000, Treadmill training in incomplete spinal cord injured rats, *Behav. Brain Res.* 115:107–113.

Frostick, S.P., Yin, Q., and Kemp, G.J., 1998, Schwann cells, neurotrophic factors, and peripheral nerve regeneration, *Microsurgery* 18:397–405.

Fu, S.Y. and Gordon, T., 1997, The cellular and molecular basis of peripheral nerve regeneration, *Mol. Neurobiol.* 14:67–116.

Garcia-Verdugo, J.M., Ferron, S., Flames, N., Collado, L., Desfilis, E., and Font, E., 2002, The proliferative ventricular zone in adult vertebrates: A comparative study using reptiles, birds, and mammals, *Brain Res. Bull.* 57:765–775.

Gianola, S. and Rossi, F., 2002, Long-term injured purkinje cells are competent for terminal arbor growth, but remain unable to sustain stem axon regeneration, *Exp. Neurol.* 176:25–40.

Goldberg, J.L. and Barres, B.A. 2000, The relationship between neuronal survival and regeneration, *Annu. Rev. Neurosci.* 23:579–612.

Goldberg, J.L., Klassen, M.P., Hua, Y., and Barres, B.A., 2002, Amacrine-signaled loss of intrinsic axon growth ability by retinal ganglion cells, *Science* 296:1860–1864.

Golding, J.P., Bird, C., McMahon, S., and Cohen, J., 1999, Behaviour of DRG sensory neurites at the intact and injured adult rat dorsal root entry zone: Postnatal neurites become paralysed, whilst injury improves the growth of embryonic neurites, *Glia* 26:309–323.

Gomez, T.M. and Letourneau, P.C., 1994, Filopodia initiate choices made by sensory neuron growth cones at laminin/fibronectin borders in vitro, *J. Neurosci.* 14:5959–5972.

Grotewiel, M.S., Beck, C.D., Wu, K.H., Zhu, X.R., and Davis, R.L. 1998, Integrin-mediated short-term memory in Drosophila, *Nature* 391:455–460.

Harkema, S.J., 2001, Neural plasticity after human spinal cord injury: Application of locomotor training to the rehabilitation of walking, *Neuroscientist* 7:455–468.

Hedgecock, E.M., Culotti, J.G., and Hall, D.H., 1990, The unc-5, unc-6, and unc-40 genes guide circumferential migrations of pioneer axons and mesodermal cells on the epidermis in *C. elegans*, *Neuron* 4:61–85.

Hermanns, S., Reiprich, P., and Muller, H.W., 2001, A reliable method to reduce collagen scar formation in the lesioned rat spinal cord, *J. Neurosci. Meth.* 110:141–146.

Hildebrand, A., Romaris, M., Rasmussen, L.M., Heinegard, D., Twardzik, D.R., Border, W.A. *et al.*, 1994, Interaction of the small interstitial proteoglycans biglycan, decorin and fibromodulin with transforming growth factor beta, *Biochem. J.* 302:527–534.

Hopker, V.H., Shewan, D., Tessier-Lavigne, M., Poo, M., and Holt, C., 1999, Growth-cone attraction to netrin-1 is converted to repulsion by laminin-1, *Nature* 401:69–73.

Ito, J., Murata, M., and Kawaguchi, S., 1999, Regeneration of the lateral vestibulospinal tract in adult rats by transplants of embryonic brain tissue, *Neurosci. Lett.* 259:67–70.

Jones, F.S. and Jones, P.L., 2000, The tenascin family of ECM glycoproteins: Structure, function, and regulation during embryonic development and tissue remodeling, *Dev. Dyn.* 218:235–259.

Jones, L.S. and Grooms, S.Y., 1997, Normal and aberrant functions of integrins in the adult central nervous system, *Neurochem. Int.* 31:587–595.

Joosten, E.A., Dijkstra, S., Brook, G.A., Veldman, H., and Bar, P.R., 2000, Collagen IV deposits do not prevent regrowing axons from penetrating the lesion site in spinal cord injury, *J. Neurosci. Res.* 62:686–691.

Jung, M., Petrausch, B., and Stuermer, C.A., 1997, Axon-regenerating retinal ganglion cells in adult rats synthesize the cell adhesion molecule L1 but not TAG-1 or SC-1, *Mol. Cell. Neurosci.* 9:116–131.

Kamiguchi, H. and Lemmon, V., 2000, Recycling of the cell adhesion molecule L1 in axonal growth cones, *J. Neurosci.* 20:3676–3686.

Kamiguchi, H., Long, K.E., Pendergast, M., Schaefer, A.W., Rapoport, I., Kirchhausen, T. *et al.*, 1998, The neural cell adhesion molecule L1 interacts with the AP-2 adaptor and is endocytosed via the clathrin-mediated pathway, *J. Neurosci.* 18:5311–5321.

Keino-Masu, K., Masu, M., Hinck, L., Leonardo, E.D., Chan, S.S., Culotti, J.G. *et al.*, 1996, Deleted in Colorectal Cancer (DCC) encodes a netrin receptor, *Cell* 87:175–185.

Kennedy, T.E., Serafini, T., de la Torre, J.R., and Tessier-Lavigne, M., 1994, Netrins are diffusible chemotropic factors for commissural axons in the embryonic spinal cord, *Cell* 78:425–435.

Kirschstein, R. and Skirboll, L.R./National Institutes of Health, 2001, *Stem Cells: Scientific Progress and Future Research Directions* (May 2, 2002); http://www.nih.gov/news/stemcell/scireport.htm.

Kolodziej, P.A., Timpe, L.C., Mitchell, K.J., Fried, S.R., Goodman, C.S., Jan, L.Y. *et al.*, 1996, Frazzled encodes a Drosophila member of the DCC immunoglobulin subfamily and is required for CNS and motor axon guidance, *Cell* 87:197–204.

Kuhn, T.B., Schmidt, M.F., and Kater, S.B., 1995, Laminin and fibronectin guideposts signal sustained but opposite effects to passing growth cones, *Neuron* 14:275–285.

Kury, P., Stoll, G., and Muller, H.W., 2001, Molecular mechanisms of cellular interactions in peripheral nerve regeneration, *Curr. Opin. Neurol.* 14:635–639.

Leonardo, E.D., Hinck, L., Masu, M., Keino-Masu, K., Ackerman, S.L., and Tessier-Lavigne, M., 1997, Vertebrate homologues of C. elegans UNC-5 are candidate netrin receptors, *Nature* 386:833–838.

Letourneau, P.C., Condic, M.L., and Snow, D.M., 1994, Interactions of developing neurons with the extracellular matrix, *J. Neurosci.* 14:915–928.

Li, M., Shibata, A., Li, C., Braun, P.E., McKerracher, L., Roder, J. *et al.*, 1996, Myelin-associated glycoprotein inhibits neurite/axon growth and causes growth cone collapse, *J. Neurosci. Res.* 46:404–414.

Lim, D.A., Flames, N., Collado, L., and Herrera, D.G., 2002, Investigating the use of primary adult subventricular zone neural precursor cells for neuronal replacement therapies, *Brain Res. Bull.* 57:759–764.

Lindvall, O., 1998, Update on fetal transplantation: The Swedish experience, *Mov. Disord.* 13:83–87.

Logan, A., Baird, A., and Berry, M., 1999, Decorin attenuates gliotic scar formation in the rat cerebral hemisphere, *Exp. Neurol.* 159:504–510.

Long, K.E., Asou, H., Snider, M.D., and Lemmon, V., 2001, The role of endocytosis in regulating L1-mediated adhesion, *J. Biol. Chem.* 276:1285–1290.

Lundberg, C., Martinez-Serrano, A., Cattaneo, E., McKay, R.D., and Bjorklund, A., 1997, Survival, integration, and differentiation of neural stem cell lines after transplantation to the adult rat striatum, *Exp. Neurol.* 145:342–360.

Mason, M.R., Campbell, G., Caroni, P., Anderson, P.N., and Lieberman, A.R., 2000, Overexpression of GAP-43 in thalamic projection neurons of transgenic mice does not enable them to regenerate axons through peripheral nerve grafts, *Exp. Neurol.* 165:143–152.

McFarlane, S., 2000, Attraction vs. repulsion: The growth cone decides, *Biochem. Cell Biol.* 78:563–568.

McKenna, M.P. and Raper, J.A., 1988, Growth cone behavior on gradients of substratum bound laminin, *Dev. Biol.* 130:232–236.

McKeon, R.J., Hoke, A., and Silver, J., 1995, Injury-induced proteoglycans inhibit the potential for laminin-mediated axon growth on astrocytic scars, *Exp. Neurol.* 136:32–43.

Meiners, S., Mercado, M.L., and Geller, H.M., 2000, The multi-domain structure of extracellular matrix molecules: Implications for nervous system regeneration, *Prog. Brain Res.* 128:23–31.

Meyer, R.L., Miotke, J.A., and Benowitz, L.I., 1994, Injury induced expression of growth-associated protein-43 in adult mouse retinal ganglion cells in vitro, *Neuroscience* 63:591–602.

Miranda, J.D., White, L.A., Marcillo, A.E., Willson, C.A., Jagid, J., and Whittemore, S.R., 1999, Induction of Eph B3 after spinal cord injury, *Exp. Neurol.* 156:218–222.

Moon, L.D., Asher, R.A., Rhodes, K.E., and Fawcett, J.W., 2001, Regeneration of CNS axons back to their target following treatment of adult rat brain with chondroitinase ABC, *Nat. Neurosci.* 4:465–466.

Moreno-Flores, M.T. and Wandosell, F., 1999, Up-regulation of Eph tyrosine kinase receptors after excitotoxic injury in adult hippocampus, *Neuroscience* 91:193–201.

Nakamura, F., Kalb, R.G., and Strittmatter, S.M., 2000, Molecular basis of semaphorin-mediated axon guidance, *J. Neurobiol.* 44:219–229.

Neugebauer, K.M. and Reichardt, L.F., 1991, Cell-surface regulation of beta 1-integrin activity on developing retinal neurons, *Nature* 350:68–71.

Neumann, S., Bradke, F., Tessier-Lavigne, M., and Basbaum, A.I., 2002, Regeneration of sensory axons within the injured spinal cord induced by intraganglionic cAMP elevation, *Neuron* 34:885–893.

Neumann, S. and Woolf, C.J., 1999, Regeneration of dorsal column fibers into and beyond the lesion site following adult spinal cord injury, *Neuron* 23:83–91.

Nishimura, S.L., Boylen, K.P., Einheber, S., Milner, T.A., Ramos, D.M., and Pytela, R., 1998, Synaptic and glial localization of the integrin alphavbeta8 in mouse and rat brain, *Brain Res.* 791:271–282.

Nogradi, A. and Vrbova, G., 1994, The use of embryonic spinal cord grafts to replace identified motoneuron pools depleted by a neurotoxic lectin, volkensin, *Exp. Neurol.* 129:130–141.

Palecek, S.P., Horwitz, A.F., and Lauffenburger, D.A., 1999, Kinetic model for integrin-mediated adhesion release during cell migration, *Ann. Biomed. Eng.* 27:219–235.

Palecek, S.P., Loftus, J.C., Ginsberg, M.H., Lauffenburger, D.A., and Horwitz, A.F., 1997, Integrin-ligand binding properties govern cell migration speed through cell-substratum adhesiveness, *Nature* 385:537–540.

Parent, J.M. and Lowenstein, D.H., 2002, Seizure-induced neurogenesis: Are more new neurons good for an adult brain? *Prog. Brain Res.* 135:121–131.

Pasterkamp, R.J., De Winter, F., Holtmaat, A.J., and Verhaagen, J., 1998, Evidence for a role of the chemorepellent semaphorin III and its receptor neuropilin-1 in the regeneration of primary olfactory axons, *J. Neurosci.* 18:9962–9976.

Pasterkamp, R.J., Giger, R.J., Ruitenberg, M.J., Holtmaat, A.J., De Wit, J., De Winter, F. *et al.*, 1999, Expression of the gene encoding the chemorepellent semaphorin III is induced in the fibroblast component of neural scar tissue formed following injuries of adult but not neonatal CNS, *Mol. Cell. Neurosci.* 13:143–166.

Pasterkamp, R.J. and Verhaagen, J., 2001, Emerging roles for semaphorins in neural regeneration, *Brain Res. Brain Res. Rev.* 35:36–54.

Pinkstaff, J.K., Detterich, J., Lynch, G., and Gall, C., 1999, Integrin subunit gene expression is regionally differentiated in adult brain, *J. Neurosci.* 19:1541–1556.

Poppel, E., Held, R., and Frost, D., 1973, Residual visual function after brain wounds involving the central visual pathways, *Nature* 243:295–296.

Qiu, J., Cai, D., Dai, H., McAtee, M., Hoffman, P.N., Bregman, B.S. *et al.*, 2002, Spinal axon regeneration induced by elevation of cyclic AMP, *Neuron* 34:895–903.

Ramon y Cajal, S., 1991, Cajal's degeneration and regeneration of the nervous system. In *History of Neuroscience* (J. Defelipe and E.G. Jones, eds.), Oxford University Press, New York, p. 769.

Reichardt, L.F. and Tomaselli, T.J., 1991, Extracellular matrix molecules and their receptors, *Ann. Rev. Neurosci.* 14:531–570.

Richardson, P.M., McGuinness, U.M., and Aguayo, A.J., 1980, Axons from CNS neurons regenerate into PNS grafts, *Nature* 284:264–265.

Romero, M.I., Rangappa, N., Li, L., Lightfoot, E., Garry, M.G., and Smith, G.M., 2000, Extensive sprouting of sensory afferents and hyperalgesia induced by conditional expression of nerve growth factor in the adult spinal cord, *J. Neurosci.* 20:4435–4445.

Rossi, F., Bravin, M., Buffo, A., Fronte, M., Savio, T., and Strata, P., 1997, Intrinsic properties and environmental factors in the regeneration of adult cerebellar axons, *Prog. Brain Res.* 114:283–296.

Rossi, F. and Cattaneo, E., 2002, Opinion: Neural stem cell therapy for neurological diseases: Dreams and reality, *Nat. Rev. Neurosci.* 3:401–409.

Ruoslahti, E., Yamaguchi, Y., Hildebrand, A., and Border, W.A., 1992, Extracellular matrix/growth factor interactions, *Cold Spring Harb. Symp. Quant. Biol.* 57:309–315.

Salinero, O., Garrido, J.J., and Wandosell, F., 1998, Amyloid precursor protein proteoglycan is increased after brain damage, *Biochim. Biophys. Acta.* 1406:237–250.

Salinero, O., Moreno-Flores, M.T., and Wandosell, F., 2000, Increasing neurite outgrowth capacity of beta-amyloid precursor protein proteoglycan in Alzheimer's disease, *J. Neurosci. Res.* 60:87–97.

Sango, K., Horie, H., Inoue, S., Takamura, Y., and Takenaka, T., 1993, Age-related changes of DRG neuronal attachment to extracellular matrix proteins in vitro, *NeuroReport* 4:663–666.

Schmalfeldt, M., Bandtlow, C.E., Dours-Zimmermann, M.T., Winterhalter, K.H., and Zimmermann, D.R., 2000, Brain derived versican V2 is a potent inhibitor of axonal growth, *J. Cell. Sci.* 113:807–816.

Schreyer, D.J. and Skene, J.H., 1991, Fate of GAP-43 in ascending spinal axons of DRG neurons after peripheral nerve injury: Delayed accumulation and correlation with regenerative potential, *J. Neurosci.* 11:3738–3751.

Shimizu, I., Oppenheim, R.W., O'Brien, M., and Shneiderman, A., 1990, Anatomical and functional recovery following spinal cord transection in the chick embryo, *J. Neurobiol.* 21:918–937.

Shirasaki, R., Katsumata, R., and Murakami, F., 1998, Change in chemo-attractant responsiveness of developing axons at an intermediate target, *Science* 279:105–107.

Sholomenko, G.N. and Delaney, K.R., 1998, Restitution of functional neural connections in chick embryos assessed in vitro after spinal cord transection in Ovo, *Exp. Neurol.* 154:430–451.

Singer, W., 1982, Recovery mechanisms in the mammalian brain. In *Repair and Regeneration of the Nervous System* (J.G. Nicholls, ed.), Springer-Verlag, New York, pp. 203–226.

Snow, D.M., Atkinson, P.B., Hassinger, T.D., Letourneau, P.C., and Kater, S.B., 1994, Chondroitin sulfate proteoglycan elevates cytoplasmic calcium in DRG neurons, *Dev. Biol.* 166:87–100.

Stein, E. and Tessier-Lavigne, M., 2001, Hierarchical organization of guidance receptors: Silencing of netrin attraction by slit through a Robo/DCC receptor complex, *Science* 291:1928–1938.

Stichel, C.C., Niermann, H., D'Urso, D., Lausberg, F., Hermanns, S., and Muller, H.W., 1999, Basal membrane-depleted scar in lesioned CNS: Characteristics and relationships with regenerating axons, *Neuroscience* 93:321–333.

Subramanian, T., 2001, Cell transplantation for the treatment of Parkinson's disease, *Semin. Neurol.* 21:103–115.

Sunderland, S., 1965, The connective tissues of peripheral nerves, *Brain* 88:841–854.

Sunderland, S., 1970, Anatomical features of nerve trunks in relation to nerve injury and nerve repair, *Clin. Neurosurg.* 17:38–62.

Sunderland, S., 1990, The anatomy and physiology of nerve injury, *Muscle Nerve* 13:771–784.

Tanelian, D.L., Barry, M.A., Johnston, S.A., Le, T., and Smith, G.M., 1997, Semaphorin III can repulse and inhibit adult sensory afferents in vivo, *Nat. Med.* 3:1398–1401.

Temple, S., 2001, The development of neural stem cells, *Nature* 414:112–117.

Terenghi, G., 1999, Peripheral nerve regeneration and neurotrophic factors, *J. Anat.* 194:1–14.

Turlejski, K. and Djavadian, R., 2002, Life-long stability of neurons: A century of research on neurogenesis, neuronal death and neuron quantification in adult CNS, *Prog. Brain Res.* 136:39–65.

Tzeng, S.F., Cheng, H., Lee, Y.S., Wu, J.P., Hoffer, B.J., and Kuo, J.S., 2001, Expression of neural cell adhesion molecule in spinal cords following a complete transection, *Life Sci.* 68:1005–1012.

Vaudano, E., Campbell, G., Anderson, P.N., Davies, A.P., Woolhead, C., Schreyer, D.J. *et al.*, 1995, The effects of a lesion or a peripheral nerve graft on GAP-43 upregulation in the adult rat brain: An in situ hybridization and immunocytochemical study, *J. Neurosci.* 15:3594–3611.

Vielmetter, J., Kayyem, J.F., Roman, J.M., and Dreyer, W.J., 1994, Neogenin, an avian cell surface protein expressed during terminal neuronal differentiation, is closely related to the human tumor suppressor molecule deleted in colorectal cancer, *J. Cell Biol.* 127:2009–2020.

Wang, X.M., Basso, D.M., Terman, J.R., Bresnahan, J.C., and Martin, G.F., 1998a, Adult opossums (Didelphis virginiana) demonstrate near normal locomotion after spinal cord transection as neonates, *Exp. Neurol.* 151:50–69.

Wang, X.M., Terman, J.R., and Martin, G.F., 1998b, Regeneration of supraspinal axons after transection of the thoracic spinal cord in the developing opossum, Didelphis virginiana, *J. Comp. Neurol.* 398:83–97.

Weidner, N., Grill, R.J., and Tuszynski, M.H., 1999, Elimination of basal lamina and the collagen "scar" after spinal cord injury fails to augment corticospinal tract regeneration, *Exp. Neurol.* 160:40–50.

Wictorin, K. and Bjorklund, A., 1992, Axon outgrowth from grafts of human embryonic spinal cord in the lesioned adult rat spinal cord, *Neuroreport* 3:1045–1048.

Wilkinson, D.G., 2001, Multiple roles of EPH receptors and ephrins in neural development, *Nat. Rev. Neurosci.* 2:155–164.

Wu, A., Pangalos, M.N., Efthimiopoulos, S., Shioi, J., and Robakis, N.K., 1997, Appican expression induces morphological changes in C6 glioma cells and promotes adhesion of neural cells to the extracellular matrix, *J. Neurosci.* 17:4987–4993.

Yick, L.W., Wu, W., So, K.F., Yip, H.K., and Shum, D.K., 2000, Chondroitinase ABC promotes axonal regeneration of Clarke's neurons after spinal cord injury, *NeuroReport* 11:1063–1067.

Yin, Q., Kemp, G.J., and Frostick, S.P., 1998, Neurotrophins, neurones and peripheral nerve regeneration, *J. Hand Surg. [Br.]* 23:433–437.

13

Developmental Mechanisms in Aging and Age-Related Diseases of the Nervous System

Mark P. Mattson and Tobi L. Limke

INTRODUCTION

This chapter provides developmental neurobiologists with an overview of cellular and molecular changes that occur in the nervous system during aging, describes the current state of understanding of how aging impacts developmental processes operative in the adult nervous system, and considers how developmental mechanisms may contribute to the pathogenesis of neurodegenerative disorders such as Alzheimer's and Parkinson's diseases. Although studies of invertebrates, particularly *Caenorhabditis elegans* and *Drosophila*, have provided vital information on the molecular regulation of development, they have not yet been tapped to study mechanisms of nervous system aging. This chapter, therefore, focuses almost exclusively on the aging of mammalian nervous systems. While many age-associated changes in the nervous system also occur in other tissues, we will focus on those that have the highest impact (such as oxidative stress and protein accumulation) and those that are relatively unique to the nervous system (such as the age-associated alterations in the Notch–Delta signaling pathway). We will then explore some of the mechanisms that not only regulate development of the nervous system, but also play a role in aging in both the normal and diseased brain.

We now know that a spectrum of developmental processes operates in the adult mammalian nervous system. The adult nervous system is not "hard-wired"; instead, neuronal circuits undergo structural remodeling in response to environmental demands. Like other tissues, there are cells in the nervous system capable of undergoing proliferation, differentiation, and programmed cell death (apoptosis), as well as a number of more subtle changes that alter neural structure and function. For example, hippocampal synapses may form, disassemble, or change their shape in response to learning, stress, and fluctuations in levels of sex steroids (McEwen, 2001). In neurogenic regions of the adult brain, there are dynamic populations of stem cells capable of dividing and differentiating into neurons or glial cells (Gage, 2000). Programmed cell death (apoptosis) also occurs in the

adult nervous system, at a low level under normal conditions, and at an accelerated pace following injury or in certain neurological disorders (Mattson, 2000). As far as is known, developmental processes in the mature nervous system are regulated by similar, if not identical, signaling mechanisms to those employed during embryonic development. Thus, members of each of the major types of signaling systems employed in embryonic development are operative in the adult. The impact of aging on these signaling pathways, and the consequences for age-related alterations in the cytoarchitecture and function of the nervous system, will therefore be given considerable attention in this chapter. In order to understand how developmental mechanisms may contribute to normal aging and age-related dysfunction and diseases in the nervous system, it is first necessary to understand the cellular and molecular changes that occur during aging.

CELLULAR AND MOLECULAR CHANGES DURING NORMAL AGING

Aging in all tissues, including the nervous system, involves a progressive loss of normal function as a result of intrinsic and extrinsic forces (Fig. 1). These processes occur during normal aging, in the absence of disease; however, as will be discussed later, many of these processes are exacerbated during age-related neurodegenerative disorders and often accelerate the damage and/or inhibit effective repair. Changes that occur in the nervous system during normal aging include increased oxidative damage to proteins and DNA, accumulation of protein and lipid byproducts (e.g., lipofuscin and advanced glycation end products), reduced metabolic activity, mitochondrial dysfunction, and cytoskeletal alterations. These processes affect terminally differentiated cells as well as proliferating and maturing stem/progenitor populations. However, there are also age-related changes that are unique to the nervous system that are likely the result of the molecular complexity of neurons and glial cells, which express approximately 50–100 times more genes than cells in other

---

Mark P. Mattson and Tobi L. Limke • Laboratory of Neurosciences, National Institute on Aging Intramural Research Program, Baltimore, MD.

Developmental Neurobiology, 4th ed., edited by Mahendra S. Rao and Marcus Jacobson. Kluwer Academic / Plenum Publishers, New York, 2005.

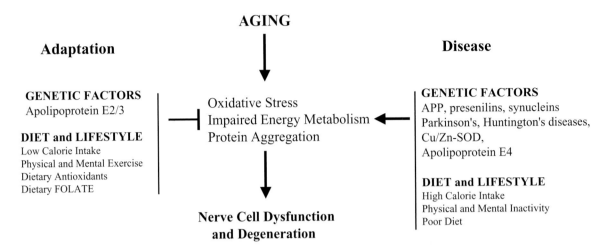

FIGURE 1. The nervous system may age successfully, or may suffer disease, depending upon its ability to adapt to adversity. Both intrinsic (genetic) and extrinsic (environmental) factors influence the outcome of aging. Successful aging of the nervous system is achieved when cells are able to adapt by enhancing their ability to resist degeneration and restore damaged neuronal circuits.

TABLE 1. Mechanisms that Regulate Successful and Unsuccessful Development and Aging in the Nervous System

Trophic factors (bFGF, BDNF)	Oxidative stress
Adhesion molecules (integrins)	Metabolic stress
Neurotransmitters (glutamate)	Diet (caloric intake)
Gases (nitric oxide)	Behavior (exercise)

tissues. The many different signal transduction pathways for neurotransmitters, trophic factors, and cytokines are examples of such complex regulatory systems that may be particularly prone to modification by aging. Many different genetic and environmental factors undoubtedly play roles in determining whether the nervous systems ages successfully by adapting to the aging process, or unsuccessfully resulting in disease. Interestingly, many of these determinant factors also play a critical role in developmental processes (Table 1).

Age-Related Cytoarchitectural Changes in the Nervous System

While the most dynamic structural changes in the cellular composition of the nervous system occur during embryonic and early postnatal development, there are similar but more subtle changes that occur throughout adult life. The changes include neurogenesis and gliogenesis, cell death, dendritic and axonal growth or retraction, synapse loss and remodeling, and glial cell reactivity. Alterations in cellular signaling pathways that control cell growth and motility may contribute to both adaptive and pathological structural changes in the aging brain. A prime example is glutamate, the major excitatory neurotransmitter in the mammalian central nervous system (CNS). Glutamate plays important roles in regulating dendritic growth cone motility and synaptogenesis during brain development (Mattson et al., 1988a, b, 1989) and in regulating synaptic plasticity in the adult (Izquierdo, 1994), but may also contribute to synaptic degeneration and cell death in

aging and age-related disorders such as Alzheimer's disease and stroke (Hugon et al., 1996; Mattson and Furukawa, 1998).

Because cellular structure is controlled by the cytoskeleton, many architectural changes in the brain with aging result from alterations in cytoskeletal proteins. The primary cytoskeletal components of cells are actin microfilaments (6 nm diameter); intermediate filaments (10–15 nm diameter), made of one or more cell type-specific intermediate filament proteins (e.g., neurofilament proteins in neurons and glial fibrillary acidic protein in astrocytes); and microtubules (25 nm in diameter), which are made of tubulin. In order to control the polymerization dynamics of cytoskeletal filaments and their interactions with other cytoskeletal components and membranes, cells express an array of cytoskeleton-associated proteins that are particularly complex in neurons. For example, several different microtubule-associated proteins (MAPs) are expressed in neurons where they are differentially distributed within the complex neuritic architecture of the cell. A well-known example is the presence of MAP-2 in dendrites and its absence in the axon, whereas an MAP called tau is present in axons but not in dendrites (Mandell and Banker, 1995). Alterations in the subcellular localization and phosphorylation state of MAPs are widely documented in aging and neurodegenerative disorders (Mandelkow and Mandelkow, 1995).

Studies of rodents and primates have revealed several changes in the cytoskeleton of neurons and glial cells during aging (Fig. 2). Overall levels of cytoskeletal proteins (tubulin, actin, and neurofilament proteins) do not change appreciably with normal aging, with a few exceptions. One cytoskeletal protein that does increase consistently during normal brain aging in humans and laboratory animals is the astrocytic intermediate filament protein glial fibrillary acidic protein (Morgan et al., 1999); this increase is characteristic of activated astrocytes and may therefore result from a reaction to subtle neurodegenerative changes. Several changes in the cytoskeletal organization and in posttranslational modifications of cytoskeletal proteins occur in the aging nervous system. Neurites may become distorted or dystrophic,

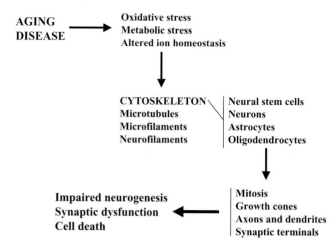

AGING
DISEASE

Oxidative stress
Metabolic stress
Altered ion homeostasis

CYTOSKELETON   Neural stem cells
Microtubules   Neurons
Microfilaments   Astrocytes
Neurofilaments   Oligodendrocytes

Impaired neurogenesis   Mitosis
Synaptic dysfunction   Growth cones
Cell death   Axons and dendrites
                                            Synaptic terminals

FIGURE 2. Roles of the cytoskeleton in aging and disorders of the nervous system. Increases in oxidative stress, impaired energy metabolism, and perturbed cellular ion homeostasis result in modifications of the cytoskeleton of neurons, glia, and neural stem cells. The modifications may include increased or decreased protein phosphorylation, oxidative modifications, and changes in polymerization state and interactions with cytoskeleton-associated proteins. The alterations in the cytoskeleton may adversely affect neurogenesis, neurite outgrowth, and synaptic plasticity, and may ultimately result in the death of neurons, glia, and neural stem cells.

while astrocytes may assume a more ramified structure. One prominent type of posttranslational alteration that occurs during aging is an increase in phosphorylation of several cytoskeletal proteins. For example, increased phosphorylation of the MAP tau occurs in neurons in some brain regions, particularly those involved in learning and memory, such as the hippocampus and basal forebrain. Increased or decreased proteolysis of cytoskeletal proteins may result in localized loss or accumulation of the proteins. Calcium-mediated proteolysis of cytoskeletal proteins, such as MAP-2 and spectrin, increases in some neuronal populations during aging (Nixon et al., 1994). On the other hand, aggregates of several proteins occur during aging in humans including tau, amyloid beta-peptide, alpha-synuclein, and ubiquitin (Johnson, 2000). As the result of increased levels of oxidative stress during aging, there is increased oxidative modification of cytoskeletal proteins which can manifest as carbonyls, glycation, and covalent binding of lipid peroxidation products such as 4-hydroxynonenal (Keller and Mattson, 1998). Cytoskeletal alterations are also a prominent feature of Parkinson's disease, with abnormal accumulations of neurofilaments, associated MAPs (particularly MAP-1b), alpha-synuclein, and actin-related proteins such as gelsolin, forming in neurons (Braak and Braak, 2000). Lower motor neurons are also vulnerable to age-related disease; in amyotrophic lateral sclerosis, motor neurons become filled with massive accumulations of neurofilaments that are concentrated in proximal regions of the axon (Julien and Beaulieu, 2000).

Synaptic remodeling occurs in the adult nervous system with the extent of remodeling depending on the particular neuronal circuits involved and the environmental demands that are placed upon those circuits. For example, synaptic connections in the

hippocampus are modified by learning and memory (Muller et al., 2000), physical activity (Cotman and Berchtold, 2002), psychosocial stress (Fuchs et al., 2001), and even changes in diet (Prolla and Mattson, 2001). Studies of synapses during the aging of rodents and humans suggest that in some brain regions there may be decreases in synaptic numbers, but that such decreases may be offset by increases in synaptic size, whereas in other brain regions, no changes in synapse numbers or size can be discerned (Bertoni-Freddari et al., 1996). There may be a preferential loss of synapses and neurons with particular neurotransmitter phenotypes during aging. For example, cholinergic synapses on dendrites of cortical layer V pyramidal neurons are reduced in numbers during aging to an extent greater than other types of synapses (Casu et al., 2002). Studies of cerebellar circuitry indicate that the numbers of synapses on Purkinje cell dendrites decrease during aging, but the size of each synapse increases (Chen and Hillman, 1999). Thus, there is considerable evidence that synaptic remodeling occurs in the CNS during aging (DeKosky et al., 1996).

Age-Related Molecular Changes in the Nervous System

Many of the molecular alterations that occur in the nervous system also occur in other tissues and can therefore be considered typical of aging. However, some age-related molecular changes may be confined to specific regions of the nervous system, or to specific neuronal circuits. For example, a progressive loss of D2 dopamine receptors occurs during aging and may contribute to age-related deficits in motor function (Roth, 1995). In humans, the protein content of the brain typically decreases with aging, which likely plays a major role in the progressive decrease in overall brain weight that occurs with aging. Insoluble aggregates of proteins accumulate in the brain during aging, with the cytoskeletal protein tau and Aβ being the two most closely linked to age-related neurodegeneration. Changes in membrane lipids during aging have been documented in numerous studies, with one prominent change being an increase in the levels of sphingomyelin (Giusto et al., 1992). A conspicuous lipid alteration during aging is the intracellular accumulation of damaged membrane lipids which form autofluorescent lipofuscin granules. Although there is little or no change in overall DNA content in the brain during aging, brain region-specific changes in RNA levels have been documented. Thus, levels of RNA decrease in the basal nucleus of Meynert, in several regions of cerebral cortex, and in some cranial nerve nuclei with advancing age, whereas RNA levels increase in the subiculum (Naber and Dahnke, 1979). While global changes in the molecular composition of the nervous system do not change dramatically during aging, numerous alterations in specific molecules have been identified.

Oxidative Damage during Aging

The most widely documented changes during aging are those resulting from increased oxidative stress. Free radicals are molecules with an unpaired electron in their outer orbital, which makes them highly reactive and capable of damaging other molecules by abstracting hydrogen ions. A prominent free radical produced in cells is the superoxide anion radical ($O_2^-\cdot$), which is generated in mitochondria during the electron transport process, as well as by the activities of various oxygenases (e.g., cyclooxygenases). Superoxide is normally eliminated from cells via the activity of manganese- and copper/zinc superoxide dismutases (MnSOD and Cu/ZnSOD), which convert $O_2^-\cdot$ to hydrogen peroxide (H_2O_2). However, hydrogen peroxide is a source of a damaging free radical called hydroxyl radical ($\cdot$OH), formed in a reaction catalyzed by Fe^{2+} and Cu^+. Because of its potential to be toxic, cells possess enzymes called glutathione peroxidases and catalases that eliminate hydrogen peroxide. Another free radical in cells of the nervous system is nitric oxide which is formed as the result of calcium-mediated activation of enzymes called nitric oxide synthases. A related reactive oxygen molecule called peroxynitrite is formed as the result of the interaction of superoxide with nitric oxide. The importance of oxyradicals in aging is emphasized by compelling evidence that there is an increase in production and accumulation of oxyradicals in essentially all tissues in the body during the aging process, including the brain (Sastre *et al.*, 2000). As a result, there is progressive oxidative damage to membrane lipids, proteins, and nucleic acids that apparently contributes to neural impairments during aging.

During aging, free radicals can attack the double bonds of membrane lipids in a process called lipid peroxidation. This process impairs the function of various types of membrane proteins in neurons and glial cells including receptors, ion-motive ATPases, glucose and glutamate transporters, and GTP-binding proteins (Mattson, 1998). This may occur as the result of covalent modification of the membrane proteins by an aldehydic product of lipid peroxidation called 4-hydroxynonenal. Lipid peroxidation-related changes may also contribute to a variety of age-related changes throughout neurons and other cells. For example, covalent modification of cytoskeletal proteins by 4-hydroxynonenal can alter protein phosphorylation resulting in abnormalities in cytoskeletal dynamics (Mattson *et al.*, 1997). In addition, functions of mitochondria and the endoplasmic reticulum can be adversely affected by lipid peroxidation. By altering the function of ion channels and ion-motive ATPases, lipid peroxidation can have a particularly damaging effect on cellular ion homeostasis (Mattson, 1998; Lu *et al.*, 2002).

Oxidative damage to nuclear and mitochondrial DNA occurs in cells of the nervous system during development and throughout adult life. In the nucleus, damaged DNA is normally repaired by highly efficient DNA repair enzyme systems, whereas in mitochondria, damaged DNA is less readily repaired. During aging, and particularly in age-related neurodegenerative disorders, DNA damage may become excessive and may trigger cell death (Rao, 1993; Mattson, 2000). DNA damage can also cause cell cycle arrest and/or death of mitotic cells including glia and neural progenitor cells (LeDoux *et al.*, 1996; Cheng *et al.*, 2001). Many age-related oxidative processes are greatly enhanced in neurodegenerative disorders. Studies of brain tissues of patients with Alzheimer's and Parkinson's diseases have revealed increased levels of protein oxidation in vulnerable brain regions and, in particular, in degenerating neurons. Two proteins shown to be heavily glycated in AD are Aβ and tau, the major components of plaques and neurofibrillary tangles, respectively.

Mitochondrial DNA damage can be extensive during normal aging, largely because mitochondria are sites where the vast majority of free radicals are generated and because cells do not possess effective systems for repair of damaged mitochondrial DNA. Damage to mitochondrial DNA can lead to failure of mitochondrial electron transport and reduced ATP production, and can impair calcium-regulating functions of mitochondria. These changes can render neurons vulnerable to excitotoxic and metabolic insults. The importance of mitochondrial oxyradical production in aging in general is underscored by recent studies of the mechanism whereby caloric restriction extends lifespan in rodents and nonhuman primates. Levels of cellular oxidative stress (as indicated by oxidation of proteins, lipids, and DNA) are decreased in many different nonneural tissues of rats and mice maintained on a calorie-restricted diet (30–40% reduction in calories). Recent studies suggest that levels of oxidative stress are also reduced in the brains of calorie-restricted rodents (Dubey *et al.*, 1996). The current dogma for the underlying mechanism is that reduced mitochondrial metabolism due to reduced energy availability results in a net decrease in mitochondrial ROS production over time, and hence less radical-mediated cellular damage. Thus, one factor contributing to brain aging is simply the constant production of oxyradicals and resultant progressive damage to cells.

Alterations in Signaling Pathways during Aging

Additional alterations of aging that may be more specific to the nervous system are impaired calcium signaling and neurotrophic factor signaling, which may promote perturbed synaptic function and neuronal degeneration. Alterations in neuronal calcium regulation and expression of certain Ca^{2+}-binding proteins are observed in aged rodents (Disterhoft *et al.*, 1994); such changes in the hippocampus are associated with age-related deficits in learning and memory. Changes in the levels of voltage-dependent calcium channels and glutamate receptors may also occur during aging (Clayton *et al.*, 2002). An age-related decrease in nerve growth factor (NGF) levels and levels of NGF receptors in the aging rodent brain apparently contributes to age-related cognitive impairment (Koh *et al.*, 1989; Nabeshima *et al.*, 1994). Brain-derived neurotrophic factor (BDNF) signaling also decreases during aging, with an associated decline in learning and memory (Lapchak *et al.*, 1993; Croll *et al.*, 1998). Similarly, neurotrophin-3 and neurotrophin-4 levels decrease in the targets of sensory neurons during aging (Bergman *et al.*, 2000), which may play a role in age-related sensory deficits. The ability of the nervous system to modulate neurotrophic factor signaling in response to stress may be compromised during aging (Smith and Cizza, 1996). Analysis of gene expression in individual neurons in the basal forebrain of young and old rats revealed significant decreases in the percentage of neurons expressing choline acetyltransferase and of neurons expressing glutamate decarboxylase (Han *et al.*, 2002), suggesting that neurons cease producing acetylcholine and GABA during aging, and/or that neurons expressing these neurotransmitters are preferentially lost during aging.

Aging and Programmed Cell Death

The programmed cell death of neurons that occurs during development is easily documented in many regions of the nervous system as relatively large numbers of cells die during a brief time window (Johnson and Oppenheim, 1994). Apoptosis is the predominant form of programmed developmental cell death; it is characterized by cell shrinkage, membrane blebbing, and nuclear chromatin condensation and fragmentation. A biochemical cascade involving pro-apoptotic Bcl-2 family members such as Bax and Bad, mitochondrial alterations resulting in the release of cytochrome *c*, and activation of death effector enzymes called caspases mediates apoptosis (Fig. 3). It is believed that one important trigger of developmental neuronal death is insufficient access to target-derived neurotrophic factors that occurs at the time synapses are being formed. Neural precursor cells may also undergo apoptosis (de la Rosa and de Pablo, 2000), but the factors that control their survival remain to be determined.

Considerable evidence suggests that many neurons die during adult life, and that such cell deaths are increased during aging and even more so in neurodegenerative disorders (Mattson, 2000). Age-related decreases in number of neurons have been documented in some brain regions, but not in others (West *et al.*, 1994). Age-related neuronal death presumably results from apoptosis or a related form of programmed cell death, but this has not been conclusively established. It is unlikely that neurons undergo necrosis because this form of cell death, which is characterized by cell swelling and rupture, usually involves large numbers of cells dying in clusters; this phenomenon has not been observed in the nervous system during normal aging. In an immunohistochemical study of the cerebellum and hippocampus of young adult and old rats, it was shown that levels of the apoptotic protein p53 are increased in Purkinje cells and hippocampal CA1 neurons of old rats (Chung *et al.*, 2000). Many neurons and glial cells may undergo adaptive responses during aging that allow them to survive. Levels of the anti-apoptotic protein Bcl-2 are increased in hippocampal and cerebellar neurons during aging, and this increase appears to be a cytoprotective response to age-related increases in levels of oxidative stress (Kaufmann

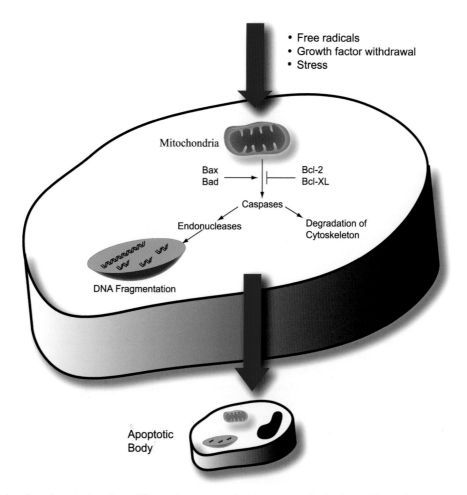

FIGURE 3. Simplified outline of apoptosis pathway. When cells are exposed to severe stress (in the form of free radicals, growth factor withdrawal, etc.), a signaling cascade is activated in which pro-apoptotic factors (such as Bax and Bad) activate caspases, which then activate other proteins which degrade the cytoskeleton and gauge fragmentation of nuclear DNA. Apoptosis is a common feature in both normal development and in several neurodegenerative disorders.

et al., 2001). The strongest evidence for neuronal apoptosis during aging comes from studies of neurodegenerative disorders in which numbers of cell deaths are greatly increased. Numerous studies of postmortem brain tissues from patients with Alzheimer's disease, Parkinson's disease, and stroke have provided evidence that neurons die by apoptosis. Hallmarks of apoptosis, including increased levels of p53, Par-4, Bax, activated caspases, are present in neurons affected in these disorders (Mattson, 2000). In addition, interventions known to prevent apoptosis, such as inhibitors of p53 and caspases, and agents that stabilize mitochondria, can prevent neuronal death in animal and cell culture models (Robertson *et al.*, 2000; Culmsee *et al.*, 2001; Liu *et al.*, 2002). The factors that trigger neuronal apoptosis during normal aging are not known, but may involve oxidative and metabolic stress, and reduced trophic support.

Neural Control of Aging

As described above, the brain undergoes profound changes during the aging process. Interestingly, there is increasing evidence suggesting that the brain also plays a role in regulating lifespan as well as health status during the aging process. The nervous system contains several signaling pathways that influence and possibly regulate lifespan in individuals. One such pathway is the insulin-like signaling pathway in mammals, in which activated plasma membrane receptor kinases phosphorylate tyrosine residues on an intracellular adapter protein termed insulin receptor substrate-1 (IRS-1). IRS-1 then activates phosphatidylinositol-3-kinase (PI3K), which activates Akt (protein kinase B), a regulator of several targets including forkhead transcription factors (van Weeren *et al.*, 1998; Tang *et al.*, 1999). In the mammalian brain, this pathway influences several aspects of neural development, including neuronal growth and differentiation, retinal axon pathfinding (Song *et al.*, 2003) and growth factor-mediated neuronal survival (Vaillant *et al.*, 1999; Gary and Mattson, 2001). Insulin-like signaling decreases in the rat brain during aging (Sonntag *et al.*, 1999), while infusion of insulin-like growth factor-1 (IGF-1) into the lateral ventricle of aged rats can ameliorate age-related deficits in brain energy metabolism (Lichtenwalner *et al.*, 2001) and memory (Markowska *et al.*, 1998). Thus, insulin-like signaling apparently plays a critical role in both neural development and age-related neural decline; additionally, it may play a role in determining the lifespan of an individual, as demonstrated by studies in nonmammalian species. Mutations in the insulin receptor (Tatar *et al.*, 2001) and the IRS homolog CHICO (Clancy *et al.*, 2001) result in an increased lifespan in *Drosophila*. In *C. elegans*, there are several homologs of members of the insulin-like signaling pathway including the insulin receptor (daf-2), PI3K (age-1), and the forkhead transcription factor (daf-16). Mutations in *daf-2* and *age-1* increase lifespan in *C. elegans*. When cell-type specific promoters are used to overexpress wild-type daf-2 or age-1 in *daf-2* or *age-1* mutants, the increased longevity of the mutants is reversed but only when overexpression occurs in the nervous system, but not when overexpression is targeted to muscle or intestinal cells (Wolkow *et al.*, 2000). Similar increases in lifespan are reported

for mutations in the *C. elegans* tryptophan hydroxylase homolog *tph-1* (Sze *et al.*, 2000), suggesting that more than one pathway regulates lifespan.

In mammals, there is indirect evidence that neural signaling pathways can influence lifespan. Dietary restriction extends lifespan in animals and causes a corresponding decrease in circulating insulin levels and increased insulin sensitivity and glucose tolerance (Kalant *et al.*, 1988; Weindruch and Sohal, 1997; Wanagat *et al.*, 1999). Dietary restriction also increases levels of BDNF in several brain regions in rodents (Duan *et al.*, 2001; Prolla and Mattson, 2001; Lee *et al.*, 2002b). BDNF interacts with the trkB receptor, whose signaling pathway is remarkably similar to the insulin pathway (Foulstone *et al.*, 1999) and is generally considered to be a neuroprotective trophic factor. Significantly, dietary restriction delays age-related deficits in learning and memory in rodents (Ingram *et al.*, 1987) and can protect neurons against dysfunction and death in rodent models of Alzheimer's disease, Parkinson's disease, and stroke (Bruce-Keller *et al.*, 1999; Duan and Mattson, 1999; Yu and Mattson, 1999; Zhu *et al.*, 1999). Dietary restriction also increases neural levels of antioxidant enzymes, stress proteins (such as HSP-70 and GRP-78), and anti-apoptotic proteins (such as Bcl-2), suggesting that the lifespan-increasing effect of dietary restriction may result from decreased oxyradical production and enhanced cellular stress resistance (Bruce-Keller *et al.*, 1999; Duan and Mattson, 1999; Yu and Mattson, 1999). The dual effect on oxidative stress and trophic factors emphasizes the point that aging is a complex process with many overlapping and converging pathways that play a role in the aging process. Further, the ability of alterations in specific signaling pathways to alter aspects of aging indicates that the nervous system is not only affected during the aging process, but may also play an active role in determining an individual's lifespan.

DEVELOPMENTAL MECHANISMS UNDERLYING AGE-RELATED ALTERATIONS IN NEUROGENESIS

As described in the previous sections, many of the mechanisms that regulate neural development are believed to play a role in the aging of the nervous system. This is especially true for neural stem cells, which continue dividing in the adult brain long after most neural cells have undergone terminal differentiation, albeit at a lower rate than in the developing brain. Neural stem cells are defined as cells that can self-renew through cell division and are multipotent (i.e., they can produce differentiated progeny of all three mature neural cells: neurons, astrocytes, and oligodendrocytes). Neural stem cells in the adult brain have several potential fates (Fig. 4). The first is to remain quiescent and not re-enter the cell cycle, thus preventing self-renewal and differentiation into mature neurons and glia. This process has the additional consequence of reducing the stem cell pool as it is not renewed by new cell divisions. Stem cells may also enter the cell cycle but undergo apoptosis and die, or they may re-enter the cell cycle and successfully produce differentiated progeny.

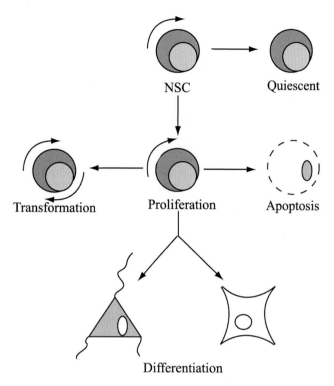

FIGURE 4. A neural stem cell has several fates. It can remain quiescent and not undergo any further cell divisions. Alternatively, it can re-enter the cell cycle and divide symmetrically (to produce more neural stem cells) or asymmetrically (to produce differentiated neurons and/or glia). If the stem cell becomes transformed, it will divide uncontrollably and may contribute to tumor formation. Finally, a stem cell and/or its progeny can undergo apoptosis and be removed from the tissue.

Finally, stem cells may successfully undergo division to produce healthy, functional daughter cells. Such division may be symmetric, to produce two identical daughter cells, or asymmetric, to produce one new stem cell and one daughter cell that will become a differentiated cell. The final outcome depends on the convergence of many intrinsic and extrinsic signals received by the cell and is influenced by factors such as cell density, receptor expression, and cross-talk between various signaling pathways (Sommer and Rao, 2002). Many of the mechanisms driving neural stem cell proliferation, differentiation, and survival in the adult and aging brain are similar to those regulating neural development and are often implicated in the pathogenesis of age-related neurodegenerative disorders.

Neurogenesis in the Developing and Adult Nervous System

The two primary populations of neural stem cells in the adult brain are located in the subventricular zone adjacent to the lateral ventricles and in the dentate gyrus of the hippocampus. Stem cells in the subventricular zone give rise to interneurons of the olfactory bulb, a population of neurons that die and are replaced throughout life. Stem cells in the dentate gyrus can form either granule cell layer neurons or astrocytes. Cells with a more restricted developmental potential than neural stem cells also exist in the CNS and can give rise to differentiated progeny. Such cells are generally restricted to a neuronal fate (neuronal

restricted progenitors) or a glial fate (glial restricted progenitors) (Fig. 5). In the developing spinal cord, it has been demonstrated that multipotent neural stem cells give rise to lineage-restricted progenitor cells, as assessed by differential expression of lineage specific markers (Mayer-Proschel et al., 1997; Kalyani and Rao, 1998; Quinn et al., 1999). The ultimate fate of these stem and progenitor cells depends on a number of factors, including but not limited to the presence or absence of trophic factors; system stress from oxidative or metabolic stress; and diet.

Many of the characteristics of adult neural stem cells are similar to those of fetal neural stem cells, including the two defining characteristics of a stem cell: the capacity for self-renewal through cell division, and the ability to produce differentiated progeny of all three types of mature neural cells. Both fetal and adult stem cells respond to a variety of growth factors and cytokines, including epidermal growth factor (EGF) and basic fibroblast growth factor (bFGF). Additionally, adult neural stem cells give rise to neurons that are integrated into existing neuronal circuitry and appear to be fully functional, as determined by electrophysiological recordings from newly formed neurons in the adult mouse hippocampus (van Praag et al., 2002). However, there is some evidence to suggest that the mechanisms that regulate stem cell processes change as the organism matures. For example, the early embryonic spinal cord is derived from multipotent neuroepithelial cells. At early developmental stages (E10.5 in the rat), neuroepithelial cells in the neural tube express neural stem cell-specific markers, such as fibroblast growth

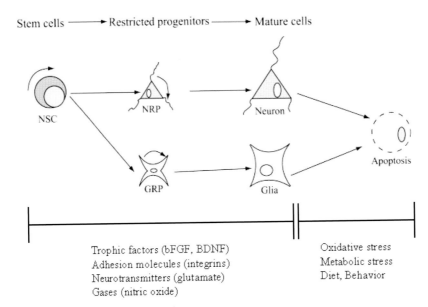

FIGURE 5. Mechanisms that regulate neurogenesis and gliogenesis in the adult nervous system. Multipotent neural stem cells can give rise to neuron-restricted progenitor cells (NRP) and glia-restricted progenitor cells (GRP). Differentiated neurons and glia may become functional and endure, or may undergo apoptosis; GRP and NRP may also undergo apoptosis. Proliferation, differentiation, and survival are regulated by a myriad of factors, including trophic support and environmental conditions (oxidative stress, metabolic stress, etc.).

factor receptor-4 (FGFR4), Frizzled 9 (Fz-9), and Sox-2 (Cai et al., 2002). Expression of Fz-9 and FGFR4 is downregulated as neuroepithelial cells become committed to a specific cell linage (neuronal or glial) and are virtually undetectable by E14 (Kalyani et al., 1999; Cai et al., 2002), suggesting they are uniquely expressed by early but not late embryonic neural stem cells. This is supported by the lack of FGFR4 expression in neural stem cells of the late embryonic and adult rat hippocampus (Limke et al., 2003). In contrast, the transcription factor Sox-1, a HMG-box protein related to SRY, is expressed in ectodermal cells fated to become neural cells (Pevny et al., 1998) and is also found in late embryonic and young adult hippocampus in proliferative cells (Limke et al., 2003), suggesting that some factors may regulate both developmental and adult stem cell populations.

Neurogenesis in the Aging Nervous System

Neural stem and progenitor cells are subject to many of the same environmental stressors other neural cells experience during aging, which can alter their capacity for self-renewal as well as their survival (Fig. 6). Stem cells appear to be affected by at least some of these factors, as there is decreased incorporation of bromodeoxyuridine in the aged rat hippocampus, suggesting a decline in the neurogenic capacity of the adult nervous system with age (Kuhn et al., 1996). Neurogenesis might be impaired as the result of reduced proliferation or differentiation of neural stem cells, increased quiescence of cells as they mature, or increased death of newly generated neurons. Even in the young adult brain, studies in which neural stem cells were labeled with bromodeoxyuridine provide evidence that most newly generated cells in the hippocampus and subventricular zone eventually die, with some of them dying before they differentiate into functional

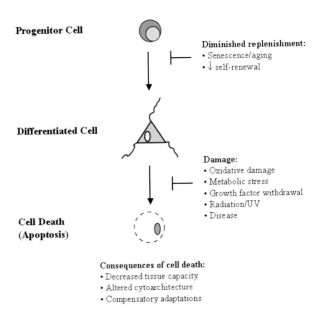

FIGURE 6. Neural stem cells and their progeny are exposed to stressors which may affect their ability to function and can ultimately lead to cellular senescence or, in severe situations, apoptosis. These stressors are present during normal aging and are often heightened during neurodegenerative disorders. As stem and mature cells are removed from the brain, there is a decreased capacity for proliferation/cell replacement, as well as alterations in the brain's structure and plasticity.

neurons or glial cells (Levison et al., 2000; Lee et al., 2002a). The decline in hippocampal neurogenesis does not appear to be caused by metabolic impairment, but may result from decreased proliferation or a decrease in the numbers of neural stem cells (Kuhn et al., 1996). Presumably, age-related increases in cellular oxidative stress or decrements in neurotrophic factor levels contribute to the decline in neurogenesis during aging (Haughey et al., 2002), although this remains to be established.

Various growth factors and cytokines drive the formation, maturation, and survival of the neural cells during development; modification of these factors may influence neurogenesis in the aged brain. Adult neural stem cells respond to several growth factors, particularly EGF and bFGF, which promote proliferation of stem cells and progenitor cells derived from the adult subventricular zone (Kuhn et al., 1997). Factor bFGF also induces the proliferation of hippocampal neural progenitor cells (Ray et al., 1993); the responsiveness of these cells to bFGF may decrease during aging (Cheng et al., 2002). EGF has a similar mitogenic effect in proliferating cells in the subventricular zone, although its effects appear to promote gliogenesis rather than neurogenesis (Kuhn et al., 1997; Gritti et al., 1999). Injection of EGF and NGF into the lateral ventricle of aged mice promotes proliferation of subventricular zone cells (Tirassa et al., 2003). Interestingly, this protocol also causes an upregulation of mRNA for BDNF, a trophic factor which promotes survival of newly born neurons. BDNF itself promotes the differentiation and survival of newly generated neurons in the hippocampus (Lee et al., 2002a, b). Age-related declines in BDNF and the BDNF receptor, TrkB, have been described in the rat and primate brain (Hayashi et al., 1997; Katoh-Semba et al., 1998; Romanczyk et al., 2002). Interestingly, when mice are maintained on dietary restriction, hippocampal neurogenesis is increased (Lee et al., 2002b), possibly as a result of a BDNF-mediated increase in survival of newly generated neurons (Lee et al., 2002a). Another growth factor that declines in the aging brain is IGF-1, which is reduced in the hippocampus of aged rats (Lai et al., 2000). Age-associated diminishment of hippocampal neurogenesis in the aged rat can be reversed by administration of IGF-1 (Lichtenwalner et al., 2001), suggesting that its receptor plays a role in the aging process.

Other molecules that drive development, including cytokines, neurotransmitters, and hormones, are also critical regulators of neurogenesis and gliogenesis during development and aging. Leukemia inhibitory factor (LIF) and ciliary neuro-trophic factor (CNTF) act through gp130 heterodimer receptors to promote maintenance of an undifferentiated state in mouse embryonic stem cells, but promote gliogenesis in the adult mouse brain (Williams et al., 1988; Conover et al., 1993; Yoshida et al., 1993). Neurotransmitter signaling may also play important roles in regulating adult neurogenesis. For example, antidepressants that enhance serotonergic signaling stimulate hippocampal neurogenesis by a mechanism that may involve upregulation of BDNF (Duman et al., 2001). Neurogenesis can also be stimulated by adrenalectomy, suggesting that endocrine signals can also modulate neurogenesis in the aged brain (Cameron and Gould, 1994; Cameron and McKay, 1999). Interestingly, sex hormones (estrogen and testosterone) may directly and/or indirectly affect neurogenesis in the aging brain. For example, estrogen levels decline abruptly in post-menopausal women not receiving hormone replacement therapy. Estrogen deprivation significantly reduces hippocampal BDNF levels in the female rat hippocampus; interestingly, exercise and/or hormone replacement therapy restore BDNF mRNA and protein content to normal levels (Berchtold et al., 2001). Estrogen alone promotes proliferation of both embryonic and adult neural stem cells (Brannvall et al., 2002). Similarly, men experience an age-related decline in testosterone levels. Testosterone promotes neurogenesis in the adult songbird neostriatum (Louissaint et al., 2002) and, like estrogen, causes an upregulation of survival-promoting BDNF (Rasika et al., 1999). Thus, age-related declines in neurogenesis may be linked to loss of hormone levels associated with normal aging.

Other manipulations which increase neurogenesis in the aged rodent hippocampus include physical exercise (van Praag et al., 1999) and enriched environments (Kempermann et al., 1998; Nilsson et al., 1999), consistent with beneficial effects of exercise and intellectual activities in preserving brain function during aging in humans. Increased hippocampal neurogenesis creates new neurons with apparently functional circuitry (Snyder et al., 2001; van Praag et al., 2002) and is associated with cognitive improvement in aged rodents (Kempermann et al., 2002). Reduced hippocampal neurogenesis is associated with loss of ability to form trace memories, which is regained when neurogenesis is restored (Shors et al., 2001). Additionally, exercise-induced neurogenesis significantly improves learning, exploratory behavior, and locomotion in aged mice (Kempermann et al., 2002). What cannot be determined from these studies is the contribution of increased neurogenesis to the observed changes, as compared to other beneficial effects of exercise (increased trophic support, etc.). Interestingly, age-related reductions in neurogenesis do not correlate with spatial memory impairment (Merrill et al., 2003), suggesting that increased neurogenesis is not a "cure-all" for all age-related hippocampal impairments.

While the level of neurogenesis can be modulated by factors such as diet, environmental stimulus, and trophic factor levels, there is little information to date regarding the intrinsic mechanisms underlying the age-related decline in neural stem function. Proliferating, non-transformed cells will undergo a certain number of cell divisions before exiting the cell cycle to become senescent. This number of divisions, termed the "Hayflick limit," is controlled by telomerase, an enzyme that adds a six-base DNA repeat onto the ends of chromosomes (telomeres) and thereby prevents their shortening during successive rounds of cell division. Telomerase levels are high in developing neural progenitor cells, but then decrease as cells differentiate into neurons and glia (Klapper et al., 2001). Telomerase has been suggested to play a role in aging because its absence in somatic cells results in telomere shortening and cell senescence. Telomeres are generally shorter in older people than in younger people, suggesting that telomere length may provide a molecular clock for measuring lifespan. Alterations in telomere length can dramatically affect the onset and maintenance of aging. For example, accelerated shortening of telomeres in disease such as Werner's syndrome and Down's syndrome is associated with early onset of aging. Recent studies have shown that telomerase promotes the survival of neurons and neuronal precursor cells (Fu et al., 2000; Lu et al., 2001) and its reduction during aging may therefore play a role in age-related neuronal loss and impaired neurogenesis.

DEVELOPMENTAL MECHANISMS IN AGE-RELATED NEURODEGENERATIVE DISORDERS

As described in the previous sections, aging involves a series of changes within the brain that are a normal part of the aging process. These include elevation of reactive oxygen species, increased oxidative damage to proteins and DNA, accumulation of protein and lipid byproducts, reduced metabolic activity, and cytoskeletal changes. Such changes are distinct from the effects of age-related neurological disorders which often exacerbate the factors contributing to the general decline observed during aging. A number of neurodegenerative disorders exist which are positively correlated with aging (Table 2). Some diseases, such as Parkinson's disease, target a distinct population of neurons (in this case, the dopaminergic neurons of the substantia nigra), while others, such as Alzheimer's disease, affect a more diffuse set of cells (in this case, primarily the cortex and hippocampus). What is of interest is that, like the normal alterations in brain physiology that accompany aging, many of the age-related neurological disorders also have a foundation in developmental processes.

Inherited Disorders with Abnormal Aging Phenotype

Inherited disorders that are characterized by premature aging are providing insight into the overlap of mechanisms of aging and development in the nervous system. Werner's syndrome is an autosomal recessive disorder caused by mutations in a DNA

helicase that manifests accelerated aging of tissues throughout the body (van Brabant et al., 2000). Age-related alterations in the brains of Werner's patients have been documented and include amyloid deposition and neurofibrillary tangles in frontal and temporal lobes (Leverenz et al., 1998). Cockayne syndrome is characterized by a defect in DNA repair (van Gool et al., 1997) and manifests widespread aging-like changes in the nervous system including retinal and cochlear degeneration, peripheral neuropathies, and neurodegenerative changes in the brain (Rapin et al., 2000). Patients with progeria exhibit a dramatic acceleration of age-related pathologies including cerebrovascular disease and neuronal degeneration (Rosman et al., 2001). A more common inherited disorder that manifests premature age- and Alzheimer-like pathologies in the brain is Down's syndrome (trisomy of chromosome 21). Patients with Down's syndrome exhibit extensive amyloid deposition in the brain with associated neurofibrillary pathology and cognitive dysfunction, as well as degeneration of cholinergic and noradrenergic systems (Coyle et al., 1986; Sawa, 1999). Although the gene(s) responsible for the phenotypes of Down's syndrome has not been clearly established, those encoding amyloid precursor protein (APP) and proteins involved in oxyradical metabolism are located on chromosome 21. In particular, a role for APP is suggested by studies showing that APP plays important roles in regulating neuronal plasticity (dendrite outgrowth and synaptic plasticity) and cell survival (Mattson, 1997). Thus, disruption of mechanisms that regulate development can result in symptoms which mimic changes observed during aging, supporting the idea that the mechanisms driving development and aging are often the same.

Developmental Mechanisms Underlying Age-Related Neurodegenerative Disorders

How might developmental mechanisms contribute to the pathogenesis of neurodegenerative disorders? Each neurodegenerative disorder is characterized by selective vulnerability of particular populations of neurons (Fig. 7). The mechanisms that regulate the survival and plasticity of neurons and glia during aging are not well understood, but studies of age-related neurodegenerative disorders have revealed novel genes and environmental factors that influence both the development of the nervous system and its susceptibility to dysfunction and degeneration during aging.

Studies of the brains of Alzheimer's disease patients have revealed several development-related processes occurring in association with amyloid plaques and neurofibrillary tangles, the major pathological lesions in this disease. For example, fetal forms of MAPs are present in dystrophic neurites (Joachim et al., 1987) and aberrant axonal sprouting occurs in some brain regions (Larner, 1995). In addition, growth factors such as bFGF and transforming growth factor-beta are present at high levels in amyloid plaques (Cummings et al., 1993; Finch et al., 1993). Damage to nuclear DNA in striatum of Huntington's disease patients, and in hippocampus and vulnerable cortical regions of Alzheimer's patients, has been documented. For example, levels

TABLE 2. Age-Related Diseases of the Nervous System

Disease	Primary symptoms
Alzheimer's disease	β-amyloid plaques and neurofibrillary tangles, primarily in hippocampus and cortex; results in memory deficits
Parkinson's disease	Loss of dopaminergic neurons in the substantia nigra and striatum; results in motor control problems
Huntington's disease	Cell death in neostriatum and cortex, with accompanying movement and cognitive dysfunction; results in severely reduced lifespan
Werner's syndrome	Amyloid deposition and neurofibrillary tangles in frontal and temporal lobes; results in accelerated aging
Cockayne syndrome	Defect in DNA repair causing retinal and cochlear degeneration, peripheral neuropathies, and neurodegenerative changes in the brain; symptoms resemble nervous system changes observed in aging
Down's syndrome	Amyloid deposition, neurofibrillary tangles, and cognitive dysfunction, degeneration of cholinergic and noradrenergic systems

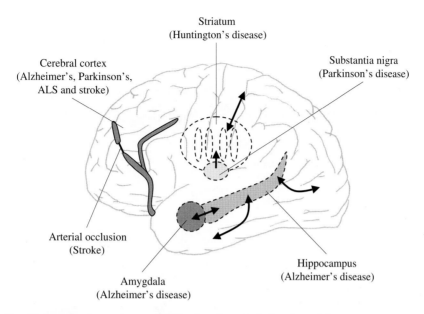

Striatum
(Huntington's disease)

Cerebral cortex
(Alzheimer's, Parkinson's,
ALS and stroke)

Substantia nigra
(Parkinson's disease)

Arterial occlusion
(Stroke)

Hippocampus
(Alzheimer's disease)

Amygdala
(Alzheimer's disease)

FIGURE 7. Brain regions affected in age-related neurodegenerative disorders. Synaptic dysfunction and degeneration and neuron death occur in the affected brain regions in the indicated disorders. Accordingly, the symptoms of each disorder are directly related to the functions of the affected brain regions. For example, brain regions involved in cognitive processes (hippocampus and cerebral cortex) and emotional behaviors (amygdala) are affected in Alzheimer's disease, while brain regions involved in controlling body movements (substantia nigra and striatum) are affected in Parkinson's disease.

of 8-hydroxyguanosine are increased suggesting DNA damage caused by reactive oxygen molecules such as hydroxyl radical and peroxynitrite. Interestingly, a dietary deficiency of folate can have striking adverse effects on the developing nervous system and may also increase the risk of Alzheimer's disease and Parkinson's disease by promoting DNA damage in neurons (Duan *et al.*, 2002; Kruman *et al.*, 2002).

The most striking links between development and neuro-degenerative disorders comes from studies of Alzheimer's disease. Although the cause of most cases of Alzheimer's disease is unknown, some cases result from genetic mutations. Three disease-causing genes have been identified; they encode the APP, presenilin-1 (PS1), and presenilin-2 (PS2). APP is a transmembrane protein that is the source of the amyloid beta-peptide that forms insoluble plaques in the brains of Alzheimer's patients (Mattson, 1997). Cleavage of APP within the amyloid beta-peptide sequence by an enzyme activity called alpha-secretase releases a soluble form of APP (sAPP) from the cell surface; this cleavage occurs normally and is stimulated by various growth factors and by electrical activity in neurons. In Alzheimer's disease, there is a decrease, in the production of sAPP; instead, APP is cleaved by enzymes that cut it at the N- (beta-secretase) and C- (gamma-secretase) termini of amyloid beta-peptide to generate the full-length amyloidogenic peptide. APP, PS1, and PS2 mutations increase the production of amyloid beta-peptide. Presenilin and APP mutations may alter neuronal plasticity and promote neuronal degeneration by perturbing cellular calcium homeostasis (Mattson, 1997).

Recent studies have revealed important roles for APP and presenilins in the development of the nervous system and in adult neuroplasticity. The secreted form of APP has been shown to regulate neurite outgrowth and cell survival in embryonic rat hippocampal neurons (Mattson *et al.*, 1993; Mattson, 1994) and can protect neurons against death in experimental models of Alzheimer's disease and stroke (Goodman and Mattson, 1994; Smith-Swintosky *et al.*, 1994). Studies of synaptic transmission in hippocampal slices showed that sAPP can enhance long-term potentiation (Ishida *et al.*, 1997), suggesting that sAPP facilitates learning and memory, a possibility consistent with *in vivo* studies (Roch *et al.*, 1994). In addition to its neurotrophic effects and roles in synaptic plasticity, sAPP may play a role in neurogenesis. When cultured embryonic cortical stem cells were exposed to sAPP, their proliferation rate increased (Hayashi *et al.*, 1994; Ohsawa *et al.*, 1999). The signal transduction pathway that mediates the biological activities of sAPP may involve cyclic GMP and the transcription factor NF-κB (Furukawa *et al.*, 1996; Guo *et al.*, 1998).

Notch is a type 1 membrane protein that, when activated by cell-associated ligands, is proteolytically processed in a manner that releases an intracellular C-terminal fragment of Notch which then translocates to the nucleus where it may regulate gene expression (Fig. 8). The developmental roles of presenilins are thought to result from a function in the Notch signaling pathway because the phenotype of PS1 null mice is essentially identical to that of Notch knockout mice (Conlon *et al.*, 1995; Shen *et al.*, 1997). In addition, the cellular expression of PS1 and Notch in the developing rodent nervous system is very similar, being high during neurogenesis and decreasing as the embryo develops. Levels of Hes5, a gene induced by activation of the Notch signaling pathway, are decreased in the ventricular zone of PS1 null mice, whereas levels of a Notch ligand are elevated. The Drosophila PS1 homolog is highly expressed in neurons during development; mutations of PS1 alter the

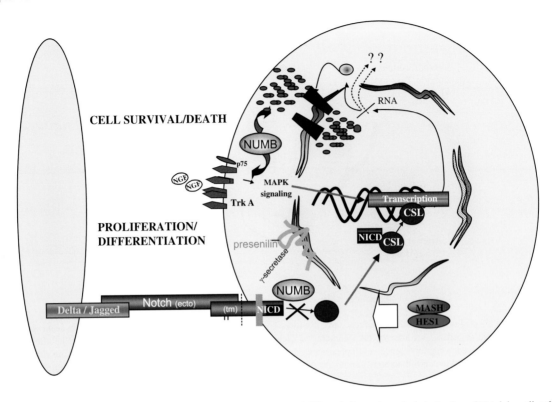

FIGURE 8. Model for the mechanisms whereby Notch and Numb regulate neuronal differentiation and survival. Activation of Notch by cell–cell interactions results in a proteolytic cleavage of an intracellular domain (NICD), which interacts with a protein called CSL and thereby regulates gene transcription. Numb can antagonize Notch signaling. Numb may also enhance NGF signaling by facilitating activation of the high-affinity receptor trkA resulting in activation of mitogen-activated protein kinases (MAPK). In neural progenitor cells, Notch signaling promotes cell proliferation, whereas Numb promotes cell differentiation. In differentiated neurons, Notch may promote cell survival, while Numb can facilitate apoptosis. Notch and Numb may play important roles in aging and neurodegenerative disorders.

subcellular localization of Notch and result in defects in eye development and neuronal differentiation. PS1 and PS2 have considerable homology to two genes in the nematode *C. elegans* called *spe*-4 and *sel*-12; *spe*-4 functions in spermatogenesis and *sel*-12 plays a role in the process of egg-laying by a mechanism involving the Notch signaling pathway. The *sel*-12 mutants can be rescued by PS1 demonstrating a conserved function for these two genes. Moreover, human PS1 can rescue defective egg-laying resulting from mutations in *sel*-12, strongly suggesting similar functions of PS1 and *sel*-12 (Levitan and Greenwald, 1995). PS1 is necessary for ligand-induced transmembrane cleavage of Notch (Hartmann *et al.*, 2001), and may thereby regulate cell fate decisions.

Numb is an evolutionarily conserved protein identified by its ability to control cell fate in the nervous system of Drosophila, wherein Numb may act by antagonizing Notch signaling (Artavanis-Tsakonas *et al.*, 1999) (Fig. 8). Mammals express four isoforms of Numb that differ in the composition of a phosphotyrosine-binding domain (PTB) and a proline-rich region (PRR). Numb regulates the sensitivity of cells to neurotrophin-induced differentiation and cell survival dependency in an isoform-specific manner (Pedersen *et al.*, 2002). Numb isoforms containing a short PTB enhance the differentiation response to NGF, and enhance apoptosis in response to NGF withdrawal by a mechanism dependent upon release of calcium from endoplasmic

reticulum stores. These findings suggest that isoform-specific modulation of neurotrophin responses by Numb may play important roles in the development and plasticity of the nervous system. Additional studies have examined the possible roles of Numb in the pathogenesis of Alzheimer's disease. Numb isoforms containing a short PTB domain increase the vulnerability of neural cells to death induced by amyloid beta-peptide (Chan *et al.*, 2002). Dysregulation of cellular calcium homeostasis occurs in cells expressing Numb isoforms with a short PTB domain, and the death-promoting effect of Numb is abolished by pharmacological inhibition of calcium release. The levels of Numb are increased in cultured primary hippocampal neurons exposed to Aβ, suggesting a role for endogenous Numb in the neuronal death process. Furthermore, higher levels of Numb were detected in the cortex of mice expressing mutant APP relative to age-matched wild-type mice. These findings suggest that the effects of Numb on cell fate decisions, both during development of the nervous system and in neurodegenertive disorders, are mediated by changes in cellular calcium homeostasis.

Deficits in neurotrophic factors may contribute to neurodegenerative processes in aging and disorders of aging. Analyses of neurotrophic factor expression in brain tissues from young and old rodents, and from patients with age-related neurodegenerative disorders, suggest that neurotrophic support of neurons declines with advancing age and more so in neurodegenerative disorders. It was reported that transgenic mice that express an antibody against NGF exhibit neuronal degeneration with features of AD including amyloid deposits and neurofibrillary tangle-like pathology in the hippocampus and cerebral cortex (Capsoni *et al.*, 2000). Although depletion of a neurotrophic factor or impaired neurotrophic signal transduction has not yet been shown to cause a neurodegenerative disorder, recent findings suggest major contributions of diminished neurotrophic support in Alzheimer's, Parkinson's, and Huntington's diseases. It was reported that the normal huntingtin protein induces the expression of BDNF, and that disease-causing mutations in huntingtin result in a marked decrease in BDNF expression (Zuccato *et al.*, 2001).

The evidence that developmental mechanisms are involved in Alzheimer's disease is now quite strong, and investigations of other age-related neurodegenerative disorders are revealing similar processes. Sprouting of nitric oxide synthase-positive neurites occurs in Parkinson's disease (Sohn *et al.*, 1999), suggesting a role for aberrant nitric oxide signaling in the pathogenesis of this disorder. Glial cell-line-derived neurotrophic factor (GDNF) can promote the survival, production of dopamine, and neurite sprouting in dopaminergic neurons in experimental models of Parkinson's disease (Gash *et al.*, 1998) and is currently being tested in clinical trials in human patients. Parkinson's disease can be caused by mutations in alpha-synuclein, and studies of songbirds and mammals have provided evidence that alpha-synuclein functions in the regulation of synaptic plasticity (Clayton *et al.*, 2002). Ischemic stroke involves a complex set of neurodegenerative and neurorestorative cellular responses. Apoptosis appears to be a prominent form of neuronal death in stroke, while neurogenesis and neurite outgrowth are compensatory responses that likely influence the extent of recovery from a stroke (Stroemer *et al.*, 1995; Jin *et al.*, 2001). Thus, there is significant evidence that developmental mechanisms play a role in many neurodegenerative disorders of the aging brain.

SUMMARY

The mechanisms driving development of the nervous system are complex and involve the integration of many intrinsic and extrinsic signals. Many of the mechanisms which regulate development, including trophic factors, cytokines, and hormones, are the same mechanisms that dysfunction during aging and contribute to the pathogenesis of neurodegenerative disorders. A better understanding of how abnormalities in developmental signaling mechanisms may contribute to the pathogenesis of neurodegenerative disorders, and how developmental mechanisms might be tapped to restore damaged neuronal circuits are important areas for future investigations.

ACKNOWLEDGMENTS

The authors would like to thank the members of their respective laboratories for insightful discussion and suggestions during the preparation of this chapter. T.L. was supported by an NIA IRTA fellowship. The authors thank Lance A. Edwards and Sic L. Chan for the contribution of several figures presented in this publication.

REFERENCES

Artavanis-Tsakonas, S., Rand, M.D. *et al.*, 1999, Notch signaling: Cell fate control and signal integration in development, *Science* 284(5415): 770–776.

Berchtold, N.C., Kesslak, J.P. *et al.*, 2001, Estrogen and exercise interact to regulate brain-derived neurotrophic factor mRNA and protein expression in the hippocampus, *Eur. J. Neurosci.* 14(12):1992–2002.

Bergman, E., Ulfhake, B. *et al.*, 2000, Regulation of NGF-family ligands and receptors in adulthood and senescence: Correlation to degenerative and regenerative changes in cutaneous innervation, *Eur. J. Neurosci.* 12(8):2694–2706.

Bertoni-Freddari, C., Fattoretti, P. *et al.*, 1996, Synaptic structural dynamics and aging, *Gerontology* 42(3):170–180.

Braak, H. and Braak, E., 2000, Pathoanatomy of Parkinson's disease, *J. Neurol.* 247(Suppl 2):II3–II10.

Brannvall, K., Korhonen, L. *et al.*, 2002, Estrogen-receptor-dependent regulation of neural stem cell proliferation and differentiation, *Mol. Cell. Neurosci.* 21(3):512–520.

Bruce-Keller, A.J., Umberger, G. *et al.*, 1999, Food restriction reduces brain damage and improves behavioral outcome following excitotoxic and metabolic insults, *Ann. Neurol.* 45(1):8–15.

Cai, J., Wu, Y. *et al.*, 2002, Properties of a fetal multipotent neural stem cell (NEP cell), *Dev. Biol.* 251(2):221–240.

Cameron, H.A. and Gould, E., 1994, Adult neurogenesis is regulated by adrenal steroids in the dentate gyrus, *Neuroscience* 61(2):203–209.

Cameron, H.A. and McKay, R.D., 1999, Restoring production of hippocampal neurons in old age, *Nat. Neurosci.* 2(10):894–897.

Capsoni, S., Ugolini, G. *et al.*, 2000, Alzheimer-like neurodegeneration in aged antinerve growth factor transgenic mice, *Proc. Natl. Acad. Sci. USA* 97(12):6826–6831.

Casu, M.A., Wong, T.P. *et al.*, 2002, Aging causes a preferential loss of cholinergic innervation of characterized neocortical pyramidal neurons, *Cereb. Cortex* 12(3):329–337.

Chan, S.L., Pedersen, W.A. *et al.*, 2002, Numb modifies neuronal vulnerability to amyloid beta-peptide in an isoform-specific manner by a mechanism involving altered calcium homeostasis: Implications for neuronal death in Alzheimer's disease, *Neuromolec. Med.* 1(1): 55–67.

Chen, S. and Hillman, D.E., 1999, Dying-back of Purkinje cell dendrites with synapse loss in aging rats, *J. Neurocytol.* 28(3):187–196.

Cheng, A., Chan, S.L. *et al.*, 2001, p38 MAP kinase mediates nitric oxide-induced apoptosis of neural progenitor cells, *J. Biol. Chem.* 276(46): 43320–43327.

Cheng, Y., Black, I.B. *et al.*, 2002, Hippocampal granule neuron production and population size are regulated by levels of bFGF, *Eur. J. Neurosci.* 15(1):3–12.

Chung, Y.H., Shin, C. *et al.*, 2000, Immunocytochemical study on the distribution of p53 in the hippocampus and cerebellum of the aged rat, *Brain Res.* 885(1):137–141.

Clancy, D.J., Gems, D. *et al.*, 2001, Extension of life-span by loss of CHICO, a Drosophila insulin receptor substrate protein, *Science* 292(5514): 104–106.

Clayton, D.A., Mesches, M.H. *et al.*, 2002, A hippocampal NR2B deficit can mimic age-related changes in long-term potentiation and spatial learning in the Fischer 344 rat, *J. Neurosci.* 22(9):3628–3637.

Conlon, R.A., Reaume, A.G. *et al.*, 1995, Notch1 is required for the coordinate segmentation of somites, *Development* 121(5):1533–1545.

Conover, J.C., Ip, N.Y. *et al.*, 1993, Ciliary neurotrophic factor maintains the pluripotentiality of embryonic stem cells, *Development* 119(3): 559–565.

Cotman, C.W. and Berchtold, N.C., 2002, Exercise: A behavioral intervention to enhance brain health and plasticity, *Trends Neurosci.* 25(6):295–301.

Coyle, J.T., Oster-Granite, M.L. *et al.*, 1986, The neurobiologic consequences of Down syndrome, *Brain Res. Bull.* 16(6):773–787.

Croll, S.D., Ip, N.Y. *et al.*, 1998, Expression of BDNF and trkB as a function of age and cognitive performance, *Brain Res.* 812(1–2):200–208.

Culmsee, C., Zhu, X. *et al.*, 2001, A synthetic inhibitor of p53 protects neurons against death induced by ischemic and excitotoxic insults, and amyloid beta-peptide, *J. Neurochem.* 77(1):220–228.

Cummings, B.J., Su, J.H. *et al.*, 1993, Neuritic involvement within bFGF immunopositive plaques of Alzheimer's disease, *Exp. Neurol.* 124(2): 315–325.

de la Rosa, E.J. and de Pablo, F. 2000, Cell death in early neural development: Beyond the neurotrophic theory, *Trends Neurosci.* 23(10):454–458.

DeKosky, S.T., Scheff, S.W. *et al.*, 1996, Structural correlates of cognition in dementia: Quantification and assessment of synapse change, *Neurodegeneration* 5(4):417–421.

Disterhoft, J.F., Moyer, J.R., Jr. *et al.*, 1994, The calcium rationale in aging and Alzheimer's disease. Evidence from an animal model of normal aging, *Ann. N Y Acad. Sci.* 747:382–406.

Duan, W., Ladenheim, B. *et al.*, 2002, Dietary folate deficiency and elevated homocysteine levels endanger dopaminergic neurons in models of Parkinson's disease, *J. Neurochem.* 80(1):101–110.

Duan, W., Lee, J. *et al.*, 2001, Dietary restriction stimulates BDNF production in the brain and thereby protects neurons against excitotoxic injury, *J. Mol. Neurosci.* 16(1):1–12.

Duan, W. and Mattson, M.P., 1999, Dietary restriction and 2-deoxyglucose administration improve behavioral outcome and reduce degeneration of dopaminergic neurons in models of Parkinson's disease, *J. Neurosci. Res.* 57(2):195–206.

Dubey, A., Forster, M.J. *et al.*, 1996, Effect of age and caloric intake on protein oxidation in different brain regions and on behavioral functions of the mouse. *Arch. Biochem. Biophys.* 333(1):189–197.

Duman, R.S., Nakagawa, S. *et al.*, 2001, Regulation of adult neurogenesis by antidepressant treatment, *Neuropsychopharmacology* 25(6): 836–844.

Finch, C.E., Laping, N.J. *et al.*, 1993, TGF-beta 1 is an organizer of responses to neurodegeneration, *J. Cell. Biochem.* 53(4):314–322.

Foulstone, E.J., Tavare, J.M. *et al.*, 1999, Sustained phosphorylation and activation of protein kinase B correlates with brain-derived neurotrophic factor and insulin stimulated survival of cerebellar granule cells, *Neurosci. Lett.* 264(1–3):125–128.

Fu, W., Killen, M. *et al.*, 2000, The catalytic subunit of telomerase is expressed in developing brain neurons and serves a cell survival-promoting function, *J. Mol. Neurosci.* 14(1–2):3–15.

Fuchs, E., Flugge, G. *et al.*, 2001, Psychosocial stress, glucocorticoids, and structural alterations in the tree shrew hippocampus, *Physiol. Behav.* 73(3):285–291.

Furukawa, K., Barger, S.W. *et al.*, 1996, Activation of K+ channels and suppression of neuronal activity by secreted beta-amyloid-precursor protein, *Nature* 379(6560):74–78.

Gage, F.H., 2000, Mammalian neural stem cells, *Science* 287(5457): 1433–1438.

Gary, D.S. and Mattson, M.P., 2001, Integrin signaling via the PI3-kinase-Akt pathway increases neuronal resistance to glutamate-induced apoptosis, *J. Neurochem.* 76(5):1485–1496.

Gash, D.M., Zhang, Z. *et al.*, 1998, Neuroprotective and neurorestorative properties of GDNF, *Ann. Neurol.* 44(3 Suppl 1):S121–S125.

Giusto, N.M., Roque, M.E. *et al.*, 1992, Effects of aging on the content, composition and synthesis of sphingomyelin in the central nervous system, *Lipids* 27(11):835–839.

Goodman, Y. and Mattson, M.P., 1994, Secreted forms of beta-amyloid precursor protein protect hippocampal neurons against amyloid beta-peptide-induced oxidative injury, *Exp. Neurol.* 128(1):1–12.

Gritti, A., Frolichsthal-Schoeller, P. *et al.*, 1999, Epidermal and fibroblast growth factors behave as mitogenic regulators for a single multipotent stem cell-like population from the subventricular region of the adult mouse forebrain, *J. Neurosci.* 19(9):3287–3297.

Guo, Q., Robinson, N. *et al.*, 1998, Secreted beta-amyloid precursor protein counteracts the proapoptotic action of mutant presenilin-1 by activation of NF-kappaB and stabilization of calcium homeostasis, *J. Biol. Chem.* 273(20):12341–12351.

Han, S.H., McCool, B.A. *et al.*, 2002, Single-cell RT-PCR detects shifts in mRNA expression profiles of basal forebrain neurons during aging, *Brain Res. Mol. Brain Res.* 98(1–2):67–80.

Hartmann, D., Tournoy, T., Saftig, P., Annaert, W., DeStrooper, B., 2001, Implication of App secretases in notch signalling *J. Mol. Neurosci.* 17:171–181.

Haughey, N.J., Liu, D. *et al.*, 2002, Disruption of neurogenesis in the subventricular zone of adult mice, and in human cortical neuronal precursor cells in culture, by amyloid beta-peptide: Implications for the pathogenesis of Alzheimer's disease, *Neuromolec. Med.* 1(2): 125–135.

Hayashi, M., Yamashita, A. *et al.*, 1997, Somatostatin and brain-derived neurotrophic factor mRNA expression in the primate brain: Decreased levels of mRNAs during aging, *Brain Res.* 749(2):283–289.

Hayashi, Y., Kashiwagi, K. *et al.*, 1994, Alzheimer amyloid protein precursor enhances proliferation of neural stem cells from fetal rat brain, *Biochem. Biophys. Res. Commun.* 205(1):936–943.

Hugon, J., Vallat, J.M. *et al.*, 1996, Role of glutamate and excitotoxicity in neurologic diseases, *Rev. Neurol. (Paris)* 152(4):239–248.

Ingram, D.K., Weindruch, R. *et al.*, 1987, Dietary restriction benefits learning and motor performance of aged mice, *J. Gerontol.* 42(1): 78–81.

Ishida, A., Furukawa, K. *et al.*, 1997, Secreted form of beta-amyloid precursor protein shifts the frequency dependency for induction of LTD, and enhances LTP in hippocampal slices, *Neuroreport* 8(9–10): 2133–2137.

Izquierdo, I., 1994, Pharmacological evidence for a role of long-term potentiation in memory, *FASEB J.* 8(14):1139–1145.

Jin, K., Minami, M. *et al.*, 2001, Neurogenesis in dentate subgranular zone and rostral subventricular zone after focal cerebral ischemia in the rat, *Proc. Natl. Acad. Sci. USA* 98(8):4710–4715.

Joachim, C.L., Morris, J.H. *et al.*, 1987, Tau epitopes are incorporated into a range of lesions in Alzheimer's disease, *J. Neuropathol. Exp. Neurol.* 46(6):611–622.

Johnson, J. and Oppenheim, R. 1994, Neurotrophins. Keeping track of changing neurotrophic theory, *Curr. Biol.* 4(7):662–665.

Johnson, W.G., 2000, Late-onset neurodegenerative diseases—the role of protein insolubility, *J. Anat.* 196(Pt 4):609–616.

Julien, J.P. and Beaulieu, J.M. 2000, Cytoskeletal abnormalities in amyotrophic lateral sclerosis: Beneficial or detrimental effects? *J. Neurol. Sci.* 180(1–2):7–14.

Kalant, N., Stewart, J. *et al.*, 1988, Effect of diet restriction on glucose metabolism and insulin responsiveness in aging rats, *Mech. Ageing Dev.* 46(1–3):89–104.

Kalyani, A.J., Mujtaba, T. *et al.*, 1999, Expression of EGF receptor and FGF receptor isoforms during neuroepithelial stem cell differentiation, *J. Neurobiol.* 38(2):207–224.

Kalyani, A.J. and Rao, M.S. 1998, Cell lineage in the developing neural tube. *Biochem. Cell. Biol.* 76(6):1051–1068.

Katoh-Semba, R., Semba, R. *et al.*, 1998, Age-related changes in levels of brain-derived neurotrophic factor in selected brain regions of rats, normal mice and senescence-accelerated mice: A comparison to those of nerve growth factor and neurotrophin-3, *Neurosci. Res.* 31(3): 227–234.

Kaufmann, J.A., Bickford, P.C. *et al.*, 2001, Oxidative-stress-dependent up-regulation of Bcl-2 expression in the central nervous system of aged Fisher-344 rats, *J. Neurochem.* 76(4):1099–1108.

Keller, J.N. and Mattson, M.P. 1998, Roles of lipid peroxidation in modulation of cellular signaling pathways, cell dysfunction, and death in the nervous system, *Rev. Neurosci.* 9(2):105–116.

Kempermann, G., Gast, D. *et al.*, 2002, Neuroplasticity in old age: Sustained fivefold induction of hippocampal neurogenesis by long-term environmental enrichment, *Ann. Neurol.* 52(2):135–143.

Kempermann, G., Kuhn, H.G. *et al.*, 1998, Experience-induced neurogenesis in the senescent dentate gyrus, *J. Neurosci.* 18(9):3206–3212.

Klapper, W., Shin, T. *et al.*, 2001, Differential regulation of telomerase activity and TERT expression during brain development in mice, *J. Neurosci. Res.* 64(3):252–260.

Koh, S., Chang, P. *et al.*, 1989, Loss of NGF receptor immunoreactivity in basal forebrain neurons of aged rats: Correlation with spatial memory impairment, *Brain Res.* 498(2):397–404.

Kruman, II, Kumaravel, T.S. *et al.*, 2002, Folic acid deficiency and homocysteine impair DNA repair in hippocampal neurons and sensitize them to amyloid toxicity in experimental models of Alzheimer's disease, *J. Neurosci.* 22(5):1752–1762.

Kuhn, H.G., Dickinson-Anson, H. *et al.*, 1996, Neurogenesis in the dentate gyrus of the adult rat: Age-related decrease of neuronal progenitor proliferation, *J. Neurosci.* 16(6):2027–2033.

Kuhn, H.G., Winkler, J. *et al.*, 1997, Epidermal growth factor and fibroblast growth factor-2 have different effects on neural progenitors in the adult rat brain, *J. Neurosci.* 17(15):5820–5829.

Lai, M., Hibberd, C.J. *et al.*, 2000, Reduced expression of insulin-like growth factor 1 messenger RNA in the hippocampus of aged rats, *Neurosci. Lett.* 288(1):66–70.

Lapchak, P.A., Araujo, D.M. *et al.*, 1993, BDNF and trkB mRNA expression in the rat hippocampus following entorhinal cortex lesions, *Neuroreport* 4(2):191–194.

Larner, A.J., 1995, The cortical neuritic dystrophy of Alzheimer's disease: Nature, significance, and possible pathogenesis, *Dementia* 6(4): 218–224.

LeDoux, S.P., Williams, B.A. *et al.*, 1996, Glial cell-specific differences in repair of O6-methylguanine, *Cancer Res.* 56(24):5615–5619.

Lee, J., Duan, W. *et al.*, 2002a, Evidence that brain-derived neurotrophic factor is required for basal neurogenesis and mediates, in part, the enhancement of neurogenesis by dietary restriction in the hippocampus of adult mice, *J. Neurochem.* 82(6):1367–1375.

Lee, J., Seroogy, K.B. *et al.*, 2002b, Dietary restriction enhances neurotrophin expression and neurogenesis in the hippocampus of adult mice, *J. Neurochem.* 80(3):539–547.

Leverenz, J.B., Yu, C.E. *et al.*, 1998, Aging-associated neuropathology in Werner syndrome, *Acta Neuropathol. (Berl)* 96(4):421–424.

Levison, S.W., Rothstein, R.P. *et al.*, 2000, Selective apoptosis within the rat subependymal zone: A plausible mechanism for determining which lineages develop from neural stem cells, *Dev. Neurosci.* 22(1–2):106–115.

Levitan, D., Greenwald, I., 1995, Facilitation of lin-12-mediated signalling by sel-12, a Caenorhabditis elegans S182 Alzheimer's disease gene. *Nature* 377:351–354.

Lichtenwalner, R.J., Forbes, M.E. *et al.*, 2001, Intracerebroventricular infusion of insulin-like growth factor-I ameliorates the age-related decline in hippocampal neurogenesis, *Neuroscience* 107(4):603–613.

Limke, T.L., Cai, J. *et al.*, 2003, Distinguishing features of progenitor cells in the late embryonic and adult hippocampus, *Dev. Neurosci.* 25:257–272.

Liu, D., Lu, C. *et al.*, 2002, Activation of mitochondrial ATP-dependent potassium channels protects neurons against ischemia-induced death by a mechanism involving suppression of Bax translocation and cytochrome c release, *J. Cereb. Blood Flow Metab.* 22(4):431–443.

Louissaint, A., Jr., Rao, S. *et al.*, 2002, Coordinated interaction of neurogenesis and angiogenesis in the adult songbird brain, *Neuron* 34(6): 945–960.

Lu, C., Chan, S.L. *et al.*, 2002, The lipid peroxidation product 4-hydroxynonenal facilitates opening of voltage-dependent Ca^{2+} channels in neurons by increasing protein tyrosine phosphorylation, *J. Biol. Chem.* 277(27):24368–24375.

Lu, C., Fu, W. *et al.*, 2001, Telomerase protects developing neurons against DNA damage-induced cell death, *Brain Res. Dev. Brain Res.* 131(1–2):167–171.

Mandelkow, E. and Mandelkow, E.M. 1995, Microtubules and microtubule-associated proteins, *Curr. Opin. Cell Biol.* 7(1):72–81.

Mandell, J.W. and Banker, G.A. 1995, The microtubule cytoskeleton and the development of neuronal polarity, *Neurobiol. Aging* 16(3):229–237; discussion 238.

Markowska, A.L., Mooney, M. *et al.*, 1998, Insulin-like growth factor-1 ameliorates age-related behavioral deficits, *Neuroscience* 87(3):559–569.

Mattson, M.P., 1994, Secreted forms of beta-amyloid precursor protein modulate dendrite outgrowth and calcium responses to glutamate in cultured embryonic hippocampal neurons, *J. Neurobiol.* 25(4):439–450.

Mattson, M.P., 1997, Cellular actions of beta-amyloid precursor protein and its soluble and fibrillogenic derivatives, *Physiol. Rev.* 77(4): 1081–1132.

Mattson, M.P., 1998, Modification of ion homeostasis by lipid peroxidation: Roles in neuronal degeneration and adaptive plasticity, *Trends Neurosci.* 21(2):53–57.

Mattson, M.P., 2000, Apoptosis in neurodegenerative disorders, *Nat. Rev. Mol. Cell. Biol.* 1(2):120–129.

Mattson, M.P., Cheng, B. *et al.*, 1993, Evidence for excitoprotective and intraneuronal calcium-regulating roles for secreted forms of the beta-amyloid precursor protein, *Neuron* 10(2):243–254.

Mattson, M.P., Dou, P. *et al.*, 1988a, Outgrowth-regulating actions of glutamate in isolated hippocampal pyramidal neurons, *J. Neurosci.* 8(6): 2087–2100.

Mattson, M.P., Fu, W. *et al.*, 1997, 4-hydroxynonenal, a product of lipid peroxidation, inhibits dephosphorylation of the microtubule-associated protein tau, *Neuroreport* 8(9–10):2275–2281.

Mattson, M.P. and Furukawa, K. 1998, Signaling events regulating the neurodevelopmental triad. Glutamate and secreted forms of beta-amyloid precursor protein as examples, *Perspect. Dev. Neurobiol.* 5(4):337–352.

Mattson, M.P., Lee, R.E. *et al.*, 1988b, Interactions between entorhinal axons and target hippocampal neurons: A role for glutamate in the development of hippocampal circuitry, *Neuron* 1(9):865–876.

Mattson, M.P., Murrain, M. *et al.*, 1989, Fibroblast growth factor and glutamate: Opposing roles in the generation and degeneration of hippocampal neuroarchitecture, *J. Neurosci.* 9(11):3728–3740.

Mayer-Proschel, M., Kalyani, A.J. *et al.*, 1997, Isolation of lineage-restricted neuronal precursors from multipotent neuroepithelial stem cells, *Neuron* 19(4):773–785.

McEwen, B.S., 2001, Plasticity of the hippocampus: Adaptation to chronic stress and allostatic load, *Ann. N Y Acad. Sci.* 933:265–277.

Merrill, D.A., Karim, R. *et al.*, 2003, Hippocampal cell genesis does not correlate with spatial learning ability in aged rats, *J. Comp. Neurol.* 459(2):201–207.

Morgan, T.E., Xie, Z. *et al.*, 1999, The mosaic of brain glial hyperactivity during normal ageing and its attenuation by food restriction, *Neuroscience* 89(3):687–699.

Muller, D., Toni, N. *et al.*, 2000, Spine changes associated with long-term potentiation, *Hippocampus* 10(5):596–604.

Naber, D. and Dahnke, H.G., 1979, Protein and nucleic acid content in the aging human brain, *Neuropathol. Appl. Neurobiol.* 5(1):17–24.

Nabeshima, T., Nitta, A. *et al.*, 1994, Oral administration of NGF synthesis stimulators recovers reduced brain NGF content in aged rats and cognitive dysfunction in basal-forebrain-lesioned rats, *Gerontology* 40(Suppl 2):46–56.

Nilsson, M., Perfilieva, E. *et al.*, 1999, Enriched environment increases neurogenesis in the adult rat dentate gyrus and improves spatial memory, *J. Neurobiol.* 39(4):569–578.

Nixon, R.A., Saito, K.I. *et al.*, 1994, Calcium-activated neutral proteinase(calpain) system in aging and Alzheimer's disease, *Ann. N Y Acad. Sci.* 747:77–91.

Ohsawa, I., Takamura, C. *et al.*, 1999, Amino-terminal region of secreted form of amyloid precursor protein stimulates proliferation of neural stem cells, *Eur. J. Neurosci.* 11(6):1907–1913.

Pedersen, W.A., Chan, S.L. *et al.*, 2002, Numb isoforms containing a short PTB domain promote neurotrophic factor-induced differentiation and neurotrophic factor withdrawal-induced death of PC12 Cells, *J. Neurochem.* 82(4):976–986.

Pevny, L.H., Sockanathan, S. *et al.*, 1998, A role for SOX1 in neural determination, *Development* 125(10):1967–1978.

Prolla, T.A. and Mattson, M.P. 2001, Molecular mechanisms of brain aging and neurodegenerative disorders: Lessons from dietary restriction, *Trends Neurosci.* 24(11 Suppl):S21–S31.

Quinn, S.M., Walters, W.M. *et al.*, 1999, Lineage restriction of neuroepithelial precursor cells from fetal human spinal cord, *J. Neurosci. Res.* 57(5):590–602.

Rao, K.S., 1993, Genomic damage and its repair in young and aging brain, *Mol. Neurobiol.* 7(1):23–48.

Rapin, I., Lindenbaum, Y. *et al.*, 2000, Cockayne syndrome and xeroderma pigmentosum, *Neurology* 55(10):1442–1449.

Rasika, S., Alvarez-Buylla, A. *et al.*, 1999, BDNF mediates the effects of testosterone on the survival of new neurons in an adult brain, *Neuron* 22(1):53–62.

Ray, J., Peterson, D.A. *et al.*, 1993, Proliferation, differentiation, and long-term culture of primary hippocampal neurons, *Proc. Natl. Acad. Sci. USA* 90(8):3602–3606.

Robertson, G.S., Crocker, S.J. *et al.*, 2000, Neuroprotection by the inhibition of apoptosis, *Brain Pathol.* 10(2):283–292.

Roch, J.M., Masliah, E. *et al.*, 1994, Increase of synaptic density and memory retention by a peptide representing the trophic domain of the amyloid beta/A4 protein precursor, *Proc. Natl. Acad. Sci. USA* 91(16): 7450–7454.

Romanczyk, T.B., Weickert, C.S. *et al.*, 2002, Alterations in trkB mRNA in the human prefrontal cortex throughout the lifespan, *Eur. J. Neurosci.* 15(2):269–280.

Rosman, N.P., Anselm, I. *et al.*, 2001, Progressive intracranial vascular disease with strokes and seizures in a boy with progeria, *J. Child Neurol.* 16(3):212–215.

Roth, G.S., 1995, Changes in tissue responsiveness to hormones and neurotransmitters during aging, *Exp. Gerontol.* 30(3–4):361–368.

Sastre, J., Pallardo, F.V. *et al.*, 2000, Mitochondria, oxidative stress and aging, *Free Radic. Res.* 32(3):189–198.

Sawa, A., 1999, Neuronal cell death in Down's syndrome, *J. Neural Transm. Suppl.* 57:87–97.

Shen, J., Bronson, R.T. *et al.*, 1997, Skeletal and CNS defects in Presenilin-1-deficient mice, *Cell* 89(4):629–639.

Shors, T.J., Miesegaes, G. *et al.*, 2001, Neurogenesis in the adult is involved in the formation of trace memories, *Nature* 410(6826):372–376.

Smith, M.A. and Cizza, G., 1996, Stress-induced changes in brain-derived neurotrophic factor expression are attenuated in aged Fischer 344/N rats, *Neurobiol. Aging* 17(6):859–864.

Smith-Swintosky, V.L., Pettigrew, L.C. *et al.*, 1994, Secreted forms of beta-amyloid precursor protein protect against ischemic brain injury, *J. Neurochem.* 63(2):781–784.

Snyder, J.S., Kee, N. *et al.*, 2001, Effects of adult neurogenesis on synaptic plasticity in the rat dentate gyrus, *J. Neurophysiol.* 85(6):2423–2431.

Sohn, Y.K., Ganju, N. *et al.*, 1999, Neuritic sprouting with aberrant expression of the nitric oxide synthase III gene in neurodegenerative diseases, *J. Neurol. Sci.* 162(2):133–151.

Sommer, L. and Rao, M. 2002, Neural stem cells and regulation of cell number, *Prog. Neurobiol.* 66(1):1–18.

Song, J., Wu, L. *et al.*, 2003, Axons guided by insulin receptor in Drosophila visual system, *Science* 300(5618):502–505.

Sonntag, W.E., Lynch, C.D. *et al.*, 1999, Alterations in insulin-like growth factor-1 gene and protein expression and type 1 insulin-like growth factor receptors in the brains of ageing rats, *Neuroscience* 88(1):269–279.

Stroemer, R.P., Kent, T.A. *et al.*, 1995, Neocortical neural sprouting, synaptogenesis, and behavioral recovery after neocortical infarction in rats, *Stroke* 26(11):2135–2144.

Sze, J.Y., Victor, M. *et al.*, 2000, Food and metabolic signalling defects in a *Caenorhabditis elegans* serotonin-synthesis mutant, *Nature* 403(6769):560–564.

Tang, E.D., Nunez, G. *et al.*, 1999, Negative regulation of the forkhead transcription factor FKHR by Akt, *J. Biol. Chem.* 274(24):16741–16746.

Tatar, M., Kopelman, A. *et al.*, 2001, A mutant Drosophila insulin receptor homolog that extends life-span and impairs neuroendocrine function, *Science* 292(5514):107–110.

Tirassa, P., Triaca, V. *et al.*, 2003, EGF and NGF injected into the brain of old mice enhance BDNF and ChAT in proliferating subventricular zone, *J. Neurosci. Res.* 72(5):557–564.

Vaillant, A.R., Mazzoni, I. *et al.*, 1999, Depolarization and neurotrophins converge on the phosphatidylinositol 3-kinase-Akt pathway to synergistically regulate neuronal survival, *J. Cell Biol.* 146(5):955–966.

van Brabant, A.J., Stan, R. *et al.*, 2000, DNA helicases, genomic instability, and human genetic disease, *Annu. Rev. Genomics Hum. Genet.* 1:409–459.

van Gool, A.J., van der Horst, T.G. *et al.*, 1997, Cockayne syndrome: Defective repair of transcription? *EMBO J.* 16(14):4155–4162.

van Praag, H., Kempermann, G. *et al.*, 1999, Running increases cell proliferation and neurogenesis in the adult mouse dentate gyrus, *Nat. Neurosci.* 2(3):266–270.

van Praag, H., Schinder, A.F. *et al.*, 2002, Functional neurogenesis in the adult hippocampus, *Nature* 415(6875):1030–1034.

van Weeren, P.C., de Bruyn, K.M. *et al.*, 1998, Essential role for protein kinase B(PKB) in insulin-induced glycogen synthase kinase 3 inactivation. Characterization of dominant-negative mutant of PKB, *J. Biol. Chem.* 273(21):13150–13156.

Wanagat, J., Allison, D.B. *et al.*, 1999, Caloric intake and aging: Mechanisms in rodents and a study in nonhuman primates, *Toxicol. Sci.* 52(2 Suppl): 35–40.

Weindruch, R. and Sohal, R.S., 1997, Seminars in medicine of the Beth Israel Deaconess Medical Center. Caloric intake and aging, *N. Engl. J. Med.* 337(14):986–994.

West, M.J., Coleman, P.D. *et al.*, 1994, Differences in the pattern of hippocampal neuronal loss in normal ageing and Alzheimer's disease, *Lancet* 344(8925):769–772.

Williams, R.L., Hilton, D.J. *et al.*, 1988, Myeloid leukaemia inhibitory factor maintains the developmental potential of embryonic stem cells, *Nature* 336(6200):684–687.

Wolkow, C.A., Kimura, K.D. *et al.*, 2000, Regulation of *C. elegans* life-span by insulinlike signaling in the nervous system, *Science* 290(5489): 147–150.

Yoshida, T., Satoh, M. *et al.*, 1993, Cytokines affecting survival and differentiation of an astrocyte progenitor cell line, *Brain Res. Dev. Brain Res.* 76(1):147–150.

Yu, Z.F. and Mattson, M.P., 1999, Dietary restriction and 2-deoxyglucose administration reduce focal ischemic brain damage and improve behavioral outcome: Evidence for a preconditioning mechanism, *J. Neurosci. Res.* 57(6):830–839.

Zhu, H., Guo, Q. *et al.*, 1999, Dietary restriction protects hippocampal neurons against the death-promoting action of a presenilin-1 mutation, *Brain Res.* 842(1):224–229.

Zuccato, C., Ciammola, A. *et al.*, 2001, Loss of huntingtin-mediated BDNF gene transcription in Huntington's disease, *Science* 293(5529):493–498.

14

Beginnings of the Nervous System

Marcus Jacobson[†][*]

BEGINNINGS OF THE NERVOUS SYSTEM

There is, it would seem, in the dimensional scale of the world, a kind of delicate meeting place between imagination and knowledge, a point, arrived at by diminishing large things and enlarging small ones, that is intrinsically artistic.

Vladimir Nabokov (1899–1977),
in *Speak Memory* (1966, revised edition)

A HISTORICAL ORIENTATION

This was a theory of trial and error—of conjectures and refutations. It made it possible to understand why our attempts to force interpretations upon the world were logically prior to the observation of similarities. Since there were logical reasons behind this procedure, I thought that it would apply in the field of science also; that scientific theories were not the digest of observations, but that they were inventions—conjectures boldly put forward for trial, to be eliminated if they clashed with observations; with observations which were rarely accidental, but as a rule undertaken with the finite intention of testing a theory by obtaining, if possible, a decisive refutation.

Karl R. Popper (1902–), *Conjectures and refutations: The Growth of Scientific Knowledge*, 1962

The Nervous System Is Made of Cells

The history of neuroscience can be viewed as a gradual improvement of techniques with which complex organisms could be analyzed and reduced to their constituent cells and molecules. This can be called the reductionist neuroscience research program. The reductionist program was found on two main assumptions: Firstly, development is the assembly of elementary units in various combinations and configurations, advancing from simple to complex, each stage caused by the conditions of the immediately preceding stage. The second assumption underlying the reductionist program is that one of the main aims of embryology is to deduce the events of development and morphogenesis from the activities of elementary units.

The idea that living organisms are reducible to elementary components, invisible corpuscles and fibers, which were already in the 17th century sometimes termed molecules, was held by Pierre Gassendi (1592–1655) and Robert Boyle (1627–1691) among others (reviewed by Hall, 1979). This idea was one of the essential parts of a research program that culminated in what I have called the mechanization of the brain picture. As Popper states in the epigraph to this section, the idea leads, and the techniques and the results follow, during the construction of a scientific research program. Atomistic and molecular theories of living organisms were proposed in the 17th and 18th centuries, long before the cell theory was advanced in the 19th—reduction of organisms to their constituent parts did not only occur in the logical order from larger to smaller components, from the top down, but also from the atoms and molecules to the macroscopic structures, from the bottom up. Modern theories of cellular structure echo the micro mechanical models of the 17th and 18th century theorists. One of the consequences was that conjectures about molecular mechanisms were made that could not be tested experimentally until centuries later. Many examples (of which Mendel's theory of inheritance is the best) can be given of such premature conjectures that were systematized and generalized to form premature theories (see Jacobson, 1993). Methods to implement the reductionist program would remain unavailable for almost a century, though it was clearly perceived as the desired goal. Koelliker prophetically states in the first English edition of his *Manual of Human Histology* (1852):

If it be possible that the molecules which constitute cell membranes, muscular fibrils, axile fibres of nerves should be discovered, and the laws … of the origin, growth and activity of the present so-called elementary parts, should be made out, then a new era will commence for histology, and the discoverer of the law of cell genesis, or, of a molecular theory, will be as much or more celebrated than the originator of the doctrine of all animal tissues out of cells.

The cell theory, introduced by Schwann in 1839, defined cells as elementary units whose division, transformation, combination, and permutation to form complex organisms could be observed with the lately improved microscopes and histological methods.

Marcus Jacobson ● Department of Neurobiology and Anatomy, University of Utah, SOM, Salt Lake City, UT 84132.
[†] Deceased.
[*] Edited and updated by Dr. T. N. Parks—Chair—Department of Neurobiology and Anatomy, University of Utah School of Medicine, Salt Lake City 84124.

Developmental Neurobiology, 4th ed., edited by Mahendra S. Rao and Marcus Jacobson. Kluwer Academic / Plenum Publishers, New York, 2005.

Because of its inclusivity, the cell theory permitted generalization from one form to all others, so that development of the nervous system was no longer considered to be governed by its own exclusive laws. The cell theory is a paradigm; the neuron theory emerged from it as an important special case. Sherrington says this in the opening sentence of *The Integrative Action of the Nervous System* (1906): "Nowhere in physiology does the cell theory reveal its presence more frequently in the very framework of the argument than at the present time in the study of the nervous reactions."

From 1828 to 1839, the concept of epigenesis was established by VonBaer as a central theory of embryology: "The general before the specific, and so on down to the smallest parts." This theory opened the way for a causal analysis of development of organs and relatively gross structures. Epigenesis was interpreted by VonBaer and most of his contemporaries in terms of homology and the unity of body plan of all animals. The theory of germ layers and the theory of segmentation were results of the interpretation of development. Embryology was still practiced as a branch of morphology not requiring reference to histological structure. That changed after 1830, when the achromatic compound microscope made it possible to see fibers and globular bodies in the central nervous system. In 1837 Purkinje discovered the flask shaped cells in the cerebellar cortex which now bear his name. He concluded that in all animals the nervous system is formed of three components, fluid, fibers, and globules. Such globules were at first thought to develop by precipitation out of a homogeneous fluid substance. This notion was consistent with the theory of epigenesis and it persisted for at least a decade after the introduction of the cell theory.

Beginning in 1839 with the publication of Schwann's book on the cell theory, cells were shown to be the basic units of all multicellular organisms, and the primary roles of cells in heredity and development were very slowly revealed. There was an initial period of confusion, during which cells were still believed to originate by precipitation from a homogeneous "cytoblasteme" (Schwann's term, 1839) in addition to their production by mitosis (Fleming's term, 1860). However, by 1850–1855 Remak could give an account of vertebrate development in terms of cells originating only from other cells, a concept that was finally generalized in the dictum *omnis cellula e cellula* (Virchow, 1858).

Nevertheless, Darwin was able to write *The Origin of the Species* over a period of about 20 years until its publication in 1859 without any reference to the cellular structure. This I find one of the most remarkable facts of the history of science. It shows that scientists do not necessarily work under the influence of the spirit of the times (zeitgeist) or within a "paradigm" as defined by Thomas Kuhn (1962, 1970, 1974).

During the second epoch, particularly as the result of the work of Remak (1850–1855), Koelliker (1852), and Virchow (1858), nerve cells and neuroglial cells were shown to be the basic units of organization of the nervous system, and nerve fibers were recognized as parts of nerve cells. By 1874 Wilhelm His could formulate a purely mechanistic theory of development of the nervous system in terms of cell division, migration, aggregation, differentiation, and changes in cell form and function.

This is an indication of the rate of progress that occurred in understanding the behavior of cells following the first glimmerings of the cell theory in the minds of Schleiden (1838) and Schwann (1839). Apart from formulating a general cell theory, these two got almost everything else wrong, so that their original ideas about cells were almost totally overturned before 1870.

Morphogenesis

The history of ideas about morphogenesis has an extensive literature (Russell, 1930; Radl, 1930; Needham, 1931, 1934; Hughes, 1959; Adelmann, 1966; Hall, 1969, to cite only well-known secondary sources). A sketch using broad strokes should be sufficient for the purposes of this historical orientation. Goethe, to whom we owe the term morphology, and his contemporaries and immediate successors, conceived of morphogenesis in terms of plastic transformations of tissues and organs driven by innate formative stimuli and molded by environmental forces. This concept was greatly advanced by Schwann and his successors, who could begin to understand morphology and morphogenesis in terms of cells. Morphogenesis of the nervous system was analyzed in terms of histogenesis, changes of cell shape, cell movements, and migrations (Remak, 1855; Koelliker, 1852; His, 1868, 1887). Wilhelm His (1874, 1894) showed how changes in cell shape are involved in folding of tissues such as the neural plate. The concept of cell migration in the developing nervous system was also worked out by His, first from his observations on the origins of the peripheral nervous system from the neural crest (His, 1868), the migration of cells from the olfactory placode (His, 1889; Koelliker, 1890), and later from the discovery that the neuroblasts migrate individually from the ventricular germinal zone to the overlying mantle layer of the neural tube (His, 1889). The revolutionary discovery of cell migration in the vertebrate central nervous system by His was at first greeted with skepticism, but the universality of this form of cellular behavior soon became apparent. In 1893 Loeb showed migration of pigment cells in *Fundulus*, a teleost, and migration of presumptive skeleton cells and mesenchyme cells in sea urchin embryos was discovered by Herbst (1894) and Driesch (1896). Mechanisms of cell migration were proposed by analogy with the locomotory movements of protozoa (Korschelt and Heider, 1903–1909), but there were no means for making progress along those lines at that time. Other cases of cell migration in the vertebrate central nervous system were also baffling. After discovery of the cerebellar granular layer (Obersteiner, 1883; Herrick, 1891), several attempts were made to follow the migration of granule cells until the problem yielded to Cajal's definitive analysis in the 1890s (Ramón y Cajal, 1911; see Jacobson, 1993).

THE GERMINAL CELL, HISTOGENESIS, AND LINEAGES OF NERVE CELLS

For all those who are enchanted by the magic of the infinitely small, hidden in the bosom of the living being are millions of palpitating cells whose only demand for the surrender of their secret, and with it the halo of fame, is a lucid and tenacious intelligence to contemplate them.

The final sentence of Ramon y Cajal's autobiography, *Recuerdos de mi vida: Historia de mi labor cientifica*, Tercera edicion, 1923

Historical Orientation

A mind historically focused will embody in its idea of what is "modern" and "contemporary" a far larger section of the past than a mind living in the myopia of the moment. "Contemporary civilization" in our sense, therefore, goes deep into the 19th century.

Johann Huizinga (1872–1945), *Homo Ludens*, 1938

From its inception by Wilhelm His in 1887, the concept of subclasses of germinal cells that are the progenitors of corresponding classes of neurons and glial cells has been opposed by the concept of multipotential progenitors, put forward by Vignal (1888), Schaper (1894a, b), and Koelliker (1897). Both concepts have continued to exert powerful heuristic effects for more than a century.

Discovery of the germinal cells of the vertebrate nervous system, by Wilhelm His in 1887, can be fully understood only in historical context—it was only one of many interlocking pieces of research out of which a coherent picture of the behavior of cells during development was rapidly assembled in the second half of the 19th century (O. Hertwig, 1893–1896; E.B. Wilson, 1896). The contributions of His must also be viewed as part of the ongoing program to reconcile comparative anatomy and embryology with the gradual improvements in understanding cell structure and function (Koelliker 1852, 1854, 1896). That program gained momentum throughout the 19th century, and the investigation of the histogenesis of the nervous system was pushed forward rapidly by the advances made by cytologists and embryologists. In many ways it was a period like our own in which powerful new research techniques produced results that challenged the assumptions of time. The cell theory put forward by Schwann in 1838 was radically modified during the next 50 years, most notably by the discovery of cell division and the cell cycle.

Wilhelm His (1831–1904) towers over the field of research on histogenesis of the nervous system in the 19th century. He casts a long shadow into the 20th century through his pupils Franklin P. Mall, who brought the science of human embryology to the United States, and Friedrich Miescher, founder of the chemistry of nucleic acids and nucleoproteins. The new concepts of cell biology were transmitted to neuroembryology by His: He was the first to recognize the significance of cell migration in development of the central and peripheral nervous systems. He discovered the neural crest and showed that cranial and spinal ganglia are formed by cells which migrate from the neural crest. He was among the earliest to give evidence that the nerve fiber is an outgrowth of the nerve cell, the first to try to show when neuronal and glial cell lineages diverge, and he discovered that nerve cells originate by mitosis of stem cells near the ventricle of the neural tube. He showed that neurons originate from specific progenitor cells, which he called germinal cells (*Keimzellen*), recognizable by their mitotic figures lying close to the lumen of the neural tube (His, 1887a, b, 1888a, b, 1889a, 1890a, b).

It should be remembered that Walther Flemming's *Zellsubstanz, Kern und Zelltheilung* (1882) was hot off the press when His discovered the germinal cells in the neural tube of human embryos. Flemming's book provided the first clear demonstration of the transformation of the resting cell nucleus into the mitotic figure, and he showed that the essential event of mitosis is the duplication and division of the chromosomes. Flemming recognize that chromatin (which is the name he gave to the material in the nucleus which he stained with azo dyes) is probably the same as the nucleic acid which Miescher (1871) had purified from the nuclei of leukocytes and had called nuclein (reviewed by Hughes, 1952, 1959). By the mid-1880s, it had become evident that chromatin, the material of the chromosomes (named by Waldeyer in 1888), is the basis of hereditary. With those discoveries the links were forged between cytology, embryology, and evolution.

The great achievements of Wilhelm His are in no way diminished by the fact that he was misled by histological artifacts which were unavoidable at that time. Artifacts led him to conclude that there are two different classes of cells in the neural tube of the early vertebrate embryo: germinal cells that are visible as mitotic figures lining the lumen, which give rise to neurons, and spongioblasts that appear to form a syncytium from which neuroglial cells originate. These observations led His to formulate four separate theories: theory 1 was concerned with the different stem cells for neurons and glial cells; theory 2 was about the syncytium; theory 3 was about the significance of the large extracellular spaces and their contents; theory 4 was about guidance of migrating neuroblasts by radically aligned spongioblasts (see Jacobson, 1993 for a full discussion).

His (1887a, b, 1888a, 1889) deduced that the germinal cell divides repeatedly: one daughter cell remains close to the lumen of the neural tube and reenters the mitotic cycle while the other daughter cell becomes a neuroblast. Then the neuroblast, which is incapable of further division, migrates away from the germinal layer and eventually develops into a neuron. Recognition of the asymmetrical division of the germinal cell was a significant conceptual advance that has become assimilated into modern theory.

The suffix "blast" derives from the Greek word *blastos*, which means a germ or bud, and indicates that a cell is capable of further division. However, neuroblasts in the vertebrates do not incorporate [^{3}H]thymidine, which indicates that they have ceased DNA synthesis and mitosis. Therefore, the cell called neuroblast by His is better referred to as an undifferentiated neuron or young neuron, and the term neuroblast is best reserved for the progenitor of neurons in the invertebrates.

Another theory of neuronal and neuroglial histogenesis, in almost total disagreement with the theory of His, was proposed by Vignal (1888) and Schaper (1894a, b) and supported by Koelliker (1897). They based their theory on essentially the same histological observations as those of His, but they interpreted them in a different way. The observation that mitotic figures occur only in the cells lining the lumen of the early neural tube had been interpreted by His to mean that the mitotic figures belong to germinal cells (*Keimzellen*), whereas he believes that the other cell nuclei of the neural tube belong to different classes of cells. He identifies some as neuroblasts giving rise to neurons and others as spongioblasts giving rise to neuroglial cells. Alternative interpretations of the histological picture were given by Schaper (1897a, b). He suggested that "*the so-called 'Keimzellen' of His lying near the central cavity of the neural tube, along the membrana limitans interna, are not to be considered as a special type of cell in contrast to the main epithelial cells in process of continuous proliferation*" (Schaper, 1897b). According to Schaper, the so-called germinal cells and spongioblasts are really cells of the same type which move to different levels on the neural tube during different phases of the mitotic cycle. The "*Keimzellen*" of His are merely cells that have rounded up close to the lumen in preparation for mitosis, after which the nuclei of the daughter cells move away from the lumen during interphase and return inward during prophase.

Schaper also showed that the young neuron, with a clear nucleus and abundant cytoplasm, can be distinguished from the neuroglial cell precursors. Some of the glial cell precursors have a small, round, densely chromatic nucleus and very scanty cytoplasm, but others may have different appearances. We would now call these cells glioblasts. Schaper showed that the young neurons and glioblasts migrate away from the lumen and form the mantle layer outside the germinal zone. Some of the cells in the mantle layer undergo mitosis. According to Schaper (1897a, p.100), these are "indifferent cells", which he thought are capable of giving rise either to neurons or to neuroglial cells. These would now be called pluripotential progenitor cells.

Schaper's theory was premature. On the authority of both His and Ramón y Cajal (1909, p.637), Schaper's theory was consigned to oblivion. It was not accepted into a research program until 50 years later when new evidence in its favor was provided by F.C. Sauer (1935a, b). He confirmed that the neural epithelium, until the time of closure of the neural tube, consists of a single type of epithelial cell in various stages of the mitotic cycle. In addition, Sauer showed that the appearance of the cell changes and its nucleus moves to different positions in the cytoplasm during the different phases of the mitotic cycle.

NEUROGLIAL ONTOGENY

On a perfectly translucent yellow field appear thin, smooth, black filaments, neatly arranged, or else thick and spiny, arising from triangular, stellate or fusiform black bodies! One might say they are like a Chinese ink drawing on transparent Japanese paper. The eye is disconcerted, so accustomed is it to the inextricable network stained with carmine and hematoxylin which always forces the mind to perform feats of critical interpretation. Here everything is simple, clear, without confusion … The technique of dreams is now reality! The metallic impregnation has made such a fine dissection, exceeding all previous hopes. This is the method of Golgi.

Ramón y Cajal (1852–1934), *Histologie du systèm nerveux*, Vol. 1 p. 29, 1909

History of Neuroglia

I have for this young researcher [Ramón y Cajal] the greatest regard, and as I have admired his great activity and initiative, I can appreciate the importance of his original observations. The small differences between his conclusions and my own cannot have an effect on my sentiments, as I am profoundly convinced that such divergence, by which one can push research forward, is always useful to science.

Camillo Golgi (1843–1926), *La rete nervosa diffusa degli organi centrali del sistema nervousa*, 1901

The original concept of neuroglia meaning "nerve glue" was based on Rudolf Virchow's assumption that there must be a mesodermal connective tissue element of the nervous system (Virchow, 1846, 1858, 1867). Even if neuroglial cells did not exist Virchow would have had to invent them as a requirement for his theory—as a bold conjecture thrown out for refutation. But techniques were inadequate to either corroborate or refute that conjecture. The mesodermal origin of neuroglial cells continued to receive corroboration (Andreizen, 1893; Weigart, 1895; W. Robertson, 1897, 1899, 1900a) in the face of strong counter evidence showing that both neurons and glial cells originate from embryonic ectoderm (His, 1889, 1901).

Virchow and his disciples were primarily interested in pathology and thus in neuroglial tumors and in the reaction of the neuroglia to disease and injury. Their theories guided practice in the direction of neuropathology, and away from normal development. Virchow's research program was aimed at showing that the causes of disease can be found in derangements of cells.

It is doubtful whether Virchow saw neuroglial cells in 1846 and there are no convincing pictures of them in Virchow's book, *Die Cellularpathologie* (1858), although he there discusses the theory of a neuroglial tissue. At that time the theory led and the facts followed. Progress was slow because techniques were inadequate in the 1840's and 1850's to provide reliable evidence. Virchow (1885) later described that period: "The great upheavals that microscopy, chemistry, and pathological anatomy had brought about were at first accompanied by the most dismal consequences. People found themselves helpless … filled with exaggerated expectations they seized on any fragment which a bold speculator might cast out." Virchow's bold theory of the neuroglia consisted of two such bold speculations: first, that neuroglial cells form a connective tissue, and second, that neuroglia develop from mesoderm in the embryo. These conjectures were not Virchow's original creations. He derived the concept of tissues from Bichat (1801), who first proposed that tissues are where the functions of life and dysfunctions of disease occur— Virchow extrapolated, saying that cells are where life and disease occur. Virchow obtained the concept of mesoderm from Remak to whom he is also indebted for the idea that all cells originate from other cells.

The problem of the ectodermal or mesodermal origins of the neuroglial cells has a complex history. Wilhelm His (1889) corrected Virchow's misconception that the neuroglia form the connective tissue elements of the nervous system by showing that neuroglia cells as well as neurons originate from neurectoderm.

The problem was resolved in the last decade of the 19th century when numerous stains for connective tissue failed to stain neuroglial cells but only stained blood vessels in the central nervous system: the Unna-Taenzer orcein method (Unna, 1890, 1891) and the Weigert (1898) resorcinol-fuschin method for elastic fibers; the van Geison (1889) acid fuschin-picric acid stain for collagen; and the Mallory (1900) aniline blue stain for connective tissue. That so many different connective tissue stains failed to stain neuroglial cells could not conclusively falsify Virchow's theory because it could be maintained that glial cells are a special form of connective tissue that is not stained by any other method. That was the argument used by Andriezen (1893) and Weigert (1895). Contrariwise, the invention of specific neuroglial stains (Ramón y Cajal, 1913; Rio-Hortego, 1919, 1921a, b, 1932) was not considered to be conclusive refutation of Virchow's theory. Cajal could still maintain in 1920 that the glial cells belonging to his "third element" are of mesodermal origin. The most compelling evidence against it was that neuroglial cells in the embryo originate from the neurectoderm (His, 1889). However, the available evidence did not exclude the possibility of subsequent entry of mesodermal cells into the central nervous system. The gradual resolution of this problem is considered below, in connection with the history of the microglial cells.

Wilhelm His misidentified the progenitors of both the neurons and the neuroglial cells in the neural tube. He conjectured that neurons originate from germinal cells but the glial cells originate from a syncytial tissue named the "*myelospongium*" or "*neurospongium*" formed of spongioblasts. The theory that glial cells remain permanently anastomosed with one another continued to be corroborated by many authorities (Hardesty, 1904; Held, 1909; Streeter, 1912). In 1912, Streeter could confidently invert the truth by pronouncing that "earlier conceptions of neuroglia cells were based on silver precipitation methods (Golgi) which failed to reveal the true wealth of their anastomosing branches, and there thus existed a false impression of neuroglia as consisting of scattered and independent cells." At that time cells were believed to be naked protoplasmic bodies, lacking a membrane, and connected by protoplasmic and fibrous bridges.

The neurospongium theory was based entirely on artifacts and could not be refuted conclusively until the 1950s when the electron microscope showed that all cells in the neurectoderm and neural tube are separated by narrow intercellular clefts from the beginning of development. In the 19th century the counter evidence to the neuroglial syncytium was already quite strong. Golgi staining of the neural tube always showed separate glial cells forming a series of stages of development from the radically aligned spongioblasts to mature astrocytes (Fig. 1) (Koelliker, 1893; Lenhossék, 1893; Ramón y Cajal, 1894). As evidence against the neuroglial syncytium Alzheimer (1910) showed that one neuroglial cells may undergo pathological change while its neighbors remain normal. By the 1920's the neurospongium theory had been abandoned, not because it could be falsified conclusively, but because it failed to predict new observations predicted by the alternative theory that glial cells are always separate (Penfield, 1928, review)

The research program of His split into four separate research programs with different goals. The first dealt with the

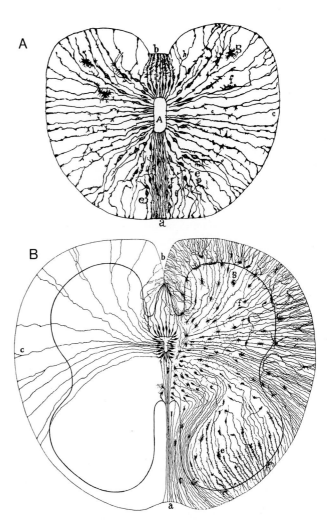

FIGURE 1. Epithelial cells and orgin of the neuroglial cells in the spinal cord of the 9-day chick embryo (A) and 14-cm humand embryo spinal cord (B). Golgi preparations. A = ependymal canal; a, b = epithelial cells of the posterior and anterior median sulci; these two types of cells maintain both peripheral and central attachments; c = ependymal cell; e = displaced epithelial cell migrating in the posterior horn; f = displaced epithelial cell migrating in te anterior horn; these two types of cells have entirely lost their central ends and conserved their peripheral ends terminating by conical boutons at the pia mater; g = epithelial cell about to become a neuroglial corpuscle; there remains no more than an outgrowth going to the pia mater. A from S. Ramón y Cajal, *Les nouvelles idees sur la structure du systeme nerveus*, 1894; B from M. von Lenhossék, Der feinere Bau des Nervensystems, 1893.

origins of neurons and glial cells from progenitors on the neural tube—His conjectured that neurons originate from germinal cells near the ventricle, and glial cells originate from spongioblasts which span the full thickness of the neural tube. The second theory dealt with the syncytial connections between cells in the neural tube. The third dealt with the enormous intercellular spaces which appeared as an artifact in sections of neural tube. This space was supposed to contain a "central ground substance" in which the cells were embedded (Boeke, 1942; Bauer, 1953). A fourth theory dealt with the role of migration of neuroblasts in these spaces, guided by the radically aligned spongioblasts

(Magini, 1888; His, 1889) (see Jacobson, 1993). These theories illustrate Cajal's dictum that "in biology theories are fragile and ephemeral constructions ... while hypotheses pass by, facts remain" (Ramón y Cajal, 1928). He was also well aware that the facts were often indistinguishable from artifacts.

The Golgi research program on neuroglia started in 1875 with Golgi's studies of glial cell tumors. Golgi's earliest papers on neuroglia were published in 1870 and 1871 using hematoxylin and carmine staining which was incapable of showing either neurons or glial cells completely. By 1873 Golgi had developed his potassium dichromate-silver technique, the first of his methods for metallic impregnation which was capable of staining entire neurons and glial cells. In Golgi's 1875 paper on gliomas stained by means of his potassium dichromate-silver technique, he was able to give the first morphological definition of neuroglial cells as a class distinct from neurons. Golgi discovered the glial cell perivascular foot and conjectured that the glial cell "protoplasmic process" mediates transfer of nutrients between the blood and brain (Golgi, 1894). This led ultimately to the modern theory of the blood–brain barrier. This is only one sign of the very progressive character of Golgi's research program, not only in terms of a great technical advance but also because of its theoretical boldness.[1] By showing that the protoplasmic processes of neuroglial cells end on blood vessels and those of nerve cells end blindly, Golgi refuted Gerlach's theory that the protoplasmic processes (later called dendrites by His, 1890) end in a diffuse nerve net by which all nerve cells were believed to be interconnected. Having refuted Gerlach's theory, Golgi conjectured that the nerve net is really between afferent axons and the axon collaterals of efferent neurons. It should be emphasized that it was a mature and progressive theory for its time and became postmature and degenerative only near the very end of the 19th century. Guided by Golgi's conjectures about the relationship between neurological cells and blood vessels and about the active role of neuroglia in mediating exchange with the blood, progressive research programs have continued to the present time (see Chapters 7 and 8).

Golgi published his mercuric chloride method in 1879 but it went unnoticed until after he published his rapid method in 1887. The Golgi techniques were widely used after 1887 and were mainly responsible for the accelerated progress in neurocytology during the 1890's. One of the first advances was recognition of different types of glial cells. Astrocytes were named by Lenhossék (1891) who recognized them as a separate subclass although Golgi had identified them as early as 1873. Koelliker

(1893) and Andriezen (1893a) subdivided them into fibrous and protoplasmic types according to the presence or absence of fibers in the cytoplasm. They noted that fibrous astrocytes predominate in white matter and protoplasmic astrocytes in grey matter.

There were two theories of astrocyte histogenesis. Schaper (1897a,b) conjectured that the germinal cells of His, positioned close to the ventricle, give rise to neuroblasts, spongioblasts, and "indifferent cells." He speculated that spongioblasts give rise to glial cells only in the embryo but indifferent cells persist in the postnatal period and give rise to both neurons and glial cells. A different theory was proposed by Koelliker (1890) and supported by Cajal (1909). They believed that spongioblasts persist into the postnatal period and are the only progenitors of all glial cells. Evidence for this was given by Lenhossék (1893), who showed a series of stages of astrocyte histogenesis, starting with detachment of radially aligned spongioblasts from the internal and external limiting membranes, followed by migration into the brain parenchyma where they continue to divide to give rise to astrocytes. Transformation of radial glial cells into astrocytes in the spinal cord was confirmed by Ramón y Cajal (1896) in the chick embryo. Almost a century later those premature discoveries were rediscovered; one might say they matured (Schmechel and Rakic, 1979a, b; Levitt and Rakic, 1980; Choi, 1981; Hajós and Bascó, 1984; Benjelloun-Touimi et al., 1985; Federoff, 1986; Munoz-Garcia and Ludwin, 1986a, b; M.Hirano and Goldman, 1988). Both Lenhossék and Cajal observed that the radial glial cells are the first glial cells to differentiate and this has also been confirmed with modern techniques (Rakic, 1972, 1981; Choi, 1981). Lenhossék and Cajal found that the peripheral expansions of some of the radial glial cells persist in the spinal cord to form the glia limitans exterior, and that too has recently been confirmed (Liuzzi and Miller, 1987).

Cajal's gold chloride-sublimate method stains astrocytes very well and this enabled him to confirm Lenhossék's theory of the origin of astrocytes from radial glial cells and to refute the theory that astrocytes originate from mesoderm (Cajal, 1913a, 1916). Cajal (1913b) also showed that astrocytes can divide in the normal brain. Mitosis of astrocytes after brain injury was demonstrated by Rio-Hortega and Penfield (1927). When those theories of glial cell origins were originally proposed, they could neither be corroborated nor falsified by means of histological evidence because no techniques for tracing cell lineages existed at that time. Those theories of glial cell lineages were bypassed for a century until more reliable techniques for tracing neuroglial cell lineages recently became available (see Chapter 8).

Oligodendroglial cells were discovered by W. Robertson (1899, 1900a) using his platinum stain. He did not understand their significance in myelination and he conjectured that they originate from mesoderm (he called them mesoglia, not to be confused with the mesoglia of Rio-Hortega, which are brain macrophages). They were rediscovered by Rio-Hortega (1919, 1921), who called them oligodendroglia because their processes are shorter and sparser than those of astrocytes. He made the distinction between perineuronal satellites in the gray matter, most of which he believed to be oligodendroglia, and interfascicular ogliodendroglia situated in rows between the myelinated fibers of

---

[1] A theory is mature when it is accepted in a research program where it operates to guide techniques to obtain new facts. A progressive theory is defined as one that leads the facts; a degenerative theory is one that falls behind the facts. A progressive theory stimulates research programs aimed at obtaining corroborative evidence or counterevidence. In a progressive research program many theories are proposed and tested. Their refutation and replacement is the basis of progress. A theory becomes degenerative when it fails to predict new facts and when the counterevidence becomes sufficient to falsify it.

the white matter. From their anatomical position, and the fact that they appear only in late embryonic and early postnatal stages during the period of myelination, Rio-Hortega (1921b, 1922) conjectured that oligodendroglia are involved in myelination in the central nervous system. Rio-Hortega (1921b) and Penfield (1924) also conjectured that a common precursor migrates into the white matter and then divides to produce astrocytes or oligodendroglia. This conjecture has only recently been possible to corroborate (see Chapter 7).

Before the introduction of good specific stains for glial cells (Ramón y Cajal, 1913; Rio-Hortega, 1919) it was not possible to differentiate, with any degree of certainty, between processes of neurons and those of neuroglia. Neuroglial cell processes were identified by a process of exclusion. Weigert understood that there was an urgent need for a specific neuroglial stain, and he spent 30 years trying to perfect one. He pioneered the use of hematoxylin for staining nervous tissue, developed a very good stain for myelinated nerve fibers, and introduced aniline dye stains (with the help of his cousin Paul Ehrlich). Weigert (1895) was the first to invent a stain (fluorochrome-methylviolet) that was specific for glial cells, but it only stained glial fibers intensely, the remainder of the glial cell weakly, did not stain neurons, and failed to stain glia in embryonic tissue. It showed glial fibers apparently outside as well as inside the glial cells. Those findings misled Weigert to conclude that glial fibers form a connective tissue in the central nervous system analogous to collagen.

Nineteenth century theories of neuroglial functions in the adult nervous system reviewed by Soury (1899) include their nutritional and supportive functions, formation of myelin, formation of a glial barrier between the nervous system and the blood and cerebrospinal fluid, their role in limiting the spread of nervous activity, their proliferation and other changes in response to degeneration of neurons, and their involvement in conscious experience, learning, and memory. These theories of neuroglial functions were sustained more by clever arguments than by the available evidence; indeed, they continued to flourish because the means to test them experimentally were not available until recent times.

Soury (1899, pp. 1615–1639) gives a masterful critique of the theories of glial function that were being debated at the end of the 19th century. The theories were weakest in dealing with the origins and early development of glial cells and with their functions in the embryo. This is not surprising because the specific methods required for identifying embryonic glial cells were not invented until much later (the glia-specific histological stains of Ramón y Cajal, 1913, and Rio-Hortega, 1919; tissue culture of glial cells in the 1920's; identification of glial cells with the electron microscope after the 1950's; glial cell-specific antibodies after the 1970s). The concept that glial cells help to guide migrating neuroblasts and outgrowing axons, first proposed by His (1887, 1889) and Magini (1888), was not possible to test experimentally, for almost a century (Mugnaini and Forströnen, 1967; Rakic, 1971a, b). Their myelinating functions were suggested by a few but rejected by most authorities in the 19th century. It was thought that myelin in the CNS is produced as a secretion of the axon and in the peripheral nerves as a secretion by either the axon ir the sheath of Schwann.

In 1913 Cajal introduced his gold chloride sublimate method for staining neuroglia. It stained astrocytes well but oligodendroglia were incompletely impregnated. Cajal (1920) mistook the latter for a new type of neuroglial cell lacking dendrites, which he called the "third element." He thought that these *celulas adendriticas* (adendroglia, Andrew and Ashworth, 1945) are responsible for myelination of fibers in the CNS, and that they are of mesodermal origin. Rio-Hortega's ammoniacal silver carbonate method (1919), which clearly stains oligodendroglial and microglial cells,, showed that these are the authentic "third element" and that Cajal's conclusions were based on incompletely stained cells. Cajal opposed this explanation, and although he reluctantly, and with reservations, acknowledged the authenticity of microglial cells, he continued to deny the existence of oligodendrocytes long after they were clearly demonstrated and their role in the central myelination had been revealed (Rio-Hortega, 1919, 1924, 1928, 1932; Penfeild, 1924). However, it is only fair to say that the definitive proof that oligodendrocytes are solely responsible for CNS myelination had to await the ultrastructural evidence (Farquhar and Hartman, 1957; R.L Schultz *et al.*, 1957; Mugnaini and Walberg, 1964; R.L. Schultz, 1964).

As Penfield (1924) put it:

Oligodendroglia has received no confirmation as yet though accepted by several writers. This is probably due to two causes; first, the difficulty of staining this element, and second, the fact that Cajal, repeating the work of his disciple, was unable to stain these cells, and, although he confirmed microglia as a group, he cast considerable doubt upon the validity of del Rio-Hortega's description of the remaining portion of the cells previously termed by Cajal "the third element".

Then comes the critical thrust: "As Cajal, the great master of neurohistology, has himself so often pointed out, it is extremely dangerous to assign value to negative results." Wilder Penfield (1977), who worked with Rio-Hortega in Madrid in 1924, describes how the disagreement resulted in an estrangement between the two great Spanish neurocytologists and may have been a factor that precipitated the older man into a state of depression.

Even Ramón y Cajal did not escape being overtaken and corrected by his intellectual progeny. In historical perspective we see that what Cajal is to the neuron, Rio-Hortega is to the neuroglia (Diaz, 1972; Albarracín, 1982). Rio-Hortega was the first to deduce the origin and functions of oligodendrocytes and microglial cells correctly, the first to show their structural transformations in relation to their functions and to emphasize the dynamic state of these cells in normal and pathological conditions. His artistic talents equaled those of his mentor, but while Cajal's drawings have the nervous vitality and intensity of vision of a Velásquez, Rio-Hortega's figures display the deliberately perfected beauty of a Murillo.

Two types of microglial cells in the mammalian CNS were first described by Rio-Hortega (1920, 1932): ameboid and ramified microglia. Ameboid microglia have shorter processes, appear to be motile and phagocytic, appear prenatally, and increase rapidly in the first few days after birth in the dog, cat, and rabbit. He concluded that these are macrophages originating from the blood, as Hatai (1902b) had observed earlier. Marinesco (1909) showed that brain macrophages ingest India ink and thus behave like macrophages elsewhere. Rio-Hortega and Asua (1921) showed that microglial cells are morphologically very similar to macrophages in other parts of the body. They conjectured that microglia and macrophages both originate from the reticuloendothelial system, which at that time was being vigorously discussed (Aschoff, 1924). The ameboid microglia appeared to Rio-Hortega (1921a) to originate in what he called "fountains" of ameboid cells at places where the pia matter contacts the white matter: beneath the pia of the cerebral penducles, from the tela choroidea of the third ventricle, and from the dorsal and ventral sulci of the spinal cord. He identified another type, ramified microglia, with long processes, apparently sedentary and nonproliferative. These appear postnatally and persist in the adult. In his 1932 paper Rio-Hortega shows a series of transitional forms between ameboid and ramified microglia and concludes that these represent normal transformations between the two types of microglial cells, thus anticipating recent findings (Perry and Gordon, 1988). In the same paper Rio-Hortega shows that microglia migrate to sites of brain injury, where they proliferate and engulf cellular debris. These are the macrophages of the nervous system, whose roles in defense against infection and injury he was the first to recognize. Confirmation of most of Rio-Hortega's conclusions had to wait until the modern epoch, when the tools were forged that have made it possible to reveal the origins and functions of neuroglial cells.

The debate, started by Rio-Hortega (1932), about whether brain macrophages are derived from the blood or from the brain has continued for 50 years (reviewed by Boya *et al.*, 1979, 1986; Adrian and Schelper, 1981; Schelper and Adrian, 1986). The currently available evidence shows that in adults both microglia and blood monocytes can contribute to brain macrophages, depending on whether the blood–brain barrier is intact or not. Present evidence shows that in the embryo the microglia originate from monocytes that enter the brain before development of the blood–brain barrier.

In modern times those who have concluded that brain macrophages are entirely hematogenous in origin include Konigsmark and Sidman (1963), S. Fugita and Kitamura (1975), Ling (1978; 1981), and Del Cerro and Mojun (1979). Those who have concluded that macrophages are derived from microglial cells include Maxwell and Kruger (1965), Mori and Leblond (1969), Vaugh and Pease (1970), Torvik and Skjörten(1971), Torvik (1975), and Boya (1976). The ultimate fate of the brain macrophages after repair of an injury is also controversial: they have been reported to degenerate (Fujita and Kitamura, 1975), return to the blood (Kreutzberg, 1966; McKeever and Balentine, 1978) or transform into microglial cells (Mori, 1972; Blakemore, 1975; Imamoto and Leblond, 1977; Ling, 1981; Kaur *et al.*, 1987), but the latter possibility is denied by Schelper and Adrian (1986). The same techniques, in the hands of skillful workers, have led to diametrically opposite conclusions. The brain macrophages may indeed originate from more than one source and have multiple fates, but the neurocytologist tends to select

his facts according to prevailing prejudices—in this he is no different from other scientists and nonscientists. The main difference between them is that the scientist, more often than the nonscientist, submits his prejudices for refutation.

THE NEURAL CREST AND ITS DERIVATIVES

> … a physiological system … is not a sum of elements to be distinguished from each other and analyzed discretely, but a pattern, that is to say a form, a structure; the element's existence does not precede the existence of the whole, it comes neither before nor after it, for the parts do not determine the pattern, but the pattern determines the parts: knowledge of the pattern and of its laws, of the set and its structure, could not possibly be derived from discrete knowledge of the elements that compose it.
>
> Georges Perec (1936–1982),
> *"La vie, mode d'emloi"* (1978)

Historical Perspective

> The principle, whereby the germinal discs or organ rudiments are represented in a planar pattern, and conversely, every single point of the germinal disc reappears in an organ, I name the principle of organ-forming germinal.
>
> Wilhelm His (1831–1904),
> *Unsere Körperform*, p. 19, 1874

In 1868 Wilhelm His discovered the neural crest and he traced the origin of spinal and cranial ganglia from the neural crest. These discoveries raised several theoretical problems—the origin and boundaries of the neural crest, the modes of cell migration, and the fates of cells of neural problems—origin, migration, and fate—have been the main conceptual platforms from which research programs on neural crest development have been launched.

The first of those research programs was aimed at finding the origins and morphological boundaries of the neural crest in the brain as well as the spinal cord. Wilhelm His thought that the neural crest originates from the *Zwischenstrang* (meaning "intermediate cord") situated between the cutaneous ectoderm and the neural plate, and he regarded it as distinct from the neural tube and the lateral ectoderm, both anatomically and with respect to the structures to which it gives rise. A priority dispute arose when Balfour (1876, 1878) announced the "discovery" of the neural ridge, part of the neural tube, as the origin of the spinal ganglia in elasmobranch fish. Wilhelm His (1879) then asserted his claim to priority of discovery of the origin of the cranial and spinal ganglia from a distinct organ-forming zone, the neural crest. Wilhelm His (1874) identified the neural crest as one of the organ-forming germinal zones as defined in the epigraph to this section, and he conjectured that it contained subsets of cells with restricted fates. During the following century that conjecture was subjected to numerous weak refutations and partial corroborations, and it has recently received further corroboration (Maxwell *et al.*, 1988; Fontaine-Perus *et al.*, 1988; Baroffio *et al.*, 1988; Smith-Thomas and Fawcett, 1989).

The problem of migration of neural crest cells was conceived by His as part of the general problem of cell migration and cell assembly during morphogenesis, in purely mechanistic terms of mechanical guidance either of coherent masses of cells or of individual cells. This mechanistic theory was greatly influenced by Carl Ludwig, professor of physiology at Leipzig, to whom His dedicated *Unsere Körperform* (1874), as a kind of mechanistic embryological manifesto. The aim of his manifesto was to refute theories which held that nonmaterial vital forces were at work during development, for example, Ernst Haeckel's "morphogenetic forces" ("*form-bildende Kräfte*") and Justus Liebig's "life forces" ("*lebens Kräfte*"). Mechanistic materialism of the kind upheld by His, Moleschott, Ludwig, and others was unable to refute vitalistic theories because there always remained some phenomena which could not be given a completely mechanistic explanation. Vitalists were unwilling or unable to disclose the conditions for refutation of their theories, and by placing them virtually beyond refutation they also place them outside the empirical sciences.

The history of research on the neural crest can be divided into three overlapping research programs. The first program started with the discovery of the neural crest by Wilhelm His in 1868 and lasted until about 1900. Its primary objectives were to describe development of the neural crest in terms of cellular activities and to fit the observations into comparative morphological and evolutionary theories (summarized in Neumayer, 1906). To achieve those objectives it was necessary to improve microscopic and histological techniques. His made important contributions to those technical advances, such as the invention in 1866 of the first microtome to have a micrometer advance (His, 1870).

This epoch may also be dated from the publication of the first significant treatise on comparative embryology (Balfour, 1881). Twenty-six years later the achievements of this research program were collected in the magisterial compendium edited by O. Hertwig (1906, over 5,000 pages dealing with all the classes of vertebrates, with chapters by such luminaries as O. and R. Hertwig, W. Waldeyer, F. Keibel, K. von Kupffer, H. Braus, E. Gaupp, W. Flemming, and F. Hochstetter, to note only the most famous). There has been no other period of 26 years in the history of embryology in which so much has been done by so few.

Wilhelm His made the distinction between two periods of development in general, and nerve cells in particular: the period of cell proliferation or "numerical growth," as he called it, and the period of "trophic growth," which is characterized by nerve cell differentiation and by outgrowth of nerve fibers (His, 1868, p. 187). He was also the first to give convincing histological evidence that the nerve fibers are outgrowths from the nerve cells, although the concept that nerve fibers are protoplasmic outgrowths of the nerve cell had been proposed earlier by Bidder and Kupffer, Remak, and Koelliker. The observation, first made by His, that nerve fibers grow out only after neural crest cells migrate to the sites at which ganglia are formed, was critical

evidence that rendered dubious the theory, then considered to be certain, that all neurons are connected by delicate fibers from the start of the migration.

One of the most significant advances during this epoch was the report by Julia Platt (1893) that she could follow "mesectoderm" from the neural crest to cartilages of the head in *Necturus*. This was the first experimental evidence challenging the dogma of origin of cartilage from mesoderm, and thus a refutation of the germ layer theory. Her findings were confirmed by von Kupffer in cyclostomes (1895), Dohrn (1902) in selachians, and Brauer (1904) in the limbless amphibian *Gymnophiona*. These findings raised fundamental questions about the origins of mesodermal segmentation of the head. The segmental pattern of the head and the relationships of the cranial nerves had been worked out by methods of comparative anatomy (Van Wijhe, 1882; O. Strong, 1895; J.B. Johnston, 1902; C.J. Herrick, 1899). The question arose about how the primary mesodermal segmentation of the head could be related to origin of cartilage cells from the apparently unsegmented cranial neural crest. The answer emerged as a result of the transplantation experiments that were the vogue during the next epoch. It was that the segmental pattern must develop primarily in the mesenchyme of the head that is not of neural crest origin; later the neural crest cells, which are not themselves segmentally determined, migrate into the predetermined metameric pattern of the mesenchyme (Landacre, 1910, 1914, 1921; Stone, 1922, 1929; Celestino da Costa, 1931; Starck, 1937; Ortmann, 1943).

The first research program terminated gradually after about 1900 with the decline of interest in comparative anatomy and with the reciprocal rise of the research program of experimental embryology. These two research programs, which overlapped in time, correspond with what Merz (1904) calls the morphological and genetic concepts of nature. The first tries to explain development in terms of rules that underlie invariant or universal patterns of morphological organization such as segmentation, polarity, and symmetry, or more generally in terms of body plans. It attempts to define homologous forms that show how morphological patterns have been conserved during development and evolution. The second tries to explain the genetic rules by which the patterns are expressed and change during development and evolution. These views are not mutually exclusive, of course, and both views were held by many workers. However, after 1900 the genetic view gained ascendancy.

The second research program can be called the Program of Experimental Embryology. It may be dated from 1888 with publication of a report by Wilhelm Roux showing that destruction of one of the first two blastomeres of the frog embryo results in development of only one-half of the embryo from which partial regeneration of the other half appears to occur. Roux's experimental method of analysis set an example for the entire research program. However, his interpretation that the egg is a self-differentiating mosaic, and even the reproducibility of his results, were later challenged by Oscar Hertwig (1894, Vol. 2 is a critique of Roux). The crucial counterexample to Roux's result was the discovery by Hans Driesch that separation of the first two blastomeres of the sea urchin embryo results in development of two complete larvae. It has been said that while Roux pointed the way to the methodology of experimental embryology, the fundamental problem of development, namely, embryonic regulation, was raised by Driesch, and that the heir to both traditions was Spemann (Horder and Weindling, 1985). Certainly, with regard to the history of the neural crest, the relative role of self-differentiation and of regulation of crest cells has been an important issue.

This program was defined by Wilhelm Roux in 1894 with his manifesto, the celebrated *"Einleitung,"* published in the first issue of his journal *Archival fur Entwickelungsmechanik.* The opening sentence states that "Developmental mechanics or casual morphology of organisms, which this Archiv is dedicated to serve, is the Doctrine of the causes of organic form, consequently the cotrine of the causes of the origin, maintenance, and involution of these forms." The experiment is the method of casual analysis, Roux declared. The comparative embryologists and anatomists have arrived where the new science of *Entwickelungsmechanik* is starting, Roux states, and he refers in a single sentence to the work of Wilhelm His. It is also strange that the name of His does not appear in the long list of those who gave their support to the new *Archiv.*

Manifestos proclaiming new theses always exaggerate their importance and originality, and Roux's was no exception. The three distinctive features of Roux's contribution were all derivative. His performationism was derived from a long tradition and most directly from Weismann and His, albeit with modification. Roux's mechanistic materialism also had a venerable tradition, but for the present purposes it is only necessary to say that Roux's outlook was less rigorously and uncompromisingly materialistic than that of His. Roux's claim that he had introduced a totally new causal-analytic experimental method was true only in that the experimental method became increasingly used in embryology after Roux. Experimental perturbation of development had been used to analyze development long before—for example, Johann Friedrich Blumenbach (1752–1840) made animal chimeras of differently colored hydra to show that the whole animal can be reconstituted by regulation of the different parts and not by regeneration of each part separately (Blumenbach, 1782). Experimental analysis of artificial fertilization of frog eggs, performed by Lazzaro Spallanzani (1784), is another early example of the causal analysis of development prior to the era of *Entwickelungsmechanik.*

Two important experimental techniques were invented at the start of this research program, namely, tissue culture and analysis of chimeras. Tissue culture was invented by Wilhelm Roux and Gustav Born (1895) but was first used for studying neurite outgrowth by Ross Harrison (1910, 1912). Surprisingly fast, a spate of publications appeared in which nervous tissue was studied *in vitro* (Lewis and Lewis, 1911, 1912a, b; Marinesco and Minea, 1912) and the technique has remained indispensable to the present time. The second important technical advance made at that time was the analysis of amphibian chimeras by Gustav Born (1876) and Ross Harrison (1903). The technique of inter-specific and intergeneric grafting was very useful in the early studies of neural crest migration in urodeles (Lehman, 1927; Raven, 1931, 1936, 1937; Detwiler, 1937). In frogs, chimeras

made up of cells from two species with different cellular markers have also been used quite often since the early experiments of Born and Harrison (G. Wagner, 1948; Thiébaud, 1983; Sadaghiani and Thiébaud, 1987; Krostoski *et al.*, 1988). Analysis of neural crest development has been greatly advanced by using the quail/chick chimera introduced by Le Douarin in 1969 (Le Douarin, 1973; Le Douarin and Teillet, 1973).

Another advance made during this period was clarification of the contribution of ectodermal placodes to cranial ganglia. This was first done by Landacre (1910, 1912, 1916) in the fishes *Ameiurus* and *Lepidosteus* and the dogfish *Squalus.* Later he was able to trace the fates of neural crest cells in the head because of their characteristic pigmentation and yolk granules in the urodele *Plethdon* (Landacre, 1921). The placodal contributions to cranial ganglia were further defined by Knouff (1927, 1935) in the frog. The vital staining method of Vogt (1925, 1929) was used to trace the fates of cranial neural crest in urodeles by Detwiler (1937a), Yntema (1943), and Hörstadius and Sellman (1946).

Studies on mammalian neural crest development were limited to descriptive observations, and the advances that were made by the methods of experimental embryology, using amphibian embryos, could not be applied at that time to mammalian embryos. The first description of neural crest in a mammal was given by Chiarugi (1894) in the guinea pig. This was followed by the extensive series of guinea pig embryos studied during the 1920s by Celestino da Costa (1931) and Adelmann's (1925) classical study of cranial neural crest development in more than 200 rat embryos. Those descriptive studies are unlikely to be repeated and are still valuable sources of information.

Perhaps the most significant achievement of the second research program on neural crest development was the demonstration that neural crest cells are pluripotent and that they can give rise to a wide range of cell types in all three classical germ layers (reviewed by Raven, 1931–1933, 1936; Starck, 1937; Hörstadius, 1950). The ablation and grafting experiments of the earlier part of this epoch have been repeated more recently using better techniques and have been extended to include avian and mammalian embryos. That program also remained incomplete, so that the mechanisms of cell migration, inductive cellular interactions, and cell determination remained to be elucidated by molecular biological methods.

In the Developmental Neurobiology Research Program, from the 1960s, the search for molecular mechanisms has dominated the field. The thrust of the new research program was articulated by Joshua Lederberg in the introduction to the first volume of *Current Topics in Developmental Biology* (1966): "The field has had enough fancy; more recently its methodology has been under enormous pressure to accommodate the inspirations of molecular biology and the models of development that can be read into microbial genetic systems. But now, as this volume amply shows, it is responding."

AXONAL DEVELOPMENT

If there exists any surface or separation at the nexus between neurone and neurone, much of what is characteristic of the conduction exhibited by the reflex-arc might be more easily explainable. ... The characters distinguishing reflex-arc conduction from nerve-trunk conduction may therefore be largely due to intercellular barriers, delicate transverse membranes, in the former.

In view, therefore, of the probable importance physiologically of this mode of nexus between neurone and neurone, it is convenient to have a term for it. The term introduced has been synapse.

Charles Scott Sherrington (1857–1952), *The Integrative Action of the Nervous System*, 1906

Historical Perspective

When you are criticizing the philosophy of an epoch, do not chiefly direct your attention to those intellectual positions which its exponents feel it necessary explicitly to defend. There will be some fundamental assumptions which adherents of all the variant systems within the epoch unconsciously presuppose. Such assumptions appear so obvious that people do not know what they are assuming because no other way of putting things has ever occurred to them.

Alfred North Whitehead (1861–1947), *Science and the Modern World*, p. 71, 1925

The history of ideas about the forms and functions of neurons shows that the conditions which permit different scientists to uphold totally opposed hypotheses are, firstly, that the evidence is contradictory and inconclusive and, secondly, that one or both hypotheses are based on erroneous assumptions. One of the major assumptions of the late 19th century was that animal cells lack a cell membrane. The cell surface was believed to be a transition between two phases, without special structure. Therefore, it was assumed that protoplasmic bridges between cells could freely appear and disappear. To recognize the significance of this fundamental assumption is to gain an entirely fresh view of the history of rival theories of formation of nerve connections. Proponents of the *neuron theory* believed that nerve cells only come into close contact and are never in direct protoplasmic continuity, whereas proponents of the *reticular theory* believed that nerve cells are directly connected by protoplasmic bridges or networks. The reticular theory is consistent with the fundamental assumption that cells lack membranes; the neuron theory is in conflict with that assumption.

The concept of the cell without a surface membrane is as old as the cell theory itself. Schwann (1839, p. 177) wrote: "Many cells do not seem to exhibit any appearance of the formation of a cell membrane, but seem to be solid, and all that can be remarked is that the external portion of the layer is somewhat more compact." No special structures or functions were attributed to the cell surface, but the "physical basis of life" that was so much discussed by 19th-century biologists was assumed to reside in minute particles and fibers in "the protoplasm" (reviewed by Hall, 1969). For example, Max Schultze (1861) defined cells as "membraneless little lumps of protoplasm with a nucleus." The concept that the cell lacks a membrane was supported by Carl Gegenbauer in his influential essay on the evolution of the egg, published in 1861. Sedgwick (1895)

regarded the embryo as a giant protoplasmic mass in which numerous cell nuclei are embedded. In discussing the structure of nerve cells, Koelliker (1896, Vol. 2, p. 45) believed that the "central cells lack a definite membrane and possess as boundaries only the tissues of the grey substance, which consists in varied proportions of nerve fibers, glial cells and blood vessels." In the 1st edition of E.B. Wilson's very influential book *The Cell in Development and Inheritance* (1896, p. 38), I came across the statement that "the cell-membrane of intercellular substance is of relatively minor importance, since it is not of constant occurrence, belongs to the lifeless products of the cell, and hence plays no direct part in the active cell-life." Wilson maintains the same opinion in the second edition (1902, p. 53), but in the third edition (1925, p. 54) he provides some evidence for the existence of a plasma membrane.

The intellectual climate before about 1920 nurtured the concept of protoplasmic connections between neurons. When the inadequate histological methods of those times failed to resolve membranes between cells, it was quite reasonable to assume that the cells are connected to form a syncytium. This flawed assumption, as much as the histological artifacts, formed the basis for reticular theories of connections between neurons.

All the reticular theories of neuronal connectivity claimed that neurons are in direct protoplasmic continuity, and form various types of networks (Gerlach, 1858, 1872; Golgi, 1882–1883, 1891; Apáthy, 1897). The outlines of these theories have so often been reviewed that they do not require repetition insofar as they narrate the sequences of events and their main ideas (Stieda, 1899; Soury, 1899; Barker, 1901; Ramón y Cajal, 1933; Van der Loos, 1967; Clarke and O'Malley, 1968). However, none of those authors seems to have recognized that the unstated assumption beneath all the variant theories was that cells normally lack a cell membrane. We can now understand more adequately how the minds of the proponents of different theories of neuronal organization were conditioned by the prevalent assumptions about cellular organization, and how far their theoretical speculations exceeded what they could have seen in their histological preparations.

The idea that nerve cells are directly interconnected through a network of fine fibers is usually attributed to Camillo Golgi (1843–1926). However, that notion was originally the brainchild of Joseph von Gerlach (1820–1896). In sections of the spinal cord stained with carmine or gold chloride, Gerlach (1872) saw a fine feltwork of fibers in the gray matter. He interpreted this to be a genuine network formed by anastomosis between branches of the dendrites, which were at that time called protoplasmic processes. Remak (1854) and Deiters (1865) have shown that the branched protoplasmic processes are different from the single, unbranched axis cylinder, but they had not been able to show how they end or form connections in either the central or peripheral nervous systems. Gerlach (1872) depicted the sensory fibers of the dorsal roots originating indirectly by branching from a diffuse nerve network formed by interconnected branches of the protoplasmic processes. Gerlach's concept was accepted by all neuroanatomists at that time because it was consistent with the prevailing belief in protoplasmic bridges connecting cells in general.

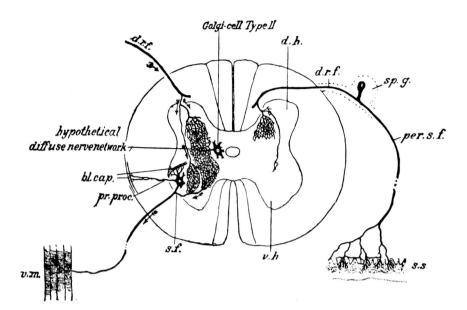

FIGURE 2. Representation of the Golgi's hypothetical diffuse nerve network in the mammalian spinal cord. The netwrok was supposed to be formed by anastomosis between collaterals of incoming dorsal root fibers (d.r.f), the much branched axons of Golgi type II cells, and collaterals (s.f.) of Golgi Type I cells (the motor neuron.) The direction of impulse traffic is shown by the arrows. The cell body and dendrites were supposed to be excluded from the conduction pathways, and the dendrites were supposed to connect with blood capillaries (bl. cap.) and to be purely mitritive in function. From L. Barker, *The Nervous System and its Constituent Neurones* (1898).

Gerlach's construct was eventually demolished by Golgi (1882–1883), using his method of impregnation with potassium bichromate and silver nitrate, which revealed the entire neuron for the first time. Golgi (1882–1883) showed that, contrary to Gerlach's construct, the protoplasmic processes end freely, without any interconnecting network, and appear to make contact only with blood vessels and glial cells (Fig. 2). Therefore, he thought that they must have nutritive functions, and he looked elsewhere for the conducting pathways between nerve cells. He believed that he had found them in the axon collaterals, which he discovered. Golgi then made the distinction between two main types of neurons: type I with a long axon and type II with a short axon. He suggested that the axon collateral of type I neurons are connected directly by a diffuse nerve network *reticola nervosa diffusa* to branches of the axons of type II neurons, thus excluding the dendrites from the conducting pathways. Golgi at first treated this as a speculative hypothesis, and the diffuse nerve network is merely discussed but never shown in any of the numerous figures in his 1882–1883 papers or in his long polemical paper of 1891 which deals specifically with the question of the functional significance of the nerve net. Pictures of the *reticola nervosa diffusa* are shown for the first time in two figures in a communication from Golgi which appears in the *Trattato di disiologia* of Luigi Luciani (1901), and the same figures are reproduced in Golgi's *Opera Omnia* (1903, Figs. 41 and 42) and his Nobel lecture (1907). One figure shows a network formed by axon collaterals of granule cells of the hippocampal dentate gyrus, and the other depicts a network in the granular layer of the cerebellar cortex formed by axonal branches of the basket cells. Neither of these curious illustrations shows the second type of cell participating in the network, as the theory requires. On this

point the text also reveals Golgi's reluctance to commit himself to specific details, saying that these figures only *give an idea of the network* (Golgi, 1907, p. 15).

The reticular theory, increasingly destitute of empirical and intellectual support, found its final refuge in structures of incredible subtlety, at the limits of resolution of light microscopy. These are aptly described by Golgi (1901) as "*di organizza-zione di meravigliosa finezza.*" These structures took the form of tenuous fibrillar networks that were depicted as direct extensions of intracellular neurofibrils (Apáthy, 1897; Bethe, 1900, 1903, 1904; Held, 1905). Such fancies had to be abandoned when it was shown that neurons are separated at some large synapses by a membrane which is not crossed by neurofibrils and that the neurofibrils do not enter all synaptic terminals (Bartelmez and Hoerr, 1933; Hoerr, 1936; Bodian, 1937, 1947). The neurofibrils that are stained with silver in light microscopic preparations of mammalian central nervous tissue were eventually shown to be clumps of neurofilaments when examined with the electron microscope (Peters, 1959; Gray and Guillery, 1961).

Evidence showing a discontinuity from neuron to neuron came from four quarters: embryology, histology, physiology, and pathological anatomy. The first embryological evidence of the individuality of neurons was reported by Wilhelm His (1886, 1887, 1889). He demonstrated that neuroblasts originate and migrate as separate cells and that nerve fibers grow out of individual neuroblasts and have free endings before they form connections. In 1887 His described with remarkable accuracy the outgrowth of the nerve fiber: "The fibers which grow out from the nerve cells advance by growing into existing interstitial spaces between other tissue elements. *In the spinal* cord and in the brain, the medullary stroma already formed, provides pathways for expansion and its structure undoubtedly determines the course of the process of extension. ... " These observations were later confirmed by Ramón y Cajal (1888, 1890a, b, c, 1907, 1908). This evidence showing free outgrowth of axons did not shake the faith of the reticularists because they could point out that it did not exclude the later development of protoplasmic continuity between neurons after they had made contact (Nissl, 1903; Bethe, 1904; Held, 1905). Of course, the fact that axons and dendrites develop as independent extensions of the neuron did not prove that the entire neuron remains an independent cell, but evidence for that gradually accumulated. Ramón y Cajal (1888, 1890a, b, c, 1907) showed that neurons stained by means of the Golgi technique were always completely isolated from others, and he followed Golgi in assuming that they were revealed in their entirety. This assumption could not be tested before the advent of the electron microscope.

During the 19th century there were three main theories of development of the axon: the cell-chain theory, the plasmodesm theory, and the outgrowth theory. According to the cell-chain theory, originated by Schwann and supported by F.M. Balfour among other excellent embryologists, the axon is formed by fusion of the cells that form the neurilemmal sheath. This theory of development was an extrapolation from interpretations of regeneration of peripheral nerves, in which Schwann cells and fibroblasts were mistaken for the precursors of the regenerating axons. Ramón y Cajal (1928, pp. 7–16) heaps scorn and derision on this theory and on its proponents who are "little acquainted with the severity and rigour of micrographic observation, and with the secrets of histological interpretation," but his rhetoric cannot conceal the fact that histological observations alone, without experimental evidence, were insufficient to disprove the cell-chain theory. It is fair to say that Cajal tended to use such rhetorical flourishes to conceal weaknesses in his arguments. The cell-chain theory was finally refuted by the elegant experiments of Harrison (1904, 1906, 1924a), when he showed that removal of the neural crest, from which the Schwann cells originate, results in the development of normal nerve fibers in the absence of Schwann cells. He also demonstrated that removal of the neural tube, which contains the developing neurons, prevents the formation of nerves, although the Schwann cells are left intact.

The plasmodesm or syncytial theory originated with Viktor Hensen (1835–1924) and was supported by Hans Held (1866–1942). According to this theory, the nerve fiber differentiates from preestablished filaments that connect all the cells of the nervous system. This theory was founded on the fundamental assumption, which has been discussed above, that cells originate as a syncytium in the embryo and retain protoplasmic bridges throughout life. This theory was consistent with much of the evidence available at that time and was widely supported. For example, as late as 1925 in the third edition of *The Cell in Development and Inheritance*, E.B. Wilson was still defining plasmodesms as "the cytoplasmic filaments or bridges by which in many tissues adjoining cells are connected." The plasmodesm theory provided explanation for several observation that were supposed to support the contact theory of the synapse; for example,

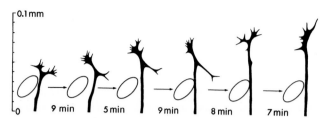

FIGURE 3. Six successive views of the end of a growing nerve fiber, showing its change of shape and rate of growth. The sketches were made with the aid of a camera lucida at the time intervals indicated. The red blood corpuscle, shown in outline marks a fixed point. The average rate of elongation of the nerve was about I μm/minute. The total length of the nerve fiber was 800 μm. The observations were made on a preparation of from embryo ectoderm, isolated in lymph, 4 days after isolation Adapted from R. G. Harrison, *J. Exp. Zool.* 9:787–846 (1910).

reflex delay was ascribed to slowing of conduction during passage of the nerve impulse across extremely thin plasmodesms.

The outgrowth of nerve fibers was eventually demonstrated in tissue culture by Ross Harrison (1907b, 1910). Harrison excised pieces of neural tube from early tailbud frog embryos, at a stage before any nerve fibers are present, and explanted the tissue into a drop of frog lymph suspended from a coverslip. Nerve fibers grew out of the explant, in some cases from single isolated cells, for distances up to 1.15 mm at rates ranging from 15.6 to 56 &m/hour (Fig. 3). Harrison's observations were very rapidly confirmed by other reports of outgrowth or regeneration of axons in tissue culture of nervous tissue of amphibians (Hertwig, 1911–1912; Legendre, 1912; Oppel, 1913), chick (Burrows, 1911; Lewis and Lewis, 1911, 1912; Ingebrigsten, 1913a,b) and mammals (Marinesco and Minea, 1912 a–d).

The locomotion and growth of epithelial cells, young neurons, and nerve fibers in tissue culture were shown to occur only when they are in contact with a surface such as fibrin fibers in a fluid medium, or are at the interface between the solid substratum and liquid medium, or are at the liquid-air interface (Loeb, 1902; Harrison, 1910, 1912; W.H. Lewis and Lewis, 1912). This phenomenon was called *stereotropism* by Loeb (1902) and Harrison (1911, 1912), *contact sensibility* by Dustin (1910), and *tactile adhesion* by Ramón y Cajal (1910, 1928). Wilhelm His was the first to recognize the importance of mechanical factors in embryonic development. This and many other important contributions of Wilhelm His to developmental neurobiology are reviewed by Picken (1956). His clearly understood and described cases of axonal guidance by the tissue substratum and his 1894 review of the mechanical basis of animal morphogenesis contain numerous aperçus of the concepts and mechanism of nerve growth later promoted by Ross Harrison and by Paul Weiss.

These results were the final confirmation of the theory of the outgrowth of nerve fibers from young neurons first stated tentatively by Koelliker in 1844: "The fine fibers arise in the ganglia ... as simple continuations of the processes of the ganglion-globules. In other words, the processes of the ganglion-globules are the beginnings of these fibers." Only 13 years later Bidder and Kupffer could assert "with the greatest degree of certainty, that ... every fiber must ... be conceived merely as a colossal 'outgrowth' of the nerve cell". In 1886 Koelliker could state quite definitely that "*I consider the primitive nerve fibers to be protoplasmic outgrowths of the central nerve cells*" (Koelliker, 1886). This was only one of five theories ultimately united around the end of the 19th century to construct a theory of organization of the nervous system. The others were that the nerve cell and fibers are parts of the same unit; that dendrites are fundamentally different from axons; that nerve cells connect by surface contact and not by cytoplasmic continuity; that the contact regions are the principal sits of functional integration and modifiability; and the nerve cells and their connection are initially formed in excess, and the redundancy is eliminated during later development. Construction of the general theory ("the neuron theory") from these components is considered in detail by Jacobson (1993).

Harrison (1910) observed the outgrowth of the axons from the Rohon–Beard cells, which are the primary sensory cells and which can be seen in the dorsal part of the neural tube just beneath the epidermis in living frog embryos. Beard (1896), in his original description of the development of Rohon–Beard neurons, had accurately depicted the outgrowth of the axon from the cell body, but he failed to draw the general conclusion. Harrison observed that as the axon grows out of the Rohon–Beard cells into the subepidermal tissue, it slowly increases in length and gives rise to many branches. The growth of the axons of Rohon–Beard cells occurs in the same way as the growth of axons in tissue culture. The tip of the initial outgrowth, as well as the end of each branch, consists of an enlargement from which ameboid terminal filaments are constantly emitted and retracted. These were the first observations of the activities of the growth cone during normal development in a living animal. They fully confirmed Cajal's descriptions of growth cones in fixed specimens, and they provided a standard by which to assess whether growth cones in histological preparations are normal or artifactual. The fact that Harrison's observations on the growth of living nerve fibers agreed with descriptions of the growth of axons in histological preparations of the developing nervous system at one stroke established the validity of histological observations of Koelliker (1886), His (1886, 1887, 1889), and Cajal (1888, 1890a–c), which were far more detailed and diverse that any that could be obtained *in vitro* at that time. Harrison's experiments had, in his own estimation, taken the mode of formation of the axon "out of the realm of inference and placed it upon the secure foundation of direct observation."

Harrison coined the term "exploratory fibers" for the nerve fiber that precedes the rest of the development of a fiber pathway. Cajal gives many vivid descriptions of these pathfinders. For instance, during the outgrowth of the dorsal spinal root from the spinal sensory ganglia, "a bundle of precocious bipolar cells strikes with its cones, like battering-rams, on the posterior basal membrane and opens a narrow breach in it. Other sensory fibers, differentiating later, make use of this opening, and assault the interior of the spinal cord along its dorsal portion."

Cajal was the most vigorous advocate of the neuron theory, and it is significant that he never doubted the objective reality of the cell membrane. The sources of his conviction are difficult to trace because he does not discuss the evidence or defend his belief

in the existence of the cell membrane. Already in the first edition of *Elementos de Histologia Normal* (1895, p. 303) he defines a "fundamental membrane" which is "a living organ of the cell which is a continuation of the protoplasm." In the fourth edition of his *Manual de Histoligical Normal* (1909, pp. 150, 154) he says: "All the cells of the central nervous system and sensory organs as well as the sympathetic possess a membrane of extreme thinness, a fundamental membrane. ... This membrane is not a peculiarity of certain neurons, it is a general property without exceptions." This statement is made *ex cathedra*, without supporting evidence, as if it were a self-evident truth, at a time when most authorities, even some who supported the neuron doctrine, held the opposite opinion, namely, that the cell membrane is either a histological artifact or a lifeless structure without significant function.

I have been unable to determine whether Cajal's belief in the existence of a cell membrane preceded or followed his adoption of the neuron theory. The two beliefs are now seen to be so obviously interdependent that it is not easy to understand how, at that time, it was possible to affirm the one while denying the other. Yet none of the other supporters of the neuron theory shared Cajal's deep conviction, not Koelliker, not Lenhossék, not van Gehuchten. Nor can one find any discussion of the authenticity of the nerve cell membrane the 19 massive volumes of *Biologische Untersuchungen* (1881–1921) of Gustav Retzius, a consistent proponent of the neuron theory. The matter was ignored, either because they did not understand the importance of the cell membrane or because they denied its very existence. For example, Koelliker (1896, Vol. 2, p. 48) states that "with reference to the envelope of the nerve cell, it can be shown with certainty that the latter apparently at all times lacks a cell membrane." Another defender of the neuron theory, Mathias Duval in his *Précis d'Histologie* (1897, p. 774) says of the nerve cell that "formerly one described it as having an envelope, by reason of artifacts produced by coagulating reagents; nowadays it is recognized that it is a naked protoplasmic body." All those who opposed the neuron doctrine were at least logically consistent in also denying the existence of a nerve cell membrane. Thus, after reviewing the evidence, Sterzi (1914, p. 19) concludes that "a cellular membrane does not exist. ... The nervous cytoplasm is in direct relationship, through fine reticular fibrils that constitute the interstitial part of the nervous tissue." Cajal's convictions were not dogmatic—he was too shrewd not to be willing to acknowledge that exceptions to the neuron theory may exist. He admits that perineural connective tissue cells but not the neurons are sometimes connected by protoplasmic bridges. He says that "these mesodermic cells form a net with meshes of variable size," and he shows axons "growing through the plasmatic interstices [of the] anastomozed fibroblasts" (1928, p. 183 and Fig. 101D). Cajal treats reports of protoplasmic connections between Schwann cells with skepticism (1928, p. 83) but finally agrees that in degenerating nerves the Schwann cells form a syncytium (1928, p. 130 and Fig. 23).

In his final statement on the evidence for the neuron theory, Cajal (1933) still find it necessary to ask: "Do the terminal nerve arborizations actually touch the nude protoplasm of the cell or do limiting membranes exist between the two synaptic factors?" Then comes the prescient conclusion: "I definitely favor this latter opinion, although with the reservation that the limiting films are occasionally so extremely thin that their thickness escapes the resolution power of the strongest apochromatic objectives." He then admits that "neuronal discontinuity, extremely evident in innumerable examples, could sustain exception … for example those existing in the glands, vessels and intestines."

The achievements of Cajal may be seen as signals rising far above the intellectual noise. But it is not always possible to see how they were generated in Cajal's mind, or even how he found the empirical stimuli for his creativity. His autobiographical account of his creative processes, *Recuerdos de mi vida* (first ed. 1917, third ed. 1923), deserves to be treated with as much skepticism as respect. His *Reglas y consejos para investigación cientifica* (1923) shows how difficult it must have been for him to discipline his unrepented romanticism (see Jacobson, 1993). In Cajal the genius of the artist and scientist were combined to a unique degree. He had the gift, usually granted only to the artist, of incorporating vague and chaotic elements of experiences into an orderly synthesis. The artist is more or less free to adopt, modify, or invent a language to represent and express his experience. The scientist is not so free and usually lacks the originality and courage that are necessary to liberate himself from the assumptions of his times.

Cajal's methods of drawing from the microscope can be inferred from his own testimony and other evidence. There is a solitary reference to his use of a camera lucida (Ramón y Cajal, 1891, legend to Fig 1), but to suggest that he used such a drawing aid habitually (De Felipe and Jones, 1988) is like saying that a life preserver is needed by a powerful swimmer within reach of the shore (see further discussion in Jacobson, 1993). Cajal describes different models of camera lucida in his *Manual de Histologia Normal*, but he also describes other instruments such as the microspectroscope and the polarizing microscope, which he probably never used. Further evidence against his habitual use of the camera lucida is that the latter is neither mentioned in Cajal's autobiography nor visible in the photographs showing him at his worktable. Cajal's line drawings of Golgi or silver preparations were evidently made with a metal pen or a goose quill, with which the width of the line can be delicately shaped by varying pressure on the point. Penfield (1954) saw Cajal writing with a good quill (but not drawing with one or wiping it on his bed sheets, as stated by De Felipe and Jones, 1988). For halftone figures, he used pencils, crayon, and fine paintbrushes (Penfield, 1977, p. 104). Cajal was well aware also of the special artistic effects obtainable with paper of different grades and textures (cf. Ramón y Cajal, 1905, p. 36). I believe that he could have used the camera lucida for laying out the picture at low magnification, but that the details were drawn freehand, keeping one eye and hand on the microscope while using the other hand and eye for drawing. This was the way in which students were trained to use the monocular microscope for making histological drawings, and it was also the principal method recommended by Cajal. He notes in his *Manual de Histologia Normal* (p. 36,

4th ed., 1905) that this method "requires a facility for copying from nature as well as artistic taste which, alas, does not always coexist in the dedicatees of the natural sciences." Cajal would undoubtedly have found this direct method no less accurate and much less cumbersome than using a camera lucida attached to this microscope, especially when a very strong source of light is required for viewing the image of a Golgi preparation 100 micrometers thick. From his own evidence it is certain that his preferred method of freehand drawing would have been inhibited and frustrated by the used of a camera lucida.

Cajal's unique gift was his ability to grasp in a novel synthesis the relationships between neurons that were seldom if ever seen in a single view through microscope. Justifying this method, he wrote:

> A histological drawing is never an impersonal copy of everything present in the preparation. If that were true our figures would be far too complicated and almost incomprehensible. By virtue of an incontestable right, the scientific artist, for the purpose of clarity and simplicity, omits many useless details. … In order to decrease the number of figures artists are sometimes forced to combine objects which are scattered in two or three successive sections (Ramón y Cajal, 1929a).

Cajal had definite presuppositions regarding the functional significance of the living structure he observed in dead, fixed specimens and was not averse to making bold inferences that went beyond anything that he could have seen (Jacobson, 1993). For example, in his drawings the conspicuous arrows pointing in the assumed direction of flow of nerve activity were meant to show an intrinsic property of the neuron to conduct in one direction only, a "dynamic polarization" of the cell (Fig. 4). His vivid description of activity of the growth cone, which he saw only as a fixed and stained structure, is also typical of the strong inductive vein in his mode of thought. Cajal's procedure was akin to that of the method of Chinese painting called *xie-yi hua*, literally "writing the meaning painting," which I have described as a combination of uninhibited fluency with deep insight (Jacobson, 1985; see also Jacobson, 1993).

The most compelling pathological–anatomical evidence of discontinuity between one neuron and the others comes from the experiments of von Gudden (1869) and Forel (1887). They observed that when axons are cut degeneration is confined to the corresponding neurons. Reactive changes occur in neighboring glial cells (Weigert, 1895; Nissl, 1894), but the neighboring, uninjured neurons remain unaffected. After the discovery of specialized nerve terminals called "endkolben" (terminal knobs) or "endfusse" (endfeet) (Held, 1897; Auerbach, 1898; Ramón y Cajal, 1903; Wolff, 1905), it became apparent that these endings are not in direct continuity with the neurons they contact. This could be deduced from the fact that injury to a neuron results in rapid degeneration of its nerve terminal structures but not of the neurons that they contact (Hoff, 1932; Foerster *et al.*, 1933). Conversely, after nerve injury resulting in retrograde degeneration there is not an immediate effect on the nerve endings in contact with the degenerating neurons (Barr, 1940; Schadewald, 1941, 1942). That the acute degenerative changes are confined to

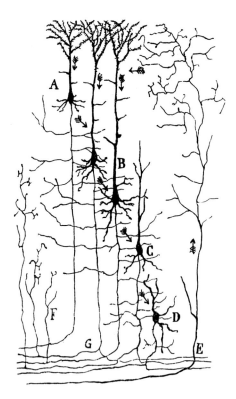

FIGURE 4. Schematic illustation of the probable currents and the nervosopro-toplastmic connections between the cells of the cerebral cortex. A = small pyramidal cell; B = large pyramidal cell; C and D = polymorphic cells; E = fiber terminal coming from other centers; F = collaterals of the white matter; G = axis cylinder bifurcating in the white matter. From S. Ramóny Cajal, Les nouvells idées sur la structure du système nerveux chez l'homme et chez les vertébrés, Reinwald Paris, 1894.

the injured neuron was much later confirmed with the electron microscope (de Robertis, 1956; reviewed by Gray and Guillery, 1966).

Physiological experiments showed that conduction in nerve fibers is bidirectional whereas conduction in reflex pathways is unidirectional. Sherrington (1900, p. 798) proposed that one-way conduction in the reflex arc is due to a valve-like property of the junction between neurons, which he named the synapse in 1897. Thus the unidirectional conduction in the central nervous system was conceived by Sherrington to be a function of one-way conduction at the synapse and not due to the "dynamic polarization" of the entire neuron from dendrites to axon, as conceived by van Gehuchten (1891) and Ramón y Cajal (1895). Additional physiological evidence supporting the concept of the synapse was provided by measurements of the reflex conduction time, which showed a delay of 1–2 msec more than could be accounted for by the conduction time in the nerve fibers (Sherrington, 1906, p. 22; Jolly, 1911; Hoffmann, 1922; Lorente de Nó, 1935, 1938).

The question of whether transmission of excitation from neuron to muscle is electrical or chemical was first discussed by du Bois-Reymond (1877, p. 700 *et seq.*), who concluded in favor of chemical transmission, largely on the basis of the effects of

curare. The accumulation of evidence in favor of chemical transmission at the vagus endings in the heart (Loewi, 1921, 1933), at the neuromuscular junction (Dale, 1935, 1938), and in autonomic ganglia (Feldberg and Gaddum, 1934; Dale, 1935) dealt directly only with transmission at synapses in the peripheral nervous system. Both Adrian (1933) and Eccles (1938) argued that the rapid speed of conduction across central synapses precludes chemical transmission. This objection was removed after it was shown that cholinesterase can act within milliseconds on the amounts of acetylcholine released at the neuromuscular junction and probably also at central synapses (Brzin et al., 1940).

The advent of the electron microscope and of intracellular microelectrode recording finally proved the identity of synapses in both central and peripheral nervous systems. Until the invention of glass knives for cutting ultrathin sections (Latta and Hartmann, 1950), the electron microscope revealed little more than the light microscope, namely, that there was what appeared to be a single membrane separating the neurons at the synapse (reviewed by Robertson, 1987). Later electron microscopic observations of synapses made with tissues fixed in osmium tetroxide, embedded in methacrylate, and sectioned with glass knives showed the presynaptic and postsynaptic and postsynaptic membranes separated by a synaptic cleft about 20 nm wide (Robertson, 1952; Palade and Palay, 1954). Synaptic vesicles, which characterize presynaptic nerve endings (Palade and Palay, 1954; de Robertis and Bennett, 1955), were immediately recognized as possible storage sites of chemical transmitters (del Castillo and Katz, 1956). After the introduction of epoxy embedding (Glauert et al., 1956, 1958) and potassium permanganate fixation (Luft, 1956), it became possible to see the structure of the presynaptic and postsynaptic membranes and to obtain more accurate measurements of the width of the synaptic clefts. Even before glutaraldehyde fixation (Sabatini et al., 1965) provided reliable pictures of cytoplasmic structure, the first attempts were made to find ultrastructural differences between excitatory and inhibitory synapses (Gray, 1959; Anderson et al., 1963; reviewed by Eccles, 1964).

The crucial physiological evidence proving the existence of chemical synaptic transmission was obtained by means of intracellular microelectrode recording at the neuromuscular junction (Fatt and Katz, 1951; Fatt, 1954), in the vertebrate spinal cord (Brock et al., 1952), and in the abdominal ganglion of *Aplysia* (Tauc, 1955). Intracellular microelectrode recording showed that the excitatory postsynaptic potential is thousands of times greater than can be accounted for by electrotonic transmission from neuron to neuron, thus making it certain that the observed amplification is mediated by chemical synaptic transmission. This evidence settled the long-standing controversy between electrical and chemical theories of synaptic transmission (reviewed by Eccles, 1959, 1964, who reversed his early position in support of electrical transmission to finally accept the general validity of chemical transmission at central synapses). It is ironic that no sooner had the dispute been resolved in favor of chemical synaptic transmission than the first report appeared of an authentic case of electrical transmission in the crayfish giant fiber to motor fiber synapses (Furshpan and

Potter, 1957). When potassium permanganate fixation allowed membrane structure to be resolved with the electron microscope, it became evident that the presynaptic and postsynaptic membranes are in close apposition at electrically coupled synapses, whereas a synaptic cleft is characteristic of chemical synapses (Robertson, 1963; 1965; Pappas and Bennett, 1966).

In addition to the preconceived ideas about the absence of cell membranes, the other main reason for the slow acceptance of the neuron theory was that the evidence in its favor arrived in bits and pieces from several different disciplines, over a period of more than 50 years. The neuron theory was built up gradually, by a process of conjecture and refutation. Such theoretical constructs can include approximations, need only have sufficient internal consistency to work, and can be continually adjusted to fix the data.

The theory of scientific revolutions, as expounded by Kuhn (1962, 1970, 1974), does not provide a satisfactory explanation of how the neuron theory ultimately replaced the reticular theory, both of which are components of different paradigms. The neuron theory replaced the reticular theory because it was a more consistent predictor of discoveries and ultimately because the electron microscopy evidence falsified the reticular theory. It is said by Kuhn (1970) that scientific paradigms are created under the influence of the *Zeitgeist* (a term invented by Goethe for events that occur "*neither by agreement nor by fiat, but self-determined under the multiplicity of climates of opinion*"). Yet the history of the neuron and reticular theories shows that both existed in the same *Zeitgeist* and people continued to believe in the reticular theory long after the passing of the *Zeitgeist* in which the theory had grown up. There were also some central figures who remained indifferent to either theory, which shows that the climate of opinion did not affect them crucially. As an example of this indifference it is instructive to quote Wilhelm Wundt (1904), the founder of physiological psychology:

> Whether the definition of the neurone in general, and whether in particular the views of the interconnexion of neurones promulgated especially by Ramón y Cajal will be tenable in all cases, cannot now be decided. Even at the present day, the theory does not want for opponents. Fortunately, the settlement of these controversies among the morphologists is not of decisive important for a physiological understanding of nervous functions.

Wundt was wrong but he had enormous influence (see Titchener, 1921; Boring, 1950). Wundt's reactionary view of the neuron theory is consistent with his archaic exposition of neuroanatomy which occupies the first volume of his very influential textbook *Principles of Physiological Psychology* (5th ed., English transp., 1904) in which the important advances in neuroanatomy and neurocytology of the previous decades are completely ignored.

The physiological psychologists, more than any other group of neuroscientists, continued to support the last vestiges of the reticular theory or variants of it. Karl Lashley (1929) was still upholding a theory of equipotentially of different regions of the cerebral cortex 50 years after Golgi had proposed a similar theory. According to Lashley (1929, 1937), learning and memory depend on the quantity of cortex, not on the specific cortical region. After destruction of large regions of cortex, the *engram* (Lashley's term for the form in which memory is stored) moves to another place without losing its essential form, as in the "harmonius equipotential system" proposed by Driesch (1908) to account for embryological regulation in his vitalistic view. Sperry's chemoaffinity theory of formation of specific nerve connections (Sperry, 1963) was a reaction to theories of equipotential or random neural networks and to theories of development of functional specificity from an initially diffuse neural network (see Jacobson, 1993).

The synapse was conceived as a theoretical necessity by Sherrington in 1897, more than 50 years before it was conclusively shown to exist and function as a physical entity. None of the 19th century proponents of the different theories of connections between neurons lived to witness the final solution of their problem. Ironically, neither the intracellular microelectrode technique not electron microscopy owed anything to the rival theories of nerve connections—the stimuli for their invention came from other sources and those techniques were first applied to other problems before being put to use in solving the problem of neuronal connections. The problems were resolved by opportunistic application of any techniques that seemed likely to work, and finally by means of microelectrode recording and electron microscopy, not merely as a result of wrangling about different theories.

This understanding of the history of neuronal connections and of the neuron theory is different from the generally prevalent view. I see the neuron theory reaching its canonical form, as we finally understand it, only gradually as a result of convergence and coalescence of several theoretical positions and research programs rather than as a revolutionary overthrow of one theory by another. The techniques that were favored by the neuronists, especially the Golgi methods, corroborated their position by showing free nerve endings. The reticularists favored techniques, for example the gold technique of Apáthy and neurofibrillar stains, which apparently revealed fine fibrils directly connecting nerve cells. The neuronists conceived of strict localization of function in the nervous system, whereas the reticularists conceived of diffusely distributed functions. These differences were, in general, also related to different values, for example, as reflected in the mechanism-versus- vitalism and in the nature-versus-nurture debates. The importance of values in neuroscience research programs is discussed further in my *Foundations of Neuroscience* (1993).

It is true that the adversarial theoretical positions occupied by the neuronists and the reticularists had a powerful heuristic effect, driving them on to seek evidence to corroborate their own position and to refute that of their adversaries. The research program progressed by means of a dialectical process of conjecture and refutation. Because refutation always lags behind conjecture, and because each side held tenaciously to its own theoretical position for as long as possible in the face of counter-evidence, both theories continued to be contested long after the reticular theory became impossible to defend. The reticular research program, with its untenable theory, selective techniques, and special world view, became a degenerating program in the sense that it was supported by artifacts and refuted by the facts.

History provides many examples to show that to gain a scientific reputation it is sufficient to apply a new technique to resolve an old problem, but it is neither necessary to be a profound thinker nor to support the correct theory. As Schopenhauer (1851) remarks in a celebrated footnote in *The Art of Controversy*, in order to win a dispute "unquestionably, the safest plan is to be right to begin with; but this in itself is not enough in the existing disposition of mankind, and, on the other hand, with the weakness of the human intellect, it is not altogether necessary."

FORMATION OF DENDRITES AND DEVELOPMENT OF SYNAPTIC CONNECTIONS

> I find the fundamental features of a theory of the central ganglion cells in the observation of Remak, that every cell makes connection exclusively with only one motor nerve cell root, and that this is a fiber chemically and physiologically different from all other central processes … . The body of the cell is continuous, without interruption, with a more or less large number of processes which branch frequently … . These processes which … must not be considered as the source of axis cylinders, or as having a nerve fiber growing from them … will hereafter be called protoplasmic processes.
>
> Otto Friedrich Karl Deiters (1834–1863),
> *Untersuchungen über Gehirn und Rückenmark*
> *des Menschen und der Saugethiere*, 1865

Historical and Theoretical Perspective

> So long as one only substitutes one theory for another, in the absence of direct proof, science gains nothing; one old theory deserves another.
>
> Claude Bernard (1813–1878), *Leçons sur la physiologie*
> *er la Pathologie du systèm nerveux*, Vol. 1, p. 4, 1858

Five cardinal theories of the organization of nervous systems originated in the second half of the 19th century and formed the basis of the neuron theory on which modern neuroscience research programs are constructed. The most important advance in our understanding of the historical development of the neuron theory is that it did not originate in the 1880s and 1890s as a single theory but was constructed over a much longer period, starting in the 1840s, by convergence of at least five different cardinal theories and several other auxiliary theories. Those cardinal theories were firstly, that the nerve cell and its fibers are parts of the same unit (Wagner, 1847; Koelliker, 1850; Remak, 1853, 1855); secondly, that nerve fibers are protoplasmic outgrowths of nerve cells (Bidder and Kupffer, 1857; His, 1886); thirdly, that dendrites and axons are fundamentally different types of nerve fibers (Wagner, 1842–1853; Remak, 1854; Deiters, 1865; Golgi,

1873, 1882–1885); fourthly, that nerve conduction occurs in only one direction—from dendrites to axons to dendrites between different nerve cells (van Gehuchten, 1891; Lenhossék, 1893; Ramón y Cajal, 1895); fifthly, that nerve cells are connected by surface contact and not by cytoplasmic continuity (Koelliker, 1887, 1890,; His, 1886; Forel, 1887). In addition to those five cardinal theories, a number of additional auxiliary theories also entered into construction of the neuron theory, especially the theory that the contact regions, named synapses by Sherrington (1897, 1906), are the principal sites of functional integration and modification. One of the consequences of understanding the historical development of the neuron theory in these terms is that priority for its discovery cannot be fairly attributed to a single individual, least of all to Ramón y Cajal, who only appeared on the scene in 1888 after the theory had been largely constructed by others.

A revolutionary advance in understanding the cellular organization of the CNS occurred in the decades before those cardinal theories were promulgated. It was necessary to advance from the elementary level of understanding that the brain is composed of separate globules and fibers suspended in a fluid or ground substance (Purkinje, 1837; Valentin, 1836, 1838, among many others), to the higher level of understanding that the globule and fiber belong to the same cell. This advance was mainly accomplished by Rudolph Wagner (1847), Albert Koelliker (1850), and Robert Remak (1853, 1855). The theory of the unity of nerve cells and fibers was advanced further by the conjecture that nerve fibers grow out of nerve cells (Bidder and Kupffer, 1857; His, 1886).

The discovery of the difference between dendrites and axons (Wagner, 1851; Remak, 1853; Deiters, 1865) intensified efforts to discover how nerve cells are connected together. The earliest theories were based on the assumption that there is a special tissue consisting of fine fibrils forming the link between neurons. Joseph Gerlach (1872) thought this linkage was made by find fibrils connecting dendrites of different neurons. Camillo Golgi (1882–1885) thought that the linkage was formed by a fiber network interposed between afferent axons and collaterals of efferent axons.

Albert Koelliker (1879, 1883, 1886) was the first to conjecture that cells are connected by contact and not by continuity. It is significant that the first good evidence in support of the contact theory was experimental and not merely histological, because histological methods were at that time not capable of resolving the problem—August Forel (1887) showed that, after eye enucleation or lesions of the visual cortex, degeneration was confined to the injured neurons and did not extend to those in contact with them.

It should be emphasized that all these theories, including those which were eventually refuted, were at first progressive, in the sense that they were ahead of the facts (there were few facts on which to base any theory). They were also mature, in the sense that they could be accepted immediately into the construction of research programs and could guide research within the constraints of available techniques. Only in the late 1880s and early 1890s were the facts accumulated to unify and consolidate the five cardinal theories into the neuron theory.

Let us start by considering the level of conceptualization that had been reached in the first half of the 19th century. The concept that the nervous system consists of separate globules and fibers prevailed until the 1840s. It was believed that globules and fibers originate separately and that when they join, which occurs only rarely, it was a secondary and perhaps an impermanent union. The early microscopic observations of nervous globules were probably artifacts created by chromatic aberration. Joseph Lister (1830), who developed the achromatic objective, concluded that virtually all previous microscopic observations of histological structure were invalid because of the gross optical aberrations produced by the available lenses. The poor methods of fixation, sectioning, and staining also set severe limitations on the accuracy of histological observation. Construction of the research program concerning nerve cells and fibers and their connections was closely linked to the progress of microscopic and histological techniques.

The theory that the nerve fiber is an outgrowth of the nerve cell was originally proposed by Bidder and Kupffer (1857, p. 116): "It can be stated with the greatest degree of certainty, that the nerve cell is endowed with the conditions for allowing the fiber to grow as a direct extension out of itself. ... every fiber must thereafter until its peripheral termination, regarded morphologically, be conceived merely as a colossal 'outgrowth' of the nerve cell." This theory was based on very flimsy evidence, but it was a mature theory in the sense that it could be accepted into a research program. Its heuristic power was tremendous, and it has continued to guide research until now. It was also a progressive theory in that it was ahead of the facts. It continued to lead the facts for the following 50 years or more, as they were slowly accumulated, culminating in Harrison's (1910) demonstration of nerve fibers growing in tissue culture.

Here we have an interesting exercise in assigning priorities. Priority of discovery of the unity of the nerve cell and fiber could be awarded to Remak for having shown the unity between fibers and sympathetic ganglion cells (1838); to Helmholtz (1842) for showing it in an invertebrate; to Koelliker (1844) for generalizing that concept to all nerve cells. Priority of discovery of outgrowth of fibers from nerve cells could be claimed by Bidder and Kupffer (1857), who first advanced the idea; by His (1886) for histological demonstration of fiber outgrowth from cells in chick and human embryos; by Cajal for discovery of the axonal growth cone, in 1890(c); and by Harrison (1910) for showing nerve fiber outgrowth in tissue culture. Is priority established by the one who pronounced the original theory, or who finally proved the theory? Or should priority be given to those who invented the techniques that made it possible to obtain the facts? No doubt they all did well and all deserve praise, but I think that priority belongs principally to the one who planted the tree of knowledge, less to those who tended it and pruned it, and least to those who marketed the fruit.

Wilhelm His was the first to obtain histological evidence showing that the axon grows out of the nerve cells in the spinal cord of the chick embryo. In 1886 His made this pregnant statement: "As a firm principle I advocate the following law: that every nerve fiber extends as an outgrowth of a single cell. That is its genetic, its nutritive and its functional center; all other connections of the fiber are either merely collaterals or are formed secondarily." From the deduction of Forel that nerve fibers end by contacting other nerve cells in the nerve centers, and those of His that the fiber is an outgrowth of the nerve cell, the neuron theory began to be constructed. His could not see the full extent of outgrowing nerve fibers in his preparations stained with carmine, gold, or hematoxylin, and he did not have the advantage of apochomatic lenses or of the Golgi technique, both of which became generally available after 1886. For example, His was unable to see the growth cones and could not at that time deal explicitly with the question of how one neuron connects with another. It is not sufficiently appreciated that in 1856 His had obtained the earliest evidence that nerve fibers end freely in preparations of cornea stained with silver nitrate and blackened by exposure to light. His suggested that nerve fibers in the center might also end freely but could not obtain evidence with the techniques at this command.

An indirect approach to this problem was taken by August Forel in 1887 which led him to obtain the first experimental evidence showing that nerve cells connect by contact and not by continuity. Forel's work was based on the discovery by Bernard von Gudden (1870, 1874) that removal of the eye of the newborn rabbit results, after a short survival period, in atrophy of the visual centers. He thus provided the first method for tracing pathways in the CNS. He extended this method in 1881 to show that lesions in the cerebral cortex result in degeneration in the corresponding subcortical structures. It was clear that acute degeneration is confined to the injured nerve cells, but von Gudden did not understand the general significance of his results. That was accomplished by Forel, who in 1887 showed that removal of one eye of a rabbit results in degeneration restricted to the optic nerve fibers without extending to nerve cells of the lateral geniculate nucleus, whereas removal of the visual cortex results in loss of the lateral geniculate neurons without apparently affecting the optic nerve fibers. From this Forel made the brilliant deduction that there are two separate neurons linking the retina to the cerebral cortex and that they make contact but do not form direct connections in the lateral geniculate nucleus. In his autobiography (published posthumously in 1935) Forel states:

> I considered the findings of Gudden's atrophy method, and above all the fact that total atrophy is always confined to the processes of the same group of ganglion-cells, and does not extend to the remoter elements merely functionally connected with them ... All the data convinced me ever more clearly of simple contact ... I decided to write a paper on the subject and risk advancing a new theory ... and sent it immediately to the Archiv für Psychiatrie in Berlin. However, this periodical was then appearing at long intervals, so my paper did not appear until January 1887. ... Without my knowledge Professor His of Leipzig had arrived at similar results, and had published them in a periodical which was issued more promptly, in October 1886, so that formally speaking the priority was his.

Cajal, in Chapter 5 of his autobiography, deals with the contributions of His and Forel in the following manner: "Two main hypotheses disputed the battlefield of science: that of the network, defended by nearly all histologists; and that of free endings, which had been timidly suggested by two lone workers, His and Forel, without rousing any echo in the schools. ... My work consisted just in providing an objective basis for the brilliant but vague suggestions of His and Forel." Their statements were certainly neither timid nor vague. Perhaps Cajal's perception of the contributions of His and Forel reflects his lack of understanding of their classical or Apollonian style in contrast to his own romantic or Dionysian style (Jacobson, 1993).

Let us now consider the history of concepts of dendritic form and function and the origins of the cardinal theory that dendrites are fundamentally different from axons. In 1851 Rudolph Wagner described the large nerve cells in the electric lobe of the brain of *Torpedo* and noted that usually only one of its several processes is continuous with the nerve fiber (Wagner, 1851, Vol. 3, p. 377). Before that time nerve cells had been described as ganglionic globules, notably by Christian Gottfried Ehrenberg (1833, 1836), Gustav Gabriel Valentin (1836, 1839), and Jan Evangelista Purkinje (1837a, b), but the relationship of globules to nerve fibers was incorrectly understood, and the dendrites had not been identified. For example, Valentin (1836, 1838) thought that the fibers approach the ganglionic globules and even loop around them without making contact. In 1837 Purkinje described the ganglionic globules (they were called cells only after 1839) that now bear his name in the cerebellar cortex and showed the cell body and proximal part of the dendrites without identifying the latter. In the first edition of Koelliker's *Mikroskipische Anatomie* (1850–1852) the dendrites are not identified as such.

Wagner's identification of two different types of nerve cell processes was confirmed in the multipolar nerve cells of the spinal cord of the ox by Robert Remak (1854), who also clearly showed that the axon is in direct continuity with the nerve cell body. Those observations were made by Ramak on tissue section sent to him by Stilling, who was the master of the freehand technique for cutting thin frozen sections. The relationship of the cell body to the two types of processes was described more clearly by Otto Deiters, who dissected single motor neurons from the spinal cord of the ox, after macerating the cord in a weak solution of potassium dichromate. He showed that the dendrites, which he named protoplasmic processes, are different from the axon, and he generously gave priority to Remak for discovery of two different nerve cell processes (see the epigraph to this chapter). Deiters also described a separate system of fine fibers originating from the dendrites, which he believed to run into the ground substance, in which he thought nerve cells are embedded. This was the source of the idea that there are two systems of fibers connecting neurons. Deiters, and later Gerlach, thought that axons connect with one another to form one system but a separate system is formed by the connections between dendrites of different nerve cells. Deiters' observations, which were published posthumously in 1865 by Max Schultze, elevated the difference between axon and dendrites to a general theory of nerve cell morphology. This theory was characterized by Henle (1871, p. 26) in his historical review of the progress of anatomy, as the single most important advance made up to that time in

understanding the nervous system. Only the proximal segments of large dendrites could be seen until Golgi, in several works published between 1873 and 1886, showed the complete dendritic trees of neurons in the spinal cord, olfactory bulb, and cerebral and cerebellar cortex. However, it was only in 1890 that they were named dendrites by His. Another important distinction was made by Nissl (1894), who showed that the basophilic granules which now bear his name extend from the cell body into dendrites but never into axons.

By 1860 a major unresolved problem was recognized to be the way in which nerve cells are connected together in the CNS. Koelliker was able to state in 1863 (p. 313): "The case is related undoubtedly to the connections of the nerve cells to one another. Many describe anastomoses and see such where others find nothing definite. I could name many well known researchers who have shown me such variation with which I could not agree." Gerlach (1872, p. 353) claimed to have discovered the necessary link in the form of a very fine feltwork of fibrils between nerve cells in the spinal cord stained with carmine and ammonia or with gold. Gerlach conceived of the feltwork arising from the tips of the fine branches of the protoplasmic processes forming one system of connections between nerve cells and conceived of a separate system of connection between axons. I have been able to find only two reports claiming to corroborate Geralach's theory. One of these is by Boll (1874), and it is of interest because it shows fine fibers linking the protoplasmic processes (dendrites) of Purkinje cells. That might possibly have been a premature discovery of the parallel fibers.

Adequate counterevidence to Gerlach's theory could not be obtained with the existing techniques and was delayed until invention of the Golgi technique and publication of Camillo Golgi's major work. Golgi showed protoplasmic processes (dendrites), fully stained for the first time, in the spinal cord, cerebellar cortex, cerebral cortex, and olfactory bulb (Fig 5). He showed that the protoplasmic processes end blindly, without any connections to one another, and thus totally demolished Gerlach's theoretical construct. We should recognize the significance of Golgi's discoveries in relation to his progress ahead of his predecessors and contemporaries and not only in relation to the later advances made by his followers. Golgi's view of the cellular structure of the CNS was as far in advance of those of his predecessors as the views of Cajal were in advance of those of Golgi. Cajal could see farther not only because he had sharp vision, but because he stood on Golgi's shoulders.

Golgi proposed that dendrites end on or close to blood vessels, and he believed that they have nutritional functions and are not in the main conducting pathways. Golgi rejected the authenticity of Gerlach's fibrillar feltwork, but he proposed that the link between nerve cells is a network of fibrils which form connections between afferent axons and the collaterals of efferent axons. Golgi did not show a picture of his conjectured network or reticulum but in 1886 he described it in very guarded terms:

Out of all these branchings of the different nerve processes there arises, of course, an extremely complicated texture

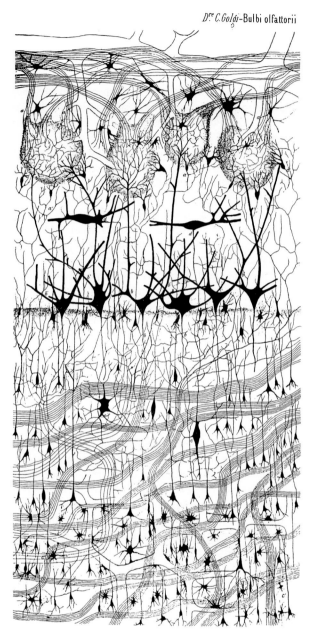

Dᶜ C. Golgi-Bulbi olfattorii

FIGURE 5. Structure of the olfactory bulb of a dog revealed by means of Golgi's newly invented technique of metallic impregnation of nerve cells, using potassium dichromate and silver nitrate. In the original figure the nerve cells and processes are shown in black and blue and the glial cells in red. Golgi states that although this figure is semischematic, it is an accurate depiction made with the aid of a *camera chiara* (same as a *camera lucida*). Golgi discovered the astrocyte perivascular endfeet (seen at the top left). At the time of publication of this figure Golgi was uncertain about the modes of connection of nerve cells, and he shows the axons as well as dendrites ending blindly. The figure was made with an achromatic objective at a total magnification of 250× and was the best that could be obtained unti apochromatic objectives became available in 1886 and enabled others to build on Golgi's work using his staining mehods. From *C. Golgi, Riv. sper. freniat. Reggio-Emilia* 1:405–425 (1875).

which extends throughout the whole grey substance. It is very probably that out of the innumerable further subdivisions there arises a network, by means of complicated anastomoses,

and not merely a feltwork; indeed one would be inclined to believe in it from some of my preparations, but the extraordinary complexity of the texture does not permit this to be stated for certain.

There followed the well-known dispute between the supporters of Golgi's reticular theory (notably Apáthy, Bethe, Held, Nissl) and the supporters of the alternative theory of neuronal connections by contact (Koelliker, 1879, 1883, 1886, 1887; Forel, 1887; His, 1886, 1889; Ramón y Cajal, 1890, 1891; Lenhossék, 1890, 1892; Retzius, 1892; Van Gehuchten, 1892; Vignal, 1889). This polemic is discussed in some detail Jacobson (1993). None of the supporters of the neuron theory actually saw the synaptic contact zone, only the free nerve endings. It should be noted that all these proponents of the theory of neuronal contact at first also admitted that there were some neurons linked by protoplasmic anastomoses. The difference between the opposing factions was that while one side asserted that contact between nerve cells is the rule and anastomoses are the exception, the reverse was asserted by the other side. This is most clearly seen in Koelliker's theoretical position, which shifted progressively from the reticular to the neuron theory. The confrontation between these opposing theories, unpleasant as it often was, had great heuristic value, resulting in efforts to obtain corroborative and refutative evidence.

Evidence showing that dendrites are in the direct conducting pathway accumulated rapidly to refute Golgi's theory that dendrites only have nutritive functions. Firstly, many bipolar neurons had been described, for example, in the spinal ganglia and cranial ganglia of fish (Wagner, 1847) and in the cochlear and vestibular ganglia. In those neurons the dendrites had to be in the conducting pathway. Max Schultze (1870, p. 174) said, "It is obvious that such a ganglion cell is only a nucleated swelling of the axis cylinder." Secondly, many neurons were found in which the axon comes off a dendrite rather than the cell body, for example, in cerebellar granule cells (Ramón y Cajal, 1888, 1890). In the invertebrate nervous system, where the cell body is outside the line of conduction, the beautiful methylene blue and Golgi preparations of Retzius (1890, 1891, 1892a) clearly showed dendrites as a necessary part of the conduction pathways and failed to show any signs of Golgi's conjectured network.

Let us now consider how concepts of the organization of specific regions of the CNS evolved historically. A number of regions were selected by Golgi for study with his metallic impregnation techniques—the spinal cord, cerebellar cortex, hippocampal formation, and cerebral neocortex, and these became the principal battlefields on which different theories of organization of nervous connections were fought (Jacobson, 1993). Briefly, I confine the discussion to the history of concepts of organization of the olfactory bulb.

The direct connection of olfactory nerve fibers to dendrites of mitral cells in the glomeruli of the olfactory bulb was discovered by Owsjannikow (1860) and Walter (1861). Their discovery that the olfactory nerve fibers connect directly with protoplasmic processes (dendrites) was contrary to all concepts at that time. Their findings were disputed by Golgi (1875), who pointed out that the olfactory nerve fibers often stain when the mitral cell dendrites fail to stain, and *vice versa*, and that he could find no connection between them. Instead, Golgi claimed to have found fiber-to-fiber connections between the olfactory and fine nerve fibers entering the olfactory glomeruli. Golgi's depiction of cellular relationships in the olfactory bulb was correct in many aspects, was a significant advance over earlier concepts, and formed the basis for all subsequent studies. For example, Golgi discovered the astrocytic perivascular endfoot, and he correctly recognized the relationship between astrocytes and brain capillaries. Golgi (1875, 1882) gave the first modern description of the olfactory mitral cells and their relationship to the glomeruli, but he failed to trace the mitral cell axons out of the olfactory bulb. He correctly showed mitral cell dendrites entering the glomeruli, but instead of contacting the olfactory afferents he showed the dendrites contacting blood vessels. Golgi did not fail to see the irregularities on the surface of dendrites but he regarded them as artifacts caused by metallic precipitates. The dendrites are depicted with perfectly smooth surfaces in Golgi's figures of olfactory bulb, spinal cord, cerebellar cortex, and cerebral cortex. He also depicted many mitral cell dendrites ending blindly in the external plexiform layer. Golgi's observations led him to the wrong conclusion that all dendrites end blindly in association with blood vessels and are not nervous conducting elements. Golgi thought that the olfactory nerve fibers connect with fine nerve fibers which enter the glomeruli from the olfactory tract. As we know now, the only fibers from the olfactory tract which end near the glomeruli, but do not enter them, are the centrifugal fibers originating from the nucleus of the horizontal limb of the diagonal band, and it is probably that Golgi erroneously traced those into the glomeruli. He correctly showed some of those fibers ending in the external plexiform layer and also correctly showed other fibers ending in the granule cell layer (these are now known to originate from the anterior olfactory nuclei of both sides). We should remember that he had then only recently invented his method of metallic impregnation of nerve cells, which he gradually improved over the following decade. Also, his observations were made before the invention of apochromatic microscope objectives in 1886. Golgi's advance over his predecessors was at least as great as the further advances that were made by Koelliker, Lenhossék, van Gehuchten, Retzius, and Cajal. To deny Golgi the full credit due to the original discoverer, because he failed to see as far as his successors, is like denying the credit to Columbus for discovering America, because he did not land at New York and failed to explore the entire continent.

Golgi's error regarding the blind ending of the mitral cell dendrites and the fiber-to-fiber connections in the olfactory glomeruli persisted until 1890. The problem could then be resolved with apochromatic objectives, which Golgi did not have when he did his pioneering neurocytological investigations. Ramón y Cajal (1891), van Gehuchten and Martin (1891), Retzius (1892b), and Koelliker (1892), using Golgi's technique, showed that the olfactory nerve fibers end by contacting mitral cell dendrites, and that the mitral cell axons extend into the olfactory tract. Axodendritic contacts were also demonstrated

between climbing fibers and cerebellar Purkinje cell dendrites (Ramón y Cajal, 1888, 1890; Retzius, 1892) and between optic nerve terminals and dendrites of neurons in the optic tectum of the chick embryo (Ram Ramón y Cajal, 1891; van Gehuchten, 1892).

The original demonstration of a specialized presynaptic ending at an identified synapse was made by Held (1897a), who showed that during development of the giant presynaptic terminals (the calyces of Held) on the cells of the trapezoid nucleus there is a clear line of demarcation between the axon and the dendrite on which it ends. However, Held incorrectly concluded that the two neurons fuse later in development. The axon terminal expansion were named: "Endfüsse" (endfeet) and "Endkolben" (Held, 1897; Auerbach, 1898). Ramón y Cajal, 1897, immediately understood the significance of these specializations as a means of increasing the contact area, and he was also the first to show that presynaptic terminal expansions occur as a rule and to demonstrate them clearly by means of his reduced silver stain. In 1903 Cajal refers to these axon endings as "mozas," "anillos," "varicosidades," "bulbos," and "botones" (knobs, rings, varicosities, bulbs, and buttons). They were called "boutons terminaux" by van Gehuchten (1904). The terminology was simplified when Sherrington (1897) introduced the term "synapse," and it became conventional to refer to presynaptic and postsynaptic structures and functions.

Confrontation between the reticular theory and neuron theory did not end with these advances—the reticular theory merely changed from a progressive theory, meaning that it was overtaken by the facts and continued to be supported only by histological artifacts. For example, in support of the reticular theory, there were several reports which claimed that fine filaments cross between neurons at the synapse and that those filaments persist even after degeneration of the presynaptic terminals (Tiegs, 1927; Boeke, 1932, Stohre, 1935). In support of the neuron theory it was shown that the end bulbs of degenerating axons swell and disappear completely within 6 days after axotomy, without apparently affecting the cell on which they terminated (Hoff, 1932). This became an effective method of tracing fibers to their terminations, and it generated a vast literature from the 1930s to the 1950s. The reticular theory was by then only a historical relic. It was finally refuted when the synapse could be studied with the electron microscope (Palade and Palay, 1954).

Many examples of axo-somatic junctions were also compelling evidence that a fine network does not form the link between neurons. For example, Ramón y Cajal showed the contacts between cerebellar basket cell axonal terminal and Purkinje cell bodies in 1888, the contacts between centrifugal optic nerve fibers and retinal cells in birds in 1889 and 1892, and the termination of cochlear nerve fibers on cell bodies in the ventral cochlear nucleus of mammals in 1896. The evidence conclusively refuted Golgi's theory, but Golgi continued to adhere to it despite the counter evidence—an outstanding case of tenacity.

A theory of synaptic receptors was first proposed by Langley (1906) from his experiments on the effects of nicotine on neuromuscular transmission in the chicken. Langley (1906) stated that the "receptive substance ... Combines with nicotine and curari [sic] and is not identical with the substance which contracts." This theory was included in a research program that had started with the simultaneous discovery by Claude Bernard and Albert Koelliker, about 1844, that curare blocks transmission at the neuromuscular junction (Bernard, 1878, pp. 237–315), and it culminated in purification of the nicotinic actylcholine receptor, its molecular cloning, and elucidation of its primary structure (reviewed by Changeux et al., 1984; Schuetze and Role, 1987).

The theory that dendrites change shape and retract or extend in response to functional demands was widely held at the end of the 19th century. The theory of ameboid movements of the dendrites was proposed by Rabl-Rückhard (1890) and Duval (1895). At first Cajal (1895) supported the theory and added to it the possibility that neuroglial cells penetrate into the space left by retraction of dendrites during sleep or anesthesia. Later Cajal (1909–1911) argued that the theory was unsupported by any evidence showing the required anatomical changes at synapses, but, as we know, lack of evidence is not a good reason for abandoning a theory—for that there must be well-corroborated counterevidence. Some counterevidence was obtained by Sherrington (1906, p. 24), who pointed out that the reflex delay ("latent period") is longer on the second occasion when a reflex is produced in two stages than when a single full-strength reflex is produced: "This argues against an amoeboid movement of the protoplasm of the cell being the step which determines its conductive communication with the next." That was a fine argument at the time, but subsequent research has shown changes in synaptic size and shape as a result of stimulation (Sotelo and Palay, 1971; Baily and Chen, 1983, 1988; Wernig and Herrara, 1986), and the molecular mechanism for rapid changes in size and shape of dendritic spines has been discovered (Coss, 1985; Fifková, 1985a, b).

We should now consider the theory that synapses initially develop in excess and are later eliminated selectively. The concept of competition and selection on the basis of "fitness," "Adaptiveness," and "competitiveness" derives from Charles Darwin. Once the selectionist idea was grasped it could be extrapolated to deal with populations of molecules, cells, nerve fibers, synapses, or any other parts of the organism. The first to do so was Wilhelm Roux in 1881 in his book *Der Kampf der Theile in Organismus* ("Struggle between the Parts of the Organism"). Charles Darwin considered this "the most important book on Evolution which has appeared for some time" and noted that its theme is "that there is a struggle going on within every organism between the organic molecules, the cell and the organs. I think that his basis is, that every cell which best performs its functions is, in consequence, at the same time best nourished and best propagates its kind" (Darwin, 1888, Vol. 3, p. 244). As Roux recognized, competition is keenest between individuals that are similar and will finally result in one type completely displacing the other. In his 1881 book, Roux introduced two other principles of biological modifiability and plasticity: Itrophische Reizung ("trophic stimulation") *and funktionelle Anpassung* ("functional adaptation"). In his autobiography Roux (1923) noted that he had shown that these are "also applicable as a partial elucidation of

adaptation during learning in the spinal cord and brain" (1881, p. 196; 1883, p. 156; 1895, Vol. 1, pp. 357, 567).

In a single theoretical construct, Roux included competition, trophic interactions, and functional adaptation as causes of plasticity. This was a premature theory in the sense that it was too far in advance of the facts to be of immediate use in constructing research programs. Starting in the 1960s, the technical methods were devised that could be used to test this theory and include it in a research program. The theory of competition and selection was then reinvented in more modern terms. This was done without acknowledging Roux's priority in spite of attention that had been drawn to his contribution in both previous editions of this book. By contrast, the significance of Roux's theoretical construct was well known to his contemporaries, but it was difficult to test the theory with techniques available during the 19th century.

Ramón y Cajal was aware of Roux's theory of cellular competition and selection. Cajal showed that overproduction of axonal and dendritic branches represents a normal phase of development in which excessive components are eliminated. He tells us: "We must therefore acknowledge that during neurogenesis there is a kind of competitive struggle among the outgrowths (and perhaps even among nerve cells) for space and nutrition … However, it is important not to exaggerate, as do certain embryologists, the extent and importance of the cellular competition to the point of likening it to the Darwinian struggle …" (Ramón y Cajal, 1929). The last sentence indicates the influence of Roux's theoretical position, which is the origin of so-called neural Darwinism (Edelman, 1988). Cajal (1892, 1910) also adopted Roux's idea of trophic agents in the mechanism of competitive interaction, survival of the fittest, and elimination of the unfit nerve terminals, synapses, and even entire nerve cells. Since then selectionist mechanism have been proposed for development of functionally validated synaptic connections (Hirsch and Jacobson, 1974; Changeux and Danchin, 1976), for development of connections between sets of neurons by various forms of competitive interaction between nerve terminals, and for development of behavior and learning (Jerne, 1967; Changeux et al., 1984; Edelman, 1988).

The first evidence of specificity of formation of synaptic connections was obtained by J.N. Langley (1895, 1897), who showed that, after cutting of the preganglionic fibers of the superior cervical ganglion, selective regeneration of presynaptic fibers occurs from different spinal cord levels to the correct postganglionic neurons. Thus, stimulation of spinal nerve T1 dilates the pupil but does not affect blood vessels of the ear whereas the opposite effect is produced by stimulation of T4; T2 and T3 have both effects, but to different degrees. Langley (1895) proposed the theory that preganglionic fibers recognized postganglionic cells by a chemotactic mechanism. Guth and Bernstein (1961) concluded that this selection was made on the basis of competition between the presynaptic terminals.

Experimental tests of competition between cells or cellular elements are very difficult to do. When one structure supplants another during development, the deduction is often made that one has been eliminated as a result of competition. However, there are cases in which one structure is replaced by another without any competition, for example, the pronephros by the mesonephros and the latter by the metanephros. In that case there is not even a causal relationship between the three kidneys that develop in succession. In general, mere succession is not evidence of causal relationship and is thus not evidence of mechanism, competitive or otherwise (M. Bunge, 1959; Mayr, 1965; Nagel, 1965). An experimental test of neuronal competition was first done by Steindler (1916) by implanting the cut ends of the normal and foreign motor nerves into a denervated muscle. Steindler found no selective advantage of the normal nerve. When two different nerves innervate a muscle, the resulting pattern is a mosaic in which individual muscle fibers are innervated at random by one nerve or the other. Steindler's observations have been repeatedly corroborated (Weiss and Hoag, 1946; Bernstein and Guth, 1961; Miledi and Stefani, 1969). Similarly, when two optic nerves are forced to connect with one optic tectum, their terminals segregate to form strips and patches in the optic tectum of the goldfish (Levine and Jacobson, 1975) and the frog (Constantine-Paton and Law, 1978).

Several theories of the possible mechanisms of competitive exclusion and elimination of synapses have been proposed. The oldest of these is the theory of formation of selective connections between neurons that have correlated activities. This is an extension of the psychological theory of association of ideas. That theory, deriving from the epistemology of John Locke and David Hume, was first given a neurological explanation by David Hartley. In his *Observations on Man* (first published in 1749), Hartley proposed that mental associations form as a result of corresponding vibrations in nerves (an idea that Newton had thrown out in the last paragraph of his *Principia*). The step from a psychological to a neurophysiological theory of association appears to have been made before the mid-19th century, as evidenced by Herbert Spencer's statement: As every student of the nervous system knows, the combination of any set of impressions, or motions, or both, implies a ganglion in which the various nerve-fibres concerned are put into connection" (*Principles of Psychology*, 1855). The hypothesis that synapses form or become altered between neurons whose electrical activities coincide has become widely accepted in approximately the way in which it was formulated by Ariëns Kappers et al. (1936): "The relationships which determine connections are synchronic or immediately successive functional activities.

The general idea that learning is predicated by selective strengthening of synapses (Ramón y Cajal, 1895) has been accepted and elaborated in various forms (Hebb, 1949, 1966; J.Z. Young, 1951; Eccles, 1964; Konorski, 1967; Beritoff, 1969; Anokhin, 1968; Stent, 1973). Neurophysiological theories of strengthening of synapses between neurons that have synchronous functional activities imply that linkages initially are extensive but become more restricted, functionally and anatomically, as a result of functional activity. In this view, the final arrangement is the result of cooperative interactions between neurons. This view has been extended to include competitive functional interactions between neurons with equal activities being able to maintain connections with a shared postsynaptic target, while functional imbalance results in the more active neuron excluding the less active neuron from a share of the postsynaptic space (Guillery, 1972a; Sherman et al., 1974; Sherman and Wilson, 1975; C. Blakemore et al., 1975; Edelman, 1987).

Theories of competitive elimination of synapses are based on the assumption that presynaptic terminals compete with one another for necessary molecules in limited supply such as trophic factors Ramón y Cajal, 1919, 1928; Changeux and Danchin, 1976; M.R. Bennett, 1983); or that synapse elimination occurs as a result of secretion of inhibitory or toxic factors (Marinesco, 1919; Aguilar et al., 1973; O'Brien et al., 1984; Connold et al., 1986). In 1919 Cajal noted that these factors could be produced by and act upon presynaptic or postsynaptic elements, or both, and that neurotrophic factors could also be secreted by glial cells. It was also recognized that the nerve cell body has a trophic influence on the axon and on the peripheral structures with which it connects (Goldscheider, 1898; Parker, 1932). It was also conjectured that a retrograde trophic stimulus travels from peripheral structures to neurons. The observation that dendrites of spinal motor neurons sprout only after their axons have grown into the muscles led to the theory that a neuron's dendritic growth is dependent on its axonal connections (Ramón y Cajal, 1909–1911, p. 611; Barron, 1943, 1946; Hamburger and Keefe, 1944). Related to this is the "modulation theory" of Paul Weiss (1936, 1947, 1952), according to which the motoneuron modulates its central synaptic connections to match the muscle with which its axon connects.

DEVELOPMENT OF NERVE CONNECTIONS WITH MUSCLES AND PERIPHERAL SENSE ORGANS

> The quest of a single neuromuscular unit has in fact had many of the dramatic features associated with the quest for a single atom, and the success achieved by the physiologist is in most respects quite as remarkable as that of the physicist.
>
> John F. Fulton (1899–1960), *Physiology of the Nervous System*, 1st ed., p. 40, 1938

Notes on the History of Ideas about the Connections made by Peripheral Nerves

> If any one offers conjectures about the truth of things from the mere possibility of hypothesis, then I do not see how any certainty can be determined in any science; for it is always possible to contrive hypotheses, one after another, which are found to lead to new difficulties.
>
> Isaac Newton, "Letter to Pardies, 10 June 1672"
> (In *The Correspondence of Isaac Newton* [H.W. Turnbull, ed.], Cambridge University Press, Cambridge, 1959)

We have considered the growth of knowledge about outgrowth of nerve fibers and evolution of ideas about peripheral nerve endings. The history of ideas about the modes of termination of peripheral nerve fibers parallels that of ideas about endings of nerve fibers in the central nervous system (CNS).

Until the 1860s it was generally believed that the peripheral nerves end by anastomosing with one another to form plexuses in the skin and muscles. It was also thought that sensory nerve fibers branch and anastomose in the skin and mucous membranes and then recombine to form fibers that return to the CNS (Beale, 1860, 1862). The concept of anastomosis between the processes of nerve cells in the CNS was supported by the evidence available at that time. Both central and peripheral nervous systems were believed to be organized on the principle of nerve networks. Microscopes could not resolve individual fine unmyelinated nerve fibers in the peripheral nerves. They revealed fascicles which were mistaken for single nerve fibers. Interlacing of such fascicles was construed as true anastomoses between fibers. As we shall see later, this misconception persisted until Ranson (1911) showed that peripheral nerves contain large numbers of unmyelinated fibers and proved that they are sensory (Ranson, 1913, 1914, 1915).

Wilhelm His (1856) was the first to discover free nerve endings in the epithelial later of the cornea stained with silver nitrate, and this was confirmed in 1867 by Julius Cohnheim, using the recently invented method of staining nerve fibers with gold chloride. The use of gold chloride made it possible to see fine peripheral nerve endings in skin, mucous membranes, and smooth muscle. Free termination of nerve fibers in smooth muscle was demonstrated by Löwit in 1875 using gold chloride followed by formic acid, which is the basis of the modern technique of gold staining. When Friedrich Merkel (1875, 1880) described the cutaneous nerve endings that now bear his name, he thought that the *Tastzellen* (touch cells) were ganglion cells from which the nerve fibers originate. That they are modified epithelial cells in contact with disklike expansions of the nerve endings was shown by methylene blue staining of nerve endings at different stages of development in the skin of the pig's snout (Szymonowicz, 1895).

The gold chloride method also led to uncertainty about whether nerve terminals penetrate into the peripheral cells, and even into the cutaneous hairs (Bonnet, 1878). The beautiful Golgi preparations of Retzius (1892, 1894) and van Gehuchten (1892) left no doubt that all the different types of nerve endings end freely in the hair follicles and adjacent skin. Very rapid progress in describing peripheral nerve endings was made after introduction of methylene blue staining (Ehrlich, 1885) and after Golgi published his rapid method in 1886. As an indication of the sudden burst of activity in this field, Kallius (1896) cites 185 papers in his review of the histology of sensory nerve endings. Lenhossék (1892–1893) and Retzius (1892) showed that nerve endings end freely among the cells of the taste bud. Prior to their reports it was believed that the taste cells give off nerve fibers which run to the CNS. The periodic varicosities of autonomic nerve endings in the mucous membranes of the bladder and esophagus were clearly demonstrated by Retzius (1892).

The concept of anastomosis between peripheral nerve fibers was not laid to rest by the evidence that they end blindly and that there are one-to-one relationships between some sensory nerve endings and some peripheral sensor cells and organelles. It seems to have passed unnoticed by historians of neuroscience that the concept of anastomosis between peripheral nerve fibers

persisted long after the neuron theory was well established. The reason for this is that until the introduction of the pyridine silver method by Walter Ranson (1911), it was not possible to stain unmyelinated nerve fibers reliably and to count them in peripheral nerves. Before Ranson's work the unmyelinated axons were seen only after dissociation of the nerve fibers by soaking pieces of peripheral nerve in weak acid solutions after fixation in alcohol (Ranvier, 1878). Ranson (1911, 1912a, 1913, 1914) discovered that the majority of small unmyelinating peripheral nerve fibers are sensory, showed that they have their cell bodies in the dorsal root ganglia, and traced their fine central processes into Lissauer's tract of the spinal cord. He also correctly conjectured that they subserve pain (Ranson, 1915). Ranson's evidence that the majority of unmyelinated peripheral axons are afferent was not accepted immediately and continued to be denied for another 20 years, for example, by Bishop *et al.* (1933), and indisputable evidence that they are afferents was finally published only in 1935 (Ranson *et al.*, 1935).

The nerves growing into the skin appear to be confronted with a large number of potential targets from which each nerve has to select one target. The situation is complicated by the fact that the density of cutaneous innervation and the number of sensory corpuscles are quite constant in each region of the skin. These aspects of the problem were first fully grasped by Ramón y Cajal, who, in a remarkable paper published in 1919, established the theoretical framework into which all subsequent contributions to the problem have ineluctably had to be fitted. He believed that both selective growth (that is, chemotropism) and selective terminal connection (that is, chemoaffinity) probably play a part in regulating the pattern of cutaneous innervation. He pointed out that the density of innervation of each region of the skin is precisely determined and that "each fiber is destined for an epithelial territory devoid of nerves, and there are no vast aneuritic spaces in some regions nor excessive collections of fibrils in others" (Ramón y Cajal, 1919). He suggested that the nerve fibers are attracted by chemicals in the epidermis, which are either used up or neutralized by the nerves as they grow into the skin, so that "after invasion of the epithelium a state of chemical equilibrium is crested, by virtue of which the innervated territories are incapable of attracting new sprouts."

In addition to the general attractive effect of the epithelium, Cajal proposed a more specific neurotropic effect to account for the specific innervation of different types of sensory organelles and muscles. He pointed out that this specificity is unlikely to be the result of mechanical guidance, because then

it becomes difficult to understand how, of the large nervous contingent arriving at the mammalian snout, some fibers travel without error to the cutaneous muscle fibers, others toward the hair follicles, others to the epidermis and finally some to the tactile apparatus of the dermis. A similar multiple specificity is found in the tongue, trigeminal fibers innervate the ordinary papillae, and facial (geniculate ganglion) and glossopharyngeal fibers go to the gustatory papillae

(Ramón y Cajal, 1919).

Motor nerves were described as ending freely in both skeletal muscle and smooth muscle, in the form of loops or plexuses on the surface of the muscle fibers. For example, Koelliker (1852) remarks that "with respect to the ultimate termination of the nerves, it may be stated that in all muscles there exist anastomoses of the smaller branches forming the so-termed plexuses." At that time the striated muscle fibers were known to be cells called primitive tubes, containing fibrils, and surrounded by a sarcolemma. Schwann (1847) showed that several mononucleate myoblasts fuse to form a multinucleate myotube. Remak (1844) and Lebert (1850) showed that striated muscle fibers differentiate by elongation of myotubes and that self-multiplication of their nuclei occurs. It should perhaps be noted, because it is not well known, that both Remak (1844) and Lebert (1845) were among the first to apply the cell theory rigorously to development and pathology, and in that respect Lebert's *Physiologie pathologique*, published in 1845, was a forerunner to Virchow's more renowned *Cellularpathologie* published in 1849.

The motor end-plate was discovered and named by Willy Kühne in 1862. He was at first unable to see whether the nerve and muscle are continuous or only contiguous. In 1869 Kühne asked the question: "In what way do nerves terminate in muscle?" He came to the wrong conclusion: "We now believe that we are able to perceive the direct continuity of the contactile with the nervous substance." He then expressed some doubts: "Yet it may still happen that, in consequence of further improvements in our means of observation, that which we regard as certain may be shown to be illusory." Sixteen years later, in his Croonian Lecture, Kühne was able to say that "nerves end blindly in the muscles … Contact of the muscle substance with the non-medullated nerve suffices to allow transfer of the excitation from the latter to the former" (Kühne, 1888). Thus, the concept of transmission of nervous excitation by contact rather than by continuity between nerve and muscle was formed before the concept of contact between neurons in the CNS. Once the concept of nervous transmission by contact between nerve and muscle was accepted, it became easier to generalize it to transmission by contact between nerve and nerve in the CNS. The theory was at that time ahead of the evidence, which was obtained remarkably quickly during the decade at the close of the 19th century.

The anatomical concept of what is now known as the motor unit originated in the late 19th century. However, the modern term was first used in 1925 by Liddell and Sherrington and later defined by Eccles and Sherrington (1930) as "an individual motor nerve fibre together with the bunch of muscle fibres it innervates." Counts of the nerves and muscle fibers were made by Tergast (1873), who showed that the ratio of nerve to muscle fibers ranges from 1:80 to 1:120 in limp muscles but is only 1:3 in the extraocular muscles of the sheep, but he did not know that nearly half of the nerves to muscle are sensory, which was discovered much later (Ranson, 1911). Nevertheless, Tergast (1873) established the principle that muscles which perform fine movements have smaller motor units than those which perform gross movements. This was eventually confirmed by counting muscles and nerve fibers and correlating the counts with the tension developed by each motor unit (Clark, 1931).

The muscle spindle was first identified and named by Willy Kühne (1863). The definitive work on muscle spindles and their sensory and motor nerve endings was accomplished by Ruffini (1892, 1898). After Ruffini there was little that others could add with the methods then available, and Ruffini's account of muscle spindles was not superseded until much later (Denny-Brown, 1929; Boyd, 1960). Sherrington (1894, 1897) proved that muscle spindles are proprioceptors of muscle, and the classical experimental analysis of muscle proprioceptive function was done by Mott and Sherrington (1895), who analyzed the effects of cutting various combinations of dorsal roots supplying the limb in monkeys.

NEURONAL DEATH AND NEUROTROPHIC FACTORS

> According to tradition, the development of the vertebrate nervous system has hitherto seemed to proceed straight on in a gradually ascending path, without turnings, temporary expedients, or regressive changes. As a consequence none were looked for and none were found.
>
> John Beard (1858–1918), *The History of a Transient Nervous Apparatus in Certain Ichthyopsida*, 1896

Prolegomena to a History of Nerve Cell Death during Development

> Men make their own history, but they do not make it just as they please; they do not make it under circumstances chosen by themselves, but under circumstances directly encountered, given and transmitted from the past. The tradition of all the dead generations weighs like a nightmare on the brain of the living.
>
> Karl Marx (1818–1883), *The Eighteenth Brumaire of Louis Bonaparte, 1852*

The tyranny of theory over the evidence is nowhere more glaringly evident than in the history of the delayed discovery of neuronal death during normal development. The tyranny in this case was imposed by the theory that both ontogeny and phylogeny are progressive, from lower and less organized to higher and more organized nervous systems. Evidence of neuronal death during normal development was reported but was ignored because it was in conflict with the idea of progressive development. Reports of neuronal death were buried in the literature, to be unearthed much later as curious historical relics. Such reports come back to haunt us as they haunted previous generations who could not accept evidence that conflicted with their cherished theories.

Neuron death during normal development was discovered and described in considerable detail by John Beard in 1896 in the Rohon–Beard cells of the skate: "This normal degeneration of ganglion-cells and of nerves is now for the first time described and figured for vertebrate animals, in which hitherto such an occurance is without precedent" (Beard, 1896a). Rohon–Beard cells had been discovered by Balfour (1878), who illustrated them in the spinal cord of elasmobranch embryos, and they were further described by Rohon (1884) in the trout and in other fish embryos by Beard (1889, 1892). Beard traced their origin from "immediately laterad to the medullary place," in other words, from the neural crest. He described their differentiation and outgrowth of their neurites, discovered their degeneration (Beard, 1896a), and tried to build a general theory on that evidence. Beard (1896b) thought that cell death occurs generally during what he called "critical periods" (the first time that term was used in neuroscience). According to Beard, the critical period represents a period of regression and reorganization of embryonic structures and a transition to the definitive structures of the adult. The quotation from Beard that forms an epigraph to this chapter shows that he recognized the prejudice against the idea of normal regressive developmental stages. He noted that his evidence contradicted the biogenetic law of Ernst Haeckel according to which the embryo simply climbs the phylogenetic tree, recapitulating the structures of its ancestors as it ascends to its appropriate level.

After Beard's definitive work on death of Rohon–Beard neurons, the few reports of neuron death during normal development were consigned to obscurity not altogether undeserved in view of their inability to relate the facts to a general theory of neuron death. Cell death during normal development of the nervous system was first reported in the chick embryo neural tube by Collin (1906a, b).

Ernst (1926) was the first to recognize that overproduction of neurons was followed by death of a significant fraction of neurons in many regions of the nervous system of vertebrates. For example, he reported death of a third of the neurons in the dorsal root ganglia. The originality of his findings may be appreciated from the following brief extracts. After discussing the report by Sánchez y Sánchez (1923) of massive cell death during metamorphosis of insects (now undeservedly forgotten, e.g., in the review by Truman and Schwartz, 1982), Ernst says:

We find ourselves in agreement with Sánchez y Sánchez. He states that he found such extensive cell death in all ganglia of appropriate stages that he at first hesitated to publish descriptions, because he could not believe that such results were not already well known. We too found such massive cell death, above all in the retina, in the trigeminal and facial ganglia, in the upper jaw, and in the anterior horn, that we were at first doubtful whether we were dealing with normal events ... We have at the same time the explanation of why ganglia of older embryos always have fewer cells than those of very young stages ... The results are in complete agreement in showing that degenerations always occur most strongly in the ganglia from which the nerves grow out to the extremities ... There remains only a group of degenerations which are always found, namely in the anterior horns of the spinal cord, the floor of the third and fourth ventricles and at the transition from the thick lateral wall of the brain to the thin roof of the ventricle ... Characteristic of all these degenerations is the timepoint of their occurrence: it consists of a striking correspondence between the vascularization of these regions and the occurrence of degenerations ... For all these cases we

must for the present be satisfied with confirmation of the facts that in the named regions a large number of cells are available for differentiation into nerve cells are available for differentiation into nerve cells but that only a fraction of them are used for that purpose whereas the remainder are destined for disintegration.

Ernst (1926) deserves credit for proposing a general theory of neuron death during normal development and for obtaining a diversity of evidence to support it. The work of Glücksmann (1940, 1951, 1965) merely confirmed the findings of Ernst and others and provided an incomplete but convenient summary in English of some of the literature in other languages. This led to the deplorable practice (e.g., Saunders, 1966, but soon followed by others, e.g., P.G. Clarke, 1985a; Hurle, 1988) of ignoring the work of Ernst and his predecessors and crediting Glücksmann with the concept of three different modes of cell death, when all he did was to give them names. As Ramon y Cajal (1923) noted: "In spite of all the flatteries of self-love, the facts associated at first with the name of a particular man end by being anonymous, lost forever in the ocean of universal science. The monograph permeated with individual human quality becomes incorporated, stripped of sentiments, in the abstract doctrine of the general theories."

Ernst (1926) had provided good evidence of his own and reviewed the previous evidence showing that there are three main types of cell death during normal development: the first occurring during regression of vestigial organs; the second occurring during cavitation, folding, or fusion of organ anlage; the third occurring as part of the process of remodeling of tissues. These were later named phylogenetic, morphogenetic, and histogenetic cell death by Glücksmann (1940, 1951, 1965).

The great neurocytologists of the 19th and early 20th century were in a position to see the death of cells in the developing nervous system but failed to discover it. Why scientists fail to see important things that are staring them in the face is notoriously difficult to understand. Cajal was fond of saying the truth is revealed to the prepared mind, and I should agree that the minds of the great neurocytologists of his time were not prepared for the truth about neuronal death during normal vertebrate development. Another important reason for their failure to see neuronal death was their reliance on Golgi and silver impregnation techniques which do not show cellular debris clearly or obscure it with metallic precipitates. The Nissl stain could have revealed neuronal death to the unprejudiced observer, but the observers were prejudiced by the idea of progressive development. Death of embryonic cells was recognized as a phenomenon of significance only during disintegration of vestigial organs and during metamorphosis, as brilliantly studied by Cajal's student Domingo Sánchez y Sánchez. It is remarkable that the concept of regression of axonal and dendritic structures was easily accepted whereas death of large number of neurons in the vertebrate nervous system was not an acceptable fact. Cajal relished the analog between regression of axonal and dendritic branches and pruning of excessive branches from trees and bushes in a formal garden, but he never conceived of uprooting and destroying large numbers of trees in the process of laying out the garden.

The long delay in accepting the evidence of developmental neuronal death has been regarded as an historical enigma. Here is how the puzzle may now be solved. Nineteenth-century biologists saw that development has an overriding *telos*, a direction and a gradual approach to completion of the embryo, and also saw a terminal regression and final dissolution of the adult; but a fallacy arose when the progression and regression, which coexist from early development, were separated in their minds. Development was conceived in terms of progressive construction, of an epigenetic program—from simple to more complex. For every event in development they attempted to find prior conditions such that, given them, nothing else could happen. The connections and interdependencies of events assure that the outcome is always the same. Such deterministic theories of development made it difficult to conceive of demolition of structures as part of normal development, and it was inconceivable that construction and destruction can occur simultaneously. It became necessary to regard regressive developmental processes as entirely purposeful and determined. For example, elimination of organs that play a role during development but are not required in the adult or regression of vestigial structures such as the tail in humans were viewed as part of the ontogenetic recapitulation of phylogeny. Regression in those cases is determined and is merely one of several fates: cellular determination may be either progressive or regressive. The idea of progress in all spheres, perhaps most of all in the evolution and development of the vertebrate nervous system, has appealed to many thinkers since the 18th century. Such ideas change more slowly than the means of scientific production; thus new facts are made to serve old ideas. That is why the history of ideas, even if it does not exactly repeat itself, does such a good job of imitation.

In the realm of ideas held by neuroscientists, the idea of progressive construction, of hierarchically ordered programs of development, has always been dominant over the idea of a plenitude of possibilities, from which orderly structure develops from disorderly initial conditions by a process of selective attrition. (M. Jacobson, 1970b, 1974b; Changeaux *et al.*, 1973; Changeaux and Danchin, 1976; Edelman, 1985). Progressive development implies increasing orderliness gained by the organism, "sucking orderliness from its environment," and by "feeding on negative entropy" (Schrödinger, 1944, p. 74). Schrödinger did not recognize that the organism can lose entropy (that is, gain orderliness) by ridding itself of internal disorder as effectively as by "attracting, as it were, a stream of negative entropy upon itself" (Schrödinger, 1944, p. 74). Cell death may be a quick way for the embryo to reduce its entropy level.

The idea of development of organiztion by means of selective cellular attrition has gained popularity since the 1970s. Before that time, the dominating idea was that matching between different nerve centers is achieved by programs of cell proliferation, migration, and differentiation in which orderly progress always prevails. But this early period of construction is now known to be followed by a period of deconstruction.

Another dominant idea from the beginning of the century until now (e.g., Cowan *et al.*, 1984) was that the number of neurons is matched to the size of their targets as a result of reciprocal interactions between nerves and their peripheral innervation fields: neuronal proliferation, migration, and survival were conceived to result from trophic influences coming from the target tissues, and a reciprocal trophic influence of nerves on the target tissues ensured the vitality of muscles and sense organs (reviewed historically by Oppenheim, 1981). For the past 200 years the nutritional functions of nerves have generally been regarded as distinct from their roles in sensation and movement. For example, Procháska (1784) states: "Sylvius, Willis, Glisson and others considered that there were two fluids in the nerves, one thick and albuminous, subservient to nutrition, the other very thin and spiritour, intimately connected with the former, and subservient to sensation and movement ..."

A century ago the word "trophic" was on everyone's lips to signify the mysterious life-giving effects of nerves on one another and on the tissues which they supply. In Foster's *Text-Book of Physiology* (7th ed., 1897), trophic action is defined as "the possibility of the nervous system having the power of directly affecting the metabolic actions of the body, apart from any irritable, contractile or secretory manifestations." The first experimental evidence of a trophic action of sensory nerves was the demonstration that taste buds degenerate after denervation and regenerate only if sensory nerves are present (von Vintschgau and Hönigschmied, 1876; von Vintschgau, 1880; Hermann, 1884). Wilhelm Roux (1881) discusses "the trophic action of functional stimuli" under which he has a section "on trophic nerves" (p. 125). There he reviews the trophic effects of nerves on the muscles and other tissues, and he makes the distinction between a direct trophic action of the nerves on these tissues and the indirect effects of lack of stimulation, disease, changes in blood flow, etc. He concludes that nerves have a trophic effect which is not entirely due to excitation. He maintains that not only are the peripheral organs provided with a trophic stimulus independently of the nervous activity, but also "the central nervous substance likewise is influenced in its nourishment by the peripheral organs with which it has formed an excitation-unity" and that "the central nervous tissue should be regarded, so to speak (practically) not as a one-sided provider but at the same time as the nutritive provider by the peripheral tissues." In 1899 L.F. Barker could write, "The more thought one gives to the subject the more he will find in the trophic relations of neurons to make him hesitate before he denies the possibility of conduction of impulses or influences in either direction throughout the neurone." Goldscheider (1898) first conjectured that materials are transported from the nerve cell body to the axon terminals. During the following decades evidence built up to support the theories that trophic factors flow from the nerve cell body to the axonal endings (Olmsted, 1920a,b, 1925; May, 1925) and that nerves release specific trophic factors into the tissues they innervate (Parker, 1932; see M. Jacobson, 1993, for a discussion of the significance of those premature theories.) Perhaps here I should say that those premature conjectures fell on deaf ears and unprepared minds. To arrive on the scene with a message prematurely might be like someone in the position of shouting "fire" in an empty theater.

Experimental analysis of the changes in the developing nervous system resulting from altering peripheral sensory and motor fields was pioneered by Braus (1905) and Shorey (1909) and followed by many others. Removal of limbs or grafting additional limbs was shown to result in hypoplasia or hyperplasia, respectively, and those results were interpreted consistently in terms of regulation of cellular proliferation, as the reader can easily verify from the general textbooks dealing with the subject, such as Samuel Detwiler's *Neuroembryology* (1936) and *Principles of Developmment* by Paul Weiss (1939). The appearance of Glücksmann's 1951 review of cell death during normal development prompted a reconsideration of the effects of limb amputation. Prior to the publication of Glücksmann's review, Hamburger and Levi-Montalcini (1949) concluded that "two basically different mechanisms operate in the control of spinal ganglion development by peripheral factors: (a) the periphery control the proliferation and initial differentiation of undifferentiated cells which have no connections of their own with the periphery; (b) the periphery proivdes the conditions for continued growth and maintenance of neurons following the first outgrowth of neurites" (Hamburger and Levi-Montalcini, 1949). After the discovery that a mouse sarcoma implanted in the chick embryo results in neuronal hyperplasia and hypertrophy (Bueker, 1948) the effect was consistently misinterpreted as a primary action of the factor on neuronal proliferation (Levi-Montcalcini and Hamburger, 1951, 1953).

Victor Hamburger (1958) was able to show that the number of motoneurons in the chick embryo decreases after limb amputation as a result of increased cell death, not because of failure of mitosis or of motoneuron differentiation. However, he did not yet recognize the significance of death during normal development. Arthur Hughes (1961) was the first in recent times to show that a large overproduction of motoneurons occurs during normal development and that motoneuron death is a major factor regulating their final numbers, and Martin Prestige (1965) was the first to demonstrate the same in spinal ganglia. Yet the belief persisted that the periphery controls cell proliferation, even after the discovery of nerve growth factor (NGF), which was at first said to have mitogenic effects (Levi-Montalcini, 1965, 1966; Levi-Montalcini and Angeletti, 1968). The confusion was resolved only after [^{3}H]thymidine autoradiography showed that changes in mitotic activity in the nervous system, following limb grafting or amputation, is confined entirely to glial cells (Carr, 1975, 1976). The path to discovery of the biological effects of NGF and other neurotrophic factors detoured around the difficulties and confusions created by surgical manipulation of limbs. Those were prologues to the biochemical identification of NGF—the rest is history, that ultimate act of imaginative reconstruction.

HISTOGENESIS AND MORPHOGENESIS OF CORTICAL STRUCTURES

That the cortex of the cerebrum, the undoubted material substratum of our intellectual activity, is not a single organ which enters into action as a whole with every physical function, but consists rather of a multitude of organs, each of which subserves definite intellectual processes, is a view presents itself to us almost with the force of an axiom …. If … definite portions of the cerebral cortex subserve definite intellectual processes, there is a possibility that we may some day attain a complete organology of the brain-surface, a science of the localization of the cerebral functions.

Alexander Ecker (1816–1887),
Die Hirnwindung des Menschen, 1869

Historical Orientation

In my opinion there are only quantitative differences, not qualitative differences, between the brain of a man and that of a mouse. Accordingly, all cortical regions which are vested with a specific structure and a specific function and are differentiated in humans are also represented—with the corresponding simplification and reduction—in the mammals and probably even in the lower vertebrates.

Ramón y Cajal (1852–1934), Estudios
sobre la corteza cerebral humana.
III. Cortez motriz. *Revista Trimestral
Micrográfica* 5: 1–11, 1890

Three important theories of nervous organization, valid for our time, emerged from the cell theory. Firstly, the demonstration that the nerve cell and fiber are parts of the same structure (first claimed by Remak, 1838) was the first step in the formulation of the neuron theory. Secondly, recognition that there are different types of nerve cells, even in the same region, was the beginning of the theory of neuronal typology. Thirdly, realization that there are regionally specific patterns of nerve cells and fibers, especially in the cerebral cortex, was the beginning of a theory of cytoarchitectonics (reviewed by Brodmann, 1909; Lorente de Nó, 1943; Kemper and Galaburda, 1984).

Those extensions of the cell theory were linked to the theory of evolution of the nervous system and, especially as seen from the viewpoint of this chapter, to the theory of evolution of the forebrain. Evolution of the telencephalon was understood as a process which exploited the neural structures—cell groups and their connecting fiber tracts-laid down during earlier stages of evolution. Telencephalization involves selective expansion and elaboration of the front end of the neural tube. This starts phylogenetically with the evolution of the floor plate which becomes the huge basal cell masses of fishes. The later phylogenetic advances may be seen as successive additions of new pallial formations: first the primordial pallium of fishes, next the primary hippocampo-pyriform fallial formation of Amphibia, thereafter the secondary hippocampal and pyriform cortices of reptiles, and finally the neopallium of mammals. Efforts were made to trace the phylogenetic order of emergence of different fields in the neopallium and to relate phylogeny to ontogeny. This research program was constructed, around the end of the 19th and beginning of the 20th centuries, by many workers, notably L. Edinger, C.J. Herrick, Elliot Smith, and Ariëns Kappers.

The two quotations standing at the head of this chapter emphasize the early historical origin of two major concepts of organization of the cerebral cortex: firstly, the concept of parcellation of the cerebral cortex into different areas which subserve specialized functions; secondly, the concept of a common organizational scheme for the entire cortex. Questions arising out of the first concept relate to how the different regional specializations develop. For example, to what extent are the specialized areas preformed from the time of their origin and to what extent do they differentiate epigenetically from a single primordial pattern to a more complex final organization? Karl Ernst von Baer recognized that "each step in development is made possible only by the immediately preceding state of the embryo … From the most general in form-relationships the less general develops, and so on, until finally the most special emerges" (Entwickelungsgeschichte der Thiere, Part 1, pp. 147, 224, 1828). Subsequent studies of brain development were made within the framework of the theory of epigenesis—from simple to more complex stages of ontogeny—and also within the framework of a theory of ontogeny recapitulating phylogeny.

The concept that the mature organization of the cortex develops from a more uniform early state and the final state emerges by addition as well as elimination of components was already well established by the beginning of this century. Korbinian Brodmann (1909, p. 226) summarized that concept of progressive versus regressive differentiation as follows: "Considered genetically it is partially new production of anatomical cortical fields, partially their regression or reversion which are combined here …. Undoubtedly both processes, that is progressive and regressive differentiation, occur concurrently during development of cortical fields." This was a premature theory which could not be substantiated until more than 70 years later.

Several questions emerged regarding the conversation of certain features of cortical organization in different regions in all mammals. For example, how have the six layers and their characteristic cell types, inputs, and outputs been conserved? Are the similarities based on homology, meaning that they share the same evolutionary ancestry, or are they based on analogy, meaning that they evolved under similar functional and adaptive pressures regardless of ancestry?

Franz Joseph Gall (1825) first theorized that different mental faculties are represented in separate regions of the surface of the human brain. Although he claimed to be able to relate the cortical representations to bumps on the cranium, he did not claim to be able to delimit separate cortical areas subserving different faculties. Before the 1860s it was generally believed that the cerebral cortex is the seat of psychic and mental functions while motor functions were believed to be controlled by the brainstem. Those beliefs were established by Jean-Pierre-Marie Flourens (1794–1867) on the basis of his surgical ablation experiments. One of his principal achievements was to demonstrate that the cerebellum functions to coordinate voluntary movements. That was then the strongest refutation of Gall's phrenological theory which localized sexual functions in the cerebellum (Gall, 1835, Vol. 3, pp. 141–239; for a brief history of concepts of cerebellar function see Dow and Moruzzi, 1958, pp. 3–6).

Flourens concluded that the cerebrum is the seat of sensation but is not directly involved in control of voluntary movements. Flourens understood that different functions are localized in different parts of the brain, but he concluded that the cerebral cortex functions as a whole, as the organ of sensory perception, intellect, the will, and the soul. (For detailed consideration of Flourens' views, which changed in the two editions of his Recherches, see R.M. Young, 1970.)

There were two opposing schools of thought about cerebral localization—we can call those "lumpers" who saw unity in diversity, and we can call those "splitters" who saw diversity in unity. The prevailing views at different moments of history have tended to oscillate between the extreme lumper and splitter positions. Flourens belonged to the school of lumpers who believed that the cerebral hemispheres function as a whole.

Those beliefs were put in doubt by the observations of Hughlings Jackson (1863) that tumors and other disease processes involving the cerebral cortex sometimes cause seizure movements that progress from distal to proximal limb muscles, often involve the facial muscles, and resemble fragments of purposeful movements. Jackson proposed that the cerebral cortex directly controls body movements is organized in terms of coordinated movements and not of individual muscles, for any muscle could be brought into play in a variety of different movements.

Experimental support for part of Jackson's theory was provided by Fritsch and Hitzig (1870), who evoked coordinated movements of body parts in the dog in response to galvanic stimulation around the cruciate sulcus of the cerebral cortex on the opposite side. Much better evidence of a somatotopic motor representation was obtained by Faradic stimulation of the cerebral cortex of the monkey (Ferrier, 1875, 1876, 1890) and higher apes (Grünbaum and Sherrington, 1902, 1903; Leyton and Sherrington, 1917). The latter also showed that the postrolandic area is inexcitable, contrary to the general belief at that time that the rolandic area is both sensory and motor (Mott, 1894; Bechteres, 1899; see Fulton, 1943, and A. Meyer, 1978, for the history of the concept of sensorimotor cortex and of the efforts to delimit sensory regions of cortex). Cushing (1909) provided the first evidence that stimulation of the postcentral gyrus in humans can result in somatic sensation without movement. It was only after it became possible to record electrical cortical responses evoked by peripheral stimulation that the somatotopic sensory projections to the cortex could be mapped physiologically in cat, dog, and monkey (Adrian, 1941; C.N. Woolsey, 1943).

The area of cerebral cortex from which body movements could be evoked with shortest latency and lowest threshold was defined physiologically as the primary motor cortex. However, it was known that movements can be elicited from widespread cortical areas by using suprathreshold electrical stimuli (Fulton, 1935; Hines, 1947a, b). Mapping those cortical area led to the discovery of the supplementary motor cortical area, which was found first on the mesial surface of the frontal lobe of the human brain (Penfield and Welch, 1951) and later confirmed in experimental animals (C.N. Woolsey, 1951) and later confirmed in experimental animals (C.N. Woolsey, 1952, 1958; G. Goldberg, 1985, review). The areas defined physiologically were correlated with the anatomical localization of giant pyramidal cells and with the origins of the pyramidal and extrapyramidal pathways (Bechterew, 1899; Brodmann, 1905). The structure–function correlations were strengthened by observation of functional deficits and the extent of nerve fiber degeneration following cortical lesions (Fulton and Kennard, 1934; Fulton, 1935; Hines 1947b).

From those studies the motor cortex appeared to be organized as a mosaic in which each body part is represented in somatotopic order. Whether fundamental units of cortical organization are movements or individual muscles (e.g. H.-T. Chang et al., 1947) is an important question that has been reviewed by Kaas (1983) and D.R. Humphrey (1986), but is beyond our scope.

Let us now briefly summarize the evolution of modern concepts regarding the cellular organization of the cerebral cortex (see also M. Jacobson, 1993). The principal concepts regarding cellular organization evolved in parallel with construction of the neuron theory as noted above. There were five crucial conceptual advances made surprisingly rapidly in the final 60 years of the 19th century: recognition that the nervous system is composed of many types of nerve cells and fibers grouped in characteristic morphological patterns; understanding that nerve fibers are outgrowths of nerve cells; making the distinction between axons and dendrites in terms of differences in structure and in the direction of transmission of nervous activity; understanding that nerve cells are linked by contact at synaptic junctions; and conceiving of function in terms of integration of excitatory and inhibitory actions mediated by different synapses. Making allowances for the inevitable overlap between them, it may be useful to consider these concepts evolving in the order given above, and as parts of a research program, advancing to progressively higher levels of understanding.

Koelliker, in the first edition of his *Handbuch der Gewebelehre*, 1852, was already able to classify nerve cells according to shape (pyriform, fusiform, etc.) And according to the number of processes emerging from the cell body (apolar, unipolar, or bipolar). Koelliker's cellular typology was originally based on the appearance of unstained neurons dissociated from fixed brain. The first evidence confirming that similar differences between cell types occurs in a regular histological pattern in section of the cerebral cortex stained with carmine was reported by Berlin (1858). The concept of structural types was linked to that of functional differentiation, termed by A. Milne-Edwards (1857, Vol. 1) the *"physiological division of labour,"* one of the dominant concepts of biology in the latter half of the 19th century (see Herbert Spencer, 1866, p. 166; Oscar Hertwig, 1893–1898, Vol. 2, p. 79). In adopting that concept, Cajal (1900) also emphasized that the "principle of division of labour, which holds sway more in the brain than in any other organ, requires that the organs which register sensations are different from those which register memories."

In addition to the principle of functional differentiation, 19th-century studies of the cerebral cortex were guided by two other principles, namely, the principle of functional and structural homology of cortical areas in different mammals, and the principle of divergent differentiation of homologous parts in relation to their use and disuse in different mammals. These three principles are discussed at length by Brodman (1909, Chapter 7), and they continue to influence our current ideas about the development and evolution of the cerebral cortex. For example, evidence that cells with similar functional properties are clustered together anatomically in the cerebral cortex is consistent with the principles of functional differentiation and of functional and structural homology. Examples in the visual cortex are the ocular dominance and orientation columns in the primary visual cortex (Hubel and Wiesel, 1962, 1968) and color clusters in the primary visual area (Livingstone and Hubel, 1984; Tootell et al., 1988c) and second visual area (Hubel and livingstone, 1987). Horizontal and corticocortical connections also link clusters or groups of neurons with similar functional specificities (Gilbert and Wiesel, 1989).

The first schemata of cortical architectonics were guided by the principle of regional structural–functional differentiation and were based on differences in sizes and shapes of cell bodies and by their horizontal layering. Those features dominate the histological picture in sections of cerebral cortex stained with carmine, which was the best method of staining then available (Berlin, 1858; Meynert, 1872; Lewis, 1878; Lewis and Clarke, 1878). Despite the limitations of the histological techniques, the architectonics of the cerebral cortex was first worked out in remarkable detail by Theodor Meynert (1867–1868, 1872). Cajal (1911, p. 601) says that Meynert's "study was so exact that, notwithstanding the imperfection of his methods, it is still the best we possess." Meynert (*Bau der Grosshirnrinde*, 1867, p.58) subdivided the cortex into two main types: one with a white surface layer and the other with a gray surface layer. The latter he subdivided into five-layered cortex ("general type" and "claustrum formation") and eight-layered cortex (e.g., calcarine cortex). The white-surface cortex he also called "defective cortex" (including Ammon's horn, uncus, septum pellucidum, and olfactory cortex).

Another guiding principle was that certain cortical regions have been conserved during evolution in all mammals and can be recognized by their functions and structures, especially with respect to layering of certain types of neurons and their afferent and efferent connections. This principle of structural and functional homology generated a terminology in which the homologies are implied. Terminology often reflects the theoretical prejudice of the users. Edinger (1908a, b) coined the term paleoencephalon to mean the phylogenetically most ancient part of the central nervous system (CNS), and the only part in most fishes, as contrasted with what he termed the neoencephalon, of which the neocortex is the most recent culmination. The concept that the CNS of modern amniotes contains a core of ancient structures that are overlaid by layers of structures that evolved at later times was originated by L. Edinger (1908a, b) and Ariëns Kappers (1909). This concept has been extended by MacLean in his theory of the triune brain. According to MacLean (1970, 1972), the brain of higher primates is formed of three systems that originated in reptiles, early mammals, and late mammals. A related concept is that the cerebral cortex enlarges during evolution simply by addition of new areas (Smart and McSherry, 1986a). But evolution does not simply add new levels of organization on top or by the side of the old levels, so to say, like strata in an archeological site. No, the old adapts to the new and they all continue to evolve. Progression of the old and new occur together. The progression is not A_AB_ABC but A_A′B_A+″B′C and so on. Terms such as paleocortex, neocortex, and archicortex imply a phylogenetic progression which is not well based on evidence, and I use those terms only with certain qualifications. Those terms were coined by Ariëns Kappers (1909) on the basis of comparative studies of lower vertebrates, and their transferral to the mammals, and especially to primates, is a questionable practice. The terms rhinencephalon and pallium were adopted by Koelliker (1896) in his monumental attempt to attach ontogenetic and phylogenetic significance to the different regions of the cerebral cortex.[2] The term "rhinencephalon," for example, was associated with the concept of macrosmatic, and anosmatic brains, that is, with the importance of the sense of smell in the evolution of the species (Broca, 1878; Turner, 1891; Retzuis, 1898). The rhinencephalon was seen as either hypertrophied or atrophied in different species, depending on their use of the sense of smell. For example, the olfactory tubercle, prepyriform area, retrosplenial area, and amygdaloid nucleus were regarded as atrophied in the primates, in which the sense of smell is relatively weak.

Finally, the principle of divergent differentiation embraces the concepts of differentiation of several cortical areas from a protocortex, of progressive adaptation of cortical differentiation as a result of natural selection during evolution and as a result of use and experience of the individual, and also includes the concept of plasticity of the cortex after injury. As Brodmann, (1909, p. 243) clearly understood, all these can occur as a result of progressive or regressive transformations.

We should also remember that a theory prevalent at the end of the 19th century held that nerve cells are all multipotential or even equivalent at early stages of development, and that nerve cell differentiation is controlled by afferent stimulation. Koelliker (1896, Vol. 2, p. 810) summed up the evidence in no uncertain terms: "So I am finally forced to the conclusion that all nerve cells at first possess the same function, and that their differentiation depends solely and entirely on the various external influences or excitations which affect them, and originates from the various possibilities that are available for them to respond to those contingencies." This concept has attained current validity

with the evidence that cerebral cortical functions can be specified by afferent nerve fibers. The concept that localization of functions in the cerebral cortex is determined by the input from the periphery and is not autonomously determined within the CNS has endured for more than a century and continues to receive support (e.g., D.M. O'Leary, 1989, review). This theory was held by Golgi and Nissl among anatomists, Flourens and Goltz among physiologists, and S. Exner, Wundt, W. James, and Lashley among psychologists (reviewed by Neuburger, 1897; Soury, 1899; Lashley, 1929; Riese and Hoff, 1950; Walker, 1957; Tizard, 1959). For example, Wundt (1904, p. 150) says, "We know, of course, that the cell territories stand, by virtue of the cell processes, in the most manifold relation. We shall accordingly expect to find that the conduction paths are nowhere strictly isolated from one another. We must suppose, in particular, that under altered functional conditions they may change their relative positions within very wide limits." As Brodmann (1909) says, "All these theories are in fundamental agreement in their concept that the ganglion cells are equivalent forms, unencumbered by their origins, their positions, or their external forms."

The concept of an organology of the cerebral cortex, as expressed by Ecker in the epigraph to this chapter, for example, attained maturity with the cytoarchitectonic and myeloarchitectonic maps, which aimed at showing the structural and presumed functional parcellation of the cortex. This concept can be traced back to the phrenological theory of Gall and Spurzheim, whose *Anatomie et Physiologie due Système Nerveux* (1810–1819), especially in Volumes 1 (1810) and 2 (1812), tried to establish a relationship between the intellectual functions and the shape of the cranium and the underlying convolutions. The phrenological theory, while incorrect in the localization of so-called intellectual and moral functions, was based on much correct anatomical observation, especially that of Gall. Its main significance was to have given an impetus to studies of the relationship between structure and function of the cerebral cortex (see E. Clarke and O'Malley, 1968; R. M. Young, 1970). Out of such studies has come the principle that the magnification of cortical representation is proportional to the functional importance of the peripheral sensory or motor fields and that the primary gyri correspond fairly well, although not precisely, with cytoarchitectonic fields and with functional representation in the cortex (Connolly, 1950, pp. 264–269; Kaas, 1983).

The cortical cytoarchitectonic map of Campbell (1905) is the prototype based on the differences in layering of the cell bodies revealed in Nissl-stained sections. Campbell's structure–function correlations had the virtues of simplicity and reasonableness and were initially communicated to the Royal Society of London by Sherrington in 1903 before publication in book form in 1905. The introduction of the Weigert stain in 1882 resulted in an efflorescence of studies of the fiber tracts of the CNS (Bechterew, 1894; Edinger, 1896) and of the cerebral cortex (Vogt, 1904; Poliak, 1932). Difference in the time of development of myelin in the cerebral cortex was another criterion that was pressed into service to demarcate different regions of the cortex (Flechsig, 1896, 1901, 1927). This direction of research led to the publication of cerebral cortical maps of increasing

---

[2] The successive editions of *Handbuch der Gewebelehre* by Koelliker (six editions from 1852 to 1896) are invaluable for tracing progress during the second half of the 19th century. A very useful single source of information, in English translation, about the mid-19th century levels of understanding of development and structure of the nervous system is the *Manual of Histology* edited by S. Stricker (English edn, 3 vols, 1870–1873). It contains chapters on research techniques, the cell theory, and embryonic development by Stricker, spinal cord by J. Gerlach, the retina by M. Schultze, and on brains of mammals by T. Meynert. In his autobiography, Cajal refers to Stricker's treatise as "a model … invaluable for the devotee of the laboratory" and notes that he acquired a copy in 1883, before he started his investigations of the histology of the nervous system, and considerably earlier than his initial use of the Golgi technique in 1887–1888.

complexity, in which the relevance to function and development tends to be inversely proportional to the number of cortical regions demarcated (Campbell, 1905, shows 20; Brodmann, 1909, shows 52; Von Economo and Koskinas, 1925, delimit more than 100). We may ask whether the trend shows an increasing departure from reality or a progressive approach to the truth. Neither fits snugly into any theory of history of neuroscience, unless it is a theory of evolution of hypertrophic species which results in the ultimate extinction of the monstrosities. The last in the line of progressively more complicated cortical maps by Von Economo and Koskinas (1925) and the myeloarchitectonic studies of the Vogts (1902, 1904, 1919, review) are probably destined to remain forever enshrined in their gigantic volumes, to be opened by the curious bibliophile, but ignored by the working scientist looking for useful information.

Distrust of the validity of cortical maps was based on what the mappers left out as much as on their excessive zeal to split fields. The reaction to these errors of omission or errors of commission took an extreme, almost nihilistic form in the statement by Lashley and Clark (1946) that in some cases, "architectonic charts of the cortex represent little more than the whim of the individual student." Their skepticism was aroused by the evidence available at that time showing lack of correspondence between electrophysiologically defined cortical areas and anatomically defined architectonic areas. The correspondence has more recently been shown to be quite precise (Kaas, 1983, review). Moreover, quantitative computerized morphometric analysis of cortical cytoarchitectonics has eliminated subjective evaluation and has largely confirmed the validity of classical cytoarchitectonic and myeloarchitectonic maps (Fleischhauer et al., 1980; Zilles et al., 1980, 1982; Wree et al., 1983). The more criteria used to distinguish between different cortical areas, the more subdivisions tend to be revealed: in addition to classical cytoarchitectonics, modern cortical maps take into considera-tion cortical inputs and outputs, intracortical connections, his-tochemical and immunocytochemical markers, electrical activity and functional properties of the constitutent neurons, overall functions, and the effects of lesions (Brak, 1980; Wise and Goschalk, 1987).

Cajal (1922) was alert to the problem of errors of omission resulting from the use of highly selective stains: "There is little value, therefore, in dealing with the differentiation of corticl areas based exclusively on the revelations of the Nissl and Weigert methods, because they show an insignificant portion of the constituent features of the gray matter." In Cajal's hands, the Golgi technique, complemented by staining with methylene blue and reduced silver nitrate, led to the discovery of the dendritic spines and many new types of neurons. The breadth of his observations on the anatomical relationships between different types of neurons in the cerebral cortex and his views on the functional subdivisions of the cortex (for example, he argued that precentral and postcentral gyri are both sensory and motor) can now be appreciated more easily in the English translation by De Felipe and Jones (1988) of a large selection of Cajal's writings on the cerebral cortex.

Koelliker was the pivotal figure connecting the epochs before and after the general use of the Golgi technique. I agree with the assessment by Lorente de Nó (1938a) that "Kölliker's account of the structure of the human cortex (1896) is one of the masterpieces of neuroanatomy and it marks the end of an historical period of research on the cerebral cortex." Golgi occupies a unique position in the earlier epoch as a solitary worker who discovered his marvelous technique prematurely in 1873 and worked with it in virtual isolation unaided by critique, until the significance of his work was first recognized by Koelliker in 1887. Cajal occupies a different position as the person uniquely capable of using the Golgi technique to build on the foundations laid by those in the earlier epoch.

The wealth of information about the cellular organization and functions of the nervous system that had been amassed by the end of the 19th century was codified in the large textbooks published at that time (Koelliker, 1896; Bechterew, 1899; Soury, 1899; Ramón y Cajal, 1899–1904; Sherrington, 1906; Van Gehuchten, 1906). It was finally possible to consider the organization of the CNS in terms of functional assemblies of different types of neurons linked together by synaptic junctions and to analyze their input and output relationships anatomically and physiologically. This synthesis is set forth in Sherrington's lectures on "The Integrative Action of the Nervous System" (1906), in which he gives the evidence showing the importance of reflex inhibition and the concept of the final common path. For a history of the concept of nervous inhibition see Fearing (1930, pp. 187–217) and Fulton (1938, p. 77). We may ask how it was possible for Fearing (1930), in his masterly history of reflex action, to make only two passing references to the work of Cajal.

It is curious that Cajal appears to have failed to grasp the significance of inhibition: he does not refer to it, and he did not accept the challenge of trying to identify the structural substrates of inhibitory functions in the CNS. The question of whether entire neurons or only parts of them had inhibitory functions seems to have eluded him. This may have been because his grasp of anatomical organization was descriptive, albeit at a very high level of insight. His concept of functional organization was dominated by the notion of polarized flow of excitation, which could be altered by changes at the synapses, but he failed to see the significance of inhibition and the final common path as they were worked out by Sherrington (1897, 1904, 1906). By contrast, Sherrington was one of the first to fully understand the significance of Cajal's discoveries.

It has often been said that achievements in neuroscience in recent years are unprecedented. Never before has so much been discovered by so many at such great expense, but I am unable to affirm that we are witnessing a golden age such as that which took place a century ago. It should be remembered that almost all of Ramon y Cajal's important work was done in the four years from the beginning of 1888 to the end of 1891. Other giants of that era, Koelliker, Lenhossek, van Gehuchten and Retzius, were also phenomenally productive in the first decade after 1887, following their use of the Golgi technique. The flow of discoveries in neuroscience has never been stronger than in the 1890's, and it must have been exhilarating to live in a time when the tide came rushing in day after day bearing new gifts. In at least one other respect it resembles our own molecular biological era: priority for discovery is determined by speed of publication as much as by the moment of discovery. The smart investigator joins the rush to publish every new find, but few can match Cajal, who published 19 of his 28 papers in 1890 and 1891, at the rate of one almost every month, in the *Gaceta Médica Catalana* and other local journals, to establish priority. Later he expanded and republished many of them in journals with international distribution. Not all his contemporaries had the advantage of rapid publication.

By the closing decades of the 19th century less that 100 workers had laid the secure foundations on which neuroscience continues to be upheld. This can be confirmed by perusal of the great textbooks that appeared at the end of that golden age. To go back in search of the golden age is not to deny, but rather to affirm, the values of the present. That more than half my references are to works published since 1980 [i.e., in the decade prior to publication of the third edition in 1991—Ed.] shows that I have chosen to put my money on the future. A final comment—we cannot simply compare the past with the present. Each has to be dealt with on its own terms. We have new problems, and even when we take up their old problems we arrive at new solutions, different from theirs.

DEPENDENCE OF THE DEVELOPING NERVOUS SYSTEM ON NUTRITION AND HORMONES

He who admits the principle of sexual selection will be led to the remarkable conclusion that the nervous system not only regulates most of the existing functions of the body, but has indirectly influenced the progressive development of various bodily structures and of certain mental qualities. Courage, pugnacity … bright colours and ornamental appendages, have all been indirectly gained by the one sex or the other, through the exertion of choice, the influence of love and jealousy … and these powers of the mind manifestly depend on the development of the brain.

Charles Darwin (1809–1882), *The Descent of Man and Selection in Relation to Sex*, 2nd Ed., 1875

Vulnerability of the Human Brain to Malnutrition

The destiny of nations depends on the manner of their nutrition.

Jean Anthelme Brillat-Savarin (1755–1826), *The Physiology of Taste, or Meditations on Transcendental Gastronomy*, 1825

A large percentage of children in the "Third World," and a significant number in the rich industrial countries, are unable to obtain food necessary for normal development, and many pregnant women suffer from malnutrition. It is therefore a question of the highest importance whether fetal or childhood malnutrition retards or otherwise alters neurological development. If so, the types of changes, their casual mechanisms, and the permanence or degrees of reversibility of the lesions are of very great moral

and social concern. Ethical values are an important component of this research program. It should have attracted great scientific interest and generous public support, yet relatively little money and effort have been expended to answer important questions about causes, prevention, and treatment of physical and functional neurological damage resulting from fetal and neonatal malnutrition. This is evident from the small number of publications in this field in the past decade compared with the efflorescence of publications from other research programs of no greater scientific importance and of lesser human significance. The pregnant phrase of Brillat-Savarin at the beginning of this section should be written on the doorposts of every national legislative assembly.

The effects of maternal malnutrition on the human fetus are not well understood but can result in placental insufficiency causing premature birth, which carries a high risk of mental retardation. Children subjected to chronic malnutrition from birth to 18 months of age, severe enough to result in growth retardation, suffer permanent deficits in emotional, cognitive, and intellectual functions. Acute episodes of neonatal malnutrition also may result in permanent brain damage. The lesions caused by malnutrition in children are not well characterized but are most likely reduction in glial cell numbers and functions, retardation of growth of dendrites and synaptogenesis, and defective myelination. These effects may be totally reversed by nutritional therapy and enriched social conditions provided before age 2 but only partially reversed if therapy is delayed to later ages.

Retardation caused by malnutrition and socioeconomic deprivation interact to produce the syndrome of physical and mental retardation and dysfunction. All the organ systems may be affected. The effects on the nervous system are both direct, due to insufficient nutrients required for growth, as well as indirect, due to lack of trophic and growth factors and hormones required for normal development.

It is necessary to conduct epidemiological studies of the populations at risk in order to discover the causal factors and to design and implement programs to prevent and treat childhood malnutrition. The biological mechanisms can also be analyzed in animal models chosen because they are believed to be relevant to human conditions. However, animals in the wild rarely, if ever, suffer from malnutrition, which is a condition created by the human species on a global scale.

At least half the world's population has suffered a period of nutritional deprivation during childhood, and at present about 300 million children throughout the world are malnourished (World Health Organization, Scientific Publication No. 251, 1972). The majority of these children suffer from chronic lack of proteins and calories punctuated by episodes of acute malnutrition caused by illness and exacerbated by wars and other adverse political and economic malnutrition on the nervous system is difficult because of the other disadvantages of poor children (Pollitt and Thomson, 1977, review).

There is no doubt that economic poverty and malnutrition interact to stunt physical and mental development. For example, in a study of more than 7,000 children in the United States, Edwards and Grossman (1980) found intelligence quotient (IQ) and scholastic achievement test scores significantly correlated with height and weight, and a history of chronic nutritional deprivation was correlated with subnormal test scores. The fact that sociocultural factors play such an important role in intellectual and scholastic development makes it necessary to analyze the variables, including the contribution of malnutrition to the retardation of impoverished children (Cravioto and DeLicardi, 1972; Herzig *et al.*, 1972; Manocha, 1972; Greene and Johnston, 1980; Brozek, 1985). The methodological problems of such a research program are thoroughly discussed by Barrett and Frank (1987).

Prenatal nutritional deprivation of the human fetus may occur in multiple pregnancy as a result of competition between the fetuses for nutrients. Birth weight of twins is lower than that of singletons, and birth weight is significantly correlated with later intelligence (Churchill *et al.*, 1966). Birth weights below 2,000 g invariably affect behavioral and intellectual development, whereas birth weights between 2,500 and 4,500 g have variable adverse effects on development of higher nervous functions, depending on other factors such as adequate infant care, disease, nutrition, and sociocultural conditions (Drillien, 1958; Weiner, 1962; Scarr, 1969). Twins average about 7 IQ points below singletons (Stott, 1960; Vandenberg, 1966; Inouye, 1970). The importance of intrauterine competition for nutrients is shown by the fact that identical twins with the same birth weight have the same IQ, but if their birth weights are unequal, the twin with the lower birth weight has the lower IQ in later life (Willerman and Churchill, 1967). The number of variables that enter into human intelligence makes these results difficult to interpret. Thus, in a review of the effect results difficult to interpret. Thus, in a review of the effect of very low birth weight (1,500 g or less) on later intelligence, Francis-Williams and Davies (1974) point out that as many of the harmful factors in the treatment of such infants

have been eliminated (excessive use of oxygen, hypothermia, prolonged starvation, infections), there has been progressive improvement of the ultimate IQ attained by children with low birth weight. When babies with low birth weight are cared for under optimal conditions, mental retardation or significant deficits in IQ do not occur (P.A. Davies and Stewart, 1975).

In humans, it is not known whether twins have a lower brain weight or fewer brain cells than singletons, as would be predicted if animal experimental results, to be described later, can be extrapolated to humans. The deficit is unlikely to be due to failure of nerve cell production because production of nerve cells ceases by 25 weeks of gestation (Dobbing and Sands, 1970), whereas differences in the weight of multiple fetuses compared with a single fetus become apparent only after 26 weeks of gestation (McKeown and Record, 1952). The deficit is more likely to be due to some failure of glial cell production plus retarded neuronal cell growth and differentiation rather than to a reduction in neuronal cell numbers. Nevertheless, there is good evidence that loss of brain cells can occur after severe malnutrition in human infants. Children who die of severe malnutrition in the first 2 years after birth have greatly reduced quantities of DNA in the cerebrum, cerebellum, and brainstem (Fig. 6) compared with well-nourished children of the same age (Winick and Rosso, 1969; Winick, 1970). It is not known whether the deficit is in the number of neurons or glial cells, but it is more likely to be a glial cell deficit which is known to occur in malnourished rats (Robain and Ponsot, 1978; Bhide and Bedi, 1984).

Assessment of the effects of malnutrition on nervous development in children is complicated by at least three main difficulties. First, the effects of malnutrition cannot be entirely separated from the effects of other harmful conditions, such as maternal neglect, environmental impoverishment, and lack of stimulation and incentive. Second, malnourished children show behavioral abnormalities which are variable and are difficult to measure accurately, such as reduced social responsiveness, increased irritability, and emotional disturbances. Third, the effects of malnutrition on the human brain can rarely be assessed directly by postmortem physical and chemical measurements. Instead, less reliable indices, such as physical status, ratio of weight to height, head circumference, and IQ, are generally used. These indices are themselves complex variable, and their interpretation is usually difficult and frequently ends in controversy. For example, general intelligence is a complex product of many variable factors, one of them being the size of the brain. As a result, authorities disagree about the relationship between brain size and intelligence—the relationship can be shown in animals (see Section 2.13) but in humans it tends to be overridden by other factors that vary in different sociocultural contexts. In those cases where IQ is found to be reduced in malnourished children, it is often difficult, if not impossible, to determine whether the reduced IQ is the result of retarded brain development due to malnutrition and associated conditions such as disease, or whether the poor performance is largely or entirely due to social and economic disadvantages. Moreover, a single IQ test has little value, particularly is if is done at an early age: IQ at 1 year of age has no correlation with the IQ at age 17 (B.S. Bloom, 1964).

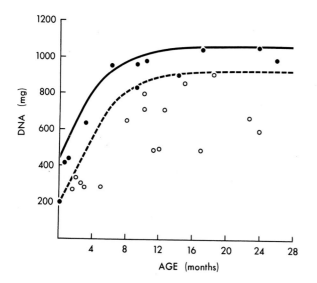

FIGURE 6. Severe childhood malnutrition can result in a reduced DNA content of the brain. The DNA content of the cerebrum of normal children (dark line, black dots) and severely malnorished children (dashed line, circles). From M. Winick, P. Rosso, and J. Waterlow, *Exp. Neurol.* 26:393–400 (1970), copyright Academic Press, Inc.

Beside, the IQ is subject to change during childhood: changing the educational level of children can result in increase or reduction of their IQ by as much as 28 points (Skeels, 1966). Obviously, the rate of development of general intelligence is a more revealing index than a single measurement.

Occipitofrontal circumference of the head is another index that is often used to assess the effects of malnutrition on brain development. However, the circumference of the head also related to the thickness of the skull and scalp, so that it has been claimed that there is a poor correlation between head size and brain size (Eichorn and Bayley, 1962). Intelligence is normal in a small percentage of microcephalic children (H.P. Martin, 1970), and reduced head circumference in infants and young children may not be irremediable (H.P. Martin, 1970; Stoch and Smythe, 1976). Nevertheless, there is a linear relationship between occipitofrontal head circumference over a range of 100–700 g from 25 weeks of gestation to 8 months postnatal (Lemons et al., 1981). Head circumference is significantly correlated with brain size in cases of intrauterine growth retardation (Battist et al., 1986) and in cases of nutritional deprivation during the first 2 years of life, when most of the growth of the brain occurs (Johnston and Lampl, 1984).

In view of criticisms of the validity of IQ tests, especially that they are racially and culturally biased, it is significant that deficits in IQ of malnourished children have been reported from countries which differ in race, culture, economic, and social conditions: India (Champakam et al., 1968), Indonesia (Liang et al., 1967), Lebanon (Botha-Antoun et al., 1968), Latin America (Pollit and Granoff, 1967; Birch, 1972; Cravioto and DeLicardi, 1975; Klein et al., 1977; Barrett and Frank, 1987), Yugoslavia (Cabak and Najdanvic, 1965; Cabak et al., 1967) and Africa

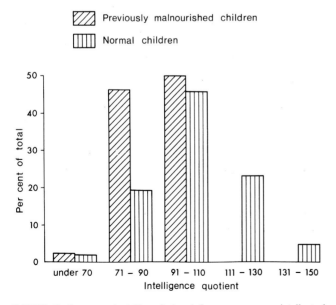

FIGURE 7. Severe malnutrition during infancy may cause intellectual impairment. Distribution of IQs of 36 children aged 7–14 years who had been severely malnourished between the ages of 4 and 24 months, compared with the IQs of normal children of the same ages. Adapted from V. Cabak and R. Najdanvic, *Arch. Dis. Child.* 40:532–534 (1965).

(Stoch and Smythe, 1963; Fisher et al., 1978). In one such study, Stoch and Smythe (1963, 1967) found that severely undernourished South African children from 1 to 8 years of age were 20 points lower in IQ than well-nourished children of similar parentage. When these children were studied 15 and 20 years later, the same severe intellectual deficits were found, and physical abnormalities of the brain were found by computerized tomography scans in some cases, showing that the changes are permanent (Stoch and Smythe, 1976; Handler et al., 1981; Stoch et al., 1982). Similar conclusions were reached in another study of the effects of severe undernutrition during infancy on subsequent intellectual functions in Yugoslavian children (Cabak and Najdanvic, 1965; Cabak et al., 1967). There were marked deficits in IQ in 36 children who had been hospitalized between the ages of 4 and 24 months because of malnutrition but had not suffered from chronic illness thereafter, and who had no significant reduction in body growth. The mean IQ of the malnourished children was 88, in contrast to 101 and 109 for two groups of comparable normal children. Significantly, none of the malnourished children had an IQ above 110 (Fig. 7). No correlation was found between the age of hospitalization and subsequent intellectual impairment. Hospitalization during the first year of life is frequently associated with sensory deprivation and separation of the child from the mother, and these contribute to subsequent emotional disturbances and intellectual impairment (Spitz and Wolf, 1946; Coleman, 1957).

Malnourished children have learning and scholastic difficulties. This is a complex syndrome in which apathy, emotional disturbances, and impaired cognitive and intellectual abilities are synergistic in producing dysfunction (Lester, 1975). Children who are severely malnourished in the first year of life subsequently show behavior disturbances and difficulties with reading, spelling, and arithmetic (Richardson et al., 1973; Barrett and Frank, 1987). Reduced ability to integrate inputs from various sensory modalities, for example, auditory–visual integration, has been found in children suffering from malnutrition before 3 years of age (Birch, 1964; Kahn and Birch, 1968; Cravioto and DeLicardie, 1975). Lester et al., (1975). studied effects of acute malnutrition in 1-year-old children in Guatemala and found much slower habituation to repeated stimulation. Comparable changes in habituation have been reported in malnourished rats (Bronstein et al., 1974; Franková and Zemanová, 1978).

Language development is a significant indication of general intellectual status in children (Cameron et al., 1967). Findings which are from different countries, and therefore unlikely to be culturally or racially biased, show retarded language development in malnourished children (Barrera-Moncada, 1963; Monckeberg, 1968; Champakam et al., 1968; Chase and Martin, 1970). Retardation of language development has a complex etiology, which includes social deprivation, illness, and maternal deficits as well as malnutrition.

Several studies have provided evidence supporting the concept of a critical period from birth to about 2 years of age during which the nervous system is most vulnerable to malnutrition therapy. The amount of reversibility of impairment following childhood malnutrition is of great practical importance in planning food supplementation programs and other forms of prevention and treatment. The time and duration of malnutrition, as well as the time at which nutritional therapy and environmental enrichment are started, may be critical.

One study shows that there can be negligible long-term effects of acute starvation—no physical or mental deficits were found in victims of the severe famine in Holland from 1944 to 1945 when they were examined 19 years later (Stein *et al.*, 1975). The famine was very severe but of relatively short duration and the amount of intrauterine or neonatal growth retardation is not known. However, it seems that any effects were later reversed by nutrition, education, and good sociocultural conditions. This should be compared with the significant intellectual impairment reported in German school children who were severely undernourished during and after World War I (Blanton, 1919).

Several studies have shown that severely malnourished children can benefit from early nutritional therapy and enhanced sociocultural conditions (Yatkin and McLaren, 1970; McKay *et al.*, 1972; Chavez *et al.*, 1974; Irwin *et al.*, 1979; Monckeberg, 1979; Mora *et al.*, 1979). Winick *et al.* (1975) found that IQ and scholastic performance between age 6 and 12 were normal in malnourished Korean children adopted into American families before the age of 2 but were significantly below normal in children adopted after 2 years of age.

Cravioto and Robles (1965) reported that severely malnourished Mexican children admitted to hospital before 6 months of age showed the slowest rate of recovery and greatest loss of intellectual ability compared with malnourished children who entered hospital between age 37 and 42 months, who recovered more rapidly and had least intellectual deficit. McKay *et al.* (1978) found that food supplements given to chronically malnourished Colombian children improved their cognitive, language, and social abilities. The greatest gains were made when treatment was started at age 3, and the gains were progressively less when treatment started at ages 4 and 5. In a study of chronically malnourished pregnant women and children in Louisiana (Hicks *et al.*, 1982), a positive relationship was found between the age at which food supplementation was started and subsequent gains of IQ. Mothers received extra food during pregnancy, and their children who continued to receive it from birth to 4 years had higher IQ scores at 6 years of age than their siblings who started receiving extra food after 1 year for 3 years.

Other studies have failed to find any evidence of a critical period although they clearly show that malnutrition and poor socioeconomic conditions result in physical and intellectual impairment. Chase and Martin (1970) reported that children in the United States who were admitted to hospital in a severely malnourished condition before 4 months of age were subsequently less impaired than children who entered hospital after 4 months of age, although both groups later showed significant retardation. This could be interpreted to mean that starting therapy early during the critical period is more effective than starting treatment later. Jamaican children who were hospitalized with severe malnutrition at ages 3–24 months were found to have no correlation between the age of admission to hospital and retardation of IQ at 5–11 years of age (Hertzig *et al.*, 1972). Presumably

severity of malnutrition and age of onset are only two variables among many, including social and educational deprivation, family instability, and maternal illiteracy, all of which combine to retard child development.

The biological mechanisms of the dysfunction and retardation caused by malnutrition in humans are not well understood, but they are likely to be complex. The nervous system may be affected both directly and indirectly. Nutrients are required for growth as well as to provide materials for synthesis of growth factors, trophic factors, and hormones required for normal development of the nervous system and the other supporting systems. Severe malnutrition directly affects the heart muscle, resulting in cardiac insufficiency and reduced blood supply to the brain. The reduced plasma proteins cause edema in all the tissues including the brain. Anemia, which is often associated with malnutrition, may diminish the oxygen supply to the brain. Neonatal malnutrition results in retarded development of the immune system and consequently in greater risk of infectious disease. Those who survive the insults of malnutrition during infancy and childhood frequently remain with physical and mental disabilities as well as social and economic disadvantages.

REFERENCES

Adelmann, H.B. 1925, The development of the neural folds and cranial ganglia of the rat, *J. Comp. Neurol.* 39:19–171.

Adelmann, H.B. 1966, Marcello malpighi and the evolution of embryology, 5 Vols. Cornell University Press, Ithaca.

Adrian, E.D. 1933, The all-or-nothing reaction. *Ergeb. Physiool.* 35:744–755.

Adrian, E.D. 1941, Afferent dischrges to the cerebral cortex from peripheral sense organs. *J. Physiol. (London).* 100:159–191.

Adrian, E.K., Jr., and Schelper, R.L. 1981, Microglia, monocytes, and macrophages. *Prog. Clin. Biol. Res.* 59A:113–124.

Aguilar, C.E., Bisby, M.A., Cooper, E., and Diamond, J. 1973, Evidence that axoplasmic transport of trophic factors is involved in the regulation of peripheral nerve fields in salamanders. *J. Physiol.* 234:449–464.

Alzheimer, A. 1910, Beitrage zur Kenntniss der pathologischen Neuroglia und ihrer Beziehungen zu den Abbauvorgangen im Nervengewebe. *Hist. u. Histopath. Arb. Nissl-Alzheimer* 3:401.

Andersen, P., Eccles, J.C., and Loyning, Y. 1963, Recurrent inhibition in the hippocampus with identification of the inhibitory cell and its synapses. *Nature* 198:540–542.

Andrew, W. and Ashworth, C.T. 1945, The adendroglia. *J. Comp. Neurol.* 82:101–127.

Andriezen, W.L. 1893a, The neuroglia elements of the brain. *Br. Med. J.* 2:227–230.

Andriezen, W.L. 1893b, On a system of fibre-cells surrounding the blood-vessels of the brain of man and its physiological significance. *Inter Monasschr. Anat. Physiol.* 10:533.

Anokhin, P.K. 1968, The biology and Neurophysiology of the Conditional Reflex (in Russian). Meditsina, Moscow.

Apathy, S. 1897, Das leitende Element des Nervensystems und seine topographischen Beziehungen zu den Zellen. *Mittheil. aus. der. zool. Station. zur. Neapal.* 12:495–748.

Ariens Kappers, C.U. 1909, The phylogenesis of the paleocortex and archicortex compared with the evolution of the visual neocortex. *Arch. Neurol. Psych.* 4:161–173.

Aschoff, P. 1924, Reticulo-endothelial system. *In* Lectures on Pathology. Hoeber, New York.

Auerbach, L. 1898, Nervenendigung in den Centralorganen. *Neurol. Zbl.* 17:445–454.

Bailey, C.H., and Chen, M. 1988, Long-term memory in Aplysia modulates the total number of varicosities of single identified sensory neurons. *Proc. Natl. Acad. Sci. U S A.* 85:2373–2377.

Balfour, F.M. 1876, The development of elsmobrance fishes. Development of the trunk. *J. Anat. Physiol. (London).* 11:128–172.

Balfour, F.M. 1878, A monograph on the development of elasmobranch fishes. Macmillan, London.

Balfour, F.M. 1881, A Treatise on Comparative Embryology. 2 vols. Macmillan, London.

Barker, T.F. 1901, The Nervous System, New York, Appleton.

Baroffio, A., Dupin, E., and Le Douarin, N.M. 1988, Clone-forming ability and differentiation potential of migratory neural crest cells. *Proc. Natl. Acad. Sci. U S A.* 85:5325–5329.

Barr, M.L. 1940, Axon reaction in motoneurones and its effect upon the endbulbs of Held-Auerbach. *Anat Rec.* 77.

Barrera-Moncada, G. 1963, Estudios Sobre Alteraciones del Crecimiento y Desarrollo Psicologico del Sindrome Pluricarencial. Editoria Grafos, Caracas, Venezuela.

Barrett, D.E., and Frank, D.A. 1987, The effects of undernutrition on children's behavior. Gordon and Breach, New York.

Barron, D.H. 1943, The early development of the motor cells and columns in the spinal cord of the sheep. *J. Comp. Neurol.* 78:1–26.

Barron, D.H. 1946, Observations on the early differentiation of the motor neuroblasts in the spinal cord of the chick. *J. Comp. Neurol.* 85:149–169.

Bartelmez, G.W., and Hoerr, N.L. 1933, The vestibular club endings in Ameiurus. Further evidence on the morphology of the synapse. *J. Comp. Neurol.* 57:401–428.

Battisti, O., Bach, A., and Gerard, P. 1986, Brain growth in sick newborn infants: a clinical and real-time ultrasound analysis. *Early Hum. Dev.* 13:13–20.

Bauer, K.F. 1953, Organisation des Nervengewebe und Neurencytiumtheorie. Urban & Schwarzenberg, Berlin.

Beale, L.S. 1860, On the distribution of nerves to the elementary fibres of striped muscle. *Phil. Trans. Roy. Soc. (London) Ser B.* 1862:611–618.

Beale, L.S. 1862, Further observations on the distribution of nerves to the elementary fibres of striped muscles. *Phil. Trans. Roy. Soc. (London) Ser B.* 1862:889–910.

Beard, J. 1889, The early development of Lepidosteus osseus. *Proc. Roy. Soc. (London) Ser B.* 46:108–118.

Beard, J. 1892, The transient ganglion cells and their nerves in Raja batis. *Anat Anz.* 7:191–206.

Beard, J. 1896, The history of a transient nervous apparatus in certain Ichthyopsida. An account of the development and degeneration of ganglion-cells and nerve fibres. *Zool. Jahrbucher. Abt. Morphol.* 9:1–106.

Bechterew, W. v. 1894, Die Leitungsbachnen im Gehirn und Ruckenmark. Translated from the Russian by R. Weinberg. Verlag von Arthur Georgi, Leipzig.

Bechterew, W. v. 1899, Die Leitungsbahnene. *In* Gehirn und ruckenmark, 2 Aufl. Verlag von Arthur Georgi, Leipzig.

Benjelloun-Touimi, S., Jacque, C.M., Derer, P., De Vitry, F., Maunoury, R., and Dupouey, P. 1985, Evidence that mouse astrocytes may be derived from the radial glia. An immunohistochemical study of the cerebellum in the normal and reeler mouse. *J. Neuroimmunol.* 9:87–97.

Bennett, M.R. 1983, Development of neuromuscular synapses. *Physiol. Rev.* 63:915–1048.

Beritoff, J.S. 1969, Structure and functions of the cerebral cortex (in Russian). Nauka, Moscow.

Berlin, R. 1858, Beitrage zur strukturlehre der grosshirnwindungen. Junge, Erlangen.

Bernard, C. 1878, Le curare. *In* La Science experimentale. Bailliere, Paris.

Bernstein, J.J., and Guth, L. 1961, Nonselectivity in establishment of neuromuscular connections following nerve regeneration in the rat. *Exp. Neurol.* 4:262–75.

Bethe, A. 1900, Ueber die Neurofibrillen in den Ganglienzellen von Wirbelthieren und ihre Beziehungen zu den Golginetzen. *Archiv. Mikroskop. Anat.* 55:513–558.

Bethe, A. 1903, Allgemeine Anatomie und Physiologie des Nervensystems. Verlag von Georg Thieme, Leipzig.

Bhide, P.G., and Bedi, K.S. 1984a, The effects of a lengthy period of environmental diversity on well-fed and previously undernourished rats. I. Neurons and glial cells. *J. Comp. Neurol.* 227:296–304.

Bhide, P.G., and Bedi, K.S. 1984b, The effects of a lengthy period of environmental diversity on well-fed and previously undernourished rats. II. Synapse-to-neuron ratios. *J. Comp. Neurol.* 227:305–10.

Bichat, X. 1801, Anatomie generale appliquee a la physiologie et a la medecine. Paris.

Bidder, F., and Kupffer, C. 1857, Untersuchungen uber die Texture des Ruckenmarks und die Entwickelung seine Formelemente. Leipzig.

Birch, H.G., and Belmont, L. 1964, Auditory-Visual Integration in Normal and Retarded Readers. *Am. J. Orthopsychiatry* 34:852–61.

Bishop, G.H., Heinbecker, P., and O'Leary, J.L. 1933, The function of the non-myelinated fibers of the dorsal roots. *Am J. Physiol.* 106: 647–669.

Blakemore, W.F. 1975, The ultrastructure of normal and reactive microglia. *Acta. Neuropathol. Suppl. (Berl).* Suppl 6:273–8.

Blanton, S. 1919, Mental and nervous changes in the children of the Volksschulen of Trier, Germany, caused by malnutrition. *Ment. Hyg. N.Y.* 3:343–386.

Bloom, B.S. 1964, Stability and change in human characteristics. Wiley, New York.

Blumenbach, J.F. 1782, Handbuch der Naturgeschichte. J. Chr. DIeterich, Gottingen.

Bodian, D. 1937, The structure of the vertebrate synapse. A study of the axon-endings on Mauthner's cell and neighboring centers in the goldfish. *J. Comp. Neurol.* 68:117–159.

Bodian, D. 1947, Nucleic acid in nerve-cell regeneration. *Symp Soc Exp Biol* 1:163–178.

Boeke, J. 1932, Nerve endings, motor and sensory. *In* Cytology and Cellular Pathology of the Nervous System (W. Penfield, Ed.), Vol. 1: pp. 243–315. Hoeber, New York.

Boeke, J. 1942, Sur les synapses a distance. Les glomerules cerebelleux, leur sturcture et leur developpement. *Schweiz. Arch. Neurol. Psychiat.* 49.

Boll, j. 1874, Die histologie und hostogenese der nervosen Centralorgane. *Arch. f. Psychiatr. u. Nervenkr.* 4.

Bonnet, R. 1878, Studien ueber die innervation der haarbalge der hausthiere. *Morphol. Jahrb Leipzig.* 4:329–398.

Boring, E.G. 1950, A history of experimental psychology. Appleton-Century-Crofts, New York.

Born, j. 1896, Ueber Verwachsungversuche mit Amphibienlarven. *Arch. Entw. Mech. Organ.* 4:349–465, 517–623.

Boya, J. 1976, An ultrastructural study of the relationship between pericytes and cerebral macrophages. *Acta. Anat. (Basel).* 95:598–608.

Boya, J., Calvo, J., Carbonell, A.L., and Garcia-Maurino, E. 1986, Nature of macrophages in rat brain. A histochemical study. *Acta. Anat. (Basel).* 127:142–145.

Boya, J., Calvo, J., and Prado, A. 1979, The origin of microglial cells. *J. Anat.* 129:177–86.

Boyd, J.D. 1960, Development of striated muscle. *In* Structure and functions of muscle (G.H. Bourne, Ed.), Vol. 1, pp. 63–85. Academic Press, New York.

Braak, H. 1980, Architectonics of the Human Telencephalic Cortex. Springer-Verlag, Berlin.

Brauer, A. 1904, Beitrage zur Kenntnis der Entwicklung und Anatomie der Gymnophionen, IV. *Zool. Jahrb. Suppl.* 7:409–428.

Braus, H. 1905, Experimentelle Beitrage zur frage nach der entwicklung peripherer nerven. *Anat. Anz.* 26:433–479.

Broca, P.P. 1878, Anatomie comparee des circonvolutions cerebrales. Le grand lobe limbique et la scissure limbique dans l serie mammiferes. *Rev. Anthrop.* 1:385–498.

Brock, L.G., Coombs, J.S., and Eccles, J.C. 1952, The recording of potentials from motoneurones with an intracellular electrode. *J. Physiol.* 117:431–460.

Brodmann, K. 1905, Beitrage zur histologischen lokalisation der grosshirnrinde 4, Mitteilung: der riesenpyramidentypus und sein verhalten zu den furchen bei den karnivoren. *J. fur. Psychologie. und. Neurologie.* 6:108–120.

Brodmann, K. 1908, Vergeichende Lokalisationslehre der Grosshirnrinde. Barth, Leipzig.

Bronstein, P.M., Neiman, H., Wolkoff, F.D., and Levine, M.J. 1974, The development of habituation in the rat. *nim. Learning. Behav.* 2:92–96.

Brzin, M., Dettbarn W.-D., Rosenberg P, Nachmansohn D 1940, Cholinesterase activity per unit surface area of conducting membrane. *Science* 92:353–364.

Bueker, E.D. 1948, Implantation of tumors in the hindlimb field of the embryonic chick nd developmental response of the lumbosacral nervous system. *Anat. Rec.* 102:369–390.

Bunge, M. 1959, Causality. Harvard Univ Press, Cambridge, Mass.

Burrows, M.T. 1911, The growth of tissues of the chick embryos outside the animal body, with special reference to the nervous system. *J. Exper. Zool.* 10:63–84.

Cabak, V. 1967, The long-term prognosis of infantile malnutrition. *In* Proceedings of the first congress of the international association for scientific study of mentl deficiency (B.W. Richards, Ed.), pp. 521–532. Michel Jackson, Montpellier, France.

Cabak, V., and Najdanvic, R. 1965, Effect of undernutrition in early life on physical and mental development. *Arch. Dis. Child.* 40:532–534.

Cameron, J., Livson, N., and Bayley, N. 1967, Infant vocalizations and their relationship to mature intelligence. *Science* 157:331–333.

Campbell, A.W. 1905, Histological Studies on the Localisation of Cerebral Function. Cambridge Univ. Press, Cambridge.

Carr, V.M. 1975, Peripheral effects on early development of chick spinal ganglia. *Neurosci. Abstr.* 1:749.

Carr, V.M. 1976, Ph.D. thesis. Northwestern Univ., Evaston, Ill,

Celestino da Costa, A. 1931, Sur la constitution et le developpement des ebauches ganglionnaires craniens chez les mamiferes. *Arch. biol. (Liege)* 42:71–105.

Champakam, S., Srikantia, S.G., and Gopalan, C. 1968, Kwashiorkor and mental development. *Am. J. Clin. Nutr.* 21:844–52.

Chang, H.T., Ruch, T.C., and Ward, A.A. 1947, Topographic respresentation of muscles in the motor cortex of monkeys. *J. Neurophysiol.* 10:39–56.

Changeux, J.P., and Danchin, A. 1976, Selective stabilisation of developing synapses as a mechanism for the specification of neuronal networks. *Nature* 264:705–12.

Changeux, J.P., Devillers-Thiery, A., and Chemouilli, P. 1984, Acetylcholine receptor: an allosteric protein. *Science* 225:1335–45.

Chase, H.P., and Martin, H.P. 1970, Undernutrition and child development. *N. Engl. J. Med.* 282:933–9.

Chavez, A., Martinez, C., and Yaschine, T. 1974, The importance of nutirtion and stimuli on child mental and social development. *In* Early nutrition and mental development (Cravioto, Hambraeus, and Vahlquist, Eds.), pp. 255–269.

Chiarugi, G. 1894, Contribuzioni allo studio dello sviluppo dei nervi encefalici nei mammiferi. Firenze.

Choi, B.H. 1981, Radial glia of developing human fetal spinal cord: Golgi, immunohistochemical and electron microscopic study. *Brain. Res.* 227:249–67.

Churchill, J.A., Neff, J.W., and Caldwell, D.F. 1966, Birth weight and intelligence. *Obstet. Gynecol.* 28:425–429.

Clark, D.A. 1931, Muscle counts of motor units: A study in innervation ratios. *Am. J. Physiol.* 96:296–304.

Clarke, E., and O'Malley, C.D. 1968, The human Brain and Spinal Cord. A Historical Study Illustrated by Writings from Antiquity to the Twentieth Century. Univ. Caif. Press, Berkeley and LosAngeles.

Clarke, P.G. 1985, Neuronal death during development in the isthmo-optic nucleus of the chick: sustaining role of afferents from the tectum. *J. Comp. Neurol.* 234:365–379.

Coleman, R.W., and Provence, S. 1957, Environmental retardation (hospitalism) in infants living in families. *Pediatrics* 19:285–292.

Collin, R. 1906, Recherches cytologique sur le developpement de la cellule nerveuse. *Nevraxe.* 8:181–308.

Collins, R. 1906, Recherches cytologiques sur le developpement de la cellule nerveuse. *Nevraxe.* 8:181–308.

Connold, A.L., Evers, J.V., and Vrbova, G. 1986, Effect of low calcium and protease inhibitors on synapse elimination during postnatal development in the rat soleus muscle. *Brain. Res.* 393:99–107.

Connolly, C.J. 1950, External Morphology of the Primate Brain. Thomas, Springfield, Ill,

Constantine-Paton, M., and Law, M.I. 1978, Eye-specific termination bands in tecta of three-eyed frogs. *Science* 202:639–641.

Coss, R.G., and Perkel, D.H. 1985, The function of dendritic spines: a review of theoretical issues. *Behav. Neural. Biol.* 44:151–185.

Cowan, W.M., Fawcett, J.W., O'Leary, D.D., and Stanfield, B.B. 1984, Regressive events in neurogenesis. *Science* 225:1258–1265.

Cravioto, J., and DeLicardie, E.R. 1972, Environmental correlates of severe clinical malnutrition, and language development in survivors from Kwashiorkor and marasmus. Pan American Health Organization, Scientific Publ. No. 251, Washington, D.C.

Cushing, H.P. 1909, A note upon the faradic stimulation of the postcentral gyrus in conscious patients. *Brain* 32:44–53.

Dale, H.H. 1935, Pharmacology and nerve-endings. *Proc. Roy. Soc. Med* 28:319–332.

Darwin, C. 1888, The life and letters of Charles Darwin including an autobiographical chapter. John Murry, London.

Davies, P.A., and Stewart, A.L. 1975, Low-birth-weight infants: neurological sequelae and later intelligence. *Br. Med. Bull.* 31:85–91.

DeFelipe, J., and Jones, E.G. 1988, Cajal on the Cerebral Cortex. Oxford Univ Press, New York.

Del Castillo, J. and Katz, B. 1965, Biophysical aspects of neuro-muscular transmission. *Progr. Biophys.* 6:121-.

del Cerro, M., and Monjan, A.A. 1979, Unequivocal demonstration of the hematogenous origin of brain macrophages in a stab wound by a double-label technique. *Neuroscience* 4:1399–404.

Denny-Brown, D. 1929a, The histological features of striped muscle in relation to its functional activity. *Proc. Roy. Soc. (London) Ser B* 104:371–411.

Denny-Brown, D. 1929b, On the nature of postural reflexes. *Proc. Roy. Soc. (London) Ser B* 104:252–301.

Detwiler, S.R. 1936, Neuroembryology: An experimental study. Macmillan, New York.

Detwiler, S.R. 1937a, Does the developing medulla influence cellular proliferation within the spinal cord? *J. Exp. Zool.* 77:109–122.

Detwiler, S.R. 1937b, An experimental study of spinal nerve segmentation in amblystoma with reference to the plurisegmental contribution to the brachial plexus. *J. Exp. Zool.* 77:395–441.

Detwiler, S.R. 1937c, Observations upon the migration of neural crest cells., and upon the development of the spinal ganglia and vertebral arches in Amblystoma. *Am J. Anat.* 61:63–94.

Dobbing, J., and Sands, J. 1970a, Growth and development of the brain and spinal cord of the guinea pig. *Brain. Res.* 17:115–23.

Dobbing, J., and Sands, J. 1970b, Timing of neuroblast multiplication in developing human brain. *Nature* 226:639–40.

Dohrn, A. 1902, Weiterer Beitrage zur Beurteilung der Occipitalregion und der Banglienleiste der Selachier. *Mittheil. aus. der. zool. Station zur Neapal* 15:555–654.

Dow, R.S., and Moruzzi, G. 1958, The physiology and pathology of the cerebellum. Univ Minnesota Press, Minneapolis.

Driesch, H. 1896, Die tactische Reizbarkeit der Mesenchymzellen von Echinus microtuberculatus. *Arch. Entw-Mech* 3.

Driesch, H. 1908, The Science and Philosophy of the Organism. Adam and Charles Black, London.

Drillien, C.M. 1958, Growth and development in a group of children of very low birth weight. *Arch. Dis. Child.* 33:10–8.

Dustin, A.P. 1910, Le role des tropismes et de l'odogenes dans la regeneration du systeme nerveus. *Arch. biol. Liege.* 25:269–388.

Duval, M. 1895, Hypothese sur la physiologie du centre nerveux: theorie histologique du sommeil. *Compts Rendu. Soc. Biol. Paris* 10:74–77.

Duval, M. 1897, Precis d'Histologie. Masson, Paris.

Eccles, J.C. 1959, The development of ideas on the synapse. *In* The historical development of physioligcal thought (C.M. Brooks and P.F. Cranefield, Eds.). Hafner, New York.

Eccles, J.C. 1964, The physiology of synapses. Springer-Verlag, Berlin.

Eccles, J.C., and Sherrington, C.S. 1930, Numbers and contraction-values of individual motor units examined in some muscles of the limb. *Proc. Roy. Soc. (London) Ser. B* 106:326–357.

Edelman, G.M. 1987, Neural Darwinism. Basic Books, New York.

Edelman, G.M. 1988a, Morphoregulatory molecules. *Biochemistry* 27:3533–3543.

Edelman, G.M. 1988b, Topobiology. Basic Books, New York.

Edinger, L. 1896, Untersuchungen uber die vergleichende Anatomie des Gehirn. III. Neue Studien uber das Vordergehirn der Reptilien. *Abh. Senckenb. Naturforsch. Ges.* 19:313–388.

Edinger, L. 1908a, The relations of comparative anatomy to comparative psychology. *J. Comp. Neurol. Psychol.* 18:437–457.

Edinger, L. 1908b, Vorlesungen uber den Bau der Nervosen Zentralorgane des Menschen und der Tiere. II. Vergieichende Anatomie des Gehirns. Vogel, F.C.W., Leipzig.

Edwards, L.N., and Grossman, M. 1981, Children's health and the family. *Adv. Health Econ. Health Serv. Res.* 2:35–84.

Ehrlich, P. 1885, Sauerstoffbedurfnis des oorganismus. Springer, Berlin.

Eichorn, D.H., and Bayley, N. 1962, Growth in head circumference from birth through young adulthood. *Child. Dev.* 33:257–71.

Ernst, M. 1926, Uber Untergang von Zellen wahrend der normalen Entwicklung bei Wirbeltieren. *Z. Anat. Entwicklungsgesch.* 79: 228–262.

Fatt, P. 1954, Biophysics of junctional transmission. *Physiol. Rev.* 34:674–710.

Fatt, P., and Katz, B. 1951, Spontaneous subthershold activity at motor nerve endings. *J. Physiol. (London).* 117:109–128.

Fearing, F. 1930, Reflex action: a study in the history of physiological psychology. Williams and Wilkins, Baltimore.

Fedoroff, S. 1986, Prenatal ontogenesis of astrocytes. *In* Cellular neurobiology: A Series. Astrocytes, development, morphology and regional specialization of astrocytes (S. Fedoroff and A. Vernadakis, Eds.), Vol. 1. Academic Press, Orlando.

Feldberg, W., and Gaddum, J.H. 1934, The chemical transmitter at synapses in a sympathetic ganglion. *J. Physiol. (London)* 81:305–319.

Ferrier, D. 1875, Experiments in the brain of monkeys. *Proc. Roy. Soc. (London). Ser. B* 23:409–430.

Ferrier, D. 1876, The function of the brain. Smith, Elder and Co., London.

Ferrier, D. 1890, The croonian lectures on cerebral localisation. Smith, Elder and Co., London.

Fifkova, E. 1985a, Actin in the nervous system. *Brain. Res.* 356:187–215.

Fifkova, E. 1985b, A possible mechanism of morphometric changes in dendritic spines induced by stimulation. *Cell. Mol. Neurobiol.* 5:47–63.

Flechsig, P.E. 1896, Gerhirn und Seele..Veit, Leipzig.

Flechsig, P.E. 1901, Ueber die entwickelungsgeschichtliche (myelogenetische) Flachengliederung der grosshirnrinde des menschen. *Arch. Ital. Biol.* 36:30–39.

Flechsig, P.E. 1927, Meine mytelogenetische hirnlehre. Mitbiographischer einleitung. Springer, Berlin.

Fleischhauer, K., Zilles, K., and Schleicher, A. 1980, A revised cytoarchitectonic map of the neocortex of the rabbit (oryctolagus cuniculus). *Anat. Embryol. (Berl)* 161:121–143.

Flemming, W. 1882, Zellsubstanz, Kern und Zelltheilung. F.C.W. Vogel, Leipzig.

Foerster, O., Gagel, O., and Sheehan, D. 1933, Veranderungen an den Endosen im Ruckenmark des Affen nach Hinterwurzeldurchschneidung. *Z. Anat. EntwGesch.* 101:553–565.

Fontaine-Perus, J., Chanconie, M., and Le Douarin, N.M. 1988, Developmental potentialities in the nonneuronal population of quail sensory ganglia. *Dev. Biol.* 128:359–375.

Forel, A. 1887, Einige hirnanatomische Betrachtungen und Ergebnisse. *Arch. f. Psychiat. u. Nervenkr.* 18:162–198.

Francis-Williams, J., and Davies, P.A. 1974, Very low birthweight and later intelligence. *Dev. Med. Child. Neurol.* 16:709–728.

Frankova, S., and Zemanova, R. 1978, Development and long term characteristics of habituation to novel environment in malnourished rats [proceedings]. *Act. Nerv. Super. (Praha)* 20:113–114.

Fritsch, G., and Hitzig, E. 1870, Ueber die electrische Erregbarkeit des Grosshirns. *Arch. Anat. Physiol. Wiss. Med.* 37:300–332.

Fujita, S., and Kitamura, T. 1975, Origin of brain macrophages and the nature of the so-called microglia. *Acta. Neuropathol. Suppl. (Berl)* Suppl 6: 291–296.

Fulton, J.F. 1934, A study of flaccid and spastic paralysis produced by lesions of the cerebral cortex. *Res. Publ. Assoc. Nerv. Ment. Dis.* 13:158–210.

Fulton, J.F. 1935, A note on the definition of motor and 'premotor' areas. *Brain* 58:311–316.

Fulton, J.F. 1938, Physiology of the Nervous System. Oxford Univ. Press, London.

Fulton, J.F. 1943, Physiology of he Nervous System, 2nd Ed. Oxford Univ Press, London.

Furshpan, E.J., and Potter, D.D. 1957, Mechanism of nerve-impulse transmission at a crayfish synapse. *Nature* 180:342–3.

Gall, F.J. 1825, Organologie ou exposition des instincts, des penchans, des sentimens et des talens, ou des qualites morales et des facultes intellectuelles fondamentales de l'homme et des animaux, et du siege de leurs organes. Bailliere, J.B., Paris.

Gall, F.J., and Spurzheim, J.C. 1835, On the functions of the brain and of each of its parts. with observations on the possibility of determining the instincts, propensities, and talents, or the moral and intellectual dispositions of men and animals, by the configuration of the brain and head. March, Capen and Lyon, Boston.

Gerlach, J. 1858, Mikroskopische studien aus dem gebiete der menschlichen morphologie. Enke, Erlangen.

Gerlach, J. 1872, Von dem Ruckenmark. *In* Stricker's Hndbuch der Lehre von den Geweben (S. Stricker, Ed.), Vol. 2:pp. 665–693. Engelmann, Leipzig.

Gilbert, C.D., and Wiesel, T.N. 1989, Columnar specificity of intrinsic horizontal and corticocortical connections in cat visual cortex. *J. Neurosci.* 9:2432–42.

Glauert, A.M., Rogers, G.E., and Glauert, R.H., 1956, A new embedding medium for electron microscopy, *Nature* 178:8–.

Glauert, A.M., and Glauert, R.H.,1958, Araldite as an embedding medium for electron microscopy, *J. Biophysic. Biochem. Cytol.*, 4:191–195.

Glucksmann, A. 1940, Development and differentiation of the tadpole eye. *Br. J. Ophthalmol.* 24:153–178.

Glucksmann, A. 1951, Cell deaths in normal vertebrate ontogeny. *Biol. Rev.* 26:59–86.

Glucksmann, A. 1965, Cell death in normal development. *Arch. biol. (Liege)* 76:419–437.

Goldberg, G. 1985, Supplementary motor are structure and function: Review and hypotheses. *Beh. Brain Sci.* 8:567–616.

Goldscheider, A. 1898, Die Bedeutung der Reize fur Pathologie und Therapie im Lichte der Neuronlehre. Leipzig.

Golgi, C. 1873, Sulla stuttura della sostanza grigia del cervello. *Gazz Med Ital Lombardo* 6:41–56.

Golgi, C. 1875a, *Riv sper freniat Reggio-Emilia Oper Omnia 1903*, 1:66–78.

Golgi, C. 1875b, Sull fina stuttura dei bulbi olfattorii. *Riv. sper. freniat. Reggio-Emilia* 1:405–425.

Golgi, C. 1882–1885, Sulla fina anatomia degli orgni centrali del sistema nervoso. *Rivist. sperimentale. di. Freniatria.* 8:165–195, 361–391 (1882); 9:1–17, 161–192, 385–402 (1883); 11:72–123, 193–220 (1885). Also in Opera Omnia 295–393, 397–536 (1903).

Golgi, C. 1883, Recherches sur l'histologie des centres nerveux. *Archives Ital. Biol.* 4:285–317; 4:92–123.

Golgi, C. 1891a, La rete nervousa diffusa degli orgni centrali del sistemo nervosa: Suo significato fisiologico. *Rendiconti R. 1st Lombardo Sci. Lett, Ser. 2* 24:595–656.

Golgi, C. 1891b, Le reseau nervus diffus des centres du systeme nerveux; ses attributs physiologiques; methode suivie dans les recherches his-tologiques. *Archives. Ital. Biol.* 15:434–463.

Golgi, C. 1894, Das diffuse nervose Netz der Centralorgane des Nervensystems. *In* Untersuchungen uber den feineren Bau des centralen und peripherischen Nevensystems (C. Golgi and R. Teuscher, Eds.). Fisher, Jena.

Golgi, C. 1901, Sulla fina organizzazione del sistema nervoso. *In* Trattato di fisiologia dell'uomo (L. Luciani, Ed.). Societa editrice libraria, Milan.

Gray, E.G. 1959, Axo-somatic and axo-dendritic synapses of the cerebral cortex:an electron microscope study. *J. Anat.* 93:420–433.

Gray, E.G., and Guillery, R.W. 1961, The basis for silver staining of synapses of the mammalian spinal cord: a light and electron microscope study. *J. Physiol.* 157:581–588.

Gray, E.G., and Guillery, R.W. 1966, Synaptic morphology in the normal and degenerating nervous system. *Int. Rev. Cytol.* 19:111–182.

Greene, L.S., and Johnston, F.E. 1980, Social and biological predictors of nutritional status, physical growth and neurological development. Academic Press, New York.

Guillery, R.W. 1972, Binocular competition in the control of geniculate cell growth. *J. Comp. Neurol.* 144:117–129.

Guth, L., and Bernstein, J.J. 1961, Selectivity in the re-establishment of synapses in the superior cervical sympathetic ganglion of the cat. *Exp. Neurol.* 4:59–69.

Hajos, F., and Basco, E. 1984, The surface-contact glia. *Adv Anat. Embryol. Cell Biol* 84:1–79.

Hall, B.K., and Tremaine, R. 1979, Ability of neural crest cells from the embryonic chick to differentiate into cartilage before their migration away from the neural tube. *Anat. Rec.* 194:469–75.

Hall, T.S. 1969, Ideas of Life and Matter. Studies in the History of General Physiology 600 B.C.-1900 A.D. Chicago, Univ. of Chicago Press.

Hamburger, V. 1944, The effects of peripheral factors on the proliferation and differentiation in the spinal cord of the chick embryo. *J. Exp. Zool.* 96:223–242.

Hamburger, V. 1949, Proliferation differentiation and degeneration in the spinal ganglia of the chick embryo under normal and experimental conditions. *J. Exp. Zool.* 111:457–501.

Hamburger, V. 1958, Regression versus peripheral control of differentiation in motor hypoplasia. *Am. J. Anat.* 102:365–409.

Handler, L.C., Stoch, M.B., and Smythe, P.M. 1981, CT brain scans: part of a 20-year development study following gross undernutrition during infancy. *Br. J. Radiol.* 54:953–954.

Hardesty, I. 1904, On the development and nature of the neuroglia. *Am. J. Anat.* 3:229–268.

Harrison, R.G. 1903, Experimentelle Untersuchungen uber die Entwicklung der Sinnesorgane der Seitenlinie bei den Amphibien. *Arch Mikroskop Anat.* 63:35–149.

Harrison, R.G. 1904, An experimental study of the relation of the nervous system to the developing musculature in the embryo of the frog. *Am. J. Anat.* 3:197–220.

Harrison, R.G. 1906, Further experiments on the development of peripheral nerves. *Am. J. Anat.* 5:121–131.

Harrison, R.G. 1907a, Experiments in transplanting limbs and their bearing upon the problem of the development of nerves. *J. Exp. Zool.* 4:239–281.

Harrison, R.G. 1907b, Observations on the living developing nerve fiber. *Anat. Rec.* 1:116–118.

Harrison, R.G. 1910, The outgrowth of the nerve fiber as a mode of proto-plasmic movement. *J. Exp. Zool.* 9:787–846.

Harrison, R.G. 1912, The cultivation of tissues in extraeous media as a method of morphogenetic study. *Anat. Rec.* 6:181–193.

Harrison, R.G. 1924a, Neuroblast versus sheath cell in the development of peripheral nerves. *J. Comp. Neurol.* 37:123–205.

Harrison, R.G. 1924b, Some unexpected results of the heteroplastic trans-plantation of limbs. *Proc. Natl. Acad. Sci. U S A.*10.

Hatai, S. 1902, On the origin of neuroglia tissue from the mesoblast. *J. Comp. Neurol.* 20:119–147.

Hebb, D.H. 1949, The organization of behavior. John Wiley, New York.

Hebb, D.H. 1966, A textbook of psychology. Saunders, Philadelphia.

Held, H. 1897a, Beitrage zur Struktur der Nervenzellen und ihre Fortsatze. Dritte Abhandlung. *Arch. Anat. Physiol. Anat. Abt. Leipzig Suppl,* 273–312.

Held, H. 1897b, Beitrage zur Struktur der Nervenzellen und ihrer Fortsatze (Zweite Abhandlung). *Arch. Anat. Physiol. Anat. Abt.* 21:204–294.

Held, H. 1905, Zur Kenntniss einer neurofibrillaren Continuitat im Centralnervensystem der Wirbelthiere. *Arch. Anat. Physiol. Leipzig,* 55–78.

Held, H. 1909, Die Entwicklung des Nervengewebes bei den Wirbeltieren. Barth, Leipzig.

Henle, J. 1871, Handbuch der systematischen Anatomie des Menschen, Bd. 3, Nervenlehre. F. Vieweg, Braunschweig.

Herbst, C. 1894, Ueber die Bedeutung der Reizphysiologie fur die kausale Auffassung von Vorgangen in die Thierische Ontogenie. *Biol. Centralbl.,* 14.

Herrick, C.L., and Herrick, C.J. 1897, Inquires regarding current tendencies in neurological nomenclature. *J. Comp. Neurol.* 7:162–168.

Hertwig, O. 1893–1898, Die Zelle und die Gewebe. Fischer, Jena.

Hertwig, O. 1906, Handbuch der vergleichenden und experimentellen Entwickelungslehre der Wirbeltiere. Verlag von Gustav Fischer, Jena.

Hertwig, O. 1911–1912, Methoden und VErsuche zur Erforschung der Vita propria abgetrennter Gewebs- und Organstuckchen von Wiebeltieren. *Arch. Mikroskop. Anat.* 79:113–120.

Hertzig, M.E., Birch, H.G., Richardson, S.A., and Tizard, J. 1972, Intellectual levels of school children severely malnourished during the first two years of life. *Pediatrics* 49:814–824.

Hicks, L.E., Langham, R.A., and Takenaka, J. 1982, Cognitive and health measures following early nutritional supplementation: a sibling study. *Am. J. Public Health* 72:1110–1118.

Hines, M. 1947a, The motor areas. *Fed. Proc.* 6:441–447.

Hines, M. 1947b, Movements elicited from precentral gyrus of adult chim-panzee by stimulation with sine wave currents. *J. Neurophysiol.* 3:442–466.

Hirano, M., and Goldman, J.E. 1988, Gliogenesis in rat spinal cord: evidence for origin of astrocytes and oligodendrocytes from radial precursors. *J. Neurosci. Res.* 21:155–167.

Hirsch, H.V., and Jacobson, M. 1973, Development and maintenance of connectivity in the visual system of the frog. II. The effects of eye removal. *Brain. Res.* 49:67–74.

His, W. 1856, Beitrage zur normalen und pathologischen Histologie der Kornea. Schweighauser, Basel.

His, W. 1868, Untersuchungen über die erste anlage des wirbeltierleibes: Die erste Entwicklung des Hühnchens im Ei. Vogel, Leipzig.

His, W. 1870, Beschreibungeines Mikrotoms. *Arch. Mikroskop. Anat.* 6:229–232.

His, W. 1874, Unserer Körperform und das Physiologische probelm ihrer entstehung. Engelmann, Leipzig.

His, W. 1879, Ueber die Anfange des peripherischen Ruckenmarks und der nerenwurzeln. *Arch. Anat. Physiol. Leipzg. Anat Abth*, 455–482.

His, W. 1886, Zur Geschichte des menschlichen Ruckenmarks und der Nervenwurzeln. *Abh. kgl. sachs. Ges. Wissensch. math. phys. Kl.* 13:147–513.

His, W. 1887a, Die Entwickelung der ersten Nervenbahnen beim menschlichen Embryo. Ubersichtliche Darstellung. *Arch. Anat. Physiol. Leipzg. Anat. Abth.* 92:368–378.

His, W. 1887b, Zur Geschichte des menschlichen Rückenmarks und der Nervenwurzeln. *In* Abh. kgl. sächs. Ges. Wissensch. math. phys. Kl. pp. 13:479–513.

His, W. 1888a, On the principles of animal morphology. *Proc. Roy. Soc. (Edin)* 15:287–297.

His, W. 1888b, Sur Geschichte des Gehirns sowie der centralen und peripherischen Nervenbahnen beim menschlichen Embryo. *Abh kgl sachs Ges Wissensch math phys Kl* 24:339–392.

His, W. 1889a, Die Formentwickelung des menschlichen Vorderhirns vom Ende des ersten bis zum Beginn des dritten Moknats. *Abh. kgl. sachs. Ges. Wissensch. math. phys. Kl* 15:673–736.

His, W. 1889b, Die Neuroblasten und deren Entstehung im embryonalen mark. *Abh. kgl. sachs. Ges. Wissensch. math. phys. Kl* 15:311–372.

His, W. 1889c, Uber die Entwicklung des Riechlappens und des Riechganglions und uber diejenige des verlangerten Markes. *In* Verhandl. der anatom. Ges. auf der 3. Versammlung zu Berlin.

His, W. 1890a, Die Entwickelung des menschlichen Rautenhirns vom Ende des ersten bis zum Beginn des dritten Monats. I. Verlangertes Mark. *Abh. kgl. sachs. Ges. Wissensch. math. phys. Kl* 29:1–74.

His, W. 1890b, Histogenese und Zusammenhang der Nervenelemente. *Arch. Anat. Physiol. Leipzg. Anat. Abt. Suppl* 95:95–119.

His, W. 1894, Ueber mechanische grundvorgänge thierischer formenbildung.

His, W. 1901, Das Princip der organbildende Keimbezirke und die Verwandschaft der Gewebe. *Zeitschr. ges. Neurol. Psychiat.* 87:167.

Hoerr, N.L. 1936, Cytological studies by the Altmann-Gersh freezing-drying method. III. The preeistence of neurofibrillae and their disposition in the nerve fibre. *Anat. Rec.* 66:81–90.

Hoff, E.C. 1932, Central nerve terminals in the mammalian spinal cord and their eamination by experimental degeneration. *Proc. Roy. Soc. (London) Ser. B* 111:175–188.

Hoffman, P. 1922, Untersuchungen uber die Eigenreflee (Sehnenreflexe) menschlicher Muskeln. Springer, Berlin.

Horder, T.J., and Weindling, P.J. 1985, Hans Spemann and the organiser. *In* A History of Embryology (T.J. Horder, J.A. Witkowski, and C.C. Wylie, Eds.), pp. 188–242. Cambridge Univ Press, Cambridge.

Horstdius, S., and Sellman, S. 1946, Experimenteler Untersuchungen uber die Determination des Knorpeligen Kopfskelettes bei Urodelen. *Nova. Acta. Regiae. Soc. Sci. Upsal.* IV 13:1–170.

Hubel, D.H., and Livingstone, M.S. 1987, Segregation of form, color, and stereopsis in primate area 18. *J. Neurosci.* 7:3378–3415.

Hubel, D.H., and Wiesel, T.N. 1962, Receptive fields, binocular interaction and functional architecture in the cat's visual cortex. *J. Physiol.* 160:106–154.

Hubel, D.H., and Wiesel, T.N. 1968, Receptive fields and functional architecture of monkey striate cortex. *J. Physiol.* 195:215–243.

Hughes, A.F. 1952, The mitotic cycle, the cytoplasm and nucleus during interphase and mitosis. Academic Press, New York.

Hughes, A.F. 1959, A history of cytology. Abelard-Schuman, New York.

Hughes, A.F. 1961, Cell degenration in the larval ventral horn of Xenopus laevis (Daudin). *J. Embryol. Exp. Morphol.* 9:269–284.

Humphrey, D.R. 1986, Representation of movements and muscles within the primate precentral motor cortex: historical and current perspectives. *Fed. Proc.* 45:2687–2699.

Hurle, J.M. 1988, Cell death in developing systems. *Methods Achiev. Exp. Pathol.* 13:55–86.

Imamoto, K., and Leblond, C.P. 1977, Presence of labeled monocytes, macrophages and microglia in a stab wound of the brain following an injection of bone marrow cells labeled with 3H-uridine into rats. *J. Comp. Neurol.* 174:255–279.

Inouye, E. 1970, Twin studies and human behavioral genetics. *Jinrui Idengaku Zasshi* 15:1–25.

Irwin, M., Klein, R.E., Townshend, J.W., Owens, W., Engle, P.L., Lechtig, A., Martorell, R., Yarbrough, C., Lasky, R.E., and Delgado, H.L. 1979, The effect of food supplementation on cognitive development and behavior among rural Guatemalan children. *In* Behavioral effects of energy and protein deficits (Brozek, Ed.), pp. 239–254. US Department of Health, Education and Welfare, NIH Publ. No. 79–1906.

Jackson, J.H. 1898, Remarks on the relations of different divisions of the central nervous system to one another and to parts of the body. *Br. Med. J.* 1:65–69.

Jacobson, M. 1970a, Development, specification and diversification of neuronal connections. *In* The neurosciences: second study program (F.O. Schmitt, Ed.). The Rockefeller Univ Press, New York.

Jacobson, M. 1970b, Developmental Neurobiology. Holt Rinehart and Winston, New York.

Jacobson, M. 1974a, Neuronal plasticity: concepts in pursuit of cellular mechanisms. *In* Plasticity and recovery of function in the central nervous system (G. Gottlieb, Ed.), pp. 31–43. Academic Press, New York.

Jacobson, M. 1974b, A plentitude or neurons. *In* Studies on the development of behavior and the nervous system (G. Gottlieb, Ed.), Vol. 2, pp. 151–166. Academic Press, New York.

Jacobson, M. 1974c, Through the jungle of the brain: neuronal specificity and typology re-explored. *Ann. N.Y. Acad. Sci.* 228:63–67.

Jacobson, M. 1985, Clonal analysis and cell lineages of the vertebrate central nervous system. *Annu. Rev. Neurosci.* 8:71–102.

Jacobson, M. 1993, Foundations of Neuroscience. Plenum, New York.

Jerne, N.K. 1967, Waiting for the end. *Cold Spring Harb. Symp. Quant. Biol.* 37:591–603.

Johnston, J.B. 1902, An attempt to define the primitive functional divisions of the central nervous system. *J. Comp. Neurol.* 12:87–106.

Jolly, W.A. 1911, The time relations of he knee-jerk and simple reflexes. *Quart J Exp Physiol* 4:67–87.

Kaas, J.H. 1983, What, if anything, is SI? Organization of first somatosensory area of cortex. *Physiol. Rev.* 63:206–231.

Kahn, D., and Birch, H.G. 1968, Development of auditory-visual integration and reading achievement. *Percept. Mot. Skills.* 27:459–460.

Kallius, E. 1896, Endigungen sensibler nerven bei wirbeltieren. *Ergebn Anat Entw Gesch* 5:55–94.

Kaur, C., Ling, E.A., and Wong, W.C. 1987a, Localisation of thiamine pyrophosphatase in the amoeboid microglial cells in the brain of postnatal rats. *J. Anat.* 152:13–22.

Kaur, C., Ling, E.A., and Wong, W.C. 1987b, Origin and fate of neural macrophages in a stab wound of the brain of the young rat. *J. Anat.* 154:215–227.

Kemper, T.L., and Galaburda, A.M. 1984, Principles of cytoarchitectonics. *In* Cerebral cortex, Vol.1, Cellular components of the cerebral cortex (A. Peters and E.G. Jones, Eds.), pp. 35–58. Plenum, New York.

Klein, R.E., Irwin, M., Engle, P.L., and Yarbrough, C. 1977, Malnutrition and mental development in rural Guatemala: An applied cross-cultural research study. *In* Warren, Advances in Cross-Cultural Psychology, pp. 92–121. Academic Press, New York.

Knouff, R.A. 1927, Origin of the cranial ganglia of Rana. *J. Comp. Neurol.* 44:259–361.

Knouff, R.A. 1935, The developmental pattern of ectomdermal placodes in Rana pipiens. *J. Comp. Neurol.* 62:17–71.

Koelliker, R.A. 1850, Mikroskopische Anatomie oder Gewebelehre des Menschen, Vol. 2: part 1. Engelmann, Leipzig.

Koelliker, R.A. 1852, Handbuch der Gewebelehre des Menschen. Wilhelm Engelmann, Leipzig.

Koelliker, R.A. 1854, Manual of Human Microscopical Anatomy. Transl. by G. Busk and T. Huxley, edited by J. da Costa. Lippincott, Grambo & Co., Philadelphia.

Koelliker, R.A. 1886, Histologische Studien an Batrachierlarven. *Z. wiss. Zool.* 43:1–40.

Koelliker, R.A. 1887, Die Untersuchungen von Golgi uber den feineren Bau des centralen Nervensystems. *Anat. Anz.* 15:480–.

Koelliker, R.A. 1890, Zur feineren anatomie des centralen nervensystems. *Z. wiss Zool* 49:663–689.

Koelliker, R.A. 1893, Handbuch der Gewebelchre des menschen. Wilhelm Engelmann, Leipzig.

Koelliker, R.A. 1896, Handbuch der Gewebelehr des menschen, Vol. Bd. 2, Nervensystem des menschen und der thiere, 6, Aufl. Wilhelm Engelmann, Leipzig.

Konigsmark, B.W., and Sidman, R.L. 1963, Origin of Brain Macrophages in the Mouse. *J. Neuropathol. Exp. Neurol.* 22:643–676.

Korschelt, E., and Heider, K. 1902–1909, Lehrbuch der vergleichenden Entwicklungsgeschichte der Wirbellosen Thiere. Gustav Fischer, Jena.

Kreutzberg, G.W. 1966, Autoradiographische Untersuchung uber die Beteiligung von Gliezellen an der aonalen Reaktion im Facilialiskern der Ratte. *Acta. Neuropathol.* 7:149–161.

Krotoski, D.M., Fraser, S.E., and Bronner-Fraser, M. 1988, Mapping of neural crest pathways in Xenopus laevis using inter- and intra-specific cell markers. *Dev. Biol.* 127:119–132.

Kuhn, T.S. 1962, The structure of Scientific Revolutions. Chicago University Press, Chicago.

Kuhn, T.S. 1970, The Structure of Scientific Revolutions. Chicago University Press, Chicago.

Kuhn, T.S. 1974, Second thoughts on paradigms. *In* The structure of scientific, pp. 459–482. Univ. Illinois Press, Urbana.

Kuhne, W. 1888, On the origin and the causation of vital movement. *Proc. Roy. Soc. (London) Ser B* 44:427–447.

Kupffer, C. von 1895, Studien zur vergleichenden Entwicklungsgeschicchte des Kopfes der Cranioten. Heft III: Die Entwicklung der Kopfnerven von Ammocoetes planeri. Lehmann, Munich and Leipzig.

Landacre, F.L. 1910, The origin of the cranial ganglia in Ameiurus. *J. Comp. Neurol.* 20:389–411.

Landacre, F.L. 1914, Embryonic cerebral ganglia and the doctrine of nerve components. *Folia. Neurobiol.* 8:601–815.

Landacre, F.L. 1916, The cerebral ganglia and early nerves of Squalus acanthias. *J. Comp. Neurol.* 27:20–55.

Landacre, F.L. 1921, The fate of the neural crest in the head of the urodeles. *J. Comp. Neurol.* 33:1–43.

Langley, J.N. 1895, Note on regeneration of pre-ganglionic fibres of the sympathetic. *J. Physiol. (London)* 18:280–284.

Langley, J.N. 1897, On the regeneration of pre-ganglionic and post-ganglionic visceral nerve fibres. *J. Physiol. (London)* 22:215–230.

Langley, J.N., and Anderson, H.K. 1904a, On the union of the fifth cervical nerve with the superior cervicl ganglion. *J. Physiol. (London)* 30:439–442.

Langley, J.N., and Anderson, H.K. 1904b, The union of different kinds of nerve fibres. *J. Physiol. (London)* 31:365–391.

Langley, J.N. 1906, On nerve-endings and on special excitable substances in cells:Croonian lecture. 78, 170–194.

Lashley, K.S. 1929, Brain mechanisms and intelligence. A quantitative study of injuries to the brain. Univ Chicago Press, Chicago.

Lashley, K.S. and Clark, G. 1946, The cytoarchitecture of the cerebral cortex of Ateles: a critical examination of architectonic studies. *J. Comp. Neurol.* 85:223–305.

Latta, H. and Hartmann, J.F. 1950, Use of a aglass edge in thin sectioning for electron microscopy. *Proc. Soc. Exp. Biol. Med.,* 74:436–.

Le Douarin, N. 1973, A biological cell labeling technique and its use in expermental embryology. *Dev. Biol.* 30:217–22.

Le Douarin, N.M., and Teillet, M.A. 1973, The migration of neural crest cells to the wall of the digestive tract in avian embryo. *J. Embryol. Exp. Morphol.* 30:31–48.

Lebert, H. 1845, Physiologie pathologique ou recherches cliniques, experimentales et mmicroscopiques sur l'inflmmation, la tuberculisation, les tumeurs, la formation dur cal etc. Bailliere, Paris.

Lehman, F.E. 1927, Further studies on the morphogenetic role of the somite in the development of the nervous system of amphibians. *J. Exp. Zool.* 49:93–132.

Lemons, J.A., Schreiner, R.L., and Gresham, E.L. 1981, Relationship of brain weight to head circumference in early infancy. *Hum. Biol.* 53:351–354.

Lenhosek, M. v. 1890, Zur Kenntniss der ersten Entstehung der Nervenzellen und Nervenfasern beim Vogelembryo. *Verhndl. des. S. internat. Med. Congresses, Berlin* 2:115.

Lenhosek, M. v. 1891, Zur Kenntnis der Neuroglia des menschlichen Ruckenmarkes. *Verh. Anat. Ges.* 5:193–221.

Lenhosek, M. v. 1892–1893, Der feinere Bau und die Nervenendigungen der Geschmacksknospen. *Anat. Anz.* 8:121–127.

Lenhosek, M. v. 1893, Der feinere Bau des Nervensystems im Lichte neuester Forschungen. *In* Fischer's Medicinische Buchhandlung (H. Kornfeld, Ed.), Berlin.

Lester, B.M., Klein, R.E., and Martinez, S.J. 1975, The use of habituation in the study of the effects of infantile malnutrition. *Dev. Psychobiol.* 8:541–546.

Levi-Montalcini, R. 1949, The development to the acoustico-vestibular centers in the chick embryo in the absence of the afferent root fibers and of descending fiber tracts. *J. Comp. Neurol.* 91:209–241, illust, incl 3 pl.

Levi-Montalcini, R. 1965, Morphological and metabolic effects of the nerve growth factor. *Arch. biol. (Liege)* 76:387–417.

Levi-Montalcini, R. 1966, The nerve growth factor: its mode of action on sensory and sympathetic nerve cells. *Harvey Lect.* 60:217–259.

Levi-Montalcini, R., and Hamburger, V. 1951, Selective growth stimulating effects of mouse sarcoma on the sensory and sympathetic nervous system of the chick embryo. *J. Exp. Zool.* 116:321–361.

Levi-Montalcini, R., and Hamburger, V. 1953, A diffusible agent of mouse sarcoma producing hyperplasia of sympathetic ganglia and hyperneurotization of the chick embryo. *J. Exp. Zool.* 123:233–288.

Levine, R., and Jacobson, M. 1975, Discontinuous mapping of retina onto tectum innervated by both eyes. *Brain. Res.* 98:172–176.

Levitt, P., and Rakic, P. 1980, Immunoperoxidase localization of glial fibrillary acidic protein in radial glial cells and astrocytes of the developing rhesus monkey brain. *J. Comp. Neurol.* 193:815–840.

Lewis, E.B. 1978, A gene complex controlling segmentation in Drosophila. *Nature* 276:565–70.

Lewis, M.R., and Lewis, W.H. 1911, The cultivation of tissues from chick embryos in solutions of NaCl, $CaCl_2$, KCl and $NaHCO_3$. *Anat. Rec.* 5:277–293.

Lewis, M.R., and Lewis, W.H. 1912, The cultivtion of chick tissues in medi of known chemicl constitution. *Anat. Rec.* 6:207–211.

Lewis, W.B., and Clarke, H. 1878, The cortical lamination of the motor are of the brain. *Proc. Roy. Soc. (London) Ser B* 27:38–49.

Leyton, A.S.F., and Sherrington, C.S. 1917, Observations on the excitable cortex of the chimpanzee, orang-utan and gorilla. *Quarterly J. Exp. Physiol.* 11:135–222.

Liang, P.H., Hie, T.T., Jan, O.H., and Gick, L.T. 1967, Evaluation of mental development in relation to early malnutrition. *Am. J. Clin. Nutr.* 20:1290–1294.

Ling, E.A. 1978, Electron microscopic studies of macrophages in Wallerian degeneration of rat optic nerve after intravenous injection of colloidal carbon. *J. Anat.* 126:111–121.

Ling, E.A. 1981, The origin and nature of microglia. *In* Advances in Cellular Neurobiology (S. Fedoroff and L. Hertz, Eds.), Vol. 2: pp. 33–82. Academic Press, New York.

Lister, J.J. 1830, On some properties in achromatic object-glasses applicable to the improvement of the microscope. *Phil. Trans. Roy. Soc. Lond.* 120:187–200.

Liuzzi, F.J., and Miller, R.H. 1987, Radially oriented astrocytes in the normal adult rat spinal cord. *Brain. Res.* 403:385–388.

Livingstone, M.S., and Hubel, D.H. 1984, Anatomy and physiology of a color system in the primate visual cortex. *J. Neurosci.* 4:309–56.

Loewi, O. 1921, Uber humorale Ubertragbarkeit der Herznervenwirkung. *Pflug Arch ges Physiol* 189:239–242.

Loewi, O. 1933, The Ferrier lecture on problems connected with the principle of humorl transmission of nervous impulses. *Proc. Roy. Soc. (London)* 118:299–316.

Lorente de No, R. 1935, The sysnaptic delay of motoneurons. *Am. J. Physiol..* 111:272–281.

Lorente de No, R. 1938a, The cerebral cortex: Archecture, intracorticl connections and motor projections. *In* Physiology of the Nervous system (J.F. Fulton, Ed.), pp. 291–325. Oxford Univ Press, London.

Lorente de No, R. 1938b, Limits of variation of the sysnaptic delay of motoneurons. *J Neurophysiology* 1.

Lorente de No, R. 1943, Cerebral cortex: architecture, intracortical connections, motor projections. *In* Physiology of the nervous system (J.F. Fulton, Ed.), pp. 274–301. Oxford Univ Press, London.

Luft, J.H. 1956, Permanganate—a new fixative for electron microscopy. *J. Biophys. Biochem. Cytol.* 2:229–232.

MacLean, P.D. 1970, The triune brain, emotion, and scientific bias, pp. 336–349. *In* The Neurosciences, 2nd Study Program (F.O. Schmitt, ed) Rockefeller Univ. Press, New York.

MacLean, P.D. 1972, Cerebral evolution and emotional processes: New findings on the striatal complex. Ann. N.Y. Acad. Sci. 193, 137–149.

Magini, G. 1888, Sur la neuroglie et les cellules nerveuses cerebrales chez les foetus. *Arch. Ital. Biol.* 9:59–60.

Manocha, S.L. 1972, Malnutrition and Retarded Human Development. Charles Thomas, Springfield, Illinois.

Marinesco, G. 1909, La cellule nerveuse. O. Doin, Paris.

Marinesco, G. 1919, Nouvelles contributions a l'etude de la regeneration nerveuse et du neurotropisme. *Phil. Trans. Roy. Soc. (London) B* 209:229.

Marinesco, G., and Minea, J. 1912a, Croissance des fibres nerveuses dans le milieu de culture, in vitro, des ganglions spinaux. *Soc. Biol. Compt Rend* 73:668–670.

Marinesco, G., and Minea, J. 1912b, Culture des ganglions spinaux des mamiferes (in vitro). Suivant le procede de M. Carrel. *Acad. nat. med. Paris. Bull. Ser.. 3* 68:37–40.

Martin, H.P. 1970, Microcephaly and mental retardation. *Am. J. Dis. Child.* 119:128–31.

Maxwell, D.S., and Kruger, L. 1965, Small Blood Vessels and the Origin of Phagocytes in the Rat Cerebral Cortex Following Heavy Particle Irradiation. *Exp. Neurol.* 12:33–54.

Maxwell, G.D., Forbes, M.E., and Christie, D.S. 1988, Analysis of the development of cellular subsets present in the neural crest using cell sorting and cell culture. *Neuron* 1:557–568.

May, R.M. 1925, The relation of nerves to degenerating and regenerating taste buds. *J. Exp. Zool.* 42:371–410.

Mayr, E. 1965, Cause and effect in biology. Free Press, New York.

McKay, H., McKay, A., and Sinisterra, L. 1972, Behavioral intervention studies with malnourished children. A review of experiences. U.S. Government Printing Office, Washington.

McKay, H., Sinisterra, L., McKay, A., Gomez, H., Lloreda, P. 1978, Improving cognitive ability in chronically deprived children. *Science* 200:270–278.

McKeever, P.E., and Balentine, J.D. 1978, Macrophages migration through the brain parenchyma to the perivascular space following particle ingestion. *Am. J. Pathol.* 93:153–164.

McKeown, T., and Record, R.G. 1952, Observations on foetal growth in multiple pregnancy in man. *J. Endocrinol.* 8:386–401.

Merkel, F. 1875, Tastzellen und Tastkorperchen bei den Hausthieren und beim Menschen. *Arch. mikr. Anat. Bonn.* 11:636–652.

Merkel, F. 1880, Ueber die Endigungen der sensiblen Nerven in der Haut der Wirbeltiere. Schmidt, Rostock.

Meynert, T. 1867, Der Bau der Grosshirnrinde und seine ortlichen Verschiedenheiten, nebst einem pathalogischanatomischen Corollarium. Engelmann, Leipzig.

Meynert, T. 1868–1869, Der Bau der Bross-Hirnrinde und seine ortlichen verschiedenheiten, nebst einem pathologischanatomischen corollarium. *Vschr Psychiat Vienna* 1:77–93.

Meynert, T. 1872, Vom Gehirne der Saugethiere. *In* Handbuch der Lehre von den Geweben des Menschen und die Thiere (S. Stricker, Ed.), Vol. 2: pp. 694–808. Engelmann, Leipzig.

Miescher, F.J. 1871, Ueber die chemische Zusammensetzung der Eiterzellen. *Med. Chem. Untersuch.* 1:441–460.

Miledi, R., and Stefani, E. 1969, Non-selective re-innervation of slow and fast muscle fibres in the rat. *Nature* 222:569–571.

Milne-Edwards, H. (1857–1881). Lecons sur la physiologie et l'anatomie comparee de l'homme et des animaux. Victor Masson, Paris.

Monckeberg, F. 1968, Effect of early marasmic malnutrition on subsequent physical and psychological development. *In* Malnutrition, learning and behavior (Scrimshaw and Gordon, Eds.), Vol. 10: pp. 269–278. MIT Press, Cambridge.

Monckeberg, F. 1979, Recovery of severely malnourished infants: effects of early sensory-affective stimulation. *In* Behavioral effects of energy and protein deficits (Brozek, Ed.), pp. 120–130. U.S. Department of Health, Education and Welfare. NIH Publ. No. (NIH) 79–1906.

Mora, J.O., Clement, J., Christiansen, N., Ortiz, N., Vuori, L., and Wagner, M. 1979, Nutritional supplementation, early stimuation and child development. *In* Proc Int Nutr Conf Behavioral Effects of Energy and Protein Deficits (Brozek, Ed.), pp. 255–269. NIH Publ. No. 79–106.

Mori, S., and Leblond, C.P. 1969, Identification of microglia in light and electron microscopy. *J. Comp. Neurol.* 135:57–80.

Mott, F.W. 1894, The sensory motor functions of the central convolutions of the cerebral cortex. *J. Physiol. (London)* 15:464–487.

Mott, F.W., and Sherington, C.S. 1895, Experiments upon the influence of sensory nerves upon movement and nutrition of the limbs. Preliminary communication. *Proc. Roy. Soc. (London) Ser. B.* 57:481–488.

Mugnaini, E. and Forstronen 1967, Ultrastructural studies on the cerebellar histogenesis. I. Differentiation of granule cells and development of glomeruli in the chick embryos. *Z. Zellforsch. Mikrosk. Anat. Abt. Histochem.* 77:115–143.

Munoz-Garcia, D., and Ludwin, S.K. 1986a, Gliogenesis in organotypic tissue culture of the spinal cord of the embryonic mouse. I. Immunocytochemical and ultrastructural studies. *J. Neurocytol.* 15:273–90.

Munoz-Garcia, D., and Ludwin, S.K. 1986b, Gliogenesis in organotypic tissue culture of the spinal cord of the embryonic mouse. II. Autoradiographic studies. *J. Neurocytol.* 15:291–302.

Nagel, E. 1965, Types of causal explanation in science. Free Press, New York.

Needham, J. 1931, Chemical embryology, 3 Vols. Macmillan, Cambridge Univ. Press, New York, Cambridge.

Neuburger, M. 1897, Die historische Entwickelung der experimentellen Gehirn und Ruckenmarksphysiologie vor Flourens. Enke, Stuttgart.

Neumayer, L. 1906, Histogenese und Morphogeneese des peripheren nerven-systems, der Spinnlganglien und des Nervus sympthicus. *In* Handbuch der vergleichenden und experimentellen Entwickelungslehre der Wirbeltiere (O. Hertwig, Ed.), Vol. 2: part 3, pp. 513–626. Gustv Fischer Verlag, Jena.

Nissl, F. 1894, Ueber die sogenannten Granula der Nervenzellen. *Neurol. Centralbl.* 13:676–685, 781–789, 810–814.

Nissl, F. 1903, DieNeuronlehre und ihre Anhanger. Fischer, Jena.

Obersteiner, H. 1883, Der feinere Bau der Kleinhirnrinde beim Menschen und bei tieren. *Biol Zentralbl* 3:145–155.

O'Brien, R.A., Ostberg, A.J., and Vrbova, G. 1984, Protease inhibitors reduce the loss of nerve terminals induced by activity and calcium in developing rat soleus muscles in vitro. *Neuroscience* 12:637–646.

Olmsted, J.M.D. 1920a, The nerve as a formative influence in the development of taste-buds. *J. Comp. Neurol.* 31:465–468.

Olmsted, J.M.D. 1920b, The results of cutting the seventh cranial nerve in Ameiurus nebulosus (Lesueur). *J. Exp. Zool.* 31:369–401.

Olmsted, J.M.D. 1925, Effects of cutting the lingual nerve of the dog. *J. Comp. Neurol.* 33:149–154.

Oppel, A. 1913, Explantation (Deckglaskultur, in vitro-Kultur). *Zentr. Zool. Allgem. Exp. Biol.* 3:209–232.

Oppenheim, R.W. 1981a, Cell death of motoneurons in the chick embryo spinal cord. V. Evidence on the role of cell death and neuromuscular function in the formation of specific peripheral connections. *J. Neurosci.* 1:141–51.

Oppenheim, R.W. 1981b, Neuronal cell death and some related regressive phenomena during neurogenesis: A selective historical review and progress report. Oxford Univ. Press, Oxford.

Palade, G.E., and Palay, S.L. 1954, Electron microscope observations of interneuronal and neuromuscular synapses. *Anat. Rec.* 118:335–336.

Pappas, G.D., and Bennett, M.V. 1966, Specialized junctions involved in electrical transmission between neurons. *Ann. N. Y. Acad. Sci.* 137: 495–508.

Parker, G.H. 1932, On the trophic impulse so-called, its rate and nature. *Am. Naturalist* 66:147–158.

Penfield, W. 1924, Oligodenroglia and its relation to classicl neuroglia. *Brain* 47:430–452.

Penfield, W. 1928, Neuroglia and microglia. The interstitial tissue of the central nervous system. *In* Special Cytology (E.V. Cowdry, Ed.), pp. 1032–1067. Hoeber, New York.

Penfield, W. 1954, Ramon y Cajal, an appreciation. *In* Neuron theory or reticular theory. Instituo Ramon y Cajal, Madrid.

Penfield, W. 1977, No man alone. A neurosurgon's life. Little Brown, Boston.

Penfield, W., and Welch, K. 1951, The supplementary motor area of the cerebral cortex; a clinical and experimental study. *AMA Arch. Neurol. Psychiatry* 66:289–317.

Perry, V.H., and Gordon, S. 1988, Macrophages and microglia in the nervous system. *Trends Neurosci.* 11:273–277.

Peters, A. 1959, Experimental studies on the staining of nervous tissue with silver proteinates. *J. Anat.* 93:177–194.

Picken, L. 1956, The fate of Wilhelm His. *Nature* 178:1162–1165.

Platt, J.B. 1893, Ectodermic origin of the cartilages of the head. *Anat. Anz.* 8:506–509.

Poliak, S. 1932, The main afferent fiber systems of the cerebral cortex of primates. *Univ. Calif. Publ. Anat.* 2:1–370.

Pollit, E., and Granoff, G. 1967, Mental and motor development of Peruvian children treated for severe malnutrition. *Revista. Interamer. Psicol.* 1:93–102.

Prestige, M.C. 1965, Cell Turnover in the Spinal Ganglia of Xenopus Laevis Tadpoles. *J. Embryol. Exp. Morphol.* 13:63–72.

Prochaska, J 1784, English translation by T. Laycock in The Principles of Physiology, by John Augustus Unzer; ad a Dissertation on the Functins of the Nervous System, by George Prochaska. The New Syndenham Society, London, 1851.

Purkinje, J.E. 1837a, Neueste Beobachtungen uber die Struktur des Gehirns. *Opera omnia* 2:88.

Purkinje, J.E. 1837b, Neueste Beobachtungen uber die Struktur des Gehirns. *Opera. ominia.* 3:45–49a.

Purkinje, J.E. 1838, Uber die gangliose Natur bestimmter Hirntheile. *Ber. Vers. dtsch. Naturf. Arzte. (Prag), 1837, 15–174 ff. in Oper omnia 1939, Vol 3: Prague, pp. 45–49.*

Rabl-Ruckhard, H. 1890, Sind die Ganglianzellen amoboid? Eine Hypothese zur Mechnik psychischer Vorgange. *Neurol. Centralbl. Leipzig.* 9:199.

Rádl, E. 1930, The history of Biological theories. Oxford University Press, London.

Rakic, P. 1971a, Guidance of neurons migrating to the fetal monkey neo-cortex. *Brain. Res.* 33:471–476.

Rakic, P. 1971b, Neuron-glia relationship during granule cell migration in developing cerebellar cortex. A Golgi and electronmicroscopic study in Macacus Rhesus. *J. Comp. Neurol.* 141:283–312.

Rakic, P. 1972a, Extrinsic cytological determinants of basket and stellate cell dendritic pattern in the cerebellar molecular layer. *J. Comp. Neurol.* 146:335–354.

Rakic, P. 1972b, Mode of cell migration to the superficial layers of fetal monkey neocortex. *J. Comp. Neurol.* 145:61–83.

Rakic, P. 1981a, Development of visual centers in the primate brain depends on binocular competition before birth. *Science* 214:928–931.

Rakic, P. 1981b, Neuronal-glial interaction during brain development. *Trends Neurosci.* 4:184–187.

Ramon y Cajal, S. 1888a, Estructura de los centros nerviosos de las aves. *Rev. trim. Histol. normal. y. Pathol.*

Ramon y Cajal, S. 1888b, Terminaciones nerviosas los husos musculares de la rana. *Rev. trim. Histol. normal y Pathol.*

Ramon y Cajal, S. 1890a, A propos de certains elements bipolaires du cervelet avec quelques details nouveaux sur l'evolution des fibres cerebelleuses. *Int. Mschr. Anat. Physiol* 7:12–31.

Ramon y Cajal, S. 1890b, Sur les fibres nerveuses de la conche granuleuse du cervelet et sur l'evolution des elements cerebelleux. *Int. Mschr. Anat. Physiol.* 7:12–31.

Ramon y Cajal, S. 1890c, Sur l'origine et les ramifications des fibres nerveuse de la moelle embryonnaire. *Anat. Anz.* 5:85–95, 111–119, 609–613, 631–639.

Ramon y Cajal, S. 1892, Nuevo concepto de la histologia de los centros nerviosos. *Rev. Cien. med. Barcelona* 18.

Ramon y Cajal, S. 1894a, Die Retina der Wirbelthiere. Bergmann-Verlag, Wiesbaden.

Ramon y Cajal, S. 1894b, Les nouvelles idees sur la structure du systeme nerveux chez l'homme et chez les vertebres. Reinwald, Paris.

Ramon y Cajal, S. 1895a, Algunas conjecturas sobre el mecanismo anatomico de la ideacion, associacion y atencion. *Rev. med. cirurg. pract.* 36: 497–508.

Ramon y Cajal, S. 1895b, Manual de Histologia Normal y de Tecnica Micrografica. N Moya, Madrid.

Ramon y Cajal, S. 1896a, Beitrag zur Studium der Medulla oblongata, des Kleinhirns und des Ursprung der Gehirnnerven. Ambrosius Barth, Leipzig.

Ramon y Cajal, S. 1896b, Las espinas colaterales de las celulas del cerebro tenidas con el azul de metileno. *Revista. trimestral. micrografica.* 1:123–136.

Ramon y Cajal, S. (1897, 1899–1904). Textura del sistema nervioso del hombre y de los vertebrados. N. Moya, Madrid.

Ramon y Cajal, S. 1900, Estudios sobre la corteza cerebral humana III. Corteza motriz. *Rev. Trim. Micrografica.* 5:1–11.

Ramon y Cajal, S. 1903, Un sencillo metodo de coloracion selectiva del reticulo protoplasmatico y sus efectos en los diversos organos nerviosos. *Trab. Lab. Invest. Biol. Univ. Madrid* 2:129–221.

Ramon y Cajal, S. 1905a, Genesis de las fibras nerviosas del embrion. *Trab. Lab. Invest. Biol. Univ. Madrid* 2:227–294.

Ramon y Cajal, S. 1905b, Mecanismo de la regeneracion de los nervios. *Trab. Lab. Invest. Biol. Univ. Madrid* 4:119–210.

Ramon y Cajal, S. 1907, Die histogenetische Beweise der Neuronentheorie von His und Forel. *Anat. Anz.* 5:113–144.

Ramon y Cajal, S. 1909–1911, 'Histologie du Systeme Nerveux de l'Homme et des Vertebres.'

Ramon y Cajal, S. 1910, Algunas observaciones favorables a la hipotesis neurotropica. *Trab. Lab. Invest. Biol. Univ. Madrid* 8:63–134.

Ramon y Cajal, S. 1913a, Contribucion al conocimiento de la neuroglia del cerebro humano. *Trab. Lab. Invest. Biol. Univ. Madrid* 11:255–315.

Ramon y Cajal, S. 1913b, Sobre un nuevo proceder de impregnacion de la neuroglia y sus resultados en los centros nerviosos del hombre y animales. *Trab. Lab. Invest. Biol. Univ. Madrid* 11:219–237.

Ramon y Cajal, S. 1916, El proceder del oro-sublimado para la coloracion de la neuroglia. *Trab. Lab. Invest. Biol. Univ. Madrid* 14:155–162.

Ramon y Cajal, S. 1919, Accion neurotropica de los epitelios (Algunas detalles sobre el mecanismo genetico de las ramificationes nerviosas intraepiteliales, sensitivas y sensoriales). *Trab. Lab. Inv. Biol. Univ. Madrid* 17:181–228.

Ramon y Cajal, S. 1920, Algunas consideraciones sobre la mesoglia de Robertson y Rio-Hortega. *Trab. Lab. Invest. Biol. Univ. Madrid* 18: 109–127.

Ramon y Cajal, S. 1922, Studies on the fine structure of the regional cortex of rodents. I: Subcortical cortex (retrosplenial cortex of Brodmann). *Trab. Lab. Invest. Biol. Univ. Madrid* 20:1–30.

Ramon y Cajal, S. 1928, 'Degeneration and Regeneration of the Nervous System.' Hafner, New York, 1959.

Ramon y Cajal, S. 1929a, Considerations critiques sur le role rophiques des dendrites et leurs pretendues relations vasculaires. *Trab. Lab. Invest. Biol. Univ. Madrid* 26:107–130.

Ramon y Cajal, S. 1929b, 'Etude sur la neruogenese de quelques vertebres.' Thomas, Springfield, Ill., 1960.

Ramon y Cajal, S. 1933a, 'Histology.' Wood, Baltimore.

Ramon y Cajal, S. 1933b, Neuronismo o reticularismo? Las pruebas objectivas de la unidad anatomica, de las cellulas nerviosas. *Arch. Neurobiol. Psicol. Madr.* 13:217–291; 579–646.

Ranson, S.W. 1911, Non-medulated nerve fibers in the spinal nerves. *Am. J. Anat.* 12:67–87.

Ranson, S.W. 1912, The structure of the spinal ganglia and of the spinal nerves. *J. Comp. Neurol.* 22:159–175.

Ranson, S.W. 1913, The course within the spinal cord of the non-medullated fibers of the dorsal roots: a study of Lissauer's tract in the cat. *J. Comp. Neurol.* 23:259–281.

Ranson, S.W. 1914, An experimental study of Lissauer's tract and the dorsal roots. *J. Comp. Neurol.* 24:531–545.

Ranson, S.W. 1915, Unmyelinated nerve-fibers as conductors of protopathic sensation. *Brain* 38:381–389.

Ranson, S.W., Droegemueller, W.H., Davenport, H.K., and Fisher, C. 1935, Number, size and myelination of the sensory fibers in the cerebrospinal nerves. Research publications. *Assoc. Res. Nerv. Ment. Dis.* 15:3–34.

Ranvier, M.L. 1878, 'Lecons sur l'histologie du systeme nerveux.' Savy, Paris.

Raven, C.P. 1931–1933, Zur Entwicklung der Ganglienleiste. *Arch. Entw. Mech. Organ* 125:210–292; 129:179–198; 130:517–561.

Raven, C.P. 1936, Zur Entwicklung der Ganglienleiste. V. Differenzierung des Rumpfganglienleistenmaterials. *Arch. Entw. Mech. Organ* 134: 122–146.

Raven, C.P. 1937, Experiments on the origin of the sheath cells and sympathetic neuroblasts in amphibia. *J. Comp. Neurol.* 67:221–240.

Remak, R. 1838a, 'Observations anatomicae et microscopicae de systematis nervosi structura.' Reimer, G., Berlin.

Remak, R. 1838b, Ueber die verrichtungen des organischen nervensystems. *Froriep's Notizen* 7:65–70.

Remak, R. 1844, 'Neurologische Erlauterungen.'

Remak, R. 1853, 'Uber gangliose Nervenfasern beim Menschen und bei den Wirbeltieren.'

Remak, R. 1854, Ueber multipolare Ganglienzellen. *Verhandlungen der Konig. Preuss. Akad. der Wissenschaften* 19:26–32.

Remak, R. 1855, 'Unterschungen uber die Entwickelung der Wirbelthiere.' G. Reimer, Berlin.

Retzius, G. 1890, Zur Kenntniss des Nervensystems der Crustaceen. *Biol. Untersuch. Neue Folge* 1:1–50.

Retzius, G. 1891a, Zur Kenntniss des centralen Nervensystems der Wurmer. *Biol. Untersuch. Neue Folge* 2:1–28.

Retzius, G. 1891b, Zur Kenntniss des centralen Nervensystems von Amphioxus lanceolatus. *Biol. Untersuch. Neue Folge* 2:29–46.

Retzius, G. 1892a, Das Nervensystem der Lumbricinen. *Biol. Untersuch. Neue Folge* 3:1–16.

Retzius, G. 1892b, Die endigungsweise des Riechnerven. *Biol. Untersuch. Neue Folge* 3:25–28.

Retzius, G. 1892c, Die Nervenendigungen in dem Geschmacksorgan der Saugetiere und Amphibien. *Biol. Untersuch. Neue Folge* 4:19–32.

Retzius, G. 1892d, Die nervosen Elemente der Kleinhirnrinde. *Biol. Untersuch. Neue Folge* 3:17–24.

Retzius, G. 1892e, Ueber die Nervenendigungen an den Haaren. *Biol. Untersuch. Neue Folge.*

Retzius, G. 1892f, Zur Kenntniss der motorischen Nervenendigungen. *Biol. Untersuch. Neue Folge* 3:41–52.

Retzius, G. 1894a, Die Neuroglia des Gehirns beim Menschen und bei Saugethieren. *Biol. Untersuch. Neue Folge* 6:1–28.

Retzius, G. 1894b, Ueber die Endigungsweise der Nerven an den Haaren des Menschen. *Biol. Untersuch. Neue Folge* 6:61–62.

Retzius, G. 1894c, Weitere Beitrage zur Kenntniss der Cajal'schen Zellen der Grosshirnrinde des Menschen. *Biol. Untersuch. Neue Folge* 6:29–36.

Retzius, G. 1898a, Zur ausseren Morphologie des Riechhirns der Saugethiere und des Menschen. *Biol. Untersuch. Neue Folge* 8:23–48.

Retzius, G. 1898b, Zur Frage von der Endigungsweise peripherischer sensibler Nerven. *Biol. Untersuch. Neue Folge* 8:114–117.

Riese, W. and Hoff, E.C. 1950, A history of the doctrine of cerebral localization. *J. Hist. Med.* 1:50–71.

Rio-Hortega, P. d. 1919, El tercer elemento de los centros nerviosos. I. La microglia normal. II. Intervencion de la microglia en los procesos patologicos. (Celulas en bastoncito y cuerpos granulo-adiposos). III. Naturaleza probable de la microglia. *Bol. Soc. Esp. Biol.* 9:69–129.

Rio-Hortega, P. d. 1920, La microglia y su transformacion en celulas en basoncito y cuerpos granulo-adiposos. *Trab. Lab. Inv. Biol. Univ. Madrid* 18:37–82.

Rio-Hortega, P. d. 1921a, Estudios sobre la neuroglia. La glia de escasas radiaciones (oligodendroglia). *Bol. Soc. Esp. Biol.* 21:64–92.

Rio-Hortega, P. d. 1921b, Histogenesis y evolucion normal exodo y distribucion regional de la microglia. *Mem. de la Real. Soc. Esp. Hist. Nat.* 11:213–268.

Rio-Hortega, P. d. 1922, Son homologables la glia de escasas radiacion es y la celula de Schwann? *Bol. Soc. Esp. Biol.* 10.

Rio-Hortega, P. d. 1924, La glie a radiations peu nombreuses et la cellule de Schwann sont elles homologables? *Compt. Rend. Soc. Biol* 91: 818–820.

Rio-Hortega, P. d. 1928, Tercera aportacion conocimiento morfologico e interpretacion functional de la oligodendroglia. *Mem. Real. Soc. Esp. Hist. Nat.* 14:5–122.

Rio-Hortega, P. d. 1932, 'Microglia.' Hoeber, New York.

Rio-Hortega, P. d. and Penfield, W. 1927, Cerebral cicatrix: the reaction of neuroglia and microglia to brain wounds. Bull. Johns Hopkins Hosp. 41, 278–303.

Robain, O., and Ponsot, G. 1978, Effects of undernutrition on glial maturation. *Brain. Res.* 149:379–397.

Robertson, J.D. 1953, Ultrastructure of two invertebrate synapses. *Proc. Soc. Exp. Biol. Med.* 82:219–223.

Robertson, J.D. 1963, The Occurrence of a Subunit Pattern in the Unit Membranes of Club Endings in Mauthner Cell Synapses in Goldfish Brains. *J. Cell. Biol.* 19:201–221.

Robertson, J.D. 1965, The synapse: Morphological and chemical correlates of function: A report of an NRP work session. *Neurosci. Res. Prog. Bull.* 3:1–79.

Robertson, M. 1987, Retinoic acid receptor. Towards a biochemistry of morphogenesis. *Nature* 330:420–421.

Robertson, W. 1897, The normal histology and pathology of neuroglia. *J. Ment. Sci.* 43:733–752.

Robertson, W. 1899, On a new method of obtaining a black reaction in certain tissue-elements of the central nervous system (platinum method). *Scottish Med. Surg. J.* 4:23.

Robertson, W.F. 1900, A microscopic demonstration of the normal and pathological histology of mesoglia cells. *J. Ment. Sci.* 46:733–752.

Rohon, V. 1884, 'Zur histogenese des Ruckenmarks der Forelle.'

Roux, W. 1881, 'Der Kampf der Theile im Oranismus.' Wilhelm Engelmann, Leipzig.

Roux, W. 1923, 'Autobiographie.' Verlag von Felix Meiner, Leipzig.

Ruffini, A. 1892, Sulla terminazione nervosa nei fusi muscolari e sul loro significato fisiologico. *R.C. Accad. Lincei ser.5, 2nd sem.*, 31–38.

Ruffini, A. 1898, On the minute anatomy of the neuromuscular spindles of the cat, and on their physiological significance. *J. Physiol. (London)* 23:190–208.

Russell, E.S. 1930, The interpretation of development and heredity. *In* A study in biological method. Clarendon press, Oxford.

Sabatini, M.T., Pellegrinodeiraldi, A., and Derobertis, E. 1965, Early Effects of Antiserum against the Nerve Growth Factor on Fine Structure of Sympathetic Neurons. *Exp. Neurol.* 12:370–83.

Sadaghiani, B., and Thiebaud, C.H. 1987, Neural crest development in the Xenopus laevis embryo, studied by interspecific transplantation and scanning electron microscopy. *Dev. Biol.* 124:91–110.

Sanchez y Sanchez, D. 1923, L'histogenese dans les centres nerveux des insectes pendent les metamorphoses. *Trab. Lab. Inv. Biol. Univ. Madrid* 23:29–52.

Sauer, F.C. 1935a, The cellular structure of the neural tube. *J. Comp. Neurol.* 63:13–23.

Sauer, F.C. 1935b, Mitosis in the neural tube. *J. Comp. Neurol.* 62:377–405.

Scarr, S. 1969, Effects of birth weight on later intelligence. *Soc. Biol.* 16:249–256.

Schadewald, M. 1941, Effects of cutting the trochlear and abducens nerves on the end-bulbs about the cells of the corresponding nuclei. *J. Comp. Neurol.* 74.

Schadewald, M. 1942, Transsynaptic effect of neonatal axon section on bouton appearance about somatic motor cells. *J. Comp. Neurol.* 77:739–746.

Schaper, A. 1846, Die morphologische und histologische Entwicklung des Kleinhirns der Teleostier. *Morphol. Jahrb.* 21:625–708.

Schaper, A. 1894, Die morphologische und histologische Entwicklung des Kleinhirns der Teleostier. *Anat. Anz.* 9:489–501.

Schaper, A. 1897a, Die fruhesten Differenzierungsvorgange im Centralnervensystem. *Arch. Entw. Mech. Organ* 5:81–132.

Schaper, A. 1897b, The earliest differentiation in the central nervous system of vertebrates. *Science* 5:430–431.

Schelper, R.L., and Adrian, E.K., Jr. 1986, Monocytes become macrophages; they do not become microglia: a light and electron microscopic autoradiographic study using 125-iododeoxyuridine. *J. Neuropathol. Exp. Neurol.* 45:1–19.

Schleiden, M.J. 1838, Müller's arch. f. anat. *Physiol. u. wissensch.*

Schmechel, D.E., and Rakic, P. 1979, A Golgi study of radial glial cells in developing monkey telencephalon: morphogenesis and transformation into astrocytes. *Anat. Embryol. (Berl)* 156:115–52.

Schopenhauer, A. 1851, Paregra and Paralipomena, tr. E. Payne, ed. D. Cartwright. Oxford, Clarendon.

Schrodinger, E. 1944, What Is Life? Cambridge Univ. Press, Cambridge.

Schuetze, S.M., and Role, L.W. 1987, Developmental regulation of nicotinic acetylcholine receptors. *Annu. Rev. Neurosci.* 10:403–57.

Schultz, M.G. 1962, Male pseudohermaphroditism diagnosed with aid of sex chromatin technique. *J. Am. Vet. Med. Assoc.* 140:241–244.

Schultz, M.J.S. 1870, 'In Manual of Human and Comparative Histology.' New Sydenham Society, London.

Schwann, T. 1839a, 'Microscopical Researches into the Accordance in the Structure and Growth of Animals and Plants.' The Sydenham Society, London.

Schwann, T. 1839b, 'Mikroskopische Untersuchungen uber die ubereinstimmung in der Struktur und dem Wachstum der Thiere und Pflanzen.' Berlin.

Schwann, T. 1839c, 'Mikroskopische Untersuchungen über die Übereinstimmung in der Struktur und dem Wachstum der Thiere und Pflanzen.' London, 1847.

Sedgwick, A. 1895, On the inadequacy of the cellular theory of development, and on the early development of nerves etc. *Quart. J. Microsc. Sci.* 37:87–101.

Sherman, S.M., Guillery, R.W., Kaas, J.H., and Sanderson, K.J. 1974, Behavioral, electrophysiological and morphological studies of binocular competition in the development of the geniculo-cortical pathways of cats. *J. Comp. Neurol.* 158:1–18.

Sherman, S.M., and Wilson, J.R. 1975, Behavioral and morphological evidence for binocular competition in the postnatal development of the dog's visual system. *J. Comp. Neurol.* 161:183–95.

Sherrington, C.S. 1894, On the anatomical constituion of nerves of skeletal muscles; with remarks on recurrent fibres in the ventral spinal nerve roots. *J. Physiol. (London)* 17:211–258.

Sherrington, C.S. 1897a, The central nervous system. Macmillan, London.

Sherrington, C.S. 1897b, Further note on sensory nerves of muscles. *Proc. Roy. Soc. (London) Ser. B.* 61:247–249.

Sherrington, C.S. 1897c, On the reciprocal innervation of antagonistic muscles. Third note. *Proc. Roy. Soc. (London) Ser. B.* 60:414–417.

Sherrington, C.S. 1900, 'The spinal cord.' Pentland, Edinburgh.

Sherrington, C.S. 1904, Correlation of reflexes and the principle of the final common path. *In* 'Rep. Br. Assoc. Adv. Sci. 74th meeting, transact., Sec.1, Physiology', pp. 728–741.

Sherrington, C.S. 1906, The Integrative Action of the Nervous System. Yale Univ. Press, New Haven.

Shorey, M.L. 1909, The effect of the destruction of peripheral areas on the differentiation of the neuroblasts. *J. Exp. Zool.* 7:25–64.

Skeels, H.M. 1966, Adult status of children with contrasting early life experiences. A follow-up study. *Monogr Soc Res Child. Dev.* 31:1–56.

Smart, I.H., and McSherry, G.M. 1986a, Gyrus formation in the cerebral cortex in the ferret. I. Description of the external changes. *J. Anat.* 146:141–152.

Smart, I.H., and McSherry, G.M. 1986b, Gyrus formation in the cerebral cortex of the ferret. II. Description of the internal histological changes. *J. Anat.* 147:27–43.

Smith-Thomas, L.C., and Fawcett, J.W. 1989, Expression of Schwann cell markers by mammalian neural crest cells in vitro. *Development* 105:251–262.

Sotelo, C., and Palay, S.L. 1971, Altered axons and axon terminals in the lateral vestibular nucleus of the rat. Possible example of axonal remodeling. *Lab Invest* 25:653–671.

Soury, J. 1899, 'Le Systeme Nerveux Central. Structure et Fonctions.' Carre et Naud, Paris.

Spallanzani, L. 1784, Memoria seconda ed ultima sopra la riproduzione della testa nelle lumache terrestri. In: Memorie di Matematica e Fisica della Societá Italiana, Tomo II, parte II. Ramazzini, Verona.

Spencer, H. 'The Principles of Biology.' Appleton, New York.

Sperry, R.W. 1963, Chemoaffinity in the Orderly Growth of Nerve Fiber Patterns and Connections. *Proc. Natl. Acad. Sci. U S A.* 50:703–10.

Spitz, R., and Wolf, A.M. 1946, Hospitalization: An inquiry into the psychiatric conditions in early childhood. *Psychoanal. Study Child* 1:53.

Starck, D. 1937, Ueber einige Entwicklungsvorgange m kopf der Urodelen. *Morphol Jahrb* 79:358–435.

Stein, Z., Susser, M., Saenger, G., and Marolla, F. 1975, 'Famine and human development. The Dutch hunger winter of 1944–45.' Oxford University Press, New York.

Steindler, A. 1916, Direct neurotization of paralyzed muscles, further study of the question of direct nerve implantation. *Am. J. Orthopedic. Surg.* 14:707–719.

Stent, G.S. 1973, A physiological mechanism for Hebb's postulate of learning. *Proc. Natl. Acad. Sci. U S A.*70:997–1001.

Sterzi, G. 1914–1915, 'Anatomia del Sistem Nervoso Centrle dell'Uomo.' Padova.

Stieda, L. 1899, Geschichte der Entwickelung der Lehre von den Nervenzellen und nervenfasern Whrend des XIX. Jahrhunderts. I. Teil: Von Sommering bis Deiters. Verlag von Gustav Fischer, Jena.

Stoch, M.B., and Smythe, P.M. 1963, Does undernutrition during infancy inhibit brain growth and subsequent intellectual development? *Arch. Dis. Child.* 38:546–552.

Stoch, M.B., and Smythe, P.M. 1967, The effect of undernutrition during infancy on subsequent brain growth and intellectual development. *S Afr. Med. J.* 41:1027–1030.

Stoch, M.B., and Smythe, P.M. 1976, 15-Year developmental study on effects of severe undernutrition during infancy on subsequent physical growth and intellectual functioning. *Arch. Dis. Child.* 51:327–336.

Stoch, M.B., Smythe, P.M., Moodie, A.D., and Bradshaw, D. 1982, Psychosocial outcome and CT findings after gross undernourishment during infancy: a 20-year developmental study. *Dev. Med. Child. Neurol.* 24:419–436.

Stohre, P. 1935, Beobachtungen und Bemerkungen uber die Endausbreitung des vegetativen Nervensystems. *Z. Anat. Entw. Gesch.* 104:133–158.

Stone, L.C. 1922, Experiments on the development of the cranial ganglia and the the lateral line sense organs in Amblystoma punctatum. *J. Exp. Zool.* 35:421–496.

Stone, L.C. 1929, Experiments showing the role of migrating neural crest (mesectoderm)in the formation of head skeleton and loose connective tissue in Rana palustris. *Arch. Entwicklungsmech.* 118:40–77.

Stott, D.H. 1960, Interaction of heredity and environment in regard to 'measured intelligence'. *Brit. J. Educ. Psychol.* 30:95–102.

Streeter, G.L. 1912, The development of the nervous system. *In* 'Manual of Human Embryology', Vol. 2: pp. 1–156. Lippincott, Philadelphia.

Strong, O.S. 1895, The cranial nerves of Ampohibia; a contribution to the morphology of the vertebrate nervous system. *J. Morphorl.* 10:101–230.

Szymonowicz, W. 1895, Beitrage zur Kenntniss der Nervenendigungen in Hautgebilden. Ueber Bau und Entwickelung der Nervenendigungen in der Schnauze des Schweines. *Arch. mikr. Anat. Bonn.* 14:624–635.

Tergast, P. 1873, Ueber das Verhaltniss von Nerve und Muskel. *Arch. f. mikr. Anat.* 9:36–46.

Tiegs, O.W. 1927, A criticl review of the evidence on which is based the theory of discontinous synapsesin the spinal cord. *Aust J Exp Biol Med Sci* 4:193–212.

Titchener, E.B. 1921, Wilhelm Wundt. *Am. J. Psychol.* 32:161–178; 575–580.

Tizard, B. 1959, Theories of brain localization from Flourens to Lashley. *Med Hist* 3:132–45.

Tootell, R.B., Silverman, M.S., Hamilton, S.L., De Valois, R.L., and Switkes, E. 1988, Functional anatomy of macaque striate cortex. III. Color. *J. Neurosci.* 8:1569–1593.

Torvik, A. 1975, The relationship between microglia and brain macrophages. Experimental investigations. *Acta. Neuropathol. Suppl. (Berl)* Suppl 6, 297–300.

Torvik, A., and Skjorten, F. 1971a, Electron microscopic observations on nerve cell regeneration and degeneration after axon lesions. I. Changes in the nerve cell cytoplasm. *Acta. Neuropathol. (Berl)* 17:248–264.

Torvik, A., and Skjorten, F. 1971b, Electron microscopic observations on nerve cell regeneration and degeneration after axon lesions. II. Changes in the glial cells. *Acta. Neuropathol. (Berl)* 17:265–282.

Turner, W. 1891, The convolutions of the brain: A study in comparative anatomy. *J. Anat. (London)* 25:105–153.

Unna, P.G. 1890, Uber die Taenzersche (Orcein-) Farbung des elstischen Gewebes. *Monatsschr. Prak. Dermatol.* 11:366–367.

Unna, P.G. 1891, Uber die Reifung unserer Farbstoffe. *Z. Wiss. Mikrosk.* 8:475–487.

Van der Loos, H. 1967, 'The history of the neuron.' Elsevier,

Van Gehuchten, A. 1891, La structure des centres nerveux la moelle epiniere et le cervelet. *Cellule* 8:1–43.

Van Gehuchten, A. 1892a, Contributions a l'etude de l'innervation des poils. *Anat. Anz.* 7:341–348.

Van Gehuchten, A. 1892b, La structure des lobes optiques chez l'embryon de poulet. *La Cellule* 8:1–43.

Van Gehuchten, A. 1904, Boutons terminaux et reseau pericellulaire. *Nevraxe* 8:81–116.

Van Gehuchten, A. 1906, 'Anatomie du systeme nerveux de l'homme.' A. Uystpruyst-Dieudonne, Louvain.

Van Gieson, J. 1889, Laboratory notes of technical methods for the nervous system. *N.Y. Med. J.* 50:57–60.

Vandenberg, S.G. 1966, Contributions of twin research to psychology. *Psychol Bull* 66:327–352.

Vaughn, J.E., and Pease, D.C. 1970, Electron microscopic studies of wallerian degeneration in rat optic nerves. II. Astrocytes, oligodendrocytes and adventitial cells. *J. Comp. Neurol.* 140:207–226.

Vignal, W. 1888, Recherches sur le developpement des elements des couches corticales du cerveau et du cervelet chez l'homme et les mamifers. *Arch. Physiol. Norm. Path. (Paris)* 4:228–254, 311–338.

Vignal, W. 1889, 'Developpement des elements du systeme nerveux cerebrospinal.' Paris.

Virchow, R. 1846, Uber das granulirte Ansehen der Wandungen der Gehirnventrikel. *Allgem. Zeitschr. Psychiat.* 3:242–250.

Virchow, R. 1858, Cellularpathologie in ihre Begrundung auf Physiologische und Pathologische Gewebelehre. A. Hirschwald, Berlin.

Virchow, R. 1867, Knogenitale encephalitis und myelitis. *Virchow's Archiv* 38:129.

Virchow, R. 1885, Cellular Pathology. *Arch. Path. Anat. Klin. Med.* 8:1–15.

Vogt, C., and O, V. (1902–1904). 'Neurobiolgische arbeiten. I. Zur Erforschungder Hirnfaserung. II Die Markreifung des Kindergehirn wahrend der ersten vier Lebensmonate und ihre methodologische bedeutung.' VErlag von Gustav Fischer, Jena.

Vogt, O., and Vogt, C. 1919, Ergebnisse unnserer Hirnforschung. *J. Psychol. Neurol. Leipzig.* 25:277–462.

Vogt, W. 1925, Gestltungsanalyse am Amphibienkeim mit ortlicher Vitalfarbung. I. Methodik und Wirkungsweise der ortlichen Vitalfarbung mit gr als Farbrager. *Arch. Entw. Mech. Organ.* 120:384–706.

von Econonmo, C., and Koskinas, G.N. 1925, Die cytoarchitektonik der hirnrinde des eruchsenen menschen. *In* 'Text und Atlas'. Springer, Berlin.

von Gudden, B.A. 1870, Experimentluntersuchungen uber ds peripherischungen uber das peripherische und centrle nervensystem. *Arch Psychiat* 2:693–723.

von Gudden, B.A. 1874, 'Experimental-Untersuchungen über das Schädelwachstum.' München

Von Vintschgau, M. 1880, Beobachtungen uber die Veranderungen der Schmeckbecher nach Durchschneidung des N. glossopharyngeus. *Arch. Ges. Physiol.* 23:1–13.

Von Vintschgau, M., and Honigschmied, J. 1876, Nervus glossopharyngeus und Schmeckbecher. *Arch. Ges. Physiol.* 14:443–448.

Waganer, G. 1949, Die Bedeutung der Neuralleiste fu die Kopfgestaltung der Amphibienlarven. Untersuchungen an Chimaeren von Triton und bombinator. *Rev. Suisse. Zool.* 56:519–619.

Wagner, R. 1842–1853, Handworterbuch der Physiologie it Rucksicht auf physiologische Pathologie. Vieweg, F., Braunschweig.

Wagner, R. 1847, 'Neue Untersuchungen uber den Bau und die Endigungen der Nerven und die struktur der Ganglienzellen.' Leipzig.

Walker, A.E. 1957, The development of the concept of cerebral localization in the nineteenth century. *Bull. Hist. Med.* 31:99–121.

Walter, G. 1861, Ueber den feineren Bau des Bulbus ofactorius. *Virchow's Archiv.* 22:241–259.

Weigert, C. 1895, Festschrift zum 50-jahrigen jubil d. arztl Vereins zu Frankfurt a. M. *In* 'Beitrage zur Kenntniss der normalen menschlichen Neuroglia'.

Weigert, C. 1898, Uber eine Methode zur Farbung elasticher Fasern. *Zentralbl. Allg. Pathol.* 9:289–292.

Weiss, P. 1936, Selectively controlling the central-peripheral relations in the nervous system. *Biol. Rev.* 11:494–531.

Weiss, P. 1939, 'Principles of Development.' Henry Holt, New York.

Weiss, P. 1947, The problem of specificity in growth and development. *Yale J. Biol. Med.* 19:235–278.

Weiss, P. 1952, Central versus peripheral factors in the development of coordination. *Res Publ Assoc. Res. Nerv. Ment. Dis.* 30:3–23.

Weiss, P., and Hoag, A. 1946, Competitive reinnervation of rat muscles by their own and foreign nerves. *J. Neurophysiol.* 9:413–418.

Wernig, A., and Herrera, A.A. 1986, Sprouting and remodelling at the nerve-muscle junction. *Prog. Neurobiol.* 27:251–91.

Willerman, L., and Churchill, J.A. 1967, Intelligence and birth weight in identical twins. *Child. Dev.* 38:623–629.

Wilson, E.B. 1896, 'The cell in development and inheritance,' New York, Macmillan.

Winick, M. 1970, Cellular growth in intrauterine malnutrition. *Pediatr. Clin. North. Am.* 17:69–78.

Winick, M., Meyer, K.K., and Harris, R.C. 1975, Malnutrition and environmental enrichment by early adoption. *Science* 190:1173–5.

Winick, M., and Rosso, P. 1969, The effect of severe early malnutrition on cellular growth of human brain. *Pediatr. Res.* 3:181–184.

Winick, M., Rosso, P., and Waterlow, J. 1970, Cellular growth of cerebrum, cerebellum, and brain stem in normal and marasmic children. *Exp. Neurol.* 26:393–400.

Wise, j. P., and Godschalk, M. 1987, Functional fractionation of frontal fields. *Trends Neurosci.* 10:449–450.

Wolff, M. 1905, Zur Kenntnis des Heldschen nervenendfusse. *Psychol. Neurol.* 4:144–157.

Woolsey, C.N. 1943, Second somatic receiving areas in the cerebral cortex of cat, dog and monkey. *Fed. Proc.* 2:55–56.

Woolsey, C.N., Settlage, P.H., Meyer, D.R., Spencer, W., Pinto-Hamuy, T., Travis, M. 1952, Pattern of localization in precentral and supplementary motor area sand their relation to the concept of a premotor cortex. *Assoc. Res. Nerv. Ment. Dis.* 30:238–264.

Woolsey, C.N. 1958, Organization of somatic sensory and motor areas of the cerebral cortex. In: Harlow, H.F. and Woolsey, C.N. (Eds.) The Biological and Biochemical Bases of Behavior, Univ. Wisc. Press, Madison.

Wree, A., Zilles, K., and Schleicher, A. 1983, A quantitative approach to cytoarchitectonics. VIII. The areal pattern of the cortex of the albino mouse. *Anat. Embryol. (Berl)* 166:333–353.

Wundt, W. 1904, 'Principles of physiolgical psychology.' Swan Sonnenschein, Macmillan, New York.

Yntema, C.L. 1943, Deficient efferent innervation of the extremities following removal of neural crest in Amblystoma. *J. Exp. Zool.* 94:319–349.

Young, J.Z. 1951, Growth and plasticity in the nervous system. *Proc. R. Soc. Lond. B. Biol. Sci.* 139:18–37.

Young, R.M. 1970, 'Mind, brain and adaptation in the nineteenth century.' Clarendon Press, Oxford.

Zilles, K., Zilles, B., and Schleicher, A. 1980, A quantitative approach to cytoarchitectonics. VI. The areal pattern of the cortex of the albino rat. *Anat. Embryol. (Berl)* 159:335–360.

Index

A3B5$^+$/PSA-NCAM$^+$ cells, 165
Ablation studies, 78
Acetylcholine (ACh), 295–296
Acetylcholine esterase (AChE), 284
Acetylcholine receptor (AChR) channels, 276
Acetylcholine receptor inducing activity (ARIA), 294
Acetylcholine receptors (AChRs), 273, 276, 278–281, 284, 290, 293, 309
  distribution, 285
  types of, 269, 271
Achaete-scute homologue *ash1*, 97
Actin, retrograde flow of, 246
Actin-binding protein filamin 1 (filamin-α), 225
Active zones (AZ), 270, 275, 279–280
Activin, 9, 10
Adhesion, tactile, 377
Adhesion proteins, synaptic, 300–305
Adhesive cell-surface signals, 251
Adrenoleukodystrophy, 173
Age-related alterations in neurogenesis, developmental mechanisms
    underlying, 354–357
Age-related cytoarchitectural changes in nervous system, 350–351
Age-related molecular changes in nervous system, 351
Age-related neurodegenerative disorders, 361
  brain regions affected in, 359
  developmental mechanisms in, 358–361
Aging
  cellular and molecular changes during normal, 349–354
  mechanisms that regulate nervous system development and, 350, 361
  neural control of, 354
Aging phenotype, inherited disorders with abnormal, 358
Agrin, 284, 291–294, 297, 298, 305–306
  structure/function, 291
Agrin hypothesis, 292
Agrin-induced signaling pathways, 293
Alcohol syndrome, fetal, 176
ALD protein, 173
Alkaline phosphatase (AP), 257
Alzheimer's disease, 358
Ameboid microglia, 371–372
4-Aminopyridine (4-AP), 180–181
Amniotes, 80–81
Amphibian embryos, 4; *see also Xenopus* embryo
Amyloid precursor protein (APP), 358, 359
Anamniotes, 81
Anastomosis between peripheral nerve fibers, 389
Anterior maintenance, 48
Anterior neural plate border, preplacodal field at, 102–104
Anterior-posterior (A-P) topography, mechanisms of, 260
Anteroposterior (AP) pattern
  early decisions, 44

first division, 44
  head neural induction and maintenance, 45–46
  trunk neural induction, 46–50
  regional patterning, 50–57
Antibodies that promote remyelination, 182
Apaf-1, 322–323
Apoptosis-inducing factor (AIF), 319
Apoptosis pathway, 353
Apoptotic cell death, 318, 319; *see also* Programmed cell death
Ara-C (cytosine arabinoside), 205
Architectonic maps, 396–397
Astrocyte development
  in cerebellum, 208
  in forebrain
    pathways of, giving rise to different types of astrocytes, 207–208
  is not uniform across different regions of CNS, 199
  in spinal cord, 208–209
Astrocyte genesis, interplay of multiple pathways contributes to, 216
Astrocyte lineages
  model of, 212
  in vitro, heterogeneity within, 211
Astrocyte precursor cells (APCs), 212–213
  contrasted with NG2 cells, 209–210
  radial glia as, 203–204
Astrocyte progenitors, emigrate from SVZ along radial glial guides, 207
Astrocyte-restricted precursors, 213
Astrocyte specification
  induced by alpha helical family of cytokines, 213
  induced by TGF-β family of cytokines, 213–214
  regulated by multiple signals, 213–216
Astrocytes, 164–166, 171, 197, 217
  cell culture studies reveal multiple lineages of, 210–211
  defined, 197
  directly descended from SVZ, 205–207
  functions, 197
  maintenance in adult brain, 209–210
    by endogenous precursors, 209
  subset that are direct descendants of ventricular zone, 200–205
  and their developmental expression, molecular markers for, 200
  types of, 197, 199, 211
    in different regions, 197–199
    type 1 lineage, 210–211
    type 2 and the O-2A lineage, 211
    in vivo, 197–199
Autism
  mercury, vaccines, and, 177
Autonomic nervous system, 95–96
  structure, 95
Autophagic cell death, 318–320; *see also* Programmed cell death
Avian embryos, 46